W9-BOB-052

Evolution

Third Edition

Monroe W. Strickberger

Museum of Vertebrate Zoology
University of California, Berkeley

JONES AND BARTLETT PUBLISHERS

Sudbury, Massachusetts

BOSTON TORONTO LONDON SINGAPORE

World Headquarters
Jones and Bartlett Publishers
40 Tall Pine Drive
Sudbury, MA 01776
978-443-5000
info@jbpub.com
www.jbpub.com

Jones and Bartlett Publishers Canada
2100 Bloor Street West
Suite 6-272
Toronto, ON M6S 5A5
CANADA

Jones and Bartlett Publishers International
Barb House, Barb Mews
London W6 7PA
UK

The creature pictured in the right-hand page corner throughout this book is an early tetrapod, a "four-footed" animal. Moving from water to land was a critical step in tetrapod evolution; fossil evidence of the animal itself and its footprints suggests that this may have occurred about 365 million years ago. Hypothesized advantages to moving to land include avoiding predators and exploiting terrestrial food sources, such as insects. Flip the pages of this book from front to back to watch the creature "evolve."

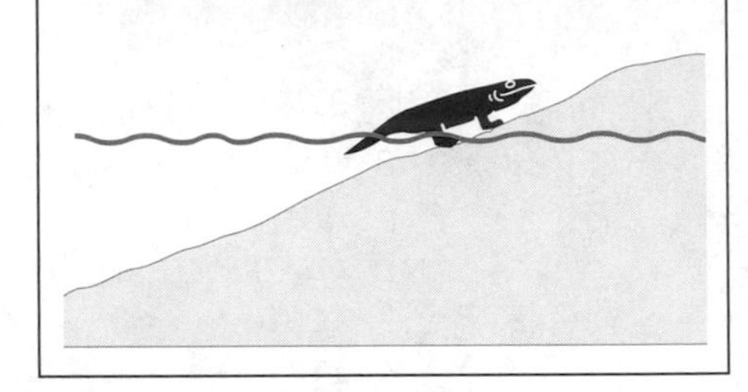

Chief Executive Officer: Clayton Jones
Chief Operating Officer: Don Jones, Jr.
President: Tom Walker
V.P., Sales and Marketing: Tom Manning
V.P., Managing Director: Judith H. Hauck
V.P., College Editorial Director: Brian L. McKean
V.P., Director of Interactive Technology: Mike Campbell
Director of Design and Production: Anne Spencer
Director of Manufacturing and Inventory Control: Therese Bräuer
Production Editor: Rebecca S. Marks
Editorial/Production Assistants: Tim Gleeson and Jennifer E. Angel
Web Designer: Dean A. Wetherbee
Interior Design: Seventeenth Street Studios
Illustrations: Elizabeth Morales
Copy Editor: Kate Scheinman
Typesetting and Composition: Carlisle Communications, Ltd.
Cover Design: Stephanie Torta
Printing and Binding: Courier Kendallville
Cover Printing: Phoenix Color

Library of Congress Cataloging-in-Publication Data

Strickberger, Monroe W.
Evolution / Monroe W. Strickberger. — 3rd ed.
p. cm.
Includes bibliographical references and indexes.
ISBN 0-7637-1066-0
1. Evolution (Biology) I. Title.
QH366.2.S78 2000
576.8—dc21

99-32072
CIP

Printed in the United States of America
03 02 01 00 99 10 9 8 7 6 5 4 3 2 1

Brief Contents

Contents

Preface

All biological phenomena derive from evolutionary relationships and past interactions. As the great evolutionary geneticist Theodosius Dobzhansky remarked, "Nothing in biology makes sense except in the light of evolution." Unifying all biology under an evolutionary theme is still difficult, although the explosive increase in molecular, organismic, and populational information makes the realization of this goal more possible now than ever before.

The purpose of this book is to bring together some prevailing knowledge and ideas about evolution in order to provide an informed evolutionary framework of thought for undergraduates. It is based on a course I have given for many years to biology majors who have had prior introductory biology courses. (Reviews of some basic biological and genetic concepts are nevertheless included.)

Academic biology is heavily partitioned, and biology majors take a variety of specialty courses such as development, ecology, genetics, microbiology, physiology, and they also concentrate in specific areas. Thus the evolutionary theme that runs through all of biology is often fragmented, or entirely ignored. In many courses and in many institutions, evolution is little more than a curricular afterthought, and biologists emerge from such programs with little grasp of the following basic questions and topics covered in this book:

- **Chapters 1, 2, 3:** What is the philosophical and historical background of evolutionary concepts, how did these concepts develop, and why were they so readily accepted by most scientists?
- **Chapter 4:** Why is evolution still considered controversial by so many? What basic issues cause conflict between believers in evolution and various believers in religion?
- **Chapters 5, 6:** How did cosmological and geological evolution lead to those features responsible for life on earth?
- **Chapter 7:** What chemical factors enabled life to originate on this planet, and what molecular

developments provided its substance and early direction? From what molecular sources did natural selection arise?

- **Chapter 8:** What are some proposed concepts on how protein synthesis and the genetic code evolved?
- **Chapter 9:** Whence came life's metabolic pathways, their functions, and relationships? What do we know about early cellular forms and their differences?
- **Chapter 10:** How does genetics provide the constancy and variability used in evolution?
- **Chapter 11:** How are organisms classified, and what problems are there in classifying them so they reflect evolutionary relationships?
- **Chapter 12:** What sources and techniques are used to obtain molecular information about evolution, and what phylogenies do these provide?
- **Chapter 13:** What do we know about how the major forms of plants evolved, and the evolutionary processes they employed?
- **Chapter 14:** What ideas do we have about the origin of multicellular animals and their basic features?
- **Chapter 15:** What are we discovering about organismic development, and the various paths it follows? What do we know about how and why development changes between generations over time?
- **Chapter 16:** What do we know about the evolution of invertebrate phyla? What characteristics and body forms did they assume, and how did they change?
- **Chapters 17, 18:** From what organisms may vertebrates have originated? What adaptations enabled vertebrate tetrapods to invade land, and what factors caused their various extinctions and replacements?
- **Chapter 19:** What were the early stages in mammalian evolution? How were these affected by their Mesozoic experiences, and how did geological changes affect their distributions?
- **Chapter 20:** From whence came our own species, and how does evolution explain our features, our mental attributes, and our behavior?
- **Chapters 21, 22:** What factors are involved in conserving or changing the genetic characteristics of populations, and how do these affect their evolutionary paths?
- **Chapter 23:** What contributions do topics such as neutral mutation; selectionism; the advantages of sex; coevolution; group selection; and adaptive landscapes offer that help understand evolutionary mechanisms?
- **Chapter 24:** What do we know about races and species in terms of their features, their adaptational patterns, and the kinds of barriers that led to their evolutionary differences?
- **Chapter 25:** From what sources did our culture arise? What mechanisms enabled it to evolve? What impact does our biology have on our culture? How does our culture affect our evolution?

Should the modern biology major have at least a modest understanding of these topics? The answer is an unequivocal *Yes*! Biologists trained to represent our science need evolutionary understanding at all levels! To "make sense" of biology demands more than a short mention of a few evolutionary events in courses primarily confined to more specialized fields, whether ecology, genetics, or population biology.

In general, I have considered evolution from a historical point of view both biologically and conceptually. On the biological level, historical information passed on by transmitted genetic material connects the biology of organisms to past events—a modern organism derives from earlier organisms. On the ideological level, present evolutionary concepts derive from previous concepts. In both these forms of transmitted information, "like" not only produces "like" but also produces "unlike" because of (1) genetic changes in hereditary material and (2) conceptual changes in the ideas of evolution. Almost every aspect of existence has an evolutionary background and framework, and knowledge of the past is essential to fully understand the present. In fact, what makes biology unique compared to chemistry or physics is that biological forms and functions, in all their many variations, originated through historical events and continue to do so—an understanding of biology is inseparable from its history, and evolutionary predictability cannot escape from contingency.

The realm of evolutionary science therefore includes both chronology and mechanisms—we seek concepts that explain both the sequence of events and their causes. For this purpose, evolutionary scientists have developed, and continue to develop, methods that provide reconstructions of evolutionary events and let us understand not only biological chronology but also its genetic connections. That is, we hold that evolution follows a sequence of logically understandable causes that provides us with rational explanatory powers and reliable knowledge of the past.

Evolution is a majestic story—certainly the longest and most encompassing the world offers. Since its grand outline covers both history and mechanism, evolution is an exciting subject to students, and I have found over the years that they respond best when the textual material is generously illustrated. I have therefore provided close to 450 figures, tables, and diagrams. To further help the student master the material, the text includes boxed reviews of special topics, end-of-chapter summaries, lists of key terms, discussion questions, and a glossary. For research and reference, complete chapter bibliographies as well as separate author and subject indexes are provided. Some added points of interest, not crucial to understanding the text, are given in footnotes.

Although the order of topics offered here has worked well for my own classes, I know there are different ways to organize this material, and the chapters have generally been written to allow considerable flexibility. I have avoided an overly theoretical treatment of the subject. Nothing beyond elementary algebra is needed to understand the mathematics used.

Since evolution is the broadest of biological fields, covering the greatest range of disciplines, even the brief survey of evolution offered here has been impossible to achieve without errors and ambiguities. To the extent that this book has been spared many such failings, I owe thanks to many reviewers, who commented on one or more sections of this new edition:

Wyatt Anderson, University of Georgia, Athens

Michael Benton, University of Bristol, United Kingdom

Peter Bowler, Queen's University, Belfast, United Kingdom

Robert Foley, University of Cambridge, United Kingdom

Michael Ghiselin, California Academy of Sciences

Thomas Givnish, University of Wisconsin, Madison

Anthony Hallam, University of Birmingham, United Kingdom

Manyuan Long, University of Chicago

Stanley Miller, University of California-San Diego

Kenneth Mowbray, American Museum of Natural History, New York

Claus Nielsen, Zoologisk Museum, Copenhagen

Karl Niklas, Cornell University

Timothy Prout, University of California-Davis

Ursula Rolfe, Children's Hospital Oakland

Trinh Thuan, University of Virginia, Charlottesville

David Wake, University of California-Berkeley

I am especially grateful to Elizabeth Morales for the extraordinarily fine artwork that graces many illustrations. I also owe many thanks to the staffs at Jones and Bartlett, and Carlisle Publishers Services, for the care and attention they devoted to producing the text.

Evolution on the Web

ones and Bartlett Publishers is pleased to present Evolution on the Web, a web site designed to accompany **Evolution, 3/e.** This site, located at www.jbpub.com/evolution, has been created to provide students with an additional learning resource and to help them take full advantage of the abundant information available on the internet. The site, prepared by Professor William A. Brindley of Utah State University, contains **Web Exercises** for each of the chapters and also **Evolution Links**, a compilation of informative links relating to the study of evolution. Evolution is a broad-ranging and interdisciplinary study. From history and theory to scientific study and fact, this web site helps students explore all aspects of evolution.

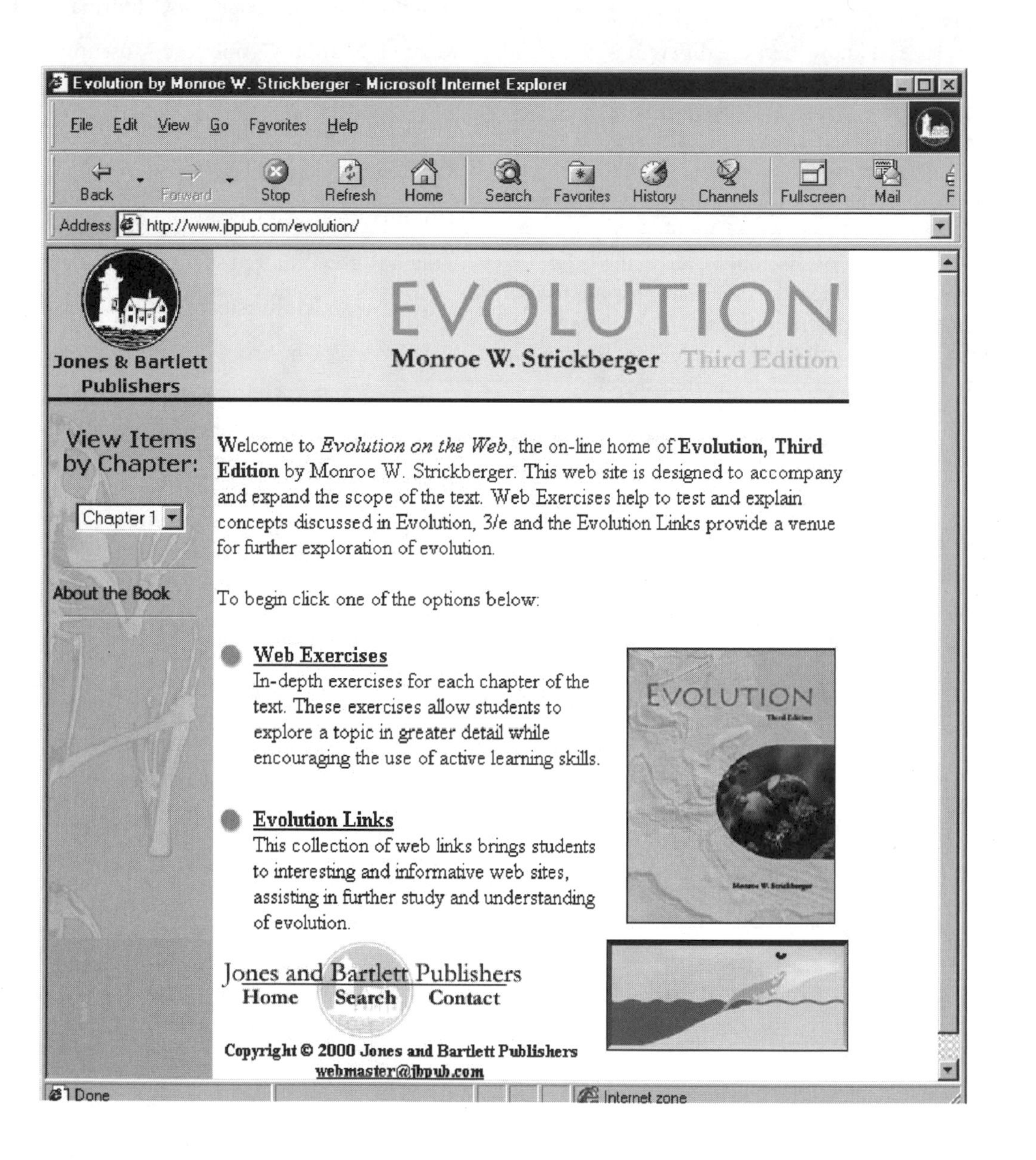

I
The Historical Framework

1 Before Darwin

Biological evolution entails inherited changes in populations of organisms, over time, that lead to differences among them. Essential to our present concept of evolution are the notions that a group of organisms is bound together by its common inheritance; that the past has been long enough for inherited changes to accumulate; and perhaps most essential of all, that discoverable natural events and relationships explain the phenomena of evolution. Although people studied and discussed each of these aspects at various times in human history, it is only since the published work of Charles Darwin in the late nineteenth century, that biological evolution became socially accepted. This acceptance was based on many changes in how people view the world and explain natural phenomena.

The purpose of this chapter and the three that follow is to review some of the underpinnings that enabled the modern Darwinian concept of evolution to unfold. For brevity and simplicity, various events and concepts are not elaborately developed. Fuller historical treatments can be found in books listed in the references, including those by Bowler, Depew and Weber, Desmond, Greene, Mayr, Richards, and Ruse.[1]

Idealism and the Species

Attempts to understand the world in a rational way—that is, by commonly accepted methods of thought and logic—began about the fifth century B.C. in Greece. Plato (428–348 B.C.), the philosopher who along with Aristotle (384–322 B.C.) had the greatest impact on Western thought, suggested that the observable world—our experience—is no more than a shadowy reflection of underlying "ideals" that are true and eternal for all time. Most things, according to Plato, were originally in the form of such eternal ideals, and any change represents disharmony. The Platonic goal for

[1]The term "evolution" actually has a seventeenth century embryological origin, defined as the "unfolding" of parts and organs in attaining a preformed body plan. It was only in the nineteenth century that people came to use evolution to mean the transformation of species.

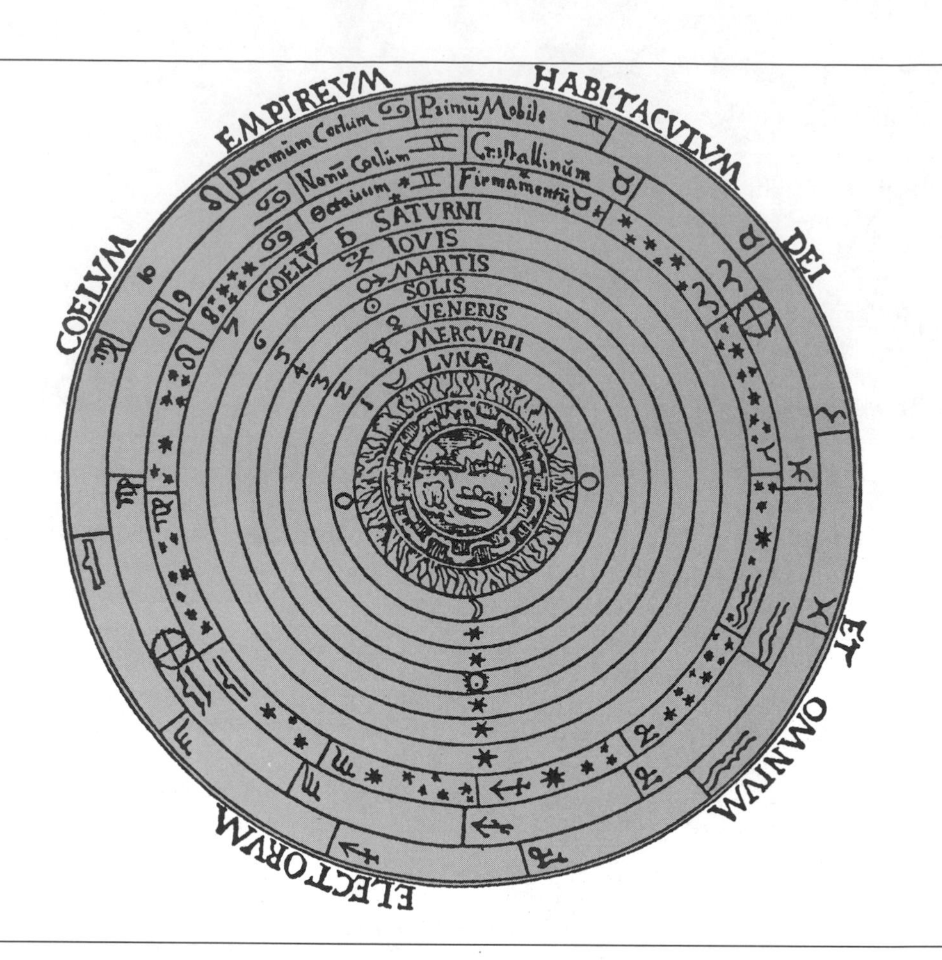

FIGURE 1–1 A medieval concept of the ten spheres of the universe with Earth and its four elements (earth, air, fire, water) at the center, according to Apian's *Cosmographia* (published 1539 in Antwerp). Surrounding Earth are transparent crystal spheres containing in succession the moon, Mercury, Venus, the sun, Mars, Jupiter, Saturn, the fixed stars, and spheres involved in the motion of the stars and of the entire universe ("Primu Mobile"). Beyond these spheres lies Heaven ("The Empire and Habitation of God and All the Elect").

human society was to analyze experience in order to understand and strive for ideal perfection. The notions of "perfect circles" to explain the motions of the heavenly bodies (Fig. 1–1), "perfect numbers" such as 6 (1 + 2 + 3) and 10 (1 + 2 + 3 + 4), and the four "elements" (earth, water, fire, and air) to which all matter could be reduced were among the results of the search for perfection.[2] What are the sources of such **idealism?**

To a large extent, idealism originates from our often used ability to abstract concepts from experience—to think, for example, of "cat" rather than one particular animal of specific size and head shape, with claws, tail, fur, and so on. Such abstraction lets us generalize our experience, to differentiate between cat and tiger, to pet the cat and run away from the tiger, and to communicate these general concepts or universals to others through our symbolic language. Despite these advantages, however, generalizations are not always reliable because our experiences may modify the generalizations: not all cats or tigers are the same.

In fact, the struggle between generalization and particularization is continual. Only by generalizing can we conceive of regularity in nature and thereby consciously adapt to its needs, and only by particularizing can we contact and observe reality. No sooner do we conceive of some new generality than we often discover further details and may thereby be forced to modify our original conception. Experience stresses continual change, and generalization stresses stability. That few of the Greek thinkers, with the notable exception of Heraclitus (540–475 B.C.), tried to incorporate change into their philosophies may indicate that the stability generalization confers is one of the prevailing comforts and prejudices of human thought.

Unfortunately, Plato and his successors assumed that only ideal generalizations are real while all else is a shadowy illusion.[3] From our present viewpoint, reality is not

[2]Variations on this theme were not uncommon. To the four elements Empedocles (c. 490–430 B.C.) added two active principles: love (which binds elements together) and hate (which separates them). An additional element, the "quintessence," was presumed by Aristotle and others to be the component of heavenly bodies. In respect to mystical numbers, Oken (1779–1851), one of the German Natural Philosophers, proposed that the highest mathematical idea is zero, and God, or the "primal idea," is therefore zero.

[3]In Plato's famous parable of the cave (in his dialogue *The Republic*), humans deprived of philosophy are pictured as cave-confined prisoners facing a wall on which are displayed their own shifting, distorted shadows and those of objects behind them that they cannot directly see; their chains prevent them from turning their heads toward the light. As a result, they interpret the deformed, shadowy aberrations they observe as reality, while the actual unchanging figures and objects are the true

so narrowly defined but represents all the interactions of the universe. Thus it is true that different cats are imperfect reflections of our universal concept of "cat," but these pluralities are not imperfect reflections of reality: they are the realities that furnish and allow the generalization. Without "cats," there is no "cat"! Biologically, this viewpoint may extend to different groups of individuals in that they can interact as a group (for example, as a population, race, or as a natural unit) with other elements of reality. For example, cats have common features in the way they interact with prey and predators. The dilemma for biologists traditionally has been to recognize the reality of differences among members of a group and yet to recognize the reality of the group itself. Idealism offered practically no means of reconciling these two aspects of reality.

Aside from its intellectual roots, an important source for Platonic idealism can be found in the underlying social structure from which it arose. The ethic of Pythagoras (c. 570–500 B.C.), which in many ways gave rise to Platonic idealism, considered the most exalted state of citizenship that of the philosopher-spectator who does not partake in activity but only contemplates it in order to understand it.[4] This contemplative ideal must have derived, at least partially, from ancient social inequities in which the upper classes maintained their position by exploiting the activities of slaves or social classes deemed inferior.

For example, the model for Plato's ideal republic with its philosophers-statesmen-guardians is often suspected to have been the city-state of Sparta with its mostly idle warrior-rulers supported by serfs, or helots. Even Athens, which pretended to democracy, restricted political participation to only a small percent of its population, since the disenfranchised majority was primarily composed of women, slaves, and foreign-born. To be a Platonic idealist meant, therefore, to live in a world of cruel exploitation in which a large portion of reality may have been distasteful but was either ignored or accepted without serious question. In later periods, especially during the rigidly structured feudalism of medieval Europe, idealistic philosophies bolstered the concepts of idealized social classes and a perfectly ordered society (as had long been the case in the caste system of India), to help maintain the status quo.

Whatever its sources and sustenance and the guises under which it is hidden, idealism has been a persistent and pervasive philosophy and has had pronounced effects on biology and the study of evolution. To Plato, the form of a structure could be understood from its function, since the function dictated the form. Thus, Plato believed that the form of the universe derived from its function of goodness and harmony imposed by an external creator. Aristotle, who may be regarded as the founder of biology (among other sciences) modified this notion to accommodate the development of organisms, pointing out that the last stage of development, the adult form, explains the changes that occur in the immature forms. This type of explanation is called **teleological** because the adult represents the "telos," or final goal, of the embryo.

To many later thinkers, teleology became associated with Platonic mystical processes by which advanced stages, in some unknown manner, influenced and affected earlier stages. Thus, because ideals implied conscious creation, it seemed as though organs and organisms were designed for some special purpose and each species was created as an ideal in anticipation of its future use. Pliny the Elder (A.D. 23–79) carried this notion to the point of claiming that all species were created for the benefit of man. Some two hundred years later, Lactantius (c. A.D. 260–340) wrote, "Why should anyone suppose that, in the contrivance of animals, God did not foresee what things were living, before giving life itself?" This view helped cast the teleological origin of species more permanently into the religious form it took in Christian Europe throughout the Middle Ages until Darwin's time.[5]

"essences" or "forms." In another of his dialogues (*Timaeus*, see Flew), Plato writes:

> . . . we must agree that one category [idealism] involves Form existing in its own right, without beginning and without end . . . unseen and unsensed in other ways; for this is the province of rational knowledge. The second province [reality] is one in which things have the same names as, and resemble, things in the first—the sphere of the sensible, of the generated, of what is always being moved about, of things originating in one place and then perishing in another; and this is accessible to opinion aided by perception. . . .

To Plato, "opinion" is the reflection of reality through the senses, and is faulty and unreliable; only "knowledge," the philosophical perception of the ideal, provides the truth. In a later dialogue (*Parmenides*), Plato elaborates on these concepts, and also questions them.

[4]One formulation of the Pythagorean ethic is quoted by Russell:

> In this life, there are three kinds of men, just as there are three sorts of people who come to the Olympic Games. The lowest class is made up of those who come to buy and sell, the next above are those who compete. Best of all, however, are those who come simply to look on. The greatest purification of all is, therefore, disinterested science, and it is the man who devotes himself to that, the true philosopher, who has most effectually released himself from the "wheel of birth."

[5]In his *Summa Theologica*, St. Thomas Aquinas, the prominent thirteenth century Christian theologian, wrote:

> Whatever lacks knowledge cannot move towards an end, unless it be directed by some being endowed with knowledge and intelligence; as the arrow is directed by the archer. Therefore some intelligent being exists by whom all natural things are directed to their end; and this being we call God.

Five centuries later, the Swedish taxonomist Linnaeus (p. 10) extended teleology even to science:

> If the Maker has furnished this globe, like a museum, with the most admirable proofs of his wisdom and power; if this splendid theater would be adorned in vain without a spectator; and if man the most perfect of all his works is alone capable of considering the wonderful economy of the whole; it follows that man is made for the purpose of studying the Creator's work that he may observe in them the evident marks of divine wisdom. (Linnaeus, 1754, *Reflections on the study of nature.*)

ERNST MAYR

Birthday:
July 5, 1904

Birthplace:
Germany

Present position:
Professor Emeritus
Harvard University
Cambridge, Massachusetts

WHAT PROMPTED YOUR INITIAL INTEREST IN EVOLUTION?

I was a born naturalist, roaming the fields and woods ever since I was a small boy. What most attracted me was the immense diversity of life. Why are there so many different kinds of species? Later on I asked, How do they originate? I looked for the solution on expeditions to New Guinea and the Solomon Islands in the 1920s.

WHAT DO YOU THINK HAS BEEN MOST VALUABLE OR INTERESTING AMONG THE DISCOVERIES YOU HAVE MADE IN SCIENCE?

Science advances both by discoveries and by the introduction of new concepts. It is to the latter that I made most of my contributions:

- The biological species concept
- The concept of sibling species
- The importance of geographic speciation
- Speciation through founder populations
- The origin of evolutionary novelties
- Species turnover on islands
- The relation between population size and evolution
- The holistic concept of the genotype
- Individuals and social groups are the targets of selection, not genes

My most interesting discovery was the great speedup of evolutionary rate in small populations isolated beyond the previous species borders.

WHAT AREAS OF RESEARCH ARE YOU (OR YOUR LABORATORY) PRESENTLY ENGAGED IN?

I am now engaged in exploring the philosophical consequences of the discovery of new evolutionary principles.

IN WHICH DIRECTIONS DO YOU THINK FUTURE WORK IN YOUR FIELD NEEDS TO BE DONE?

The internal structure of the genotype, the workings of the central nervous system, and the interaction of species in the ecosystem are now the most exciting frontiers of biology.

WHAT ADVICE WOULD YOU OFFER TO STUDENTS WHO ARE INTERESTED IN A CAREER IN YOUR FIELD OF EVOLUTION?

My advice to beginners is to become thoroughly familiar with one particular group of organisms in order to be able to test any new ideas or theories against the background of that set of solid facts. Pure speculation and model building without a solid factual basis rarely leads to sound advances.

The Great Chain of Being

Through idealism the concept of a species became strongly tied to its use in explaining the divine origin and design of nature. Plato had defined the species as representing the initial mold for all later replicates of that species: "The Deity wishing to make this world like the fairest and most perfect of intelligible beings, framed one visible living being containing within itself all other living beings of like nature." Aristotle expanded this view to a chainlike series of forms, each form representing a link in the progression from most imperfect to most perfect (Fig. 1–2). He called this the Scale of Nature, a concept that continued long in the history of European thought and merged with other ideas into the **Ladder of Nature** and the **Great Chain of Being.**

Philosophically satisfying as it was, the concept of the Great Chain of Being did not necessarily put humans on the highest, or even near the highest, rung of the Ladder of Nature. Many who contemplated the innumerable steps between humans and perfection (God) felt the despair of occupying a relatively lowly position and only consoled themselves with the thought that there were even more lowly organisms. However, even such consolations were unable to quell troubled feelings about a concept that suggested that the evils of nature are also part of the universal fabric and that the special divine creation of everything might allow nothing to change. Nevertheless, despite its discomforts, the Great Chain of Being was generally accepted well into the eighteenth century.

In Germany this notion was fostered by Herder (1744–1803) and soon adopted by Goethe (1749–1832) and others of the Natural Philosophy (Naturphilosophie) school who tied it in strongly with an idealistic concept of biological forms. According to Goethe, the creation of each level of organisms was based on a fundamental primitive plan—an **archetype** or ***Bauplan.*** Goethe conceived the morphology of plants, for example, to be founded on an "Urpflanze," or original plant,

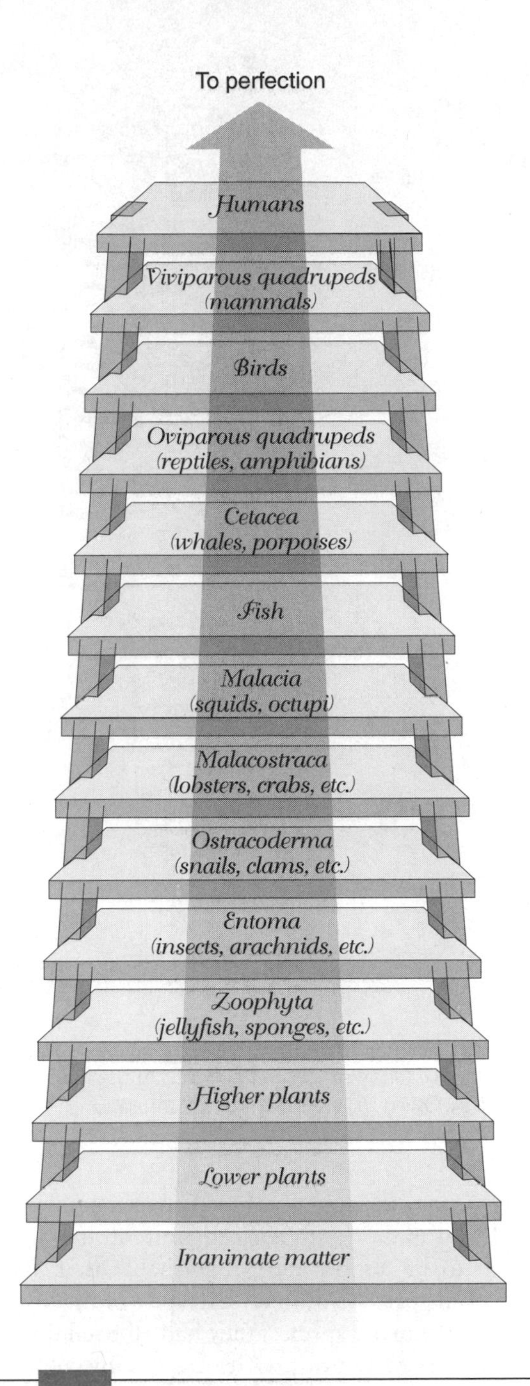

FIGURE 1-2 Aristotle's Scale of Nature. (*Adapted from descriptions in Guyénot.*)

that had only one main organ, the leaf, from which the stem, root, and flower parts derived as variations (Fig. 1–3*a*). Similarly, the bones of the skull were supposed to be merely modifications of the vertebrae of an animal archetype, or "Urskeleton," composed of only vertebrae and ribs (Fig. 1–3*b*).

To most of its exponents, the Ladder of Nature had the comforting quality of stressing a precisely ordered regularity of relationships among organisms and could also be used to support and justify the prevailing social and political orders. As expressed by Soame Jenyns (1757):

> The universe resembles a large and well-regulated family, in which all the officers and servants, and even the domestic animals, are subservient to each other in a proper subordination; each enjoys the privileges and perquisites peculiar to his place, and at the same time contributes, by that just subordination, to the magnificence and happiness of the whole.

Among the relatively few who at first disputed this concept, Voltaire (1694–1778) incisively pointed to its earthly model:

> This hierarchy pleases those good folks who fancy they see in it the Pope and his cardinals followed by archbishops and bishops; after whom come the curates, the vicars, the simple priests, the deacons, the subdeacons; then the monks appear, and the line is ended by the Capuchins.

Voltaire also addressed the question of the many observed gaps among species, an observation that did not seem to be in accord with the expected innumerable steps in the continuous progression from imperfect to perfect. He proposed that although there were no living species to fill these gaps, such gaps were real, perhaps caused by the extinction of species. In this respect Voltaire essentially echoed the thoughts of the philosophers Descartes (1596–1650) and Leibniz (1646–1716). Leibniz had even proposed evolutionary changes to account for these gaps, suggesting that many species had become extinct, others had become transformed, and different species that presently share common features may at one time have been a single race.

To Leibniz, evolution of species was tied in with the perfection toward which the universe continually progressed, and his philosophy thus represented a major shift from a perfectly created universe to one in the process of becoming perfect. Progress toward the perfection of species was also expressed by biologists such as Bonnet (1720–1793), who maintained that the development of any organism from its "seed" was an unfolding of a preconceived plan inherent in the seeds of previous generations.[6] This notion of progress was thus fitted into a teleological framework—that it was "necessary" and directed toward some particular end.

[6]Bonnet predicted that in the future, humans might reach the level of angels and animals might reach the level of humans (see Lovejoy):

> Man—who will then have been transported to another dwelling place more suitable to the superiority of his faculties—will leave to the monkey or the elephant that primacy which he, at present, holds among the animals of our planet. In this universal restoration of animals, there may be found a Leibniz or a Newton among the monkeys or the elephants, a Perrault or a Vauban among the beavers.

Although this concept may seem quite advanced and evolutionary for the period, we should also keep in mind that Bonnet conceived of this process as a perfectibility of souls—that is, a series of progressive reincarnations.

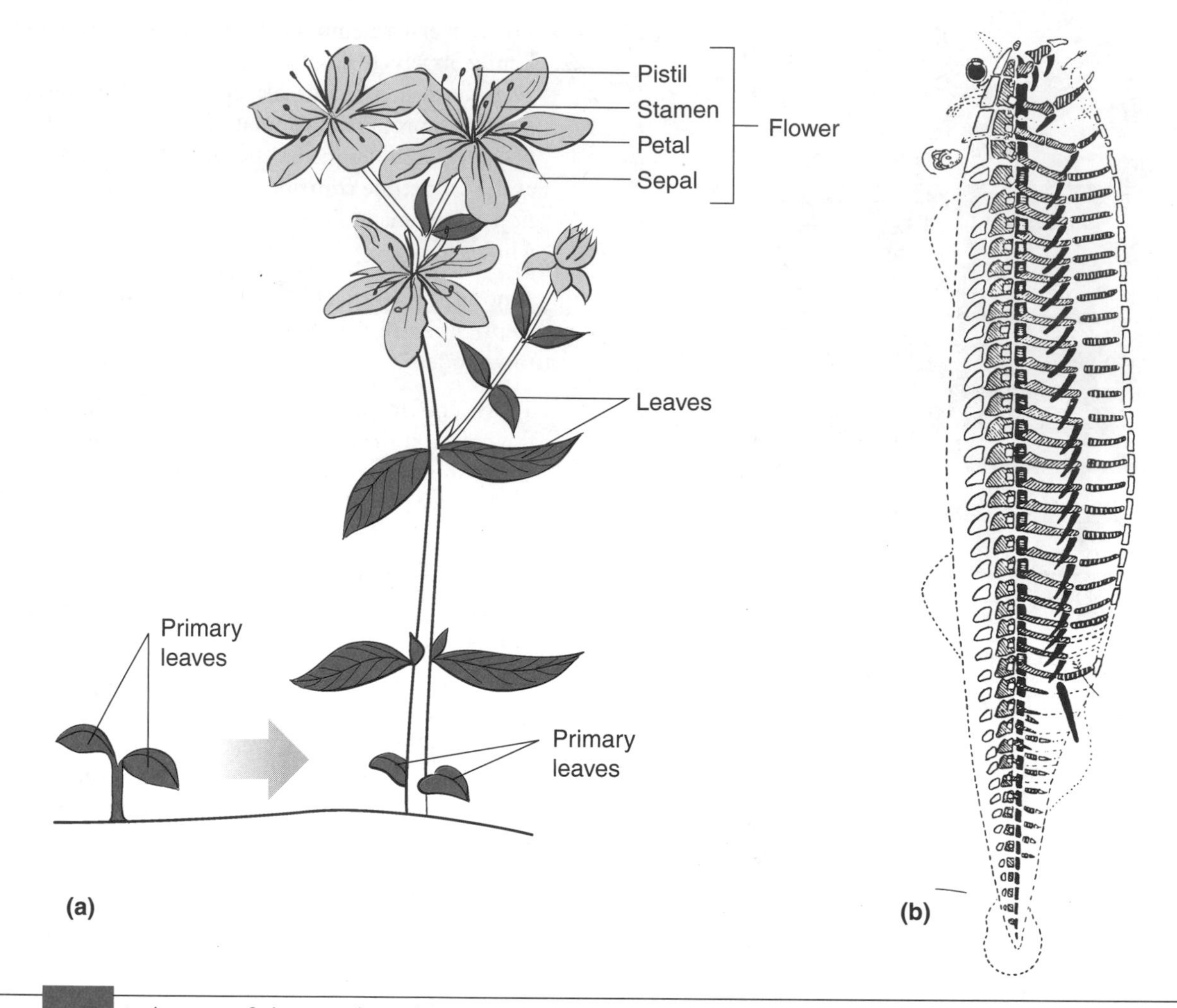

FIGURE 1–3 Archetypes of plants and vertebrate animals. The idealized plant (*a*) shows Goethe's concept of the derivation of all plant parts from the leaf. The segments in the vertebrate skeleton pictured by Owen (*b*) are alike from cranium to tail. (*Adapted from Wardlaw; Owen.*)

Along with other changes in thought during the eighteenth century, these evolutionary forebodings were probably associated with some of the major changes then being undergone by society. That is, the progressive weakening of feudalism, which had begun in the fourteenth century with the rise of commerce and the new power of the merchant classes, was now accelerating because of rapid advances in technology and the Industrial Revolution. The old, rigid, land-based class structures were breaking up, and both social institutions and the ideas expressed by many thinkers were becoming more mobile and flexible.

The Great Chain of Being also had important effects on plant and animal classification, which derived partly from the search for that multitude of living organisms that many felt would be found to occupy all the various rungs of the Ladder of Nature. There were proposals that even humans could be linked to other species through the "wild man" (orangutan), which, according to some writers was of the human species (Fig. 1–4). Other authors thought the link between humans and animals was in the South African Hottentots, who were believed to be almost indistinguishable in reasoning power from apes and monkeys. In spite of the observed gaps between many species, they had all been linked by the principle of continuity, expressed by Leibniz as "Nature makes no leaps." Although not espousing the evolution of species as such, the philosopher Kant expressed this same idea as "the principle of affinity of all concepts, which requires continuous transition from every species to every other species by a gradual increase of diversity."

Thus, in spite of its idealistic nature, the Great Chain of Being led almost directly to the idea that the perfection of organisms may demand multiple intermediary stages. By the eighteenth century the basic concept of evolution, the actual transformation of one species into another, can be said to have been merely awaiting the philosophical acceptance of actual change between the innumerable steps in the Great Chain of Being.

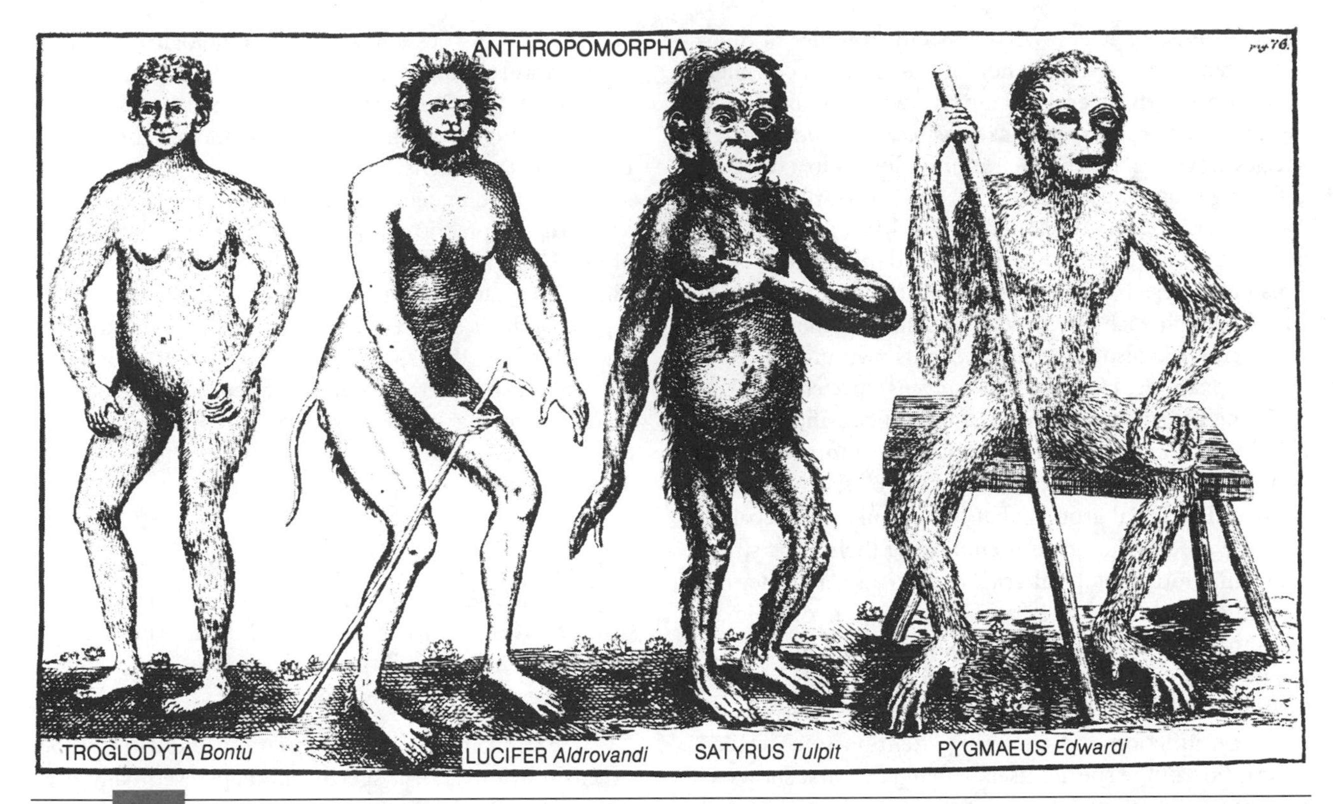

FIGURE 1-4 Presumed "missing links" between apes and humans in the Ladder of Nature. These individuals received binomial species designations, and Linnaeus made attempts to place them in his *Systema Naturae.* This figure is reproduced from an eighteenth century work by Linnaeus's student, C. E. Hoppius, who also noted the close similarity between humans and apes, "So near are some among the genera of Men and Apes as to structure of body: face, ears, mouth, teeth, hands, breasts; food imitation, gestures, especially in those species which walk erect and are properly called Anthropomorpha, so that marks sufficient for the genera are found with great difficulty." Social institutions, however, often greeted such conjectured relationships with horror or derision: the 1770 suggestion by DeLisle de Sales that the orangutan was the human ancestor led to a prison sentence.

The Origin of Systematics

From the biological viewpoint, however, considerable difficulties still existed in respect to how species were to be defined and classified, that is, distinguished one from the other, and placed into groups that reflected their most significant features. As discussed in Chapter 11, without a rational system of classification, evolutionary relationships between most species would probably have been impossible to establish. But the recognition of the biological importance of species took considerable time. During the Middle Ages of Europe, species were generally collected and described on the basis of their culinary or medical properties. When the expansion of worldwide exploration and trade occurred in the sixteenth and seventeenth centuries, the discovery of many new species of plants and animals greatly increased the problems of classifying them. For example, Moufet (1553–1604), attempting to describe grasshoppers and locusts, writes:

> Some are green, some black, some blue. Some fly with one pair of wings, others with more; those that have no wings they leap, those that cannot either fly or leap, they walk; some have longer shanks, some shorter. Some there are that sing, others are silent. And as there are many kinds of them in nature, so their names were almost infinite, which through the neglect of naturalists are grown out of use.[7]

Early attempts at classification were usually made in Aristotelian fashion by postulating a broad category (for example, "substance") and then subdividing this into subsidiary categories (for example, "body," "animal") until an individual species could be placed into a particular

---

[7]Plants also were not exempt from difficulties in classification. In Al-Dinawari's (820–895) *Book of Plants,* whose fame lasted through the Middle Ages:

> Plants are divided into three groups: in one, root and stem survive the winter; in the second the winter kills the stem, but the root survives and the plant develops anew from this surviving rootstock; in the third group both root and stem are killed by the winter, and the new plant develops from seeds scattered in the earth. All plants may also be arranged in three other groups: some rise without help in one stem, others rise also but need the help of some object to climb, whilst the plants of the third group do not rise above the soil, but creep along its surface and spread upon it.

subdivision. Linnaeus (1707–1778), the founder of modern systematics, used a method of classification considerably more advanced, beginning with as precise a description of each species as possible. He then grouped species closely related by their morphology into a category called a **genus** (plural **genera**) [as had been foreshadowed more than a century earlier by Bauhin (1560–1624)], grouped related genera into **orders,** and these into **classes.** This helped establish the system of **binomial nomenclature** in which each species name defines its membership in a genus and also provides it with its own unique identity, for example, *Homo* (genus) *sapiens* (species).

Using the species as the basic unit of classification enabled Linnaeus to arrive at groupings far more "natural" in their interrelationships than many of the previously proposed artificial groups: that is, he could now separate or unite species into groups employing their basic structural and morphological features. To use a somewhat simplified example, one early classification of animals was into those that can fly and those that cannot fly. Flying fish were therefore considered to be hybrids between birds and fish. However, by ignoring these "ideal" classes based on function and confining attention to a detailed description of the species itself, a flying fish first shows its fishlike relationships and then the change in its fins that enables it to glide. Therefore, except for those patterns all vertebrate groups share, there are obviously no special birdlike structures in such fish at all. Linnaeus's contribution to classification was thus an essential step leading to the discovery of natural evolutionary relationships among organisms.

For much of his career, however, Linnaeus conceived of the species as a fixed entity, deriving his concept essentially from John Ray (1627–1705) who defined a species on the basis of its common descent: "The specific identity of the bull and the cow, of the man and the woman, originate from the fact that they are born of the same parents." Ray had therefore attempted to separate different species on the basis of whether they could be traced to different ancestors: "A species is never born from the seed of another species." Thus, a species, with only rare exceptions, could never change, and its ultimate ancestor could only be divinely created. Linnaeus essentially adopted this view, with the proviso that varieties within a species may show considerable nonheritable differences among themselves.

Under Linnaeus the art of systematics developed rapidly, and many species were described mainly on the basis of their reproductive parts and classified into groupings still valid today. Generally, however, classification was almost always based on appearance and not on observations of ancestry, since the classifiers (taxonomists) usually described preserved specimens whose natural behavior and origins were often unknown. In accord with idealist concepts, each species was believed to possess a unique "essence" that determined all its specific characters. This "essentialist" or "typological" view of species was reinforced by taxonomists who practiced the method of depositing "type" specimens in museums or herbaria, which were then used as the models for classifying further specimens.[8]

Although Linnaeus placed special emphasis on the species as the practical unit of classification, it was Buffon (1707–1788) who codified the notion that species are the only biological units that have a natural existence ("Les espèces sont les seuls êtres de la nature."). Buffon introduced the idea that species distinctions should be made on the basis of whether there were reproductive barriers to crossbreeding between groups ("reproductive isolation") as indicated by whether fertile or sterile hybrids were produced:

> We should regard two animals as belonging to the same species if, by means of copulation, they can perpetuate themselves and preserve the likeness of the species; and we should regard them as belonging to different species if they are incapable of producing progeny by the same means.

To Buffon, considerable variation could occur between individuals of a species, perhaps eventually even producing completely new varieties through time (for example, different kinds of dogs). However, despite such variation, a species itself remained permanently distinguished from other species, although at times Buffon seemed to indicate the possibility that significant species changes could occur.[9]

Strangely enough, the eighteenth century barrier to the acceptance of evolution seemed to rest mostly on the reality of species. If species were real, then they seemed inevitably fixed. How then could new species arise? Buffon, who had proposed evolutionary events on both the cosmological and geological levels, had, at the same time, actually established three basic arguments against

---

[8]Mayr (1976) points out, "The ultimate conclusions of the population thinker and of the typologist are precisely the opposite. For the typologist the type (*eidos*) is real and the variation an illusion, while for the populationist, the type (average) is an abstraction and only the variation is real. No two ways of looking at nature could be more different."

[9]In the fourth volume of his *Natural History* (1753), Buffon wrote:

> Not only the ass and the horse, but also man, the apes, the quadruped, and all the animals, might be regarded as constituting but a single family. . . . If it were admitted that the ass is of the family of the horse, and differs from the horse only because it has varied from the original form, one could equally well say that the ape is of the family of man, that he is degenerate man, that man and ape have a common origin; that, in fact, all the families, among plants as well as animals, have come from a single stock, and that all animals are descended from a single animal, from which have sprung in the course of time, as a result of progress or of degeneration, all the other races of animals.

However, in spite of this clear statement of an evolutionary view, Buffon felt forced to reject it because it is contrary to religion (". . . all animals have participated equally in the grace of direct creation . . .") and because of the further arguments he offered.

biological evolution which were used by antievolutionists well into the nineteenth century:

- New species have not appeared during recorded history.
- Although matings between different species lead to the inviability or sterility of hybrids, this mechanism could certainly not apply to matings between individuals of the same species. How then could individuals of a single species be separated from others of the same kind and become transformed into a new species?
- Where are all the missing links between existing species if transformation from one to the other has taken place? (Numerous missing links had been imagined—refer back to Fig. 1–4—but none had actually been found.)

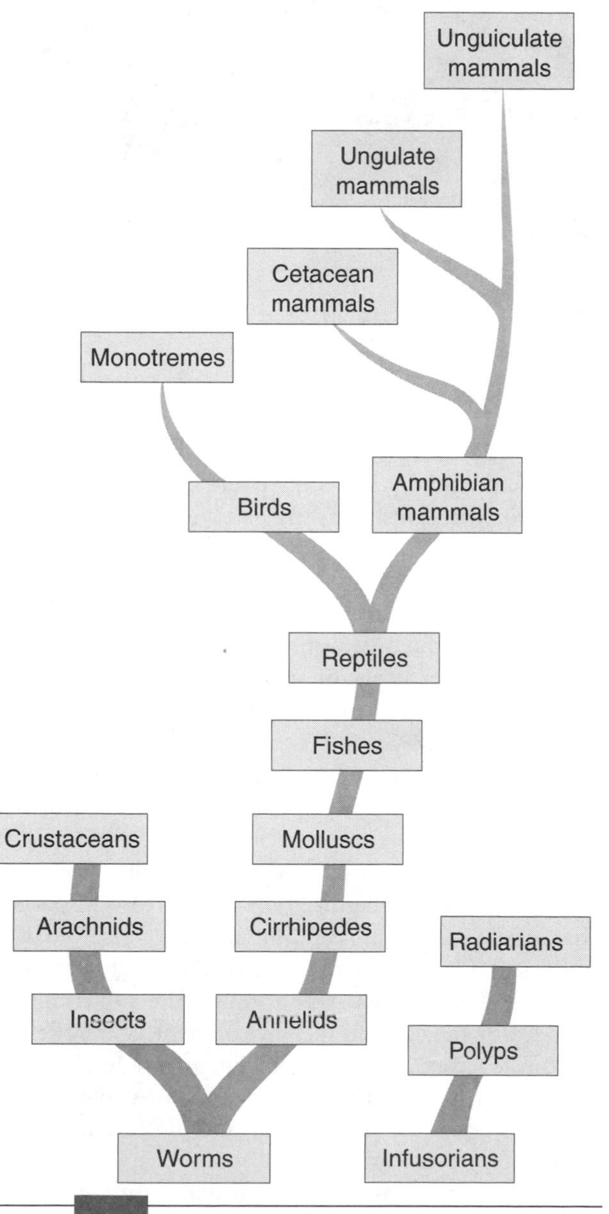

FIGURE 1–5 Evolutionary relationships among animals according to Lamarck.

Because these arguments were not refuted until after Darwin, it is no surprise that one of the first serious pre-Darwinian proponents of biological evolution, Jean-Baptiste de Lamarck (1744–1829), felt that one must do away with the reality of species to establish the possibility of evolution. Lamarck proposed that all organisms are tied together by intermediate evolutionary forms and that species distinctions are artificial and arbitrary, although they may be helpful in classification. The observable gaps between species, genera, families, and so on, according to Lamarck, were only apparent, not real, since all intermediate forms existed someplace on earth as varieties or races, although they were not necessarily easy to discover.

Thus, although Lamarck shared the Great Chain of Being concept that species do not become extinct, he did not believe these forms were separately created but instead proposed that they had evolved from each other. In fact, his branching classification of animals (Fig. 1–5) introduced a direct challenge to the venerable doctrine of a Scale of Nature, which goes in only one direction, from imperfect to perfect: "In my opinion, the animal scale begins with at least two separate branches and . . . along its course, several ramifications seem to bring it to an end in specific places."

As discussed later (Chapter 2), the mechanisms Lamarck offered to account for these evolutionary changes were inadequate. However, even if Lamarck's explanations had seemed reasonable, a most serious impediment to evolutionary thought concerned the question of life itself: Is continuity between different generations of a species necessary at all?

Spontaneous Generation

Until perhaps the middle of the nineteenth century, it had been common to believe that although most large organisms reproduced by sexual means, smaller organisms could arise spontaneously from mud or organic matter. Some folklore suggested that larger organisms decomposed into smaller ones when they died, and there were even common legends that magical transitions could change a living member of one species into another (for example, werewolves). About 300 years ago, Van Helmont (1577–1644) offered a classic expression of spontaneous generation:

> If you press a piece of underwear soiled with sweat together with some wheat in an open mouth jar, after about 21 days the odor changes and the ferment, coming out of the underwear and penetrating through the husks of wheat, changes the wheat into mice. But what is more remarkable is that mice of

FIGURE 1-6 Pasteur's flasks used in demonstrating that spontaneous generation of bacteria does not occur even when nutrient broth is exposed to air. Because airborne bacteria are trapped in the lower bends of the curved necks of these flasks, the broth remains clear, and no fermentation occurs. Once the neck of a flask has been snipped off, bacteria enter directly into the broth, enabling them to multiply, thus causing fermentation. (*From Taylor.*)

> both sexes emerge, and these mice successfully reproduce with mice born naturally from parents. . . . But what is even more remarkable is that the mice which come out of the wheat and underwear are not small mice, not even miniature adults or aborted mice, but adult mice emerge!

Two serious and somewhat contradictory obstacles to the development of evolutionary concepts therefore prevailed almost simultaneously. The Linnaean contribution of species constancy helped raise the question of the origin of species, but, by insisting on species fixity, prevented consideration of any evolutionary transformations. Belief in spontaneous generation, in contrast, seemed contrary to species fixity, but at the same time cast doubt on any permanent continuity between organisms. If species could arise *de novo* at any time or be capriciously changed into other species, could there ever be a rational mechanism to explain their origin or the sequence of their appearance?

Fortunately, in the late seventeenth century, use of the experimental method had begun in biology, and a number of the new scientists were able to show that, at least for insects, spontaneous generation was not taking place. In 1668, Redi (1621–1697) demonstrated that maggots (larvae) arise only from the eggs laid by flies, and flies arise only from maggots. If meat is protected so that adult flies cannot lay their eggs, then maggots and flies are not produced. A year later, Swammerdam (1637–1680) showed that the insect larvae found in plant galls arise from eggs also laid by adult insects.

Within a century, further experiments demonstrated that even appearance of the microscopic "beasties" observed by Van Leeuwenhoek (1632–1723) in decaying or fermenting solutions and broth could be explained as deriving from previously existing particles. The Abbé Spallanzani (1729–1799) heated various types of broth in sealed containers and observed no growth of microscopic organisms. Only when the containers were open to airborne particles did organisms grow.

Although the theory of spontaneous generation was not generally abandoned until the crucial experiments of Pasteur (1822–1895) and Tyndall (1820–1893) in the nineteenth century (Fig. 1–6), serious attempts to replace it with a theory called **preformationism** were begun much earlier (Pinto-Correia). In the words of Swammerdam, preformation embodied the idea that "there is never generation in nature, only an increase in parts." That is, when each embryonic organism is conceived, it is preformed as a perfect replicate of the adult structure, and then gradually enlarges through the nourishment provided by the egg and the environment. Some preformationists proposed that this miniature adult is contributed by the maternal egg (ovists) while others suggested that it was contained within the paternal seminal fluid (spermists or animalculists).

In its most extreme form, preformationism led to the ***emboitement*** (encasement) theory espoused by Bonnet and others in which the initial member of a species encapsulates within it the preformed "germs" of all future generations. As Bonnet's critics Dumas (1800–1884) and Prévost (1790–1850) pointed out, "it seems easier [for preformationists] to imagine a time when nature, as it were, labored and gave birth all at once to the whole of creation, present and future, than to imagine continual activity." Thus, although preformation had the satisfying quality of explaining the many different plans of organismic growth and disputed the idea of spontaneous generation, it led once again to the fixity of species and brought the question of the origin of species back to a mystical, unknowable creation.

By the nineteenth century, however, development of improved experimental techniques and microscopic observations led biologists to replace preformationism with the theory of **epigenesis.** According to epigenesis, an embryo develops by gradually differentiating uniform, undifferentiated tissues into organs that were not themselves present at conception. At first this differentiation of undifferentiated tissue was believed to occur because of mystical, nonphysical forces, such as Aristotle's suggestion of the contribution of "form" by the seminal fluid, or Harvey's (1578–1657) "*aura seminalis,*" or Wolff's (1738–1794) "*vis essentialis.*" These explanations were **vitalistic:** they ascribe to living beings a vital force that cannot be explained by any underlying physical or chemical principles. Fortunately, by the time of von Baer (1792–1876), the prevailing view of epigenesis had changed so that biologists could accept differentiation and growth as natural and as explainable a set of processes as any others. In addition, Wohler's (1800–1882) 1828 biochemical synthesis of an organic compound (urea), the first such extraorganismic synthesis, showed there was no mystical essence in organic molecules that had to be understood outside the laws of chemistry. Such ideas of rational biology helped cultivate the climate in which evolutionary concepts could develop further (Box 1–1).

Fossils

An essential basis for understanding evolutionary relationships among organisms of the past, and for appreciating their lengthy history, was a study of their fossil remains. People had long noted the fossilized bones of animals that did not resemble existing species, and had even found strange seashells in the most unlikely places, such as mountaintops. The ancient Greeks were aware of such **fossils,** and a number of ancient writers, including Herodotus (484–425 B.C.), suggested that they could be explained by changes in the positions of sea and land. To Aristotle, there was no question that these changes occurred over considerable periods of time:

> The whole vital process of the earth takes place so gradually and in periods of time which are so immense compared with the length of our life, that these changes are not observed; and before their course can be recorded from the beginning to end, whole nations perish and are destroyed.

However, with the ascendancy of Christianity in Europe, many influential church authorities measured the age of the world by the number of generations since Adam in the biblical book of Genesis, and calculated its origin no earlier than perhaps 4000 to 7000 B.C. (Dalrymple). Limited to such a relatively short period, fossils could hardly be ascribed to a long historical process, and they were therefore commonly called *lusi naturae,* or "jokes of nature." Serious consideration of fossils as representing the remains of real organisms began only after the medieval "Dark Ages" had ended.

The breakup of feudalism and the expansion of trade and exploration led, as we have seen, to a number of important changes and challenges. Foremost among these was the challenge posed to the Great Chain of Being concept by the discovery of fossils in exposed riverbanks, mines, and eroded surfaces. Among the many who engaged in fossil hunting was Thomas Jefferson (1743–1826), third president of the United States, who was a discoverer of the extinct clawed giant sloth *Megalonix jeffersoni* (Fig. 1–7), which he mistakenly thought to be a giant lion.

Did the fossils indicate possible errors in the plan of nature, causing some species to become extinct? Were there gaps in the Ladder of Nature caused by the loss of these extinct species? Jefferson, like many others who addressed themselves to these questions, proposed that these species were not truly extinct, only rare: "Such is the economy of nature, that no instance can be produced of her having permitted any one race of her animals to become extinct; of her having formed any link in her great works so weak to be broken." Other theories sought to explain fossils as caused by the Noachian flood described in Genesis or having purposely been implanted into the earth at the time of creation in order to test humanity's faith in religion.

Contrary arguments, proposing the reality of fossil species, were offered by Hooke (1635–1703) and Steno (1638–1686) and led to more naturalistic attempts to understand fossil origins. Such views helped place fossils in a historical sequence: when arranged by stratigraphic age (deeper strata signifying older age than superimposed strata), older fossils showed greater differences from modern species than later fossils (see Fig. 6–5), indicating changes over time.

Given the reality of fossils, one of the commonly held theories among biologists during the late 1700s and early

BOX 1–1 CULTURE, SCIENCE, AND PHILOSOPHY

It has become common to acknowledge that our views of the world are strongly influenced by the culture in which we grow up. That is, different cultures place different emphases on how people perceive various events and relationships and on how they explain these perceptions. Thus, in one culture death is a recycling of a person's spirit into another organismic form; another culture believes death is a state of reward and punishment for an individual's behavior; while still another culture regards death as the end of a person's existence.

What we consider science is also culturally dependent in the sense that large differences can exist between cultures as to whether or how we should apply scientific concepts to natural events. Explanations that many of us accept as scientific—analyses based on rational, understandable, nonmysterious principles and laws—others do not necessarily accept, or accept to varying degrees. Thus, some people consider that human behavior and interactions can be explained by natural processes; others believe these are predestined actions produced by one or more godlike creators; and still others propose events are determined by constellations of planets, stars, and phases of the moon that have mysterious powers and properties.

In general, nonliving phenomena have been considered more acceptable to scientific analyses in western European culture than matters that touch on life itself, or on human life in particular. For example, physics and chemistry were well established as sciences by the nineteenth century, whereas biology, especially evolutionary events, has continually been subjected to vitalistic interpretations by various religious groups even to this day. As discussed in Chapter 4, modern creationists hold strongly to religious concepts that the origin and evolution of life and its species are mystical and miraculous and cannot be analyzed or explained by scientific method. Gaining freedom from cultural constraints has therefore been more difficult for biology, and especially evolution, than for physics and chemistry.

However, not all objections to evolution are cultural or religious, although they may be prompted by such viewpoints. One philosophical criticism is that evolutionary explanations (hypotheses) cannot be tested and supported in the same fashion as hypotheses in physics and chemistry. The claim is made that since evolutionary studies deal with events that occurred in the past, which are generally impossible to repeat in a laboratory, evolutionary biology can never reach the status of a science such as physics and chemistry. Some critics even extend these arguments to geology and to astronomy, fields of study that also deal with the past and with matters on such a large scale that they too cannot be repeated experimentally in the laboratory.

Further objections to evolution are that many studies in this area cannot be properly evaluated by scientific method. That is, rejection or acceptance of a scientific hypothesis is generally based on whether events relating to ("testing") that hypothesis refute it or not. Hypotheses constructed so they can never be refuted ("falsified" according to the philosopher Karl Popper) are not considered scientific. Thus, concepts that invisible angels are responsible for the birth or death of an organism, or that God created the universe, are not scientific because any events that seem to conflict with such a concept can always be reinterpreted to support it ("Sperm-egg movements may seem random but are really guided by angels," "supernatural entities control apparent gravitational relationships among stars and galaxies," "any event is ultimately, although undetectably, caused by God's will," and so on). Some people claim that since evolutionary concepts derive from history, these concepts seem to be irrefutable and unscientific because unknown past events might always be recruited to support a hypothesis.

Nevertheless, crucial as these philosophical objections appear, they have not much influenced the practice of evolutionists. Like studies in geology and astronomy, biologists continue to undertake evolutionary investigations

1800s was called **catastrophism,** popularized largely by followers of Cuvier (1769–1832), the French comparative anatomist. According to catastrophism, the sharp discontinuities in the geological record—the stratifications of rocks, the layering of fossils, the transition from marine fossils to freshwater fossils—indicated sudden upheavals caused by catastrophes, glaciations, floods, and so on. Fossils were recognized as extinct species "whose place those which exist today have filled, perhaps to be themselves destroyed and replaced by others." To some of the upholders of the biblical account, catastrophism had the advantage of explaining at least some catastrophes as obvious departures from "natural" laws that could be ascribed to divine intervention. Some writers, such as Agassiz (1807–1873), conjectured that there may have been as many as 50 to 100 successive special divine creations. This approach justified both the prior existence of fossil species and the biblical flood and made it possible to conceive that all present organisms arose within the time span the Judeo-Christian Bible provides, although preceded by many geological ages.

In contrast to Cuvier's catastrophist position, Lamarck (p. 11) proposed that the geological discontinuities represented gradual changes in the environment and

and continue to propose hypotheses. Part of the reason for this attitude is simply the profound recognition by "curious" humans that the past has influenced the present, and that an understanding of the past is a highly desirable and satisfying goal, whether the methodology is philosophically correct or not.

More to the philosophical point, however, is the realization that rationally we can explain historical events as we explain other "scientific" events if the explanations we use are consistent with observations. Thus, the sequence of hominid primatelike fossils extending from the far past to the present supports the hypothesis that humans have a primate origin (Chapter 20). Similarly, the correspondence in the amino acid sequences of myoglobin and hemoglobin supports the evolutionary relationship between them (Chapter 12). Since either hypothesis can be disproved by finding, for example, fossilized horselike or birdlike hominid ancestors, such as "centaurs" or "gargoyles," or noting an absence of any amino acid sequence similarities between myoglobin and hemoglobin, these hypotheses bear the earmarks of science without requiring laboratory repetition. On the whole, therefore, "historical" sciences concerned with the past, such as astronomy, geology, and evolution, can make use of observations to refute or support hypotheses.[10]

Of special interest in historical sciences is the emphasis on understanding a particular sequence of events rather than on primarily discovering general laws such as those of physics and chemistry. For example, subjects of historical sciences are events that led to our solar system, to the separation of South America from Africa, and to the origin of humans; recognizing at the same time that many of these events are singular and do not apply to all stars, or to all continental separations, or to the evolution of all species. Thus although historical sciences make use of general laws such as those of gravity, mechanics, and biochemistry, their aim is to discover the causes of diversity and uniqueness rather than to discover causes that apply uniformly to all matter.

[10]According to Popper (1973), the process by which reliable knowledge is attained—"conjecture and refutation"—follows a sequence similar to evolution by natural selection: ". . . our knowledge consists, at every moment, of those hypotheses which have shown their (comparative) fitness by surviving so far in their struggle for existence; a competitive struggle which eliminates those hypotheses which are unfit." The crucial issue in the survival of hypotheses is "testing," which permits their rejection or acceptance, and leads to further proposals and refinements.

Put more simply, the properties of different hydrogen atoms, for example, can be explained on a common physical basis, whereas the properties of different organisms—the organization and function of their component parts—can only be explained by their organismic histories. There is an obvious distinction between attributes such as temperature (or entropy) that have "the same meaning for all physical systems," from a biological attribute such as fitness that, "although measured by a uniform method, is qualitatively different for every different organism" (Fisher).

Nevertheless, when historical conditions are repeated, and different organisms are subjected to similar selective evolutionary forces, some common features can be predicted. For example, animals adapted to cave conditions generally show loss of pigment, rudimentary eyes, and improved chemosensory organs, among other common attributes (Culver).

For our present purposes, an appropriate way to deal with evolutionary biology is to ask, "What happened?" and "Why did it happen this way?" and to seek rational explanations that observations can support.

Books by Hull, Mayr (1997), Rosenberg, and Sober discuss these concerns and many others relevant to the philosophy of biology.

climate to which species were exposed, and through effects on organisms these changes led to species transformation. This **uniformitarian** concept, that the steady, uniform action of the forces of nature could account for the earth's features, had been foreshadowed by Buffon and others[11] and was strongly developed in the work of the geologist Hutton (1726–1797). Later, Lyell (1797–1875), a geologist and contemporary of Charles Darwin, offered the uniformitarian reply to catastrophism through the following arguments:

1. Sharp, catastrophic discontinuities are absent if geological strata are examined over widespread geographical areas. In fact, any widely distributed stratum often shows considerable regularity in its structure and composition. Only in specific localities do rapid shifts seem to appear and then because of local changes.
2. When changes occur in the geological record, they arise from the action of erosive but natural forces

[11]In physics, Isaac Newton had pointed out, "We are to admit no more causes of natural things than such are both true and sufficient to explain their appearance."

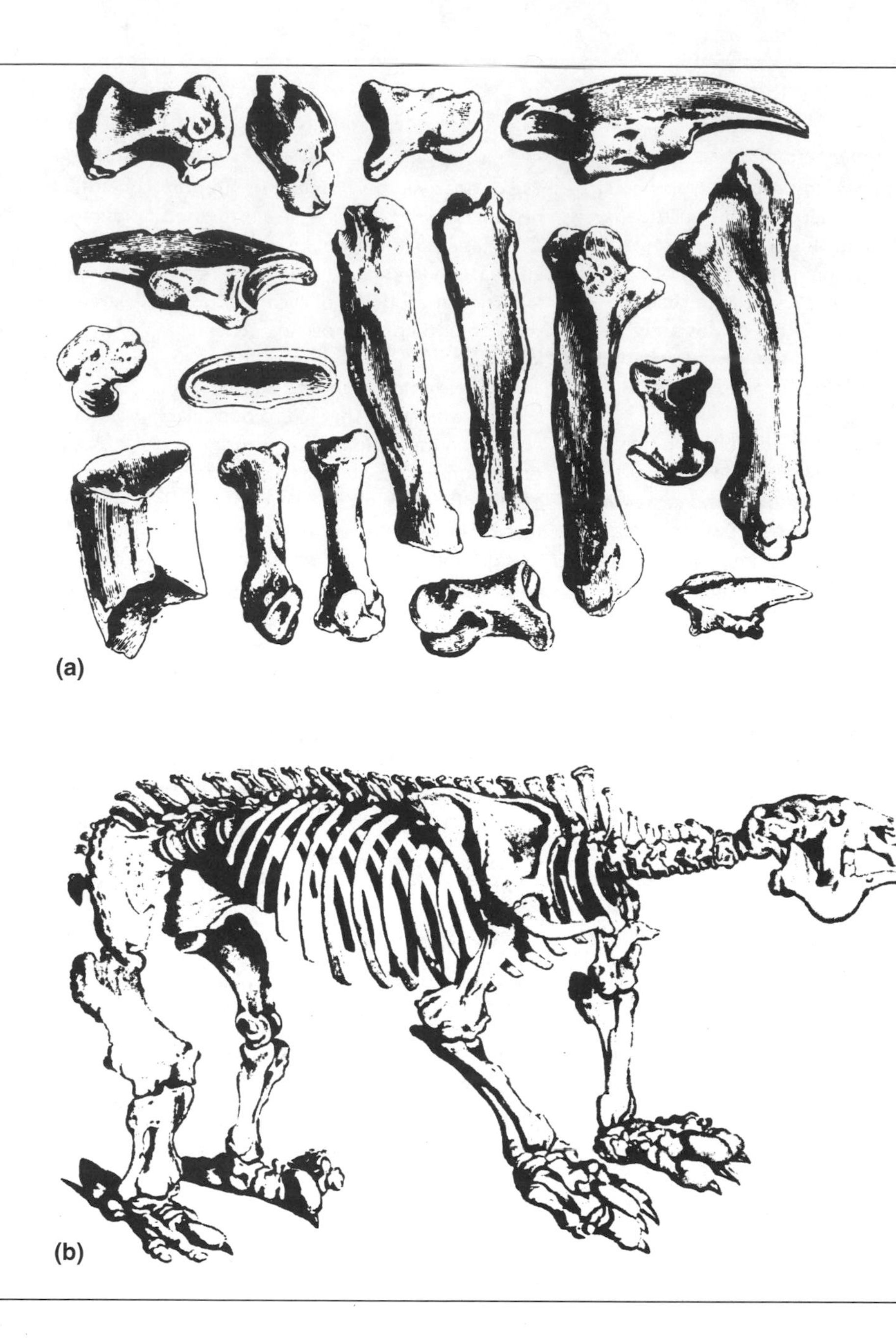

FIGURE 1–7 (*a*) Some of the bones of the extinct giant sloth, *Megalonix jeffersoni,* discovered in western Virginia in 1796. (*b*) Reconstruction by Cuvier of the skeleton of a similar extinct South American giant sloth, *Megatherium.* Both sloths were edentates, clawed mammals without cutting teeth. (*From Greene.*)

such as rain and wind, as well as from volcanic upthrusts and flood deposits. The laws of motion and gravity that govern natural events are constant through time. Thus, phenomena that occurred in the past were essentially caused by the same forces that produce phenomena at present, although the extent to which phenomena like volcanism occurred may have fluctuated in the past. This means that all natural causes for phenomena should be investigated before supernatural causes are used to explain them.

3. The earth must be very old for its many geological changes to have taken place by such gradual processes.

Thus, although uniformitarianism did not exclude sudden geological changes such as floods, volcanic eruptions, and meteorite impacts—events that were of common or recorded knowledge—it led to the view that even such "catastrophes" could be naturally caused and rationally explained.[12] The transition from catastrophism to

[12]The frontispiece of Lyell's *Principles of Geology* is a portrait of the three remaining columns of the ruined "Temple of Serapis" in Pozzuoli, Italy, showing that they had been historically subjected to both rise and fall in sea level. (A 9-foot section of these columns shows the borings of a molluskan bivalve, indicating that these columns had once been partially submerged.) He used this portrait through 12 editions of his book as an example of gradual geological change. Although as a strict uniformitarian Lyell did not elaborate further on this issue, we should recognize that

uniformitarianism had profound effects, because it helped liberate scientific thinking from the concept of a static universe powered by capricious, unexplainable changes to one that is perpetually dynamic and more historically understandable. In biology, it was Charles Darwin who first offered an acceptable explanation for historical changes among organisms and thereby helped tie all organisms together by a community of descent—evolution.

SUMMARY

Many intellectual threads led to the modern theory of evolution, which requires recognition that the earth is ancient, that there is a common inheritance within a biological group, and that natural events can be explained by discoverable natural laws. But it took a long time before these threads were brought together into an evolutionary concept.

Plato's idealistic concepts, according to which all natural phenomena are imperfect representations of the true essence of an ideal unseen world, was a long prevailing philosophy in western Europe that profoundly inhibited the development of evolutionary ideas. Because the world of essences is perfect, all change is illusory. Following Platonic ideas, Aristotle suggested that not only were species immutable but that there was a hierarchical order of species from most imperfect to most perfect, a concept refined over the centuries as the Great Chain of Being. This unchanging order remained unquestioned until inexplicable gaps in the chain of nature prompted philosophers such as Leibniz to propose that the universe was not perfect, only becoming so, and that it might go through successive intermediary stages on the pathway to perfection.

By the seventeenth and eighteenth centuries, new attention to living creatures and far-flung explorations led to an increasing interest in classifying organisms within the natural chain. Linnaeus revolutionized systematics by using the species as the basic unit and building his system from the species upward to larger taxonomic categories. The naturalist Buffon went farther, implying that the species is not just a category in classification but the only natural grouping. But he remained wedded to the Platonic ideal of a species as a "real" unit, thereby precluding change or the formation of new species. Lamarck evaded this problem by proposing that species are arbitrary, not "real" and that there could and must be forms intermediate between species.

The idea that organisms could arise from nonliving materials by spontaneous generation or that organisms did not change during development but were already "preformed" in their ancestors further hindered the development of evolutionary thought. Not until the nineteenth century was spontaneous generation finally disproved and the idea established that organisms develop epigenetically, by differentiating from undifferentiated tissues. At last biological phenomena became amenable to rational explanation.

The most severe blow to antievolutionary ideas was struck by the fossil record. The discovery of fossils of unknown types of organisms and the apparently inappropriate location of some fossils suggested that the surface of the earth and the organisms on it had existed for a long time. This, however, conflicted with the Judeo-Christian view of a recent origin, and fossil data were interpreted to accord with biblical catastrophes such as the Noachian flood or as "jokes of nature." Geologists asserted that fossil evidence was only explicable if the earth were indeed old and forces of nature had shaped its surface. Changes on the earth's surface would then have led to alterations in the organisms that lived on it, and these changes would be reflected in their fossil remains. Charles Lyell, a contemporary of Darwin, invalidated the idea that capricious catastrophic and miraculous events had influenced the geological structure of the earth, and so he helped establish the validity of a world that was comprehensible and rational.

KEY TERMS

archetype
binomial nomenclature
Bauplan
catastrophism
classes
emboitement
epigenesis
fossils
genera (plural of genus)
Great Chain of Being
idealism
Ladder of Nature
orders
preformationism
species
spontaneous generation
teleological
uniformitarian
vitalistic

DISCUSSION QUESTIONS

1. What is Platonic idealism?
2. Why did idealism become such an important approach to how people looked at nature?
3. What is the connection between idealism and the description and classification of organisms?
4. Is the concept of the Great Chain of Being (the Ladder of Nature) idealistic? Why?
5. Can the concept of biological evolution arise in people who believe firmly in idealism?

what is uniformitarian at one level can be considered catastrophic on another: the recorded evidence for the rise and fall in sea level that Lyell used to support uniformitarianism may well have been catastrophic to terrestrial organisms that became submerged.

6. Why has freedom from cultural constraints and prejudices been more difficult for evolutionary studies than for physics and chemistry?
7. If the concept of the spontaneous generation of species contradicts the concept that each species is created individually, why didn't people who believed in spontaneous generation become evolutionists?
8. What obstacles made it difficult for people to consider the reality of fossil species?
9. a. Is there a difference between the concepts of catastrophism and uniformitarianism in offering explanations for evolutionary changes?
   b. Can uniformitarianism be defined to include occasional catastrophic changes?

EVOLUTION ON THE WEB

Explore evolution on the web! Visit the accompanying web site for *Evolution,* 3/e at www.jbpub.com/evolution for web exercises and links relating to topics covered in this chapter.

REFERENCES

Bowler, P. J., 1984. *Evolution: The History of an Idea.* University of California Press, Berkeley.

Culver, D. C., 1982. *Cave Life: Evolution and Ecology.* Harvard University Press, Cambridge, MA.

Dalrymple, C. B., 1991. *The Age of the Earth.* Stanford University Press, Stanford, CA.

Depew, D. J., and B. H. Weber, 1995. *Darwinism Evolving: Systems Dynamics and the Genealogy of Natural Selection.* MIT Press, Cambridge, MA.

Desmond, A., 1982. *Archetypes and Ancestors: Palaeontology in Victorian London 1850–1875.* Blond and Briggs, London.

———, 1989. *The Politics of Evolution: Morphology, Medicine, and Reform in Radical London.* University of Chicago Press, Chicago.

Fisher, R. A., 1958. *The Genetical Theory of Natural Selection,* 2d ed. Dover, New York.

Flew, A., 1989. *Introduction to Western Philosophy: Ideas and Arguments from Plato to Popper.* Thames and Hudson, London.

Gasking, E., 1967. *Investigations into Generation: 1651–1828.* Hutchinson, London.

Glass, B., O. Temkin, and W. L. Straus, Jr. (eds.), 1959. *Forerunners of Darwin: 1745–1859.* Johns Hopkins University Press, Baltimore.

Greene, J. C., 1959. *The Death of Adam.* Iowa State University Press, Ames.

Guyénot, E., 1941. *Les Sciences de la Vie: L'Idée d'Evolution.* Albin Michel, Paris.

Hull, D. L., 1974. *Philosophy of Biological Science.* Prentice Hall, Englewood Cliffs, NJ.

Lamarck, J. B., 1809. *Zoological Philosophy.* Translated into English by H. Elliot, 1914, Macmillan, New York.

Lovejoy, A. O., 1936. *The Great Chain of Being.* Harvard University Press, Cambridge, MA.

Lyell, C., 1830ff. *Principles of Geology, Being an Attempt to Explain the Former Changes of the Earth's Surface by References to Causes Now in Operation.* J. Murray, London (many editions).

Mayr, E., 1976. *Evolution and the Diversity of Life: Selected Essays.* Harvard University Press, Cambridge, MA.

———, 1982. *The Growth of Biological Thought: Diversity, Evolution, and Inheritance.* Harvard University Press, Cambridge, MA.

———, E., 1997. *This Is Biology: The Science of the Living World.* Harvard University Press, Cambridge, MA.

Newton, I., 1687. *Principia Mathematica.* Reprinted 1965, University of Chicago Press, Chicago.

Nordenskiold, E., 1928. *The History of Biology.* Knopf, New York.

Owen, R., 1848. *On the Archetype and Homologies of the Vertebrate Skeleton.* Voorst, London.

Pinto-Correia, C., 1997. *The Ovary of Eve: Egg and Sperm and Preformation.* University of Chicago Press, Chicago.

Popper, K., 1959. *The Logic of Scientific Discovery.* Hutchinson, London.

———, 1973. *Objective Knowledge.* Oxford University Press, Oxford, England.

Richards, R. J., 1992. *The Meaning of Evolution: The Morphological Construction and Ideological Reconstruction of Darwin's Theory.* University of Chicago Press, Chicago.

Rosenberg, A., 1985. *The Structure of Biological Science.* Cambridge University Press, Cambridge, England.

Rudwick, M. J., 1972. *The Meaning of Fossils: Episodes in the History of Paleontology.* Macdonald, London.

Ruse, M., 1996. *From Monad to Man: The Concept of Progress in Evolutionary Biology.* Harvard University Press, Cambridge, MA.

Russell, B., 1945. *A History of Western Philosophy.* Simon & Schuster, New York.

Singer, C., 1959. *A History of Biology,* 3d ed. Abelard Schuman, London.

Sirks, M. J., and C. Zirkle, 1964. *The Evolution of Biology.* Ronald Press, New York.

Smith, C. U. M., 1976. *The Problem of Life.* Wiley, New York.

Sober, E., 1993. *Philosophy of Biology.* Westview Press, Boulder, CO.

Taylor, G. R., 1963. *The Science of Life.* McGraw-Hill, New York.

Toulmin, S., and J. Goodfield, 1965. *The Discovery of Time.* Harper & Row, New York.

Wardlaw, C. W., 1965. *Organization and Evolution in Plants.* Longmans Greens, London.

Young, D., 1992. *The Discovery of Evolution.* Cambridge University Press, Cambridge, England.

Darwin

By the early nineteenth century, many of the basic concepts necessary to develop a belief in organic evolution were already present: (1) geologists such as Hutton conceived the age of the earth to be in the range of millions of years ("We find no vestige of a beginning—no prospect of an end"); (2) people had accepted the reality of previously extinct fossil species; (3) systematists, comparative anatomists, and embryologists had noted the close similarities among many different species; and (4) most scientists, if not all, believed organisms had descended through inheritance from previously existing organisms.

The notion of a divine "common plan," which had been proposed to account for the relationships among species by supernatural acts of creation, was therefore only one step away from the materialist evolutionary notion that species relationships derive from their common ancestry: a change "from archetypes to ancestors." Nevertheless, the materialist still had to face at least two important questions: What natural cause or mechanism could explain why organisms change? What hereditary mechanism could enable organisms to change?

Charles Darwin (1809–1882) answered the first question in 1859, and he thereby helped transform biology into an evolutionary science. An acceptable answer to the second question had to wait for the twentieth century, although soon after Darwin, Gregor Mendel (1822–1884) provided the essential basis for understanding materialist biological inheritance (see Chapter 10).

Charles Darwin

Many biographies have been written of Charles Darwin (Fig. 2–1), a man who defied his own social and religious background not only by espousing a radical concept but also by becoming the instrument that made it acceptable to many of his compatriots. The enigma, as described by Desmond and Moore, was:

> How could an ambitious thirty-year-old gentleman open a secret notebook [in 1837] and, with a

FIGURE 2–1 Portrait of Charles Darwin in 1849 at the age of 40, by T. H. Maguire. (*Copyright British Museum.*)

> devil-may-care sweep, suggest that headless hermaphrodite mollusks were the ancestors of mankind? A squire's son, moreover, Cambridge-trained and once destined for the cloth. A man whose whole family hated the "fierce & licentious" radical hooligans.

In its barest outlines, Darwin's life is the history of a genial, curious, and intellectually creative man who was courageous yet fearful, living in a society undergoing considerable change. Briefly, he was born to an English middle-class family whose fortunes derived largely from his father, Robert Darwin (1766–1848), and his paternal grandfather, Erasmus Darwin (1731–1802), both prosperous physicians. Erasmus Darwin was, in fact, an early popularizer of evolution, but most of his contemporaries judged him wildly speculative in science. Even Charles himself, when he was searching for evolutionary explanations in the 1830s, did not seriously consider many of his grandfather's suppositions. In any case, historians believe that evolutionary ideas probably had little effect on Charles's early development.

At age 16 Charles left grammar school in Shrewsbury and was sent to Edinburgh University to study medicine. Because surgical procedures at that time were quite brutal, Darwin found the experience distasteful and after two years transferred to Cambridge University with the intention of becoming a minister in the Church of England. However, his interests were not in academic or ministerial pursuits but in hunting, collecting, natural history, botany, and geology. He despised formal classical education and was usually no more than a mediocre student. His father apparently felt that Charles had betrayed the family trust of industrious professionalism and castigated him: "You care for nothing but shooting, dogs, and rat-catching, and you will be a disgrace to yourself and all your family."

In 1831, through the recommendation of John Henslow (1796–1861), a botany professor at Cambridge, and the intercession of an uncle, Josiah Wedgwood (1769–1843), Darwin was able to put off further study for the ministry and instead undertook his now famous voyage around the world on the **H.M.S. *Beagle*** (Fig. 2–2). His post on the *Beagle* was that of naturalist, a special unpaid position created by the British Admiralty on naval ships making broad geographical surveys.

The *Beagle* voyage lasted approximately 5 years, and during this interval Darwin was transformed from a casual amateur to a dedicated geologist and biologist. His letters to Henslow on many of the observations made during the voyage, along with his collections of plants, animals, fossils, and minerals, excited considerable scientific interest even before his return to England. Darwin's account of the voyage, published later as his *Journal of Researches* and reprinted many times under the title *Voyage of the Beagle,* remains one of the most interesting and perceptive chronicles of exploration in the nineteenth century.

On his return to England, a substantial income and inheritance enabled Darwin to forgo financial pursuits and dedicate himself entirely to biology. He married his cousin, Emma Wedgwood (1808–1896), and in 1842 settled near the village of Down in the Kent countryside, 16 miles from London. There he and his wife, aided by servants, began to raise a large family.

For 40 years, to the time of his death in 1882, Darwin lived at home, mostly as a semi-invalid subject to heart palpitations, rashes, and gastric discomfort. The cause for his disability is not known, and conjectures have ranged from parasitic infection (trypanosomes that cause Chagas disease) and heavy metal (arsenic) poisoning via some of the "cures" of his time, to psychosomatic illness (Colp) involving severe symptoms of "panic disorder" (Barloon and Noyes). Whatever the cause, his illness isolated him from most of the world about him except through letters and publications. Despite his physical discomforts, Darwin appears to have lived a harmonious life, probably because of his own warm personality and behavior and the sympathetic concern of his wife. It has often been said that Darwin was the perfect patient and his wife the perfect nurse.

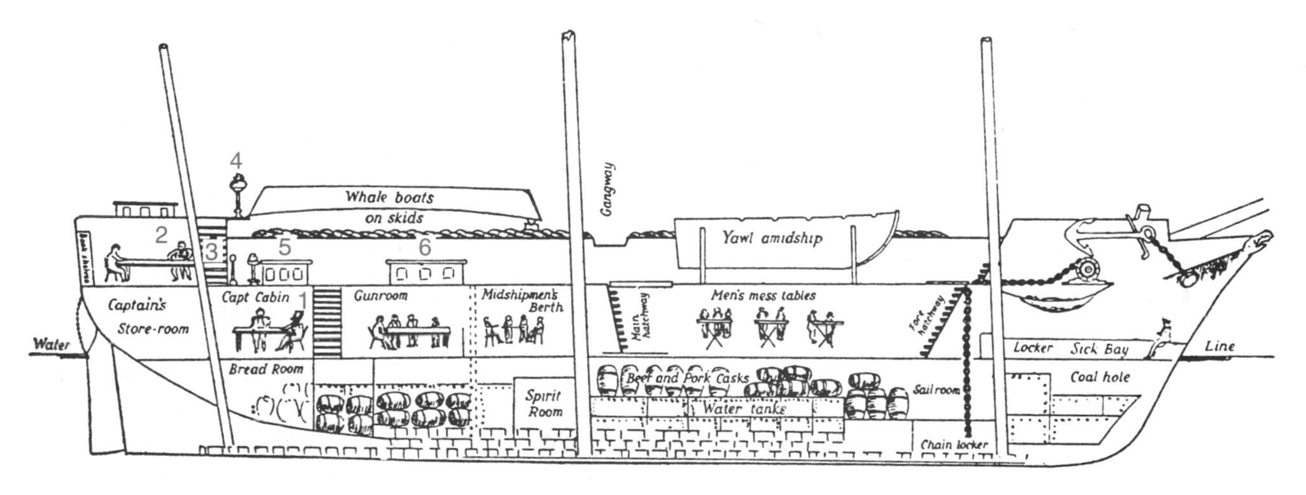

H. M. S. Beagle 1832

1 Mr Darwin's seat in Captain's Cabin   2 Mr Darwin's seat in Poop Cabin   3 Mr. Darwin's drawers in Poop Cabin
4 Azimuth Compass   5 Captain's skylight   6 Gunroom skylight

FIGURE 2–2 *Top:* H.M.S. *Beagle* in the Strait of Magellan at the southern tip of South America. The ship was a 10-gun brig, 90 feet long, weighing 240 tons. In 1831, the year Darwin began his voyage, it had been refitted for circumnavigation in order to fix world longitudinal markings and chart the coast of South America. *Bottom:* Side elevation of the *Beagle,* based on a drawing by one of Darwin's shipmates showing the general plan of the ship and the cramped quarters that held a crew of about 70. (In a recent account of the ship's voyages, Thompson notes: "To say that the *Beagle* was extremely cramped, even given the expectations of the time, would be a supreme understatement. The ship was, after all, no longer than the distance between two bases on a baseball field.") Darwin slept in the poop cabin at the stern of the ship, which he shared with two officers. This cabin also held a 10-by-6-foot chart table and various chart lockers, as well as drawers for his own equipment and specimens. He wrote, "I have just room to turn around and that is all."

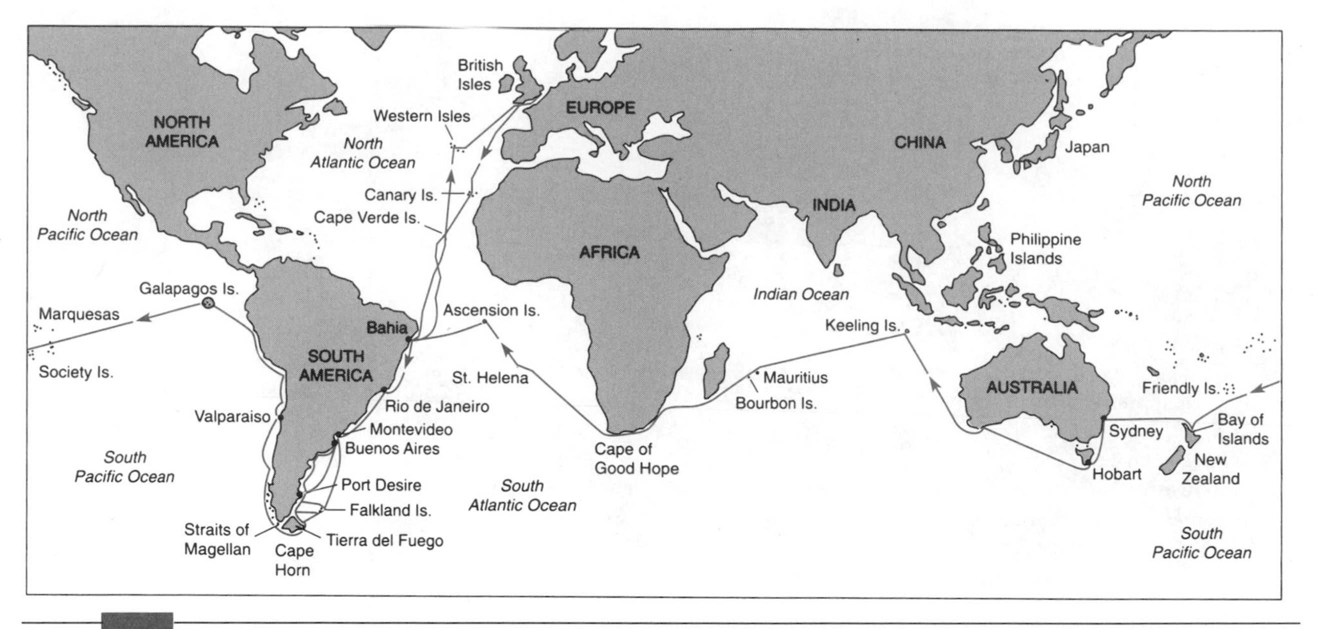

FIGURE 2-3 Route of the 5-year voyage of the *Beagle,* beginning at Plymouth, England, in December 1831, and ending in Falmouth, England, in October 1836. Almost four years were spent in South America, including one month in the Galapagos Islands (September–October 1835).

The Voyage of the *Beagle*

The 5-year voyage on the *Beagle* (Fig. 2–3) enabled Darwin to observe and think about a relatively wide range of organisms and geological formations. He collected birds, insects, spiders, and plants in the Brazilian tropical forests. At Punta Alta, on the coast of Argentina, he unearthed fossil bones of the 20-foot-high giant sloth, *Megatherium,* the hippopotamuslike *Toxodon,* the giant armadillo *Glyptodon,* and other animals resembling present species yet recognizably different. The primitiveness and wildness of the Tierra del Fuego Indians at the southern tip of South America impressed him with the severity of their struggle for subsistence in a meager and unrelenting environment.

During the voyage, Darwin carried with him Lyell's *Principles of Geology* and assiduously noted the geological features of many terrains he covered. To explain some of the geological uplifting processes that shaped the South American landscape, he gathered evidence showing the distribution of marine shells at various places above sea level, the loss of pigment in the older shells found at higher elevations, and the terracing of land by erosion as it was lifted upward. At the Bay of Concepción, along the coast of Chile, he experienced a severe earthquake that raised the level of the land in some places from about 2 or 3 feet above sea level to as much as 10 feet. This experience had a deep effect on him:

> A bad earthquake at once destroys our associations: the earth, the very emblem of solidity, has moved beneath our feet like a thin crust over a fluid; one second of time has created in the mind a strange idea of insecurity, which hours of reflection would not have produced.

An experience that, years later, had great impact on Darwin's thinking about evolution was the month he spent in the bleak, lava-ridden **Galapagos Islands** off the coast of Ecuador. Here, 500 miles[1] from the mainland, was a strange collection of organisms: giant tortoises, yard-long marine and land iguanas, as well as many unusual plants, insects, lizards, and seashells. As he had already noted on the mainland, different geographical localities, although possessing some environmentally similar habitats, were not always occupied by similar species.

This situation was most striking in the Galapagos, where insect-eating warblers and woodpeckers were absent but various species of finches, usually seed eating, now assumed the insect-eating patterns of the missing species (Fig. 2–4). Also, the observation that each island appeared to have its own unique, closely related constellation of species raised the important question: What could account for this distribution of organisms? In Darwin's words:

> It is the circumstance that several of the islands possess their own species of the tortoise, mocking-thrush, finches, and numerous plants, these species having the same general habits, occupying

[1]The following American units of measurement are often used in this text, but can be easily converted to the metric system. For length, 1 inch = 2.54 centimeters, 1 foot = 30.48 centimeters, and 1 mile = 1.6093 kilometers. For weight, 1 pound = 453.6 grams. For volume, 1 cubic foot = 28.32 liters.

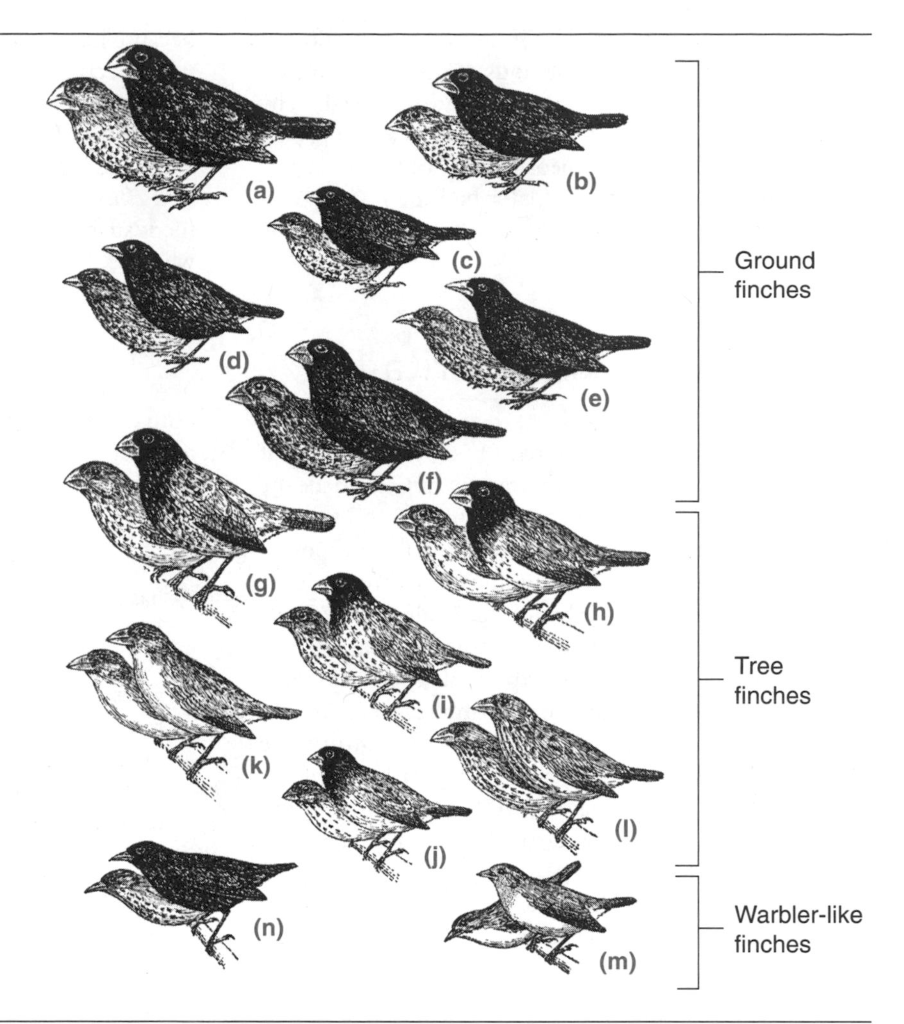

FIGURE 2-4 Species of finches (male on left, female on right; about 20 percent of actual size) that Darwin observed in the Galapagos Islands. (*a*) *Geospiza magnirostris* (large ground finch), (*b*) *Geospiza fortis* (medium ground finch), (*c*) *Geospiza fuliginosa* (small ground finch), (*d*) *Geospiza difficilis* (sharp-beaked ground finch), (*e*) *Geospiza scandens* (cactus ground finch), (*f*) *Geospiza conirostris* (large cactus ground finch), (*g*) *Camarhynchus crassirostris* (vegetarian tree finch), (*h*) *Camarhynchus psittacula* (large insectivorous tree finch), (*i*) *Camarhynchus pauper* (large insectivorous tree finch on Charles Island), (*j*) *Camarhynchus parvulus* (small insectivorous tree finch), (*k*) *Camarhynchus pallidus* (woodpecker finch), (*l*) *Camarhynchus heliobates* (mangrove finch), (*m*) *Certhidea olivacea* (warbler finch), and (*n*) *Pinaroloxias inornata* (cocos finch). Evolutionary relationships among these finches are illustrated in Fig. 3-3. (*From* Darwin's Finches: An Essay on the General Biological Theory of Evolution, *1947 by D. Lack. Reprinted by permission of Cambridge University Press.*)

> analogous situations, and obviously filling the same place in the natural economy of this archipelago, that strikes me with wonder.

Did separate and different creations make one species in one place slightly different from another species in another place? Why?[2]

The *Beagle* voyage stirred the seed of evolutionary thought in Darwin, leading him to begin his first notebook on the *Transmutation of Species* in 1837. He adopted the view that only changes among species could reasonably explain observations that present species resemble past species and that different species share similar structures: "The only cause of similarity in individuals we know of is relationship." The differences between the flora and fauna of different geographical areas, he thought, must have arisen because not all plants or animals are universally distributed.

For the Galapagos Islands, for example, Darwin raised the question:

> Why on these small points of land, which within a late geological period must have been covered with ocean, which are formed of basaltic lava, and therefore differ in geological character from the American continent, and which are placed under a peculiar climate,—why were their aboriginal inhabitants . . . created on American types of organization?

It seemed clear to Darwin that islands such as the Galapagos will contain only those organisms able to reach them, and evolution can transform only those species that are available:

> Seeing this gradation and diversity of structure in one small, intimately-related group of birds, one might really fancy that from an original paucity of birds in this archipelago one species had been taken and modified for different ends.

[2]It should be noted that Darwin's account of his 1831–1836 voyage on the *Beagle* was first published in 1838 and revised some years later. The ornithologist David Lack and historians such as Sulloway have pointed out that although Darwin's *Journal of Researches* expresses these and other evolutionary forethoughts, the significance of his observations on the Galapagos Islands and elsewhere did not become apparent to Darwin until after his return to England. This was especially true for the various Galapagos finches, which were first classified in England by John Gould, a British ornithologist whom Darwin met in 1837.

However, the mechanism for the transformation of species was by no means as obvious as the reasonable assumption that such transformation had occurred. Why do species change? In seeking an answer, Darwin apparently explored a variety of theories. One of the most persistent concepts, a theory that later had many adherents in France and the United States, was put forth by Lamarck.[3]

The Lamarckian Heritage

Jean-Baptiste de Lamarck (1744–1829), the first biologist to actively advocate evolution, made the important leap from what appeared to others as species extinction (as evidenced by fossils) to proposing their continuity by gradual modification through time (see p. 11). The means by which these evolutionary modifications occurred and the exquisite relationships through which organisms exploited their environments—**adaptations**—were areas Lamarck began to explore at the beginning of the nineteenth century. He proposed that the variations among organisms originate because of response to the needs of the environment, and this ability to respond in a particular direction accounts for a trait's adaptation.

For example, Lamarck suggested that the long legs of water birds such as herons and egrets have arisen through the following mechanism:

> We find . . . that the bird of the waterside which does not like swimming and yet is in need of going to the water's edge to secure its prey, is continually liable to sink in the mud. Now this bird tries to act in such a way that its body should not be immersed in the liquid, and hence makes its best efforts to stretch and lengthen its legs. The long-established habit acquired by this bird and all its race of continually stretching and lengthening its legs, results in the individuals of this race becoming raised as though on stilts, and gradually obtaining long bare legs, denuded of feathers up to the thighs and often higher still.

Implicit in this process is that organisms learned new habits and could change appropriately by exercising an unknown, inner "perfecting principle." This special power could sense the needs of the environment and respond by developing new traits in appropriate adaptive directions, mostly from simple to complex. The source for such directional orientation was not clear to Lamarck, because he was not aware, as Darwin later was, that natural selection is the device that leads to continued improvement of adaptive mechanisms.

At times Lamarck ascribed his belief in evolutionary "progress" to an inner, mystical, vitalistic property of life (*feu éthéré,* ethereal fire), whereas at other times he denied such supernatural causes. However, no matter whether their direction was caused by natural or supernatural events, the origin of organic changes and their transmission to further generations was believed by Lamarck to be aided by two universal mechanisms that he codified into two basic "Laws of Nature" (although both these concepts can be traced back to the folklore of antiquity and were also incorporated into *Zoonomia,* a popular work by Charles Darwin's grandfather, Erasmus Darwin):

> *1. Principle of Use and Disuse:* In every animal which has not passed the limit of its development, a more frequent and continuous use of any organ gradually strengthens, develops and enlarges that organ, and gives it a power proportional to the length of time it has been so used; while the permanent disuse of any organ imperceptibly weakens and deteriorates it, and progressively diminishes its functional capacity, until it finally disappears.
>
> *2. The Inheritance of Acquired Characters:* All the acquisition or losses wrought by nature on individuals, through the influence of the environment in which their race has long been placed, and hence through the influence of the predominant use or permanent disuse of any organ; all these are preserved by reproduction to the new individuals which arise, provided that the acquired modifications are common to both sexes, or at least to the individuals which produce the young.

This remarkable hereditary plasticity by which organisms could adapt to their environments led Lamarck (as we have seen on p. 11), to the notion that species exist in name only, since what is called a *species* must be merely a continuum between organisms that are at different points in the process of change. Thus, fossil species, according to Lamarck, were not truly extinct but had become modified in time and thereby evolved into later, more complex organisms.

It was Cuvier (1769–1832) who marshaled what seemed at the time the most telling arguments against Lamarck's evolutionary proposals. Cuvier pointed out that no intermediate forms were found, either alive or as fossils, that bridged the gaps between different species. Also, when a species hybrid was occasionally formed, such as the mule, it was always sterile. The Lamarckian concept that organisms strive for perfection seemed ludicrous: What elements of consciousness could one ascribe to plants and lower organisms? Even among animals,

---

[3]Although abandoned by practically all biologists in modern countries, a form of Lamarckianism was adopted by the Soviet Union as official policy during the 1948–1963 period as a result of political demagoguery and experimental fabrications by the Russian agronomist T. D. Lysenko and his supporters. (For a review of this episode, see Joravsky.)

how could new habits of swimming or flying produce organs enabling such habits without these organs being already present? Furthermore, Cuvier argued, in spite of 4,000 years of recorded history, no new species had evolved. Why assume they *can* evolve? Like begets like!

Although Lamarck's theories fell into disfavor, it is important to note that the attitudes for which he was denounced during the nineteenth century were often attitudes that were eventually accepted. For example, Lyell, in his *Principles of Geology,* wrote of Lamarck:

> His speculations know no definite bounds: He gives the rein to conjecture, and fancies that the outward form, internal structure, instinctive faculties, nay, that reason itself, may have been gradually developed from some of the simple states of existence,—that all animals, that man himself, and the irrational beings, may have had one common origin; that all may be parts of one continuous and progressive scheme of development from the most imperfect to the more complex; in fine, he renounces his belief in the high genealogy of his species, and looks forward, as if in compensation, to the future perfectibility of man in his physical, intellectual, and moral attributes.

Although it was by no means obvious at the time, we can, with hindsight, summarize Lamarck's contribution as the concept that evolution depends on natural processes. These processes, the inheritance of acquired characters and the effects of use and disuse, were later proved incorrect but nevertheless had the important advantage that they were uniformitarian in principle and did not immediately rely on supernatural or catastrophic events. Lamarck thus helped in developing the climate of opinion in which evolution could be understood in the same fashion as any other natural event.

If we accept the basic idea that organisms can change through time and discard the Lamarckian explanation for this, the question is then: Why do they change?

Natural Selection

It remained for Darwin to elaborate a mechanism for evolution more acceptable to biologists than that of Lamarck. He briefly defined the mechanism he proposed, natural selection, as follows:

> As many more individuals of each species are born than can possibly survive; and as, consequently, there is a frequently recurring struggle for existence, it follows that any being, if it vary however slightly in any manner profitable to itself, under the complex and sometimes varying conditions of life, will have a better chance of surviving, and thus be *naturally selected.* From the strong principle of inheritance, any selected variety will tend to propagate its new and modified form.

Behind this simple explanation is a complex set of causative events that Darwin spent most of his life investigating, although some aspects of a selective process in nature had been previously argued by others. In antiquity, Empedocles (c. 490–430 B.C.) had suggested that the initial appearance of life was in the form of parts and organs floating freely and combining together to form whole organisms. Those organisms that were adapted to "some purpose" survived and those which did not "perish and still perish." From this original selective act, Empedocles proposed, all present organisms stem.

Aristotle disputed Empedocles' concept of the randomness on which such selection acts with an argument often used since. He said the Scale of Nature, like any other teleological process, obviously arises through a fixed progression of steps from lowest to highest stages. There cannot, therefore, be anything arbitrary or random in this progression that would require selection.

In the eighteenth century, Buffon saw natural selection (as well as selection by humans) as the agent responsible for the extinction of species: "All the bodies imperfectly organized and all the defective species would vanish, and there would remain, as there remain today, only the most powerful and complete forms, whether plants or animals." However, Buffon did not see natural selection as responsible for the generation of new species. He believed that new species could arise by spontaneous generation and that differences in the conditions under which spontaneous generation occurred probably caused differences between species.

In the Lamarckian view, environmental effects initiated variations that occurred only in an adaptive direction. So there could be no extinction of "imperfect" or "defective" species, because organisms could always adapt themselves to changing environments by inheriting acquired characteristics. To Lamarck, variations were not separate from evolution, and therefore they could not be random. Thus, selection was not needed to choose adaptive traits.

In the early nineteenth century, a number of authors, including Wells (1757–1817) and Matthew (1790–1874), separated the origin of variations from the forces responsible for preserving them and used the principle of natural selection to explain changes within species. Unfortunately, their ideas seemed highly speculative, since they did not provide sufficient support, and their works were recorded in obscure publications that did not come to the general attention of biologists.

Note that Darwin ascribes his notion for the tendency of a species to produce more members than resources can sustain—the primary populational pressure which leads to competition and selection—not to biological literature but to the sociology of his time. During the Victorian period in England, a variety of social and economic problems had become apparent because of the rapid increase in the poor resulting from the Industrial Revolution. This tide of poverty had begun with the impoverishment of small handicrafts establishments and was continually fed by small farmers pushed from their lands (the "commons") by the Enclosure Acts.

Among British economists, one attitude, expressed by Rev. Thomas **Malthus** (1766–1834), was that the fate of the poor is inescapable; their reproductive powers will always exhaust their means of subsistence. Food supplies, Malthus pointed out, can at best increase arithmetically (1 → 2 → 3 → 4 → 5 . . .) by the gradual accretion of land and improvement of agriculture, whereas the number of poor people will increase geometrically (1 → 2 → 4 → 8 → 16 . . .), because the children of each family are usually more numerous than the parents. Thus, famine, war, and disease inevitably become major factors among the controls that limit population growth.

The only hope that Malthus held out for the poor was self-restraint: delay marriage and refrain from sexual activity. He held all other solutions—such as the Poor Laws (welfare) or the redistribution of wealth and the improvement of living conditions—to be inadequate, because such measures would stimulate a further increase in the number of poor people and begin again the cycle of famine, war, and disease.

Like many others of the time, Darwin was deeply impressed by the Malthusian argument, although Malthus was not an evolutionist. In fact, Malthus believed that limiting population growth would prevent evolutionary change because individuals who departed from the population norm would be more susceptible to extinction. To Darwin, however, the importance of Malthus lay in revealing the conflict between a population's limited natural resources and its continued reproductive pressure.[4] In contrast to Malthus's proposals for alleviating the impact of population increase, Darwin pointed out that plants and animals had no such alternatives: "There can be no artificial increase in food, and no prudential restraint from marriage."

Under such pressure of continuously limited resources, selection could act by choosing for reproduction those individuals or types with increased chances of survival and could therefore change the composition of the population. From his inquiries on breeding domesticated species, Darwin had obtained clear evidence that selection (in this case, human or **artificial selection**) could have marked hereditary effects. Although Darwin's autobiography, completed one year before he died, did not always faithfully reconstruct all earlier conceptual events, there is little question that Malthus played an important role in developing the idea of natural selection. Darwin writes:

> I soon perceived that selection was the keystone of man's success in making useful races of animals and plants. But how selection could be applied to organisms living in a state of nature remained for some time a mystery to me.
>
> In October 1838, that is, fifteen months after I had begun my systematic enquiry, I happened to read for amusement 'Malthus on Population' and being well prepared to appreciate the struggle for existence which everywhere goes on from long-continued observation of the habits of animals and plants, it at once struck me that under these circumstances favourable variations would tend to be preserved, and unfavourable ones to be destroyed. The result of this would be the formation of new species. Here then I had at last got a theory by which to work; but I was so anxious to avoid prejudice, that I determined not for some time to write even the briefest sketch of it. In June 1842 I first allowed myself the satisfaction of writing a very brief abstract of my theory in pencil in 35 pages; and this was enlarged during the summer of 1844 into one of 230 pages, which I had fairly copied out and still possess.

Because he needed to gather supporting evidence and feared that his theory was not ready to be accepted, Darwin withheld publication for a long time.[5] But the idea was "in the air." In 1858 Alfred Russel Wallace (1823–1913), a naturalist then collecting mainly birds, insects, and mammals in the islands of Southeast Asia, sent Darwin a paper to be published in which he described the theory of natural selection in the essential form Darwin had envisaged. Wallace too had puzzled about a mechanism for evolution and had also read Malthus. Remarkably, Malthus performed the same function for Wallace as he had for Darwin twenty years earlier. Wallace wrote:

> At that time [February 1858] I was suffering from a rather severe attack of intermittent fever at Ternate in the Moluccas . . . and something led me to think of the positive checks described by Malthus in his 'Essay on Population', a work I had read several

[4]The importance of Malthus to Darwin's theory has been disputed by some historians (see, for example, Gordon).

[5]In 1844, a Scots writer, Robert Chambers, had published a book, *Vestiges of the Natural History of Creation,* that elaborated the idea that all matter, inorganic and organic, evolved out of inorganic dust. The mechanism of biological evolution that Chambers proposed was an accumulation of accidental mutations caused somehow by changes in nutrition or environment. Chambers also espoused many questionable notions such as the spontaneous generation of organisms by electrical currents and phrenology (study of bumps on the head) in understanding how the mind works. Although popular for a time (there were 14 editions in Britain, and many others in the United States), considerable religious and scientific denunciation focused on Chambers's work, which made Darwin very fearful of exposing his own ideas to ridicule.

years before, and which had made a deep and permanent impression on my mind. These checks—war, disease, famine, and the like—must, it occurred to me, act on animals as well as on man. Then I thought of the enormously rapid multiplication of animals, causing these checks to be much more effective in them than in the case of man; and, while pondering vaguely on this fact there suddenly flashed upon me the *idea* of the survival of the fittest—that the individuals removed by these checks must be on the whole inferior to those that survived. In the two hours that elapsed before my ague fit was over I had thought out almost the whole of the theory, and the same evening I sketched the draft of my paper, and in the two succeeding evenings wrote it out in full, and sent it by the next post to Mr. Darwin. (Introductory note to Chapter II of *Natural Selection and Tropical Nature*, revised edition, 1891.)

To prevent Darwin losing his priority in applying natural selection to evolution, his friends Lyell and Hooker arranged for short papers on the topic by both authors to be published in 1858 in *The Journal of the Linnaean Society*.[6] Surprisingly, the scientific and nonscientific communities made little response at that time, indicating perhaps that the theory itself, without supporting evidence and without enlisting large-scale evolutionary phenomena, did not invite serious interest and did not threaten established opinion. Only with the publication of Darwin's expanded and heavily documented *On the Origin of Species* in November 1859 did the world take notice.

The evolutionary principle expressed by Darwin and Wallace is briefly outlined in Figure 2–5. Note that the evolutionary process is a continual one; the achievement of an adaptation by individuals leads to enhanced reproductive ability relative to other individuals, followed by further competition for the limited resources and further natural selection. Since each evolutionary stage builds on the one before, the process spirals in the direction of improved adaptation for any particular environment.

At the base of the process is the mutational "fuel" of evolution, the continual introduction of new heritable variations on which selection can act. Darwin did not know the biological basis for heredity or its variations, and his arguments were weakest in these areas. At times, he proposed that either environmental changes or a large

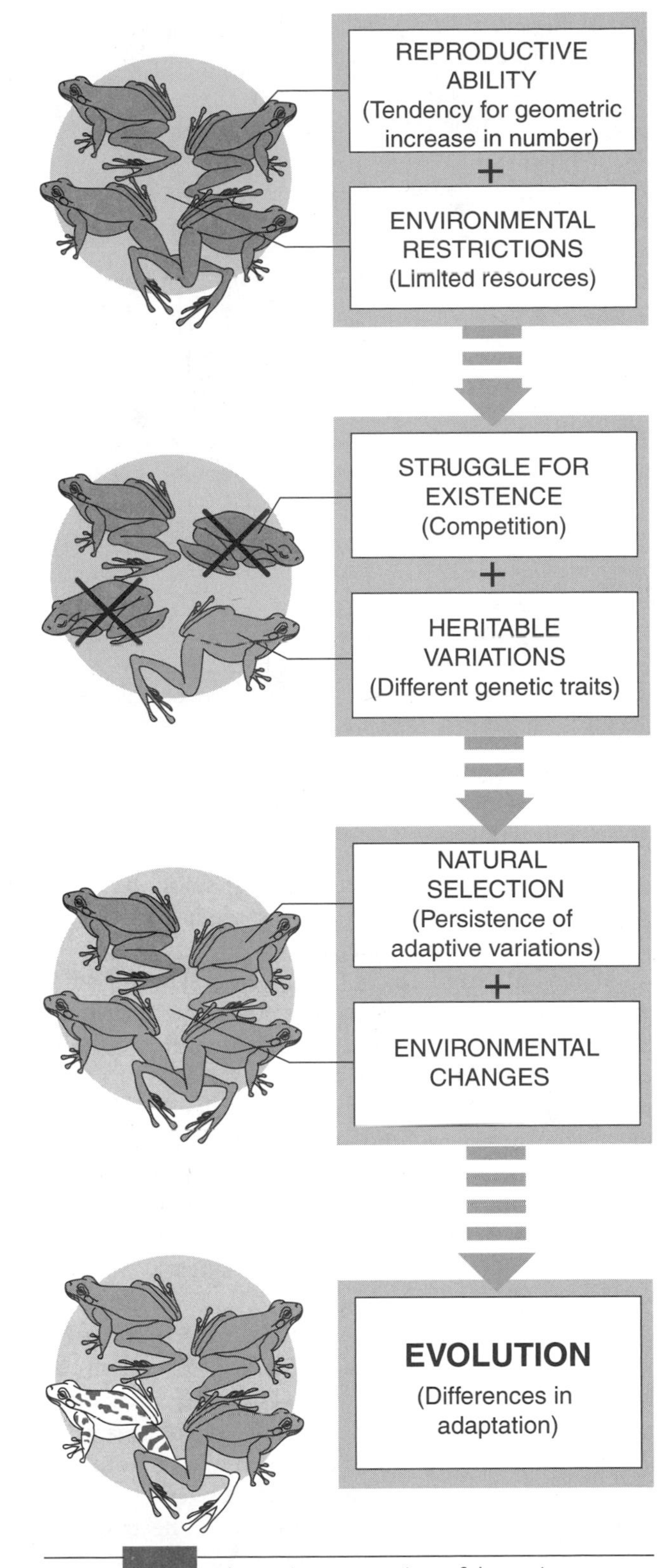

FIGURE 2-5 Schematic presentation of the main conceptual arguments for evolution by natural selection given by Charles Darwin and Alfred Russel Wallace. For modern versions of these arguments, see Chapters 10 and 21–24. (*Based on a table by Wallace.*)

[6]Although some Wallace biographers (for example, Brackman, Brooks) have suggested that Darwin either "borrowed" or "stole" Wallace's ideas, most historians agree there is no evidence for this (see, for example, Beddall). As noted from the writings of Wells and Matthew, natural selection and the evolutionary divergence of species were "in the air," and both Darwin and Wallace, although separated by many thousands of miles, were obviously original thinkers who could grasp these concepts independently. Their cordial personal relationship, which began when Wallace returned to England in 1862 and continued throughout their lives, testifies to the absence of any hostility or rancor that would have accompanied plagiarism or deception.

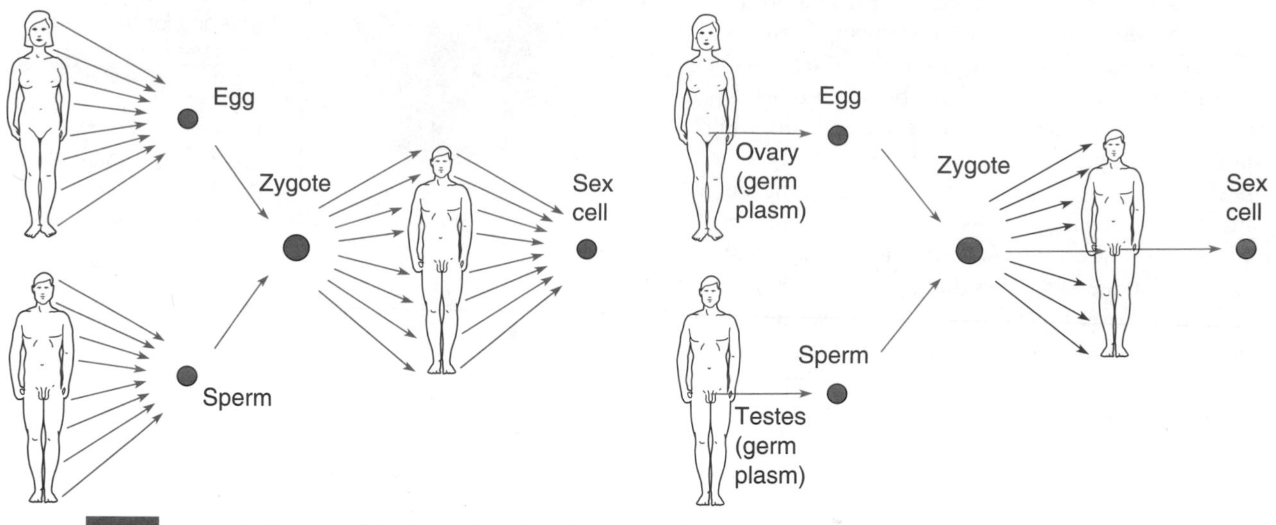

FIGURE 2-6 Comparison between (*a*) pangenesis and (*b*) germ plasm theories in the formation of a human. In pangenesis all structures and organs throughout the body contribute copies of themselves to a sex cell. In the germ plasm theory the plans for the entire body are contributed only by the sex organs. (*From* Genetics Third Edition *by Monroe W. Strickberger. Copyright © 1985 by Monroe W. Strickberger. Reprinted by permission of Prentice Hall, Inc., Upper Saddle River, NJ.*)

increase in numbers may enhance variability among a population of individuals; at other times, he adopted the Lamarckian view of the inheritance of use and disuse.

At the bottom of Darwin's difficulties, as with other biologists of the time, was the commonly accepted theory that heredity is mostly a blend of the heredity of both parents, much as diluting red paint with white produces pink. As Chapter 3 shows, the idea of **blending inheritance** confronted evolutionary theory with the serious enigma of trying to explain how adaptive variations can be preserved by natural selection if they are blended out by the mating of their carriers with other members of the population.

Instead of blending inheritance, Darwin reinstituted an old theory called **pangenesis.** He suggested that small atomic "gemmules" or "pangenes" derived from all the tissues of a parent are incorporated into the parental gametes. When fertilization occurs and parental gametes unite, these gemmules would then spread out to form the tissues of the offspring (Fig. 2–6*a*). Pangenesis would therefore help account for the presumed effects of use and disuse, by suggesting that changes can arise in the frequencies of particular gemmules, and for the observation that not all traits become blended, by postulating that the structure of gemmules can remain constant.

However, there was no evidence for pangenesis, and Weismann (1834–1914) effectively disproved it some years later. Weismann cut off the tails of 22 generations of mice and showed that the tail length was not affected by the presumed loss of tail gemmules in each generation. For pangenesis, Weismann substituted the modern **germ plasm theory** of inheritance, in which only the reproductive tissues (testes and ovaries) transmit the heredity factors of the entire organism, and changes that occur in nonreproductive somatic tissues are not transmitted (Fig. 2–6*b*). Thus changes in heredity cannot be simply explained by inheritance of acquired characters or by use and disuse.[7] With the development of the modern science of genetics, many of the difficulties Darwin faced were resolved. Several views on variation and heredity are compared in Table 2–1.

[7]Nineteenth century Lamarckians generally understood "acquired characters" to be somatic traits that can be reproduced each generation without instruction from germ-line tissue. There was presumably no difference between somatic tissue and germinal tissue in ability to transmit hereditary traits. If we use later terms (p. 113) and distinguish between *genotype* (nucleotide coding sequences) and *phenotype* (developmental product of the genotype), Lamarckianism can be treated as a proposal for phenotypic self-reproduction based entirely on nongenotypic sources of information. Since material elements that can consistently produce nongenetic transmission of biological information are not identified (see Chapter 15), Lamarckianism has an inherent notion that organisms possess unspecified, and perhaps mystical, agents that can sense environmental needs and respond appropriately. In cases where authors claim a material basis for Lamarckian inheritance and evolution (for example, Jablonka and Lamb), the source seems to be a form of chromosomal change that others would consider genotypic rather than phenotypic, such as the methylation of cytosine nucleotides in DNA (see p. 270). Most biologists would agree, however, that Lamarckian inheritance does occur in learning and culture (Chapter 25) where transmission and change are determined by conscious agents (for example, humans with brains) rather than by nucleotide sequences.

TABLE 2-1 Comparison of views on variation and heredity

	Creationist	Lamarck	Darwin	Present Biology
What accounts for the similarity among many species?	The divine plan of creation (purpose unknown) produced the basic "kinds" of organisms.	Descent from a common ancestor.	Descent from a common ancestor.	Descent from a common ancestor.
What accounts for the origin of variations among members of a species?	Although they can be environmentally caused,[a] they are part of the divine plan of creation.	Environmentally caused.	At times: unknown causation. At times: environmental changes cause new variations, although the variations may be in any direction.	Heritable differences are caused by random changes (mutation) in the genetic material. Noninheritable differences are caused by the environment.
What accounts for the presence of particular organs and structures through time?	They were initially designed so by the creator. Many present creationists believe that organ defects, diseases, etc., are caused by the fall of humans from divine grace and/or intervention by a devil.	Use enhances the development of adaptive variations, and disuse eliminates nonadaptive ones.	Natural selection perpetuates only adaptive traits and eliminates nonadaptive traits. At times: use and disuse.	Primarily natural selection but other forces may be involved, as discussed in Chapter 22.
What accounts for the variation among species?	The separate creation of each species. Many present creationists believe that the original "kinds" of organisms were perfect, and variations leading to species differences have been degenerative.	Each species has responded to different environmental needs by developing new organs or discarding old ones.	At times: selective differences among species account for their changed inheritance. At times: differences in the use and disuse of particular organs has caused changed inheritance.	Changes occur in the genetic material of each species through the process of mutation and the various forces that change gene frequencies.
What accounts for the resemblance of organisms to their parents?	Mechanisms unknown, but acquired characters are inherited as part of the divine plan.[b]	Those characters acquired through use and disuse are inherited through a pangenesislike process.	At times: unknown. At times: pangenesis.	Transmission of genetic material through the germ plasm.

[a]See the story of Jacob and the sheep in Genesis 30: 37–39.

[b]"Visiting the iniquity of the fathers upon the children and children's children unto the third and fourth generation." (Exodus 34: 7.)

SUMMARY

The basic ideas essential to evolutionary theory were already present by the time Darwin made his famous voyage on the H.M.S. *Beagle.* The most important ideas were that the earth was ancient, that fossils represented the remains of extinct species, that many species showed close similarities, and that organisms descended from previously existing organisms. However, the mechanism for evolutionary change and the agents that allowed organisms to change remained undiscovered.

The information Darwin accumulated on his 5-year voyage engendered his evolutionary ideas and suggested to him a mechanism by which evolution might proceed. A keen observer of both geology and natural history,

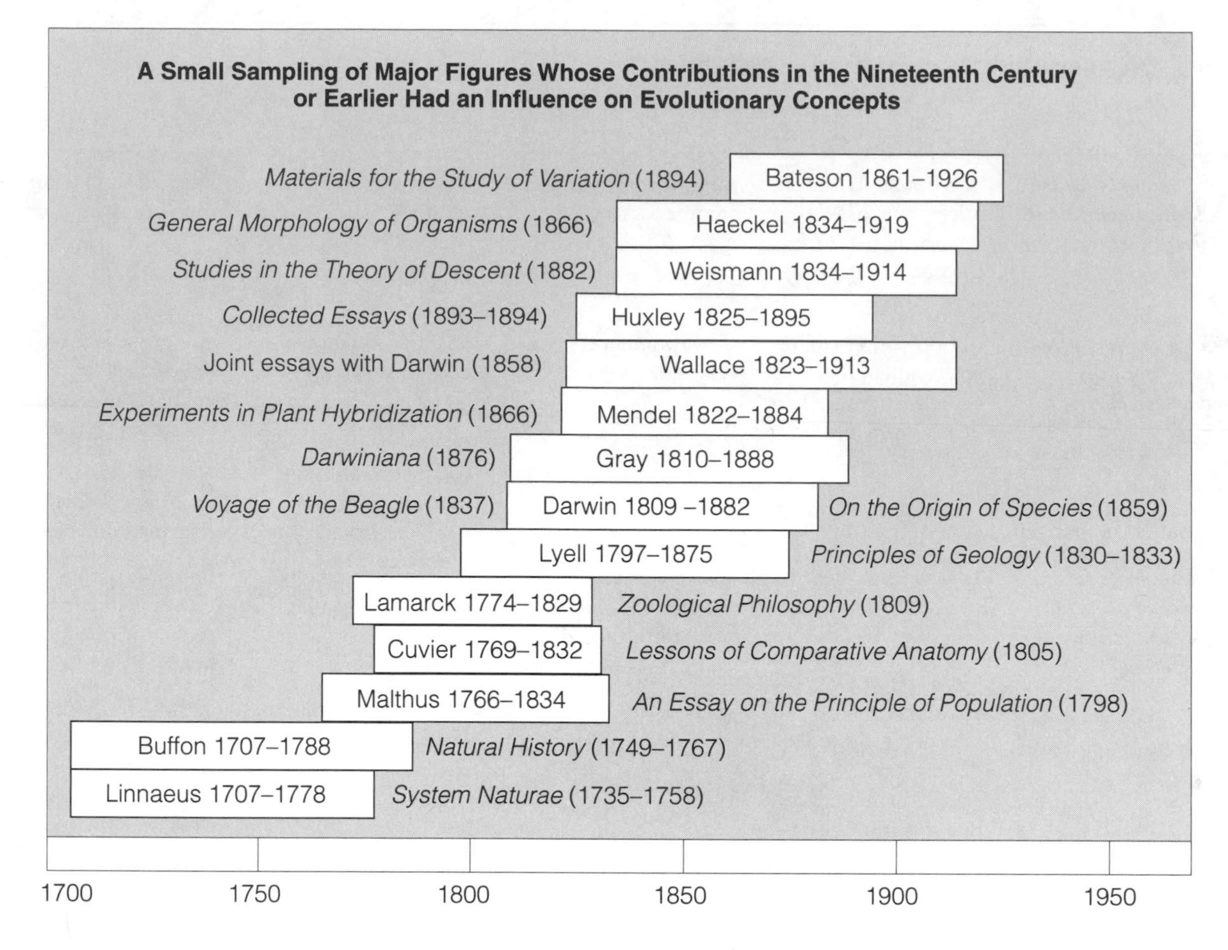

Darwin noted geological formations that gave evidence of historical transformation, as well as the peculiar geographical distribution of organisms and the close similarities of species. He recognized that the only rational explanation for these phenomena must be that species could be transformed. At first, however, he could find no mechanism by which transformation might occur.

Darwin rejected Lamarck's contention that structures survived or deteriorated through use or disuse and, furthermore, that traits so acquired could be inherited. It was left to Darwin to propose a more acceptable alternative—natural selection. From reading Malthus he derived the idea of a superabundance of progeny competing for limited resources, and this "struggle" provided Darwin with a scenario for changing the composition of a population. Organisms that had traits better suiting them to their environment ("adaptations") would tend to reproduce more prolifically than others, and their traits could then be passed on in higher proportion to future generations. Thus populations could continually improve their adaptations to environments to which they were subjected, and populations with inadequate adaptations would become extinct. Limited environmental resources, eventually faced by all organisms, made reproductive success the ultimate judge of survival. By coincidence, the naturalist Alfred Russel Wallace simultaneously proposed the same mechanism. The papers of both men were presented in 1858, and Darwin's *On the Origin of Species* was published in 1859.

However, the question of *how* organisms might change remained unresolved. The theory of natural selection depended on the presence of inheritable variations on which selection could act. Neither Darwin nor Wallace knew how such variants might be produced. Not until the science of genetics developed was this difficulty resolved.

KEY TERMS

adaptations
artificial selection
blending inheritance
Galapagos Islands
germ plasm theory
H.M.S. *Beagle*
inheritance of Acquired Characters
Lamarckianism
Malthus
natural selection
pangenesis
principle of use and disuse

DISCUSSION QUESTIONS

1. How did his experiences in the Galapagos Islands affect Charles Darwin's thinking in searching for evolutionary explanations?
2. a. What were the Lamarckian explanations for evolutionary change?

b. Why are these explanations now unacceptable to most biologists?

3. a. What is the concept of natural selection?
   b. From what sources did the concept of natural selection develop?
   c. What are the differences between Lamarckian explanations for evolution and the Darwin-Wallace concept of natural selection?

4. If Darwin's concept of natural selection explained how giraffes attained longer necks—by reaching higher branches during times of drought and intense competition—what explanation could he have offered to explain why sheep do not also develop long necks?

5. What is the pangenesis theory, and why did Darwin espouse it?

EVOLUTION ON THE WEB

Explore evolution on the web! Visit the accompanying web site for *Evolution,* 3/e at www.jbpub.com/evolution for web exercises and links relating to topics covered in this chapter.

REFERENCES

Barloon, T. J., and R. Noyes, Jr., 1997. Charles Darwin and panic disorder. *JAMA,* **277,** 138–141.

Beddall, B. G., 1968. Wallace, Darwin, and the theory of natural selection. *J. Hist. Biol.* **1,** 261–323. 1988.

———, Darwin and divergence: The Wallace connection. *J. Hist. Biol.,* **21,** 1–68.

Bowlby, J., 1990. *Charles Darwin.* Hutchinson, London.

Brackman, A. C., 1980. *A Delicate Arrangement: The Strange Case of Charles Darwin and Alfred Russel Wallace.* Times Books, New York.

Brent, P., 1981. *Charles Darwin: A Man of Enlarged Curiosity.* Harper & Row, New York.

Brooks, J. L., 1984. *Just Before the Origin: Alfred Russel Wallace's Theory of Evolution.* Columbia University Press, New York.

Burkhardt, R. W., Jr., 1977. *The Spirit of System: Lamarck and Evolutionary Biology.* Harvard University Press, Cambridge, MA.

Clark, R. W., 1984. *The Survival of Charles Darwin: A Biography of Man and Idea.* Weidenfeld & Nicolson, London.

Colp, R., 1998. To be an invalid, redux. *J. Hist. Biol.,* **31,** 211–240.

Darwin, C., 1845. *The Voyage of the Beagle.* (Originally published as *Journal of Researches,* it has now appeared in numerous editions.)

———, 1859. *On the Origin of Species by Means of Natural Selection or the Preservation of Favoured Races in the Struggle for Life.* Murray, London.

Darwin, F., 1887. *The Life and Letters of Charles Darwin.* Appleton, New York.

De Beer, G., 1963. *Charles Darwin.* Nelson, London.

Desmond, A., and J. Moore, 1991. *Darwin.* Warner Books, New York.

Eiseley, L. C., 1958. *Darwin's Century: Evolution and the Men Who Discovered It.* Doubleday, New York.

Gayon, J., 1998. *Darwinism's Struggle for Survival: Heredity and the Hypothesis of Natural Selection.* Cambridge University Press, Cambridge, England.

Gordon, S., 1989. Darwin and political economy: The connection reconsidered. *J. Hist. Biol.,* **22,** 433–459.

Greene, J. C., 1959. *The Death of Adam.* Iowa State University Press, Ames.

Irvine, W., 1955. *Apes, Angels, and Victorians.* McGraw-Hill, New York.

Jablonka, E., and M. J. Lamb, 1995. *Epigenetic Inheritance and Evolution: The Lamarckian Dimension.* Oxford University, Oxford, England.

Joravsky, D., 1970. *The Lysenko Affair.* Harvard University Press, Cambridge, MA.

Keynes, R. D. (ed.), 1979. *The Beagle Record.* Cambridge University Press, Cambridge, England.

Kohn, D. (ed.), 1985. *The Darwinian Heritage.* Princeton University Press, Princeton, NJ.

Lack, D., 1947. *Darwin's Finches: An Essay on the General Biological Theory of Evolution.* Cambridge University Press, Cambridge, England.

Lamarck, J. B., 1809. *Zoological Philosophy.* Translated into English by H. Elliott, 1914, Macmillan, New York.

Lerner, I. M., 1959. The concept of natural selection: A centennial view. *Proc. Amer. Phil. Soc.,* **103,** 173–182.

McKinney, H. L. (ed.), 1971. *Lamarck to Darwin: Contributions to Evolutionary Biology, 1809–1859.* (Contains short excerpts from original writings of J. B. Lamarck, W. C. Wells, P. Matthew, C. Lyell, E. Blyth, R. Chambers, A. R. Wallace, and C. Darwin.) Coronado Press, Lawrence, KS.

———, 1972. *Wallace and Natural Selection.* Yale University Press, New Haven, CT.

Millhauser, M., 1959. *Just Before Darwin: Robert Chambers and Vestiges.* Wesleyan University Press, Middletown, CT.

Moorehead, A., 1969. *Darwin and the Beagle.* Hamilton, London.

Ospovat, D., 1981. *The Development of Darwin's Theory: Natural History, Natural Theology, and Natural Selection, 1838–1859.* Cambridge University Press, Cambridge, England.

Richards, R. J., 1987. *Darwin and the Emergence of Evolutionary Theories of Mind and Behavior.* University of Chicago Press, Chicago.

Strickberger, M. W., 1985. *Genetics,* 3d ed. Macmillan, New York.

Sulloway, F. J., 1982. Darwin and his finches: The evolution of a legend. *J. Hist. Biol.,* **15,** 1–53.

Thompson, K. S., 1995. *HMS Beagle: The Story of Darwin's Ship.* W. W. Norton, New York.

Zirkle, C., 1946. The early history of the idea of the inheritance of acquired characters and of pangenesis. *Trans. Amer. Phil. Soc.,* **35,** 91–151.

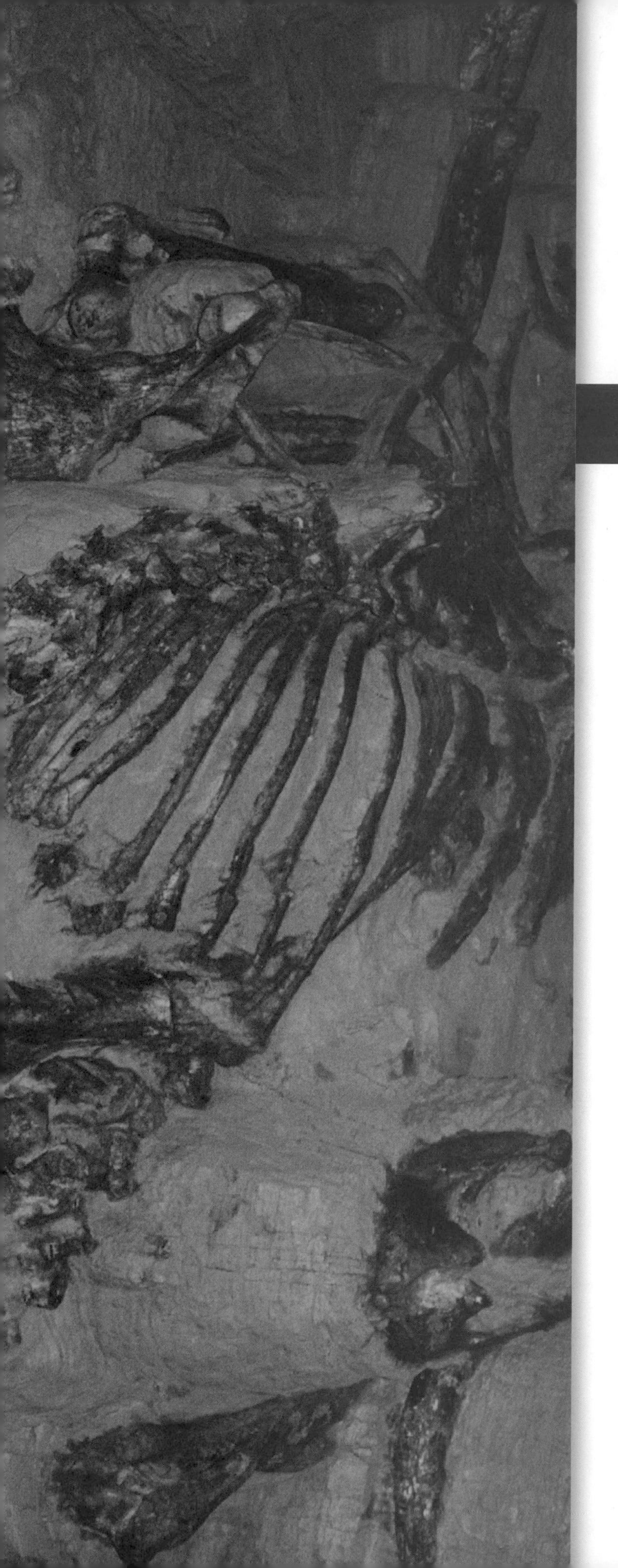

3 The Arguments and the Evidence

Although the public quietly received the brief natural selection papers by Darwin and Wallace in 1858, Darwin's book, *On the Origin of Species,* published the following year, had profound effects. Many biologists found Darwin's detailed exposition of his theory, supported by 20 years of thought and documentation, impossible to overlook. They rapidly recognized evolution as the cause for the diversity of species, accepting to various degrees that natural selection was an important if not primary mechanism for evolution. Thomas Huxley (1825–1895), who later became Darwin's main public defender, is said to have exclaimed, "How extremely stupid not to have thought of that."

Scientific Objections

Objections, however, came rapidly in the nineteenth century and took various forms, including religious attacks that will be discussed in Chapter 4. On the scientific level, opponents raised a variety of major questions about Darwin's theory, which he and his supporters answered in various ways at different times.

BLENDING INHERITANCE

Since the prevailing concept of inheritance was that maternal and paternal contributions blended in their offspring, a number of critics objected that new adaptations would be successively diluted with each generation of interbreeding. According to this argument natural selection would be incapable of maintaining a trait for more than a few generations. Darwin had early been aware of this objection and made various replies, among them the following:

1. A beneficial trait could maintain itself if those who had it were isolated from the rest of the population. Darwin pointed to the familiar practice among animal breeders of isolating newly appearing "sports"—called *mutations* by Hugo de Vries

(1848–1935)—and their offspring. This mechanism was commonly used to develop new stocks.

2. Some traits are "prepotent," or dominant, and appear undiluted in later generations.

3. An adaptive trait does not appear only once in a population; rather, such traits must arise fairly often—witness the large amount of variability present in most populations. Since variability is common, it cannot dilute out as easily as if it were rare. Moreover, Darwin believed, some forms that carry a particular variation pass on to future generations the tendency for the same variation to arise again.

4. Natural selection not only enhances the reproductive success of favorable variants but also diminishes the reproductive success of unfavorable ones. Thus the frequency of favorable variations increases when unfavorable ones die out, and less chance arises of diluting out favorable variations.

5. As explained earlier (p. 28), Darwin developed the concept of pangenesis by gemmules to help explain the inheritance of traits that he believed were affected by use and disuse and also to provide constancy for the determining agents of inheritance. There were presumably many gemmules for each particular trait, and their numbers could vary during passage from one generation to another; that is, gemmules could be lost but were not changed by "blending."

VARIABILITY

In *On the Origin of Species,* Darwin explicitly confined evolution by natural selection to small, continuous variations and (in the earlier editions of his book) excluded larger variations as not being useful. He had, in fact, literally adopted the Leibnizian dictum that "nature makes no leaps" (p. 8). A number of objections followed almost immediately.

The first objection, raised by various critics and emphasized by Fleeming Jenkin (1833–1885) in a review of 1867, concerned the limits of variability on which selection could act. Except for monstrosities that were highly abnormal or sterile, most observed variations were only of small changes and did not depart from the species pattern. How then could new species arise? To this Darwin replied that no limits really apply to variability because each stage in evolution of a species entails further variability, on which selection then acts. Darwin maintained that the succession of changes through time, rather than a single simultaneous set of changes, leads to species differences.

A second, more common objection was the difficulty in determining how selection would recognize each of the very small modifications Darwin proposed. Certainly, in many instances, such as size, a very small modification might hardly be enough to confer significant advantage on an organism, whereas a large modification might well be selected. Darwin could not successfully reply to this argument, yet he doggedly held to his concept of gradual accretion of small modifications, and the findings of modern genetics (Chapter 10) later added considerable support to his position.

Many traits have been discovered that stem from small heritable changes ascribed to many different genes, each with small effect, called *polygenes.* For example, size differences are often distributed in populations so that some large individuals possess many genes that lead to an increase in size while others have relatively few such genes and are therefore smaller. Thus, although differences in size may stem from genetic differences of small effect, selection may nevertheless act on accumulations of such differences in various individuals (Chapters 10 and 22).

A further aspect of this same problem was the question of determining the initial adaptive level that a trait or organ would have to reach for selection to favor it. If the trait already existed before selection acted on it, perhaps some quantitative expression of the trait would suffice for further evolution; that is, a larger eye might function better than a smaller eye. But if the trait did not exist, or only barely existed, how could selection act on it? For example, many critics felt that the earliest incipient stages of complex organs such as the eye, brain, and liver would have no function at all and could hardly be selected. Can one conceive of an appropriate adaptive function in only one cell of an eye, brain, or leg?

An evolutionary answer to this question of the origin of new traits came from the concept of preadaptation, a principle of which Darwin was aware. According to **preadaptation,** a new organ need not arise *de novo* but may already be present in an organism that is using it for a purpose other than that for which it is later selected. For example, in his monograph on barnacles, Darwin suggested that the cementing mechanism by which present-day barnacles attach to their substrate is related to the cementing mechanism by which the barnacle oviduct coats its eggs in order to attach them to solid objects. That is, only after its earlier evolution in oviducts was this mechanism adapted for attaching the barnacle itself. Similarly, Darwin pointed out that the evolution of lungs from swim bladders in fish illustrated "that an organ originally constructed for one purpose, namely flotation, may be converted into one for a wholly different purpose, respiration" (see also p. 404). Among modern evidence for this notion are findings that optical neural pathways, no longer needed in blind cave animals, may be used to enhance new olfactory and tactile functions (Voneida and Fish).

Given continued environmental pressure and selection, such evolutionary transitions are a commonly expected feature. A highly specialized organ like the vertebrate eye did not arise all at once but probably represents a succession of further evolutionary adaptations of a previous light-gathering organ and its ancillary tissues that may

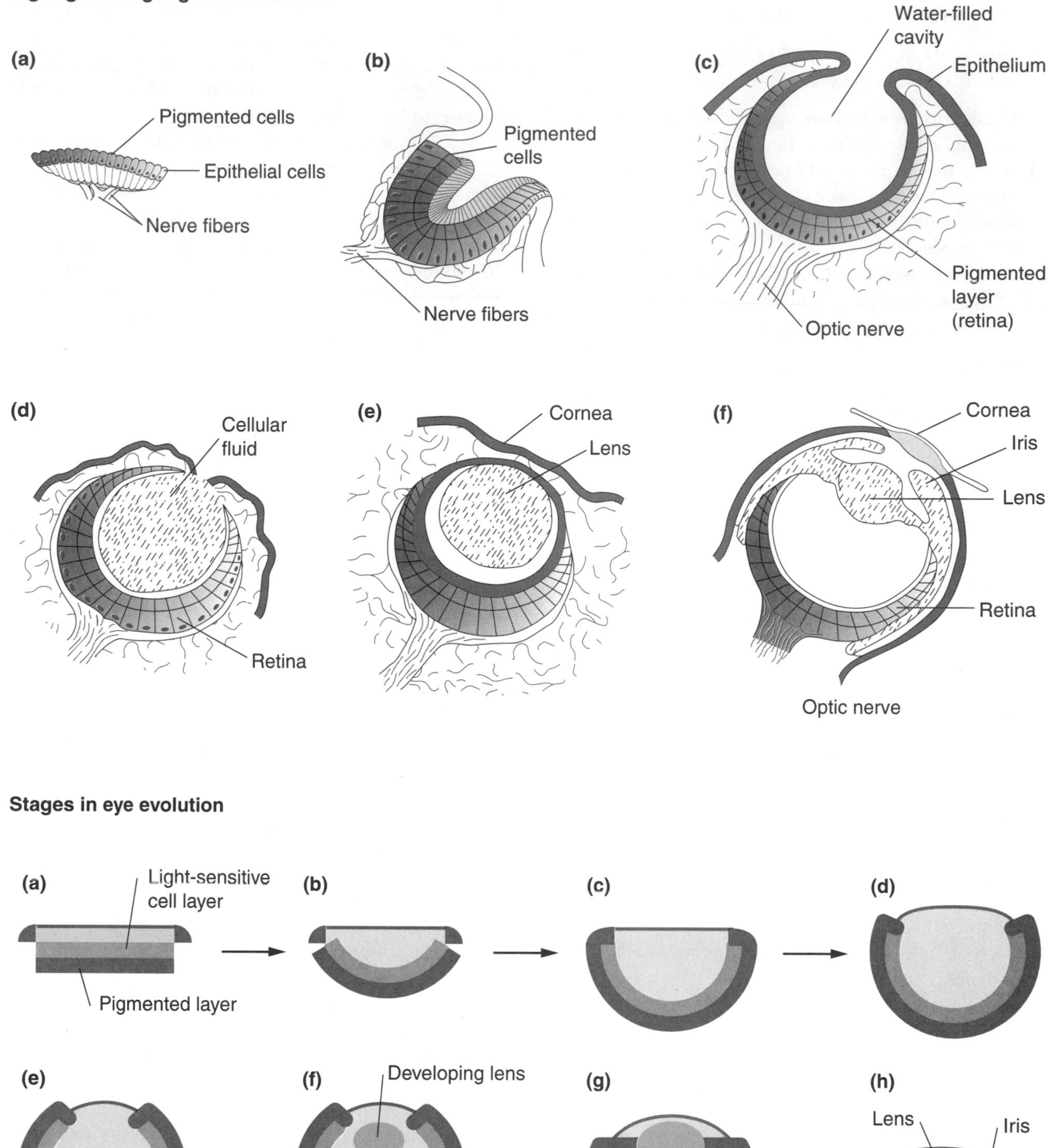
Light-gathering organs in mollusks
(a)
Pigmented cells
Epithelial cells
Nerve fibers
(b)
Pigmented cells
Nerve fibers
(c)
Water-filled cavity
Epithelium
Pigmented layer (retina)
Optic nerve
(d)
Cellular fluid
Retina
(e)
Cornea
Lens
(f)
Cornea
Iris
Lens
Retina
Optic nerve
Stages in eye evolution
(a)
Light-sensitive cell layer
Pigmented layer
(b)
(c)
(d)

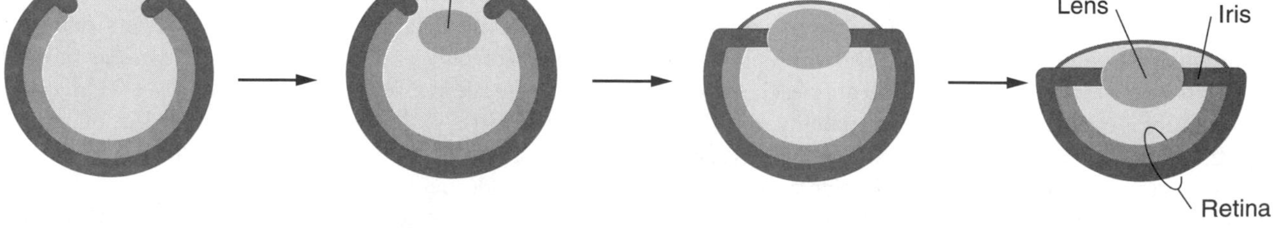
(e)
(f)
Developing lens
(g)
(h)
Lens
Iris
Retina

FIGURE 3-1

have originally involved only a few cells. A turn-of-the-century illustration of one such progression is shown in Figure 3–1 *top*. More recently, Nilsson and Pelger have calculated that even a small one percent per generation change in eye anatomy produces a marked effect in a very brief geological period. A flat patch of light-sensitive cells can change to a complex squidlike eye with a focused refractive lens in less than half a million years (Fig. 3–1 *bottom*).

Unfortunately, Darwin's search for small modifications led him to place less emphasis on the fact that many traits often show distinct steps and differences—such as different colors, presence and absence of structures, and different numbers of structures. These large variations may also be important for selection, as various biologists, including Huxley, suggested. Interestingly, such traits, showing large observable differences, enabled Mendel to develop the basic laws that explain inheritance (Chapter 10). Until the twentieth century, Darwinists did not resolve the problem of where, how, and to what extent variations originated, and it remained the most often attacked element in Darwin's theory.

ISOLATION

Critics also singled out Darwin's almost complete emphasis on the transformation of a single species into another single species (phyletic evolution; see Chapter 11). They pointed out that although Darwin's approach accounted for the evolution of a particular species in time, it did not easily account for the multiplication of species in geographical space. What explains the origin of many new species rather than the transformation of one old species?

Furthermore, argued Moritz Wagner (1813–1887), among others, Darwin did not even fully explain the evolution of a single species into a single new species, because how could a new species possibly evolve in the same locality as its parents?

> Free crossing of a new variety with the old unaltered stock will always cause it to revert to the original type. . . . Free crossing, as the artificial selection of animals and plants uncontestably teaches, not only renders the formation of new races impossible, but invariably destroys newly formed individual varieties.

Missing in Darwin's 1859 argument was a strong emphasis on the barriers that prevent exchange of hereditary material between different groups that would let each such isolated group follow its own evolutionary path.

Because Darwin did not emphasize isolation among groups as a primary cause for evolution, he also dismissed the notion that sterility among separately evolved groups might be beneficial. That is, as Wallace showed, it would be advantageous for isolated populations, each with its special adaptations, to produce sterile hybrids when they meet, because sterility would let each group maintain its unique adaptations without dilution. Darwin insisted that sterility was primarily accidental. This view blocked him from explaining the almost universal prevalence of sterility among species and from using this important isolating barrier to account for the divergent evolution of closely related species.

On the whole, although Darwin knew isolation could be important in helping a population evolve, he apparently

FIGURE 3–1 *Top*: Some stages in the evolution of eyes as found in mollusks, a phylum whose various groups show different needs for vision and a wide range of light-gathering organs. (*a*) A pigment spot with neural connections that light can stimulate. (*b*) Folding of pigment cells concentrates their activity, thus providing improved light detection. (*c*) A partly closed, water-filled cavity of pigment cells that allows images to form on the pigmented layer as in a pinhole camera. (*d*) Secreted transparent cellular fluid, instead of water, forms a barrier that protects the pigmented layer (retina) from external injury. (*e*) A thin film or transparent skin covers the entire eye apparatus, adding further protection. Also, some of the fluid within the eye hardens into a convex lens that improves the focusing of light on the retina. (*f*) A complex eye found in squids, which has an adjustable iris diaphragm and focusing lens. (*Adapted from Conn.*) Of course, not all mollusks need visual devices with focusing refractive lenses, and selection may even go in an opposite direction in cave animals, from more to less visual acuity. For recent discussions of the evolution of eyes and photoreceptor pigments, see Goldsmith. Interestingly, genetic studies indicate that similar inherited factors (*pax*-6 gene sequences) regulate development of anterior sense organ patterns in both invertebrates and vertebrates (p. 354). These factors probably extend back to a wormlike common ancestor of both groups (Loosli et al.). Nevertheless, despite some common regulatory features, specific cellular pathways in embryonic eye development differ noticeably between squids (*f*) and vertebrates: squid photoreceptor cells derive from the epidermis, whereas vertebrate retinae derive from the central nervous system (Harris). As explained by the process of "convergent evolution" (pp. 37, 242), the structural similarity of their eyes does not come from a common ancestral visual structure, but rather from similar selective pressures leading to similar organs that enhance visual acuity. Such morphological convergences may have also arisen independently in numerous other animal lineages subject to similar selective visual pressures (Salvini-Plawen and Mayr).
Bottom: Stages in eye evolution displayed by a computerized model in which random changes in eye structure are followed by selection for visual acuity. Beginning with a light-sensitive middle layer of skin backed by pigment (*a*), successive selective steps for improved optical properties lead to a concave buckling that enhances light gathering (*b–e*), a focusing lens (*f–g*), and an eye with a flattened iris in which the focal length of the lens equals the distance between lens and retina (*h*). (*Adapted from Nilsson and Pelger.*)

felt that it was more essential for him to establish that speciation could occur without isolation. It therefore remained the task of others to explore the role of isolation in forming species (Chapter 24).

AGE OF THE EARTH

Essential to Darwin's argument was a belief that the age of the Earth extended beyond anything ever proposed before. As Darwin pointed out in an 1844 essay:

> The mind cannot grasp the full meaning of the term of a million or hundred million years, and cannot consequently add up and perceive the full effects of small successive variations accumulated during almost infinitely many generations.

This emphasis on evolution taking a long time ran counter to the time spans usually given. The heliocentric theory tied the Earth's origin to the sun, and Newton had calculated that a sphere the size of the Earth would take about 50,000 years to cool down to its present temperature. Since even such a short period contradicted the 5,000 or so years of history allowed in the Judeo-Christian Bible, Newton piously rejected these calculations. Buffon, in contrast, calculated approximately 75,000 years of age for the Earth, reconciling this with the biblical time scale by interpreting each of the seven days of creation in Genesis as a separate geological epoch, varying in length from 3,000 to 3,500 years.

In Darwin's time, William Thomson (Lord Kelvin, 1824–1907) reassessed the temperature gradients observed in mine shafts, the conductivity of rocks, and the presumed temperature and cooling rate of the sun. He then calculated the total age of the Earth's crust at about 100 million years. Of this duration, however, Thomson suggested that only the last 20 to 40 million years could have been sufficiently cool for life to exist. This figure, although large by previous estimates, was still too small to account for the Darwinian evolution of organisms.

Darwin had no answer to these various calculations; their underestimate of terrestrial age and age of the sun came from unawareness of the radioactivity that accounts for the Earth's interior heat and of the atomic fusion reactions that account for the sun's continued radiation. Geological age was not accurately measured until twentieth-century scientists developed radioactive dating techniques (Chapter 6).

Support of Darwin

Although critics opposed Darwin, and he and his supporters did not always have the knowledge and the data to answer each objection satisfactorily, Darwin's works made the evolution of species an acceptable concept. One important reason for this success was that although Darwin presented many hypotheses—such as the struggle for existence, natural selection, the divergence between species, and the improvement of adaptations—each of these mechanisms relied on natural processes and could be supported by observations. These natural processes were in strong contrast to previous, more speculative, evolutionary theories such as Lamarck's, which were tied to nonmaterial agents impossible to observe.

Another attractive feature was Darwinism's expansion of the role of biology to include the study of relationships among all living creatures, including humans, who were formerly thought to be divine and separate. In every area of biology, from anthropology to botany to paleontology to zoology, Darwinism opened new lines of thought and new areas of investigation. What are the relationships among different kinds of cells? different parts of cells? different flowers? How did orchids evolve? How did bone structures change? What are the steps in the evolution of circulatory systems? nervous systems? Why do some species mimic others? How did sterile insect castes such as worker bees evolve? What accounts for the geographical distribution of specific organisms? Although each topic demanded separate techniques and study, all sprang from an evolutionary source that rational, understandable mechanisms could explain. By offering an overall view of adaptation and evolution, Darwin helped start the process by which scientists could eventually bind all of biology together.

Of considerable importance to biologists of the time were also the many lines of both direct and indirect evidence that rapidly began to accumulate in support of an evolutionary view. Some of these are briefly outlined as follows:

SYSTEMATICS

Although the evidence was indirect, after Darwin it seemed clear that the gradation of different organisms observed in classification procedures, whether from simple to complex or from one type to another, could most easily be explained by evolutionary relationships (Fig. 3–2).

GEOGRAPHICAL DISTRIBUTION

Many biologists became aware that groups of organisms that are evolutionarily related are usually, as expected, geographically connected. Large geographical barriers such as oceans and mountain ranges isolate groups from one another and lead to considerable differences among the separated groups. Colonizers that transcend such barriers often become the ancestors of entirely new groups. This showed up most graphically in the wide evolutionary radiation of species that descended from the finches that reached the Galapagos Islands (p. 22). Beginning with

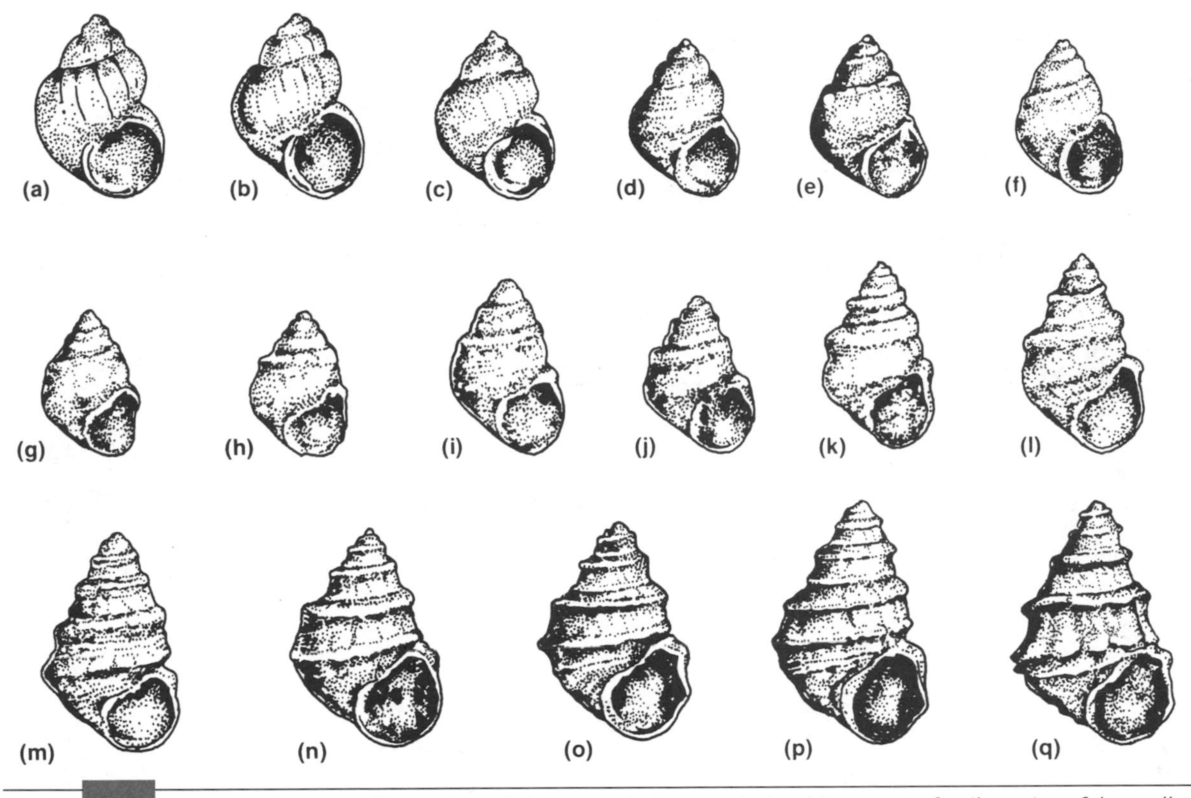

FIGURE 3-2 A nineteenth century illustration of obvious evolutionary relationships among fossil species of the mollusk *Paludina,* ranging from the oldest form, *P. neumayri* (*a*), to the youngest form, *P. hoenesi* (*q*). To Darwin and many others, differences of this kind "blend into each other in an insensible series; and a series impresses the mind with the idea of an actual passage." (*From Romanes.*)

what was probably an ordinary mainland finch, new kinds of finches evolved in the Galapagos that could function in habitats that other bird species left vacant (Fig. 3–3).

The name given to this process, **adaptive radiation,** signifies the rapid evolution of one or a few forms into many different species that occupy different habitats within a new geographical area. The marsupial radiation in Australia (Fig. 3–4) shows how, when marsupials are protected from competition with placental mammals by the isolation of a continent, this process can lead to an entire array of species with widely divergent functions, from herbivores to carnivores.

COMPARATIVE ANATOMY

The comparative anatomy area of biology—the study of comparative relationships among anatomical structures in different species—became for a period of time after Darwin the most popular biological discipline. A search for evolutionary relationships made it possible to trace, especially in vertebrates, many stepwise changes in bones, muscles, nerves, organs, and blood vessels (Fig. 3–5). Such studies made clear that as each species and group of species evolved, previously inherited structures could become modified in entirely new ways.

Derived from terminology Richard Owen (1804–1892) introduced in the 1840s, organs that related to each other through common descent, although now perhaps functioning differently, were called **homologous.** For example, a study of bones and muscles showed that evolutionary homology could explain the forelimbs of widely different vertebrates (Fig. 3–6). In contrast, **analogous** organs that performed the same function in different groups, such as the wings of bats and the wings of insects, do not show a common underlying plan of structure, since these organs did not evolve from the same organ in a common ancestor. Even when analogous organs seemed strikingly similar, such as the eye of an octopus and the eye of a mammal, biologists could demonstrate that they differed in retinal position.

The evolution of different organisms, or parts of organisms, in such similar directions was called **convergent evolution,** indicating that selection for similar habitats in different evolutionary lineages could occasionally lead to functionally similar (although not identical) anatomical structures (Fig. 3–7). However, with the general exception of such events whose genetic distinctions were not yet understood, comparative anatomy followed the logic that organisms with shared structures derived from a common group of ancestors, whereas organisms with unlike structures represented divergent

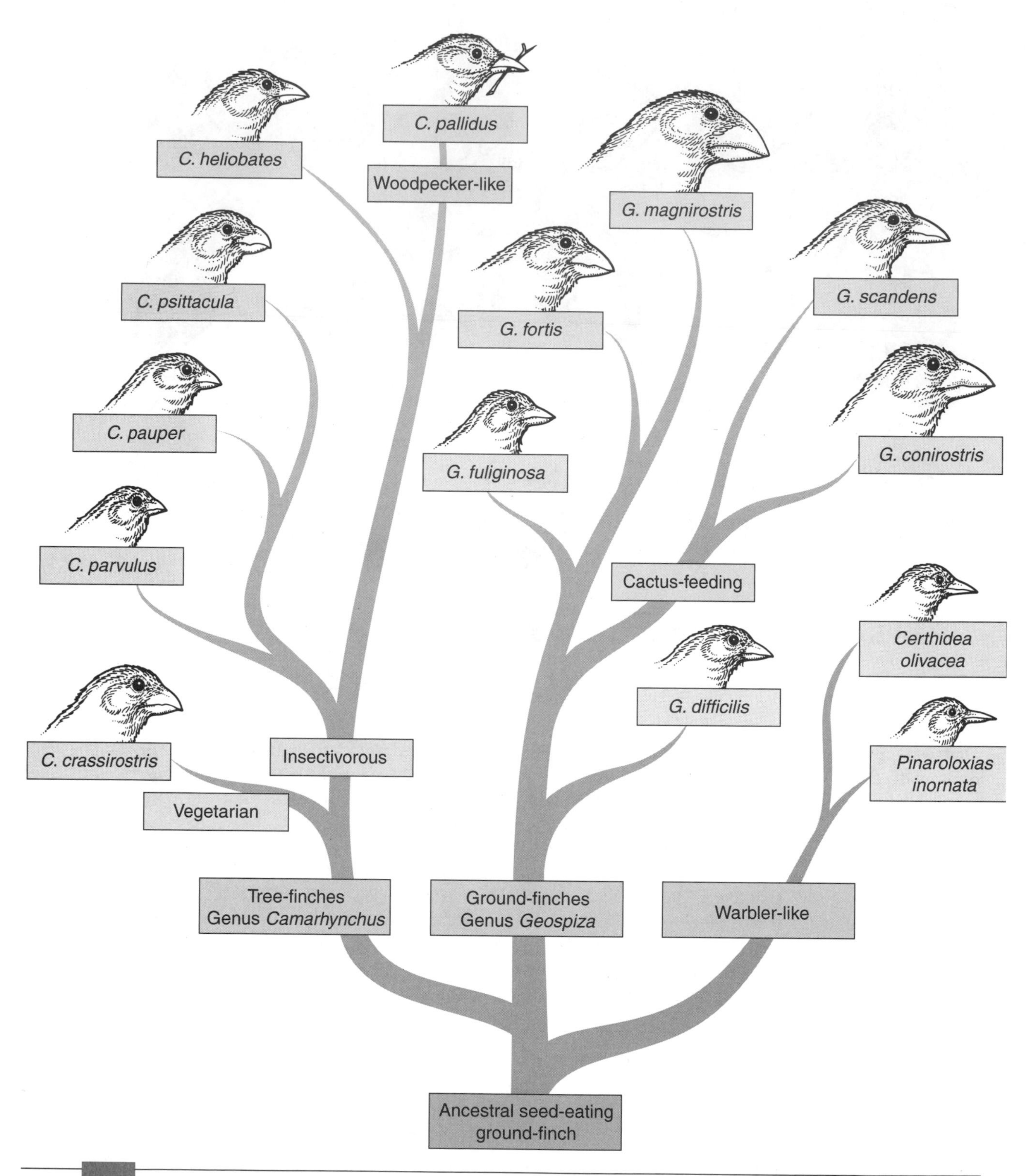

FIGURE 3-3 Evolutionary tree of Darwin's finches showing beak adaptations of the individual species. (*Adapted from Lack.*) A recent molecular study by Sato and coworkers supports some aspects of this phylogeny. Using comparisons among mitochondrial DNA sequences (see Chapter 12), the molecular study distinguishes between tree finches and ground finches, but shows that distinctions among members within each group have not yet been firmly established. ("[T]he incomplete species differentiation within the ground and tree finch groups is probably a sign of an adaptive radiation in progress.") This molecular study also indicates that these finches are all descended from a single species, now identified as a warbler-type "dull-colored grassquit."

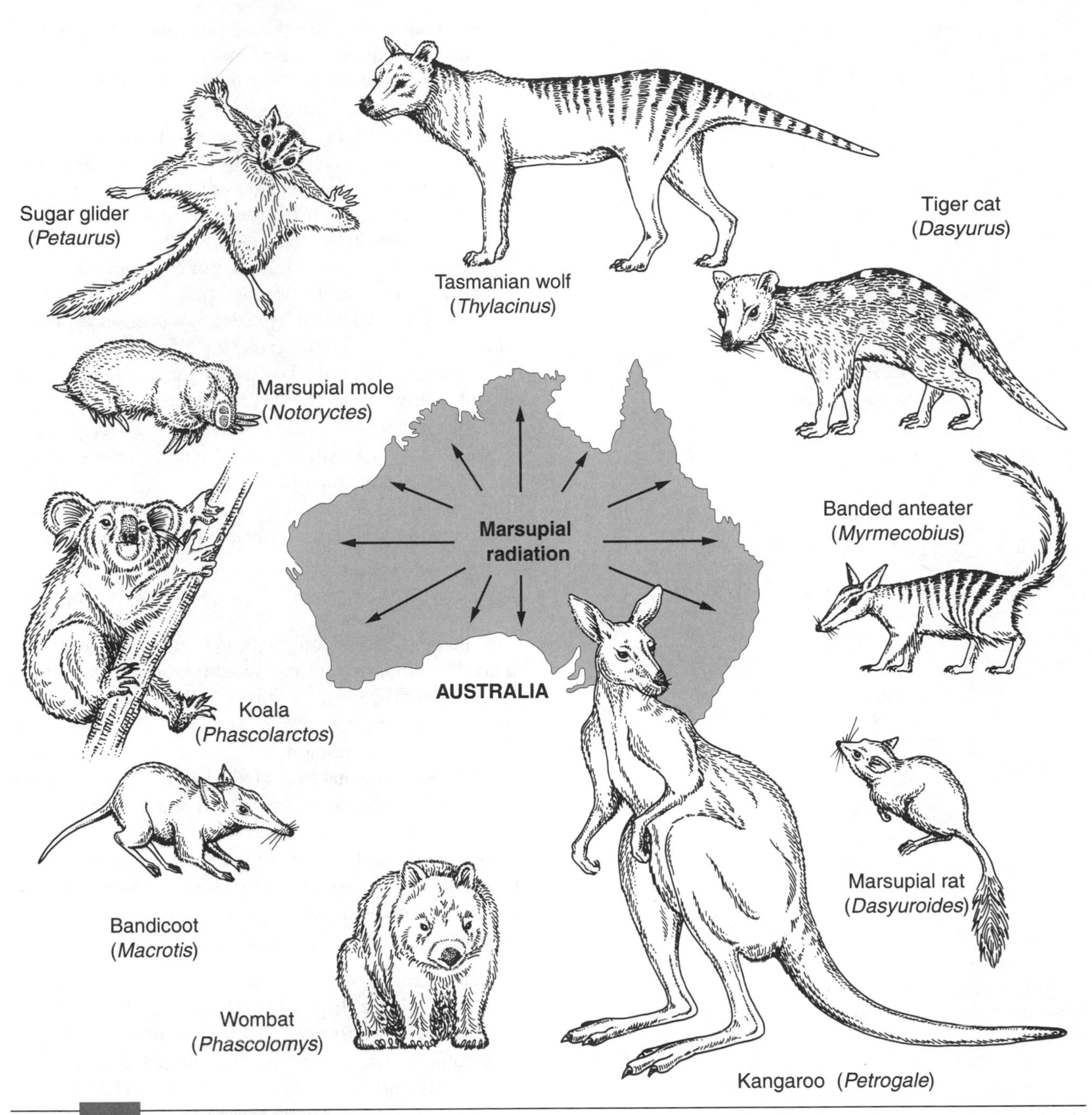

FIGURE 3-4 Adaptive radiation of Australian marsupial mammals showing the many divergent forms that evolved independently of changes occurring among placental mammals on other continents. The striking similarity between some of these marsupial mammals and placental mammals arises because selection for survival in similar habitats can lead to similar adaptations—that is, parallel or convergent evolution has taken place (see also Fig. 3-7 and p. 242). (*Adapted from Simpson and Beck, with additions.*)

evolutionary pathways. Careful anatomical dissections and comparisons provided the criteria for constructing detailed evolutionary trees (Chapter 11).

Of considerable interest to comparative anatomists was finding structures that seemed to have lost some or all of the function they had in earlier ancestors. From an evolutionary viewpoint, biologists could explain the presence of such rudimentary or **vestigial organs** as arising because an organism adapting to a new environment usually carries along some previously evolved structures that are now no longer necessary. According to the principle of natural selection, individuals that devote less energy to the specific elaboration and maintenance of such extraneous structures would be more reproductively successful than individuals that maintain them. Moreover, some structures that were no longer necessary might well interfere with the functioning of new adaptations, or, in some cases, even give rise to an entirely new evolutionary function. Among such latter innovations are vestigial rudiments of the reptilian jaw apparatus that evolved into

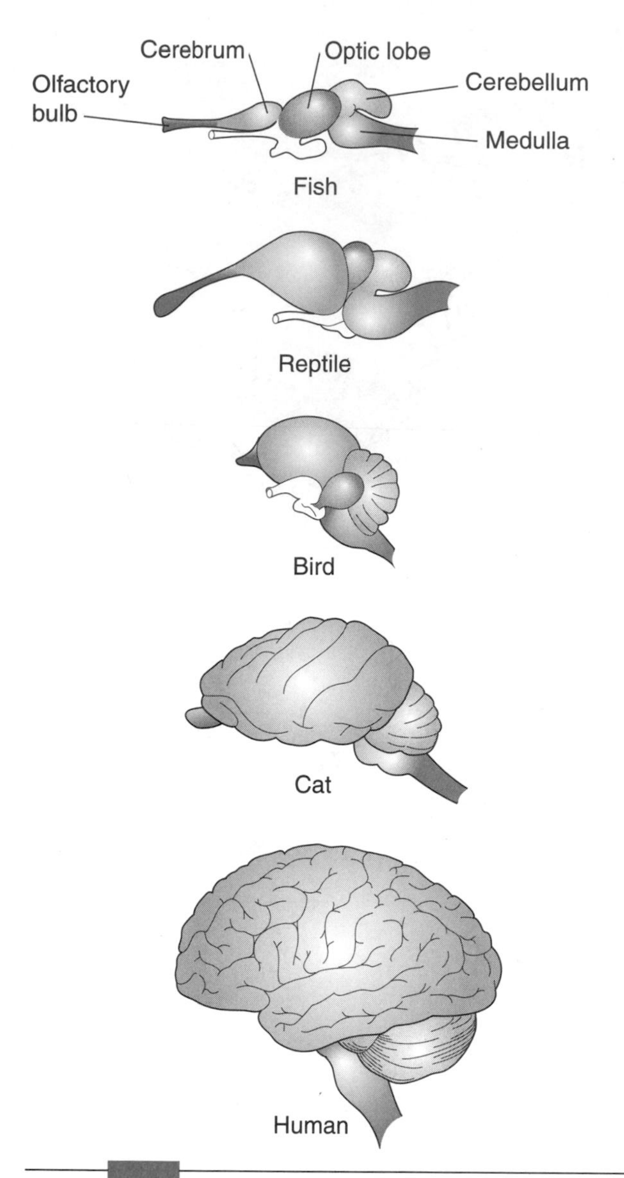

FIGURE 3–5 Redrawn nineteenth century diagrams of vertebrate brains. Homologous structures shared by all these groups are the cerebellum, cerebrum, medulla, olfactory bulbs, and optic lobes. In humans, overgrowth of the cerebrum covers the olfactory bulbs and optic lobes. (*After LeConte.*)

the three mammalian ossicles that carry sound to the inner ear (Fig. 19–3). Similarly, new olfactory and tactile functions were assumed by former optical neural pathways of blind cave animals (p. 33).

Thus, as time went on, obsolete structures, when not adopting new roles, would tend to diminish, showing only traces of their former size and function. Examples of these were found in the rudimentary bones that were all that remained of the former hind limbs in the whale and snake species shown in Figure 3–8. The presence of reduced eye stalks in blind, cave-dwelling crustaceans also indicated the evolutionary process by which obsolete structures gradually became rudimentary.

In humans, a number of vestigial structures indicate not only the obsolescence of organs but also a relationship to other mammals and primates. For example, muscles of the external ear, as well as scalp muscles, are rudimentary in humans and often nonfunctional, but are common to many mammals. The inflection of the feet for grasping, along with extension of the great toe, is a primate trait that is often expressed in human infants but that degenerates in human adults.

The marvelous gripping power of the hands of human infants is also a vestigial trait, because other primates develop it even further for brachiating purposes (grasping of tree branches) and to permit the infant to cling tightly to its mother. Obvious vestigial organs in humans and the other great apes are the reduced tail bones (*os coccyx*) and the remnants of a few tail muscles. To these one can probably add the nictitating membrane of the eye, the appendix of the cecum, rudimentary body hair, and wisdom teeth (Fig. 3–9). All these are apparent vestiges of more developed structures present in earlier mammals.

EMBRYOLOGY

Early in the nineteenth century, von Baer had noticed the remarkable similarity among vertebrate embryos that are quite different from each other as adults. He generalized such observations into a "law" that early embryos of related species bear more common features than do later, more specialized developmental stages ("the increasing individuality of the growing animal"). Darwin and others therefore considered the early stages of development more conservative or evolutionarily stable than the adult stages. Embryonic comparisons among vertebrates showed even remote evolutionary relationships in the persistence of a common stage called the **pharyngula** (Fig. 3–10 *top row*). As Darwin stated:

> In two groups of animals, however much they may at present differ from each other in structure and habits, if they pass through the same or similar embryonic stages, we may feel assured that they have both descended from the same or nearly similar parents, and are therefore in that degree closely related. Thus community in embryonic structure reveals community of descent.

Ernst Haeckel (1834–1919), the main propagandist for evolution in Germany, further developed and popularized this concept into the **biogenetic law**:

> Ontogeny [development of the individual] is a short rapid recapitulation of phylogeny [the ancestral sequence]. . . . The organic individual repeats during the swift brief course of its individual development the most important of the form-changes which its ancestors traversed during the slow protracted course of their paleontological evolution according to the laws of heredity and adaptation (1866).

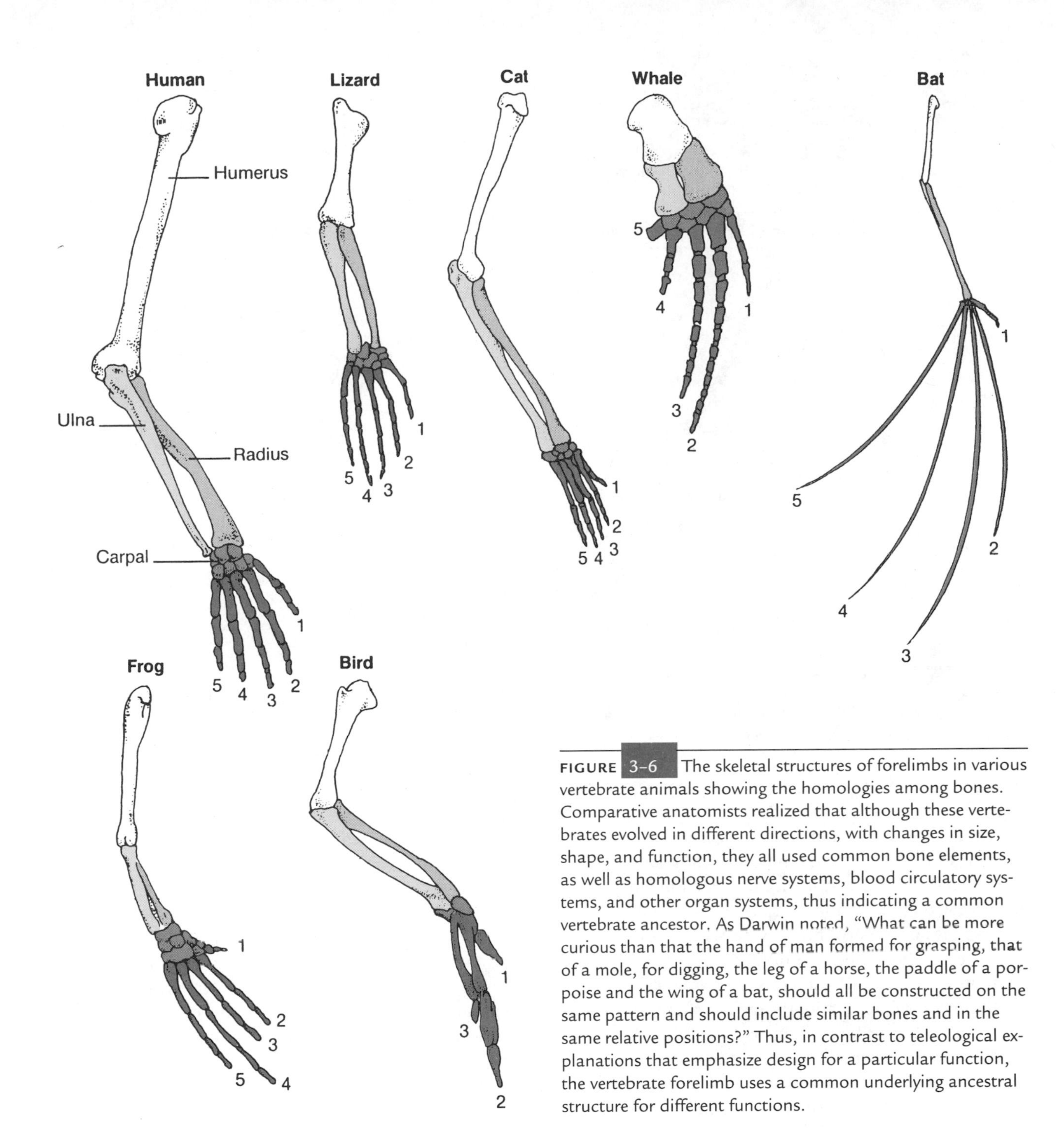

FIGURE 3–6 The skeletal structures of forelimbs in various vertebrate animals showing the homologies among bones. Comparative anatomists realized that although these vertebrates evolved in different directions, with changes in size, shape, and function, they all used common bone elements, as well as homologous nerve systems, blood circulatory systems, and other organ systems, thus indicating a common vertebrate ancestor. As Darwin noted, "What can be more curious than that the hand of man formed for grasping, that of a mole, for digging, the leg of a horse, the paddle of a porpoise and the wing of a bat, should all be constructed on the same pattern and should include similar bones and in the same relative positions?" Thus, in contrast to teleological explanations that emphasize design for a particular function, the vertebrate forelimb uses a common underlying ancestral structure for different functions.

To Haeckel this apparently meant that early stages of development may also recapitulate the adult ancestral forms. Scientists have widely disputed this point, and most biologists consider Haeckel's law an oversimplification.[1] According to present views, early stages of development

[1]Difficulties in applying the concept that ontogeny recapitulates phylogeny are obvious for a number of developmental processes, especially those that show progressive but imperfect functional changes during the life cycle of an individual. In humans, for example, infantile stages of stumbling, falling, and sitting precede walking. It is difficult to imagine that these stages recapitulate prehuman stages in which populations fell or stumbled about. Admittedly, bipedal adaptations for walking took many generations to evolve, but those primate populations in which they were evolving must have had functional mobility. Moreover, the discovery that juvenile stages of ancestral organisms can be retained in the adult forms of their descendants—such as the preservation of juvenile ape features in adult humans (**neoteny,** see Montagu, and also Fig. 20–26)—directly contradicted the Haeckelian notion that descendants retain ancestral *adult* features. More often retained in closely related species are therefore *embryological* processes, because related organisms generally rely on common genetic agents that can produce characteristic developmental stages, such as those associated with the vertebrate "phylotype" (Fig. 3–10).

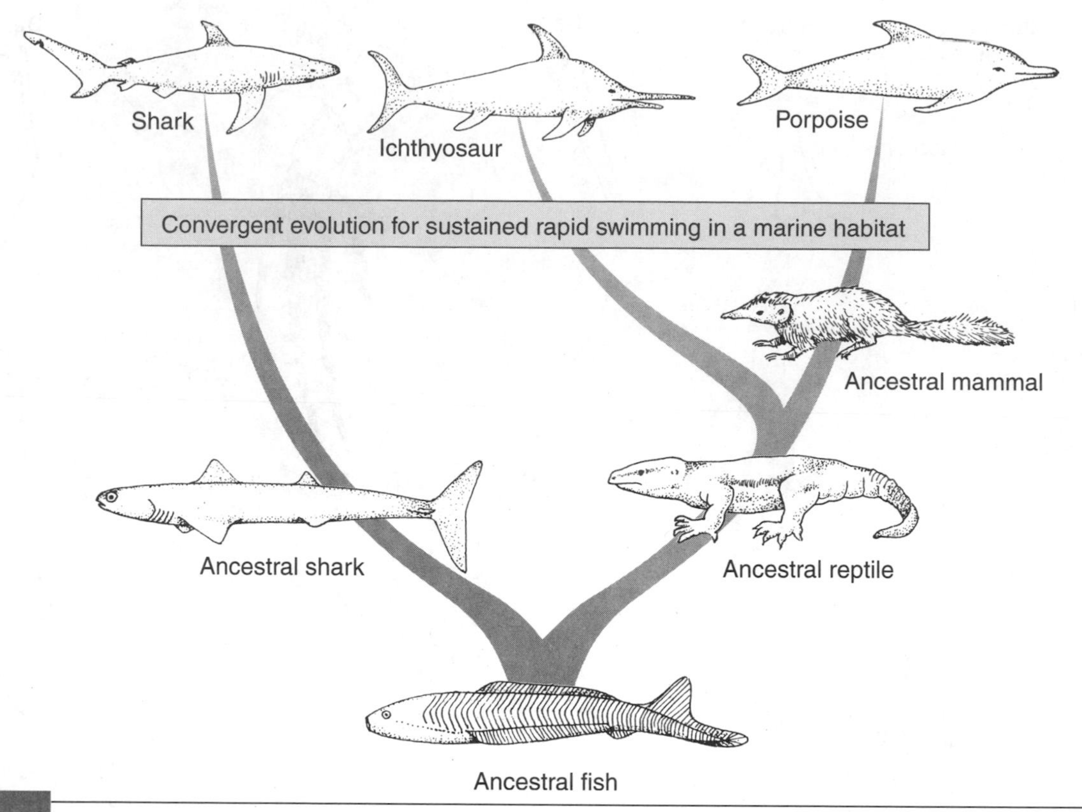

FIGURE 3–7 Convergent evolution in three marine predators that have different ancestries: shark (fish), ichthyosaur (extinct reptile), and porpoise (placental mammal). This illustrates that similar adaptations for rapid movement through water have been independently selected in each of these three lineages.

probably recapitulate only early ancestral developmental stages. It is as though organisms that share common descent make use of common underlying embryological patterns on which to build later but different adult patterns. Genetic and molecular evidence discussed in Chapter 15 provides strong support for this view.

Examples of the evolutionary persistence of underlying patterns often appear in human development where gill arches (see Fig. 3–10) serve as the basis for the further development of head and thoracic structures, yet no functional gills ever form. Similarly, the human embryo, like the embryo of the chick and other advanced vertebrates, goes through the stage of having a two-chambered heart like a fish, although the final functional human heart is four-chambered.

Even the anatomical positions of some adult nerves, blood vessels, and other structures are intelligible only owing to their evolutionary developmental patterns. For example, as Figure 3–11*a* shows, each branch of the vagus nerve in fish runs through an arterial arch pierced by a gill slit. From studies in many vertebrates, it is clear that two of these vagal nerve branches eventually evolved in mammals for stimulating the larynx. The most anterior of these, called the *anterior laryngeal nerve,* loops around the third arterial arch (now the carotid artery), and a posterior nerve branch, called the *recurrent laryngeal nerve,* loops around the sixth arterial arch (Fig. 3–11*b*).

However, in contrast to its function in fish, the left side of the sixth arterial arch in mammals is the *ductus arteriosis,* which is used embryonically to carry blood to the placenta until birth but then atrophies to become a pulmonary artery ligament. Since the ligamentous remnant of this old sixth arterial arch is close to the heart in adult mammals, to complete its circuit the left side of the recurrent laryngeal nerve must travel from the cranium to the thoracic cavity and back to the larynx. In mammals with long necks, such as giraffes, this extra loop is obviously many feet longer than it would be if nerve development were independent of evolutionary pattern.

All these embryonic stages and anomalies make sense to biologists only if we consider that humans and other terrestrial vertebrates had fishlike ancestors that provided them with some of their basic developmental patterns.

FOSSILS

In Darwin's day the fossil record was even more spotty than today. Fossil remains predominantly appear in sedimentary rocks originally laid down by a succession of deposits in seas, lakes, riverbeds, deserts, and so on and occur in some areas but not in others (Fig. 3–12). Even in appropriate sedimentary environments, many dead organisms decompose before they fossilize or are later destroyed by the erosion of sedimentary rocks even when they have fossilized.

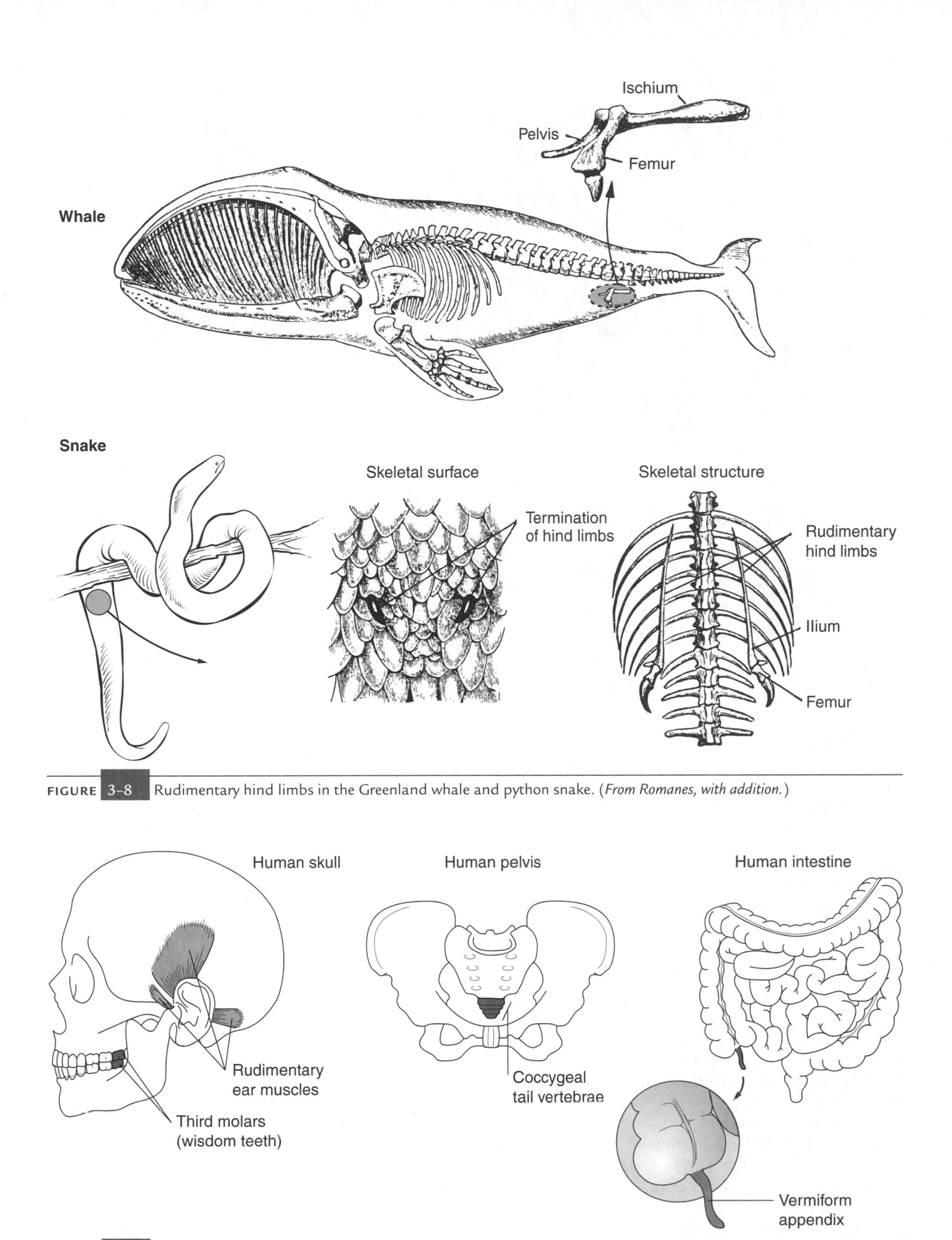

FIGURE 3-8 Rudimentary hind limbs in the Greenland whale and python snake. (*From Romanes, with addition.*)

FIGURE 3-9 Some vestigial structures found in humans. (*After Romanes, modified.*)

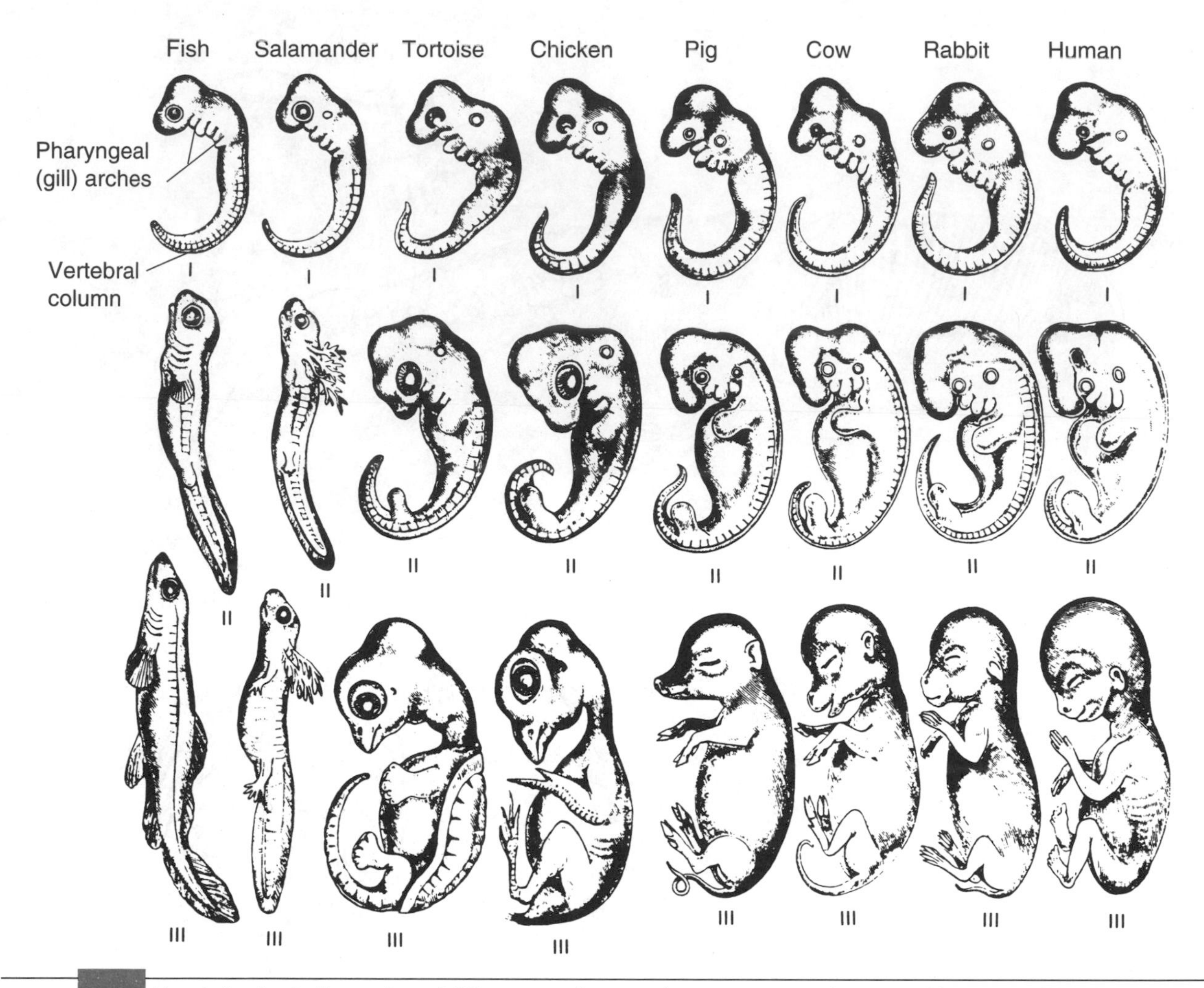

FIGURE 3–10 Haeckel's classic illustration of different vertebrate embryos at proposed comparable stages of development. Although Haeckel took some liberties in drawing these figures, it was apparent that the earlier the stage of development, the more strikingly similar are the different groups. Note that each of the embryos begins with a similar number of pharyngeal (gill) arches (pouches below the head) and a similar vertebral column. Some biologists call this a *phylotypic* stage (see also Fig. 15-9), representing the distinctive developmental substructure for a phylum's body plan, called a *pharyngula* in vertebrates. In later stages of development, these and other structures are modified to yield each vertebrate group's characteristic form. However, modifications of such patterns may vary considerably even within groups, so the terms *phylotype* and *pharyngula* really describe comparable (rather than identical) stages. (The embryos in the different groups have been scaled to the same approximate size so that comparisons can be made among them.) (*From Romanes, adapted from Haeckel.*)

Also, since isolation between populations encourages and sustains their evolutionary differences, researchers rarely find transitional forms in the same place as the original forms. Thus, a complete evolutionary progression of fossils from most ancient to most modern has never been found in a single locality. In spite of these difficulties, the search for fossils engendered considerable interest, because biologists could use fossils to provide hard evidence for evolution.

Fortunately, in Darwin's lifetime a few paleontological findings came to light that strongly supported the Darwinian position. One was the discovery in 1861 of a true "missing link"—in this case, an animal that stood approximately midway between reptiles and birds. As shown in Figure 3–13, this fossil, *Archaeopteryx,* had a number of reptilian features, including teeth and a tail of 21 vertebrae, but also had a number of birdlike features such as wishbone and feathers. Huxley argued convincingly that *Archaeopteryx* was probably a "cousin" to the lineage running from reptiles (dinosaurs) to birds—that such primitive forms were predictable consequences of evolution that helped prove the theory.

By the 1870s, paleontologists such as Marsh (1831–1899) were able to use fossils of both North American and European horses to provide the first classic example of a stepwise evolutionary tree among vertebrates, showing various transitional stages. One year after the publication of *On the Origin of Species,* Owen had described the earliest known horse-like mammal, first called *Hyracotherium* (also *Eohippus*). It was about 20

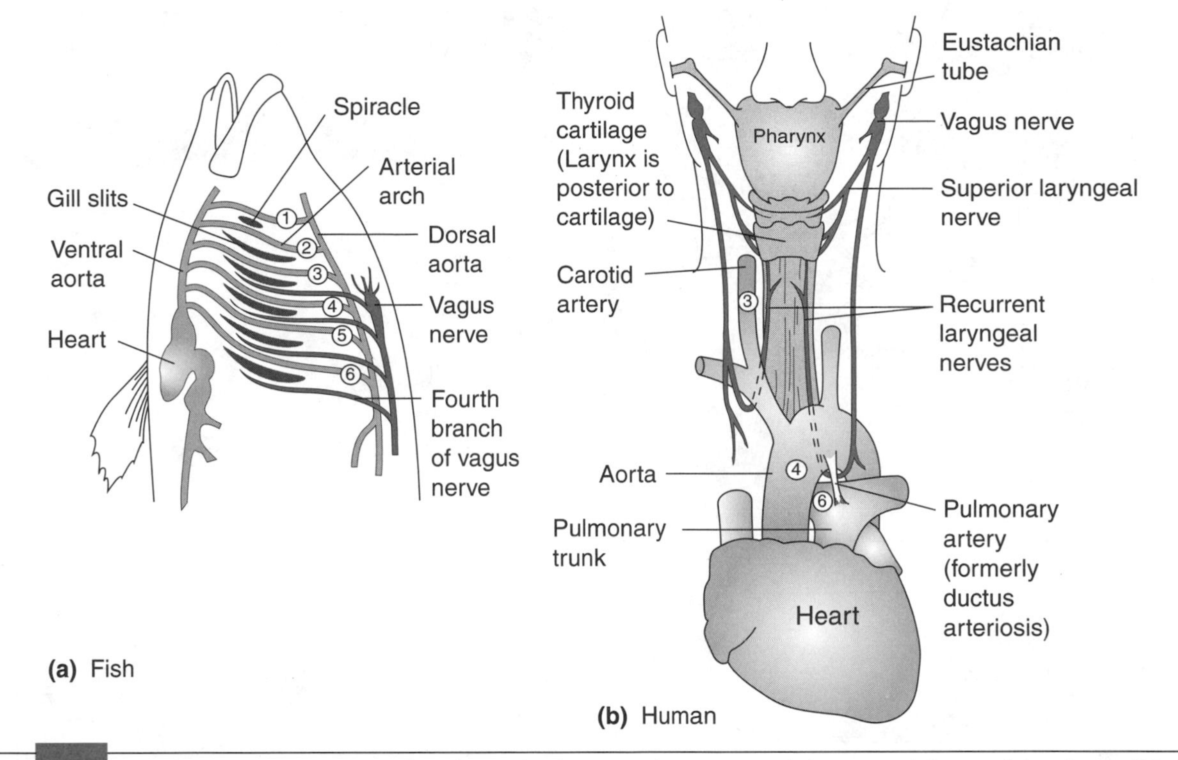

FIGURE 3–11 Schematic diagram showing the relationship between the vagus cranial nerve and the arterial arches in fish (*a*) and human (*b*). Only the third, fourth, and part of the sixth arterial arches remain in placental mammals, the sixth acting only during fetal development to carry blood to the placenta. The fourth vagal nerve in mammals (the recurrent laryngeal nerve) loops around the sixth arterial arch just as it did in the original fishlike ancestor, but must now travel a greater distance since the remnant of the sixth arch is in the thorax.

inches high, weighing about 50 pounds, with four toes on its pad-footed front legs and three on its hind legs, and with simple teeth adapted for browsing on soft vegetation. Later fossil finds indicate that *Hyracotherium* ranged widely from North America to Europe, and the genus now includes a variety of fossil herbivores, some no larger than an average-sized house cat (MacFadden).

In the approximately 60 million years since *Hyracotherium,* horses have changed radically. They now run on hard ground; chew hard, silica-containing grasses; and show special adaptations for this particular environment. Their elongated legs are built for speed, bearing most of the limb muscles in the upper part of the legs, enabling a powerful, rapid swing. They now have the distinction of being the only vertebrates with a single toe on each foot, which, together with a special set of ligaments, provides them with a Pogo-Stick-like springing action while running on hard ground. Their teeth also show unique qualities adapted for chewing hard, abrasive grasses. The molars and premolars look identical in shape and are very long, continuing growth for the first 8 years of life until the roots form. The high crowns of these grinding teeth have vertical layers of enamel and cement. As the tooth wears, the cement breaks down more rapidly than the enamel, letting sharp grinding edges of enamel stay above the cement.

Remarkably, we now know almost all the intermediate stages between *Hyracotherium* and the modern horse, *Equus*—from low-crowned to high-crowned, from browsers to grazers, from pad-footed to spring-footed, and from small-brained to large-brained forms. However, as Figure 3–14 shows, it is now also clear that evolutionary changes among these forms did not go in only a single direction (**orthogenesis**). Horses evidently evolved adaptations for their habitats in different ways, with some individual branches maintaining fairly distinct structures until they became extinct.[2]

The rate of evolution for any particular trait among the various branches was also not constant. Size, for example, underwent relatively few changes for both the first 30 million years and for the last few million years (MacFadden). Even when horse evolution was proceeding

[2]MacFadden and coworkers point out that reversion from grazing to browsing occurred in some Florida species. Although they all occupied the same general area, they made use of different environmental resources ("resource partitioning," see also p. 572). Some species became browsers, feeding on shrubs and trees; other remained as grazers, feeding on grasses; and still others both grazed and browsed. Although all had high-crowned molars, indicating a grazing ancestry, differences in feeding habits are evident because of differences in the carbon isotope ratios ($^{12}C/^{13}C$) of grasses and shrubs. Different diets produced different $^{12}C/^{13}C$ ratios in teeth, in addition to dental scratches caused by grazing and dental pits caused by browsing.

FIGURE 3–12 Fossilization process in which an animal dies in a watery environment that protects it from scavengers. Reduced oxygen levels at lower aqueous depths also help resist deterioration. The remains are gradually silted over and eventually covered by successive layers of soil that compact it into sedimentary rock. In time, because of erosion, the fossil surface may become exposed. (*From Kardong.*)

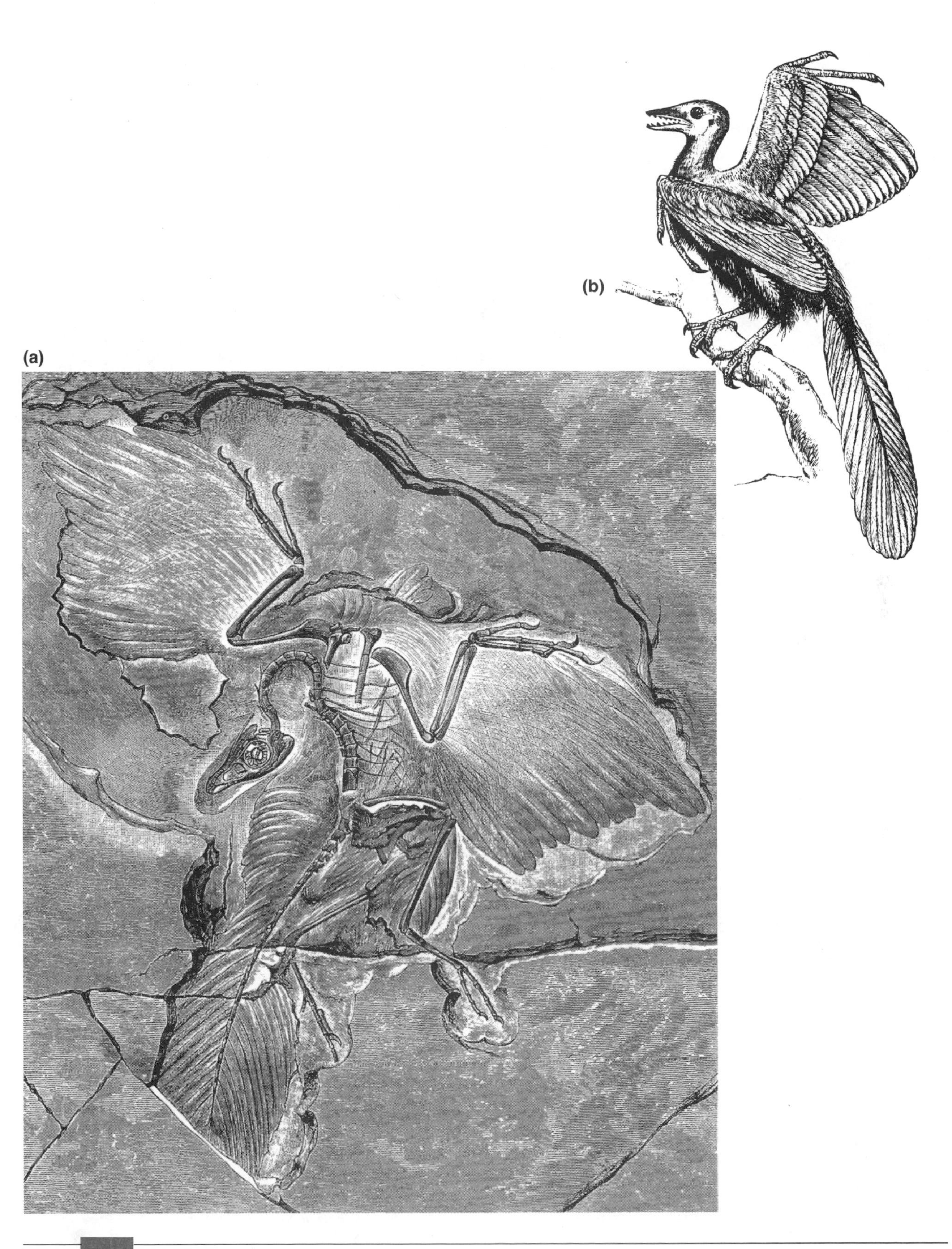

FIGURE 3-13 (*a*) The Berlin example of *Archaeopteryx* found in the Upper Jurassic limestones of Bavaria. (*b*) Nineteenth century restoration of the bird's appearance during life. (*Adapted from Romanes.*)

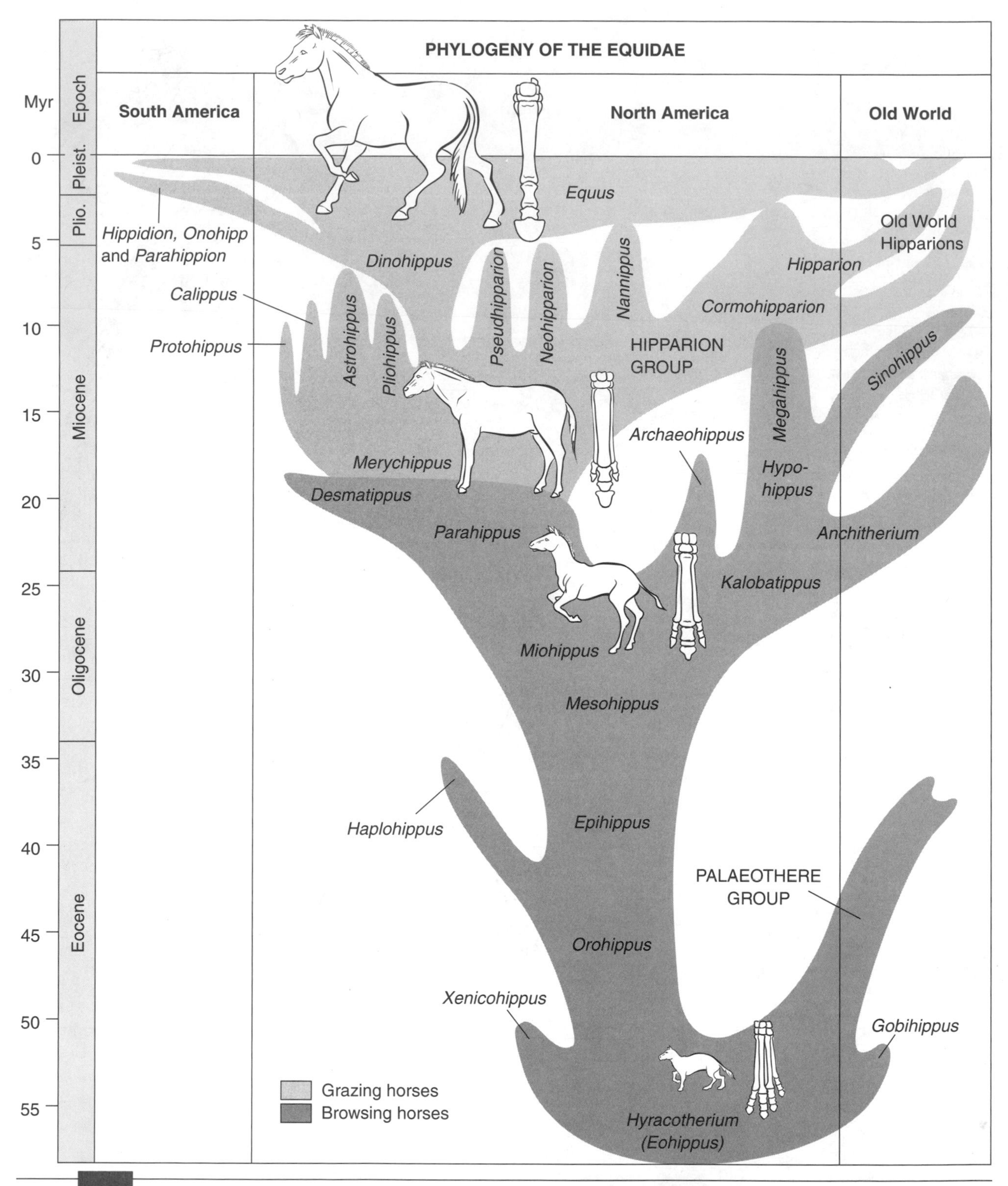

FIGURE 3-14 Evolutionary relationships among various lineages of horses, with emphasis on North American and Old World groups. Sample reconstruction of some fossil horses and modern *Equus* are shown, including diagrams of their hind foot bone structure. Using terms described in Chapter 11, horse lineages obviously followed two different evolutionary patterns: "cladogenetic" (branching) and "anagenetic" (nonbranching). (*Adapted from MacFadden, with additions.*)

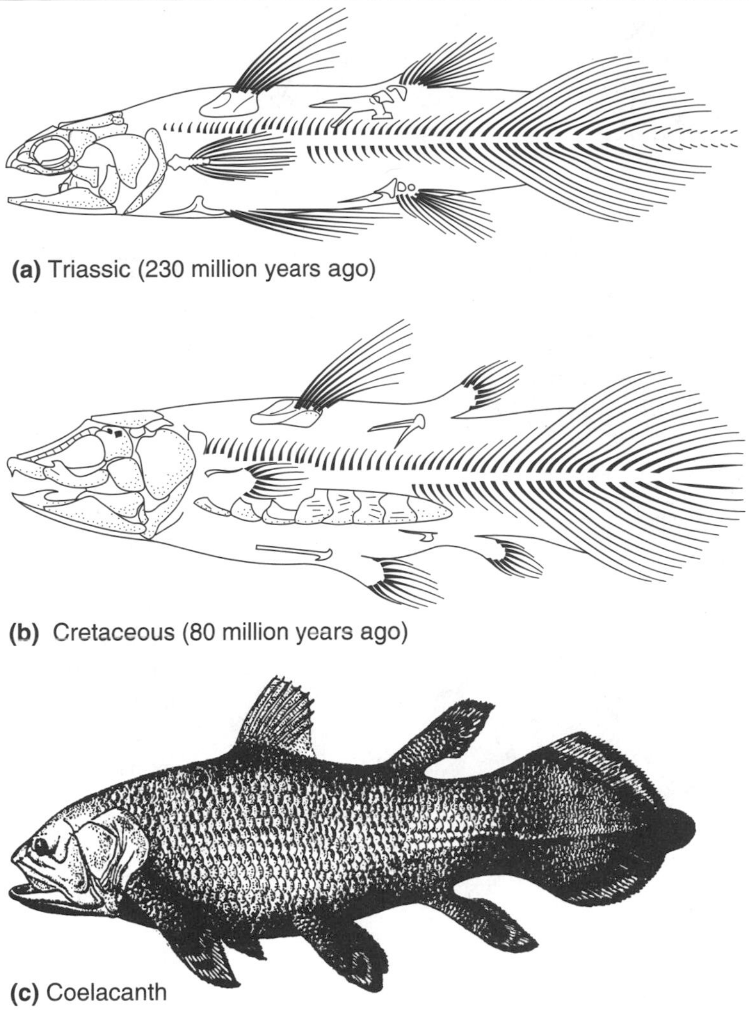

FIGURE 3–15 (*a*) Fossil lobe-finned coelacanth (*Laugia groenlandica*) from the Lower Triassic period, about 230 million years ago. (*b*) *Macropoma mantelli* from the Upper Cretaceous period about 80 million years ago. (*From P. L. Forey, 1988. Golden jubilee for coelacanth* Latimeria chalumnae. Nature, *336, 727–732. Reprinted by permission.)* (*c*) Modern coelacanth (*Latimeria chalumnae*) found off the eastern coast of South Africa.

(a) Triassic (230 million years ago)

(b) Cretaceous (80 million years ago)

(c) Coelacanth

rapidly, during the Miocene epoch, considerable size diversity evolved for both small- and large-sized species. No unidirectional orthogenetic trend in size applied to all lineages. According to paleontologists, this finely detailed phylogeny encompasses hundreds of fossil species and is one of the best illustrations of some of the realities and complexities of evolution.

Interestingly, ancient organisms may persist to modern times without further evolving morphologically. Such **living fossils** include opossums, alligators, sturgeons, lungfish, horseshoe crabs, *Lingula* brachiopods, and ginkgo trees. Occasionally, biologists discover species remarkably similar to organisms believed to have become extinct many ages ago. For example, coelacanths are ancient lobe-finned fishes (Fig. 3–15) related to those which evolved into terrestrial vertebrates about 200 million years ago (see Chapter 17). Although the fossil record of coelacanths seemed to have begun in the Devonian period about 380 million years ago and ended 80 to 100 million years ago, fishermen have found live coelacanths (*Latimeria chalumnae*) in deep waters off the eastern coast of South Africa. Similarly, a very ancient form of segmental mollusk (*Neopilina*), believed extinct since the Devonian period, has been found in deep sea trenches off the coast of lower California, Costa Rica, and Peru.

Although these findings support the validity of paleontological claims that fossils indicate the existence of real organisms, they also show the inadequacy of the paleontological record; that is, the disappearance of fossils of a particular type in the fossil record does not necessarily mean that this type immediately became extinct. However, aside from such rare "living relics," fossils in almost every instance differ from present forms, and the more recent geological strata generally show forms more like the present than do the older strata. Taken as a whole, fossil evidence provides strong support for evolution.

ARTIFICIAL SELECTION

To support his concept of evolutionary change, Darwin discussed a number of examples of evolution by selection, although the selection method involved human choice rather than natural events. In artificial selection as practiced before Darwin's time and later, the breeder

FIGURE 3-16 Heads of some different breeds selected by dog fanciers. Molecular evidence from mitochondrial DNA sequencing (see Chapter 12) indicates that the domestic dog is more closely related to the gray wolf than to any other species of the canid family (such as the jackal and fox). This evidence also shows that the domestication event probably took place a number of times over the last 10,000–15,000 years.

(whether of dogs, cats, pigeons, cattle, horses, or whatever) selects the parents deemed desirable for each generation and culls, or destroys, the undesirable types. Since the selected parents may produce a variety of different offspring, the breeder can usually continue to select in a particular direction until the traits in which he or she is interested are considerably different from their initial appearance. For example, humans have for thousands of years selected dogs, which now range in size from the St. Bernard to the Chihuahua, and in features from the greyhound to the bulldog (Fig. 3–16). Fanciers have long bred pigeons, which now show a wide variety of beaks, shapes, and feathers. The same is true for sheep, cattle, and all the many agricultural species of plants and animals.

Artificial selection demonstrated to Darwin and his colleagues that continued selection was powerful enough to cause observable changes in almost any species. The Darwinian claim that natural selection for particular environments could accomplish even greater changes than artificial selection and lead to speciation therefore seemed reasonable, given the much longer periods of time in evolutionary history and the "unrelenting vigilance" of natural selection.

SUMMARY

The publication of *On the Origin of Species* in 1859 aroused a variety of objections against Darwin's theory of evolution by natural selection. The prevalence of the idea that an individual's traits were a blend of those of the parents' made it difficult to see how traits could be selected for, since successive generations of mating with non-adapted individuals would dilute any trait. Darwin's insistence that evolutionary forces act on small and continuous variations led to criticism that selection could not recognize such slight variations, which could thus not lead to the formation of new species. An additional question concerned the appearance of new traits, which Darwin apparently thought might form by preadaptation (the conversion of some structure already present to a new use). While selection might account for the evolution of one species into another (phyletic evolution), biologists found it difficult to understand how many species might arise from a single one. To account for this factor it is necessary to emphasize the isolation of groups from one another, which Darwin did not do. Finally there did not seem to be sufficient time for evolution to occur, since geologists had not yet established the great antiquity of the Earth.

Nevertheless, Darwin's theory had the advantages of relying on understandable mechanisms, being supported by much data, and of successfully explaining many features of natural systems. The transitional forms observed by taxonomists could be explained on this basis. Evolutionary theory could also explain why organisms with similar characteristics appear geographically close to each other, whereas groups separated by geographical barriers have fewer characteristics in common. In comparative anatomy, structures found in different organisms but having an underlying similarity of plan could be explained by their relationship through a common ancestor (homology). Vestigial structures such as the rudiments of limb bones in snakes and whales would be the remnants of organs that had become obsolete. The similarities among vertebrate embryos during early developmental stages also suggested a common evolutionary past. Fossil evidence was particularly important. Some fossil forms intermediate between living groups (such as the bird-reptile *Archaeopteryx*) were found. But the horse provided the most complete and continual fossil record of evolution of a group. It had evolved from a small, four-toed, browsing animal to a large, single-toed creature with hard teeth adapted to chewing tough grasses. Finally, Darwin recognized that humans, in seeking to perpetuate favorable traits in domestic plants and animals, could select (over a relatively short period of geological time) radical alterations in organisms—albeit artificially rather than naturally. The heavy accumulation of data made Darwin's claims for evolution appear reasonable to many scientists.

KEY TERMS

adaptive radiation
analogous
biogenetic law
comparative anatomy
convergent evolution
geographical distribution
homologous
isolation
living fossils
neoteny
orthogenesis
pharyngula
preadaptation
variability
vestigial organs

DISCUSSION QUESTIONS

1. Why did the concept of blending inheritance conflict with the concept of natural selection, and how did Darwin attempt to deal with this problem?
2. Since Darwin believed that evolution was a very gradual process, resulting from the accumulation of only small differences over time, how did he seek to deal with the following objections?
   a. How can new species arise when the variations that organisms produce are only small?
   b. How can small variations be recognized by natural selection?
   c. How can organs with new functions arise?
3. In what way did critics use the following points as objections against Darwin's theory, and how were these objections later resolved?
   a. The multiplication of species
   b. The age of the Earth
4. Nineteenth century support for Darwinism came from various sources and studies:
   a. Systematics
   b. Geographical distribution of organisms
   c. Comparative anatomy
   d. Embryology
   e. Fossils
   f. Artificial selection among plants and animals

   How did these areas of biology support Darwinism, and what examples of such support did they offer? For example, how were concepts such as adaptive radiation, homology, analogy, convergent evolution, vestigial structures, and living fossils used in supporting Darwinism?
5. Did these studies indicate that
   a. Evolution always went in a single direction?
   b. There was always an increase in the number of structures or organs in a lineage?
6. What are the differences between explaining vestigial structures by the Lamarckian principle of use and disuse and explaining their presence by the Darwinian principle of natural selection?
7. What is Haeckel's biogenetic law, and how do biologists presently regard it?

EVOLUTION ON THE WEB

Explore evolution on the web! Visit the accompanying web site for *Evolution,* 3/e at www.jbpub.com/evolution for web exercises and links relating to topics covered in this chapter.

REFERENCES

Ali, M. A., 1984. *Photoreception and Vision in Invertebrates.* Plenum Press, New York.

Appleman, P. (ed.), 1970. *Darwin: A Norton Critical Edition.* Norton, New York.

Badash, L., 1989. The age-of-the-earth debate. *Sci. Amer.,* **261**(2), 90–96.

Barnett, S. A. (ed.), 1958. *A Century of Darwin.* Heinemann, London.

Bowler, P. J., 1996. *Life's Splendid Drama: Evolutionary Biology and the Reconstruction of Life's Ancestry, 1860–1940.* University of Chicago Press, Chicago.

Conn, H. W., 1900. *The Method of Evolution.* Putnam's, New York.

Darwin, C., 1859. *On the Origin of Species by Means of Natural Selection or the Preservation of Favoured Races in the Struggle for Life.* Murray, London.

Forey, P. L., 1988. Golden jubilee for the coelacanth *Latimeria chalumnae. Nature,* **336,** 727–732.

Ghiselin, M. T., 1969. *The Triumph of the Darwinian Method.* University of California Press, Berkeley.

Goldsmith, T. H., 1990. Optimization, constraint, and history in the evolution of the eyes. *Q. Rev. Biol.,* **65,** 281–322.

Gould, S. J., 1977. *Ontogeny and Phylogeny.* Harvard University Press, Cambridge, MA.

Haber, F. C., 1959. *The Age of the World.* Johns Hopkins University Press, Baltimore.

Haeckel, E., 1866. *Naturliche Schöpfungsgeschichte.* Reimer, Berlin.

———, 1905. *The Evolution of Man.* Translated from the 5th German ed. by J. McCabe, Watts, London.

Harris, W. A., 1997. *Pax-6*: Where to be conserved is not conservative. *Proc. Nat. Acad. Sci.,* **94,** 2098–2100.

Hull, D. L., 1973. *Darwin and His Critics.* Harvard University Press, Cambridge, MA.

Kardong, K. V., 1998. *Vertebrates,* 2d ed. WCB/McGraw-Hill, Boston.

Lack, D., 1947. *Darwin's Finches.* Cambridge University Press, Cambridge, England.

LeConte, A., 1888. *Evolution, Its Nature, Its Evidences, and Its Relation to Religious Thought.* Appleton, New York.

Loosli, F., M. Kmita-Cunisse, and W. J. Gehring, 1996. Isolation of a *Pax-6* homolog from the ribbonworm *Lineus sanguineus. Proc. Nat. Acad. Sci.,* **93,** 2658–2663.

MacFadden, B. J., 1992. *Fossil Horses: Systematics, Paleobiology, and Evolution of the Family Equidae.* Cambridge University Press, Cambridge, England.

MacFadden, B. J., N. Solounias, and T. E. Cerling, 1999. Ancient diets, ecology and extinction of 5-million-year-old horses from Florida. *Science,* **283,** 824–827.

Montagu, M. F. A., 1962. Time, morphology, and neoteny in the evolution of man. In *Culture and the Evolution of Man,* M. F. A. Montagu (ed.). Oxford University Press, Oxford, pp. 324–342.

Nilsson, D.-E., and S. Pelger, 1994. A pessimistic estimate of the time required for an eye to evolve. *Proc. Roy. Soc. Lond.* (B), **256,** 53–58.

Romanes, G. J., 1910. *Darwin, and After Darwin.* Open Court, Chicago.

Rudwick, M. J. S., 1973. *The Meaning of Fossils.* Macdonald, London.

Ruse, M., 1979. *The Darwinian Revolution.* University of Chicago Press, Chicago.

Salvini-Plawen, L. V., and E. Mayr, 1977. On the evolution of photoreceptors and eyes. *Evol. Biol.,* **10,** 207–263.

Sato, A., et al., 1999. Phylogeny of Darwin's finches as revealed by mtDNA sequences. *Proc. Nat. Acad. Sci.,* **96,** 5101–5106.

Simpson, G. G., 1951. *Horses.* Oxford University Press, New York.

Simpson, G. G., and W. Beck, 1965. *Life,* 2d ed. Harcourt, Brace, and World, New York.

Voneida, T. J., and S. E. Fish, 1984. Central nervous system changes related to the reduction of visual input in a naturally blind fish (*Astyanax hubbsi*). *Amer. Zool.,* **24,** 775–782.

Vorzimmer, P. J., 1970. *Charles Darwin: The Years of Controversy.* Temple University Press, Philadelphia.

Wayne, R. K., 1993. Molecular evolution of the dog family. *Trends in Genet.,* **9,** 218–224.

Wolken, J. T., 1986. *Light and Life Processes.* Van Nostrand Reinhold, New York.

4 The Darwinian Impact: Evolution and Religion

By making evolution an acceptable concept, Darwin's impact was profound. Darwinian evolution offered a vast historical framework in which people could understand biological change; it made clear that species fixity was not at all "natural"; and it proposed that the form and function of living organisms did not arise by creation but by selection. These were revolutionary ideas in the nineteenth century, and they not only revolutionized biology but also affected fields such as sociology (Herbert Spencer), anthropology (Lewis Henry Morgan), economics (Karl Marx, Thorstein Veblen), politics (Walter Bagehot), women's rights (Charlotte Perkins Gilman, Elizabeth Cady Stanton), fiction (Joseph Conrad, George Eliot, Thomas Hardy, Jack London, Jules Verne, Theodore Dreiser, H. G. Wells), poetry (Robert Browning, Alfred Tennyson, Walt Whitman), linguistics (William Dwight Whitney), philosophy (Charles S. Peirce, John Dewey, Henri Bergson), and psychology (William James, Sigmund Freud).

The impact of Darwinism, however, was most dramatic in respect to religion. To many of Darwin's religious contemporaries and to others since, *On the Origin of Species* as well as *The Descent of Man,* which Darwin published in 1871, raised controversial matters of vast proportion. One of the chief popular issues of the late nineteenth and early twentieth centuries was the struggle over the acceptance of evolution, with many scientists arrayed on one side and religionists on the other. As pointed out by Ellegård, "To the general public, Darwinism was at least as much a religious as a scientific question."

The stressful relationship between evolution and religion stemmed not only from the vulnerability of religion, as we shall see later, but also from the recognition that evolutionary concepts were not impregnable. Darwin, for example, pointed out that at least two phenomena could refute his theory:

1. Discovery of an inexplicable reversion in the evolutionary sequence, such as evidence of the presence of humans in the Paleozoic or Mesozoic eras
2. The finding of exactly the same species in two separated geographical locations whose presence was not caused by migration between these areas

If such incidents appeared, Darwin recognized that he would have to abandon his view of evolution, and its major alternative seemed only a religious supernatural doctrine (in Darwin's words, "the common view of actual creation"). Although it is now almost a century and a half after Darwin, this dichotomy of views, evolution versus religion, still presents itself as a popular controversy: the alternative proposed by antievolution groups and individuals is almost always religious (creationist).[1] (For a bibliography of more than 1,800 antievolution publications, see McIver.) Understanding how these two concepts interact is important to comprehend and accept the role that evolution can play in modern life.

The Religious Attack

An early model of the battles to come took place at Oxford soon after publication of *On the Origin of Species.* In this debate Bishop Samuel Wilberforce (1805–1873) of the Anglican Church (Fig. 4–1) attacked Darwinian theory as incompatible with the Bible, and, coached by Richard Owen, a former student of Cuvier, attempted to destroy it through scientific arguments. Wilberforce's final point was made directly to the Darwinian defender, Thomas Huxley, when Wilberforce asked whether it was through Huxley's grandfather or grandmother that Huxley claimed descent from a monkey. The wit of Huxley's response, recounted in a letter to a friend, has often been quoted:

> If, then, said I, the question is put to me would I rather have a miserable ape for a grandfather, or a man highly endowed by nature and possessed of great means and influence, and yet who employs these faculties and that influence for the mere purpose of introducing ridicule into a grave scientific discussion, I unhesitatingly affirm my preference for the ape.

Many theologians, however, were unimpressed with scientific arguments and continually hammered away at the heresy of evolution. Wilberforce accused Darwin of "a tendency to limit God's glory in creation." Cardinal Manning, a leader of English Catholicism, called Darwinism "a brutal philosophy—to wit, there is no God, and the ape is our Adam." The religious attacks were worldwide, frequent, harsh, and almost always focused on the same points. That is, religious opponents accused Darwinists of seeking "to do away with all idea of God," "to produce in their readers a disbelief of the Bible," "to displace God by the unerring action of vagary," and "destroy humanity's unique status."

FIGURE 4–1 Caricatures of Bishop Samuel Wilberforce (*left*) and Thomas Huxley (*right*) that appeared in the British magazine *Vanity Fair* some years after their 1860 debate at Oxford University.

If these claims were true—and there was very reasonable cause for religious alarm—then one could well ask how such a heretical doctrine could have developed in a religious European country and become acceptable to so many of Darwin's learned compatriots. Among the answers is one that helps show the struggle between evolution and religion to be not accidental but part of the historical framework of the time.

From a social point of view, we have already observed that the development of evolutionary theory was one aspect of that all-pervasive political and economic revolution in social behavior and thought that began with the overthrow of the rigidly ordered feudal class structures that had prevailed in Europe until the rise of capitalism. The economic challenges posed by capitalism and its new monied classes in many ways allowed ideological challenges to the prevailing religious and philosophical systems that had supported the old social order. The divine right of kings, for example, was being overthrown socially, philosophically, and religiously. Without such

[1]Many Americans are "creationist." According to a 1997 Gallup poll, 44 percent of Americans agreed with the statement that "God created humankind in its present form almost 10,000 years ago," 39 percent agreed with the statement that "Humans have evolved over millions of years from less-advanced forms of life, but God guided this process," 10 percent agreed that "Humankind has developed over millions of years from less-advanced forms of life but God had no part in this process," and 9 percent had "no opinion." Similar findings come from a 1994 Louis Harris survey: 46 percent of American adults do not believe that humans evolved from earlier species of animals, and an additional 9 percent are "unsure."

changes, European science could not have flourished as it did. Science became a way of asking interesting and challenging questions to be answered by relevant nonauthoritarian, nonreligious explanations.

However, not too surprisingly, capitalism in its triumph also sought ideological justifications for its power. As in feudalism, many of these justifications rested on religious concepts, which commonly appealed to an inherent hierarchy of nature and the wisdom of the creator. In the words of Alexander Pope (1688–1744):

> Order is Heaven's first law, and this confest,
> Some are, and must be, greater than the rest,
> More rich, more wise.

Or, as stated in Darwin's century by C. F. Alexander in her hymn "All Things Bright and Beautiful" (1848):

> The rich man in his castle,
> The poor man at his gate,
> God made them, high or lowly,
> And order'd their estate.

Looking at evolution historically, we can see that by threatening basic religious concepts of a fixed universal order, evolution seemed to have exceeded the "game plan" for permissible ideological challenges. Nevertheless, there was no sustained political attempt to suppress evolutionary ideas, although they were attacked far and wide and even outlawed in some American states. One reason for the relative freedom that evolution enjoyed is probably its close ties with all other aspects of science. Science and technology were, after all, the mainstays of economic expansion. Many scientists were or became evolutionists, and most social leaders of the time must have felt that evolution was a scientific foible that would have to be tolerated.

To many religionists, however, evolution was a deep, abiding threat. They had considerable cause for concern, and to understand their hostility it is valuable to review those aspects of religion most directly threatened by evolution.

The Basis of Religious Belief

At present, different levels of religious development appear in various cultures that provide clues to the evolution of religion itself. Through the efforts of anthropologists and psychologists we know that religion first develops in a culture as a form of behavior through which humans seek to deal with aspects of their experience they cannot control or understand. People have apparently sustained religions from the common feeling that what is outside our control may nevertheless be humanlike and therefore subject to appeal and thus indirectly to control. In the earliest stages of religion, people directly endow the forces of nature with the spirits of animals and humans.[2]

In the attempt to control these forces by sympathetic magic, people enact particular events by imitation—hunting dances, rain dances, and so on—to guide these events toward desired ends. Ritual develops when people repeat magical ceremonies to help ensure their efficacy. Ritualized behavior seems to have become especially important in the transition from hunting to agricultural societies in which crops have to be planted and harvested at appropriate seasons each year and where one's efforts could be either rewarded or damned by forces that remained mysterious and arbitrary. In his advice to farmers, Hesiod (800 B.C.) wrote:

> Everyone praises a different day, but few fully understand their nature. One day may be like a stepmother, another like a mother. A man will be happy and lucky if, having an eye to all of these things, he completes his work without offending the gods, reads the omens of the birds, and avoids all transgressions.

People of today can ill afford to laugh at such beliefs. The uncertainty, hope, and wooing of good fortune Hesiod expressed are basic feelings that many share. Hesiod more baldly states them here than in the more sophisticated forms in which we usually acknowledge them.[3]

---

[2]Boulding points out:

> It is not surprising that the first theologies tended to be animistic. We observe that we can change the behavior of our neighbors by talking to them. Why cannot we similarly change the behavior of the skies, or the plants, or the animals? Furthermore, it is quite easy to accumulate evidence for animism: if we say something to our neighbor and he does not respond, we conclude that we said the wrong thing and should say something else. If we say something to the mysterious forces of nature and they do not respond, then we decide we said the wrong thing, and will do better next time. It is hard to avoid selective memory of response; we can remember the successes and forget the failures. Even today the gambler kisses the dice he is about to throw. We even sacrifice a bottle of champagne when launching a ship, no doubt to cheer up its spirits. And we make huge economic and human sacrifices in the name of national defense to appease the spirit of the nation. . . .
>
> Animism, however, seems to lead to the gods and to the great pantheons of Greece and Rome, India and Japan. These must surely begin as fantasy and poetry, arising out of the extraordinary human capacity for creating a vast inner landscape in which we can personify and satisfy desires, exorcise our sense of impotence, and magnify our capacities both for good and for evil. We cannot resist imagining that which is larger than life, and the fantasies, the legends, and the myths then take on a reality of their own.

[3]It was common to believe that danger arose from mystical and magical sources that surrounded daily life. For example, in describing early Israelite peasant life, Coote points out:

> An entire cast of village gods and goddesses, spirits, sprites, demons, goblins, and genies were thought to participate in the processes of production and reproduction. Every nook and cranny of house and field were inhabited by evil sprites who emerged at dark and returned to their dens at break of day. Their noises—shuffle and bustle,

Underlying many human anxieties about the world, and often at the root of the difficulty in treating nature as an independent object of study, was the feeling that nature reflected a supernatural evaluation of human affairs (by imposing punishment and reward—that is, calamity and prosperity). Since people saw the forces of nature as humanlike, religion sustained and encouraged the hope that those appeals that humans understood—submission, supplication, gifts, sacrifice, obedience, and loyalty—could appease nature's judgment and recrimination.

Religion's universal attractiveness is supported by the universality of many human emotions and behaviors, such as fear, insecurity, dependency, aggression, and guilt (see also p. 499). To alleviate and control the effects of such emotions, religion preserved and also often provided primary belief systems (myths) that helped explain and guide peoples' relationships to the world about them. Where did we and our society come from, and why? Each culture's mythological world usually offered comprehensive accounts of how and for what purpose the world was created; models that upheld its views of human society, history, and behavior—often made relevant to personal human experience.[4]

Priestly leadership mostly reinforced such cultural myths through rituals, pageants, festivals, holidays, symbolic adornments, statuary, and temple observances that included prayers, incantations, and sacrifices. By these means and many prescribed rules and regimens, religions granted their followers favor with supernatural ruling deities and aid in correcting personal and social ills that departed from mythological ideals and purity.

Soul and God

In support of a supernatural view of events, religion has relied on two basic concepts that probably arose early in history, the **soul** and **God.** Most religions consider the soul, or the spirit, or what many would call the **personality** or *self,* to be a humanlike, conscious entity without physical attachments or properties. According to various psychologists and anthropologists, the idea of the separation of the soul from the body probably originated in the separation of mind from reality in dreams. The restrictions of space, time, and even death vanish in dreams, and people can then suppose that one aspect of human life, the soul, is immune to such restrictions.[5]

This attitude is reinforced by the fact that people find it hard to conceive that the personality has only a limited existence. That is, it is a formidable task to realize that the mental organ of sensation and feelings that we use to perceive our relationships to the environment, and through which we integrate, evaluate, and interact with the world around us, has a beginning and an end. Although one can intellectually conceive of death, related to other creatures or phenomena or even to one's own body, it is impossible to "feel" death as one may feel and anticipate other events and sensations. To a human, and perhaps to other creatures as well, this organ of awareness and feelings, the personality, seems to be eternally present, since there is usually no self-knowledge of how it arrived or of how it ends.

Theologians have often attempted to provide a "scientific" argument for the nonmaterial nature of the soul or the personality, claiming that intellectual processes cannot have a material origin. According to their argument, freedom of choice, ethics, and all the many complex ideas about which humans can think must be considered the fruits of nonbiological matter, since they are so obviously separate from purely physical processes. However, the biological view is quite different: although we do not yet know the precise relationship between the matter of the brain (neurons, synapses, and so on) and the thoughts and feelings it produces, that such a relationship exists is no mystery. Many creatures have thoughts, some have personalities, and some have dreams. Although probably none of the thoughts in most creatures are as complex as in primates—and among primates, none are apparently as complex as in humans—we have every reason to believe that the complexity of thinking and feeling has evolved like any other trait.

Like the concept of the soul, the concept of God has many qualities that reflect human experience. Each of the

whispering and shrieking, hissing and cackling—were often difficult to distinguish from the sounds of animals, insects, and wind. . . . [They] had to be coddled, cozened, evaded, and thwarted, treated simultaneously with reverence and suspicion.

[4]For example, Gaster points out that the Jewish Old Testament myths and stories:

> . . . are paradigms of the continuing human situation; we are involved in them. Adam and Eve, characters in an ancient tale, are at the same time Man and Woman in general; we are all expelled from our Edens and sacrifice our happiness to the ambitions of our intellects. All of us metaphorically flee our Egypts, receive our revelation, and trek through our deserts to a promised land which only our children or our children's children may eventually enjoy. In every generation God fights the Dragon, and David fells Goliath. We all wrestle with angels through a dark night.

[5]One of the nineteenth century founders of anthropology, E. B. Tylor, explained the soul concept among primitive tribes as follows (see also La Barre):

> As it is well known by experience that men's bodies do not go on these excursions, the natural explanation is that every man's living self or soul is his phantom or image, which can go out of his body and see and be seen itself in dreams. Even waking men in broad daylight sometimes see these human phantoms, in what are called visions or hallucinations. They are further led to believe that the soul does not die with the body, but lives on after quitting it, for, although a man may be dead and buried, his phantom-figure continues to appear to the survivors in dreams and visions.

A somewhat different concept of the soul was as an entity that would obtain reward and punishment after death. In Old Testament biblical sources, the Book of Daniel first makes this concept explicit; it was written in the second century B.C., and became popular in various Jewish sects by the beginning of the Christian era (Cohen).

various gods personifies, often in human form, forces or tasks that seem beyond the capabilities of humans, like a powerful parent seen through the eyes of a child. How better for primitive people to alleviate anxiety and fears about harvests, thunder, fertility, woods, and rivers, than to believe that these elements of nature embody extensions of human behavior? The progression from individual gods for each element of nature to gods that people can rank in respect to their power and then to a God of gods, such as Jove, Jehovah, Allah, Brahma, and others, was only a succession of steps.[6]

Major religions also generally moved from visualizing God in material forms to more abstract concepts: from a god who literally walks in the Garden of Eden to an invisible god who rules from a distant heaven (Friedman, 1997). Whether gods were material or abstract, faithful believers could approach these deities through prayer, obedience, loyalty, and sacrifice,[7] in search of divine aid and comfort, whether in terms of future compensation on earth, resurrection after death, or attainment of a state of bliss, such as heaven, paradise, or nirvana.

At the heart of religion is the feeling of reliance and dependence on what is held to be—and what is certainly desired to be—a wise, responsive, and caring nature. That not all events seem justified or understandable only reflects human inability to fathom God's behavior. To echo St. Augustine (354–430), an early Christian theologian: "If you understand it, it is not God; if it is God, you do not understand it." Different religions endow these dependent feelings with structures derived from their own societies, often with considerable imagination. For example, reality could probably never produce pleasures and torments as great and enduring as those provided by the imaginations of believers in heaven and hell.

Challenges to Religion: The Question of Design

Clearly, as long as people needed to believe that active intervention by a divine power was necessary to explain most or all observed events and to allow the continued maintenance of the universe, that belief inhibited the search for natural laws to account for changes in the present, past, or future. The first significant cracks in the theological armor of continued divine intervention in nature were made by Copernicus (1473–1543), Galileo (1564–1642), and Kepler (1571–1630) in their discoveries of natural laws regulating the motion of the solar system.

These openings were considerably widened by the mechanistic explanations offered by Newton (1642–1727) on the motion of the solar system through the force of gravity, and the postulate of an unbounded infinite universe in which our world was proportionately smaller than even a grain of sand. Later, geologists such as Lyell extended this mechanistic approach by proposing how natural forces could mold the earth's surface. Although these scholars were not atheists, their findings about natural processes made through sciences such as mechanics, optics, and chemistry helped reduce God from a continually active, intervening agent to a prime force more like a

[6]Even the Jews, whose monotheism helped found Christianity, the major western religion, went through a succession of polytheistic stages. According to various studies (see, for example, Albertz, Armstrong, Baron, Cohen), the Jewish primary god, Jehovah (Yahweh), began as a provincial god, perhaps in Midian (now Jordan), competing with other provincial gods such as those of Canaan and Babylon, whom they also worshipped (for example, Anath, Ashtoreth, Baal, Dagan, El, Moloch, Resheph, Shemesh). By the time the first five books of the Old Testament were brought together (from "Jahvist," "Elohist," "Deuteronomist," and "Priestly" sources), codified, and edited (about the sixth and seventh centuries B.C.), Jehovah had generally been given a more powerful status as the supreme but not exclusive god. During the beginnings of the Second Temple (sixth century B.C.), Jewish prophets finally established the concept that there was only a single true god, although they often lamented widespread polytheistic recidivism in the populace. To make monotheism more acceptable to people who needed closer religious ties to their personal lives, the use of angels to account for human events dramatically increased during the Second Temple period, which ended in A.D. 70. Angels were godlike creatures who were mostly guards and attendants of the major deity, and could serve as messengers between the deity and humans. Some angels, such as Michael the "Prince" of Israel, were protective and benevolent, while others such as Satan were responsible for wickedness, treachery, disease, and death, thus helping to displace blame for evil from a benevolent God. In a sense, polytheism with its humanlike gods went underground and crept back into religion via angels and saints, each with distinctive supernatural powers that allowed the comfort of "miracles"—that a deity can actively intervene in normal events to deflect or modify unfavorable natural laws or restraints.

Such concepts of friendly supernatural agents opposed by hostile counterparts helped cast differences and conflicts between individuals and groups into a dualistic struggle between "good" and "evil," often described religiously as "angelic" and "satanic." Among the early Christians, St. Paul put this as follows:

> Our contest is not against flesh and blood, but against powers, against principalities, against the world-rulers of this present darkness, against spiritual forces of evil in heavenly places. (Eph. 6:12)

Interestingly, some monotheistic thinkers at the time, such as Celsus (ca. A.D. 180), objected to this view, claiming it is "as if there were opposing factions within the divine, including one that is hostile to God" (Pagels).

[7]Although different kinds of sacrifice were demanded at different times and for different purposes, many early religions practiced human as well as animal sacrifice. The sacrificial victim served not only as a gift to a deity, but also often enabled the celebrants to identify with the offering and open a mystical portal allowing entry into a spiritual world. The story of Abraham and his aborted sacrifice of Isaac in the Old Testament (Gen. 22:1–18) probably marked the end of human sacrifice among forebears of the Jewish religion. Nevertheless, the power of using human sacrifice to achieve reward or forgiveness long remained a potent concept in many cultures, and was adopted as a mainstay of Christianity by St. Paul (1 Cor. 5:7), The Gospel of John, and other early Christian patriarchs in the form "Jesus Christ died for our sins." The practice of congregation members consuming the "flesh" and "blood" of Christ in the Eucharist ritual became a common mode of partaking in Christ's sacrifice, and thereby attaining a state of grace and salvation.

master artisan who has designed logically contrived, self-functioning machines.[8]

This development helped modify the human attitude toward the divine establishment from simple fear and subjection to the more comfortable attitudes of admiration and respect, but it also led to a closer examination of the nature of God, with many ensuing contradictions. If God functioned only as prime initiator of the universe, what was the need for God (and therefore for religion) at the self-functioning levels that followed the world's origin? Moreover, if the universe is logical, what was the logical purpose in creating it? An arbitrary God such as Jehovah obviously needs no understandable reason for creation, whereas the motivation of a logical and rational God is implicitly questionable. Was the motivation the pride of reaping adoration from human subjects as they admired God's works? Could God be vain? Was the motivation to create perfection? Did God need testimony for God's perfect attributes? How could imperfection and evil arise from perfection? And, most pernicious of all paradoxes, if the world was not created perfect, what moral good could there have been in its creation?

However much these philosophical questions challenged religion, they were not as damaging as the frontal attack Darwin and his evolutionists made. In the most sensitive area of all—life itself—Darwinian evolution offered different answers to religious claims *why* life's important events occur. Darwin's works made clear that people no longer needed to believe that only the actions of a supernatural creator could explain biological relationships. God was neither "The Great Speciator," nor "The Great Sculptor of Nature." Instead, Darwin presented the concept that nature entails continual change, unpredictable chance events, an unrelenting struggle for survival among living creatures, and no obvious guidance.

To religious believers who found such ideas distasteful, Darwin's natural selection also appeared to substitute waste for economy by viewing life as a continually expendable commodity rather than a divinely premeditated and consecrated goal. He thus replaced what many had seen as an understandable view of nature—that is, the creativity of a humanlike God—by the most heretical concepts (randomness and uncertainty) or the fear that now no one could really understand the source and purpose of any natural event or design. An example of how essential it was to believe that each design had a creative purpose is reflected in Paley's (1743–1805) statement:

> There cannot be design without a designer; contrivance without a contriver; order without choice; arrangement without anything capable of arranging. . . . Arrangement, disposition of parts, subservience of means to an end, relation of instruments to a use, imply the presence of intelligence and mind.[9]

In terms of the origin of humans, of course, this simply means that the "silk purse," which is the human being (or his or her soul or personality) could not have been made out of a "sow's ear," that is, out of the rest of our animal kin, without the active intervention at some point, early or late, of an intelligent, sympathetic deity. This assurance gave people a reason for their own creation—they were not born in vain. It also helped mollify the most unkind cut of all to intelligent, sensitive creatures—that they might die in vain—because the designer who gave them life was also responsible for their death and for the immortal preservation of their souls.

What people conceived as simple common sense also seemed to support **supernatural design.** Can there be a watch without a watchmaker, and by extension, can there be a person without a personmaker, laws without a lawmaker? To evolutionary theory, the essential challenge that religion poses has always been, "How from the disorder of random variability can nature achieve the beauty of adaptation without intelligent intervention?" Darwin's contribution was to answer this question by means of a phenomenon that no one had thoroughly explored before—natural selection. We can illustrate the evolutionary view of selection in simple form as follows.

Although chance events arise in evolution, it is primarily a historical process. That is, what evolves depends on what has evolved before. Thus, no complex structure arises all at once by a lucky combination of events, but rather evolution builds new structures from old ones. For example, if one had a large bowl full of ten different letters (*A, C, E, I, L, N, O, T, U, V*) with each letter present in equal frequency (Fig. 4–2), nine letters randomly drawn from this bowl can be arranged in many millions of ways. The chances of getting the exact word *EVOLUTION* from a random draw of nine letters is obviously small $[(1/10)^9 = .000000001]$.

If we assume, however, that a selection mechanism exists that will perpetuate certain adaptive combinations,

[8]In the words of Bernal, "God had, in fact, like his anointed ones on earth, become a constitutional monarch." Some writers, even some scientists, complained that such views replaced simple beliefs in moral innocence with hard unfeeling laws of science. Schrödinger, an esteemed physicist, claimed that a "fateful division" between "the heart [religion] and pure reason [science] has hampered us for centuries and become unendurable in our days." More recently, Kauffman argued similarly, "Paradise has been lost, not to sin, but to science." Such claims uphold pretensions that humans historically lived in a state of blissfully veiled religious ignorance until religious mysteries were brutally crushed by scientific knowledge. In contrast, all evidence indicates that pre-science human history is primarily an account of unredeemed hardship, high mortality, continuous social conflicts, and barbaric practices that often included, at times, cannibalism and human sacrifice. Anthropologists have long disproved the myth of "The Noble Savage."

[9]Erasmus Darwin, Charles's grandfather, expressed this view as follows:

> Dull atheist, could a giddy dance
> Of atoms lawlessly hurl'd
> Construct so wonderful, so wise,
> So harmonised a world?

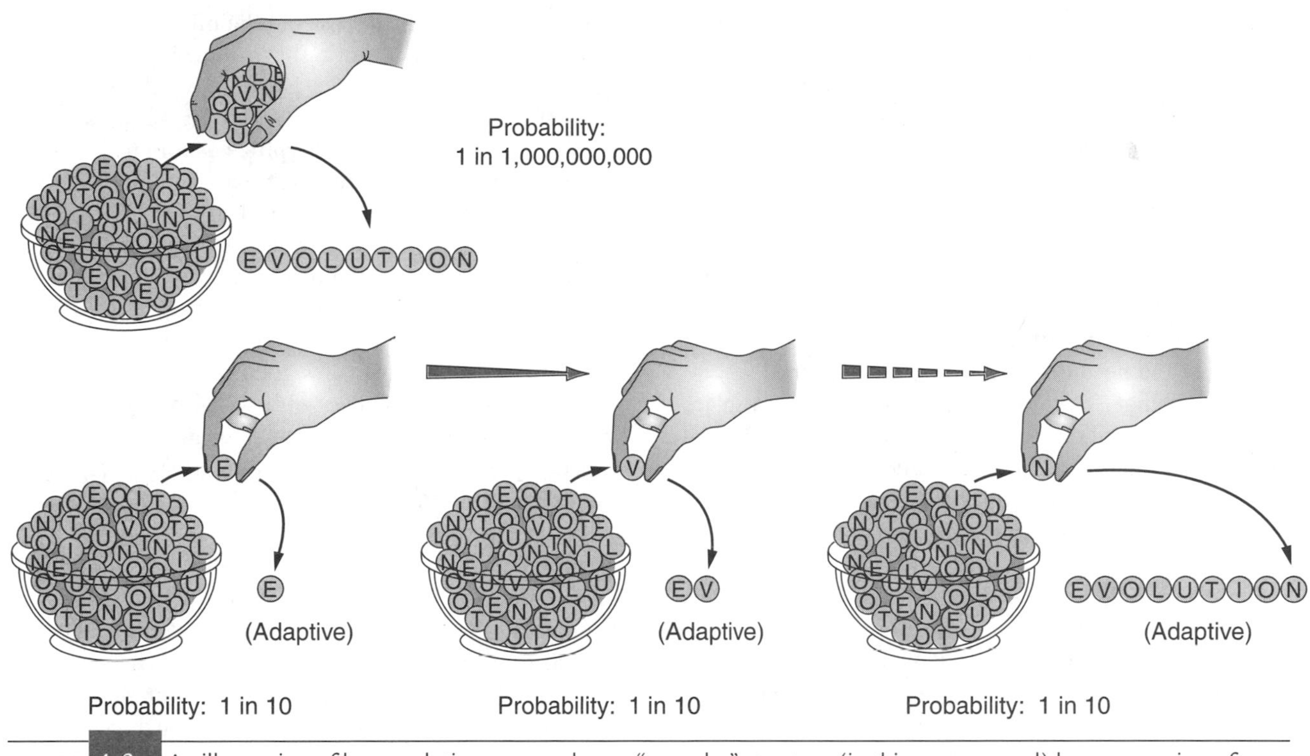

FIGURE 4–2 An illustration of how evolution can produce a "complex" structure (in this case, a word) by a succession of selective steps. If we assume a bowlful of 10 different letters (*A, C, E, I, L, N, O, T, U, V*) with each letter present (or generated by mutation) in equal frequency, there is a considerable chance of obtaining the word *EVOLUTION* within a short period, if each letter is chosen one at a time and each new sequence of letters (for example, *E, EV, EVO, EVOL*, and so on) is adaptive. That is, a *succession* of adaptive evolutionary events occurs, each stage having a reasonable probability (1/10).

then the "evolutionary" attainment of the word *EVOLUTION* is far greater than on first appearance. Let us assume that *E* is the only letter that can survive by itself. The chance of drawing an *E* from the bowl is therefore, on average, 1/10. Let us assume next that a *V* in combination with the already adaptive *E* has additional survival value; then the chance of achieving this particularly adaptive combination *EV* is again 1/10. (Note that once *E* has been drawn and survives, the chance of getting the *EV* combination is the 1/10 chance of drawing *V* and *not* the 1/100 chance of drawing *EV* together.) Similarly, if we assume that the next adaptive combination consists of adding an *O* to *EV*, and that further successive combinations are selected because of their adaptive value, the entire word, *EVOLUTION*, can eventually be selected with relatively high probability without the intervention of any agent other than the strictly opportunistic one of what is adaptive at each separate stage. Since there may be other adaptive combinations in this bowlful of letters, similar selective mechanisms can also lead to words such as *EVOCATION*, and *ELEVATION*. Like *EVOLUTION*, the chances are extremely small for such words to arise by choosing nine letters at random in a single selective event, yet they can easily be produced by *successive* selection of adaptive combinations.

Evolution can thus create complex organs such as the eye and the brain by naturally selecting successively improved adaptations for preservation. The variability on which the choice is exercised is random (the letters, or mutations, are of different kinds, both adaptive and nonadaptive), but the structure that is built over many generations of selection has been historically molded and created and is not at all random. Thus, in the words of the population geneticist R. A. Fisher, "natural selection is a mechanism for generating an exceedingly high degree of improbability." Or to paraphrase Monod, mutation provides the random noise from which selection draws out the nonrandom music.[10]

To repeat this in a somewhat different way: The biological changes produced by selection can occur whether humans are the selective agents (artificial selection) or nature is the selective agent (natural selection). Artificial selection, for example, has given us the many forms of dogs created by humans over a relatively short period of time from the variability present in that particular species. Natural selection, in contrast, acting over much longer periods of time, has given us the various species we see, all created from the variability that accumulated in many different prior species. These prior forms, in turn, were selected from previous forms going back to those

[10]Note that the only way to circumvent selection would be to allow individuals carrying every different hereditary variation to survive and reproduce—an obvious impossibility in a world of limited resources.

first primeval organisms of billions of years ago that could transmit their characteristics to their progeny.

To Darwinians, all biology has had an accidental origin in the sense that hereditary variables arose at first randomly without purposeful foresight. Yet most, if not all, biological features that survived were adaptive in the sense that only adaptive combinations of these random variables could perpetuate themselves in the face of selection. It was the stepwise process of selection that led from randomness to adaptation. These arguments made clear that purposeful function can result from naturally selective forces and need not be purposefully designed.

The fear that Darwinism was an attempt to displace God in the sphere of creation was therefore quite justified. To the question, "Is there a divine purpose for the creation of humans?" evolution answers no. To the question "Is there a divine purpose for the creation of any living species?" evolution answers no. According to evolution, the adaptations of species and the adaptations of humans come from natural selection and not from design. The properties that species have do not exist as ideas or teleological causes prior to or outside of their evolution (except, of course, for those types of organisms for which humanity itself has now become the designer).[11]

Surprisingly, the conflict between these seemingly irreconcilable points of view on the same subjects, the origin of species and the origin of humans, did not result in either the defeat of religion or the defeat of evolution. Evolution suffered almost no loss at all. In fact, tied to the scientific method and supported by it, evolution became the main unifying force in biology and led to an expansion of research in almost every area. Although religious critics continued to argue that evolution could not be "seen," the evolutionary view became as strongly entrenched in biology as the atomic, molecular, and gravitational theories were in the physical sciences. None of these principles were seen, yet they were all working models indispensable to a consistent and scientific comprehension of events.

Sources for the Preservation of Religion

One might think that, through the closely reasoned arguments offered by biologists, evolution had successfully undermined one of religion's prime justifications for itself—the special creation of humans. The demise of religion should therefore have been just a matter of a short time. This demise did not occur, and some of the reasons are worth looking at.

First of all, the change from regarding nature as operating in terms of **anthropomorphic wisdom** to operating in terms of what evolutionists viewed as **evolutionary opportunism** (expressed by the survival of the fittest) was certainly difficult for people raised in the bosom of paternalistic social structures and religious beliefs. For many centuries it was, after all, easy to follow a pattern of simple parallels between society and nature: the father rules, the Lord rules, the King rules, the Pope rules, God rules. It is hoped that each rules with wisdom, concern, and sympathy, but in all cases humans depend on the beneficences of their rulers and try to influence their judgments by special pleas, sacrifices, and rituals.

Even for those who questioned the legitimacy of kingly rule, it must have seemed far too audacious a leap to question the rule of nature by the wisdom of God. Even the philosopher Kant, who had an evolutionary approach to cosmology and to the origin of the solar system, found it at times abhorrent to admit that species could evolve; in one place he described such notions as "ideas so monstrous that the reason shrinks before them." Darwin too is reported to have felt quite uncomfortable about his role in proposing the evolution of species and wrote in an 1844 letter to Joseph Hooker that "it is like confessing a murder."

We must remember that one source for these uncomfortably guilty attitudes was the staid, conservative milieu that held strongly to the notion that man was created in the image of God and endowed with the rule over other biological and social groups.[12] Kinship to those below, whether ape or servant, was a repugnant idea. An oft-reported example of this repugnance is the response of the wife of the Bishop of Worcester when informed that Huxley had announced that man was descended from apes: "Descended from apes! My dear, let us hope that it is not true, but if it is, let us pray that it will not become generally known."[13]

[11]Whether the "final cause" is given as God or a mystical aspect of nature, scientists reject explanations outside the sequence of physical causes that determine events. The Dutch philosopher Baruch Spinoza (1632–1677) put the matter succinctly in a letter commenting on Robert Boyle's (1627–1691) proposal that nature designed birds for flying and fish for swimming, "He seeks the cause in the purpose."

[12]The Judeo-Christian Bible expresses these attitudes in various places. For example:

> God created man in his own image, in the image of God he created him. (Gen. 1:27)
>
> Thou hast made him only a little lower than angels, and hast crowned him with glory and honor. . . . Thou madest him to have dominion over the works of thy hand, thou hast put all things beneath his feet. (Ps. 8)

[13]It is ironic that many Western religionists were affronted by the concept of humanity's ascent from apes because it gave us a lowly ancestor, but were not offended by the concept of our fall from the "state of blessedness" in the biblical Garden of Eden, although that meant that humans live in a state of sin and relative degradation. Certainly, to arise from the beasts is a nobler attainment than to fall from the gods. What was at issue, of course, was not which of these positions was relatively higher or lower but which one allowed us the comfort of thinking we were a sort of angel, although fallen. Better, it seemed, to be a fallen angel than a risen beast!

Reinforcing these attitudes was the reaction of many of the European middle classes to the upheavals of the French Revolution and to the atheism of many of its leaders. (According to the critic St. George Mivart, Darwinism led to "horrors worse than the Paris Commune.") These reactions probably helped to strengthen the religious climate among the middle classes in the nineteenth century and led, in various groups, at least for a time, to a more literal interpretation of the Bible.

Essential to the preservation of religion in the midst of the evolutionary bombardment was also the fact, as pointed out previously, that religion answers a series of strong emotional needs. We can identify these as the need to feel that one's life has a purpose; the need to feel that there is a powerful force ("someone up there") who is protective and on whom one can depend, but to whom, in any case, one can appeal; and the need, in some way, to fill in the important personal mysteries of birth and death (the question of identity, "Who am I?" and "Why me?"). Furthermore, religion has traditionally served as the main repository for the ethics and morals of society (what is right and wrong) and fostered the use of penance to alleviate guilt. Religion has also often served to maintain confidence in the social order and to unify nationalistic and provincially chauvinist sentiments ("God is on our side").[14] In fact, one can claim that religion serves an important role in helping provide a common social identity to groups of individuals, and by increasing group power, confers "biological advantage" (Wilson).

Evolution, in contrast, deals with many basic questions of life that are of concern to religion, but as a science it did little to meet emotional needs until Freud and others began to develop psychotherapeutic methods. It is important to realize that as knowledge diminishes fear, the emotional needs that spring from such fears usually also diminish. Unfortunately, the feelings of humans arise from interaction among so many genetic, developmental, and social factors that fulfilling human emotional needs is still quite difficult.

The "Truce"

Society has therefore held on to both evolution and religion with a sort of armed truce. With the exception of some fundamentalists, religion essentially withdrew from the domain of biological evolution, leaving both the origin of species and the origin of humans in the hands of the evolutionists. To help make evolution acceptable among some of the main religious groups, various theologians then placed more emphasis on reinterpreting the Judeo-Christian Bible by either ignoring the creation story in Genesis or describing it as allegorical or mythical. This enabled scientists and intellectuals who maintained religious affiliations to insist that one could believe in both evolution and religion ("theistic evolutionists").

However, such reinterpretations evoked considerable discomfort in orthodox believers since, like any other religious document, the Bible is an attempt to explain the unknown in a religious framework. To concede that parts of the Bible can be known and understood outside the religious structure easily opens the door to further loss of religious credibility, which generated considerable nineteenth century dissension among theologians (see, for example, Roberts). Among various compromises attempted was that by Asa Gray (1810–1888), the American evolutionist, who proposed that the variability on which natural selection acts during evolution was itself specially created. To this Darwin replied that since not all variations are useful, it is inconceivable that evolution could be "designed" by such means.[15]

Some nineteenth century writers, such as Owen, Campbell, and Mivart, suggested that although natural selection may have caused some adaptations, design must have caused many basic generalized patterns such as the vertebrate archetype and the parallel appearance of similar organs (such as eyes) in different groups. Essentially, these authors moved the hand of the "designer" from specifying minor adaptations for species to devising major plans for higher taxonomic groups.[16] Most biologists maintained

[14]Some authors, such as the psychologist Erich Fromm, put the role of religion more strongly:

> Religion has a threefold function: for all mankind, consolation for the privations exacted by life; for the great majority of men, encouragement to accept emotionally their class situation; and for the dominant minority, relief from guilt feelings caused by the suffering of those whom they oppress.

[15]In a letter to Lyell, Darwin wrote:

> If you say that God ordained that at some time and place a dozen slight variations should arise, and that one of them alone should be preserved in the struggle for life and the other eleven should perish in the first or few generations, then the saying seems to me mere verbiage. It comes to merely saying that everything that is, is ordained.

In Dobzhansky's (1967) words, if evolution is "ordained," the ordination is evil:

> If evolution follows a path which is predestined (orthogenesis), or if it is propelled and guided toward some goal by divine interventions (finalism), then its meaning becomes a tantalizing, and even distressing, puzzle. If the universe was designed to advance toward some state of absolute beauty and goodness, the design was incredibly faulty. Why, indeed, should many billions of years be needed to achieve the consummation? The universe could have been created in the state of perfection. Why so many false starts, extinctions, disasters, misery, anguish, and finally the greatest of evils—death? The God of love and mercy could not have planned all this. Any doctrine which regards evolution as predetermined or guided collides head-on with the ineluctable fact of the existence of evil.

[16]Such proposals emphasizing major "designs" rather than selection appear similar to more recent claims that many developmental patterns are independent of genetic control because they are primarily caused by nonselected developmental constraints or noninherited environmental forces (Chapter 15).

that evolution operated on all levels and its mechanisms were enough to account for similarities and differences among major groups. By the beginning of the twentieth century, biologists rarely used arguments of supernatural design in biology, and such viewpoints were considerably less popular among other intellectuals as well.

Although religion retreated in various places, some groups held strongly to the concept that a creator must be directly involved in human relationships (Sarna):

> A creator-god who withdraws from his creation and leaves his creatures entirely to their own devices is a functionless deity, an inactive being, remote and aloof from the world of men and women. He represents no ideal, makes no demands, enjoins no obligations, provides no moral governance of the world, imposes no moral law. Human strivings rest in no assurance of being other than unreality and futility, and the human race is bereft of ultimate destiny.
>
> Not so the Creator-God of the Bible. He is vitally concerned with the welfare of His creatures, intensely involved in their fate and fortune. An unqualifiedly moral Being, He insistently demands human imitation of His moral attributes. He imposes His law on the human race, and He judges the world in righteousness. History, therefore, is the arena of divine activity, and the weal and woe of the individual and of the nation is the product of God's providence, conditioned by human response to His demands.

Other denominations, more willing to compromise, felt that how humans should behave (ethics and morality), and the sanctity and the meaning of what has evolved (such as human personality, identity, and social relationships), can be considered separately from how human biology came into being. To varying extents, such religionists share some of the reasonable claims that cultural evolutionists made, that the means and goals of cultural evolution can differ from those involved in biological evolution (Chapter 25).

Challenges to religion have continued; once people removed the notion of fixity and design from such a sacred concept as species and their creation, people needed to take only one further step to question and investigate the fixity and design of religion itself. The awe and mystery that surround an institution usually diminish in direct proportion to "earthly" knowledge of its origin and development. Religion is no exception. The evolutionary approach expressed through various natural and social sciences has seriously undermined its power, which depends on awe and mystery.[17]

To evolution, nothing is necessarily permanent, even society and morality. In fact, since the time of Darwin, sociologists and anthropologists have often pointed out that different standards of morality develop in different societies because of historical social reasons. Evolutionists such as Huxley, Dewey, and others extended these concepts to show that the types of morality considered desirable in Western industrialized societies could be derived from human rational thinking and do not at all depend on religious beliefs. Such proposals emphasize moral codes based on mutually beneficial social goals that respect human rights of all individuals and do not detract from their capacity for education, knowledge, and personal fulfillment. Thus, we can answer logically the ethical and moral questions of how humans should live without enlisting the common religious justification—"God willed it so!" The active social, scientific, and technological changes that have accelerated rapidly and radically in this century have fed this skeptical evolutionary climate.

Religious Fundamentalism and "Creation Science"

Fundamentalist religious groups have generally not accepted the uneasy truce between evolution and religious institutions in the Western world. In the United States, many people who believe in the Judeo-Christian account of creation in Genesis have banded together to form political pressure groups to impose their beliefs on public education. Their origins date back at least to the early 1900s when many fundamentalists were part of an anti-intellectual and antiestablishment populist movement with strong roots among economically threatened tenant farmers and small landholders, especially in the South and Southwest. Sociologists have suggested that believing in the literal truth of the Bible, revivalism, and other aspects of fundamentalist religion popularized in a series of pamphlets called *The Fundamentals* helped many of these rural groups defend their way of life against domination and control by the more intellectual but exploitative northern and eastern social and economic establishments.

[17]For example, historians have shown that the biblical Old and New Testaments, formerly considered of divine origin and inspiration, date considerably later than the depicted events, and were edited by various groups in the interim (see footnote 6 on p. 57). Thus, the Five Books of Moses in the Old Testament were not codified in written form until the time of Ezra, at least 400 years later than many of their described events (Friedman 1987), and the earliest surviving copies of New Testament manuscripts date 175 years after the death of Jesus Christ (Funk et al.). Quotations attributed to prominent biblical figures, such as Moses and Jesus, have also been seriously questioned. In their analysis of the Fourth Gospel (John) in the New Testament, Funk and his colleagues point out they "were unable to find a single saying they could with certainty trace back to the historical Jesus." Such information does not deny the personal value of religious beliefs, but it does show how ancient religious writers imposed their theology on history in order to cast their religion in distinctive directions.

Whatever their initial motivations, fundamentalist groups were occasionally successful in pursuing their antievolution goals in the South and Southwest during the first few decades of this century. By the end of the 1920s, fundamentalists had introduced antievolution bills into a majority of U.S. State legislatures, and had passed some in various southern states. Probably the most famous confrontation between evolution and biblical creationism during that period was the 1925 trial of a schoolteacher, John Scopes, who was convicted of ignoring the ban against teaching evolution in Tennessee schools.

Although many evolutionists felt that the Scopes trial essentially defeated the intellectual validity of the creationist position,[18] creationists apparently lost little ground in these regions and managed to have an impact on public education far beyond the South and Southwest. As Nelkin points out, by influencing textbook adoption procedures in various local and state school boards, creationists successfully minimized evolutionary explanations in secondary school textbooks for a long time. In accommodating its texts to creationist pressures, one publisher baldly admitted "Creation has no place in biology books, but after all we are in the business of selling textbooks"—a view reflected in the common downgrading of evolution by other textbook publishers. By 1942, more than 50 percent of high school biology teachers throughout the country excluded any discussion of evolution from their courses.

The impetus for an increase of evolutionary teaching in secondary schools was the result of a movement to reform the science curriculum in the late 1950s and early 1960s, when it was realized that science education in the United States lagged behind that of other countries (specifically the Soviet Union, which had launched the first space satellite, *Sputnik,* in 1957). Among these innovations were new high school textbooks in both biological and social sciences ("Biological Science Curriculum Study," "Man as a Course of Study") that discussed evolution and analyzed changes in human social relationships. By the end of the 1960s antievolution laws were either repealed or declared unconstitutional.

The fundamentalist response to these challenges was to intensify attacks on the teaching of evolution and to adopt a new strategy claiming that creation was as much a science as evolution and therefore should be given equal time whenever evolution was taught. Within the last few decades, a number of societies and institutes established by fundamentalists for the propagation of **creation science** have entered the fray to further this approach.[19]

In spite of the name, there is little, if any, recognizable science in creation science. One can ask, How or why did the divine creation of different species occur? What are the scientific mechanisms of creation science? How are proposed mechanisms for creation described, compared, and evaluated? How, for example, can a "creation scientist" determine which of the following ten **creation myths** are correct?[20]

1. God arose from the depths of the ocean, created dry land, and then created all creatures on the hill at Heliopolis at the center of the universe (Egypt).
2. God made sky and earth by splitting the powers of evil in half, and then produced humans for purposes of worship (Mesopotamia).
3. God creates all that is good and struggles with an evil being that creates all that is bad. Each struggle lasts about 3,000 years and will continue until evil is vanquished, at which time creation will be complete and perfect (Iran).
4. God, a female, divided the sky from the sea, and produced a serpent with whom she copulated. She then laid a giant egg out of which came the earth, its creatures, and all the heavenly bodies, as well as the subsidiary powers to rule these various entities (Greece).
5. God created himself from a golden egg, and from the various parts of his body everything was born. After a time, life is destroyed and the cycle begins again (India).
6. God created the universe in 6 days ending with the creation of humans, according to Genesis 1; or, God first created Adam in the Garden of Eden and then created animals and birds and eventually Eve, according to Genesis 2 (Israel).
7. God was a woman who produced twins—the sun and the moon. During various eclipses, the twins came

[18]The verbal interchanges between the creationist, William Jennings Bryan, and the defending evolutionist attorney, Clarence Darrow, were widely disseminated and made the subject of a popular play and film, *Inherit the Wind.* Although Scopes was fined by the court, the Tennessee Supreme Court later reversed the decision on the grounds that the fine had been wrongly imposed. The antievolution law in Tennessee remained until 1967, when the state legislature repealed it.

[19]For historical accounts of the creation science movement, see the books by Larson, Numbers, and Webb.

[20]This small sample of creation myths, in which "God" designates the creator, derives from various religions, past and present. (Many additional myths can be found in collections by Farmer, Hamilton, Leeming and Leeming, Leach, Sproul, and Van Wolde.) In general, creationists ignore questions of how divine creation occurred, and profess creationism to be a "science" without any data on creation! To "creation scientists," a supernatural creator transgresses "natural" laws that can be analyzed and explained by science! Their " scientific" arguments are mostly limited to attempts at denigrating evolutionary findings, and to proposing that the only alternative to intelligent creation that evolution offers is randomness or chance. As Paley did almost two centuries ago, they argue that since organized complex structures have an extremely low probability of arising purely by chance, only "godlike" design can effectively account for complexity. Although such views may have seemed valid before people understood Darwinian selection, to present such arguments now is unconscionably misleading.

FIGURE 4–3 Creationist view of evolution as "the root of the tree of evil" that affects society. "Evolution is the taproot which is feeding the oppressive, murderous and infidel directions we see gaining acceptance around us" (Bartz). A similar "Evolution tree," published by the Pittsburgh Creation Society, shows 21 evil evolutionary fruits including terrorism, racism, sex education, inflation, hard rock music, suicide, and women's liberation (Toumey). It should be obvious that such fanaticism is not open to rational scientific discussion.

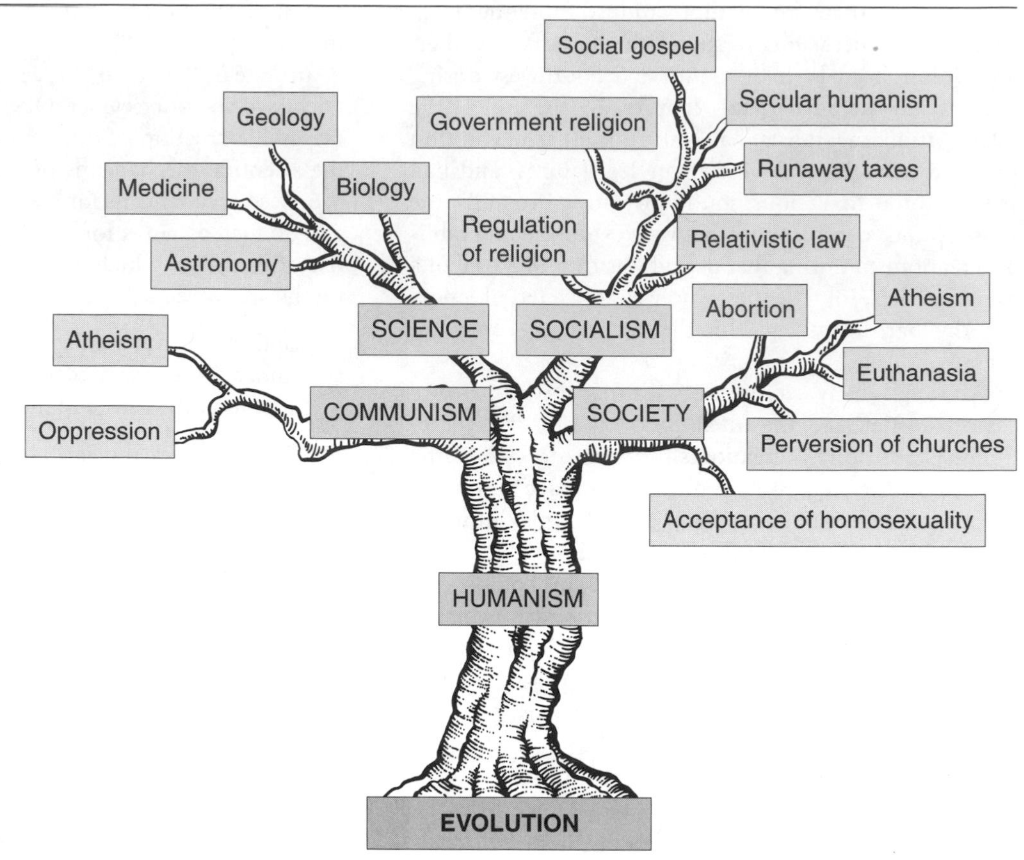

together to create the various gods and spirits of earth and sky that rule over humans (Benin, Africa).

8. God created the world in four distinct periods, each separated by a flood (Yucatán).
9. God created the earth and its creatures from mud gathered in the webbed feet of ducks who swam on a primeval ocean (American Crow Indians).
10. The universe was originally in the shape of a hen's egg, out of which God emerged and chiseled its main physical features. After 18,000 years God died and the remainder of the world was derived from his body: the dome of the sky from his skull, rocks from his bones, soil from his flesh, rain from his sweat, plant life from his hair, and humans from his fleas (China).

Fundamentalists have used "creation science" as a pseudonym for the religious creation myths in the Judeo-Christian Bible; they make no pretension at all to study this "science." As stated by Henry Morris, a founder of the Institute for Creation Research:

> Since nothing in the world has been created since the end of the creation period [in Genesis], everything must then have been created by means of processes which are no longer in operation and which we therefore cannot study by any of the means of science. We are limited exclusively to divine revelation as to the date of creation, the duration of creation, the method of creation, and every other question concerning the creation.

Clearly, evolutionary discoveries and methodologies can have little or no persuasive power with creationists who hold such positions. In fact, to many creationists, evolutionary concepts are the source of all "evil" (Fig. 4–3).[21] Although considerable literature deals with creationist attacks on evolution (see, for example, books by Futuyma and Kitcher and that edited by Godfrey), the implacable hostility of fundamentalist creationists toward evolution shows no promise of ever being resolved. The simple "black and white" theology of creationists—good versus evil; salvation versus damnation; creation

[21]From Morris (p. 83):

> The origin of all the evil in the universe must have been coincident with the origin of the idea of evolution, both stemming from Satan's rejection of God's revelation of himself as Creator and Ruler of the universe. This primal act of unbelief and pride later led to the fall of man. Similarly, unbelief in God's Word and man's pride in his own ability to rule his own destiny have yielded the bitter fruits of these thousands of years of human sin and suffering on the earth. And today, this God-rejecting, man-exalting philosophy of evolution spills its evil progeny—materialism, modernism, humanism, socialism, Fascism, communism, and ultimately Satanism—in terrifying profusion all over the world.

versus anarchy; God's people versus God's enemies, God versus the Devil—may offer comfort to believers but does not allow for other views. Casting differences in such a dichotomous manner helps religious believers to dehumanize their opponents as "instruments of Satan," a practice long used to condemn adversaries and their ideas as "heretics" and "heresies" (Pagels). By contrast, views in evolution, whether conforming or dissenting, are evaluated, as in all science, by evidence and not by unverifiable decree (see Box 4–1).

BOX 4–1 RESPONSES TO CREATIONIST ARGUMENTS

The following are responses by evolutionists to some issues raised by creationists opposed to evolution. Information on these and other topics are fully discussed throughout this text. For those who have access to the Internet, a list of more than 50 creationist misinterpretations and distortions is discussed by Mark Vuletic at the site *http://www2.uic.edu/~vuletic/cefec.html.* Another extensive Internet source is the Talk.Origins Archive at *http://www.talkorigins.org/origins/faqs-qa.html.*

- **Creationist claim:** *Evolution is a controversial subject even among evolutionists, since they often engage in major disputes.*
- **Evolutionist response (Stephen J. Gould):** "Scientists regard debates on fundamental issues of theory as a sign of intellectual health and a source of excitement. Science—and how else can I say it—is most fun when it plays with interesting ideas, examines their implications, and recognizes that old information might be explained in surprisingly new ways. Evolutionary theory is now enjoying this uncommon vigor. Yet, amidst all this turmoil, no biologist has been led to doubt the fact that evolution occurred; we are debating how it happened. We are all trying to explain the same thing: the tree of evolutionary descent linking all organisms by ties of genealogy. Creationists pervert and caricature this debate by conveniently neglecting the common conviction that underlies it, and by falsely suggesting that evolutionists now doubt the very phenomenon we are struggling to understand."
- **Creationist claim:** *Evolutionists' belief in evolution is as much a matter of blind faith as belief in the Bible is to creationists.*
- **Evolutionist response (Kenneth E. Boulding):** "The world scientific community has both many similarities to a world religion, but also very important differences. It has, in the first place, a very distinctive ethic of its own, which I am almost tempted to call "the four-fold way," as it has four essential components. The first is a high value on curiosity, which not all cultures possess. The second is a high value on veracity—that is, on not telling lies—which many other cultures also do not possess. The one thing that can get a scientist excommunicated from the scientific community is to be caught deliberately falsifying his [or her] results—that is, in telling lies. Error is often pardonable, but lying is the sin that cannot be forgiven. The third ethical principle is the high value on the testing of images of the world against the external world that they are supposed to map. Mere internal consistency is not enough, for there may be views of the world which are internally consistent, but which are, nevertheless, not true, in the sense that the real world does not conform to them. There are many methods of testing. Experiment is an important method where it is appropriate, though only perhaps a third to a half of the testing activities of science consist of experiment. Careful observation and recording, coupled with systematic analysis, is another important method, such as we have in celestial mechanics and in national economic statistics. Comparative studies of systems which are alike in many respects but differ in others is another important method; for instance, in medical research and the social sciences. Underlying all these, however, is a profound belief that the real world will speak for itself if it is asked the right questions. The fourth principle of scientific ethics is abstention from threat, embodied in the principle that people should be persuaded only by evidence and never by threat. This, of course, is in striking opposition to the ethics of many religious organizations and of all political organizations."
  To Boulding's response we can also emphasize the concept, discussed in Box 1-1, that analyses and explanations of historical events can be accepted by scientists even without experimental recapitulation. We regard geology, astronomy and evolutionary biology as historical sciences because their methodology has become sufficiently precise to allow explanations of past events that are "consistent with observations" of other past events and with present events.
- **Creationist claim:** *Since no one has really seen all the events evolutionists claim, the most they can say is that evolution is a "theory" and not a "fact."*
- **Evolutionist response:** Like previous points, this creationist argument shows a deep misunderstanding of science. There are many "theories" in science, from the atomic theory, to the molecular theory, to the theories of relativity and gravity, all based on explanations that account for observable events ("facts"). For example, the atomic theory was accepted by scientists because it explained chemical reactions,

although the atoms themselves were invisible. Similarly, the theory of evolution accounts for the historical sequence of past and present organisms ("facts") by explaining their existence in terms of factors that caused changes in the genetic inheritance of organisms over time ("theory"). The facts of evolution come from the anatomical similarities and differences among organisms, the places they are geographically found, the metabolic pathways they use, the embryological stages in which they develop, the fossil forms they leave behind, and the genetic, chromosomal, and molecular features that relate and separate them. These observable phenomena—all subjects of this book—show conclusively that "nothing in biology makes sense except in the light of evolution" (Dobzhansky 1973). The theory of evolution (genetic changes over time) explains the historical course of biology (facts) in terms of natural processes, such as mutation, selection, genetic drift, and migration (Chapter 22). These explanations are consistent with all observations so far. Evolution occurred—it is a "fact." Evolutionists also present the "theory" that explains this fact by the rules of science. Such approach is not different from the "fact" that an apple dropped on Newton's head (or thereabouts), and he explained it with the "theory" of gravity.

- **Creationist claim:** *"Creation science" is as much a biological science as evolution.*

- **Evolutionist response:** An important part of biological science is the search for understandable explanations for the history of organisms: to seek answers to the questions What happened? and Why did it happen this way? (Box 1-1). "Creation science" cannot be considered science, since creationists do not believe this search should be pursued scientifically, and instead offer biblical or religious explanations for the history of organisms. The following is a quotation from *Evolution?—The Fossils Say No!* by Duane T. Gish, Associate Director of the Institute for Creation Research, the leading creationist organization:

> By creation we mean the bringing into being of the basic kinds of plants and animals by the process of sudden, or fiat, creation described in the first two chapters of Genesis. Here we find the creation by God of the plants and animals, each commanded to reproduce after its own kind using processes which were essentially instantaneous. We do not know how God created, what processes He used, for *God used processes which are not now operating anywhere in the natural universe.* This is why we refer to divine creation as special creation. We cannot discover by scientific investigations anything about the creative processes used by God.

As the evolutionist Michael Ruse points out:

> In science an explanation must explain more than that for which it was invented. Saying God created the eye of a dog, for example, does not explain anything about why it is structured the way it is, why it works the way it does, and why it is similar to the eye of a cat in some respects and different in others.

- **Creationist claim:** *Since the second law of thermodynamics holds that entropy (disorder and disorganization) increases, organized life is thermodynamically impossible without intervention of a creator who can circumvent this law.*

- **Evolutionist response:** The second law of thermodynamics applies to closed systems isolated from external energy. It does not apply to open systems such as living organisms that obtain energy from external sources, which they convert into negative entropy—organized forms. Many complex natural phenomena, such as snowflakes, tornadoes, and stalactites, are produced by the conversion of external energy into organized structures. Furthermore, as Hugo Franzen points out:

> If thermodynamics requires the intervention of a supernatural agent to originate life, it requires the continued intervention of that agent to sustain life. There would then be a sustained and repeated violation of the laws of thermodynamics in life processes [to produce all necessary catalytic molecules]. It is, of course, possible to accept this notion. It is not possible to accept it and thermodynamics as well.

Unlike creation, evolution is a natural event in perfect accord with all physical and chemical laws.

- **Creationist claim:** *If evolution were true, we would find fossils of all transitional organisms, such as between fish and amphibians, reptiles and birds, apes and humans. According to creationist authority Gary Parker, "Famous paleontologists at Harvard, the American Museum, and even the British Museum say that we have not a single example of evolutionary transition at all."*

  Evolutionist response: Paleontologists have firmly identified thousands of fossil intermediates, such as between fish and amphibians, reptiles and birds, reptiles and mammals, and apes and hominids. For example, Roger J. Cuffey provides long lists of paleontological sources describing fossil intermediates between species of the same genus as well as transitional forms between different genera, orders, and classes. These fossil intermediates are found in Algae, Ginkgophytes, Angiosperms, Foraminiferans, Corals, Bryozoans, Brachiopods, Gastropods, Pelecypods, Ammonoids, Trilobites, Crustaceans, Echinoids, Carpoids, Blastoids, Graptolites, Conodonts,

Fishes, Fish-Tetrapods, Amphibians, Amphibian-Reptiles, Reptiles, Reptile-Birds, Reptile-Mammals, Mammals, Hominids.

That paleontologists have not found all transitional fossils does not mean such organisms were absent: dead organisms are rarely preserved (p. 42) and transitional forms even less so because they are often in small populations that survived for relatively short time periods (p. 379). Even when preserved, most fossils remain imbedded in rocks and inaccessible to paleontologists. The admission that we have not discovered everything does not mean we know nothing. Also, the absence of evidence is not evidence of absence—that the sun is not visible at night does not imply that it vanishes in the evening and is recreated in the morning.

In respect to the obviously false claim by creationist Gary Parker, Niles Eldredge, a paleontologist at the American Museum of Natural History writes:

> A prominent creationist interviewed a number of paleontologists at those institutions and elsewhere (actually, he never did get to Harvard). I was one of them. Some of us candidly admitted that there are some procedural difficulties in recognizing ancestors and that, yes, the fossil record is rather full of gaps. Nothing new there. This creationist then wrote letters to various newspapers, and even testified at hearings that the paleontologists he interviewed admitted that there are no intermediates in the fossil record. Thus, the lie has been perpetuated by Parker. All of the paleontologists interviewed have told me that they did cite examples of intermediates to the interviewer. The statement is an outright distortion of the willing admission by paleontologists concerned with accuracy that, to be sure, there are gaps in the fossil record. Such is creationist "scholarship."

Eldredge's complaint can be extended to many other such matters, from creationist literature to their practices in debates. Creationists consistently misrepresent evolutionary findings and often quote evolutionists completely out of context. They repeat this tactic in debates: by the time the evolutionist has dealt with one distortion or misquotation, the creationist has introduced others. Their fundamental strategy is to deny any event, such as a transitional fossil, that questions creationist doctrine.

Thus, since creationists insist new species cannot appear after the initial creation period, new types are not novel species to them but only differences in "kind." That is, australopithecines who walked upright were just a "kind" of ape, and dog-sized four-toed leaf-eating *Eohippus* (*Hyracotherium*) of the Eocene epoch was just a "kind" of horse. Transitional forms are therefore deemed not transitional at all. For example, *Archaeopteryx* with its reptilian teeth and skeleton was just a "bird," and *Ichthyostega* with its fish-like shape, vertebrae, and fin-rayed tail was just an "amphibian." Such distorted concepts obviously contribute nothing to how organismic distinctions and relationships are determined and how they can be understood and explained (Chapters 11 and 24)—basic elements of biology.

- **Creationist claim:** *The earth is really young—perhaps 10,000 years old—and evolutionists' proposals for an older earth measured by decay rates of radioactive elements and other techniques are faulty and inaccurate.*

- **Evolutionist response (Joel Cracraft):** "Geologists have established, virtually beyond scientific doubt, that the earth is approximately 4.5 to 4.6 billion years old. That the stratigraphic record of sediments can be sequentially dated by radiometric decay rates is not now a matter of question among geologists who study dating techniques. Each radioactive element decays at a unique and constant rate, and these rates are not influenced by external factors such as extremes of temperature or pressure. The creationists simply assert that these rates are not constant. Yet, at the [1981] Arkansas creation trial, every one of the creationists' geological witnesses—including Robert Gentry, their chief expert on radioactivity—testified that no scientific evidence exists which questions the constancy of these decay rates.

The creationists sometimes invoke a "singularity" at about 6,000 years corresponding to their suggested time for the Noachian "Flood"; at this singularity, the decay rates slowed down significantly—more or less to their present level. Prior to that time, the rates were much higher, thus giving the appearance that the earth is billions of years old, when actually it is only thousands. James Hopson of the University of Chicago has suggested to me a simple response to this supposition of a supernatural event: if the creationists are correct in believing that the earth is only thousands of years old, and that the decay rates at one time accelerated, then the amount of heat released from that amount of radioactive decay would have been sufficient—by a large margin—to have vaporized the earth.

By their own admission, the creationists cannot provide any scientific evidence for such dramatic changes in radioactive decay rates or their assertion that these rates do not measure the true age of the earth. The creationists turn to a supernatural "singularity," which is a belief derived from religion, not from the evidence of science."

- **Another Evolutionist response (Niles Eldredge):** "Perhaps the most dramatic demonstration of the validity and accuracy of modern geologic dating comes from the deep-sea cores stored by the thousands in various oceanographic institutions. The direct sequence is preserved in these drill cores, of course, and the microscopic fossils in them allow the usual "this-is-older-than-that" sort of relative dating to be done. We can also trace the pattern of changes in the orientation of the earth's magnetic field: as you go up a core, portions are positively charged, while others are negative. Major magnetic events, reflecting a flipping of the earth's magnetic poles, are recognizable, and the sequence of fossils, the same from core to core, always matches up with the magnetic history in the same fashion from core to core. *Then,* when we obtain absolute dates from the cores (usually by using oxygen isotopes), we *always* find that the date of the base of the "Jaramillo event"—one of the pole-switching episodes—always yields a date of about 980,000 years ago. The dates are *always* the same (again, with a minor plus-or-minus factor). They are *always* in the right order. They are *always* in the tens or hundreds of thousands of years for the most recent dates, and in the millions of years further down the cores."

- **Creationist claim:** *Evolutionists say they have found transitional forms between apes and humans, but these have turned out to be misclassified apes or misclassified humans. "The number of known human fossils would barely fill an average-sized coffin."*

- **Evolutionist response:** Although primate fossil hunting has been a difficult pursuit, large numbers of different kinds of hominid fossils are known, from *Ardipithecus* to *Homo* (Chapter 20). Kenneth E. Nahigian notes that even creationists who do not regard such fossils as evolutionary intermediates are impressed by the size of these findings:

  > Creationist Michael J. Ord, in his book review of *Bones of Contention,* had this admission: "I was surprised to find that instead of enough fossils barely to fit into a coffin, there were over 4,000 hominid fossils as of 1976. Over 200 specimens have been classified as Neanderthal and about one hundred as *Homo erectus.* More of these fossils have been found since 1976." Marvin L. Lubenow, the author of *Bones of Contention,* wrote to the editor of the same creationist journal: "The current figures are even more impressive: over 220 *Homo erectus* fossil individuals to date and well over 300 Neanderthal fossil individuals discovered to date [1994]."

- **Creationist claim:** *Evolutionists claim that a complex living structure arose purely by chance—an event with such low probability as to be impossible in our universe.*

- **Evolutionist response:** As explained on p. 59, it is important to recognize that random chance events are not the same as evolutionary events based on natural selection. Chance can be defined as events whose causes are independent of each other, so that a succession of such events need bear no mutual relationship, and therefore often leads to disorganization ("chaos"). Evolution by natural selection, in contrast, marks events whose occurrence depends on previous events, so that a succession of such events can lead to organized structures and increased complexity. The nonorganismic (abiotic) synthesis of amino acids and other basic biological molecules (Chapter 7) also indicates that chemical reactions are biased to produce such molecules, and natural selection can subsequently operate to increase their complexity and organization. Although the exact steps to the origin of life are not yet known, this does not detract from our understanding of many basic mechanisms that allowed the present diversity of life to evolve, nor affect the evidence that such evolution occurred. Moreover, since the origin of life is a molecular biology question, and we are just learning molecular biology techniques, it is not surprising that we don't yet know exactly how life on Earth originated. Even so, we are learning how enzymatic reactions evolve, how metabolic pathways evolve, how developmental controls evolve, and how genetic systems evolve. That's a lot of learning about early life questions in a very short time!

  Always missing in creationist arguments is an understanding of selection—that the alternative to design by a "creator" is not random chance, but selection. Selection is a sequential process that ties individual chance events into a "creative" sequence because particular steps in the sequence are "adaptive" and allowed to persist. An adaptation is not a sudden event but the result of a succession of selective events, each with reasonable probability.

  Thus, selection for improved sight leads to improved visual apparatus (Fig. 3–1), selection for improved vertebral support leads to improved spinal apparatus (Fig. 18–8), selection for improved hearing leads to improved auditory apparatus (Fig. 19–3), and so forth—steps along the way are distinctively adaptive.

- **Creationist claim:** *Natural selection can only eliminate misfits or produce minor changes, but cannot produce new adaptations. If selection was potent and evolution continuous, we should be bombarded with new species all the time. Where are they?*

- **Evolutionist response:** First, selection, as explained above, is a "creative" force in producing adaptations by a succession of selective steps that improve fitness. Second, the answer to where new species can be found is simple—they are always about! New species have continually evolved over the last 3.5 billion years and have continually replaced older species. Most organisms we see before us today are different from ancestors who may have lived only hundreds or thousands of generations ago, and even more different from ancestors who lived many hundreds of thousands of generations ago. We do not have film to show these changes, but we do have fossils and genetic techniques that indicate how different ancestral species were, and how novel present species are.

The answer to why we don't commonly observe new species emerge is also simple—speciation takes time! In many groups, species formation is generally recognized when reproductive isolation occurs between two populations, allowing each to embark on its own distinctive evolutionary path. Such events can take many thousands of years, because they involve many genetic and selective (adaptive) changes. We can nevertheless document that intense selective differences have caused rapid genetic differentiation between populations, as have also certain large chromosomal mutations. For example, we know that cultivated corn evolved in only about 7,000 years from the wild grass, teosinte, because it was continuously selected for food by Central American Indians (Fig. 21-1). Other selective effects, observed in both laboratory experiments and nature, include changes in temperature, food sources, habitats, and environmental toxins. Among such events investigated by population geneticists are fly populations that became resistant to insecticides (Fig. 10-37), plants that developed new characters (p. 313), *Drosophila* strains that developed unique sexual behaviors (p. 595), cichlid fishes that rapidly adapted to different conditions in African lakes (Fig. 12-18), and parasitic insects that adapted to different plant hosts (p. 597). Although we have not yet discovered the exact genes necessary to produce speciation, we have identified some, and know they vary in different groups and circumstances (p. 287 and Chapter 24).

We also know that certain chromosomal changes in structure and number can be more radical in their effects, and lead to spontaneous speciation events. Among such examples are translocations responsible for new species of *Oenothera* (p. 516) and doubled chromosome numbers responsible for new hybrid species of *Nicotiana* (Fig. 10-14).

- **Creationist claim:** *Evolutionists cannot account for anomalies that contradict evolutionary concepts, such as:*

  a) The sun has much less angular (rotational) momentum than would be expected for its large mass.

  b) There are deviations from the evolutionary concept that more recent geological strata should always be positioned over older strata.

  c) Human and dinosaur footprints have been found together in Cretaceous rocks in Texas.

  d) The finding of sudden large deposits of fossils is more in accord with catastrophic events such as the biblical Noachian flood than with slow evolutionary sedimentation processes.

  e) Since mutations are deleterious and not beneficial, evolutionary selection can never lead to adaptation but only go from bad to worse.

- **Evolutionist response:** Scientists have demonstrated many times that these and other so-called anomalies are really "pseudoanomalies," yet they nevertheless keep reappearing in creationist literature.

  a) The angular momentum of the sun was transferred to the planets by the solar wind during condensation of the solar system (p. 86).

  b) Older geological strata can be boosted over younger strata in some places by "overthrusting," especially in mountain-building areas (Fig. 6–15). They are nevertheless recognized as "older" by the kinds of fossils they contain, and by structural features that show continuity with nonoverturned deposits elsewhere.

  c) The Paluxy River anomalies (human footprints in Cretaceous deposits) have been demonstrated to be dinosaurian, and the human footprints were recently carved by humans. (The earliest known fossil hominid footprints were made by australopithecines and date back about 3.7 million years (p. 472).

  d) Organismic fossils are not concentrated in any particular geological stratum, but are found even in Precambrian deposits (Figs. 9–11 and 14–2). Geological strata are multilayered, some many kilometers thick, bearing sediments and fossils deposited over very long time periods. There is no geological or paleontological support for a biblical 1-year worldwide flood.

  e) Although most mutations are deleterious—organisms are generally so well-adapted that a random genetic change will cause malfunction—some mutations, such as resistance to toxins or parasites (p. 575) can obviously improve fitness. Other mutations, even those not immediately adaptive, persist in populations through various mechanisms (Chapter 23), and provide the genetic variability allowing populations to evolve and improve fitness when environmental conditions change (pp. 229–230).

- **Creationist claim:** *In fairness to both sides of the argument, we need laws that creationism be taught along with evolution.*

- **Evolutionist response (Richard D. Alexander):** "When a creationist, Darwinist, Marxist, or supporter of any other theory defends his or her views publicly, he or she does everyone a service. But when anyone attempts to establish laws or rules requiring that certain theories be taught or not be taught, he or she invites us to take a step toward totalitarianism. Whether a law is to prevent the teaching of a theory or to require it is immaterial. No laws were ever passed saying that evolution had to be taught in biology courses. The prestige of evolutionary theory has been built by its impact on the thousands of biologists who have learned its power and usefulness in the study of living things. No laws need to be passed for creationists to do the same thing. When creation theorists strive to introduce creation into the classroom as an alternative biological theory to evolution they must recognize that they are required to give creation the status of a falsifiable idea—that is, an idea that loses any special exemption from scrutiny, that is accepted as conceivably being false, and that must be continually tested until the question is settled. A science classroom is not the place for an idea that is revered as holy. The greatest threat to society and to our children is not whether students are exposed to wrong ideas—after all, many high school biology students are legally adults with voting privileges, and all high school biology students have already been exposed to many wrong ideas. What is important is whether each has been taught how and given the freedom to test new ideas, evaluate them, and respond appropriately."

To Alexander's statement, we can add the following: If "fairness" is a matter of teaching proposed alternate versions of science in schools, then other "believers" can rightfully claim that their concepts should also be given "equal time": astrology with astronomy, alchemy with chemistry, flat Earth with spherical Earth, geocentric solar system with heliocentric, phrenology (study of personality in cranial bumps) with psychology, Christian Science disease theory with germ theory, and so forth. Obviously, subjects to be taught in schools as science should be based on scientific concepts and judged by scientific rules of evidence rather than include nonscientific or pseudoscientific matter mandated by religious and special interest groups.

SUMMARY

Darwinian evolution had a profound impact not only on biology but also on many other fields, religion in particular. Its extraordinary influence was due to social, economic, and technological developments that had helped overthrow the old social order of feudalism and monarchy and had brought about the rise of capitalism. The traditional religious rationale for social and biological systems was that the universe followed a designed order established by an intelligent deity. Thus evolutionary theory, by denying that a god purposely designed biological creatures, was perceived as a great threat to religious interests.

The roots of religious beliefs lie in human attempts to appeal to and control the forces of nature, which were long incomprehensible and thought to be supernatural and humanlike. From these roots arose the concept of God and soul, both of which were supposed to be eternal and immaterial. To the insecure human, the instability of nature could only be managed and controlled by engaging divine support and warding off divine wrath. Such propitiation was essential in establishing and maintaining all religious cults—the altar of sacrifice often being the primary means by which offerings to the divine power could restore order, provide benefits, and ameliorate guilt.

Until Copernicus and Galileo in the sixteenth century, no one seriously challenged the idea of a powerful deity controlling the physical universe. In the new worldview they and others brought about, God appeared as an initial creator rather than as an incessant manipulator of the solar system.

The advent of Darwinism posed even greater threats to religion by suggesting that biological relationships, including the origin of humans and of all species, could be explained by natural selection without the intervention of a god. Many felt that evolutionary randomness and uncertainty had replaced a deity having conscious, purposeful, human characteristics. The Darwinian view that evolution is a historical process and present-type organisms were not created spontaneously but formed in a succession of selective events that occurred in the past, contradicted the common religious view that there could be no design, biological or otherwise, without an intelligent designer.

According to evolution, interaction between organisms and their environment selects successful traits and further selective events then enhance these traits. In this way adaptation to the environment can continuously modify organs and structures over long periods, and complexities that seem unlikely as singular spontaneous events become evolutionarily probable events. The variability on which selection depends may be random, but adaptations are not; they arise because selection chooses and perfects only what is adaptive. In this scheme a god of design and purpose is not necessary.

Neither religion nor science has irrevocably conquered. Religion has been bolstered by paternalistic social systems in which individuals depend on the beneficences of those more powerful than they, as well as by the comforting idea that humanity was created in the image of a god to rule over the world and its creatures. Religion provided emotional solace, a set of ethical and moral values, and support for the established social system. Many Judeo-Christian denominations evaded interpretation of evolutionary biological events or attempted compromises between traditional religious explanations and scientific ones. Nevertheless, faith in religious dogma has been eroded by natural explanations of its mysteries, by a deeper understanding of the sources of human emotional needs, and by the recognition that ethics and morality can change among different societies and that acceptance of such values need not depend on religion.

In the United States fundamentalist religious groups who oppose evolution were quite successful in preventing its teaching in various states and reducing or eliminating evolution in many biology textbooks. However, in the 1950s revulsion arose against this attitude, and a wave of reform began in American science education. Fundamentalists countered with the "creation science" movement, which, despite its name, does not use scientific method. The positions of the creationists and the scientific world appear irreconcilable.

KEY TERMS

anthropomorphic wisdom	God
creation myths	personality
creation science	religious fundamentalism
evolution of religion	soul
evolutionary opportunism	supernatural design

DISCUSSION QUESTIONS

1. From the sixteenth century onward, what relationship developed between the introduction of new scientific concepts and changes in Western society?
2. How and why did the concepts of soul and god become established in religious institutions?
3. The question of design:
   a. Why do religious institutions want to believe that a designer creates natural events?
   b. How does Darwinism, using the concept of natural selection, explain the design of organisms?
4. Why have religion and Darwinism existed side-by-side in Western society for more than 100 years despite conflicting explanations for natural events?
5. "Creation science"
   a. Which scientific principles and laws should we apply in analyzing creation?
   b. Which of the many creation stories should we use in science texts and for teaching purposes?

EVOLUTION ON THE WEB

Explore evolution on the web! Visit the accompanying web site for *Evolution*, 3/e at www.jbpub.com/evolution for web exercises and links relating to topics covered in this chapter.

REFERENCES*

Albertz, R., 1994. *A History of Israelite Religion in the Old Testament Period. Volume I: From the Beginnings to the End of the Monarchy.* Westminster/John Knox Press, Louisville, KY.

Alexander, R. D., 1983. Evolution, creation, and biology teaching. In *Evolution Versus Creationism: The Public Education Controversy*, J. P. Zetterberg (ed.), Oryx Press, pp. 90–91.

Armstrong, K., 1993. *A History of God: The 4,000-Year Quest of Judaism, Christianity, and Islam.* Knopf, New York.

Baron, S. W., 1952–1983. *A Social and Religious History of the Jews*, 2d ed. (18 volumes). Columbia University Press, New York.

Bartz, P. A., 1984. *Bible Science Newsletter*, **22,** 2.

Bernal, J. D., 1969. *Science in History.* MIT Press, Cambridge, MA.

Boulding, K. E., 1984. Towards an evolutionary theology. In *Science and Creationism*, A. Montagu (ed.). Oxford University Press, Oxford, pp. 142–158.

Bowler, P. J., 1977. Darwinism and the argument from design: Suggestions for a reevaluation. *J. Hist. Biol.*, **10,** 29–43.

Cohen, S. J. D., 1987. *From the Maccabees to the Mishna (Library of Early Christianity).* Westminster Press, Philadelphia.

Coote, R. B., 1990. *Early Israel: A New Horizon.* Fortress Press, Minneapolis.

Cracraft, J., 1983. The scientific response to creationism. In *Creationism, Science, and the Law: The Arkansas Case*, M. C. La Follette (ed.). MIT Press, Cambridge, MA, pp. 138–149.

Cuffey, R. J., 1972. Paleontological evidence and organic evolution. In *Science and Creationism*, A. Montagu (ed.). Oxford University Press, Oxford, England, pp. 255–281.

Dobzhansky, Th., 1967. *The Biology of Ultimate Concern.* New American Library, New York.

———, 1973. Nothing in biology makes sense except in the light of evolution. *Amer. Biol. Teacher*, **35,** 125–129.

Eldredge, N., 1982. *The Monkey Business: A Scientist Looks at Creationism.* Washington Square Press, New York.

Ellegård, A., 1958. *Darwin and the General Reader.* Göteborgs Universitets Årsskrift, Gothenburg, Sweden. (Republished 1990 by University of Chicago Press, Chicago.)

Farmer, P. (ed.), 1979. *Beginnings: Creation Myths of the World.* Atheneum, New York.

Franzen, H. F., 1983. Thermodynamics: The red herring. In *Did the Devil Make Darwin Do It? Modern Perspectives on the Creation-Evolution Controversy*, D. B. Wilson (ed.). Iowa State University Press, Ames, IA, pp. 127–135.

Friedman, R. E., 1987. *Who Wrote the Bible?* Harper and Row, New York.

———, 1997. *The Hidden Face of God.* Harper, San Francisco.

*Extensive portions of this chapter originally appeared in *Bioscience*, **23,** 417–421 (1973).

Fromm, E., 1963. *The Dogma of Christ and Other Essays on Religion, Psychology and Culture.* Holt, Rinehart and Winston, New York.

Funk, R. W., R. W. Hoover, and the Jesus Seminar, 1993. *The Five Gospels: The Search for the Authentic Words of Jesus.* Polebridge Press (Macmillan), New York.

Futuyma, D. J., 1983. *Science on Trial: The Case for Evolution.* Pantheon Books, New York.

Gaster, T. H., 1969. *Myth, Legend and Custom in the Old Testament.* Harper & Row, New York.

Gillespie, N., 1979. *Charles Darwin and the Problem of Creation.* University of Chicago Press, Chicago.

Gillispie, C. C., 1951. *Genesis and Geology.* Harvard University Press, Cambridge, MA.

Gish, D. T., 1972. *Evolution?—The Fossils Say No!* Creation-Life Publishers, San Diego, CA.

Glick, T. F. (ed.), 1974. *The Comparative Reception of Darwinism.* University of Texas Press, Austin.

Godfrey, L. R. (ed.), 1983. *Scientists Confront Creationism.* Norton, New York.

Gould, S. J., 1981. Evolution as fact and theory. In *Science and Creationism,* A. Montagu (ed.). Oxford University, Oxford, England, pp. 117–125.

Greene, J. C., 1961. *Darwin and the Modern World View.* Louisiana State University Press, Baton Rouge.

Hamilton, V., 1988. *In the Beginning: Creation Stories from Around the World.* Harcourt Brace Jovanovich, San Diego.

Kauffman, S. A., 1995. *At Home in the Universe: The Search for Laws of Self-Organization and Complexity.* Oxford University Press, Oxford, England.

Kitcher, P., 1982. *Abusing Science: The Case Against Creationism.* MIT Press, Cambridge, MA.

La Barre, W., 1970. *The Ghost Dance: Origins of Religion.* Doubleday, New York.

Larson, E. J., 1985. *Trial and Error: The American Controversy over Creation and Evolution.* Oxford University Press, Oxford, England.

Leach, M., 1992. *Guide to the Gods.* ABC-CLIO, Santa Barbara, CA.

Leeming, D., and M. Leeming, 1994. *A Dictionary of Creation Myths.* Oxford University Press, New York.

Lovejoy, A. O., 1959. Kant and evolution. In B. Glass, O. Temkin, and W. L. Straus, Jr. (eds.). *Forerunners of Darwin.* Johns Hopkins Press, Baltimore, pp. 173–206.

Lubenow, M. L., 1992. *Bones of Contention.* Baker, Ada, MI.

McIver, T., 1992. *Anti-Evolution: A Reader's Guide to Writings Before and After Darwin.* Johns Hopkins University Press, Baltimore.

Monod, J., 1971. *Chance and Necessity.* Knopf, New York.

Moore, J. R., 1979. *The Post-Darwinian Controversies.* Cambridge University Press, Cambridge, England.

Morris, H. M., 1963. *The Twilight of Evolution.* Baker, Grand Rapids, MI.

Nahigian, K. E., 1997. Impressions: An evening with Dr. Hugh Ross. *Reports of the National Center for Science Education,* **17,** 27–29.

Nelkin, D., 1982. *The Creation Controversy: Science or Scripture in the Schools.* Norton, New York.

Numbers, R. L., 1992. *The Creationists: The Evolution of Scientific Creationism.* Knopf, New York.

Oldroyd, D. R., 1980. *Darwinian Impacts: An Introduction to the Darwinian Revolution.* Open University Press, Milton Keynes, Oxford, England.

Pagels, E., 1995. *The Origin of Satan.* Random House, New York.

Paley, W., 1802. *Natural Theology: or Evidences of the Existence and Attributes of the Deity, Collected from the Appearances of Nature.* Faulder, London.

Parker, G., 1980. *Creation: The Facts of Life.* Creation-Life Publishers, San Diego.

Roberts, J. H., 1988. *Darwinism and the Divine in America.* University of Wisconsin Press, Madison.

Ruse, M., 1988. *But Is It Science? The Philosophical Question in the Creation/Evolution Controversy.* Prometheus Books, Buffalo, NY.

Russell, C. A. (ed.), 1973. *Science and Christian Belief: A Selection of Recent Historical Studies.* University of London Press, London.

Russett, C. E., 1976. *Darwin in America: The Intellectual Response 1865–1912.* Freeman, San Francisco.

Sarna, N. M., 1986. *Exploring Exodus: The Heritage of Biblical Israel.* Schocken Books, New York.

Schrödinger, E., 1954. *Nature and the Greeks.* Cambridge University Press, Cambridge, England.

Sober, E., 1993. *Philosophy of Biology.* Westview Press, Boulder, CO.

Sproul, B. C., 1979. *Primal Myths: Creation Myths Around the World.* Harper, San Francisco.

Tawney, R. H., 1926. *Religion and the Rise of Capitalism.* Harcourt, Brace, New York.

Toumey, C. P., 1994. *God's Own Scientists: Creationists in a Secular World.* Rutgers University Press, New Brunswick, NJ.

Tylor, E. B., 1881. *Anthropology: An Introduction to the Study of Man and Civilization.* Appleton, New York.

Van Wolde, E., 1996. *Stories of the Beginning: Genesis 1–11 and Other Creation Stories.* Morehouse Publishing, Ridgefield, CT.

Wallace, A. C. F., 1966. *Religion: An Anthropological View.* Random House, New York.

Webb, G. E., 1994. *The Evolution Controversy in America.* University Press of Kentucky, Lexington, KY.

Weber, M., 1963. *The Sociology of Religion.* Beacon Press, Boston.

White, A. D., 1896. *A History of the Warfare of Science with Theology in Christendom.* Appleton, New York.

Wilson, E. O., 1978. *On Human Nature.* Harvard University Press, Cambridge, MA.

II
The Physical and Chemical Framework

5

The Beginning

People find it difficult to appreciate the enormity of time that surrounds us. We are almost always focusing on the here and now, and our own mortal histories limit our direct experience of the past. In the last five or six thousand years, written genealogies and historical accounts somewhat lifted these limitations, but even then our notion of history usually extended no further than the dim legends of our culture. To explore even a small portion of the past beyond our culture—let us say 10 or 100 million years—long seemed an unnatural feat. What could have existed before our memories and traditions? The author of Ecclesiastes said, "There is nothing new under the sun." (Eccles. 1:9)

To conceive of a reality past our own records demands considerable evidence. Fortunately, as we have seen, the evidence of fossils and of long-term geological changes ended the historical isolation of humans from the world around them. However, the question remains: What, and how long ago, was the beginning?

The Origin of the Universe

Theories of the origin of the universe that astronomers have developed in recent years have sought to deal with a few main observations:

1. Hydrogen is the basic fuel stars use when they begin radiating the large amounts of energy that makes them visible to us.
2. As stars burn hydrogen, other elements such as helium and carbon accumulate through various fusion reactions.
3. Further elements are produced as stars undergo aging processes, including, in some cases, cataclysmic events such as supernovae.
4. The continual transformation of hydrogen indicates that this fundamental element will eventually diminish until no more stars can come into being unless a new source of matter is present.

To many scientists, these considerations all point to a time when the universe must have consisted mostly or entirely of hydrogen, and our present universe is only a stage in the evolution of this primordial mass. Among many proposed cosmological theories, one former concept, called the **steady-state hypothesis,** suggested that matter in the universe never really disappears, but as hydrogen diminishes, new matter from an unknown source replaces it. The amount of new matter that must enter according to this hypothesis would have to be astronomically large, requiring the birth of approximately 50,000 new stars per second.

The presently accepted view that Gamow and others proposed suggests that at a distant time in the past the whole universe was a small sphere of concentrated energy/matter. This substance then exploded in a big bang to form hydrogen at first and then eventually the galaxies and stars. In the dispute with proponents of the steady-state hypothesis, considerable evidence came to light that helped reconstruct various aspects of the origin of the universe. Much of this evidence supports the **big bang theory** and includes the following:

- GALACTIC EXPANSION About 70 years ago, the American astronomer Slipher discovered that light waves emitted from distant galaxies of stars shifted toward the red end of the spectrum, indicating that these galaxies are rapidly receding from us. This effect, called the *Doppler shift,* arises because emitted wavelengths from any source appear longer (red) or shorter (blue) if an observer is moving, respectively, away from the source or toward it. As shown in Figure 5–1, absorption lines of a known element (calcium in this case) shift toward the red as the distance from the earth increases. This indicates that some of the farthest galaxies may be receding from us at speeds that approach 25 to 90 percent of the speed of light. Also, the rate at which this recession occurs has probably been changing: the universe seems to have expanded more rapidly in the past than at present. The velocity of early cosmic expansion in this "inflationary universe" may even have exceeded the speed of light (Linde).
- BLACK BODY "FOSSIL" RADIATION Although visible energy seems to be radiating primarily from the galaxies and their stars, there is evidence from radio telescopes that a background of fairly uniform low-temperature radiation, about 3°K[1], pervades the entire universe. This black body radiation is predictable if the original big bang occurred between 10 to 20 billion years ago and began with an initial temperature of about 10^{32} degrees Kelvin. As time went on and the universe expanded, its temperature decreased, reaching the present 2.726°K.
- RADIO WAVES Evidence suggests that celestial radio wave sources are basically associated with the presence of galaxies. These sources greatly increase in number when observed at light-year distances that reflect the time period when galaxies first came into being. That is, observations of galaxies that are about 7 or 8 billion light-years from us indicate an increase in radio sources and point to a time when the universe was only about one-third to one-fifth its present age and was considerably more compact than today. Before this period, few or no radio sources are clearly discerned, indicating a time when galaxies may not yet have formed.
- HYDROGEN AND HELIUM PROPORTIONS The proportion of helium in the universe (about 23 percent in mass) is far greater than the amounts synthesized by galactic stars in which thermonuclear burning of hydrogen produces helium. Moreover, the relative proportions of hydrogen (about 75 percent in mass) and helium appear about the same in all stars and galaxies in contrast to later-formed heavier elements (2 percent in mass) whose compositions can vary by factors of more than 100 (pp. 82–83). Most helium and hydrogen seems to have been synthesized before galaxies formed, and big bang theory calculations support the observed abundances of these two fundamental elements in stars and galaxies.

Although these findings are strong evidence for the big bang theory, a number of astronomers have suggested that they do not exclude more than one big bang. That is, the universe may oscillate between an expanding state caused by a big bang to a contracting state (the "big crunch") followed by a subsequent big bang, perhaps *ad infinitum* (Fig. 5–2, **oscillating big bang theory**). Among other implications, this theory would mean that enough matter exists in the universe for gravity to halt its expansion and reverse galactic dispersion.

Researchers have therefore focused on measuring the amount of matter by measuring its average density. So far, calculations indicate that the universe can contract if the density is 10^{-29} gm/cm^3 (one proton per 10 cubic feet). Since the actual average density of all luminous matter in stars and galaxies is only 10^{-31} gm/cm^3, or about 100 times less than the density necessary for contraction, the oscillating universe at first seemed questionable. However, cosmologists propose that the uneven distribution of galaxies follows along gravitational paths primarily influenced by "dark matter," which provides

[1]Convert degrees Kelvin (°K) into degrees centigrade or Celsius (°C) by subtracting 273. Thus 3°K is −270°C or −454°F (Fahrenheit). The freezing and boiling points of water are respectively 273°K (0°C = 32°F) and 373°K (100°C = 212°F). Zero degrees Kelvin signifies absolute zero, at which all molecular motion ceases.

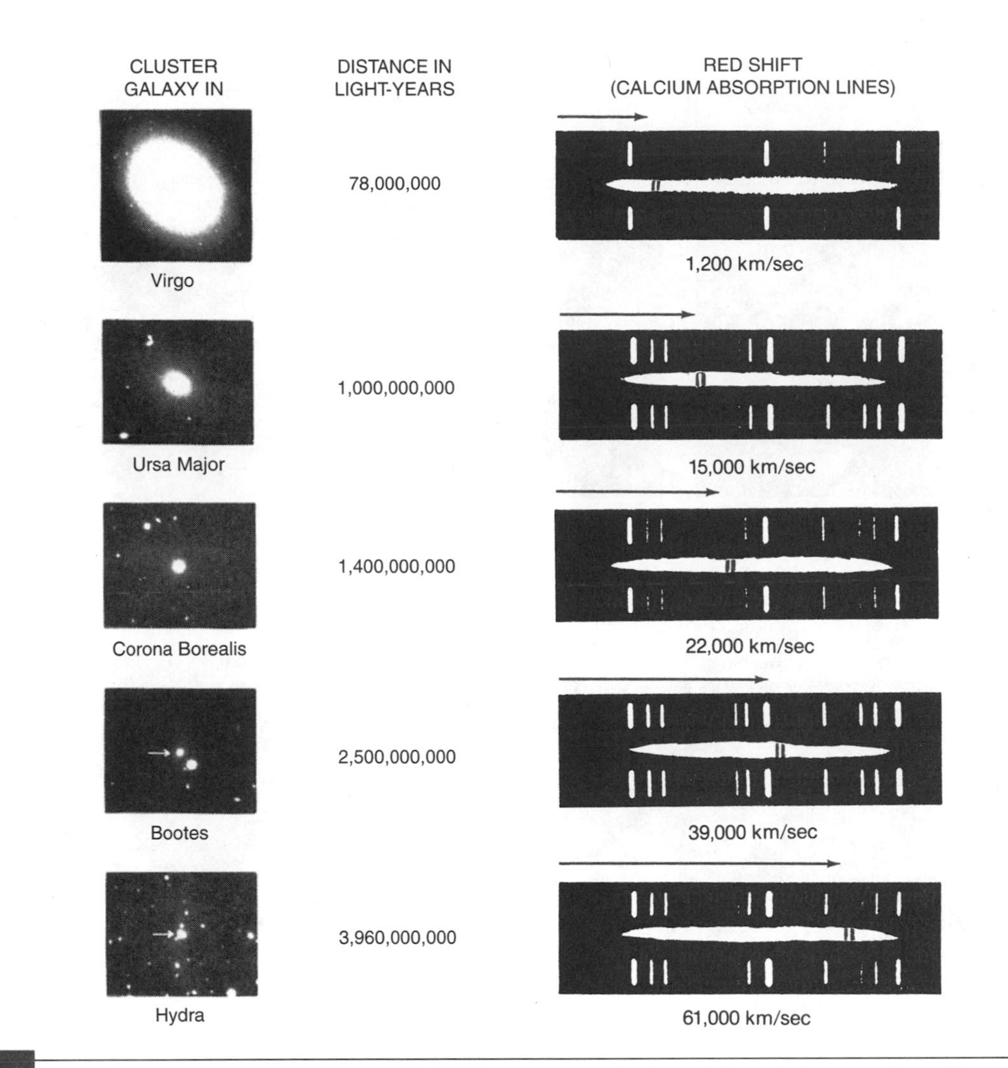

FIGURE 5-1 The shift (*arrows*) in calcium absorption line spectra towards the red for galaxies at various distances from the earth, indicating that the velocity at which a galaxy moves away from the observer (recessional velocity) is proportional to its distance, a relationship first described by Hubble. Because the ratio of speed of recession to distance from the observer is considered the same between all galactic clusters, it is called the **Hubble constant.** Values of the Hubble constant reflect the expansion rate of the universe and enable the time of its origin to be calculated. Low values of the Hubble constant signify a long time interval needed to achieve present intergalactic distances, and therefore an older universe. The exact value of the Hubble constant, however, is in dispute, and former values that provided estimates of a universe 10 to 20 billion years old have been challenged by estimates suggesting an age of only 7 or 8 billion years (Freedman et al.). As yet, the issue is not resolved although new methods of determining distance may help provide more exact estimates of the Hubble constant (Sperger et al., Freedman). The observation that many stars appear older than the age estimate derived from the Hubble constant is also being reevaluated (Watson). (*Adapted from Jastrow and Thompson, with distances according to Kutter.*)

considerably more mass to density calculations.[2] To date, the problem of deciding between continued expansion or a series of big bangs followed by big crunches remains one of measuring the exact amount of dark matter, and understanding its nature, although sufficient evidence now firmly excludes the steady-state theory and some of its variations. Beginning with a big bang origin, possible future scenarios include an "open universe" in which distance between galaxies continues to expand, a "flat universe" in which intergalactic distances reach a constant level, and a "closed universe" in which all galaxies condense back into a ball of energy/matter.

Whatever its fate, the question remains: What sequence of events followed the (last) big bang? In seeking an answer, astronomers have helped provide us with an understanding of the origin of the elements, the development of

[2]This form of matter is believed to constitute most of the gravitational mass of the universe, at least ten times that of luminous matter, but is invisible because it does not emit any type of radiation. Various presently investigated particles, such as neutrinos and WIMPS (weakly interacting massive particles), have been suggested as dark matter components (Rosenberg).

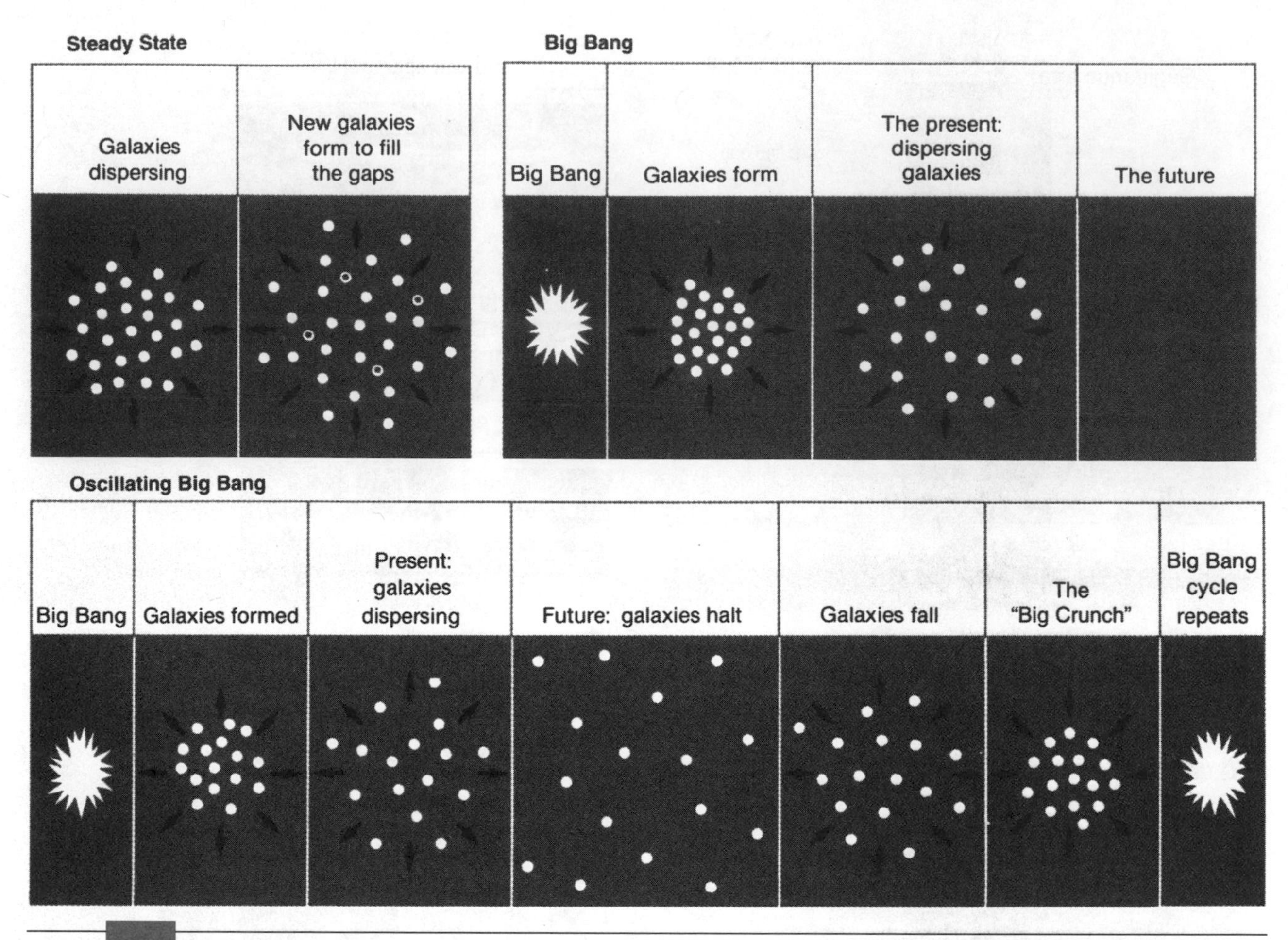

FIGURE 5–2 Schematic diagrams of three major cosmological theories. Note that these are two-dimensional oversimplifications, since it is the universe that is expanding, and not galaxies within a constant space. The dimensions of space and its relationships to time and matter are complex research topics, broadly discussed in books by Barrow and by Thorne. (*Adapted from Jastrow and Thompson.*)

the solar system, and a concept of the immense reaches of time that were essential to our existence.

According to most modern reconstructions of the event, the big bang event occurred about 15 billion years ago at a time when all the matter and energy in the universe was indistinguishable, compacted at an infinitely high temperature and density into a point, called a *singularity.* What preceded the singularity remains unknown, but once it appeared, its rapid expansion led to the introduction of space, time, and energy. Various cosmological models are proposed for these initial stages (see Barrow), but all agree that as the expansion proceeded, the temperature decreased, starting the long ascent of cosmological complexity. Within a short period, protons and neutrons formed, followed by atoms (mostly hydrogen and some helium).

When the universe reached an age of perhaps 100 million years, large masses of hydrogen gas separated out to form **protogalaxies.** Because of gravity, these bodies of matter then gradually started to collapse inwardly to produce the giant galaxies in which individual stars evolved. About 100 billion galaxies are believed to exist, clustered in space in the form of galactic sheets, filaments, and even knots. Because of their distances from each other the night sky appears quite dark, although each galaxy may contain perhaps 100 billion or more stars.

Our own galaxy, the Milky Way, for example, is a member of a "local group" containing the equally large Andromeda galaxy and about 20 smaller ones. This group, in turn, is a member of the Virgo Supercluster comprising many thousands of galaxies. Like the Andromeda, our galaxy may have 150 billion stars or more and is organized in the shape of a flattened disc with spiral arms, about 100,000 light-years in diameter (Fig. 5–3). Its stars rotate around the galactic center at fairly rapid speeds. Thus our sun, which is approximately 28,000 light-years from the galactic center, moves at a rate of more than 200 kilometers per second through space to complete its galactic orbit in about 250 million years. Spherical and elliptical galaxies are also common, the ex-

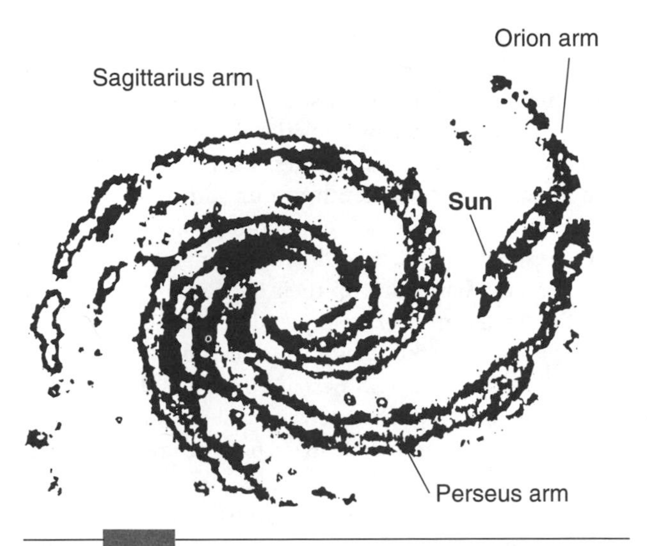

FIGURE 5-3 A sketch of our Milky Way galaxy showing its spiral structure and the position of our sun in relation to three nearby arms. The high speeds at which stars orbit around the center indicates an unseen mass of unknown composition that constitutes about 90 percent of the mass of the galaxy. *(Adapted from Taylor.)*

act shape probably arising from the initial distribution of matter in the galaxy and the degree of spin imposed on this mass when it was formed.

Evolution of a Star and Its Elements

There are four fundamental forces that act on matter and participate in the evolution of a star, such as our sun:

- A "strong" nuclear force that pulls particles of the atomic nucleus (protons, neutrons, etc.) together into densities of 1 billion tons per cubic inch. This force acts only over very small distances, no greater than 1 ten-trillionth of an inch.
- Another nuclear force, one million times weaker that the strong nuclear force, is called the "weak" nuclear force, and is responsible for the manner in which neutrons can eject electrons and transform into protons during radioactive decay.
- An electromagnetic force that is 100 to 1,000 times weaker than the strong nuclear force, electromagnetism binds electrons to nuclei-forming atoms but weakens with distance. "Weak" though it is, by linking atomic nuclei, electromagnetism introduces "chemistry" into the universe, advancing the complexity of matter to the molecular level.
- Gravity is a force that is 10^{38} times weaker than electricity and can aggregate matter into structures and patterns. Gravity acts over long distances such as between the earth and its moon, the sun and its planets, and the galaxy and its suns.

Initially, the first step in the evolution of a star is the gravitational condensation of fragments derived from the galactic cloud of gas and dust. Some of these condensing masses, or **protostars,** can be seen in the heavens as dark globules that have not yet reached the temperature necessary to emit intense light of their own. With the passage of time, the protostar continues to contract by gravity, releasing energy in the form of heat and light. This process arises because atoms moving inward from gravitational attraction pick up speed as the center increases in mass and density, and the greater the speed, the higher the temperature: gravitational potential energy transforms into atomic kinetic energy.

At first, the larger molecules in the protostar dissociate into atoms, and at about 1,800°K this dissociation occurs even for hydrogen molecules. As the protostar becomes smaller, its temperature increases further, and within about 500,000 years its interior may reach a temperature of 100,000°K. A temperature of this magnitude causes the ionization of atoms (loss of outer electrons) but does not yet enable the **thermonuclear reactions** that are necessary for this mass to start burning its own material. The mass needs a much greater temperature, 10 million°K, to overcome the repulsive forces between hydrogen nuclei (protons) and to let fusion reactions occur.

Continued contraction of the initial gravitational mass keeps increasing the protostar's temperature. A critical stage arrives when the 10-million-degree temperature for hydrogen fusion arrives and the subsequent movement of particles in the interior of the protostar is fast enough to prevent further contraction. At that point, approximately 10 million years from its origin (for a star the mass of our sun), the radiant energy of the star is maintained by thermonuclear reactions leading to the conversion of hydrogen (H) into helium (^{4}He):

Proton (H nucleus) + proton (H nucleus) $\rightarrow$ deuteron (proton + neutron) + positron (positively charged electron) + neutron

Deuteron + proton $\rightarrow$ ^{3}He (2 protons + 1 neutron) + γ rays

^{3}He + ^{3}He $\rightarrow$ ^{4}He (2 protons + 2 neutrons) + 2 protons [3]

[3] ^{3}He and ^{4}He are **isotopes** of the element helium, having the same number of protons in their nucleus, but differing in the number of neutrons.

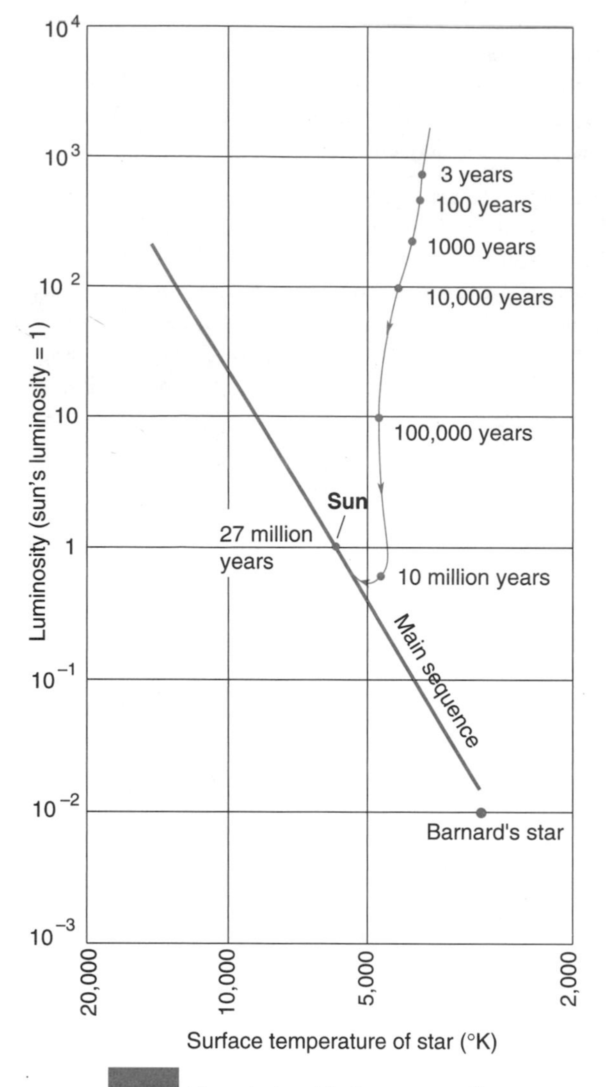

FIGURE 5–4 The relationship between surface temperature and luminosity for hydrogen-burning stars, shown as a diagonal line (the main sequence). The dashed lines and arrows show the path of a protostar about the size of our sun as it approaches the main sequence, and the number of years indicates the time from its origin until it reaches a particular stage. Our sun has been positioned on the "main sequence" for about 4.6 billion years and has enough hydrogen to keep burning at such luminosities and temperatures for another equal period. (*From Jastrow and Thompson.*)

Hydrogen burning enables the star to become hotter and more luminous, and it subsequently moves into what is called the **main sequence** or **main line** of stellar evolution.

As shown in Figure 5–4, when researchers measure many stars in terms of their external temperature and degree of luminosity, they can generate a graph called the **Hertzsprung-Russell diagram,** in which hydrogen-burning stars occupy positions along the main sequence (diagonal line). A star's position on this main sequence depends primarily on its mass. Large stars with masses many times greater than our sun occupy positions on the main sequence at high luminosities and high surface temperatures (*upper left*). Small stars, such as Barnard's star, with a mass about one-tenth of our sun, have relatively low luminosity and surface temperatures (*lower right*).

Interestingly, because of the high temperatures generated, the large stars burn up their hydrogen more rapidly than the small stars. For example, the main sequence lifetime of a star 20 to 30 times greater than our sun is only a few million years, whereas Barnard's star has the potential to continue burning hydrogen for trillions of years. Our sun, which is of intermediate mass, has already spent about half its lifetime, or 4.6 billion years, on the main sequence.

When a star has burned up a considerable portion of its hydrogen, a helium core accumulates and nuclear fusion diminishes. Because the interior of the star now has fewer thermonuclear reactions to prevent gravitational collapse, the star begins to shrink. This compression acts to heat the core rapidly and increase the rate of burning of the hydrogen shell around it. The collapse thus leads to an increase in the release of nuclear energy from the interior of the star, which expands its outer layers. Expansion of the outer mass, in turn, absorbs most of the star's increased energy until the surface temperature drops to between 3,000°K and 4,000°K, and appears to have a reddish glow.

The contraction of the helium core meanwhile continues, increasing the core temperature and increasing the burning rate of the enveloping hydrogen shell. This increases the luminosity of the star so that it now appears as a red giant, located in the upper right of the Hertzsprung-Russell diagram, far off the main line. At the time that our sun becomes a red giant, it will be approximately 100 times its present size, and its radius will reach Earth's orbit.

The increased heating of the core of this expanded star eventually reaches the 100-million-degree temperature at which helium nuclei can fuse. This leads to an expansion of the core (the **helium flash**) followed by a further contraction, a process that may repeat itself a few times, during which various atomic products of helium-burning are produced, including beryllium (^{8}Be) and carbon (^{12}C):

$$^4\text{He} + {}^4\text{He} \rightarrow {}^8\text{Be (unstable)}$$

$$^4\text{He} + {}^8\text{Be} \rightarrow {}^{12}\text{C} + \gamma \text{ rays, or } 3\,{}^4\text{He} \rightarrow {}^{12}\text{C} + \gamma \text{ rays}$$

The burning of helium, however, cannot continue for too long. Carbon gradually accumulates in the core, which then slows down the burning of helium just as the helium core, with its higher fusion temperature, had previously slowed down hydrogen burning. With the reduction in thermonuclear reactions, the star again begins to collapse. Should the star have a relatively small- or medium-size mass like our sun, the heat of gravitational collapse will be insufficient for the carbon core to reach

VIRGINIA TRIMBLE

Birthday:
November 15, 1943

Birthplace:
Los Angeles, California

Undergraduate degree:
University of California–Los Angeles, 1964

Graduate degrees:
M.S., California Institute of Technology, 1965
Ph.D., California Institute of Technology, 1968
M.A., University of Cambridge, England, 1969

Present position:
Professor of Physics
University of California–Irvine

■

WHAT PROMPTED YOUR INITIAL INTEREST IN EVOLUTION?

My interest in biological evolution is that of an enthusiastic amateur, and dates back to reading, while I was still in grade school, a marvelous book called *You and Heredity,* by Amram Scheinfeld. That stars and galaxies also evolve, as individuals and as populations, I did not discover until college. Some aspects of stellar and galactic structure and evolution that I continue to find fascinating are (a) that they can be described by exactly the same principles of physics—gravitation, electromagnetism, thermodynamics, and the rest—that we study in terrestrial laboratories; (b) that they all fit together to make a consistent pattern, in which changing populations of stars add up to make the galaxies we see both here and now and very long ago and far away; and (c) that if any of a number of things had been different (not much carbon built by helium fusion; only massive, short-lived stars formed, etc.) we could not be here to worry about scientific problems.

WHAT DO YOU THINK HAS BEEN MOST VALUABLE OR INTERESTING AMONG THE DISCOVERIES YOU HAVE MADE IN SCIENCE?

My Ph.D. dissertation was a study of the Crab Nebula, remnant of a stellar explosion seen in the year 1054. 1 was able to show that the remnant indeed started expanding about then; that it is being "pushed on" by magnetic fields and high energy particles and so speeding up; and that the amount of matter in it is consistent with the evolution of the giant stars that we think ought to give rise to such supernova explosions. Other areas where I have published original papers include (a) the determination of some of the properties of white dwarfs—the stars left behind when small, long-lived stars like the sun die; (b) studies of populations of binary stars (gravitationally bound pairs) suggesting that their formation is the last stage in a more general problem of star formation; and (c) investigations of some of the short-lived and rare phases of stellar evolution, showing that they, too, fit into the pattern, though they are things our sun will never do.

WHAT AREAS OF RESEARCH ARE YOU (OR YOUR LABORATORY) PRESENTLY ENGAGED IN?

In recent years, I have focussed increasingly on the field called *scientometrics* and on history of science (that is, in effect, on structure and evolution of the astronomical and physics communities, rather than astronomical objects). This has resulted in information about how productive different telescopes are, what kinds of careers astronomers and others of different ages can expect, and other items useful for the community in planning ahead.

IN WHICH DIRECTIONS DO YOU THINK FUTURE WORK IN YOUR FIELD NEEDS TO BE DONE?

I think most people in the field would agree that the single most important unsolved problem in modern astrophysics is the formation of galaxies. The basic problem is for the matter to gather together into large, complex agglomerations, while leaving the cosmic microwave radiation (which also comes to us out of a hot, dense big bang) very smooth throughout the observable universe. Understanding galaxy formation requires information about dark matter, particle physics, plasmas, and a number of other topics.

WHAT ADVICE WOULD YOU OFFER TO STUDENTS WHO ARE INTERESTED IN A CAREER IN YOUR FIELD OF EVOLUTION?

Two generations ago, a wonderful woman astronomer named Cecilia Payne Gaposchkin said, "A woman should do astronomy only if nothing else will satisfy her; because nothing else is what she will get." At the present time, I think the main modification that should be made in this advice is to replace "A woman" by "you" and "astronomy" by "scientific research." That is, it applies to both genders, all races, and so forth, and to all the sciences. You are unlikely to get rich, unlikely to become famous, unlikely to be understood by your family and friends. The rewards of finding out things that no one has ever known before (that is what is meant by research) are enormous, but they are to be achieved only by exceedingly hard work, and do not bring much recognition or other secondary rewards.

the 600-million-degree K temperature at which carbon nuclei fuse (**carbon burning**). The star's outer envelope, however, greatly heats up and expands. It may even separate from the core to form a ghostly planetary nebula.

In such medium-size stars, contraction of the core will eventually diminish, although the core will continue to emit considerable heat. At that point, the contracted state of the star is such that if it were initially the size of our sun—that is, 1 million miles in diameter—it would now occupy a diameter of only 20,000 miles. Further collapse is hindered by repulsion between electrons that cannot be further compressed. The gravitational force at the surface of this compacted mass is high, 1 million times that on Earth, and this body is now called a "white dwarf." In time, the white dwarf gradually cools, forming a dead, cold lump of matter.

However, if the star is originally massive, or if it has accumulated large amounts of material from a companion star, its collapse (when helium burning begins to decline) may enable the carbon core to reach the critical 600-million-degree K temperature. The burning of carbon would then take place, halting further collapse and producing various new elements, among which are oxygen (^{16}O), neon (^{20}Ne), sodium (^{23}Na), magnesium (^{24}Mg), silicon (^{28}Si) and sulfur (^{32}S):

$$^{12}C + {}^{12}C \rightarrow {}^{24}Mg + \gamma \text{ rays}$$

$$^{12}C + {}^{12}C \rightarrow {}^{23}Na + \text{proton}$$

$$^{12}C + {}^{12}C \rightarrow {}^{20}Ne + {}^{4}He$$

$$^{4}He + {}^{12}C \rightarrow {}^{16}O + \gamma \text{ rays}$$

$$^{12}C + {}^{16}O \rightarrow {}^{28}Si$$

$$^{4}He + {}^{28}Si \rightarrow {}^{32}S + \gamma \text{ rays}$$

As occurred previously in the helium core, the exhaustion of carbon burning then leads to a further gravitational collapse and further contraction. New nuclear fuels begin to burn, and new elements continue to come into being until an iron (^{56}Fe) core appears.

In contrast to previously synthesized elements, the nuclear fusion of iron absorbs energy rather than radiates it.

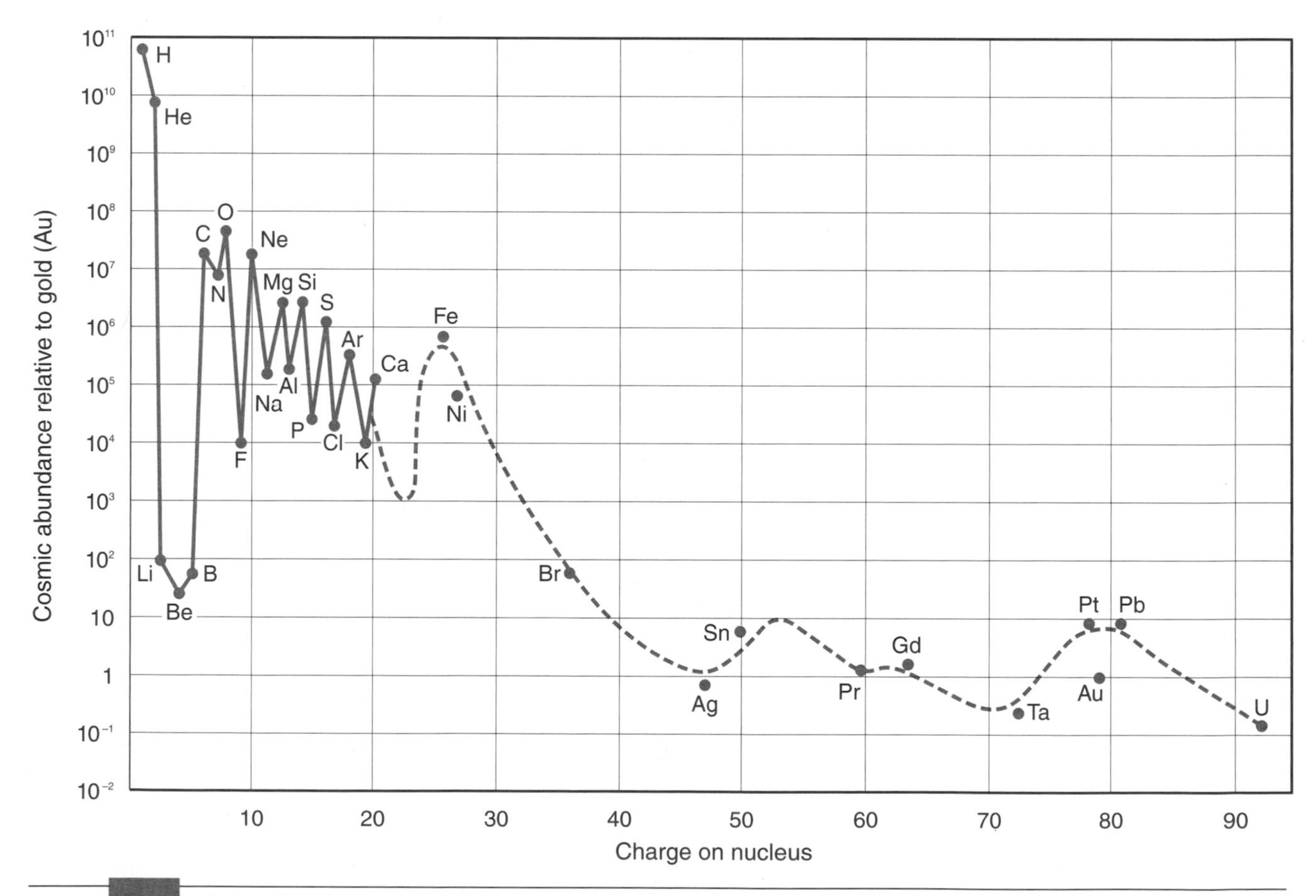

FIGURE 5-5 Relative abundance of elements in the universe, using gold (Au) as a standard of 1.00. Note that elements with atomic weights less than iron are relatively abundant since they are produced in many stars. Elements on the periodic table higher than iron, such as silver (Ag), lead (Pb), and uranium (U), are about 100,000 times less common because they are produced in the very brief interval of a supernova explosion. The abundances of additional unspecified elements, such as rare earths, fall below the dashed line. *(Adapted from Jastrow and Thompson.)*

The **iron core** thus acts as a "heat sink," leading to a gradual extinction of thermonuclear reactions. As nuclear fusion stops, collapse of the core begins again, but because no further reactions are taking place, the collapse of this large mass continues. According to one view of subsequent events, the combination of heat and pressure makes this collapsing material rebound rapidly, like a giant spring, in an explosion of immense proportions. A large part of the star's material ejects into space, and the sky may then light up to form an object as bright as a whole galaxy, a **supernova.**

For suns whose masses ranges from 1.4 to 5 times that of our own, the remaining core then contracts to form a "neutron star" in which all protons, neutrons, and electrons combine into a pure ball of neutrons about 10 miles in diameter. Further collapse is prevented by neutron incompressibility, and the star reaches a density of one billion metric tons per cubic centimeter. Astronomers believe these bodies are the "pulsars" that emit radio waves in two narrow beams from their surface as they spin on their axes from less than a few times a second to several hundred times a second. For suns with masses greater than five times our own, collapse continues even beyond neutron density, until the core's gravitational force is strong enough to prevent electromagnetic waves such as light from leaving its surface. When it reaches this enormous density, the collapsing star becomes a "black hole" in space.

Such cataclysmic supernovae events now occur at a frequency of about one per 100 years in our galaxy but were probably more common in the past when star formation was more frequent. An example of the remnants of a supernova is the Crab Nebula, a large, rapidly expanding cloud of matter in the constellation of Taurus, which arose from a supernova explosion in 1054, that was visible on Earth during daylight. Astronomers estimate that about one billion supernovae have occurred during our galactic history.

The supernova material, distributed widely through space, has important effects on later-born stars that use this material as part of their own formation. One consequence of a supernova is the creation of very high fluxes of free neutrons, which various atomic nuclei then capture to form elements such as gold and uranium. This is the primary means by which elements heavier than iron (the 26th element in the periodic table) appear. Also, the relatively short duration of the supernova effect accounts for the fact that these heavier elements are rare (Fig. 5–5). Our solar system, which contains samples of all elements, must have used the remnants of at least one supernova explosion in its formation, and some astronomers have already mapped likely positions where such nearby supernovae may have occurred. Our sun may therefore be a second- or even third-generation star.

SUMMARY

To understand evolution, we had to extend the history of the Earth to times far earlier than most people previously envisaged. Although we now know that the universe is immensely old, when and how it came into being is still highly speculative. People have proposed two views of the origin of the universe, both based on the universal use of hydrogen as the stellar fuel. The first, and less probable, steady-state theory is that hydrogen continually replenishes from outside the universe. The second, more probable, big bang theory is that the universe originated about 15 billion years ago in an explosion of a small volume of extremely dense energy/matter. Support for this theory comes from the apparent continuing expansion of the universe, the constant abundances of hydrogen and helium in celestial bodies, and the remnant radiation from the early universe. There may have been only one big bang, or the universe may oscillate between expansions and contractions.

After the big bang, the temperature of matter was enormously high. As it cooled, hydrogen and helium atoms formed. Perhaps 100 million years after the formative explosion, masses of hydrogen began to condense into galaxies, each galaxy producing many billions of stars in a manner still occurring today. Stars forming from galactic matter condense and increase in temperature, enabling the fusion of hydrogen atoms to form helium. The combustion of hydrogen raises the temperature still more, and the star becomes a main sequence star. The larger the star, the more rapidly it consumes its hydrogen, and nuclear fusion declines. It undergoes a complex series of events that cause it to expand and become more luminous, at which point astronomers call it a red giant. The internal temperature continues to increase until helium nuclei can fuse. Eventually a medium-size star contracts and cools, forming a white dwarf, which is the end of its evolution. Large stars may attain extremely high core temperatures, enabling carbon to burn and producing many new elements. This process continues until the core of the star consists of iron, the combustion of which requires energy, and eventually thermonuclear reactions stop. The star collapses, heats further, and explodes into a supernova while its core collapses to a neutron star (or pulsar) or black hole. When supernovae occur, heavy elements form. All these elements exist in our solar system and appear to be remnants of nearby supernovae formations.

KEY TERMS

big bang theory
carbon burning
helium flash
Hertzsprung-Russell diagram
Hubble constant
hydrogen burning
iron core
isotopes
main line

main sequence
oscillating big bang theory
protogalaxies
protostars
steady-state hypothesis
supernova
thermonuclear reactions

DISCUSSION QUESTIONS

1. Steady-state and big bang theories of the universe
   a. What is the difference between these theories?
   b. Which of these theories has gathered most support?
2. Stars
   a. How do stars originate?
   b. What sequence do they follow in their evolution?
   c. What information does the Hertzsprung-Russell diagram offer about the size of stars and the fuel they burn?
   d. How do the various elements form in stars?
   e. Why do some astronomers consider our sun a second- or third-generation star?

EVOLUTION ON THE WEB

Explore evolution on the web! Visit the accompanying web site for *Evolution,* 3/e at www.jbpub.com/evolution for web exercises and links relating to topics covered in this chapter.

REFERENCES

Barrow, J. D., 1994. *The Origin of the Universe.* Basic Books, New York.

Emiliani, C., 1992. *Planet Earth: Cosmology, Geology, and the Evolution of Life and Environment.* Cambridge University Press, Cambridge, England.

Field, G. B., G. L. Verschuur, and C. Ponnamperuma, 1978. *Cosmic Evolution: An Introduction to Astronomy.* Houghton Mifflin, Boston.

Finkbeiner, A., 1995. Closing in on cosmic expansion. *Science,* **270,** 1295–1296.

Freedman, W. L., 1998. Measuring cosmological parameters. *Proc. Nat. Acad. Sci.,* **95,** 2–7.

Freedman, W. L., et al., 1994. Distance to the Virgo cluster galaxy M100 from Hubble Space Telescope observations of Cepheids. *Nature,* **371,** 757–762.

Gamow, G., 1952. *The Creation of the Universe.* Viking Press, New York.

Hawking, S. W., 1988. *A Brief History of Time: From the Big Bang to Black Holes.* Bantam Books, Toronto.

Jastrow, R., and M. H. Thompson, 1972. *Astronomy: Fundamentals and Frontiers.* Wiley, New York.

Kragh, H., 1996. *Cosmology and Controversy: The Historical Development of Two Theories of the Universe.* Princeton University Press, Princeton, NJ.

Kutter, G. S., 1987. *The Universe and Life: Origins and Evolution.* Jones and Bartlett, Boston.

Linde, A., 1994. The self-reproducing inflationary universe. *Sci. Amer.,* **271** (5), 48–55.

Morris, R., 1993. *Cosmic Questions.* Wiley, New York.

Rosenberg, L. J. , 1998. Direct searches for dark matter: Recent results. *Proc. Nat. Acad. Sci.,* **95,** 59–66.

Silk, J., 1989. *The Big Bang: The Creation and Evolution of the Universe,* 2d ed. Freeman, New York.

Sperger, D. N., M. Bolte, and W. L. Freedman, 1997. The age of the universe. *Proc. Nat. Acad. Sci.,* **94,** 6579–6584.

Taylor, S. R., 1992. *Solar System Evolution: A New Perspective.* Cambridge University Press, Cambridge, England.

Thorne, K. S., 1994. *Black Holes and Time Warps: Einstein's Outrageous Legacy.* W. W. Norton, New York.

Thuan, T. X., 1995. *The Secret Melody.* Oxford University Press, New York.

Watson, A., 1998. The universe shows its age. *Nature,* **279,** 981–983.

Weinberg, S., 1977. *The First Three Minutes.* Basic Books, New York.

6 The Earth

Astronomers have proposed two main theories for the evolution of planets in our solar system. We can call the first, proposed originally by Buffon and later by Jeans and Jeffries, the **collision theory.** It suggests that another star passing close to our sun pulled out, through gravity, material that became the planets. The most serious of many objections to this theory is the extreme rarity of such events: astronomers estimate collisions or near collisions between stars to have occurred in our galaxy only ten times in the last 5 billion years, whereas more than a billion stars in our galaxy appear to have planets.

In its various forms, most astronomers today hold the second theory, which was first suggested by Kant and later by Laplace. This **condensation theory,** or nebular hypothesis, proposes that the large, whirling mass of matter out of which our solar system initially condensed about 4.6 billion years ago did not have a uniform distribution of material. According to this theory, the large condensing mass at the center of this cloud became the sun when it reached thermonuclear reaction temperatures, whereas the smaller peripheral condensations never reached such critical temperatures and therefore became the **protoplanets** (Fig. 6–1). These peripheral masses remained tied to the solar orbit and formed the planets, although some captured subplanets or moons of their own.[1]

[1] In the case of Earth's moon, present thinking leans toward its origin from one very large impact on Earth early on in its history by a planetisimal perhaps the size of Mars (Taylor). After impact, a major part of this planetesimal rebounded into space, clumped together, and remained gravitationally tied to Earth as its moon. This process was followed by intense meteorite bombardment of both Earth and moon, ending about 3.9 billion years ago. The massive cratering of the Moon, marked by about 50 basins, each more than 300 kilometers across, dates back to this period. This period also indicates that the origin of present life probably began no earlier, since such intense bombardments may well have sterilized the Earth's surface (see also p. 172 and Fig. 9–13). Nevertheless, such impacts probably brought to Earth large amounts of water, hydrogen, nitrogen, other elements, and organic compounds (Fig. 7–8 and p. 122) that could be used for biological purposes (Chyba et al.). Wetherill (1995) suggests that subsequent catastrophic impact events greatly diminished in planets of the inner solar system because large amounts of cometary impact material were swept up by the giant planet Jupiter. The formation of Jupiter thus allowed Earth's orbit and climate to stabilize.

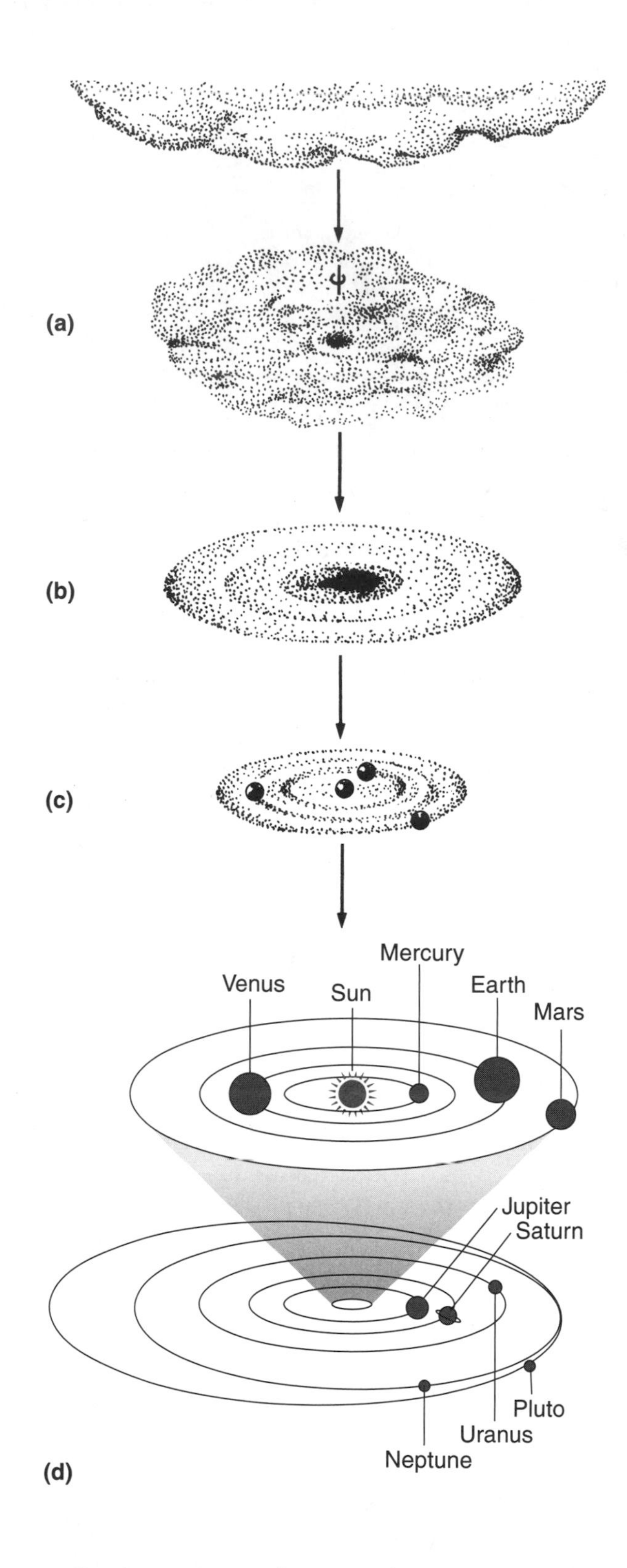

FIGURE 6–1 Stages during the condensation of the solar nebula into the solar planetary system. (*a*) Fragmentation of an interstellar cloud. (*b*) Contraction and flattening of the solar nebula. (*c*) Condensation of nebular material into meteorites and protoplanetary bodies. (*d*) Solidification of planets, with an indication of present orbits. An "asteroid belt," consisting of many thousands of bodies with sizes ranging up to about 1,000 kilometers in diameter, lies between Mars and Jupiter. Its fragmented nature is probably the result of the massive proximity of Jupiter, which swept up or ejected objects that would ordinarily have aggregated into a planet of their own. Along with comets whose orbits cross that of Earth, the asteroid belt and its subsidiaries provide nearly all the 20,000 meteorites that annually enter Earth's gravitational field, most in the range of 1–10 kilograms. (*Adapted from Field et al., with additions.*)

Solar Condensation

The condensation theory helps explain both the motion and location of the planets, although a number of questions are still being answered. For example, the rotation of the sun is about 100 times slower than we would expect from its large mass and the relatively smaller mass of its planets. (The sun has only 2 percent of the angular momentum of the solar system, rotating around its axis only once every 26 days, yet has 99.9 percent of the mass of the solar system.) Although this reduced rotation rate would seem to occur only if the mass of planetary material were several hundred times greater than observed, theorists offer a reasonable explanation for this anomaly: as the early solar system condensed, the solar wind carried off most of the angular momentum from the central body. Another peculiarity, but so far unexplained, is the source for the difference in angle between the equatorial plane of the sun and the planes along which the planets revolve (Hughes).

In spite of these difficulties, most astronomers have little doubt that gravitational condensations must have occurred in the formation of the planets. In fact, the condensation theory implies that each time a star condenses out of the gaseous matter of space, the opportunity, or even likelihood, exists that planets will form. Unfortunately, aside from our own solar system, planets elsewhere cannot be seen directly, because they emit no radiation of their own and reflect only the light of their suns. Even if a planet of a nearby star were relatively large, such reflections would be much too feeble to identify with available earth-bound telescopes. So astronomers have attempted indirect observations by noting whether interaction with unseen planets affects the motion of a star through the galaxy. Through such means, a fair number of extrasolar planetary systems have been discovered, with differently sized planets following varied orbits around their respective suns (see Glanz, Lunine). The presence of other planetary solar systems underscores a long-standing question—Does life exist elsewhere in the universe (Box 6–1)?

BOX 6–1 IS THERE LIFE ELSEWHERE IN THE UNIVERSE?

Our concept of ourselves in relation to the universe we live in has changed radically, especially in this last century. From an imagined center of the universe, we have been ignominiously moved to its periphery and to cosmic diminutiveness. Not only are we on a planet on the fringes of a galaxy containing more than 100 billion other stars, but we are in a universe containing billions of other galaxies. There are even serious proposals among astrophysicists that our universe may be only one of many (for example, Linde). Inevitably, this raises the question whether the evolution of life has been repeated on planets of other stars, albeit taking different forms.

From our knowledge so far, life on Earth and its myriad adaptations and forms depend on the presence of special features:

- Appropriate atomic elements and available reactive molecules
- A sun of moderate size (between .8 and 1.5 solar masses) located on the Hertzsprung-Russell "main sequence" (Fig. 5-4), providing radiant energy for many hundred million years
- A planet properly distant from its sun, following an orbit that eliminates extreme temperatures
- A protective yet reactive atmosphere
- The presence of liquid water—a solvent that allows essential biochemical reactions[2]

The presence of life then becomes a question of how unique are these features, and could they exist elsewhere?

As mentioned in the text, the recent recognition of planets in other solar systems indicates that planet-formation must be common in our galaxy and in most, or all, others. Most astronomers feel that many features supporting life could undoubtedly have developed throughout the universe. Given the more than 100 billion stars in our galaxy, even a 1 percent chance for the origin of an Earthlike planet would provide more than a billion opportunities for the evolution of life. Furthermore (although still very controversial), a report of ancient microscopic fossils in a Martian-derived meteor raises the possibility that the evolution of primitive life is not confined to Earth even within our own solar system (D. S. McKay et al.).[3]

However rational these expectations, the immense distances of extraterrestrial space make it impossible to observe life directly outside our own planetary world. To obtain a material sample of life from another solar system, or send observers there, would not only entail enormous expense to build a spaceship, but would require immense traveling times. For example, at present rocket speeds, it would take years to leave our own solar system, and centuries more to reach even the nearest star system, Alpha Centauri, four light-years (25 trillion miles) distant.

Evidence for extrasolar life is therefore presently restricted to detecting electromagnetic signals emitted by intelligent creatures.[4] Such signals, used in radar or in radio and television communication, can carry considerable information through pulsed or modulated frequencies which move through space at the fastest rate possible, the speed of light.

Although humans have only engaged in producing these signals for less than a century, this may not be true for civilizations in other systems. Planets elsewhere may have possessed intelligent and technologically advanced life forms for many thousands or millions of years, and their coded electromagnetic transmissions may now be reaching Earth although we are very many light-years away. The technical problem astronomers on Earth face is where to look for such signals, and how to detect them.

At present, the SETI Institute (Search for Extra Terrestrial Intelligence: *http://www.seti-inst.edu*) is the largest research institution engaged in searching the skies and analyzing spatial radio waves for intelligent communication. It uses both very large (1,000 foot diameter) and small radio telescopes focused on a variety of stars, including those that appear to have solar-type planetary systems. SETI investigations cover a range of frequencies, concentrating on those that can carry signals with low noise levels for long cosmic distances.

One frequency band receiving special attention lies between 1.4 and 1.8 gigahertz (wavelengths between 18 and 21 centimeters). This band contains

[2] Among proposals for an alternative to our carbon-based water-solvent biochemistry are silicon-based systems and an ammonia solvent. Such proposals have not been investigated.

[3] According to C. P. McKay:

> Geomorphological evidence suggests that liquid water existed on the surface of Mars at approximately the time that the first life appeared on Earth, between 3.8 and 3.5 billion years ago. The possibility of the origin of life on Mars is based on analogy with Earth. All the major habitats and microenvironments that would have existed on Earth during the formation of life would have been expected on early Mars as well: hot springs, salt pools, rivers, lakes, volcanos, and so forth. Even tidal pools would have existed on Mars, albeit at a much reduced level because there would have been only solar tides. The possible nonbiological sources of organic material would have supplied both planets. Perhaps the major unknown is the duration of time that Mars had Earthlike environments compared with the time required for the origin of life. The length of neither of these times is known precisely, but current theories suggest that the lengths may be comparable.

[4] Given that living forms face continually challenging environments over very long periods, we can ask the reasonable question: Is the evolution of intelligence inevitable? We do know that selection for improved behavioral strategy is a ubiquitous feature of life in the "arms race" between prey and predator (pp. 336 and 426) and for many organisms competing with others for reproductive success. Such adaptations involve improved neurological structures, including increased brain size (Fig. 19-10), that may well lead to higher forms of intelligence in groups with complex social systems (Chapter 20).

the radiation wavelength for hydrogen (21 cm), the most common element, and has been proposed as the "cosmic water hole" where interplanetary intelligent creatures would seek to communicate. SETI techniques are sufficiently sensitive to identify extraterrestrial signals whether continuous, pulsed, or modulated, and to exclude signals from our own planet and its artificial satellites. Deciphering extraterrestrial signals will, of course, be another problem. (In Carl Sagan's 1985 novel, *Contact,* a message from the Vega star system is first detected as 21-cm frequency pulses that code for a repeated series of prime numbers.)

Although many efforts have been made since Drake's pioneering work in the 1960s,[5] not a single intelligent message has yet been confirmed. To some scientists, such negative results indicate that the many environmental contingencies that led to human intelligence are most probably unique to Earth's history, and the evolution of technologically advanced creatures elsewhere is of immensely low probability (Conway Morris, Mayr, Tipler). Others, such as Horowitz, point out that our cost for detecting extraterrestrial signals is sufficiently low to warrant a continued search in helping resolve basic human questions: How unique are we? If we are not unique, what can we learn elsewhere?

Our curiosity remains, and the search continues. In Drake's (1961) words, "Those who feel that the goal justifies the great amount of effort required will continue to carry on this research, sustained by the possibility that sometime in the future, perhaps a hundred years from now, or perhaps next week, the search will be successful."

[5] Among Drake's contributions was a famous "equation" that offered seven factors for estimating the number of extraterrestrial technological civilizations emitting detectable signals (N):

- Number of sunlike stars (N_*)
- Fraction of such stars with planets (f_p)
- Number of above planets that are habitable maintaining liquid water (n_e)
- Fraction of above planets that evolve life (f_l)
- Fraction of above planets that evolve intelligent creatures (f_i)
- Fraction of above planets that develop civilization and technology (f_c)
- The lifetime of such civilizations (L)

where $N = N_* \times f_p \times n_e \times f_l \times f_i \times f_c \times L$

Although the values of some fractions may be quite small, there are so many stars in our galaxy and universe (estimates are about 10^{20}) that conjectures for N range from many hundreds of millions of extraterrestrial civilizations downward (Hart, Harrison). It is interesting to note that these estimates show a correspondence in the number of communicating civilizations and their lifetime in years (for example, for L = 100 years only 100 technical civilizations persist, for L = 1,000 years, there are 1,000, and so forth). The chance for finding a communicating civilization among all these billion of stars is therefore a matter of social survival. "So to listen for a signal is, in a sense, the expression of faith in science and technology. It evinces the belief that 'intelligent' creatures—here defined, again, as those with big radio sets—generally manage to survive, rather than blowing themselves up" (Ferris).

The Earth's Atmosphere

The planetary distribution of elements as our solar system condensed was not apparently uniform. According to some astronomers, a density gradient established, with many heavier elements condensing into the "Earthlike" planets nearest the sun (Mercury, Venus, Earth, Mars) and relatively large amounts of the lighter, more volatile elements condensing into planets farthest from the sun (Jupiter, Saturn, Uranus, Neptune). These differences in condensation, as well as the pressure of heat and solar radiation on nearby planetary atmospheres, caused the Earthlike planets to lose their initial hydrogen and helium atmospheres. Loss of this primary atmosphere then left these planets with the rocky materials so characteristic of them today.

Various gases remained in the interior of the Earthlike planets and gradually escaped to form a secondary atmosphere. On Earth, the outgassing of hydrogen, the most prevalent of cosmic elements, enabled three essential hydrogen-bearing compounds to form: methane (CH_4), water (H_2O), and ammonia (NH_3). Other gases present at the time probably included carbon monoxide (CO), nitrogen (N_2), and some that even now issue from volcanoes and hot springs, such as carbon dioxide (CO_2), hydrochloric acid (HCl), and hydrogen sulfide (H_2S). As time went on, and Earth's surface temperature cooled, liquid water formed enabling CO_2 to react with silicates to produce carbonates, thus reducing CO_2 in the atmosphere. A number of the noble gases such as neon, argon, and xenon may also have been prevalent and should have persisted to this day in relatively high quantities since they are chemically inert. Their almost complete absence is so far unexplained.

In any case, there is now little question that the early atmosphere of the Earth was either strongly or mildly **reducing** because of the prevalence of hydrogen compounds capable of providing electrons to **oxidizing** agents capable of accepting them. Evidence for this view exists in deposits laid down in South Africa and other places 2 or more billion years ago, which became inaccessible to Earth's later

TABLE 6–1 Present composition of the Earth's atmosphere

Gas	Percent by Volume
Nitrogen (N_2)	78.09
Oxygen (O_2)	20.95
Argon (Ar)	0.93
Water (H_2O)	Variable (up to 1.00)
Carbon dioxide (CO_2)	0.03
Neon (Ne)	0.002
Helium (He), methane (CH_4), carbon monoxide (CO), krypton (Kr), nitrous oxide (N_2O), hydrogen (H), ozone (O_3), Xenon (Xe)	Less than 0.001

atmosphere. Such deposits include sulfides of iron (FeS), lead (PbS), and zinc (ZnS), compounds that are highly unstable in the presence of oxygen. If oxygen were present in the atmosphere at the time these compounds formed, they would have deposited in the form of sulfates (for example, $FeSO_4$) rather than sulfides.

Where, then, did our present oxygen come from (Table 6–1)? The answer to this question is not clear, although geochemists seem to generally agree that the proportion of free oxygen in the atmosphere began to increase about 2 or 3 billion years ago. Ultraviolet irradiation of water in the upper atmosphere may produce free hydrogen ($2H_2O \rightarrow 2H_2 + O_2$), which can then escape the Earth's gravity and leave behind increasing amounts of molecular oxygen. A more popular proposal relies on the apparent correlation between increase in oxygen and increased domination of the Earth's surface by plant life. As we discuss in Chapter 9, electron transfer in the plant photosynthetic process involves removing of hydrogen atoms from water molecules, producing free oxygen that then diffuses to the atmosphere. In whatever manner it first appeared, geochemists generally agree that the proportion of atmospheric oxygen is now related to photosynthesis in plants.

The Earth's Structure

Geologists believe the formation of the Earth from the wide band of material in its original orbit was a process in which many subsidiary condensations first occurred. These subsidiary planetesimals were then drawn into the condensing Earth, probably along with uncondensed orbital material, to form a structure that probably had some degree of differentiation. That is, different compounds and minerals probably occupied different positions in the Earth depending on the temperature at which they condensed, the temperature of the condensing Earth, and other variables. Within the first billion years of the Earth's history, geologists believe, differentiation of its structure proceeded at a fairly steady rate until relationships developed similar to those which exist today. Radioactive elements trapped within the Earth during its condensation gave off small but incremental amounts of heat which gradually increased the temperature of surrounding material. Along with the heat of condensation and pressure, the center of the Earth probably soon developed temperatures high enough to melt iron.[6]

Present information concerning the interior of the Earth primarily derives from vibrational waves that earthquakes generate. Researchers can detect these **seismic waves** with sensitive seismographs, and their paths and velocities can be shown to depend on the composition, fluidity, and thickness of the materials through which they travel. Combined with studies of the Earth's magnetic, electric, and gravity fields, seismic information indicates that the interior of the Earth has a number of concentric layers that differ in temperature, pressure, composition, and degree of crystallization (Fig. 6–2). At the center is a **core,** a solid iron mass (with some nickel) about 800 miles in radius surrounded by a liquid iron envelope mixed with sulfur or silicon about 1,300 miles thick. Shifts in the molten iron core are believed responsible for changes in the Earth's magnetic field.

Surrounding the iron core is a hot **mantle** layer of rock, about 1,800 miles thick, that comprises approximately four-fifths of the Earth's volume. Because of radioactivity, pressure, and localized heating or cooling, the mantle has experienced repeated melting and crystallization, and geologists now characterize it as a partly molten plastic structure whose density increases with its nearness to the core. Floating on the surface of the mantle is a thin **crust** of rock with a thickness of about 20 miles for the less dense continental crust and about 5 to 7 miles for the heavier oceanic basins (Fig. 6–3). We know most about the crust, and can distinguish three basic types of crustal rocks:

- IGNEOUS ROCKS **Igneous rocks** crystallize out of the molten liquid magma pushed up through cracks in the crust by the mantle. When deposited under existing rocks, igneous intrusions may be detected by the erosion of covering strata. Magma may also be deposited directly on the surface in the form of lava.

[6] Some theorists believe the newborn Earth was superheated because of bombardment by planetesimals as large as the moon or Mars. High temperatures of this kind may have effected the distribution of elements such as gold and iridium in both the core and mantle.

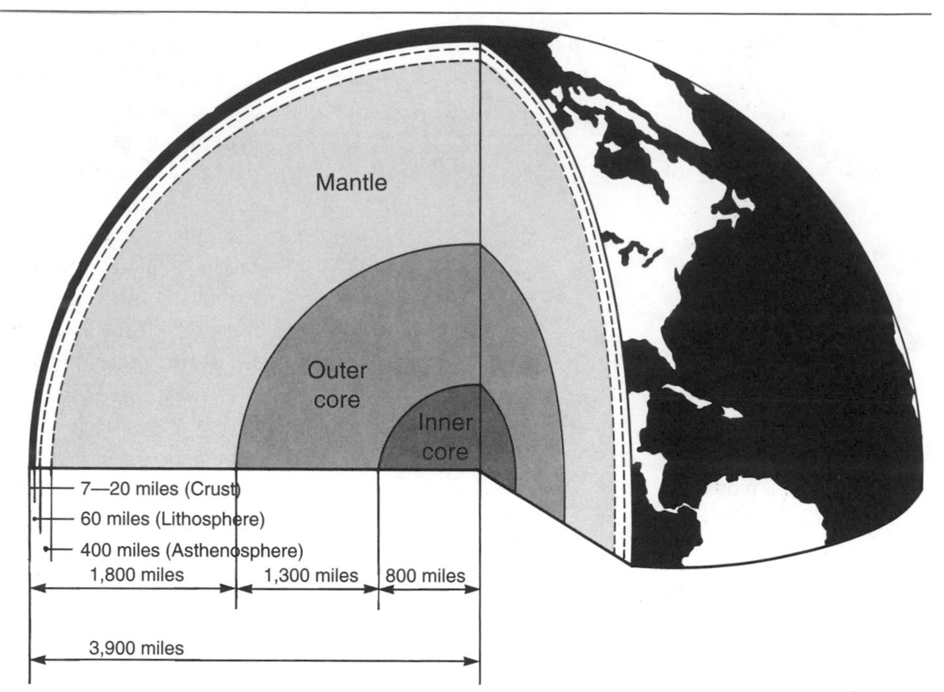

FIGURE 6-2 Section through the Earth's interior, which possesses a radius of 3,948 miles (6,357 kilometers) at the poles and 3,960 miles (6,378 kilometers) at the equator. The **lithosphere** consists of relatively rigid plates composed of the rock-like crust plus a portion of the underlying mantle that reaches to a depth of about 50 miles at the oceanic basins and 60 to 90 miles at the continents. Below the lithosphere is a more fluid, deformable material, the **asthenosphere,** that allows the lithospheric plates to move about. Further distinctions between these and other layers in the outer Earth are reviewed by Rogers. (*Adapted from Wyllie.*)

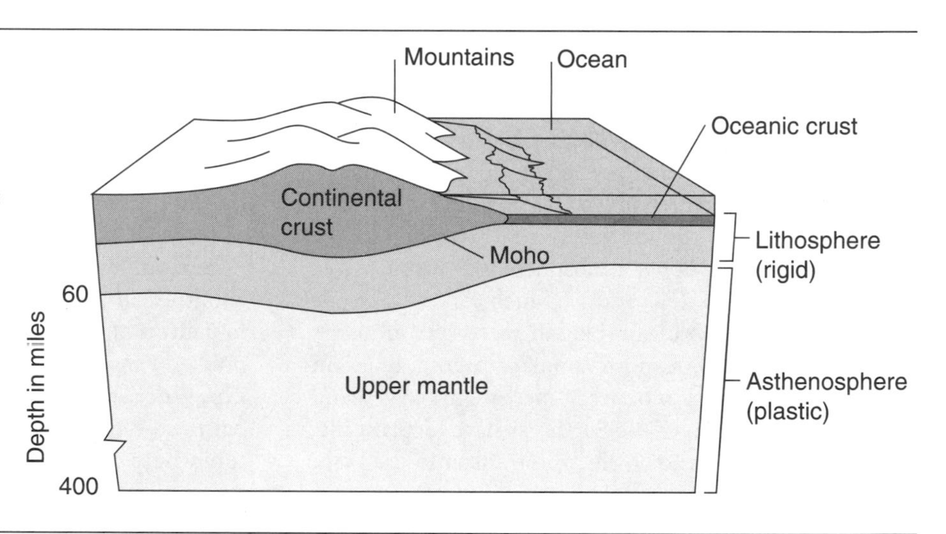

FIGURE 6-3 Section of the Earth's crust, showing differences in thickness. Since the crust has less density than the Earth's subsurface material, it floats on the mantle, and different thicknesses of the crust float at different levels. Thus the thicker, and therefore more buoyant, continental landmasses float higher compared to the thinner, less buoyant, oceanic basins. At the Moho discontinuity, a sharp change occurs in the velocity of certain seismic waves, and geologists consider this area to represent the boundary between crust and mantle.

The granites are a common example of igneous rocks, as are the dark, fine-grained basalts that often appear as the solidified lava of volcanoes.

- SEDIMENTARY ROCKS The erosion of igneous rocks by water, wind, and chemical reactions, as well as the dust and effluent thrown up by volcanic activity, produces particles that can then be transported and reformed into new arrangements. Thus, a stream may deposit its sediments at the bottom of a lake; wind, waves, and ice can shift sand, pebbles, and other geological debris into layers that settle out on various surfaces. Should such layers harden, either through the pressure of other layers above them or by chemical means, **sedimentary rocks** form. In this process, gravity is the primary force accounting for the settling and layering that geologists observe. Sandstone (sand origin), shale (mud origin), and limestone (calcium carbonate) are examples of sedimentary rocks. Limestone is most often found associated with the remains of organisms such as corals, mollusks, and other organisms that lived in marine reefs and shallow seas and used calcium carbonate for their skeletal and habitat structures.

- METAMORPHIC ROCKS **Metamorphic rocks** were originally either igneous or sedimentary and later underwent significant changes because of heat, pres-

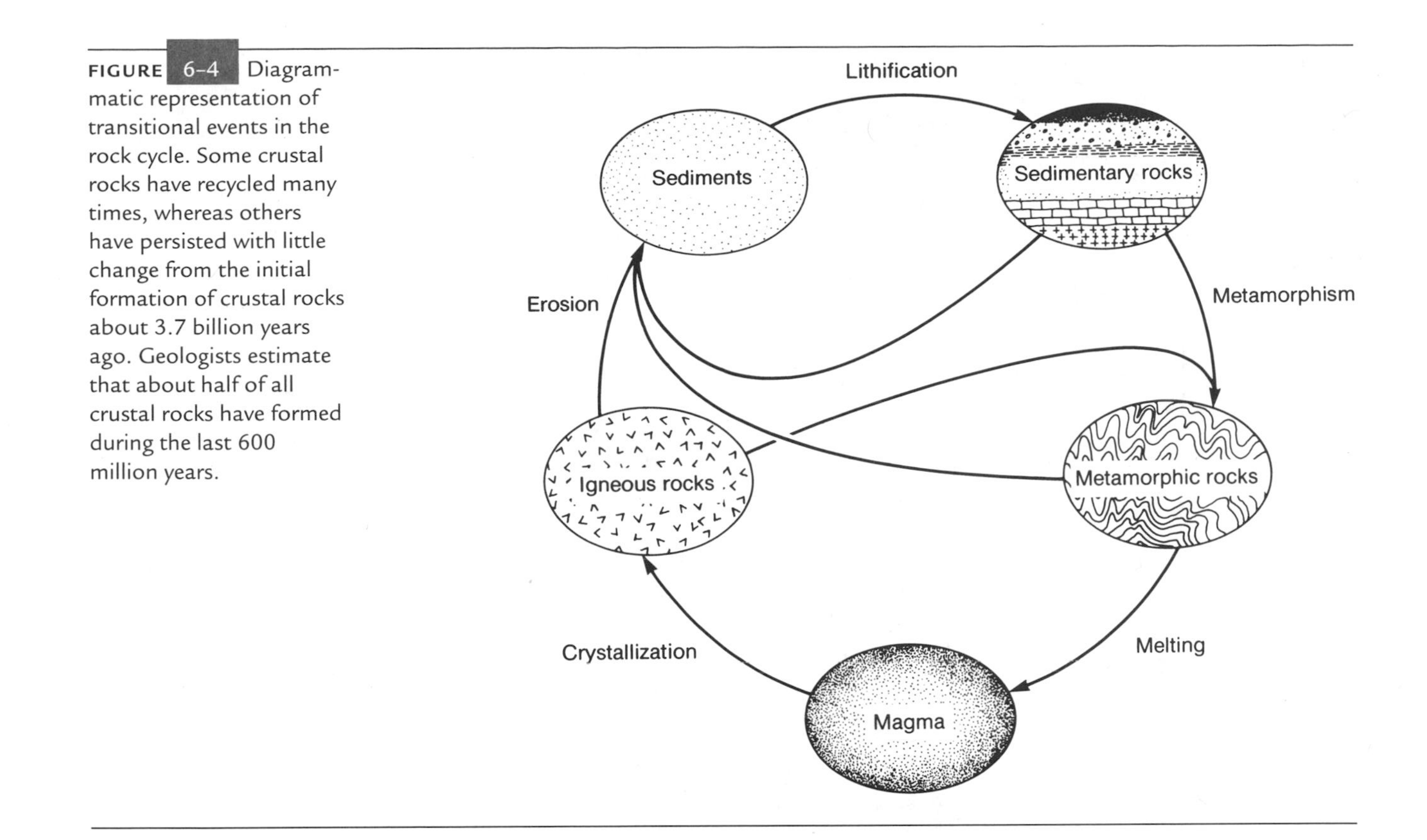

FIGURE 6-4 Diagrammatic representation of transitional events in the rock cycle. Some crustal rocks have recycled many times, whereas others have persisted with little change from the initial formation of crustal rocks about 3.7 billion years ago. Geologists estimate that about half of all crustal rocks have formed during the last 600 million years.

sure, and/or chemical interactions. Marble, for example, is a metamorphic rock that was originally limestone, and slate is a metamorphic rock that was originally shale. According to some geologists, some forms of granite are also metamorphic rocks.

As Figure 6–4 shows, a rock cycle exists in which these three major types of rock, given enough time, transform from one to the other, although not necessarily in equal proportions. At present, geologists think the Earth's crust consists, by volume, of 65 percent igneous rocks, 8 percent sedimentary rocks, and 27 percent metamorphic rocks. A layer of sedimentary rocks covers most of the surfaces of continental landmasses.

Geological Dating

Beginning in the seventeenth and eighteenth centuries, geologists became aware that they could use the relative positions of different rocks to determine their relative ages. Steno, an early proponent of the validity of fossils, was among the first to establish the **law of superposition,** which states that if a series of sedimentary rocks has not been overturned, the oldest layers or strata are at the bottom of the series and the youngest stratum is at the top. More than a century later, William Smith (1769–1839) discovered how to identify different strata by the unique kinds of fossils found within them. As Cuvier and others showed, the relative ages of the fossils seemed to correspond closely to the relative ages of the strata in which fossil hunters discovered them. That is, fossils from the uppermost strata seemed more like modern organisms than fossils from lower strata (Fig. 6–5).

Fossils became a primary means by which scientists could trace a particular geological stratum or group of strata (system) in various localities. For example, the Cambrian system (named after a Welsh tribe by Sedgwick in 1835) represents strata in which many marine invertebrate skeletons such as trilobites, brachiopods, and simple mollusks first appear. Cambrian strata exist on all continents and occupy the same relative positions; that is, they lie above Precambrian strata (absence of fossil shells) and below Ordovician and Silurian strata (true corals, echinoderms, small primitive fishes, and so on).

Unfortunately, fossils are infrequent in all geological strata, since they represent only a partial sampling of organisms, mostly those with shells, skeletons, or hard parts deposited in appropriate sediments (see Fig. 3–12). Soft-bodied organisms, which could perhaps also identify strata, are extremely rare in the fossil record. Furthermore, the same fossils are not always present in all locations of a stratum, since they may have lived only in restricted habitats or areas. Nevertheless, fossils can usually identify a particular stratum because all its areas generally contain at least some fossils characteristic of that period.

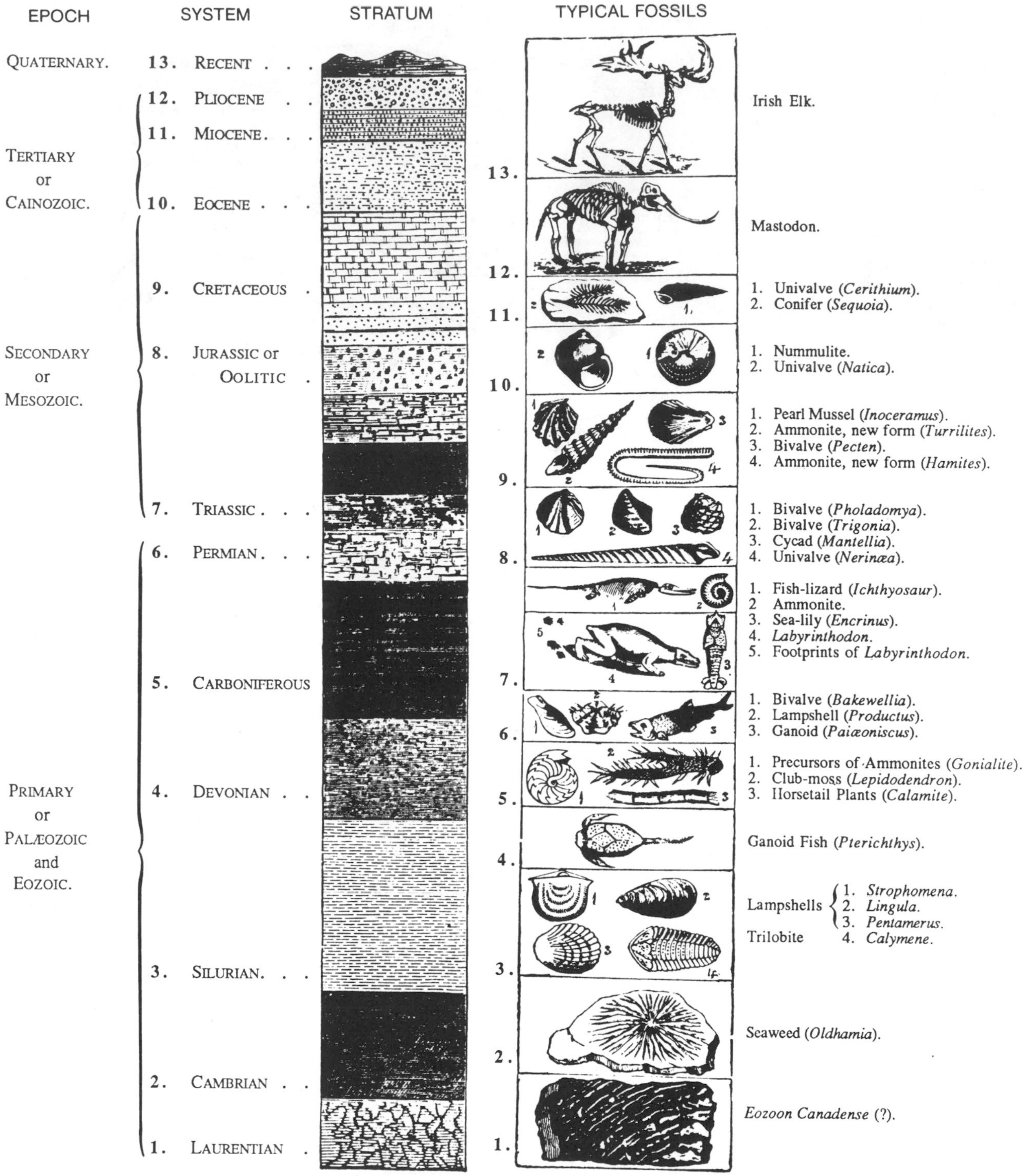

EPOCH	SYSTEM
QUATERNARY.	13. RECENT . . .
TERTIARY or CAINOZOIC.	12. PLIOCENE . .
	11. MIOCENE. . .
	10. EOCENE . . .
SECONDARY or MESOZOIC.	9. CRETACEOUS .
	8. JURASSIC or OOLITIC .
	7. TRIASSIC . . .
PRIMARY or PALÆOZOIC and EOZOIC.	6. PERMIAN. . .
	5. CARBONIFEROUS
	4. DEVONIAN . .
	3. SILURIAN. . .
	2. CAMBRIAN . .
	1. LAURENTIAN .

STRATUM

TYPICAL FOSSILS

13. Irish Elk.

Mastodon.

12. 1. Univalve (*Cerithium*). 2. Conifer (*Sequoia*).

11. 1. Nummulite. 2. Univalve (*Natica*).

10. 1. Pearl Mussel (*Inoceramus*). 2. Ammonite, new form (*Turrilites*). 3. Bivalve (*Pecten*). 4. Ammonite, new form (*Hamites*).

9. 1. Bivalve (*Pholadomya*). 2. Bivalve (*Trigonia*). 3. Cycad (*Mantellia*). 4. Univalve (*Nerinæa*).

8. 1. Fish-lizard (*Ichthyosaur*). 2 Ammonite. 3. Sea-lily (*Encrinus*). 4. *Labyrinthodon*. 5. Footprints of *Labyrinthodon*.

7. 1. Bivalve (*Bakewellia*). 2. Lampshell (*Productus*). 3. Ganoid (*Palæoniscus*).

6. 1. Precursors of Ammonites (*Gonialite*). 2. Club-moss (*Lepidodendron*). 3. Horsetail Plants (*Calamite*).

5. Ganoid Fish (*Pterichthys*).

4. Lampshells {1. *Strophomena*. 2. *Lingula*. 3. *Pentamerus*. Trilobite 4. *Calymene*.

3. Seaweed (*Oldhamia*).

2. *Eozoon Canadense* (?).

1.

FIGURE 6-5 Nineteenth century illustration of a table of stratified rocks that classifies geological strata according to their relative age and shows some of the fossils associated with each period. (*From Clodd*).

TABLE 6-2 Geological ages and associated organic events

Time Scale (eon)	Era	Period	Epoch	Millions of Years Before Present (approx.)	Duration in Millions of Years (approx.)	Some Major Organic Events
Phanerozoic	Cenozoic	Quaternary	Recent (last 5,000 years)		1.6	Appearance of humans
			Pleistocene	1.64		
		Tertiary	Pliocene	5.2	3.5	Dominance of mammals and birds
			Miocene	23.5	18.3	Proliferation of bony fishes (teleosts)
			Oligocene	34	10.5	Rise of modern groups of mammals and invertebrates
			Eocene	55	21	Dominance of flowering plants
			Paleocene	65	10	Radiation of primitive mammals
	Mesozoic	Cretaceous		146	81	First flowering plants Extinction of dinosaurs
		Jurassic		208	62	Rise of giant dinosaurs Appearance of first birds
		Triassic		245	37	Development of conifer plants
	Paleozoic	Permian		290	45	Proliferation of reptiles Extinction of many early forms (invertebrates)
		Carboniferous	Pennsylvania	320	30	Appearance of early reptiles
			Mississippian	363	43	Development of amphibians and insects
		Devonian		409	46	Rise of fishes First land vertebrates
		Silurian		459	30	First land plants and land invertebrates
		Ordovician		505	66	Dominance of invertebrates First vertebrates
		Cambrian		545	40	Sharp increase in fossils of invertebrate phyla
Precambrian	Proterozoic	Upper		900	355	Appearance of multicellular organisms
		Middle		1,600	700	Appearance of eukaryotic cells
		Lower		2,500	900	Appearance of planktonic prokaryotes
	Archean			3,900	1,400	Appearance of sedimentary rocks, stromatolites, and benthic prokaryotes
	Hadean			4,500	600	From the formation of Earth until first appearance of sedimentary rocks; no observable fossil organisms

*Note: Dates derived mostly from Harland et al. Some geologists divide the Precambrian eon into two major eras, Proterozoic and Archean, and then denote the Hadean as the first Archean period (Fig. 9-13). However, the exact dates that mark each geological period are often only approximate, and other authors provide somewhat different time spans.

By these means, geologists have defined a **Phanerozoic time scale** (or eon) as the period in which abundant visible (*phanero*) life (*zoon*) appears. It consists of three major eras of geological strata, beginning with the Paleozoic—the first in which significant numbers of hard-bodied fossils are found. As shown in Table 6–2, each era contains a number of subsidiary systems or periods, often further subdivided into series or epochs.[7]

Although relative dating by stratigraphic methods usually establishes a sequential relationship between different rocks and between different fossils, stratigraphy does not offer information on the time lengths involved. Sediments do not deposit in identical thicknesses from time to time or from place to place. Furthermore, in all localities large sections of the geological record have been worn away by erosions or destroyed by new rock formations and Earth movement. Nowhere does the geological record offer a complete sequence that we can trace continuously, year by year, to the present time.

Dating with Radioactive Elements

Fortunately, geologists have discovered dating methods using radioactivity that permit them to date rocks even billions of years old with a fair degree of accuracy. All these methods of **radioactive dating** rely on three main factors:

1. The ease with which researchers can detect many radioactive elements
2. The known isotopes into which their atoms disintegrate
3. The known rates at which this disintegration occurs

[7] The system of geological classification adopted in the eighteenth and early nineteenth centuries (initially suggested by the Italian geologist Arduino) followed the practice of designating *primary* rocks as those without fossils. These were believed to date from the origin of the Earth's crust and appeared as typical nonstratified, ore-bearing outcroppings in mountainous areas. Geologists called stratified fossiliferous rocks, such as sandstone and limestone, *secondary*, and believers in the Judeo-Christian Bible attributed their origin to the Noachian deluge. These secondary strata contained obviously ancient molluskan fossils such as ammonites and belemnites (Chapter 15) as well as early fish and reptiles that differed considerably from present forms. Geologists believed tertiary sedimentary rocks to be derived from secondary strata by flooding, erosion, volcanic action, and so on and contained ancient representatives of more recent forms such as mammals. Quaternary rocks represented the glacial and alluvial deposits of relatively recent times. Since not all mountains nor all strata were of the same age, scientist found these divisions difficult to apply universally, and eventually abandoned the terms, with the exception of Tertiary and Quaternary. *Tertiary* came to mean the period of preglacial deposits corresponding to most of the Cenozoic; *Quaternary* means the period dating from the Pleistocene ice age deposits to the present.

For example, the radioactive element uranium 238 (^{238}U) is present in the mineral zircon found in most igneous rocks and disintegrates to form the lead isotope ^{206}Pb at a rate that transforms half the uranium into lead over a period (the half-life) of about 4.5 billion years (Fig. 6–6). Thus, after we make allowances for the presence of lead that the uranium disintegration (^{204}Pb) did not produce, and assuming these two isotopes have fully persisted, their relative amounts in a particular rock provide a fairly accurate dating method for older rocks.

A somewhat simplified formula that scientists can use for this purpose is

$$t = 1/\lambda \ln (^{206}Pb/^{238}U + 1)$$

where t is time in years, λ is decay rate per year (1.537×10^{-10} for ^{238}U), and ln is the natural logarithm (base e). Thus, a $^{206}Pb/^{238}U$ ratio of 0.360 in a particular sample would indicate that

$$t = 1/(1.537 \times 10^{-10}) \ln (1.360) = (6.508 \times 10^{9}) (.307) = 1.998 \times 10^{9}$$

that is, approximately 2 billion years have elapsed since the ^{238}U was first incorporated into this sample. Researchers can check dates determined in this fashion by the disintegration rates of other radioactive elements present in the same material such as the decay of ^{235}U to ^{207}Pb (half-life of about 0.7 billion years). As shown in Table 6–3, additional radioactive elements that researchers use in dating include rubidium 87 (which disintegrates to strontium 87, with a half-life of 48.8 billion years) and potassium 40 (which disintegrates to argon 40) with a half-life of 1.3 billion years. For dating fairly recent events, geologists commonly use carbon 14 (which disintegrates into nitrogen 14 with a half-life of only 5,730 years). Another method for dating young volcanic rocks (as well as ceramic artifacts) is to count the fission tracks they have incorporated over time because of the steady decay of uranium atoms.

So far, geologists have mostly applied radioactive dating methods to igneous rocks and have extended the dates to sedimentary rocks by the relative positions of the two kinds of rock (Fig. 6–7). Thus, igneous rocks that coincide with the age of the Cambrian sediments are approximately 540 million years old. Later sedimentary rocks, as shown in Table 6–3, can be dated fairly precisely up to the recent period. As we go further back in time, the oldest terrestrial rocks are somewhat more than 3.5 billion years old, whereas estimates based on the combined isotope composition of lead in all Earth materials (the $^{206}Pb/^{204}Pb$ ratio) point to an overall terrestrial age of about 4.6 billion years. This 4.6 billion-year estimate accords with the ages of moon rocks brought to Earth by the *Apollo* lunar missions, as well as with similar estimates made for meteorites that astronomers believe originated at the birth of the solar system.

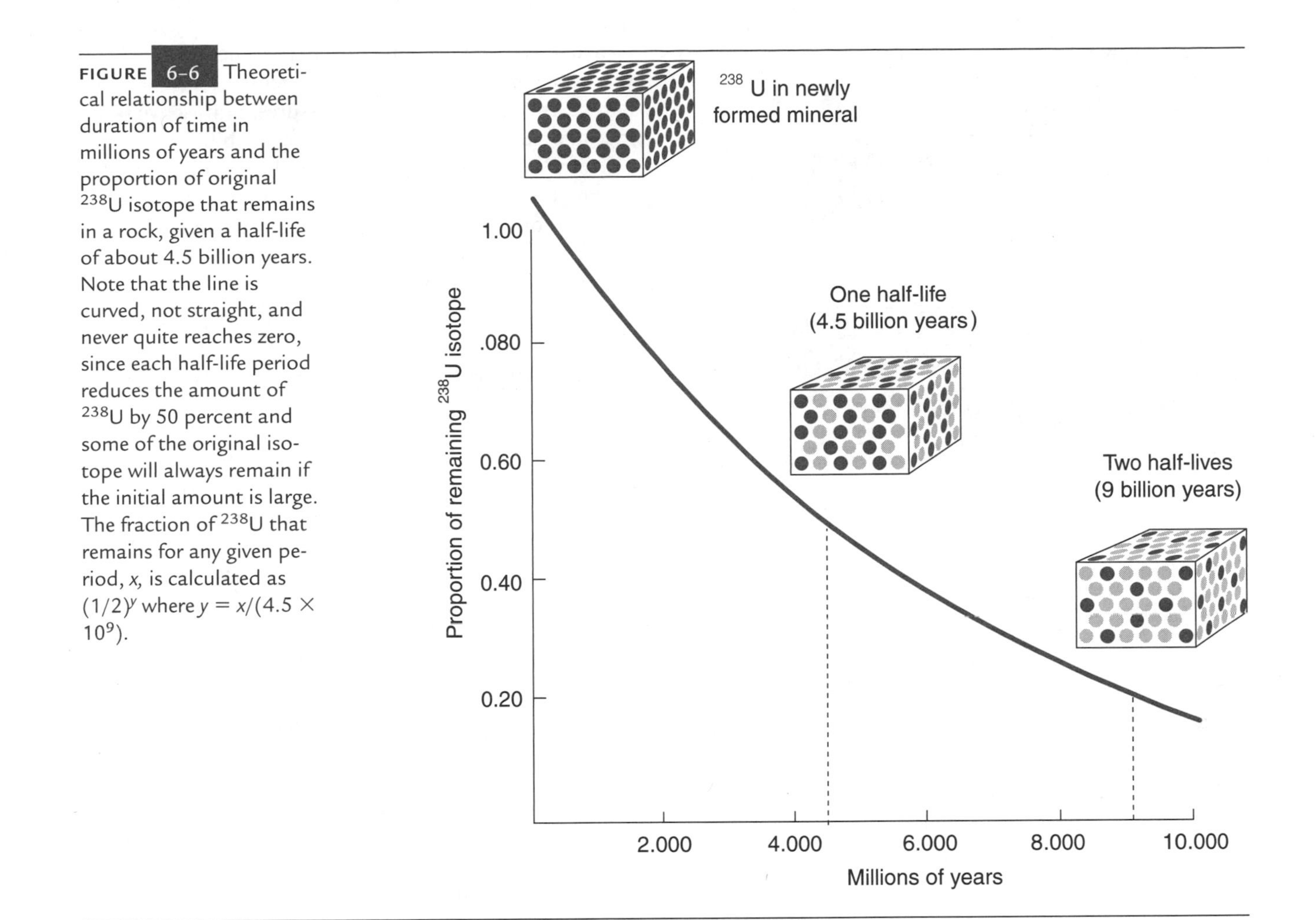

FIGURE 6-6 Theoretical relationship between duration of time in millions of years and the proportion of original ^{238}U isotope that remains in a rock, given a half-life of about 4.5 billion years. Note that the line is curved, not straight, and never quite reaches zero, since each half-life period reduces the amount of ^{238}U by 50 percent and some of the original isotope will always remain if the initial amount is large. The fraction of ^{238}U that remains for any given period, x, is calculated as $(1/2)^y$ where $y = x/(4.5 \times 10^9)$.

TABLE 6-3 Some radioactive isotopes used in dating

Parent		Daughter	Half-Life	Usable Range
Samarium 147	→	Neodymium 147	110 billion years	>1 billion years
Rubidium 87	→	Strontium 87	49 billion years	>100 million years
Thorim 232	→	Lead 208	14 billion years	>300 million years
Uranium 238	→	Lead 206	4.5 billion years	>100 million years
Potassium 40	→	Argon 40	1.3 billion years	>100 thousand years
Uranium 235	→	Lead 207	0.7 billion years	>100 million years
Uranium 234	→	Thorium 230	0.25 million years	>1 million years
Carbon 14	→	Nitrogen 14	5,730 years	<50,000 years

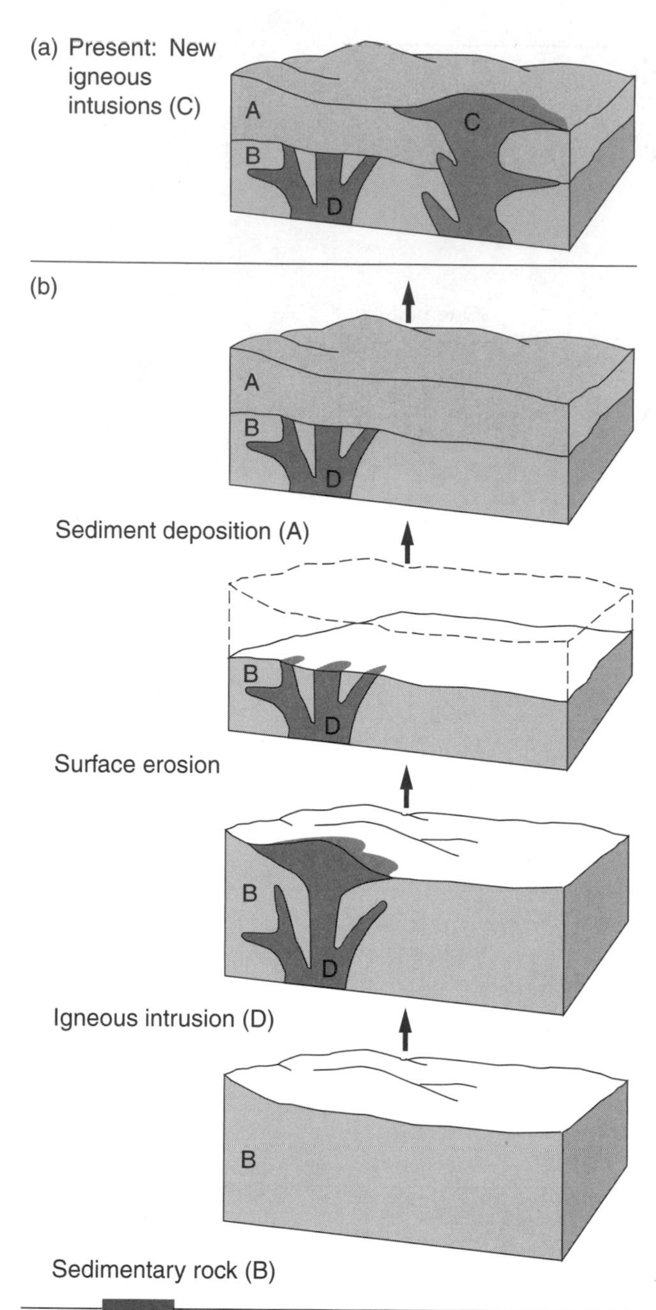

FIGURE 6–7 Use of relative and absolute dating in determining the ages of sedimentary and igneous rocks. (*a*) Diagram showing the observed relationships among four geological assemblies, two sedimentary layers (A and B) and two igneous intrusions or "dikes" (C and D). (*b*) Historical interpretation of the arrangement of these rocks based on the rules of superposition (younger sediments lie above older sediments) and crosscutting relationships (igneous rocks are younger than the rocks through which they cut across). According to these principles, the B sediments are the oldest of rocks, and D represents a later igneous intrusion into B. Erosion then occurred, removing part of the intrusive igneous rock B, followed by the later deposition of A sediments. The last geological event was a new igneous intrusion of rock C into both A and B. The age relationships are therefore B>D>A>C. Thus, if the absolute ages of the two intrusions, C and D, can be determined by radioactive dating techniques, the upper and lower limits for the age of the A sediments can be determined.

Scientists therefore now generally accept that the Earth was probably formed at low temperatures 4.6 billion years ago, and that high pressures, radioactive heating, and surface cooling took about a billion years to generate continental masses and their igneous rocks. During the Archean era, 3.5 to 3.7 billion years ago, the presence of water and other weathering conditions was sufficient to enable the first of the presently observed sedimentary rocks to appear.

Continental Drift

In the period between 1912 and 1930 a meteorologist, Alfred Wegener (1880–1930), developed the concept that all the continents were at one time a single land mass that he called **Pangaea.**[8] He suggested that fissures occurred within this mass and the resulting fragments drifted apart to form the present continents. According to Wegener, drifting was caused by gravitational forces moving the continents through the viscous sea floor material.

For the next few decades most geologists considered Wegener's theory little more than an imaginative fantasy until the evidence for continental drift became so overwhelming that they could no longer ignore it. This evidence includes observations made of the fit between continents; similarity of rocks, fossils, and glaciation; paleomagnetism; and ocean floor spreading.

FIT OF THE CONTINENTS

As shown in Figure 6–8, one of the most striking geographic correlations is the exact match between the east coast of South America and the west coast of Africa. Not quite so obvious but nevertheless observable is the match between the east coast of North America and the northwest coast of Africa. These and other geographical juxtapositions indicate that the continents at one time either joined together or were extremely close.

SIMILARITY OF ROCKS, FOSSILS, AND GLACIATIONS

A group of rock strata in India, called the Gondwana system, dates from the late Carboniferous to the early Cretaceous period. Formations of extremely similar nature and composition exist in South Africa, South America, Antarctica, the Falkland Islands, and Madagascar. As Figure 6–9 shows, associated with a few of these Gondwana formations are unique types of fossil plants (*Glos-*

[8] Francis Bacon (1561–1626) had noted such a relationship long ago; he proposed that a continent called Atlantis once exactly fitted the mid-Atlantic, and later sunk beneath the ocean.

FIGURE 6–8 Matched fit between the offshore continental shelves at 500 fathoms deep on opposite sides of the Atlantic Ocean. (*From Eicher, D.L., and A.L. McAlester, 1980.* History of the Earth. *Reprinted by permission of Prentice Hall.*)

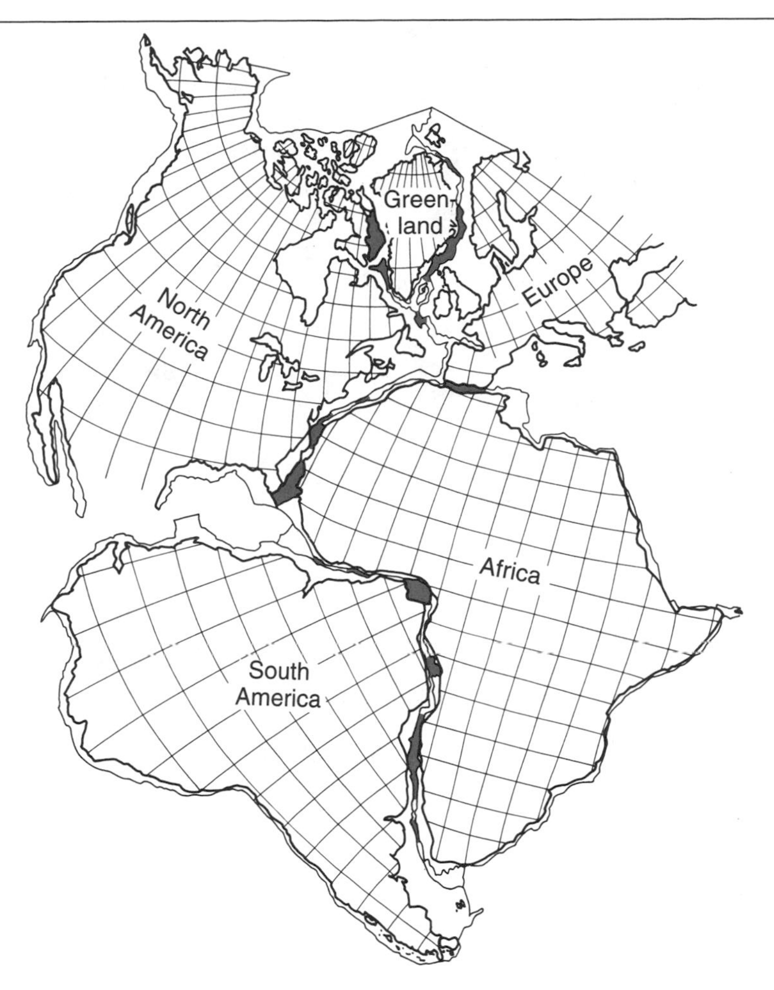

sopteris) and animals (*Mesosaurus, Lystrosaurus, Cynognathus*). Furthermore, all the areas bearing Gondwana formations, along with Australia, were apparently covered by the same glaciation event during a Paleozoic ice age. To account for these observations, geologists have suggested the existence of a massive southern continent, **Gondwana,** which included the areas that now carry the Gondwana formations and Australia. These land areas centered much closer to the South Pole than their present tropical locations and therefore glaciers developed on them more easily.

PALEOMAGNETISM

As new rocks arise from the cooling of magma, ferrous material within them (for example, magnetite, Fe_3O_4) magnetizes in a direction that depends on the location and strength of the Earth's magnetic field prevailing at the time. Should this magnetic field change for any reason, the magnetic field of newly formed rocks would also be expected to change. Thus geologists can study rocks from all eras and all continents for their fossilized magnetism, or paleomagnetism, and can then deduce the direction and distance of the Earth's magnetic poles relative to these rocks.

Although we would expect paleomagnetic studies to show slight shifts in the magnetic poles, it was strange to find that these poles had shifted during past ages over thousands of miles and that the magnetic poles of different continents did not coincide for long periods of time. For example, although magnetite deposits in recent igneous rocks from South America and Africa show the same magnetic orientation , this is not true for older Paleozoic rocks. As shown in Figure 6–10*a*, the magnetic poles derived from analyzing continental rocks that date between the Silurian and Permian periods indicate seemingly independent positions for each continent. Since different magnetic poles could not exist simultaneously, we can best explain these different polar wanderings as arising from the movement of continents relative to each other as well as from their movements relative to the poles.

Figure 6–10*b* shows that the South American and African poles coincide for the Silurian–Permian period if we juxtapose the positions of these two continents. Both continents were united during the Paleozoic era so the magnetic orientation of magnetite deposits during that

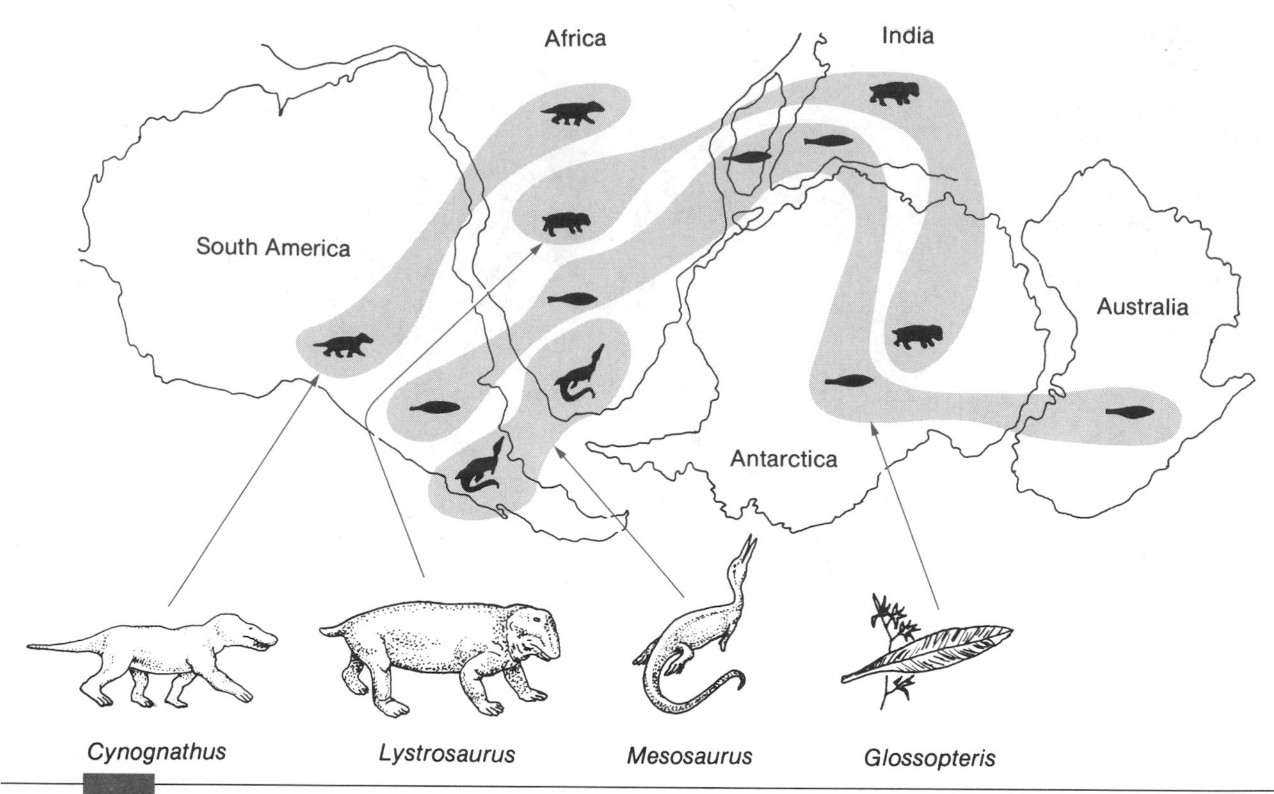

FIGURE 6-9 Distribution of various fossil plants and animals throughout Gondwana continents. The presumed fit of the continental margins during the Permian–Triassic period is also shown. *Cynognathus* was a carnivorous mammal-like reptile, (therapsid, Chapter 18) with a distinctive doglike skull, found in Triassic-period deposits in South America and Africa. *Lystrosaurus* was also a Triassic mammal-like reptile but larger than *Cynognathus* and probably herbivorous, with beaklike jaws and two large tusks. The genus *Mesosaurus* represents a fossil order of freshwater reptiles restricted to Permian deposits in Brazil and South America. This reptile was about 1 1/2 feet in length with distinctive features of skull and limbs. *Glossopteris* was a fossil plant with many features similar to seed ferns (pteridosperms), bearing also large tongue-shaped leaves patterned with many reticulate veins. These fossil leaves appear in all the Gondwana formations and date back to the early Permian period. (*Adapted from Colbert.*)

period all pointed to the same geographic position for the South magnetic pole. As the continents separated in the Mesozoic era, the "fossilized" magnetic orientations of these deposits now pointed to apparently different South magnetic pole positions, showing the anomalies in Fig. 6–10*a*. What changed was not the Paleozoic magnetic pole position, but the geographic location of Paleozoic magnetized deposits.

THE OCEAN FLOOR

Geologists made a puzzling observation soon after ocean floor samplings became common—the relative youth of the ocean floor. Its sediments seemed no older than 100 to 200 million years, and about 50 percent of its rock composition was no older than the beginning of the Tertiary period. Also, in contrast to the often folded and compressed sedimentary rocks in continental mountains, the oceanic mountains consisted almost exclusively of igneous basalts. Since considerable evidence shows that oceans have existed since early geological history, the relative youth of the present ocean basins clearly indicated that they must have replaced older ocean floors.

Another unusual oceanic feature was the existence of magnetized belts that parallel the long midoceanic ridges found in almost all ocean basins. Measurement of the magnetic direction on both sides of such ocean ridges showed that each belt symmetrically paired with a belt of approximately equal width and of the same magnetic orientation on the other side of the ridge. Belts adjacent to each other on the same slope, however, usually magnetized differently.[9] Radioactive dating showed that the youngest belts were closest to the crest of the ridge and the older belts were farther away.

[9] Geologists call changes in magnetic orientation between adjacent belts *reversals*, since they are caused by a 180° reversal in the polarity of the Earth's magnetic pole. That is, the degree of magnetism weakens with time, and at some point it reverses so that the south-pointing needle on a compass now points north. Using data from ocean floors and other sources, geologists have shown that the duration of a particular magnetic polarity before it reverses may vary from several thousand to 700,000 years.

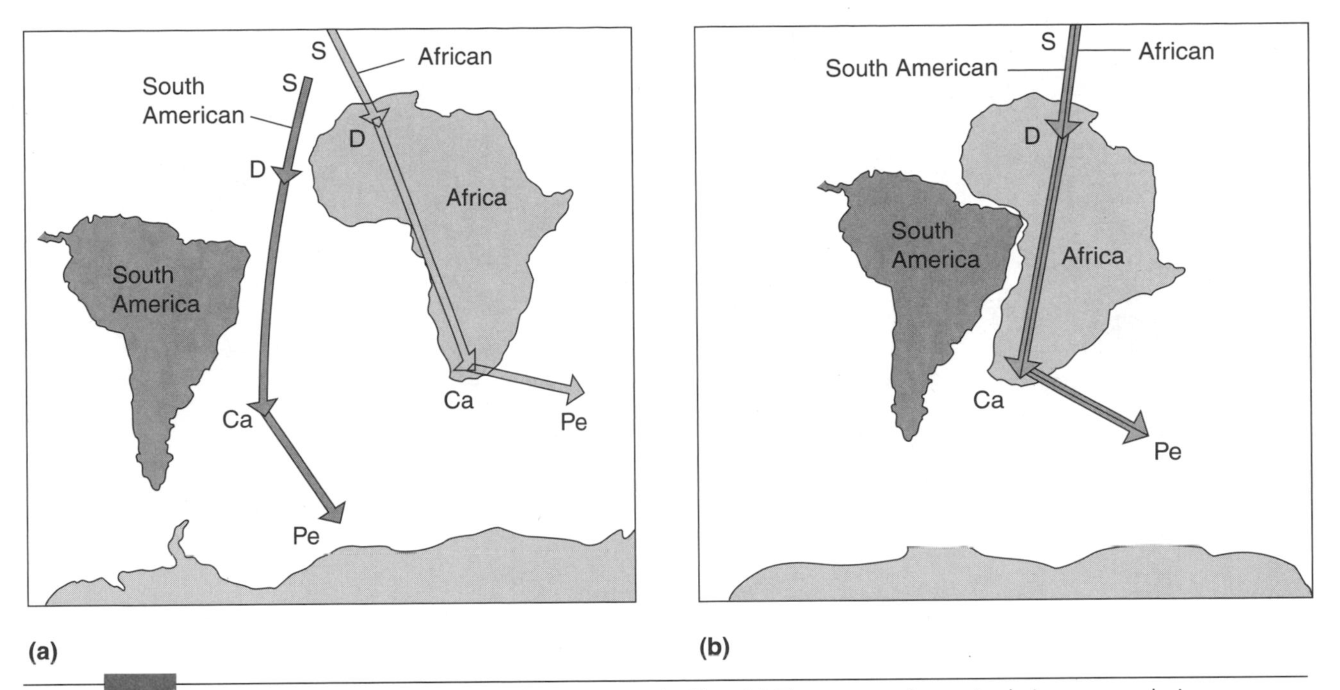

FIGURE 6–10 Magnetic pole wanderings of South America and Africa. (*a*) The two continents in their present relative positions, showing paleomagnetically determined locations of the south magnetic pole for the Silurian (S), Devonian (D), Carboniferous (Ca), and Permian (Pe) periods. The magnetic pole for these four periods is apparently in a different location for each continent, an anomaly that would be difficult to explain if each continent had always occupied its present relative position. (*b*) This reconstruction demonstrates that the two seemingly independent polar pathways shown in (*a*) coincide when the two continents are fitted together according to Fig. 6.8. The paleomagnetic data indicate that the continents moved as a unit across the South Pole through the Paleozoic era and began to separate during the Mesozoic (Fig. 6–12). (*Adapted from Cox and Moore.*)

We can best explain all these observations if we assume that the midoceanic ridges represent fissures out of which new ocean floor emerges and then spreads to either side. Thus, molten rock spouting from the oceanic ridge magnetizes upon cooling and is then displaced from the ridge by later emerging material (Fig. 6–11). Somewhat like the annual growth rings of a tree, the ocean floor therefore retains its history in a series of parallel bands of rocks marked by magnetic fields prevailing at the time of their origin. However, this **sea floor spreading** is not uniform, and its annual rates vary from about 1 centimeter in the North Atlantic Ridge to 3 centimeters in the South Atlantic Ridge to as much as 9 centimeters in some portions of the Eastern Pacific Ridge.

To and From Pangaea

One view of events that emerges from these studies is shown in Figure 6–12. It begins with a Devonian geography indicating separation between the Gondwana group of continents and a North American–Eurasian group called Laurasia (Fig. 6–12*a*). Most geologists now agree that by the end of the Paleozoic era, these two major continental groups had united to form the giant landmass Pangaea (Fig. 6–12*b*, *c*). According to these reconstructions, Pangaea began to break up during the Triassic period, about 225 million years ago (Fig. 6–12*d*).

The fragmentation of Pangaea began with an oceanic rift that developed between Western Gondwana (South America and Africa) and Eastern Gondwana (India, Antarctica, and Australia), and a further rift that separated Laurasia from Western Gondwana (Fig. 6–12*e*). By the Late Jurassic period, sea floor spreading began to separate North America from Europe, and by the Cretaceous period, South America from Africa (Fig. 6–12*f*, *g*). The Indian subcontinent, moving independently from about the mid-Cretaceous period on, continued northward from the Antarctic–Australian mass until it reached Southern Asia in the Cenozoic era about 40 million years ago. The Himalayan Mountains, in which mountain building still seems to be going on, now mark compressions caused by this Indian–Asian collision.

In the Western Hemisphere, the rapid drift of South America away from Africa, which began about 100 million years ago, led eventually to a reunion with North America approximately 4 or 5 million years ago. In the Southern Hemisphere, although New Zealand had already drifted away from the Australian–Antarctican–South American landmass before the end of the Cretaceous period, the

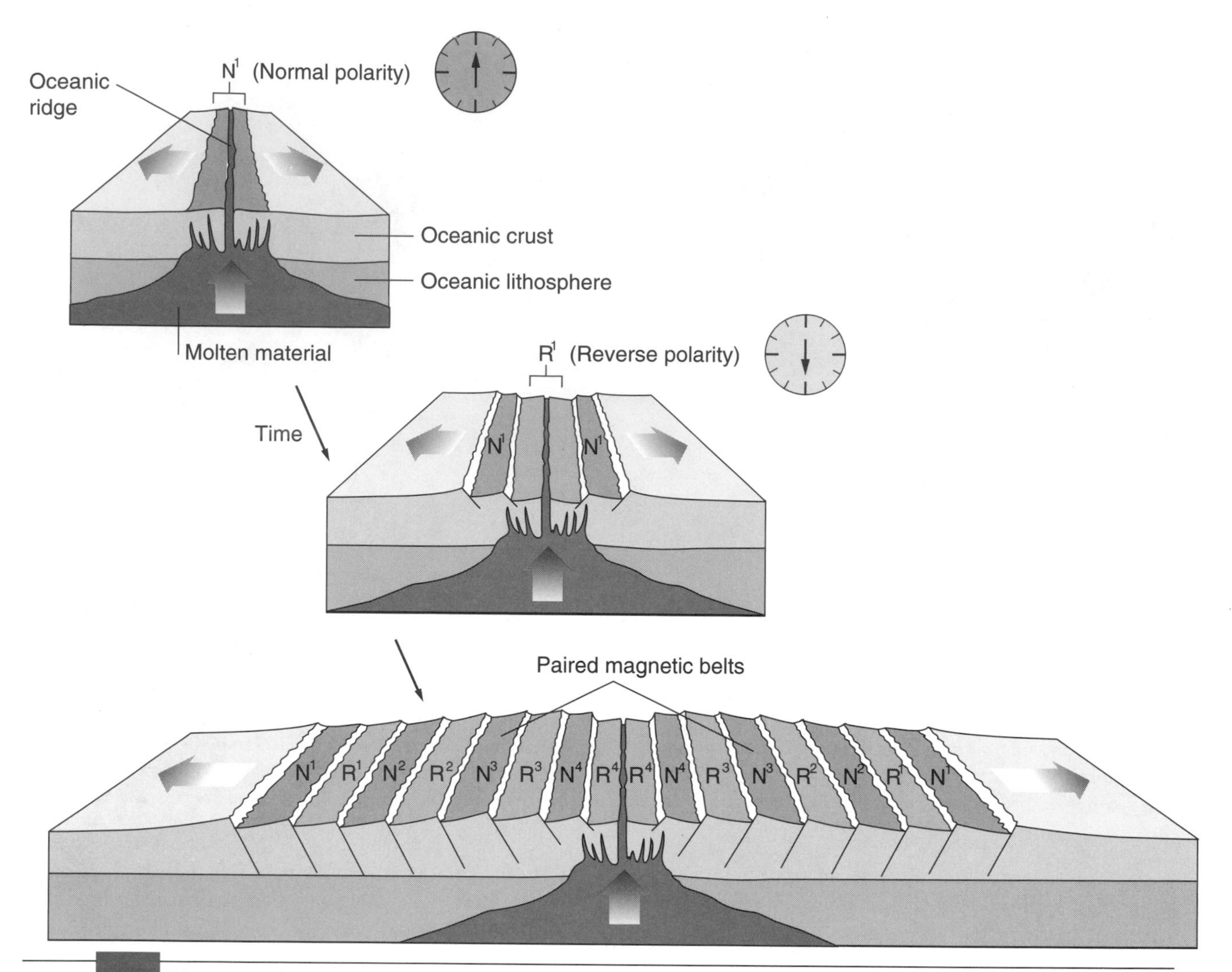

FIGURE 6-11 Diagrammatic sections through an oceanic ridge showing how sea floor spreading produces differently magnetized belts. Hot molten material adds to the ridge from the mantle, falls away on both sides, and magnetizes in the orientation of the prevailing magnetic field as it cools. With time, the magnetic field changes in strength and/or direction, and new material added to the ridge forms a pair of belts distinctly different from adjacent belts.

other Southern continents were joined until the beginning of the Tertiary period, about 65 million years ago. However, by the Eocene epoch, 20 million years later, Australia had begun its northward journey that will eventually lead to its union with Asia.

The process of drifting and colliding continents has extended even to Precambrian times. For example, geologists can show that the magnetic poles of the North American continent and the Gondwana group shared a common pathway for a period lasting more than a billion years during the Proterozoic era (Fig. 6–13). Although these events indicate the existence of a giant continent containing most of the Earth's surface, Northern Europe may have remained independent of North America until the middle of the Paleozoic era. Some geologists have also suggested that a number of Asian subsections did not actually unite with each other and with Europe until the Mesozoic era (Ziegler et al).

Tectonic Plates

Scientists now know these seemingly varied and intricate continental movements are based on movements (tectonics) of gigantic plates that compose the Earth's lithosphere. So far, researchers have delineated eight major plates as well as several minor ones, marked mostly by Earthquake belts that accompany plate movements (Fig. 6–14). In general, they have observed three major types of plate boundary events:

- Plates can separate from each other by the addition of new lava material to their adjoining boundary. Such events occur in the oceanic ridges and account for sea floor spreading and an increase in size of some oceanic basins (Fig. 6–11).

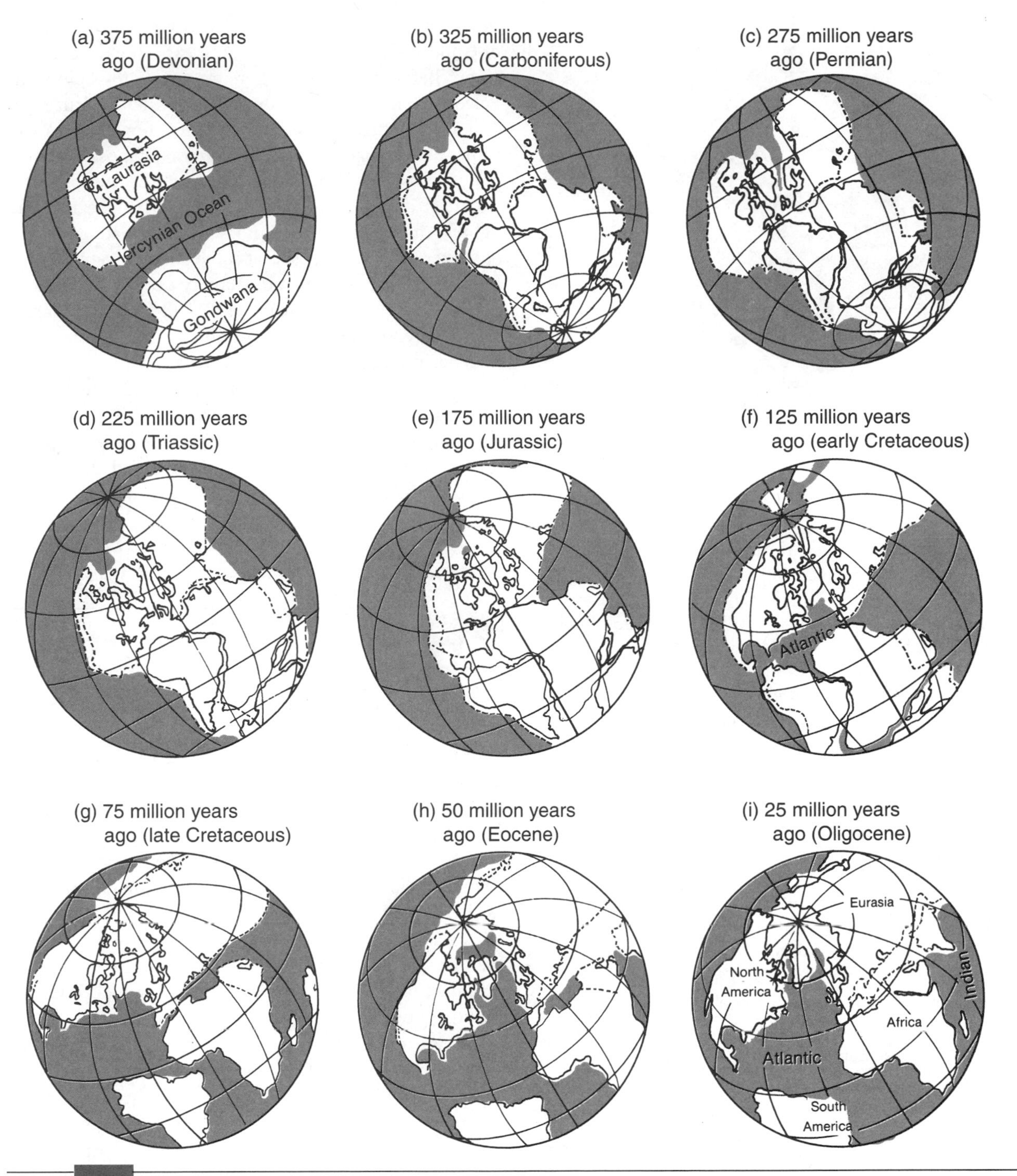

FIGURE 6–12 Some of the major geographical changes caused by continental drift from the Devonian period onward. (*Drift of the Major Continental Blocks since Devonia, by E. Irving,* Nature *270: 304–309, 1977. Reprinted by permission.*)

- Two adjoining plates can slide past each other at a common boundary, or fault, without any significant change in size. An example is the motion of the Pacific plate carrying a section of western California past the North American plate at the San Andreas fault. The speed at which this is occurring (6 centimeters per year) will bring Los Angeles to the same latitude as San Francisco in about 10 million years and to the Aleutian Islands near Alaska in about 60 million years.

- One plate can move toward another, causing a convergent boundary. When one such plate carries oceanic crust, the convergent event is often marked by the loss of plate material as this crustal mass plunges into the mantle. For example, the Pacific

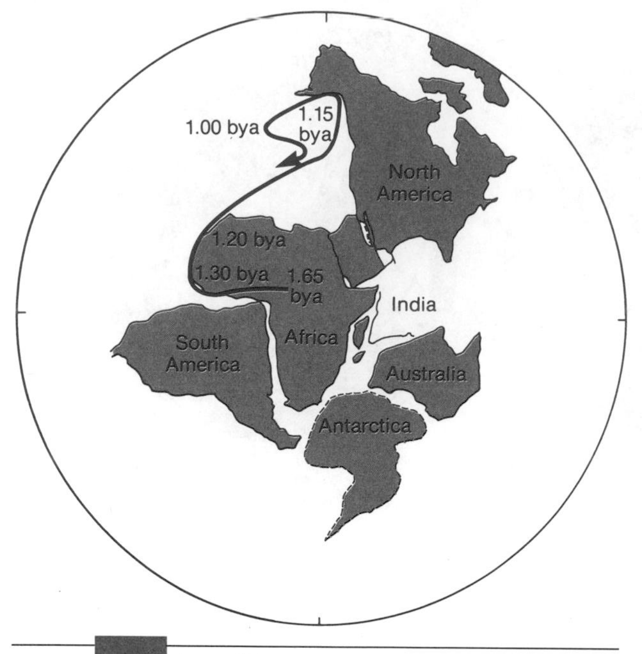

FIGURE 6–13 A possible reconstruction of the Gondwana–North American landmass for the approximate period between 1.0 and 2.2 billion years ago (bya). The heavy dark line shows the pathway of the magnetic pole determined for this landmass for given years (for example, the magnetic pole was at the northwest tip of Africa 1.2 billion years ago). The exact position of Antarctica is uncertain. Dalziel presents a more detailed scenario showing movements of the North American continents between 750 and 250 million years ago. *(Adapted from Piper, 1974.)*

plate in its motion northward meets a border of the North American plate at the Aleutian Islands and is then pushed (subducted) under this trench (Fig. 6–15*a*). The descending plate changes or deforms the mantle, which then produces volcanic activity and mountain formation in the crustal region above. Such processes account for some belts of volcanic islands, such as those near the Aleutian and Java trenches, and also explain the origin of the South American Andes Mountains that lie near the Chilean trench, where the Pacific (Nazca) plate undercuts the westward-moving American plate (Fig. 6–15*b*). Once subduction has begun, a plate can continue to descend into the mantle, pushing oceanic sediments along with it until a continental mass meets the convergent boundary. Because the rocks of the continents are lighter and thicker than the ocean floor, they cannot apparently be forced under another plate, and, as a consequence, mountains such as the Himalayas form through the foldings and pressures of these colliding landmasses (Fig. 6–15*c*).

Although we do not know the exact causes for plate tectonics, plate tectonics has helped, more than any other geological theory, to explain the relative motion of continents since the Mesozoic era and clarified the localization of Gondwana deposits, island arcs, earthquake belts, and so on. Before the Mesozoic era, and certainly before the Paleozoic, mountain building and continental drift must also have occurred, but it is not yet clear how far back tectonic plates formed. Some authors propose tectonic plates must have originated during the period after chemical differentiation of the mantle had begun, and lighter materials (aluminum silicates) had surfaced. Some researchers suggest that convection currents arising from heat produced by radioactivity during the Hadean period (4.5–3.9 billion years ago) were four times larger than at present, and led to motion of the slaglike crust, thus beginning a tectonic-like process.

However tectonic movements began, the separation and joining of landmasses from the Proterozoic onward had important biological effects, because they often determined the distribution of organisms and thus influenced subsequent evolution.

Biological Effects of Drift

One of the most prominent examples of the effect of continental drift on the distribution and evolution of organisms is the unique collection of primitive mammals found in Australia and South America. If we divide existing mammals (hairy skin, mammary glands, special auditory skull bones; Chapter 19) into three groups, **Prototheria** (egg-laying monotremes such as the duckbilled platypus, Fig. 6–16*a*); **Metatheria** (marsupials, which undergo part of their development in an external female pouch, Fig. 6–16*b*); and **Eutheria** (placental mammals, which have their entire fetal development *in utero*, Fig. 6–16*c*); then we find that by the Pliocene epoch the placentals had replaced the more primitive monotremes and marsupials in all localities except these two continents.

In Australia there still exist two families of monotremes and 13 of marsupials, but, with the possible exception of bats, researchers found no placentals on the continent until relatively recent times. The South American mammalian fauna seems to have been somewhat more advanced than that of Australia, and included, until the mid-Tertiary period, a number of placental families in addition to 5 families of marsupials. However, even the native South American placentals were generally primitive, as evidenced by mammals that persist there such as armadillos, anteaters, and tree sloths.

The picture of mammalian evolution that emerges from these studies is that primitive prototherians and metatherians had probably entered southern parts of Pangaea by the Upper Jurassic and Lower Cretaceous periods (Fig. 6–17*a*). The subsequent rifting of Australia isolated its primitive mammalian fauna from later com-

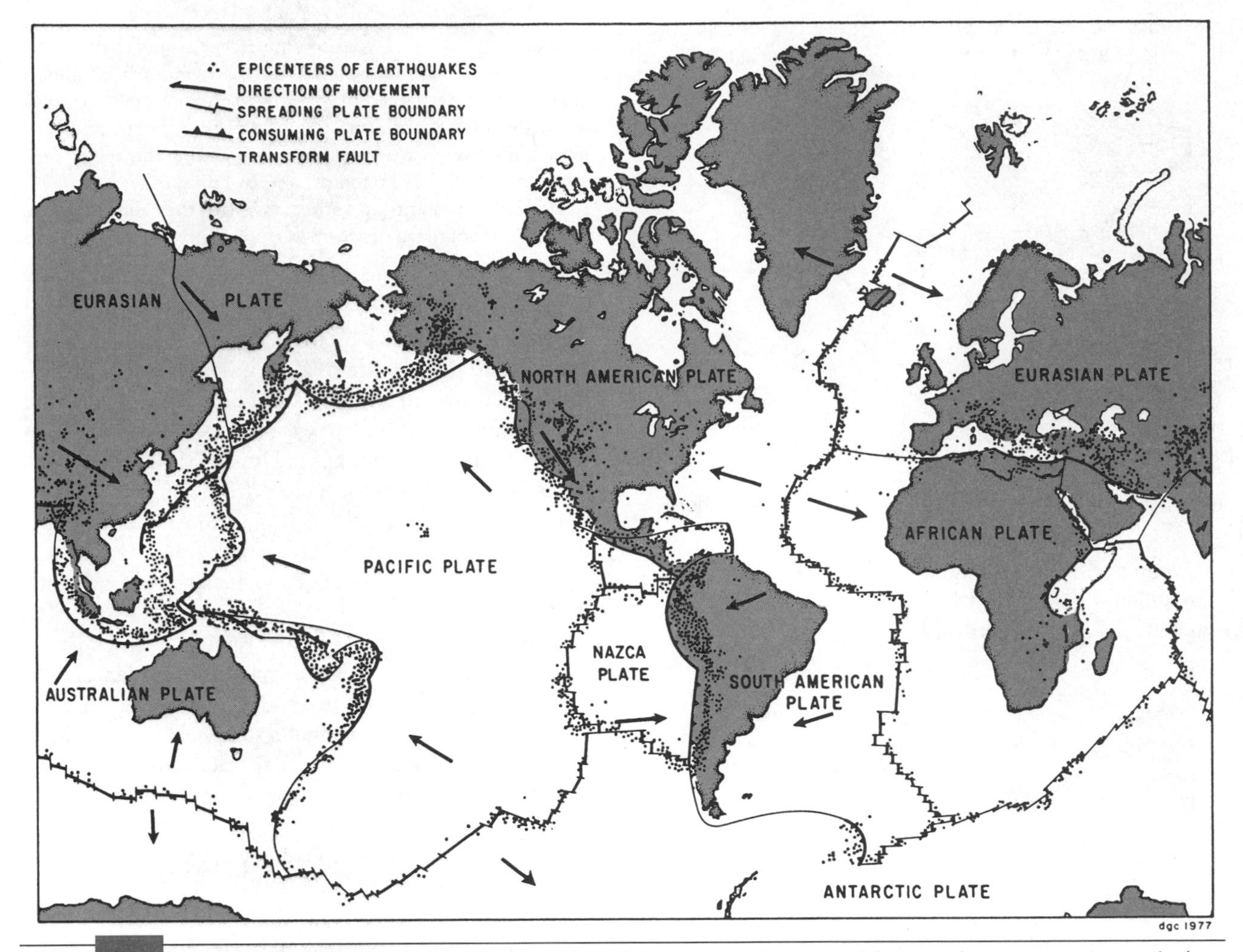

FIGURE 6–14 The major geological plates (and their boundaries) that account for many of the crustal movements. Interestingly, in an 1838 meeting of London's Geological Society, Charles Darwin also pointed to the relationship between volcanic action and mountain building: "The contemplation of volcanic phaenomena in South America has induced the author to infer, that the crust of the globe in Chile rests on a lake of molten stone, undergoing some slow but great change . . . that mountain building and volcanos are due to the same cause, and may be considered as mere subsidiary phaenomena, attendant on continental elevations; that continental elevations, and the action of volcanos, are phaenomena now in progress, caused by some slow but great change in the interior of the earth; and, therefore, that it might be anticipated that the formation of mountain-chains is likewise in progress; and at a rate which may be judged of, by either actions, but most clearly by the growth of volcanoes." (*From* Cosmos, Earth and Man: A Short History of the Universe *by P. Cloud, 1978, Fig. 11, p. 82. Reprinted by permission.*)

petition with more advanced eutherian groups that evolved in western Pangaea during the late Cretaceous and early Tertiary periods (Fig. 6–17*b*). In South America, the metatherians and primitive eutherians had probably replaced the prototherian groups by the early and mid-Tertiary period, but by that time the South American continent had drifted considerably from Africa and also had separated from North America (Fig. 6–17*c*).

The evolution of mammals on the isolated island of South America was therefore largely independent of mammalian evolution elsewhere,[10] until South America rejoined North America via the Panama Isthmus during the Pliocene epoch (Fig. 6–17*d*). By the Pliocene epoch, however, considerable evolution toward more advanced eutherian forms had occurred either in Africa or Laurasia (North America–Eurasia), whereas most of the South American mammalian fauna were still relatively primitive. When the Pleistocene epoch began, massive invasions of advanced northern eutherians were making their way south across the Central American land bridge, causing the rapid extinction of many South American mammalian families (Chapter 19). Only rarely, as in the case of opossums, did primitive South American mammals manage to successfully invade North America.

Supporting this view of continental drift and mammalian evolution is abundant fossil evidence of South

[10] During the Oligocene epoch, however, some island hopping combined perhaps with transport on floating debris ("rafting") seems to have occurred, and various monkeys and caviomorph rodents made their way to South America either from Africa or North America (see also p. 468).

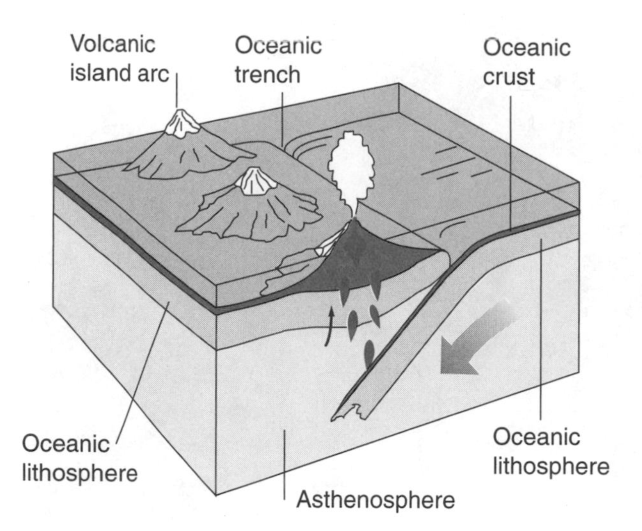

(**a**) Collision of two plates, both with oceanic lithosphere

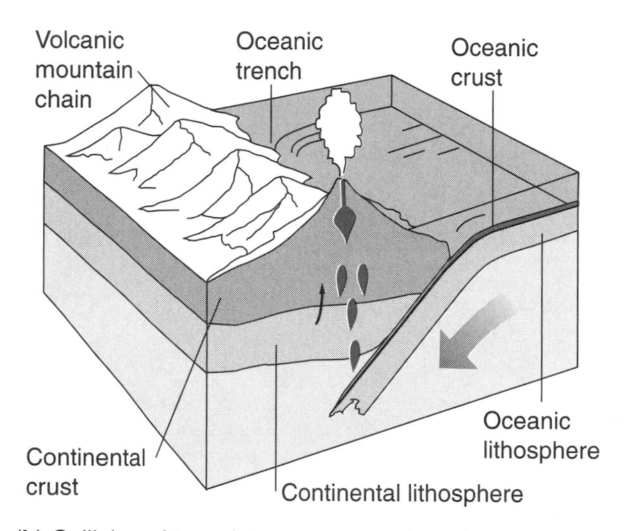

(**b**) Collision of two plates, one oceanic and one continental

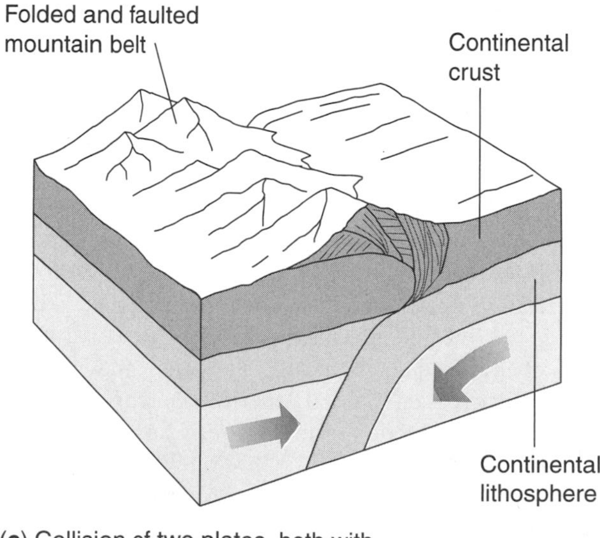

(**c**) Collision of two plates, both with continental lithosphere

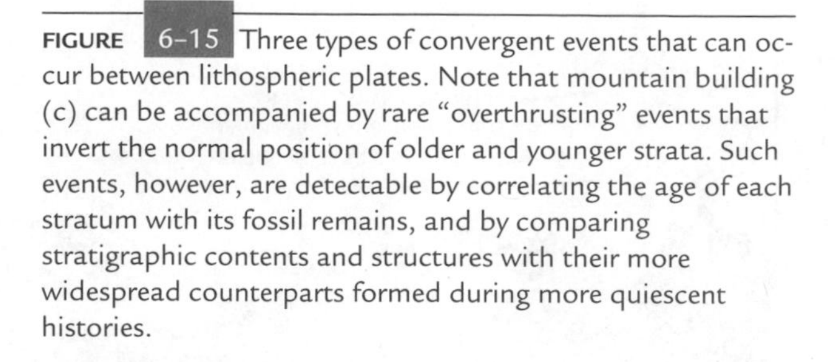

FIGURE 6-15 Three types of convergent events that can occur between lithospheric plates. Note that mountain building (c) can be accompanied by rare "overthrusting" events that invert the normal position of older and younger strata. Such events, however, are detectable by correlating the age of each stratum with its fossil remains, and by comparing stratigraphic contents and structures with their more widespread counterparts formed during more quiescent histories.

American extinctions in the Pliocene and Pleistocene epochs, and the absence of any eutherian fossils in Australia up to recent times. As Fooden pointed out, this view would also lead us to expect fossil remains of prototherians and metatherians in Upper Triassic deposits of other landmasses that separated from Pangaea concurrently with Australia, such as India and Antarctica. If paleontologists discovered such fossils, they would add considerably to the predictive value of the continental drift theory.

In summary, continental drift must have had profound biological effects. It subjected moving landmasses to new climatic conditions and geographical relationships, enabling their inhabitants to be selected for different evolutionary adaptations. Because of the breakup of land masses, drift must have also separated groups of organisms that were formerly associated, setting each such isolated species or group on its own evolutionary pathway. On the other hand, the joining of landmasses because of continental drift led to competition among previously separated groups of plants and animals that had evolved unique adaptations. These could then interact to produce increased complexity as well as extinction.

An important lesson that emerges from these events is the close connection between biological evolution and historical circumstances of all kinds. The preceding illustrations show evolutionary changes to be allied to geological changes, which can cause environmental effects of many varieties and descriptions, causing, in turn, many kinds of organismic interaction (pp. 451–452) and different evolutionary consequences. It is the tie between the development of biological systems and the impact of diverse historical circumstances that has essentially molded biology into a historical science.

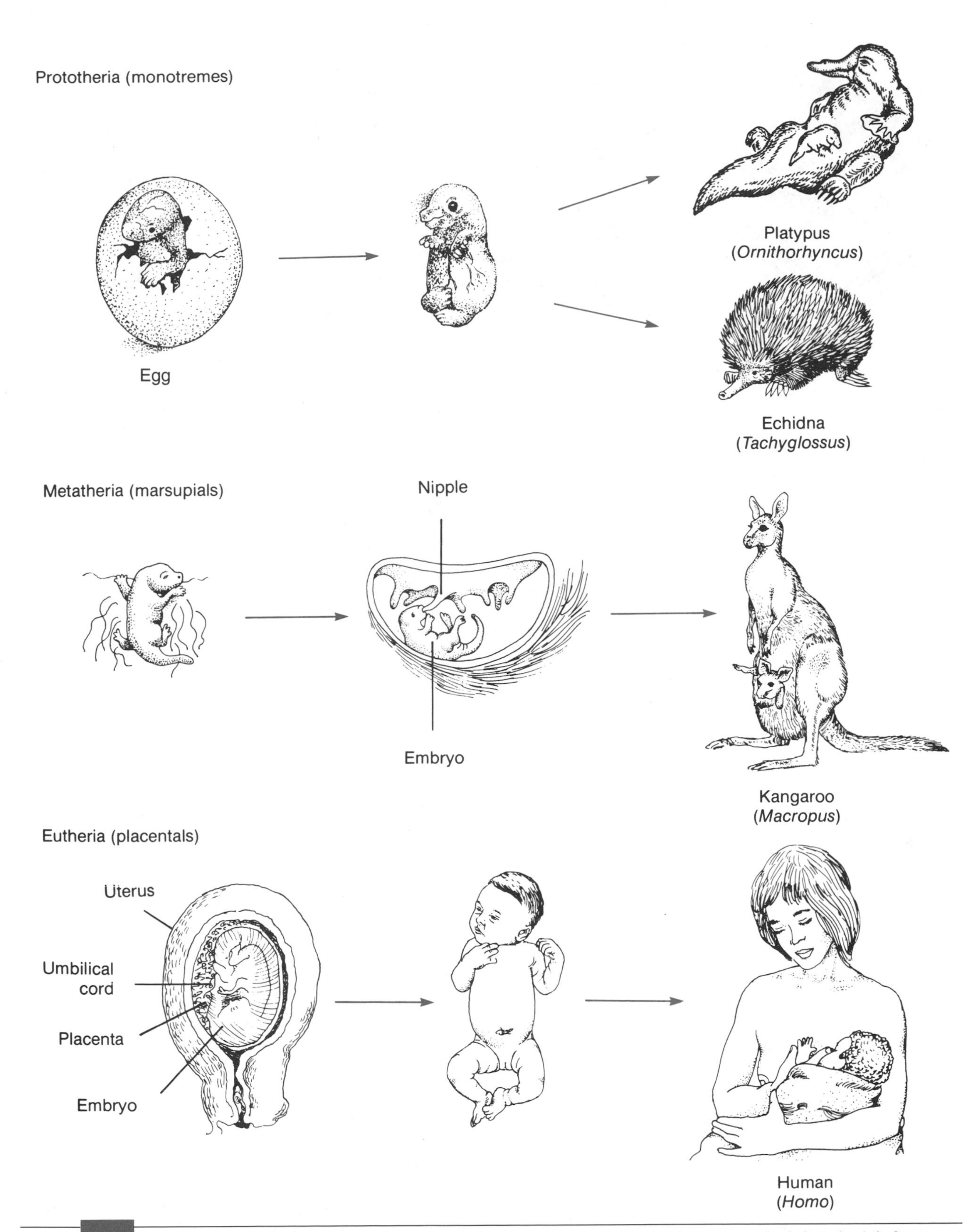

FIGURE 6-16 The three major groups of existing mammals, showing early developmental stages on the left and adult forms on the right. They all have hair, mammary glands, and other features that distinguish them from reptiles but differ from each other in their prenursing development. Prototheria (*a*) lay eggs whose embryos hatch out to attach onto maternal abdominal hairs connected to mammary glands. In Metatheria (*b*), developing eggs remain in utero, and the emerging offspring then climb into the maternal pouch and attach themselves to mammary nipples. Eutherians (*c*) maintain the fetus in utero until a relatively late stage of development.

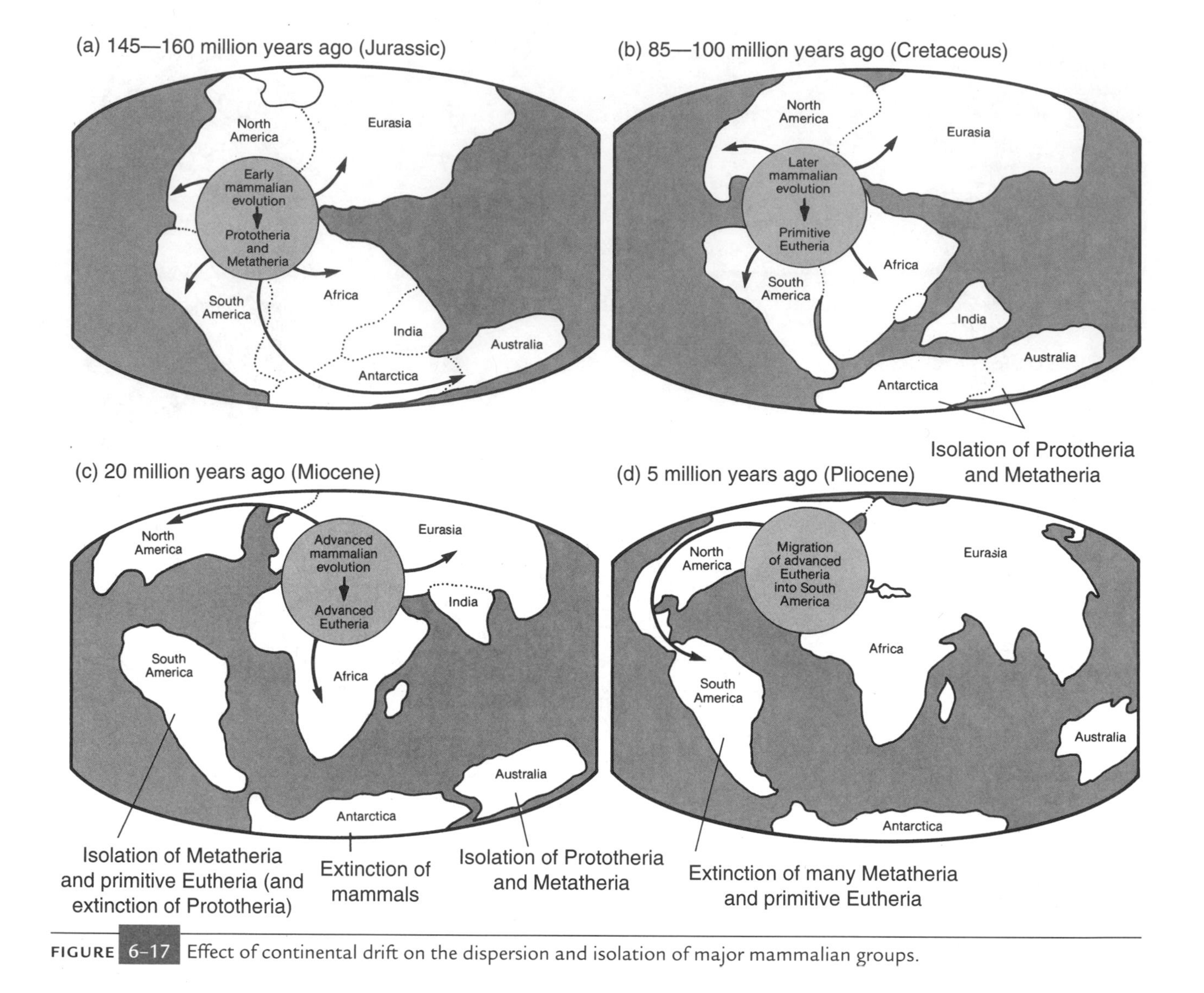

FIGURE 6-17 Effect of continental drift on the dispersion and isolation of major mammalian groups.

SUMMARY

Our solar system most probably originated by condensation from a rotating mass of gas about 4.6 million years ago. The matter from which it condensed was unevenly distributed, and the central mass became the sun; the peripheral masses, the planets.

During this condensation, elements scattered according to a density gradient, with the outer planets receiving the lighter elements and the inner ones capturing the heavier ones. On Earth an atmosphere arose through the emission of hydrogen from the interior. Subsequently many gaseous compounds such as ammonia, water, methane, carbon dioxide, and nitrogen appeared, resulting in a reducing atmosphere with little or no oxygen. The high proportions of oxygen in the Earth's present atmosphere probably arose from the actions of photosynthetic organisms.

The interior of the Earth consists of several concentric layers differing in composition and physical properties. The core is mainly molten iron and nickel and is surrounded by a thick, partially molten mantle. A thin crust of rock blankets the mantle and is composed of various types of rock: igneous, sedimentary, and metamorphic. The distribution of rocks in layers or strata accords with the age of the rock, and in the eighteenth and nineteenth centuries geologists found that they could recognize different strata by the types of fossils each contains. However, exact dating of strata had to await the discovery of radioactive elements, such as uranium, whose disintegration rates could be used for this purpose. Using such techniques, geologists have determined the age of the Earth to be 4.6 billion years.

Wegener's proposal, early in the twentieth century, that the present land masses of the Earth had once been united as one huge continent, Pangaea, has since garnered much support in the form of the compatible profiles of continents, the similarity of rocks and fossils in previously conjoined areas, the direction of magnetism in rocks, and the youthful structure of the ocean floor. Pangaea began to disintegrate about 225 million years ago when one section, Gondwana, broke into South America–Africa and India–Antarctica–Australia, and both separated from Laurasia (North America–Eurasia). Other rifts and fusions created the present-day continents. We can explain the movements of the continents and many features of the Earth's surface, such as earthquake belts, by the movement of at least eight "plates" comprising the Earth's crust. These plates can separate from each other, slide past each other, or converge on each other. Researchers call such activity *plate tectonics,* and explain the distribution of many organisms on the basis of the historical fusion and separation of these huge landmasses. Most notable is the way in which the separation of part of Gondwana from the rest of Pangaea restricted monotreme mammals to Australia and marsupial mammals to both Australia and South America, while elsewhere placental mammals replaced them. Many other examples also indicate the biological effects of alterations in the Earth's surface on evolutionary patterns.

KEY TERMS

asthenosphere
collision theory
condensation theory
continental drift
core
crust
Eutheria
Gondwana
igneous rocks
law of superposition
lithosphere
mantle
metamorphic rocks
Metatheria
oxidizing
paleomagnetism
Pangaea
Phanerozoic time scale
protoplanets
Prototheria
radioactive dating
reducing
sea floor spreading
sedimentary rocks
seismic waves
tectonic plates

DISCUSSION QUESTIONS

1. What is the prevailing theory for the origin of the solar system?
2. Reduction and oxidation
   a. How do these two modes of chemical reaction differ?
   b. Was the Earth's early atmosphere primarily reducing or oxidizing? What evidence can you offer?
3. Geological structures
   a. What are the general compositions of the major types of rocks?
   b. How do rocks transform from one type to another?
4. Geological dating
   a. How did scientists date geological strata in the nineteenth century?
   b. How do geologists perform radioactive dating?
   c. How can geologists use igneous rocks to date sedimentary rocks? (How do they use the rules of superposition and crosscutting relationships for this purpose?)
5. Geological time scales
   a. What is the sequence of eras in the Phanerozoic time scale?
   b. What sequence of geological periods do scientists ascribe to each of these eras?
6. Continental drift
   a. What evidence supports continental drift? (Explain terms such as *continental fit, Gondwana, Pangaea, paleomagnetism,* and *sea floor spreading.*)
   b. How do geologists delineate tectonic plates, and how do these plates interact?
   c. What major changes in the positions of continents occurred during the Mesozoic and Cenozoic eras? (North America, Eurasia, Africa, South America, India, Antarctica, and Australia)
   d. How did continental drift affect the distribution and evolution of the major mammalian subgroups?

EVOLUTION ON THE WEB

Explore evolution on the web! Visit the accompanying web site for *Evolution,* 3/e at www.jbpub.com/evolution for web exercises and links relating to topics covered in this chapter.

REFERENCES

Allégre, C., 1992. *From Stone to Star: A View of Modern Geology.* Harvard University Press, Cambridge, MA.

Bullard, E., J. E. Everett, and A. G. Smith, 1965. The fit of the continents around the Atlantic. *Proc Trans. Roy. Soc. Lond.,* **A258,** 41–51.

Chyba, C. F., T. C. Owen, and W.-H. Ip, 1995. Impact delivery of volatiles and organic molecules to Earth. In *Hazards Due to Comets and Asteroids,* T. Gehrels (ed.). University of Arizona Press, Tucson, pp. 9–58.

Clodd, E., 1988. *Story of Creation.* Longmans Green, London.

Cloud, P., 1978. *Cosmos, Earth, and Man: A Short History of the Universe.* Yale University Press, New Haven, CT.

Colbert, E. H., 1973. *Wandering Lands and Animals.* Hutchinson, London.

Condie, K. C., 1997. *Plate Tectonics and Crustal Evolution,* 4th ed. Pergamon, New York.

Condie, K. C., and R. E. Sloan, 1998. *Origin and Evolution of Earth: Principles of Historical Geology.* Prentice Hall, Upper Saddle River, NJ.

Conway Morris, S., 1998. *The Crucible of Creation.* Oxford University Press, New York.

Cox, C. B., and P. D. Moore, 1980. *Biogeography,* 3d ed. Blackwell, Oxford, England.

Dalrymple, C. B., 1991. *The Age of the Earth.* Stanford University Press, Stanford, CA.

Dalziel, I. W. D., 1995. Earth before Pangea. *Sci. Amer.*, **272**(1), 58–63.

Darwin, C., 1838. On the connexion of certain volcanic phaenomena, and on the formation of mountain-chains and volcanos, as the effects of continental elevations. *Proc. Geol. Soc. Lond.*, **2**(56), 654–660.

Dietz, R. S., and J. C. Holden, 1970. Reconstruction of Pangaea: Breakup and dispersion of continents, Permian to present. *J. Geophys. Res.*, **75**, 4939–4956.

Drake, F., 1961. Project Ozma. *Phys. Today,* **14**(4), 40–46

———, 1962. *Intelligent Life in Space.* Macmillan, New York.

———, 1965. The radio search for intelligent extraterrestrial life. In *Current Aspects of Exobiology*, G. Mamikunian and M. H. Briggs (eds.). Pergamon Press, Oxford, England, pp. 323–345.

Eicher, D. L., and A. L. McAlester, 1980. *History of the Earth.* Prentice Hall, Englewood Cliffs, NJ.

Emiliani, C., 1992. *Planet Earth: Cosmology, Geology, and the Evolution of Life and Environment.* Cambridge University Press, Cambridge, England.

Ferris, T., 1997. *The Whole Shebang: A State of the Universe(s) Report.* Simon and Schuster, New York.

Field, G. B., G. L. Verschuur, and C. Ponnamperuma, 1978. *Cosmic Evolution: An Introduction to Astronomy.* Houghton Mifflin, Boston.

Fooden, J., 1972. Breakup of Pangaea and isolation of relict mammals in Australia, South America, and Madagascar. *Science,* **175**, 894–898.

Glanz, J., 1997. Worlds around other stars shake planet birth theory. *Science,* **276,** 1336–1339.

Hallam, A., 1989. *Great Geological Controversies,* 2d ed. Oxford University Press, Oxford, England.

———, 1994. *An Outline of Phanerozoic Biogeography.* Oxford University Press, Oxford, England.

Harland, W. B., A. V. Cox, P. G. Llewellyn, C. A. G. Pickton, A. G. Smith, and R. Walters, 1982. *A Geologic Time Scale.* Cambridge University Press, Cambridge, England.

Harrison, A. A., 1997. *After Contact: The Human Response to Extraterrestrial Life.* Plenum Press, New York.

Hart, M. H., 1995. Atmospheric evolution, the Drake equation and DNA: Sparse life in an infinite universe. In *Etraterrestrials, Where Are They?* 2d ed., B. Zuckerman and M. H. Hart (eds.). Cambridge University Press, Cambridge, England, pp. 215–225.

Horowitz, P., 1997. Extraterrestrial intelligence: The search programs. In *Carl Sagan's Universe,* Y. Terzian and E. Bilson (eds.). Cambridge University Press, Cambridge, England, pp. 98–120.

Hughes, D. W., 1989. Planetary planarity. *Nature,* **337,** 113.

Irving, E., 1977. Drift of the major continental blocks since the Devonian. *Nature,* **270,** 304–309.

Jardine, N., and D. McKenzie, 1972. Continental drift and the dispersal and evolution of organisms. *Nature,* **234,** 20–24.

LeGrand, H. E., 1988. *Drifting Continents and Shifting Theories: The Modern Revolution in Geology and Scientific Change.* Cambridge University Press, Cambridge, England.

Lunine, J. I., 1999. In search of planets and life around other stars. *Proc. Nat. Acad. Sci.,* **96,** 5353–5355.

Mayr, E., 1995. The search for extraterrestrial intelligence. In *Extraterrestrials, Where Are They?* 2d ed., B. Zuckerman and M. H. Hart (eds.). Cambridge University Press, Cambridge, England, pp. 215–225.

McKay, C. P., 1996. The origin and evolution of life in the universe. In *The Origin and Evolution of the Universe,* B. Zuckerman and M. A. Malkan (eds.). Jones and Bartlett, Sudbury, MA, pp. 924–930.

McKay, D. S., et al., 1996. Search for past life on Mars: Possible relic biogenic activity in Martian meteorite ALH84001. *Science,* **273,** 924–930.

Parker, B., 1998. *Alien Life: The Search for Extraterrestrials and Beyond.* Plenum Press, New York.

Piper, J. D. A., 1974. Proterozoic crustal distribution, mobile belts and apparent polar movements. *Nature,* **251,** 381–384.

———, 1987. *Paleomagnetism and the Continental Crust.* Wiley, New York.

Raven, P. H., and D. I. Axelrod, 1972. Plate tectonics and Australasian paleobiogeography. *Science,* **176,** 1379–1386.

Rogers, J. J. W., 1993. *A History of the Earth.* Cambridge University Press, Cambridge, England.

Scientific American, 1976. Introduction by J. T. Wilson. *Continents Adrift and Continents Aground.* Freeman, San Francisco.

Taylor, S. R., 1998. *Destiny or Chance: Our Solar System Evolution and Its Place in the Cosmos.* Cambridge University Press, Cambridge, England.

Tipler, F. J., 1985. Extraterrestrial beings do not exist. In *Extraterrestrials: Science and Alien Intelligence,* E. Regis, Jr. (ed.). Cambridge University Press, Cambridge, England.

Van Andel, T. H., 1994. *New Views on an Old Planet,* 2d ed. Cambridge University Press, Cambridge, England.

Wetherill, G. W., 1988. Formation of the Earth. In *Origins and Extinctions,* D. E. Osterbrock and P. H. Raven (eds.). Yale University Press, New Haven, CT, pp. 43–82.

———, 1995. How special is Jupiter? *Nature,* **373,** 470.

Windley, B. F., 1995. *The Evolving Continents,* 3d ed. Wiley, London.

Wyllie, P. J., 1976. The Earth's mantle. In *Continents Adrift and Continents Aground,* Scientific American. Freeman, San Francisco, pp. 46–57.

Ziegler, A. M., C. R. Scotese, W. S. McKerrow, M. E. Johnson, and R. K. Bombach, 1979. Paleozoic paleogeography. *Ann. Rev. Earth Planet. Sci.,* **7,** 473–502.

7 Molecules and the Origin of Life

The problem of how life could have originated by ordinary chemical means long seemed insuperable. This is hardly surprising if we consider the present complexity of even the most elementary unit of life, the cell. At its simplest, a modern cell (Fig. 7–1) is surrounded by a highly selective permeable membrane composed of lipids and proteins that regulates the kinds of substances that pass through. Within the cell, the cytoplasm consists of a multitude of structures and substructures involved in the synthesis, storage, and breakdown of a large variety of chemical compounds.

Amino Acids

Foremost among the metabolic agents that enable the cell to function are the many different proteins that catalyze and regulate practically all living chemical reactions. In basic structure, proteins consist of subunits, called **amino acids,** that have the following features:

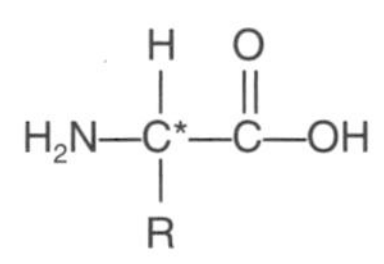

- An alpha carbon atom (C*) to which all other parts are attached
- An amino NH_2 group with a potential positive charge (NH_3^+)
- A carboxyl COOH group with a potential negative charge (COO^-)
- An H atom
- An R side chain that varies in structure among the different amino acids (Fig. 7–2)

These amino acids link together by chemical bonds called **peptide linkages** (Fig. 7–11) into linear **polypeptide chains** that are the constituents of proteins. The

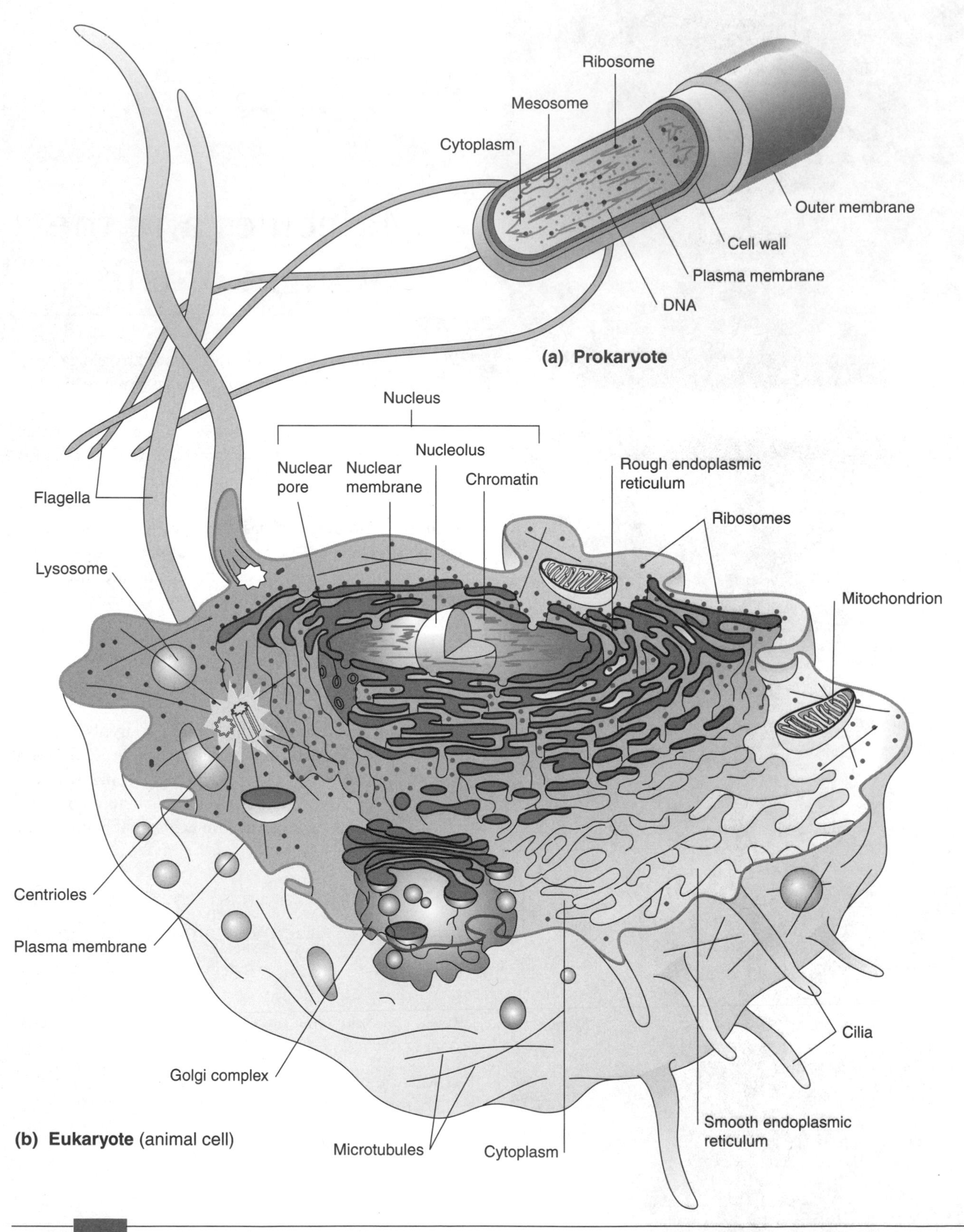

FIGURE 7–1 Diagrammatic representation of generalized prokaryotic and eukaryotic cells showing cross sections through various important cellular organelles.

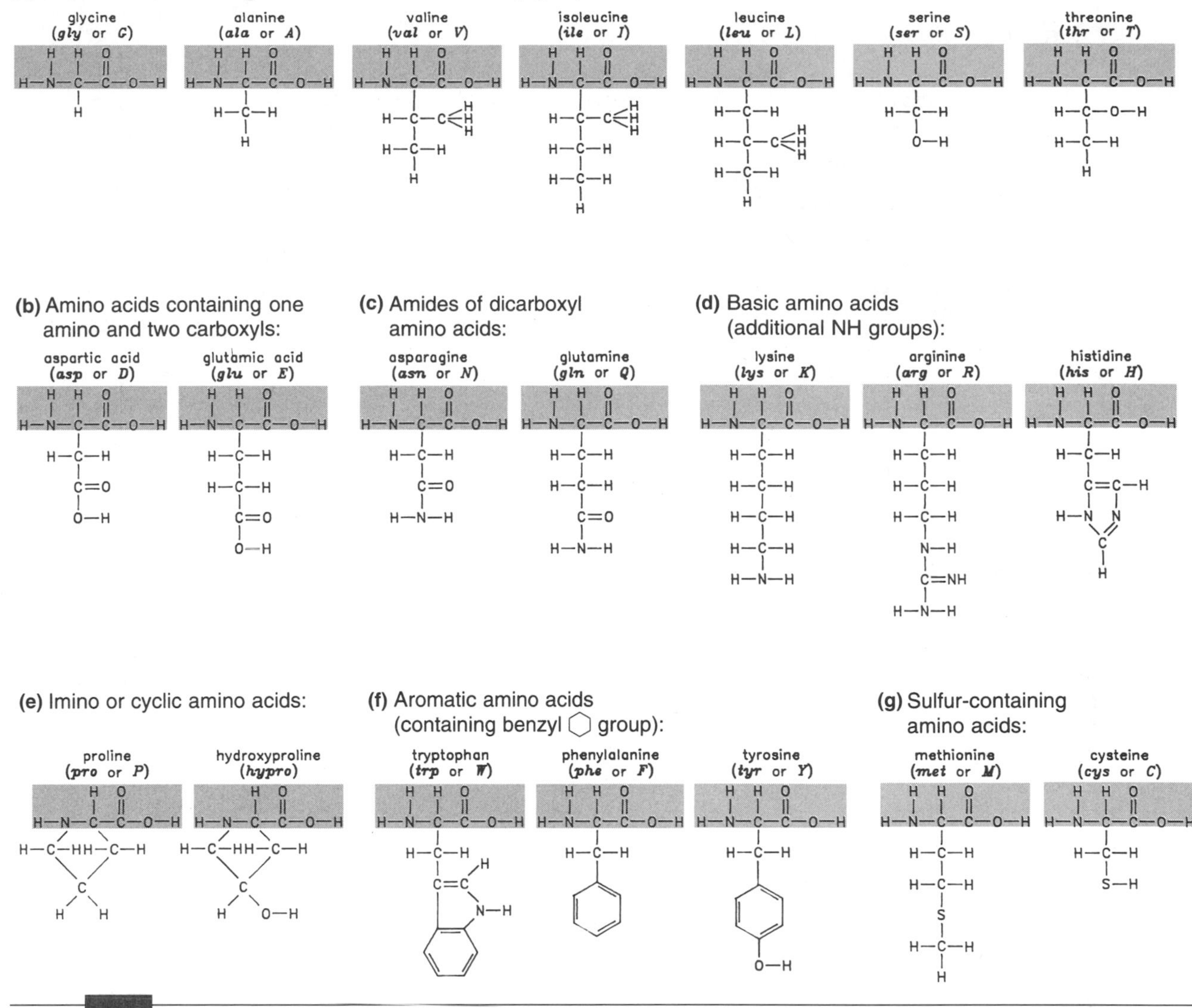

FIGURE 7–2 Structures of the common amino acids found in proteins.

highly specific structure of any protein molecule, whether it functions as an **enzyme** (catalyst) or for some other purpose, derives from the exact linear placement of its various component amino acids. These specific amino acid sequences enable polypeptide chains to fold into specific three-dimensional forms that confer specific properties on proteins. It is reasonable to claim that most present living phenomena, whether absorption, sensation, motion, structure, or whatever, derive from the enzymatic and regulatory activities of these long sequences of amino acids. Complexity, however, is not limited to proteins, since the amino acid sequences of proteins are actually determined by the nucleotide sequences in another group of basic molecules, nucleic acids.

Nucleic Acids

Nucleic acids, as Figure 7–3 shows, are long-chained molecules composed of **nucleotide** subunits, each containing a pentose (5-carbon) sugar, a monophosphate group, and a nitrogenous base. The two kinds of sugar used in nucleic acids, ribose (hydroxylated at the 2' carbon position) and deoxyribose sugar (lacks 2' hydroxyl), provide the names for the two kinds of nucleic acids, ribonucleic acid (RNA) and deoxyribonucleic acid (DNA). In both these nucleic acids, the phosphate groups occupy the same position, tying the 3' carbon of one sugar to the 5' carbon of its neighbor via a phosphodiester bond. Connected to the 1'carbon

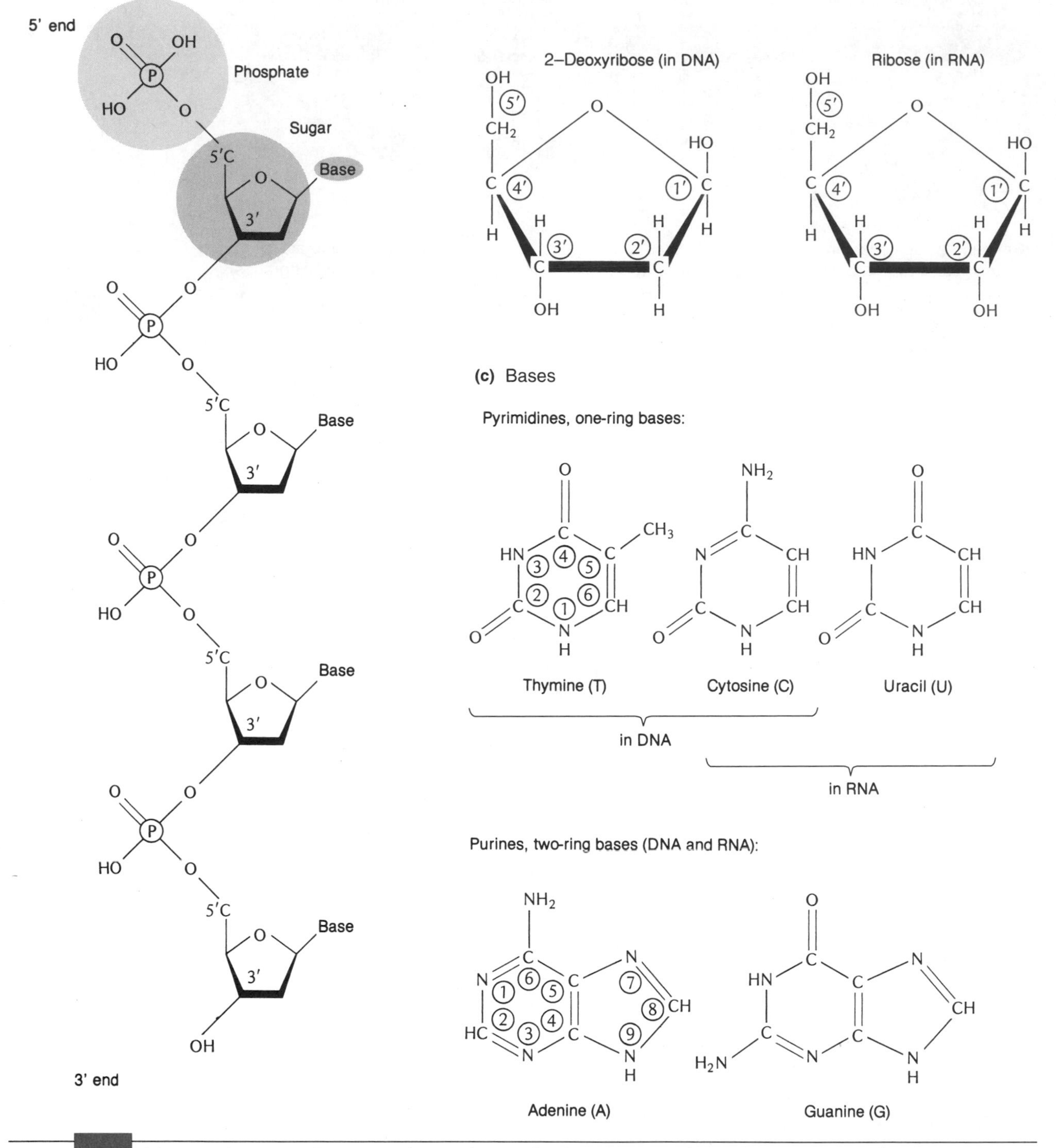

FIGURE 7-3 (*a*) General structure for DNA and RNA chains. The chains, or strands, may be of considerable length and are composed of a linear sequence of nucleotides. Each nucleotide consists of a phosphate group, a sugar, and a nitrogenous base, linked together in the manner shown. (*b*) Differences between the sugars found in DNA (deoxyribose) and RNA (ribose). (*c*) The basic kinds of nitrogenous bases found in DNA (T, C, A, G) and RNA (C, U, A, G).

of each sugar is one of four kinds of nitrogenous heterocyclic bases, of which two are purines [A (adenine), and G (guanine), in both DNA and RNA] and two are pyrimidines [C (cytosine), and T (thymine), in DNA, and C (cytosine), and U (uracil), in RNA].

Since the complexity of proteins derives from the complexity of nucleic acids, you might think that the restriction of nucleic acid composition to only four different kinds of bases would limit the message-bearing capacity of these molecules to only four kinds of messages, but it doesn't. The fact that nucleic acid molecules may be many thousands or millions of nucleotides long, and that each message can be encoded by a unique linear sequence of nucleotides, endow these molecules with the capacity to carry an immense variety of highly complex messages. For any one nucleotide position, four different messages are possible (A, G, C, or T); for two nucleotides in tandem, 4^2 or 16 different messages are possible (AA, AG, AC, AT, GG, GC, . . .); and so on: the rule being simply that for a linear sequence of n nucleotides, 4^n different possible messages can be encoded. Thus a linear sequence of only 10 nucleotides can discriminate among more than 1 million (4^{10}) potentially different messages.

All this helps explain the information-carrying role of nucleic acids but does not explain how they replicate and transmit this information. The model presently accepted for nucleic acid replication derives from the now-familiar **double helix** structure that Watson and Crick first offered. The DNA double helix consists of two antiparallel strands coiled around each other in the form of a right-hand screw, with complementary pairing between purine bases on one strand and pyrimidine bases on the other (A-T, G-C). In the familiar B form of DNA, diagrammed in Figure 7–4, there are approximately 10 base pairs for each complete turn of the helix, and the bases are stacked almost perpendicularly to the helical axis.[1]

The replication power of the DNA double helix derives from the ability of each of the two strands to serve as a template for a newly complementary strand, so that two new double helices can form bearing nucleotide sequences that are identical to each other as well as to that of the parental molecule (Fig. 7–5). This unique quality of exact molecular replication, enabling similar messages to be transmitted from generation to generation, confers on nucleic acids their function as "genetic material."

Fundamental to our understanding of the relationship between genetic material and protein is an important concept: the three-dimensional structure of a protein—its form, shape, and subsequent function—is primarily determined by the linear sequence of amino acids of which it consists. This linear sequence of amino acids in turn derives from the linear sequence of bases in nucleic acids by means of a protein-synthesizing apparatus involving three different kinds of RNA.

In brief, as diagrammed in Figure 7–5, the genetic material, through the process of transcription, produces a molecule of **messenger RNA (mRNA)** that is, base for base, a complement to the bases on one of the DNA strands. Through the mediation of ribosomes, which themselves consist of **ribosomal RNA (rRNA)** and protein, a sequence of bases in mRNA then translates into a sequence of amino acids. This translation follows the triplet rule that a sequence of three mRNA bases designates 1 of the 20 different kinds of amino acids used in protein synthesis.

Note that during translation, no physical material is actually inserted by the mRNA into the protein; only information is transferred. That is, the mRNA only designates the linear position in which each amino acid is to be placed by special molecules of **transfer RNA (tRNA)** that bring the amino acids to the messenger (Fig. 7–6). With the aid of the ribosome and various enzymes, the amino acids then connect in sequence through peptide linkages. A polypeptide chain forms in which the genetic material has ultimately designated the precise position of each component amino acid. DNA (or RNA in some viruses) provides the **genotype,** or genetic endowment of an organism. The expression of this nucleic acid information, via transcription and/or translation, provides the various aspects of an organism's appearance, or **phenotype.**

The presently observed circular interdependency of all these events, such as the replication of nucleotides because of the presence of appropriate enzymes and the determination of enzyme structure because of the presence of appropriate nucleotide sequences, points to certain difficulties in finding a reasonable explanation for the origin of life. Which of the many biochemical agents came first? How did they arise? How could they have functioned before an entire cellular structure formed? Many proposals exist for the origin of life, and we can generally divide such proposals into two broad categories: life on Earth developed from previous life; life on Earth originated by chemical means.

Life Only from Prior Life

The concept that life did not originally arise on Earth is embodied in almost all human creation myths. These myths usually presume the special creation of life on Earth by one or more superior, intelligent, and all-powerful beings who themselves possess attributes of life such as sensation, thought, and purposive movement (Chapter 4). This concept explains the origin of terrestrial life in a simple fashion that is especially attractive to those who believe that conscious agents govern natural events. It does not explain the source of the initial creator and therefore does not explain life's origin.

Another ancient concept is that life can arise spontaneously at any time (Chapter 1), such as the presumed origin of insects from sweat and crocodiles from mud. Strangely enough, this view was often held simultaneously

[1]DNA takes various structural forms depending on relative humidity, salt concentration, and so on. Rosalind Franklin, an X-ray crystallographer whose photographs provided Watson and Crick essential information in devising their model, gave the first two DNA forms investigated, crystalline (dry) and wet (hydrated), the respective names A and B. In the A form, each turn of the helix has 11 bases, and the bases are noticeably tilted (20°) relative to the helical axis.

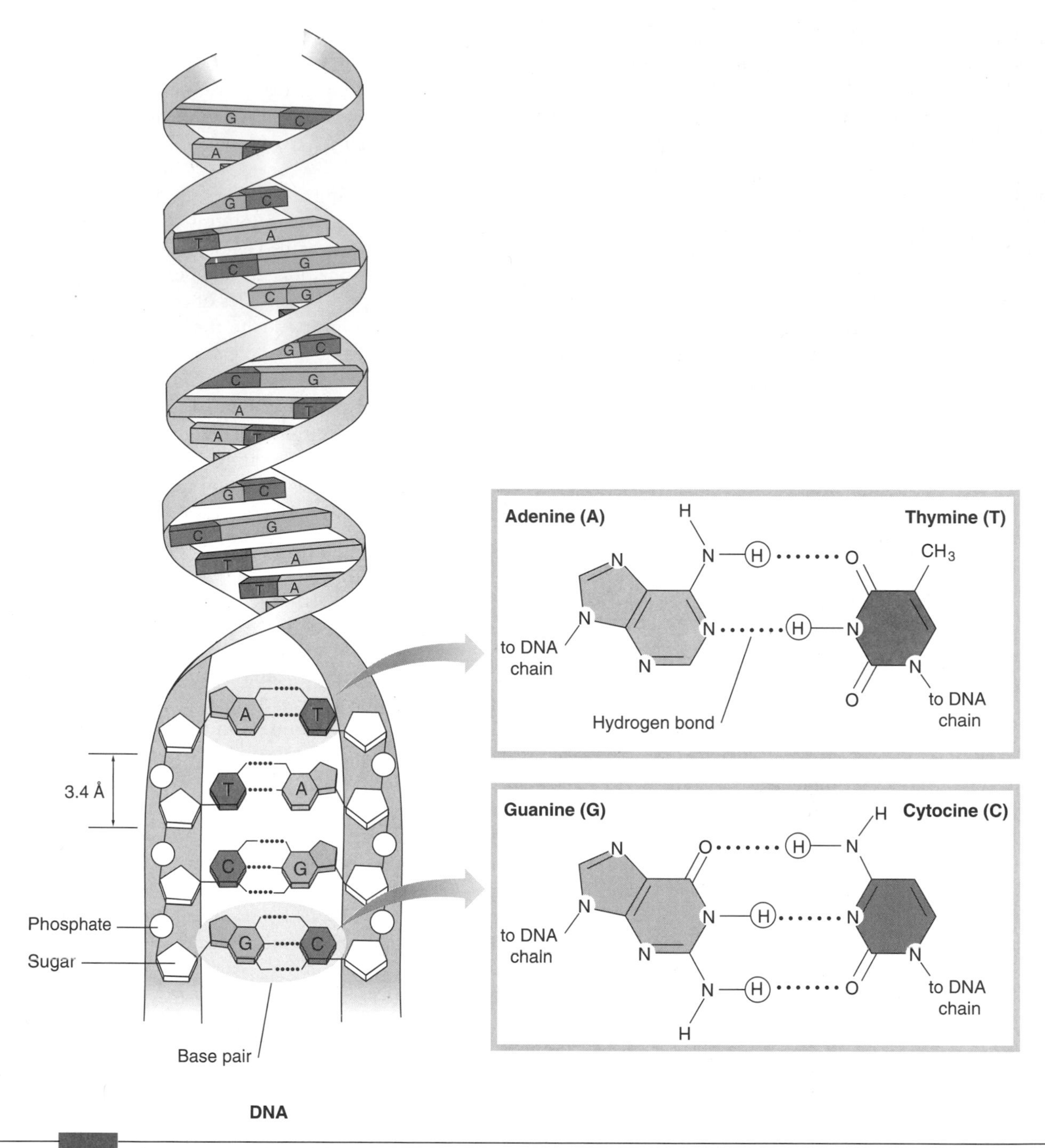

FIGURE 7-4 The Watson-Crick model of the standard (B form) DNA double helix on the left, with examples of hydrogen-bond pairing between bases on the right (A–T, G–C). The two strands of the double helix are bridged by parallel rows of such paired nucleotide bases stacked at regular 3.4 angstrom intervals—1 angstrom = 0.0001 micrometer (μm) = 0.0000001 millimeter (mm).

with the view that life derives from a conscious creator, and was popular in Europe throughout medieval times. Pasteur and others put spontaneous generation theory to rest in the 1860s and it has not since been revived in its original form. As we shall see later, the modern concept of the spontaneous origin of life on Earth does not include such simple means as the immediate action of sunlight on liquid or clay but proposes instead the past existence of more complex yet more understandable biochemical processes.

One variation of the theory that life comes only from life is the proposal that life is somehow ingrained in all matter, and the creation of matter by whatever cause is responsible for the creation of life. Although this notion offers the advantage of ascribing life to a natural event, its origin would seem difficult if not impossible to understand. Obviously, many aspects of matter show no evidence of life if we define life to include those attributes possessed by terrestrial organisms, such as metabolism and reproduction. How did these attributes arise?

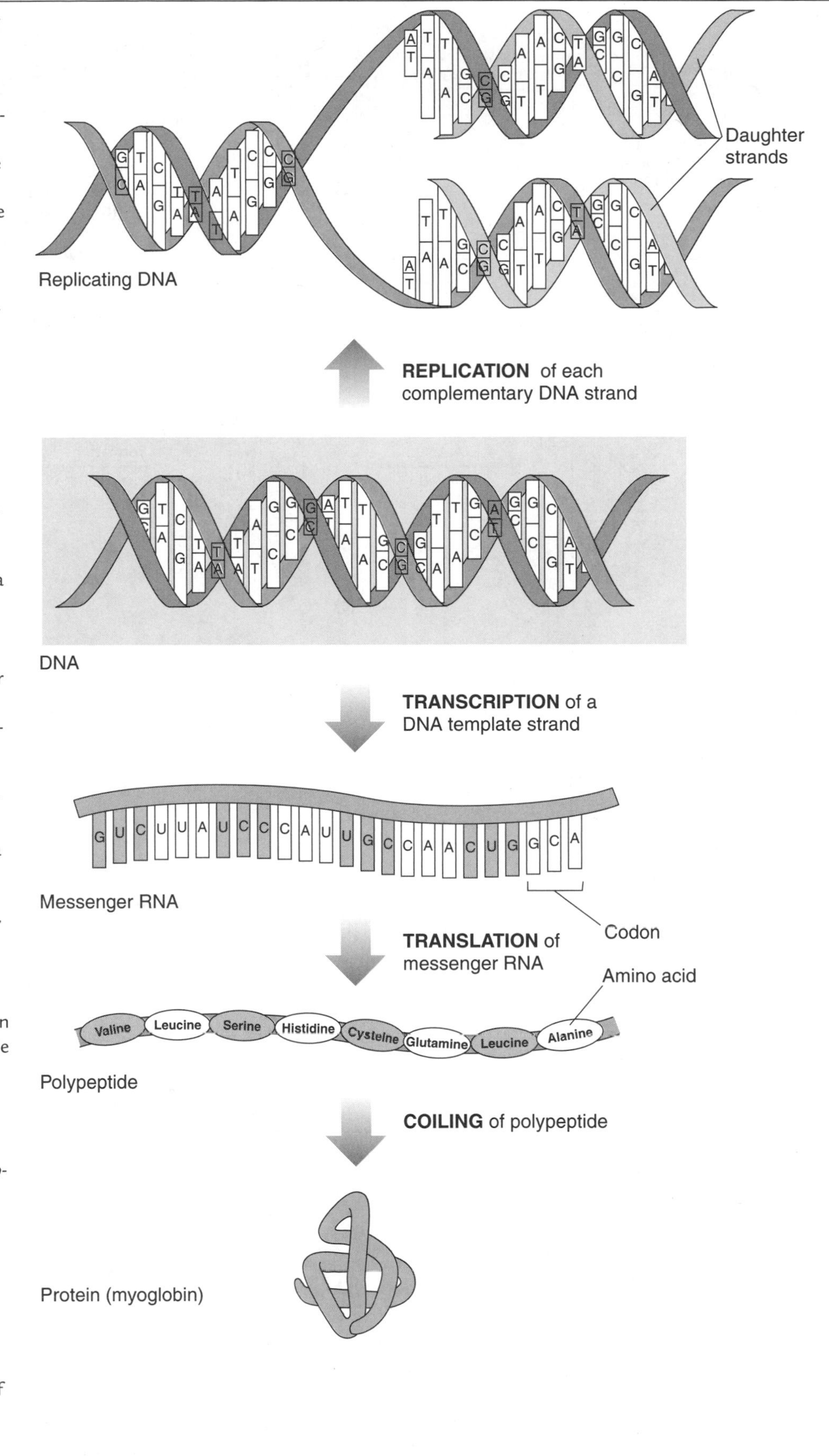

FIGURE 7-5 Diagrammatic illustration of how DNA replicates and how information transfers from DNA to RNA to protein. In DNA replication, special proteins break the hydrogen bonds between paired bases, allowing the two strands to unwind. Each unwound strand then acts as a template producing a new complementary strand through base pairing catalyzed by a DNA polymerase enzyme. As a result, two double-stranded DNA molecules form, each an exact replica of the original parental double helix. In transcription, one of the two DNA strands (*dark gray color*) serves as a template upon which a molecule of messenger RNA is transcribed (*light gray color*). This messenger then serves in turn as a template on which a molecule of protein is translated. Note that the messenger RNA is exactly complementary to its DNA template and that a sequence of three nucleotides (a triplet codon) on the messenger specifies one amino acid (for example, 24 nucleotides = 8 amino acids). The amino acids in the illustrated polypeptide chain coil into a right handed α-helix that enables this particular molecule to fold into a special protein called *myoglobin* used in the cellular transport of oxygen. Other proteins have, of course, different amino acid sequences, which may form different kinds of polypeptide structures such as β-pleated sheets and which may consist of two or more polypeptide chains.

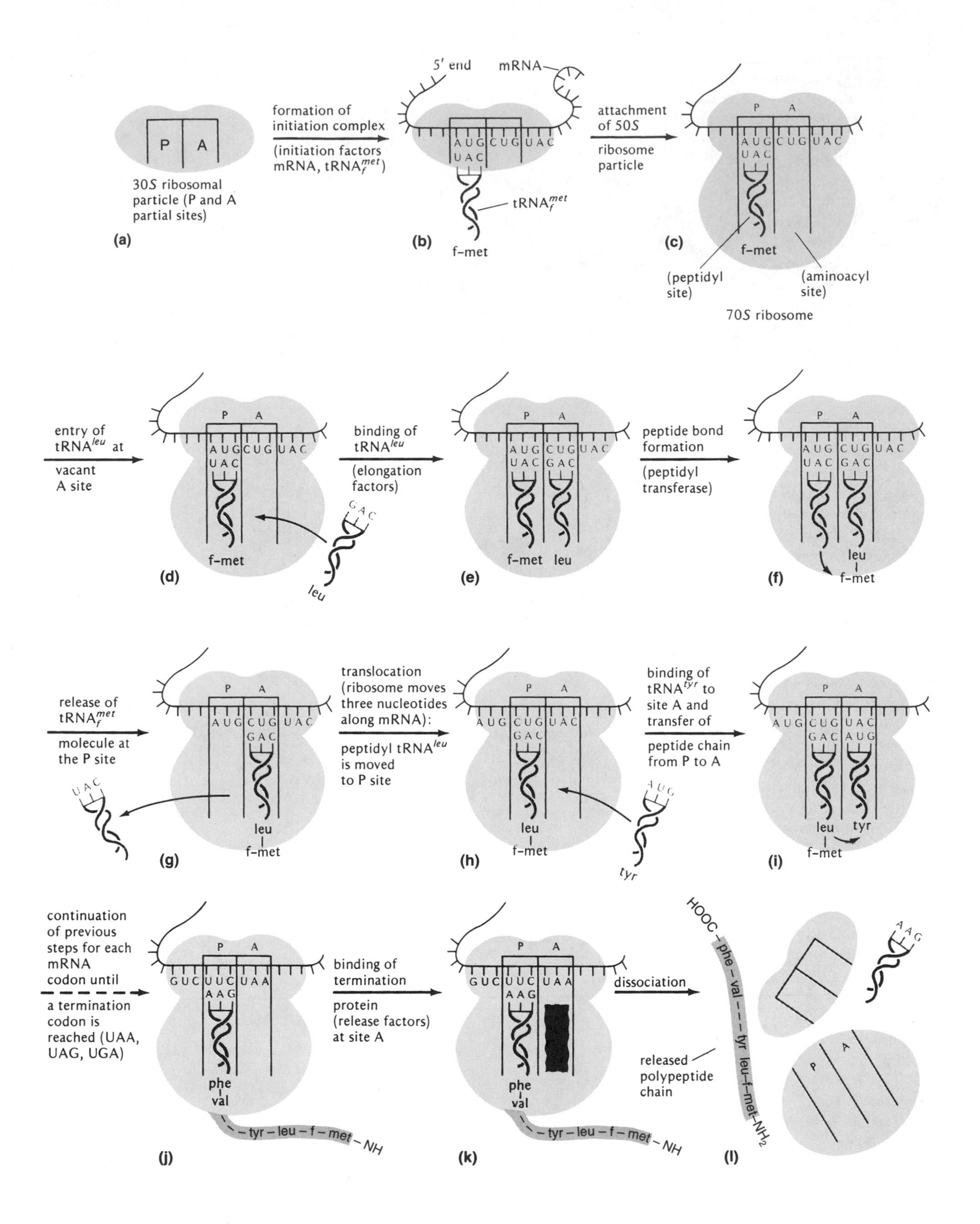
30S ribosomal particle (P and A partial sites)
(a)
formation of initiation complex
(initiation factors mRNA, tRNA$_f^{met}$)
5′ end
mRNA
tRNA$_f^{met}$
f-met
(b)
attachment of 50S
ribosome particle
(c)
f-met
(peptidyl site)
(aminoacyl site)
70S ribosome
entry of tRNA^{leu} at
vacant A site
(d)
binding of tRNA^{leu}
(elongation factors)
(e)
f-met leu
peptide bond formation
(peptidyl transferase)
(f)
release of tRNA$_f^{met}$
molecule at the P site
(g)
translocation (ribosome moves three nucleotides along mRNA):
peptidyl tRNA^{leu} is moved to P site
(h)
binding of tRNA^{tyr} to site A and transfer of
peptide chain from P to A
(i)
continuation of previous steps for each mRNA codon until
a termination codon is reached (UAA, UAG, UGA)
(j)
binding of termination
protein (release factors) at site A
(k)
dissociation
released polypeptide chain
HOOC – phe – val – – – tyr leu–f–met–NH$_2$
(l)

FIGURE 7-6 General scheme of protein synthesis in the bacterium *Escherichia coli.* The ribosome consists of two subunits designated by their rates of sedimentation in a centrifuge as 30*S* and 50*S*. Each subunit in turn consists of various proteins and molecules of ribosomal RNA. (*a*) The 30*S* ribosomal subunit bears two partial sites, peptidyl (P) and acceptor (A), which become functionally complete only when combined with the larger 50*S* subunit. (*b*) In the presence of mRNA and necessary protein initiation factors, only the special initiator transfer RNA, $tRNA_f^{met}$, bound to a formyl-methionine amino acid (f-met) occupies the partial P site opposite the mRNA initiation codon. (*c*) The 30*S* and 50*S* subunits then join, accompanied by cleavage of the phosphate-bond energy donor, guanosine triphasphate (GTP), releasing guanosine diphosphate (GDP) and inorganic phosphate (P_i). (*d*) Once the 70*S* ribosome forms, the completed A site of the ribosome can then be entered by aminocyl-charged tRNAs [tRNAs that carry amino acids such as leucine (leu) and tyronsine (tyr)] whose anticodons can match the mRNA codon at that site. (*e*) Special protein elongation factors allow the binding of the appropriate aminoacyl-charged tRNA to the A site, and this step also accompanies cleavage of GTP to GDP and P_i. (*f*) The amino acid (or peptide) at the P site transfers to the amino acid at the A site through peptide bond formation (Fig. 7-11*a*) catalyzed by a peptidyl transferase enzyme located on the 50*S* subunit. As a result, the lowermost amino acid in this diagram is at the amino (NH_2) end of the chain, and the uppermost amino acid is at the carboxyl (COOH) end. (*g, h*) Once the peptide bond forms, the tRNA molecule that has donated its amino acid or peptide to the A site releases from the P site. Simultaneously, the ribosome translocates along the mRNA molecule for a distance of three nucleotides, thereby placing the former A-site tRNA (with its newly elongated peptide chain) in the P site. An additional protein elongation factor is needed for this to occur, and the reaction $GTP \rightarrow GDP + P_i$ again provides energy. This translocation step lets a new mRNA codon appear at the now vacant A site. (*i*) Transfer of the peptide chain from P-site tRNA to A-site tRNA again follows binding of aminoacyl-charged tRNA to the A site. (*j*) These steps are repeated as the ribosome moves along the mRNA molecule until it reaches a termination codon. (*k*) A protein release factor recognizes the termination codon at the A site and prevents further translocation of the ribosome along the mRNA molecule. (*l*) A termination reaction then releases the peptide chain from the P-site tRNA and expels the tRNA molecule from the ribosome. The ribosome then separates from the mRNA strand and dissociates into 30*S* and 50*S* subunits. (*From* Genetics Third Edition *by Monroe W. Strickberger. Copyright © 1985 by Monroe W. Strickberger. Reprinted by permission of Prentice Hall, Inc., Upper Saddle River, NJ.*)

Another variation suggests that life on Earth arose elsewhere, perhaps on a distant planetary body circling a distant star, and was then transported to Earth by radiation-resistant spores or other means. This notion, called **panspermia,** was fostered by the chemist Arrhenius (1859–1927) in the early part of this century and still has some adherents among scientists today (see, for example, Brooks and Shaw; Crick; and Hoyle and Wickramasinghe). It overcomes the difficulty of seeking a chemical explanation for the origin of life on Earth but does not, of course, explain the origin of life elsewhere.

Some proponents of the panspermia hypothesis point to the discovery of a number of different organic compounds in carbon-containing meteorites (**carbonaceous chondrites**), ranging from carbohydrates to amino acids. Looked at closely, however, the structural forms or isomers of amino acids in these carbonaceous chondrites possess the two different kinds of **optical activity** (dextro- and levorotary) in approximately equal amounts, therefore comprising a **racemic mixture** that shows little if any optical activity. In contrast, the amino acids of living forms generally show only one type of optical activity, levorotary (Fig. 7–7). Furthermore, a number of the amino acids in meteorites do not appear in proteins. Along with other observations, these findings indicate that organic compounds probably formed through random chemical reactions in the meteorite itself or in its parent body rather than through ordered living processes.

Opponents to the panspermia hypothesis also note the difficulty of envisaging how spores or "bugs" from outer space get to Earth without the help of conscious agents in spaceships. If the bug is too large, it cannot be easily

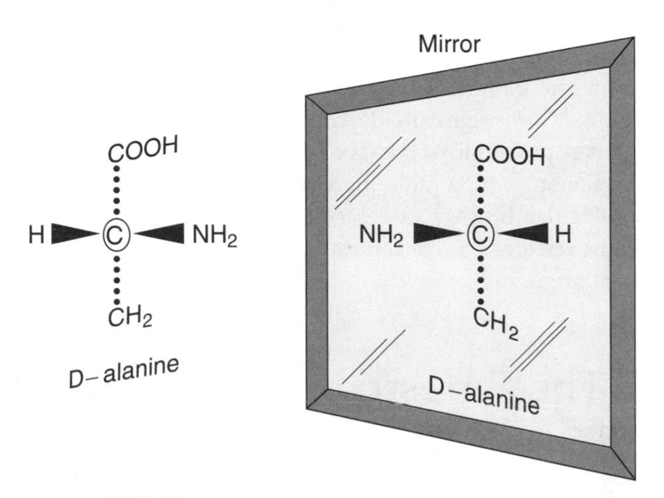

FIGURE 7-7 Structures of an L-form and D-form amino acid. The dark, wedge-shaped bonds indicate that the attached NH_2 and H groups project above the plane of the paper. Note that although D-alanine is a mirror image of L-alanine, they are not identical, since the two molecules cannot be superimposed on each other. Each stereoisomer has optical activity that can be measured by the extent to which it rotates polarized light. However, the two forms rotate light in opposite directions, and an equimolar mixture of the two is not optically active.

ejected from its home planet nor subsequently pushed out by sun radiation in its particular solar system. If it is too small, it will be kept from entering Earth's field by radiation pressure from our sun. Shklovskii and Sagan suggest that bugs larger than 0.6 μm cannot escape from Earth, while bugs smaller than 0.6 μm would be pushed away from Earth. So the donor planet must have been capable of ejecting bugs that we cannot eject (for example, about 1 μm); that is, the sun of the donor planet must have been very hot (high radiation pressure). But if such a sun were hot enough for this purpose, it would destroy ejected particles by radiation.

In addition, the hazards of interstellar travel are many. For example, ultraviolet radiation (UV) from our sun will kill ejected particles from Earth within about one day of interplanetary travel. If the bugs were shielded from UV, they would be too heavy to be ejected. Among other space hazards are the hot, ionized gases that surround early-type stars, the presence of cosmic rays, and the absorption of bugs into passing suns by gravitational attraction.

Although some writers assume that bugs could have survived such obstacles (see Parsons), difficulties still remain. A serious obstacle is the vastness of space that would disperse these spores so widely that their chance of reaching Earth seems infinitesimally small. Shklovskii and Sagan have calculated that 100 million life-bearing planets in our galaxy would each have had to eject about 1,000 tons of spores in order for Earth to have received a single microorganism during its first billion years of history. Such difficulties make questionable whether an event as seemingly rare as panspermia is more probable than the chemical origin of life on this planet: Why should the origin of life on other bodies have had a greater probability than its origin on Earth? Such misgivings, along with a rapid increase in our understanding of molecular biology and biochemistry, have influenced most scientists to concentrate their attention on a terrestrial origin of life.

The Terrestrial Origin of Life

The difficulty in visualizing life originating on Earth is essentially one of visualizing the molecular environment and events that occurred in a long-distant past. Unfortunately, we have as yet no certainties about the details of our molecular past, and we may never have such knowledge because molecular fossils are indeed sparse. At best, we can try to deduce the general nature of some of the original molecular events from present living structures and reactions and try to reconstruct such events experimentally under controlled conditions. However, before undertaking such a molecular review, it is important to consider the framework in which people usually pose the question of the origin of life from a terrestrial source. That is, we must attempt to deal with the problem of whether a highly complex, ordered phenomenon such as life could have arisen at all from the molecular chaos assumed to have existed during the Earth's early history. To recapitulate a question discussed in Chapter 4: How can order arise from disorder?

The probability for a modern, self-reproducing cell to arise from complete disorder is tiny. To use an oft-quoted example: Can a monkey, given billions of years, produce the works of Shakespeare by randomly pressing the keys of a typewriter? Even if we restrict Shakespeare's writings to 1 million (10^6) alphabetical letters and limit the typewriter to 26 keys, the chance for such an event is $(1/26)^{10^6}$. This means that even if a monkey could type 1 million words a second, we could expect such an event to occur only once in $7 \times 10^{1,414,965}$ years!

By similar reasoning, the chances for most complex organic structures to arise spontaneously are infinitesimally small. Even a small enzymatic sequence of 100 amino acids would have only one chance in 20^{100} ($= 10^{130}$) to arise randomly, since there are 20 possible kinds of different amino acids for each position in the sequence. Thus, if we randomly generated a new 100-amino-acid-long sequence each second, we could expect such a given enzyme to appear only once in 4×10^{122} years! In terms of the volume necessary to generate all such possibilities, the difficulty appears just as immense: if an entire universe 10 billion light-years in diameter were densely packed with randomly produced polypeptides, each 100 amino acids long, the number of such molecules—10^{103}—would not equal their 10^{130} possibilities.[2]

These arguments long seemed formidable and were further strengthened by the suggestion that nature itself would deteriorate any complex organization of matter even if such complexity were to arise accidentally. Theorists often pointed out that according to the **second law of thermodynamics,** the energy in a system tends toward diffusion rather than concentration. That is, in an *isolated* system, in which events occur in the absence of outside sources of energy, conditions go from a more ordered to a less ordered state because **entropy** (disorder) increases rather than decreases. Thus, there appeared to be only a negative answer to the question that Pasteur had posed in the nineteenth century: "Can matter organize itself?" It seemed either that the living organization of matter must be explained as arising from a mystical nonnatural source, or that this event, if it were of natural origin, was so improbable that any attempt at comprehension or reconstruction would be meaningless.

In answer to these apparent difficulties, many scientists today point to three important considerations:

1. The likelihood that primeval "living" organisms did not have many of their present complexities.

[2]Similar "enigmas" are often presented by creationists and others to show that chance events are inadequate to explain complex structures, such as the improbability that a monkey could create Leonardo Da Vinci's *Mona Lisa* when given simply paint and canvas. As explained in Chapter 4 and p. 136, the creative force is not chance but successive acts of selection that act as the creative filter to channel random variations into adaptive forms and patterns, whether the selected components are elements such as letters, colors or amino acids.

2. The formation of organic molecules and subsequent organic structures was not the result of completely random events, although such events were nevertheless natural in the sense that they followed chemical and physical laws.

3. Life is not an "isolated" system unable to maintain organized structures, but continually receives energy and materials from outside the organism that enables it to preserve "order" until death.

The first consideration, that early organisms were more primitive than those of today, is extremely important. Perhaps the most basic quality of life we would recognize in even a primitive living organism is its ability to perform those reactions necessary for it to grow and replicate. Certainly the endless loop of metabolism and information transfer that we now see embodied in the intricate relationship between proteins and nucleic acids need not always have been of the same complexity. As we shall see later, proteinlike compounds may arise in reaction mixtures of amino acids without intervention by nucleic acids and, although formed randomly, these compounds may then function in a variety of enzymatic ways.

Although it is true that chemically generated proteins do not have the repeatable and precisely ordered sequences of amino acids in cellular proteins, it is probably also true that the metabolic functions necessary for survival and growth were much simpler in the past. The relative simplicity of early "life" and its precursors, especially in the absence of competition with the more sophisticated later forms, seems a reasonable assumption to make.

The second and third considerations, that life did not arise from absolute chaos, and was not isolated from external forces which decrease entropy, have been supported in many ways. Scientists have pointed out that the evolution of our solar system offered a number of essential prerequisites that enabled the development and sustenance of life:

1. Our planct possessed a sun of moderate size that was on the main sequence of stellar evolution (Fig. 5-4). This sun provided a steady rate of emitted radiation over a long enough period of time for life to develop. As shown in Table 7–1, the amount of energy available from solar radiation seems to have always been far greater than any other source, although energy from other sources may also have been important in initiating particular chemical reactions.

2. A fairly large sampling of different elements existed, such as H, O, C, N, S, P, and Ca. These elements must have provided considerable chemical diversity, enabling reactions to occur that were necessary to form organic molecules involving carbon. The important chemical attributes of carbon, its ability to effect four covalent bonds and the tetrahedral arrangement of its outer electrons, provided the opportunity for the formation of a large number of different kinds of stable molecules with considerable three-dimensional variety and complexity. Interestingly, the terrestrial presence of such molecules is not unique: by means of spectroscopy, researchers can now observe a variety of organic molecules in the dense interstellar clouds that give rise to stars and planets (Fig. 7–8). Such observations indicate that a number of compounds necessary for the origin of life were already present before and during the formation of our solar system, and their synthesis derived from **abiotic** chemical interactions—independent of living systems.

TABLE 7–1 Present energy sources (calories per square centimeter per year) that were probably available for organic synthesis on the primitive earth

Source	Energy
Total solar radiation (all wavelengths)	260,000
Ultraviolet light wavelengths (in angstroms)	
Below 3,000	3,400
Below 2,500	563
Below 2,000	41
Below 1,500	1.7
Electrical discharges (lightning, corona discharges)	4
Shock waves (meteorite impacts, lightning bolt pressure waves)	1.1
Radioactivity (to depth of 1 km)	0.8
Volcanoes (heat)	0.13
Cosmic rays	0.0015

Source: From *The Origins of Life on the Earth* by S. L. Miller and L. E. Orgel, 1974. Reprinted by permission.

3. The Earth followed a nearly circular orbit at a fairly uniform distance from the sun. Such an even orbit would eliminate temperature extremes preventing organic molecules from forming, surviving, and functioning.

4. There was present on Earth large amounts of an excellent solvent, water, which is stable in liquid form over a relatively wide range of temperatures and enables both acids and bases to ionize and react. Water also has the advantage that it floats in its crystalline frozen form (ice), so that bodies of water containing organic matter may remain liquid under a surface of ice, rather than freezing because of the subsurface accumulation of ice. Geochemists now believe that water must have been present early in Earth's history in the cold planetesimal condensations

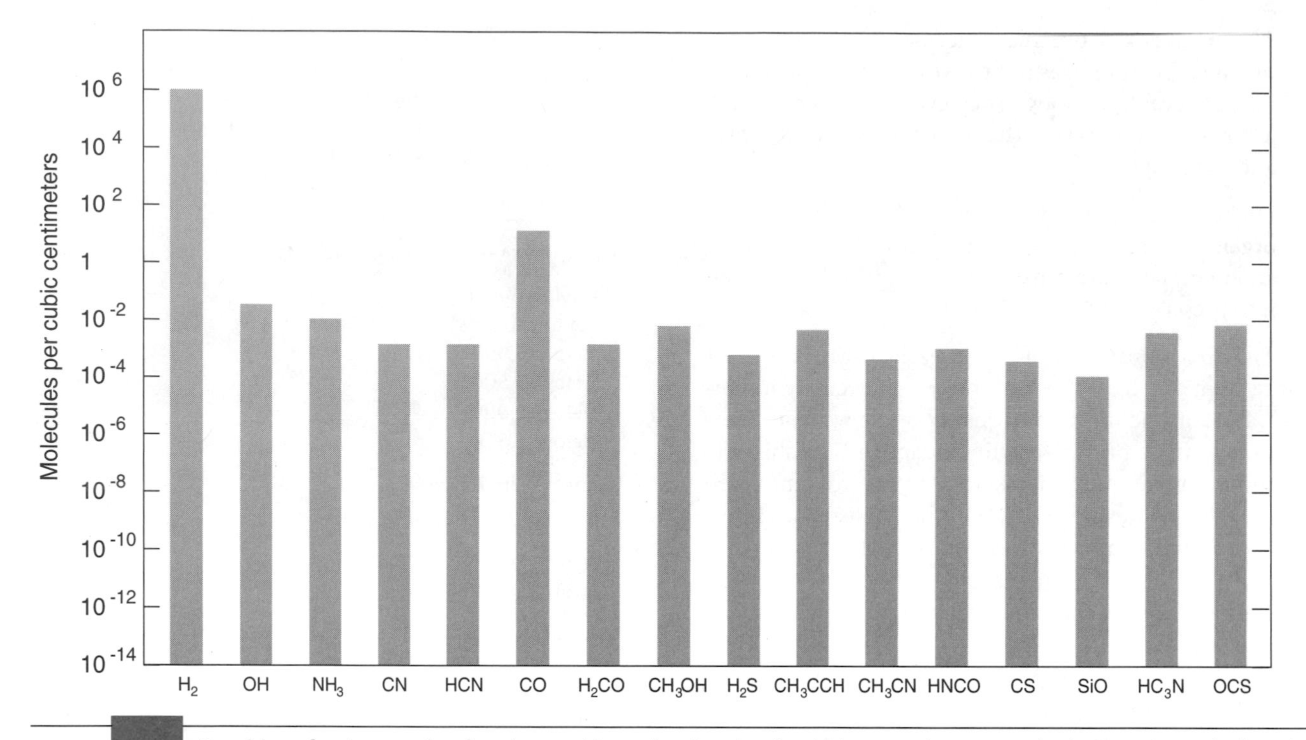

FIGURE 7-8 Densities of various molecules observed in molecular clouds within our galaxy. H_2 = hydrogen, OH = hydroxyl, NH_3 = ammonia, CN = cyanogen, HCN = hydrogen cyanide, CO = carbon monoxide, H_2CO = formaldehyde, CH_3OH = methyl alcohol, H_2S = hydrogen sulfide, CH_3CCH = methylacetylene, CH_3CN = methyl cyanide, HNCO = isocyanic acid, CS = carbon monosulfide, SiO = silicon monoxide, HC_3N = cyanoacetylene, and OCS = carbonyl sulfide. *(Adapted from Buhl.)*

and appeared in liquid form as soon as the lithosphere reached appropriate temperatures. Additional water has been continually emitted into the atmosphere through volcanic activity, which was probably greater in the past than at present. Since water causes crustal erosion that leads to sedimentary rock formation, the presence of significant amounts of water dates back to the beginning of the geological record.

5. Hydrogen-containing gases existed for a long initial period in the Earth's history, derived from the high cosmic abundance of hydrogen and the consequent abundance of its compounds. Even though free hydrogen was probably lost early in the Earth's history, outgassing from the Earth's interior would have led to at least a partially reducing atmosphere in some or many localities in which hydrogen atoms could be donated to a variety of hydrogen-accepting elements, especially carbon.[3] Such gases, along with the energy from solar and ultraviolet radiation, enabled the formation of a variety of organic molecules that could, in turn, provide both structure and energy for living processes (for example, amino acids, sugars, fatty acids, purines, and pyrimidines).

In summary, we can say that there was considerable **molecular preadaptation** for the biochemical events leading to the formation of life even before life appeared. There were appropriate energy sources, chemicals, temperature, and solvent—the foundations of a "universal organic chemistry." What kinds of reactions would then have taken place?

[3]Theories about the composition of the Earth's early atmosphere range from strongly reducing to mildly reducing to neutral and nonreducing (see Chang et al.). However, there is no dispute that gases presently emitted from the Earth's interior include hydrogen (H_2) and methane (CH_4), and it seems reasonable to assume that such gases were also emitted in the past into a more strongly reducing atmosphere than we see now. Support for this view can also be gained from heating meteorite-like materials that model the composition of the planet. The primary gases produced are methane, nitrogen, hydrogen, ammonia, and water vapor, with small amounts of H_2S, CO, and CO_2. Hydrogen was probably also produced by the reaction of ferrous oxide or hydroxide—both more plentiful in the past than today—with water to form ferric compounds and hydrogen: for example, $2\ FeO + H_2O \rightarrow Fe_2O_3 + H_2$ (or $2\ Fe(OH)_2 + 2\ H_2O \rightarrow 2\ Fe(OH)_3 + H_2$). (As shown later on page 172, the ferrous-ferric oxidation reaction takes place in the presence of oxygen which, at some stage, was supplied by photosynthesizers that split water molecules and release O_2.) Hydrothermal deep-sea vents, with water temperatures of 350°C can also produce very large amounts of ammonia by combining nitrogen or its oxides with overheated water in the presence of iron and other mineral catalysts (Brandes et al.).

FIGURE 7-9 Diagrammatic representation of apparatus Miller (1953) used to demonstrate the synthesis of organic compounds by electrical discharge in a reducing atmosphere.

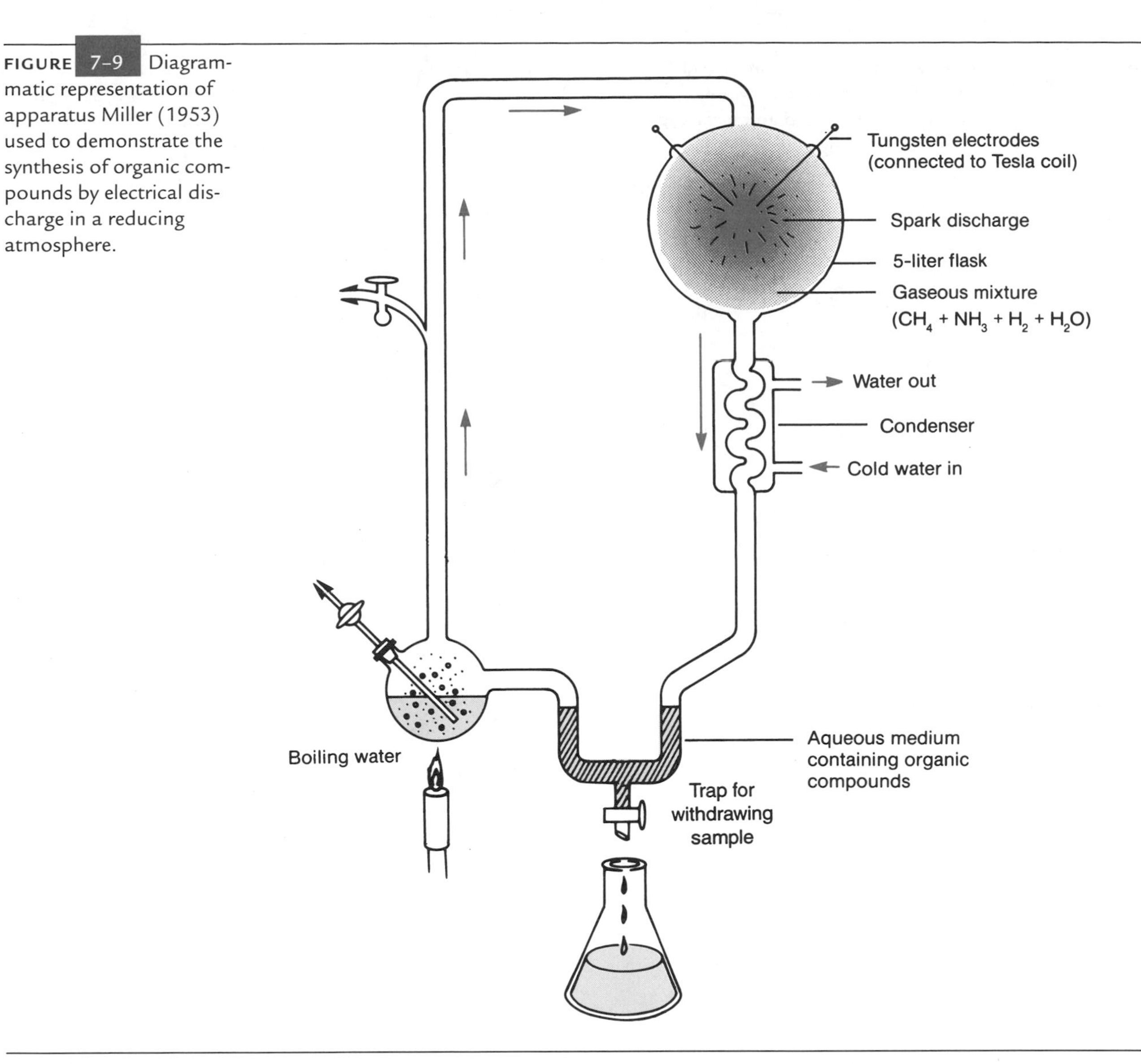

Origin of Basic Biological Molecules

In the 1920s, Oparin (1894–1980), a Russian biochemist, and Haldane (1892–1964), an English geneticist, independently suggested that the primitive atmosphere of the Earth was reducing and that organic compounds formed in such an atmosphere might be similar to those presently used by living organisms. However, almost 30 years elapsed before someone undertook an experimental test of this hypothesis. In 1953, Miller placed together in a glass apparatus (Fig. 7–9) methane, ammonia, and hydrogen gases. He generated an electric spark in a large 5-liter flask, and boiled water in a smaller flask to provide vapor to the spark as well as to circulate the gases. Compounds formed by sparking were then condensed, or recirculated if they were volatile. After one week of continuous electrical discharge, he chromatographed and analyzed the products accumulated in the aqueous phase (Table 7–2).

Note that many of these compounds are molecularly simple, including both amino acids and other substances, such as urea, found in living organisms. In fact, of the wide array of possible complex molecules that such apparently random chemical reactions could have produced, it is remarkable that significant amounts of such relatively simple compounds essential to life actually formed. These experiments and others that followed[4] point strongly to the likelihood that the chemical environment that existed

[4]In addition to electrical discharges, other energy sources such as β-rays, γ-rays, x-rays, thermal heating, and ultraviolet light produce amino acids and other organic compounds from simple gases. For example, the Fischer–Tropsch Type reaction (originally discovered by passing steam over heated charcoal) can produce a variety of organic compounds when a heated mixture of carbon monoxide and hydrogen is passed over a catalyst. Adding ammonia produces purine and pyrimidine nucleotide bases that the Miller reactions do not produce.

TABLE 7-2 Yields of various organic compounds obtained from a mixture of water, hydrogen, methane, and ammonia exposed to electrical sparking

Compound	Yield (percent)[a]
Glycine	2.1%
Glycolic acid	1.9
Sarcosine	0.25
Alanine	1.7
Lactic acid	1.6
N-methylalanine	0.07
α-amino-n-butyric acid	0.34
α-aminoisobutyric acid	0.007
α-hydroxybutyric acid	0.34
β-alanine	0.76
Succinic acid	0.27
Aspartic acid	0.024
Glutamic acid	0.051
Iminodiacetic acid	0.37
Iminoaceticpropionic acid	0.13
Formic acid	4.0
Acetic acid	0.51
Propionic acid	0.66
Urea	0.034
N-methyl urea	0.051
TOTAL	15.2

Note: These products represented only about 15 percent of the carbon that had been added to the apparatus. The remaining carbon products were mostly in the form of polymerized tarlike substances that were not analyzed.

[a]The percent yields are based on the amount of carbon that was added to the mixture as methane.

Source: From *The Origins of Life on the Earth* by S. L. Miller and L. E. Orgel, 1974. Reprinted by permission.

TABLE 7-3 Comparison between the relative abundances (asterisks) of amino acids in the Murchison meteorite and in electric discharge synthesis

Amino Acid[a]	Murchison Meteorite	Electric Discharge
Glycine	****	****
Alanine	****	****
α-amino-n-butyric acid	***	****
α-aminoisobutyric acid	****	**
Valine	***	**
Norvaline	***	***
Isovaline	**	**
Proline	***	*
Pipecolic acid	*	<*
Aspartic acid	***	***
Glutamic acid	***	**
β-alanine	**	**
β-amino-n-butyric acid	*	*
β-aminoisobutyric acid	*	*
γ-aminobutyric acid	*	**
Sarcosine	**	***
N-ethyglycine	**	***
N-methylalanine	**	**

Note: Analysts did not observe purine and pyrimidine compounds found in the nucleic acids of living organisms in the meteorite, although they did note traces of nonbiological pyrimidines (for example, 4-hydroxypyrimidine).

[a]Strongly indicative of their nonbiological origin was the finding that the meteorite amino acids were optically inactive (racemic) mixtures of both D and L forms, rather than consisting exclusively of the levorotary forms produced biologically on Earth.

Source: From *Cold Spring Harbor Symposia on Quantitative Biology*, Volume LII, 1987 by Cold Spring Harbor Laboratory.

before the origin of life was probably not "chaos." Rather, the Earth had a significant amount of simple organic molecules that could participate in forming living organisms.

Moreover, astronomers can see such compounds in interstellar clouds in our galaxy (Fig. 7–8) and also in various carbonaceous meteorites that they believe represent material remaining in space from the original solar condensation 4.6 billion years ago. One such example, the Murchison meteorite that fell in Australia in 1969, contains more than 80 kilograms of carbonaceous material, of which about 1 percent is organic carbon (Fig. 7–10). Since a number of different laboratories analyzed the meteorite almost immediately after it landed and the results were consistent overall, researchers doubt that the protein-type amino acids could have originated from terrestrial contamination. Also remarkable is the fact that practically all the amino acids found in the meteorite, both protein and nonprotein, are similar to amino acids researchers have produced by sparking mixtures in laboratory experiments (Table 7–3).

FIGURE 7-10 An approximately 5-inch-long fragment of the Murchison carbonaceous chondrite that fell near Murchison, Victoria, Australia in 1969. Frictional heating produced the charred crust as the meteorite passed through the Earth's atmosphere. Geochemists have suggested that major meteorite impacts during the Hadean era (Fig. 9-13) contributed significant amounts of water and organic compounds used in origin of life reactions. (*From Miller 1992.*)

These meteorite observations strongly indicate that the laboratory experiments reflect actual chemical processes that also occurred in the synthesis of prebiotic organic compounds on the primitive Earth. However, the cause for these impressive correlations is not clear: What chemical mechanisms restrict or bias the production of organic compounds to those that are observed?

It seems likely that the amino acids synthesized under primitive Earth conditions arise primarily from the formation of aldehydes,

$$R{-}\overset{\overset{\displaystyle O}{\|}}{C}{-}H$$

(where R may represent any group), which then interact with ammonia and cyanide compounds. These reactive chemicals may have arisen from a variety of simple gases or from their further interactions:

$$N_2 + H_2,\; N_2 + H_2O \longrightarrow NH_3 \text{ (ammonia)}$$

$$CH_4 \longrightarrow C_2H_2 \text{ (HC}{\equiv}\text{CH, acetylene)}$$

$$CH_4 + N_2,\; CH_4 + NH_4,\; CO + NH_3,\; C_2H_2 + N_2 \longrightarrow HCHN \text{ (HC}{\equiv}\text{N, hydrogen cyanide)}$$

$$CH_4 + H_2O,\; CH_4 + CO_2,\; CO_2 + H_2,\; CO_2 + H_2O \longrightarrow HCHO \text{ (H}\overset{\overset{\displaystyle O}{\|}}{C}\text{H, formaldehyde)}$$

$$CH_4 + H_2O \longrightarrow CH_3CHO \text{ (H}_3\text{C}{-}\overset{\overset{\displaystyle O}{\|}}{C}\text{H, acetaldehyde)}$$

According to one of the possible pathways of amino acid synthesis (the Strecker synthesis), subsequent steps are as follows:

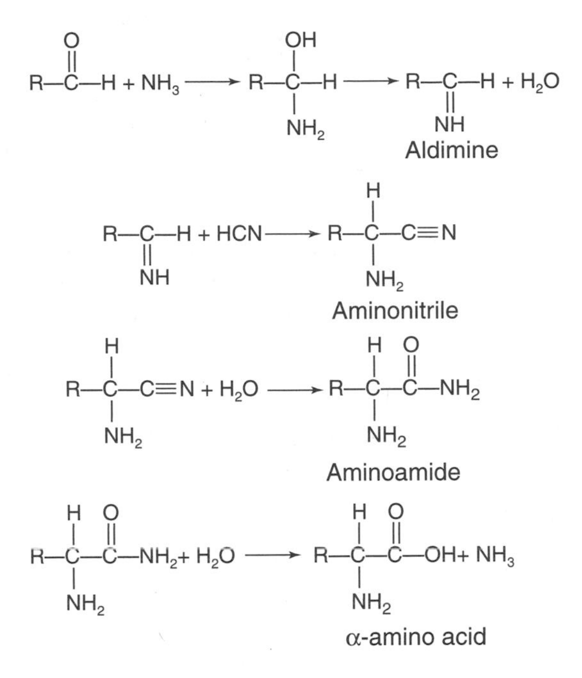

If R in the preceding reactions is a hydrogen atom—that is, if the initial molecule is formaldehyde (HCHO)

$$H{-}\overset{\overset{\displaystyle O}{\|}}{C}{-}H$$

—then the resulting amino acid is glycine. Glycine can also result from adding water (hydrolysis) to cyanide polymers:

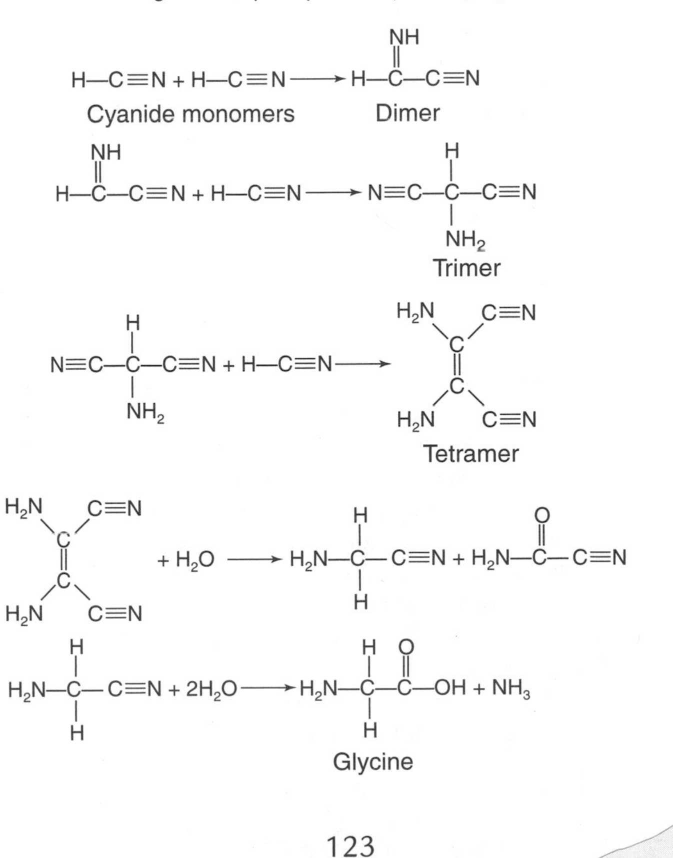

Adding formaldehyde to glycine under alkaline conditions can then produce serine:

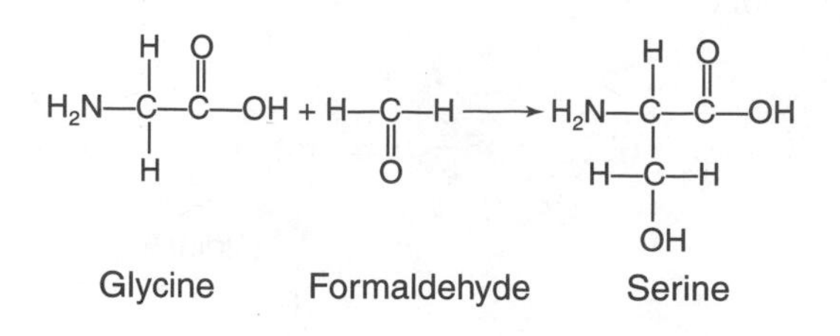

All 20 different amino acids now used in protein synthesis have a similar structural pattern, as Figure 7–2 shows, although synthesized by different biochemical pathways.

Among other basic organic molecules that could easily be synthesized under fairly simple conditions are the **sugars.** In the "formose reaction," small but significant yields of glucose and ribose, for example, occur from condensing formaldehyde:

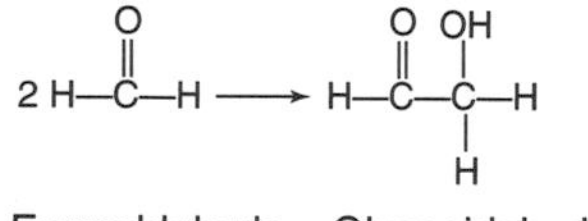

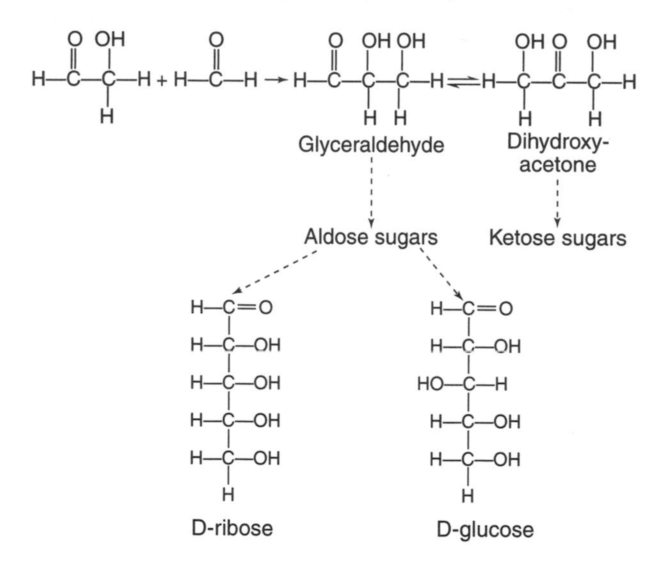

Among all possible 6-carbon sugars that could have been used as the common metabolic fuel, the dominance of glucose is most probably related to its environmental stability. It may therefore have accumulated in greater quantities than other sugars, establishing its basic role in early organismic reactions.

The **purine** and **pyrimidine** bases that are essential components of nucleic acids can also be synthesized under prebiotic conditions. For example, Oró and coworkers have shown that heating aqueous solutions of ammonium cyanide (prepared by the reaction of HCN with NH_4OH) produces up to 0.5 percent yield of adenine. Similarly, ultraviolet radiation acting on hydrogen cyanide solution produces a number of purines, including adenine and guanine. Condensation reactions in forming adenine have been studied in some detail, and researchers have suggested that one sequence may be as follows:

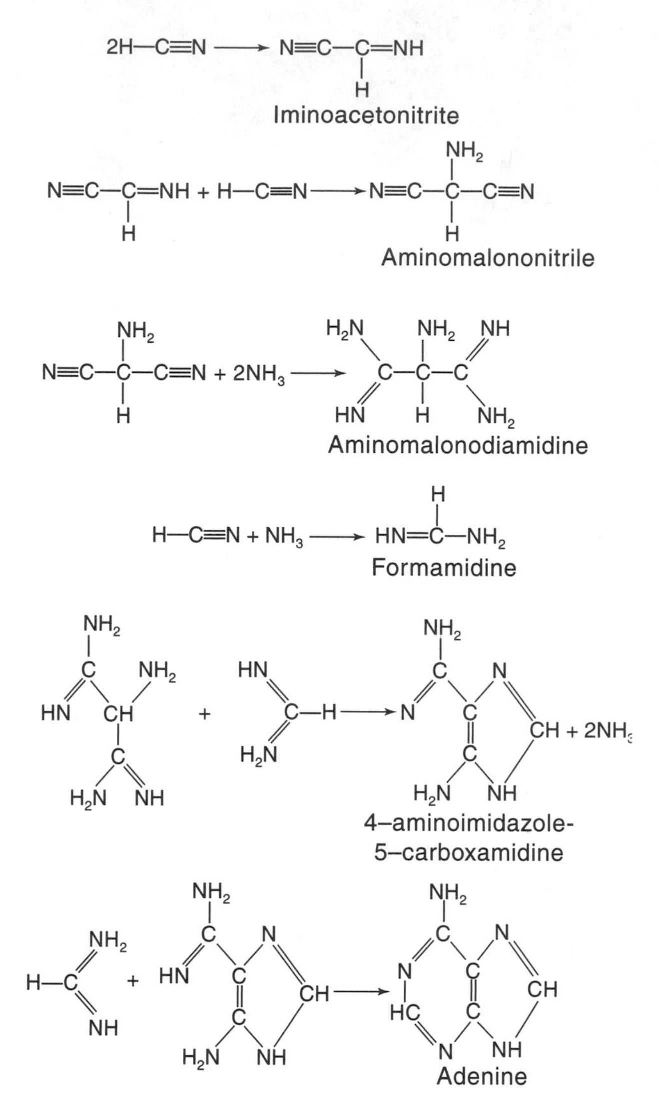

The reaction of cyanoacetylene with cyanates such as urea has produced the pyrimidine cytosine, as shown next, and researchers have proposed similar synthetic procedures for the other pyrimidines uracil and thymine:[5]

[5]Ferris and Joshi have shown that orotic acid (which is a pyrimidine) produced biologically as a precursor to uracil and also produced abiologically by polymerization of HCN, can be relatively efficiently decarboxylated to uracil by ultraviolet light. Because the reactions are so alike in both these systems, they suggest that early biological synthesis of uracil derivatives could easily have followed the pattern of abiological synthesis, but using catalytic enzymes rather than ultraviolet light.

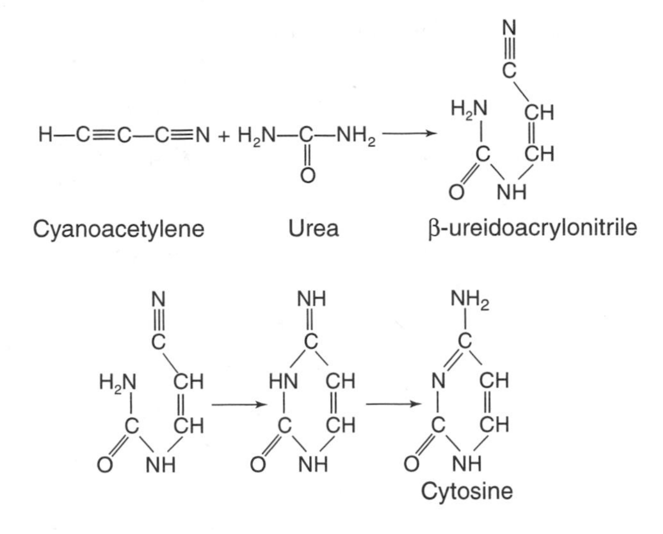

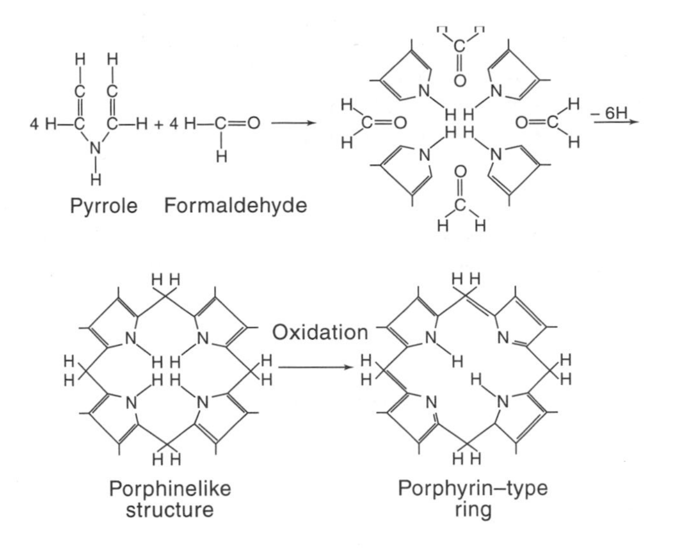

Fatty acids, now used in membranes and storage tissues of living organisms, are among other basic molecules that have been synthesized under high atmospheric pressures, with γ-rays as an energy source:

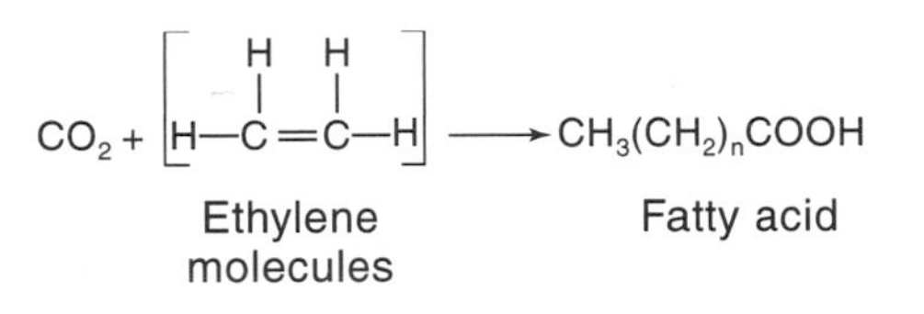

For evidence of prebiotic fatty acid synthesis, we can look to carbonaceous chondrites that contain compounds of the kind synthesized in the early solar system. Interior portions of the Murchison meteorite, for example, have been shown to possess fatty acids up to eight carbons long (Lawless and Yuen). Moreover, experiments by Deamer indicate that a portion of uncontaminated Murchison meteorite compounds can produce fatty-like structures and boundaried vesicles that resemble membranes. Some researchers suggest that high temperatures and pressures found in hydrothermal plumes, common along midoceanic ridges, could have provided conditions appropriate for fatty acid synthesis on Earth.

Pyrroles, which are precursors of porphinelike compounds, can be synthesized in mixtures of CH_4, NH_3, and H_2O and can then react with formaldehyde (also benzaldehyde) to form porphine structures. Although not considered to be a prebiotic sequence, oxidation then yields the **porphyrin** rings found in heme, chlorophyll, and other pigments:

The porphyrin structure has alternating double and single bonds that can "resonate" by assuming a variety of different configurations without changing the position of their constituent atoms. Such resonance confers stability on porphyrins, enabling them to hold extra electrons and thus to function as electron acceptors (oxidation) or electron donors (reduction). Similar oxidative and reductive functions can be performed by nucleotide derivatives such as nicotinamide adenine dinucleotide (NAD), called **coenzymes** because they act in union with protein enzymes to catalyze a wide variety of chemical reactions.

We can thus see that many of the basic organic molecules used in living organisms form relatively easily in many reactions. The amounts per reaction are usually small, but the overall quantities of such substances may have been quite large if sufficient reactive compounds such as methane were present. Even if we limit reactions to available energy, significant concentrations of organic material could have been produced. Shklovskii and Sagan, for example, point out that 1 photon of ultraviolet radiation produces a quantum yield of about 1/100,000 to 1/1,000,000 of a simple organic molecule. If we take 10^{-22} grams as the average mass of such a molecule, then the quantum yield per photon is about $10^{-5} \times 10^{-22} = 10^{-27}$ grams.

Shklovskii and Sagan estimate the number of photons at the top of the Earth's atmosphere in primitive times at 3×10^{14} photons/cm^2/sec, so the quantum yield per square centimeter per second may have been $10^{-27} \times (3 \times 10^{14}) = 3 \times 10^{-13}$ grams. Thus, if the reducing atmosphere lasted for 300 million years (about 10^{16} seconds), enough energy would have been available to produce $(3 \times 10^{-13}) \times 10^{16} = 3 \times 10^{3}$ grams of matter per square centimeter of the Earth's surface. Furthermore, even if this material were diluted in as deep an ocean as the present (3×10^5 centimeters), the concentration of the solution would still be significant: (3×10^3)gm/(3×10^5) cc, or .01 gram per cubic centimeter.

Of course, ultraviolet radiation and heat also decompose organic material, and such degradative effects may have been considerable. Nevertheless, once organic material formed, it would undoubtedly have had many opportunities to accumulate in relatively cool, protected localities, such as the fissures of rocks, oceanic depths, and pools inaccessible to decomposition by ultraviolet rays. In such places, the concentrations of organic materials may have been quite high.

Condensation and Polymerization

Given localized concentrations of amino acids, sugars, and other organic molecules, further chemical evolution would depend on the polymerization or condensation of these **monomers** into peptides, polysaccharides, and so on. Such events are not spontaneous: to obtain one small polypeptide (molecular weight 12,000) from a one molar aqueous amino acid solution, in the absence of any other chemical forces, has been said to require a volume of amino acids 10^{50} times that of the Earth. How then could such polymerizations occur?

As Figure 7–11 shows, most polymerizations depend on the removal of water molecules from the monomers to be condensed. Amino acids, for example, are ordinarily bonded into peptides on cellular ribosomes through **phosphate-bond** energy (Fig. 7–6). Outside the cell, the task is more difficult, but bonds can nevertheless form either in aqueous media or under anhydrous conditions.

So far, compounds identified as condensing agents that could have existed early in the Earth's history include the following:

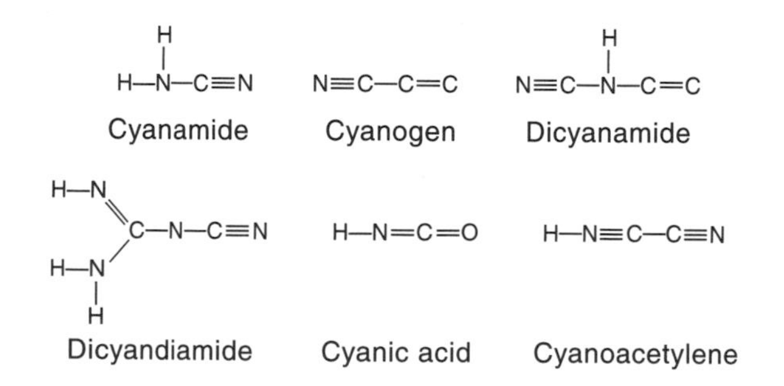

In each case, the unsaturated cyano–carbon–nitrogen bonds enable the condensing agent to combine with water and release energy during this hydration. For example,

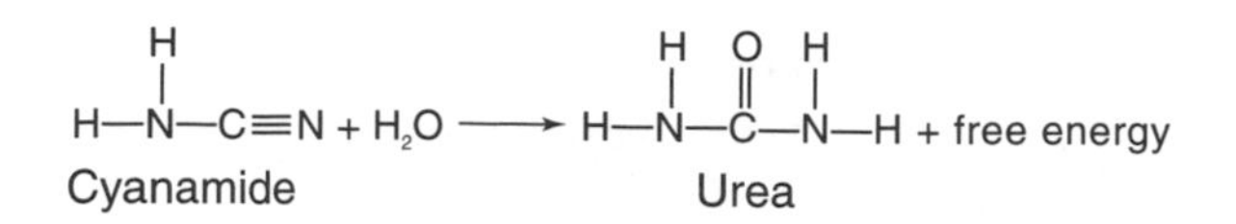

Thus, the condensation of two amino acids into a dipeptide can couple to the **hydrolysis** of cyanamide:

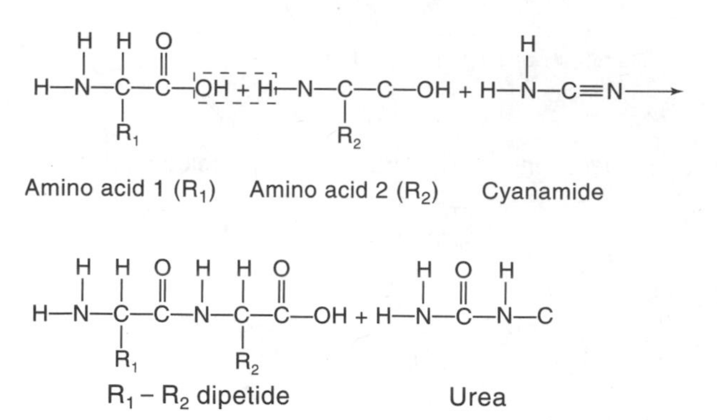

Because of their apparent preference for reacting with organic molecules carrying anions (for example, phosphate HPO_4^-), many of the cyanic condensing agents produce peptide bonds between amino acids even in aqueous solutions. Some, such as cyanogen and cyanamides, also cause nucleotides to form by the phosphorylation of adenosine, uridine, and cytosine; for example,

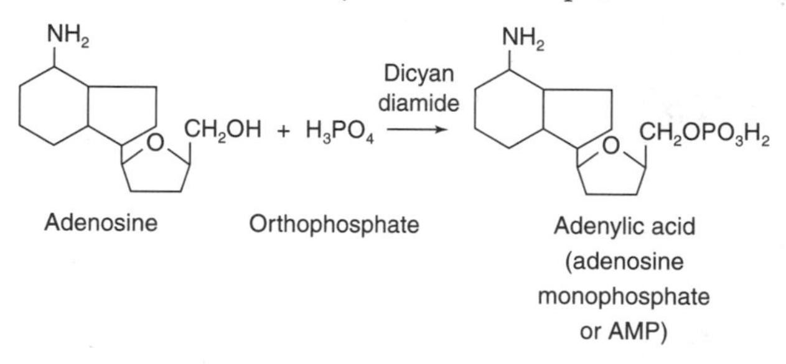

Under anhydrous conditions, with no or few water molecules, heat can promote condensation and polymerization by causing extraction and loss of water molecules from chemical substrates even in the absence of specific condensing agents. One such reaction, accomplished by heat in the laboratory and by enzymes in living organisms, is the formation of high-energy phosphate bonds from orthophosphate:

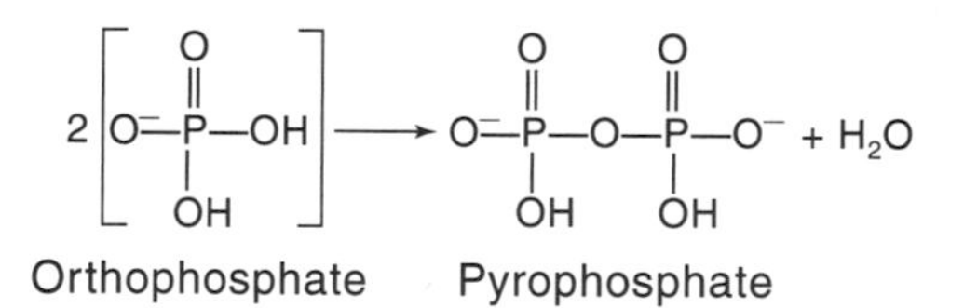

High yields of inorganic pyrophosphate (PPi) can also be synthesized by the condensing agent cyanic acid (cyanate) reacting on precipitated hydroxyapatite [$Ca_{10}(PO_4)_6(OH)_2$], a major phosphate mineral. Such pyrophosphates can then be made available to form adenosine diphosphate (ADP) and **adenosine triphosphate (ATP),** reactions that can then be reversed by hydrolysis to yield energy:

$$ATP + H_2O \rightarrow ADP + \text{orthophosphate} + 7.30 \text{ kilocalories per mole}$$

FIGURE 7-11 Examples of condensation reactions leading to the formation of peptides, polysaccharides, lipids, and nucleic acids. In (*d*) the sugar unit is ribose and the nucleic acid produced is ribonucleic acid (RNA), whereas the sugar unit in deoxyribonucleic acid (DNA) lacks the oxygen atom at the 2′ carbon position. (*Adapted from Calvin.*)

(a) Proteins

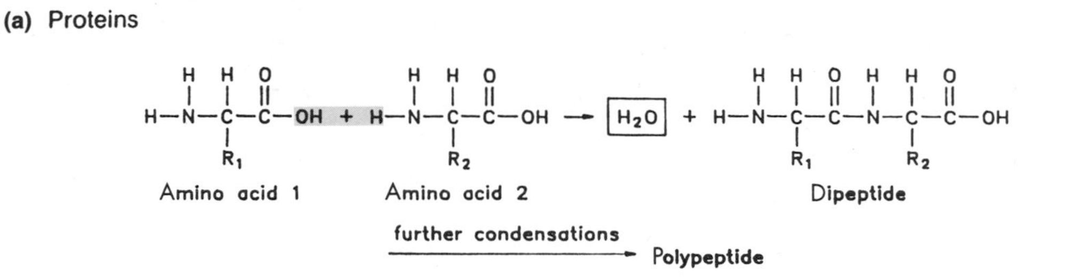

(b) Polysaccharides

Glucose + Glucose → H_2O + Maltose (disaccharide)

further condensations → Starch (polysaccharide)

(c) Lipids

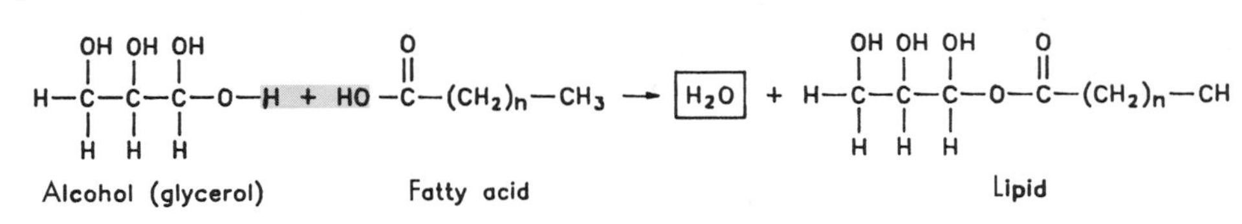

(d) Nucleic acids

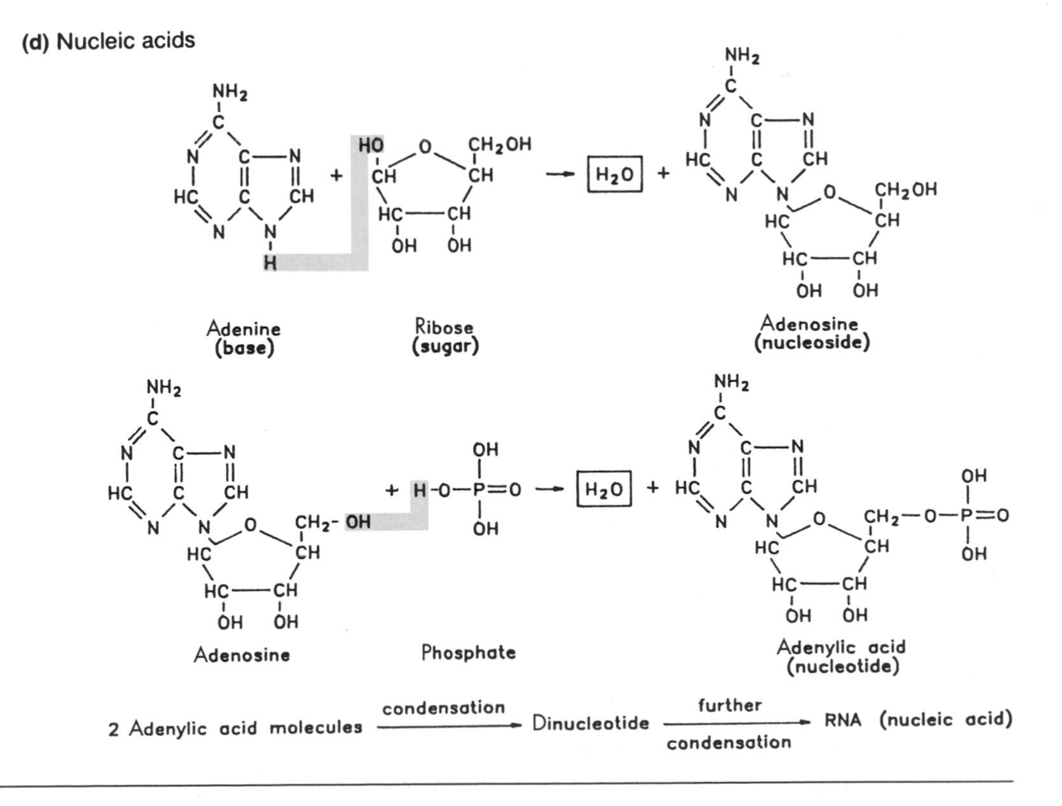

$ADP + H_2O \rightarrow AMP + \text{orthophosphate} + 7.30$ kcal/mole

$AMP + H_2O \rightarrow \text{adenosine} + \text{orthophosphate} +$ 3.40 kcal/mole

When one or more steps in this hydrolytic sequence occur, the cell gets its main source of energy and the primary means of removing further water molecules during condensation reactions. Biochemists have suggested that polyphosphate chains may have provided some of the first organismic energy sources, and the adenosine component in ATP added later to act as a label that would allow enzymatic recognition.

Proteinoids

In the 1950s, Fox and coworkers developed a technique in which heat could also be used to produce peptides from dry mixtures of amino acids. Depending on the kinds of amino acids in the mixture, they found that temperatures of 150° to 180°C could produce as much as 40 percent yield of peptidelike products with molecular weights between 4,000 and 10,000 daltons. Fox called these polymers *proteinoids* (also *thermal proteins*), and he and his group proposed that these compounds bear proteinlike features. According to their analyses, the proteinoids possess nonrandom proportions of amino acids; that is, their compositions are not simply based on the frequency of the different amino acids in the initial mixture (Table 7–4).

They also suggest that the positions of the amino acids in the polymer are not based on their overall frequencies in the chain, since some amino acids preferentially occupy the N- and C-terminals of the proteinoids.[6] The nonrandomness of proteinoid structure also seems supported by the finding that these polymers all show similar properties as tested by sedimentation rates, electrophoretic techniques, column fractionation, and other measurements. Thus, some preferential interaction be-

[6]Since the interior amino acid sequences have not yet been analyzed, this point can be disputed.

TABLE 7–4 Amino acid compositions in molar percentages of two proteinoids compared to the initial reaction mixtures

	2:2:1 Proteinoid		2:2:3 Proteinoid	
Amino Acid	Initial Mixture	Proteinoid Product	Initial Mixture	Proteinoid Product
Aspartic acid	42.0	66.0	30.0	51.1
Glutamic acid	38.0	15.8	27.0	12.0
Alanine	1.25	2.36	2.72	5.46
Lysine	1.25	1.64	2.72	5.38
Semicystine	1.25	0.94	2.72	3.37
Glycine	1.25	1.32	2.72	2.79
Arginine	1.25	1.32	2.72	2.44
Histidine	1.25	0.95	2.72	2.03
Methionine	1.25	0.94	2.72	1.73
Tyrosine	1.25	0.94	2.72	1.66
Phenylalanine	1.25	1.84	2.72	1.48
Valine	1.25	0.85	2.72	1.16
Leucine	1.25	0.88	2.72	1.06
Isoleucine	1.25	0.86	2.72	0.90
Proline	1.25	0.28	2.72	0.59
Serine[a]	1.25	0.6	0.0	0.0
Threonine[a]	1.25	0.1	0.0	0.0

[a]Serine and threonine were omitted from the 2:2:3 proteinoid. Tryptophan was present in the 2:2:1 proteinoid.

Source: From "Amino Acid Compositions in Molar Percentages of Two Proteinoids Compared to the Initial Reaction Mixtures," by Fox, et al, in *Arch. Biochem. & Biophysics* 102:439, 1963. Reprinted by permission.

J. WILLIAM SCHOPF

Birthday:
September 27, 1941

Birthplace:
Urbana, Illinois

Undergraduate degree:
Oberlin College (Ohio), 1963

Graduate degrees:
A.M. Harvard University, 1965 (Biology)
Ph.D. Harvard University, 1968

Present position:
Professor of Paleobiology
Director, Center for the Study of Evolution and the Origin of Life
University of California–Los Angeles

■

WHAT PROMPTED YOUR INITIAL INTEREST IN EVOLUTION?

My father was a paleobotanist and professor in the Department of Geological Sciences at Ohio State University. As a high school student, I was encouraged both by my parents and by my teachers to pursue a career in the natural sciences.

WHAT DO YOU THINK HAS BEEN MOST VALUABLE OR INTERESTING AMONG THE DISCOVERIES YOU HAVE MADE IN SCIENCE?

- Helping to establish and set up the interdisciplinary format of a new area of science, Precambrian paleobiology
- Encouraging international activity in this area of science
- On the basis of direct fossil evidence, posing and attempting to answer new questions fundamental to science regarding, for example, the time, mode, and environmental and evolutionary impact of the origin of oxygen-producing photoautotrophy; the time of origin of eukaryotic cells and the evolutionary impact of eukaryotic sexuality; and the differences in both tempo and mode between Precambrian (prokaryotic) and Phanerozoic (eukaryotic) evolution.
- Discovery of the oldest fossils now known—approximately 3,465 million-year-old cyanobacterium-like cellular filaments from the Apex chert of northwestern Western Australia.

WHAT AREAS OF RESEARCH ARE YOU (OR YOUR LABORATORY) PRESENTLY ENGAGED IN?

We are continuing work on the last two items just listed.

IN WHICH DIRECTIONS DO YOU THINK FUTURE WORK IN YOUR FIELD NEEDS TO BE DONE?

- Increased studies are needed of the evolutionary impact of major, long-term environmental change. For example, over the past approximately 4 billion years, we need to know more about what changes occurred in day length, atmospheric composition, solar luminosity, and ambient surface temperature.
- More studies are needed of the extent to which the concept of "molecular clocks" (revealed by the biochemistry of extant organisms) can be applied to determining the rate of ancient evolutionary change.

WHAT ADVICE WOULD YOU OFFER TO STUDENTS WHO ARE INTERESTED IN A CAREER IN YOUR FIELD OF EVOLUTION?

Early in their undergraduate years, students should obtain a sound background in organic chemistry or biochemistry, biology (including molecular biology), and geology. Nature, in all its evolutionary splendor, is not compartmentalized into discrete disciplines, which means that major advances in understanding nature will come from soundly based interdisciplinary studies.

tween amino acids in proteinoid formation seems to dictate their position and frequency and lead to some degree of uniformity in the kinds of molecules produced.

Although not all the amino acid bonds formed in such proteinoids are of the usual peptide variety, nor do the shapes of these molecules follow the familiar α-helix of protein structure, there still seem to be enough peptide linkages to characterize them as proteins in many tests. Thus, proteinoids give positive color tests with the same reagents that proteins do; their solubilities resemble proteins; they are precipitable with similar reagents; and Fox and Dose propose they have other proteinlike traits listed in Table 7–5. They therefore suggest that some proteinoid reactions, combined into a particular sequence, may have served as the beginnings of later metabolic systems. Thus, decarboxylation of oxaloacetic acid can be followed by decarboxylation of its product, pyruvic acid, leading to acetic acid and carbon dioxide; or amination of pyruvic acid can lead to alanine (Fig. 7–12). Furthermore, some proteinoids even show relatively sophisticated hormonal activity and can stimulate the production of melanin-producing cells.

Although researchers have debated whether the thermal synthesis of proteins could occur extensively in present natural surroundings (see, for example, Miller and Orgel), the exact conditions encountered on the primitive Earth are certainly not known. Surfaces near some volcanic regions, or upwellings from shallow marine hydrothermal plumes, may have maintained appropriate temperatures for the condensation of amino acids.

TABLE 7-5 Properties common to thermally produced proteinoids and to biologically produced proteins

Qualitative amino acid composition
Range of quantitative amino acid composition (except serine and threonine)
Limited heterogeneity
Range of molecular weights (4,000–10,000)
Reaction in color tests (including biuret reaction)
Inclusion of non-amino acid groups (iron, heme)
Range of solubilities
Lipid quality
Salting-in and salting-out properties
Precipitability by protein reagents
Some optical activity (for polymers of L amino acids)
Hypochromicity
Infrared absorption patterns
Recoverability of amino acids with mineral acid hydrolysis
Susceptibility to proteolytic enzymes
Various enzymelike properties
Inactivation of catalysis by heating in aqueous buffer
Nutritive qualities
Hormonal activity (melanocyte stimulation)
Tendency to assemble into microparticle systems

Source: From *Molecular Evolution and the Origin of Life* by S. W. Fox and K. Dose, W. H. Freeman, 1972.

Cooling rains or currents may then have dispersed such thermally produced proteinoids to places where further interactions could take place.

The unusual living conditions of many recently discovered organisms—oceanic sea vents, boiling hot springs—show that life can be maintained and probably arise even under the most stringent conditions. Thermophilic (heat-loving) organisms have now been found at temperatures above the boiling point of water, some living more than a mile below the surface (Kerr). Proposals that much of Earth's very early landmass was composed of volcanic islands, and that high global "greenhouse" temperatures persisted for more than a billion years because of high CO_2 atmospheric pressures (Lowe), support such rigorous scenarios.

In any case, a wide-enough array of condensation mechanisms have been established, some of which could have functioned in the past. Among these are layered clays such as montmorillonite, which some researchers

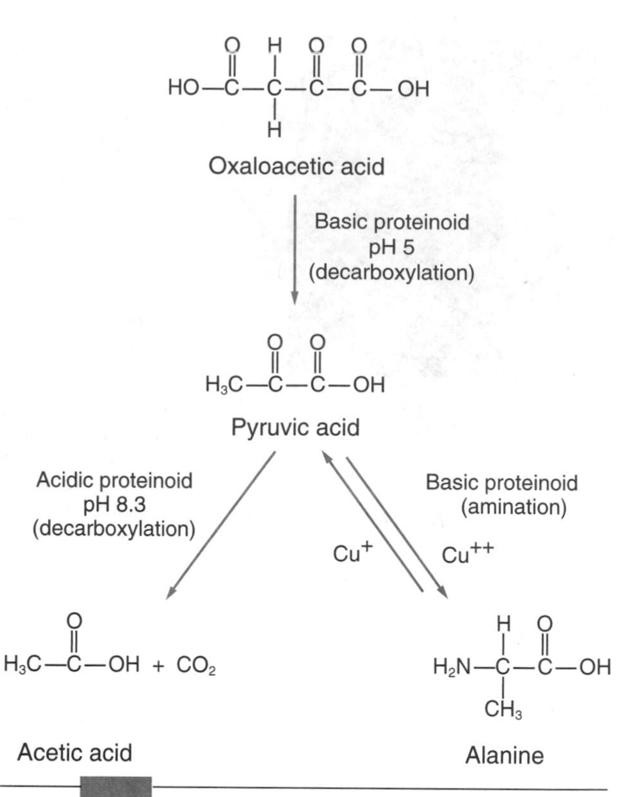

FIGURE 7-12 Some sequential reactions believed to be catalyzed by different proteinoids or proteinoid complexes. (*Adapted from Fox and Dose.*)

propose served as polymerizing templates on which condensation occurred. Paecht-Horowitz and coworkers, for example, used such clays to report that phosphate-activated amino acids such as aminoacyl adenylates will condense to form high yields of polypeptide chains:

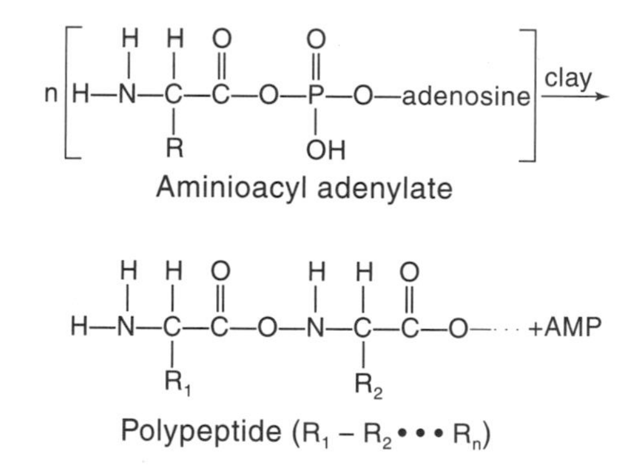

The amino acid ends of the adenylates apparently penetrate the narrow layers of the clay, and the condensation reactions take place there. Studies by Ferris and coworkers report that nucleotides can be joined into 55-length chains on montmorillonite clays, as can amino acids polymerizing into peptides on other mineral surfaces (such as hydroxyapatite, illite). The superheated plumes arising from hydrothermal vents can produce similar polymers as shown by Imai and coworkers who polymer-

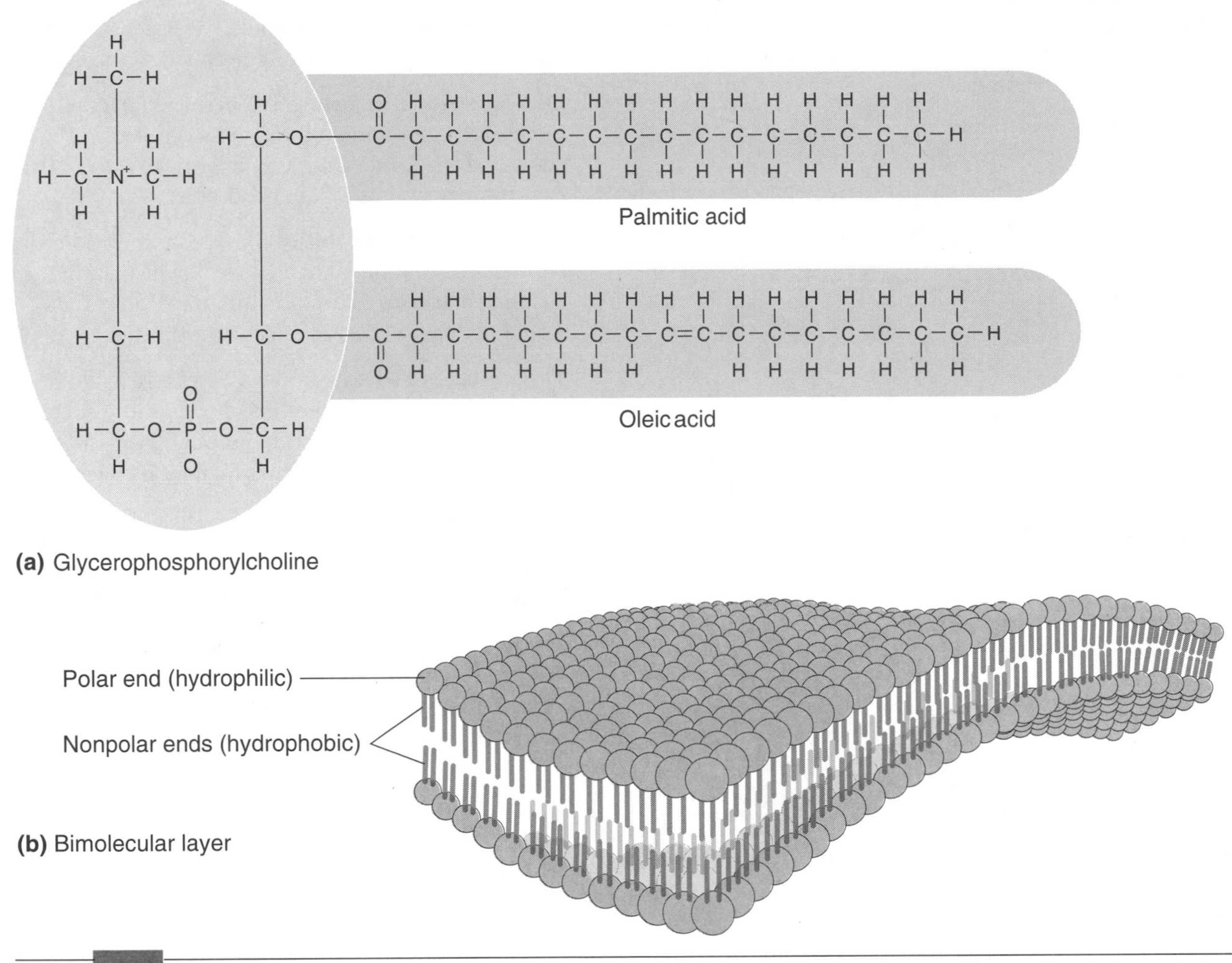

FIGURE 7–13 (*a*) A phospholipid molecule, lecithin. (*b*) A diagrammatic view of a bimolecular sheetlike double layer of phospholipid molecules that have self-assembled with their hydrophilic phosphate heads (*colored circles*) facing the water solvent, and their hydrophobic hydrocarbon tails facing each other. Such polar–nonpolar molecules are called *amphipathic* or *amphiphilic,* and characterize the plasma membrane that circumscribes the cell.

ized glycine amino acids in a flow reactor system that simulates submarine hydrothermals. Even volcanic sediments would have produced spongelike minerals ("zeolites") that can retain and catalyze organic compounds (Smith). Thus, the early availability of cyanamides, heat, clays, and other condensing agents makes it highly probable that polypeptides, polysaccharides, lipids, and perhaps even polynucleotides were present early in the Earth's history, and could have been used for primitive organismlike reactions and structures.

The Origin of Organized Structures

The presence of appropriate organic monomers and polymers is only a first step in the origin of life. Living processes of metabolism and function occur because the materials of which organisms consist are highly organized. How did such organization come about?

At its earliest, interactions among molecules must have led them to assume relative positions based on forces such as hydrogen bonding, ionization, solubility, adhesion, and surface tension. Phospholipids, for example, are organic molecules with a phosphorus-containing polar group at one end and nonpolar fatty acid groups at the other end (Fig. 7–13*a*). In water, a polar solvent, the polar ends of these molecules are oriented toward water (hydrophilic), while their nonpolar ends are oriented toward each other, away from water (hydrophobic). As a result, phospholipid membranous structures can form quickly, yielding **vesicles** composed of bimolecular layers in which the nonpolar surfaces of each of the two layers "dissolve" in each other (Fig. 7–13*b*). Carried further, such vesicles can encapsulate inclusions in tide pools that undergo drying and wetting cycles (Deamer 1993).

Membranous droplets or vesicles composed of lipids, polypeptides, or other molecules undoubtedly formed in great quantities, produced by the mechanical agitation of molecular films on liquid surfaces (Fig. 7–14) or even spontaneously (Deamer 1986). The attainment of such droplet levels of organization would have been an important step in the origin of life for a number of reasons:

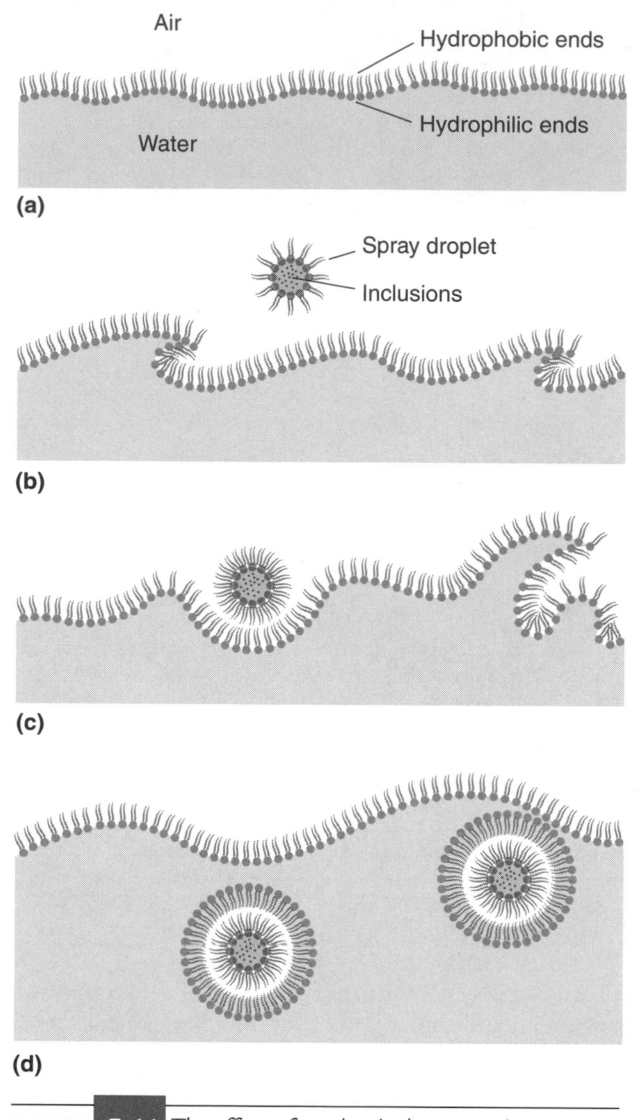

FIGURE 7-14 The effect of mechanical wave action on a surface film (e.g., "foaming") containing molecules oriented with one end pointed away from water (hydrophobic) and the other end toward the water (hydrophilic). In organic molecules such as lipids, the hydrocarbon chain is hydrophobic and the carboxyl end is hydrophilic. Wave action causes the formation of droplets and the bilayered vesicles shown in (*d*). Surface aggregates of peptide chains, nucleic acid sequences, or their combinations may have been among the various vesicle inclusions. Appearance of cell-like double-membraned structures may also have occurred through incorporation of one bilayered vesicle within another (Kaler et al.). Other accretions, composed at least partially of hydrophobic amino acids (Table 8-3), may have persisted within a membrane's layer of hydrophobic tails, acting as selective channels to help transport materials, molecules, and ions between exterior and interior.

1. Depending on its structure and permeability, the membrane surrounding the droplet can selectively choose which compounds can enter from the environment and exit from the droplet.
2. Such **selective permeability** allows concentrations of particular compounds to differ across the membrane, enabling reactions to occur within the droplet that would not have occurred outside the droplet.
3. The presence of a basic protein causes a 100-fold increase in the entrapment of nucleic acids into such droplets. As Jay and Gilbert point out, "Protein-mediated encapsulation creates high local concentrations of protein and nucleic acids within the vesicular volume.... This would enhance the interaction of molecules with low affinities, potentiating the formation of aggregates with biological function."
4. The small size of the droplet can permit a chain or network of reactions to occur, the products of one reaction being available to serve as substrates for another reaction.
5. Both the small size of the droplet and the concentration of various materials within it would permit localized precipitation to occur as well as the organization of compartments and substructures enabling reaction specificity.
6. Droplet membranes could easily have incorporated "amphipathic" peptides (proteins that span phospholipid bilayers), and thus channel ionic and molecular transfers. Once in place, some channels could further evolve into "proton pumps," enabling hydrogen ion (H^+) gradients for energy transport and accumulation (p. 164 and Fig. 9-6).
7. We can think of "advanced" droplets of this kind as unique subsystems that were able to preserve their organizational framework by partially separating themselves from the entropy or disorder in their surrounding environment. That is, although it is true that entropy tends to increase in the universe according to the second law of thermodynamics, it can nevertheless decrease in such subsystems during their life spans. Because of their semipermeable membranes they are not isolated, and can acquire external energy and matter to retain, and even enhance, their organizational and informational structures as long as they can continue to perform biological processes (see Van Holde).[7]

[7]One way of defining life is therefore as a system that prevents attaining a mass action equilibrium that has increased positive entropy. Biological organisms accomplish this by acquiring energy from external

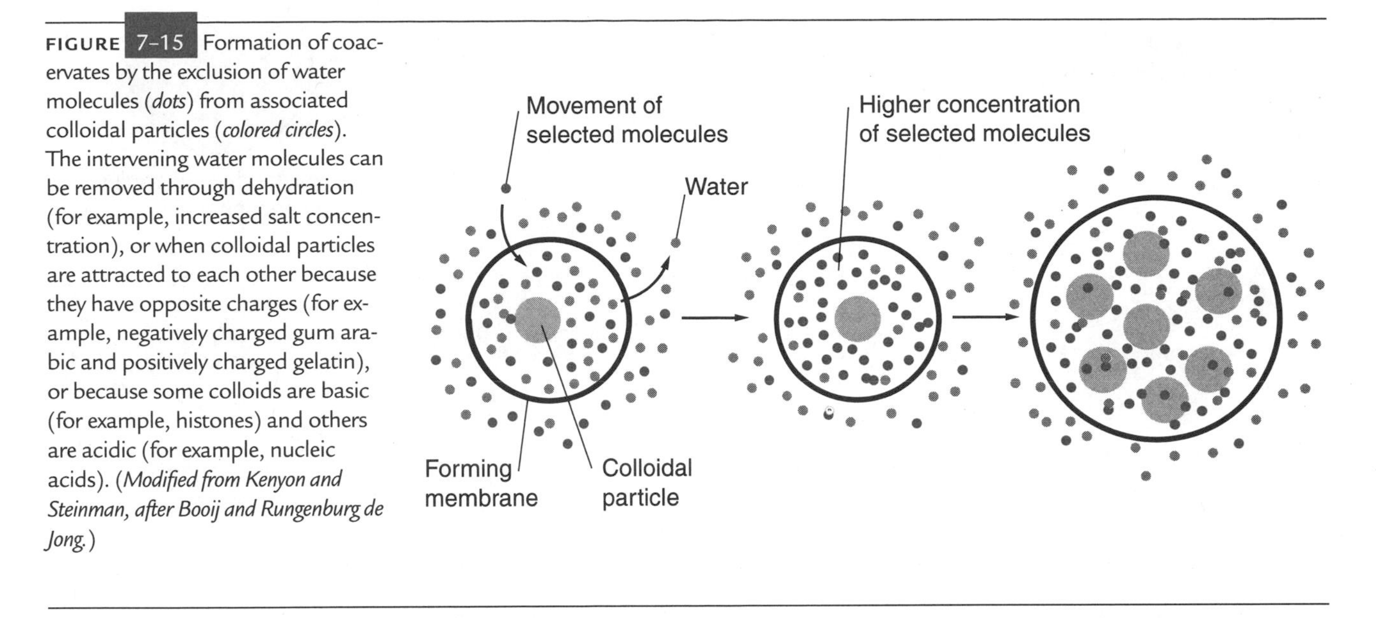

FIGURE 7-15 Formation of coacervates by the exclusion of water molecules (*dots*) from associated colloidal particles (*colored circles*). The intervening water molecules can be removed through dehydration (for example, increased salt concentration), or when colloidal particles are attracted to each other because they have opposite charges (for example, negatively charged gum arabic and positively charged gelatin), or because some colloids are basic (for example, histones) and others are acidic (for example, nucleic acids). (*Modified from Kenyon and Steinman, after Booij and Rungenburg de Jong.*)

It seems presumptuous and unrealistic to assume that the only kinds of organization capable of growth, metabolism, and reproduction are the structures found in present-day organisms. Although present forms are highly efficient, early forms could have functioned at a much lower level of efficiency, because they were not then competing with the more advanced forms. For a primitive form to show some (but certainly not all) "living" attributes, it would have been sufficient if it could merely grow (for example, increase in size), maintain its individuality, and divide. It is therefore interesting to note that some authors ascribe such properties to bimolecular vesicles (for example, Morowitz), and there are, in addition, at least two types of fairly simple laboratory-produced structures that seem to possess some aspects of these basic prerequisites: Oparin's coacervates and Fox's microspheres. Although artificial, they point to the likelihood that nonbiological membraned enclosures could have sustained reactive systems for at least short periods of time.

COACERVATES

Coacervates have been known to occur when dispersed colloidal particles separate spontaneously out of solution into droplets because of special conditions of acidity, temperature, and so on (Fig. 7–15). If there is more than one type of macromolecular particle in the colloid, complex coacervates can form that show a number of interesting properties:

- They possess a simple but persistent organization.
- Although they are mostly unstable, some coacervates can maintain themselves in solution for extended periods.
- They can increase in size.

Oparin, the first to draw serious attention to these droplets, developed artificial coacervate systems that could incorporate enzymes that performed functions such as the synthesis and hydrolysis of starch (Fig. 7–16) as well as synthesizing polynucleotides. In a coacervate system containing chlorophyll irradiated with visible light, Oparin and coworkers showed that there can be a constant inflow of reduced ascorbic acid and oxidized methylene red, which then converts into a constant outflow of oxidized ascorbic acid and reduced methylene red. The chlorophyll picks up electrons from the ascorbic acid and then supplies these for the methylene red reduction—a process similar to common noncyclic photosynthesis in which water molecules supply electrons for reducing the coenzyme $NADP^+$ to NADPH (Chapter 9).

MICROSPHERES

Fox showed that these small spheres formed when the thermally produced proteinoids were boiled in water and allowed to cool. The microspheres are uniform in size, stable, bounded by double membranes that appear somewhat cell-like, and can undergo fission and budding (Fig. 7–17). They appear in large numbers: 1 gram of proteinoid material can produce 10^8 or 10^9 microspheres. Among microsphere qualities indicating active internal processes are their selective absorption and diffusion of certain chemicals but not others, their growth in size and mass, and observations demonstrating

sources that they convert by metabolic processes into negative entropy (biological structures that sustain metabolism) for a period of time. Death for an organism is simply the end of its ability to continue its metabolic functions, accompanied by a slide into mass action equilibrium, or positive entropy.

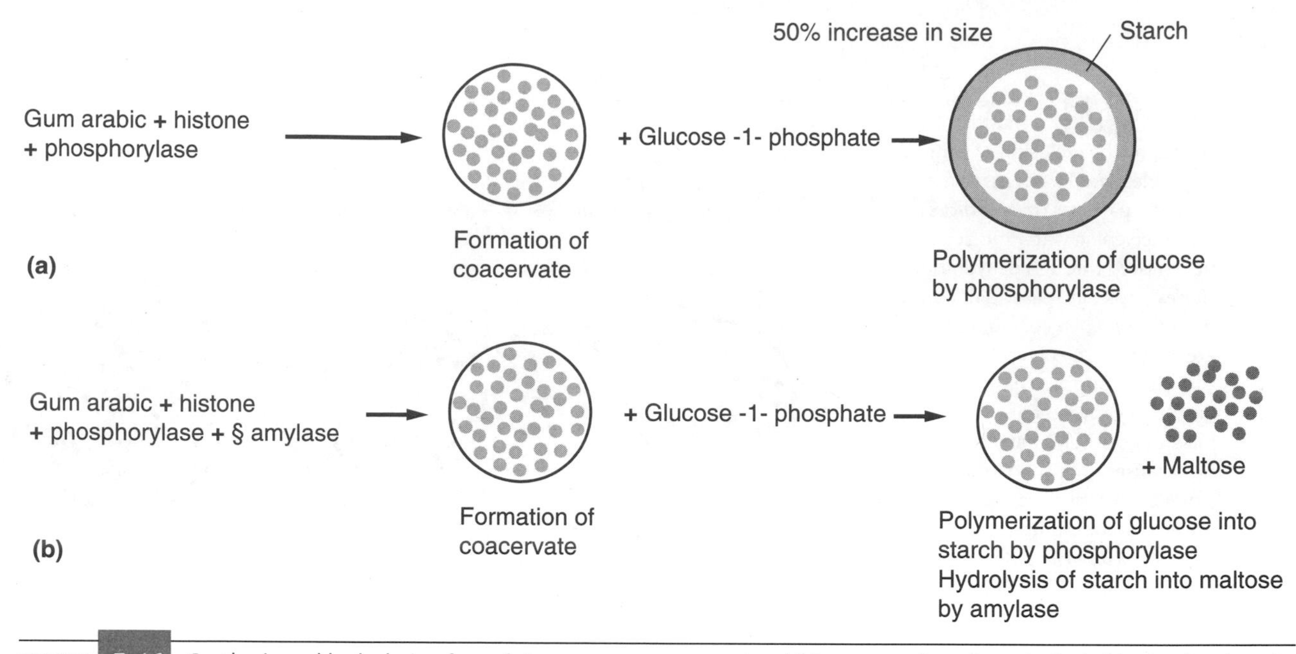

FIGURE 7-16 Synthesis and hydrolysis of starch in coacervate systems in which enzymes have been included in the droplets. In (*a*) the phosphorylase enzyme acts to polymerize phosphorylated glucose into starch, while in (*b*) the starch formed this way is hydrolyzed into maltose by the amylase enzyme.

osmosis, movement, and rotation. Moreover, microspheres show the potential for transferring information, in that proteinoid particles pass through junctions between them.

* * * * *

The spontaneous **self-assembly** of macromolecules into vesicles, coacervates, and microspheres indicates that the occurrence of similar entities under primitive conditions would probably not have been an unusual event. Such entities are not cells, of course, and considerable time may well have elapsed before more elegant structures with more complex metabolic capabilities could develop. Nevertheless, substantial evidence shows that the component materials of even more complex structures can self-assemble without the immediate presence of a prior pattern.

For example, the protein and nucleic acid components of tobacco mosaic virus spontaneously aggregate into the exact configuration needed to produce an active virus. A more complex virus such as T4 (see Fig. 15-1), containing many different kinds of protein, also has a significant number of steps in which self-assembly occurs. Nomura and coworkers have shown even cellular organelles such as ribosomes to be capable of forming by self-assembly from component materials. Each level of self-assembly, from monomers to polymers to coacervates to various cellular organelles, may thus have derived from nonrandom events, in that certain combinations form more quickly and easily than others.

Nonrandom self-assembly, however, is hardly sufficient to account for more than a few complexities of life. In their most essential aspects, such as the precisely ordered monomer sequences found in proteins and nucleic acids, present forms of life certainly did not result from mere chemical attractions between component amino acids or nucleotides. At the same time, we also have good reason to believe that, because of their specificity, the positioning of monomers in these precisely ordered sequences may nevertheless have had a nonrandom basis. To recapitulate a theme raised previously: How can the nonrandom biological order of amino acid and nucleotide sequences arise from disorder? The answer lies in selection.

The Origin of Selection

Primitive structures, whether coacervates, microspheres, or other localized organizations, would have had one important evolutionary feature: they would serve as the first distinctive, multichemical **individuals,** or membrane-enclosed **protocells,** that could interact as units with their environment. Together with their various neighbors and progenies, such individuals would form a group or **population** on which selection could act. That is, protocells incorporating those organizations and metabolic activities most successful in growth and division would increase most in relative frequency or in area occupied.

We can thus say selection arises when the following conditions are reached:

- A population of individuals exists.
- The properties of these individuals are governed by reactions in which they absorb and transform environmental material into their own material.

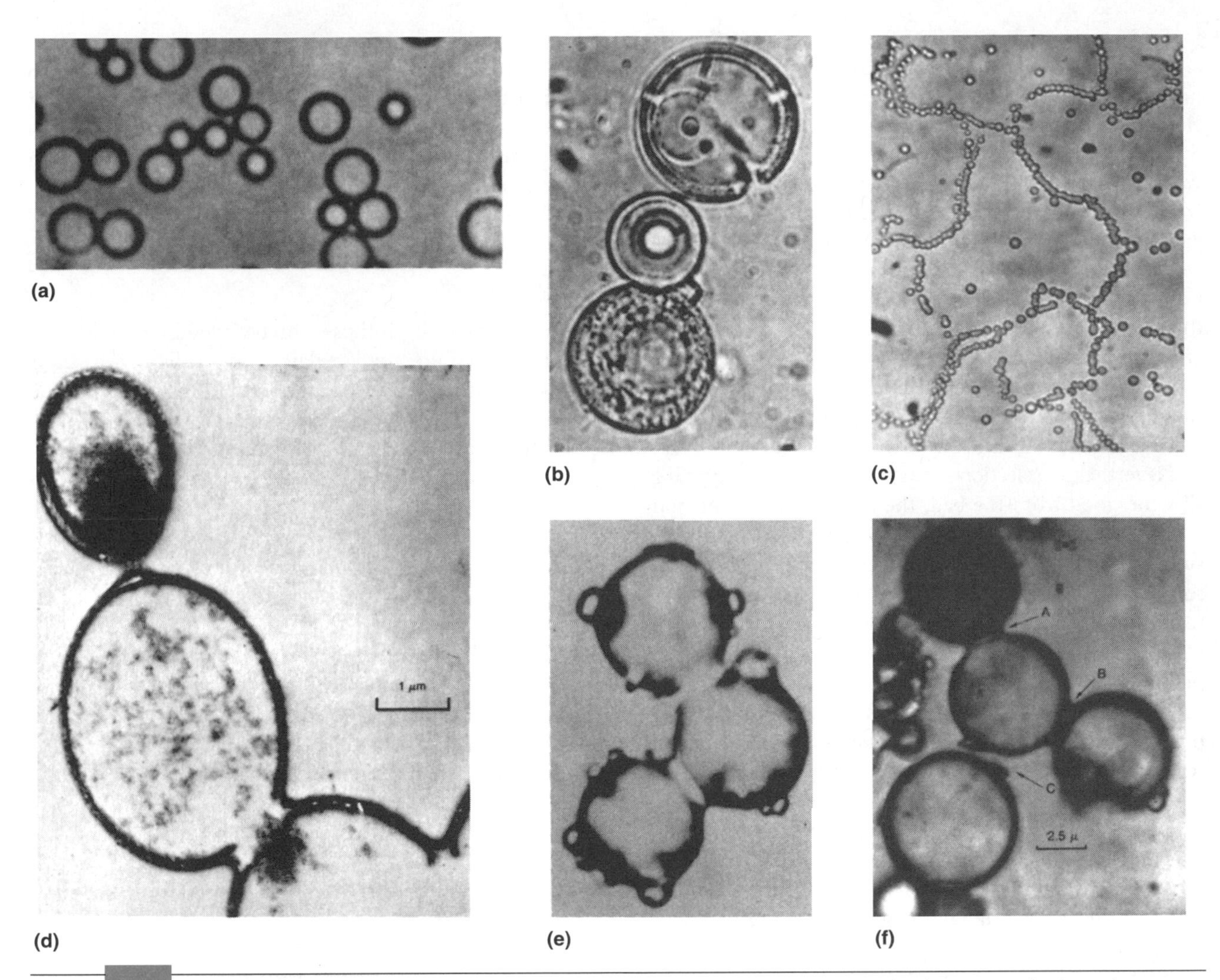

FIGURE 7-17 Various forms of proteinoid microspheres. (*a*) From a chilled solution of proteinoids. (*b*) Structured microspheres. (*c*) Streptococcus-like assemblages. (*d*) Electron micrograph showing double-layer membrane structure. (*e*) Budding. (*f*) Junctions form between pairs of microspheres (*arrows*). (*From* Molecular Evolution and the Origin of Life *by S. W. Fox and K. Dose, W. H. Freeman, 1972.*)

- Individuals differ in the efficiency with which these processes take place.
- Availability of materials and energy is limited so that not all types of individuals can form, nor can all types of individuals formed survive.

The mechanism that enabled the formation of protocells is a crucial issue in understanding selection itself. If protocells could have originated only by self-replication, this would indicate that selection had always operated on the same efficient basis as it does now, with advantageous traits rapidly transmitted to succeeding generations by fairly exact replicative mechanisms. However, if protocells initially formed only through acts of prevailing environmental chemistry, then selection was probably not very efficient in the past and would itself have undergone evolution from chemical nonreproductive selection to the more modern biological natural selection.

That is, early selection would have been confined to the survival of nonreproductive individuals who could wrest the most material from their environment and transform it for their own benefit with the least expenditure of energy. Although differences among such individuals could not be precisely transmitted, the fact that some such individuals survived and others did not would undoubtedly have affected the composition and further interactions of succeeding groups.[8]

Inheritance would therefore, at first, have been mostly a matter of transmitting molecular "things" that permit survival, rather than transmitting exact nucleic acid patterns that produce the "things." As discussed in the next chapter, the earliest forms of "living" individuals may well have replicated themselves poorly, yet they passed on some of their metabolic and enzymatic properties, which continued to be selected and improved.

[8]Kauffman offers mathematical models showing that a "chaotic" system of interacting components can spontaneously become an ordered system on which selection can act. That such spontaneous assembly can *completely* self-reproduce without transmittable genetic information is more difficult to accept.

From a materialistic point of view, unless we postulate an accident of immense proportions and infinitely low probability, selection must have bridged the gap between chemical evolution (changes in the composition of nonreproductive or poorly reproductive molecules, coacervates, microspheres, and so on) and biological evolution (changes in inherited differences among reproductive organisms). So far, selection is the only natural mechanism we know that can account for the creative changes among nonreproductive individuals that could have led them in the direction of living organisms: from "molecules"—to "function"—to "message." Although the events may be complex, the device is simple: organisms that react to their environment with improved useful information replace those that lack such information.

Nevertheless, selection is not merely a passive agent that sifts the good from the bad, the adaptive from the nonadaptive, but, because of its historical continuity, enables a succession of adaptations to accumulate that leads to something entirely new. Selection thus acts as a creative force that has made possible biological organizations that would otherwise have been highly improbable. To use a previous example, a polypeptide chain consisting of a specific sequence of 100 amino acids has an extremely low probability (10^{-130}) of occurring spontaneously without selection. However, as explained in Chapter 4, if each step in the growth of the chain attains a selective advantage when the correct amino acid inserts, then the probability of achieving a functionally advantageous polypeptide is almost immeasurably increased.[9]

Once the game of life has begun, the evolutionary replacement of players bearing information, whether they are genes, organisms, races, species, or other entities, becomes inextricably bound to their ability to play the game further—an ability that selection has previously molded and is now in turn measured anew by selection. Some random replacement of the players certainly occurs by accident rather than by selection, but participation in life is selective by its very nature since the resources of life are always limited in one way or another.

Given reproductive expansion, limited resources, transmitted variation, and environmental change—the organismic condition—selection is inevitable and constant. Thus, it is not merely their origin by selection that characterizes living systems but their continued ability to subject themselves to selection. As discussed in the next chapter, this ability has led to a coupling between function and reproduction that provides living forms with their relatively rapid evolutionary rates.

[9]For example, assuming that each of the 20 different amino acids are present in the surrounding medium in equal frequency, there is a probability of 1/20 that random chance will supply the correct, functionally advantageous amino acid to any position of the chain. Thus $100 \times 1/20 = 5$ positions that will be correctly occupied by chance alone without selection. For the remaining 95 positions, selection will operate so that each position, once occupied by its correct amino acid, enables further selection to occur at the next succeeding position. This stepwise procedure would entail perhaps 20 trials to achieve the correct amino acid at one position, another 20 trials to achieve the correct amino acid at one position, and so on. In sum, a succession of only $95 \times 20 = 1{,}900$ trials may be necessary for selection to provide a functional amino acid sequence for a chain 100 amino acids long—a probability of $1/1{,}900 \approx 2 \times 10^{-3}$ compared to the 10^{-130} probability in the absence of selection.

SUMMARY

For many years even those who believed in the reality of evolution could not explain how cells, with their presently complex membranes and compartmentalized systems, could have arisen. Recently, however, biochemists have identified chemical pathways that may have led to the formation of the most important molecules in cells.

Proteins, composed of specific linear sequences of amino acids linked together by peptide bonds, are crucial to organisms because of their catalytic functions. Their structure is specified by the sequence of nucleotides in nucleic acids, which act as information storage molecules in cells. The problem is to explain the origin of these now totally interdependent molecules.

If life arose on Earth by chemical means, as seems more likely than by special creation or panspermia ("seeding" from outer space), the origin of complex organic molecules from simple substances present on the primitive Earth must be explained. The probability of this occurring randomly is almost incalculably small, especially since the universe increases in entropy. The probability of organic syntheses increases, however, if the first organic molecules were not as complex as they are at present and if chemical reactions were biased to produce such molecules. Conditions on the Earth several billion years ago probably offered a most favorable chemical environment: sufficient and continuous energy, availability of carbon and other important elements, an abundance of water (an excellent solvent), and vast amounts of hydrogen (and its compounds) providing a reducing or partially reducing atmosphere.

Some years ago Miller demonstrated that amino acids and other organic compounds could spontaneously synthesize from hydrogen, ammonia, methane, and water in the presence of an energy source. Amino acids can also be synthesized in the interaction of aldehydes with nitrogenous compounds and by cyanide-cyanide reactions. Furthermore, formaldehyde can condense to form the pentose sugars used in nucleic acids. The nitrogenous bases of nucleic acids arise from cyanide compounds; fatty acids from simple chemicals. The removal of water can polymerize these simple subunits into complex molecules, and heat or condensing agents can expedite polymerization. Fox showed that amino acid mixtures subjected to high temperatures polymerize to form proteinoids, structures that have some peptidelike properties such as nonrandom amino acid frequencies and low-level catalytic activity. Amino acids and even nucleotides can also polymerize in certain types of clays.

If complex molecules were produced from simpler ones long ago, how might they have become organized into cells? Intermolecular forces can bind macromolecules into membrane-enclosed droplets (coacervates and microspheres), which exhibit some features of living systems such as organization, selective permeability, and energy use. That these droplets, like primitive biological systems such as viruses, can self-assemble in nonrandom ways suggests that similar structures might have occurred on the pathway to living organisms. The droplet systems, or protocells, that could best maintain themselves would perpetuate. Later, protocells that could reproduce, however inefficiently, would be acted on by natural selection, which would act as a creative force successively enhancing the probability of advantageous molecules, reactions, and structures.

KEY TERMS

abiotic
adenosine triphosphate (ATP)
amino acids
carbonaceous chondrites
coacervates
coenzymes
condensation
double helix
entropy
enzyme
fatty acids
genotype
hydrolysis
individuals
membranous droplets
messenger RNA (mRNA)
microspheres
molecular preadaptation
monomers
nucleic acids
nucleotides
optical activity
panspermia
peptide linkages
phenotype
phosphate bond
polymerization
polypeptide chains
population
porphyrins
proteinoids
protocells
purines
pyrimidines
racemic mixture
ribosomal RNA (rRNA)
second law of thermodynamics
selective permeability
self-assembly
sugars
transfer RNA (tRNA)
vesicles

DISCUSSION QUESTIONS

1. DNA
   a. What explains the information-carrying capacity of DNA?
   b. How does DNA replicate?
   c. How does information transfer from DNA to protein?
2. What is the difference between genotype and phenotype?
3. The origin of life from previous life
   a. What proposals suggest that life originated extraterrestrially?
   b. Why have scientists not generally accepted these proposals?
4. The terrestrial origin of life
   a. What prevailing environmental conditions would have enabled the origin of life on Earth?
   b. What chemical compounds and reactions may explain the early origin of basic biological molecules? in what possible quantities?
   c. What reactions would have condensed such basic organic molecules into macromolecules?
5. Proteinoids
   a. How are proteinoids formed?
   b. What biological properties do they possess?
6. Membranous droplets
   a. How do membranous droplets permit increased control and organization of biological molecules, structures, and reactions?
   b. Why does entropy decrease in such systems?
   c. What are coacervates, and what lifelike properties do they have?
   d. What are microspheres, and what lifelike properties do they have?
7. Selection
   a. What conditions allow selection to occur?
   b. Can selection occur among nonreproductive entities?
   c. Can selection produce biological compounds and structures that, most probably, would not have arisen spontaneously?

EVOLUTION ON THE WEB

Explore evolution on the web! Visit the accompanying web site for *Evolution,* 3/e at www.jbpub.com/evolution for web exercises and links relating to topics covered in this chapter.

REFERENCES*

Booij, H. L., and H. G. Bungenberg de Jong, 1956. *Biocolloids and Their Interactions.* Springer-Verlag, Vienna.

Brandes, J. A., N. Z. Boctor, G. D. Cody, B. A. Cooper, R. Hazen, and H. S. Yoder, Jr., 1998. Abiotic nitrogen reduction on early Earth. *Nature,* **395,** 365–367.

Brooks, J., and G. Shaw, 1973. *Origin and Development of Living Systems.* Academic Press, New York.

Buhl, D., 1974. Galactic clouds of organic molecules. *Origins of Life,* **5,** 29–40.

*Many seminal conceptual papers covering different topics of the origin of life have been brought together and republished in the collection by Deamer and Fleischaker, listed here.

Cairns-Smith, A. G., 1982. *Genetic Takeover and the Mineral Origins of Life.* Cambridge University Press, Cambridge, England.

Calvin, M., 1969. *Chemical Evolution.* Oxford University Press, Oxford, England.

Chang, S., D. DesMarais, R. Mack, S. L. Miller, and G. E. Strathearn, 1983. Prebiotic organic synthesis and the origin of life. In *Earth's Earliest Biosphere: Its Origin and Evolution,* J. W. Schopf (ed.). Princeton University Press, Princeton, NJ, pp. 53–92.

Crick, F., 1981. *Life Itself: Its Origin and Nature.* Simon and Schuster, New York.

Deamer, D. W., 1986. Role of amphiphilic compounds in the evolution of membrane structure on the early Earth. *Origins of Life,* **17,** 3–25.

———, 1993. Prebiotic conditions and the first living cells. In *Fossil Prokaryotes and Protists,* J. H. Lipps (ed.). Blackwell Scientific, Boston, pp. 11–18.

Deamer, D. W., and G. R. Fleischaker, 1994. *Origins of Life: The Central Concepts.* Jones and Bartlett, Boston. (Contains classic papers by Haldane, Oparin, Urey, Miller, and many others, covering areas such as early environmental conditions, prebiotic chemistry, self-assembly, biochemical energetics, and informational molecules.)

Eigen, M., 1971. Self-organization of matter and the evolution of biological macromolecules. *Naturwiss.,* **58,** 465–523.

———, 1983. Self-replication and molecular evolution. In *Evolution from Molecules to Man,* D. S. Bendall (ed.). Cambridge University Press, Cambridge, England, pp. 105–130.

Eigen, M., and P. Schuster, 1979. *The Hypercycle.* Springer-Verlag, Berlin.

Emiliani, C., 1992. *Planet Earth: Cosmology, Geology, and the Evolution of Life and Environment.* Cambridge University Press, Cambridge, England.

Ferris, J. P., and P. C. Joshi, 1978. Chemical evolution from hydrogen cyanide: Photochemical decarboxylation of orotic acid and orotate derivatives. *Science,* **201,** 361–362.

Ferris, J. P., A. R. Hill, Jr., R. Liu, and L. E. Orgel, 1996. Synthesis of long prebiotic oligomers on mineral surfaces. *Nature,* **381,** 59–61.

Fox, S. W., 1984. Proteinoid experiments and evolutionary theory. In *Beyond Neo-Darwinism,* M.-W. Ho and P. T. Saunders (eds.). Academic Press, London, pp. 15–60.

Fox, S. W., and K. Dose, 1972. *Molecular Evolution and the Origin of Life.* Freeman, San Francisco.

Haldane, J. B. S., 1929. The origin of life. *The Rationalist Annual,* **148,** 3–10.

Hoyle, F., and N. C. Wickramasinghe, 1993. *Our Place in the Cosmos.* Dent, London.

Imai, E., H. Honda, K. Hatori, A. Brack, and K. Matsuno, 1999. Elongation of oligopeptides in a simulated submarine hydrothermal system. *Science,* **283,** 831–833.

Jay, D. G., and W. Gilbert, 1987. Basic protein enhances the incorporation of DNA into lipid vesicles: Model for the formation of primordial cells. *Proc. Nat. Acad. Sci.,* **84,** 1978–1980.

Kaler, E. W., A. K. Murthy, B. E. Rodriguez, and J. A. N. Zasadzinski, 1989. Spontaneous vesicle formation in aqueous mixtures of single-tailed surfactants. *Science,* **245,** 1371–1374.

Kauffman, S. A., 1993. *The Origins of Order: Self-organization and Selection in Evolution.* Oxford University Press, New York.

Kenyon, D. H., and G. Steinman, 1969. *Biochemical Predestination.* McGraw-Hill, New York.

Kerr, R. A., 1997. Life goes to extremes in the deep earth—and elsewhere? *Science,* **276,** 703–704.

Lawless, J. G., and G. U. Yuen, 1979. Quantification of monocarboxylic acids in the Murchison carbonaceous meteorite. *Nature,* **251,** 40–42.

Lowe, D. R., 1994. Early environments: Constraints and opportunities for early evolution. In *Early Life on Earth* (Nobel Symposium No. 84), S. Bengtson (ed.). Columbia University Press, New York, pp. 24–35.

Matsuno, K., K. Dose, K. Harada, and D. L. Rohlfing (eds.), 1984. *Molecular Evolution and Protobiology.* Plenum Press, New York.

Miller, S. L., 1953. A production of amino acids under possible primitive Earth conditions. *Science,* **117,** 528–529.

———, 1974. The atmosphere of the primitive Earth and the prebiotic synthesis of amino acids. *Origins of Life,* **5,** 139–151.

———, 1992. The prebiotic synthesis of organic compounds as a step toward the origin of life. In *Major Events in the History of Life,* J. W. Schopf (ed.). Jones and Bartlett, Boston, pp. 1–28.

Miller, S. L., and L. E. Orgel, 1974. *The Origins of Life on the Earth.* Prentice Hall, Englewood Cliffs, NJ.

Morowitz, H. J., 1992. *Beginning of Cellular Life: Metabolism Recapitulates Biogenesis.* Yale University Press, New Haven, CT.

Nomura, M., 1973. Assembly of bacterial ribosomes. *Science,* **179,** 864–873.

Oparin, A. I., 1924. *Proiskhozhdenie Zhizny* ("The Origin of Life"). Moscovsky Robotschii, Moscow. (Original Russian edition of Oparin's theory; a revised edition was published in English 1938, and reprinted 1953 by Dover Publications.)

Oró, J., and A. P. Kimball, 1961. Synthesis of purines under possible primitive Earth conditions. I. Adenine from hydrogen cyanide. *Arch. Biochem. Biophys.,* **94,** 217–227.

Paecht-Horowitz, M., J. Berger, and A. Katchalsky, 1970. Prebiotic synthesis of polypeptides by heterogeneous polycondensation of amino acid adenylates. *Nature,* **228,** 636–639.

Parsons, P., 1996. Dusting off panspermia. *Nature,* **383,** 221–222.

Schopf, J. W. (ed.), 1983. *Earth's Earliest Biosphere: Its Origin and Evolution.* Princeton University Press, Princeton, NJ.

Shklovskii, I. S., and C. Sagan, 1966. *Intelligent Life in the Universe.* Holden-Day, San Francisco.

Smith, J. V., 1998. Biochemical evolution. I. Polymerization on internal organophilic silica surfaces of dealuminated zeolites and feldspars. *Proc. Nat. Acad. Sci.,* **95,** 3370–3375.

Strickberger, M. W., 1985. *Genetics,* 3d ed. Macmillan, New York.

Van Holde, K. E., 1980. The origin of life: A thermodynamic critique. In *The Origins of Life and Evolution,* H. O. Halvorson and K. E. Van Holde (eds.). Liss, New York, pp. 31–46.

Wald, G., 1974. Fitness in the universe: Choices and necessities. *Origins of Life,* **5,** 7–27.

8 Proteins and the Genetic Code

In general, function in living organisms depends on transforming material and energy outside the organism into processes that take place within it. These living processes originally must have existed on a fairly simple level, not much different from some of the processes we have seen in coacervates and microspheres (Chapter 7). For a long time these processes may have depended at least partly on heat to provide some of the reactions that produce various cellular constituents. However, thermal energy, which may vary in place and in time, would hardly have been reliable and consistent enough for cellular needs. At best, heat from the environment or from rapid oxidations would have provided only an explosive, uncontrolled release of energy. More valuable to the cell was the development of chemical energy providers, such as adenosine triphosphate (ATP), which can release small but significant amounts of phosphate-bond energy that the enzyme apparatus of the cell can both control and localize to specific reactions.[1]

The shift to chemical systems of energy meant the elaboration of **organic catalysts** (enzymes), which could restrict chemical reactions to the most opportune times and places. Accompanying this shift must have been a remarkable increase in the efficiency with which particular reactions could take place. We know that catalysts function by lowering the energy level necessary for a reaction, thereby increasing its frequency. Enzyme proteins add even more speed to this process by providing specific sites at which potential reacting molecules can be localized and manipulated to enhance the reaction.

[1] Phosphorylated nucleotides such as adenosine (ATP, ADP) or guanosine (GTP, GDP) are not the sole agents capable of transferring energy-rich phosphate bonds. In a reaction that Siu and Wood discovered, phosphorylated enolpyruvate served as a phosphate donor and inorganic phosphate (P_i) as an acceptor, yielding inorganic pyrophosphate (PP_i): phosphoenolpyruvate (PEP) + CO_2 + $P_i \leftrightarrows$ oxaloacetate + PP_i. Lipmann (1965) suggested that this reaction represents the metabolic fossil of a primitive form of energy transfer, $2P_i \leftrightarrows PP_i + H_2O$ (see also p. 126). Other reactions support this view, and Baltscheffsky and Baltscheffsky point out:

> The suggested role of PP_i in early biological conversion of energy is based on its formation in bacterial photophosphorylation, its capabilities as biological energy and phosphate donors, its comparatively uncomplicated structure, its occurrence as mineral, and its formation from hot volcanic magma.

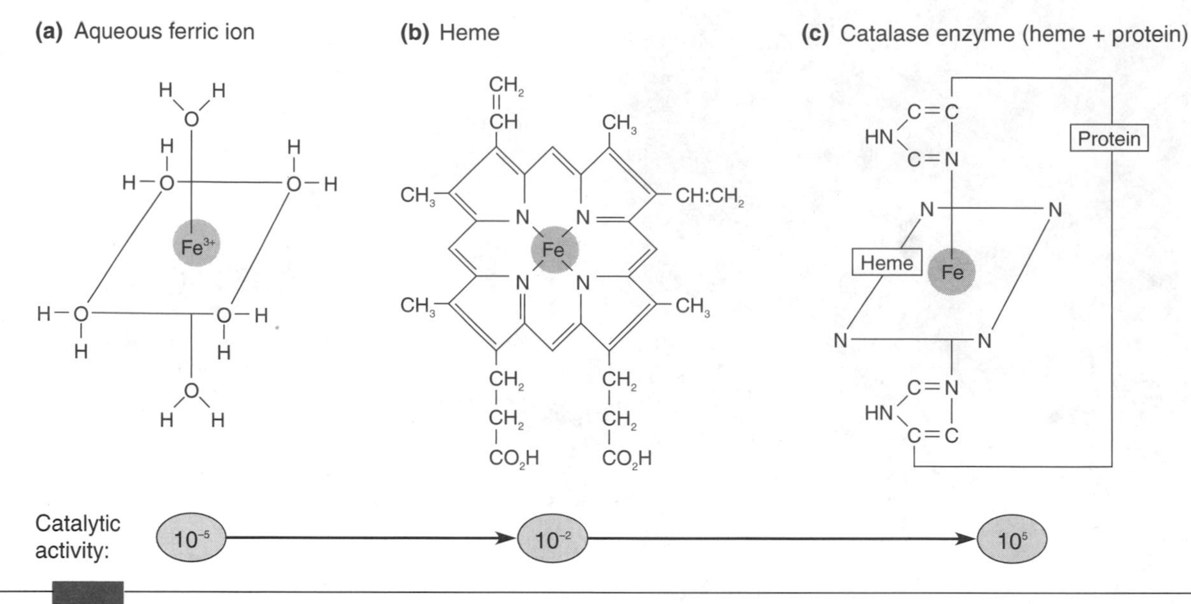

FIGURE 8-1 Change in catalytic activity for the reaction $2H_2O_2 \rightarrow 2H_2O + O_2$ when the iron atom is used by itself (*a*), or in different molecular combinations (*b, c*). The catalase protein in (*c*) provides an enzymatic advantage to the reaction because it binds rapidly to hydrogen peroxide molecules and distorts them so their decomposition proceeds at a lower "activation energy" than without the enzyme. (*Adapted from Calvin.*)

For example, inorganic ferric ion (Fe^{3+}) shows some catalytic activity in a variety of reactions, including the decomposition of hydrogen peroxide into water and oxygen (Fig. 8–1*a*). However, when such ions incorporate into porphyrin molecules to form **heme** (Fig. 8–1*b*), the molecules are about a thousand times more effective than Fe^{3+} alone. If the protein component of the enzyme catalase then adds to the heme unit, catalytic efficiency increases by a further factor of 1 billion (Fig. 8–1*c*). How did such proteins evolve?

Proteins or Nucleic Acids First?

At present, the amino acid sequences of the enzymes that serve as catalysts in the cell derive entirely from the nucleotide sequences in ribonucleic acid (RNA), which in turn derive from the nucleotide sequences of deoxyribonucleic acid (DNA), as discussed in Chapter 7. The fact that the entire chain of information transfer—replication (DNA → DNA), transcription (DNA → RNA), translation (RNA → protein)—depends on appropriate enzymes (Fig. 8–2) poses the serious question of how these functional and informational systems could possibly have evolved independently of each other.

One answer to this problem has been that nucleic acids arose first, and their self-replicating power then enabled selection to develop protein systems that would support further **self-replication.** The geneticist Hermann Muller long ago suggested that since the basic appearance, or phenotype, of an organism derives essentially from its genetic material, or genotype, this relationship must also have existed in the past. That is, the genotype was probably first in the evolutionary sequence. **Protein synthesis** might then have evolved through specific amino acids directly interacting with specific nucleotide sequences, or perhaps through indirect placement of amino acids into such sequences by intermediary **adaptor molecules** that brought amino acids to the nucleotide chain.

Supporters of the view that nucleic acid replication arose first have generally argued that only a self-replicating system can provide the basis on which selection can build a cooperative functional unit. In the absence of self-replication, function presumably would quickly disappear and the "organism" could not maintain itself. Scientists have sought for an **autocatalytic** process to explain the origin of a "naked gene" that could replicate itself without the help of proteins.

The discovery that some RNA molecules do possess catalytic properties even in the absence of proteins (Cech and Bass) has given some support to the possibility of autocatalytic **nucleic acid replication.** Presumably, by using such **RNA catalysts** or **ribozymes,** short RNA sequences theoretically could replicate themselves without protein enzymes by forming templates for complementary RNA sequences (Loomis). Some researchers have therefore proposed the very early existence of an **RNA world** dependent on such self-replication RNA nucleotide sequences. Opposing these proposals, Joyce and coworkers have pointed out that such RNA replication strongly would have been

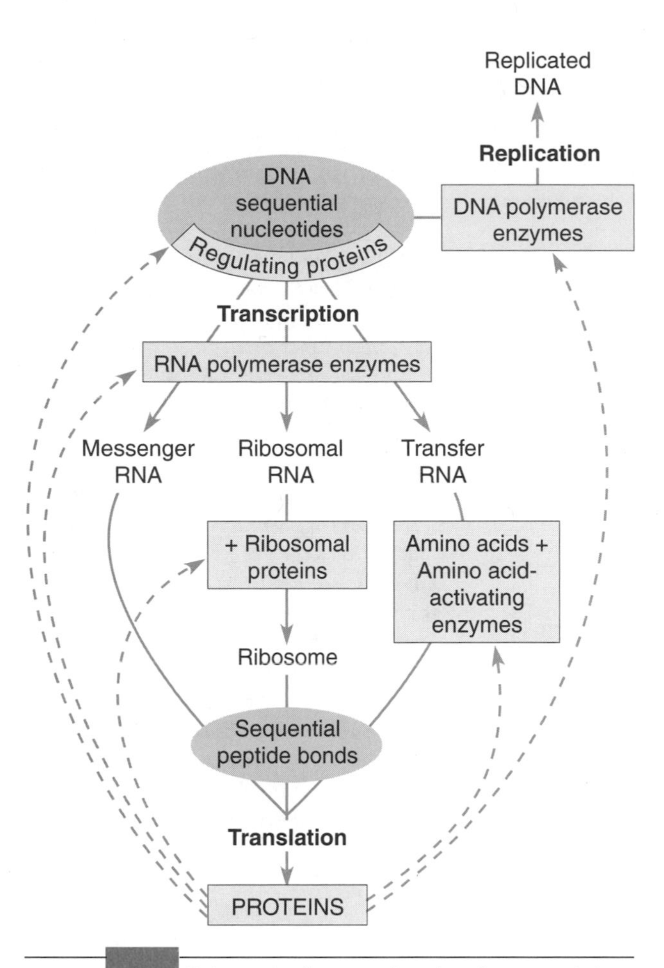

FIGURE 8–2 Schematic diagram showing the mutual dependence of information carried by nucleotide sequences and function governed by proteins. Solid lines indicate the general directions of information transfer, and *dashed lines* point to proteins that this process synthesizes. Clearly, the nucleotide sequence information determines the amino acid sequences of proteins, and proteins in turn regulate and catalyze the transfer of nucleotide information. How could one process have developed without the other?

inhibited in prebiological times because different stereoisomers (Fig. 7–7) in the ribose–sugar backbone would have prevented complementary base pairing. The consistent prebiological supply of ribose needed for such backbone formation has also been questioned because, like many sugars, ribose is extremely unstable (Larralde et al.).

Joyce and coworkers therefore suggest that early nucleic acid genetic material was not based on ribose-containing nucleotides but on riboselike analogues in which such pairing difficulties were minimized and that were more easily synthesized than ribose. One such system has been experimentally demonstrated by Zielinski and Orgel. Nevertheless, numerous questions remain as to the range of reactions that RNA could catalyze (its present range is quite limited) and how such systems, even if they occurred, would have made the transition to RNA synthesis and replication. Some of the views concerning the existence of an RNA world are presented in Box 8–1.

Instead of nucleic acids, Cairns-Smith suggests that the primitive self-replicating unit may have consisted of organic, claylike silicate crystals or layered clay mixtures (for example, montmorillonite, see p. 130). Crystals can grow by adding subunits into their highly ordered structures, and thus we can view them as having some self-replicatory powers. According to Cairns-Smith, nucleic acids such as RNA could have incorporated into such an assembly, followed by the formation of peptides and the evolution of a protein-synthesizing system.

Although mineral surfaces are gaining more attention as a locale for biosynthetic reactions (p. 131), their ability to develop a highly complex metabolic sequence extending from self-replicating RNA to protein synthesis is still difficult to visualize.[2] Some biologists have therefore emphasized the possibility that either proteins themselves or protein-nucleic acid combinations were the first self-replicating systems. Black, for example, suggests that some early peptides might have served as templates for the aggregation of nucleotides that may then have bonded together to form a precise mold for the replication of these same peptides. This view has the advantage of offering a mechanism by which the replication of both peptide chains and nucleotide chains would have been interdependent from the start. Also, as one chain lengthens during further evolution, so does the other, each gradually improving its replicatory role until some of the present features of nucleic acid replication and protein synthesis evolve. Unfortunately, as with naked genes, we still find it difficult to imagine the spontaneous origin of such a precisely organized nucleic acid template.

Various theories propose that protein systems probably developed diverse functional properties before they coupled to nucleic acid replicative systems. Foremost among the evidence for this view is the likelihood that proteins always had more functionally different forms than nucleic acids. Part of this functional variety arises from the almost inexhaustible array of permuted amino acid sequences that proteins can achieve.

For example, since 20 different amino acid alternatives exist for each position in a polypeptide, a sequence of 5 amino acids has 20^5, or more than 3 million possible arrangements. By contrast, a sequence of 5 nucleotides in a nucleic acid has only 4^5 or 1,024 possibilities. Moreover, many amino acids are quite different in structure (Fig. 7–2), enabling them to interact in many different ways both with each other and with molecules such as water, metal ions, and various monomers and polymers. These

[2] Among other problems, ribose is a lesser product in synthesis of sugars from formaldehyde (p. 124), and ribonucleotides can easily be degraded or functionally inhibited by other chemical structures. According to Joyce and coworkers, "the accumulation of substantial quantities of relatively pure mononucleotides on the primitive Earth is high implausible."

BOX 8-1 THE RNA WORLD

Because it is difficult to conceive how informational nucleotide sequences could have evolved simultaneously with the polypeptide sequences necessary to replicate them, researchers place emphasis on discoveries that some RNAs possess catalytic activity that might have enable self-replication. The possibility that such an early RNA world antedated cellular enzymatic proteins comes from various findings:

- The "intron" portion of transcribed RNA (p. 175) in the protozoan ciliate *Tetrahymena* splices itself out of the RNA molecule and helps form a chemical bond between the protein-coding RNA sections ("exons") on either side (Cech).
- Some cellular RNA molecules catalyze reactions that include binding to ATP, the common energy transfer molecule (Sassanfar and Szostak).
- Researchers can design synthetic RNA molecules to perform precise catalytic reactions (Haseloff and Gerlach).
- Selection experiments in the laboratory have led to the evolution of RNAs into new kinds of molecules that show catalytic activities many orders of magnitude greater than in the initial mixture (Lehman and Joyce). Some such selected ribozymes can even combine a ribose sugar with a thiouracil base to make nucleotides at a rate more than ten million times greater than the uncatalyized reaction (Unrau and Bartel).
- Some RNA molecules function as gene regulators by binding to nucleic acids and affecting gene expression (Wicken and Takayama).
- RNA "fragments" appear in coenzymes used in various metabolic reactions (for example, coenzyme A, FAD, NAD). Since other chemicals could have assumed the function of these RNA fragments, researchers consider their presence "historical"—a presumed remnant of the earlier RNA world (Benner et al.).
- The protein translation system that all cells use depends on RNA—messenger RNA, ribosomal RNA, and transfer RNA. In fact, peptide formation catalyzed by ribosomal enzymatic action, called "peptidyl transferase," apparently depends more on ribosomal RNA than on ribosomal proteins (Noller et al.). Zhang and Cech demonstrate that even a selected noncellular ribozyme can perform this same amino acid-binding function.

These findings emphasize the question whether RNA could have originally served as a self-replicating molecule by acting as both template and catalyst; that is, did it "translate" itself into a sequence of RNA nucleotides or ribozymes whose enzymatic activity included RNA replication? The answer is not simple. In general, as the fidelity of replication increases, longer RNA molecules can persist even when their survival relative to other molecules is low.

For example, we can calculate that 90 percent fidelity of replication may conserve a molecule no longer than 12 nucleotides (at some given level of relative survival), whereas 95 percent fidelity of replication will conserve a molecule about twice as long. Similarly, superiority of survival influences length: the greater its relative superiority over other molecules, the longer the sequence of nucleotides that can be maintained for some given fidelity of replication. But fidelity of replication has an essential requirement: that an RNA molecule be long enough to act as a replicase enzyme. So far this length is unknown, but Joyce and Orgel suggest that the minimum length for RNA catalytic activity is possibly "a triple stem-loop containing 40-60 nucleotides (Fig. 8-3)." Such an RNA replicase could hardly have arisen by chance, nor can we visualize how intermediary steps in its evolution would have been functional.

Other problems of RNA replication fidelity arise because known RNA polymerase enzymes lack the "proofreading" attributes of DNA polymerases, and thus produce relatively high mutation rates (pp. 146 and 224). Although high mutation rates might seem an advantage in a rapidly evolving world, they would cause increasingly inefficient enzymes with each mutational generation. Moreover, RNA sequences, especially guanine nucleotides, coil back on themselves to produce tangled strands impossible to replicate without intervention of proteins, such as the Qβ virus "replicase" described on p. 290. How could RNA sequences duplicate by RNA alone?

Furthermore, synthesis of an enzymatic protein that can replicate RNA, even without proofreading, probably requires no less than 1,000 nucleotides, coding for at least 300 or more amino acids—a number too large for RNA itself to replicate without an already existing replicase enzyme. According to one proposal, this dilemma ("enzymes need coding by long RNA sequences, and long RNA sequences need replicating by long enzymes") may have been overcome by smaller functional polypeptide subunits, each coded by smaller, more easily replicated RNA sequences.

In a system called a **hypercycle** (Eigen and Schuster, see also Eigen), these subunits enter into a symbiotic relationship conferring advantages to the overall system by successively coupling their individual effects—one subunit aiding replication of the other—thus forming a network that can produce a much longer, more accurately

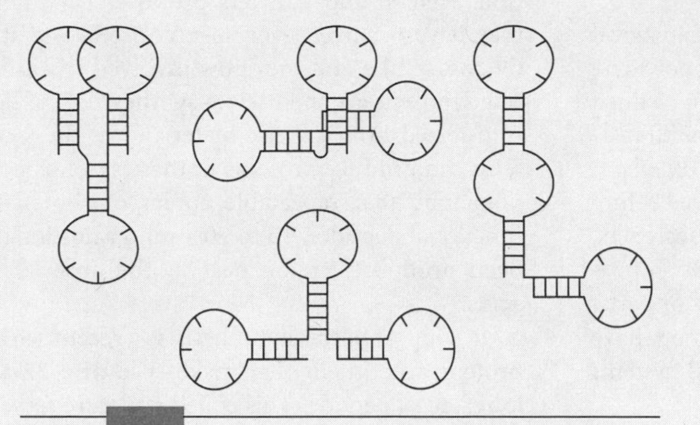

FIGURE 8-3 Some triple stem-loop structures that RNA molecules can assume. (*From Joyce, G. F., and L. E. Orgel, 1993. Prospects for understanding the origin of the RNA world. In* The RNA World, *R. F. Gesteland and J. F. Atkins (Eds.). Reprinted by permission.*)

replicated, sequence. If such system is compartmentalized, the success of the compartment ("cell") and the success of its symbiotic components become mutually dependent—a form of "group selection" discussed later (Chapter 23). To these proposals, Maynard Smith and Szathmáry add other systems that could also have accounted for an increase in genetic information.

Despite its attractiveness, however, further doubts about an RNA world stem from what we know of protein synthesis. Although homologies between ribosomal RNA sequences from all cellular organisms indicate they share a common ancestral function, homologies between ribosomal proteins also show that proteins took part in this ancestral role. Furthermore, making proteins via ribosomes is hardly evidence of an RNA world devoted only to making more RNA. As Moore puts it, "Why would a device for making polypeptides evolve in an organism that had no use for protein?" Also, even if some proteins were of value to RNA, we can still question whether or how RNA replication alone could initiate the separate highly intricate and energy-expensive translation process needed to produce such proteins. Translation is probably the most complex biochemical cellular process, needing more than 120 different molecular elements ranging from messenger RNA to ribosomes and their many protein and RNA accessories.[3]

RNA-protein questions still remain: Why is RNA still here? Why is RNA still necessary for translation? Why haven't proteins assumed all RNA catalytic functions? What were the functions of RNA in the origin-of-life period?

From current knowledge, we can say that RNA still exists in cells because no successful alternative presented itself in the past as an intermediary between genetic material and protein translation. Through base pairing, RNA serves as a "messenger" to ribosomes (messenger RNA), and through codon-anticodon pairing it serves as an "adaptor" transferring amino acids to polypeptides (transfer RNA). We do not fully understand its role in the ribosome itself, but judging from the complexity of ribosomal RNA, probably various types of nucleotide pairing and interaction are needed. For example, Purohit and Stern demonstrated that a nucleotide sequence in ribosomal RNA mediates base pairing between the codons and anticodons used in translation. Perhaps once pathways evolved using RNA for these or other purposes (for example, intron splicing), substituting a different molecular mechanism would have entailed too many widespread changes, causing lethality (see "frozen accidents," p. 150).

Although we have not resolved many questions, there may well have been a world in which RNA was the only genetic material, in which it was the primary or sole constituent of ribosomes, and in which it performed various catalytic functions, even more than at present. Illangasekare and coworkers have shown that RNA could have substituted for at least one enzyme in protein synthesis by finding (through selection) an RNA molecule that could catalyze its own attachment to an amino acid, a function normally assumed by special amino acid activating enzymes. In fact, Ribas de Pouplana and coworkers show that an amino acid activating enzyme used in protein synthesis was likely preceded during evolution by a transfer RNA. Furthermore, even RNA polymerization by RNA alone has been shown possible by Ekland and Bartel who selected an RNA ribozyme that can copy a 6-ribonucleotide-long template.

Encouraging as these findings are, an RNA world capable of catalyzing complex reactions and replicating long RNA molecules efficiently and faithfully has not been demonstrated. Also, if an RNA world did occur, serious questions remain: Was such world the earliest of self-replicating organic worlds or only one step in a progression? Did the RNA world include or exclude proteins (Orgel)? Answers, if possible, depend on further molecular investigations and reconstructions.

[3] According to Fraser and coworkers, even the smallest of cellular organisms (*Mycoplasma genitalium*) need a minimum of 90 different proteins for translation and about 30 for DNA replication.

differences confer an astronomical variety of possible three-dimensional configurations on a protein, in contrast to the relatively more rigid shapes assumed by many nucleic acids.

Chemical experiments under presumed prebiological conditions also show the difficulty of producing polymerized nucleic acids spontaneously, whereas long-chained polypeptides are produced in such experiments with relative ease. A number of authors (for example, Fox) claim that a protein catalytic system must have developed before a nucleic acid replicative system. Many other researchers, however, feel this view has a serious shortcoming: If proteins arose first and were used by primitive cells or particles for functional purposes, how could they have replicated without a nucleic acid translational system? Could proteins alone have synthesized proteins?

If we consider only present organisms, it is difficult to conceive of protein synthesis independent of nucleic acids since no apparent complementary relationships exist between amino acids as do between nucleotides. That is, the precise **stereochemical fit** that occurs between the base pairs of complementary nucleotide chains (adenine–thymine, adenine–uracil, cytosine–guanine) and that accounts for the replicative, transcriptional, and translational properties of nucleic acids nowhere echoes in a similar complementary stereochemical fit between amino acids in polypeptide chains. Nevertheless, a process does exist in which proteins make proteins.

As Lipmann (1971) and others showed, spore-forming *Bacillus brevis* bacteria produces at least two antibiotics, gramicidin S and tyrocidin, that are formed exclusively by enzymes in the absence of messenger RNA (mRNA). Both these antibiotic molecules are circular **oligopeptides** ten amino acids long and are synthesized by the sequential addition of amino acids. Should one amino acid be omitted, peptide synthesis ceases, indicating that the enzyme involved functions as a precisely ordered template for the amino acid sequence. An unfilled position on the template prevents bonding between amino acids on either side.

Of special interest is the form in which peptides elongate among these antibiotics. Single amino acids bind to sulfhydryl (–SH) groups on the enzyme before they join the peptide chain and are then connected by the sequential removal of their sulfur (thiol) groups and the formation of peptide bonds. The chain maintains a thiol at its "head" end to furnish the connection for sequential growth. Lipmann called this process **head–growth polymerization** and pointed out its striking similarity to the polymerization of carbon groups during fatty acid synthesis (also polymerized by use of sulfhydryl bonds) and to the polymerization of amino acids during ribosomal peptide synthesis (polymerized by phosphate bonds).[4]

Lipmann suggested that these polymerizing similarities indicate a common underlying polymerization process that may have arisen early during chemical evolution. Heinen and Lauwers propose that thiol synthesis may have occurred even in a nonreducing atmosphere. Of course, the enzymes now involved in antibiotic peptide synthesis are themselves synthesized via information transferred from genetic material, but the fact that proteins can produce proteins in these systems points to the possibility that repeatable copies of short-chained but functional peptides, 15 to 20 amino acids long, may have been produced in the past in the absence of nucleic acids.[5]

In support of this hypothesis, is a recent self-replicating protein experiment that makes use of a 32-amino acid long helical peptide. This coiled structure serves as a template to bond 17- and 15-amino acid fragments, which then become templates for further replication (Lee et al.). Although dependent on laboratory conditions and components, such results imply that small α-helical catalytic structures can self-replicate once they have evolved. Matrices on which such reactions could have initially organized may have been the clays and zeolites mentioned in the previous chapter, or mineral surfaces such as iron sulfides discussed by Edwards.

Evolution of Protein Synthesis

Whatever the early composition of the templates used in condensing amino acids, the polymerization process itself must have been one of the first functions for which selection occurred, since only in peptide form do amino acids attain their catalytic properties. These early templates were probably inefficient, producing peptide chains with only vague similarities to one another.

As time went on, selection for improved polymerization must have led to the production of more efficient templates, which were perhaps themselves polymerized by more efficient **polymerization enzymes.** Evolution of a template capable of producing an enzyme whose amino acid chain was long enough to polymerize the template

[4] For a proposal that thiol (–SH) groups reacting with carboxylic acids to form thioester bonds –S–CO–) furnished the first mechanism for polymerizing amino acids, see De Duve.

[5] Perhaps relevant to this view are the "prion diseases" responsible for neurological degeneration such as Creutzfeldt-Jakob disease in humans and "mad cow" disease in cattle. The prion protein, normally harmless, can undergo a pathogenic change in its three-dimensional shape, which then converts other such proteins to a similar form. Through such "domino effect," a prion disease, acting remarkably like a non-nucleic acid infectious agent, increasingly develops as prion proteins increasingly change to pathogenic form (Prusiner). Thus, although prion protein sequences are genetically determined, the interactions that change their conformation can be transmitted and reproduced between cells and individuals without further genetic information.

itself, was probably a most critical step in the transition to a self-replicating molecular life form. However this selection occurred, whether for successively lengthened templates or improved enzymatic activity or through both processes (see, for example, Szathmáry 1989), the impact of increasing the amount of coded genetic material would have been profound. It seems reasonable that the circular, autocatalytic tautology of life—to make more of those substances that can interact with the environment to make more such substances—was firmly established during this early period.[6]

Given a template for the synthesis of peptides and a polymerization enzyme, various sequences of events might have followed. The value of considering any of these is simply to show that enough molecular information has been accumulating to permit the development of different step-by-step hypotheses that can be used to solve the origin-of-life puzzle. If we continue with the notion that the first templates were probably not pure nucleic acids but perhaps some combination of peptides and other materials, one possible sequence is as follows:

1. Once peptide polymerization established itself, selection for its improvement must have led to an increased number of polymerase enzymes per protocellular unit, as well as improved template efficiency of these polymerases and their ability to increase the rapidity of amino acid condensation.

2. Improvements in the mechanism that brought amino acids to the template probably accompanied improvement in polymerase function. That is, adaptor molecules evolved that could bind to amino acids in the medium, and then could bind these amino acids to the polymerase enzyme or the polymerizing template. These adaptor molecules were probably not very large, and their evolution might simply have meant new uses for some of the templates that had previously produced polymerase enzymes.

3. Further improvement probably involved selection for small, nucleotidelike coenzymes used in binding and releasing phosphate energy groups, similar perhaps to present coenzymes such as nicotinamide adenine dinucleotide (NAD) and flavin adenine dinucleotide (FAD), discussed in the next chapter. Such nucleotides were not necessarily RNA or DNA but could have consisted of combinations of various sugars, nucleic acid bases, and phosphates.

4. Both the template and the adaptor molecules may have incorporated nucleotide sequences, which could then recognize each other by complementary base pairing via hydrogen bonding similar to the type presently seen between adenine and uracil or guanine and cytosine. These new adaptors could therefore bring their amino acid to the nucleotide portions of the polymerization complex rather than to the peptide portions. Protein synthesis would gain the advantage of precise stereochemical pairing between complementary nucleotide bases as compared to the more inefficient amino acid–polypeptide interactions previously used. Such events would lead directly to improved placement of different amino acids into specific positions on polypeptide chains and to a much increased number of amino acids that could be incorporated into such chains.

5. If we consider the polymerization enzyme complex to be a primitive ribosome of sorts, some of its nucleotide components began to act as a template specifying the sequence of amino acid incorporation by base pairing with adaptor molecules. At the same time, these nucleotide sequences themselves could replicate with considerable precision by base pairing with available nucleotides.

6. Replicating nucleotides led to forming master nucleic acid molecules that could be stored as genetic material, and that could also serve in translating their sequences into amino acid sequences or produce complementary messenger strands for this purpose. Three separate functional classes of nucleotide sequences eventually arose: storage, messenger, and translational (ribosomal and transfer)—all probably RNA, since this nucleic acid still fills two of these purposes in all organisms (messenger and adaptor) and also fills genetic storage purposes in some viruses (see also Box 8–1).[7]

7. Since RNA was probably used primitively for both information storage and protein translation, difficulties in separating these two functions must have offered advantages to organisms that could use a different nucleic

[6] Black suggests that the original force responsible for polymerizing organic molecules arose because of hydrophobic interactions between various amino acids ("a search by organic compounds for a means to separate from water"). The most stable of such interactions that would dissipate hydrophobic forces and produce the lowest free energy level presumably lies in the folded globular organization of polymerized protein. Black purposes that the energy to attain these polymerizations derived from simultaneous degradation of other organic compounds. The coupling of these two processes, degradation and polymerization, would then provide the framework enabling the selection of polymers that could catalyze their own polymerization. Once such autocatalytic polymers arose, evolutionary refinements would have emphasized selection for other attributes, such as a code for polymer replication, metabolic pathways, response to environmental substances, homeostasis, and development.

[7] Dyson suggests two separate origins of life: one based on proteins and the other based on some form of replicating RNA nucleotides. According to him, RNA first entered protein-based cells as a replicating "infection," which later became incorporated as a more helpful symbiont that improved host cell replication. The likelihood that eukaryotic mitochondria and chloroplasts originated from symbiotic infections by prokaryotes (Chapter 9) indicates to him that such symbioses have a long history that may extend back to nucleic acids themselves.

acid, DNA, for storage purposes. The enzymes that translate RNA into protein do not function with DNA, thus restricting the more uniformly structured, double-helix DNA exclusively to the storage of information and to transcribing one of its strands to form mRNA.[8] The observation that deoxyribonucleotides are synthesized from ribonucleotides in cellular pathways supports the notion that DNA arose later in cellular evolution than RNA.

8. A question often raised is why both RNA and DNA are limited to four different bases, each matched with a single other base during complementary base pairing (adenine–uracil and guanine–cytosine in RNA, and adenine–thymine and guanine–cytosine in DNA). Among possible answers is the vulnerability of base pairing to mutation frequency. It seems likely that the greater the kinds of bases involved in genetic replication, the greater the chances for mismatching during the pairing process and the greater the loss of replication accuracy (Szathmáry 1991). Because RNA most likely preceded DNA as genetic material, and RNA polymerases, then as now, could probably not proofread or easily repair mismatching errors, high mutation frequency would have confined genetic material to very short sequences, producing short and inefficient enzymes. To lengthen and improve enzyme function, genetic material would have had to increase in both length and replication accuracy—a process that probably became dependent on restricting the number of different kinds of bases to a genetic alphabet of four code letters replicating via only two complementary base pairings.

The evolution of protein synthesis may thus have had its start in a primitive polymerase enzyme that could replicate inefficiently on a template of polypeptides and other materials. Evolution then proceeded to a self-contained ribosome that carried its own stored genetic information along with mRNA committed exclusively to translating a limited number of polypeptides. In later stages these genetic and translational functions sequestered to different parts of the cell, with mRNA moving from its new site of transcription, where genetic information was now stored, to the ribosome for translation.

[8] Some researchers have also proposed that DNA genetic material would offer a more easily protected molecule against mutation than RNA because the 2' hydroxyl group in the ribosome sugar of RNA causes it to undergo more rapid hydrolytic cleavage than the 2' deoxyribose of DNA. Moreover, the 2' hydroxyl group in RNA apparently provides it with catalytic activity, whereas DNA is relatively inactive—no DNA molecule so far appears to act as a catalyst. In addition, the transition from RNA to DNA as the genetic material may have simply entailed the presence of special **reverse transcriptase** enzymes that can transcribe RNA sequences into DNA sequences. Such enzymes, perhaps originally only involved in RNA replication (they can function as RNA polymerases), could later have served to transfer genetic information from RNA to DNA.

The evolution of new kinds of ribosomes, no longer committed to the production of particular proteins, enabled the same ribosome to translate different mRNAs. This transferred the burden of regulating which proteins to make from the ribosome to the transcriptional process. That is, the particular proteins to produce could now be selected by regulating which mRNA molecules to transcribe from the stored genetic material, a process that eventually gave rise to sophisticated regulatory systems such as that shown in Fig. 10–25.

Accompanying these innovations must have been changes from depending on only few enzymes and proteins, each with multiple functions and lower coupling specificities, to employing greater numbers with restricted functions and higher binding specificity. Such transitions would have followed a simple increase in the number of mRNA molecules by gene duplication. Mutations within duplicated genes and subsequent selection among them would allow individual gene products to diverge and evolve in new and more specific directions (see also pp. 260–261).

Once gene numbers increased, their linkage into chromosomes would have been advantageous by ensuring their collective presence in each cell. Chromosomes also enable linked genes to replicate as a synchronized unit rather than as individual competitors. Linking genes into such cooperative networks would have been similar to the way a long protein could be synthesized by a coordinated group of RNA sequences (Box 8–1). By such means, some of the basic modern features of protein synthesis and its control may have come into being. (For examples of other possible scenarios, see Loomis, and Maynard Smith and Szathmáry.)

Evolution of the Genetic Code

Information transfer between nucleic acids and proteins follows a genetic code that determines the placement of a particular type of amino acid in a protein from the placement of a particular trinucleotide sequence in mRNA. Table 8–1 lists the terminology and characteristics of the code, and Table 8–2 gives the code itself.

As for any other biological trait, researchers believe that this code must have evolved from a more primitive form, although they have so far discovered no ancestral codes. Attempts at an evolutionary reconstruction of the code have generally relied on a detailed analysis of the features that characterize the present code. Ten basic features are as follows:

1. Messenger RNA molecules consist of only four kinds of nucleotide bases, adenine (A), guanine (G), uracil

TABLE 8-1 Definitions of common terms used in describing the present genetic code

Term	Meaning
Code letter	Nucleotide, e.g., A, U, G, C (in mRNA) or A, T, G, C (in DNA). Note that there are four code letters in each nucleic acid "alphabet," forming two kinds of complementary base pairs (A–U and G–C in RNA, and A–T and G–C in DNA)
Codon or **code word**	Sequence of nucleotides specifying an amino acid, e.g., the RNA codon for leucine = CUG (or GAC in DNA)
Anticodon	Sequence of nucleotides on transfer RNA that complements the codon, e.g., GAC = anticodon for leucine (see Fig. 8–6)
Genetic code or **coding dictionary**	Table of all the codons, each designating the specific amino acid into which it translates (Table 8–2)
Codon length, or **word size**	Number of letters in a codon, e.g., three letters in a **triplet code** (these are the same as coding ratio in a non-overlapping code)
Nonoverlapping code	Code in which only as many amino acids are coded as there are codons in end-to-end sequence, e.g., for a triplet code, UUUCCC = phenylalanine (UUU) + proline (CCC)
Redundant or **Degenerate code**	Presence of more than one codon for a particular amino acid, e.g., UUU, UUC = phenylalanine. 20 different amino acids are therefore coded by a total of more than 20 codons
Synonymous codons	Different codons that specify the same amino acid in the redundant code, e.g., UUU = UUC = phenylalanine
Ambiguous code	Circumstances when one codon can code for more than one amino acid, e.g., GGA = glycine, glutamic acid. No ambiguities exist in the present code although such ambiguities may have existed in the past.
Commaless code	Absence of intermediary nucleotides (spacers) between codons, e.g., UUUCCC = two amino acids in triplet non-overlapping code
Reading frame	Particular nucleotide sequence coding for a polypeptide that starts at a specific point and then partitions into codons until it reaches the final codon of that sequence
Frameshift mutation	Change in the reading frame because of the insertion or deletion of nucleotides in numbers other than multiples of the codon length. This changes the previous partitioning of codons in the reading frame and causes a new sequence of codons to be read.
Sense word	Codon that specifies an amino acid normally present at that position in a protein
Replacement mutation	Change in nucleotide sequence, either by deletion, insertion, or substitution, resulting in the appearance of a codon that produces a different amino acid (**missense mutation**) in a particular protein, e.g., UUU (phenylalanine) mutates to UGU (cysteine)
Stop mutation	Mutation that results in a codon that does not produce an amino acid, e.g., UAG (also called a **chain-terminating codon** or **nonsense codon**)
Universal code	Use of the same genetic code in all organisms, e.g., UUU = phenylalanine in bacteria, mice, humans, and tobacco (with some exceptions, e.g., mitochondria, see Table 8–2)

Source: From *Genetics Third Edition* by Monroe W. Strickberger.

TABLE 8-2 Nucleotide sequences in mRNA codons specifying particular amino acids: The "standard" genetic code[a]

UUU, UUC } *phe* UUA, UUG } *leu*	UCU, UCC, UCA, UCG } *ser*	UAU, UAC } *tyr* UAA, UAG } STOP[b]	UGU, UGC } *cys* UGA STOP[b] UGG *trp*
CUU, CUC, CUA, CUG } *leu*	CCU, CCC, CCA, CCG } *pro*	CAU, CAC } *his* CAA, CAG } *gln*	CGU, CGC, CGA, CGG } *arg*
AUU, AUC, AUA } *ile* AUG[c] *met*	ACU, ACC, ACA, ACG } *thr*	AAU, AAC } *asn* AAA, AAG } *lys*	AGU, AGC } *ser* AGA, AGG } *arg*
GUU, GUC, GUA, GUG } *val*	GCU, GCC, GCA, GCG } *ala*	GAU, GAC } *asp* GAA, GAG } *glu*	GGU, GGC, GGA, GGG } *gly*

[a]Rare exceptions to this code occur in various animal mitochondria in which AUA also specifies methionine, and AGA specifies serine or serves as a stop codon (Jukes and Osawa). Such mitochondrial changes seem to be in the direction of economizing in the kinds of transfer RNA produced in a small organelle that makes relatively few polypeptides. Some other exceptions are found in several organisms displayed in Fig. 8-7.

[b]These codons are also called *chain-terminating codons* or, in the past, *nonsense codons.*

[c]This is the common codon used to initiate protein synthesis.

(U), and cytosine (C). These compose chains of varying lengths and varying sequence.

2. An mRNA codon that specifies a particular amino acid is a triplet consisting of a chain of three nucleotides.
3. The code is commaless and nonoverlapping: each codon translates in a continuous, uninterrupted sequence, three successive nucleotides at a time, from one end of an mRNA reading frame to the other.
4. Reading frames in messenger RNA generally begin with the codon AUG, and terminate at the stop codons UAA, UAG, or UGA.
5. The codon sequence complements an anticodon sequence on the adaptor or transfer (tRNA) molecule (Fig. 8–4) that carries a particular amino acid to the mRNA codon.
6. All living organisms share the ("universal") coding dictionary of Table 8–2 with some codon differences in mycoplasmas (bacteria lacking polysaccharide cell walls) and ciliate protistans. Mitochondrial organelles also show a few codon differences from those cellular nuclei use.
7. Ambiguities have not been found in the code; that is, the same codon does not specify two or more different amino acids.
8. With the exception of methionine and tryptophan, more than one codon designates each amino acid.
9. The pattern of code redundancy is mostly in the third codon position. For example, eight amino acids (including valine, threonine, and alanine) use quartets of codons, each member of a quartet varying only at the third position.
10. When an amino acid uses only a duet (two) of the codons in a quartet, the third codon positions in this duet are both pyrimidines (U and C) or both purines (A and G), never one pyrimidine and one purine.

Explanations for some features of code redundancy have not been difficult to find. In part, redundancy derives from the presence of more than one kind of tRNA for a single amino acid. The tRNAs used for leucine, for

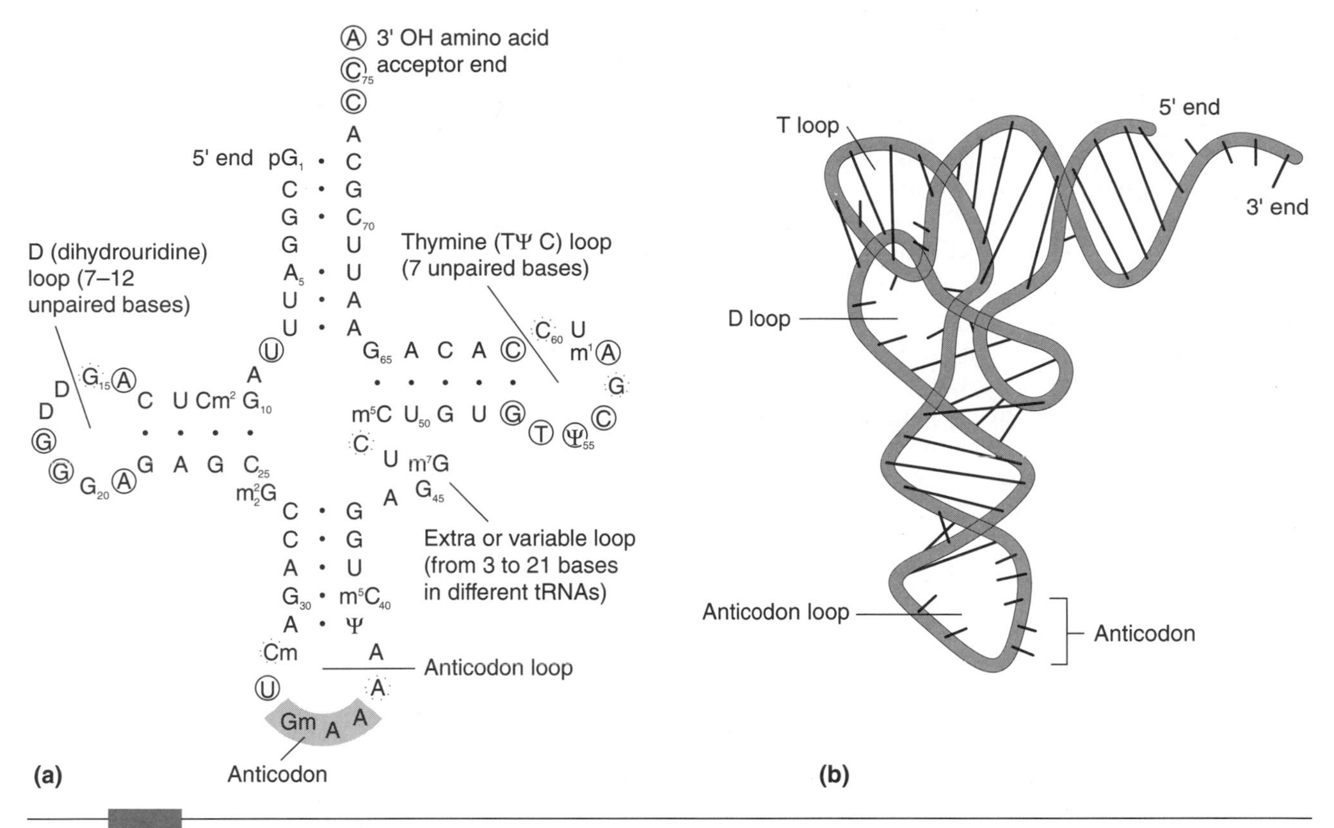

FIGURE 8-4 (*a*) Sequence of the 76 nucleotides in phenylalanine tRNA of yeast shown in the commonly portrayed two-dimensional cloverleaf form. Hundreds of different kinds of tRNA molecules sequenced in a variety of organisms can fit into this same cloverleaf pattern, offering a maximum of pairing (*dots*) between complementary bases. The four major tRNA loops are indicated, including the anticodon loop, which contains, in this case, the special sequence AAG that matches the mRNA codon for phenylalanine UUC (and also UUU because of wobble pairing; see also Fig. 8-6). Transfer RNA molecules that are specific for other amino acids bear, of course, different anticodon sequences as well as different nucleotides in some of the other positions. In the present diagram, bases encircled with *solid lines* occupy the same positions in all the different tRNAs examined so far, whereas those encircled with *dots* are more variable, indicting base pair positions that are always occupied by either purines or pyrimidines. (*b*) Three-dimensional L-shaped structure showing the molecule's functional form as two distinct domains at right angles to each other: an acceptor domain for attaching specific amino acids, and an anticodon domain for binding to specific mRNA codons. Some researchers propose that the acceptor domain evolved first, acting as a tag to attract RNA-replicating ribozymes in the RNA world (Maizels and Weiner). As evolution proceeded toward protein synthesis, some sequences within these tags served as "codons" that also coupled with ribozymes carrying amino acids. These coupling sequences then duplicated in the anticodon domain enabling them to bind to mRNA. Although we cannot exclude such scenarios (Schimmel, DiGiulio), it is still difficult to be confident about the many complex events that must have occurred during this presumed early period of RNA-protein transition. Numbering of the bases, shown as subscripts, begins at the 5′ end of the nucleotide sequence. Unusual nucleotides found in tRNA include D = dihydrouridine, ψ = pseudouridine, mX = methylated nucleotide, and T = thymine.

example, may have the anticodons AAU, AAC, and GAG. Crick explained that the redundancy pattern mostly attaches to the third codon position because of **wobble pairing** between certain bases of the tRNA anticodon and the mRNA codon in this position.

Crick's wobble hypothesis, since proved correct, was that anticodons bearing inosine (I) at this third position can pair with either U, A, or C;[9] anticodons bearing G can pair with either of the pyrimidines U and C; and anticodons bearing U can pair with either of the purines A and G. Third codon position redundancy thus points to the importance of the first two codon positions in specifying amino acids; these positions suffice to code for the eight amino acids that use codon quartets.

The striking universality of the code—its common features in all independent living organisms and its restriction to the same 20 amino acids—makes it reasonable to ask: Why this particular code? Since at least 1,070 possible different codes use 64 codons to code for 21 entities (20 amino acids + chain termination), either the exclusive use of this particular code must have derived

[9] Inosinic acid is a nucleotide that uses the purine hypoxanthine as a base. Hypoxanthine in turn derives from the loss of an amino group from adenine. Other modified tRNA nucleotides that show wobble pairing have also been found (for example, 5-methoxyuridine pairs with both A and G and less efficiently with U).

from accidental causes, or some relationship between amino acids and their codons (or anticodons) must exclude large numbers of other possible codes—or perhaps both factors operated.

Some authors have suggested that the amino acids originally associated with their codons or anticodons through stereochemical fitting or by sharing other complementary characters such as hydrophilic and hydrophobic properties. For example, some kind of pairing may have occurred between an amino acid such as phenylalanine and a codon such as UUC or its anticodon AAG (Jungck).

However, other authors maintain that, despite considerable search, there are few, if any, examples of preferential affinity between amino acids and their codons or anticodons. Furthermore, if we consistently applied such a hypothesis, it would be grossly inadequate in explaining why serine uses two such markedly dissimilar sets of codons as the UCN quartet (where N designates any base) and the AGU–AGC duet.

Theorists therefore have offered an alternative hypothesis suggesting that the universality of the code derives from the survival of only one of the possible codes that may actually have existed in the past. That is, early amino acid–codon relationships arose largely by chance rather than by restricted stereochemical pairing and may therefore have produced a number of different primordial genetic codes, each used by different groups. As time went on, however, only one group carrying a particular code succeeded enough to continue evolving, and the others became extinct.[10] The code thus reached its present form.

Further evolution of the code was then restricted as a **frozen accident** because protein synthesis in its mature form precisely positions each particular amino acid in all the various long-chain polypeptides in which it is found. So any change in the genetic code for a widely used amino acid would significantly change many different proteins carrying that amino acid. For example, if the code for phenylalanine (UUU) extended to include the serine codon (UCU), tRNA molecules carrying phenylalanine would then insert into polypeptide positions formerly occupied by serine. Since these two amino acids differ considerably in structure and function, the effect of such a sudden change would undoubtedly be lethal.[11]

Before "freezing," the genetic code must itself have evolved to accompany some of the changes taking place in protein synthesis. That is, early proteins were probably much shorter than present proteins, consisting of fewer kinds of amino acids, and produced by a much less accurate translation mechanism. These proteins may have had only minimal functional specificity, and those that a single mRNA molecule produced could best be described as statistically alike rather than exactly alike. Thus, there would likely have been room for changes between some codons of early amino acids, and not all the present amino acids would have been incorporated into the primeval code(s).

A very early hypothesis on the evolution of the code was that it derived from a prior code that used fewer than three bases per codon. That the first two nucleotide bases in each codon still primarily specify the amino acids using quartets seemed to support this idea. Thus it seemed possible to suggest that an even earlier code may have been a **singlet code,** with four mononucleotide codons (A, U, G, C) specifying perhaps four kinds of amino acids. This code evolved later into a **doublet code** with 16 dinucleotide codons (AA, AU, AG, . . .) specifying a maximum of 15 amino acids and one chain-termination codon. Only after undergoing this doublet experience did the code finally achieve its present trinucleotide form of 64 codons.

Although the idea is superficially attractive, evolution of the code in this fashion seems extremely doubtful, because each change in codon size would change the meaning of practically all former codons. For example, an mRNA sequence translated by a doublet code, AU CG UU GU AG CG. . ., would produce entirely different numbers and kinds of amino acids when translated by a triplet code, AUC GUU GUA GCG. . . . In the face of a radical transition of this kind an organism could hardly retain the function of most, if not all, of its genetic material.

Researchers therefore believe the genetic code was triplet even during its beginnings, or perhaps doublet with single **nucleotide spacers.** Mechanical considerations support this view, since anything less than a triplet codon would probably not provide a stable pairing relationship between a tRNA anticodon and an mRNA codon. At the same time, quadruplet or quintuplet codons are probably too "sticky," given selection for an optimum turnover rate that would allow rapid dissociation between codons and anticodons. Researchers have also suggested that the trinucleotide width of a triplet an-

[10] According to Wong, individuals carrying the successful code may have had one or many unique advantages that would have given them important competitive superiority, such as phased coupling of DNA replication and cell division.

[11] Frozen accidents could also help explain the optical rotations found in organic molecules that existing organisms produce (p. 117):

- All proteins consist of L-amino acids.
- All nucleic acids consist of nucleotides with D-sugars.
- DNA double helices coil in a righthanded rather than a lefthanded direction.

Perhaps once organisms developed enzymes that used subunits with optical specificity, it would have been detrimental to change these specificities. Although researchers have made various proposals (see Bonner's review), the cause(s) for these initial rotational biases still remains a mystery. Among intriguing (but probably unanswerable questions) is whether evolutionary consequences would have been different if enzymes had not been restricted to function only with these optical rotated molecules.

ticodon provides a minimal space, enabling tRNA molecules to lie close enough together for peptide bonding between their amino acids.

Given a small group of amino acids coded by such triplets, further evolution of the code probably would have proceeded under three selective conditions:

- Nucleotide substitutions caused by errors in replication (mutations) should produce as few amino acid changes as possible.
- The number of different codons per amino acid should generally be proportional to the frequency in which the amino acid occurs in proteins.
- Errors occurring during the mRNA–tRNA translation process should lead to as little drastic protein changes as possible.

In respect to the first condition, selection would tend to expand the number of different codons used by an amino acid so that random base changes would still produce the same amino acid. Redundancy—increasing the number of different codons for any particular amino acid—would also be advantageous in diminishing the number of stop or nonsense codons that do not code for any amino acid, thereby ensuring that most random mutations do not terminate protein production. In contrast, there would be a limit on the number of different codons that a single amino acid could use since the coding dictionary must accommodate a variety of amino acids. Unfortunately, the extent to which these selective forces operated in the past is now hard to determine.

In respect to the numbers of codon assignments, a proportional relationship does seem to exist between the relative number of codons possessed by most amino acids and their frequencies in proteins (Fig. 8–5). However, the source for this relationship is still not clear. It may indicate a selective relationship (more codons are selected for the use of those amino acids that occur more frequently) or an accidental relationship (the overall composition of proteins derives from the frequency of amino acid codons), or perhaps both factors prevailed.[12]

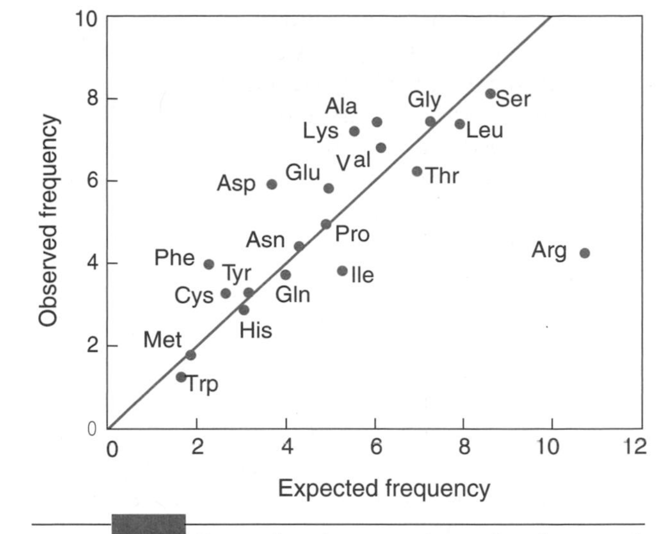

FIGURE 8–5 Comparison between observed and expected frequencies of amino acids at 5,492 positions in 53 different vertebrate polypeptides. If we exclude arginine, the correlation coefficient for these data is 0.89, which represents a high positive correlation. (The expected frequency of each amino acid derives from a technique of calculating the nucleotide composition of the mRNA from the frequency of bases in the first two codon positions used by the various amino acids that compose the proteins: U = 0.220, A = 0.303, C = 0.217, G = 0.261. Randomized triplet codons for this nucleotide composition then furnish the expected amino acids.) *(Reprinted with permission from "Non Darwinian Evolution" by King, L. L. in* Science, *16 May 1978. Copyright © 1978 American Association for the Advancement of* Science.*)*

The third level of codon evolution, selection for minimizing translational errors, may have considerably effected codon assignments. For example, the prevailing redundancy at the third codon position is precisely what we would expect if this position were the one most involved in translational errors. That is, the various quartets (for example, GUU, GUC, GUA, GUG) and duets (for example, UUU, UUC) arise from the selective advantage of assigning to the same amino acid codons that could most easily be mistaken for one another. The next most error-prone translational event occurs at the first codon position. At this position the code again shows some redundancy [for example, UUA (leucine) → CUA (leucine)] or is so constructed that an error may occasionally enable the substitution of an amino acid with related function [for example, UUA (leucine) GUA → (valine)].

[12] One anomalous feature of the code is the disproportionate number of arginine codons in the genetic code (6 codons = 9.8 percent of the total) compared to the actual frequency of this amino acid in most proteins (usually less than 5 percent). By contrast, lysine has only two codons (3.3 percent of the total) but appears more frequently (6.6 percent) in proteins. One reason for this difference may be that mutations leading to the substitution of arginine for other amino acids are usually more disadvantageous than mutations leading to the substitution of lysine. Jukes (1974) suggests that arginine is a relatively new amino acid in contrast to most others, although it has so many codons. According to him, ornithine was probably the amino acid that originally used the codons presently assigned to arginine. Presumably, the evolution of the urea cycle for nitrogen excretion led to the production of arginine, which had a strong affinity for the tRNA molecules that ornithine used. Because arginine is a more basic amino acid than ornithine, it may have conferred new advantageous properties on some proteins. In general, when arginine usurped the ornithine codons, previous ornithine amino acid functions were probably assumed by lysine, which is somewhat like ornithine. Jukes claims that lysine with only two codons now presumably occupies many amino acid positions formerly occupied by ornithine, which had perhaps six codons.

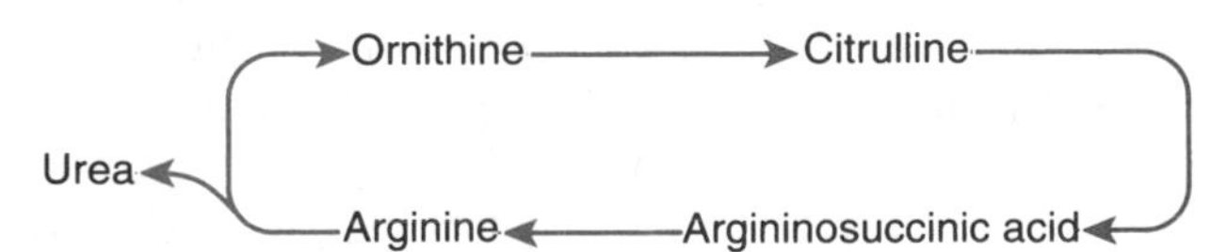

In general the first codon position is considerably less error prone than the third, apparently because of modifications of the 40th tRNA nucleotide immediately adjacent to this position. These modifications include the addition of methyl or even bulkier groups to nucleotide 40, thereby preventing wobble at the first codon position. For example, one such modified nucleotide is threonyl-6-adenine, which prevents the wobble pairing of U in the adjacent position with G in the first codon position.

Researchers have suggested that the second codon position, the least error prone during translation, may at one time have separated entire classes of amino acids with unique functions. This system would offer a selective advantage by ensuring that amino acids are rarely substituted between classes. A possible remnant of such a grouping may be the assignment of U to the second codon position for leucine, isoleucine, and valine, all of which are **hydrophobic amino acids** that exist mostly in the interior of proteins, and assignment of A to the second codon position for glutamic acid, histidine, aspartic acid, and other **hydrophilic amino acids** that commonly exist on the protein surface (Table 8–3).

TABLE 8–3 Classification of 18 amino acids according to the nucleotide found at their second codon positions and known hydration potentials of their side chains[a]

Amino Acid	Second Codon Letter	Hydration Potential (kcal/mole)	
Gly	G	+2.39	**Most hydrophobic**
Leu	U	+2.28	↑
Ile	U	+2.15	
Val	U	+1.99	
Ala	C	+1.94	
Phe	U	−0.76	
Cys	G	−1.24	
Met	U	−1.48	
Thr	C	−4.88	
Ser	C(G)	−5.06	
Trp	G	−5.89	
Tyr	A	−6.11	
Gln	A	−9.38	
Lys	A	−9.52	
Asn	A	−9.68	
Glu	A	−10.19	
His	A	−10.23	↓
Asp	A	−19.92	**Most hydrophilic**

[a]Other proposals suggest that the first codon position specifies amino acids made through similar biosynthetic pathways, or distinguishes between small and large amino acids (see Maynard Smith and Szathmáry).
Source: Data from Wolfenden et al.

In summary, it is possible to claim that the triplet genetic code may have initially coded for relatively few amino acids, or, based primarily on the second codon position, may have been able to distinguish among general classes of amino acids. As time went on, the first codon position also came into use for amino acid positioning because of the modifications that could be made to the immediately adjacent tRNA nucleotide. With the advent of two accurately translated codon positions, the genetic code could then expand its repertoire of amino acids. Thus, until translation accuracy was firmly achieved, some shuffling of codons may have taken place between different amino acids, or entirely new amino acids may have entered the process, using some or all of the codons of older amino acids. The retention of third codon position redundancy or wobble may have economized on the number of tRNA molecules needed to translate amino acids such as valine, threonine, and alanine.

Accordingly, Jukes (1983) has proposed that an early stage in the evolution of the genetic code may have been somewhat like Figure 8–6*a*. Each **codon quartet** or **codon family**, of four codons specified perhaps one amino acid, and each amino acid was represented by only one kind of tRNA molecule whose anticodon could pair with all four codons in the family. (This type of code commonly appears in present-day mitochondrial organelles that, because of their small size and limited function, economize in the kinds of both tRNA and proteins they produce.) Such extreme wobble limited this code to no more than 15 or 16 amino acids. One family or partial family of codons terminated translation since there were no amino acid-bearing tRNA molecules whose anticodons could pair with these stop codons. In subsequent evolution (Fig. 8–6*b*, *c*), tRNA gene duplications enabled new anticodons to evolve, some of which could now mutate so that new amino acids activated them.

For example, the gene for an early tRNA molecule with anticodon AAU (that specifies phenylalanine) could form an additional tRNA gene by duplication, which then mutates to produce anticodon AAG. The AAG anticodon becomes restricted to pairing with codons UUU and UUC, whereas the AAU anticodon pairs only with UUA and UUG. Although both tRNAs originally specify phenylalanine, the gene producing the tRNA with anticodon AAU may now mutate so that its tRNA product now binds to leucine rather than to phenylalanine. Thus

(a) An early code: 16 anticodons for perhaps 15 amino acids

Codons	Anticodon	Amino acid	Codons	Anticodon	Amino acid	Codons	Anticodon	Amino acid	Codons	Anticodon	Amino acid
UUU, UUC, UUA, UUG	AAU	phe?	UCU, UCC, UCA, UCG	AGU	ser	UAU, UAC	AUG	tyr	UGU, UGC, UGA, UGG	ACU	cys?
						UAA, UAG	STOP				
CUU, CUC, CUA, CUG	GAU	leu	CCU, CCC, CCA, CCG	GGU	pro	CAU, CAC, CAA, CAG	GUU	his?	CGU, CGC, CGA, CGG	GCU	arg
AUU, AUC, AUA, AUG	UAU	ile	ACU, ACC, ACA, ACG	UGU	thr	AAU, AAC, AAA, AAG	UUU	asn?	AGU, AGC, AGA, AGG	UCU	ser?
GUU, GUC, GUA, GUG	CAU	val	GCU, GCC, GCA, GCG	CGU	ala	GAU, GAC, GAA, GAG	CUU	asp?	GGU, GGC, GGA, GGG	CCU	gly

tRNA gene duplications and mutations

– – – – – →

Evolution of new anticodons

(b) A later code: 31 anticodons for perhaps 18 amino acids

Codons	Anticodon	Amino acid	Codons	Anticodon	Amino acid	Codons	Anticodon	Amino acid	Codons	Anticodon	Amino acid
UUU, UUC	AAG	phe	UCU, UCC	AGG	ser	UAU, UAC	AUG	tyr	UGU, UGC	ACG	cys
UUA, UUG	AAU	leu	UCA, UCG	AGU	ser	UAA, UAG	STOP		UGA, UGG	ACU	trp
CUU, CUC	GAG	leu	CCU, CCC	GGG	pro	CAU, CAC	GUG	his	CGU, CGC	GCG	arg
CUA, CUG	GAU	leu	CCA, CCG	GGU	pro	CAA, CAG	GUU	glu	CGA, CGG	GCU	arg
AUU, AUC	UAG	ile	ACU, ACC	UGG	thr	AAU, AAC	UUG	asn	AGU, AGC	UCG	ser
AUA, AUG	UAU	ile	ACA, ACG	UGU	thr	AAA, AAG	UUU	lys	AGA, AGG	UCU	arg
GUU, GUC	CAG	val	GCU, GCC	CGG	ala	GAU, GAC	CUG	asp	GGU, GGC	CCG	gly
GUA, GUG	CAU	val	GCA, GCG	CGU	ala	GAA, GAG	CUU	glu	GGA, GGG	CCU	gly

Further tRNA gene duplications and mutations

– – – – – →

Evolution of new anticodons (and deletion of ACU anticodon to produce a UGA stop codon)

(c) The modern code: 43 known anticodons for 20 amino acids

Codon	Anticodon	Amino acid	Codon	Anticodon	Amino acid	Codon	Anticodon	Amino acid	Codon	Anticodon	Amino acid
UUU	AAG	phe	UCU	AGI	ser	UAU	AUG	tyr	UGU	ACG	cys
UUC		phe	UCC		ser	UAC		tyr	UGC		cys
UUA	AAU	leu	UCA	AGU	ser	UAA	STOP		UGA	STOP	
UUG	AAC	leu	UCG	AGC	ser	UAG			UGG	ACC	trp
CUU	GAG	leu	CCU	GGI	pro	CAU	GUG	his	CGU	GCI	arg
CUC		leu	CCC		pro	CAC		his	CGC	GCG	arg
CUA	GAU	leu	CCA	GGU	pro	CAA	GUU	gln	CGA		arg
CUG	GAC	leu	CCG		pro	CAG	GUC	gln	CGG		arg
AUU	UAI	ile	ACU	UGI	thr	AAU	UUG	asn	AGU	UCG	ser
AUC	UAG	ile	ACC	UGG	thr	AAC		asn	AGC		ser
AUA	UAC*	ile	ACA	UGU	thr	AAA	UUU	lys	AGA	UCU	arg
AUG	UAC	met	ACG		thr	AAG	UUC	lys	AGG		arg
GUU	CAI	val	GCU	CGI	ala	GAU	CUG	asp	GGU	CCG	gly
GUC	CAG	val	GCC	CGU	ala	GAC		asp	GGC		gly
GUA	CAU	val	GCA		ala	GAA	CUU	glu	GGA	CCU	gly
GUG	CAC	val	GCG		ala	GAG	CUC	glu	GGG	CCC	gly

FIGURE 8-6 Some possible stages in the evolution of the genetic code based on a scheme that Jukes suggested. The mRNA codons are at the left of each box, and the tRNA anticodons are in shaded capital letters to their right. The cytidine nucleotide in the UAC anticodon marked with an asterisk is acetylated, restricting this tRNA molecule to AUA (isoleucine) codons on mRNA. Osawa provides lists of known anticodons in eukaryotes, prokaryotes, chloroplasts, and mitochondria. *(From* Genetics Third Edition *by Monroe W. Strickberger. Copyright © 1985 by Monroe W. Strickberger. Reprinted by permission of Prentice Hall, Inc., Upper Saddle River, NJ.)*

UUA and UUG codons would now specify leucine.[13] Jukes points out that since such aminoacylation changes have arisen in the genetic codes of mitochondria, which produce relatively few proteins, they also could have occurred in early primitive organisms.

[13] The process by which particular amino acids bind to particular tRNA molecules depends on special **amino acid–activating enzymes** that recognize unique features of both the amino acids and the tRNAs. For example, some mutations of a tRNA gene may allow an amino acid–activating enzyme that had exclusively attached amino acid X to $tRNA^X$ to attach amino acid X to this new tRNA, for example, $tRNA^Y$. A mutation of the amino acid–activating enzyme itself may then enable a new amino acid, Z, to attach to $tRNA^Y$, which can be considered as $tRNA^Z$ when it loses its prior function with amino acid X. According to Jukes and coworkers (see Osawa et al.):

> the occurrence of unassigned codons implies that some life forms use fewer than 64 codons, although the genetic code is remarkably conserved among the vast majority of organisms. The number of usable codons may vary among organisms to some extent, and can decrease during evolution, e.g., by directional mutation pressure or can increase up to 64 by capture of unassigned codons.

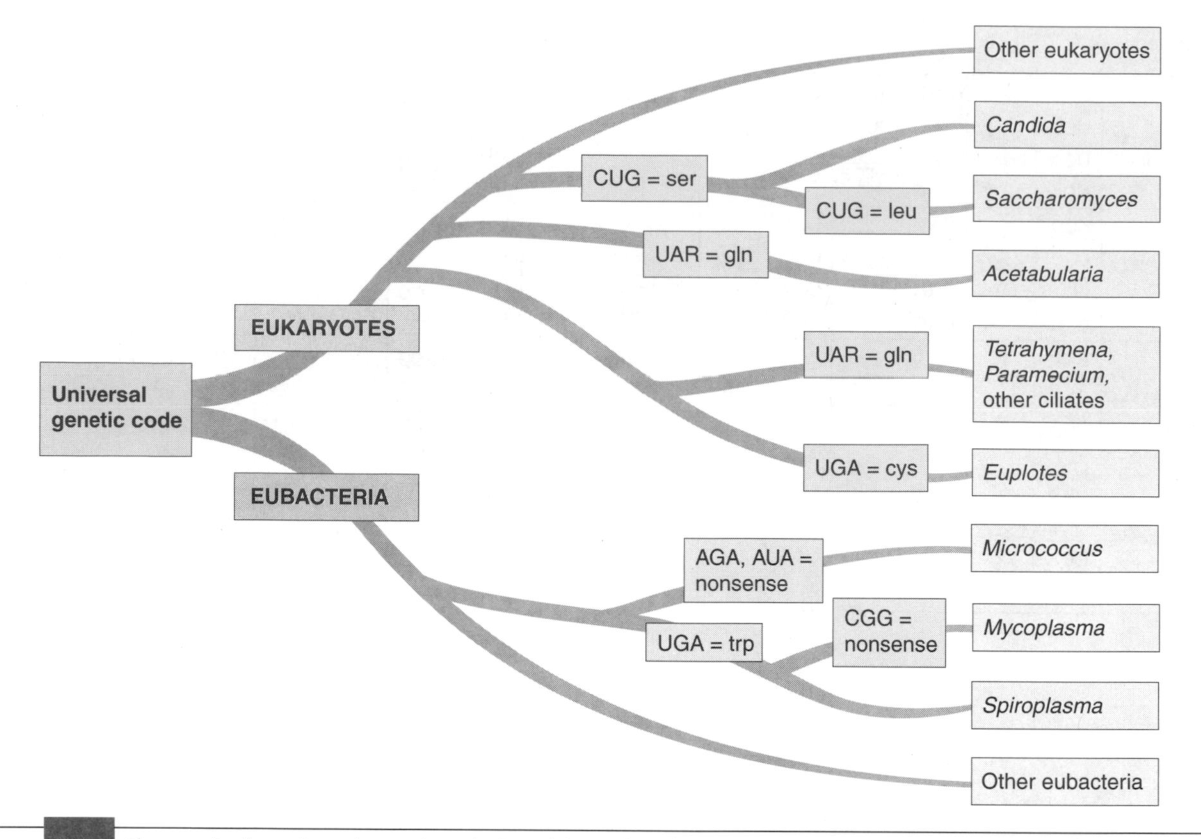

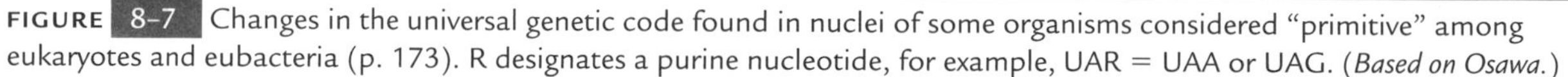

FIGURE 8–7 Changes in the universal genetic code found in nuclei of some organisms considered "primitive" among eukaryotes and eubacteria (p. 173). R designates a purine nucleotide, for example, UAR = UAA or UAG. (*Based on Osawa.*)

By such means the kinds of tRNA molecules could increase and new amino acids could add to the code. Apparently, when 20 different amino acids had incorporated into the code, these ancestral organisms were producing a large-enough number of proteins so that codon changes necessary to include any further amino acids would lead to widespread protein malfunction and widespread lethality. At that point the code "froze," limiting its codon assignments to the prevailing amino acids. The universality of the genetic code (that it is common to all organisms with a few rare codon exceptions) indicates that only the ancestral bearers of this particular code successfully survived the early evolutionary period.

Genetic code exceptions, found not only in mitochondria but also in the nuclei of a few primitive organisms (Fig. 8–7), indicate that some proteins in these entities tolerated some variation in amino acid composition or chain termination without ill effect. Osawa proposes that these exceptions are not ancient relics but are of recent origin, and that some codon changes may still be evolving in such organisms. He suggests that mutation or selection can favor the use of particular codons, thereby allowing unused codons to lose their former assignment and be "captured" to code for a new or different amino acid.

SUMMARY

At present, cells depend totally on protein catalysts for the production of both proteins and nucleic acid templates. Although some types of nucleic acids may have appeared early in biotic history in the form of an RNA world, others suggest that proteins were the earliest functioning polymers. Both types of molecules would have offered the advantage of catalytic activity, and their functions may have also extended to self-replication, a process presumably simpler for nucleic acids than for polypeptides. (Nevertheless, there are reports of a few bacterial antibiotics synthesized by using the polymerizing enzyme as a template.)

Because of the importance of lengthened molecular chains for both structure and function, polymerization was probably one of the earliest cellular processes to evolve. If we assume the existence of some kind of template (perhaps a peptide–nucleic acid chimera) as well as a polymerizing enzyme, selection could have led to the origin of adaptor molecules carrying amino acids to the template, coenzymes for the use of high-energy molecules, a base-pairing system by which adaptor and nucleotide portions of the template could recognize one another, and some primitive type of ribosome encompassing all these functions. Subsequently a separate nucleic acid molecule developed for information storage, and this added to the panoply of other nucleic acid sequences acting as messengers, adaptors, and regulators. Finally, storage and production functions separated into distinct molecules—that is, DNA and RNA.

At present, RNA in the form of three-nucleotide units known as *codons* carries the coded message from the storage molecule, DNA, to the protein synthesizing machinery. From an evolutionary perspective, the code's most interesting features are its universality and its redundancy or "degeneracy" (the use of several codons for the same amino acid). This particular code probably became fixed in its present form because the organisms bearing it survived and the organisms carrying alternative codes disappeared.

An early code, even then composed of three nucleotides, probably provided information for fewer amino acids than at present or for general classes of amino acids. To minimize errors in replicating the master code or in translating it into protein, the third nucleotide in each codon was relatively unspecific (redundant). The first and second nucleotides are less flexible and perhaps specified particular amino acids or discriminated among different groups of amino acids on the basis of certain features, such as their hydrophobic or hydrophilic properties. Having two accurately translated nucleotides increased the number of amino acids that could be determined and at the same time regularized the primary structures of proteins. The code stopped evolving once 20 amino acids were specified and a large number and variety of proteins synthesized. At that time the code "froze" in our ancestral group, and it has remained virtually invariable in every cellular organism.

KEY TERMS

adaptor molecules
ambiguous code
amino acid–activating enzymes
anticodon
autocatalytic
chain-terminating codon
code letter
coding dictionary
codon (or code word)
codon family
codon length (or word size)
codon quartet
commaless code
degenerate code
doublet code
evolution of protein synthesis
frameshift mutation
frozen accident
genetic code
head–growth polymerization
heme
hydrophilic amino acids
hydrophobic amino acids
hypercycle
missense mutation
nonsense codon
nonoverlapping code
nucleic acid replication
nucleotide spacers
oligopeptides
organic catalysts
polymerization enzymes
protein synthesis
reading frame
redundant code
replacement mutation
reverse transcriptase
ribozymes
RNA catalysts
RNA world
self-replication
sense word
singlet code
stereochemical fit
stop mutation
synonymous codons
triplet code
universal code
wobble pairing

DISCUSSION QUESTIONS

1. What is the present relationship among cellular proteins, RNA, and DNA?
2. Nucleic acids, clays, proteins
   a. Why have researchers suggested each of these substances as the earliest self-replicating genetic material or earliest cellular material?
   b. What objections have other workers raised against each of these proposals?
3. Provide a scenario of how protein synthesis could have evolved.
4. Genetic code
   a. What are the major features of the present genetic code?
   b. How does wobble pairing account for redundancy of the code?
   c. What is a "frozen accident," and how would it explain the universality of the present genetic code?
   d. Why is the present genetic code believed to have derived from a code based on triplet codons rather than from a code based on doublets or singlets?
   e. What selective factors have researchers proposed that would have influenced the assignment of specific codons to specific amino acids?
   f. How could a smaller number of different amino acids specified by an earlier code have increased so that the present genetic code could specify a larger number (20)?

EVOLUTION ON THE WEB

Explore evolution on the web! Visit the accompanying web site for *Evolution*, 3/e at www.jbpub.com/evolution for web exercises and links relating to topics covered in this chapter.

REFERENCES

Baltscheffsky, H., and M. Baltscheffsky, 1994. Molecular origin and evolution of early biological energy conversion. In *Early Life on Earth* (Nobel Symposium No. 84), S. Bengtson, (ed.). Columbia University Press, New York, pp. 81–90.

Benner, S. A, A. D. Ellington, and A. Tauer, 1989. Modern metabolism as a palimpsest of the RNA world. *Proc. Nat. Acad. Sci.*, **86,** 7054–7058.

Black, S., 1973. A theory on the origin of life. *Adv. in Enzymol.*, **38,** 193–234.

Bonner, W. A., 1991. The origin and amplification of biomolecular chirality. *Origins of Life*, **21**, 59–111.
Cairns-Smith, A. G., 1982. *Genetic Takeover and the Mineral Origins of Life.* Cambridge University Press, Cambridge, England.
Calvin, M., 1969. *Chemical Evolution.* Oxford University Press, Oxford, England.
Cech, T. R., 1987. The chemistry of self-splicing RNA and RNA enzymes. *Science*, **236**, 1532–1539.
Cech, T. R., and B. L. Bass, 1986. Biological catalysis by RNA. *Ann. Rev. Biochem.*, **55**, 599–629.
Crick, F. H. C., 1968. The origin of the genetic code. *J. Mol. Biol.*, **38**, 367–379.
De Duve, C., 1995. *Vital Dust: Life as a Cosmic Imperative.* Basic Books, New York.
DiGiulio, M., 1998. Reflections on the origin of the genetic code: A hypothesis. *J. Theoret. Biol.*, **191**, 191–196.
Dyson, F., 1985. *Origins of Life.* Cambridge University Press, Cambridge, England.
Edwards, M. R., 1998. From a soup or a seed? Pyritic metabolic complexes in the origin of life. *Trends in Ecol. and Evol.*, **13**, 178–181.
Eigen, M., 1992. *Steps Towards Life: A Perspective on Evolution.* Oxford University Press, Oxford, England.
Eigen, M., W. Gardiner, P. Schuster, and R. Winkler-Oswatitsch, 1981. The origin of genetic information. *Sci. Amer.*, **244**, 88–118.
Eigen, M., and P. Schuster, 1979. *The Hypercycle: A Principle of Natural Self-Organization.* Springer-Verlag, Berlin.
Ekland, E. H., and D. P. Bartel, 1996. RNA-catalysed RNA polymerization using nucleoside triphosphates. *Nature*, **382**, 373–376.
Fox, S. W., 1978. The origin and nature of protolife. In *The Nature of Life*, W. H. Heidecamp (ed.). University Park Press, Baltimore, pp. 23–92.
Fraser, C. M., et al., 1995. The minimal gene complement of *Mycoplasma genitalium. Science*, **270**, 397–403.
Gerbi, S. A., 1985. Evolution of ribosomal RNA. In *Molecular Evolutionary Genetics*, R. J. MacIntyre (ed.). Plenum Press, New York, pp. 419–517.
Haseloff, J., and W. Gerlach, 1988. Simple RNA enzymes with new and highly specific endoribonuclease activities. *Nature*, **334**, 585–591.
Heinen, W., and A. M. Lauwers, 1996. Organic sulfur compounds resulting from FeS, H_2S, or HCl and CO_2. *Origins of Life and Evol. of the Biosphere*, **26**, 131–150.
Illangasekare, M., G. Sanchez, T. Nickles, and M. Yarus, 1995. Aminoacyl-RNA synthesis catalyzed by an RNA. *Science*, **267**, 643–647.
Joyce, G. F., 1989. RNA evolution and the origins of life. *Nature*, **338**, 217–224.
Joyce, G. F., and L. E. Orgel, 1993. Prospects for understanding the origin of the RNA world. In *The RNA World*, R. F. Gesteland and J. F. Atkins (eds.). Cold Spring Harbor Laboratory Press, Cold Spring Harbor, New York, pp. 1–25.
Joyce, G. F., A. W. Schwartz, S. L. Miller, and L. E. Orgel, 1987. The case for an ancestral genetic system involving simple analogues of the nucleotides. *Proc. Nat. Acad. Sci.*, **84**, 4398–4402.
Jukes, T. H., 1974. On the possible origin and evolution of the genetic code. *Origins of Life*, **5**, 331–350.
——, 1983. Evolution of the amino acid code. In *Evolution of Genes and Proteins*, M. Nei and R. K. Koehn (eds.). Sinauer Associates, Sunderland, MA, pp. 191–207.
Jukes, T. H., and S. Osawa, 1991. Recent evidence for evolution of the genetic code. In *Evolution of Life: Fossils, Molecules, and Culture*, S. Osawa and T. Honjo (eds.). Springer, Tokyo, pp. 79–95.
Jungck, J. R., 1978. The genetic code as a periodic table. *J Mol. Evol.*, **11**, 211–224.
King, J. L., and T. H. Jukes, 1969. Non-Darwinian evolution. *Science*, **164**, 788–798.
Küppers, B-O., 1983. *Molecular Theory of Evolution.* Springer-Verlag, Berlin.
Larralde, R., M. P. Robertson, and S. L. Miller, 1995. Rates of decomposition of ribose and other sugars: Implications for chemical evolution. *Proc. Nat. Acad. Sci.*, **92**, 8158–8160.
Lee, D. H., J. R. Granja, J. A. Martinez, K. Severin, and M. R. Ghadri, 1996. A self-replicating peptide. *Nature*, **382**, 525–528.
Lehman, N., and G. F. Joyce, 1993. Evolution in vitro: Analysis of a lineage of ribozymes. *Current Bio.*, **3**, 723–734.
Lipmann, F., 1965. Projecting backward from the present stage of evolution of biosynthesis. In *The Origins of Prebiological Systems*, S. W. Fox (ed.). Academic Press, New York, pp. 259–280.
——, 1971. Attempts to map a process evolution of peptide biosynthesis. *Science*, **173**, 875–884.
Loomis, W. F., 1988. *Four Billion Years: An Essay on the Evolution of Genes and Organisms.* Sinauer Associates, Sunderland, MA.
Maizels, N., and A. M. Weiner, 1994. Phylogeny from function: Evidence from the molecular fossil record that tRNA originated in replication, not translation. *Proc. Nat. Acad. Sci.*, **91**, 6729–6734.
Maynard Smith, J., and E. Szathmáry, 1995. *The Major Transitions in Evolution.* Freeman, Oxford, England.
Moore, P. B., 1993. Ribosomes and the RNA world. In *The RNA World*, R. F. Gesteland and J. F. Atkins (eds.). Cold Spring Harbor Laboratory Press, Cold Spring Harbor, New York, pp. 119–135.
Noller, H. F., V. Hottarth, and L. Zimniak, 1992. Unusual resistance of peptidyl transferase to protein extraction procedures. *Science*, **256**, 1416–1419.
Orgel, L. E., 1994. The origin of life on earth. *Sci. Amer.*, **271**, 77–83.
Osawa, S., 1995. *Evolution of the Genetic Code.* Oxford University Press, Oxford, England.
Osawa, S., A. Muto, T. H. Jukes, and T. Ohama, 1990. Evolutionary changes in the genetic code. *Proc. Roy. Soc. Lond. (B)*, **241**, 19–28.
Prusiner, S. B., 1995. The prion diseases. *Sci. Amer.*, **272**(1), 48–57.
Purohit, P., and S. Stern, 1994. Interactions of a small RNA with antibiotic and RNA ligands of the 30*S* subunit. *Nature*, **370**, 659–662.
Ribas de Pouplana, L., R. J. Turner, B. A. Steer, and P. Schimmel, 1998. Genetic code origins: tRNAs older than their synthetases? *Proc. Nat. Acad. Sci.*, **95**, 11295–11300.
Sassanfar, M., and J. W. Szostak, 1993. An RNA motif that binds ATP. *Nature*, **364**, 550–553.
Schimmel, P., 1996. Origin of the genetic code: A needle in the haystack of tRNA sequences. *Proc. Nat. Acad. Sci.*, **93**, 4521–4522.

Siu, P. M. L., and H. G. Wood, 1962. Phosphoenolpyruvic carboxytransphorylase, a carbon-dioxide fixation enzyme from propionic acid bacteria. *J. Biol. Chem.,* **237,** 3044–3051.

Strickberger, M. W., 1985. *Genetics,* 3d ed. Macmillan, New York.

Szathmáry, E., 1989. The emergence, maintenance and transitions of the earliest evolutionary units. *Oxford Surv. Evol. Biol.,* **6,** 169–205.

——, 1991. Four letters in the genetic alphabet: A frozen evolutionary optimum? *Proc. Roy. Soc. Lond.* (*B*), **245,** 91–99.

Unrau, P. J., and D. P. Bartel, 1998. RNA-catalysed nucleotide synthesis. *Nature,* **395,** 260–263.

Wicken, M., and K. Takayama, 1994. Deviants—or emissaries. *Nature,* **367,** 17–18.

Woese, C. R., 1973. Evolution of the genetic code. *Naturwiss.,* **60,** 447–459.

——, 1980. Just So Stories and Rube Goldberg machines: Speculations on the origin of the protein synthetic machinery. In *Ribosomes, Structure, Function, and Genetics,* G. Chamblis, G. R. Craven, J. Davies, K. Davis, L. Kahan, and M. Nomura (eds.). University Park Press, Baltimore, pp. 357–373.

Wolfenden, R. V., P. M. Cullis, and C. C. F. Southgate, 1979. Water, protein folding and the genetic code. *Science,* **206,** 575–577.

Wong, J. T. F., 1976. The evolution of a universal genetic code. *Proc. Nat. Acad. Sci.,* **73,** 2336–2340.

Zhang, B., and T. R. Cech, 1997. Peptide bond formation by in vitro selected ribozymes. *Nature,* **390,** 96–100.

Zielinski, W. S., and L. E. Orgel, 1987. Autocatalytic synthesis of a tetranucleotide analogue. *Nature,* **327,** 346–347.

9 From Metabolism to Cells

For cellular protein and nucleic acid synthesis to evolve, biochemical pathways must also have been selected in which components of these and other polymers could be produced and chemical energy used. We now see the results of such biochemical selection everywhere: cellular metabolism is highly organized in time and space so that each metabolic step in a sequence occurs in a fairly exact repeatable order. Moreover, different metabolic sequences are often precisely coupled and regulated so that the products of one sequence (for example, ATP) are available for use in other sequences at the appropriate time.

Since there are no existing relics of ancient uncoordinated metabolic pathways, we have so far not been able to get direct evidence of precellular or early cellular metabolism. One approach toward understanding metabolic evolution has become available, however, through **comparative biochemistry;** that is, to discover **metabolic pathways** or sequences within such pathways that different organisms share and to consider such shared pathways as ancestral to these organisms.

Anaerobic Metabolism

Anaerobic glycolysis, the breakdown of glucose in the absence of oxygen, is perhaps the most elemental metabolic pathway, and all living creatures share various sections of this pathway. This universality seems to depend on the fact that all existing organisms derive their free energy from the chemical breakdown of such monosaccharides. In **heterotrophic** organisms, monosaccharides or organic materials that can convert to monosaccharides derive from sources outside the organism, whereas **autotrophic** organisms make such organic materials within themselves, usually by reducing carbon dioxide.

In both types of organisms glycolytic pathways may begin directly with glucose or with almost any organic material (such as sugars, fats, or amino acids) that can be converted into glucose. As Figure 9–1 shows, the **Embden-**

Meyerhof glycolytic pathway leads to pyruvic acid (pyruvate), providing a net yield of two high-energy phosphate bonds in ATP, the basic currency for cellular chemical energy:

$$C_6H_{12}O_6 + 2\ ADP + 2\ P_i + 2\ NAD^+ \rightarrow$$
$$2\ C_3H_4O_3 + 2\ ATP + 2\ NADH + 2\ H^+ + 2\ H_2O$$

During this process two molecules of the coenzyme NAD^+ [nicotinamide adenine dinucleotide, a carrier of protons (H^+) and electrons (e^-)] reduce to NADH, but these can then reoxidize in reactions forming either lactic acid (lactate) or ethanol.

Since only two of the reactions furnish ATP, we can consider the other steps in the pathway preparatory to these primary reactions. The length of the pathway seems quite extended, and researchers have suggested that some of these steps may partially recapitulate the succession of biochemical events that furnished energy to organisms in the past. For example, enough smaller molecules such as glyceraldehyde, or its phosphorylated form, may have been available in the past to allow relatively simple energy conversion in only a few steps. As these molecules depleted, organisms that could metabolize some of the more available larger molecules to the glyceraldehyde stage could continue to use the same, but now extended, pathway.[1] By these means, succeeding organisms would need only to add or modify one or a few enzymes for each additional step as it occurred rather than to continually elaborate entirely new metabolic pathways.

Horowitz first formally proposed the hypothesis that the gradual depletion of a necessary molecule causes biochemical pathways to lengthen to allow use of an available related compound. Researchers have called this process **retrograde** or **backward evolution** and believe it accounts for many intermediate steps in biochemical pathways that lead to the synthesis of compounds such as amino acids. For example, let us assume that A is a product essential for cellular function, B is a molecular precursor of A, and C is a molecular precursor of B. Obviously, there is no need for the organism to develop a pathway for the synthesis of A when A itself is present in the environment. But as A depletes, a considerable selective advantage occurs for catalysis of the available B precursor molecules into A (for example, by action of enzyme 1). Similarly, as the supply of B is exhausted, selection occurs for the conversion of precursor C into B (for example, by enzyme 2). By these means a chain of metabolic reactions becomes established, $\ldots \xrightarrow{3} C \xrightarrow{2} B \xrightarrow{1} A$, that represents the evolution of enzyme 1 first, 2 second, 3 third, and so on, each new enzyme catalyzing a single step of a pathway that extends backward from a relatively complex molecule to its more simple precursors.

Even if the development of metabolic pathways did not depend on the opportune appearance of appropriate enzymes or on the depletion of necessary compounds, but evolved forward rather than backward (Granick), or by a patchwork accretion of independent reactions, such sequential pathways clearly result from the chemical relatedness among compounds and the convertibility of one compound into another. This is most apparent in the simple stepwise changes noted in the glycolytic pathway and indicates that biochemical pathways (as is true of many other biological chemical phenomena) did not arise randomly, but followed rules of chemistry in pathways selected for purposes of biology.

Once such pathways developed, their survival must have depended on their ability to cope with persistent chemical problems. For example, although glucose may not have been the first energy-yielding compound, it is now a common sugar whose stability and ready availability in plants and animals lead to the importance of glycolysis in virtually all organisms. [Each enzyme in the glycolytic pathway is found throughout living multicellular organisms and in most single-celled organisms.] An interesting testimonial to evolutionary economy is that seven of these glycolytic enzymes also function in glucose *biosynthesis*, exactly reversing the direction of glycolysis.

Furthermore, the amino acid sequences in many of these enzymes are remarkably similar in organisms that have been evolving separately for at least a billion years. For example, the amino acid sequence of a catalytically **active site** in the enzyme triosephosphate isomerase has apparently persisted in organisms ranging from bacteria (*Escherichia coli*) to corn (*Zea mays*). (See Fig. 7–2 for a listing of amino acids and their letter codes.)

E. coli	QGAAAFEGAVIAYEPVWAIGTGKSATPAQ
Yeast	EEVKDWTNVVVAYEPVWAIGTGLAATPED
Fish	DDVKDWSKVVLAYEPVWAIGTGKTASPQQ
Chicken	DNVKDWSKVVLAYEPVWAIGTGKTATPQQ
Rabbit	DNVKDWSKVVLAYEPVWAIGTGKTATPQQ
Corn	EKIKDWSNVVVATEPVWAIGTGKVATPAQ

Since it is highly improbable for so many amino acid sequences from different sources to have become similar by accident, most likely a single ancestral sequence existed for this catalytic purpose in the common progenitor of all these organisms. Continuous selection for the same enzymatic function in these various lineages (an essential step in glycolysis) then preserved sequence similarity. Gest and Schopf suggest that only "sugar-based cellular systems . . .

[1] By similar reasoning, the glycolytic pathway may have lengthened by beginning with large molecules and then proceeding to smaller ones. That is, the chemical energy first provided by the molecular breakdown of large molecules such as monosaccharides or polysaccharides was successfully increased by further metabolizing their smaller component products such as glyceraldehyde. The aerobic metabolic pathways discussed later may have offered, among other qualities, this type of advantage.

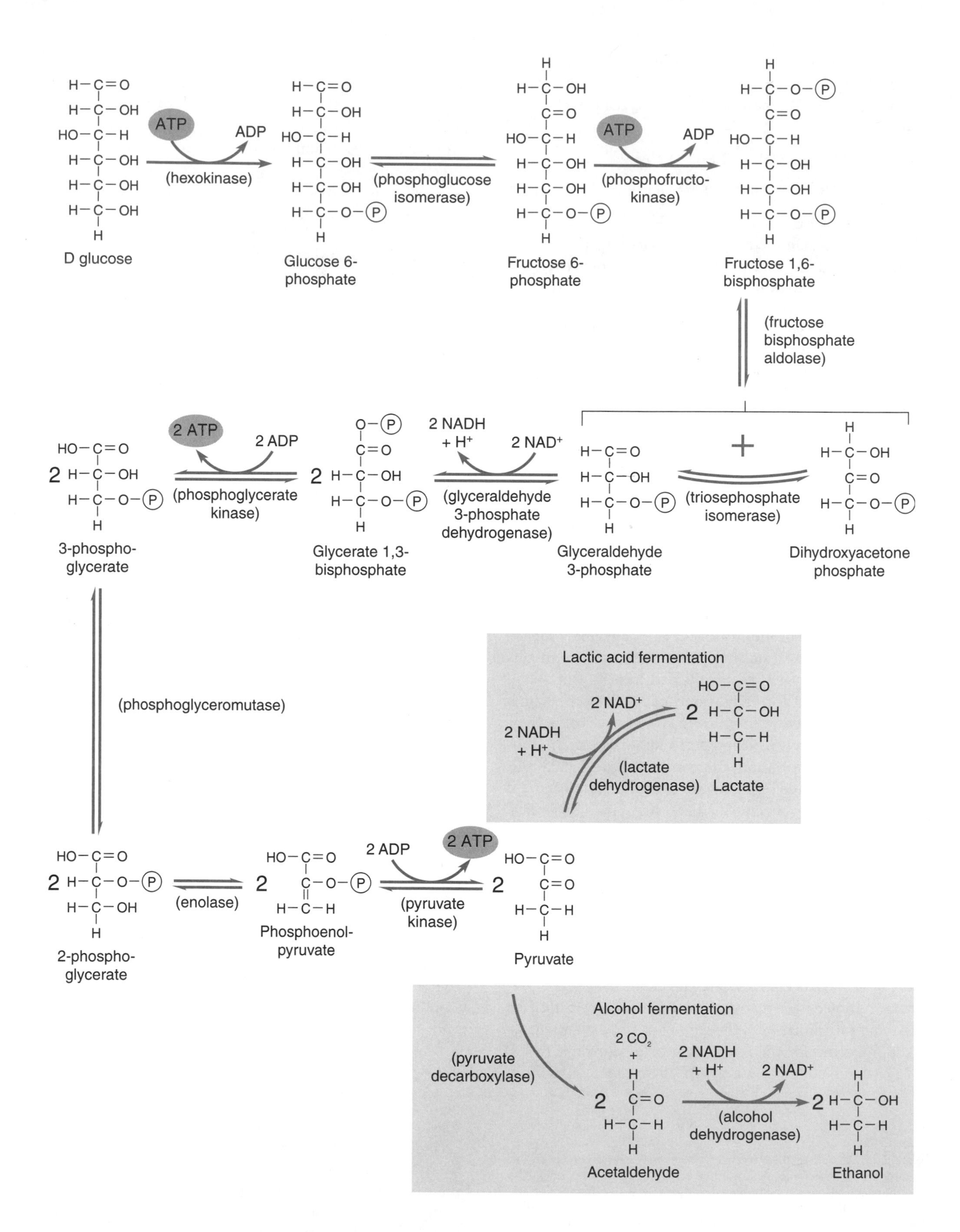
D glucose
ATP
ADP
(hexokinase)
Glucose 6-phosphate
(phosphoglucose isomerase)
Fructose 6-phosphate
ATP
ADP
(phosphofructo-kinase)
Fructose 1,6-bisphosphate
(fructose bisphosphate aldolase)
Glyceraldehyde 3-phosphate
+
(triosephosphate isomerase)
Dihydroxyacetone phosphate
2 NADH + H+
2 NAD+
(glyceraldehyde 3-phosphate dehydrogenase)
Glycerate 1,3-bisphosphate
2 ATP
2 ADP
(phosphoglycerate kinase)
3-phospho-glycerate
(phosphoglyceromutase)
2-phospho-glycerate
(enolase)
Phosphoenol-pyruvate
2 ADP
2 ATP
(pyruvate kinase)
Pyruvate
Lactic acid fermentation
2 NAD+
2 NADH + H+
(lactate dehydrogenase)
Lactate
Alcohol fermentation
(pyruvate decarboxylase)
2 CO2 +
Acetaldehyde
2 NADH + H+
2 NAD+
(alcohol dehydrogenase)
Ethanol

FIGURE 9–1 Steps in the anaerobic Embden-Meyerhof glycolytic pathway from glucose to pyruvate. Beginning with one molecule of glucose, the pathway degrades two ATP molecules to ADP but phosphorylates four ADP molecules to ATP. The overall advantage of this pathway to the cell derives from the net formation of two high-energy phosphate bonds. Also indicated is the reduction of the pyridine nucleotide coenzyme, NAD^+. The reduced form of this compound (NADH) can then be oxidized to regenerate NAD^+ by reactions that donate hydrogens and electrons to form either lactic acid or ethanol:

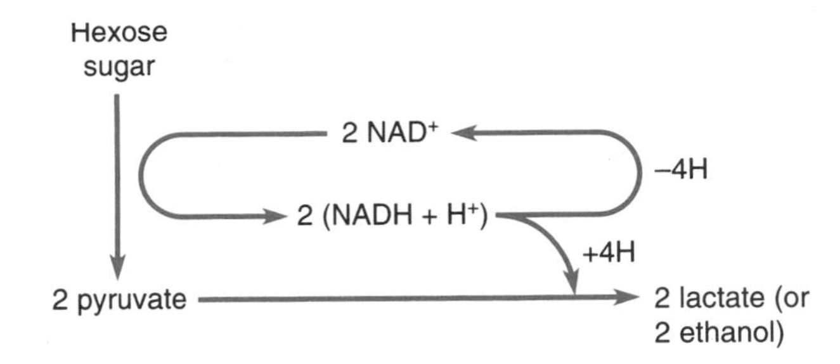

The pathway is actually more complex than illustrated here since the intermediate compounds formed during glycolysis can also function as substrates in synthesizing amino acids and nucleic acids. For example, 3-phosphoglycerate can serve as a substrate leading to the synthesis of serine, glycine, or cysteine amino acids, or purine bases.

The illustrations in this chapter show the various carboxylic acids in protonated form (HO−C=O or COOH) but are usually named as though they were unprotonated ($^-$O−C=O or COO^-), for example, pyruvic acid=pyruvate. *Circled P's* indicate phosphate groups (H_2PO_3), and *the specific enzyme for each reaction is in parentheses.*

provided the successful starting point for biochemical and cellular evolution."

Researchers have discovered other anaerobic pathways that probably are also of ancient lineage, dating back to the early anaerobic atmosphere. One example is the Entner-Doudoroff pathway, which circumvents some early steps of the Embden-Meyerhof pathway by allowing a more direct conversion of glucose-6-phosphate to three-carbon compounds, such as pyruvate, but produces a net gain of one less ATP molecule per metabolized glucose:

$$\text{Glucose} + 2\,NAD^+ + ADP + P_i \rightarrow 2\ \text{pyruvate} + 2\,NADPH + ATP$$

By whatever means anaerobic metabolism evolved, organisms that relied on it eventually must have faced a depletion of reduced carbon compounds to use as substrates. A change from such **organotrophic nutrition** to an ability to use simpler and more readily available carbon sources such as CO_2 **(lithotrophic nutrition)** must have offered considerable evolutionary advantage. However, to be effectively used, CO_2 must be reduced, or fixed, by a process that provides electrons and hydrogen ions. H_2, H_2S, NH_3, and others were probably among the reducing compounds available for this purpose. Researchers believe that early organisms, including some alive today (Vetter), broke down these inorganic compounds with the aid of catalysts (for example, hydrogenase enzyme):

$$H_2(S) \xrightarrow{\textit{hydrogenase}} 2\,H^+ + 2\,e^- + (S)$$

This process furnished electrons and hydrogens that could then be used for producing energy and hydrogenating carbon.[2]

There is considerable agreement that many of these early reactions took place with the aid of membrane-bound enzymatic systems that allowed a successive chain of **reduction–oxidations** to occur (Jones). In the energy-production pathways that evolved, the electrons are picked up by acceptor molecules, which then become transformed from the oxidized to the reduced form. These **electron acceptors** can then act as **electron donors,** transferring electrons farther down the chain until they reach the ultimate electron acceptor. Among electron carriers in these pathways are the iron-containing polypeptides known as *ferredoxins* (Fig. 9–2). These carriers are sufficiently small and widespread in nature for researchers to believe that the ferredoxins, along with the porphyrins (p. 125), were among the first cellular oxidative-reductive agents.

[2]Wächtershäuser has proposed that the energy source for early life came through the formation of pyrite (FeS_2) from hydrogen sulfide by the reaction, $FeS + H_2S \rightarrow FeS_2 + 2H^+ + 2\,e^-$, providing effects similar to those just modeled by the hydrogenase enzyme. Subsequent organic reactions presumably took place on the pyrite surface and led to various basic metabolic pathways. Although Wächtershäuser and his colleagues offer evidence for the pyrite-forming reaction (Drobner et al.), opponents raise many objections, including the difficulty of developing a complex pathway on a pyrite surface (De Duve and Miller). Nevertheless, the fact that mineral surfaces can enhance polymerization of both amino acids and nucleotides (pp. 130 and 144) makes them valuable resources for exploring molecular evolution.

1 ... 28

A Ala-Phe-Val-Ile- Asn-Asp-Ser-Cys-Val-Ser-Cys-Gly-Ala-Cys-Ala-Gly-Glu-Cys-Pro-Val-Ser-Ala-Ile-Thr-Gin-Gly-Asp-Thr-

B Ala-Leu-Tyr-Ile- Thr-Glu-Glu-Cys-Thr-Tyr-Cys-Gly-Ala-Cys-Glu-Pro-Glu-Cys-Pro-Val-Thr-Ala-Ile-Ser-Ala-Gly-Asp-Asp-

C Ala-Leu-Met-Ile- Thr-Asp-Gin-Cys-Ala-Asn-Cys-Asn-Val-Cys-Gin-Pro-Glu-Cys-Pro-Asn-Gly-Ala-Ile-Ser-Gin-Gly-Asp-Glu-

D Pro-Ile-Gin- Val-Asp-Asn-Cys-Met-Ala-Cys-Gin-Ala-Cys-Ile-Asn-Glu-Cys-Pro-Val-Asp-Val-Phe-Gin-Met-Asp-Glu-Gin-

29 ... 55

A Gin-Phe-Val-Ile-Asp-Ala-Asp-Thr-Cys-Ile-Asp-Cys-Gly-Asn-Cys-Ala-Asn-Val-Cys-Pro-Val-Gly-Ala-Pro-Asn-Gin-Glu

B Ile-Tyr-Val-Ile-Asp-Ala-Asn-Thr-Cys-Asn-Glu-Cys-Ala Ala-Cys-Val-Ala-Val-Cys-Pro-Ala-Glu-Cys-Ile-Val-Gin-Gly (60)
Gly—Leu-Asp—Glu-Gin

C Thr-Tyr-Val-Ile-Glu-Pro-Ser-Leu-Cys-Thr-Glu-Cys-Val Asp-Cys-Val-Glu-Val-Cys-Pro-Ile-Lys-Asp-Pro-Ser-His-Glu-...-Gly (81)
Val-Cys
Gly-His-Tyr-Glu-Thr-Ser-Glu

D Gly-Asp-Lys-Ala-Val-Asn-Ile-Pro-Asn-Ser-Asn-Leu-Asp-Asp-Glu-Cys-Val-Glu-Ala-Ile-Gin-Ser-Cys-Pro-Ala-Ala-Ile-Arg-Ser (56)

(a)

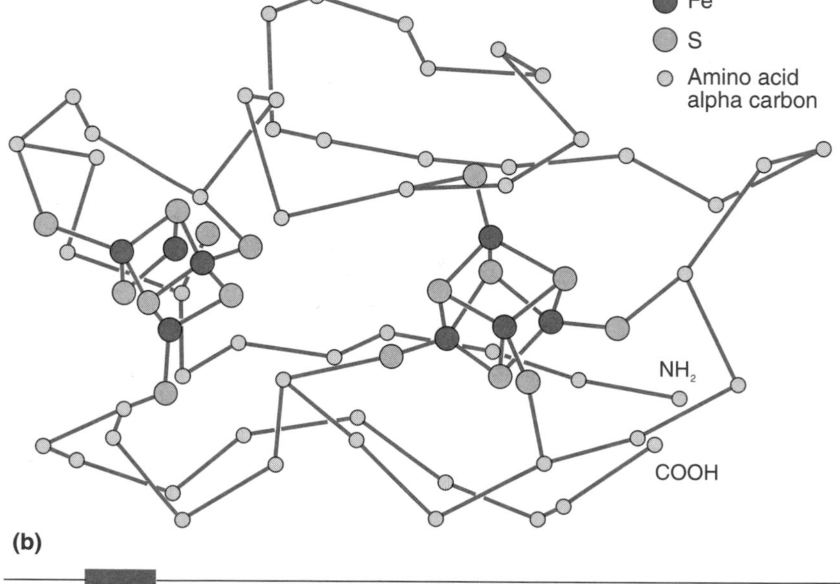

(b)

FIGURE 9-2 (*a*) Amino acid sequences of ferredoxin in four bacterial species: A, *Clostridium butyricum,* an anaerobic fermenting bacterium; B, *Chlorobium limicola,* a green photosynthetic bacterium; C, *Chromatium vinosum,* a purple photosynthetic bacterium; D, *Desulfovibrio gigas,* a sulfate-reducing bacterium. (*From Hall et al.*) Note considerable similarity between the amino acid sequence in the first halves of these molecules (nos. 1–28) and the sequence in the second halves, an observation that initially prompted Eck and Dayhoff to suggest that this protein originated from a genetic duplication. (*b*) Folding of a ferredoxin molecule for *Peptococcus aerogenes* as revealed by X-ray analysis. There are two identically distorted cubes, each containing four iron atoms that are held by sulfur bonds arising from the cysteine residues enclosed in *shaded boxes* in part (*a*). (*Adapted from Adman et al.*)

During the process of electron transfer, which now involves other agents such as cytochromes, the transfer of protons across the cell membrane consumes energy released in specific oxidation–reduction steps, thereby creating a **proton gradient** (see Fig. 9–6). The potential energy available in this gradient can convert into chemical energy by **phosphorylating** ADP to ATP. Using the proton gradient (or low-potential reductants such as H_2) as an energy source, a hydride ion (H^-) can also transfer to the coenzyme NAD^+ (or in some cases, to another coenzyme, FAD, flavin adenine dinucleotide), which then becomes reduced NADH. By means of the reduced coenzyme, a hydrogen then transfers to a carbon recipient, leading to a reduced carbon compound that can then be used either structurally or metabolically.

Photosynthesis

In spite of advantages offered by electron transport systems in these early stages, reliance on chemical energy sources undoubtedly restricted organisms to those specific localities or conditions where these organic and inorganic compounds existed. Perhaps the most important step toward environmental independence occurred when a mechanism evolved letting light-absorbing porphyrins move H^+ ions across membranes to generate ATP (**photophosphorylation**).[3]

[3]Some authors have suggested that the photosynthetic production of ATP occurred before heterotrophic phosphorylations of the kind ener-

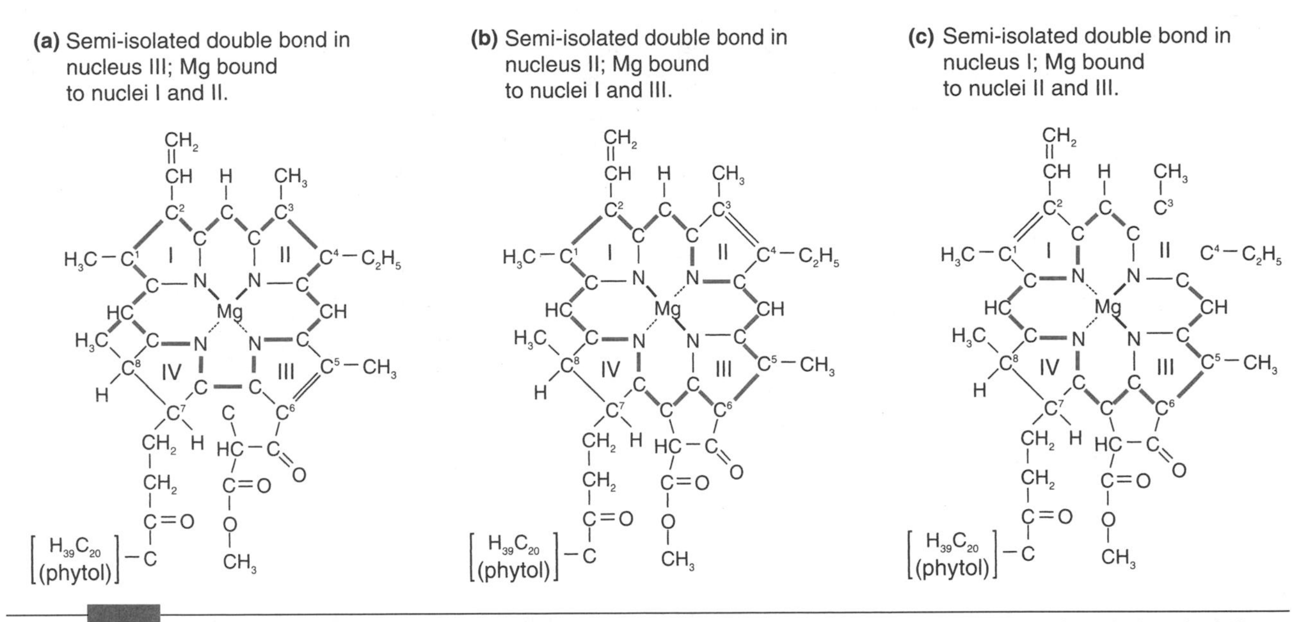

FIGURE 9-3 Three resonance forms of chlorophyll a, showing stability of the molecule as its double and single bonds shift around the ring system (C^1–C^8) in various ways (*colored lines*). This ability to resonate enables chlorophyll to temporarily retain a high electron energy level resulting either from the excitation of electrons exposed to appropriate wavelengths of light or from electrons transferred to chlorophyll from other pigments such as carotenoids, phycobilins, or flavines. Chlorophyll can also transmit such energy to other molecules used in photosynthetic reactions. (*Adapted from Wald.*)

The order in which such light-sensitive mechanisms appeared is not known, but probably porphyrinlike molecules used for oxidation–reduction were already present and could then be further selected to act as photosensitive pigments. Virtually all photosynthetic mechanisms now depend on **chlorophyll,** a porphyrin derivative, and such usage was probably also true in the past. In fact, Woese and coworkers have pointed out that five of the major bacterial groups (Gram-positive, purple, green sulfur, green nonsulfur, and cyanobacteria) possess photosynthetic species, and they therefore suggest that most, perhaps all, modern bacterial groups may have derived from a common photosynthetic ancestor.

As with other porphyrins, chlorophyll has a number of resonance forms in which double and single bonds shift while the molecule remains rigid and stable (Fig. 9–3). This resonating structure enables chlorophyll to maintain light-absorbed energy, as well as transfer it to similar molecules or receive it from pigments such as carotenoids, which absorb light energy at other wavelengths.

A photosynthetic pathway believed to be quite ancient is shown in Figure 9–4. It is **cyclic photosynthesis,** in which solar energy acting on light-sensitive chlorophyll excites the molecule to a high-energy state, allowing an electron to pass on to other electron transfer agents. At a

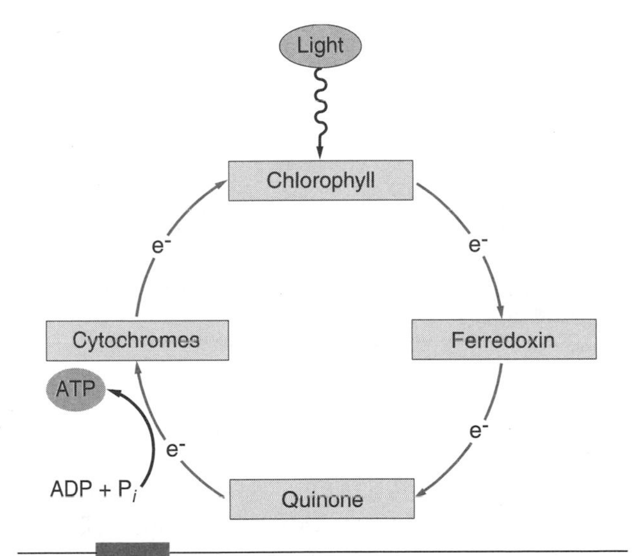

FIGURE 9–4 Diagrammatic view of a cyclic photophosphorylation pathway. Excited chlorophyll molecules force electrons to flow toward the more electronegative direction occupied by the electron acceptor, ferredoxin. Electrons then flow back to the now positively charged chlorophyll through electron carriers that include quinones and various cytochrome pigment proteins. During the transfer of electrons toward electropositive components, a proton gradient forms across the membrane, and this gradient is used to supply energy for phosphorylation of ADP to ATP (Fig. 9–6).

gized by anaerobic glycolysis (Broda). Photophosphorylation, however, appears more complex in terms of membrane organization than does anaerobic phosphorylation, and researchers generally believe the electron transfer mechanisms associated with ATP production began on the nonphotosynthetic level.

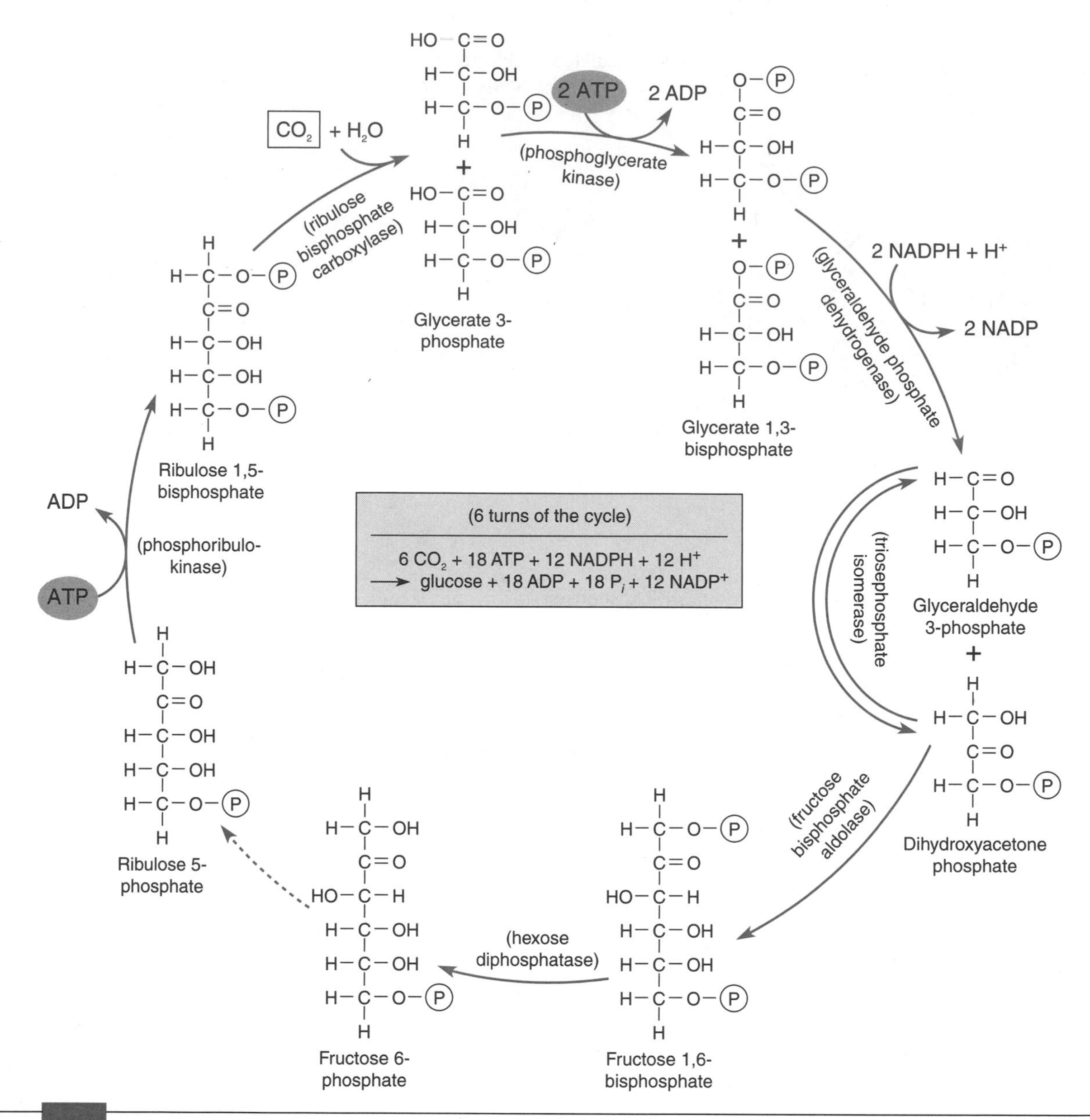

FIGURE 9-5 Simplified diagram of the Calvin cycle, the main metabolic pathway for carbon dioxide fixation in photosynthetic organisms. It relies on reducing power contributed by NADPH formed during photosynthetic reactions and on energy provided by ATP. ($NADP^+$ is the oxidized form of the coenzyme nicotinamide adenine dinucleotide phosphate, and NADPH is the reduced form.) Between glyceraldehyde 3-phosphate and ribulose 5-phosphate in the pathway are a number of reactions, only two products of which are illustrated here (fructose 1, 6-bisphosphate and fructose 6-phosphate).

very early evolutionary stage, this system must have bound to the membrane of a primitive cell or droplet, allowing photoactive energy to couple to existing membrane-contained systems that could phosphorylate ADP to ATP.[4] ATP generated by this new coupled system would be available for metabolic needs and would enable growth to occur independently of environmental chemical energy sources such as glucose intake.

The source of carbon, however, is as important to autotrophs as it is to heterotrophs. Some bacterial photophosphorylators such as the purple nonsulfur bacteria (*Rhodospirillum*) seem to depend at least partly on complex organic molecules for their carbon sources and also to use organic substrates as electron donors (for example,

[4]One such system probably originated as a membrane complex able to deacidify the cell interior by transporting protons (H^+) out of the cell using the energy provided from ATP → ADP breakdown. It is thought that incorporation of anaerobic oxidation-reduction enzymes and electron transfer components into the cell membrane provided an opportunity for evolving a system that could also reverse the ATP → ADP reaction; that is, evolution of an ATP-synthesizing system energized by re-entry of protons into the cell. The association of chlorophyll with such membrane components offered a pathway for powering proton gradient formation—the "proton pump" (p. 132)—by photoactivity.

the oxidation of succinic acid to fumaric acid). By contrast, others, such as the purple sulfur bacteria (*Chromatium*) use CO_2 exclusively and can obtain electrons from inorganic material such as H_2S.

Although it is not clear as yet which condition was more primitive, the ability to use CO_2 as a source of carbon must have offered early photosynthesizers an important opportunity to expand their distribution. In fact, the **Calvin cycle,** which is the most common pathway for the reduction of CO_2, appears in practically all observed photosynthetic organisms. As Figure 9–5 shows, one carbon dioxide molecule incorporates for each turn of this cycle, and one molecule of ribulose bisphosphate regenerates for each CO_2 molecule incorporated. Six turns of the cycle are necessary to produce one glucose molecule, and the overall reaction is:

$$6\ CO_2 + 18\ ATP + 12\ NADPH + 12\ H^+ \rightarrow \text{glucose} + 18\ ADP + 18\ P_i + 12\ NADP^+$$

Despite the advantage of using readily available CO_2 and easily obtainable photosynthetic energy, the distribution of early photosynthesizers probably remained restricted because they depended on compounds such as H_2S for hydrogen sources. Such dependence can be noted today among some bacterial photosynthesizers (purple and green sulfur bacteria) in which hydrogen sulfide provides the electrons that allow the hydrogenation of carbon:

$$2\ H_2S + CO_2 \xrightarrow{\text{light}} \underset{\text{carbohydrate}}{(CH_2O)} + H_2O + 2\ S$$

A primary revolution in the distribution of living organisms occurred when photosynthetic mechanisms evolved that could derive their electrons from readily available water molecules. This process now involves two chlorophyll systems **(noncyclic photosynthesis)** that most probably originated from fusion of two kinds of photosynthetic bacteria, each possessing a somewhat different protein reaction center for transferring electrons. The union of these two photosystems (I and II) is believed to have aided carbon dioxide reduction, especially with the advent of an aerobic atmosphere. They are localized within distinctively specialized photosynthetic membranes **(thylakoids)** found today only in prokaryotic cyanobacteria and the chloroplasts of eukaryotic algae and plants.

As shown in Figure 9–6, the source of electrons for the photosystem II chlorophyll component is the oxidation (dissociation) of water into electrons and H^+ ions and the release of molecular oxygen: $2\ H_2O \rightarrow 4e^- + 4H^+ + O_2\uparrow$. The light-excited, positively charged chlorophyll component of photosystem II readily accepts electrons released during this dissociation, and the coenzyme $NADP^+$ (nicotinamide adenine dinucleotide phosphate) is readily reduced to NADPH by hydrogens and electrons passed on to it through the electron transfer chain. The NADPH and ATP formed in the solar-illuminated **light reactions** are then used in **dark reactions** to reduce carbon by the Calvin cycle: $6\ CO_2 + 6\ H_2O \rightarrow C_6H_{12}O_6 + 6\ O_2\uparrow$.

Oxygen

The consequences of using water as an electron and hydrogen donor in photosynthesis were profound, since the liberation of molecular oxygen began to produce an oxidizing, aerobic environment whose chemical effects were quite different from the relatively more reducing environment previously encountered. We don't know the speed at which oxygen accumulated because of photosynthesis, but organisms that show up in 3-billion-year-old South African Bulawayan limestone and gunflint strata seem to have been oxygen-generating cyanobacteria, as are also the 3.5-billion-year-old filamentous cells found at Warrawoona, Australia (Fig. 9–12).

Researchers have suggested that oxygen concentration in the atmosphere may have remained at 1 percent of the present level until about 2 billion years ago and then gradually increased to its present concentration with increased success of photosynthetic forms. Although these estimates are conjectural, the initiation of an oxygen atmosphere most probably led to an increase in the number and kinds of organisms capable of utilizing aerobic metabolic pathways. We do know that by the Cambrian period or somewhat earlier, oxygen levels had apparently risen high enough to permit rapid evolution of large aerophilic multicellular organisms (Chapter 14).

Since oxygen is a highly reactive element that can rapidly "burn" (oxidize) organic material, one immediate selective effect of an oxygen atmosphere is to increase the frequency of cellular **antioxidant compounds.**[5] These antioxidants were probably not much different from some isoprenoid derivatives that can be found today, such as vitamin K, coenzyme Q, the phytol groups attached to chlorophyll, the carotenes, and vitamin E (Fig. 9–7).

Enzymes that neutralize superoxide radicals ($O_2^{\cdot}$) and peroxide products (H_2O_2) formed by oxygen within cells must also have evolved at that time, such as the

[5]**Ozone** (O_3), commonly formed in the stratosphere by breakdown of diatomic molecular oxygen ($O_2 \rightarrow 2\ O$; $O + O_2 \rightarrow O_3$), protects terrestrial organisms by absorbing lethal solar ultraviolet radiation (p. 170), but is also a biologically destructive oxidant in cells. Nitric oxide (NO), however, impedes ozone's oxidation effect, and Feelisch and Martin therefore suggest that its early prevalence may have made it the first antioxidant. They also propose that once nitric oxide incorporated into cellular metabolism, its use in other cellular activities followed, such as vasodilation, endocrine secretion, and neuronal cell communication. Unfortunately, nitric oxide is also a common atmospheric pollutant, produced by automobile exhausts and industrial plants, that reduces the protective ozone screen.

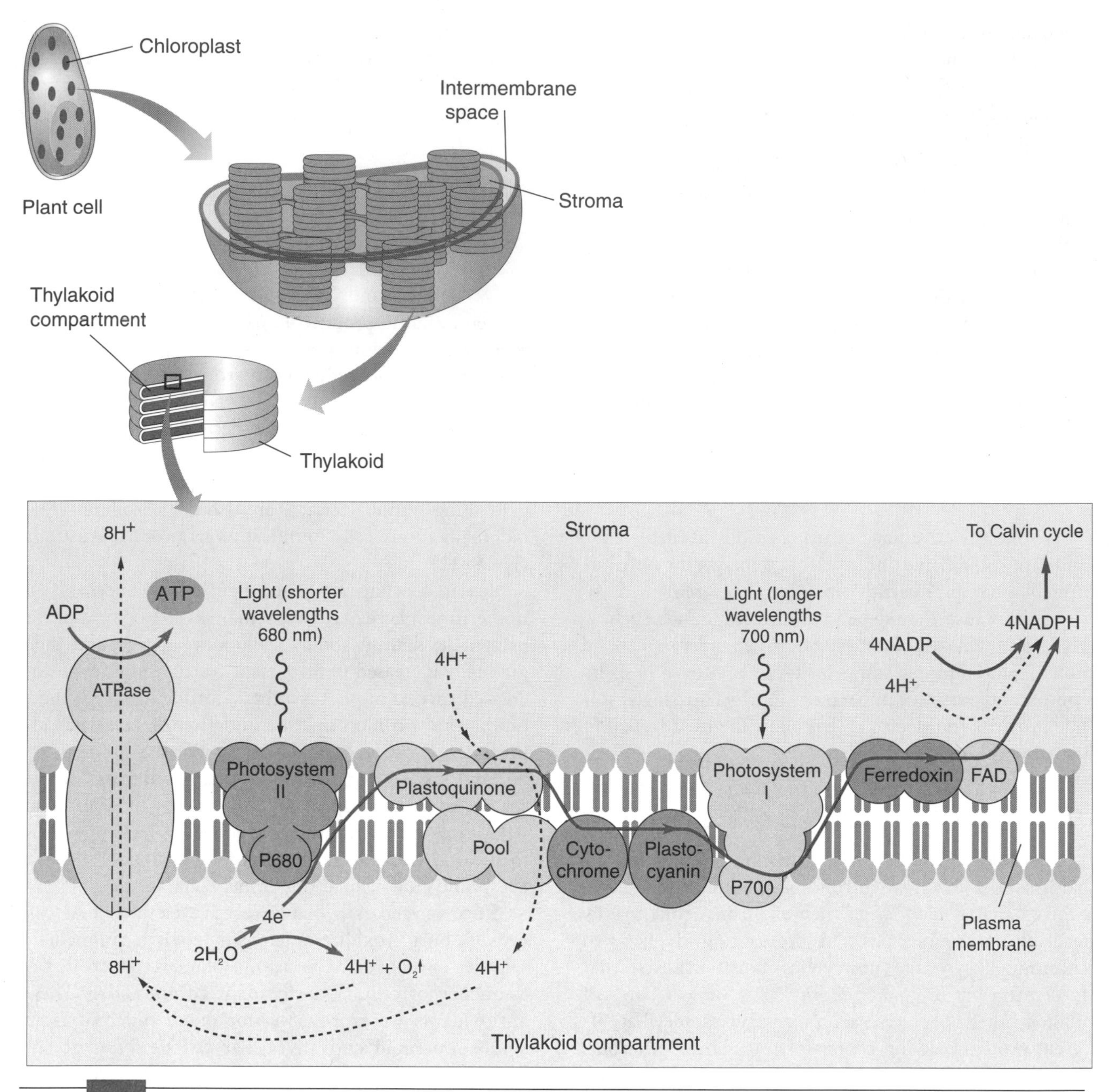

FIGURE 9-6 Diagram of noncyclic electron flow in which electrons obtained from water molecules transfer to the electron acceptor $NADP^+$. Two photosystems, I and II, each sensitive to slightly different wavelengths of light, can be activated to high energy levels so that electrons pass along the chain. The transfer of electrons from photosystem II compensates for the loss of electrons to $NADP^+$ by photosystem I. In turn, electrons derived from the photo-oxidation of water molecules replace photosystem II electrons. As in cyclic photophosphorylation, the flow of electrons produces a proton gradient that can supply energy for the phosphorylation of ADP to ATP. This appears more clearly in the thylakoid membrane structure where the ATP-synthesizing enzyme, ATPase, acts as a **proton pump** generating ATP by tapping energy from the gradient of H^+ ions flowing across the membrane into the thylakoid compartment (**chemiosmosis**). Also shown are the general locations of the two photosystem components and some of the electron transport chain proteins. An *unbroken line* indicates the presumed flow of electrons, and *dashed lines* indicate the flow of H^+ ions. (*Membrane sequence from Wolfe and also Zubay; see also Barber and Andersson.*)

superoxide dismutases, catalases, and peroxidases (Fridovich). Interestingly, researchers believe that some present organisms such as luminescent bacteria use one possible early mechanism to detoxify oxygen, namely localizing oxidation to special reactions that emit fluorescent light. Oxidation in such organisms forms the peroxides that react with luciferase enzymes to produce organic acids and water, radiating light in the process.

Among its other effects, atmospheric oxygen also inhibits nitrogenase enzyme complex activity that reduces

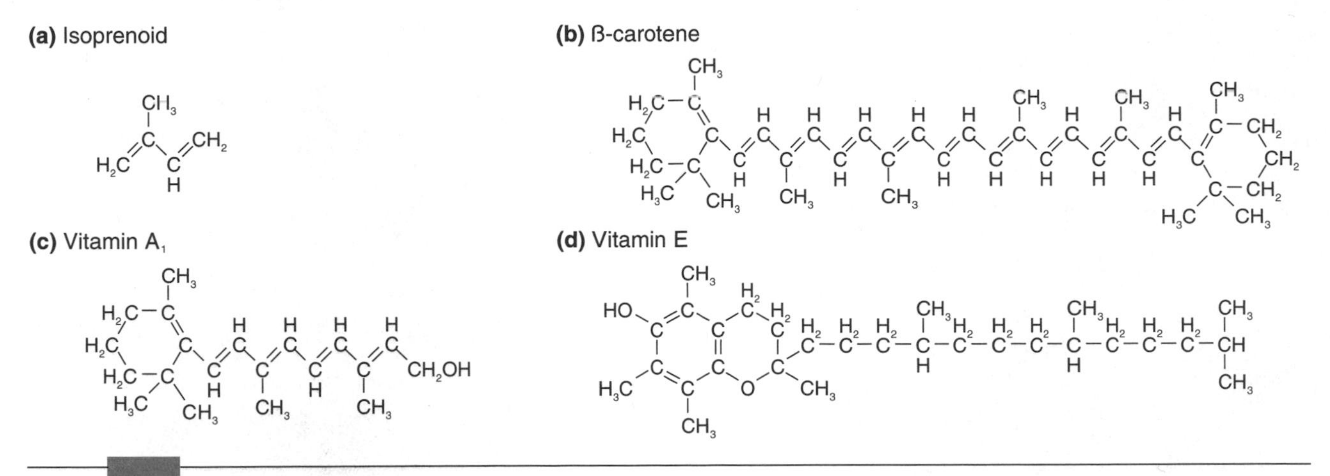

FIGURE 9-7 Examples of isoprenoid compounds. (*a*) The basic five-carbon isoprenoid unit. (*b*) β-carotene, a plant isoprenoid offering protection against oxidation in visible light. (*c*) Vitamin A_1 (retinol 1), a fat-soluble vitamin that exists only in animals, formed by the cleavage of carotenes. (*d*) Vitamin E (α-tocopherol), a plant antioxidant on which rodents are nutritionally dependent.

("fixes") molecular nitrogen (N_2) to ammonia (for example, NH_4^+) used in synthesizing amino acids, nucleotides, and coenzymes. Although this mechanism involves considerable ATP expense ($N_2 + 6\,e^- + 12\,ATP + 12\,H_2O \rightarrow 2\,NH_4^+ + 12\,ADP + 12\,P_i + 4\,H^+$), such reductive reactions evolved because of the selective value of cells that could restore metabolic ammonia depleted by early ultraviolet radiation. However, since oxygen inhibits nitrogen fixation, the nitrogenase complex is mainly found among anaerobic bacteria. Nitrogenase enzymes in aerobic cyanobacterial cells are generally protected from oxygen contact by thick walls ("heterocysts").

Aerobic Metabolism

Perhaps the most significant change in metabolism that accompanied the new aerobic environment was the evolution of a respiratory pathway by which oxygen is used to produce much more energy from the breakdown of glucose than anaerobic glycolysis can produce. This system involved the elaboration of a series of enzymes that transform pyruvic acid into an activated acetic acid group (acetyl-coenzyme A or one of its evolutionary precursors), then carry these small acetyl groups along a special cycle in which they convert into carbon dioxide and lose their hydrogens:

$$CH_3COOH\text{–coenzyme A} + 2\,H_2O \rightarrow 2\,CO_2 + 8\,(e^- + H^+) + \text{coenzyme A}$$

This part of the pathway, variously called the **Krebs cycle,** *citric acid cycle,* or *tricarboxylic acid cycle* (Fig. 9–8), has an interesting self-catalytic feature (as does the Calvin cycle) in that the cycle itself continuously generates intermediate products necessary for the cycle to occur. Oxaloacetic acid, for example, combines with acetic acid to begin the cycle and regenerates from malic acid at the end of the cycle. One or a few molecules of oxaloacetate can therefore function continuously to permit an infinite number of acetate molecules to enter the cycle. Although tied to aerobic metabolism, the Krebs cycle itself does not immediately depend on the presence of molecular oxygen, and researchers have suggested that it originates from anaerobic pathways.

One proposed evolutionary sequence Weitzman offered is that the cycle evolved from an earlier stage in anaerobes in which it split into two metabolically different arms. One arm followed a reductive pathway (see also Gest):

Pyruvate → oxaloacetate → malate → fumarate → succinate → succinyl-coenzyme A

The other arm also began with an initial pyruvate but engaged in oxidative metabolism:

Pyruvate → acetyl-coenzyme A → citrate → cis-iconotate → isocitrate → α-oxoglutarate

Coupled to reactions in the reductional arm was the oxidation of NADH to NAD^+, necessary for replacing NAD^+ used in glycolysis, which, in turn, was necessary for ATP formation. (The final product of the reductional arm, succinyl-coenzyme A, could also be used in synthesizing porphyrins and various amino acids.) Reactions in the oxidative arm had the obvious function of providing reduced nucleotides (NADH, NADPH) for carbohydrate synthesis as well as providing compounds such as α-oxoglutarate, a precursor of glutamic amino acid. Both these arms apparently still function in some organisms, such as cyanobacteria and anaerobically grown *E. coli* bacteria.

According to Weitzman, an enzyme complex (α-oxoglutarate dehydrogenase) that we can consider a

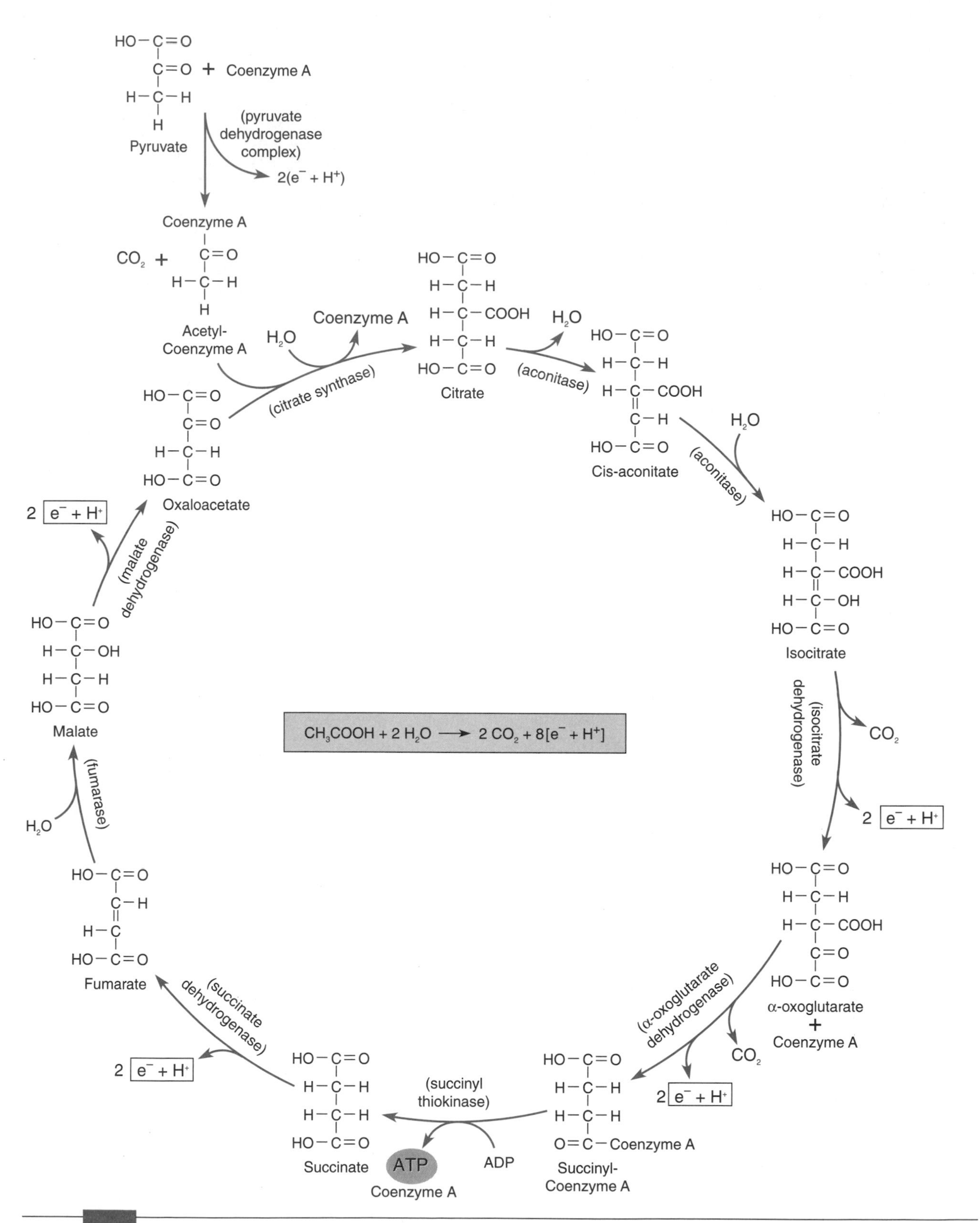

FIGURE 9-8 Diagram of Krebs cycle reactions for the conversion of a 2-carbon acetate molecule (entering the cycle as acetyl-coenzyme A) into two carbon dioxides and eight $[e^- + H^+]$ units used for reduction. In three instances, the cofactor NAD^+ functions as the hydrogen acceptor ($NADP^+$ can function in one of these steps) and FAD functions in the succinate → fumarate reaction. These reduced coenzymes then pass on to the respiratory chain illustrated in Figure 9-9, where they reoxidize by the transfer of electrons to molecular oxygen. This reoxidation process releases large amounts of energy that can help form phosphate bonds of ATP molecules. The Krebs cycle itself has only one reaction, succinyl-coenzyme A → succinic acid, in which a component molecular bond transfers its energy directly to a phosphate bond.

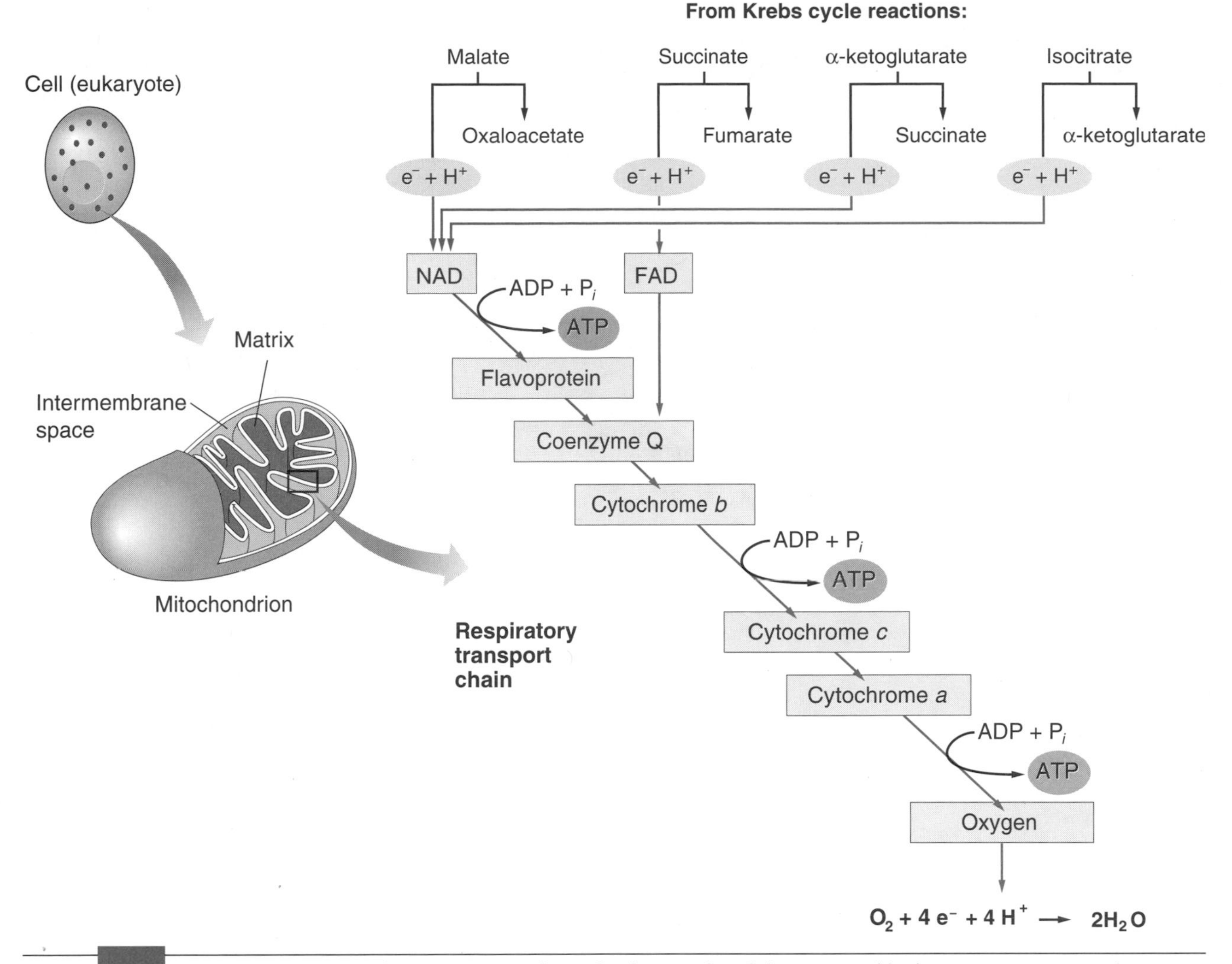

FIGURE 9–9 A simplified diagram of a respiratory pathway for the transfer of electrons and hydrogen protons occurring on the inner membrane of mitochondria. Electrons and protons from coenzymes NAD^+ and FAD pass down the respiratory transport chain, leading ultimately to oxygen and the production of water. Although different microoganisms use different electron carriers, the general sequence of oxidative phosphorylation is the same: as electrons transfer down the chain, hydrogen ions pump across the membrane, and their return flow drives ATP synthesis. For simplicity, the diagram provides coupling sites at which ATP generates, but these sites have not been precisely localized. Depending on which set of reactions bring electrons from NADH to the respiratory chain, as many as 38 ATPs can be formed during complete oxidation of glucose (Stryer).

genetic variant of the pyruvate dehydrogenase complex that helps oxidize pyruvate to acetyl-coenzyme A bridged the gap between these two arms (between α-oxoglutarate coming from the oxidative arm to succinyl-coenzyme A coming from the reductive arm). He suggests that selection to change the remaining reductive steps of the cycle in an oxidative direction associated with the increase of atmospheric oxygen. Thus, although the Krebs cycle itself does not use molecular oxygen, its evolution and adoption by aerobic organisms seems based on a membrane-bound electron transport system where oxygen serves as the final electron acceptor in the chain: one more stage in the evolution of membrane-bound systems.

In this electron transport process, the pyridine nucleotide coenzyme NAD^+ picks up electrons and associated protons (e^- and H^+) and transfers them into a respiratory chain consisting of various electron carriers. The final electron transfer to oxygen, the ultimate electron acceptor, produces water: $O_2 + 4\,e^- + 4\,H^+ \rightarrow 2\,H_2O$.[6] As diagrammed in Figure 9–9, such electron transfers occur along an electrical potential gradient that provides sufficient energy exchange at three steps (coupling sites) to let an ADP molecule phosphorylate to ATP. In sum, complete aerobic oxidation of a molecule of glucose,

[6]Dickerson has suggested that the terminal components of aerobic respiratory mechanisms must have evolved more than once in bacteria by adapting different heme proteins to carry electrons from cytochromes to oxygen. Earlier in evolutionary history, membrane-bound anaerobic respiratory chains apparently used fumarate as a terminal electron acceptor (Jones).

including **oxidative phosphorylation,** produces maximally about 38 molecules of ATP (compared to only 2 molecules of ATP formed by anaerobic glycolysis):

$$C_6H_{12}O_6 + 6\ H_2O + 6\ O_2 + 38\ ADP + 38\ P_i \rightarrow 6\ CO_2 + 12\ H_2O + 38\ ATP$$

If we consider that the hydrolysis of a mole of ATP (507 gm) to ADP probably provides at least 7 kilocalories, then cellular oxidation of a mole of glucose (180 gm) can produce at its theoretical maximum about $38 \times 7 = 266$ kilocalories. This amount is 39 percent of the 686 kilocalories produced by burning a mole of glucose in air and indicates that the efficiency of oxidative phosphorylation can be perhaps as high or higher than many human-designed mechanical energy conversion systems.

In addition to its metabolic effects, the oxygen atmosphere enabled a stratospheric ozone (O_3) layer to form that screens out short-wave ultraviolet radiation from reaching the earth's surface. Since absorption of such wavelengths by organic ring structures (for example, nucleotides) can cause lethality by deteriorating or modifying vital molecules, ozone screening must have been essential in enabling the expansion of life to ocean surfaces as well as to land.

Early Fossilized Cells

From what we can discern so far, metabolic evolutionary stages follow a progression from simple anaerobic systems dependent on energy sources in the primeval "soup," to autotrophic systems capable of generating chemical phosphate bond energy from sunlight, to aerobic systems that derive en-

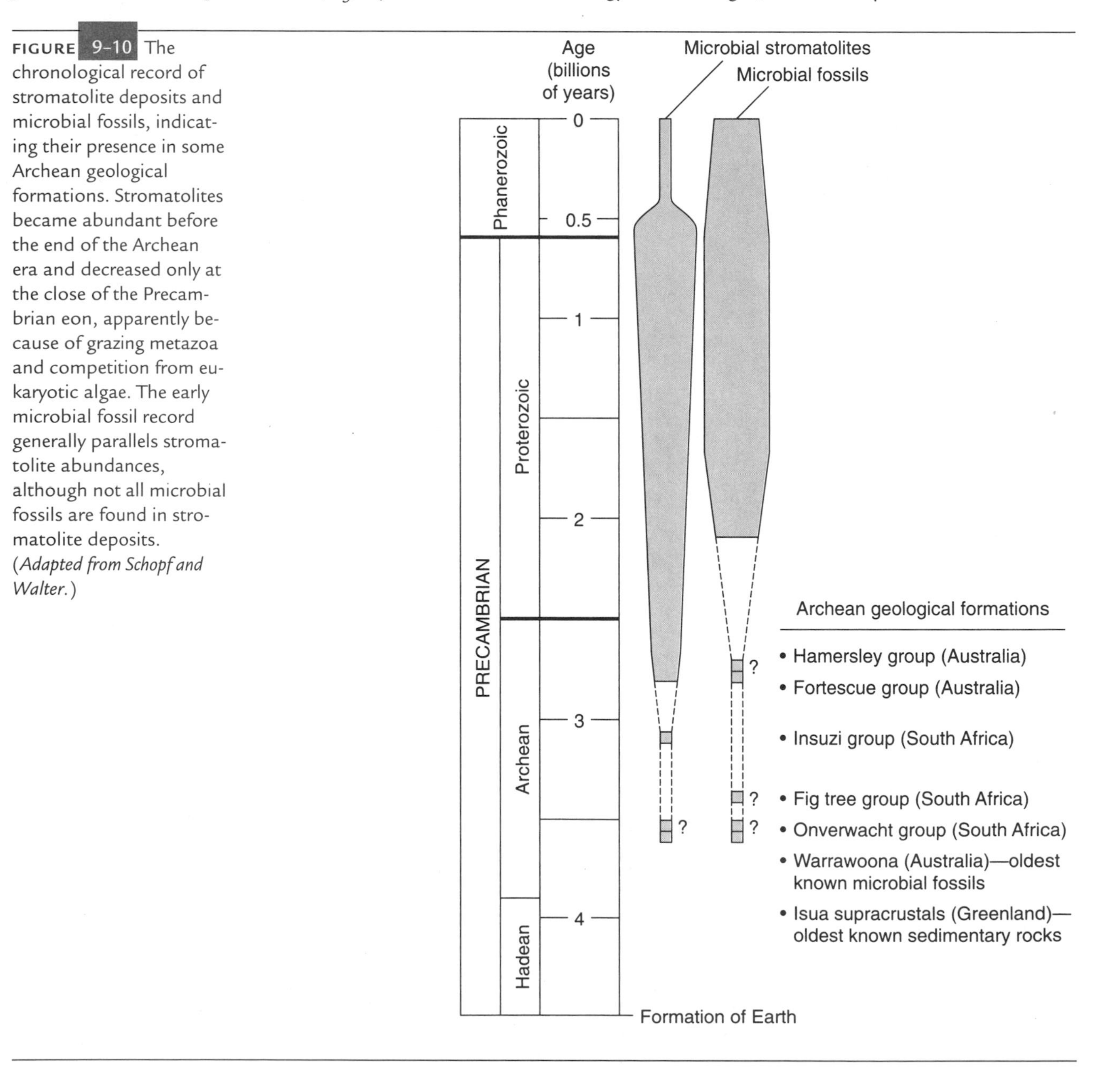

FIGURE 9-10 The chronological record of stromatolite deposits and microbial fossils, indicating their presence in some Archean geological formations. Stromatolites became abundant before the end of the Archean era and decreased only at the close of the Precambrian eon, apparently because of grazing metazoa and competition from eukaryotic algae. The early microbial fossil record generally parallels stromatolite abundances, although not all microbial fossils are found in stromatolite deposits. (*Adapted from Schopf and Walter.*)

ergy from the transfer of electrons to oxygen. Fig. 12–19 shows a proposed phylogeny of prokaryotes that includes such information, based on sequencing amino acids in proteins and nucleotides in RNA.

The exact periods during which these evolutionary steps took place are not known, but various cell-like fossils appear in geological strata that date back as far as 3.5 billion years ago. These strata are mostly unmetamorphosed rocks called **cherts,** which are dark or black because of their high carbon content and also bear considerable silicon deposits. The oldest of these cherts are in the Warrawoona group of western Australia (Fig. 9–10) which, like many other Archean cherts, associate with layered organic deposits called **stromatolites.**

In their modern form, stromatolites consist of mats of microorganisms that trap various aqueous sediments, which then cement together to form characteristic laminated structures shaped like giant knobs.[7] As shown in Figure 9–11, these modern structures are remarkably similar to ancient stromatolites, which must also have arisen from the deposition of biological organisms.

Modern and ancient forms nevertheless differ in distribution, since modern stromatolites appear only in extremely inhospitable environments protected from grazing metazoans such as snails and sea urchins (for example, they occur in salinities ten times that of sea water and at temperatures greater than 65°C), whereas ancient stromatolites undoubtedly dispersed more widely because such herbivores were absent. Interestingly, many of the fossil organisms found in stromatolite deposits are remarkably similar to modern prokaryotes (Fig. 9–12). In the words of Schopf and Walter,

> It seems reasonable to conclude that:
>
> 1. shallow water and intermittently exposed environments (and possibly also, land surfaces and open oceanic waters) were habitable by prokaryotic microorganisms at least as early as 3.5 billion years ago;
>
> 2. such organisms comprised finely laminated, multi-component, stromatolitic communities of the sediment-water interface, biocenoses where the principal surficial mat-building taxa were

(a) Fossil stromatolites (Great Slave Lake, Canada)

(b) Modern stromatolites (Shark Bay, Australia)

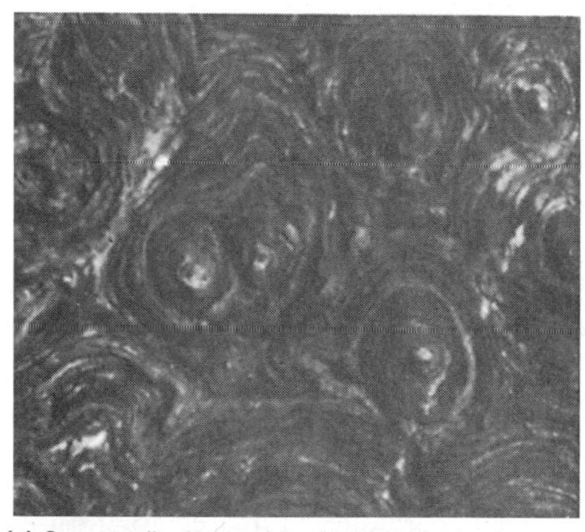

(c) Stromatolite internal laminated structure

FIGURE 9–11 Comparison between (*a*) fossil stromatolites approximately 2 billion years old and (*b*) modern stromatolites. (*c*) A cross section of a stromatolitic knob, which can have a diameter of several feet across, showing laminated layers of algal mats. (*Reprinted with permission of the Minister of Supply and Services Canada, 1989 and Paul Hoffman.*)

[7] According to Golubic and Knoll:

> Stromatolites . . . are initiated by the establishment of a thin mat of microbes on a sediment surface. As sediment particles accumulate on top of mats, they are trapped and bound into a coherent layer by the microorganisms. Microbially mediated precipitation of calcium carbonate can also contribute to sediment accumulation and stabilization. . . . Through time, commonly, a laminated structure accretes, each lamina marking a former position of the living mat community. . . . On the present-day Earth, filamentous cyanobacteria are predominant mat-builders, but coccoid cyanobacteria, other types of bacteria, and a variety of eukaryotic algae produce well-defined mats.

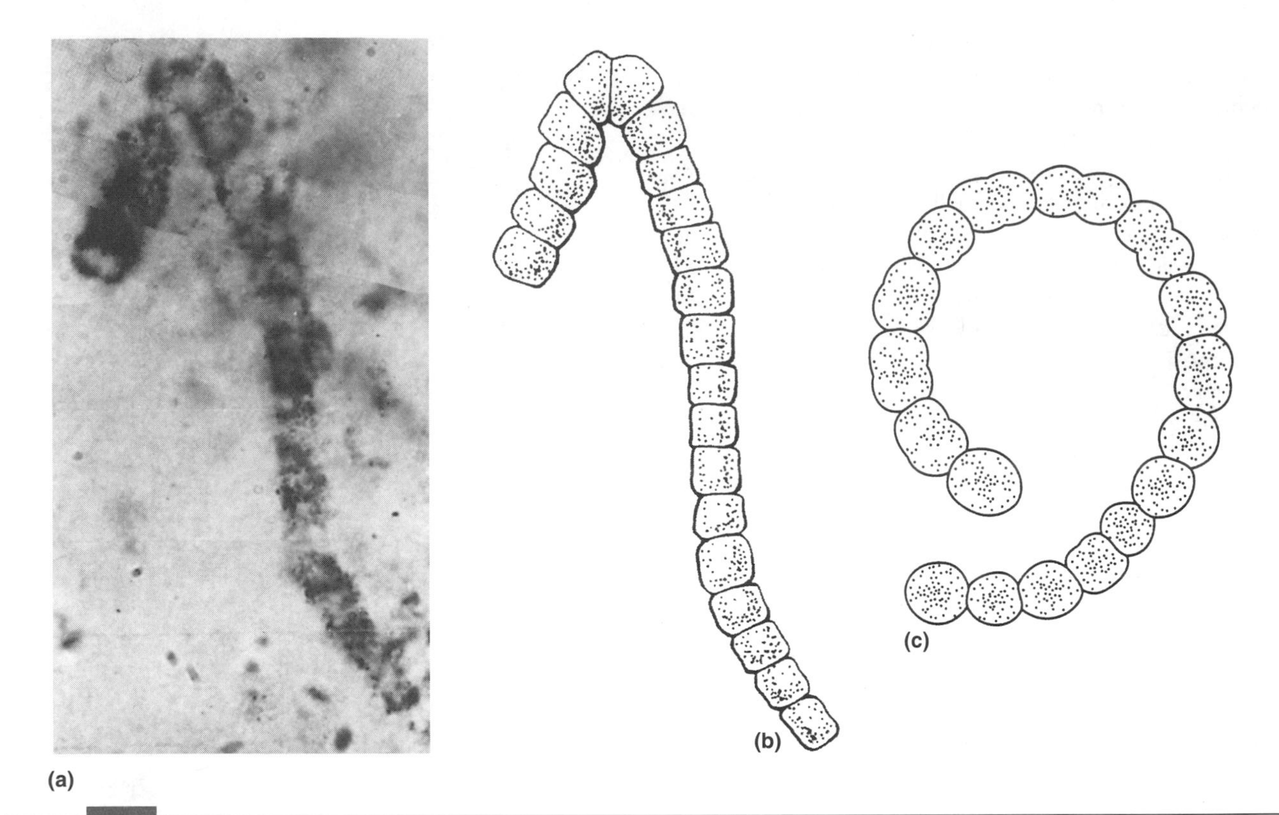

FIGURE 9-12 (*a*) Thin section of a filamentous prokaryotic fossil found in stromatolite chert in the 3.5 billion-year-old Warrawoona formation in western Australia. (*b*) Reconstruction of this fossil. (*c*) Diagram of a phase-contrast microphotograph of a filament of cells of a modern cyanobacterium (*Anabaena* species). Among other modern prokaryotic groups that appear similar to the microfossil in (*a*) are colorless sulfur-gliding bacteria (for example, *Thioplaca schmidlei*) and green sulfur bacteria (*Oscillochloris chrysea*). [*Parts (a) and (b) from J. William Schopf, ed.,* Earth's Earliest Biosphere: Its Origin and Evolution. *Copyright © 1983 Princeton University Press.*]

probably filamentous and photo-responsive, forms that may have been capable of phototactic, gliding motility;

3. such communities probably included anaerobes and both autotrophic and heterotrophic microorganisms.[8]

Using carbon isotope ratios ($^{13}C/^{12}C$), we can surmise that many of these ancient cell-like fossils had a biological origin. These two isotopes differ in respect to their participation in cellular metabolism, and these differences lead to ratios that are unique compared to those found in nonbiological material. From such analyses, almost all the stromatolite deposits dating from 3.5 billion years ago and later have carbon isotope ratios similar to rocks from the Carboniferous period and other strata in which living forms appear. Such biologically derived isotope carbon ratios have also been found in some sedimentary rocks of the Isua group (see Fig. 9–10), supporting the concept that life on Earth existed more than 3.8 billion years ago (Holland).[9]

[8]Schopf (1996) points out that aerobic cyanobacteria and nonaerobic bacteria can coexist in stromatolites by using different light-gathering pigments that enable them to occupy different subhabitats:

> The oxygen-producing cyanobacteria live in the uppermost layers of stromatolites, with the non-oxygen-producing photosynthesizers just beneath. Much of the light energy is absorbed by the cyanobacteria . . . but this does not snuff out the anoxygenic photosynthesis of the green sulfur and purple bacteria that live below because these more primitive photosynthetic anaerobes are literally able to see through the cyanobacterial layer—their pigments absorb light unused by the cyanobacteria above.

[9]Cloud suggested that **banded iron formations** (BIF), the oldest of which dates to 3.76 billion years ago, may also have formed biologically. According to him, internal sources of oxygen were responsible for the change from ferrous to ferric ion, and these sources were probably living protocyanobacteria that were splitting water molecules and releasing O_2:

$$\underset{\text{(ferrous oxide)}}{4\,FeO} + \underset{\text{(from photoautotrophs)}}{O_2} \rightarrow \underset{\text{(ferric oxide)}}{2\,Fe_2O_3}$$

The ferrous ion served as a "sink" for molecular oxygen generated during photosynthesis and would have protected the anaerobic metabolic systems of these early photoautotrophs. The oxidized BIF bands may have resulted from episodic supplies of ferrous ion, which enabled growing anaerobic photoautotrophs to precipitate iron in the form of ferric oxide.

TABLE 9-1 Some molecular biological characteristics that distinguish eubacteria, archaebacteria, and eukaryotes

	Prokaryotes		Eukaryotes
	Eubacteria	Archaebacteria	
Cell wall	All incorporate muramic acid	None incorporate muramic acid	Absent (except in plants—but no muramic acid): cell structure based mainly on internal filamentous and microtubular cytoskelton
Genome organization	One circular chromosome	Single chromosome?	Usually several linear chromosomes
Transcription-translation	Simultaneous processes: translation immediately follows transcription	Simultaneous processes (as in eubacteria)	Separate processes: transcription in the nucleus, translation in the cytophasm
Introns (transcribed intervening sequences)	None	In tRNA genes	Yes
RNA polymerases	One	Probably one	Three, and tRNA)
Capping of 5′-mRNA terminal	No	Probably absent	Yes
3′-mRNA poly-A tails	Absent	Present	Present
Ribosome unit sizes	$30S + 50S = 70S$	$30S + 50S = 70S$	$40S + 60S = 80S$
Ribosome RNA sizes	$23S$, $16\ S$, $5S$	$23S$, $16S$, $5S$	$28S$, $18S$, $5.8S$, $5S$
Ribosome sensitivity to			
Anisomycin	No	Yes	Yes
Chloramphenicol	Yes	No	No
Kanamycin	Yes	No	No
Protein translation sensitivity to diphtheria toxin	No	Yes	Yes
Protein initiator tRNA amino acid	Formylmethionine	Methionine	Methionine

Source: Adapted from Woese (1983), and Doolittle and Daniels. Cavalier-Smith provides a long list of other distinctions.

Prokaryotes and Eukaryotes

Figure 9–13 summarizes some of the major biological and geological evolutionary events from the lowermost Hadean division to the present Phanerozoic eon that originated with the Cambrian period about 545 million years ago. Aside from the origin of photosynthesis, perhaps the most significant biological change is the difference in cellular complexity that marks the "superkingdom" division between organisms classified as prokaryotes and eukaryotes. **Prokaryotes,** a term used to include all bacteria, now generally distinguishes two kingdoms (Table 9–1):

- **Eubacteria** encompass the major forms of bacteria as well as the cyanobacteria, practically all possessing unique peptidoglycan or murein cell walls (chains of sugars cross-linked with short peptides, some of which contain D-amino acids).
- **Archaebacteria** use other materials for their cell walls and live under more rigorous environmental conditions than eubacteria, such as hot sulfur springs and extreme salt concentrations.[10]

The term **eukaryote** includes many single-celled **protistan** organisms such as photosynthetic algae and nonphotosynthetic protozoans, as well as multicellular plants (metaphyta), animals (metazoa), and fungi.

The difference between prokaryotes and eukaryotes is most obvious in the absence of a **nuclear envelope** in the generally smaller prokaryotic cells (1–10 μm) and its presence in the generally larger eukaryote cells (10–100 μm). In addition, prokaryotes divide mostly by **binary fission** rather than by the **mitotic mechanisms** of nearly all eukaryotes. Prokaryotes consequently lack organelles such as

[10]Kingdom and superkingdom divisions are disputed. The traditional prokaryote-eukaryote division of life advocated by Mayr is disputed by Woese (1998b) who proposed a tripartite classification of archaebacteria-eubacteria-eukaryotes (Figure 9–16). In addition to eubacteria and archaebacteria, Lake and coworkers have suggested subdividing the archaebacteria into two groups, and considering one of these a third prokaryotic kingdom called *Eocyta.* The eocyte bacteria depend on sulfur and, according to Lake, their ribosomes are more similar to those of eukaryotes than to the other prokaryotes.

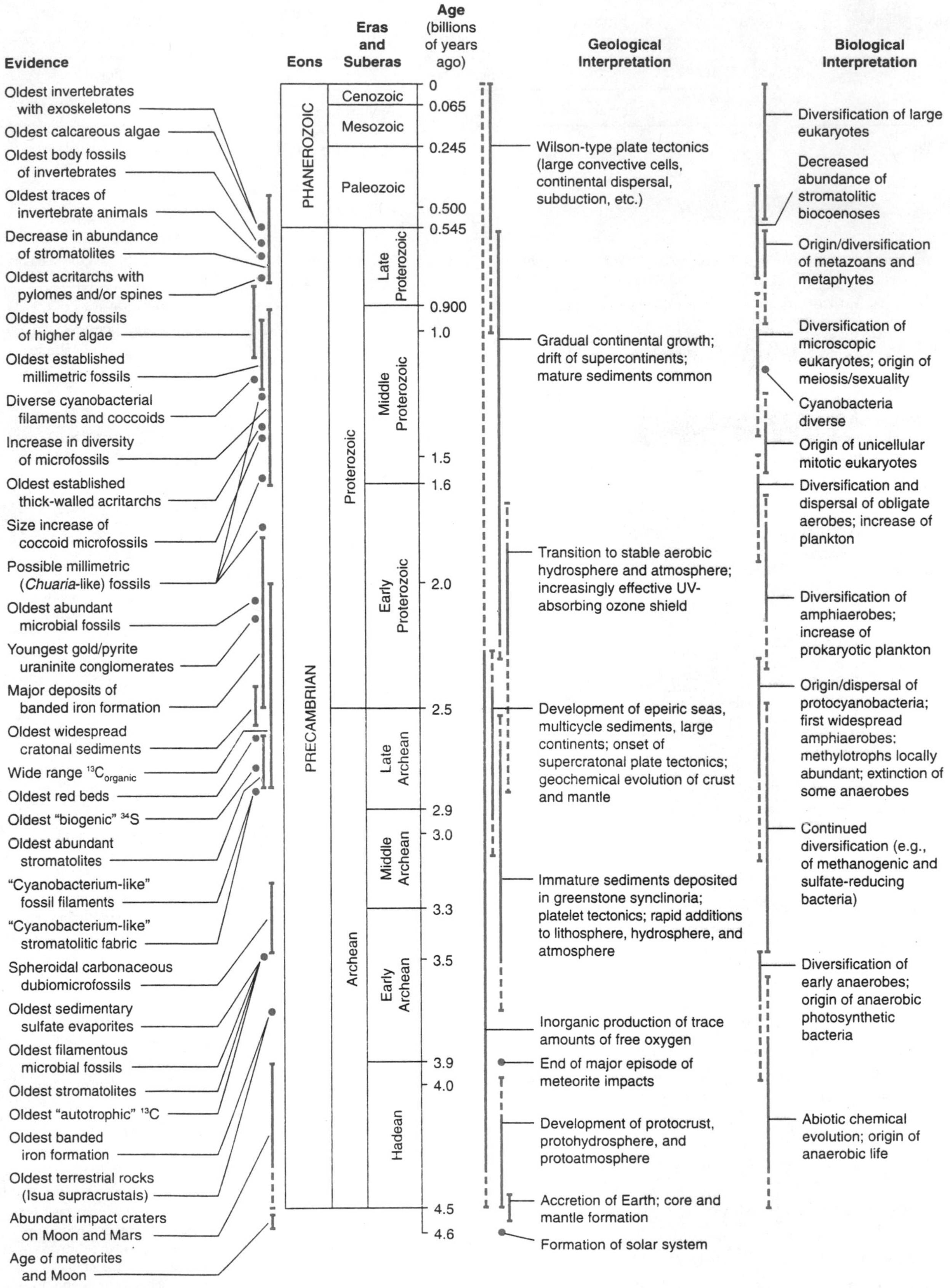

Evidence

- Oldest invertebrates with exoskeletons
- Oldest calcareous algae
- Oldest body fossils of invertebrates
- Oldest traces of invertebrate animals
- Decrease in abundance of stromatolites
- Oldest acritarchs with pylomes and/or spines
- Oldest body fossils of higher algae
- Oldest established millimetric fossils
- Diverse cyanobacterial filaments and coccoids
- Increase in diversity of microfossils
- Oldest established thick-walled acritarchs
- Size increase of coccoid microfossils
- Possible millimetric (*Chuaria*-like) fossils
- Oldest abundant microbial fossils
- Youngest gold/pyrite uraninite conglomerates
- Major deposits of banded iron formation
- Oldest widespread cratonal sediments
- Wide range $^{13}C_{organic}$
- Oldest red beds
- Oldest "biogenic" ^{34}S
- Oldest abundant stromatolites
- "Cyanobacterium-like" fossil filaments
- "Cyanobacterium-like" stromatolitic fabric
- Spheroidal carbonaceous dubiomicrofossils
- Oldest sedimentary sulfate evaporites
- Oldest filamentous microbial fossils
- Oldest stromatolites
- Oldest "autotrophic" ^{13}C
- Oldest banded iron formation
- Oldest terrestrial rocks (Isua supracrustals)
- Abundant impact craters on Moon and Mars
- Age of meteorites and Moon

Eons | **Eras and Suberas** | **Age (billions of years ago)**

- PHANEROZOIC: Cenozoic (0–0.065); Mesozoic (0.065–0.245); Paleozoic (0.245–0.545; 0.500)
- PRECAMBRIAN:
  - Proterozoic: Late Proterozoic (0.545–0.900); Middle Proterozoic (0.900–1.6; 1.0, 1.5); Early Proterozoic (1.6–2.5; 2.0)
  - Archean: Late Archean (2.5–2.9); Middle Archean (2.9–3.3; 3.0); Early Archean (3.3–3.9; 3.5)
  - Hadean (3.9–4.5; 4.0)
- 4.6

Geological Interpretation

- Wilson-type plate tectonics (large convective cells, continental dispersal, subduction, etc.)
- Gradual continental growth; drift of supercontinents; mature sediments common
- Transition to stable aerobic hydrosphere and atmosphere; increasingly effective UV-absorbing ozone shield
- Development of epeiric seas, multicycle sediments, large continents; onset of supercratonal plate tectonics; geochemical evolution of crust and mantle
- Immature sediments deposited in greenstone synclinoria; platelet tectonics; rapid additions to lithosphere, hydrosphere, and atmosphere
- Inorganic production of trace amounts of free oxygen
- End of major episode of meteorite impacts
- Development of protocrust, protohydrosphere, and protoatmosphere
- Accretion of Earth; core and mantle formation
- Formation of solar system

Biological Interpretation

- Diversification of large eukaryotes
- Decreased abundance of stromatolitic biocoenoses
- Origin/diversification of metazoans and metaphytes
- Diversification of microscopic eukaryotes; origin of meiosis/sexuality
- Cyanobacteria diverse
- Origin of unicellular mitotic eukaryotes
- Diversification and dispersal of obligate aerobes; increase of plankton
- Diversification of amphiaerobes; increase of prokaryotic plankton
- Origin/dispersal of protocyanobacteria; first widespread amphiaerobes; methylotrophs locally abundant; extinction of some anaerobes
- Continued diversification (e.g., of methanogenic and sulfate-reducing bacteria)
- Diversification of early anaerobes; origin of anaerobic photosynthetic bacteria
- Abiotic chemical evolution; origin of anaerobic life

FIGURE 9–13 A summary of the evidence for Precambrian evolution along with the geological and biological interpretations of these observations. A series of major meteorite impacts occurred between 4.1 and 3.9 billion years ago as indicated in the moon rocks retrieved during the *Apollo* mission. Geologists believe these collisions produced enough heat to sterilize Earth's surface. As a result, some proponents believe that the origin of life most probably occurred somewhat later, perhaps 3.8 billion years ago, and according to Miller, during an interval that may have been as short as 10 million years or less. (*J. William Schopf, ed.,* Earth's Earliest Biosphere: Its Origin and Evolution. *Copyright © 1983 Princeton University Press.*)

mitotic spindles and centrioles and do not have the histone proteins that structurally organize the relatively larger and more numerous eukaryotic chromosomes.

Prokaryotes also lack some of the membrane structures and **organelles** that commonly appear in eukaryotes, such as endoplasmic reticulum (usually associated with ribosomes in protein synthesis), Golgi apparatus (secretory bodies), and mitochondria (used in oxidative phosphorylation). Their membrane-enclosed compartments allow eukaryotic cells to isolate enzymes for specific reaction sequences—confining transcription to the nucleus, translation to the cytoplasm, aerobic metabolism to mitochondria, and so forth. Eukaryotes thus share mitotic and protein-synthesizing mechanisms and are almost all aerobic, with the exception of some forms such as yeast that can also function anaerobically (amphiaerobes) and some amoeban-type protists such as *Pelomyxa* (members of a lineage that may always have been anaerobic).

Among the further attributes of eukaryotes almost entirely absent in prokaryotes are **split genes** in which amino acid-coding nucleotide sequences, separated by hundreds of bases in DNA, are combined in the final messenger RNA to translate into a single polypeptide product. The nucleotide sequences that code for amino acids in such polypeptides are called **exons** ("expressed sequences"), and the intermediate noncoding nucleotide sequences are called **introns** ("intervening sequences"). These split genes transcribe their nucleotide sequences from DNA to RNA, but the RNA is then processed so that the introns are removed and the exons spliced together (Fig. 9–14). Such processed mature mRNA molecules are then transported through pores of the nuclear envelope to the cytoplasm, to be translated into polypeptides.

Intron–exon structures seem to be sufficiently advantageous to account for the finding that split genes occur in almost all vertebrate protein-coding genes that researchers have so far examined, as well as in many similar genes in other eukaryotes. The reasons for these advantages, however, are not entirely clear. Gilbert suggested that originally each exon may have coded for a single **polypeptide domain** that had a specific function, and others such as Blake suggested a relationship between the average size of an exon (coding for about 20 to 80 amino acids) and the size of the smallest polypeptide sequence that could fold into a stable structure (about 20 to 40 amino acids).

Because exon arrangement and intron removal is flexible, the exons coding for these polypeptide subunits could then act as modules ("domains") to combine in various ways to form genes that produce optimally functional polypeptides. Even some single genes can produce several different functional proteins by arranging their exons in several different ways through alternate splicing patterns (Andreadis et al.).[11] Doolittle (1978) and others favored an "introns early" hypothesis: cellular organisms ancestral to both prokaryotes and eukaryotes probably had such intron–exon structures, but these were mostly abandoned in prokaryotic lines because increased intensity of selection streamlined DNA replication and improved transcription efficiency.

An "introns late" hypothesis suggests they first inserted into full-length eukaryotic genes *after* eukaryotes had separated from prokaryotes; that is, gene assembly did not depend on the initial use of introns to connect and shuffle the modular exons. Among the evidence for this view are tests by Stoltzfus and coworkers showing lack of correspondence between exons and functional protein units, as well as indications that several introns are of recent origin in eukaryotes (Logsdon et al.).

A decision between the two hypotheses is yet to come, and some researchers propose both are correct: 30 to 40 percent of introns were "early" and the remainder "late" (De Souza et al.). However, for introns that came late, new questions arise: How did introns enter into eukaryotic genes? Why did they become so prevalent? Why are they maintained? Answers vary, but many raise the possibility or even likelihood that introns are mobile DNA sequences that can splice themselves out of, as well as into, specific "target sites" (Palmer and Logsdon), acting like mobile transposon-like elements (p. 225). Because the insertion of such elements would add excess nucleotides to messenger RNA and disrupt normal gene expression, their survival would rely on splicing themselves out of RNA transcripts before translation—that is, by acting as introns. Other suggestions for intron function and persistence include a possible regulatory role, that the increase in the number of genes from a few thousand or so in prokaryotes to tens of thousands in eukaryotes

[11]Dorit and coworkers suggest that the number of exons needed to construct all proteins may range between 1,000 and 7,000. Interestingly, researchers estimate a figure close to 1,000 as the number of ancestral amino acid sequences shared by all organisms. However, as Claverie points out, many such sequences may not correspond to exons.

FIGURE 9–14 The intron-exon structure for nucleic acid sequences involved in the production of the human β-globin polypeptide chain, 146 amino acids long, one of the components of normal adult hemoglobin. The *top portion of the diagram* indicates the approximate 1,500 nucleotide base-pair length in the β-globin DNA sequence and the structure of this sequence in terms of two introns (in1, in2) and three exons (ex1, ex2, ex3). Intron 1 is 130 nucleotides long and separates codons that will later translate into amino acids numbered 30 and 31 on the β-globin chain. Intron 2 is 850 nucleotides long and separates codons for β-globin amino acids numbered 104 and 105. Although the entire DNA sequence transcribes into mRNA, the introns are precisely removed by special RNA processing in the nucleus, and the exon sequences splice together and then translate into a continuous β-globin amino acid sequence. At the *bottom of this illustration,* a single β-globin polypeptide chain has been diagrammatically split to indicate the three subcomponents (domains) respectively produced by the three exon nucleotide sequences. (*Adapted from Strickberger.*)

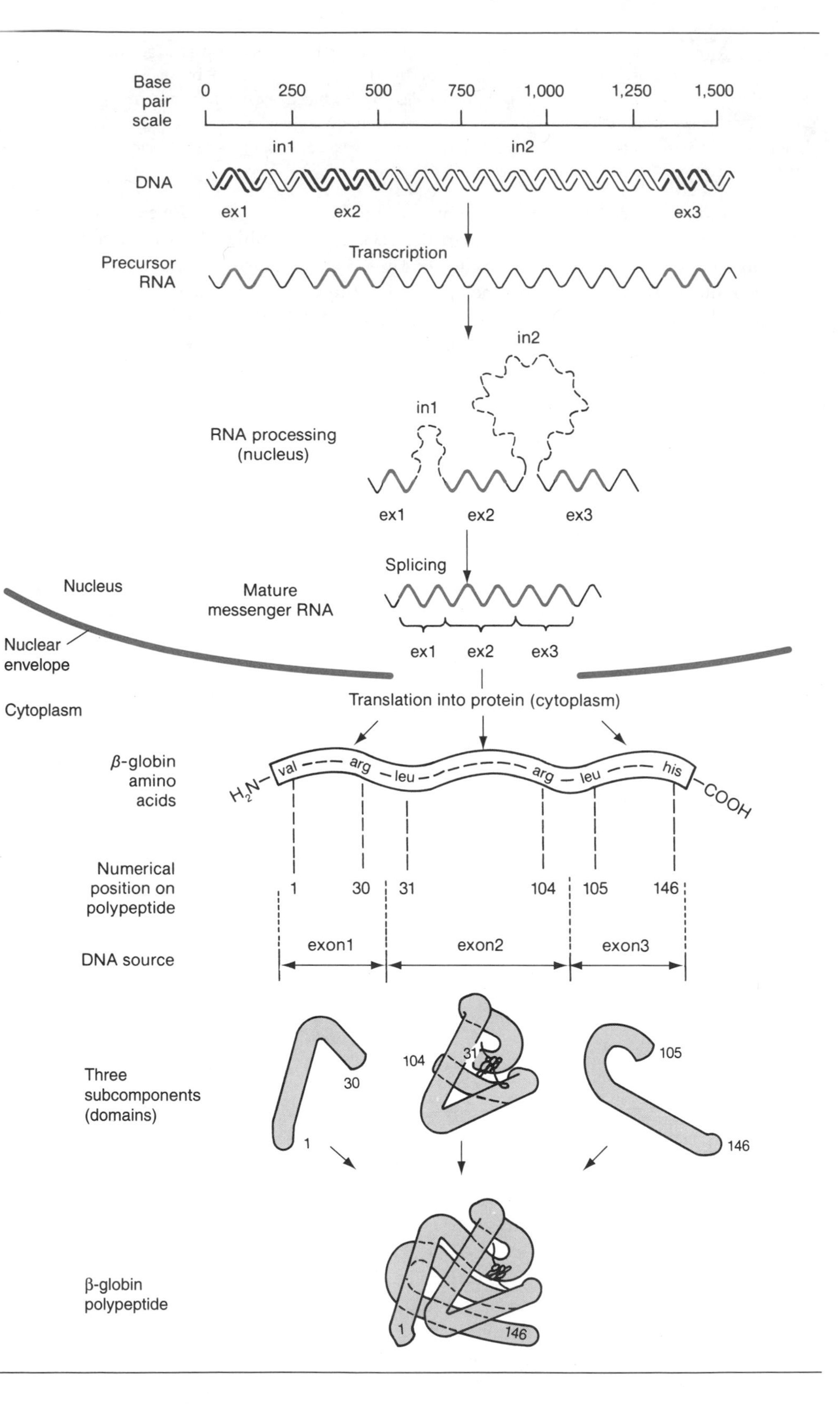

may have depended on adding introns to the gene regulatory system.[12]

Other characteristics that distinguish eubacteria, archaebacteria, and eukaryotes are shown in Table 9–1. Woese and coworkers have also catalogued the presence and frequencies of various sequences in the 16*S* RNA component of ribosomes in the three groups and shown that they have distinctive differences (Table 9–2).

Although these tables show that archaebacteria occupy an intermediate evolutionary position between eubacteria and eukaryotes, these data are not sufficient to determine the exact source of their common ancestor. Theorists have therefore suggested that no one group derived from any other: rather, they all diverged from a common cellular ancestor [or from a community of early cells in which mutation rates were high and genes were being freely exchanged—a population of **progenotes** (Woese, 1998a)]. In support, researchers observe that all three of these groups have common basic attributes such as a similar mode of DNA replication (new nucleotides add to the 3′ end of the molecule), a common genetic code, a similar protein-synthesizing system, many similar metabolic pathways, a similar cell membrane structure consisting of a phospholipid bilayer, and a similar mechanism of molecular transfer across membranes (**active transport**). For example, Figure 9–15 shows that this ancestral organism probably had enzymes involved in glycolysis, Krebs cycle, and the urea cycle. The enzymatic fidelity needed by such processes indicates that DNA with its much lower mutation rate must have already replaced RNA as the cellular genetic material.

However, subsequent events are still unclear: researchers have proposed that eubacteria and eukaryotes arose from archaebacteria (Woese), that prokaryotes arose from eukaryotes (Darnell, Poole et al.), and that eukaryotes arose directly from prokaryotes (for example, Cavalier-Smith). Because of their very ancient origins and the uncertainties caused by **horizontal gene transfer,** in which DNA from one organism incorporates into DNA of another (p. 225), many relationships are being questioned. Connections among these three major groups are often depicted as an "unrooted" tree (Fig. 9–16) without designating a particular domain as the ancestral source. Although we do not know the "root," and are uncertain about some branches (Pennisi), we can still discern general relationships. Molecular studies do point to a closer connection between eukaryotes and archaebacteria than between eukaryotes and eubacteria (Rowlands et al.):

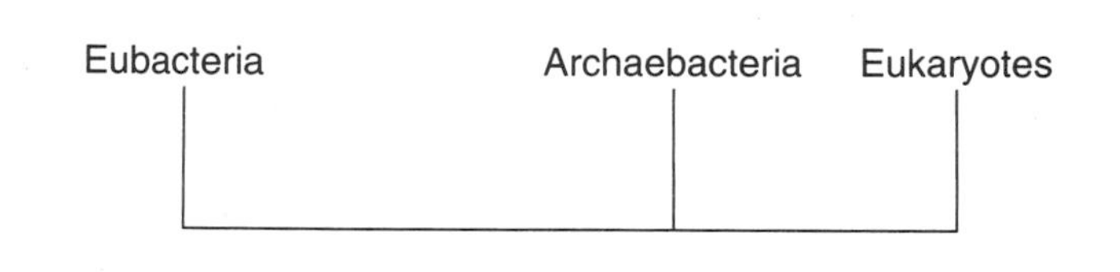

TABLE 9–2 Presence and frequencies of some oligonucleotide sequences found in the 16*S* ribosomal RNA component of eubacteria, archaebacteria, and eukaryotes

	Percent Occurrence		
Sequence[a]	Eubacteria	Archaebacteria	Eukaryotes
CYUAAYACAUG	83%	0%	0%
AYUAAG	1	62	100
ACUCCUACG	97	0	0
CCCUACG	0	97	0
ACNUCYANG	0	0	100
YYUAAAG	3	97	0
AUACCCYG	93	3	0
CAACCYUYR	91	0	0
CCCCG	0	100	0
UCCCUG	0	97	100
AUCACCUC	91	100	0

[a]R designates either purine (adenine or guanine);
Y designates either pyrimidine (uracil or cytosine);
N designates any of the four bases.
Source: Abridged from Woese (1985).

Evolution of Eukaryotic Organelles

Information has been accumulating in respect to the origin of the mitochondria and chloroplast organelles found within eukaryotic cells. The most popular hypothesis, which Stanier, Margulis, and others offer in its modern form, proposes that eukaryotic

[12]In prokaryotes, protein products of other genes regulate gene expression (see, for example, Fig. 10–25), whereas some researchers propose that nucleotide sequences in eukaryotic introns also serve as regulatory agents. Since prokaryotes do not have nuclei, and ribosome attachment with consequent protein synthesis occurs simultaneously with the transcription of messenger RNA, introns—should they be common in prokaryotes—would translate as part of messenger RNA and would result in nonfunctional proteins. In eukaryotes, however, RNA processing sequesters in the nucleus and intronless messenger RNA then translates in the cytoplasm, allowing the nuclear introns to function in a regulatory role without affecting messenger RNA translation. According to Mattick, this proposed intron-regulatory advantage may have enabled the evolution of more complex gene systems in eukaryotes and their advance to multicellularity.

Large numbers of introns found in some genes cast doubt on the presumed regulatory function. For example, the gene for dystrophin protein, whose absence or malfunction causes muscular dystrophy, has more than 65 introns totaling more than 2 million base pairs, whereas its exons total only 14,000 base pairs. Most of these introns seem excessive if their role was confined to regulating gene expression (Evans and Goyne). Transposable element duplication has also been used to explain such high intron numbers. As discussed later (p. 227), transposons can accumulate in large quantities as "junk DNA" because they replicate independently and more frequently than their host chromosomes. Fitting in to the introns-late hypothesis, frequent mobile transposon activity rather than regulation may account for the large number of introns in certain genes.

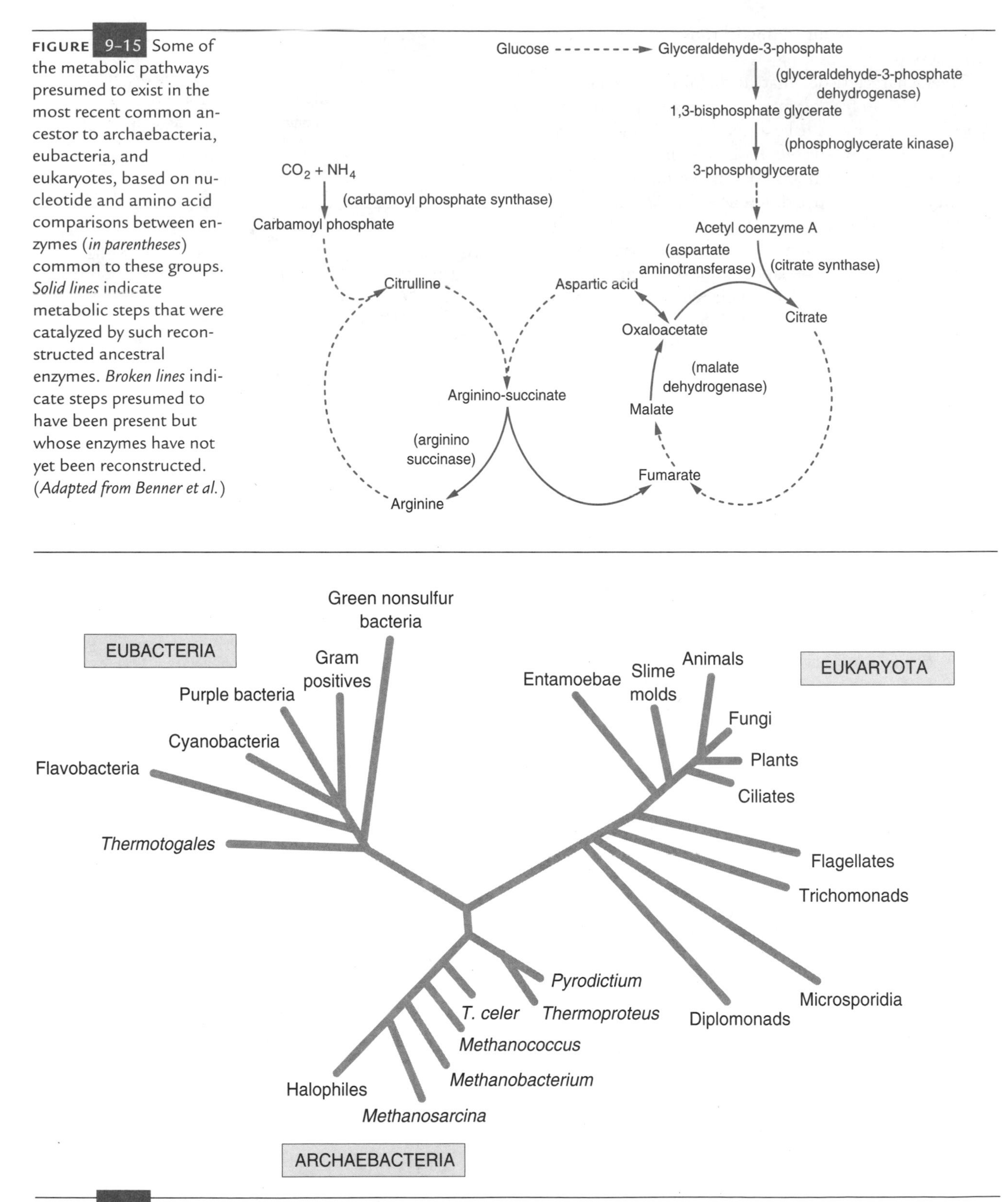

FIGURE 9-15 Some of the metabolic pathways presumed to exist in the most recent common ancestor to archaebacteria, eubacteria, and eukaryotes, based on nucleotide and amino acid comparisons between enzymes (*in parentheses*) common to these groups. *Solid lines* indicate metabolic steps that were catalyzed by such reconstructed ancestral enzymes. *Broken lines* indicate steps presumed to have been present but whose enzymes have not yet been reconstructed. (*Adapted from Benner et al.*)

FIGURE 9-16 One possible phylogenetic tree for all cellular life ("the universal tree") based on nucleotide sequence similarities among some ribosomal components (small-submit RNA, 16*S*-type RNA). In this illustration, the tree is "unrooted," meaning that no source ("root") is indicated for the origin of the common ancestor. The tree shows a number of interesting features, including the close tie between animals and fungi, and widely separate origins for some eukaryotic groups (for example, flagellates, entamoeba, diplomonads, ciliates, microsporidia) that have traditionally been classified together into the single kingdom Protista. The diplomonads are represented by *Giardia,* a protozoan whose positions as the earliest eukaryotic offshoot is supported by both nucleotide and amino acid sequence analyses (Hashimoto et al.) Although *Giardia* lacks mitochondria, we should keep in mind that its ancestors might nevertheless have maintained bacterial symbionts. Henze and coworkers point out that *Giardia,* as well as mitochondrionless *Entamoeba histolytica* possess nuclear genes that must have originated by nuclear transfer from a prokaryotic symbiont (p. 180) which was later lost from the cytoplasm—"cryptic endosymbiosis" (see also Roger et al.). (*Adapted from Sogin and from Wheelis et al.*)

LYNN MARGULIS

Birthday:
March 5, 1938

Birthplace:
Chicago, Illinois

Undergraduate degree:
University of Chicago, 1957

Graduate degree:
Ph.D. University of California-Berkeley, 1965

Postdoctoral training:
Brandeis University, Waltham, Massachusetts, 1964–1966

Present position:
Distinguished University Professor
Department of Geosciences
University of Massachusetts at Amherst

WHAT PROMPTED YOUR INITIAL INTEREST IN EVOLUTION?

I have always loved nature and the diversity of life, especially the woods. Although my undergraduate training at the University of Chicago was primarily in the liberal arts, my interest in science was piqued by prescribed readings of original scientific papers. After graduation, the master's degree program at the University of Wisconsin at Madison afforded me the opportunity to do protist research. I became fascinated by studies of the genetics of cytoplasmic particles. These particles—chloroplasts, mitochondria, and motility organelles such as ciliary kinetosomes and mitotic centrioles—were not usually investigated by geneticists, but questions about their inheritance patterns, origin, and evolution intrigued me. Although my doctoral work at the University of California at Berkeley centered on inducible mutations in *Euglena*—the loss of their photosynthetic plastids—the irreversibility of some of these mutations led me to the relationship of nonnuclear genetic systems to the history of eukaryotic cells. It became clear to me that symbiotic relationships were at the heart of the nucleated cell, whose major organelles probably had bacterial origins. The ideas and investigations that followed led to various papers and, in 1970, to publication of my first large major work on the problem, *Origin of Eukaryotic Cells* (Yale University Press).

WHAT AREAS OF RESEARCH ARE YOU (OR YOUR LABORATORY) PRESENTLY ENGAGED IN?

The area of my research that I feel most important right now concerns our discovery that spirochetes form motility associations and undergo developmental cycles that are analogous and may even be homologous to cilia development. Allied to this question is a strong interest in understanding the genetic basis of microtubular structures.

IN WHICH DIRECTIONS DO YOU THINK FUTURE WORK IN YOUR FIELD NEEDS TO BE DONE?

A great need exists in my opinion to understand more about cell division (mitotic and meiotic) in smaller organisms. Protoctists (algae, protozoans, slime molds, etc.), bacteria, and yeasts have often been ignored or misclassified as "small" invertebrates or "lower" plants, but they really represent unique diverse groups. Protists that evolved from bacterial symbiose are ancestral to all other eukaryotes. An essential part of their evolutionary importance lies in the fact that it is this group in which the fundamentals of mitosis and meiosis first appeared—patterns of growth and sexuality that radiated outward and were preserved in protoctists (single or few-celled protists) and their multicellular descendants.

WHAT ADVICE WOULD YOU OFFER TO STUDENTS WHO ARE INTERESTED IN A CAREER IN YOUR FIELD OF EVOLUTION?

Paraphrasing Ernst Mayr, "In order to do any original science one must specialize, but in order to do evolution one must be far broader than a specialist." Although there are many aspects of knowledge evolutionists can and should utilize—natural history, chemistry, atmospheric science, geology—evolutionary understanding demands integration rather than separation. Above all, I believe evolutionists should learn as much as possible about live organisms, since through the windows of the living important clues can be seen about the past.

cells evolved by physically incorporating prokaryotic organisms into their cytoplasms. According to this theory, the ancestral anaerobic eukaryotic cell achieved the ability to perform **endocytosis,** to ingest supramolecular particles because of new surface membrane properties that included loss of the rigid cell wall. An internal cytoskeleton, composed of filaments and microtubules, then helped maintain cell shape.

Provided with a new flexible coat, some members of this eukaryotic lineage became actively predaceous, so that selection led to increased predatory abilities, such as an increase in cell size, as well as to other innovations that affect movement, capture of prey, and digestion. Among the various prokaryotes on which this microbe fed were bacteria-like aerobes capable of oxidative phosphorylation as well as cyanobacteria capable of photosynthesis. At various times such ingested photosynthetic and aerobic prokaryotes assumed mutually advantageous **symbiotic** relationships with their hosts, similar to those which even now occasionally appear (**Fig. 9–17**).

FIGURE 9-17 Symbiotic relationships between a eukaryote and its photosynthetic organelles. The protozoan ciliate *Paramecium bursaria* (*left*) harbors hundreds of symbiotic algae (*right*) that may be released from the cell and cultured independently. (*From* Symbiosis in Cell Evolution *by Lynn Margulis. New York: W. H. Freeman, 1981. Reprinted by permission.*)

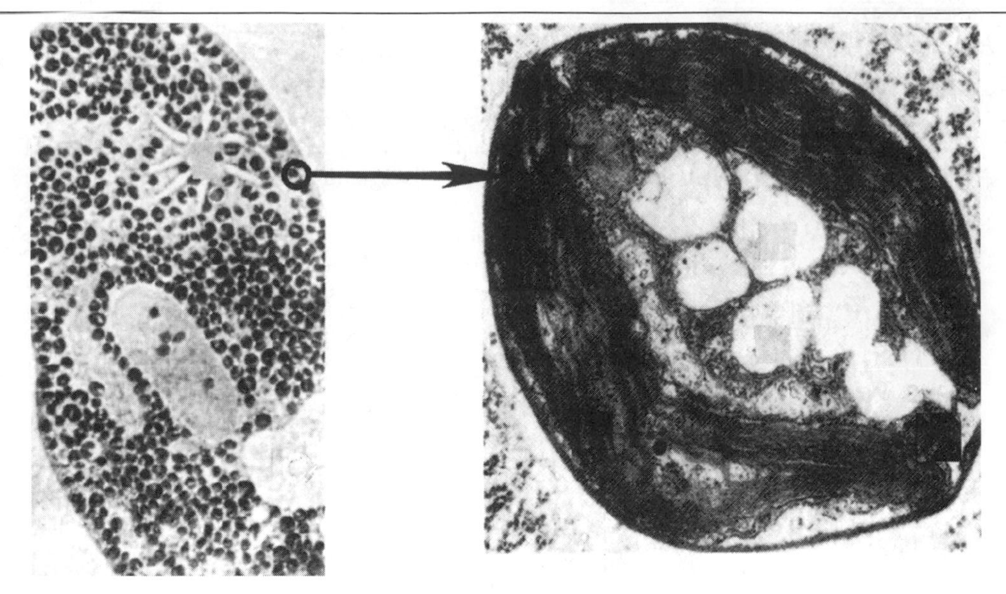

According to one scheme, eukaryotes first established symbiotic relationships with mitochondrion-like aerobic bacteria that improved eukaryotic metabolism and broadened eukaryotic predatory activity. Later, one or more lines of these new aerobic eukaryotes began similar symbiotic relationships with photosynthetic cyanobacteria, eventually evolving into the various eukaryotic algae and plants (Fig. 9–18).[13] To account for their evolutionary success, both kinds of symbiosis must have conferred selective advantages on their eukaryotic hosts, increasing their frequency compared to their nonsymbiotic competitors.

Supporting this view is the finding that mitochondria and chloroplasts have their own genetic material (DNA),[14] and their ribosomes are more like those of prokaryotes than like host ribosomes in size, sensitivity to antibiotics, and nucleotide sequences of ribosomal RNA components. Such similarities are obvious in the mitochondrion of a freshwater protozoan, *Reclinomonas.* This organelle bears the largest collection of mitochondrial genes so far, and is therefore considered more primitive than other such organelles which lost genes as they evolved in concert with more advanced eukaryotes. According to Andersson and coworkers, many genes remaining in *Reclinomonas* mitochondria are strikingly similar to a prokaryote causing epidemic typhus, *Rickettsia prowazekii,* an obligate intracellular parasite. Apparently an aerobic rickettsial ancestor, adapted to parasitizing eukaryotic cells, gave rise to mitochondria.[15]

Once incorporated in the eukaryotic cytoplasm, these symbiotic organelles transferred many of their genes to the eukaryotic nucleus (see also p. 281). They lost genes that could be provided by host nuclei, such as those for anaerobic glycolysis and amino acid biosynthesis. Such transfers and deletions had the advantage of maintaining and replicating only two copies of a symbiotic gene in a diploid host nucleus instead of sustaining a separate gene copy within each of many cellular organelles. Since some cells carry enormous numbers of organelles—more than 8,000 mitochondrial genomes in some human cells and even more copies of chloroplast genomes in some plant cells—the advantage of reducing organelle gene number by deletion or nuclear incorporation must have been highly selected. Nuclear genes that code for symbiotic organelle proteins have been widely identified (Gillham). The proteins are made by the common cellular cytoplasmic translation process, and transported to the correct organelle position using special "signal" and "transit" peptides.

[13]Since flagella, cilia, centrioles, and similar eukaryotic organelles consist of microtubules and seem structurally homologous, some proponents of the symbiotic theory suggest that these also arose by eukaryotes capturing prokaryotes and then establishing symbiosis. According to Margulis (see also Margulis and Sagan), an original cilium-type symbiont was a spirochete-like organism as now appears in the protistan *Myxotricha paradoxa.* She proposes that all eukaryotic mitotic mechanisms owe their origin to the microtubular proteins initially involved in flagellar movements.

[14]Persistence of organelle DNA within host cytoplasm can lead to undesirable intracellular competition among different mutant organelle genomes, causing replacement of functional organelles by those less functional but more successfully reproductive. Among ways this problem was apparently resolved was by restricting organelle diversity through **uniparental inheritance;** that is, by transmitting mitochondria and chloroplasts through only a single parent, commonly the female (p. 202).

[15]Comprehensive reviews of evidence for the origins of both mitochondria and chloroplasts from prokaryotic endosymbionts appear in papers by Gray (1992, 1993), who also points out that some eukaryotic algal groups probably acquired chloroplasts through secondary endosymbiosis from a eukaryote rather than from a prokaryote. Such secondary invasions by chloroplast-carrying eukaryotes may account for the retention of an algal-type cell within some protozoan parasites (Köhler et al.).

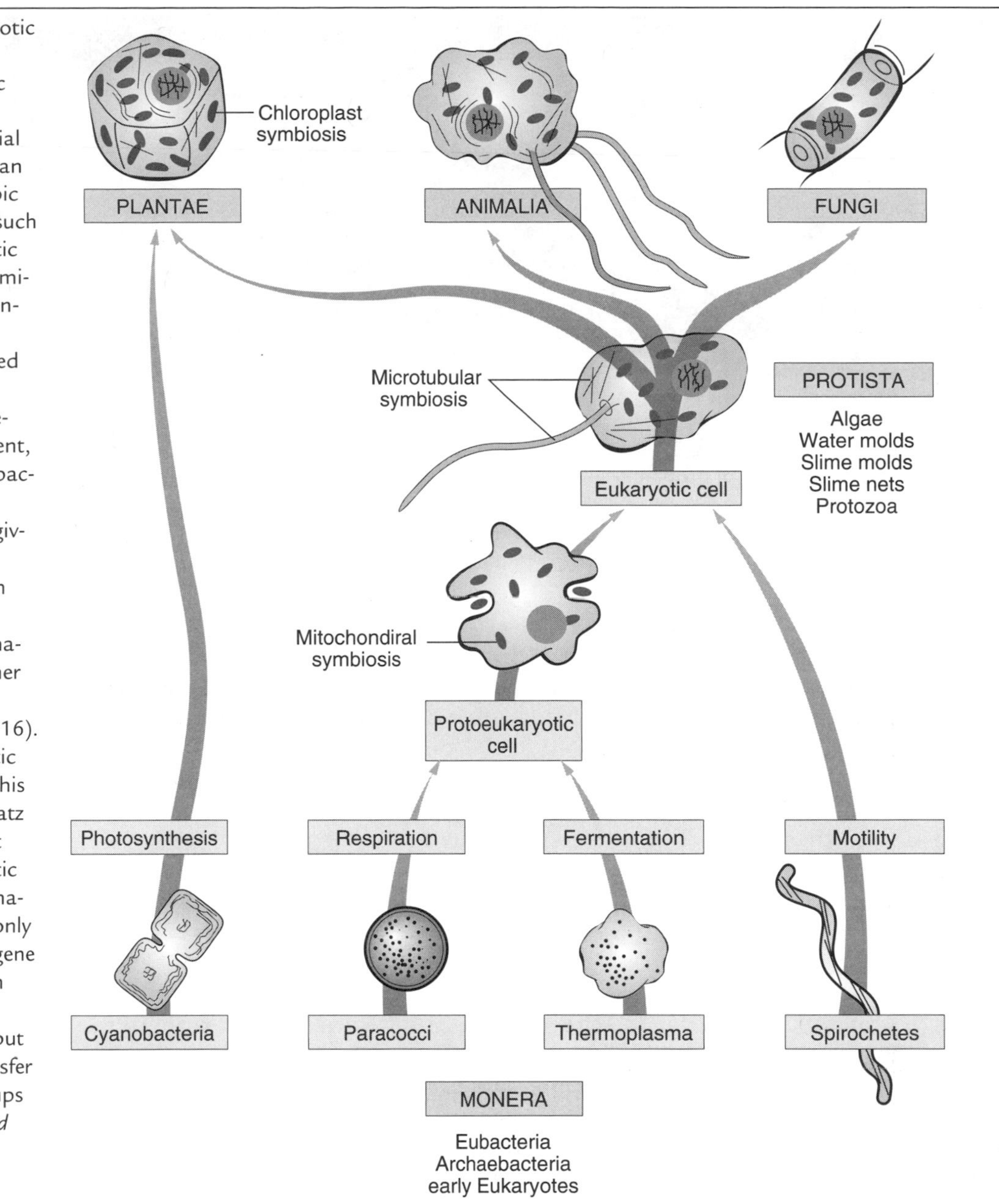

FIGURE 9-18 Symbiotic events during the evolution of eukaryotic cells according to Margulis. Mitochondrial symbioses arose from an early invasion of aerobic respiring prokaryotes such as *Paracoccus*. Eukaryotic motility as well as the microtubular structures involved in mitosis and meiosis was then gained by symbiosis with a prokaryotic spirochete-like form. In a later event, photosynthetic cyanobacteria invaded some eukaryotic cells, thus giving rise to plant chloroplasts. Although Margulis divides living forms into these five major groups, various other workers now use only three domains (Fig. 9-16). In a review of eukaryotic origin that compares this scheme with others, Katz points out that a most characteristic eukaryotic feature is its chimeric nature. Apparently, not only did normal "vertical" gene transfer occur between generations in early eukaryotic evolution, but "horizontal" gene transfer between different groups also occurred. (*Adapted from Margulis.*)

As may be expected, the eukaryotic nucleus also has collected hypotheses suggesting its origin via endosymbiosis. Thus Lake and Rivera suggest that nuclear membranes were derived from a captured prokaryotic cell which provided a portion of the eukaryotic genetic material. Using molecular sequencing techniques (Chapter 12), Gupta and coworkers presented support for this view: significant similarities exist among endoplasmic reticulum genes in the ancient eukaryote *Giardia* (Fig. 9–16) and very similar genes in certain types of "Gram-negative" bacteria detected by H. C. Gram's differential staining procedures. Since the eukaryotic double-membraned endoplasmic reticulum is continuous with the eukaryotic nuclear membrane (Fig. 7–1), these researchers suggest that the original eukaryote formed by fusion between an archaebacterium and a Gram-negative eubacterium. More specifically, an archaebacterium could have received energy and carbon through symbiosis with a bacterium that excreted hydrogen and carbon dioxide—an ancestor of the **hydrogenosome** organelle common in eukaryotes (Martin and Müller).[16]

[16]The genes analyzed for these studies produce a family of universally conserved "heat shock" proteins called hsp 70 that, among other features, act as "chaperones" to help transfer other proteins from one intracellular locality to another. According to Gupta and Singh, their sequence analysis shows a strong relationship between archaebacteria and Gram-positive eubacteria, each group having diverged from a common ancestral group more than once ("polyphyletic" evolution, see Chapter 11). They suggest

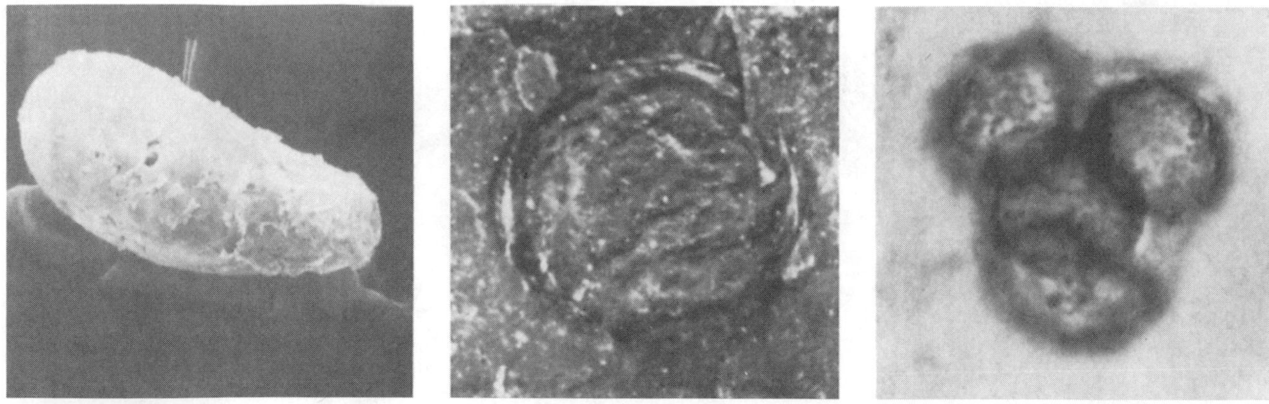

(a) Grand Canyon shales (800 million years old)

(b) Utah shales (950 million years old)

(c) Central Australian sediments (850 million years old)

FIGURE 9–19 Microfossils of probable eukaryotic cells that date back to the Proterozoic era. The cells in (*a*) and (*b*) are many times larger than any known prokaryotic cells and are considerably more complex. The group of cells in (*c*) are in a characteristic tetrahedral arrangement, suggesting they formed through either mitotic or meiotic eukaryotic cell division mechanisms. (*From "The Evolution of the Earliest Cells" by J. W. Schopf,* Scientific American *239, 1978. Reprinted by permission; see also Martin.*)

Whatever the origin of their membranes and organelles, by 1.5 billion years ago or even earlier, eukaryotic cells had clearly appeared. The approximate 2-billion-year interval between the age of these fossils and the earliest prokaryotes which date back to 3.5 billion years (p. 171) and Fig. 9–13) may have been caused by the length of time needed to incorporate and coordinate the many profound changes in cell structure and function ("coadaptive mutations," p. 451). Gaining a foothold in a world dominated by prokaryotes may also have been a protracted struggle (see p. 325).

Although early eukaryotes were all single-celled organisms (Fig. 9–19), their affinities to particular protistan groups are uncertain, and we therefore call them **acritarchs** (Greek *akritos,* "undecided"). Because of their relatively large size and complexity, these protistans had much more genetic material (DNA) than traditional among prokaryotes. The difficulty of manipulating and replicating one or more large DNA molecules having circular prokaryotic forms, each with only a single point (origin) of replication, apparently led to the division of eukaryotic DNA into one or more linear chromosomes, each with multiple origins of replication—the more such origins, the more rapid the replication.

The near-universal presence of cytoskeletal microtubules in eukaryotes enables nuclear chromosomal division to supplant the usual prokaryotic method, which depends on separating dividing chromosomes by their individual attachment to a lengthening cell membrane. Instead, each of the relatively larger, more complex, and often more numerous eukaryotic chromosomes has a **centromere** (or **kinetochore**) that can attach to a microtubular network of sliding spindle fibers during division, letting the daughter chromosomes move to opposite poles fairly efficiently. So far, we do not know the exact sequence of steps in the evolution of eukaryotic chromosome structure and cell division, although researchers have offered some hypotheses (Cavalier-Smith).

Very likely, once eukaryotic mitosis evolved, sex cell division, or **meiosis,** must have quickly followed, since meiosis and sexuality almost universally appear among the major eukaryotic taxonomic groups and practically all asexual multicellular eukaryotes seem derived from sexual forms. Some advantages of sexual reproduction, and the need for the meiotic cell divisions to reduce the number of chromosomes before cross-fertilization, are discussed in Chapters 10 and 13.

For the present we can note that although some very small multicellular forms may have evolved from single-celled protistans as long as one billion or more years ago, larger visible multicellular fossils are only apparent just before the beginning of the Phanerozoic eon. When the Cambrian, the first of the Phanerozoic periods, began about 545 million years ago, numerous new forms of multicellular invertebrates made a sudden marked appearance. In an interval of perhaps 100 million years or less, an explosive radiation of eukaryotes occurred in which a large number of animal phyla appeared in the fossil record. Evolutionary biologists have long made the story of their derivations and relationships a major focus.

However, before we deal with these phenomena, it is important to gain a clear concept of some basic genetic mechanisms that provide fuel and substance to evolu-

that Gram-negative bacteria evolved from Gram-positive bacteria at a later period and—still later—one or more fusion events united Gram-negative bacteria with archaebacteria to produce the first protoeukaryotic cells. Subsequent endosymbioses with various Gram-negative bacterial groups then furnished eukaryotes with mitochondria and chloroplasts. The "chimeric" nature of the eukaryotic nucleus, with its array of genes from many sources, thus derives from its original archaebacterial-eubacterial ancestry, along with genes transferred from endosymbiotic organelles, and perhaps also from genes of ingested cells (Doolittle 1998).

tion; that is, to understand the origin of hereditary variability among organisms and its transmission among generations. What kinds of genetic variability are there? What are their causes, and how are they transmitted? The next chapter briefly reviews this topic.

SUMMARY

The prevalence of metabolic pathways common to many organisms suggests that comparative biochemistry would be a fruitful approach to investigate the evolution of metabolism. Anaerobic glycolysis is an almost universal pathway in which energy released as glucose degrades to pyruvic acid, and the reactions comprising it may illuminate its evolutionary past. For instance, when small, molecular reactants were abundant, they would have been the first to be used for energy. After these small molecules were depleted, metabolizing larger compounds could have extended the original pathway. Whatever the correct explanation, metabolic pathways, because of their sequential nature, could not have arisen randomly but were "predetermined" by available preexisting compounds. Many pathways (and enzymes controlling individual reactions) have survived because of their selective advantages, and these commonly appear in living systems. Chance could not have caused this astonishing similarity, but evolution from a common ancestral system could.

Eventually, high-energy carbon compounds became less plentiful and some organisms switched to reducing more copious carbon compounds, such as CO_2, by means of a membrane-bound oxidation-reduction system. A revolution occurred when some organisms began to reduce carbon by using light as an energy source (photosynthesis). Such reduction requires a source of electrons, for which substances such as H_2S were first used. It later became advantageous to split water, which was plentiful. Oxygen, a by-product of this process, increased so much that it was exploited to enhance energy yields from glucose breakdown. A new pathway, the Krebs cycle, arose from anaerobic pathways, the entire process being contingent on the presence of oxygen. Electrons removed from oxidized compounds in the cycle transfer to a membrane-bound electron transport system for which oxygen acts as the ultimate electron acceptor molecule. Much more energy releases in this process than releases in anaerobic glycolysis.

Prokaryote-type cells arose about 3.5 billion years ago, and about 2 billion years later some had diversified into eukaryotes. Present prokaryotes, which include eubacteria and archaebacteria, are small, contain no nuclear membrane or complex organelles, and divide by binary fission. Eukaryotes are generally aerobic and more complex, as they contain many organelles and a microtubular apparatus for mitotic cell division. Eukaryotic genes, unlike those of prokaryotes, often have nucleotide sequences within them (introns) that do not translate into peptide sequences. In many respects the archaebacteria seem intermediate between eubacteria and eukaryotic cells, but maybe all three types of cells arose from a single common ancestor, the progenote.

Because of the cell-like characteristics of mitochondria and chloroplasts, many biologists now explain the presence of the eukaryotic organelles as remnants of an ancient symbiotic relationship. The large amount of DNA in eukaryotic cells led to its fission into several chromosomes and to the appearance of microtubule-dependent cell division. Meiosis must also have arisen early, since it is almost universal in eukaryotes. Somewhat later, perhaps just prior to the Phanerozoic era, multicellular eukaryotes appeared and underwent several explosive radiations.

KEY TERMS

acritarchs
active site
active transport
aerobic metabolism
anaerobic glycolysis
antioxidant compounds
archaebacteria
autotrophic
backward evolution
banded iron formations
binary fission
Calvin cycle
centromere
chemiosmosis
cherts
chlorophyll
comparative biochemistry
cyclic photosynthesis
dark reactions
electron acceptors
electron donors
Embden-Meyerhof glycolytic pathway
endocytosis
eubacteria
eukaryotes
exons
heterotrophic
horizontal gene transfer
hydrogenosome
introns
kinetochore
Krebs cycle
light reactions
lithotrophic nutrition
meiosis
metabolic pathways
mitotic mechanisms
noncyclic photosynthesis
nuclear envelope
organelles
organotrophic nutrition
oxidative phosphorylation
ozone
phosphorylation
photophosphorylation
polypeptide domain
progenote
prokaryotes
protistans
proton gradient
proton pump
reduction–oxidation
retrograde evolution
split genes
stromatolites
symbiotic
thylakoids
uniparental inheritance

DISCUSSION QUESTIONS

1. How does Horowitz's hypothesis of retrograde evolution explain evolution of the Embden-Meyerhof glycolytic pathway?
2. What evidence exists for the evolutionary conservation of active sites of enzymes in such pathways?

3. By which means could carbon sources such as CO_2 reduce to cellular organic compounds in the early stages of evolution?

4. Electron transport systems and protein gradients
   a. Why do researchers consider ferredoxin an early electron carrier?
   b. What roles do electron transport systems and protein gradients play in cellular energy production?

5. Photosynthesis
   a. What advantage does chlorophyll offer in photosynthesis?
   b. How do cyclic and noncyclic photosynthesis differ? Which probably evolved first?
   c. What is the relationship between photosynthesis and the Calvin cycle?
   d. What effect did noncyclic photosynthesis have on the proportions of oxygen and ozone in the atmosphere?
   e. Why do cellular antioxidants offer a selective advantage, and what types are there?

6. Aerobic metabolism
   a. According to Weitzman, how did the Krebs cycle evolve?
   b. What is the connection between the Krebs cycle and oxidative phosphorylation?
   c. How can one compare the efficiency of aerobic metabolism to anaerobic metabolism?

7. What accounts for the formation and structure of stromatolites? Their age?

8. Prokaryotes and eukaryotes
   a. What are some major differences between prokaryotes and eukaryotes?
   b. Approximately how many years ago did each type of cell originate?
   c. What are "split genes," and how do they function?
   d. Is there support for the concept of a "progenote"? If so, describe.
   e. What hypotheses have researchers offered to explain the origin of eukaryotic organelles? Which hypothesis has gathered the most support, and what kinds of evidence have been used?

9. If genes can be transferred from symbiotic organelles to the eukaryotic nucleus (p. 180), what might account for the persistence of genes in the eukaryotic mitochondrion?

EVOLUTION ON THE WEB

Explore evolution on the web! Visit the accompanying web site for *Evolution*, 3/e at www.jbpub.com/evolution for web exercises and links relating to topics covered in this chapter.

REFERENCES

Adman, E. T., L. C. Sieker, and L. H. Jensen, 1973. The structure of a bacterial ferredoxin. *J. Biol. Chem.*, **248,** 3987–3996.

Andersson, S. G. E., et al., 1998. The genome sequence of Rickettsia prowazekii and the origin of mitochondria. *Nature,* **396,** 133–140.

Andreadis A., M. E. Gallago, and B. Nadal-Ginard, 1987. Generation of protein isoform diversity by alternative splicing: Mechanistic and biological implications. *Ann. Rev. Cell Biol.,* **3,** 207–242.

Barber, J., and B. Andersson, 1994. Revealing the blueprint of photosynthesis. *Nature,* **370,** 31–34.

Barker, H. A., 1972. ATP formation by anaerobic bacteria. In *Horizons of Bioenergetics,* A. San Pietro and H. Gest (eds.). Academic Press, New York, pp. 7–31.

Benner, S. A., M. A. Cohen, G. H. Gonnet, D. B. Berkowitz, and K. P. Johnsson, 1993. Reading the palimpsest: Contemporary biochemical data and the RNA world. In *The RNA World,* R. F. Gesteland and J. F. Atkins (eds.). Cold Spring Harbor Laboratory Press, Cold Spring Harbor, New York, pp. 27–70.

Blake, C. C. F., 1985. Exons and the evolution of proteins. *Int. Rev. Cytol.,* **93,** 149–185.

Broda, E., 1975. *The Evolution of the Bioenergetic Processes.* Pergamon Press, Oxford, England.

Cavalier-Smith, T., 1991. The evolution of the cells. In *Evolution of Life: Fossils, Molecules, and Culture,* S. Osawa and T. Honjo (eds.). Springer-Verlag, Tokyo, pp. 271–304.

Claverie, J. M., 1993. Database of ancient sequences. *Nature,* **364,** 19–20.

Cloud, P., 1974. Evolution of ecosystems. *Amer. Sci.,* **62,** 54–66.

Darnell, J. E., 1978. Implications of RNA—RNA splicing in evolution of eukaryotic cells. *Science,* **202,** 1257–1260.

De Duve, C., and S. L. Miller, 1991. Two-dimensional life? *Proc. Nat. Acad. Sci.,* **88,** 10014–10017.

De Souza, S. J., M. Long, R. J. Klein, S. Roy, W. Gilbert, 1998. Towards a resolution of the introns early/late debate: Only phase zero introns are correlated with the structure of ancient proteins. *Proc. Nat. Acad. Sci.,* **95,** 5094–5099.

Dickerson, R. E., 1980. The cytochromes: An exercise in scientific serendipity. In *The Evolution of Protein Structure and Function,* D. S. Sigma and M. A. B. Brazier (eds.). Academic Press, New York.

Doolittle, W. F., 1978. Genes-in-pieces: Were they ever together? *Nature,* **272,** 581.

———, 1998. You are what you eat: A gene transfer ratchet could account for bacterial genes in eukaryotic nuclear genomes. *Trends in Genet.,* **14,** 307–311.

Doolittle, W. F., and C. J. Daniels, 1985. Prokaryotic genome evolution: What we may learn from the archaebacteria. In *Evolution of Prokaryotes,* K. H. Schleifer and E. Stackebrandt (eds.). Academic Press, London, pp. 31–44.

Dorit, R. L., L. Schoenbach, and W. Gilbert, 1990. How big is the universe of exons? *Science,* **250,** 1377–1382.

Drobner, E., H. Huber, G. Wächtershäuser, D. Rose, and K. O. Stetter, 1990. Pyrite formation linked with hydrogen evolution under anaerobic conditions. *Nature,* **346,** 742–744.

Eck, R. V., and M. O. Dayhoff, 1966. Evolution of the structure of ferredoxin based on living relics of primitive amino acid sequences. *Science,* **152,** 363–366.

Evans, B. K., and C. Goyne, 1991. Duchenne's muscular dystrophy: Review and recent scientific findings. *Amer. Med. Sci.,* **302,** 118–123.

Feelisch, M., and J. F. Martin, 1995. The early role of nitric oxide in evolution. *Trends in Ecol. and Evol.,* **10,** 496–499.

Fridovich, I., 1975. Oxygen: Boon and bane. *Amer. Sci.,* **63,** 54–59.

Gest, H., 1981. Evolution of the citric acid cycle and respiratory energy conversion in prokaryotes. *FEMS Microbiol. Lett.,* **12,** 209–215.

Gest, H., and J. W. Schopf, 1983. Biochemical evolution of anaerobic energy conversion: The transition from fermentation to anoxygenic photosynthesis. In *Earth's Earliest Biosphere: Its Origin and Evolution,* J. W. Schopf (ed.). Princeton University Press, Princeton, NJ, pp. 135–148.

Gilbert, W., 1979. Introns and exons: Playgrounds of evolution. In *Eucaryotic Gene Regulation,* R. Axel, T. Maniatis, and C. F. Fox (eds.). Academic Press, New York, pp. 1–12.

Gillham, N. W., 1994. *Organelle Genes and Genomes.* Oxford University Press, Oxford, England.

Golubic, S., and A. H. Knoll, 1993. Prokaryotes. In *Fossil Prokaryotes and Protists,* J. H. Lipps (ed.). Blackwell Scientific, Boston, pp. 51–76.

Granick, S., 1965. Evolution of heme and chlorophyll. In *Evolving Genes and Proteins,* V. Bryson and H. J. Vogel (eds.). Academic Press, New York, pp. 67–88.

Gray, M. W., 1992. The endosymbiont hypothesis revisited. *Int. Rev. Cytol,* **141,** 233–257.

———, 1993. Origin and evolution of organelle genomes. *Current Biol.,* **3,** 884–890.

Gupta, R. S., and B. Singh, 1994. Phylogenetic analysis of 70kD heat shock protein sequences suggests a chimeric origin for the eukaryotic cell nucleus. *Current Biol.,* **4,** 1104–1114.

Gupta, R. S., K. Aitken, M. Falah, and B. Singh, 1994. Cloning of *Giardia lamblia* heat shock protein HSP70 homologs: Implications regarding origin of eukaryotic cells and of endoplasmic reticulum. *Proc. Nat. Acad. Sci.,* **91,** 2895–2899.

Hall, D. O., J. Lumsden, and E. Tel-Or, 1977. Iron-sulfur proteins and superoxide dismutases in the evolution of photosynthetic bacteria and algae. In *Chemical Evolution of the Early Precambrian,* C. Ponnamperuma (ed.). Academic Press, New York, pp. 191–210.

Hashimoto, T., Y. Nakamura, F. Nakamura, T. Shirakura, J. Adachi, N. Goto, K. Okamoto, and M. Hasegawa, 1994. Protein phylogeny gives a robust estimation for early divergence of eukaryotes: Phylogenetic place of a mitochondria-lacking protozoan, *Giardia lamblia. Mol. Biol. and Evol.,* **11,** 65–71.

Henze, K., A. Badr, M. Wettern, R. Cerff, and W. Martin, 1995. A nuclear gene of eubacterial origin in *Euglena gracilis* reflects cryptic endosymbioses during protist evolution. *Proc. Nat. Acad. Sci.,* **92,** 9122–9126.

Holland, H. D., 1997. Evidence for life on Earth more than 3850 million years ago. *Science,* **275,** 38–39.

Horowitz, N. H., 1945. On the evolution of biochemical synthesis. *Proc. Nat. Acad. Sci.,* **31,** 153–157.

Jones, C. W., 1985. The evolution of bacterial respiration. In *Evolution of Prokaryotes,* K. H. Schleifer and E. Stackebrandt (eds.). Academic Press, London, pp. 175–204.

Katz, L. A., 1998. Changing perspectives on the origin of eukaryotes. *Trends in Ecol. and Evol.,* **13,** 493–497.

Knoll, A. H., and E. S. Barghoorn, 1977. Archean microfossils showing cell division from the Swaziland system of South Africa. *Science,* **198,** 396–398.

Köhler, S., et al., 1977. A plastid of probable green algal origin in apicomplexan parasites. *Science,* **275,** 1485–1489.

Krebs, H., 1981. The evolution of metabolic pathways. In *32nd Symposium of the Society for General Microbiology,* M. J. Carlile, J. F. Collins, and B. E. B. Moseley (eds.). Cambridge University Press, Cambridge, England, pp. 215–228.

Lake, J. A., and M. C. Rivera, 1994. Was the nucleus the first endosymbiont? *Proc. Nat. Acad. Sci.,* **91,** 2880–2881.

Logsdon, J. M., Jr., A. Stoltzfus, and W. F. Doolittle, 1998. Molecular evolution: Recent cases of spliceosomal intron gain? *Current Biol.,* **8,** R560–R563.

Margulis, L., 1993. *Symbiosis in Cell Evolution,* 2d ed. Freeman, New York.

Margulis, L., and D. Sagan, 1986. *Origins of Sex: Three Billion Years of Genetic Recombination.* Yale University Press, New Haven, CT.

Martin, F., 1993. Acritarchs: A review. *Biol. Rev.,* **68,** 475–538.

Martin, W., and M. Müller, 1998. The hydrogen hypothesis for the first eukaryote. *Nature,* **392,** 37–41.

Mattick, J., 1994. Intron evolution and function. *Current Opinion Genet. Devel.,* **4,** 823–831.

Mayr, E., 1998. Two empires or three? *Proc. Nat. Acad. Sci.,* **95,** 9720–9723.

Miller, S. L., 1992. The prebiotic synthesis of organic compounds as a step toward the origin of life. In *Major Events in the History of Life.* J. W. Schopf (ed.). Jones and Bartlett, Boston, pp. 1–28.

Mitchell, P., 1979. Keilin's respiratory chain concept and its chemiosmotic consequences. *Science,* **206,** 1148–1149.

Molecular Origins and Evolution of Photosynthesis, 1981. A special issue of *BioSystems,* **14** (1).

Nagy, L. A., 1974. Transvaal stromatolite: First evidence for the diversification of cells about 2.2×10^9 years ago. *Science,* **183,** 514–515.

Nes, W. R., and W. D. Nes, 1980. *Lipids in Evolution.* Plenum Press, New York.

Palmer, J. D., and J. M. Logsdon, Jr., 1991. The recent origin of introns. *Current Opinion Genet. Devel.,* **1,** 470–477.

Pennisi, E., 1998. Genome data shake the tree of life. *Science,* **280,** 672–674.

Poole, A. M., D. C. Jeffares, and D. Penny, 1998. The path from the RNA world. *J. Mol. Evol.,* **46,** 1–17.

Rao, K. K., D. O. Hall, and R. Cammack, 1981. The photosynthetic apparatus. In *Biochemical Evolution,* H. Gutfreund (ed.). Cambridge University Press, Cambridge, England, pp. 150–202.

Roger, A. J., et al., 1998. A mitochondrial-like chaperonin 60 gene in *Giardia lamblia*: Evidence that diplomonads once harbored an endosymbiont related to the progenitor of mitochondria. *Proc. Nat. Acad. Sci.*, **95,** 229–234.

Rowlands, T., P. Baumann, and S. P. Jackson, 1994. The TATA-binding protein: A general transcription factor in eukaryotes and archaebacteria. *Science,* **264,** 1326–1329.

Schopf, J. W., 1978. The evolution of the earliest cells. *Sci. Amer.,* **239**(3), 110–134.

———, 1996. Metabolic memories of Earth's earliest biosphere. In *Evolution and the Molecular Revolution,* C. R. Marshall and J. W. Schopf (eds.). Jones and Bartlett, Sudbury, MA, pp. 73–107.

Schopf, J. W., J. M. Hayes, and M. R. Walter, 1983. Evolution of earth's earliest ecosystems: Recent progress and unsolved problems. In *Earth's Earliest Biosphere,* J. W. Schopf (ed.). Princeton University Press, Princeton, NJ, pp. 361–384.

Schopf, J. W., and M. R. Walter, 1983. Archean microfossils: New evidence of ancient microbes. In *Earth's Earliest Biosphere: Its Origin and Evolution,* J. W. Schopf (ed.). Princeton University Press, Princeton, NJ, pp. 214–239.

Sogin, M. L., 1991. Early evolution and the origin of eukaryotes. *Current Opinion Genet. Devel.,* **1,** 457–463.

Stanier, R. Y., 1970. Some aspects of the biology of cells and their possible evolutionary significance. *Symp. Soc. Gen. Microbiol.,* **20,** 1–38.

Stoltzfus, A., D. F. Spencer, M. Zuker, J. M. Logsdon, Jr., and W. F. Doolittle, 1994. Testing the exon theory of genes: The evidence from protein structure. *Science,* **265,** 202–207.

Strickberger, M. W., 1985. *Genetics,* 3d ed. Macmillan, New York.

Stryer, L., 1988. Biochemistry, 3d ed. Freeman, New York.

Vetter, R. D., 1991. Symbiosis and the evolution of novel trophic strategies: Thiotrophic organisms at hydrothermal vents. In *Symbiosis As a Source of Evolutionary Innovation,* L. Margulis and R. Fester (eds.). MIT Press, Cambridge, MA, pp. 219–245.

Wächtershäuser, G., 1994. Life in a ligand sphere. *Proc. Nat. Acad. Sci.,* **91,** 4283–4287.

Wald, G., 1974. Fitness in the universe: Choices and necessities. In *Cosmochemical Evolution and the Origins of Life,* J. Oró, S. L. Miller, C. Ponnamperuma, and R. S. Young (eds.). D. Reidel, Dordrecht, Netherlands, pp. 7–27.

Weitzman, P. D. J., 1985. Evolution in the citric acid cycle. In *Evolution of Prokaryotes,* K. H. Schleifer and E. Stackebrandt (eds.). Academic Press, London, pp. 253–275.

Wheelis, M. L., O. Kandler, and C. R. Woese, 1992. On the nature of global classification. *Proc. Nat. Acad. Sci.,* **89,** 2930–2934.

Woese, C. R., 1983. The primary lines of descent and the universal ancestor. In *Evolution from Molecules to Men,* D. S. Bendall (ed.). Cambridge University Press, Cambridge, England, pp. 209–233.

———, 1985. Why study evolutionary relationships among bacteria? In *Evolution of Prokaryotes,* K. H. Schleifer and E. Stackebrandt (eds.). Academic Press, London, pp. 1–30.

———, 1987. Bacterial evolution. *Microbiol. Rev.,* **51,** 221–271.

———, 1998a. The universal ancestor. *Proc. Nat. Acad. Sci.,* **95,** 6854–6859.

———, 1998b. Default taxonomy: Ernst Mayr's view of the microbial world. *Proc. Nat. Acad. Sci.,* **95,** 11043–11046.

Woese, C. R., B. A. Debrunner-Vossbrinck, H. Oyaizu, E. Stackebrandt, and W. Ludwig, 1985. Gram-positive bacteria: Possible photosynthetic ancestry. *Science,* **229,** 762–765.

Wolfe, S. L., 1981. *Biology of the Cell,* 2d ed. Wadsworth, Belmont, CA.

Zubay, G., 1988. *Biochemistry,* 2d ed. Macmillan, New York.

III
The Organic Framework

10 Genetic Constancy and Variability

Organismic evolution relies on two fundamental aspects of genetics or biological inheritance: **constancy** and **variability.** Constancy resides in the genetic observation that like produces like and derives from the ability of nucleic acid macromolecules to replicate (Chapter 7). The evolutionary significance of constancy is that all living processes, from biochemistry to behavior, depend on the transmission of constant or reliable information from previous generations. In contrast, variability resides in the genetic observation that like also produces unlike—the replication of biological information is not always constant or exact—thus producing inherited biological changes (**mutations**). The obvious evolutionary significance of genetic variability is that it fuels evolution by allowing organisms to differ from their ancestors. Constancy and variability are indelibly embodied in evolution through hereditary information (**genotype**) that primarily determines organismic features (**phenotype**). This chapter briefly reviews some fundamental concepts of genetic constancy and variability—the genetic system that enables organisms to both change and preserve their biological attributes.

Cell Division

In modern organisms the transmission of biological information coordinates with cellular division so that both parental and daughter cells carry copies of the same information. Such cell division processes must have originated early in the history of life as a solution to the problem of how membrane-enclosed organisms could grow and expand without enlarging themselves to the point of endangering their existence.[1] Since biological information is coded as long-chain nucleic acid molecules, it is basic to cell division that the

[1]As a cell grows in the absence of cell division, its volume (which is based on the cube of its radius) increases proportionately greater than its surface area (which depends on the square of the radius). The inner constituents of nondividing cells with increasing radius would develop considerable difficulties in both obtaining and excreting metabolic substances because they have relatively less surface area for each increase in volume.

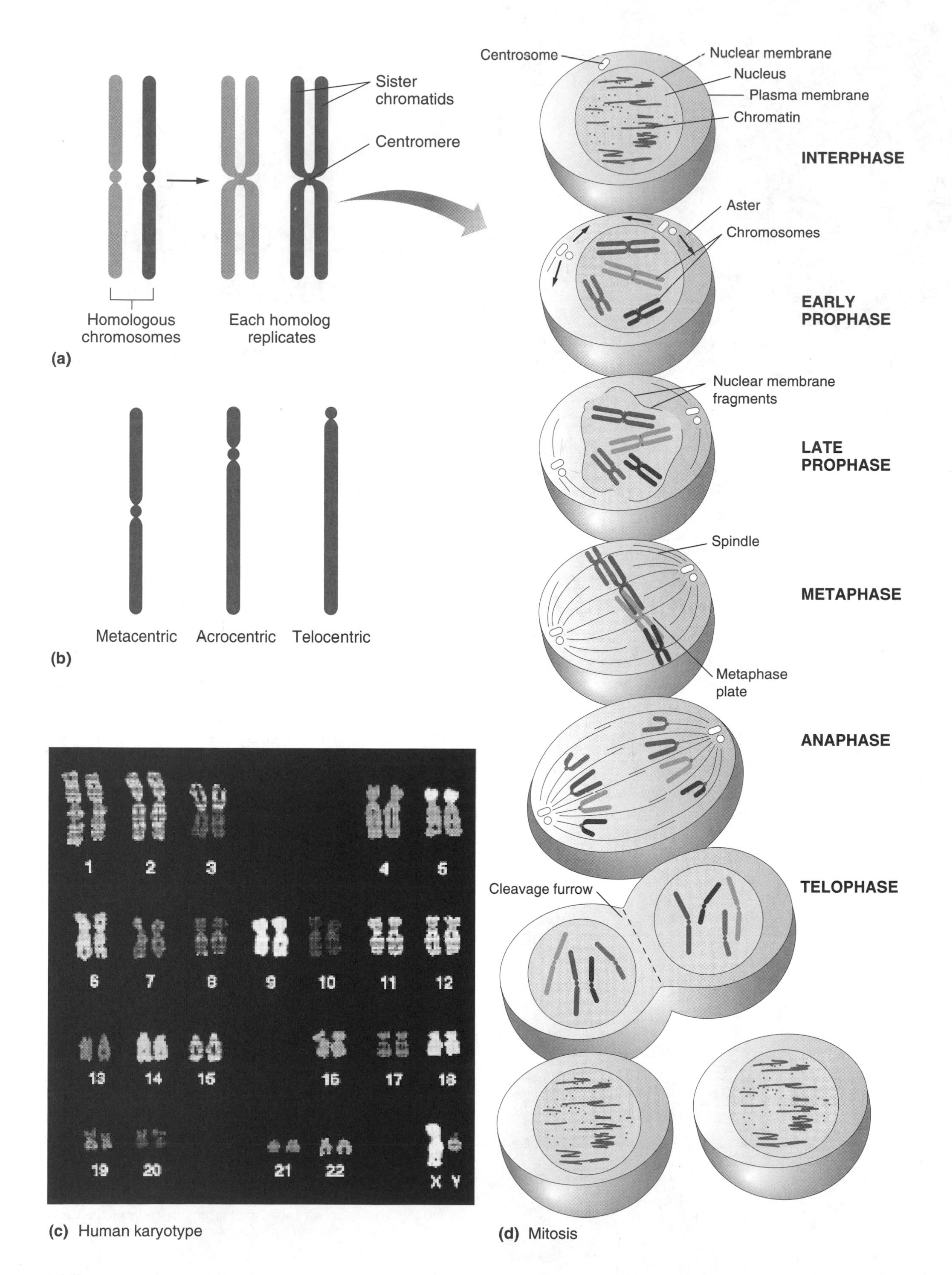

(c) Human karyotype

(d) Mitosis

FIGURE 10–1 Diagrammatic presentation of various stages of mitosis in a somatic cell. During early, medium, and late prophase stages chromosomes thicken and condense. Each chromosome with its attached replicate then lines up on a "metaphase" plate. During anaphase, the replicates (daughter chromosomes) separate, going to opposite poles. During the telophase stages a nuclear membrane re-forms around each polar group of daughter chromosomes, and these chromosomes then revert to the more extended interphase state. Division of the cytoplasm (cytokinesis) also completes during this final mitotic stage. Note that each chromosome's centromere occupies a specific position along its length: at the center for **metacentric** chromosomes, off center for **acrocentric** chromosomes, and at or near the tip for **telocentric** chromosomes. (*Part (c) from Hartl and Jones.*)

structures that bear these molecules, the **chromosomes,** must replicate and divide.

In prokaryotes that have a single chromosome, the two products of chromosomal replication attach to different points on the cell membrane. As the cell elongates to form two daughter cells, the two chromosome products separate, each becoming enclosed in a separate daughter cell. In eukaryotes, more than one chromosome is usually present, and more complex cell division processes occur, divided into **mitosis** (cell division of somatic or body tissues) and **meiosis** (cell division of gamete-producing tissues).

In mitosis, the parental and daughter cells have exactly the same numbers and kinds of chromosomes, and this form of cell division provides all the body cells of an organism with the same chromosome constitution, or **karyotype.** As Figure 10–1 illustrates, mitosis first becomes obvious during the **prophase** stage in which the gradually condensing chromosomes have replicated so that each chromosome now consists of two **chromatids** connected at their centromeres. These chromosomes next arrange themselves on a **metaphase** plate in which the two members of a pair of chromatids connect to spindle fibers that go to opposite poles.

When chromatids separate at the **anaphase** stage, each daughter cell receives one chromatid replicate (which is now an individual chromosome) derived from each of the original chromosomes in the parental cell. For example, if a parental cell has four chromosomes, A^1, A^2, B^1, and B^2, mitosis divides these into chromatids A^1–A^1, A^2–A^2, B^1–B^1, and B^2–B^2, to produce two daughter cells, each of the chromosome constitution (karyotype) A^1, A^2, B^1, and B^2.

In meiosis, the cell division process performed by eukaryotes engaged in sexual reproduction, gametes form that contain only one representative of each kind of chromosome. For example, a gamete containing two kinds of chromosomes, A^1 and B^1, fertilizes a gamete containing the same two kinds of chromosomes, A^2 and B^2, to form the initial cell (**zygote**) of an offspring that now has one pair of each, A^1A^2 B^1B^2. If these events occur in an organism whose life cycle is primarily **diploid** (cells possessing two sets of chromosomes), mitosis replicates the A^1A^2 B^1B^2 karyotype in every somatic cell, until meiosis forms gametes in the male and female gonadal tissues.

Each such meiotic gamete now contains one **haploid** karyotype (only one set of chromosomes), either A^1B^1, A^1B^2, A^2B^1, or A^2B^2. If the zygote forms in an organism whose life cycle is primarily haploid, the meiotic process produces nongametic cells carrying one of these four haploid karyotypes.

Whether haploid or diploid, meiosis generally follows the stages that Figure 10–2 illustrates and is primarily based on close **homologous pairing** between similar kinds of chromosomes called *homologues,* such as between $A^1 \leftrightarrow A^2$ and between $B^1 \leftrightarrow B^2$. Pairing between such homologues lets them separate (disjoin) from each other at the end of the first meiotic division and gives each gamete a representative of each kind of chromosome when **disjunction** is normal. Homologous pairing therefore ensures equal division between chromosomes that carry similar, although not necessarily identical, genes.

In most eukaryotes, each homologous chromosome replicates before the pairing process begins and, as in mitosis, produces two sister chromatids tied to a common centromere (Fig. 10–3).[2] A probable cause (or at least an accompaniment) of pairing between homologous chromosomes are microscopically observed crosses between paired chromatids called *chiasmata* (singular *chiasma*), believed to mark actual transfers of genetic material. Such events are usually distinctive of meiosis, whereas mitotic chromatid replication and division occurs without pairing between homologues.

Sexual reproduction and its accompanying meiotic divisions provide two important sources of variability—an obviously adaptive advantage for organisms that need new genetic innovations to survive perilous environmental challenges. First, because of the phenomenon of **recombination** or **genetic exchange** (also called **crossing over**), sections of homologous chromosomes can exchange material, thereby forming different linear arrays of nucleotides. For example, chromosome A^1, bearing the hypothetical nucleotide sequence AGC·AGC·AGC. . . , can exchange material during meiosis with chromosome A^2, AGC·TCG·TCG. . . , to yield products that are part A^1 and part A^2, such as AGC·TCG·AGC. . . . Since nucleotide sequences constitute genes (Chapter 8), recombination can produce different combinations of genes along a

[2]In some organisms such as dinoflagellates, possibly standing midway between prokaryotes and eukaryotes, homologous chromosomes pair without prior chromatid replication. That is, their homologous chromosomes pair when two cells unite sexually, but then disjoin to opposite poles in a single meiotic division to form two haploid cells (Himes and Beam). Meiosis of this kind, consisting of only a single division, may have been a stage before eukaryotes were able to use chromatid replication for gene exchange, which necessitated two meiotic divisions.

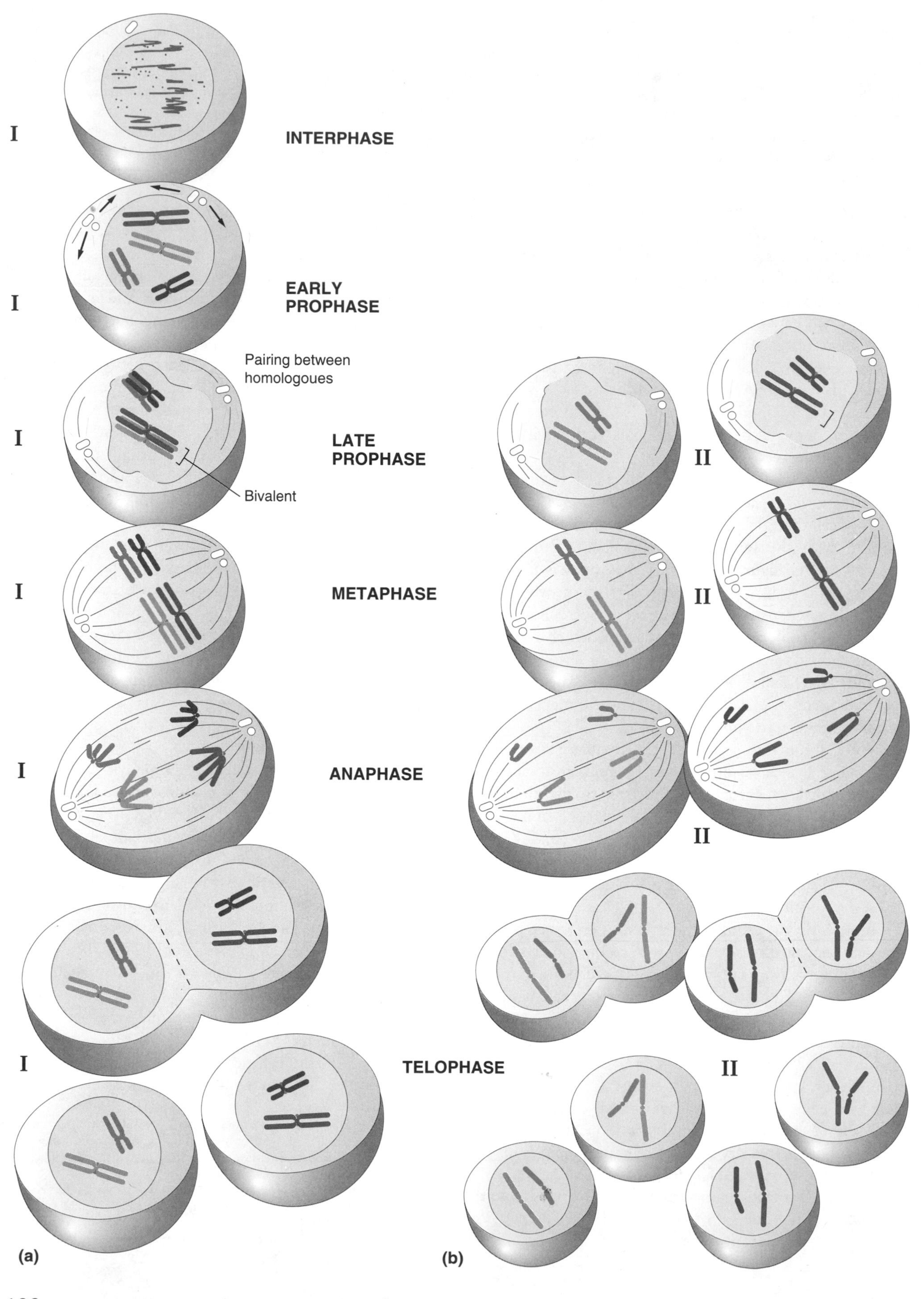
I
INTERPHASE
I
EARLY
PROPHASE
Pairing between
homologoues
I
LATE
PROPHASE
Bivalent
II
I
METAPHASE
II
I
ANAPHASE
II
I
TELOPHASE
II
(a)
(b)

FIGURE 10-2 Principal stages of meiosis, showing the first (I) and second (II) meiotic divisions. Pairing between homologous chromosomes occurs during the prophase stages. By then, each chromosome has replicated to form two sister chromatids connected at a single centromere, which then pair with the two sister chromatids of their homologue to form a group of four chromatids, or **bivalent.** Although not shown in this illustration, some exchanges between nonsister chromatids can occur or have already occurred (Fig. 10-3). At metaphase I these paired groups arrange themselves on an equatorial plate preparatory to their separation. At anaphase I the two chromosomes of a homologous pair separate, each taking their chromatids with them to opposite poles. A telophase stage may then follow, which involves some form of cytokinesis. In many organisms this first meiotic division is a reduction division because it reduces the number of homologues in a nucleus from two to one. Thus, if homologous chromosomes A^1 and A^2 differ from each other, the resultant nuclei after meiosis I will not be identical, since each will have only one of these chromosomes. The next meiotic division (II) that follows in such organisms separates the two chromatids of each chromosome, yielding two nuclei for each nucleus formed during division I. A diploid cell undergoing meiosis will therefore produce four haploid cells, each of which carries one representative of a homologous pair of chromosomes. The four haploid products of a sperm-producing cell (spermatogonium) can all function as gametes, while only one of the haploid products of an egg-producing cell (oogonium) functions as a gamete.

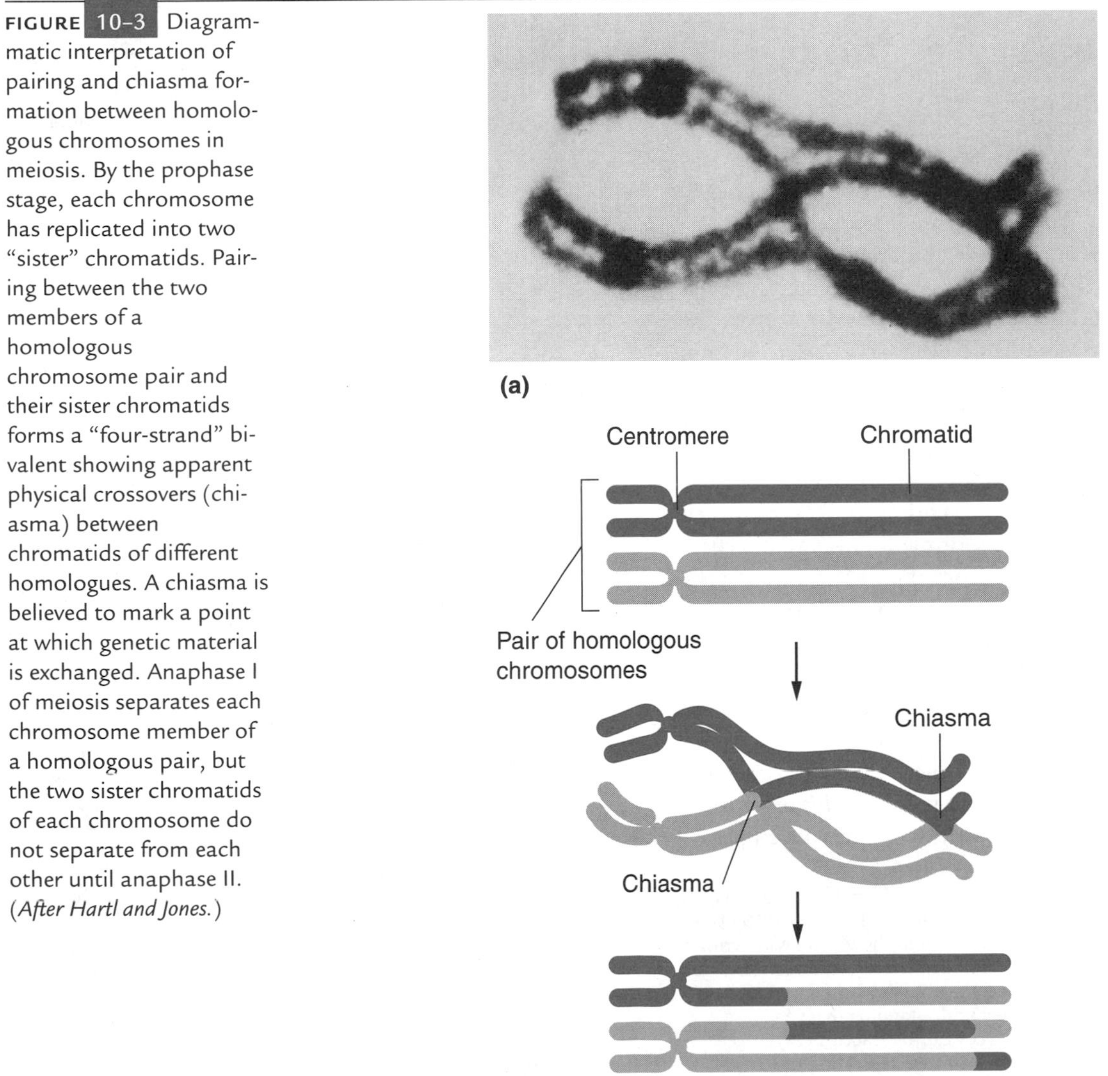

FIGURE 10-3 Diagrammatic interpretation of pairing and chiasma formation between homologous chromosomes in meiosis. By the prophase stage, each chromosome has replicated into two "sister" chromatids. Pairing between the two members of a homologous chromosome pair and their sister chromatids forms a "four-strand" bivalent showing apparent physical crossovers (chiasma) between chromatids of different homologues. A chiasma is believed to mark a point at which genetic material is exchanged. Anaphase I of meiosis separates each chromosome member of a homologous pair, but the two sister chromatids of each chromosome do not separate from each other until anaphase II. (*After Hartl and Jones.*)

chromosome, and also different kinds of genes when crossovers occur within their nucleotide sequences. In addition, as the number of pairs of homologous chromosomes increases, the meiotic process assorts them into increasingly numerous varieties of chromosome combinations.

Two homologous pairs of chromosomes ($A^1 A^2$, $B^1 B^2$) segregating in a cross can yield four different kinds of gametes (A^1B^1, A^1B^2, A^2B^1, A^2B^2). Segregation of three pairs of homologous chromosomes ($A^1 A^2$, $B^1 B^2$, $C^1 C^2$) can yield eight different gametic combinations ($A^1B^1C^1$, $A^1B^1C^2$, $A^1B^2C^1$, $A^1B^2C^2$, $A^2B^1C^1$, $A^2B^1C^2$, $A^2B^2C^1$, $A^2B^2C^2$). If we extend this to four pairs of chromosomes, 16 different gametic combinations are possible, the rule being that the number of possible different kinds of gametes equals 2^n, where n represents the number of pairs of chromosomes undergoing meiosis. In humans with normal karyotypes of 23 chromosome pairs, the different kinds of gametes that can form in this simple chromosome assortment process alone is 2^{23}, or more than 8 million varieties.

Mendelian Segregation and Assortment

Because the molecular basis of genetics long remained unknown, fundamental genetic principles primarily derived from observations on transmission of more visible and obvious biological characteristics. Gregor Mendel in Brno, Czechoslovakia, in the early 1860s, discovered the earliest genetic laws derived from such observations although biologists did not generally know his work until the beginning of the twentieth century.[3]

As Chapter 3 describes, the pre-Mendelian view of heredity prevailing throughout the nineteenth century was that heredity followed a blending process in which offspring inherited a dilution, or blend, of parental characteristics derived primarily from their appearance (phenotype). Mendel's exceptional contribution was proving that organisms had a distinct hereditary system (genotype) which transmitted biological characteristics through discrete units, later called **genes,** that remained undiluted in the presence of other genes.

Mendel demonstrated this concept by experimenting with a number of characters in the pea plant, *Pisum sativum,* in which each character possessed two alternative appearances, or traits, such as smooth or wrinkled seeds, yellow or green seeds, and tall or short plants. In his experiments, he counted the appearance of each different trait among the individuals in every generation, then analyzed these numerical results in terms of ratios that led to two fundamental principles of heredity.

- THE PRINCIPLE OF SEGREGATION When two pure-breeding parental stocks are intercrossed and they differ in respect to a character such as seed shape—for example, smooth × wrinkled seeds—the first filial generation, or F_1, carries the genes for each of these traits. Since the pea plant is diploid, an individual has a pair of such genes for seed shape, as well as pairs of genes that govern other characters such as seed color and plant size. In modern terminology, the two members of a particular gene pair are known as **alleles.** For simplicity, we can designate alleles by italic alphabetical letters, for example, the smooth and wrinkled alleles are, respectively, S and s. These alleles then segregate in the

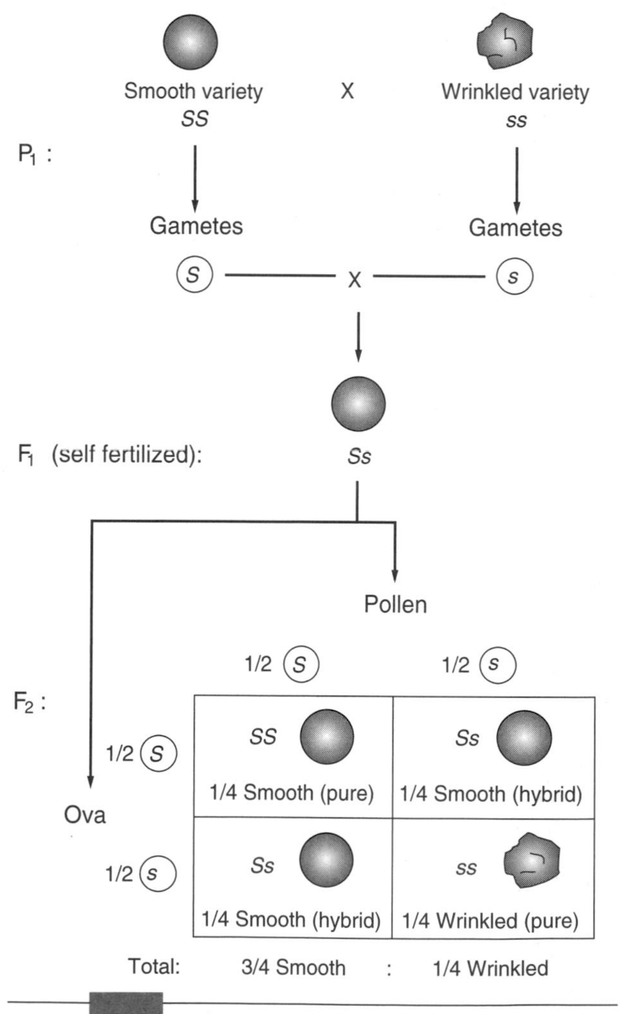

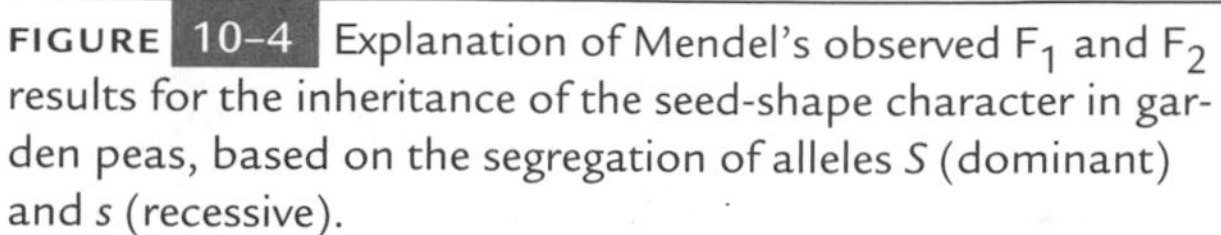
FIGURE 10-4 Explanation of Mendel's observed F_1 and F_2 results for the inheritance of the seed-shape character in garden peas, based on the segregation of alleles S (dominant) and s (recessive).

[3]Bishop suggests that Mendel's famous 1866 publication on the genetics of peas was really an antievolutionary response to Darwin's 1859 *On the Origin of Species.* That is, Mendel attempted to show that hybrid crosses do not produce new species, but rather their hereditary factors combine in later generations to restore parental types. Unwittingly, Mendel provided evolutionists with the essential understanding that genetic factors can be identified and their differences can be traced between generations. This led inevitably to research on the causes for genetic differences and to investigations of how and why such differences accumulate during evolutionary change—fundamental elements in the Darwinian "modern synthesis" (Chapter 21).

gametes of the F_1 hybrid, which then unite to form a second filial generation (F_2) in predictable proportions. As Figure 10–4 shows, these F_2 proportions arise because the allele for one trait (for example, *S*) has **dominant** effects over the other **recessive** allele (*s*) when both are present together in *Ss* individuals. That is, *Ss* peas appear smooth (as are *SS* peas), and the only wrinkled peas are *ss*. Individual organisms that carry two different alleles for any particular character (for example, *Ss*) are **heterozygotes,** and those which carry two identical alleles (for example, *SS, ss*) are **homozygotes.** Mendel's experiments showed that these alleles obviously had not been changed or blended in the heterozygote but segregated from each other to be transmitted as discrete and constant particles between generations.

- THE PRINCIPLE OF INDEPENDENT ASSORTMENT In crosses involving two different characters, such as seed shape and seed color, Mendel found that the results were predictable if genes that determined one character (that is, seed shape) had no effect on the segregation of genes for the other character (that is, seed color). That is, genes for different characters segregated independently of each other. As Figure 10–5 shows, the F_1 of a cross differing in two such

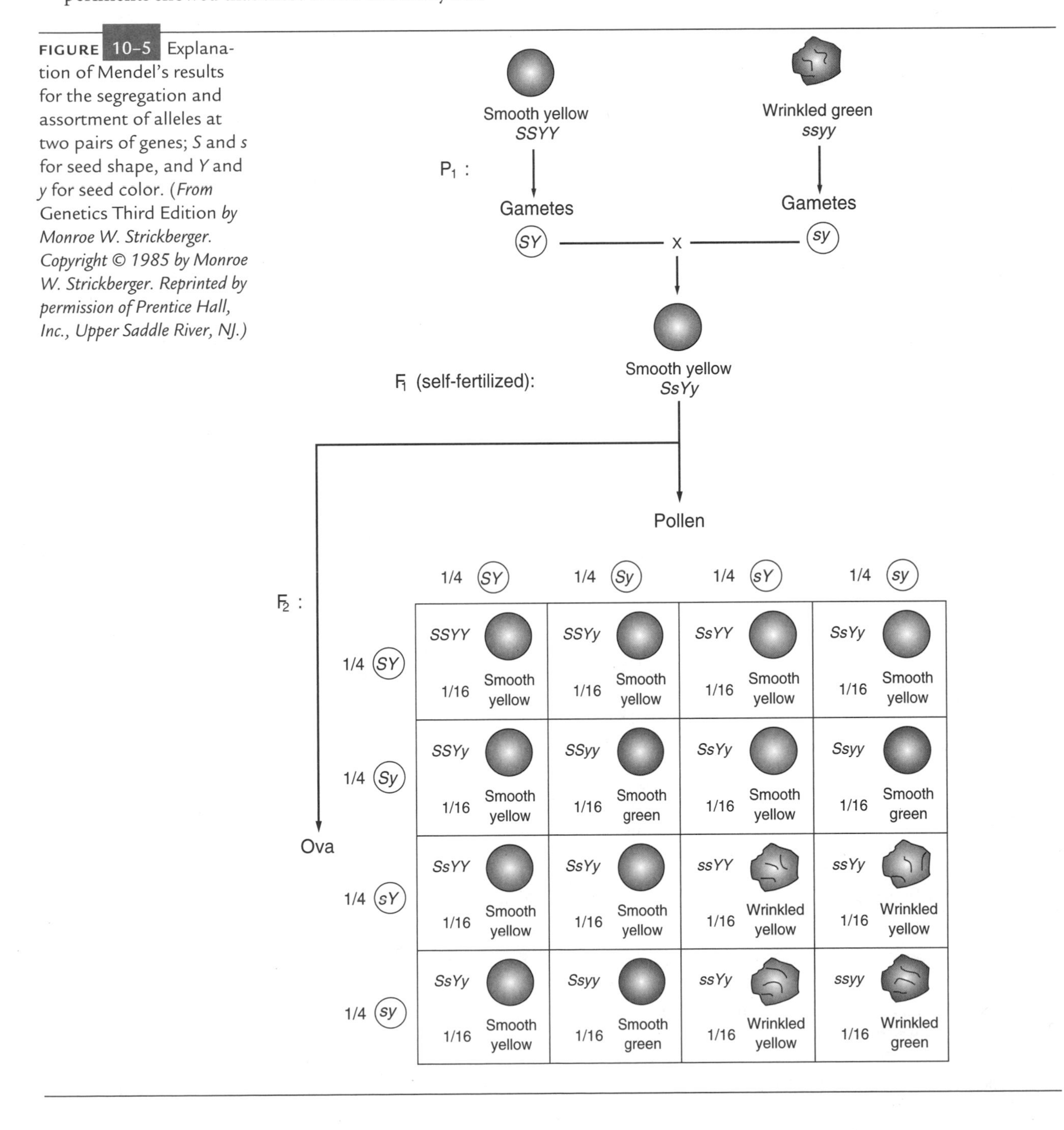

FIGURE 10–5 Explanation of Mendel's results for the segregation and assortment of alleles at two pairs of genes; *S* and *s* for seed shape, and *Y* and *y* for seed color. (*From* Genetics Third Edition *by Monroe W. Strickberger. Copyright © 1985 by Monroe W. Strickberger. Reprinted by permission of Prentice Hall, Inc., Upper Saddle River, NJ.)*

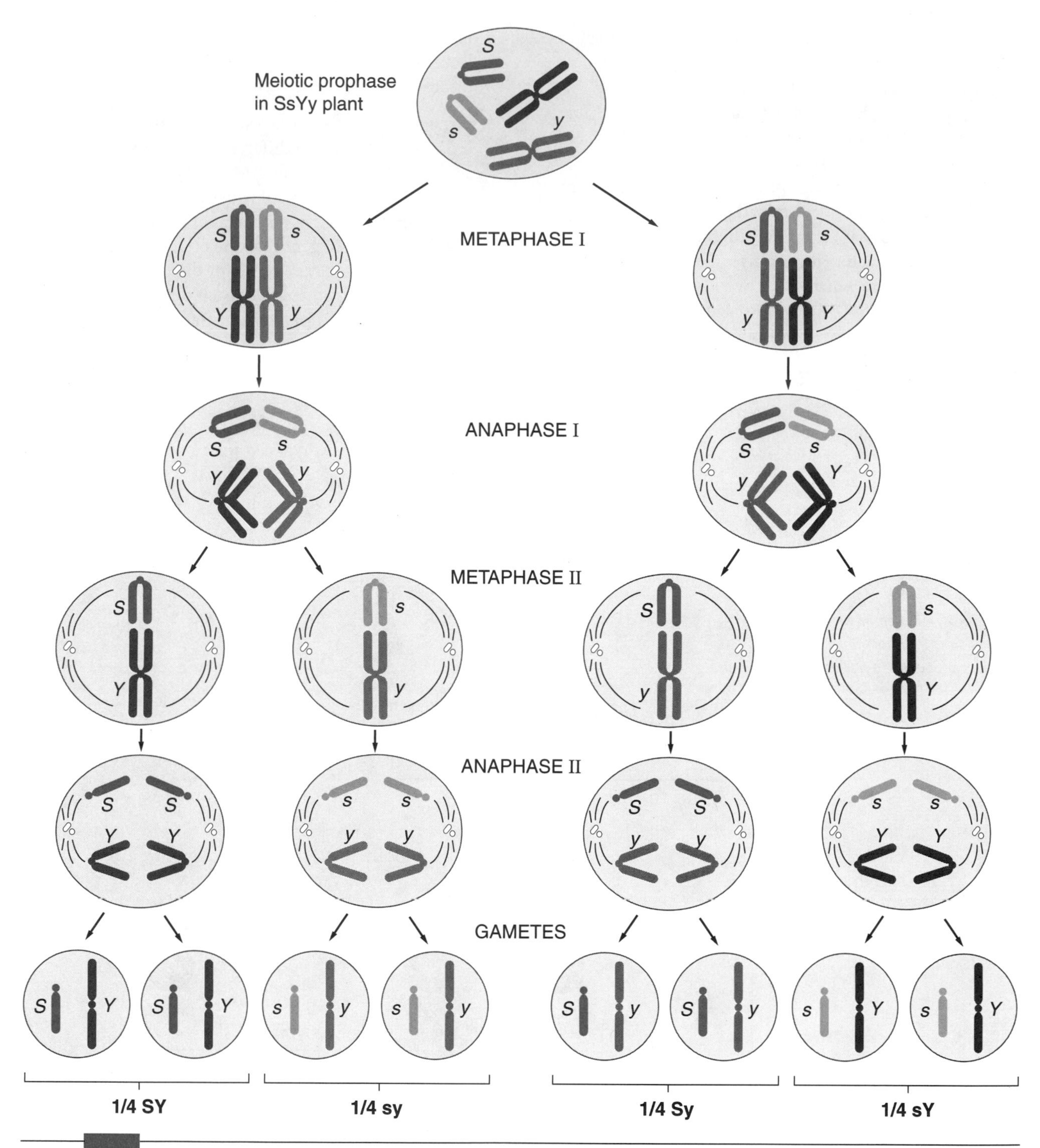

FIGURE 10-6 Explanation for the segregation and independent assortment of seed shape and seed color in Mendel's experiments in terms of factors or genes (*S, s,* and *Y, y*) localized on different chromosomes. Because of independent assortment, all four possible combinations of chromosomes in the gametes correspond to all four possible combinations of genetic factors. (*Adapted from Strickberger.*)

characters—smooth yellow × wrinkled green—shows only the dominant phenotypes, but the F_2 proportions are predictably 9/16 smooth yellow, 3/16 smooth green, 3/16 wrinkled yellow, and 1/16 wrinkled green. These predictions derive from the independent association between a seed shape allele (probability 1/2) and a seed color allele (probability 1/2) in each of the four kinds of gametes produced by the F_1, so that each different gamete has a frequency of $1/2 \times 1/2 = 1/4$. The cellular explanation for such independent assortment turned out to be the localization of the genes for each of the two characters to a different nonhomologous pair of chromosomes (Fig. 10–6): how one pair of homologous chromosomes assort themselves toward opposite poles during meiosis has no effect on how a nonhomologous pair assort themselves.

Dominance Relations and Multiple Alleles

The particular ratios Mendel observed in his diploid pea plants (3:1, 9:3:3:1, and so on) held true only as long as the genes involved showed complete dominance and recessiveness and had only two alleles for each character. After 1900, researchers quickly realized that dominance relations between alleles could range quite widely, and a gene determining a particular character could be represented by three or more alleles. For example, the condition of **incomplete dominance** can be expressed in flower color of some plants in which the homozygote G^1G^1 produces red flowers, the homozygote G^2G^2 produces white flowers, and the heterozygote G^1G^2 produces an intermediate shade of pink flowers. Or, in contrast, both alleles G^1 and G^2 are considered **codominant** when they have distinguishable effects in the heterozygote; that is, they each produce a uniquely recognizable substance, such as a special sugar compound, so that the heterozygote G^1G^2 has both compounds.

Any gene may also have more than two alleles, G^1, G^2, G^3, . . . producing, for example, a different color or different compound in each different kind of homozygote, G^1G^1, G^2G^2, G^3G^3, ... The source for such **multiple allelic** systems arises from the fact that a gene consists of a linear array of hundreds or thousands of nucleotides, and an allelic difference may be caused by only a single nucleotide change at one or many positions along its length. Moreover, there may be different dominance relationships between the alleles in such a system so that G^1, for example, produces a codominant effect with G^2, but alleles G^1 and G^2 act as though they are each completely dominant over G^3. In addition, interactions between genes in different allelic systems may occur so that the expression of allele G^1, for example, changes because certain alleles are present at a different gene pair, such as H^1 or H^2. Among such interactions are those which can modify the dominance relations of a particular allele so that the effect of G^1, for example, is dominant over G^2 in certain genotypes (for example, $H^1I^2J^1K^3$) but is recessive when the background genotype changes (for example, $H^2I^1J^3K^2$). Complications of this kind have been given the name **epistatic interactions,** to describe when one or more gene pairs change the effect caused by other gene pairs. Another term, **modifiers,** is used for quantitatively measurable effects of genes on other genes.

In general, most common, or **wild type,** alleles have been evolutionarily selected to be dominant in diploid organisms because they produce advantageous products in the presence of other alleles whose products may not be as advantageous or may even be deleterious. On the molecular level, such wild-type allelic products are mostly functional proteins such as enzymes, whereas the products of mutant alleles often appear nonfunctional, only partly functional, or even absent (**null alleles**). For example, the recessive allele causing Mendel's wrinkled peas prevents normal expression of an enzyme that makes branched starch molecules (Bhattacharyya et al.).[4]

Because so many allelic differences, as well as different dominance and interaction effects, are possible, the variability generated is far beyond that described previously for the independent assortment of chromosomes during meiosis. To take a simple example, a gene with four alleles (G^1, G^2, G^3, G^4) can produce ten different possible diploid genotypes (G^1G^1, G^1G^2, G^1G^3,G^1G^4, G^2G^1, G^2G^2, and so on), and 100 such genes can produce 10^{100} possible genotypic combinations—a figure larger than the estimated number of protons and neutrons in the universe. Since any cellular organism carries more than 100 genes, each with many possible alleles, its potential genetic variability probably is greater numerically than any conceivable aspect of reality.

Exceptions to Mendelism

Although Mendel was not aware of the cause, his principles were based (as we can see) on the meiotic process of chromosome segregation and assortment that occurs during gamete formation. When scientists discovered his contributions in 1900, they soon recognized the association between Mendelian factors and meiotic chromosome distribution, which provided the foundation for the new science of genetics. Within a short period, many other studies echoed Mendelian ratios and their variations, and—because eukaryotes shared meiotic division processes—Mendelism generally seemed applicable to most, if not all, eukaryotic organisms showing common features of sex and sex determination (Box 10–1). However, two important deviations from Mendelism became apparent in the decades that followed; both circumvent

[4]Kacser and Burns point out that "null" alleles that cannot produce an enzyme are generally recessive, since a heterozygote for such an allele with 50 percent enzymatic activity would probably not appear phenotypically different from a dominant allele homozygote with 100 percent enzymatic activity. They view dominance as an inevitable result of "the kinetic structure of enzyme networks," and suggest that there is little, if any, selection for dominance itself. However, the observation that dominance can break down on crossing certain populations (p. 546) indicates that enzymatic networks may incorporate factors that can allow changes in dominance.

BOX 10–1 EVOLUTION OF SEX-DETERMINING SYSTEMS

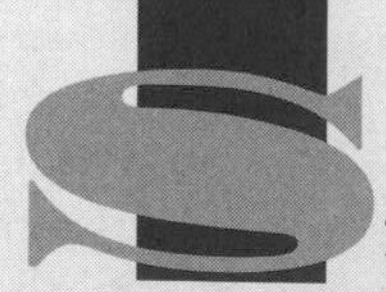

Sexual reproduction generally involves differentiation into sexual forms in which reproduction occurs only as a result of union between sex cells of different individuals to form an offspring's initial cell (zygote). As a rule the two parental gametes that unite during fertilization are physically distinct from each other (for example, sperm and egg) and form from different organs (for example, testes and ovary). This division of gametic labor in which one sex was selected to produce a large stationary egg and the other to produce small mobile sperm, has the adaptive advantage of yielding a large zygote that provides an embryo with increased nutrition, yet still receives similar nuclear genetic contributions from each parent.

In most plants and many lower animals, the different sexual organs combine within single **hermaphroditic** or **monoecious** individuals, each capable of producing both types of sexual gametes. In contrast, some plant species and most higher animals consist of separate male and female (**dioecious**) individuals that produce sperm or eggs but not both. As discussed on pp. 563–564 despite the sacrifices that such differentiation demands, the genetic and phenotypic variation produced by sex appears essential for the long-term survival of many groups that face ever-changing environments.

For many organisms familiar to us, especially mammals, sex determination is associated with chromosomal differences between the two sexes, XX females and XY males, with the Y chromosome often smaller and mostly inactive (except for male-determining and male-fertility genes). Inactive though it may be, the Y often serves an important role in meiosis by pairing with the X chromosome so that the two chromosomes can separate during the anaphase stage and go to opposite poles. Hence, the heterogametic XY individual can potentially produce two kinds of gametes (X-bearing and Y-bearing) in equal proportions, accounting for an approximately equal sex ratio among offspring. There are, however, numerous sex chromosome variations in other organisms: XX females and XO males; XX hermaphrodites and XO males; heterogametic (ZW) females and homogametic (ZZ) males; XXXX females and XXY males, and so forth.

The almost universal presence of sex chromosomes among sexual species does not necessarily mean these are the only chromosomes that affect sexual development. Since sex is a complex developmental character, it is usually affected by numerous nonsex-chromosome (**autosomal**) genes as well. For example, in *Drosophila* the X/A ratio of X chromosomes to sets of autosomes determines sex: in diploids (two sets of autosomes), females have an X/A ratio of 1 (2X/2A) and males have an X/A ratio of 1/2 (1X/2A). Ratios that depart from these can produce sexual abnormalities, for example, in triploids (three sets of autosomes) an X/A ratio of 2/3 (2X/3A) produces "intersexes." Clearly, in many cases the potential for both male and female development exists at the time of fertilization no matter what the sex chromosome complement. Thus, gene mutations such as *transformer* in *Drosophila* can turn XX zygotes into males, and the gene that causes *testicular feminization syndrome* in humans can turn XY zygotes into females. The function of sex chromosomes or the X/A ratio is to act as part of the "switch" mechanism that directs development into one of the possible paths the organism is capable of following.

In contrast to what seem to be purely genetic mechanisms, more environmentally dependent sex-determining mechanisms exist in many animals as well as in plants. In the echiuroid sea worm, *Bonellia,* larvae that are free-swimming and settle on the sea bottom develop into females with a body the size of a walnut and a proboscis about a meter long. Larvae that land on the female proboscis develop into tiny 1-millimeter-long males that lack digestive organs and exist in parasitic fashion in the genital ducts of the female.

Egg size appears to be the sex-determining mechanism in the sea worm *Dinophilus:* large eggs produce females, small eggs males. In some reptiles, high egg incubation temperatures produce mostly males (for example, the lizard *Agama agama*), whereas warm temperatures in others produce mostly females (for example, the turtle *Chrysema picta*). In certain fish, social behavior can influence sex so that loss of socially dominant males from a group causes sex conversion of the dominant female in the group. In some coral reef fishes, sex changes seem to occur as a result of visual stimulation: females become males when the surrounding fish in the group are relatively small. In the horsetail plant, *Equisetum,* female characteristics appear when the plant lives in good growth conditions and male characteristics in poor conditions.

Among the many questions these various findings raise are the following:

- How did sex determination become tied to sex chromosomes?
- What led to the different varieties of sex chromosome karyotypes?
- Since the heterogametic XY or XO (or ZW) sex carries only one copy of X (or Z) chromosome genes, and the homogametic XX (or ZZ) sex carries two such copies, are there mechanisms that compensate for this difference in gene dosage and how did they evolve?
- When only one sex is sterile or nonviable among the offspring of a cross between two species, why is it almost always the heterogametic sex ("Haldane's rule")?
- Is there some advantage for the commonly observed male/female sex ratio of 1/1?
- How and why did some species abandon chromosomal sex determination and adopt environmental sex determination?

Various workers have offered answers to these questions (see, for example, Bull, Charlesworth, Hodgkin, Maynard Smith), and we can briefly summarize some reasonable concepts as follows.

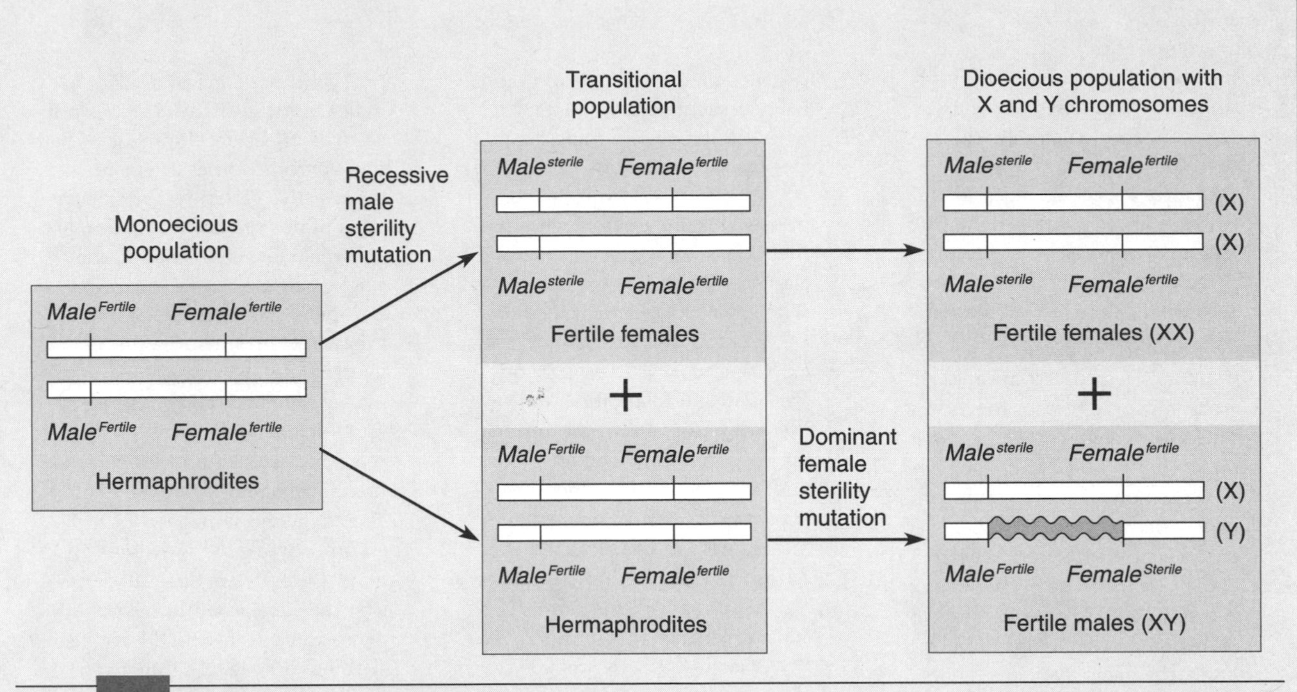

FIGURE 10–7 Sequence of nuclear mutational events leading to a transition from a monoecious to a dioecious population. Each horizontal bar represents a sex chromosome. In this example the Y chromosome determines male sex development because of dominant $Male^{Fertile}$ and $Female^{Sterile}$ alleles, whereas the X chromosome carries recessives at these loci ($Male^{sterile}$, $Female^{fertile}$)—the absence of the Y chromosome produces females. (Note that other mutational events, such as females produced by initially dominant $Male^{Sterile}$ and $Female^{Fertile}$ mutations, can lead to heterogametic ZW females and homogametic ZZ males.) The wavy section between *Male* and *Female* genes on the Y chromosome indicates a recombination deficient interval (caused either by close linkage or by crossover suppressors such as inversions, p. 209) that interferes with crossovers that might otherwise lead to sterile individuals carrying both $Male^{sterile}$ and $Female^{Sterile}$ alleles, or back to the hermaphroditic condition, $Male^{Fertile}$ $Female^{fertile}$. *(Based on Charlesworth, with modifications.)*

For chromosomal sex determination to occur, monoecy (hermaphroditism) must first change to dioecy, most probably fueled by the gain in genetic diversity from cross-fertilization (outbreeding) compared to its relative absence in self-fertilization (inbreeding). Given such benefit, there is selective advantage for a monoecious population to produce females that the male gametes of other members of the population can fertilize. As the top sequence in Figure 10–7 shows, in monoecious organisms the transition to such females requires little more than a mutation causing male sterility.

The presence of females, in turn, then offers advantages to individuals who specialize in producing male gametes, and these too can arise from monoecious organisms, but the mutations are those that cause female sterility (bottom sequence, Fig. 10–7). The result, in this example, is a dioecious population bearing two kinds of chromosomes devoted to sex determination: the X distinguished by recessive male sterility, and the Y by dominant female sterility.

Since individuals who are both male- and female-sterile cannot reproduce, and (as shown in Fig. 10–7) such individuals can arise from crossing over between these genes, a cascade of consequences follows:

1. Selection to prevent crossing over promotes genes or chromosome arrangements that interfere with recombination between the X and Y, and leads to preserving very tight linkage between genes on the Y chromosome important for the heterogametic (for example, male) sex. Thus, **sexually antagonistic genes** that benefit males but harm females, such as those that are exclusively for male mating success, can be selected to localize on the Y chromosome, and can therefore be passed on only to Y-bearing offspring. To counter effects that can harm females, Rice (1998) proposes that selection of X-linked and autosomal genes occurs which favors female fitness. The result is therefore an evolutionary see-saw of male-female "antagonistic coevolution."
2. Once X–Y recombination diminishes, its absence lets harmful mutations on the Y chromosome accumulate; for example, by a process known as Muller's ratchet. As explained on p. 564, Muller's ratchet is the concept that, because of sampling errors, populations more easily lose that class of individuals

bearing the fewest harmful mutations, so that classes with increasing numbers of such mutations tend to increase with time. Whatever the cause, Rice (1994) has now experimentally demonstrated an increase in Y-chromosome deleterious mutations associated with lack of recombination.

3. To eliminate such deleterious mutant effects, inactivation of the Y chromosome (except for its sex-determining genes) then becomes a selective advantage and results in the common X–Y sex chromosome karyotype. We most commonly see chromosome inactivation when an entire chromosome or sections of it assume a state called **heterochromatin** that stains differently under the microscope from normally active, nonmodified sections, **euchromatin.** The Y chromosome is often heterochromatic and the X euchromatic. Among other consequences of Y chromosome inactivation are increased numbers of transposable elements (p. 225) and repetitive DNA sequences (p. 267) which can persist in chromosome sections not being actively selected for viability and where recombination cannot eject them.
4. As larger sections of the Y chromosome become inactive because of reduced recombination with the X, the Y can decrease in size because selection no longer operates to preserve its length—another feature common to many XX–XY species.
5. At some point in this progression, the Y chromosome has few if any functional genes other than the sex-determining genes, and if these are translocated elsewhere (p. 209) or their functions assumed by genes in another chromosome, the result can be an XX female/XO male species, in which sex determination depends on a *Drosophila*-like X/A ratio. In *Drosophila melanogaster* sex determination is based on a key regulatory gene called *Sxl (Sexlethal)* that produces female development when turned on and male development when turned off. Turning *Sxl* on and off derives from the relationship among certain genes on the X chromosome ("numerator elements") and autosomes ("denominator element"), so that an X/A ratio of 1 turns *Sxl* on and an X/A ratio of 1/2 turns *Sxl* off (see Cline, and Parkhurst and Meneely). Conversely, a translocation between the Y and an autosome produces a "neo-Y" chromosome, which now pairs in meiosis with its translocated autosome (that can now be called X_2) as well as with the former X (now called X_1), leading to an X_1X_2Y male/$X_1X_1X_2X_2$ female sex chromosome constitution. Figure 10-8 provides an example of *Drosophila* karyotypes showing a succession of X and Y chromosome changes.
6. In some cases, only a single copy of a sex-determining gene is sufficient to embark on a pathway of sexually distinctive development. In mammals, the *SRY* gene on the Y chromosome serves such purpose for male development. Other vertebrates, from fish to reptiles, do not share *SRY*'s mammalian sex-chromosome locus, indicating that its mammalian role may have been derived from a special vertebrate function (Tiersch et al.). The books of Bull and White provide many other examples of sex chromosome varieties that arise from diverse genetic and chromosomal changes.
7. The difference in X (or Z) chromosome gene dosage between the heterogametic and homogametic sexes can be compensated in various ways. One device is simply to ignore the problem if the X (or Z) chromosomes are small with few active genes; for example, the ZW females in birds and butterflies generally produce half the amount of Z gene products of the ZZ males but this difference does not seem to affect development. When the X chromosomes are large, with many active genes, two major systems of **dosage compensation** are known so far: in *Drosophila,* the single X chromosome in the male is about twice as active as each of the two X chromosomes in the female, and in placental mammals, only one X chromosome of an XX female is functional in each cell. The result of both systems is therefore that an XX female has essentially the same level of X chromosome gene activity as an XY male. Researchers have suggested that the *Drosophila* dosage compensation system could have been selected directly to compensate the XY male as the Y chromosome was inactivated, whereas the mammalian system would have been selected because an increase in X chromosome activity could not be restricted to males, and only inactivation of the extra X chromosome in females could equalize the dosage. Whatever the sequence of events, some of the genes involved associate with sex determination, of which many genes have now been identified, especially in *Drosophila* and *Caenorhabditis* nematodes (Charlesworth 1996; Cline and Meyer; Lucchesi 1998).
8. **Haldane's rule,** often used in describing defective hybridization between animal species, states, "When in the F_1 offspring of two different animal races one sex is absent, rare, or sterile, that sex is the heterozygous [heterogametic—XY, XO, ZW, or ZO] sex." Geneticists have most often ascribed the cause for this rule to sex-linked deleterious recessive alleles. That is, interspecific hybrids carrying autosomal alleles from one species that are incompatible with sex-linked recessive alleles from the other species show

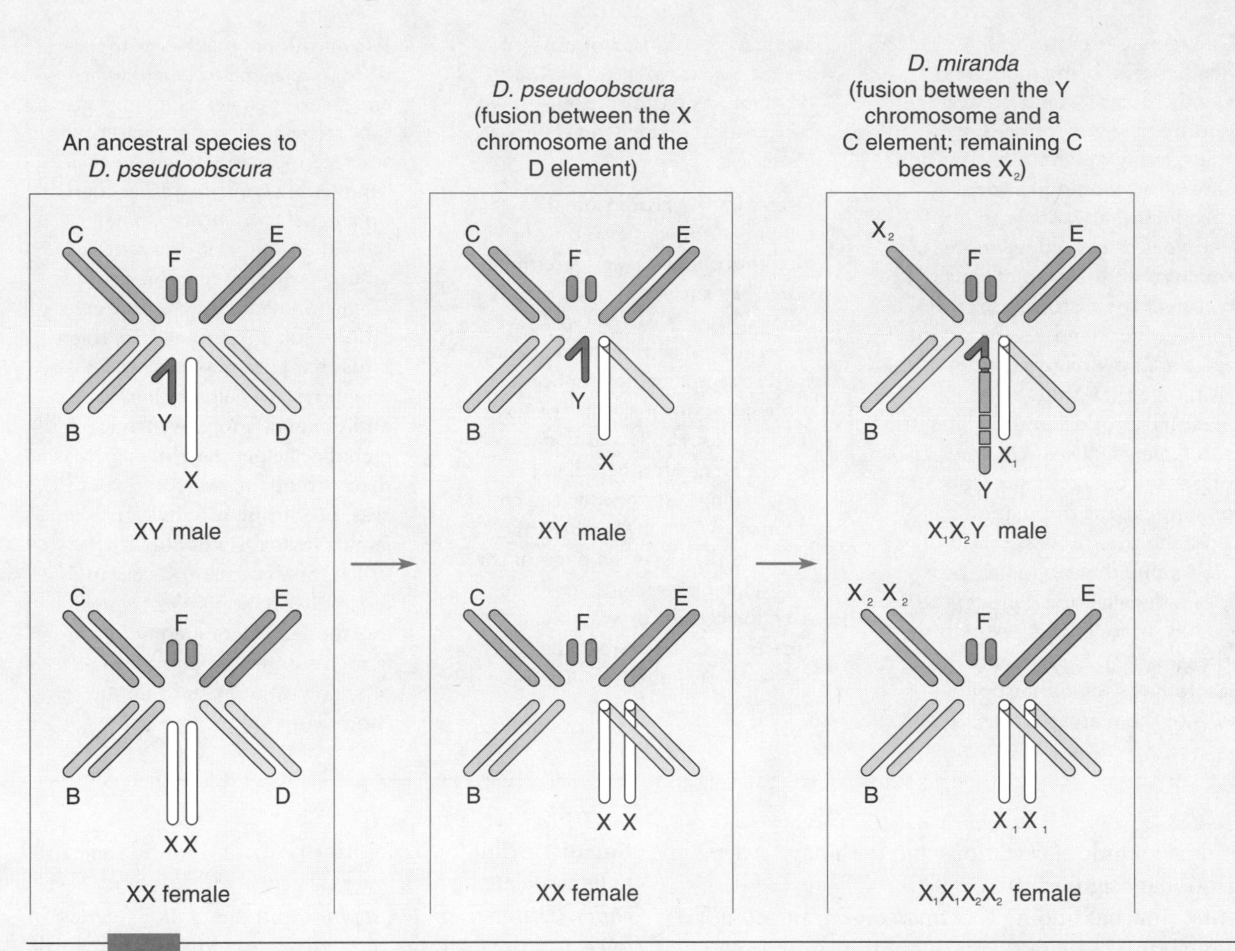

FIGURE 10-8 Diagrammatic view of sex chromosome changes in a lineage leading to *Drosophila miranda*. As shown here, the original chromosomal elements in the genus *Drosophila* are often designated by the six letters A–F, with the A element serving as the sex chromosome. The species on the left represents a hypothetical ancestor in which the common XY chromosome karyotype has been established. This was then followed by fusion of the X with the D element in *D. pseudoobscura* and fusion of the Y with the C element in *D. miranda*. The sex chromosome constitution of *D. miranda* is therefore X_1X_2Y males and $X_1X_1X_2X_2$ females, each sex also carrying three pairs of autosomes (elements B, E, and F). (*Adapted from Lucchesi (1994), with modifications.*)

deleterious effects more easily in the heterogametic sex. Thus, X-linked deleterious recessive alleles in XY males, for example, are not masked by dominant alleles on the other X chromosome as they are in homogametic XX females. To this explanation, which Muller proposed, has now been added the concept that the sexual traits of males are produced by unique sets of genes that evolve rapidly, making some or many such males more sensitive to hybrid sterility (see Turelli).

9. Although the heterogametic sex can theoretically produce equal numbers of male- and female-determining gametes, equal sex ratios need not necessarily follow. Some nuclear genes, as well as endosymbionts that cause the "sex ratio" condition in *Drosophila* (Hurst), may affect the viability of male or female zygotes. Other genes may alter the ability of the X or Y chromosomes to segregate normally during meiosis ("segregation distorters," p. 202). Either type of abnormality, whether in viability or gametic segregation, can affect male or female frequencies.[5] One way of accounting for the prevalence of an equal male:female sex ratio in many species is to consider what happens when a particular sex is rare. To use a simplified example, should a population have a scarcity of females, for example, a

[5]According to Hamilton, Y-chromosome inactivation has the advantage of inactivating segregation distortion genes that would cause more Y- than X-carrying gametes, and therefore lead to more males than females.

male:female sex ratio of 5:1, then females, being more frequently mated, would on average produce more offspring than males, many of whom cannot find mates. This would provide a reproductive advantage to genotypes that produce more females than males, and thus tend to correct the distorted sex ratio. On the other hand, assuming that males are now relatively rare, the advantage then shifts to genotypes that produce more males than females, and again tends to correct the distorted sex ratio. At some point, the population will approach the stable sex ratio of 1:1, a value that no longer provides a benefit for a genotype to produce more of one sex than the other (Fisher). Experimentally, Basolo has shown that populations of the platyfish *Xiphophorus maculatus,* composed of different sex ratio genotypes, evolve in this direction, as has also been shown by Carvalho and coworkers in *Drosophila.*

10. Interestingly, chromosomal sex determination can also transform into environmental sex determination when the sex of one or both of the XY and XX karyotypes reverses because of sensitivity to agents such as temperature and hormones. Various instances of sex reversal exist (Bull) and might well lead to an environmental sex-determining system when the environment is "patchy" rather than uniform, that is, when parts of the environment favor one sex more than the other. For example, a greater food supply in some patches may allow larger females to be produced, in contrast to other, less nutritional patches that produce smaller males. Such differences can then act as an important selective factor in establishing sex determination based on environmental sensitivity rather than on a genetic or chromosomal basis that does not distinguish among environments. An example of environmental impact on sex ratio occurs in Seychelles warblers, birds that commonly use their daughters as "helpers" in raising additional offspring. When food is plentiful, helper daughters increase their parents' reproductive success, producing broods with a female:male ratio of about 6:1. When food is scarce, such daughters hinder their parents' reproductive success by competing for the limited supply, and female:male offspring ratio drops to about 1:3 (Komdeur et al).

conventional meiotic expectations and both have genetic and evolutionary significance.

The first unusual finding was that some traits do not follow a nuclear inheritance pattern but transmit through egg cytoplasm. Such **extranuclear inheritance** (also called **cytoplasmic** or **maternal inheritance**) arises because cytoplasmic organelles such as mitochondria and chloroplasts have their own DNA genetic material. During fertilization of the zygote, practically all the egg cytoplasm including these organelles is maternally derived, while the sperm contributes primarily nuclear genetic material and, at most, very few mitochondria. It was, in fact, through cytoplasmic inheritance that genes were first identified in mitochondria and chloroplasts, supporting proposals that these organelles had an independent evolutionary origin (see p. 180).

The second departure from Mendelism was the finding that some nuclear genes do not segregate as expected in heterozygotes. As mentioned previously, the four haploid products produced by the meiotic process in heterozygotes should lead to four gametes, of which two carry one allele and two the other allele (for example, $Rr \rightarrow 2R$: $2r$). In a few cases, however, segregation is not normal, and the heterozygote for certain alleles produces more of one allele than the other (for example, $Uu \rightarrow 4U$: $1u$). Geneticists called such biased transmission **segregation distortion** (also **meiotic drive**), and found the cause to be associated with a few rare genes, most notably *Segregation Distorter* in *Drosophila* and the *tailless* genes in mice (Lyttle). Although relatively uncommon, meiotic drive can lead to the persistence of a segregation distorter gene even though the gene may have deleterious effects (p. 566).

Perhaps another deviation from Mendelism—really an elaboration rather than a departure—was the finding that genes do not necessarily assort independently of each other if they are linked together on the same chromosome. These investigations began with inferences that some genes were localized to the X (sex) chromosome, sex-linked genes, as discussed next.

Sex Linkage

Early in the twentieth century geneticists discovered that various genes could be localized to a particular chromosome associated with sex determination. A prominent example of this is the "bleeder" disease, hemophilia, localized to the X chromosome of humans and other mammals. In these organisms males and females differ in respect to a pair of sex chro-

TABLE 10–1 Some gene products and diseases linked to the X chromosome in both humans and other mammals

X Linked in Humans	X Linked in Other Mammals
α-galactosidase deficiency	More than 25 species including chimpanzee, gorilla, sheep, cattle, pig, rabbit, hamster, mouse, cat, dog, kangaroo
Anhidrotic ectodermal dysplasia	Cattle, dog, mouse
Bruton-type agammaglobulinemia	Mouse, cattle, horse
Copper transport deficiency	Mouse, hamster
Duchenne/Becker muscular dystrophy	Mouse, dog
Glucose-6-phosphate dehydrogenase deficiency	More than 30 species, including chimpanzee, gorilla, sheep, cattle, pig, horse, donkey, hare, hamster, cat, mouse, kangaroo, opossum
Hemophilia A (factor VIII deficiency)	Dog, cat, horse
Hemophilia B (factor IX deficiency)	Dog, cat, mouse
Hypoxanthine-guanine phosphoribosyl transferase (Lesch-Nyhan syndrome)	More than 25 species, including horse, hamster, dog, cat, mouse, cattle, gibbon, pig, rabbit, kangaroo
Ornithine transcarbamylase deficiency	Mouse, rat
Phosphoglycerate kinase	More than 30 species, including chimpanzee, gorilla, cattle, horse, hamster, mouse, kangaroo, opossum
Steroid sulfatase deficiency (ichthyosis)	Mouse, wood lemming
Testicular feminization syndrome	Cattle, dog, mouse, rat, chimpanzee
Vitamin D-resistant rickets	Mouse
Xg blood cell antigen	Gibbon

Source: Adapted from Strickberger with modifications and additions. Further listings can be found in Miller.

mosomes so that males are XY and females are XX. Thus males (the **heterogametic sex**) produce sperm that contain either X or Y chromosomes, whereas females (the **homogametic sex**) produce only X-bearing eggs. If the proportions of X- and Y-bearing sperm are equal, fertilization restores the two sexes in equal frequency: 1/2 X-bearing sperm × X-bearing eggs → 1/2 XX females; 1/2 Y-bearing sperm × X-bearing eggs → 1/2 XY males. However, since males carry only a single X chromosome, and the Y chromosome is mostly inactive, alleles present on the X chromosome express their effects in males, although such alleles may be recessive when in females. Thus a single hemophilia-producing allele on the X chromosome of a **hemizygous** male causes the classic hemophilia disease, whereas in XX females two such alleles are necessary to cause hemophilia.

Interestingly, of the more than 100 sex-linked genes geneticists have now identified in humans, many are also sex-linked in other mammals (Table 10–1). The conservation of the same genes on different mammalian X chromosomes indicates that a large part of this chromosome has persisted throughout mammalian evolution, at least for 90 million years. Ohno proposed that we can expect any sex-linked gene found in one mammalian species to be sex linked in other mammals as well. Box 10–1 briefly discusses the evolution of sex-determining mechanisms and some of their effects.

Linkage and Recombination

The localization of different genes to the X chromosome, which geneticists first demonstrated in the fruit fly, *Drosophila melanogaster,* provided the opportunity for establishing distances between such genes, or **linkage** relationships. Such determinations arose from crossing over or recombinational events in which exchanges occurred between the chromatids of paired homologous chromosomes. For example, the experiment diagrammed in Figure 10–9 showed that two *Drosophila* sex-linked genes, those involved in *white* eyes (w^+ and w) and *miniature* wings (m^+ and m), recombined with a frequency of about 38 percent to produce new chromosomal combinations.[6]

This frequency of recombination provided a measure of the **linkage distance** between genes on the same chromosome: the greater the recombination frequency, the greater

[6]These recombination events take place in females, since genetic crossing over does not normally occur in male *Drosophila,* and males have also only a single X chromosome. (In general, recombination is often absent or reduced in the heterogametic sex.) Interestingly, certain crosses between *Drosophila* strains can produce hybrid anomalies ("hybrid dysgenesis") in which considerable male recombination takes place. The intrusion of *P* transposable elements in some populations causes such events and is discussed on p. 226.

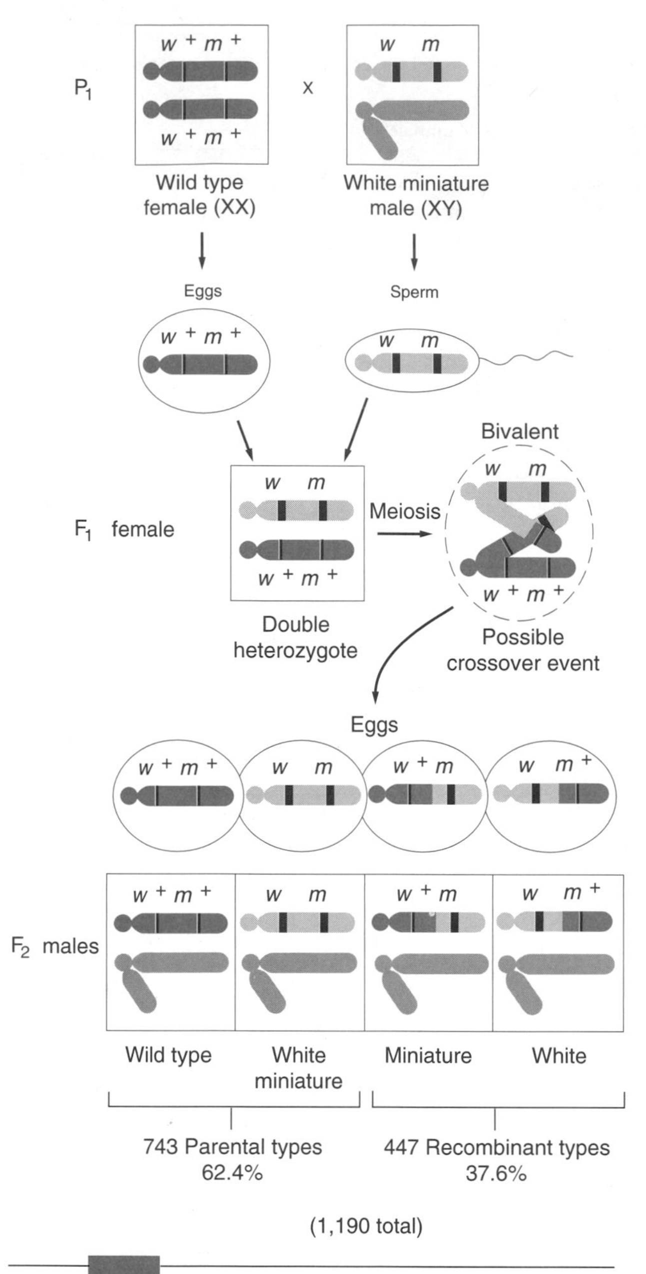

FIGURE 10-9 Recombination between the sex-linked genes *white* and *miniature* in *D. melanogaster* as evidenced by recombinant classes among F_2 males. Since the only sex chromosome the paternal parent contributes to male offspring is the Y, the meiotic events in the double heterozygous F_1 female strictly determine the phenotypes of F_2 males in respect to *white* and *miniature*. As shown in this illustration, these meiotic events involve crossovers between chromatids at the four-chromatid, or bivalent, stage (Fig. 10-3). The frequency of observed recombinants for any two given linked genes thus depends on the frequency in which such meiotic crossovers occur between the two genes. (*Adapted from Strickberger.*)

the distance. Thus, since the recombination frequency between the genes for *white* eyes and *cut* wings was about half the frequency of that between *white* and *miniature,* geneticists could assume the linkage distance between *white* and *miniature* to be about twice that between *white* and *cut.* Such experiments with both sex-linked and nonsex-linked (autosomal) genes provided linkage maps such as Figure 10–10 shows, indicating that chromosomes consist of linear arrays of genes, or **loci,** whose relative positions are additive (if three genes link in the order *H–G–I,* then the *H–I* linkage distance is the sum of *H–G* + *G–I* distances).

In humans and other mammals, geneticists have used techniques involving somatic cell fusions rather than mating between sex cells to obtain linkage relationships. These somatic hybridization techniques have been extraordinarily successful, and a large number of genes are now localized to many of the 23 human chromosomes. Moreover, as with similar sex-linked genes in mammals, genes localized to particular human chromosomes (identified by special banding patterns; see Fig. 10–17) can be localized to similar chromosomes in other mammals (for example, sheep, cattle, cats, rats, and mice)—a phenomenon known as **synteny** (O'Brien). In addition, as Figure 10–11 shows, linkage relationships among genes with similar functions in these mammals also persist.

In bacteria and viruses, recombinational detection techniques are considerably more advanced than in eukaryotes because of the very small size of bacteria and viruses, their rapid generation times, and the ease with which we can identify biochemical mutant genes. These advantages have led to elaborately detailed linkage maps such as Figure 10–12 shows. In addition, new methods that permit exact determinations of nucleotide sequences in both prokaryotes and eukaryotes are now providing information beyond anything conceived even a decade ago. All these studies indicate the immense variability that can generate through recombination: any chromosome may come to differ from its homologues by carrying a distinctive combination of alleles, for example, . . . $G^1H^2I^1J^3K^2$ . . . versus . . . $G^2H^1I^3J^2K^1$ . . . versus . . . $G^1H^1I^3J^3K^1$ . . . . Other recombinational advantages are discussed on pp. 563–564.

Chromosmal Variations in Number

Forms of variation that are often microscopically observable are changes in the number or structure of chromosomes. The numerical chromosome variations are of two major kinds: changes in the number of entire sets of chromosomes, **euploid variations**; and changes in the number of single chromo-

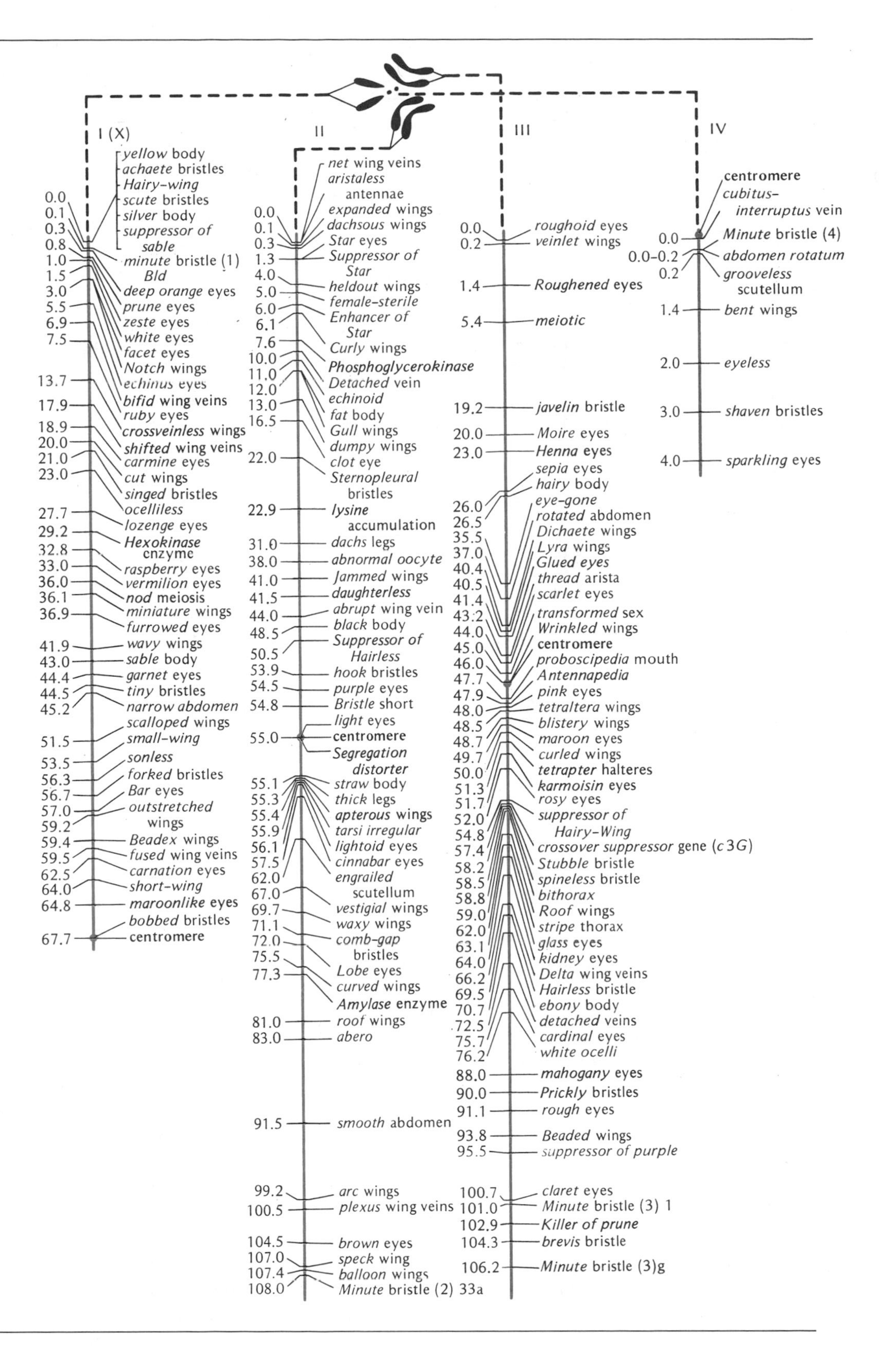

FIGURE 10-10 Linkage map of some of the important genes in the four chromosomes of *D. melanogaster*. Note that a variety of different genes may affect a specific character such as eye color, wing shape, and bristles, indicating that many steps exist in the development of a particular function, each step governed or capable of being modified by separate and different genes. (*From* Genetics Third Edition *by Monroe W. Strickberger. Copyright © 1985 by Monroe W. Strickberger. Reprinted by permission of Prentice Hall, Inc., Upper Saddle River, NJ.*)

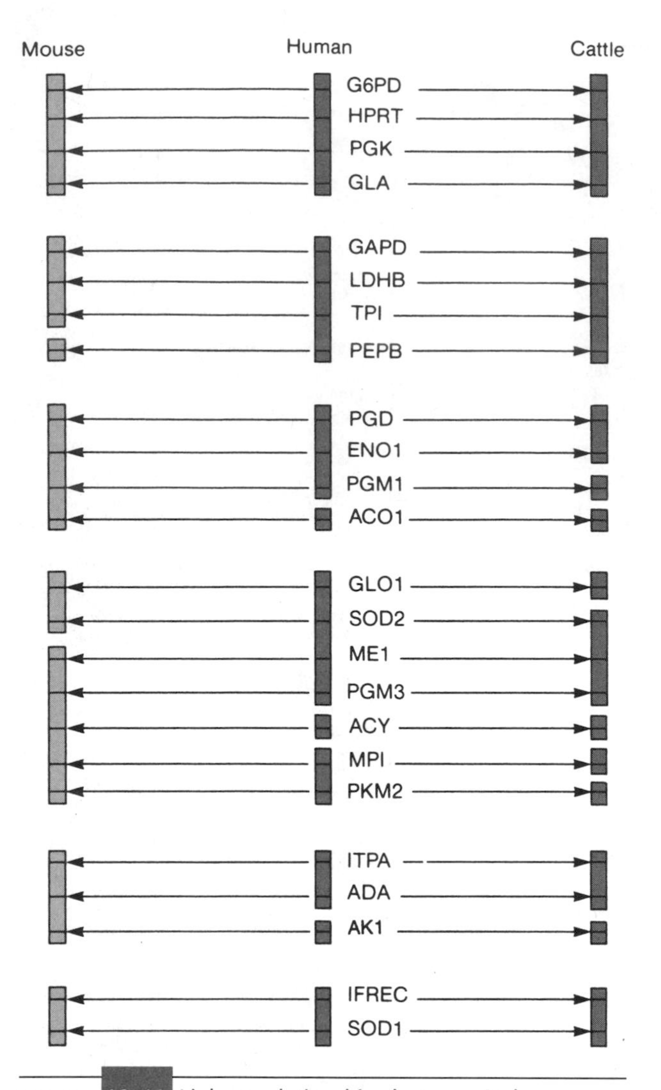

FIGURE 10-11 Linkage relationships between 24 human gene loci (*center column*) compared to those of similar genes in mice (*left*) and cattle (*right*). Each vertical rectangular block of genes indicates a chromosome in the respective species. For example, the top rectangles in each column give the relative positions of four genes on the X chromosome of each species. Note that the linkage order for many gene loci is the same in each of the three species, indicating that a large number of linkage relationships have been evolutionarily conserved. (*Adapted from Womack and Moll.*)

somes within a set, **aneuploid variations.** Since most sexually reproducing eukaryotes are diploids with two sets of chromosomes (2n), euploid variations may extend from the haploid or monoploid condition (1n) to various levels of **polyploidy** (3n, 4n, . . .) as Table 10–2 shows.

Such events may be caused by more than one sperm fertilizing an egg, or by cell division failures in which, for example, a diploid rather than haploid gamete results.

The appearance of extra sets of chromosomes within a species itself, **autopolyploidy,** seems to be a common mode of evolution in many plant groups such as mosses, apples, pears, bananas, tomatoes, and corn. Polyploids that originate from crossing between different species, **allopolyploids,** also appear in some plants such as wheat. For example, a gamete from species *A* may fertilize species *B* to produce the diploid hybrid *AB,* which can then undergo polyploidy to form the allotetraploid *AABB,* or other variations shown in Figure 10–13. Researchers have produced both auto- and allopolyploids with chemicals such as colchicine that break down spindle fiber microtubules and thereby interfere with chromosome segregation during cell division.

Interestingly, the first laboratory-created species was the product of a cross between two tobacco plants, *Nicotiana tabacum* (a diploid with 48 chromosomes whose haploid gametic number, n, was 24) and *N. glutinosa* (24 chromosomes, n = 12). As Figure 10–14 shows, the hybrid of this cross was sterile although it had two sets of chromosomes, one from each species. In such cases, sterility arises because an even division of chromosomes during meiosis depends on pairing between homologous chromosomes, but these homologues are absent when the two cross-fertilizing species have evolved chromosomal differences between them. Nevertheless, although sterile, the hybrid plant could continue to grow and increase by vegetative cuttings.

Eventually, a chromosome-doubling event produced an allopolyploid (also called amphidiploid) that was fully fertile because each chromosome now had a pairing mate, and normal meiosis could take place. Thus, at one stroke, a new species with 72 chromosomes, *N. digluta,* was created that was only fertile when crossed with itself (all its chromosomes could undergo homologous pairing) and could not produce fertile offspring when crossed to either of the parental species (only some of its chromosomes could pair homologously with those of either parent).

In animals, polyploidy is much rarer than in plants because most animals show much greater developmental sensitivity to even a small change in chromosome number. This sensitivity extends also to chromosomal sex-determining mechanisms, so that animal polyploids face difficulties in maintaining the same proportions of X and Y chromosomes present in normal diploids. For example, polyploidy in a species in which males are XY and females XX could lead to establishment of XXYY males and XXXX females. The subsequent combinations of gametes produced by these individuals (XY sperm + XX eggs) might well produce XXXY individuals that are not completely male or female. Therefore, when animal polyploid species do occur, they usually have some form of **asexual reproduction** such as **parthenogenesis** (embryonic development of eggs without fertilization).

Nevertheless, polyploidy probably has some advantages. In both plants and animals, the extra polyploid chromosomes may act as multiple buffers in various organismic processes, improving vigor and enabling such individuals to face new and drastic conditions. Under such stressful circumstances, polyploid fecundity may be

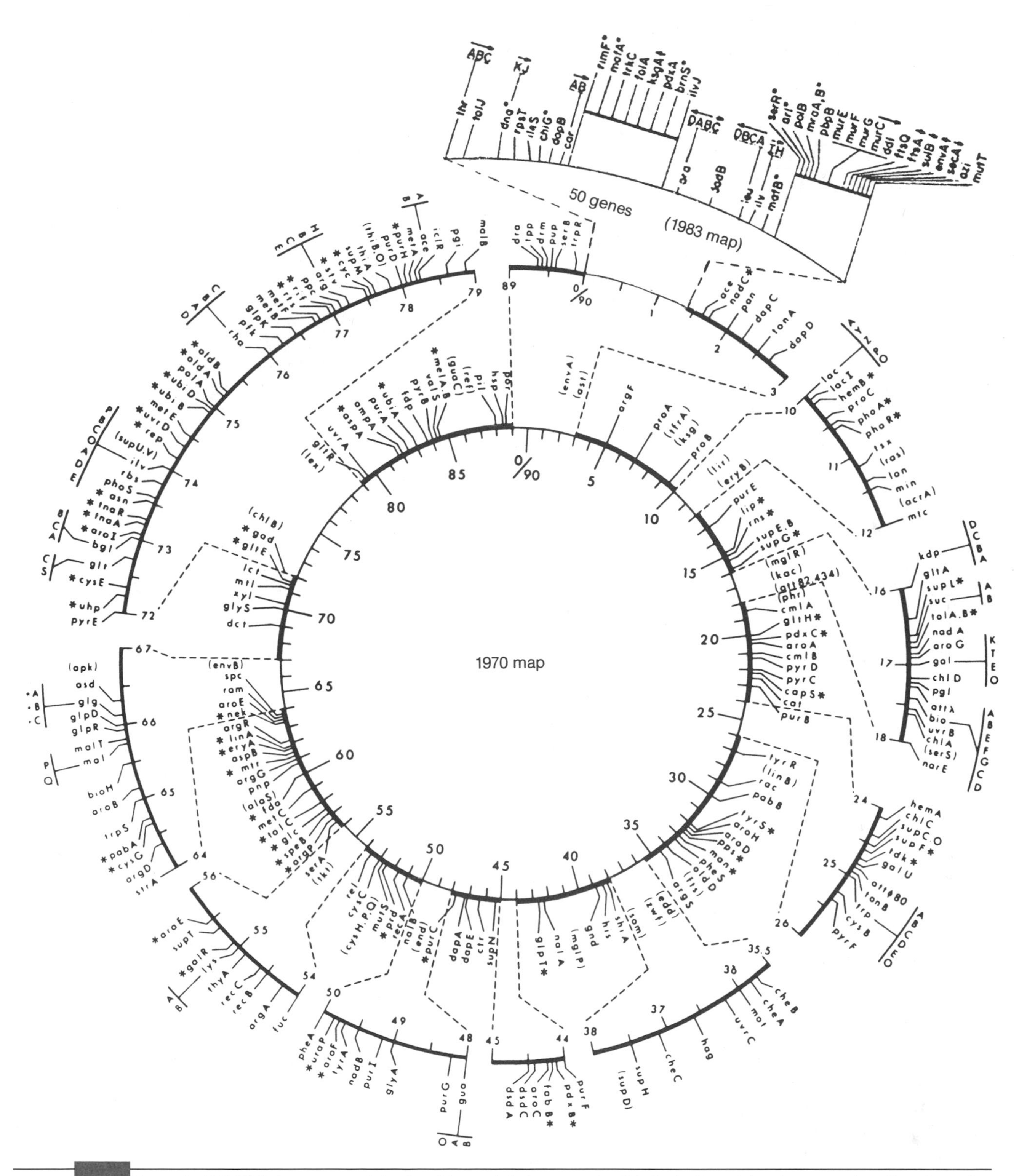

FIGURE 10-12 An early (1970) linkage map of the *E. coli* bacterial chromosome (Taylor). Geneticists have continually enlarged this map as they have discovered and localized new genes. For example, in the upper right corner is a small section, about one-fiftieth of the genome, to which 18 genes were localized in 1970 and 50 genes localized in 1983. (*Linkage Map of E. Coli K-12 ed. 7* Microbiological Reviews *47: 180–230, 1983 by B. J. Bachmann. Reprinted by permission.*)

TABLE 10–2 Euploid variations, involving entire sets of chromosomes

Euploid Type	Number of Homologues Present for Each Chromosome	Example
Haploid or monoploid	One (1n)	*A B C*
Diploid	Two (2n)	*AA BB CC*
Polyploid	More than 2	
Triploid	Three (3n)	*AAA BBB CCC*
Tetraploid	Four (4n)	*AAAA BBBB CCCC*
Pentaploid	Five (5n)	*AAAAA BBBBB CCCCC*
Hexaploid	Six (6n)	*AAAAAA BBBBBB CCCCCC*
Heptaploid	Seven (7n)	*AAAAAAA BBBBBBB CCCCCCC*
Octaploid, etc.	Eight (8n), etc.	*AAAAAAAA BBBBBBBB CCCCCCCC*, etc.

Source: Adapted from Strickberger.

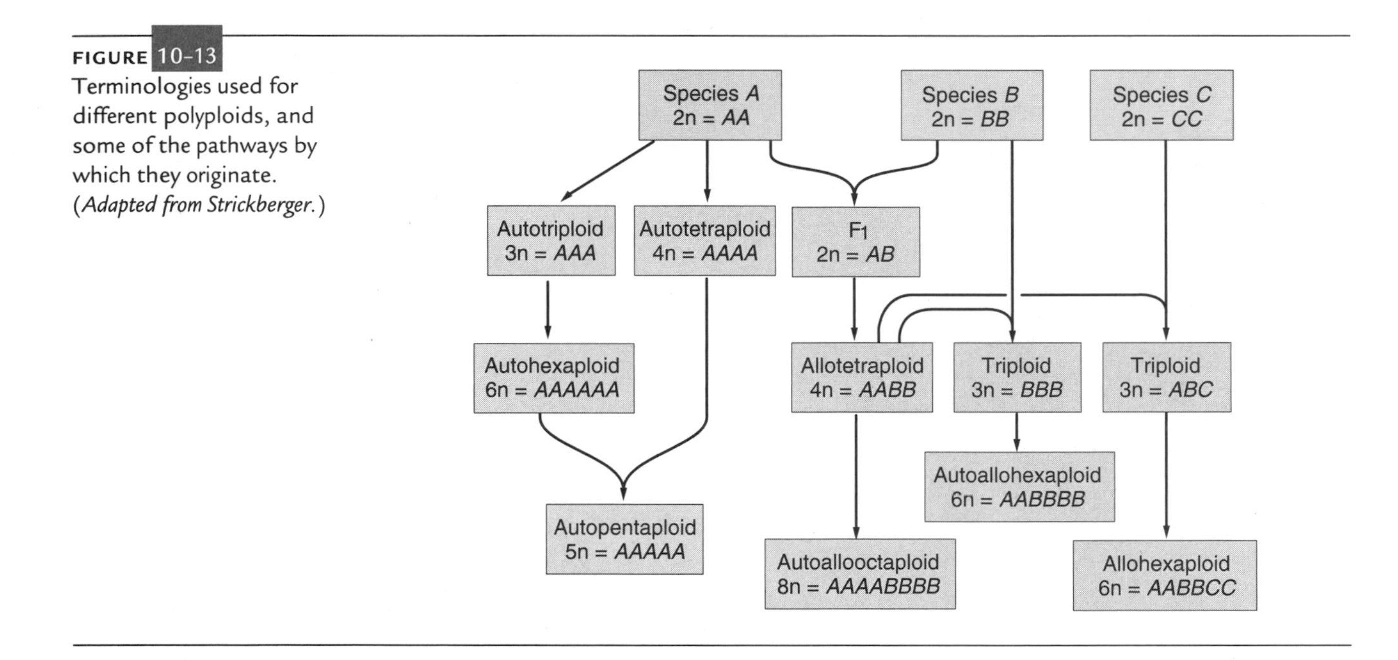

FIGURE 10–13 Terminologies used for different polyploids, and some of the pathways by which they originate. (*Adapted from Strickberger.*)

sufficient to replace previously normal diploids. Also the additional chromosomes may provide the chance to evolve new functions for their extra sets of genes leading, among other features, to increased diversity and heterozygosity (Soltis and Soltis).

In aneuploids, as Table 10–3 shows, a wide range of variations may result when chromosomes are either added or subtracted from a normal set. Such events may occur because meiotic disjunction between homologous chromosomes is abnormal (**nondisjunction**). Again, plants seem to be more tolerant of such chromosomal variations than animals, but aneuploids are found in both groups.

Chromosomal Variations in Structure

A variety of changes in chromosome structure are diagrammed in Figure 10–15.

DELETIONS OR DEFICIENCIES

The terms **deletions** and **deficiencies** describe losses of chromosomal material (Fig. 10–15*a*). In general, the

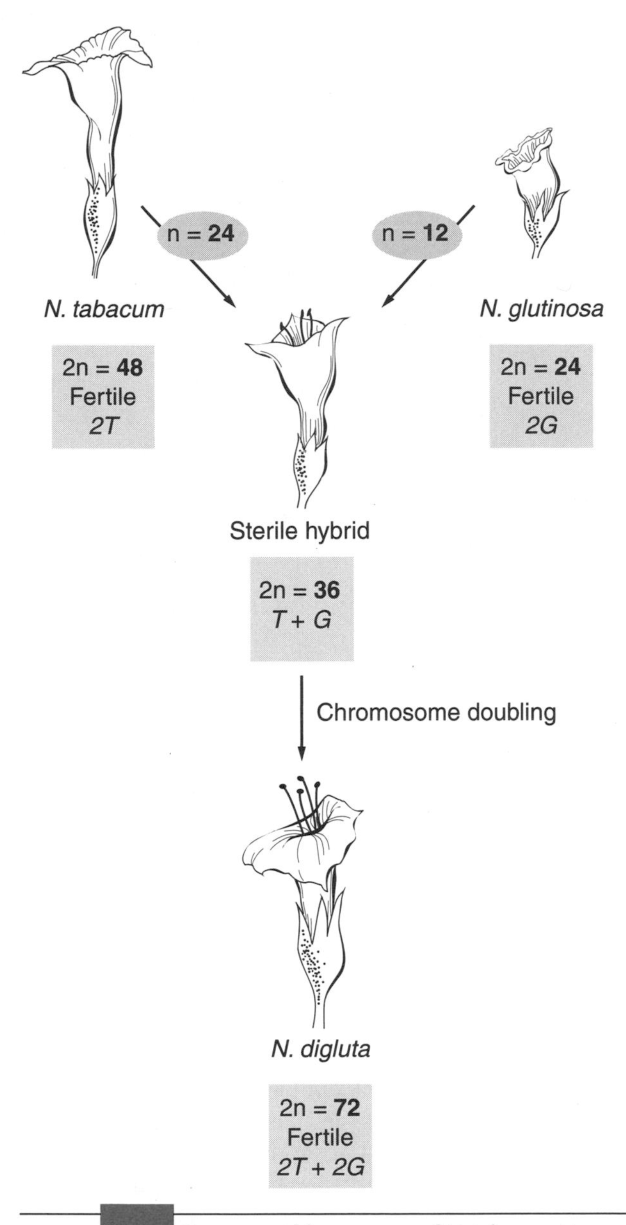

FIGURE 10–14 Flowers and karyotypes of *N. tabacum, N. glutinosa,* and their sterile deploid hybrid, and the new allotetraploid (amphidiploid) species formed by chromosome doubling in the sterile hybrid.

severity of a deletion depends on how extensive it is and on what nucleotides or genes are missing. If functional genes are involved, deletions can be quite harmful in both diploids and haploids but not necessarily harmful in polyploids or aneuploids, where such genes may be present in extra chromosomes. Researchers have detected various deletions in populations, and many of these produce "buckles" meiotically observed when the deleted and nondeleted chromosomes pair up closely in heterozygotes.

DUPLICATIONS

Duplications are either short or long segments of extra chromosome material originating from duplicated sequences within a genome (Fig. 10–15*b*). Since numerous **gene families** of similar or identical genes occur in many species, duplications show evidence of being common during evolution. For example, various eukaryotes produce the RNA components of ribosomes in clusters of long tandem arrays of duplicated genes. Other instances, such as the genes involved in producing different hemoglobin-type proteins (see, for example, Figs. 12–3 and 12–5) show that duplicated genes have evolved in different pathways, enabling them to perform different adaptive functions. As in deletions, researchers can detect many chromosome duplications in heterozygotes by the buckles formed during meiotic pairing.[7]

INVERSIONS

Inversions are reversals in chromosomal gene order that researchers can sometimes observe in the formation of loops during meiotic pairing in heterozygotes (Fig. 10–15*c*, *d*). Inversions generally lower the recombination frequency within the inverted sequence, because crossing over within such sequences in heterozygotes can lead to chromosomal abnormalities. As a result, the genes included within an inversion tend to remain together as a nonrecombinant block, called a *supergene* by some workers.

Documented in various eukaryotes are two major kinds of inversions, **paracentric** that do not include the centromere (Fig. 10–15*c*); and **pericentric** that include the centromere (Fig. 10–15*d*). Among the possible effects of pericentric inversions is a shift in the relative position of the centromere, as Figure 10–15*d* shows. In some cases, this shift may be drastic, moving the centromere from the chromosomal center (metacentric position) to one end (acrocentric). In the deer mouse genus, *Peromyscus*, there are species whose 48 chromosomes are entirely metacentric (for example, *P. collatus*) and species whose 48 chromosomes are almost entirely acrocentric (for example, *P. boylei*).

TRANSLOCATIONS

The term **translocation** primarily refers to the transfer of material from one chromosome to a nonhomologous chromosome. When the exchange of such material is mutual, it can result in the kind of **reciprocal translocation** that Figure 10–15*e* diagrams. Such translocations are

[7]Among the mechanisms that produce duplications is **unequal crossing over,** in which homologous pairing during recombination is slightly askew, thereby producing a crossover product that contains extra chromosomal material (Fig. 12–4).

TABLE 10-3 Aneuploid variations, involving individual chromosomes within a diploid set

Type	Number of Chromosomes Present	Example
Disomic (normal diploid)	2n	*AA BB CC*
Monosomic	2n − 1	*AA BB C*
Nullisomic	2n − 2	*AA BB*
Polysomic	Extra chromosomes	
Trisomic	2n + 1	*AA BB CCC*
Double trisomic	2n + 1 + 1	*AA BBB CCC*
Tetrasomic	2n + 2	*AA BB CCCC*
Pentasomic	2n + 3	*AA BB CCCCC*
Hexasomic	2n + 4	*AA BB CCCCCC*
Septasomic	2n + 5	*AA BB CCCCCCC*
Octasomic, etc.	2n + 6, etc.	*AA BB CCCCCCCC*, etc.

Source: Adapted from Strickberger.

recognizable by a cross-shaped configuration between translocated and nontranslocated chromosomes during meiotic pairing in the heterozygote. Since such meioses can produce gametes containing duplications and deficiencies, translocation heterozygotes often show sterility.[8]

The fertile and viable gametes that translocation heterozygotes produce result primarily from **alternate segregation,** in which the translocated chromosomes segregate separately from the nontranslocated chromosomes (these are the diagonal combinations of chromosomes in Fig. 10–15*e*). Thus genes in the translocated and nontranslocated chromosomes of such heterozygotes tend to be inherited as separate blocs, behaving as though all genes on each bloc were linked together.

Because translocations cause various sterility and fertility problems in heterozygotes, there is an advantage for individuals homozygous for translocations to be isolated from those homozygous for nontranslocated chromosomes. According to some proponents of "chromosome speciation" (see, for example, King), selection for separation between these groups can easily lead to speciation, and examples among plants such as *Clarkia* (Lewis) indicate that such events have occurred. Other researchers contest this view and point out that chromosomal changes have served only as a sporadic accompaniment to other causes for speciation. As yet, the matter is not resolved, although most would agree that some chromosomal changes might lower the reproductive success of hybrids between groups homozygous for different arrangements.

In terms of their cytological effects, translocations can cause changes in both the number and structure of chromosomes. For example, translocations may combine two different nonhomologous chromosomes into one larger chromosome (Fig. 10–16*a*) or cause chromosomes to significantly change in shape (Fig. 10–16*b*), or fission events can occur that increase the number of chromosomes (Fig. 10–16*c*).

Thus, in some European wild mouse populations of *Mus musculus* reductions occur from the standard number of 20 pairs of chromosomes to as few as 11 pairs. In Asiatic muntjac deer, one species has only 3 pairs of very large chromosomes compared to the normal number of 25 pairs (Fig. 10–17). According to Liming and coworkers, the relative amounts of DNA in the two species are about the same, and the very large muntjac chromosomes are derived from successive translocations that combined the smaller muntjac chromosomes. Chromosomal conservatism, in contrast, may be a strong feature in other mammalian groups. For example, present species of the 30-million-year-old lineage that includes Asian (Bactrian) and African (Dromedary) camels as well as the South American guanaco, vicuña, llama, and alpaca all have the exact same number of morphologically similar chromosomes, 74. Qumsiyeh proposes that the number of chromosome pairs reflects a species' adaptive capability: larger numbers allow increased genetic recombination, thereby producing more variability and environmental flexibility; whereas smaller numbers allow genetic combinations to persist, which lead to greater specializations for specific habitats.

Chromosomal Evolution in *Drosophila* and Primates

In those instances where researchers can identify linear sections of chromosomes through distinctive bandings, we can chart chromosomal evolution in great detail. In *Drosophila* species and other

[8]Note that the gametes produced by the translocation heterozygote illustrated in Figure 10–15e would have duplications and deficiencies if the two upper chromosomes went to one pole and the two lower chromosomes went to the other pole.

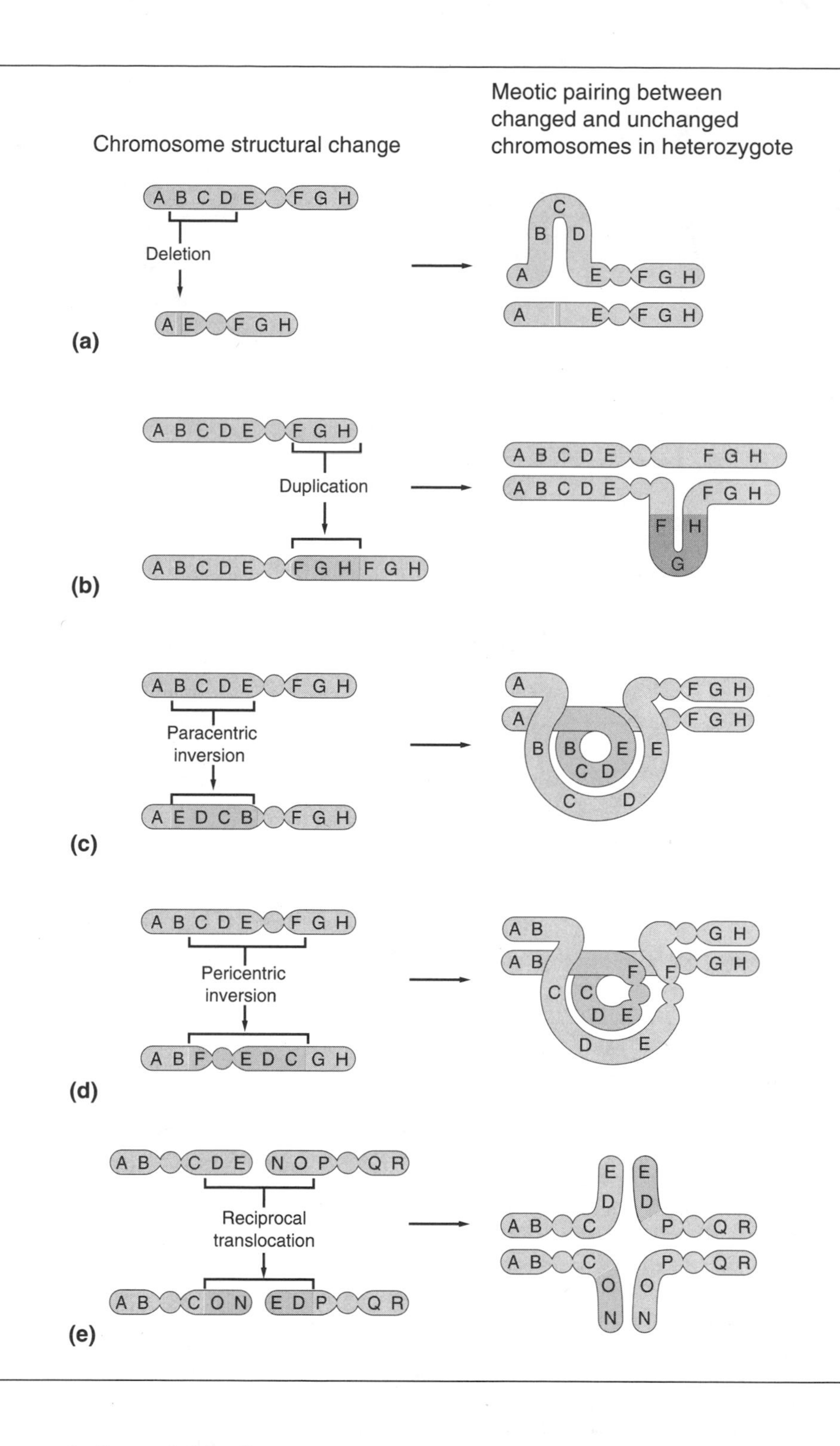

FIGURE 10–15 Major kinds of structural chromosomal changes and their effects on meiotic pairing in heterozygotes who carry both changed and unchanged homologues. (For diagrammatic simplicity, the double chromatid structure of each chromosome is omitted, although meiotic pairing between two homologues actually involves four chromatids.)

dipteran insects, the chromosomes of salivary gland cells and other tissues have replicated many times over, yet each replicate still remains closely apposed to other such replicates within the same nucleus. As a result, such **polytene chromosomes** are tremendously enlarged and show highly detailed banding configurations along their lengths. These distinctive banding arrangements enable even minor chromosomal changes to be traced. Geneticists have described practically all the chromosomal changes in the evolution of hundreds of these fly species, a sample of which Figure 10–18 shows.

Although polytene chromosomes are absent in many organisms, new chromosomal staining techniques have been devised that enable detailed comparisons between even relatively small mammalian chromosomes. For example, the G-banding technique that Figure 10–19 illustrates allows a comparison of human, chimpanzee, gorilla, and orangutan chromosomes. These bandings

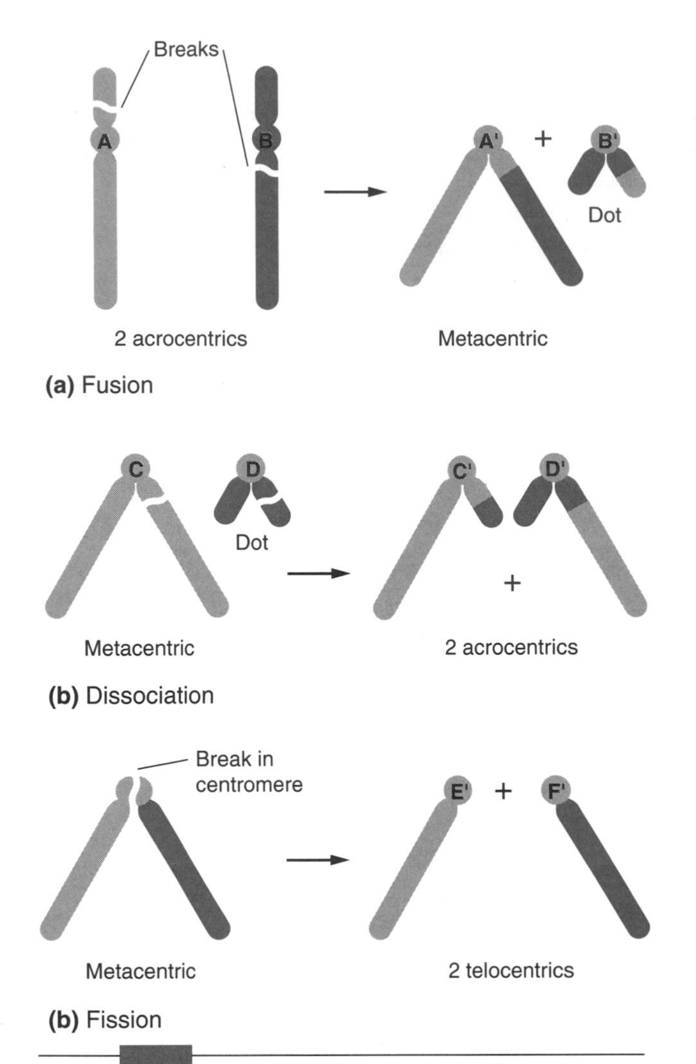

FIGURE 10–16 (*a*) Translocations that lead to **fusion** between the arms of two acrocentric nonhomologous chromosomes (A and B) to form one metacentric chromosome (A′) and a "dot" chromosome (B′) that carries only a small amount of chromosomal material. In diploids, loss of the dot chromosome will reduce the chromosome number by 2, since homozygotes for the metacentric now carry the chromosome material formerly present in the two acrocentrics. (*b*) The reverse process of **dissociation** involves reciprocal translocations between the metacentric (C) and dot (D) chromosomes leading to C′ and D′ acrocentrics. (*c*) The **fission** mechanism proposed to explain the origin of presumed single-armed telocentric chromosomes (E′, F′) from two-armed metacentrics or acrocentrics. (*From* Genetics Third Edition *by Monroe W. Strickberger. Copyright © 1985 by Monroe W. Strickberger. Reprinted by permission of Prentice Hall, Inc., Upper Saddle River, NJ.*)

Chinese muntjac
Muntiacus reevesi

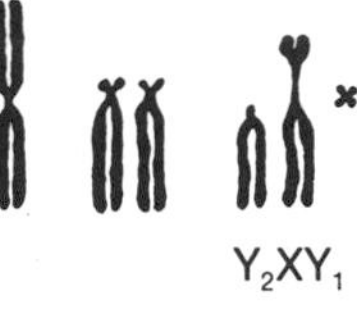

Indian muntjac
Muntiacus muntjak

FIGURE 10–17 The Chinese and Indian muntjac deer and their karyotypes. The Indian muntjacs, with two pairs of autosomes and three sex chromosomes, have the lowest known chromosome number of any mammal. (*Adapted from Austin and Short.*)

show that some chromosomes (nos. 6, 13, 19, 21, 22, and X) are practically identical in all four species, and various arms or sections of other chromosomes are homologous throughout.

We can account for the changes that have occurred during the evolution of these primates by the simple chromosomal variations just described. Thus the difference in number between humans ($n = 23$) and apes ($n = 24$) derives from a fusion event that combined the two indicated chimpanzee-type chromosomes to form the no. 2 human-type chromosome. More recent techniques involving DNA sequencing (p. 276ff) have sketched an interesting evolutionary odyssey of a human chromosome (no. 3) starting from its probable origin in an early rodentlike population perhaps 90 million years ago (Fig. 10–20).

Gene Mutations

ene mutations, or **point mutations,** are mutations not observable at the chromosomal levels discussed so far; that is, they presumably affect the nucleotide structure of the gene itself. Researchers can discern many gene mutations either directly (with modern procedures of nucleotide sequencing) or indirectly (from their effects on the amino acid sequences of proteins).

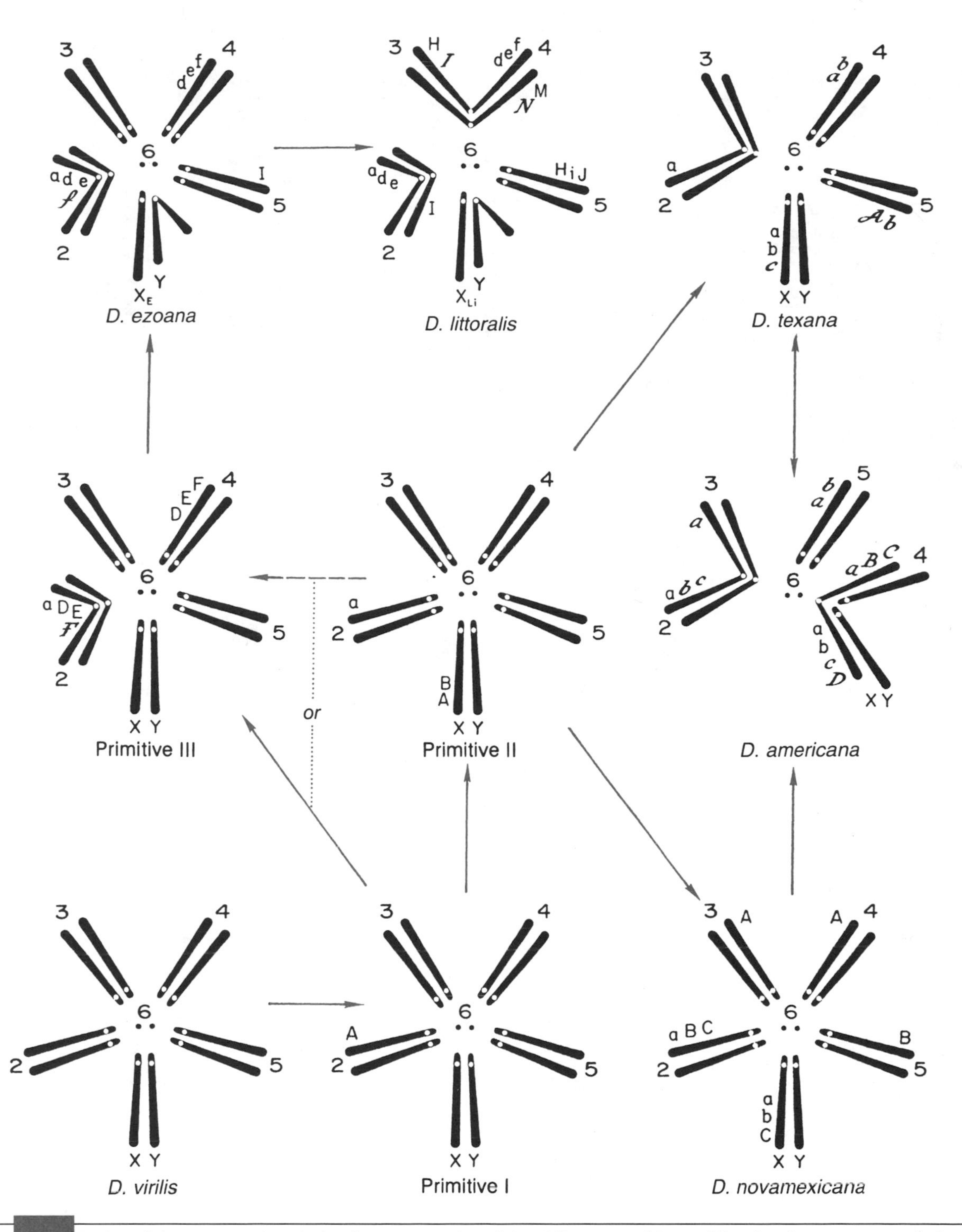

FIGURE 10-18 Paths of chromosomal evolution in some species of the *virilis* group of *Drosophila*. The chromosomes of what was probably the original karyotype of the genus *Drosophila* (*lower left*) are numbered from 1(X) to 6, and specific chromosomal banding arrangements are indicated by letters. (*Adapted from Stone.*)

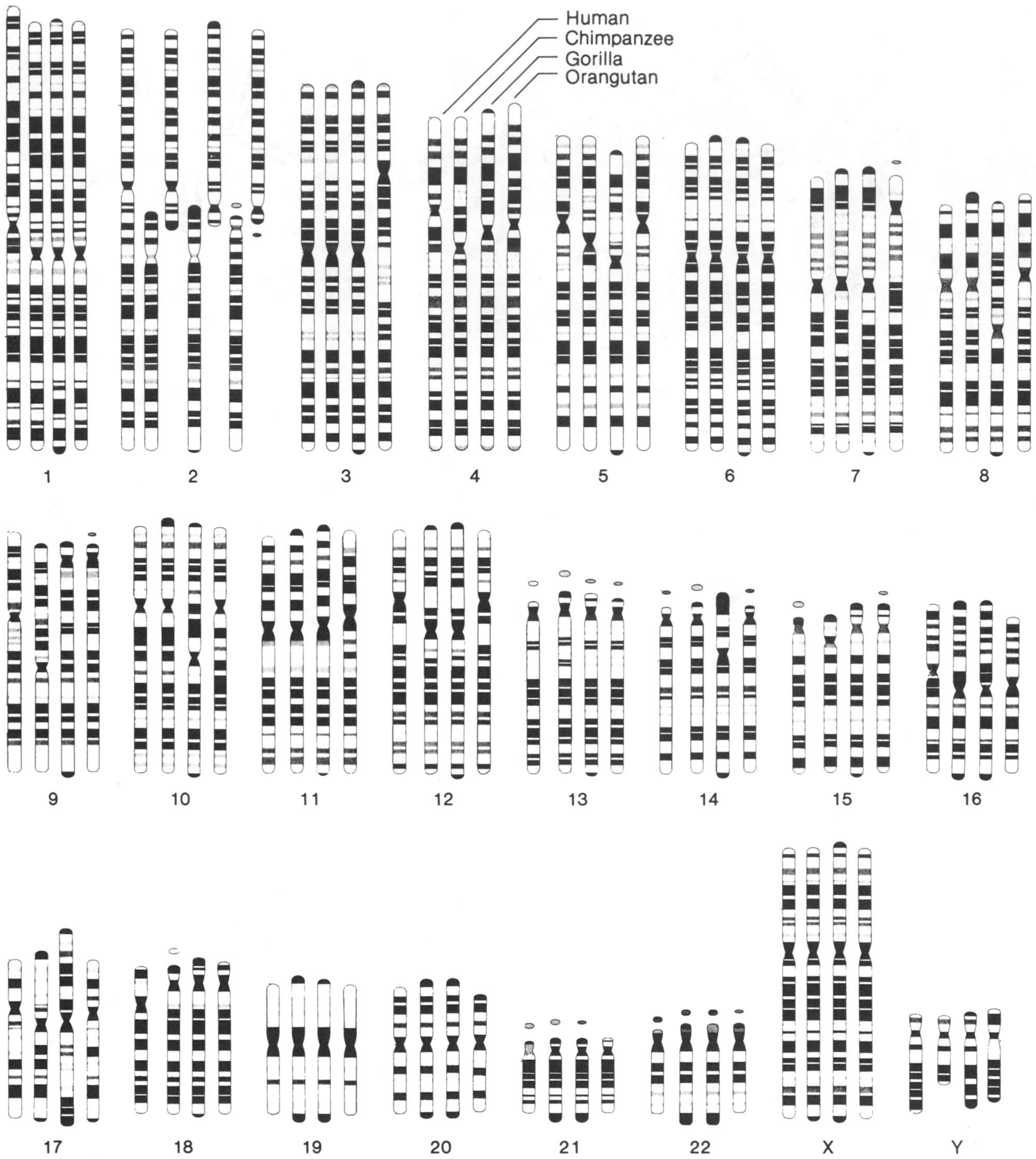

FIGURE 10–19 Banding arrangements of the chromosomes of humans, chimpanzees, gorillas, and orangutans, in respective order from left to right for each chromosome. Note that the 24 pairs of chromosomes in the great apes reduce to 23 pairs in humans (number 1 to 22 + XY) because two different chromosomes fuse into a single no. 2 human chromosome. This fusion, along with other changes (for example, inversions in chromosomes 1 and 18), must have occurred some time after the human line separated from a human-chimpanzee common ancestor. On the whole, these banding arrangements indicate that humans have a closer evolutionary relationship with chimpanzees than with gorillas and a more distant one with orangutans. (*Reprinted with permission from "The Origin of Man: A Chromosomal Pictoral Legacy" by Yunis, J. J. Copyright © 1982 American Association for the Advancement of Science.*)

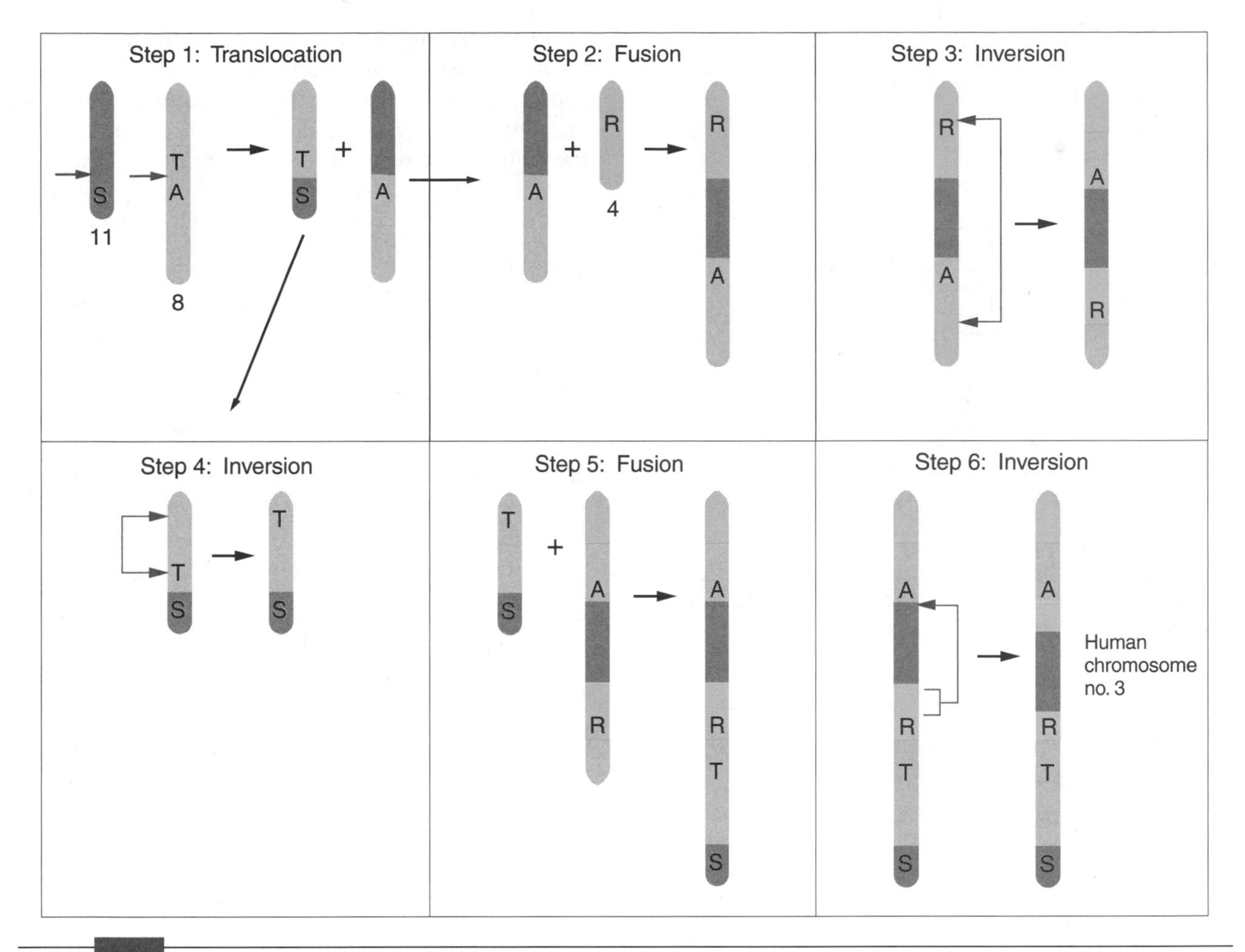

FIGURE 10–20 Six proposed major steps in the evolution of human chromosome no. 3, beginning some time during the Cretaceous period with a translocation between the two rodentlike chromosomes nos. 8 and 11. This was followed by fusion of one translocation product to the no. 4 rodent chromosome. As shown, further steps involved lengthening of the chromosome through fusion (step 5) with the remaining translocation product of step 1, and two inversions. The letters A, R, S, and T, indicate four of the marker genes whose loci we know in rodents, humans, and other mammals: A = aminoacylase-1, R = rhodopsin, S = somatostatin, T = transferrin. For simplicity, we have omitted other known chromosomal changes from this illustration, including the insertion of small segments from rodentlike chromosomes 2, 14, 15, and 16. (*Modified from Hino et al.*)

Among the various kinds of mutational changes at the molecular level are **base substitutions,** nucleotide changes that involve substituting of one base for another. As Figure 10–21 shows, researchers call these **transitions** when exchanges occur either between purines (A↔G) or between pyrimidines (T↔C) and **transversions** when purines exchange for pyrimidines or vice versa (A, G↔T, C).

Substitutions may occur spontaneously through copying errors caused, for example, by rare **tautomeric** nucleotide base changes that enable complementary pairing between adenine and cytosine or between guanine and thymine (Fig. 10–22). Other base substitutions may arise from the action of mutagenic agents such as hydroxylamine, nitrous acid, and nitrogen mustards. Also, in some instances specific nucleotide sequences cause increased mutations among adjacent nucleotides. Such "hot spots of mutation" may act by coiling the DNA molecule in ways that influence DNA polymerase enzymes to produce replication errors. The replication accuracies of polymerase enzymes also differ; some strains carry enzymes that are apparently more prone to produce mutational errors than others.

Other mutational events at the molecular level include nucleotide deletions, duplications, and insertions, which can arise spontaneously within the cell or be evoked by externally applied mutagenic agents such as acridine dyes. Rearrangements of nucleotides also occur, such as inversions (reversals in nucleotide order) or transpositions (movement of nucleotide sequences to new positions).

In general, mutational effects caused by these mechanisms may express at two levels of gene activity:

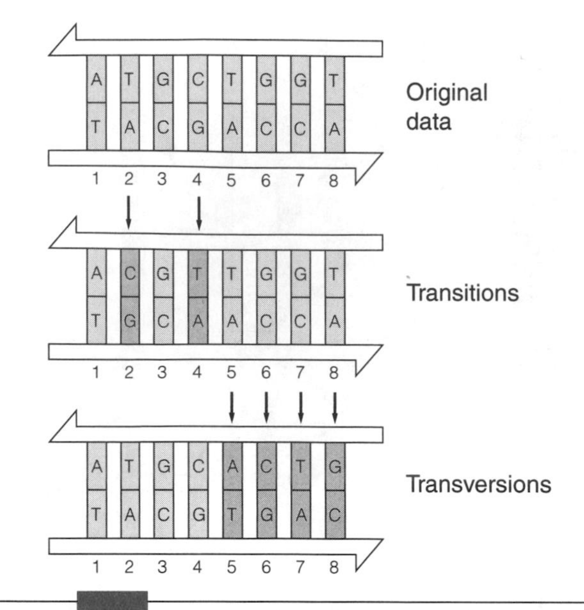

FIGURE 10-21 Examples of specific base pair changes in a section of double-stranded DNA.

(1) changes within the gene product itself—for example, in the amino acid constitution of a particular protein; and (2) changes in the regulation of a gene product—meaning that the gene product is not itself affected, but the timing of its appearance is different from normal.

Mutational effects that result in a changed gene product may arise because of nucleotide changes that cause

- A substitution for one or more of the amino acids in a protein (missense mutations, Fig. 10–23*a* and *b*).
- Changes that insert protein termination ("stop") codons in the middle of a gene sequence, thus causing premature termination of polypeptide chain synthesis (nonsense mutations, Fig. 10–23*c*).
- Nucleotide insertions or deletions that modify the messenger RNA (mRNA) protein translation reading frame so that a new and different sequence of codons appears (frameshift mutations, Fig. 10–23*d*).
- In addition, since practically all amino acids are coded by more than one kind of codon,

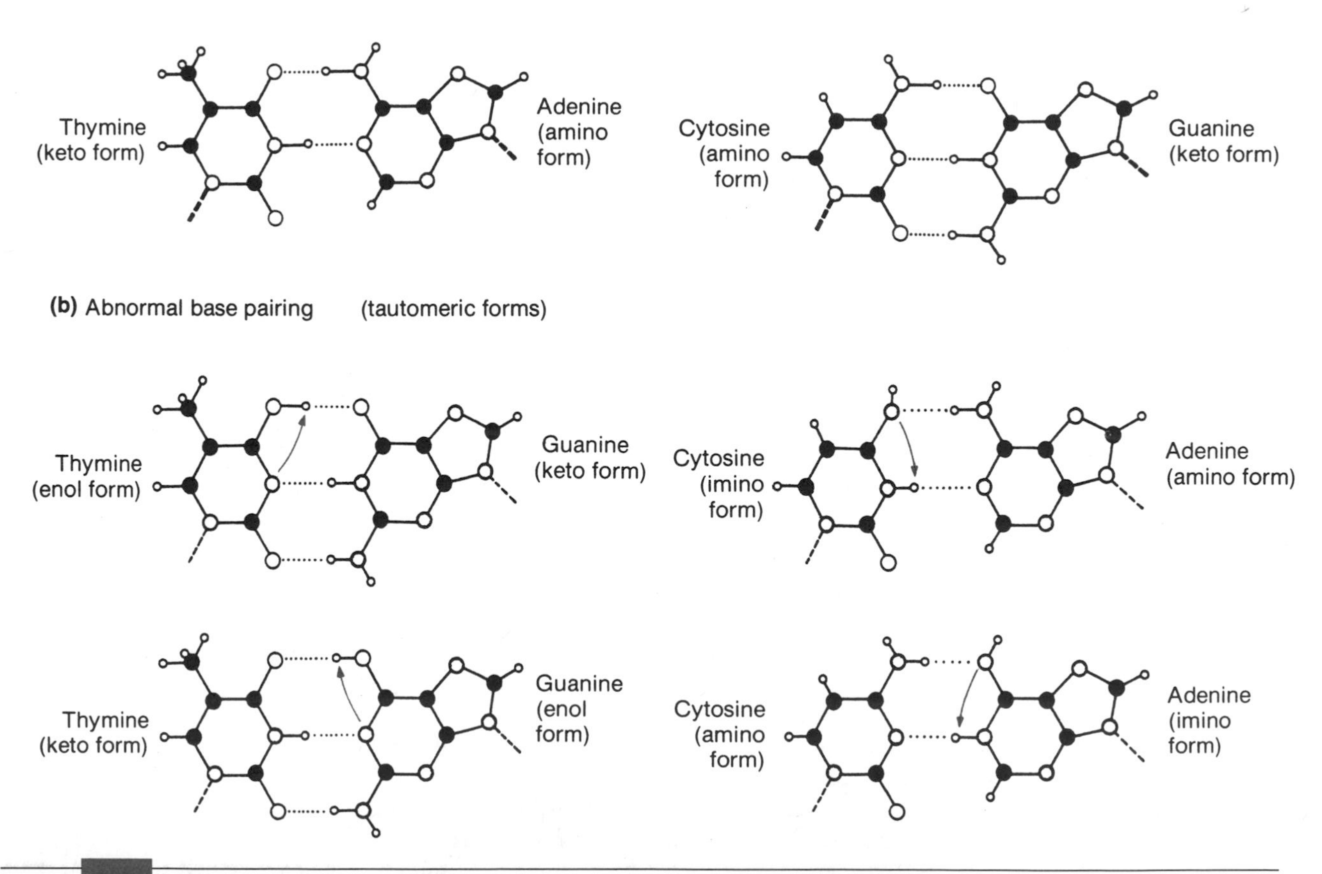

FIGURE 10-22 (*a*) Normal complementary pairing between nucleotide bases during DNA replication. (*b*) Modified base-pairing relationships that result from *tautomeric* molecular changes. Because of such changes, base substitutions can occur that produce, for example, transitions from T–A base pairs to C–G base pairs. (*Adapted from Drake 1970.*)

synonymous mutations can change an amino acid codon without producing an amino acid substitution (Fig. 10–23*e*).

The wide range of mutational possibilities has important consequences for the organism, since both too many and too few mutations can interfere with adaptation: too many generate continued errors in already adapted organisms, and too few reduce adaptive opportunity. Optimal mutation rates are therefore advantageous (p. 225), and evidence discussed on p. 291 indicates that such rates can be selected.

However, depending on its position, even one nucleotide mutation may have important consequences for the organism although it causes only a single amino acid substitution in a long-chain protein. A prominent example of such effect is that caused by the **sickle cell** mutation in humans, a gene that, in the United States, is almost entirely confined to blacks. Normally, the adult hemoglobin molecule in human blood cells consists of four polypeptide globin chains, two α's and two β's, each about 140 amino acids long with its own specific sequence. However, in homozygotes for the sickle cell gene (Hb^S/Hb^S) all β-globin chains differ from normal β's at the no. 6 position because of a transversion that changed the glutamic acid codon GAA to the sickle cell valine codon GUA.

As Figure 10–24 shows, the effects of this single genetic mutation are profound, causing a variety of phenotypic changes that often lead to inviability (death). (**Pleiotropy** is the name given to multiple phenotypic effects of a single gene.) Sickle cell disease is known to kill more than 10 percent of American black homozygotes before the age of 20, and probably has even more lethal effects in Africa, where medical facilities are limited. The high frequency of this gene in black populations is related to the selective advantage of sickle cell heterozygotes in malarial regions, a topic discussed in Chapter 22).

Regulatory Mutations

Regulatory mutations are those that affect the rates at which gene products are produced, although the products themselves may be unaffected. Among such examples are the **thalassemias,** genetic diseases in which the production of either α or β hemoglobin chains is absent or diminished.

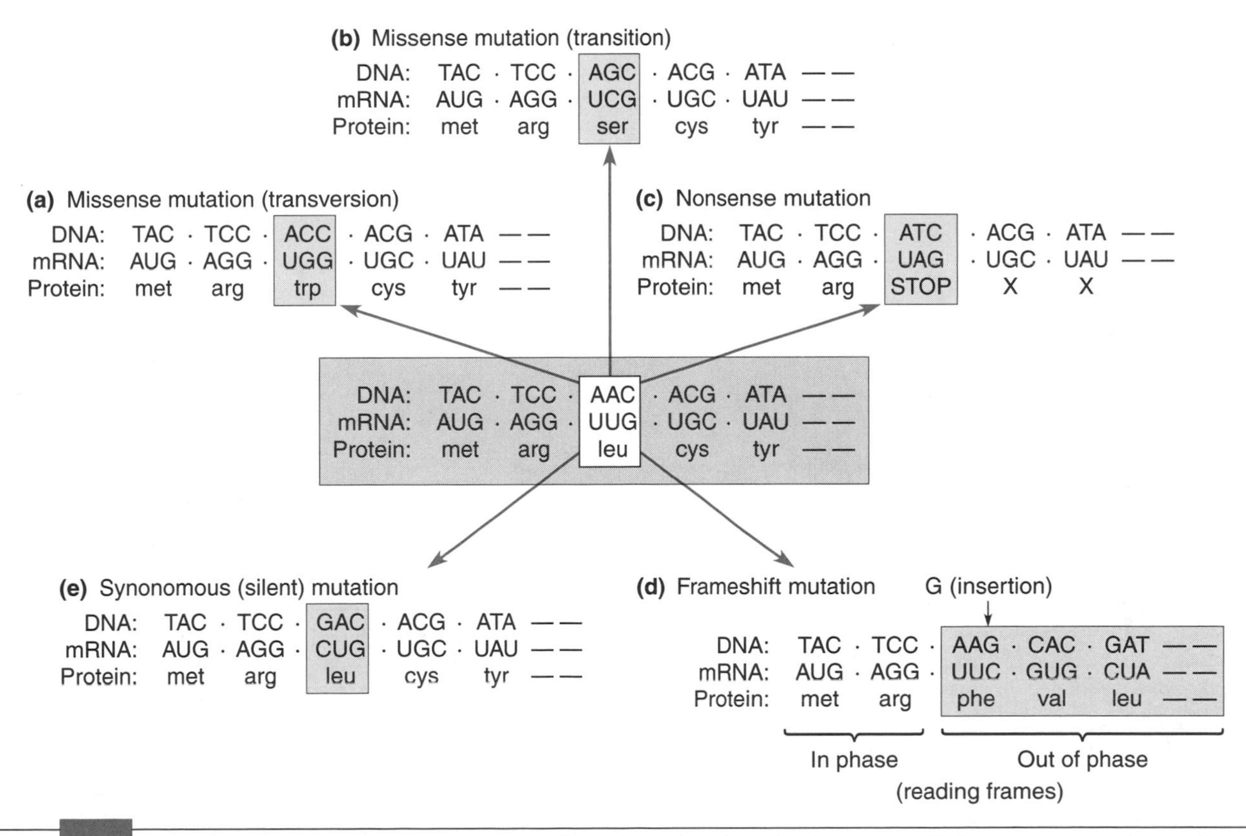

FIGURE 10–23 Different kinds of mutations produced by the indicated nucleotide changes. In all these sequences, the AUG codon on messenger RNA intiates translation into amino acids. Note that despite mutations, translation proceeds unimpeded except in (*c*), which bears a stop codon that prevents further reading of messenger RNA. Translatable sequences of nucleotides (for example, *a, b, d, e*), free of stop codons, are called **open reading frames.**

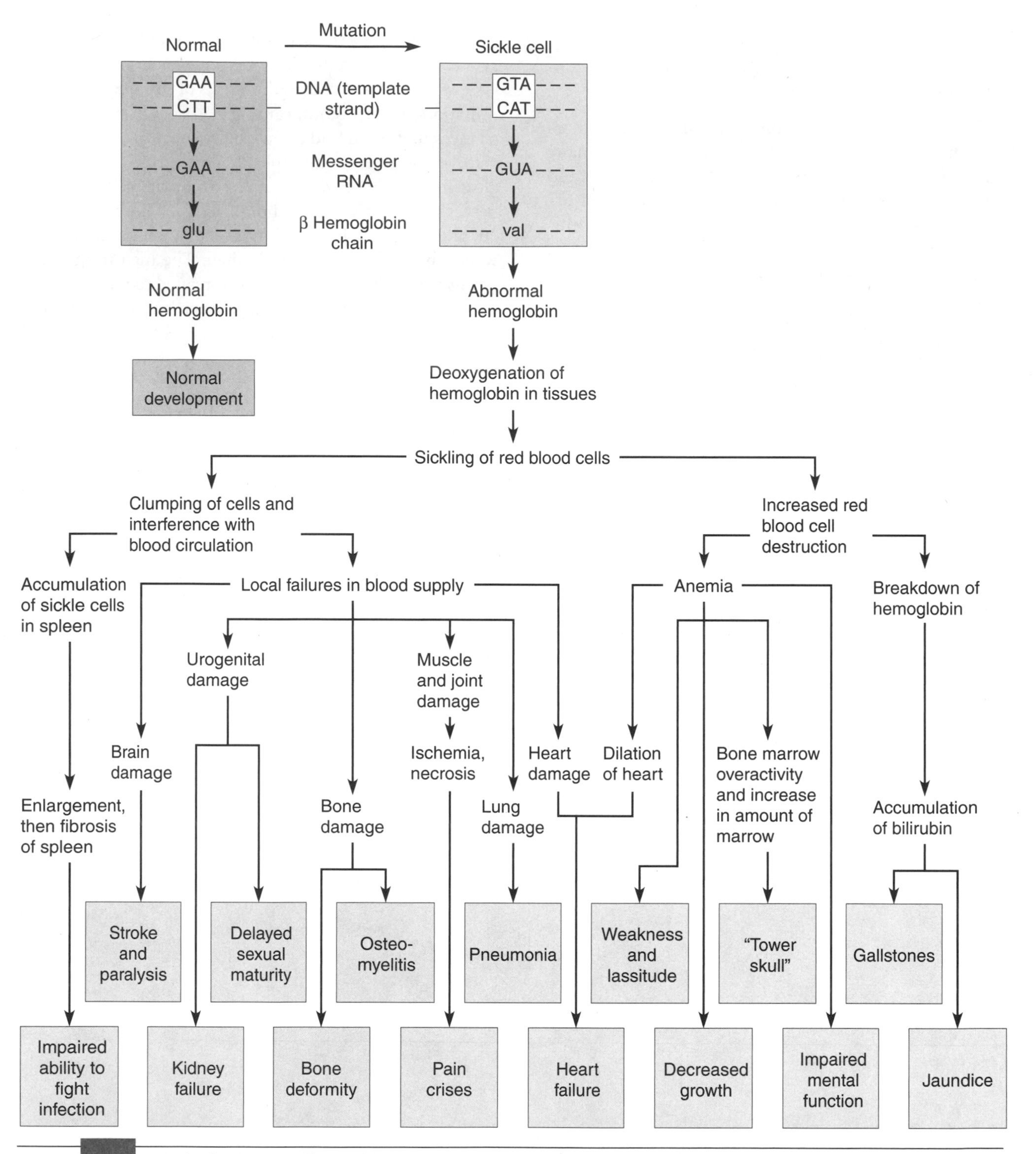

FIGURE 10-24 Varied (pleiotropic) effects of the sickle cell mutation, beginning with the transversion that changed a thymine nucleotide to an adenine nucleotide on the DNA template strand of the β-hemoglobin gene. The resultant GUA trinucleotide (triplet) coding sequence on messenger RNA then translates into a valine amino acid instead of the normal glutamic acid, producing developmental consequences that can seriously affect sickle cell homozygotes. (*From* Genetics Third Edition *by Monroe W. Strickberger. Copyright © 1985 by Monroe W. Strickberger. Reprinted by permission of Prentice Hall, Inc., Upper Saddle River, NJ.*)

Although such mutations, like the sickle cell allele, often cause lethality in homozygotes, they apparently also offer protection against the dread malarial parasites. Presumably this special advantage of some thalassemia heterozygotes explains the frequency of these genes in human populations.

Viruses and prokaryotes have, so far, provided much more information on regulatory mechanisms than have eukaryotes. The prokaryotic system in *Escherichia coli* bacteria (Fig. 10–25) governs the production of enzymes involved in lactose sugar metabolism (Fig. 10–25*a*). Since *E. coli* bacteria do not commonly encounter lactose, a **repressor protein** that occupies a specific regulatory **operator site** normally prevents the genes used in *lac* enzyme synthesis (**structural genes**) from being transcribed into mRNA. Molecular binding between the repressor and operator DNA prevents the RNA polymerase enzyme from attaching to its **promoter site,** near which transcription normally begins. As a result, transcription of *lac* enzyme genes into mRNA is prevented (Fig. 10–25*b*).

However, when bacteria encounter lactose sugar in the medium, some lactose molecules convert to a form called *allolactose,* which acts as an **inducer** that binds with the repressor. This combination releases the repressor from the operator site, thus allowing the RNA polymerase to transcribe the genes necessary to metabolize lactose (Fig. 10–25*c*).

Because of the complexity of regulatory systems, various kinds of mutations can affect the quantity and timing of gene productivity. For example, some mutations in the *I* regulator gene that produces the *lac* repressor can prevent it from binding to the operator, thereby causing *lac* enzyme synthesis to occur even in the absence of inducer (Fig. 10–25*d*). Conversely, other *I* mutations produce repressor proteins that cannot bind to inducer molecules, thus persistently preventing *lac* enzyme synthesis by keeping the repressor attached to the operator site even in the presence of inducer (Fig. 10–25*e*). In addition, various mutations of DNA at the promoter site may either increase or decrease the rate of transcription by preferentially enhancing or diminishing the attachment of RNA polymerase enzymes.

In eukaryotes, regulatory systems involve special sites on DNA sequences, called CAAT and TATA boxes, to which special proteins attach that allow transcription (Fig. 10–26). Some investigators have also demonstrated that the DNA double helix at some eukaryotic regulatory sites changes from a right-hand to a left-hand form, accompanied by a zigzag placement of phosphate groups (Fig. 10–27).

Since regulation plays an essential role in the timing and placement of all metabolic reactions, regulatory mutations can easily affect both the morphology and function of any organism, a topic that will be more thoroughly discussed in Chapter 15. For example, the *bithorax* locus in *Drosophila melanogaster,* which governs the placement of structures in various segments, may have mutations that produce an extra set of wings (Fig. 10–28). Researchers have long suggested that simple regulatory changes that affect developmental growth coordinates (Fig. 10–29) can produce changes in the shapes of various related fishes, as well as in many other species groups.

Quantitative Variation

Although large regulatory changes can explain some major differences between groups (Chapters 12, 15), the extent to which they account for most other evolutionary events is still unclear. Beginning with Darwin himself, many evolutionists suggested that rather small heritable changes provide most of the variation on which natural selection acts. In Darwin's words (*On the Origin of Species*):

> Extremely slight modifications in the structure and habits of one species would often give it an advantage over others; and still further modifications of the same kind would often still further increase the advantage. . . . Under nature, the slightest differences of structure or constitution may well turn the nicely-balanced scale in the struggle for life, and so be preserved.

Opinion as to the relative importance of small or large changes was, for a long period, a major source of contention among evolutionists. As discussed in Chapter 21, this issue became less divisive once the kinds of genetic differences could be understood in terms of their frequencies in populations and the forces affecting them. Whatever the differences between genes, it was recognized that they furnish the basic elements of evolutionary change, and their frequencies and distributions provide means for identifying causes for evolutionary change.

Nevertheless, for many measurable traits such as size and yield, researchers usually focused more narrowly on small changes or **continuous variation.** These are often seen for characters distributed in bell-shaped curves (normal distributions), such as human heights (Fig. 10–30). From an evolutionary view, it is clear that such small differences can accumulate through selection to give large quantitative differences. For example, Figure 10–31 shows that selecting for the presence or absence of white spotting in Dutch rabbits can lead to completely colored or completely white strains. The genetic causes for these changes are genes with small phenotypic effect, called **multiple factors, polygenes,** or **quantitative trait loci** (**QTLs**), which can produce the familiar normal distributions when they assort independently.

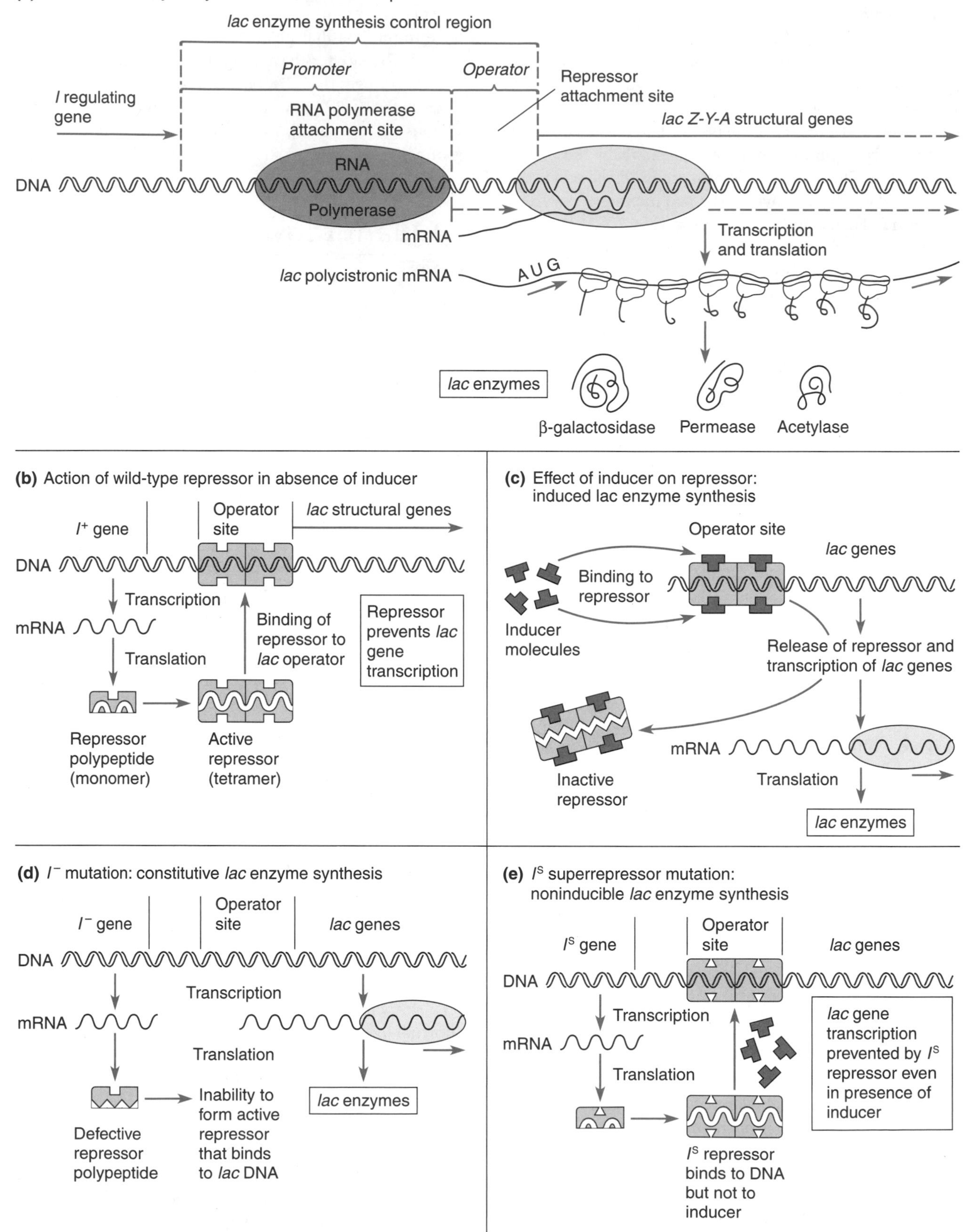
(a) Mode of lac enzyme synthesis in absence of repressor
lac enzyme synthesis control region
Promoter
Operator
Repressor attachment site
I regulating gene
RNA polymerase attachment site
lac Z-Y-A structural genes
RNA
DNA
Polymerase
mRNA
Transcription and translation
lac polycistronic mRNA
AUG
lac enzymes
β-galactosidase
Permease
Acetylase
(b) Action of wild-type repressor in absence of inducer
I^+ gene
Operator site
lac structural genes
DNA
Transcription
mRNA
Binding of repressor to lac operator
Repressor prevents lac gene transcription
Translation
Repressor polypeptide (monomer)
Active repressor (tetramer)
(c) Effect of inducer on repressor: induced lac enzyme synthesis
Operator site
lac genes
Binding to repressor
Inducer molecules
Release of repressor and transcription of lac genes
mRNA
Inactive repressor
Translation
lac enzymes
(d) I^- mutation: constitutive lac enzyme synthesis
I^- gene
Operator site
lac genes
DNA
Transcription
mRNA
Translation
Defective repressor polypeptide
Inability to form active repressor that binds to lac DNA
lac enzymes
(e) I^S superrepressor mutation: noninducible lac enzyme synthesis
I^S gene
Operator site
lac genes
DNA
Transcription
mRNA
Translation
lac gene transcription prevented by I^S repressor even in presence of inducer
I^S repressor binds to DNA but not to inducer

FIGURE 10-25 General scheme of *lac* enzyme synthesis in *E. coli* and the effects of repressor function or dysfunction on this inducible system. (*a*) The DNA region involved in controlling transcription of the *lac* structural genes *Z, Y,* and *A* consists of two major regulatory sites, the operator and promoter, each of which serves to bind specific proteins. In the absence of the repressor protein, RNA polymerase begins transcribing at the operator, which is also the site to which the repressor attaches. (The *lac* repressor itself is coded at a regulator locus, *I,* adjacent to the *lac* locus, but such proximity is not necessarily true for all systems controlled by repressor genes.) Transcription of the *lac* genes, coupled with translation, leads, as shown, to synthesis of the three *lac* enzymes. (*b*) Transcription and translation of the I^+ normal gene produces a normal repressor protein that binds to the operator site of the *lac* locus, blocking the transcription of *lac* genes by RNA polymerase. This repressed state appears in normal *E. coli* cells that are not grown on a lactose medium. (*c*) Transfer of cells to a lactose medium leads to the introduction of allolactose inducer molecules, which causes the repressor to dissociate from its DNA binding site on the operator. This allows transcription of the *lac* structural genes to proceed, followed by their translation and synthesis of *lac* enzymes. As shown diagrammatically, the repressor is a tetramer (a molecule composed of four polypeptide chains) that acts as an **allosteric protein,** signifying that it has more than one binding site: in this case, one for DNA and one for the inducer. Binding of the inducer to the repressor changes the form or steric configuration of the DNA binding site, making the repressor inactive. (*d*) When an I^- mutation produces an inactive repressor that cannot bind preferentially to the *lac* operator, the repressor does not impede transcription of *lac* genes. The synthesis of *lac* enzymes thus proceeds "constitutively" in the absence of inducers, that is, even when grown on a nonlactose medium. (*e*) A superrepressor mutation at the *I* locus causes the production of a *lac* repressor that no longer recognizes inducers but maintains its site for normal *lac* operator attachment. The result is repression of *lac* enzyme synthesis even in the presence of inducer. (*From* Genetics Third Edition *by Monroe W. Strickberger. Copyright © 1985 by Monroe W. Strickberger. Reprinted by permission of Prentice Hall, Inc., Upper Saddle River, NJ.*)

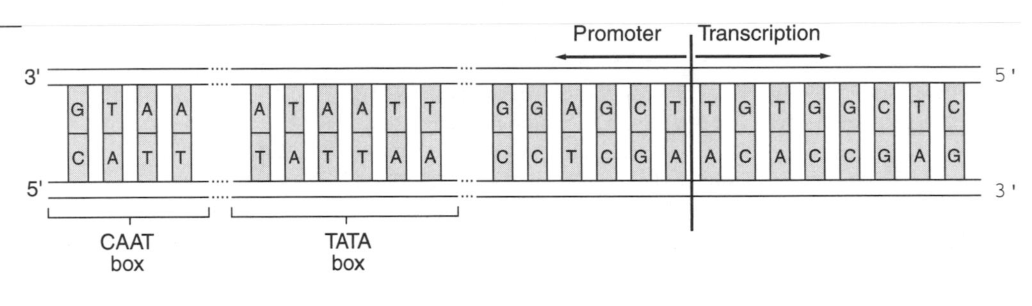

FIGURE 10-26 A eukaryotic nucleotide sequence regulating the gene that produces the thymidine kinase enzyme. The promoter region of this gene (used for attachment of RNA polymerase and other transcription-assisting proteins) contains two short sequences or boxes, called *CAAT* and *TATA,* which have also been found in promoter regions of other genes. Altering these sequences reduces the level of gene expression (Felsenfeld). In addition to promoters, other effects on transcription occur through "enhancers," sites that are often at some distance from promoters yet also affect RNA polymerase attachment. For transcription effects caused by extracellular signals (signal transduction), see Figure 15-3.

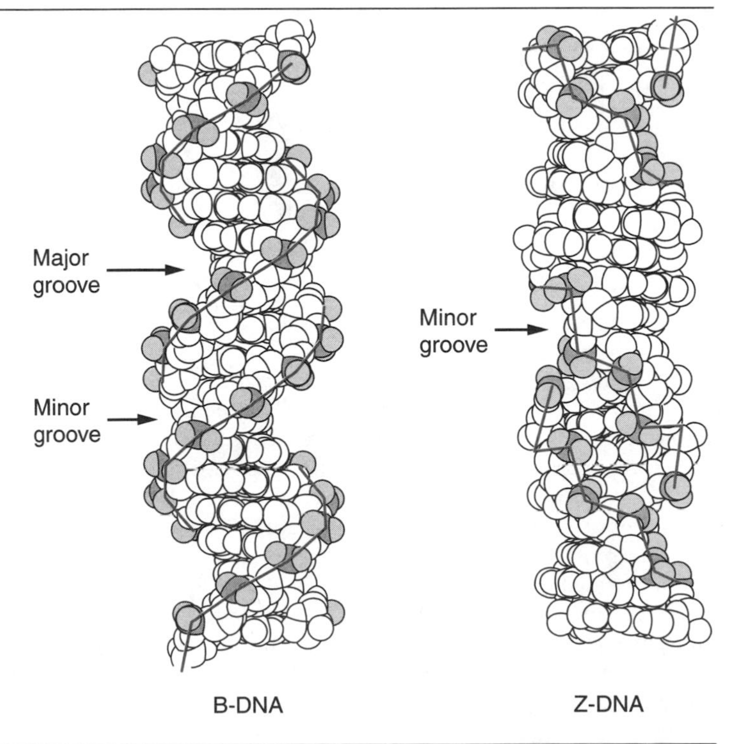

FIGURE 10-27 Space-filling models of B-DNA and Z-DNA double-helix molecules. The *lines* used to connect the phosphate groups (*color-shaded*) in each chain show the zigzag placement of phosphates in Z-DNA in contrast to the smoother curve of their relationship in B-DNA. The major and minor grooves in B-DNA differ in depth but do not extend to the central axis of the molecule, whereas the indicated Z-DNA groove penetrates the axis of the double helix. (*Reproduced with permission from the* Annual Review of Biochemistry, *Volume 53, © 1984 by Annual Reviews Inc.*)

FIGURE 10–28 A four-winged *Drosophila,* caused by mutations at the *bithorax* locus. Normally, as in all dipteran insects, *Drosophila* has only a single pair of wings, which arise from the second of the three thoracic segments. (The second dipteran wing pair evolved into balancing organs, called *halteres.*) As shown here, certain *bithorax* mutations cause the third thoracic segment to produce its own pair of wings, a condition reverting to the ancestral four-winged fly. These mutations, and others of this kind occur in genes that regulate development, a topic discussed in Chapter 15. (*Figure from E. B. Lewis. Four-winged* bithorax *mutant* Drosophila melanogaster. *Reprinted by permission.*)

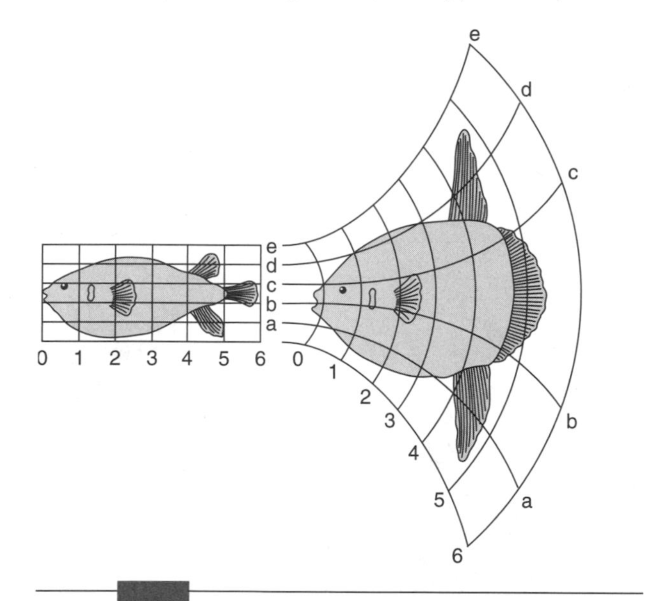

FIGURE 10–29 D'Arcy Thompson's demonstration how completely different organismic shapes can be generated by simple developmental changes in geometric coordinates. If the vertical coordinates of the puffer fish, *Diodon* (*left*), change into concentric circles, and its horizontal coordinates into hyperbolas, the resultant animal is shaped like the sunfish, *Orthagoriscus* (*right*). Such changes may be caused by a new or mutated developmental "morphogen" (see Chapter 15) that increases cell production as it moves along a gradient from anterior to posterior, and dorsally and ventrally from the midline. (*Adapted from Thompson.*)

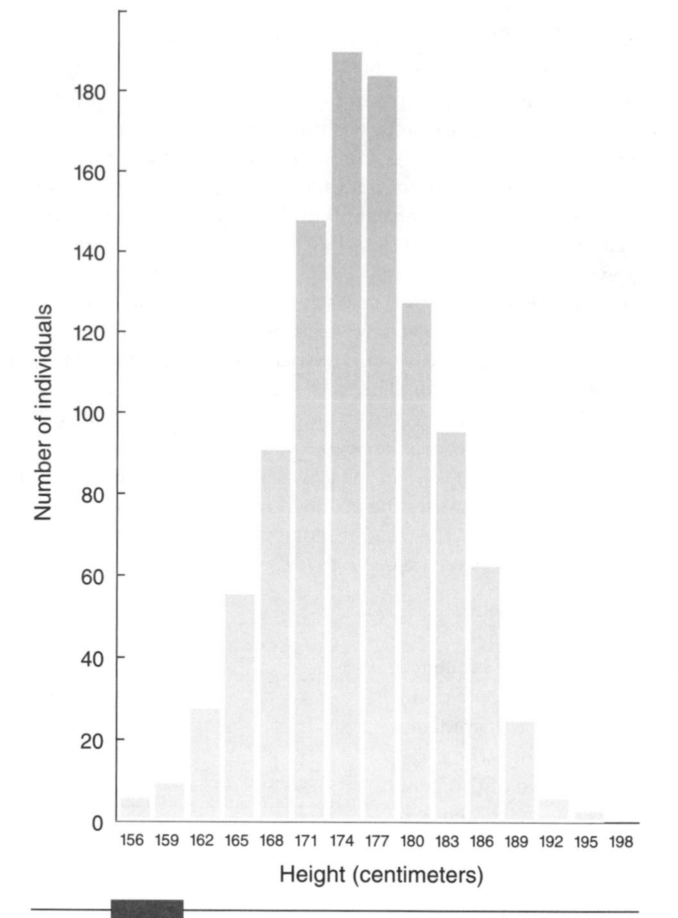

FIGURE 10–30 Distribution of the heights of 1,000 Harvard students aged 18 to 25. (*Adapted from Castle.*)

A mating between heterozygotes for three pairs of genes, each with two alleles—one colored, one white—produces the phenotypes that Figure 10–32 shows, ranging from all colored to all white. Experiments and analyses of this kind demonstrate that we can explain selection for quantitative characters on the basis of the segregation and assortment of simple mendelian genes whose individual small effects may add up to large phenotypic differences.

Detecting the numbers and chromosomal positions of such quantitative genes has become a reasonable exercise because of the availability of distinctive DNA sequences to which such genes may be linked. Differences among DNA sequences of the kinds discussed later (simple tandem repeat polymorphisms, p. 228 and restriction fragment length polymorphisms, p. 275), serve as molecularly identifiable **gene markers.** These markers, located at specific linkage positions, can be correlated with specific measurements of a trait, and thus indicate numbers and approximate positions of quantitative loci.

Quantitative inheritance, of course, does not exclude factors with large effect: highly analyzed characters such

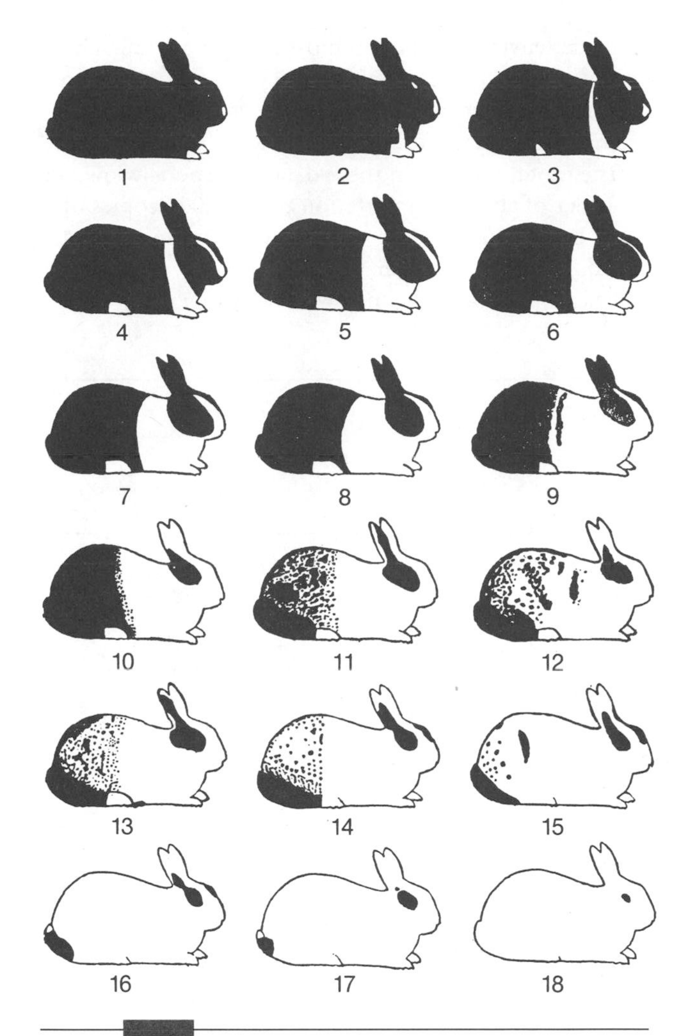

FIGURE 10-31 White spotting in Dutch rabbits, ranging from almost no spotting (grade 1) to complete spotting (grade 18). Selection experiments on animals with intermediate spotting (for example, grades 7 to 12) showed that spotting can be increased or decreased. *(Reprinted by permission of the publishers from* Genetics and Eugenics *by W. E. Castle, Cambridge, Mass.: Harvard University Press, Copyright © 1916, 1920, 1924 by Harvard University Press.)*

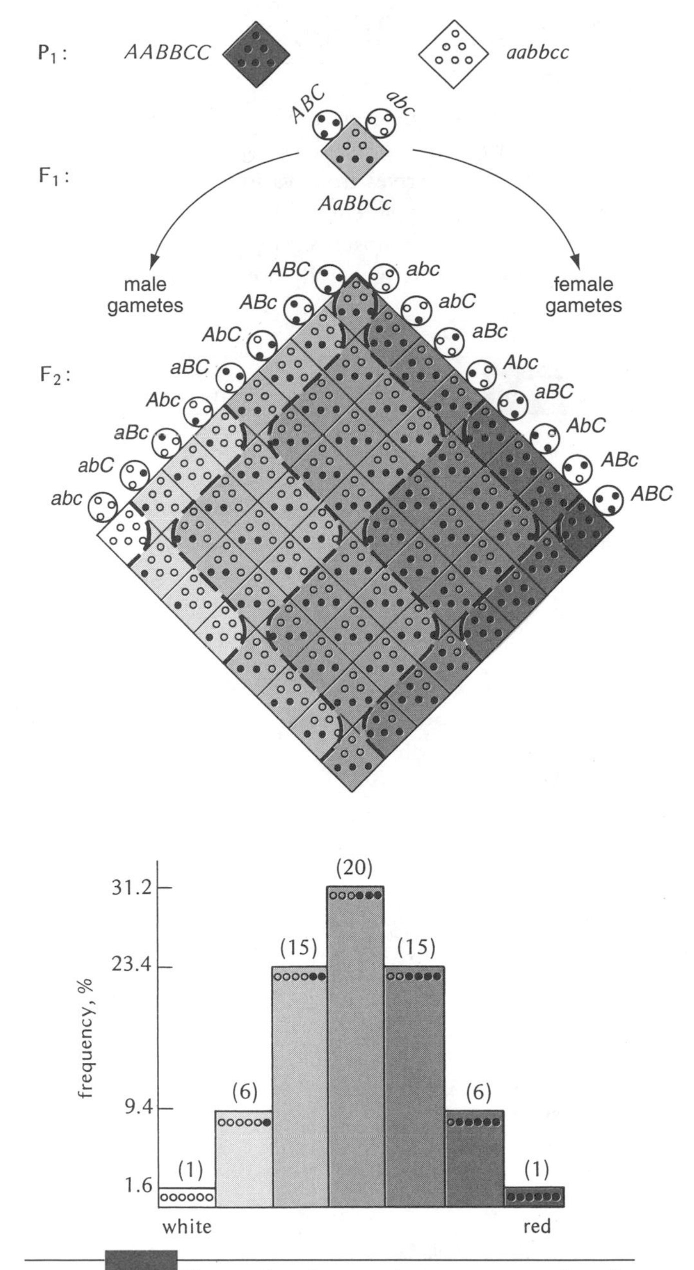

FIGURE 10-32 The results of crosses between two strains of wheat differing in three gene pairs that determine grain color. Each gene pair assorts independently of the others, and the alleles at each gene pair lack dominance, so that *Aa,* for example, has a color intermediate between *AA* and *aa.* The F_1, carrying three color (*ABC*) and three noncolor alleles (*abc*) is therefore intermediate in color to the parental stocks, and the F_2 produces a range of colors in the frequencies shown in the histogram. *(From* Genetics Third Edition *by Monroe W. Strickberger. Copyright © 1985 by Monroe W. Strickberger. Reprinted by permission of Prentice Hall, Inc., Upper Saddle River, NJ.)*

as *Drosophila* bristle number are the result of genes with varying influence (Mackay). In addition, environmental effects can play a role, such as external conditions (available food, light, moisture, and so forth) affecting size and yield. Sorting out these various influences and weighing the importance of genes, environment, effects of different genes in a particular environment, and of particular genes in different environments, involves sophisticated statistical analysis, worthy of separately dedicated texts (see Falconer and Mackay; Lynch and Walsh; Roff). In general, the emphasis of quantitative genetics has been mostly on evaluating resemblances among relatives and measuring shifts in such resemblances. One common measure emerging from such studies is **heritability,** signifying the extent to which genetic differences affect a character; that is, the degree to which a character can be modified by selection. The various kinds of selection discussed in Chapter 22, depend on heritable variation.

Mutation Rates

Considering everything discussed so far, we see that the opportunity for all kinds of mutations derives from various sources and may have various phenotypic effects. For newly arisen mutations, these effects will most likely be harmful because prevailing genotypes are generally well adapted for their particular environments, and most changes are unlikely to improve them further. Detecting new mutations in most organisms usually goes along with observing newly inherited harmful effects, some of which Table 10–4 gives.

The mutation rates in these data are generally low, on the order of about one mutation per 100,000 copies of a gene. In organisms such as humans, carrying an estimated 100,000 genes per haploid genome, this means that each sperm and egg may well carry one newly arisen

TABLE 10–4 Spontaneous mutation rates at specific loci for various organisms

Organism	Trait	Mutation per 100,000 Gametes[a]
DNA virus[b] : T4 bacteriophage	Rapid lysis ($r^+ \rightarrow r$)	7.0
	New host range ($h^+ \rightarrow h$)	0.001
Bacteria: *E. coli*	Streptomycin resistance	0.00004
	Phage T1 resistance	0.003
	Leucine independence	0.00007
	Arginine independence	0.0004
	Arabinose dependence	0.2
Salmonella typhimurium	Threonine resistance	0.41
	Histidine dependence	0.2
	Tryptophan independence	0.005
Fungus: *Neurospora crassa*	Adenine independence	0.0008–0.029
	Inositol independence	0.001-0.010
Insect: *D. melanogaster*	y^+ to *yellow*	12.0
	bw^+ to *brown*	3.0
	e^+ to *ebony*	2.0
	ey^+ to *eyeless*	6.0
Plant: corn (*Zea mays*)	*Sh* to *shrunken*	0.12
	C to *colorless*	0.23
	Su to *sugary*	0.24
	Pr to *purple*	1.10
	I to *i*	10.60
Rodent: *Mus musculus*	a^+ to *nonagouti*	2.97
	b^+ to *brown*	0.39
	c^+ to *albino*	1.02
	d^+ to *dilute*	1.25
Primate: *Homo sapiens*	Achondroplasia	0.6–1.3
	Aniridia	0.3–0.5
	Dystrophia myotonica	0.8–1.1
	Epiloia	0.4–1.0
	Huntington chorea	0.5
	Intestinal polyposis	1.3
	Neurofibromatosis	5.0–10.0
	Retinoblastoma	0.5–1.2

[a]Analysts base mutation rate estimates in viruses, bacteria, and fungi on particle or cell counts rather than on gametes.

[b]Mutation rates for RNA viruses are generally much higher than for DNA viruses, reaching in some cases one mutation per genome per replication (Drake 1993). Reasons for these high RNA viral mutation rates probably include the absence of proofreading functions in RNA polymerase enzymes and the lack of repair mechanisms such as those used to correct DNA base pair mismatches.

Source: Adapted from Strickberger.

mutation, or an average of two such mutations in a diploid fertilized zygote. However, if we extend our search for new mutations to larger numbers of genes and many possible nucleotide changes (p. 216), human mutation rates are probably even higher.

In a major study of human mutation rates, Eyre-Walker and Keightley compared amino acid compositions of 46 different proteins between humans and chimpanzees to discover changes that occurred during the time these lineages diverged. Based on a genome size of 60,000 genes, they calculate an overall rate of 4.2 amino acid-changing "missense" mutations each generation, of which at least 1.6 (38 percent) were eliminated by natural selection because they were deleterious. Since the human genome is probably larger, and mutations also occur in non-protein genes, 1.6 deleterious mutations per generation is a minimal estimate. Although such numbers may seem high, modern humans probably circumvent many deleterious effects by medical treatment and highly improved environment. Cumulative amounts of such mutations, however, may still be harmful, and researchers express concern that their impact may eventually significantly affect our health and lifestyles (Crow 1997; see also Chapter 25).

However they occur, mutation rates are not necessarily constant. Among the causes that can modify mutation rates are genes for polymerase enzymes that replicate DNA. Some alleles of these genes act as **mutator genes** that can increase mutation rates manyfold, whereas other alleles act as **antimutators** to decrease mutation rates. Mutation rates, like other essential traits, seem mostly selected for optimum values, balancing on the delicate adaptive line that stands between not undoing prevailing adaptive features yet allowing new ones to occur.

External causes may considerably affect mutation rates, including, surprisingly, infectious elements that can be transmitted from other individuals. For example, some viruses, such as herpes simplex, rubella (German measles), and chicken pox, can cause breaks and deletions in chromosomes because they release nuclease enzymes that attack host DNA.

Important factors that act to correct nuclear damage from these and other influences, including ultraviolet radiation, are a variety of **DNA repair mechanisms.** These enzyme systems can excise DNA molecular distortions and replace mutant nucleotide sequences by inserting normal sequences from specially defined DNA strands such as those introduced during recombination. Although mostly known from *E. coli* bacteria, DNA repair mechanisms appear to exist in practically all cells, indicating that all forms of life have faced common problems of DNA damage.[9]

[9]A remarkable way in which some organisms accumulate mutations without experiencing their immediate effects is to bind their gene products with heat shock proteins that normally chaperone and protect other proteins (p. 181, footnote 16). When such protective heat shock proteins are disabled, the assemblage of masked mutational products can cause significant developmental changes, perhaps leading to new evolutionary opportunities (Rutherford and Lindquist).

According to some views, improved DNA replication and repair mechanisms, especially those accompanied by diploidy (p. 302) and sex (pp. 563–564), allowed transition from small prokaryotic genomes to much larger eukaryotic genomes. No matter how it arose, its value still persists, since the inability to repair DNA damage can be lethal: in humans, such deficiencies appear in genetic diseases such as xeroderma pigmentosum, which often causes death because it increases the incidence of cancer.

On the other hand, defective DNA repair systems can be as important as polymerase enzymes in providing a wide array of mutations, some of which can let an organism face new environmental challenges. Among such examples are strains of *E. coli* and *Salmonella enterica,* in which faulty DNA repair systems caused increased mutation rates (LeClerc et al.), allowing some mutants to circumvent antibiotic challenges. These increased bacterial infectivity and pathogenicity, causing several serious epidemics of food-related illnesses.

Transposons, Repeated Sequences, and Selfish DNA

Other sources of mutational change in both prokaryotes and eukaryotes are **transposons:** nucleotide sequences that can promote their own transposition among different genetic loci. The transposon produces special transposase enzymes that let copies of the transposon insert into various target sites. For example, the IS1 transposon illustrated in Figure 10–33 makes staggered cuts at each side of a nine-nucleotide base pair sequence, and a copy of IS1 inserts within the gap these cuts produce. Depending on where transposons insert, mutations of all kinds may arise, marked by target site repeats, in which similar sequences of nucleotide bases appear at each end of the insertion but in inverted order.

Researchers have therefore used inverted repeats to detect the presence of transposable elements in various species, and such repeats indicate that a transposon can pick up DNA sequences and transfer them to other DNA locations by recombination (see Mizuuchi). Antibiotic resistance genes, for example, can be transmitted between bacterial strains by small, circular DNA particles called **plasmids** that have received transposon insertions. Such instances indicate that some hereditary traits that transposons carry may have passed "horizontally" or "laterally" between individuals of the same generation, rather than through normal "vertical" (gametic) transmission between generations. Once established in their new hosts, **horizontally transmitted genes** can undergo further mutation, leading to entirely new characteristics. Thus, as an example of rapid adaptive evolution, various pathogenic

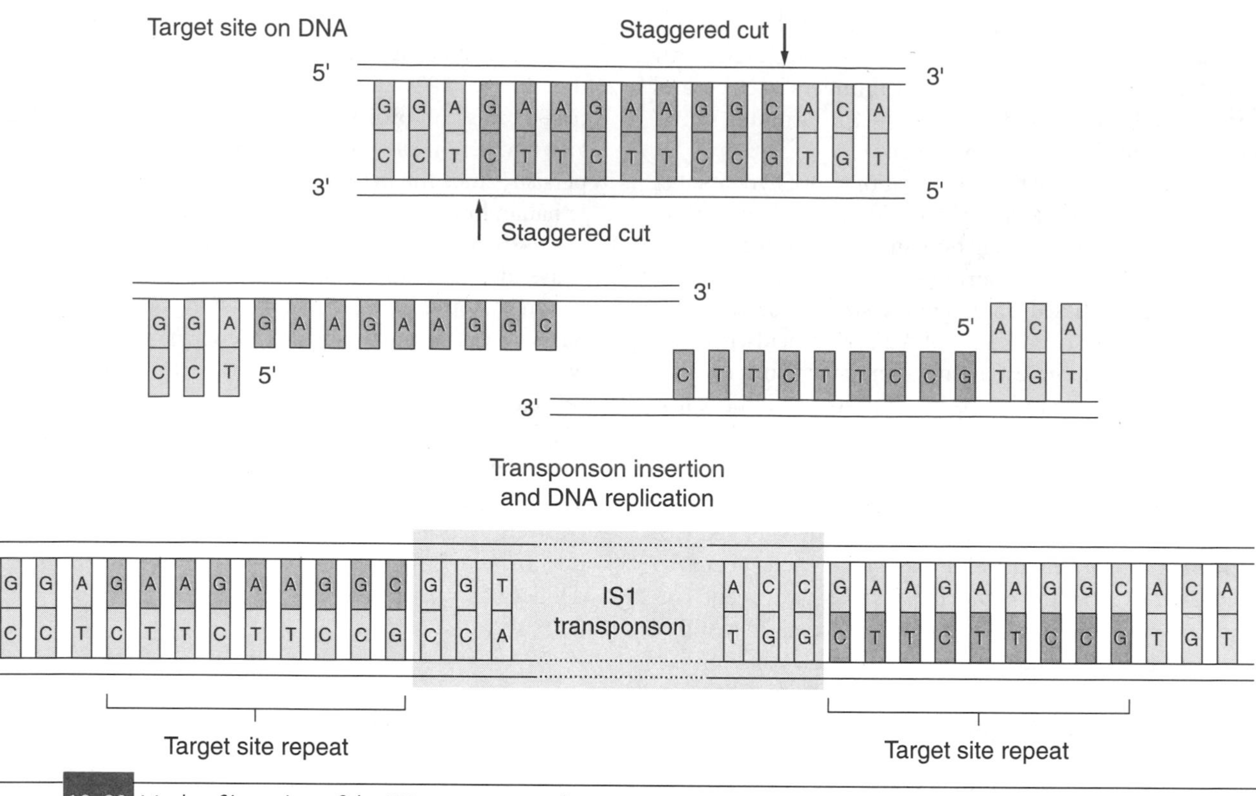

FIGURE 10-33 Mode of insertion of the IS1 transposon. The transposon recognizes the nine-base pair DNA sequence at the top of the diagram as a target site and cleaves it at the indicated *arrows.* IS1 then inserts into the resulting gap, and DNA sequences are synthesized complementary to the former single-strand sections of the target site. This process produces identical but inverted nine-base pair repeats at each end of the transposon. (*Adapted from Strickberger.*)

bacteria have widened their spectrum of resistance to many different antibiotic drugs in a decade or less (Davies).

Over longer evolutionary periods, a large number of lateral gene transfers occurred in some bacterial lines. According to Lawrence and Ochman, *E. coli* must have experienced more than 200 such events from the time this prokaryote diverged from the *Salmonella* lineage 100 million years ago.

Among eukaryotes, such horizontal transmissions have also been demonstrated in a variety of organisms. The *P* transposon, originally a transposable element common to *Drosophila willistoni,* first spread to *Drosophila melanogaster* about 50 years ago and now appears in all wild populations of that species. The likelihood that the original *willistoni–melanogaster* event involved horizontal transfer is supported by the finding that the two species differ significantly in their gene nucleotide sequences indicating a separation of about 20 million years, yet their *P* transposons differ by only a single nucleotide (Daniels et al.). Although discriminating between horizontal and vertical transfer can be difficult (Cummings), other transposon elements (such as *Hobo*) also show evidence of horizontal transmission, and the phenomenon may be more frequent than most geneticists anticipated. As Kidwell and others point out, one important advantage transposons gain by horizontal transmission is to circumvent the barriers of reproductive isolation among species, and escape inevitable extinction in vertical lineages when a species dies out.

Surprisingly, although we might expect a large increase in transposon numbers within a genome because of their simple transfer mechanisms, this is not always so. For example, the *Drosophila* genome carries only about 30–50 copies of the *P* element and a similar number of *copia* transposons. Regulatory agents within these transposons apparently control their number and thereby limit their mutagenic effects—a feature that may have been selected to ensure survival of their hosts, and therefore their own survival.

Restricted numbers, however, do not extend to all transposons. In primates, for example, a 300 base pair sequence with transposon-like features called *Alu* is represented by perhaps more than one million copies per human diploid cell. Smaller repetitive sequences of the type discussed in Chapter 12 are also widely prevalent in various eukaryotes. For example, only 3 percent of human DNA codes for proteins, whereas the remainder consists of noncoding sequences comprising many kinds of repetitive elements (Table 12–4, p. 270). What explains the widespread distribution and persistence of what appears to be extraneous DNA?

According to some writers (Orgel and Crick, Doolittle and Sapienza) many transposable elements and other forms of **repeated sequences** contribute little, if any, function to their host cells. Since the DNA replication process cannot discriminate between functional and nonfunctional sequences, it replicates any introduced DNA sequence. Transposon DNA and repeated sequences may therefore perpetuate parasitically as either "junk" or "selfish" DNA. Other explanations propose that some such sequences may function as essential elements in regulating gene activity, or help maintain the structure and integrity of the chromosome, or serve as origins of DNA replication, or act as mutator genes that occasionally provide new adaptive mutations. No agreement exists on how to weigh the selfishness or unselfishness of such sequences, and possibly some repeated sequences fulfill different roles.[10]

The Randomness of Mutation

Until the 1950s many bacteriologists felt that bacteria had a unique "plastic heredity" in which appropriate mutations arise as an immediate response to the needs of the environment. This concept seemed supported by observations in which bacteria exposed to some virus or antibiotic would quickly develop a resistant form. The explanation offered was that mutations do not originate on a random basis before exposure to some selective agent (**preadaptive** mutations), but rather that appropriate mutations are stimulated to arise only after bacteria have encountered a selective agent (**postadaptive** mutations). As with Lamarck, the postadaptive mutation concept had to rely on some unknown, perhaps mystical, agency that allowed the environment to directly cause the appearance of new adaptive hereditary factors instead of employing Darwin's natural selection to choose among preadaptive hereditary factors already present.

J. and E. Lederberg performed an impressive test that evaluated these pre- and postadaptive models, using a novel technique called *replica-plating.* They transferred samples from a "lawn" of bacteria growing on a petri dish (master plate) to other petri dishes (replica plates), which contained selective media such as viruses (bacteriophages) or antibiotics (streptomycin). They made these transfers in a manner enabling bacteria derived from specific clones on the master plate to be localized to the same positions on the replica plates (Fig. 10–34). The finding that resistant bacteria occupied identical positions on various replica plates indicated that they arose from the same clone on the master plate, a clone that must have been present before exposure to the selective medium on the replica plates. In other words, these mutants shared a common preadaptive origin on the master plate and did not arise postadaptively because of a Lamarckian response to a stimulus by the selective medium.[11]

Genetic Polymorphism: The Widespread Nature of Variability

New mutations that have an immediate beneficial effect on the organism seem generally to be quite rare, although some mutations are either neutral in their effect or harmful only when they occur in relatively rare homozygotes. Such neutral and deleterious but recessive mutations can therefore often accumulate in a population without any

[10]Although "selfishness" indicates that a unit of life is primarily concerned with its own replication, it is debatable whether, as argued by Dawkins, such selfishness is the exclusive property of DNA simply because DNA replicates itself so well. According to this argument, all biological entities such as cells and organisms are merely "vehicles" for DNA "replicators" to make more of themselves—"A chicken is DNA's way of making more DNA." But one can also invert these roles by claiming that "DNA is the chicken's way of making more chickens." Moreover, attempts to reduce explanation of life to such relationships could lead easily to a meaningless chain of arguments: organisms are the means enabling cells to replicate; cells enable chromosomes to replicate; chromosomes enable gene replication; genes enable codon replication; codons enable nucleotide replication; nucleotides enable the replication of nitrogenous bases, sugars, phosphates, and so on; until statements can be developed asserting that all organisms are the means enabling the perpetuation of various atomic or even subatomic particles. Selfishness may have some meaning in terms of one component of life competing with similar components at the same level of organization (selfish organisms, selfish cells, selfish DNA, and so on), but little seems gained by trying to establish which level is most selfish. Nevertheless, because competition is an abiding feature of evolutionary selection and survival, it is hard to ignore the claim that biological entities always act selfishly, although we should recognize that even selfishness can produce "unselfishness" such as altruist behavior in social organisms (pp. 386 and 566).

[11]Both Cairns and coworkers and Hall performed experiments suggesting that some selective environments seem to induce adaptive bacterial mutations at a frequency higher than would be expected were such mutations strictly random. Because they seemed to support a Lamarckian interpretation that adaptations arise in direct response to the environment, these experiments generated considerable controversy (see, for example, Symonds, and Lenski et al.). However, more recent work indicates that such experiments produce deleterious and neutral mutations as well as beneficial ones (Rosenberg): evolutionary direction is not caused by mystical postadaptive mutational trends unrelated to selection, such as **orthogenesis** (pp. 45, 242, and 516). Among models offered to explain such "adaptive mutation" events is **hypermutation**—that certain environmentally stressful challenges (for example, starvation, antibiotic exposure) increase the rate of mutation in selected genes, letting some favorable mutations survive and increase. How this increased mutability occurs is still unclear, but we do know that increased mutation rates arise from mutator genes and those affecting DNA repair systems (p. 225). Their effects and other causes for "hot spots" are discussed by Rosenberg and coworkers.

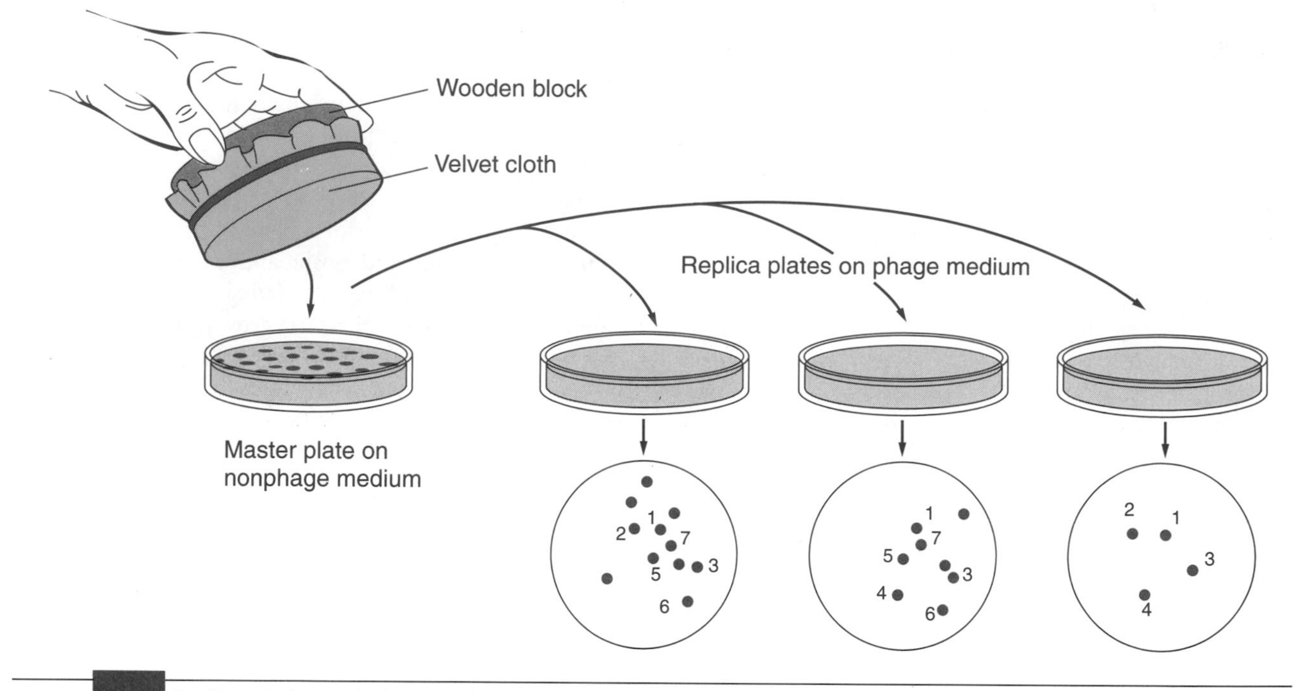

FIGURE 10-34 Replica-plating technique used to test the location of clones of *E. coli* resistant to T1 bacteriophage (virus). The test begins with a master plate that shows diffuse bacterial growth of phage-sensitive *E. coli* on a nonphage medium. Replicas are then made by pressing a velvet-covered wooden block against the master plate, then pressing this, oriented in the same direction, to the surface of petri dishes containing culture medium mixed with phage T1. One master plate has sufficient bacteria to start colonies on a number of replica plates. The replica plates show occurrence of resistant colonies (for example, 1, 3, 4) in identical locations, indicating that resistance to phage T1 must have been present at each of these positions in the master plate. (*Based on Strickberger, adapted from Lederberg and Lederberg.*)

immediately serious unfavorable effects, thereby furnishing a reservoir of genetic variability.

Genetic variability, expressed in a population by the existence of two or more genetically distinct forms—**polymorphism**—may also include the maintenance of different kinds of chromosomal anomalies such as inversions, translocations, and extra chromosomes. In *Drosophila pseudoobscura,* for example, populations in different localities in the western United States are polymorphic for a wide variety of third chromosome gene arrangements (Fig. 10–35*a*), and the frequencies of such arrangements may change seasonally (Fig. 10–35*b*). This indicates not only that chromosomal polymorphism is generally adaptive in this species, but also that certain polymorphic variations are preferentially adaptive in helping populations adjust to their specific environment at specific times.

Variation in chromosome number between very similar species can also associate with adaptive features. According to Nevo and coworkers, stressful ecological conditions such as periodic aridity and other unpredictable hardships appear to correlate with chromosome numbers in Israeli and Turkish populations of the mole rat *Spalax.* They suggest that the chromosomal increase caused by dissociation or fission (Fig. 10–16*b, c*) leads to increased genetic diversity by increasing the numbers of different possible chromosome combinations, and thus allows species in such localities to specialize for a rise in opportune ecological variation.

On the DNA level, diversity can be identified by molecular techniques (Chapter 12) which determine the presence of particular sequences or the number of repeating nucleotide strings such as CACA, CACACA, . . . CA_n. These latter repeats, known as **short tandem repeat polymorphism (STRPs),** are distributed throughout the genome. Differences among them in number of repeats, serve as gene markers that can identify particular chromosomal locations.

On the gene level, researchers can discern the magnitude of polymorphism by techniques that make allelic differences visible. Among such studies are electrophoretic methods that measure the mobility of a protein in an electric field, distinguishing even slight variations in conformation and electric charge (Fig. 10–36). Since each different electrophoretic form of a protein usually indicates a different amino acid sequence (and therefore a different nucleotide sequence in the gene that produced it), they are each considered to signify an allelic difference, or **allozyme.** Applying this technique to natural populations, beginning in 1966 (Harris, Lewontin and Hubby), led to the surprising result that populations maintain considerably more genetic variability than

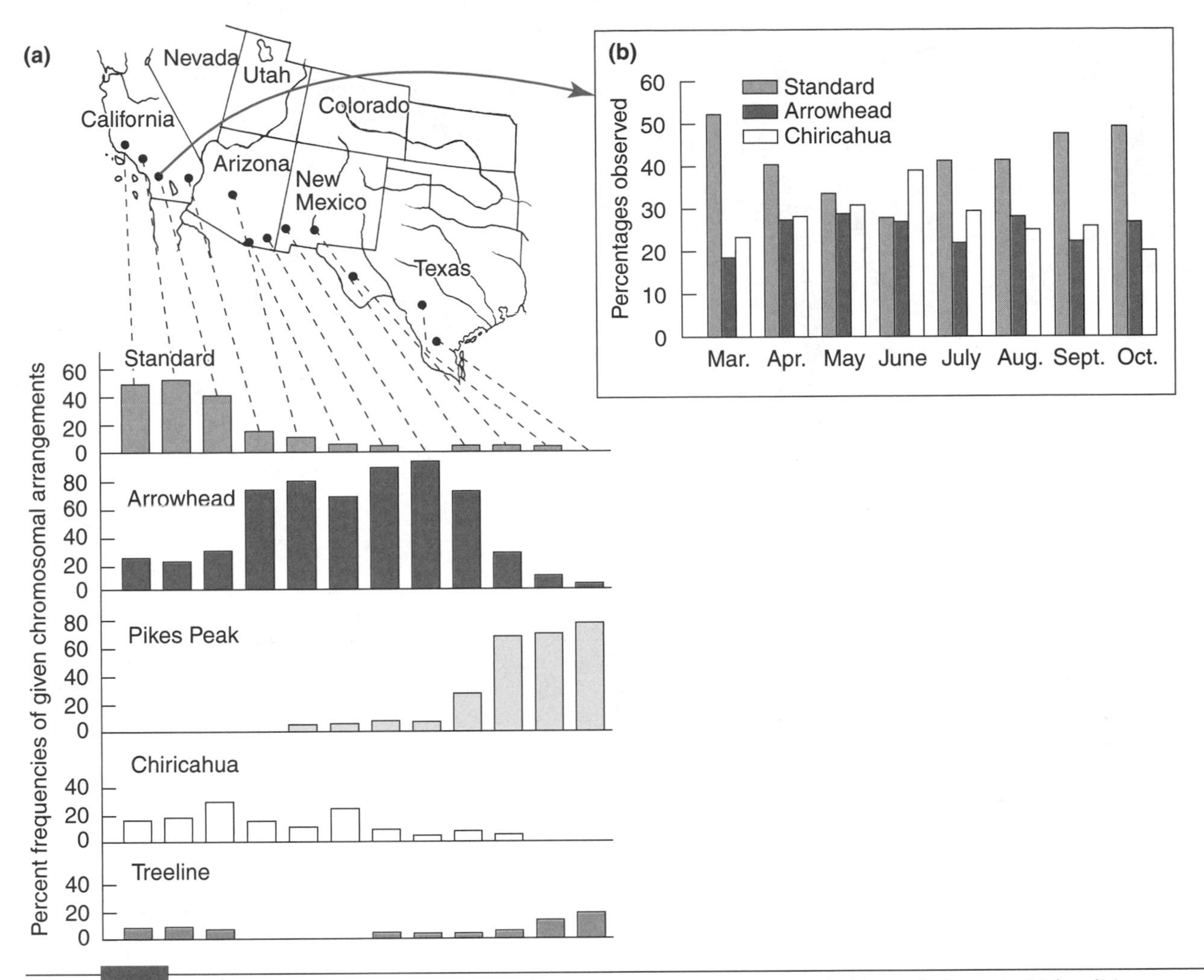

FIGURE 10-35 (*a*) Frequencies of five different third chromosome gene arrangements in *D. pseudoobscura* in 12 localities on an east-west transect along the United States-Mexican border. Each kind of arrangement consists of inversions observable as a unique system of chromosomal banding patterns. (*Adapted from Dobzhansky 1944.*) (*b*) Percentages of different third chromosomal arrangements in *D. pseudoobscura* found at different months during the year in one of these localities, Mount San Jacinto, California. It is generally believed that each of these chromosome arrangements maintains a specific gene combination that enables adaptive interactions between component alleles (epistasis). Because inversions inhibit recombination (p. 209), such allelic combinations can be maintained without disruption (see also p. 564). (*Adapted from Dobzhansky 1947.*)

previously estimated. As Table 10–5 shows, a large number of species display allozymic differences at an average of about one quarter of all loci tested, indicating that the chances for an individual to be heterozygous for any particular tested locus is more than 7 percent.

Since it is estimated that only perhaps one-third of amino acid changes in proteins are detectable by electrophoretic techniques, these observed values should probably be tripled; about two-thirds or three-quarters of all loci in many species are polymorphic, and the average individual may be heterozygous for as much as one-quarter to one-third of all its loci. This means that in *Drosophila* species with an estimated 10,000 gene loci, an individual can be a heterozygote for about 2,500 genes or more; and in humans, with an estimated 100,000 gene loci, individuals may be heterozygous for as many as 25,000 genes! Britten, for example, estimates that of the 3 billion nucleotides in the haploid human genome, one human differs from another at an average of about 5 million sites. Clearly, such past accumulations provide a much greater amount of genetic variability than do the relatively few new mutations that arise each generation.

Such genetic polymorphisms let many populations confront new environmental challenges with a large variety of mutations, some of which may be preadaptively advantageous. For example, the exposure of insect populations to DDT pesticides has caused a widespread increase in the frequency of various DDT-resistant mechanisms such as

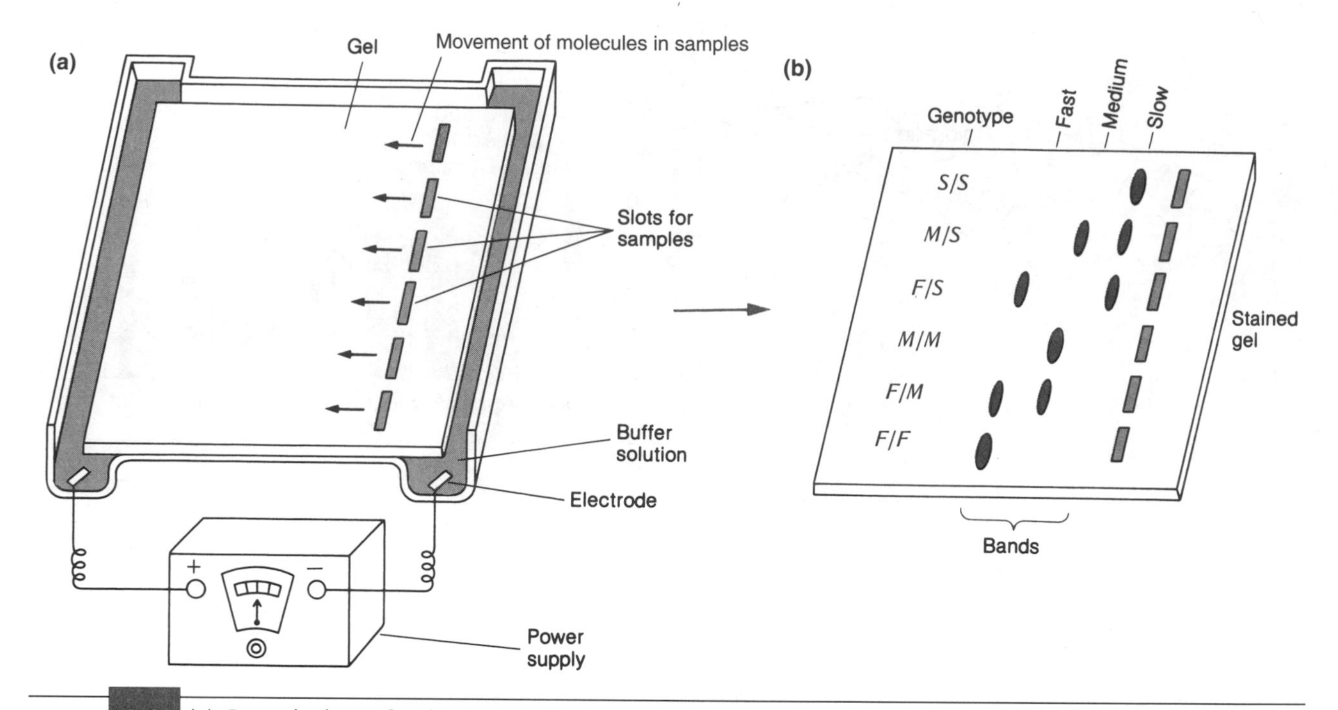

FIGURE 10–36 (*a*) General scheme for electrophoresis, using a gel (starch or polyacrylamide) in which researchers place samples along a row and subject them to an electric current carried through an aqueous buffered solution. Depending on their size and electrical charge, molecules in the samples separate, going toward either the negative or positive pole. The position they occupy on the electrical gradient can be identified as bands when the gel is treated with agents that can assay molecular or enzymatic activity or is exposed to ultraviolet light. (*b*) Treating the gel with dyes sensitive to a specific enzyme activity shows that the enzyme on this gel has three different forms, each migrating at a distinct rate (slow, medium, and fast) toward the positive pole. Since each of these three enzymatic forms is produced by a single allele of the gene for this protein, *S, M, F,* it is often called an *allozyme.* Thus an individual may possess one of six genotypes, either homozygous (*S/S, M.M, F/F*) or heterozygous (*M/S, F/S, F/M*), each different genotype producing an identifiable electrophoretic pattern of allozymes. (Some terminologies use the more general term *isozyme* for any distinct electrophoretic form of a protein, whether its uniqueness arises from genetic or nongenetic causes.) (*Adapted from Strickberger.*)

- An increase in lipid content that lets the fat-soluble DDT separate from other parts of the organism
- The presence of enzymes that break down DDT into relatively less toxic products
- A reduced toxic response of the nervous system to DDT
- Changes in the permeability of the insect cuticle to DDT absorption
- A behavioral response that reduces contact with DDT

It is therefore not surprising that insecticide resistance is associated with numerous genes. For example, the genes responsible for DDT resistance in *Drosophila* are located on all major chromosomes, each gene acting as a polygene with a small incremental effect (Fig. 10–37). Similar anti-pesticide selective events are common, and Roush and McKenzie list more than 400 such cases.

On the whole, it seems clear that most populations do not await the lucky arrival of new favorable mutations to provide for their evolutionary needs. Instead, populations tend to use their reservoir of genetic variability, consisting of many historically accumulated mutations. We can consider evolutionary potential and genetic variability as two sides of the same evolutionary coin.

TABLE 10-5 Estimates of genetic variability found in natural populations, based on electrophoretic studies

Organisms Tested	Number of Species Examined	Average Number of Loci (proteins) Studied per Species	Proportion of Polymorphic Loci	Heterozygosity per Locus
Plants	15	18	.259	.071
Invertebrates				
Various groups except insects	27	25	.399	.100
Various insects except Drosophilidae	23	18	.329	.074
Drosophila species	43	22	.431	.140
Vertebrates				
Fish	51	22	.152	.051
Amphibia	13	22	.269	.079
Reptiles	17	23	.219	.047
Birds	7	21	.150	.047
Mammals (except primates)	43	26	.148	.036
Primates				
Humans	1	71	.28	.067
Chimpanzees	1	43	.05	—
Macaque monkeys	1	29	.10	.014
Totals and Averages	242	23	.263	.074

Source: From Strickberger, derived from data collected by Nevo.

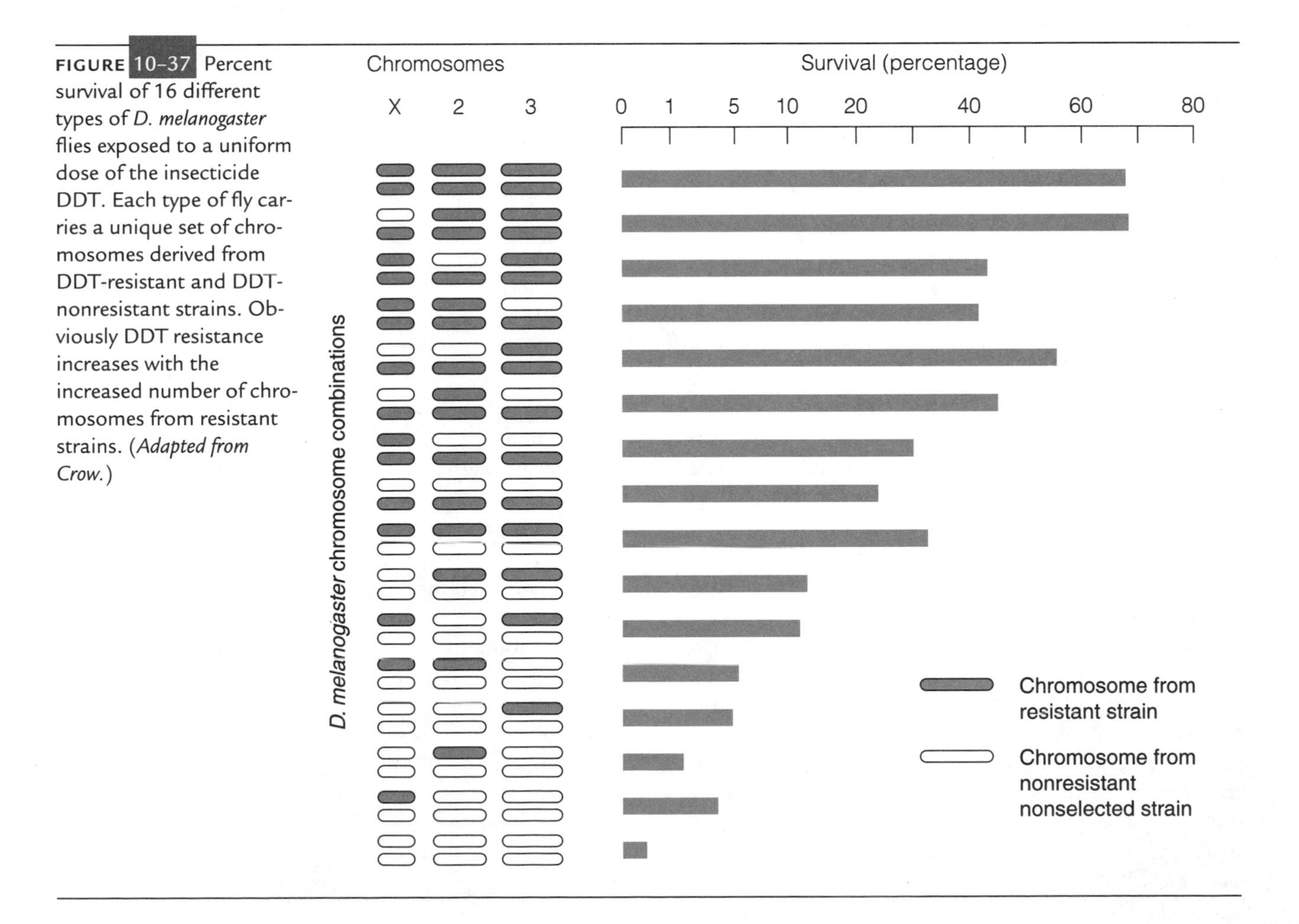

FIGURE 10-37 Percent survival of 16 different types of *D. melanogaster* flies exposed to a uniform dose of the insecticide DDT. Each type of fly carries a unique set of chromosomes derived from DDT-resistant and DDT-nonresistant strains. Obviously DDT resistance increases with the increased number of chromosomes from resistant strains. (*Adapted from Crow.*)

SUMMARY

Life depends on the genetic constancy with which organisms transmit information to their offspring, but evolution cannot occur without genetic variability. Genetic traits, changed or unchanged, are transmitted from one generation to the next by some type of cell division: by binary fission in prokaryotes and by mitosis (somatic cell division) or meiosis (gamete-producing division) in eukaryotes. In mitosis, all the daughter cells are genetically equivalent, while in meiosis variability is provided by recombination among homologous chromosomes and by random assortment of chromosomes into the gametes.

In the nineteenth century most people believed in blending inheritance. However, in the 1860s, Gregor Mendel broached two principles of inheritance that contradicted this idea. (1) The principle of segregation states that alleles of a single gene will segregate from each other into the gametes as discrete particles. (2) The principle of independent assortment involves the independent segregation into gametes of genes on different chromosomes. Although the phenotypic effects of some genes may blend with those of other genes, the genes themselves do not blend with each other.

Among the genetic sources of variability are incomplete dominance or codominance of alleles, the presence of multiple alleles for a gene, and the many instances when one gene locus affects the phenotypic expression of another (epistasis). Mendelian ratios result from genes segregating on separate chromosomes, but departures can occur because of biased transmission of particular genes (segregation distortion), or because of genes located on cytoplasmic chromosomes in mitochondria and chloroplasts, or because of linkage between genes on the same chromosome. In sex linkage, the heterogametic sex (XY) expresses all alleles lying on the X chromosome without regard to dominance relationships. In linkage, genes located on the same chromosome are constrained to varying degrees (recombination frequencies, linkage distances) to remain together during meiosis.

Variability is enhanced when there are alterations in chromosome number, either changes in entire sets (euploidy) or in individual chromosomes (aneuploidy), or when chromosomes undergo modifications in their structure, such as deletions, duplications, inversions, or translocations. Translocations can modify the size, composition, and even the number of chromosomes. Small localized chromosomal changes (mutations) are due to alterations in nucleotides or nucleotide sequences and can lead to changes in the gene product. Regulatory mutations affect the systems controlling genetic activity, as, for example, mutations in the regulatory gene that governs the expression of genes coding for lactose-using enzymes in bacteria. Mutations within a single gene generally occur rarely and usually have a characteristic frequency. But mutator genes, environment, and special movable DNA sequences called *transposons* may affect mutation rates.

Although some biologists previously thought that genes could mutate in response to environmental demand, it is now clear that selection chooses primarily from variants that are already present. Most populations are quite polymorphic, with up to 3/4 of gene loci having more than one allele. This great reservoir of variability allows populations to respond to evolutionary pressures without having to wait for new variants to arise by mutation.

KEY TERMS

acrocentric
alleles
allopolyploids
allosteric protein
allozyme
alternate segregation
anaphase
aneuploid variations
antimutators
asexual reproduction
autopolyploidy
autosomal
base substitutions
bivalent
chromatids
chromosomes
codominant
constancy
continuous variation
crossing over
cytoplasmic inheritance
deficiencies
deletions
dioecious
diploid
disjunction
dissociation
DNA repair mechanisms
dominant
dosage compensation
duplications
epistatic interactions
euchromatin
euploid variations
extranuclear inheritance
fission
fusion
gene families
gene marker
gene mutations
genes
genetic exchange
genotype
Haldane's rule
haploid
hemizygous
heritability
hermaphrodite
heterochromatin
heterogametic sex
heterozygotes
homogametic sex
homologous pairing
homozygotes
horizontal (lateral) gene transmission
hypermutation
incomplete dominance
independent assortment
inducer
inversions
karyotype
linkage
linkage distance
loci (plural; locus, singular)
maternal inheritance
meiosis
meiotic drive
metacentric
metaphase
mitosis
modifiers
monoecious
multiple alleles
multiple factors
mutations
mutator genes
nondisjunction
null alleles
open reading frames
operator site
orthogenesis
paracentric inversions

parthenogenesis
pericentric inversions
phenotype
plasmids
pleiotropy
point mutations
polygenes
polymorphism
polyploidy
polytene chromosomes
postadaptive
preadaptive
promoter site
prophase
quantitative trait loci (QTLs)
quantitative variation
recessive
reciprocal translocations
recombination
regulatory mutations
repeated sequences
repressor protein
segregation
segregation distortion
selfish DNA
sex linkage
sexually antagonistic genes
short tandem repeat polymorphisms (STRPs)
sickle cell disease
structural genes
synteny
tautomer
telocentric
thalassemia
transitions
translocations
transposons
transversions
unequal crossing over
variability
wild type
zygote

DISCUSSION QUESTIONS

1. Why are both constancy and variability essential in genetic material?
2. Cell division
   a. What role do homologous chromosomes play in producing the difference between mitosis and meiosis?
   b. Which of these modes of cell division generates variability, and how is this accomplished?
3. How do the Mendelian principles of segregation and independent assortment differ?
4. How do dominance relations, multiple allelic systems, and gene interactions generate variability?
5. What findings support Ohno's proposal that a sex-linked gene in one mammalian species will also be sex linked in other mammals?
6. Is Ohno's proposal also supported for genes that are not sex linked?
7. What are the major forms of variation in chromosome number? types of euploidy? types of aneuploidy?
8. How can a new allopolyploid (amphidiploid) species arise rapidly in nature, or be created in the laboratory?
9. What are the major forms of variation in chromosome structure?
10. What structural chromosomal changes can lead to changes in chromosomal number?
11. What kinds of chromosomal changes have occurred among some hominoids (humans and great apes)?
12. What are the different types of mutational changes on the nucleotide level, and how do they arise?
13. What is pleiotropy? Give an example.
14. How are the genes that produce the enzymes used for lactose metabolism regulated in prokaryotes? (Explain the terms *inducers, repressors, operators,* and *promoters.*)
15. What are some examples of gene regulation in eukaryotes?
16. How can multiple factors or polygenes explain continuous quantitative variation?
17. Can gene mutation rates account for the prevalence of gene mutations in populations?
18. Transposons
    a. How do transposons generate mutations?
    b. Why do transposons persist in a genome?
19. How can one discriminate whether mutations arise randomly without regard to their adaptive or nonadaptive effects (preadaptively) or whether they arise as adaptive responses to specific selective environments (postadaptively)?
20. Polymorphism
    a. What examples are there of chromosomal and gene polymorphisms?
    b. What evolutionary value do such polymorphisms offer?

EVOLUTION ON THE WEB

Explore evolution on the web! Visit the accompanying web site for *Evolution,* 3/e at www.jbpub.com/evolution for web exercises and links relating to topics covered in this chapter.

REFERENCES

Austin, C. R., and R. V. Short, 1976. *The Evolution of Reproduction.* Cambridge University Press, Cambridge, England.

Bachmann, B. J., 1983. Linkage map of *Escherichia coli* K–12, ed. 7. *Microbiol. Rev.,* **47,** 180–230.

Basolo, A. L., 1994. The dynamics of Fisherian sex-ratio evolution: Theoretical and experimental investigations. *Amer. Nat.,* **144,** 473–490.

Bender, W., M. Akam, F. Karch, P. A. Beachy, M. Peifer, P. Spierer, E. B. Lewis, and D. S. Hogness, 1983. Molecular genetics of the bithorax complex in *Drosophila melanogaster. Science,* **221,** 23–29.

Bhattacharyya, M. K., A. M. Smith, T. H. N. Ellis, C. Hedley, and C. Martin, 1990. The wrinkled-seed character of pea described by Mendel is caused by a transposon-like insertion in a gene encoding starch-branching enzyme. *Cell,* **60,** 115–122.

Bishop, B. E., 1996. Mendel's opposition to evolution and to Darwin. *Hered.,* **87,** 205–213.
Britten, R. J., 1986. Rates of DNA sequence evolution differ between taxonomic groups. *Science,* **231,** 1393–1398.
Bull, J. J., 1983. *Evolution of Sex Determining Mechanisms.* Benjamin/Cummings, Menlo Park, CA.
Cairns, J., J. Overbaugh, and S. Miller, 1988. The origin of mutants. *Nature,* **335,** 142–145.
Carvalho, A. B., M. C. Sampaio, F. R. Varandas, and L. B. Klaczko, 1998. An experimental demonstration of Fisher's principle: Evolution of sexual proportion by natural selection. *Genetics,* **148,** 719–731.
Castle, W. E., 1932. *Genetics and Eugenics,* 4th ed. Harvard University Press, Cambridge, MA.
Charlesworth, B., 1991. The evolution of chromosomal sex determination and dosage compensation. *Current Biol.,* **6,** 149–162.
Cline, T. W., 1993. The *Drosophila* sex determination signal: How do flies count to two? *Trends in Genet.,* **9,** 385–390.
Cline, T. W., and B. J. Meyer, 1996. Vive la différence: Males vs females in flies vs worms *Ann. Rev. Genet.,* **30,** 637–702.
Crow, J. F., 1957. Genetics of insect resistance to chemicals. *Ann. Rev. Entomol.,* **2,** 227–246.
———, 1997. The high spontaneous mutation rate: Is it a health risk? *Proc. Nat. Acad. Sci.,* **94,** 8380–8386.
Cummings, M. P., 1994. Transmission patterns of eukaryotic transposable elements: Arguments for and against horizontal transfer. *Trends in Ecol. and Evol.,* **9,** 141–145.
Daniels, S. B., K. R. Peterson, L. D. Strausbaugh, M. G. Kidwell, and A. Chovnick, 1990. Evidence for horizontal transmission of the P transposable element between *Drosophila* species. *Genetics,* **124,** 339–355.
Davies, J., 1994. Inactivation of antibiotics and the dissemination of resistance genes. *Science,* **264,** 375–382.
Dawkins, R., 1976. *The Selfish Gene.* Oxford University Press, New York.
Dobzhansky, Th., 1944. Chromosomal races in *Drosophila pseudoobscura* and *D. persimilis. Carnegie Inst. Wash. Publ. No. 554,* Washington, DC., pp. 47–144.
———, 1947. A directional change in the genetic constitution of a natural population of *Drosophila pseudoobscura. Heredity,* **1,** 53–64.
Doolittle, W. F., and C. Sapienza, 1980. Selfish genes, the phenotype paradigm, and genome evolution. *Nature,* **284,** 601–603.
Drake, J. W., 1970. *The Molecular Basis of Mutation.* Holden-Day, San Francisco.
———, 1993. Rates of spontaneous mutation among RNA viruses. *Proc. Nat. Acad. Sci.,* **90,** 4171–4175.
Eyre-Walker, A., and P. D. Keightley, 1999. High genomic deleterious mutation rates in hominids. *Nature,* **397,** 344–347.
Falconer, D. S., and T. F. C. Mackay, 1996. *Introduction to Quantitative Genetics,* 4th ed. Longman, Harlow, Essex, England.
Felsenfeld, G., 1985. DNA. *Sci. Amer.,* **253**(4), 58–67.
Fisher, R. A., 1958. *The Genetical Theory of Natural Selection,* 2d ed. Dover, New York.
Haldane, J. B. S., 1922. Sex-ratio and unisexual sterility in hybrid animals. *Genet.,* **12,** 101–109.
Hall, B. G., 1988. Adaptive evolution that requires multiple spontaneous mutations. I. Mutations involving an insertion sequence. *Genetics,* **120,** 887–897.
Hamilton, W. D., 1967. Extraordinary sex ratios. *Science,* **156,** 477–488.
Harris, H., 1966. Enzyme polymorphisms in man. *Proc. Roy. Soc. Lond. (B),* **164,** 298–310.
Hartl, D. L., and E. W. Jones, 1998. *Genetics: Principles and Analysis.* Jones and Bartlett, Sudbury, MA.
Himes, M., and C. A. Beam, 1975. Genetic analysis in the dinoflagellate *Crypthecodinium* (*Gyrodinium*) *cohnii*: Evidence for unusual meiosis. *Proc. Nat. Acad. Sci.,* **72,** 4546–4549.
Hino, O., et al., 1993. Universal mapping probes and the origin of human chromosome 3. *Proc. Nat. Acad. Sci.,* **90,** 730–734.
Hodgkin, J., 1992. Genetic sex determination mechanisms and evolution. *BioEssays,* **14,** 253–261.
Hurst, L. D., 1993. The incidences, mechanisms and evolution of cytoplasmic sex ratio distorters in animals. *Biol. Rev.,* **68,** 121–194.
Kacser, H., and J. A. Burns, 1981. The molecular basis of dominance. *Genetics,* **97,** 639–666.
Kidwell, M. G., 1994. The evolutionary history of the *P* family of transposable elements. *J. Hered.,* **85,** 339–346.
King, M., 1993. *Species Evolution: The Role of Chromosome Change.* Cambridge University Press, Cambridge, England.
Komdeur, J., S. Daan, J. Tinbergen, and C. Mateman, 1997. Extreme adaptive modification in sex ratio of the Seychelles warbler's eggs. *Nature,* **385,** 522–525.
Lawrence, J. G., and H. Ochman, 1998. Molecular archaeology of the *Escherichia coli* genome. *Proc. Nat. Acad. Sci.,* **95,** 9413–9417.
LeClerc, J. E., B. Li, W. L. Payne, and T. A. Cebula, 1996. High mutation frequency among *Escherichia coli* and *Salmonella* pathogens. *Science,* **274,** 1208–1211.
Lederberg, J., and E. M. Lederberg, 1952. Replica plating and indirect selection of bacterial mutants. *J. Bact.,* **63,** 399–406.
Lenski, R. E., M. Slatkin, and F. J. Ayala, 1989. Another alternative to directed mutation. *Nature,* **337,** 123–124.
Lewis, H., 1973. The origin of diploid neospecies in *Clarkia. Amer. Nat.,* **107,** 161–170.
Lewontin, R. C., and J. L. Hubby, 1966. A molecular approach to the study of genic heterozygosity in natural populations. II. Amount of variation and degree of heterozygosity in natural populations of *Drosophila pseudoobscura. Genetics,* **54,** 595–609.
Liming, S., Y. Yingying, and D. Xingsheng, 1980. Comparative cytogenetic studies on the red muntjac, Chinese muntjac, and their F_1 hybrids. *Cytogenet. and Cell Genet.,* **26,** 22–27.
Lucchesi, J. C., 1994. The evolution of heteromorphic sex chromosomes. *BioEssays,* **16,** 81–83.
Lucchesi, J. C., 1998. Dosage compensation in flies and worms: The ups and downs of X-chromosome regulation. *Curr. Opinion Genet. Devel.,* **8,** 179–184.
Lynch, M., and B. Walsh, 1998. *Genetics and Analysis of Quantitative Traits.* Sinauer Associates, Sunderland, MA.
Lyttle, T. W., 1991. Segregation distorters. *Ann. Rev. Genet.,* **25,** 511–557.

Mackay, T. F. C., 1996. The nature of quantitative genetic variation revisited: Lessons from Drosophila bristles. *BioEssays,* **18,** 113–121.

Maynard Smith, J., 1978. *The Evolution of Sex.* Cambridge University Press, Cambridge, England.

Mendel, G., 1866. Versuch über Pflanzen-Hybriden. (This is Mendel's classic paper, originally published in the *Proceedings of the Brünn Natural History Society.* It has been translated into English and reprinted under the title *Experiments in Plant Hybridization.*)

Miller, J. R., 1990. *X-Linked Traits: A Catalog of Loci in Nonhuman Animals.* Cambridge University Press, Cambridge, England.

Mizuuchi, K., 1992. Transpositional recombination: Mechanistic insights from studies of Mμ and other elements. *Ann. Rev. Biochem.,* **61,** 1011–1051.

Muller, H. J., 1942. Isolating mechanisms, evolution, and temperature. *Biol. Symp.,* **6,** 71–125.

Nevo, E., 1978. Genetic variation in natural populations: Patterns and theory. *Theor. Pop. Biol.,* **13,** 121–177.

Nevo, E., M. Filippucci, C. Redi, A. Korol, and A. Beiles, 1994. Chromosomal speciation and adaptive radiation of mole rats in Asia Minor correlated with increased ecological stress. *Proc. Nat. Acad. Sci.,* **91,** 8160–8164.

O'Brien, S. J., 1993. The genomics generation. *Current Biol.,* **3,** 395–397.

Ohno, S., 1979. *Major Sex Determining Genes.* Springer-Verlag, Berlin.

Orgel, L. E., and F. H. C. Crick, 1980. Selfish DNA: The ultimate parasite. *Nature,* **284,** 604–607.

Orr, H. A., 1997. Haldane's rule. *Ann. Rev. Ecol. Syst.,* **28,** 195–218.

Parkhurst, S. M., and P. M. Meneely, 1994. Sex determination and dosage compensation: Lessons from flies and worms. *Science,* **264,** 924–932.

Qumsiyeh, M. B., 1994. Evolution of number and morphology of mammalian chromosomes. *Hered.,* **85,** 455–465.

Rice, W. R., 1994. Degeneration of a nonrecombining chromosome. *Science,* **263,** 230–232.

———, 1998. Male fitness increases when females are eliminated from gene pool: Implications for the Y chromosome. *Proc. Nat. Acad. Sci.,* **95,** 6217–6221.

Rich, A., A. Nordheim, and A. H.-J. Wang. 1984. The chemistry and biology of left-handed Z-DNA. *Ann. Rev. Biochem.,* **53,** 791–846.

Roff, D. A., 1997. *Evolutionary Quantitative Genetics.* Chapman and Hall, New York.

Rosenberg, S. M., 1997. Mutation for survival. *Current Opinion Genet. Devel.,* **7,** 829–834.

Rosenberg, S. M., C. Thulin, and R. S. Harris, 1998. Transient and heritable mutators in adaptive evolution in the lab and in nature. *Genetics,* **148,** 1559–1566.

Roush, R., and D. R. McKenzie, 1987. Ecological genetics of insecticide and acaricide resistance. *Ann. Rev. Entomol.,* **32,** 361–380.

Rutherford, S. L., and S. Lindquist, 1998. Hsp90 as a capacitor for morphological evolution. *Nature,* **396,** 336–342.

Soltis, D. E., and P. S. Soltis, 1995. The dynamic nature of polyploid genomes. *Proc. Nat. Acad. Sci.,* **92,** 8089–8091.

Stone, W. S., 1962. The dominance of natural selection and the reality of superspecies (species groups) in the evolution of *Drosophila. Univ. of Texas Publi.,* **6205,** 507–537.

Strickberger, M. W., 1985. *Genetics,* 3d ed. Macmillan, New York.

Symonds, N., 1989. Anticipatory mutagenesis? *Nature,* **337,** 119–120.

Taylor, A. L., 1970. Current linkage map of *Escherichia coli. Bacteriol. Rev.,* **34,** 155–175.

Thaler, D. S., 1994. The evolution of genetic intelligence. *Science,* **264,** 224–225.

Thompson, D. W., 1942. *On Growth and Form,* 2d ed. Cambridge University Press, Cambridge, England.

Tiersch, T. R., M. J. Mitchell, and S. S. Wachtel, 1991. Studies on the phylogenetic conservation of the *SRY* gene. *Hum. Genet.,* **87,** 571–573.

Turelli, M., 1998. The causes of Haldane's rule. *Science,* **282,** 889–891.

White, M. J. D., 1973. *Animal Cytology and Evolution,* 3d ed. Cambridge University Press, Cambridge, England.

Womack, J. E., and Y. D. Moll, 1986. Gene map of the cow: Conservation of linkage with mouse and man. *J. Hered.,* **77,** 2–7.

Wu, C-I., N. A. Johnson, and M. F. Palopoli, 1996. Haldane's rule and its legacy: Why are there so many sterile males? *Trends in Ecol. and Evol.,* **11,** 281–284.

Yunis, J. J., and O. Prakash, 1982. The origin of man: A chromosomal pictorial legacy. *Science,* **215,** 1525–1529.

11 Systematics and Classification

The chapters that follow seek to describe, in a general way, the probable course of events that took place in the evolution of various groups of organisms. Much of this effort derives from morphological and functional descriptions of these organisms and is based on enumerating and comparing their similarities and differences. These areas are the traditional province of **classification** and **systematics**—the arts of identifying distinctions among organisms and placing them into groups that reflect their most significant features and relationships.[1]

As we have seen in Chapter 1, scientists formulated techniques of classifying organisms much earlier than they accepted the concept that their similarities and differences arose from evolutionary causes. It is, in fact, fairly easy to classify organisms in a variety of ways that can distinguish among them yet obscure their common origins. For example, we can classify fish and whales in one group, flies and birds in another, frogs and crocodiles in a third, and squirrels and monkeys in a fourth. Of course, by the eighteenth and nineteenth centuries the criteria of classification were less arbitrary, and taxonomists such as Linnaeus used a multitude of features in their descriptions, rather than the single character of whether an organism swims in water, flies in air, crawls in mud, or climbs trees. Nevertheless, scientists still dispute how many characters to compare to obtain a "natural" classification and which characters are to receive greater consideration (weighting) than others.

With the advent of the Darwinian revolution an additional consideration entered into the thinking of at least some systematists: whether they could or should use classification to reflect evolutionary relationships. There were certainly strong indications that many organisms grouped together because they possessed a large number

[1]Authors often use the terms *systematics, classification,* and *taxonomy* interchangeably, although some taxonomists such as Simpson consider systematics a much broader field—the study of the diversity of organisms and all their comparative and evolutionary relationships, including such topics as comparative anatomy, comparative ecology, comparative physiology, and comparative biochemistry. Simpson (1961) defined classification as a subtopic of systematics, as the ordering of organisms into groups, and taxonomy as the study of the principles and procedures of classification.

of similar features could also be said to descend from a common ancestor (p. 39ff). But such determinations revealed only one aspect of evolutionary classification. Another aspect was to discern lines of descent among groups that perhaps shared only a few features. As Darwin put it,

> Our classification will come to be, as far as they can be so made, genealogies . . . we have to discover and trace the many diverging lines of descent in our natural genealogies, by characters of any kind which have long been inherited.

This question of exact genealogy, or **phylogeny,** among the many different groups of organisms was not, and is often still not, easily soluble.

A primary reason for the difficulty in determining phylogenetic relationships is the difficulty in finding an unbroken line of ancestors that connects different groups. The fossil record may be fairly complete for some groups such as horses (pp. 49–50), but it is quite meager for birds, whales, insects, early angiosperms (flowering plants), and many other organisms. These fossil record inadequacies arise from a number of causes:

- The organisms themselves may be destroyed before, as well as after, their deposition in a fossil-bearing sedimentary layer, which is often localized to a former aqueous region such as a river, lake, or ocean shoreline (Fig. 3–12).[2]
- Strong winds, heavy wave action, or other potent forces can disturb the formation of sedimentary layers.
- Even after sedimentary layers form, various geological events may later erode or move them about, causing discontinuities in the record.
- Only a small portion of fossil-bearing sedimentary rocks are accessible to paleontologists.

Despite their rarity, some fossils may closely resemble ancestral forms, and their geological sequence can offer the important advantage of showing the historical order in which different phylogenetic characters were acquired or lost. For the most part, evolutionists consider and weigh all known relationships among the different groups, existing and fossil, in order to hypothesize phylogenies. Also, many evolutionists feel that how they group and classify organisms should somehow coincide with phylogenetic relationships.

Unfortunately, taxonomists do not always agree on methods of classification and therefore on which groups of organisms to classify together. As Figure 11–1 shows, entomologists have offered five different schemes of classification for the same groups of insects. Moreover, even when workers in the field agree on similarities among organisms, they are not always clear on how many subgroups to propose. For example, the single genus *Rubus* (blackberries, raspberries, loganberries) has been divided into 500 species by one botanist, 200 by another, and 25 by a third. To the evolutionist, taxonomic problems seem to be at least twofold:

- How to recognize the basic unit of classification, the species, and then to identify this unit, if possible, with a fundamental evolutionary unit
- How to order species into systems that will connect them all into a reasonably accurate phylogenetic scheme

Species

Among the variety of species definitions that have been offered, taxonomists have generally used morphological criteria, since this is how most individuals have been compared. Thus, Davis and Heywood define species as "assemblages of individuals with morphological features in common and separable from other such assemblages by correlated morphological discontinuities in a number of features."

Simple as this morphological procedure may seem, it relies heavily on the personal predilections of various taxonomists, and as described earlier, often leads to different numbers of species for the same groups. "Lumpers" tend to combine populations into single species or groups, whereas "splitters" tend to separate the same populations into different species or groups. The distinction between extreme proponents of these two points of view seems to lie in whether to use taxonomy to unite those organisms that share any features at all (lumpers), or to separate organisms that differ in any way at all (splitters).

To eliminate at least some arbitrary distinctions, some taxonomists have proposed numerical methods in which taxonomic distinctions depend on the size of the statistical correlation for as large a number of characters as possible. These characters, all given equal numerical weight, are presumably the ultimate distinctive qualities of the organisms involved and are not further divisible. A high statistical correlation among individuals for a large number of such characters indicates their membership in the same species or groups, and low correlation points to their separation into different species or groups. This method, called **numerical** or **phenetic taxonomy,** largely

[2]Newell estimates that, after organic decomposition, only the remains of individuals from 10 or 15 species that died on a tropical river bank can be identified out of the 10,000 or so species that live in the area. The proportion of preserved species is greater in some ocean environments such as limestone reefs, but even then the proportion of identifiably preserved species is no greater than 1 or 2 percent.

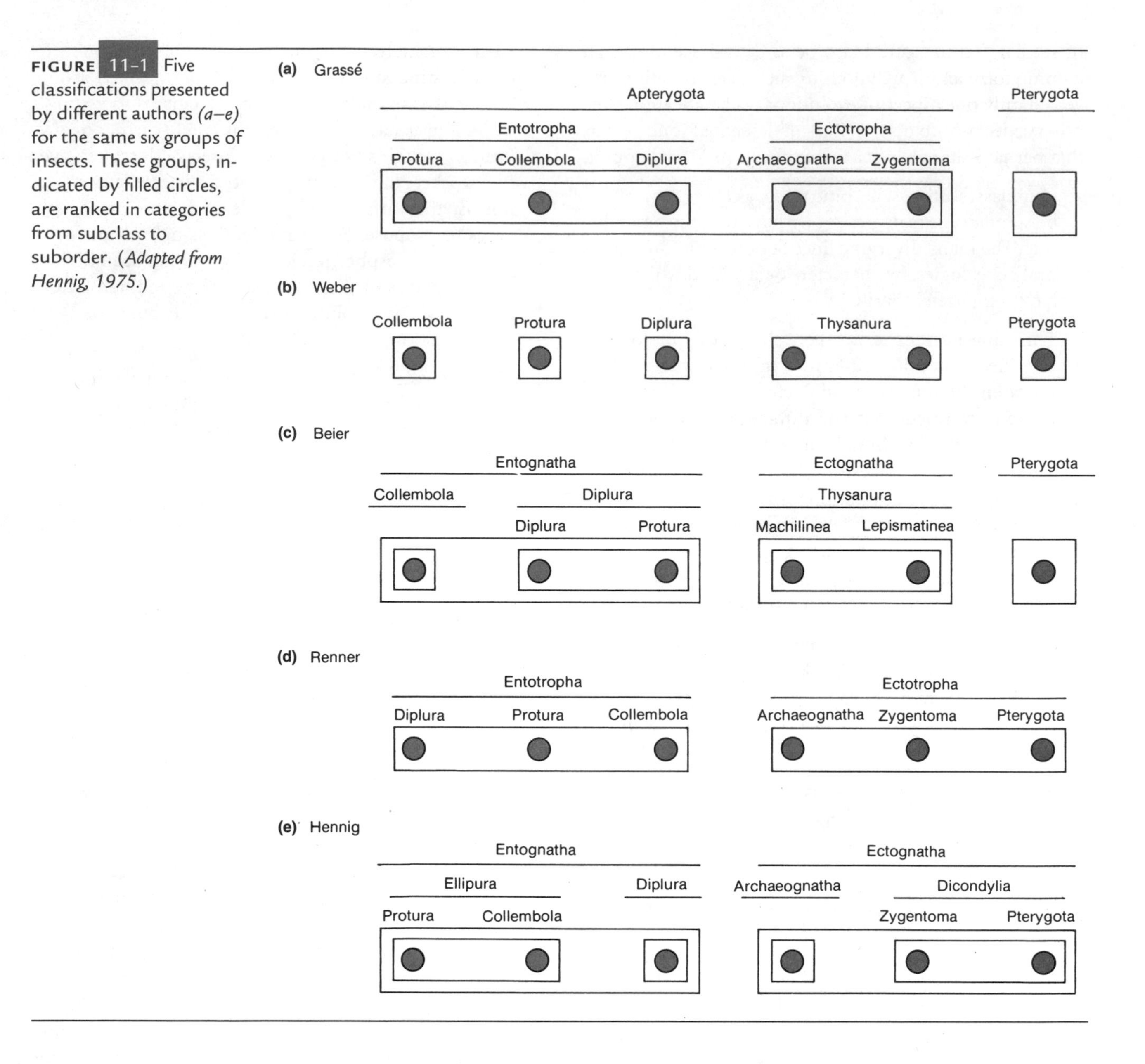

FIGURE 11-1 Five classifications presented by different authors *(a–e)* for the same six groups of insects. These groups, indicated by filled circles, are ranked in categories from subclass to suborder. (*Adapted from Hennig, 1975.*)

formalizes some of the processes taxonomists use intuitively but adds a degree of quantitative numerical consistency, which pheneticists present as a more exact alternative to the usual **classical taxonomy.**

Although there is now a wide literature in which numerical studies have helped clarify some taxonomic problems (Sneath and Sokal), and enabled many different characters to be evaluated simultaneously, taxonomists do not commonly endorse phenetic classification. Many consider the presumption in numerical taxonomy that measurable unit characters are not further divisible to be a serious difficulty. To taxonomists, characters such as wing or body length can certainly be divided into subcomponents such as wing veins, leg lengths, and so forth, whose measurements can lead to different statistics and different relationships than scoring whole, undivided units. Also, because of epistatic and pleiotropic effects (p. 197, Fig.10–24), appearance and measurement of a phenotypic character is often dependent on or correlated with other characters. Taxonomists, whether followers of phenetics or other persuasions, can greatly overestimate the number of actually independent phenotypes they use in evaluating relationships.

To give equal weight in classification to every character, no matter what its complexity, also may not be appropriate for characters that have greater evolutionary importance than others. For example, differences in chromosomal structure or in homologous pairing, characteristics that may easily isolate different groups because of cell division abnormalities, more often reflect major changes in evolutionary patterns than do differences in a character such as petal color in plants. Polyploidy (Fig. 10–13), aneuploidy (Table 10–3), and translocations (p. 210) are known to produce obvious speciation effects. Similarly, a deeply integrated genetic and developmental trait in animals, such as brain form and structure, has greater significance in differentiating organisms than a more environmentally plastic trait such as body weight. Some taxonomists also point out that the statistical results of numerical methods still depend on subjective choices as to how to apply them taxonomically. For example, which value of a correlation coefficient should taxonomists use to classify individuals into a single species?

We can see that whether the approach is traditionally morphological or more quantitatively and statistically numerical, both demand some degree of subjective judgment. That is, they both exemplify arbitrary choice, or, as John Locke commented long ago, "the boundaries of the species, whereby men sort them, are made by men."[3] To Darwin, the attempt to find "the essence of the term species" was "a vain search."

Many biologists have not been satisfied to let species definitions rest primarily on a subjective or quantitative morphological approach and have instead adopted a **biological species concept.** Derived from Buffon (p. 10) and others, this concept defines a species as a sexually interbreeding or potentially interbreeding group of individuals normally separated from other species by the absence of genetic exchange, that is, by **reproductive isolation.** The obvious advantage of this definition is that species distinctions can be objectively tested by two relatively simple criteria:

- Do populations in the same locality normally fertilize each other?
- Should cross-fertilization occur, are the hybrids viable and fertile?

If the answer to either question is no, then we consider the evaluated populations as species separated by reproductive isolation barriers (Chapter 24). [4]

[3]From "An Essay Concerning Human Understanding" (1689).

[4]Instead of defining a species by negative attributes—for example, inability to cross-fertilize with other species—some authors propose a more positive approach emphasizing inclusive qualities that tie species members together. Thus, Paterson uses the "recognition concept" by which species members are defined by their "common fertilization system" (see also p. 590); Van Valen asserts that a species occupies a unique "adaptive [ecological] zone;" and Templeton defines a species as possessing "intrinsic cohesion mechanisms" by essentially adding to Van Valen's definition of a uniform ecological niche ("demographic exchangeability") the ability to transfer genetic material between species members ("genetic exchangeability"). Nevertheless, all these species definitions present problems when assigning individuals to species, since taxonomic distinctions among individuals have to be made mostly by noting "negative" attributes primarily derived from the biological species concept: the absence of the same fertilization system, the lack of a common adaptive zone, and a failure of genetic exchangeability. Moreover, "positive" attributes are not necessarily distinctive: two species that produce sterile hybrids may still have a "common fertilization system," and two species may both occupy the same "adaptive zone." In addition, ecological and biogeographical divergence can occur not only between species but also within species. (For a brief discussion and evaluation of these various species concepts, see Endler.)

Such biological criteria have let taxonomists make species distinctions between similar-appearing populations that they could not separate on the basis of the usual morphological taxonomic criteria. Thus, the fruit fly species *Drosophila pseudoobscura* and *D. persimilis,* called **sibling species** because they are almost identical in appearance, do not normally cross-fertilize; this is also true for some leafy-stemmed sibling species in the phlox family, *Gilia tricolor* and *G. angelensis.* Such biological tests have also led researchers to unify different groups into single species that morphological and geographical criteria had separated into distinct species (for example, the union of various species of North American sparrows into one **polytypic species** consisting of multiple geographic races or subspecies, the song sparrow, *Passarella melodia*). Similarly, groups of *Achillea* plants that show distinct ecological adaptations—restricting their growth to particular environments (Figs. 24–1 and 24–2)—are generally identified as races rather than species because they are potentially able to exchange genes along a continuous geographical gradient.

Basing species distinctions mainly on reproductive isolation has also led some researchers to the opinion that the actual proportion of species that do not fit well into the biological species concept may be quite small, even among plants. A study made of 838 named plant species in the Concord, Massachusetts, area showed that 93 percent could be distinguished according to the biological species concept, and only 7 percent were problematical.

Despite these advantages, attempts to universally apply the biological species concept face considerable difficulties:

1. Although it may be possible to observe reproductive barriers between groups found in the same locality (**sympatric populations,** Chapter 24), many practical limitations block crosses between groups that are ordinarily separate (**allopatric populations**). To provide space and appropriate environments for the enormous numbers of possible crosses between all allopatric combinations of similar organisms appears beyond the present capability of biologists.[5]
2. Even when researchers can cross allopatric populations, they must still make some arbitrary decisions. For example, results from interbreeding experiments may range from no genetic exchange at all for certain attempted crosses to a fairly large degree of genetic exchange for others (Fig. 11–2). The

[5]For plants, Baker writes,

> The number of hybridization attempts which must be made, the number of plants which must be raised in the first hybrid generation and the number of subsequent generations which must be grown to see whether fertility is maintained and whether segregation occurs, all place limitations on the comprehensiveness of the experiments. . . . The total task, for naturally occurring plants, is beyond human capacity for achievement.

FIGURE 11-2
Differences in fertility observed for F_1 hybrids derived from crosses between 13 species of the plant genus *Geum,* a perennial herb. (*Evolution in the Genus Geum,* Evolution *13:378–388 by W. Gajewski. Reprinted by permission.*)

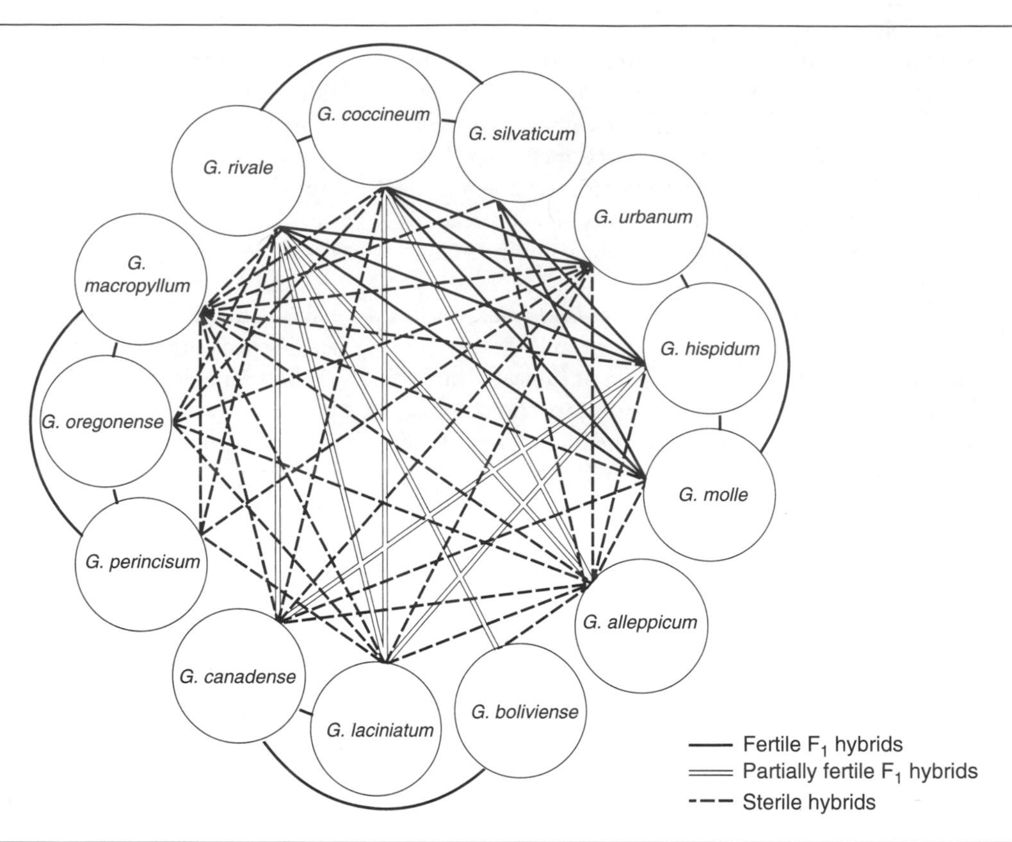

question then arises: At which point on this scale of interbreeding values shall we separate species?[6]

3. We cannot test fossil populations as to whether they can or cannot exchange genes either among themselves or with present populations.

4. In asexual organisms, no matter whether reproduction occurs by fission or by parthenogenesis, each clone of individuals is essentially genetically isolated from every other clone, yet few, if any, biologists would consider describing each clone as a separate species. [Mayr (1987) refers to asexual groups as "paraspecies," and Ghiselin (1987) calls them "pseudospecies."]

5. Plasmid-mediated horizontal gene transmission between different species (p. 225), although presumably rare, can transcend reproductive barriers and cast doubt on species (and even clonal) distinctions. For example, many bacterial clones and species groups, such as *Escherichia, Salmonella,* and *Shigella,* have very similar genes and gene sequences, indicating gene exchange ("horizontal gene transfer") has occurred among supposedly unrelated or very distantly related groups (p. 226). Similarly, viruses (bacteriophages) that infect these bacteria demonstrate "access, by a horizontal exchange, to a large common gene pool" (Hendrix et al.).

These difficulties derive from trying to apply a single definition (reproductive isolation) to different phenomena: either to situations where the definition cannot be applied at all (for example, fossils), or to traditionally accepted species that do not really satisfy the definition (for example, the plants in Fig. 11–2), or to asexual clonal groups that may partially satisfy the definition but are difficult to accept as species.

To circumvent some of these problems by broadening and reorienting the definition, various authors have proposed an **evolutionary species concept.** In this concept, species are defined in terms of differences that are not dependent on sexual isolation but rather on their "evolutionary" isolation, of which sexual isolation is only one aspect. In Simpson's (1961) words, "an evolutionary species is a lineage (an ancestor-descendant sequence of populations) evolving separately from others and with its own unitary evolutionary role and tendencies." Thus, for

[6]Interestingly, even when genetic exchange is completely uninhibited between some allopatric populations, it may still seem desirable to consider them as separate species since they do not hybridize under normal conditions. One well-known example is the discovery of two widely separate populations of trees occupying similar habitats, one in China *(Catalpa ovata)* and the other in the eastern United States (*C. bignoides).* Although they can cross with each other to produce hybrids that are as viable and fertile as the parents, these populations have probably been separate for many millions of years, and botanists have therefore generally agreed to continue their separate species identifications.

the first time, a species concept incorporates change (evolution), and also lays the groundwork for those changes that result from competition and interaction among species: the existence of separate evolutionary lineages implies that an important factor affecting their success and survival may be the success and survival of other such lineages.

The problem of an evolutionary species definition is that we may find such distinctions difficult to make in practice; taxonomists facing a large variety of specimens often have few techniques to distinguish among them, other than the purely morphological. Also, since evolutionary speciation is a process, defining the point at which groups of organisms have reached complete separation still has, of necessity, some arbitrary elements of choice. (To define a species as an "individual" clearly separated from other "individuals" can be a gross oversimplification: see Ruse's discussion on whether species are individuals, classes, or populations.) Nevertheless, the evolutionary species concept clearly justifies using ecological, behavioral, genetic, and morphological evidence to help judge evolutionary separation or distance. After all, evolutionary separation among populations accounts for species differences: without evolution, there are no biological species.

In summary, the difficulties in species taxonomy are, to a large extent, inherent in the process of speciation itself. That is, the differences among populations that makes some of them hard to classify as varieties, subspecies, or species arise from the fact that they undergo evolutionary changes that can differ in intensity and sequence, varying in different times, places, and circumstances (Chapters 22–24). Phenotypic and genotypic distinctions do not evolve uniformly among groups, but rather comparisons among them generally show different degrees of change in various characters. In sexual forms, the overall result of such differences is different degrees of reproductive isolation and morphological distinctiveness, whereas in asexual forms evolutionary distinctions are reflected in differences other than reproductive isolation.[7]

In both sexual and asexual cases, members of a species share a community of descent that explains many of their common features, yet these individuals cannot be described as members of fixed and unchanging entities. That is, members of a species are identified by their similarity but their relationship derives from their shared history, which does not necessarily lead to identity: a species name indicates singularity, but its constituents often vary. Classification and evolution emphasize different aspects of organisms. Thus, although an evolutionary explanation does not always make the taxonomist's task easier, it does offer an explanation for some difficulties in determining species.

Phylogeny

Given the existence of a uniquely different ancestor, the question of the origin of species is basically the question of the origin of new species. Darwin mostly devoted himself to explaining how, under natural selection, a single species can change through time. Figure 11–3 illustrates such an example of successional changes within a single lineage, called **phyletic evolution** or **anagenesis.** However, an important concern of many evolutionists was the problem of species multiplication (p. 35): how to explain the splits and divisions within an ancestral line that cause the appearance of more than one species—a cluster of species or **clade.** This pattern, known as **phylogenetic branching,** or **cladogenesis** (Fig. 11–4), was first offered by Lamarck (Fig. 1–5), but it was Haeckel, beginning in the 1860s, who first popularized this form of evolutionary description.

Viewed through time, as in the phylogeny of horses (Fig. 3–14), some speciation events are obviously cladogenetic and others anagenetic. However, as explained previously, the determination of phylogenetic trees is often difficult because the common ancestors of different groups of organisms are usually long extinct and the fossil record is usually inadequate. The absence of complete fossil information thus puts a lot of emphasis on constructing phylogenies by comparisons between known organisms, whether existing or fossil.

In general, the more a group of species shares common inherited attributes, the more likely their descent from a common ancestor. Taxonomists therefore enlist all available heritable characteristics in making comparisons between species: morphological, embryological, behavioral, physiological, biochemical, genetic, and chromosomal. Classifiers assume that the more two species are alike in these respects, the more they share common hereditary characteristics and the closer at hand their common ancestor. Nevertheless, even when species share common features, the genotypic basis for this may derive from different evolutionary causes, as shown in Figure 11–5:

- HOMOLOGY The same feature occurs in different species because it derives directly from a common ancestor that bore the same characteristic. For example, the many similar features in the forelimb structure of vertebrates indicate derivation from a common vertebrate ancestor (Fig. 3–6). However, since organismic "features" are not only morphological or structural but also physiological,

[7]According to some authors, we should also use the magnitude of the differences among recognized species in sexual forms, other than reproductive isolation, to distinguish species in asexual forms.

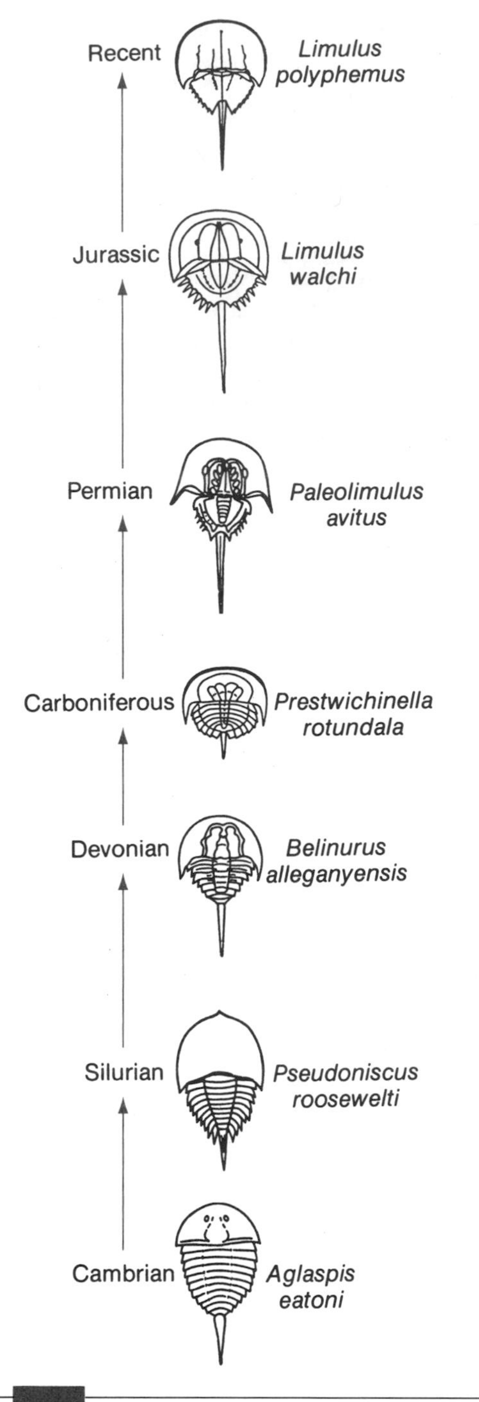

FIGURE 11-3 Phyletic evolution (anagenesis) in merostome arthropods (horseshoe crabs), indicating relatively small phenotypic changes over long periods of time. That evolution proceeds *only* in such a linear direction is, of course, a diagrammatic illusion. As in other such sequences (Fig. 3-14), there were variations within and between each of these stages. The now discarded concept that evolution occurs linearly without selection, **orthogenesis,** was quite popular among paleontologists before geneticists developed an explanation of how variation was generated and maintained (Chapters 10 and 21). (*Adapted from Newell.*)

developmental, and genetic, what is homologous at one level may not appear so at another (Chapter 15). As described later, genetic homologies provide more information on evolutionary relationships than other features that may only appear to be homologous.[8]

- PARALLELISM A similar feature occurs in different species, but their immediate common ancestor was different. For example, anteater-like features have appeared in different lines of mammals that each descended from non-anteater mammalian groups (Fig. 11–6).

- CONVERGENCE A similar feature arose independently in different species whose ancestral lineages had been separated for a considerable time. We have seen an example of convergence in the similarity of marine hydrodynamic forms among the widely separated fish, reptile, and mammalian classes of vertebrates (Fig. 3–7). As in parallelism, the cause for convergence for a character that formerly differed among species is mostly presumed to derive from their exposure to similar environmental factors evincing similar selective forces. Common adaptive features can thus be attained in each group through independent genetic changes.[9]

---

[8]Some biologists use the term homology strictly as a qualitative feature indicating presence of a common ancestral relationship, and others quantify it to measure the degree of relationship. Where homology can be quantified, especially on the molecular level (Chapter 12), it can evaluate relationships among multiple species: for example, two species are more related to each other than to a third species if they share more ancestrally derived similarities—are homologous for more numerous and longer sequences of amino acids or nucleotides. However, as explained on p. 240, we should keep in mind that horizontal gene transmission can cloud phylogenetic relationships when completely unrelated organisms share a gene or sequence obtained through lateral gene transfer rather than through common ancestry. Phylogenies ideally should be based on comparing sequences among many different genes and proteins.

Among other usages is the term **serial** ("iterative") **homology** to describe similarities among parts of the same organism, such as homologies among the various vertebrae in a vertebrate, or among the different feathers in a bird, or among the different kinds of hemoglobin molecules (α, β, γ, and so on) that a particular individual produces. The genetic basis for serial homology often comes from the duplication of a gene responsible for producing or affecting a particular structure. Such duplicates may originally have similar features, but they may also evolve differently from each other (pp. 260–262), a point also made by Darwin:

> We have already seen that parts many times repeated are eminently liable to vary in number and structure; consequently it is quite probable that natural selection, during a long continued course of modification, should have seized on a certain number of primordially similar elements, many time repeated, and have adapted them to divers purposes.

Thus, by their functional divergence, duplications can offer an important evolutionary advantage compared to new genes formed with entirely novel and untried nucleotide sequences.

[9]When the genetic variation upon which selection acts is itself similar—when similar organisms are exposed to similar environments in different localities—evolution can produce strikingly convergent results. Losos

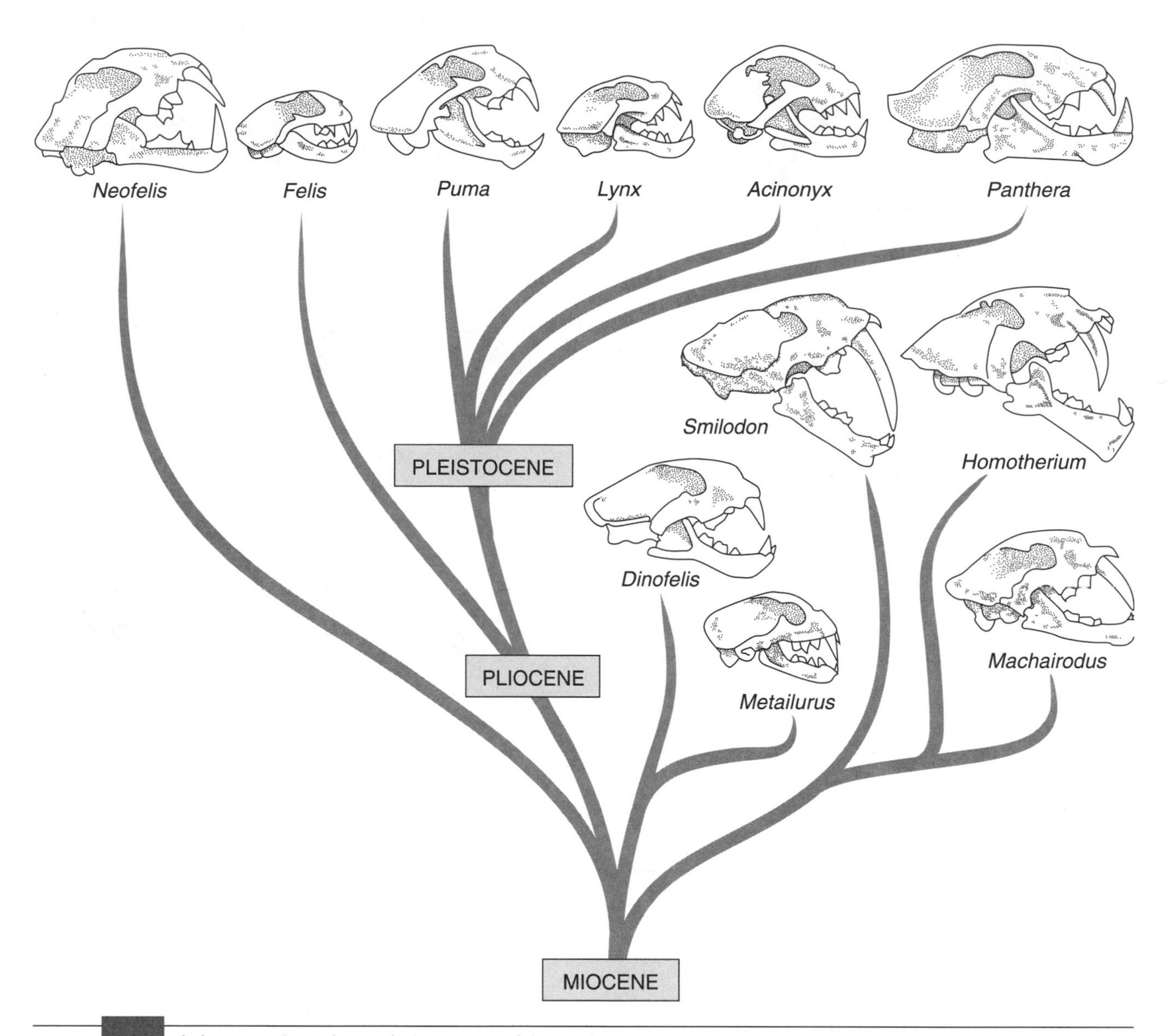

FIGURE 11–4 Phylogenetic branching (cladogenesis) of the modern cat family beginning with its origin in the Miocene epoch about 20 million years ago. The six groups along the *top row* still survive, but the other five groups (*lower right*) are extinct. *(After Benton, modified.)*

Valuable though it may seem, establishing homologous relationships is not a simple task, mostly because parallelism and convergence (**homoplasy**) can always lead to false connections, and the genetic basis for homology is not always apparent when comparing characters and adaptations confined only to structure and development. Given the chain of organismic events that extends from information to function:

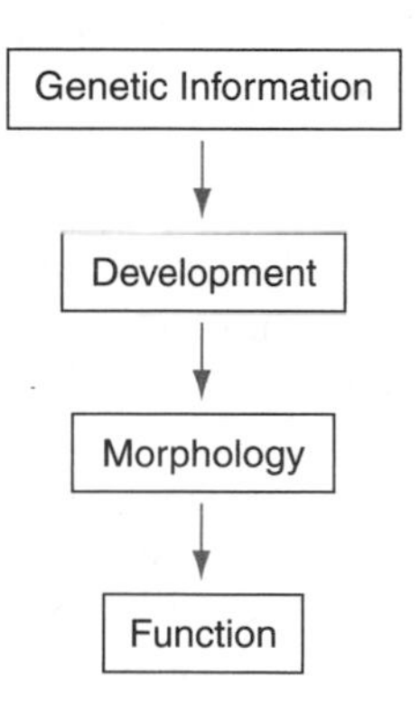

and coworkers show that anole lizards that colonized four different Caribbean islands diverged from their ancestors and evolved independently into phenotypically similar species occupying similar ecological zones on each island. In these cases, evolution seems to repeat itself fairly exactly, given similar genetic backgrounds and similar environmentally selective events. Identical behavioral breeding patterns found in widely separated groups of fiddler crabs, also seem to arise from such comparable conditions (Sturmbauer et al.). Nevertheless, *identical* convergences producing *indistinguishable phenotypes* are rarely expected since organisms colonizing different areas do not ordinarily carry identical alleles nor do they ordinarily experience identical environmental variables in identical historical order.

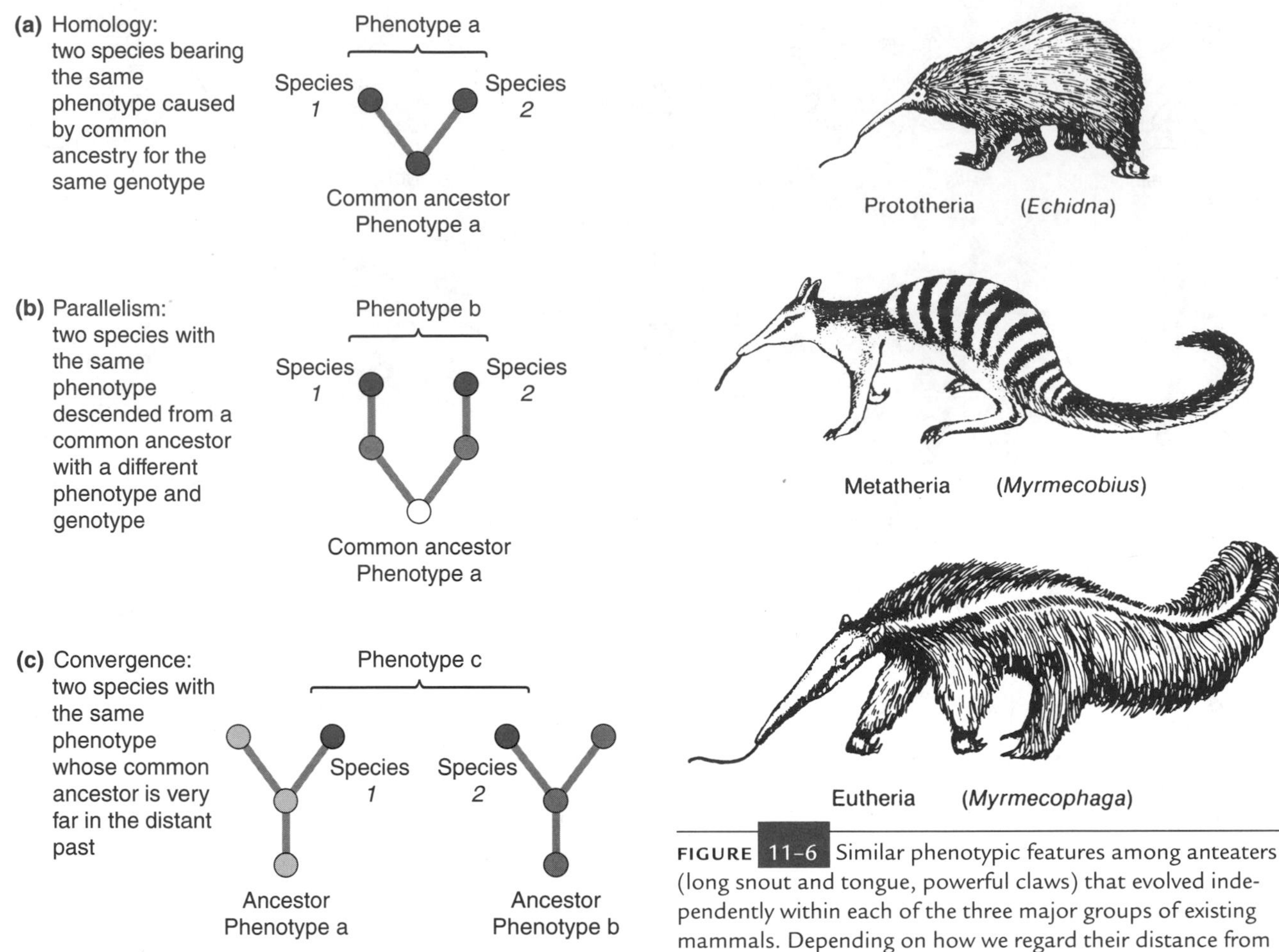

FIGURE 11-5 The phenomena of homology, parallelism, and convergence diagrammed for two species, labeled *1* and *2* that are phenotypically similar for the state in which a particular character appears (for example, long, short, round, colored). Note that the distinction between parallelism and convergence may be arbitrary, because there are no rules that specify how far in the past one can establish a common ancestor for parallel evolution, and even convergent lineages have common, albeit distant, ancestors.

FIGURE 11-6 Similar phenotypic features among anteaters (long snout and tongue, powerful claws) that evolved independently within each of the three major groups of existing mammals. Depending on how we regard their distance from a common ancestor, these forms can be considered either parallel (sharing mammalian ancestry) or convergent (descended from different mammalian groups).

it seems that the more distant a factor is from its informational source in this sequence, the greater the impact of environmental nonhereditary agents, and therefore the greater the opportunity for convergent influences to select for a similar form in unrelated organisms.

Ideally homology should be evaluated on the molecular genetic level, although convergent events can occur even there, as discussed in the next chapter (p. 265, and Box 12–3). Moreover, as already indicated, there are different organismic levels produced by a variety of metabolic and developmental pathways that are often complex, consisting of information from many genes. Homology for one gene in a pathway does not necessarily extend to all others (Fig. 3–1), nor does homology for one segment in a gene sequence necessarily extend to all other segments.

However determined, homology is the basis for establishing phylogenetic lineages, since the sharing of any character because of common descent signifies a closer relationship between organisms than any other cause for such similarity. Convergences should therefore be minimized so that acceptable phylogenies relate taxa that have fewer mutational convergences and more ancestrally derived homologies. In general, the observation that some shared characters between species may derive from very distant ancestors and others from more recent ancestors, provides a means for creating a phylogenetic tree.

For example, Figure 11–7 shows a hypothetical tree constructed from information on four characters. The eight modern descendants (species *H–O*) show they share a character (4) derived from a distant common ancestor (*A*). Although this primitive character unites all eight species, detailed branching of the tree depends on information from less primitive characters derived from more immediate ancestors. Thus, species *H–K* are related to a more recent common ancestor *B* because they share character 2 in addition to character 4. Similarly, species *L–O* are related to the more recent common ancestor *C* by

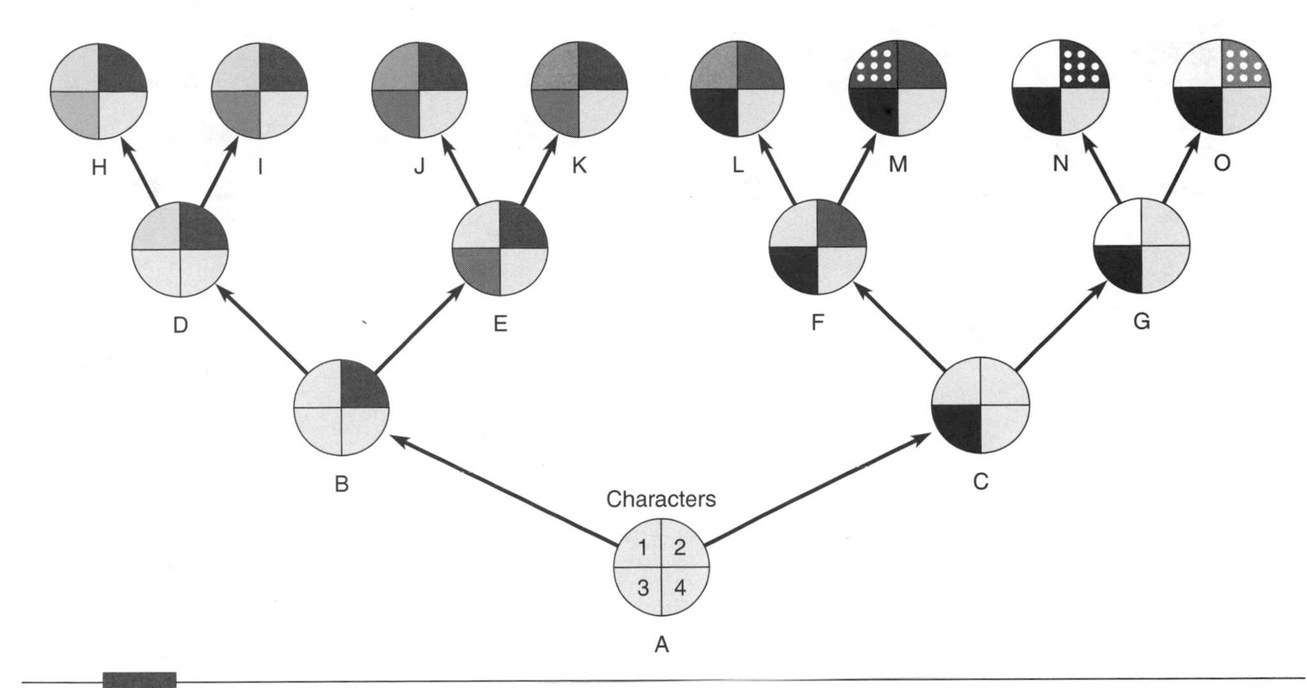

FIGURE 11-7 A simplified phylogenetic tree based on tracing four characters among 15 species (8 present and 7 ancestral), each bearing variations of these characters ("character states") that allow connections to be established. As the text explains, characters that species share indicate evolutionary relationship; for example, they all share character 4, species *H–K* share character 2, species *L–O* share character 3, species *H–I* also share 1, and so forth. The shared characters indicate common descent from species in which these variations first arose: ancestral species *A, B–C, D–G.* Note, however, that a character may be shared because of parallelism or convergence rather than because of common ancestry. In this illustration, species *K* and *L* share character 1, although they apparently derived from different lineages (ancestors *E* and *F*) in respect to characters 2 and 3.

sharing character 3. Using these characters, species *H* and *I* can be derived from an even more recent common ancestor *D* because character 1 is shared by both. Similarly: *J* and *K* derive from *E* (shared character 2); *L* and *M* from *F* (shared character 2); and *N* and *O* from *G* (shared character 1).

Again, in comparing organisms, biologists consider many characters, and it is necessary to realize that some similarities may be caused by homology (common genetic origin) and others by parallelism or convergence (independent genetic origin). Perhaps an extreme example is to note that the similarity between the shape of the pectoral flipper in mammalian whales and the shape of the pectoral flipper in ancient reptilian ichthyosaurs undoubtedly arises from convergence caused by independent selection for swimming efficiency, yet many structural features of both appendages are homologous because they derive from the forelimb of a common land vertebrate ancestor. Determining the phylogenies of these organisms depends on separating the homologies (they are both vertebrates) from the convergences or parallelisms (they are not members of the same vertebrate class).[10]

In some related lines, however, such distinctions are not always easy to make. We can define mammalian fossils, for example, by various criteria [lower jaw consisting of a single bone (dentary), a dentary-squamosal jaw joint, and ear with three ossicles, Chapter 19] and can trace their ancestry back to an ancient order of reptiles, the Therapsida. Using these criteria, some paleontologists claim that different groups among the therapsids may have independently evolved the mammalian level, or **grade,** of organization and function because of parallel evolution (Kermack and Kermack; see also Miao); that is, mammalian features that some lines of early mammals share do not result from homology but from parallelism. Biologists use the label **polyphyletic evolution** for such cases, in which different sets of organisms arrived independently at a particular grade of organization, in contrast to **monophyletic evolution,** in which sets of organisms derive from only a single ancestral population (Fig. 11–8). According to Simpson (1961), one extreme polyphyletic view suggested that each mammalian species arose separately from a single ancestral protozoan species.

[10]Some authors have introduced "partial homology" to describe instances where only a portion of either a structure or molecular sequence shared between two or more species can be traced to a common ancestor. That is, homology is the *extent* to which different species share ancestral features. Other disputants mostly ignore ancestry, and restrict the term homology to morphological and developmental *similarities.* (For a discussion of these definitions, see Donoghue.) In general, the term *homology* most often indicates both similarity *and* common ancestry, the former arising from the latter.

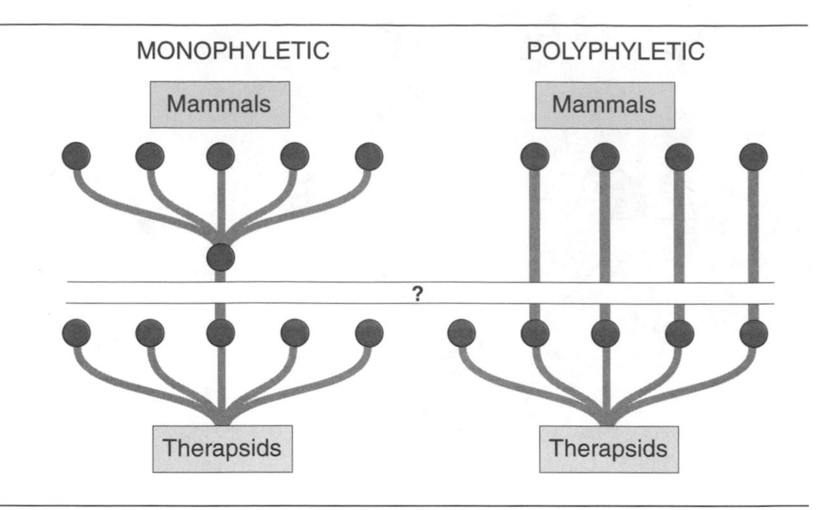

FIGURE 11-8 Monophyletic and polyphyletic schemes that can be used to explain the evolution of mammals from thereapsid reptiles. In monophyletic evolution only a single therapsid group served as ancestor to the mammalian radiation, whereas two or more therapsid groups gave rise to mammals in the polyphyletic scheme. (Instead of monophyletic, some authors use the term *holophyletic* to describe that portion of a phylogenetic tree that includes a common ancestor and all its descendants.)

Phylogenetic Classification Problems

> The fundamental difference is that phylogeny is something that happened and classification is an arrangement of its results. Although phylogeny cannot be observed as such over periods long enough to be really significant, it existed as a sequence of factual events among real things (organisms) and in a philosophical or logical sense it is objective or realistic in nature. Classification is not. It is an artifice with no objective reality. It arises and exists only in the minds of its devisers, learners, and users (Simpson 1980).

Ideally, the most descriptive phylogenetic picture of a particular population of organisms would be a portion of a multi-limbed tree that has branched connections to all present and ancestral populations and that indicates, through these connections, its degree of relationship to all other populations (Fig. 11–9). Since as many as 10 to 30 million species of organisms may exist, both known and unknown, and undoubtedly some hundreds of millions of extinct species existed in the past, mostly unknown, a complete phylogenetic tree is impossible to create, nor could we even hope to describe or compute differences among the many possible trees. For 50 existing taxa alone, 3×10^{76} possible different phylogenies would connect them to a common ancestor, exceeding the estimated number of protons and neutrons in the universe (2.4×10^{70}). Nevertheless, a picture of some sort is desirable, and evolutionists have usually placed much of this descriptive burden on how organisms are classified.

Traditional classification, as in Linnaeus's system (p. 10), did not depend on evolutionary criteria but on what seemed to be "natural." The attempt to use natural relationships to organize the many groups of organisms being discovered into simpler but fewer categories, prompted hierarchical classifications in which biologists placed an organism not only in a particular species but

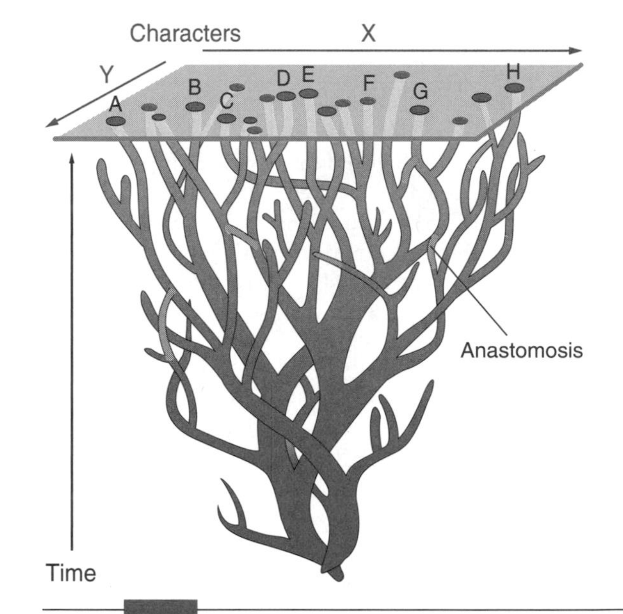

FIGURE 11-9 Diagram of a phylogenetic tree of related populations shown as continuous branches undergoing evolutionary changes through time. Some populations have become extinct, and others have merged ("reticulate" evolution) or diverged to produce new and different forms. If we consider time as the vertical *(Z)* axis in this illustration, distances along the *X* and *Y* axes might indicate measurements of different genetic traits or groups of traits. Thus, the differences between some populations, such as A and H at the present time level (*top of illustration*), may be sufficiently great to warrant separate taxonomic designations, whereas others (for example, D and E) may not yet be taxonomically distinct. Note also that convergences between two separate lineages (for example, B and C) in respect to the measured traits can conceal their evolutionary separation. (*Adapted from Levin.*)

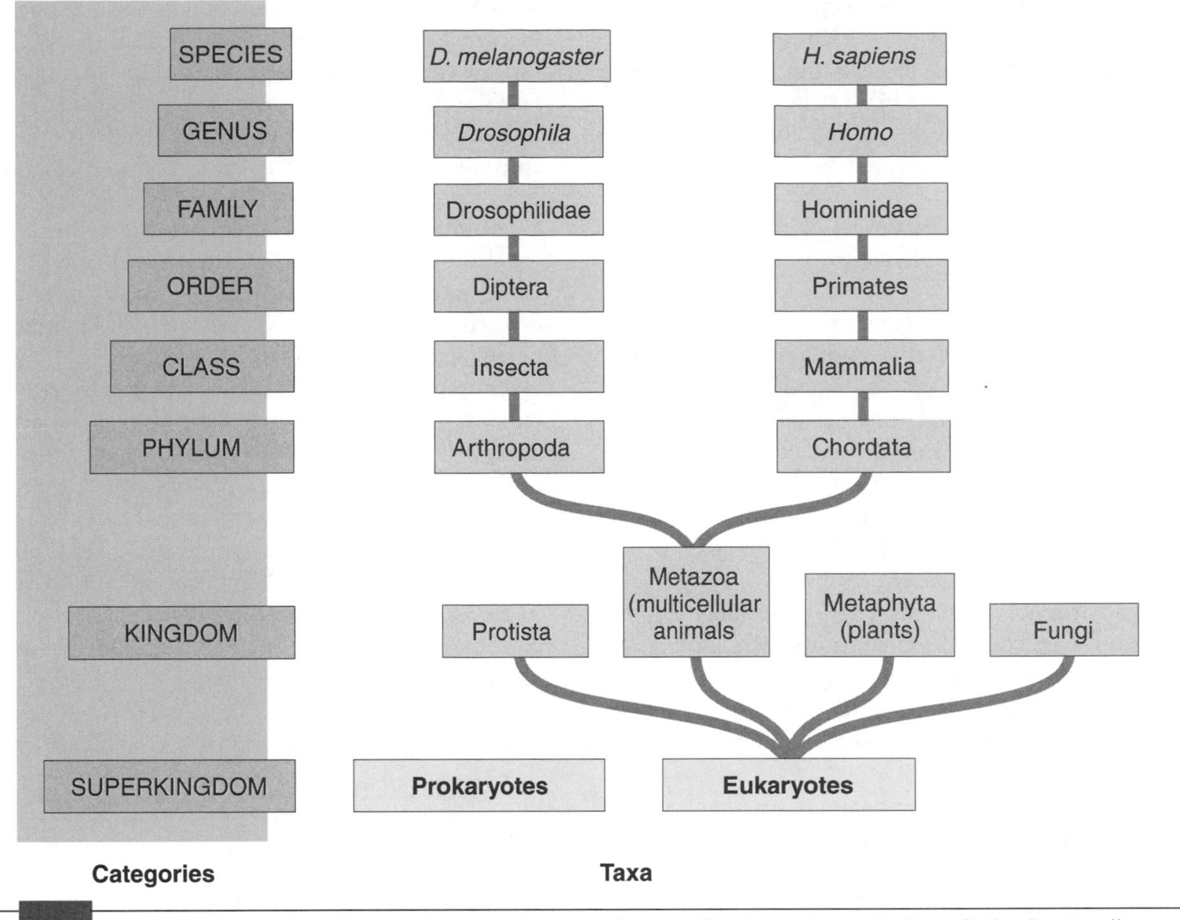

FIGURE 11–10 A classification of fruit flies and humans. Among other classification schemes is that of Wheelis, Kandler, and Woese, who substitute three "domains"—Archaea, Bacteria, and Eucarya—for the prokaryote and eukaryote "superkingdoms" illustrated here (see also Fig. 9-16.) Revisions in the "kingdom" category have also been suggested, numbers ranging from 2 ("plants" and "animals") to as many as 13. A popular classification system first proposed by Whittaker (see Margulis) used five major kingdoms: Monera (prokaryotes), Protista (unicellular eukaryotes such as protozoa, algae, slime molds, and some other groups), and the multicellular Fungi, Plantae, and Animalia. However, the wide separation among many prokaryote and eukaryote branches shown in Fig. 9-16 indicates the inadequacy of usual "kingdom" classifications when evaluated on a molecular level: plants, for example, are less separated from fungi or from animals than are microsporidia from diplomonad protistans.

also in ranked categories that included other species (the genus), other genera (the family, the order), and so forth. We still use this system today, although somewhat extended since the time of Linnaeus.

Taxonomists call each unit of classification, whether it be a particular species, genus, order, or whatever, a **taxon,** and give it a distinctive name. They arrange taxa in nested categories so that a taxon in a "higher" category includes one or more taxa in "lower" categories. For example, Figure 11–10 shows some of the categories and taxa often used in classifying humans (*Homo sapiens*) and fruit flies (*Drosophila melanogaster*).[11]

[11]Some classifications introduce additional categories to those shown in Figure 11–10, either by adding new terms (for example, *tribe,* a rank between family and genus), or by adding the prefixes *super-*, *sub-*, and *infra-*, to the given categories. Thus, the class Mammalia is usually included within the subphylum Vertebrata and the order Primates in the infraclasss Eutheria.

This mode of classification offers a simple scheme for identifying and cataloguing large numbers of species. For example, we can use phyla such as Arthropoda and Chordata to distinguish many animals and use mammalian orders such as Primates and Rodentia to distinguish many mammals. From an evolutionary point of view, this classification also has much to commend it, since there are obvious homologies between species within taxa such as Diptera or Primates. To a significant extent, these classification schemes reflect an underlying phylogenetic pattern in which each taxon seems to have originated from the one in which it is included.

Unfortunately, each traditional taxon is not always clearly monophyletic or derived from a single common ancestor. Mammals, as explained previously, may have had a polyphyletic origin. Also, some biologists believe that the arthropod grade of organization, characterized by an exoskeleton with jointed appendages, was achieved by several segmented wormlike organisms undergoing parallel evolution; some gave rise to the ancient trilobites, while others served as ancestors to groups such as crustaceans and insects (Chapter 16). Traditional monophyly assumed for some smaller taxonomic groups such as the mammalian order of insectivores, has also been refuted (Chapter 19, p. 454).

Some botanists have also claimed that major plant taxa—bryophytes, tracheophytes, gymnosperms, and angiosperms—may have had multiple origins (see, for example, discussions in Stewart and in Thomas and Spicer). If these taxa, as well as others, are polyphyletic, then the correlation between classification and phylogeny can hardly be exact: the taxon Arthropoda is not really a single phylogenetic lineage, nor is this necessarily true for the taxon Mammalia, or some others.[12]

A further serious problem is that classification at this level, like species concepts, has arbitrary elements. For example, different taxonomists working with organisms such as arthropods or angiosperms commonly dispute which groups are to be classified into which families. As with species, there are also "splitters" and "lumpers"; for example, some taxonomists split the various catlike mammals into 28 genera while others combine them into one or two genera.

Different opinions also exist about whether to consider a particular group to have attained one category or a higher one. For example, some authors consider nematodes as a class within a phylum Aschelminthes, and others consider them an independent phylum. The existence of so many possible arbitrary decisions for classifying organisms above the species level has led to a variety of proposals designed to make classification more objective.

Phenetics

One proposal, usually called *phenetic classification*, applies numerical taxonomy to arranging groups into genera and higher ranks. It has the advantage, previously described, of evaluating similarities and differences on an objective numerical basis, thereby overcoming the personal prejudices of individual taxonomists. Unfortunately, phenetics can obscure phylogenetic relationships among different taxa if some or many of the characters used in the numerical analysis have become similar because of parallel or convergent evolution. In such cases, the numerical correlations are considerably enhanced among taxa that from an evolutionary point of view should really be separated.

Also, the significance of correlations among taxa depends on presuming that the analyzed characters have evolved independently; yet it is often difficult, if not impossible, to show that all the characters used in numerical taxonomy are independent of each other, since a character such as wing length may depend on a character such as body length (p. 238). Also, some characters may have evolved at a different pace from others if they are tied to unique functions or to particular parts of the life cycle. For example, Michener has reported that a numerical taxonomic analysis of allodapine bees (a group that raises its larvae in common burrows rather than isolated cells) gives one classification for larval characteristics and quite another for adult characteristics.

Cladistics

An approach to classification that relies more on branching than does numerical taxonomy is **cladistics** (also called **phylogenetic systematics**). In this system (whose major proponent was Hennig) every significant evolutionary step marks a dichotomous branching that produces two genetically separated **sister taxa** equal to each other in rank. Since the ranking of such sister groups is below the rank of the parental group that gave rise to them, the hierarchy or ranking of groups derives logically from their genealogical position.

For example, birds and crocodiles derive from a common reptilian stem ancestor (Hedges) and cladists consider them to be sister groups of equal rank in the taxon Archosauromorpha. This taxon, in turn, ranks lower than the class of reptiles from which it arose. That is, cladists do not follow the usual classification of ranking birds (Aves) as a class separate from the class of reptiles (Reptilia), since that would imply that birds arose from a primitive prereptilian stem ancestor rather than from the reptiles themselves.[13] To cladists, groups that do not include *all* the descendants of a common ancestor (for example, the class Reptilia does not include their

[12]Simpson (1961) tried to deal with this difficulty by defining monophyly as derivation of a group from an ancestral taxon of equal or lower rank, even if more than one population in this taxon are the direct ancestors. Thus, he considered the class Mammalia as monophyletic since the different reptilian populations from which it derived were all in the same order (Therapsida). Ghiselin (1984) points out that consistently applying this view could lead to the absurdity that the vertebrate phylum is monophyletic even if birds came from echinoids and mammals from crinoids—because both these hypothetical ancestral groups are members of the same echinoderm phylum. Perhaps an alternative solution for eliminating the problem of polyphyletic origin of mammals would be to remove the taxon Synapsida (a group that includes therapsids) from the class Reptilia and place it in the class Mammalia. Adoption of this approach depends, of course, on how prone taxonomists are to accept the early, obviously reptilelike synapsids as mammals.

[13]To some cladists (Padian and Chiappe), birds, which are related to dinosaurs (p. 431), "are not only *descended* from dinosaurs, they *are* dinosaurs (and reptiles)." Following such reductive logic, many, if not all, major class distinctions can be homogenized into a conglomerate mass: mammals are reptiles, reptiles are amphibians, and so forth.

mammalian or avian offshoots) are **paraphyletic** and are not valid taxa. Similarly, cladistic schemes exclude polyphyletic groups such as Arthropoda and others that consist of convergent or parallel lineages (homoplasy). A group of taxa or clade can achieve taxonomic status only if it is strictly monophyletic and there are no ambiguities: the group must include its common ancestor and *all* of the common ancestor's descendants.[14]

Cladists stress separating ancestral (**plesiomorphic**) characters from newly derived (**apomorphic**) characters and emphasize the latter to establish phylogenies. The sharing of derived characters (**synapomorphy**) dictates the phylogeny. For example, how species A, B, and C relate can be determined by noting which share a newly derived character. Thus, unique character X shared by species A and C indicates a closer relationship than to species B; A and C branched off together from a common ancestor, whereas species B branched off separately. This emphasis on using shared derived characters provides more precise branching information than phenetic classification which offers only general estimates of similarity but no such direct method of generating branching dichotomies.

When biologists extend this approach to many taxa and more than one character, they can infer fairly complex phylogenies by various mathematical methods (for example, Felsenstein). One popular technique is to choose that phylogenetic tree for the data that minimizes the number of changes necessary to explain its evolutionary history (the "parsimony method"; see also Chapter 12).

To use the simplified example shown in Figure 11–11, if one phylogenetic tree requires that a particular character have evolved on more separate occasions than it would have had to evolve in another tree, the latter tree is preferred because it minimizes extraneous evolutionary explanations. Thus, because only a single mutation is necessary to produce a shared derived character, synapomorphy lets us attribute parsimony to cladogram *a*, which has a total of two such mutational events. Cladogram *b*, in contrast, is based on a shared ancestral character (**symplesiomorphy**), and is obviously not as parsimonious, bearing a total of three mutational events.

Because of its formal terminology, precise rules, and strict genealogical consistency in assigning branching points and patterns cladistic classification has become highly popular among taxonomists (see, for example, Ax, Forey et al., Minelli, Ridley, Wiley 1981).[15] This popularity has brought terms such as "sister group," "synapomorphy," "paraphyly," and "homoplasy" into common use. Nevertheless, cladism has long excited controversy, and the following represent some objections (see also Mayr and Ashlock, Sokal 1975, Ghiselin 1984, Panchen, and the historical account by Hull):

1. It seems unwarranted to assign a completely new taxonomic designation to each branch if one branch remains identical to the previous ancestral population (Fig. 11–12). Certainly, a single ancestral taxon giving rise to new offshoot taxa may be no different after the split from before the split. Moreover, declaring that an ancestral species has been suddenly transformed into a new species because it produced a new taxon, imposes a biologically meaningless criterion for speciation that seems more like "instantaneous creation" than evolution.

2. Dichotomous branching oversimplifies the evolutionary process, since lineages can also be "reticulate." A lineage can connect genetically with related lineages through hybridization (Chapter 24), or through "horizontal gene transmission" from even more distant

[14]Many noncladists would claim that any partition of taxa into groups, whether paraphyletic, polyphyletic, or monophyletic, is done for the convenience of taxonomists, and not because one partition is more "real" than another. Although there is continuity between an ancestor and its descendants, and taxonomy must take into account evolutionary relationships, schemes for assembling descendants into groups often seem mostly a "mental" construct. Thus, the urge to define taxa strictly according to their cladistic relationships has also led some cladists to attempt replacing the biological species concept (p. 239) with a "phylogenetic species concept." Cracraft, for example, proposes that a "phylogenetic species" is a monophyletic group composed of "the smallest diagnosable cluster of individual organisms within which there is a parental pattern of ancestry and descent." As is obvious, the "diagnosable" element in this definition is based on similar, if not identical, characters and measurements used in numerical taxonomy, and leads to similar problems of defining such characters and evaluating such measurements. In addition, emphasis on the "*smallest* diagnosable cluster" can easily result in declaring species subunits, such as races, varieties, and even smaller genetically distinct groups, as separate species—an approach that many biologists would consider an unwarranted and unrealistic multiplication of species. For example, although the many varieties of North American brown bears are generally classified into a single species, it would seem folly to separate them now into 90 species because a "splitter" taxonomist once proposed they constitute that number of subspecies (see Talbot and Shields).

[15]Not all cladists agree that there is either a necessary or desirable correspondence between cladograms and evolutionary relationships. To some, called "transformed cladists" or "pattern cladists," the goal of systematics is to discover and delineate the patterns of nature, and taxonomists can best achieve this by restricting attention to the shared derived characters of organisms however they evolved. According to this view, evolutionary mechanisms and processes only divert attention from the essential core of systematics: to clearly define taxa through the characters used in cladograms and place these taxa into a "natural order . . . basically independent of evolutionary theory" (Rieppel). In other words, to associate and compare taxa without explaining their connections. Since cladograms depend on distinguishing derived, ancestral, and convergent characters—all of which arise from evolutionary phenomena—the proposed "independence" of cladograms from evolution seems unreal. To many cladists, as Ghiselin (1984) argues, cladograms "are nested sets of characters, not descriptions of the underlying historical order." A cladistic taxon cannot evolve from another taxon of the same category, but must be "nested" within a taxon of higher category. Thus, the class Mammalia is not a major vertebrate group evolved from the class Reptilia, but is a branch within the Therapsid taxon, which is a branch within the Synapsid taxon (Chapters 18–19), which branches from higher taxa that nest successively into amphibianlike taxa, fishlike taxa, and so forth. Rather than trying to understand relationships among organisms, cladistic methods for obtaining phylogenies seem to have become ends in themselves.

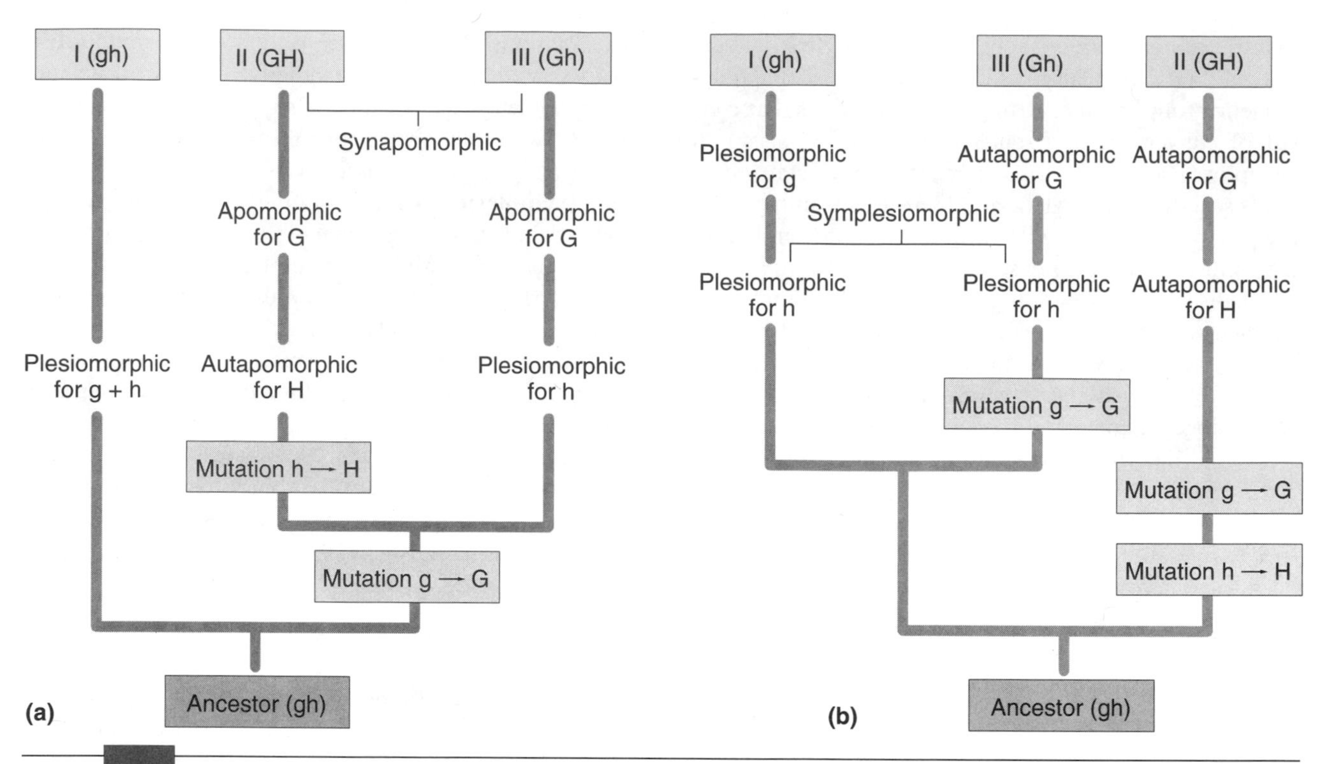

FIGURE 11-11 Cladograms for variations of two characters g and h in hypothetical taxa, I, II, III, and their presumed common ancestor. Taxon I has the ancestral (plesiomorphic) characters g and h, taxon II has derived (apomorphic) characters G and H, and taxon III is apomorphic for G and plesiomorphic for h. As shown, two alternative cladograms are possible. In (*a*) the assumption is made that character G shared by taxa II and III evolved only once by mutation of g → G, and character H evolved in the lineage of taxon II also by a single mutation of h → H. Thus taxa II and III share the derived G character (synapomorphy) and are sister taxa, while taxon I, in turn, is the sister group of (II + III). In (*b*) the assumption is made that I and III are sister taxa, because they share ancestral character h (symplesiomorphy), and taxon II is the sister group of (I + III). Note, however, for the (*b*) tree to be accepted two mutations of g → G are necessary (as well as one of h → H). Thus, by the principle of parsimony, the (*a*) cladogram based on synapomorphy, requiring a total of only two mutations, is preferable to the (*b*) cladogram based on symplesiomorphy, which requires three mutations. Derived character states that are considered unique rather than shared with other lineages are called **autapomorphies,** and do not indicate relationship.

lineages (p. 240). Such phylogenies should therefore be considered as networks rather than as simple dichotomous branching trees.

3. A parental taxon that produces a new taxon becomes the sister group of its offspring, essentially annulling its ancestral relationship. This leads to considerable difficulties in acknowledging ancestral taxa. In fact, some cladists suggest that because fossils are relatively rare and often morphologically incomplete, we should ignore the fossil record and base cladograms entirely on living forms. Such cladograms arbitrarily dismiss the hard evidence for evolution and can overlook phylogenetic sequences: Would we have noted the relationship between dinosaurs and birds without fossils such as *Archaeopteryx* (p. 431)?

4. Although anagenetic changes in a species can occur through time (see Figs. 3–2 and 11–3), cladistic classification proposes to use the same taxon name throughout evolution as long as a taxon has produced no discernible branches. Obviously, this method of classification does not recognize phyletic evolution.

5. Since cladists recognize only derived (apomorphic) characters as adaptations, they ignore the likelihood that ancestral (plesiomorphic) traits can also be adaptive—maintained by selection.

6. The cladistic dichotomous scheme is restricted to only two new species at each fork, yet a taxon may produce more than two offshoot taxa (for example, by multiple budding at the periphery of its distribution).

7. Various cladists have tied the ranking of taxa into groupings (such as families, orders, and classes) according to their chronological age. That is, branching during the Precambrian eon produced phyla; classes arose during the Cambrian–Devonian period; orders stem from the Mississippian–Permian, and so on. According to this reckoning, taxa that arose at an early age but were limited to one or very few species (such as the brachiopod genus *Lingula,* a liv-

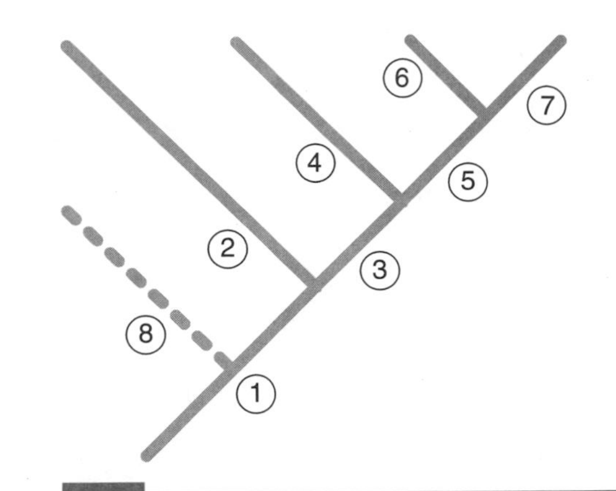

FIGURE 11-12 Illustration of the kind of diagram commonly used in cladistic classification (called a *Hennigian comb*), where each lineage that extends from the node of a branched fork represents a distinct taxon. Thus populations 1, 3, 5, and 7 are given different taxonomic designations although they may not be recognizably different genetically. If a new lineage of organisms *(indicated by the dotted line 8)* is discovered to have split from population 1, cladists will change the taxonomic designation for 1 so that it differs before and after this split. Such proceedings are based on the concept that any split produces two new sister groups of equal rank, neither of which is ancestral to the other, and the taxon that gave rise to these sister groups has become extinct.

ing fossil dating back to pre-Devonian times) should rank higher than some taxa that were produced at a later stage but include large numbers of genera and species (such as mammals and birds).[16]

8. Phylogenies that include only synapomorphic characters can omit evolutionary information when a cladistic tree does not accord with its genetic basis. For example, formal cladograms that distinguish groups because of nonhomologous differences in eye development (for example, mollusks and mammals, Fig. 3–1) overlook their control by homologous genes such as *pax-6* (p. 354). Conversely, cladistic phylogenies that associate groups because they are synapomorphic for lens crystalline proteins overlook their production by nonhomologous genes (Table 12–2, p. 262). Homology (synapomorphy) and convergence (homoplasy) may thus apply to different stages (levels) of character resolution—a point ignored in cladistic cladograms.

9. The rule that each genealogical branching point or node must rank lower than the previous one means that cladistic taxonomists must offer an unwieldy proliferation of categories for previously simple classifications: each node requires a distinctive rank. For example, most noncladist classifications rank the order Primates below the class Mammalia by two intermediate categories, the subclass Theria (mammals that produce live young) and the infraclass Eutheria (placental mammals). In contrast, one cladistic classification of primates offers ten intermediate categories (Table 11–1), which may well expand further if paleoanthropologists recognize more genealogical branches between primates and early mammals. Thus although this method can offer genealogical accuracy, it clearly can lose informational simplicity.

[16]Such cladistic proponents seem to hold an underlying assumption that new branches occur at a fairly constant rate and therefore that an older group has necessarily branched into a larger number of taxa than a younger group. This concept seems supported by the burst of new major multicellular animal phyla that appeared more than 500 million years ago during the Cambrian period (Chapter 14) and subsequently radiated into diverse subcategories such as classes and orders. However, this age-dependent diversification is not always reliable, since some groups classified as major animal phyla probably evolved later. Moreover, even modern groups can branch extensively over relatively short periods, as evidenced by the large numbers of Hawaiian Drosophilidae (more than 800 species) that are descended from perhaps one or two ancestral species that arrived in those islands no earlier than perhaps 30 or so million years ago (Fig. 22–15). Similarly, more than 40 species of Hawaiian avian honeycreepers, sporting many differently shaped beaks for extracting floral nectar, descended from a single species of finch that emigrated to these islands early in their history.

Evolutionary Classification

Evolutionary classification, a long-standing alternative to the taxonomic systems just described, dates from the post-Darwin period. Its major modern proponents (Simpson, Mayr) use the traditional taxonomic categories but give special consideration to evolutionary relationships and biological attributes rather than to strict morphological relationships. That is, incorporated into group taxonomy are factors such as functional and morphological innovations, adaptive range, and numbers of species. Evolutionary classification seeks to provide both the genealogical relationship between groups and the amount of evolutionary change or distance between them.

Unfortunately, attempts to provide a classification based on evolutionary distances can distort phylogenetic relationships and vice versa. Thus, although biologists designate birds and reptiles as distinct vertebrate classes because they have become evolutionarily widely separated, it is also clear, as we have seen, that such classification is phylogenetically misleading: the classification does not always reflect phylogeny because it includes paraphyletic groups such as Reptiles, and some might even include polyphyletic groups such as Arthropoda.

Moreover, biologists do not agree on how to evaluate evolutionary distance and how to relate it to classification. That is, one can debate which characters to measure (heads? legs? wings? tibiae? fibulae?) and what evolutionary

TABLE 11-1 Classification of primates according to one cladistic proposal

- Class MAMMALIA
  - Subclass THERIA
    - Superlegion TRECHNOTHERIA
      - Legion CLADOTHERIA
        - Sublegion ZATHERIA
          - Infraclass TRIBOSPHENIDA
            - Supercohort EUTHERIA
              - Cohort EPITHERIA
                - Magnorder PREPTOTHERIA
                  - Superorder TOKOTHERIA
                    - Grandorder ARCHONTA
                      - Order PRIMATES

Source: Adapted from McKenna.

significance to assign to differences between measurements. Measurement differences are also not consistent for different characters, and many such characters appear only in relatively few groups, which narrows comparisons of evolutionary distance. Obviously there are arbitrary elements in this system of classifying organisms, and its phylogenies may be less reproducible than those of cladists who follow more rigorous guidelines.

Concluding Remarks

From the various critiques of all these taxonomic systems we can see that, to many biologists, the desirable goals of classification include both (1) the arrangement of groups into a pattern that accurately reflects their evolutionary relationships and (2) the placement of groups into a reference system so their major features are easily and efficiently described and identified (information storage and retrieval). No single classification system reviewed so far fully accomplishes these two purposes.

Traditional morphological classification and numerical taxonomy may simplify the placement of organisms into a classification scheme but can ignore their evolutionary relationships. Cladistic classification may offer advantages in clarifying some evolutionary patterns but overlooks other patterns or seriously complicates the use of taxonomic information. Evolutionary classification can be more arbitrary than cladistics by classifying a group such as birds outside its strict phylogenetic sequence, but it can also offer descriptive and evolutionary information in a more useful form.

The demand for flexibility and change in classification has become apparent as new molecular information challenges formerly accepted phylogenies. For example, some researchers have recently shown that cetaceans (whales, dolphins, porpoises), which were formerly considered a separate mammalian order, are linked to an artiodactyl subgroup, either that of cows (Graur and Higgins) or hippopotami (Gatesy); others propose to reclassify the former monophyletic phylum Chordata as polyphyletic (Turbeville et al.).

Also complicating strict genealogies are horizontal gene transfers between species of distinctly different lineages, producing "chimeras" (pp. 180–182, 225–226, 240). Because such events may have been common, especially in early evolutionary periods, Doolittle points out:

> The integrity of organismal lineage surely has been violated by gene transfers, endosymbiosis, and "genome fusions," large and small in their consequence. . . . No single philosophy of systematics will give us the "right" answer about species history because there is no such right answer. But there will be reasonable compromises and generalizations that allow us to talk usefully about the history of life on this planet.

The value of evolutionary classification probably lies in that it synthesizes some essential attributes of other systems. The flexibility provided by the eclecticism of this system lets it offer more evolutionary features than traditional or numerical taxonomy, as well as more simplicity, and even, at times, more evolutionary information than cladistic classification (see, for example, Carpenter).[17] To many of its proponents, the irregularity of evolutionary classification is not a failing, but stems from the irregularity of evolution itself. An ideal biological classification system may be as elusive as fitting all organisms into a single ideal species concept.

[17] In Mayr's summary (1995):

> The Hennigian reference system [cladism] and the traditional classification are specially suited for different objectives. If mere genealogy, mere line of descent, is the information that is wanted, the Hennigian system is superior. However, if the student wants relatively homogeneous taxa, largely based on similarity and on the degree of genetic relationship, also reflecting their niche occupation, a traditional evolutionary classification is preferable.

SUMMARY

In its earliest stages, classification involved observations of similarities and differences among organisms, without regard to their origins. Since Darwin's time, many taxonomists have sought to construct a system that would reflect phylogenetic (genealogical) relationships, but fossil evidence of exact ancestral relationships, the ideal basis for such schemes, is often lacking. The main goals of the evolutionary taxonomist are to recognize the basic unit of classification—species—and to order them into as realistic a phylogenetic scheme as possible.

Taxonomists have used various methods to define species. Until recently they used morphology almost exclusively, but to increase exactitude, they have devised newer systems. In phenetics or numerical taxonomy, researchers assign numerical values to characters and use a cluster of these values to define a species. Systematists intended the biological species concept to overcome the subjectivity inherent in the preceding methods and used the ability of populations to interbreed and produce viable offspring as the primary criterion for determining species boundaries. However, because of the enormous number of species, their geographic dispersal, and limitations in space and human resource, and because most species are extinct, it is impossible to differentiate many species by these criteria.

The evolutionary species concept represents an effort to resolve these difficulties by defining species based on their evolutionary isolation from each other. Ideally, this method uses morphological, genetic, behavioral, and ecological variables, although it too does not resolve all the problems intrinsic to species taxonomy, since not all traits evolve at the same rate or in the same sequence.

Reconstructing phylogenies is even more difficult than defining species. Phenotypes may be alike because of common origin (homology), because of similar evolutionary patterns arising separately in different lines from not-too-distant ancestors (parallelism), or because of the development of similar characteristics in groups originating from completely different ancestors (convergence). (Because it can be difficult to distinguish between "not-too-distant" and "completely different," parallelism and convergence are commonly included in the term *homoplasy.*)

Traditional classification systems use hierarchical schemes based on natural criteria to order organisms into taxa. But these taxa are not necessarily monophyletic (coming from a common ancestor), and their designation may be arbitrary. Proponents of phenetics seek to make classification more objective by applying numerical methods to taxa, while cladists assume that evolution occurs in a series of dichotomous branchings, each evolutionary branch point giving rise to taxa of equal rank. Major difficulties arise with phenetics if convergent or parallel evolution has occurred, and cladistics does not deal adequately with phyletic evolution or multiple branching.

Evolutionary classification uses traditional taxonomic categories and morphological criteria, and incorporates many other biological factors as well. In this way biologists hope to depict genealogical relationships as well as evolutionary distance. However, resolving these issues within a single classification system remains elusive.

KEY TERMS

allopatric populations
anagenesis
apomorphic
autapomorphic
biological species concept
clade
cladistics
cladogenesis
classical taxonomy
classification
convergence
evolutionary classification
evolutionary species concept
grade
homology
homoplasy
monophyletic evolution
numerical taxonomy
orthogenesis
parallelism
paraphyletic
phenetic taxonomy
phyletic evolution
phylogenetic branching
phylogenetic systematics
phylogeny
plesiomorphic
polyphyletic evolution
polytypic species
reproductive isolation
serial homology
sibling species
sister taxa
species
sympatric populations
symplesiomorphy
synapomorphy
systematics
taxon

DISCUSSION QUESTIONS

1. Define systematics and phylogeny. Are these concepts connected? (Explain.)
2. Why can't the fossil record provide phylogenies for all organisms?
3. What difficulties do we encounter in defining a species in a manner that lets us apply this definition universally? (Discuss such species concepts as morphological species, numerical species, biological species, and evolutionary species.)
4. Which proposals for different numbers of superkingdoms (two, three, or more) would you support? (See Fig. 11–10.)
5. Would you agree with Darwin that the attempt to find "the essence of the term species" is a "vain search" (p. 239)?
6. How do parallel and convergent evolutionary events affect phylogenetic determinations?
7. Would you include instances of polyphyletic evolution in phylogenies? Why or why not?

8. What difficulties arise in applying traditional morphological classification, phenetic classification, cladistic classification, and evolutionary classification to devise phylogenies?

EVOLUTION ON THE WEB

Explore evolution on the web! Visit the accompanying web site for *Evolution,* 3/e at www.jbpub.com/evolution for web exercises and links relating to topics covered in this chapter.

REFERENCES

Ax, P., 1987. *The Phylogenetic System: The Systematization of Organisms on the Basis of Their Phylogenesis.* Wiley, Chichester, England.

Baker, H. G., 1970. Taxonomy and the biological species concept in cultivated plants. In *Genetic Resources in Plants,* O. H. Frankel and E. Bennett (eds.). Blackwell, Oxford, England, pp. 47–68.

Benton, M. J., 1991. *The Rise of the Mammals.* Apple Press, London.

Brooks, D. R., and E. O. Wiley, 1985. Theories and methods in different approaches to phylogenetic systematics. *Cladistics,* **1,** 1–11.

Carpenter, K. E., 1993. Optimal cladistic and quantitative evolutionary classifications as illustrated by fusilier fishes (Teleostei: Caesionidae). *Systematic Biol.,* **42,** 142–154.

Cracraft, J., 1983. Species concepts and speciation analysis. *Current Ornithol.,* **1,** 159–187.

Davis, P. H., and V. H. Heywood, 1963. *Principles of Angiosperm Taxonomy.* Van Nostrand, Princeton, NJ.

Donoghue, M. J., 1992. Homology. In *Keywords in Evolutionary Biology,* E. F. Keller and E. A. Lloyd (eds.). Harvard University Press, Cambridge, MA, pp. 170–179.

Doolittle, W. F., 1996. At the core of the Archaea. *Proc. Nat. Acad. Sci.,* **93,** 8797–8799.

Eldredge, N., and J. Cracraft, 1980. *Phylogenetic Patterns and the Evolutionary Process.* Columbia University Press, New York.

Endler, J. A., 1989. Conceptual and other problems in speciation. In *Speciation and Its Consequences,* D. Otte and J. A. Endler (eds.). Sinauer Associates, Sunderland, MA, pp. 625–648.

Ereshefsky, M. (ed.), 1992. *The Units of Evolution: Essays on the Nature of Species.* Massachusetts Institute of Technology Press, Cambridge, MA. This source contains a wide range of articles on species concepts by authors such as Mayr, Paterson, Hull, Sober, Kitcher, Ghiselin, Templeton, and others.

Felsenstein, J., 1982. Numerical methods for inferring phylogenetic trees. *Q. Rev. Biol.,* **57,** 379–404.

Forey, P. L., C. J. Humphries, I. L. Kitching, R. W. Scotland, D. J. Siebert, and D. M. Williams, 1992. *Cladistics: A Practical Course in Systematics.* Oxford University Press, Oxford, England.

Gajewski, W., 1959. Evolution in the genus *Geum. Evolution,* **13,** 378–388.

Gatesy, J., 1997. More DNA support for a Cetacea/Hippopotamidae clade: The blood-clotting protein gene γ-fibrinogen. *Mol. Biol. Evol.,* **14,** 537–543.

Ghiselin, M. T., 1984. Narrow approaches to phylogeny: A review of nine books on cladism. In *Oxford Surveys in Evolutionary Biology,* R. Dawkins and M. Ridley (eds.). Oxford University Press, Oxford, England, pp. 209–222.

———, 1987. Species concepts, individuality, and objectivity. *Biol. and Phil.,* **2,** 127–143.

Graur, D., and D. G. Higgins, 1994. Molecular evidence for the inclusion of cetaceans within the order Artiodactyla. *Mol. Biol. and Evol.,* **11,** 357–364.

Hedges, S. B., 1994. Molecular evidence for the origin of birds. *Proc. Nat. Acad. Sci.,* **91,** 2621–2624.

Hendrix, R. W., M. C. M. Smith, R. N. Burns, M. E. Ford, and G. F. Hatfull, 1999. Evolutionary relationships among diverse bacteriophages and prophages: All the world's a phage. *Proc. Nat. Acad. Sci.,* **96,** 2192–2197.

Hennig, W., 1966. *Phylogenetic Systematics.* University of Illinois Press, Urbana.

———, 1975. "Cladistic analysis or cladistic classification?": A reply to Ernst Mayr. *Systematic Zool.,* **24,** 244–256.

Hull, D. L., 1988. *Science as a Process: An Evolutionary Account of the Social and Conceptual Development of Science.* University of Chicago Press, Chicago.

Janvier, P., 1984. Cladistics: Theory, purpose, and evolutionary implications. In *Evolutionary Theory: Paths into the Future,* J. W. Pollard (ed.). Wiley, Chichester, England, pp. 39–75.

Kermack, D. M., and K. A. Kermack, 1984. *The Evolution of Mammalian Characters.* Croom Helm, London.

Levin, L., 1983. *The Earth Through Time.* Saunders, Philadelphia.

Losos, J. B., T. R. Jackman, A. Larson, K. de Queiroz, and L. Rodriguez-Schettino, 1998. Contingency and determinism in replicated adaptive radiations of island lizards. *Science,* **279,** 2115–2118.

Margulis, L., 1993. *Symbiosis in Cell Evolution: Microbial Communities in the Archean and Proterozoic Eons,* 2d ed. Freeman, New York.

Mayr, E., 1969. *Principles of Systematic Zoology.* McGraw-Hill, New York.

———, 1981. Biological classification: Toward a synthesis of methodologies. *Science,* **214,** 510–516.

———, 1987. The ethological status of species: Scientific progress and philosophical terminology. *Biol. and Phil.,* **2,** 145–166.

———, 1995. Systems of ordering data. *Biol. and Phil.,* **10,** 419–434.

Mayr, E., and P. D. Ashlock, 1991. *Principles of Systematics,* 2d ed. McGraw-Hill, New York.

McKenna, M. C., 1975. Towards a phylogenetic classification of the mammalia. In *Phylogeny of the Primates,* W. P. Luckett and F. S. Szalay (eds.). Plenum Press, New York, pp. 21–46.

Miao, D., 1991. On the origins of mammals. In *Origins of the Higher Groups of Tetrapods,* H.-P. Schultze and L. Trueb (eds.). Cornell University Press, Ithaca, NY, pp. 579–597.

Michener, C. D., 1970. Diverse approaches to systematics. *Evol. Biol.,* **4,** 1–38.

———, 1977. Discordant evolution and the classification of allodapine bees. *Systematic Zool.,* **26,** 32–56.

Minelli, A., 1993. *Biological Systematics: The State of the Art.* Chapman & Hall, London.

Nelson, G., and N. Platnick, 1981. *Systematics and Biogeography: Cladistics and Vicariance.* Columbia University Press, New York.

Newell, N. D., 1959. The nature of fossil record. *Proc. Amer. Phil. Soc.*, **103,** 264–285.

Padian, K., and L. M. Chiappe, 1998. The origin of birds and their flight. *Sci. Amer.*, **278**(2), 38–47.

Panchen, A. L., 1992. *Classification, Evolution, and the Nature of Biology.* Cambridge University Press, Cambridge, England.

Paterson, H. E. H., 1985. The recognition concept of species. In *Species and Speciation,* E. Verba (ed.). Transvaal Museum Monograph 4. Pretoria, South Africa, pp. 21–29.

Patterson, C., 1981. Significance of fossils in determining evolutionary relationships. *Ann. Rev. Ecol. Syst.*, **12,** 195–223.

Ridley, M., 1986. *Evolution and Classification: The Reformation of Cladism.* Longman, London.

Rieppel, O.,1984. Atomism, transformism, and the fossil record. *Zool. J. Linnean Soc.*, **82,** 17–32.

Ross, H. H., 1974. *Biological Systematics.* Addison-Wesley, Reading, MA.

Ruse, M. (ed.), 1987. *Biol. and Phil.*, **2** (2), 127–225. (An issue devoted to species concepts.)

Simpson, G. G., 1961. *Principles of Animal Taxonomy.* Columbia University Press, New York.

———, 1980. *Why and How: Some Problems and Methods in Historical Biology.* Pergamon Press, Oxford, England.

Slobodchikoff, C. N. (ed.), 1976. *Concepts of Species.* Dowden, Hutchison & Ross, Stroudsberg, PA.

Sneath, P. H., and R. R. Sokal, 1973. *Numerical Taxonomy.* Freeman, San Francisco.

Sokal, R. R., 1975. Mayr on cladism—And his critics. *Systematic Zool.*, **24,** 257–262.

———, 1985. The continuing search for order. *Amer. Nat.*, **126,** 729–749.

———, 1986. Phenetic taxonomy: Theory and methods. *Ann. Rev. Ecol. Syst.*, **17,** 423–442.

Stewart, W. N., 1983. *Paleobotany and the Evolution of Plants.* Cambridge University Press, Cambridge, England.

Sturmbauer, C., J. S. Levinton, and J. Christy, 1996. Molecular phylogeny analysis of fiddler crabs: Test of the hypothesis of increasing behavioral complexity in evolution. *Proc. Nat. Acad. Sci.*, **93,** 10855–10857.

Systematic Biology, 1969. (Proceedings of an International Conference sponsored by the National Research Council.) National Academy of Sciences, Washington, DC.

Talbot, S. L., and G. F. Shields, 1996. Phylogeography of Brown Bears (*Ursus arctos*) of Alaska and paraphyly within the Ursidae. *Mol. Phylogenet. and Evol.*, **5,** 477–494.

Templeton, A. R., 1989. The meaning of species and speciation: A genetic perspective. In *Speciation and Its Consequences,* D. Otte, and J. A. Endler (eds.). Sinauer Associates, Sunderland, MA, pp. 3–27.

Thomas, B. A., and R. A. Spicer, 1986. *The Evolution and Palaeobiology of Land Plants.* Croom Helm, London.

Turbeville, J. M., J. R. Schulz, and R. A. Raff, 1994. Deuterostome phylogeny and the sister groups of the chordates. *Mol. Biol. and Evol.*, **11,** 648–655.

Van Valen, L., 1976. Ecological species, multispecies, and oaks. *Taxon,* **25,** 233–239.

Wheelis, M. L., O. Kandler, and C. R. Woese, 1992. On the nature of global classification. *Proc. Nat. Acad. Sci.*, **89,** 2930–2934.

Wiley, E. O., 1978. The evolutionary species concept reconsidered. *Systematic Zool.*, **27,** 17–26.

———, 1981. *Phylogenetics: Theory and Practice of Phylogenetic Systematics.* Wiley, New York.

12 Molecular Phylogenies and Evolution

Modern efforts to overcome some of the usual phylogenetic problems include the use of molecular information rather than exclusive reliance on morphological information. On the molecular level, we can get information by comparing sequences of nucleotides in various DNA and RNA molecules, as well as by comparing sequences of amino acids (and their molecular configurations) in different proteins. These techniques are obviously restricted to organisms from which we can extract such compounds, and the information gathered so far generally has been limited to only some proteins and some nucleic acid sequences, although in ever-increasing number.

Nevertheless, the advantage of a molecular approach offsets many of its sampling limitations, since evolutionary changes marked by substituting amino acids and nucleotides can be measured and compared between existing organisms no matter how greatly they differ in other phenotypic features. That is, we can compare differences between sampled molecules from different organisms on a unit scale of amino acids or nucleotides when organisms have no simple, comparable units of morphology, behavior, ecology, physiology, and so on. Molecular comparisons transcend barriers among organisms whose relationships cannot be evaluated by traditional experimental techniques, such as genetic exchange. For example, as Figure 12–1 shows, we can compare an amino acid sequence in a particular protein among humans, pigs, horses, rats, chickens, alfalfa plants, yeast, and bacteria. (See also p. 159).

Although most sequence relationships are still unknown, this close tie to molecular biology has transformed evolution from a "theoretical" explanation of historical events to an observable and continuous link among all living forms. That is, organisms, past and present, *all* can be bound together through the genetics of evolution by innumerable threads of inherited molecular sequences and their variations. Such molecular records can encompass all basic genetic information transferred between organisms, affecting all attributes, from transcribed to nontranscribed nucleotide sequences, whereas more environmentally plastic morphological relationships derive from only a small fraction of transmitted information.

Bacteria	P L F D F A Y Q G F A R G – L E E D A E G L R A F A A M H K E L I V A S S Y S K N F G L Y N E R V G
Yeast	A L F D T A Y Q G F A T G D L D K D A Y A V R X X L S T V S P V F V C Q S F A K N A G M Y G E R V G
Alfalfa	P F F D S A Y Q G F A S G S L D A D A Q P V R L F V A D G G E L L V A Q S Y A K N M G L Y G E R V G
Chicken	P F F D S A Y Q G F A S G S L D K D A W A V R Y F V S E G F E L F C A Q S F S K N F G L Y N E R V G
Rat	P F F D S A Y Q G F A S G D L E K D A W A I R Y F V S E G F E L F C P Q S F S K N F G L Y N E R V G
Horse	P F F D S A Y Q G F A S G N L D R D A W A V R Y F V S E G F E L F C A Q S F S K N F G L Y N E R V G
Pig	P F F D S A Y Q G F A S G N L E K D A W A I R Y F V S E G F E L F C A Q S F S K N F G L Y N E R V G
Human	P F F D S A Y Q G F A S G N L E R D A W A I R Y F V S E G F E F F C A Q S F S K N F G L Y N E R V G

FIGURE 12–1 A comparison of eight organisms for a 50-amino-acid long sequence of the enzyme aspartate transaminase. For the amino acid abbreviations, see Fig. 7-2 or Table 12-3. (*Adapted from Benner et al.*)

Immunological Techniques

The earliest, and still among the simplest, comparative molecular methods uses immunological techniques in which antibodies produced in a particular host (usually a rabbit) against proteins (antigens) of one species are measured for their activity against proteins of other species. For example, if antibodies against species A precipitate much of the protein in species C but little of the protein in species B, then researchers presume the A and C proteins have more similar molecular configurations (antigenic components) and are more evolutionarily alike (smaller **antigenic distance**) than those of A and B.

Using a variety of antibodies, researchers can analyze immunological data by various mathematical rules (algorithms) to construct a phylogenetic tree that best correlates the antigenic distance between species with the length of time since they shared a common ancestor. If species A and C are antigenically closer to each other than to species B, we can presume species B broke off earlier from the common stem that all three originally shared. Successive comparisons are then made between all possible combinations of species until the entire phylogenetic tree is obtained.

Figure 12–2 shows a picture of such a tree for anthropoids (humans, apes, monkeys). In contrast to traditional taxonomy of the time, this technique showed that humans (*Homo*), gorillas (*Gorilla*), and chimpanzees (*Pan*) are antigenically closer to each other than to the Asian orangutan (*Pongo*). Using this method, we can separate the first three genera from the orangutans and place them in a distinct group.

Other immunological techniques, such as the **microcomplement fixation** that Sarich and Wilson used, involve measuring antibodies produced against specific proteins found in blood serum (albumin and transferrins) or enzymes such as lysozyme. Antigenic distances which researchers can detect by measuring the amount of antigen-antibody reactions[1] provide data that generally support the phylogenies we can obtain by other taxonomic methods, although some differences occur.

Amino Acid Sequences

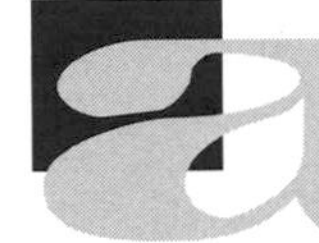

A more direct, and now more popular, approach to molecular phylogeny is sequencing the amino acids in proteins by biochemical methods and comparing such sequences for the same protein in different species. Among the first proteins to yield its amino acid sequence was **hemoglobin** (Ingram), and it probably still remains one of the most investigated of all proteins. The basic unit of hemoglobin consists of an iron-containing porphyrin (heme) that reversibly can bind oxygen attached to a globin polypeptide chain that is usually no less than 140 amino acids long. The demonstration that hemoglobin-like molecules appear in a wide range of organisms, from invertebrates to vertebrates, and even in plants, fungi, and bacteria (Hardison) indicates their origin far back in the history of life. In vertebrates, hemoglobins are usually the primary protein of red blood cells, making them relatively easy to isolate and purify in large amounts.

As explained in Chapter 10, red blood cell hemoglobin of normal human adults is a four-chain molecule or tetramer, consisting of two pairs of polypeptide chains, one pair bearing the α sequence and the other pair mostly bearing the β sequence ($\alpha_2\beta_2$). Some adult hemoglobin uses δ chains instead of β's ($\alpha_2\delta_2$), and a common form of embryonic hemoglobin has two α's and two γ's

[1] In microcomplement fixation, rabbits immunized with a protein antigen from one species produce antiserum that gives a strong reaction against that antigen (homologous antigen) but not against the same protein from another species (heterologous antigen). The researchers then measure the degree of antigenic difference between the two species by the concentration to which the antiserum must increase for the heterologous antigen to react like the homologous antigen.

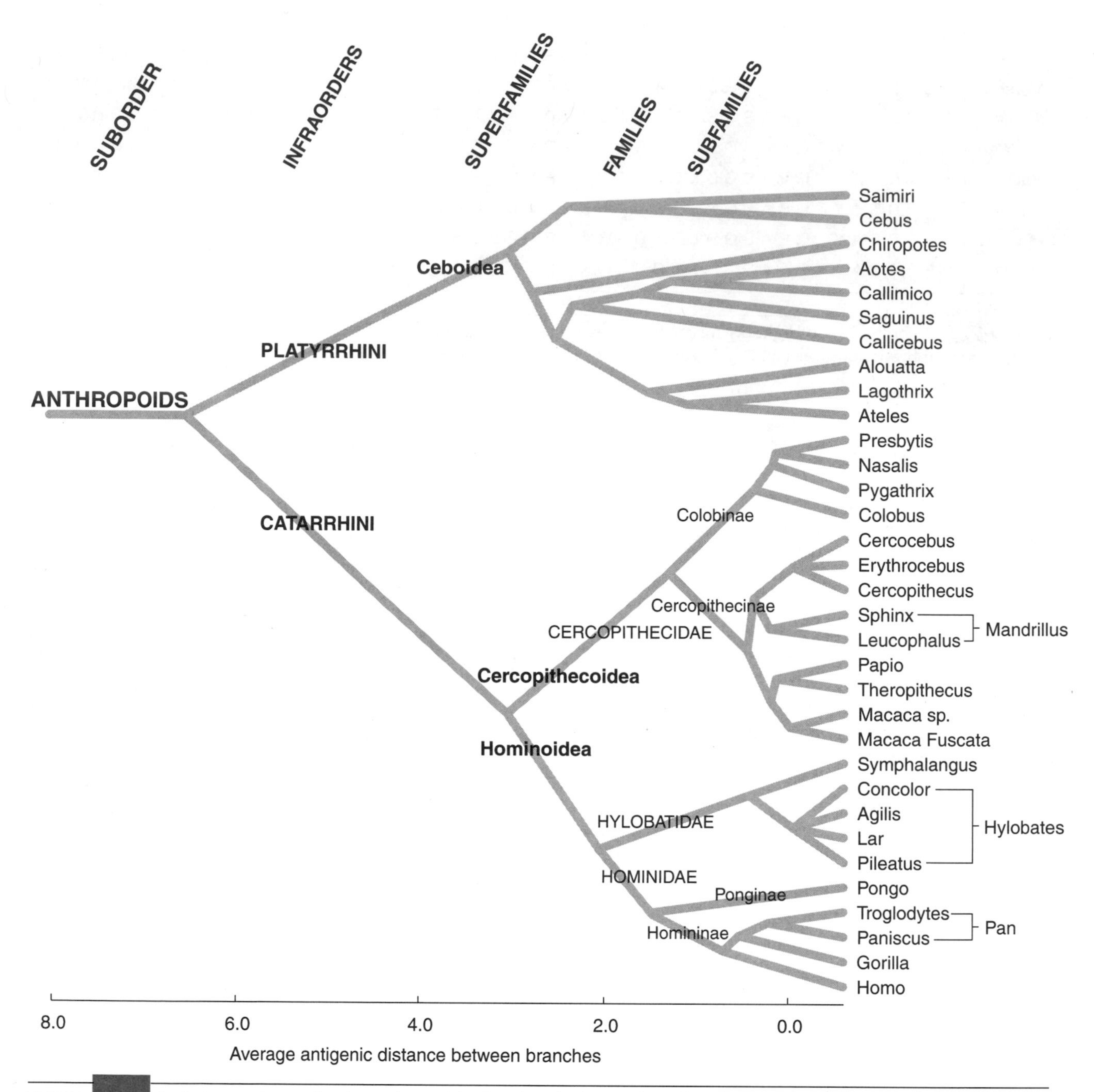

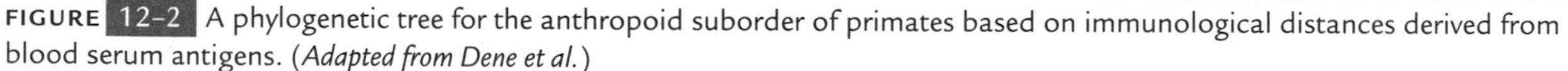

FIGURE 12-2 A phylogenetic tree for the anthropoid suborder of primates based on immunological distances derived from blood serum antigens. (*Adapted from Dene et al.*)

($\alpha_2\gamma_2$). Other types of hemoglobin chains also occur (for example, ϵ), and hemoglobin-like molecules such as **myoglobin** appear in other tissues. All these chains are distinguished by their somewhat different properties and different amino acid sequences.

That a species can possess different kinds of globin molecules and each such molecule can differ among different species, points to two major kinds of globin evolution:

- Different kinds of globin chains arose during evolution, producing the variety carried by a particular vertebrate (α differs from β, which differs from γ, and so on).
- Each particular globin chain followed its own evolutionary path, leading to changes in its amino acid sequence in different species (α chains are different in different species, as are β chains, and so on).

As an example of the first kind, Table 12–1 shows the amino acid sequences for five different human globin chains. They are all variations on a single homologous theme: the chains are all about the same length; they

TABLE 12-1 Amino acid sequences for human myoglobin and four human hemoglobin chains (α, β, γ, and δ)

	1		2	3	4	5	6	7	8	9	10	11	12	13	14	15	16	17	18	19
Myoglobin	G	—	L	S	D	G	E	W	Q	L	V	L	N	V	W	C	K	V	E	A
α chain	V	—	L	S	P	A	D	K	T	N	V	K	A	A	W	G	K	V	G	A
β chain	V	H	L	T	P	E	E	K	S	A	V	T	A	L	W	G	K	V	—	—
γ chain	G	H	F	T	E	E	D	K	A	T	I	T	S	L	W	G	K	V	—	—
δ chain	V	H	L	T	P	E	E	K	T	A	V	N	A	L	W	G	K	V	—	—

	20	21	22	23	24	25	26	27	28	29	30	31	32	33	34	35	36	37	38	39
Myoglobin	D	I	P	G	H	G	Q	E	V	L	I	R	L	F	K	G	H	P	E	T
α chain	H	A	G	E	Y	G	A	E	A	L	E	R	M	F	L	S	F	P	T	T
β chain	N	V	D	E	V	G	G	E	A	L	G	R	L	L	V	V	Y	P	W	T
γ chain	N	V	E	D	A	G	G	E	T	L	G	R	L	L	V	V	Y	P	W	T
δ chain	N	V	D	A	V	G	G	R	A	L	G	R	L	L	V	V	Y	P	W	T

	40	41	42	43	44	45	46	47	48	49	50	51	52	53	54	55	56	57	58
Myoglobin	L	E	K	F	D	K	F	K	H	L	K	S	E	D	E	M	K	A	S
α chain	K	T	Y	F	P	H	F	—	D	L	S	H	—	—	—	—	—	G	S
β chain	Q	R	F	F	E	S	F	G	D	L	S	T	P	D	A	V	M	G	N
γ chain	Q	R	F	F	D	S	F	G	N	L	S	S	A	S	A	I	M	G	N
δ chain	Q	R	F	F	E	S	F	G	D	L	S	S	P	D	A	V	M	G	N

	59	60	61	62	63	64	65	66	67	68	69	70	71	72	73	74	75	76	77
Myoglobin	E	D	L	K	K	H	G	A	T	V	L	T	A	L	G	G	I	L	K
α chain	A	Q	V	K	G	H	G	K	K	V	A	D	A	L	T	N	A	V	A
β chain	P	K	V	K	A	H	G	K	K	V	L	G	A	F	S	D	G	L	A
γ chain	P	K	V	K	A	H	G	K	K	V	L	T	S	L	G	D	A	I	K
δ chain	P	K	V	K	A	H	G	K	K	V	L	G	A	F	S	D	G	L	A

	78	79	80	81	82	83	84	85	86	87	88	89	90	91	92	93	94	95	96
Myoglobin	K	K	G	H	H	E	A	E	I	K	P	L	A	Q	S	H	A	T	K
α chain	H	V	D	D	M	P	N	A	L	S	A	L	S	D	L	H	A	H	K
β chain	H	L	D	N	L	K	G	T	F	A	T	L	S	E	L	H	C	D	K
γ chain	H	L	D	D	L	K	G	T	F	A	Q	L	S	E	L	H	C	D	K
δ chain	H	L	D	N	L	K	G	T	F	S	Q	L	S	E	L	H	C	D	K

	97	98	99	100		102		104		106		108		110		112		114	
Myoglobin	H	K	I	P	V	K	Y	L	E	F	I	S	E	C	I	I	Q	V	L
α chain	L	R	V	D	P	V	N	F	K	L	L	S	H	C	L	L	V	T	L
β chain	L	H	V	D	P	E	N	F	R	L	L	G	N	V	L	V	C	V	L
γ chain	L	H	V	D	P	E	N	F	K	L	L	G	N	V	L	V	T	V	L
δ chain	L	H	V	D	P	E	N	F	R	L	L	G	N	V	L	V	C	V	L

	116		118		120		122		124		126		128		130		132		134
Myoglobin	Q	S	K	H	P	G	D	F	G	A	D	A	Q	G	A	M	N	K	A
α chain	A	A	H	L	P	A	E	F	T	P	A	V	H	A	S	L	D	K	F
β chain	A	H	H	F	G	K	E	F	T	P	P	V	Q	A	A	Y	Q	K	V
γ chain	A	I	H	F	G	K	E	E	T	P	E	V	Q	A	S	W	Q	K	M
δ chain	A	R	N	F	G	K	E	F	T	P	Q	M	Q	A	A	Y	Q	K	V

		136		138		140		142		144		146		148		150		152	
Myoglobin	L	E	L	F	R	K	D	M	A	S	N	Y	K	E	L	G	F	Q	G
α chain	L	A	S	V	S	T	V	L	T	S	K	Y	R						
β chain	V	A	G	V	A	N	A	L	A	H	K	Y	H						
γ chain	V	T	G	V	A	S	A	L	S	S	R	Y	H						
δ chain	V	A	G	V	A	N	A	L	A	H	K	Y	H						

Note: The amino acids are abbreviated by single capital letters as shown in Figure 7-2 and Table 12-3. The chains are aligned with the 153 amino acids in myoglobin, and the boxes indicate identical amino acids found in all chains at the designated numbered positions.

possess identical amino acids at a significant number of positions; and all functioning human globin genes share a similar exon–intron structure—three exons separated by two introns (Fig. 9–14). The three-dimensional structures that amino acid sequences dictate are also similar, leading to their similar physiological functions. In addition, the genes for the β, γ, and δ are closely linked (chromosome 11), although the gene for the α chain lies on a different chromosome (16).

Gene Duplication and Divergence

Observations of multiple hemoglobin chains, especially their sequence similarities, strongly suggest that rather than arising from different genes that accidentally converged in sequence and function, all the different globin chains arose as **gene duplications** of an original globin-type gene. Once such duplicated globin genes appeared, each could then undergo its own subsequent evolutionary changes. The temporal order in which the duplications occurred can be discerned from amino acid differences on the basis that the greater the amino acid differences between any two chains, the further back in time was their common ancestor. Thus, we can order a number of events from the following three observations:

1. The myoglobin chain differs most from all others because it has distinctive amino acids at more than 100 sites.
2. The α chain differs from β at 77 sites.
3. The β chain differs from γ at 39 sites but differs from δ at only 10 sites.

These findings mean that the gene for myoglobin must have formed from a very early duplication, which was then followed by a later duplication that separated the α and β genes. Because they differ least, the separation between the β and δ chains derives from a fairly recent duplication.

Figure 12–3 portrays the genetic phylogeny of the five globins in terms of the numbers of nucleotides necessary to account for the amino acid differences, along with the chronological periods in which evolutionists presume each duplication occurred. We can see that duplication events led to the early coexistence of myoglobin with an α-like chain, the former probably assuming (or maintaining) an intracellular function and the latter probably assuming a circulatory function.

When a duplication of the α-like gene further evolved into a β-like gene, the advantage of having two pairs of

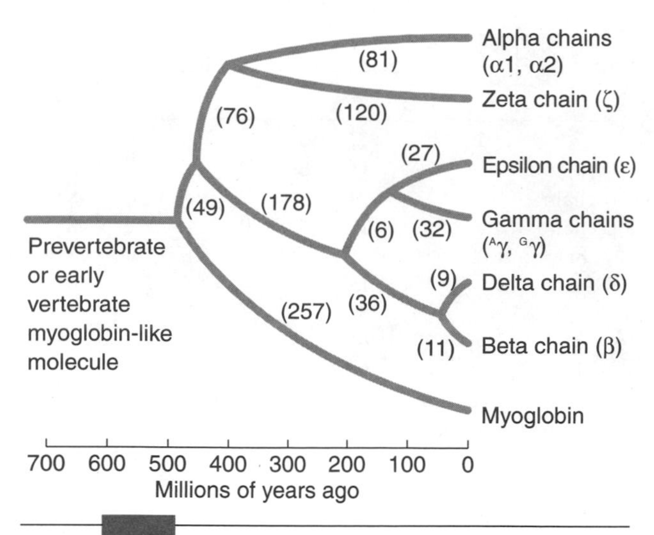

FIGURE 12–3 Phylogenetic relationships between globin-type proteins found in humans, showing the estimated times at which they originally diverged from each other. The estimated number of nucleotide replacements necessary to cause the observed amino acid changed in each branch of the lineage is given in parentheses. (*From* Genetics Third Edition *by Monroe W. Strickberger. Copyright © 1985 by Monroe W. Strickberger. Reprinted by permission of Prentice Hall, Inc., Upper Saddle River, NJ.*)

different chains in a tetramer hemoglobin molecule must have been sufficiently great to account for preserving tetramer organization in the circulating blood of most vertebrates. After the β-like gene formed, a translocation separated it from α and transferred it to a different chromosome. Duplications then occurred in the β-like gene, eventually yielding the modern β, γ, and δ genes.

It is now clear that such gene duplications are not unusual and easily can arise from an **unequal crossing over** event that produces a recombinant product possessing increased chromosome material (Fig. 12–4).[2] The human α gene, for example, is known to have two side-by-side (tandem) duplicates, α1 and α2, and the β **gene cluster,** also called a **gene family** or **multigene family,** consists of a sequence of seven such genes (Fig. 12–5).

In some cases, such as globin chains, serine proteases (chymotrypsin, trypsin), and lactate dehydrogenases, duplicated genes have preserved similar although not identical functions. In other cases, duplications retained as **pseudogenes** lost function completely because of mutations that prevent transcription or translation. Most important from an evolutionary view, are duplicated genes that have evolved in completely different functional directions, although they share enough common amino acid sequences to indicate their relationship.

[2] Unequal crossing over may also cause gene fusion by eliminating chromosomal material between two formerly separate genes. Researchers believe that the union in fungi between the A and B components of the tryptophan synthetase enzyme, normally separate in bacteria, occurred through such fusion events.

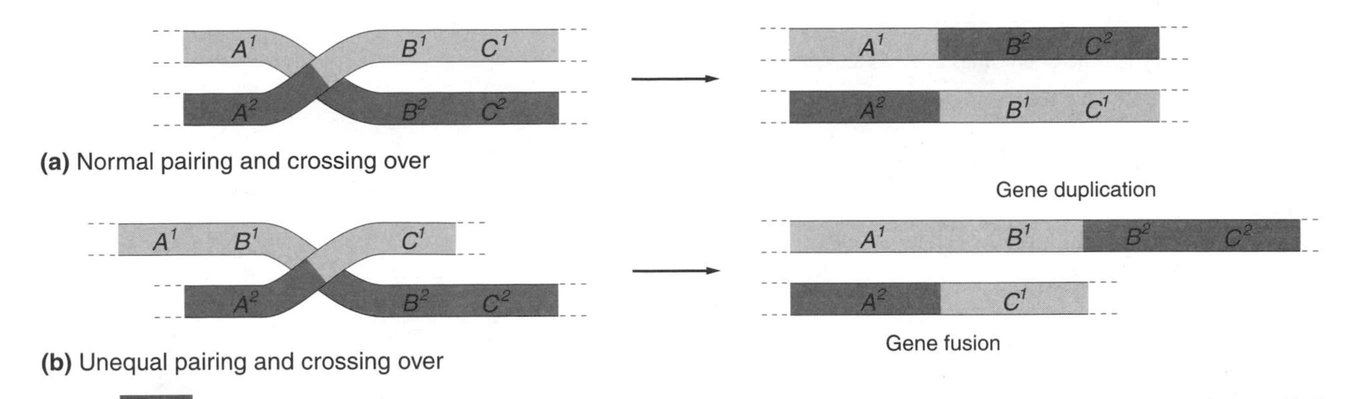

FIGURE 12-4 The results of equal and unequal crossing over for three gene segments on a chromosome. (*a*) When pairing between homologous sections on two chromosomes is equal, the crossover products have the same amounts of chromosomal material (for example, they both have an *A, B,* and *C* gene segment). (*b*) When pairing between the two chromosomes is unequal, one of the crossover products carries a gene duplication (the *B* gene segment in this illustration), and the other product shows a fusion between gene segments (*A–C*) that were formerly separated by the intervening *B* segment.

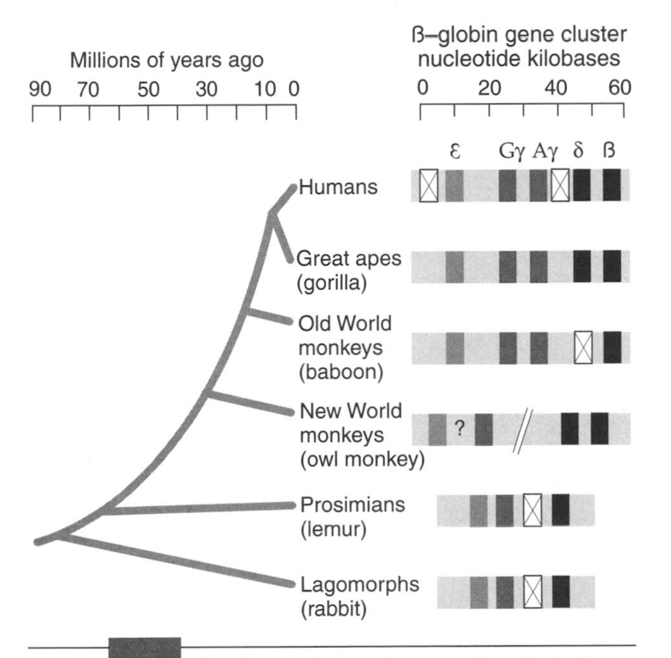

FIGURE 12-5 The clustered organization of the β-globin-type genes in five different primates and in the rabbit (*right side*), along with a proposed evolutionary tree (*left side*). The genes in each of these clusters are linked together on the same chromosome; for example, the human β-globin cluster is localized to a span of 60,000 nucleotides on chromosome 11. Each gene, denoted by a small rectangle, is transcribed from left to right, with the genes responsible for embryonic and fetal development on the left (*lighter shadings*) and the genes that produce adult β-globins on the right (*darker shading*). Genes marked with crosses indicate pseudogenes, duplicates that have become nonfunctional for various reasons. Nadeau and Sankoff estimate that deleterious mutations in gene duplicates responsible for causing pseudogenes were as likely to occur among early vertebrates as were mutations allowing duplicates to diverge functionally. (*From Strickberger, adapted from Jeffreys et al.*)

Thus the vertebrate nerve growth factor protein that enhances the outgrowth of neural cells from sympathetic and sensory ganglia has amino acid sequences similar to insulin and may share some functional similarities as well (Frazier et al.). Also, the α-lactalbumin protein, which is part of an enzyme used in the synthesis of lactose in mammalian milk, has an amino acid sequence remarkably similar to that of the lysozyme enzyme found in tears that degrades the mucopolysaccharides of bacterial cell walls (Hill et al.). This similarity is further reflected in evidence that both proteins are the products of tissues (mammary, tear duct) that were at one time probably sebaceous glands, and they both use sugar molecules as their substrates.

That natural selection can use an enzyme for more than one adaptive purpose appears to be common for transparent crystallin proteins in eye structures. Although at least ten such lens crystallins are found among organisms ranging from vertebrates to mollusks, each appears related to a functional enzyme such as alcohol dehydrogenase, α-enolase, and glutathione S-transferase (Table 12–2). In fact, both ε-crystallin and lactate dehydrogenase enzyme in crocodiles and some birds derive from a single gene product identical in both groups (Piatagorsky and Wistow), indicating that a molecule or structure may be preadapted to a new function (lens crystallin) yet retain its former enzymatic catalytic function (see also p. 353).[3]

How rapidly duplications diverge in function depends, of course, on how many amino acid substitutions

[3] Such cases, in which the same gene product performs two entirely different functions, may be considered a form of *pleiotrophy* (pp. 217 and 651) characterizing the multiple phenotypic effects caused by a gene whose product(s) enter into more than one developmental pathway. (Piatagorsky and Wistow prefer the term *gene sharing*.)

TABLE 12-2 A sample of enzymes used as lens crystallins in various taxa

Source	Crystallin Type	Enzyme
Cephalopods	S	glutathione S-transferases
Birds, Crocodiles	ϵ	lactate dehydrogenase B
Cavies, Camels	ζ	alcohol dehydrogenases
Rabbits, Hares	λ	coenzyme A derivative dehydrogenases
Many species	τ	α-enolase
Mammalian cornea	BCP54	aldehyde dehydrogenase III

Abridged from Wistow.

are necessary. A most striking example of such rapidity is the single amino acid mutation that can convert lactate dehydrogenase (LDH) to malate dehydrogenase (MDH). LDH is used in glycolysis for the lactate-pyruvate reaction (Fig. 9–1) and MDH is used in the Krebs cycle for the malate-oxaloacetate reaction (Fig. 9–8). According to Wilks and coworkers, this change can be engineered by simply substituting arginine for glutamine at the 102nd polypeptide position. Other adaptive conversions undoubtedly involve more amino acid changes, and the histories of some are now being reconstructed through "paleomolecular biochemistry." These methods trace past amino acid substitutions by noting the effects of purposely changing amino acids at particular sites using a technique called site-directed mutagenesis (Golding and Dean).

Duplications also exist within genes themselves, as shown by the presence of three homologous amino acid regions (polypeptide domains; p. 175) within the γ heavy chain of the immunoglobulin G antibody. In the human haptoglobin α-2 blood serum protein, a segment of 59 amino acids (positions 13 to 71) almost exactly repeats an adjacent segment (positions 72 to 130). Researchers have also found repeated amino acid sequences within the ferredoxin protein (used as an electron carrier in various biochemical processes, Fig. 9–2,[4] in the glutamate dehydrogenase enzyme, and various other proteins (Li 1983).

In spite of these and other fairly conspicuous examples, ancestral homologies based on similarities between amino acid sequences are not always obvious. Researchers have devised various tests for detecting amino acid sequence similarity, some more sensitive in distinguishing homologous relationships than others (Dayhoff et al., Nei, McClure et al.). In general, the most sensitive of these tests depend on comparing fairly long amino acid sequences, since comparisons between short sequences can be misleading; for example, similar sequences only three or four amino acids long may often arise in nonrelated proteins. Workers in the field have also proposed methods to align molecular sequences that overcome difficulties caused by gaps or insertions (for example, Goldman).

Phylogenies involving a large number of organisms come from the second aspect of protein evolution—comparisons between amino acid sequences in different species. To clarify such comparisons, Fitch suggested that we use different terms for two major kinds of homologous genes; duplicates within a species (for example, α, β, and γ hemoglobins in humans), and duplicates of the same gene in different species (for example, α hemoglobins in horses and humans). He called the former **paralogous,** and the latter **orthologous.** It is comparisons between orthologous genes that are valid for determining the phylogenetic relationships of the species that carry them.

Determining Molecular Phylogenies

One basic technique for detecting orthologous amino acid sequences essentially does not differ from that used in determining phylogenies based on immunological distances, except that it can furnish more precise phylogenetic positioning because it begins with an alignment of amino acid sequences from the same known protein in different species. Once aligned, the minimum number of mutations necessary to transform one amino acid in one sequence to that of a different amino acid in the same position in the other sequence can provide an estimate of mutational distances, such as those Table 12–3 shows.

[4] The small size of ferredoxin, its limited sampling of amino acids, and the simple positioning of its iron atoms indicate that it is one of the most primitive electron transport agents in the cell. These features, along with its apparent duplicated structure, have suggested that ferredoxin may owe its origin to an even earlier peptide formed on one of the then-primitive templates. (The length of ferredoxin's base amino acid sequence before duplication is only about 28 amino acids long.) Perhaps only when organisms could achieve greater reproductive and translation accuracy could such small early enzymes increase in size and thereby improve their reaction specificity, catalytic activity, and structural stability.

TABLE 12–3 Matrix of minimum number of nucleotide substitutions necessary to convert a codon for one amino acid (rows) into a codon for another amino acid (columns)

	Abbreviations		Minimum Number of Nucleotide Substitutions																			
Amino Acid	3-letter	1-letter	A	C	D	E	F	G	H	I	K	L	M	N	P	Q	R	S	T	V	W	Y
Alanine	Ala	A	0	2	1	1	2	1	2	2	2	2	2	2	1	2	2	1	1	1	2	2
Cysteine	Cys	C	2	0	2	3	1	1	2	2	3	2	3	2	2	3	1	1	2	2	1	1
Aspartic acid	Asp	D	1	2	0	1	2	1	1	2	2	2	3	1	2	2	2	2	2	1	3	1
Glutamic acid	Glu	E	1	3	1	0	3	1	2	2	1	2	2	2	2	1	2	2	2	1	2	2
Phenylalanine	Phe	F	2	1	2	3	0	2	2	1	3	1	2	2	2	3	2	1	2	1	2	1
Glycine	Gly	G	1	1	1	1	2	0	2	2	2	2	2	2	2	2	1	1	2	1	1	2
Histidine	His	H	2	2	1	2	2	2	0	2	2	1	3	1	1	1	1	2	2	2	3	1
Isoleucine	Ile	I	2	2	2	2	1	2	2	0	1	1	1	1	2	2	1	1	1	1	3	2
Lysine	Lys	K	2	3	2	1	3	2	2	1	0	2	1	1	2	1	1	2	1	2	2	2
Leucine	Leu	L	2	2	2	2	1	2	1	1	2	0	1	2	1	1	1	1	2	1	1	2
Methionine	Met	M	2	3	3	2	2	2	3	1	1	1	0	2	2	2	1	2	1	1	2	3
Asparagine	Asn	N	2	2	1	2	2	2	1	1	1	2	2	0	2	2	2	1	1	2	3	1
Proline	Pro	P	1	2	2	2	2	2	1	2	2	1	2	2	0	1	1	1	1	2	2	2
Glutamine	Gln	Q	2	3	2	1	3	2	1	2	1	1	2	2	1	0	1	2	2	2	2	2
Arginine	Arg	R	2	1	2	2	2	1	1	1	1	1	1	2	1	1	0	1	1	2	1	2
Serine	Ser	S	1	1	2	2	1	1	2	1	2	1	2	1	1	2	1	0	1	2	1	1
Threonine	Thr	T	1	2	2	2	2	2	2	1	1	2	1	1	1	2	1	1	0	2	2	2
Valine	Val	V	1	2	1	1	1	1	2	1	2	1	1	2	2	2	2	2	2	0	2	2
Tryptophan	Trp	W	2	1	3	2	2	1	3	3	2	1	2	3	2	2	1	1	2	2	0	2
Tyrosine	Tyr	Y	2	1	1	2	1	2	1	2	2	2	3	1	2	2	2	1	2	2	2	0

Source: Adapted from Fitch and Margoliash (1967).

This method of choosing a hypothesis that minimizes explanations to account for an observation invokes the **parsimony principle,** also known as "Occam's razor." It is widely used in science "to admit no more causes of natural things than such as are both true and sufficient to explain their appearances" (pp. 15, 249). Parsimony has the advantage that a true hypothesis generally involves fewer assumptions than a false hypothesis; for example, we can better explain a phenylalanine codon (UUU) as deriving from a single nucleotide substitution in a serine codon (UCU → UUU) than as deriving from a triple nucleotide substitution in a glutamic acid codon (GAA → UUU). Note that "unparsimonious" explanations are not necessarily false, but, *in the absence of further information,* choices are made on the basis of parsimony.

Given parsimoniously determined evolutionary distances between species, resolving their phylogenetic relationships can follow. For example, if the minimum or most parsimonious mutational distance for a particular protein comparison between species A and B is 25, between A and C is 20, and between B and C is 30, then the two most closely related species are obviously A and C (see also Fig. 11–11).[5] If we assign legs x, y, and z to represent the numbers of mutations responsible for their divergence, we can portray the phylogenetic relationship among the three species as follows:

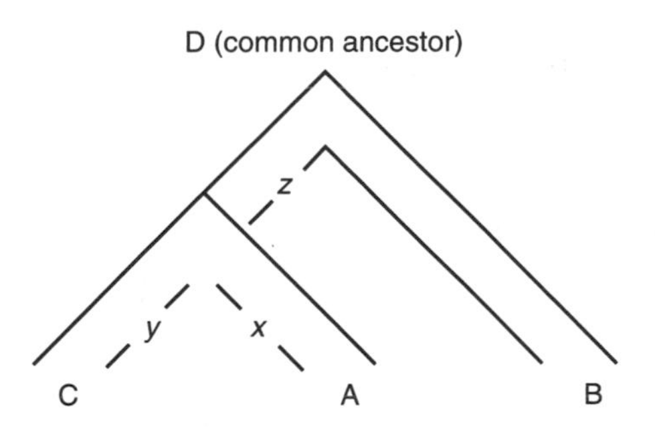

We can determine the lengths of these legs by noting that the A–B distance (25) is less than the C–B distance (30), hence the leg x must be 5 mutations less than y. Since $x + y = 20$ and $y - x = 5$, we can subtract one equation from the other and solve for x:

$$\begin{array}{r} x + \ y = 20 \\ \underline{-(-x + \ y = \ 5)} \\ 2x = 15 \\ x = 7.5, \quad y = 12.5 \end{array}$$

[5] According to Edwards, although parsimony is now generally justification for minimizing the number of phylogenetic steps in a tree, such usage was initiated earlier by "maximum likelihood" methods, a more sophisticated statistical procedure.

The leg z must therefore equal the A–B distance minus x (or the C–B distance minus y); that is, $z = 25 - 7.5$ (or $30 - 12.5$) $= 17.5$.

These mutational distances therefore yield

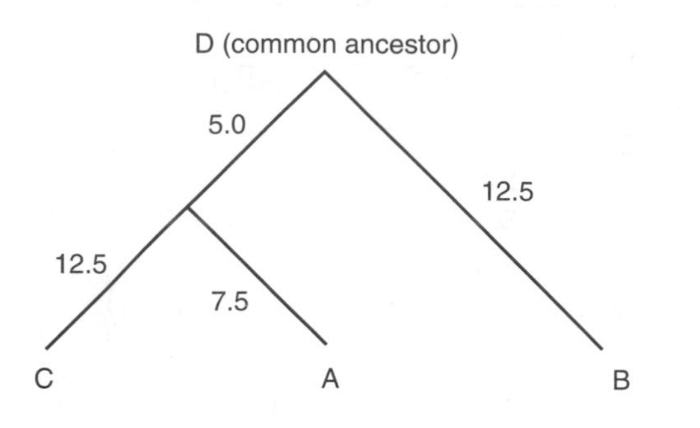

Note that an important condition of the parsimony method is that the observed mutational distance between two species must be less than or equal to the sum of the distances from their common ancestor to each species in the projected phylogenetic tree. In this example, the observed distance between A and B is 25, and the distances from their hypothesized common ancestor at the triangular apex (D) to these two species must total at least 25, if not more. The sum of the (D–A) + (D–B) legs may be more than 25 because undiscovered **back mutations** (for example, adenine → guanine → adenine) may exist in these two legs, or **parallel mutations** may have occurred in each leg (for example, cytosine → adenine; guanine → adenine), so that the difference observed between A and B is really an underestimate of the actual mutational distance.

This triangle inequality condition signifies the evolutionary presence of back and parallel mutations, and researchers often take it into account by proportionally augmenting the number of mutations in an ancestral leg in accord with its length. That is, not only do they increase numerically (for example, the A–D–B distance over the measured A–B distance) but further augment longer distances. Other approaches concentrate on taxa with smaller distances between them, thereby attempting to reduce or eliminate back and parallel mutations.

By using increased numbers of species to provide mutational data, researchers can establish a **phylogenetic tree** by various numerical and algorithmic methods (Felsenstein, Nei, Swofford et al.). Important as these methods are, they remain beyond the scope of this book, involving sophisticated mathematical techniques in choosing among many phylogenetic possibilities that increase manyfold with increasing number of taxa. For example, although about 100 different phylogenies are possible for 5 taxa, there are more than 30 million possible phylogenies for 10 taxa and more than 200 billion billion (8.2×10^{21}) for 20 taxa. On reaching 50 taxa, more phylogenetic possibilities exist than estimated atoms in the universe (10^{74}). Considering more than small sections of a tree at a time can become mathematically cumbersome and impractical.

Considering that there are many possible trees but only one true phylogeny, sampled data may not reflect it. Statisticians have searched for methods to gauge how much confidence to place on proposed phylogenies. At present, a common method is **bootstrapping,** which gives the proportion (percentage) of acceptable trees in which a branch point ("node" or "clade") appears when data is repeatedly sampled and replaced. For example, ten amino acid differences used in constructing species relationships can be resampled 100 times so that some differences are omitted and some appear more than once. Each resampling generates a tree in which a particular branch may or may not appear. The bootstrap value is the frequency in which the same branch appears. Using this method, the phylogenetic tree of cichlid fishes in Fig. 12–18 indicates that, in 99 percent of 2,000 repeated samples of the data, the six Malawi species derive from a common branch.

Unfortunately, bootstrapping values are not infallible predictors, since they are limited by the kinds and amounts of available data—how well the data represent the phylogeny. For example, when researchers change the species that represents the arthropod phylum from brine shrimp to spider, the arthropod relationship to mollusks and echinoderms also changes (Maley and Marshall). Aside from other complications that can confuse any tree, such as homoplasy (p. 243) and horizontal gene transfer (pp. 225–226), high bootstrapping values increase confidence but do not guarantee a true phylogeny.

Nevertheless, parsimony and other techniques are considered to offer reliable choices when taxonomic comparisons between related groups make use of many differences, such as nucleotide or amino acid changes comparably aligned in lengthy sequences (Russo et al.). Figure 12–6 shows one such phylogeny for the α-hemoglobin sequence in 29 different vertebrates based on analyzing a total of 630 nucleotide replacements.

Because hemoglobin phylogenies exclude invertebrates, plants, and fungi, researchers also compare more ubiquitous proteins present in widely varied groups of organisms. One such protein, cytochrome *c*, also contains heme but generally functions in aerobic organisms as part of the respiratory electron transport chain (Fig. 9–9). Figure 12–7 shows a phylogenetic tree based on 53 different amino acid sequences of cytochrome *c*.

We can hardly expect phylogenies derived from studies of one or two proteins to always reflect true evolutionary history nor discriminate accurately even between different species. That no differences show up between human and chimpanzee α-hemoglobins, marks such lack of discrimination. Similarly, that camels and whales share the same cytochrome *c* amino acid sequence hardly reflects their evolutionary divergence. (Note also, for example, the similar sequence in cows, pigs, and sheep.) Nevertheless, it is remarkable how closely the information obtained from this limited sample of only two pro-

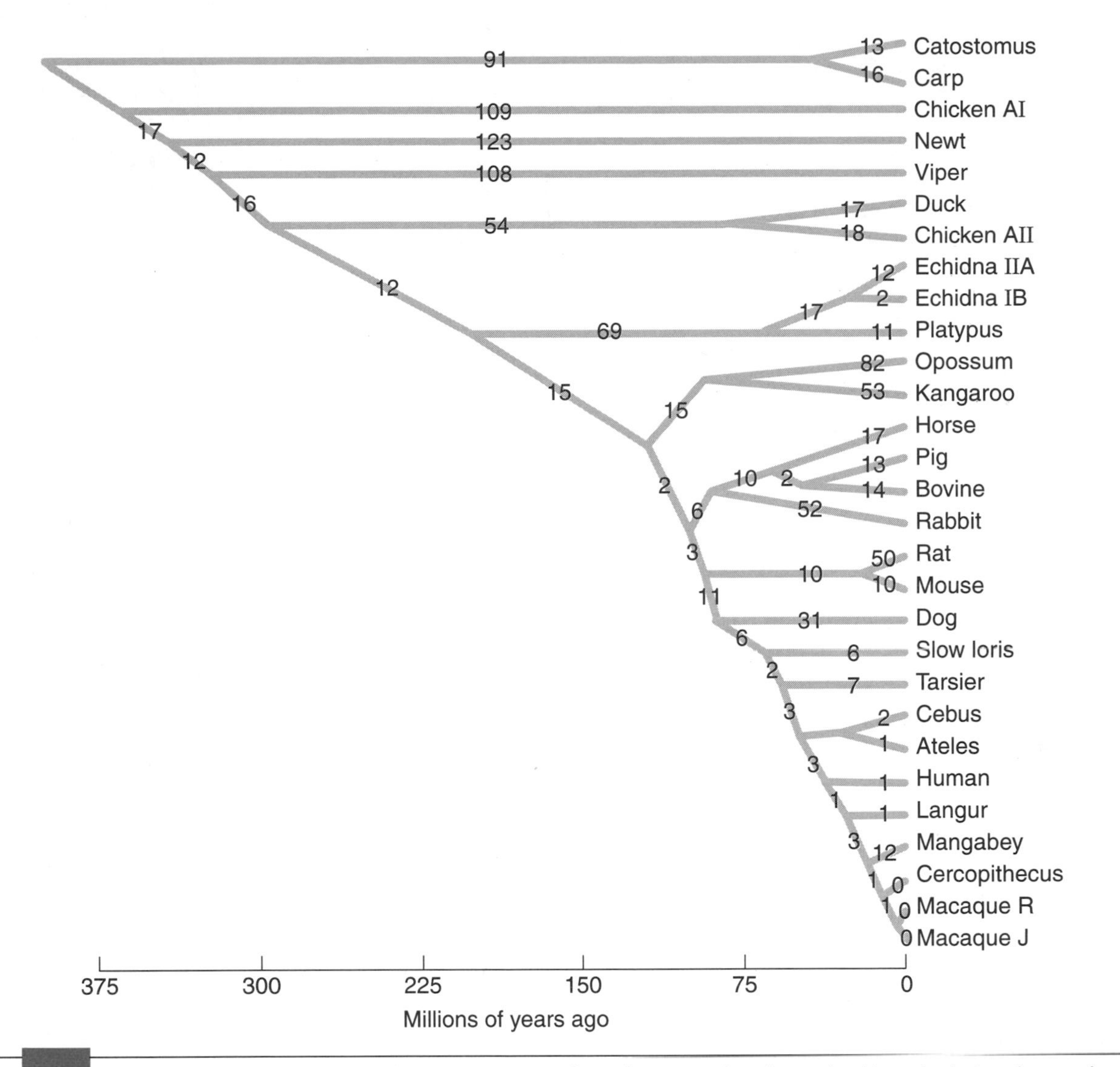

FIGURE 12–6 A phylogeny of α-hemoglobin chains in a variety of vertebrate species, determined by calculating the number of nucleotide replacements that account for the number of observed amino acid substitutions presumed to have occurred during each evolutionary interval. The horizontal scale, given in millions of years, is based on paleontological estimates for the age of the common ancestor of each branch. (*Adapted from Goodman 1976.*)

teins out of many thousands approximates the phylogenetic relationships obtained from extensive studies in comparative anatomy and paleontology.

The persistence of similarity in a protein's amino acid sequence through many branches of a long lineage is related to persistence of selection for the same function (see also Fig. 12–1). Interestingly, such selection may also be strong enough to produce similar amino acid sequences in different branches for a protein that may have originally served a different function—**convergent molecular evolution.** For example, artiodactyl ruminants (for example, cows) and langur monkeys, unique among mammals in fermenting vegetable matter in a foregut, are also unique in bearing the same five amino acid changes in the lysozyme enzyme that breaks down cell walls of fermenting bacteria (Stewart and Wilson). Some of these lysozyme changes occur even in a foregut-fermenting bird, the hoatzin (Kornegay et al.). In each case, these common protein modifications must have arisen independently, since their ancestral ties show no specific modifications.[6]

For a more faithful reflection of evolutionary history, it is preferable to identify homologies and reduce the effect of

[6] Among other molecular convergences is a mammalian protein that binds to blood vessel fibrins and increases the risk of coronary dysfunction and cerebral stroke. Although most mammals lack this protein, apolipoprotein (a), it is found in both humans and hedgehogs, having arisen independently from distinctively separate duplications of a plasminogen gene (Lawn et al.). Perhaps an even more distant molecular convergence, crossing kingdoms between fungi and animals, are the highly similar cell wall proteins of certain dipteran insects (*Chironomus*) and filamentous fungi (*Trichoderma*). Although Rey and coworkers ascribe this similarity to common usage of a 39-base-long repeating nucleotide sequence, independent emergence of the same unique protein in both groups indicates selection for convergent molecular function. Lee reviews other such cases of molecular convergence.

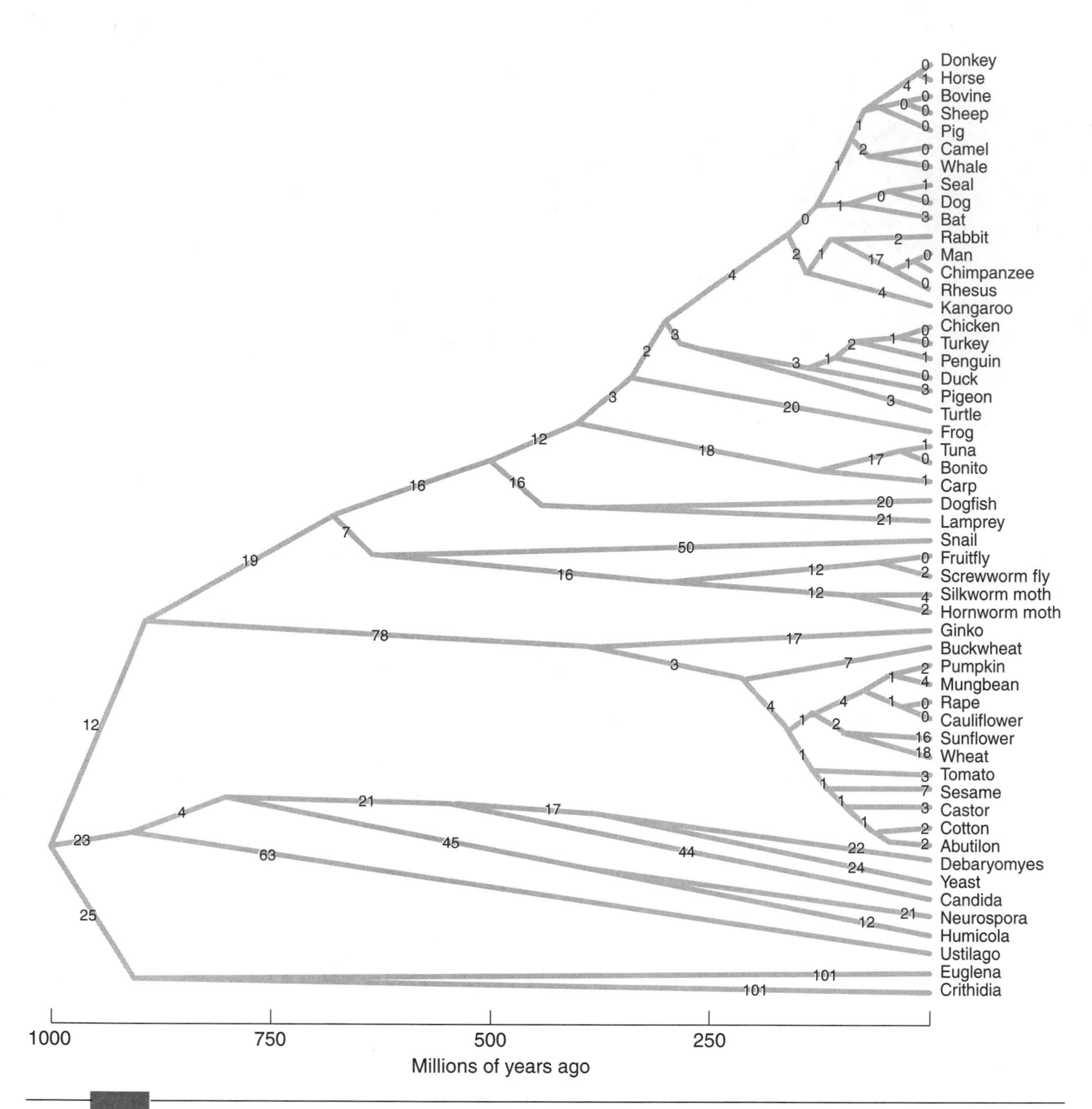

FIGURE 12–7 A cytochrome *c* phylogeny showing relationships among plants, fungi, and a variety of animals. The estimated ages of branching points appear on the horizontal scale, and the estimated number of nucleotide replacements necessary for the evolution of each branch appears within each interval. When the number of nucleotide replacements exceeds five, researchers believe that some undetectable back, or parallel, mutations have occurred, and proportionally augment that number. (*Adapted from Goodman 1976.*)

convergences by combining studies from many different proteins, allowing comparisons among many different amino acid positions. Although such procedures present difficulties, some attempts have been made, and Figure 12–8 diagrams the results of one such study. (Another example, involving both proteins and nucleic acids, is discussed later.)

DNA and Its Repetitive Sequences

The first self-replicating organisms likely had no more than a small amount of genetic material, enough to maintain those

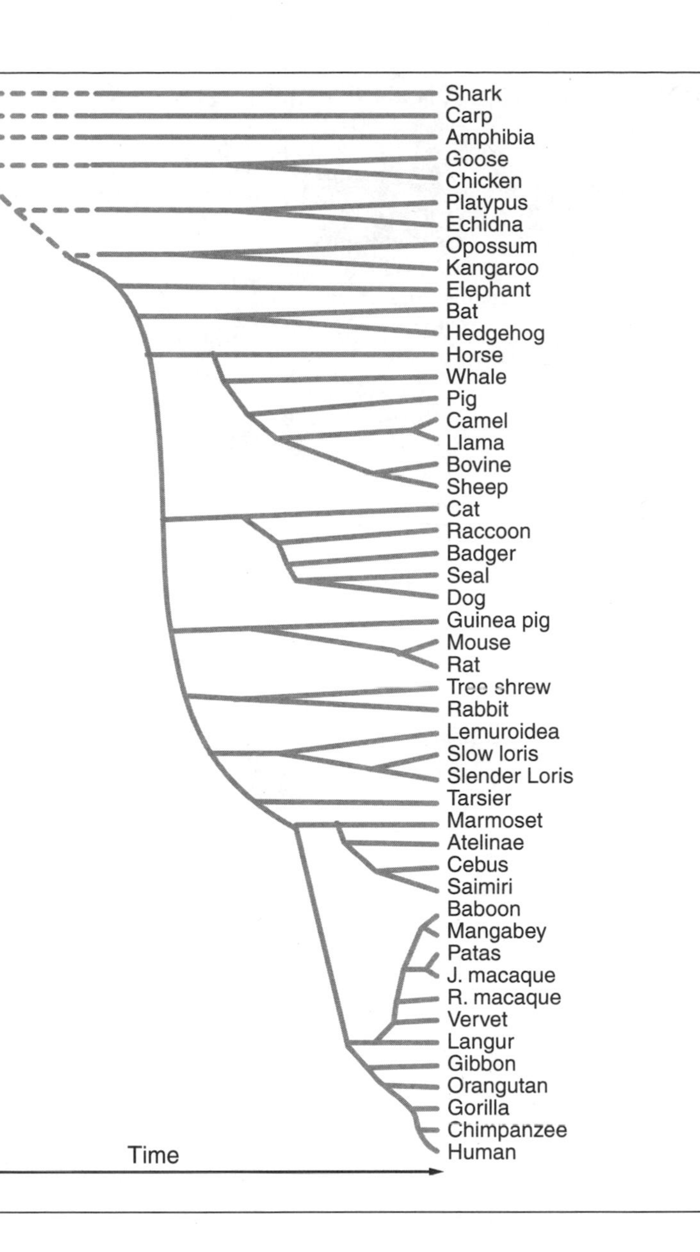

FIGURE 12–8 A phylogenetic tree for 49 vertebrate taxa derived from aligning the amino acid sequences in up to seven different polypeptide chains: α-hemoglobin, β-hemoglobin, myoglobin, lens α crystallin A, fibrinopeptide A, fibrinopeptide B, and cytochrome *c*. (*Adapted from Goodman et al.*)

relatively few functions necessary for their preservation in a mostly noncompetitive environment. As time passed and these organisms continually faced more challenges, increased amounts of genetic material provided greater selective advantages by increasing the number of functions and their regulation. Competition among organisms for successful adaptation to their environments rapidly became dependent on the numbers and kinds of genes they had, and evolutionary changes have proceeded on both levels. One technique for evaluating some consequences of these evolutionary factors was to measure the amounts of genetic material in different organisms (Box 12–1).

A further approach toward analyzing DNA in greater detail has been to determine differences among the kinds of DNA present in a single organism. These studies began with Britten's discovery that DNA sheared to specific sizes and then separated into single strands would reassociate into double-strand molecules at rates based on the nature of their nucleotide sequences. For example, a single strand of DNA bearing a sequence repeated many times over throughout a genome would find a complementary "mate" and form a double-strand molecule much more rapidly than a rare complex sequence. (Researchers can quantify reassociation rate by measuring optical changes that occur in the transition from single- to double-strand DNA.) By these means, Britten and coworkers were able to classify DNA as either **repetitive** or **unique,** referring to sequences that occurred frequently and those that occurred only as single copies.

As Table 12–4 shows, a significant fraction of DNA in tested eukaryotic organisms is repetitive, some sequences present in 200 copies or less and others repeated more than 1 million times.

BOX 12-1 QUANTITATIVE DNA MEASUREMENTS

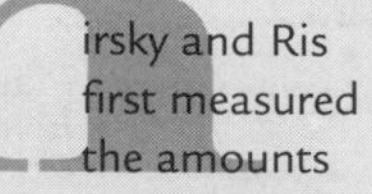irsky and Ris first measured the amounts of DNA in different organisms in the early 1950s and these measurements presently extend to more than 1,000 species. The techniques include:

- Measuring the amount of stained DNA in cells (Feulgen staining)
- Isolating DNA chemically and deriving an average DNA amount from a known number of cells
- Indirectly estimating DNA content from chromosomal size or nuclear volume

Based on results from these techniques, Figure 12-9 shows the observed range of cellular DNA content in

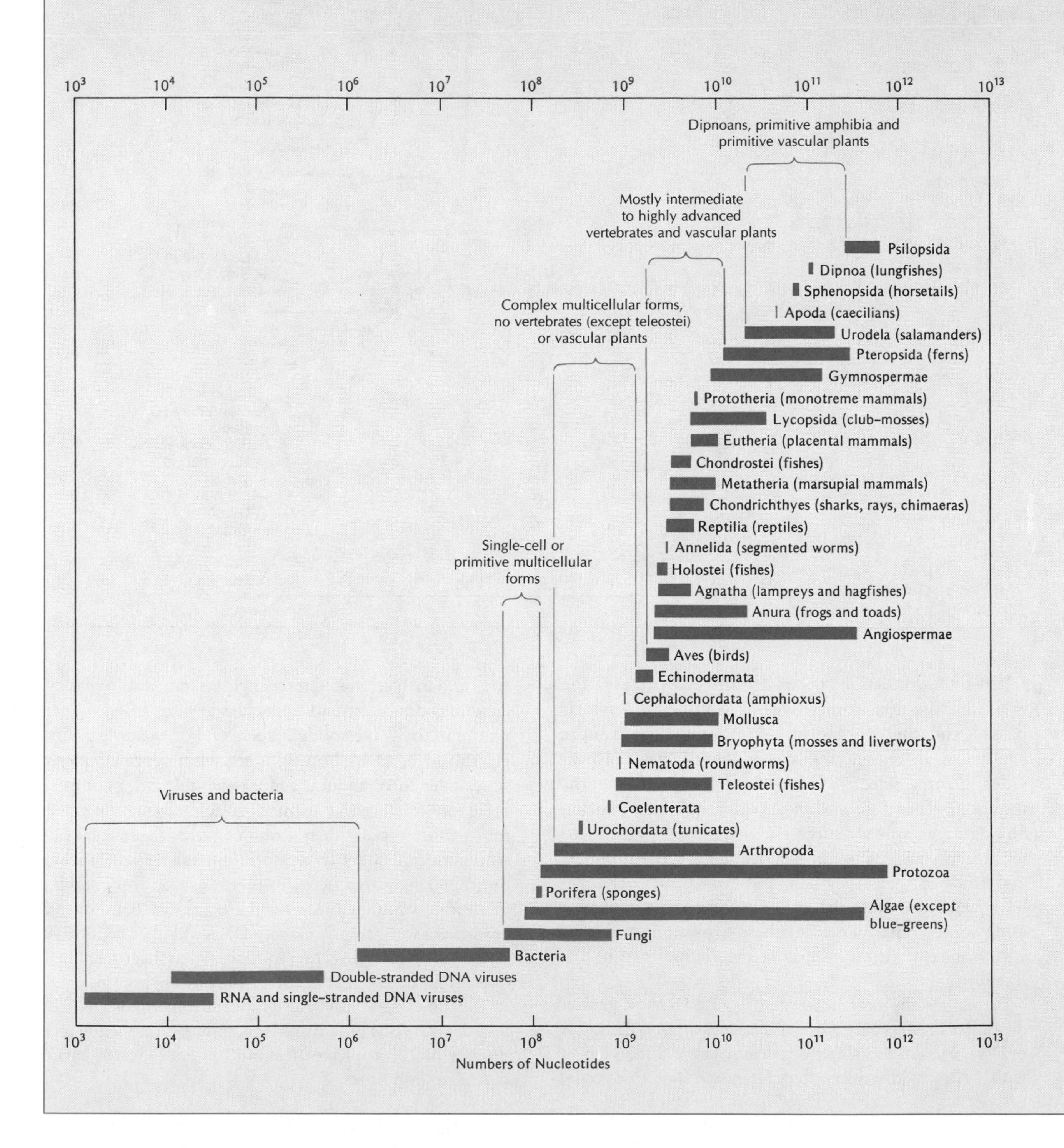

FIGURE 12-9 A comparison of the numbers of nucleotides in the genetic material of different types of organisms. Each bar in the illustration represents the range of nucleotide numbers found among the species sampled in a designated group. The nucleotide values given are largely derived from estimates of the weight of nucleic acid in the haploid complement of an organism according to the formula, 1 gm of nucleic acid = 2.0×10^{21} nucleotides. Thus the DNA in the diploid complement of human chromosomes (6.4×10^{-12} gm) contains $(2.0 \times 10^{21}) \times (6.4 \times 10^{-12}) = 12.8 \times 10^{9}$ nucleotides, or 6.4×10^{9} nucleotide pairs of double-stranded DNA. This equals a length of more than 2 meters (2.9×10^{6}) nucleotide pairs = 1 millimeter). (*From Strickberger, adapted from Sparrow et al.*)

numbers of nucleotides (1 picogram = 10^{-12} gm = 2.01×10^{9} nucleotides). Interestingly, these data indicate a relationship between the amount of DNA and general evolutionary status, going from small amounts in viruses and bacteria to relatively large amounts in the higher eukaryotes. However, this progression is not uniform, since some fairly "primitive" representatives of various groups (for example, ferns, psilopsids, lungfishes, and salamanders) show relatively large amounts of DNA. (The lungfish *Protopterus* has more than 40 times the amount of DNA in humans!) Comparative amounts of DNA are not much of a clue to complexity or to phylogenetic relationship, especially since the cells of some very different organisms (such as mammals and gymnosperms) have exactly the same DNA content, whereas obviously related species among amphibians and other groups vary widely in DNA content.[7]

Nevertheless, some authors have attempted to derive a few generalities from these data. Hinegardner has pointed out that specialized species within some groups have less DNA than the more generalized species. For example, the more generalized fish species such as salmon and cod have cellular DNA contents ranging from 1.2 to 4.4 picograms, whereas the DNA of specialized forms such as sea horses and angler fish range from 0.45 to 0.80 picograms. At the same time, however, various species of algae, protozoa, ferns, and amphibia are quite specialized yet contain relatively large amounts of DNA.

Relationship between DNA amount and chromosome number is another feature whose consistency varies: DNA content correlates well with chromosome number in plants and fishes but poorly in mammals. Even mammalian phenotypes that seem very much alike, such as Chinese and Indian muntjac deer, may have similar amounts of DNA distributed in chromosomes that differ widely in number (Fig. 10-17). In general, knowledge of DNA amount alone is insufficient to derive the fine textural patterns of evolutionary history, and a more complete analysis is necessary.

Rather than measuring the nucleotide number in different organisms, another approach toward obtaining information on evolutionary status has been to compare gene numbers. Unfortunately, many or most gene sequences that produce transcriptional products are embedded in various kinds of nontranscribed nucleotide sequences that obscure their functional distinctions. Nevertheless, some gene number estimates have been offered based on:

- Recognizing translatable nucleotide sequences that possess an initiation codon at their beginning and a stop codon at their end ("open reading frames," Fig. 10-23)
- Identifying DNA sequences that match those of genes whose sequences are already known
- Counting the number of highly repetitive cytosine-guanine sequences ("CpG islands") that usually mark the beginning of a vertebrate gene sequence
- Estimating the number of transcribed DNA genes by isolating messenger RNA molecules and using these to reassociate with their template DNA gene sequences, or to make DNA copies ("cDNA") that can then reassociate with and identify chromosome gene sequences

Using such means, generally estimated gene numbers from various present sources are:

	Organism	Estimated Number of Genes
(a)	*Escherichia coli* (bacterium)	4,000
(b)	*Saccharomyces cerevisiae* (yeast)	6,000
(c)	*Oxytricha similis* (ciliated protozoan)	12,000
(d)	*Caenorhabditis elegans* (nematode)	12,000–17,000
(e)	*Drosophila melanogaster* (fruit fly)	10,000–16,000
(f)	*Mus musculus* (house mouse)	80,000
(g)	*Homo sapiens* (human)	80,000

[7] Inconsistencies between genome size and phenotypic complexity in multicellular organisms have been called the "C-value paradox," where C-value designates genome size in terms of numbers or weight of DNA base pairs.

Although there are still too few organisms with known gene numbers to draw binding conclusions, Bird conjectures that gene number differences among prokaryotes (a), protozoan or invertebrate eukaryotes (b–e), and vertebrate eukaryotes (f, g) are caused by differences in mechanisms that restrict inefficient gene production ("noise reduction"). Thus, he suggests that prokaryotes cannot contain more than a few thousand genes on average because bacterial transcriptional controls (Fig. 10-25) become inefficient in the presence of greater numbers of genes, and would therefore cause some genes to be "turned on" even when they are not functionally necessary.

According to Bird, eukaryotes were able to circumvent such "noise" by a nuclear membrane that separates transcription from protein translation, allowing only translatable messenger RNA sequences to filter into the cytoplasm, and by tightly folding the DNA of functionally unnecessary genes into nontranscribable conformations, using nucleosomes and their histones. To these transcription-repressing mechanisms, Bird claims that vertebrates added DNA cytosine methylation, formerly used mostly to suppress genomic parasites such as transposons. Each major step in reducing wasteful transcription and translation of genes whose products were not required for immediate purposes allowed organisms to possess more genes that could be more appropriately expressed. However, whether such mechanism was a cause of accompaniment of improved systems of replication and mutation repair (p. 225) is still unknown.

TABLE 12-4 Estimates of the frequencies of nonrepetitive DNA sequences and three classes of repetitive DNA sequences in various genomes

Organism	Nonrepetitive (single copy)	Partially Repetitive (to about 200 copies per genome)	Intermediate Repetitive (250–60,000 copies per genome)	Highly Repetitive (70,000 to 1,000,000 or more copies per genome)
Chlamydomonas reinhardtii (algae)	.70		.30	
Physarum polycephalum (fungus)	.58		.42	
Ascaris lumbricoides (nematode)	.77		.23	
Drosophila melanogaster (fruit fly)	.78	.15	.07	
Strongylocentrotus purpuratus (sea urchin)	.38	.25	.34	.03
Xenopus laevis (clawed toad)	.54	.06	.37	.03
Gallus domesticus (chicken)	.70	.24		.06
Bos taurus (cattle)	.55		.38	.05
Homo sapiens (human)	.64	.13	.12	.10

Source: Data from Straus.

The lengths of these repetitive sequences may vary from less than 100 to more than 2,000 nucleotides long. With the exception of *Drosophila*, the shorter-length repetitive sequences seem to be interspersed among sequences of nonrepetitive unique DNA, although the function of this arrangement is unknown. According to some authors (Davidson and Britten) these interspersed sequences act as regulatory genes similar to prokaryotic operators and promoters (Chapter 10), whereas others have suggested that the sequences are involved in packaging long, transcribed strands of heterogeneous nuclear RNA (HnRNA) so they can be further processed into messenger RNA. The fact that many eukaryotic mRNA molecules are shorter than their parental HnRNA molecules certainly indicates that some, or even many, DNA nucleotides in a gene are not translated into protein:

some are at the gene termini, and others, known as *introns* (Fig. 9–14), are distributed throughout eukaryotic protein-coding genes.

Among the long-length repetitive DNA sequences, some undoubtedly code for multigene families such as the many duplicate, tandemly arranged ribosomal RNA genes, transfer RNA genes, and histone genes.[8] The function of the more repetitive but shorter nucleotide sequences in **satellite DNA** (recognized by its separation from the main portion of DNA after centrifugation) seems more obscure, although these have often been localized to distinctively staining chromosome sections and centromere regions (**heterochromatin**).

In the house mouse (*Mus musculus*), a satellite DNA that comprises about 10 percent of the genome interestingly shows no homology to the DNA of related rodents such as rat, field mouse, and hamster. In contrast, a number of satellite sequences are widely conserved in different species groups such as crustaceans (crabs) and insects (*Drosophila*). Such studies and others indicate that at least some satellite DNA can arise quite rapidly during the evolution of a species by adding many copies of a new DNA sequence or amplifying ancestral DNA sequences. The different kinds of satellite DNA, their different amounts, and their possible different origins signify different functions, or perhaps no function at all (Miklos). As discussed in Chapter 10, various biologists have been tempted to consider some or many such sequences as forms of "selfish DNA" (Charlesworth et al.)

Nucleic Acid Phylogenies Based on DNA-DNA Hybridizations

To estimate the extent of homology between nucleic acids of different sources, researchers can measure the degree to which homologous nucleotide sequences in different single strands pair up to form double-strand sections. In one technique presently used, DNA molecules extracted from two organisms, X and Y, are dissociated into single strands, then allowed to reassociate into X–Y hybrid double-strands by incubating them together at appropriate temperatures.

The technique depends on separating interspecific X–Y DNA from intraspecific X–X or Y–Y DNA by radioactively labeling the nonrepetitive DNA of one species, X, and using only relatively small amounts of it in the incubation mixture. Because of its rarity, the DNA of X will have very little chance of forming X–X double strands, and we can assume all radioactively labeled double-strand DNA is X–Y. We can then extract this double-strand DNA (on hydroxyapatite crystals) and examine its properties. If the DNA from X and Y are perfectly homologous and have no nucleotide differences at all (that is, they are of the same species), then the **melting temperature** at which the hybrid X–Y DNA dissociates into single strands (Fig. 12–10) will, of course, be the same as either X–X or Y–Y. However, should X and Y sequences differ, then the X–Y hybrid will dissociate more easily because of nucleotide mismatching; that is, its stability reduces and its melting temperature lowers. Various experiments show that for each 1 percent difference in nucleotide composition between X and Y, the thermal stability of the X–Y hybrid DNA molecule lowers by about 1°C.

Sibley and Ahlquist point out that such techniques enable comparisons among perhaps a billion or more nucleotides simultaneously and can provide considerably more information than usually obtained from comparing a few characters at a time. For example, they have produced detailed bird order phylogenies that once seemed difficult or impossible to determine. They also suggest that primate phylogenies based on such DNA–DNA comparisons can specify relationships that formerly seemed obscure, such as human–chimpanzee–gorilla

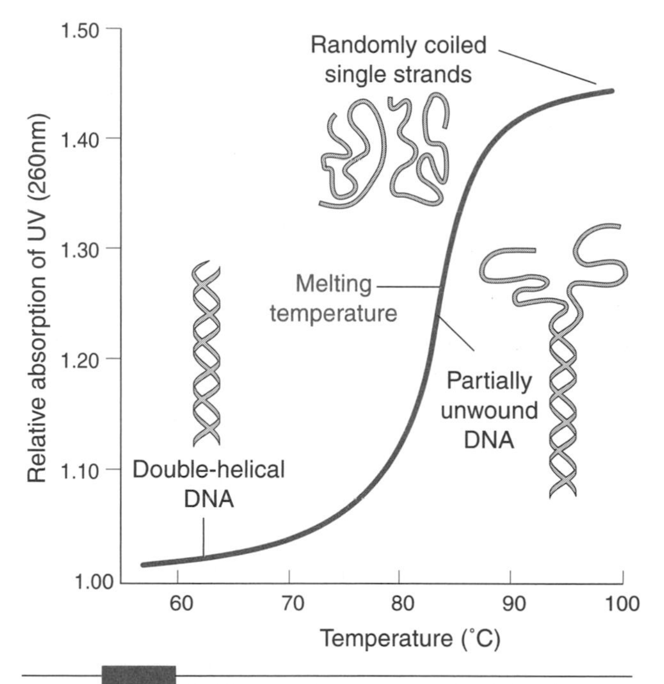

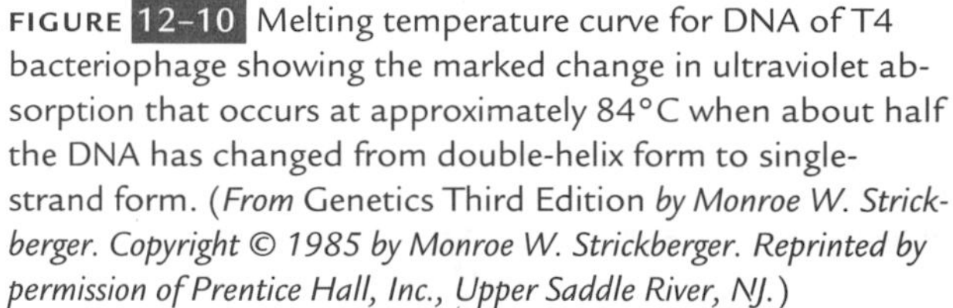

FIGURE 12–10 Melting temperature curve for DNA of T4 bacteriophage showing the marked change in ultraviolet absorption that occurs at approximately 84°C when about half the DNA has changed from double-helix form to single-strand form. (*From* Genetics Third Edition *by Monroe W. Strickberger. Copyright © 1985 by Monroe W. Strickberger. Reprinted by permission of Prentice Hall, Inc., Upper Saddle River, NJ.*)

[8] One source accounting for sequence homogeneity is an error-correcting mechanism called **gene conversion** used in some multigene families such as those coding for chromosome structural histone proteins. This mechanism involves matching the nucleotides between DNA molecules, thereby enabling two or more gene sequences to correspond. Among agents that account for gene similarity in a multigene family, gene conversion occupies a prominent position and helps explain why such genes appear to evolve in unison—**concerted evolution.**

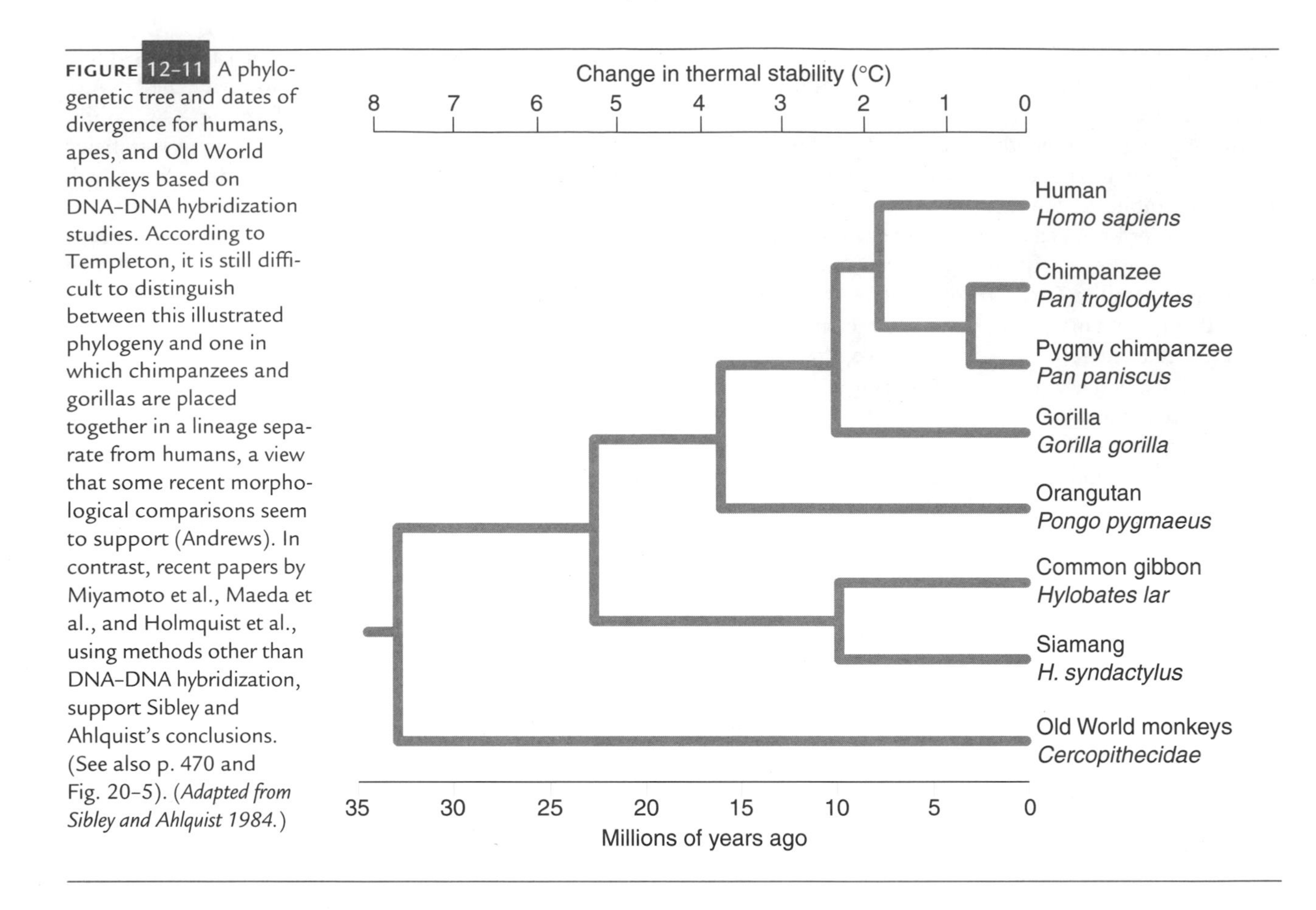

FIGURE 12–11 A phylogenetic tree and dates of divergence for humans, apes, and Old World monkeys based on DNA–DNA hybridization studies. According to Templeton, it is still difficult to distinguish between this illustrated phylogeny and one in which chimpanzees and gorillas are placed together in a lineage separate from humans, a view that some recent morphological comparisons seem to support (Andrews). In contrast, recent papers by Miyamoto et al., Maeda et al., and Holmquist et al., using methods other than DNA–DNA hybridization, support Sibley and Ahlquist's conclusions. (See also p. 470 and Fig. 20–5). (*Adapted from Sibley and Ahlquist 1984.*)

(Fig. 12–11). Moreover, Sibley and Ahlquist propose that since billions of nucleotides are involved in these determinations and changes accrue over millions of years, they can quite confidently estimate average rates of nucleotide change over time.

Assuming from paleontological evidence that the divergence between the lineages of Old World monkeys and apes–humans occurred about 33 million years ago (Chapter 20), and observing an approximate 7.7°C change in thermal stability between these groups and their common ancestor, there is an average of 1°C change for each 33/7.7 = 4.3 million-year interval. Thus the lower scale of Figure 12–11 provides estimates of the dates at which these various primate taxa diverged.

These techniques also make possible comparisons among **nucleotide substitution rates** obtained from DNA–DNA hybridization data and substitution rates from amino acid changes in known proteins. When researchers undertake such comparisons, as Figure 12–12 shows, the rate of change in DNA seems more rapid than the rate of change from most proteins except for fibrinopeptides.[9] Therefore considerable portions of these tested DNA sequences probably do not code for essential proteins such as cytochrome and insulin, and perhaps code for no proteins at all. Such sequences, perhaps largely "junk" DNA or "selfish" DNA (see p. 227) accumulate changes more rapidly than can genes that code for stringently selected proteins.[10] Thus, although we would expect the number of mutations to increase over time, we also expect that the number of mutations allowing amino acid substitutions strongly correlates with protein function.

The relationship between protein function and mutational change is supported by comparing the numbers of **fixed mutations** in amino acid codons incorporated into organisms. As Table 12–5 shows, the more distant the taxonomic relationship between the listed organisms, the more evolutionary time elapsed from their common ancestor, and the greater the number of syn-

[9] The fact that the fibrinopeptides accumulate many changes over relatively short periods of time relates to their function: they are sections of fibrinogen molecules removed during blood clot formation, and most amino acid changes in this sequence have relatively little effect on fibrinogen performance.

[10] The proportion of such noncoding "junk" DNA in many eukaryotes can be quite large. Nowak, for example, estimates 97 percent of human DNA codes neither for proteins nor functional RNA sequences. Among these presumed extraneous DNA sequences, he includes introns (Fig. 9–14), repetitive sequences such a satellites (p. 271), microsatellites (p. 280), and "interspersed elements" such as the *Alu* sequence (p. 226).

TABLE 12-5 Frequency comparisons of synonymous (silent) and replacement (amino acid substitution) mutations in homologous proteins from various organisms

Homologous Protein	Organisms Compared	Relationship	Approximate Time (millions of years) to Common Ancestor	Changes per 100 Codons per 100 Million Years: Synonymous Mutations (silent changes)	Replacement Mutations (amino acid substitutions)
β hemoglobin	Rabbit:mouse	Same class (mammal), different order (lagomorph:rodent)	80	34	25
Cytochrome *c*	Rodent:chicken	Same phylum (chordate), different class (mammal:bird)	250	43	8
Histone 3	Sea urchin:trout	Different phylum (echinoderm:chordate)	650	50	0.8

Source: Modified from a table in Jukes, with additions.

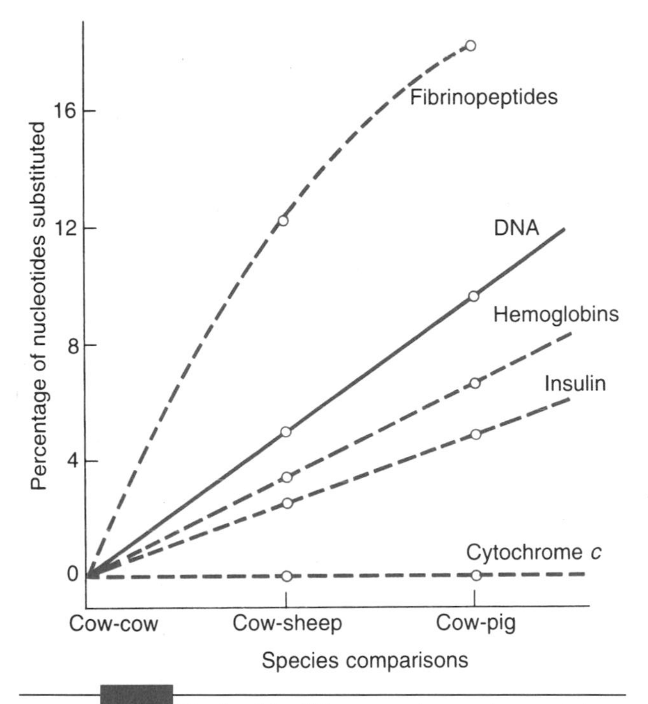

FIGURE 12-12 Nucleotide differences observed among three species of artiodactyls, using the DNA hybridization technique described in the text (*solid line*) and estimates of nucleotide substitutions derived from amino acid analysis of various proteins (*dashed lines*). *(Adapted from McCarthy and Farquhar.)*

onymous ("silent") mutations that do not cause amino acid substitutions.

In contrast, the number of fixed amino acid substitution mutations follows a pattern based on function, from greater numbers of acceptable changes in some proteins to fewer changes in others, irrespective of evolutionary time. For example, the function of histone 3 is to bind and fold DNA molecules in identical fashion in all eukaryotic organisms. This provides a common basic chromosome structure enabling common transcriptional processes as well as common chromosome replication mechanisms. Thus, because of strong selection for such uniformity, histone 3 shows practically no amino acid replacements compared to hemoglobin and even cytochrome *c*. Since the mutation process appears random, as indicated by the fairly constant rate at which silent mutations occur in these proteins, we assume that a highly critical selective process is the primary agent restricting or permitting particular amino acid replacements.

In general, despite the value of DNA–DNA hybridization techniques in detecting divergences between entire genomes, some critics still object to them. One repeated objection (for example, Templeton) is that DNA hybridization experiments compress all divergence information into a single distance measurement, thereby losing information on specific nucleotide sequence changes that could provide a more statistically supported choice among the different possible phylogenetic trees. Researchers have also made DNA comparisons with other techniques, discussed as follows.

Nucleic Acid Phylogenies Based on Restriction Enzyme Sites

One approach to comparative DNA analysis is to use **restriction enzymes** that recognize specific short nucleotide sequences and cleave the molecule at these sites. For example, the enzyme *Eco*RI, isolated from *E. coli* bacteria, recognizes

WALTER M. FITCH

Birthday:
May 21, 1929

Birthplace:
San Diego, California

Undergraduate degree:
University of California–Berkeley

Graduate degree:
Ph.D. University of California–Berkeley, 1958

Postdoctoral training:
Stanford University, 1959–1961
University College, London, 1961–1962

Present position:
Chairman, Department of Ecology and Evolutionary Biology
School of Biological Sciences
University of California–Irvine

WHAT PROMPTED YOUR INITIAL INTEREST IN EVOLUTION?

It is startling to be told that one might be related by ancestry to a fish, a fly, a worm, a plant, and a mushroom. To find evidence for or against such a possibility seemed like a wonderfully stimulating way to spend one's life.

WHAT DO YOU THINK HAS BEEN MOST VALUABLE OR INTERESTING AMONG THE DISCOVERIES YOU HAVE MADE IN SCIENCE?

My most exciting experience was developing a method to analyze sequences of amino acids in proteins, then applying that method to 20 cytochromes *c* and seeing produced, from one small protein, a wonderful tree spanning most of the eukaryotic kingdom with considerable accuracy. (This was published in a 1967 paper in *Science* with E. Margoliash, and is my most cited work.)

WHAT AREAS OF RESEARCH ARE YOU (OR YOUR LABORATORY) PRESENTLY ENGAGED IN?

I am currently developing improved methods for

- Multiple-sequence alignment
- Reconstructing molecular trees
- Assigning events (including gene conversions) to trees
- Detecting and accounting for reticulate evolution (networks rather than trees)
- Predicting the future course of human influenza evolution

I am currently applying such methods in many areas, but especially to viruses (flu, HIV, and vesicular stomatitis). I love the molecular clock problem too.

IN WHICH DIRECTIONS DO YOU THINK FUTURE WORK IN YOUR FIELD NEEDS TO BE DONE?

See my list above. Every time a result appears ambiguous, one should ask whether the problem is in the data or in the method. And every time you say "the method" you have just recognized a worthwhile problem that, if you solve it, permits you to be the first to apply a new and/or more powerful technique to many areas and discover new things as well as answer other people's questions.

WHAT ADVICE WOULD YOU OFFER TO STUDENTS WHO ARE INTERESTED IN A CAREER IN YOUR FIELD OF EVOLUTION?

You can't know everything but, in evolution, it pays to be broadly rather than narrowly trained. New insights frequently arise when different concepts come together to provide an illuminating spark of understanding, and that flash comes more readily by crossing disciplinary boundaries. And the more problems you understand, the more likely your observations will suggest a solution. In Pasteur's words, "Chance favors the prepared mind." The corollary to that is "Treasure [and understand] your exceptions."

the GAATTC/CTTAAG hexanucleotide sequence in double-strand DNA (N represents nonspecified nucleotides):

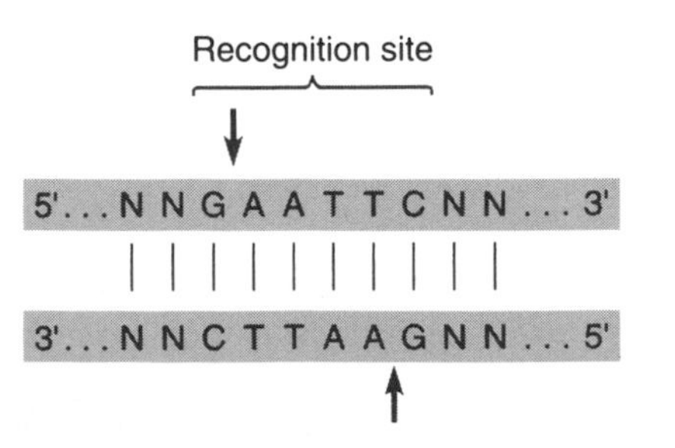

and produces cleavage products at the points between G and A on both strands indicated by the arrows:

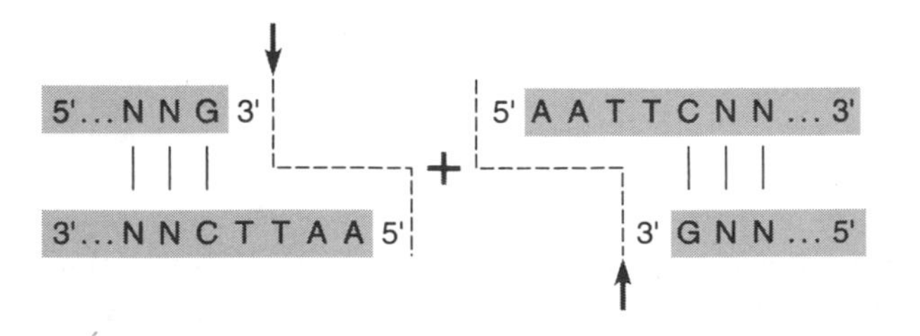

Since DNA molecules can differ from each other in nucleotide sequence, and therefore differ in the number

Human (*Homo sapiens*)

Chimpanzee (*Pan troglodytes*)

Gorilla (*Gorilla gorilla*)

Orangutan (*Pongo pygmaeus*)

Gibbon (*Hylobates lar*)

FIGURE 12-13 Cleavage maps of mitochondrial DNA from humans and four other species of higher primates, derived from the use of 19 restriction enzymes. Cleavage sites for each enzyme are designated by small letters: a, *Eco*RI; b, *Hin*dIII; c, *Hpa*I; d, *Bgl*II; e, *Xba*I; f, *Bam*HI; g, *Pst*I; h, *Pvu*II; i, *Sal*I; j, *Sac* I; k, *Kpn*I; l, *Xho*I; m, *Ava*I; n, *Sma*I; o, *Hin*cII; w, *Bst*EII; x, *Bcl*I; y, *Bgl*I; and z, *Fnu*DII. The position at the left of each map is the replication origin of the mitochondrial chromosome. (*Adapted from Ferris et al.*)

and placement of sites recognized by *Eco*RI, each particular kind of DNA will have fragments of characteristic length when subjected to the enzyme. Also, since there are different kinds of restriction enzymes, many of which recognize target sites different from those recognized by other such enzymes, a DNA molecule subjected to a battery of different enzymes will produce cleavage products unique for that particular kind of DNA. Scoring differences in these inherited fragmentation patterns between individuals (**restriction fragment length polymorphisms,** also called RFLPs) then enables estimates of genetic variation in populations (Nei), with each unique pattern designated as a **haplotype.**[11]

The distinct sites at which restriction enzymes fragment DNA also provide **restriction site maps** for sequences ranging from relatively small repetitive DNA sequences to mitochondrial DNA and the DNA of even larger chromosomes. In one example that Ferris and coworkers analyzed, mitochondrial DNA from humans and apes was subjected to 19 different restriction enzymes. As Figure 12–13 shows, these enzymes cleaved approximately 50 sites in each mitochondrial chromosome, allowing a detailed comparison of target site sequences among the five species. Thus, in accord with previously determined phylogenies, humans share more such sites with chimpanzees and gorillas than with orangutans and gibbons. Unfortunately, the exact branching order among humans, chimpanzees, and gorillas is not obvious from these data (Smouse and Li), and researchers using other techniques (for example, Maeda et al.) are testing the conclusions reached by Sibley and Ahlquist.

[11] Although more expensive to perform, restriction enzyme analysis has an advantage over allozyme variation studies detected by protein electrophoresis (p. 228) because restriction enzymes can be used for sections of DNA throughout the genome and can detect many nucleotide differences, even those that cause no amino acid substitutions (synonymous codons and noncoding DNA). By contrast, electrophoresis can only detect electrical charge differences caused by amino acid substitutions, and the sampled allozymes may not be characteristic of variation in other parts of the genome.

Nucleic Acid Phylogenies Based on Nucleotide Sequence Comparisons and Homologies

more precise method of phylogenetic determination is to compare known nucleotide sequences from different organisms rather than to infer relationships from

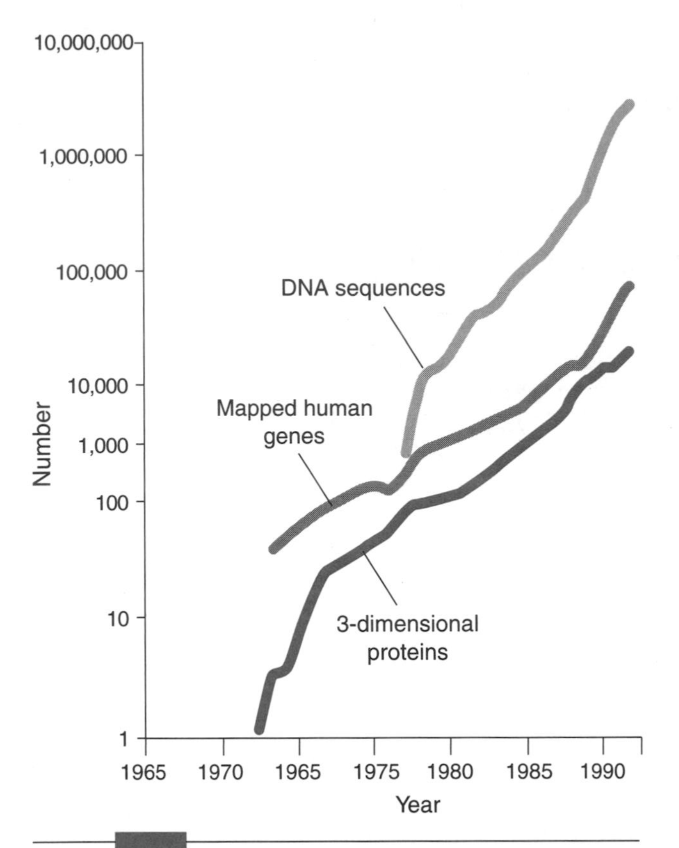

FIGURE 12–14 A growth chart of biomolecular information obtained during the last few decades. More than 2.5 million different DNA sequences are now available in the GenBank facility, an amount doubling every 18 to 24 months. 30,000 human genes are now mapped, and about 7,500 three-dimensional protein structures are recorded in the Protein Data Bank. (*After Boguski.*)

hybridization studies or restriction enzyme maps. This procedure offers advantages in comparing changes between protein-coding and noncoding DNA sequences and in determining the extent of synonymous and nonsynonymous nucleotide substitutions in the amino acid coding regions (Li et al.).

Although the gathering of nucleotide sequence information began fairly recently, data are available on 2.5 million DNA sequences, some millions of base pairs long, from organisms ranging from viruses to eukaryotes (Fig. 12–14). Complete genome sequences are now known for a variety of microbial organisms, including yeast (*Saccharomyces*)—a simple eukaryote (R. F. Doolittle)—as well as for a more complex multicellular eukaryote, the nematode *Caenorhabditis elegans.* Sequencing of other genomes, such as insects (*Drosophila*), plants (*Arabidopsis*), and mice (*Mus*) is in process, and some may have appeared by the time this book is in print. Perhaps the most ambitious project is sequencing the 3 billion base pairs of the human genome, scheduled to be completed in 2005.

From an evolutionary view, the speed in which sequencing technology advances and data accumulate is opening a flood of molecular information, offering opportunities for a wide range of comparative genetic research. For example, among the nine sequenced prokaryotic genomes so far, interesting evolutionary features appear showing:

- Extensive horizontal gene transfer between genomes
- Considerable amounts of gene duplication—as high as 25 percent in the *Bacillus subtilis* genome
- Greater similarity of archaebacterial protein sequences to eubacterial proteins rather than to eukaryotic proteins
- Proteins used in replication, transcription, and translation show a reverse relationship: greater similarity between archaebacteria and eukaryotes
- As much as 50 percent or more of genes in some genomes are "orphans" with no known function
- Based on the 480 genes in *Mycoplasma genitalium,* that number or even smaller may represent the minimal set of genes necessary for cellular life

As Box 12–2 describes, advances in technology enable sequence information to be obtained from the DNA of some fossil organisms.

Concentrating on specific genes, Figure (12–5, p. 261) shows one phylogeny derived from characterizing DNA sequences of β-globin gene clusters in various primates. Some other examples come from nucleic acid structures that are more widely distributed, such as 5*S* RNA, a component of the larger of the two ribosomal subunits that function in ribosome binding of the various transfer RNA molecules that carry the different amino acids. This cornerstone of the basic protein-synthesizing apparatus, once evolved, appears difficult, if not impossible, to change, and has been conserved evolutionarily in all organisms. The secondary structure of 5*S* RNA seems universally the same (Fig. 12–16), a feature that enables all the various 5*S* RNAs to be aligned for every nucleotide position. When such alignments are effected, researchers can use differences among 5*S* RNAs to generate a phylogenetic tree in which nucleotide changes measure evolutionary distance.

As Figure 12–17 shows, this tree illustrates a divergence between primitive prokaryotes and primitive eukaryotes at a time perhaps 50 percent earlier than the divergence between fungi (for example, yeast) and plants and animals (see also Gouy and Li). Since biologists estimate the latter divergence to have occurred about 1.2 billion years ago, the earlier prokaryote–eukaryote separation may well have occurred 1.8 billion years ago, a

BOX 12–2 ANCIENT DNA

The successful extraction and analysis of DNA sequences from dead (ancient) organisms began in 1984 with DNA taken from muscle tissue of a 140-year-old museum specimen of the quagga, a now extinct member of the horse family. The **molecular cloning** technique that Higuchi and coworkers used for this purpose was that commonly used to maintain "libraries" of DNA, in which researchers introduce sequences extracted from an organism into carrier microorganisms, such as specifically designed bacteria (also viruses or yeast cells, as discussed in Box 25–1). The workers amplify the DNA by growing the carriers in a favorable medium where they can replicate the introduced sequences, identify the "clones" that contain desired sequences, and then extract and analyze those DNA sequences (see, for example, Watson et al.).

Higuchi and coworkers confined the quagga analysis to a total of 229 nucleotide base pairs of mitochondrial DNA that showed 12 base substitutions causing only two amino acid replacements, when compared to a corresponding mitochondrial DNA sequence from zebra (*Equus zebra*). These data indicated common ancestry with the horse family as well as little if any modification of the DNA sequences after the quagga died.

There were, however, serious limitations. First, the researchers could only extract relatively short DNA sequences, none longer than 100–200 nucleotides, and the same was true for the DNA analysis Pääbo did on *Alu* repeat sequences (p. 226) in Egyptian mummies. Second, and most important, were the difficulties in obtaining sufficient numbers and kinds of these short ancient DNA sequences by bacterial cloning—the carriers had difficulties in cloning such small sequences. Fortunately, by the middle 1980s Mullis had invented the **polymerase chain reaction (PCR)** technique which let researchers amplify even small traces of DNA with great success. By 1990 thousands of laboratories were using PCR for many purposes, including analysis of ancient DNA (Mullis et al.).

Briefly described, PCR is a test-tube process that involves placing special small "primer" sequences at each end of a target DNA sequence (for example, ancient DNA), and then subjecting these to many replicating cycles (Fig. 12-15). Each such cycle exponentially doubles the number of target sequences so that, after 30 to 40 cycles, there are many millions of replicates of the original ancient DNA sequence. By obtaining such large amounts of DNA, experimenters can achieve nucleotide sequencing of the target with accuracy and confidence.

The PCR method has revolutionized molecular genetics by permitting accurate nucleotide sequencing of any extracted nucleic acid from whatever tissues are available (see the above table). DNA sequences have been analyzed by this method from many fossil organisms: kangaroo rats; the marsupial (Tasmanian) wolf; amber-embedded insects, plants, and vertebrates; human remains from the Arctic zones and from peat bogs in Florida; New Zealand flightless birds (ratites); fossilized plant material and seeds; and even fungal spores (see the collection that Herrmann and Hummel edited).

Unfortunately, because of fossil deterioration, ancient DNA appears in relatively few relics, and even when found, is often highly fragmented and modified.[12] As expected, the confluence of degradative cellular and environmental factors, normally held in abeyance during life, is the major source of damage to ancient DNA. Often, within a short time after an organism dies, nuclease enzymes and oxidizing agents such as superoxide radicals (p. 165) react with nucleotide

Biological tissues from which ancient DNA has been extracted

Type of Material	Reported Maximum Age
Human material:	
Mummies	5,000 years
"Bog" bodies	7,500 years
Bones and teeth	10,000+ years
Animal material:	
Feathers	130 years
Museum skins	140 years
Naturally preserved skins	13,000 years
Bones	25,000+ years[a]
Amber speciments[b]	???
Plant material:	
Herbaria specimens	118 years
Charred seeds and cobs	4,500 years
Mummified seeds and embryos	44,600 years

[a] Hagelberg suggests a 150,000-year-old-date for a Siberian mammoth.

[b] Austin and coworkers report that attempts to reproduce findings of ancient insect DNA in amber are not successful.

Source: Abridged and modified from Brown and Brown.

[12] Cells in bony tissues of vertebrates (osteocytes, osteoclasts, and so on) generally fare better over the long term than cells in soft tissues, due to better protection against physical damage and bacterial decay, less subjection to a damaging water environment, and absence of soft-tissue degradative enzymes. Moreover, DNA binds well to the mineral component of bone (hydroxyapatite, p. 271), also offering protection.

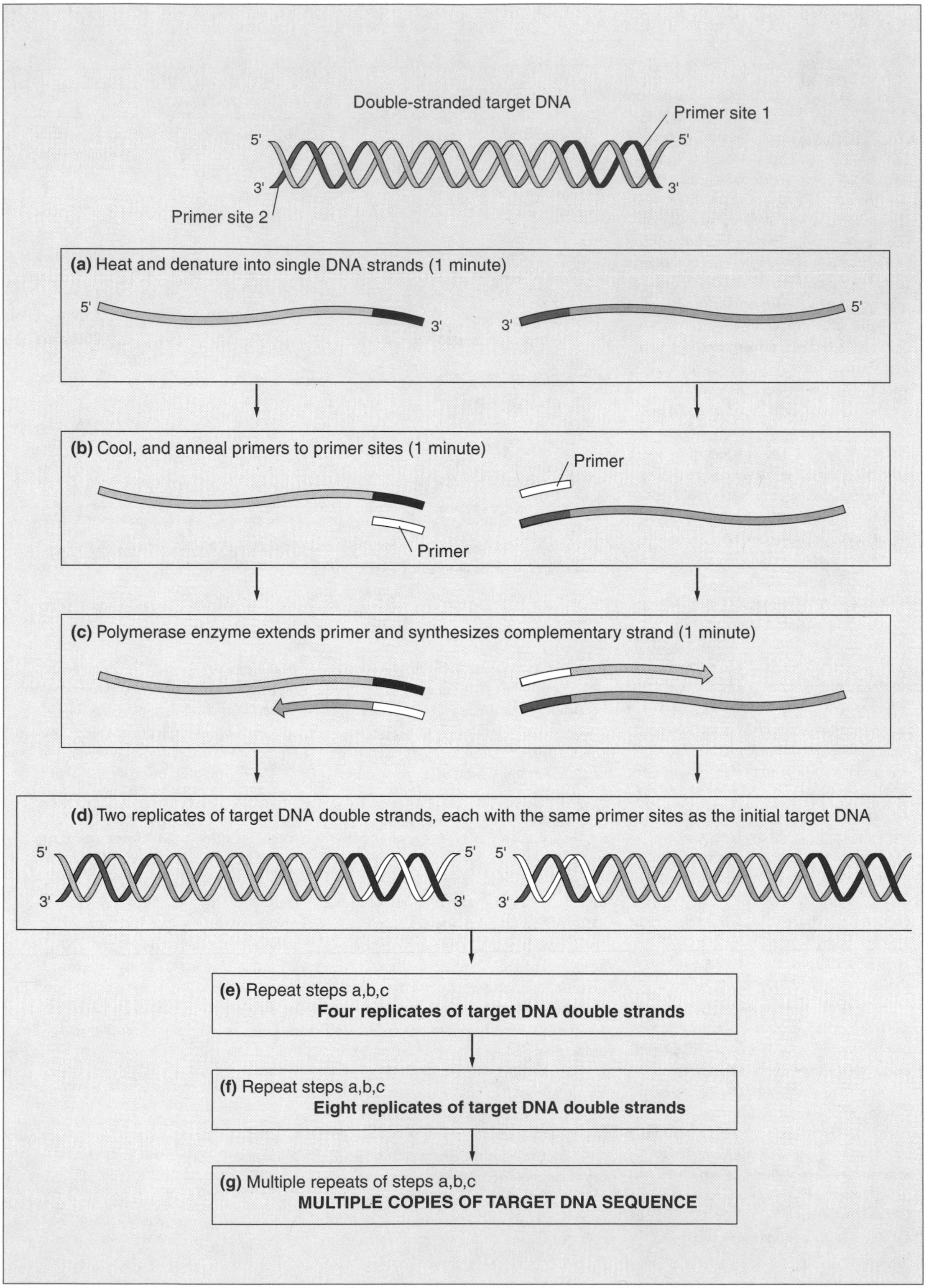

Double-stranded target DNA

Primer site 1

5'
5'
3'
3'

Primer site 2

(a) Heat and denature into single DNA strands (1 minute)

5' 3' 3' 5'

(b) Cool, and anneal primers to primer sites (1 minute)

Primer

Primer

(c) Polymerase enzyme extends primer and synthesizes complementary strand (1 minute)

(d) Two replicates of target DNA double strands, each with the same primer sites as the initial target DNA

5' 5' 5'
3' 3' 3'

(e) Repeat steps a,b,c
Four replicates of target DNA double strands

(f) Repeat steps a,b,c
Eight replicates of target DNA double strands

(g) Multiple repeats of steps a,b,c
MULTIPLE COPIES OF TARGET DNA SEQUENCE

FIGURE 12–15 Simplified diagram of the polymerase chain reaction (PCR) technique showing basic steps in replicating a DNA target sequence that may be as long as 2,000 nucleotide base pairs. Short primer sites at each end of the target are identified (*a*), and oligonucleotide sequences are synthesized (usually about 20 base pairs long) that can pair with the primer sites when "melting" the DNA into single strands (*b*). A heat-resistant DNA polymerase enzyme is used to extend nucleotide synthesis from the primers along each complementary strand (*c*), forming two double-stranded replicates of the original target DNA sequence (*d*). By alternately heating and cooling the mixture, each cycle (*a–c*) exponentially replicates the DNA target (*e–g*), so that an original sequence can potentially be amplified more than one million times ($2^{25} = 4 \times 10^6$) in 25 cycles. The duration for steps *a–c* may vary in different experimental protocols, depending on lengths and compositions of primers and target DNA segments (see Palumbi).

bases and deoxyribose sugars to break as well as cross-link DNA molecules. This is frequently followed by the degradative effects of moisture, acidity and alkalinity changes, mechanical stresses, high temperatures, and ultraviolet radiation that cause additional fragmentation as well as nucleotide base losses and alterations.[13] The consequences for RNA are even more drastic: since RNA has a single strand, it lacks the molecular stability and protection a double-strand structure offers.

Despite these limitations, research in ancient DNA is growing rapidly, and is helping to answer archaeological controversies and to resolve problematic phylogenetic relationships. Ancient DNA from human bones on Easter Island indicates that its settlers were Polynesians from other Pacific islands, and not the South American Indians that the Danish explorer Thor Heyerdahl suggested. Also, Cooper and coworkers' study of flightless New Zealand birds shows that kiwis and extinct moas were much more distantly related than previously proposed, and the kiwis were actually more closely related to the Australian emus and cassowaries.

A most interesting discovery is the distinctive "ancient" mitochondrial DNA isolated from a Neanderthal skeleton 30,000 to 100,000 years old. In answer to persistent questions on the relationship between Neanderthals and modern humans (Chapter 20), Krings and coworkers show that the Neanderthal sequence is significantly more unique than would be expected if it were a sample of normal human variation. According to their findings, the Neanderthal lineage diverged from modern humans about 600,000 years ago, and "went extinct without contributing mitochondrial DNA to modern humans."

Although the results of these and other ongoing investigations are encouraging for evolutionary understanding, their oft-popularized value in re-creating older life-forms is limited if not entirely absent. The difficulty of finding appropriate ancient DNA to analyze, and the poor fragmented form in which such DNA is found, makes it unlikely that we can resurrect entire ancient organisms from ancient DNA. Considering that single-celled organisms have thousands of genes, and most multicelled organisms depend on tens of thousands more:

> we have no idea how to piece together the millions of DNA fragments that we extract from an animal into chromosomes in a functional cell, nor can we set in motion the thousands of genes that regulate development (Pääbo).

[13] To investigators in this field, an additional problem comes from contamination, either from the DNA of bacteria, parasites, and symbionts that infected the original host, or from the DNA of later organisms that helped decompose the body, or even from the DNA of organisms present in the laboratory. Fortunately, there are means of detecting most such intrusions (see Hummel and Herrmann).

finding supported by the existence of fossil eukaryotic-type cells of Proterozic age (Chapter 9).[14]

[14] Because 5*S* RNA molecules have not been found in animal mitochondria and some experimenters consider them too small for obtaining complex phylogenies (for example, Hasegawa et al.), other ribosomal RNA sequences are used for broad-range nucleotide comparisons. For example, Figure 14–7 offers a phylogeny of eukaryotes based on sequence analysis of a larger (18*S*) ribosomal RNA component, and Figure 9–16 ("the universal tree") extends this analysis in differentiating among eubacteria, archaebacteria, and eukaryotes (see also Table 9–2). Another study of this kind by Gray and coworkers (1984) uses mitochondrial and chloroplast RNAs to support the endosymbiotic theory of organelle evolution (Chapter 9) by showing that we can trace both these organelles to a eubacterial origin. Interestingly, this study also proposes that animal and fungal mitochondria originated from nonphotosynthetic aerobic bacteria, whereas plant mitochondria originated separately from cyanobacteria. In a later study (1989) these authors suggest that plant mitochondria may be mosaics resulting from two different symbiotic events. That mitochondria derive from a bacterial origin possessing a genetic code shared by all other organisms (the universal code) supports the concept that amino acid associations with mitochondrial codons (Table 8–2) changed after the universal code was established.

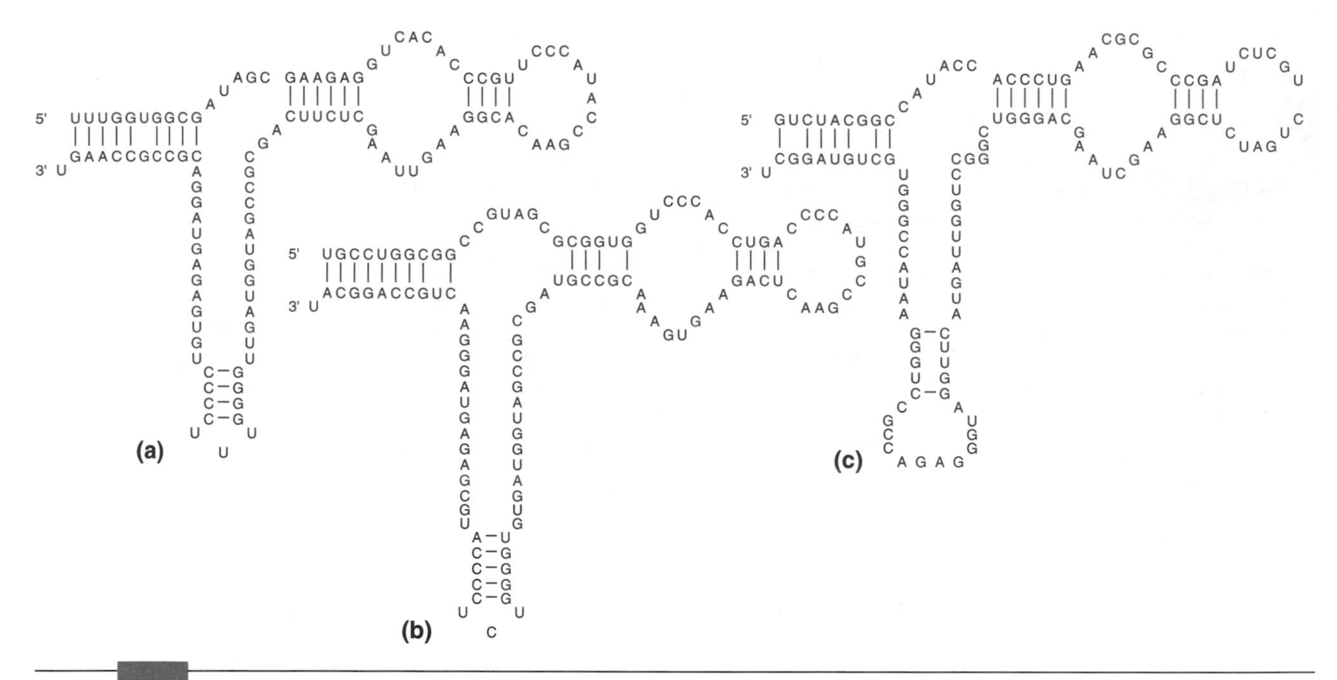

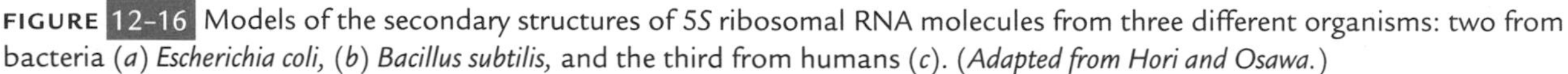

FIGURE 12-16 Models of the secondary structures of 5*S* ribosomal RNA molecules from three different organisms: two from bacteria (*a*) *Escherichia coli,* (*b*) *Bacillus subtilis,* and the third from humans (*c*). (*Adapted from Hori and Osawa.*)

In instances where biologists compare a narrower spectrum of organisms, such as a group of vertebrates, mammals, or humans, they commonly use mitochondrial DNA sequencing (see Avise). Among its advantages, mitochondrial DNA is easily isolated and it also evolves at a sufficiently rapid rate to allow recognition of distinctions and similarities among organisms that have only recently diverged (pp. 483–484). In one example, mitochondrial DNA sequence analysis indicates that the many morphologically diverse species of cichlid fishes in southern Africa have a common monophyletic origin (Meyer et al.), but their similarities in different lakes are caused by parallelism or convergence rather than common ancestry (Fig. 12–18).

The obvious lesson emerging from these and other molecular studies is that morphology alone is not always sufficient to establish phylogeny. For example, convergence can cause what seem to be morphological homologies between various aschelminth phyla (p. 374), but molecular findings show distinctly different ancestries (Winnepenninckx et al.).

In addition to procedures using enzymatic alleles (see allozymes, p. 228) researchers are analyzing relationships closer than between species, such as relationships among races or individuals within a race, using nucleic acid techniques. For example, as Chapter 20 discusses, mitochondrial DNA studies are illuminating relationships among the various present human races.

Other types of DNA analysis have begun using **microsatellites,** tandem repeats of very short nucleotide sequences such as cytosine–adenine–adenine (CAACAACAA . . .), which may vary between individuals in numbers of repeats at some particular chromosomal position. Through techniques involving the isolation and analysis of microsatellites (see Queller et al.), researchers can distinguish an individual with ten CAA repeats at a particular locus, for example, from individuals with greater or lesser numbers of repeats. Because there are many microsatellite loci (humans have at least 50,000 such loci per haploid genome), and they mutate at a relatively high rate compared to protein-coding sequences, opportunities for tracking individual differences and relationships abound. In general, molecular data from all sources can be analyzed by various elegant mathematical techniques to reconstruct phylogenetic relationships (for example, Swofford et al).

Combined Nucleic Acid–Amino Acid Phylogenies

Barnabas and coworkers have offered a comprehensive phylogeny that takes into account both nucleotide sequences from 5*S* RNA[1] and amino acid sequences from ferredoxin and the *c*-type cytochromes (diagrammed in Figure 12–19). Since ferredoxin is a very primitive iron-containing protein used in a number of basic oxidative–reductive pathways, the doubling event that occurred early in its evolution provides a baseline for the phylogenetic tree. Organisms whose ferredoxins most resemble the inferred

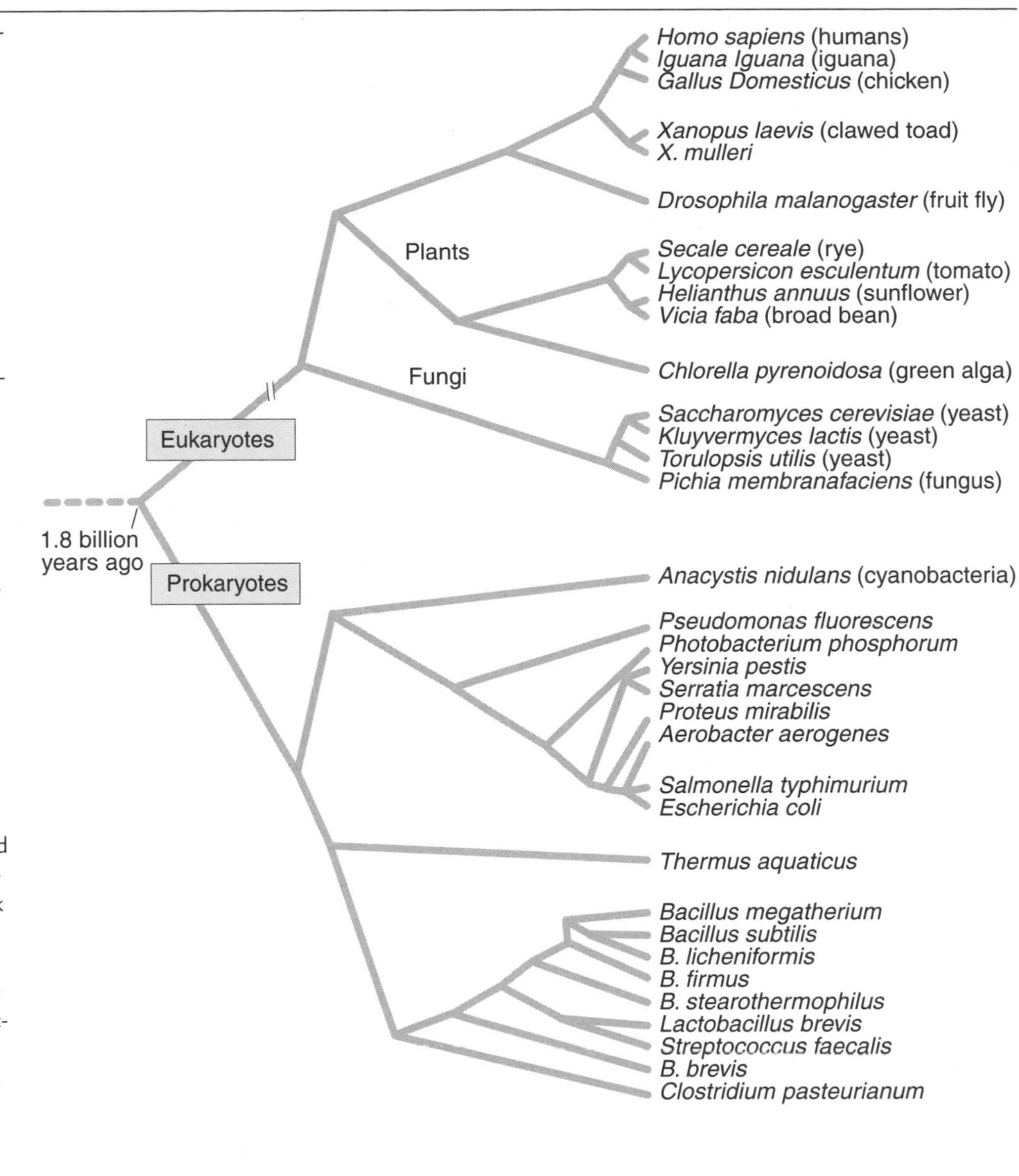

FIGURE 12-17 A phylogenetic tree that best explains the data gathered from comparing 5*S* RNA nucleotide sequences among many different species. The separation between eukaryotes and prokaryotes appears to be close to about 2 billion years ago. Other studies, comparing amino acid sequences from 57 prokaryotic and eukaryotic enzymes, also indicate the two groups shared a common ancestor about 2 billion years ago (Feng et al.). Among additional problems (W. F. Doolittle), such estimates leave in question the kinds of complex cells found in stromatolites that appear much earlier in the fossil record (Chapter 9). Were these cells variations of prokaryotes and eukaryotes, or entirely different? As discussed later (pp. 284–286), applying a single molecular clock to all data causes difficulties. (*Hori, H. and S. Osawa, 1979,* Evolutionary Change in 5S RNA Secondary Structure and a Phylogenic Tree of 54 5s RNA Species, *Proc. Nat. Acad. of Science 76: 381–385. Reprinted by permission.*)

sequence of the primitive duplicated molecule are anaerobic and heterotrophic bacteria such as *Clostridium.* They later diverged, developing into anaerobic photosynthetic bacteria such as *Chromatium* and into the main line of aerobic respiratory organisms.

The cytochrome *c* analysis that Barnabas and coworkers used shows a phylogeny in which the eukaryotic sequences are most similar to the cytochrome *c*2 sequences of the nonsulfur purple photosynthetic bacteria (Rhodospirillaceae). Since cytochrome *c* functions exclusively in the eukaryotic mitochondrion (although coded by DNA in the nucleus), most likely it is the mitochondrion organelle itself (rather than the eukaryotic cell) that derives from this bacterial line. Studies on 5*S* RNA support this view (see Fig. 12–19), showing that eukaryotes actually diverged from an earlier prokaryotic form and not from the later *Rhodopseudomonas* line. Apparently the gene for cytochrome *c* incorporated into the eukaryotic nucleus after the inclusion of the mitochondrion organelle into the eukaryotic cell.

Similarly, the evidence of strong homologies between cyanobacteria and plant chloroplasts for their cytochrome and ferredoxin sequences indicates that eukaryotic plant cells must have incorporated a chloroplast organelle that may at one time have been a prokaryotic cyanobacterium. This phylogeny offers very strong support for the symbiotic theory of organelle function (Chapter 9).[15]

[15] Such endosymbiotic events may have occurred more than once, with some eukaryotic algae receiving their chloroplasts through secondary transfer from other eukaryotes (see p. 180, footnote 15). In any case,

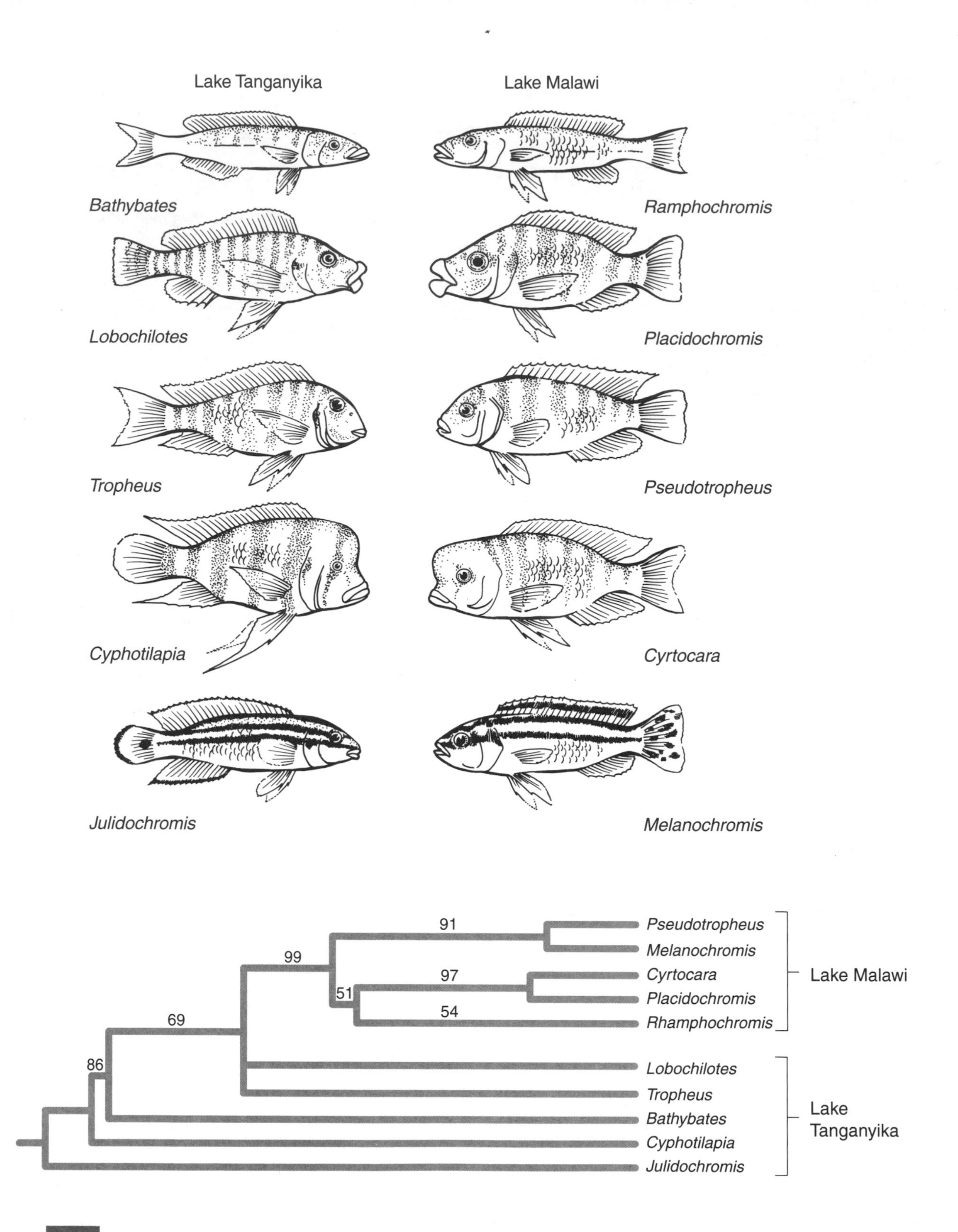

FIGURE 12-18 (*Above*): Phenotypic comparisons among some of the cichlid species from 12 genera found in Lake Tanganyika and Lake Malawi. (*Below*): Phylogenetic tree showing the separate origins of these species, indicating that the similarities among them are convergent rather than homologous. Numbers represent percent bootstrap values that are over 50 percent (p. 264) based on 2,000 samples of the data. Note that these species are only a small sample of the thousand or so different cichlid species found in various East African lakes and rivers. Having originated no later than about 3–4 million years ago, these cichlids represent a prime example of explosive radiation among vertebrates, mostly because variations in a unique pharyngeal jaw apparatus lets individual groups specialize on different prey, and because their breeding, sheltering, and feeding behaviors can restrict them to extremely localized habitats. Perhaps the most dramatic of such speciation events occurred in Lake Victoria, the youngest of the East African lakes. It is known to have dried completely during the last Ice Age 12,400 years ago, yet it has produced about 500 new cichlid species since that time (Galis and Metz). (*After Kocher et al.*)

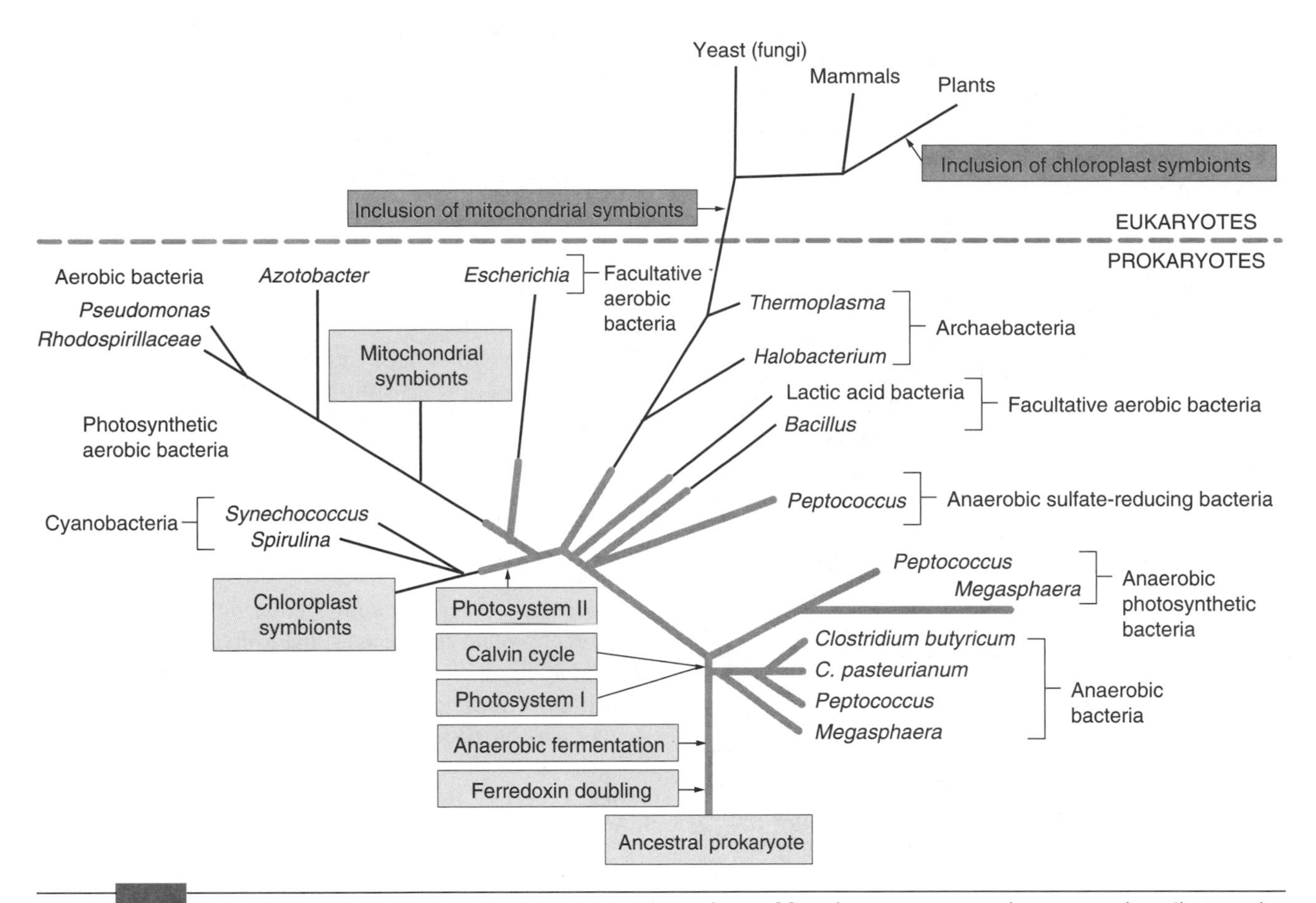

FIGURE 12–19 A composite evolutionary tree based on sequence analyses of ferredoxin, *c*-type cytochromes, and 5S ribosomal RNAs. The *colored lines* indicate segments of the tree dependent on anaerobic metabolism, and the *thin black lines* indicate groups using aerobic respiration. Not all branching relations have been clearly resolved. Note that photosynthesis evolved fairly early in prokaryotic phylogeny and is presumed to have been independently lost in a number of derived lines such as *Bacillus* and the eukaryotes but was regained later in eukaryotic plants that ingested chloroplasts from a lineage close to the cyanobacteria. According to this tree, early eukaryotic-type cells were probably facultatively aerobic, since they stemmed from a line close to *Escherichia* and *Bacillus* bacteria, and apparently only later did they replace these pathways with a more advanced aerobic system introduced by the mitochondrial symbiosis. *The dashed line* across the illustration separates the eukaryotes (*above*) from the prokaryotes (*below*). (*Adapted from Barnabas et al.*)

symbiosis was not only cytoplasmic but also extended to nuclear incorporation of organelle genes. Since chloroplasts synthesize only a small portion of the proteins they use, geneticists have often pointed out that many of the genes that the original cyanobacterial endosymbionts introduced were transferred to the nucleus, and their protein products are now being reimported into the chloroplast. Among the evidence for this is the finding that the chloroplast enzyme, glyceraldehyde-3-phosphate dehydrogenase (GAPDH) used in the Calvin cycle (Fig. 9–5) is made by a nuclear gene whose nucleotide sequence is very similar to one of the three GAPDH genes in the cyanobacterium *Anabaena variabilis*. However, even more interesting is Martin and coworkers' finding that another *Anabaena* GAPDH gene has a similar nucleotide sequence to the GAPDH gene that all plants, animals, and fungi use in gycolysis (Fig. 9–1). Thus prokaryotic endosymbionts may have transferred to eukaryotes a number of genes other than those used in mitochondria and chloroplasts. Eukaryotes can therefore be considered "chimeras," consisting partially of various prokaryotic constituents (see also p. 181).

Rates of Molecular Change: Evolutionary Clocks

Inherent in all the various phylogenies is the concept that evolutionary differences between organisms arise from mutational differences, and therefore, in general, the greater the number of mutational differences between organisms, the greater their evolutionary distance. In some of the phylogenies considered so far (for example, Fig. 12–11), evolutionists have used this notion to provide evolutionary time scales; that is, they have assumed that mutations are incorporated (or fixed) at fairly regular rates over time, and the degree of mutational distance for a phylogenetic interval correlates with the length of time in which such phylogenetic evolution took place. In other words, regarding changes in a specific gene, this assumption suggests that an **evolutionary clock** on the molecular level determines the rates at which many mutations become fixed. Since fixation of these mutations mostly depends on the clock rather than on their adaptive or selective value, Kimura and other geneticists (as discussed in Chapter 23) have proposed that mutations are primarily "neutral" in their effect.

One prominent finding that supports the concept of an evolutionary clock is the constant number of differences in amino acid sequence for the same hemoglobin chain derived from different vertebrates. Specifically, if we look at comparisons with the shark sequence for α-hemoglobin, other vertebrates differ from it by similar numbers of amino acid changes: carp 85, salamander 84, chicken 83, mouse 79, and human 79. This finding indicates that although considerable morphological changes have occurred in these different lineages over a 400 million-year period, constant rates of mutation may have been occurring for at least some proteins (see also p. 560).

Even more obvious clocklike effects are seen in data showing increasing numbers of mutational differences between pairs of organisms separated by increasing time spans (Table 12–6). Thus, the amount of β hemoglobin chain differences between human and monkey lineages, which separated from a common ancestor about 30 million years ago (*a*), increases more than three times on comparing differences between humans and artiodactyls (*b*), more than five times on comparing marsupials and placental mammals (*c*), and more than twelve times on comparing sharks and bony vertebrates (*g*).

Therefore, if evolutionary clocks exist, two consequences can be expected:

1. The lines of descent leading from a common ancestor to all contemporary descendants should have similar rates of fixed mutations because they have experienced similar durations.
2. The proportional rate of fixation that occurs in one gene relative to the rates of fixation in other genes stays the same throughout any line of descent.

In a classical study, Fitch and Langley tested attributes of the evolutionary clock hypothesis for seven proteins whose amino acid sequences they examined in 18 vertebrate taxa. Using commonly accepted dates of divergence for the various common ancestors of these taxa to "calibrate" their evolutionary clock, the researchers obtained temporal lengths for each separate line of descent. This let them compare the number of nucleotide substitutions that occurred within a given time period for all proteins together and for each protein individually. The results of the test showed that the rate at which all proteins have changed together varies significantly among the branches in the different lines of descent, indicating that molecular changes are not uniform for these geological periods.

Moreover, we cannot simply explain these differences in rates of protein change as arising from different generation times in different lines of descent, because the rate at which individual proteins changed relative to other proteins differs significantly within single branches. If changes in generation time caused molecular rate changes, we would expect all proteins to behave similarly within any particular branch, and their individual relative rates to remain unchanged. This analysis indicates that the ticking of the molecular evolutionary clock in each of these seven proteins is not constant in each branch of this phylogeny, whether scored in respect to time or generation.

Nevertheless, when we average the nucleotide substitutions over all seven proteins for each branching point in the phylogeny (rather than sum them or consider them individually), we find a marked uniformity in the rate of molecular change over time. As Figure 12–20*a* shows, a mammalian phylogeny derived from the mutational distance data provides an average number of nucleotide substitutions at each branching point that corresponds

TABLE 12–6 Evidence for progressive increases in amino acid substitutions over time (evolutionary clock) for vertebrate β-hemoglobin chains

Organisms Compared	Amino Acid Changes per 100 Codons	Approximate Time (millions of years) to Common Ancestor
(a) Human/monkey	5	30
(b) Human/cattle	18	90
(c) Marsupial/placental mammal	27	130
(d) Bird/mammal	32	250
(e) Amphibian/amniote vertebrate	49	320
(f) Teleost fish/tetrapod vertebrate	50	400
(g) Shark/bony vertebrate	65	500

Source: Abridged and modified from Jukes.

with a linear relationship to time of divergence (Fig. 12–20*b*). That is, this procedure "calibrates" an evolutionary molecular clock for these proteins, linking change to time.

Given this linear correlation, we can use the following calculations to derive the overall rate at which nucleotide substitutions occur that lead to amino acid changes. Since an average of 98.17 nucleotide substitutions occurs at the farthest point of this linear slope (number 16) for a time period of 120 million years, and a total of 1,734 nucleotide positions in the seven proteins (578 codons $\times$ 3 nucleotides) exist, the rate of nucleotide change over 120 million years is $98.17/1734 = 0.057$; that is, about 6 out of 100 nucleotides caused amino acid substitutions during this interval. The annual rate of amino-acid-changing nucleotide substitutions in this lineage is therefore $0.057/(120 \times 10^6) = 0.47 \times 10^{-9}$.

The need to average changes among different genes to obtain an annual rate of nucleotide substitutions indicates that no single evolutionary molecular clock applies to every nucleotide sequence. The most probable reason is that selection intensity varies in different parts of the genome, fixing mutations at different rates. In addition, as Britten points out (see also Li et al.), significant differences appear between taxonomic groups in nucleotide substitutions that have neutral effects on the phenotype, such as synonymous codon changes, for example, UUU (phenylalanine) → UUC (phenylalanine).

The data so far show two different rates at which such substitutions have incorporated: a slow rate of divergence for humans, apes, and birds, and a faster rate for rodents, lower primates, *Drosophila*, and sea urchins. Britten offers a possible reason for slower rates of divergence in the lower mutation rate that would result from improved DNA repair systems (p. 225): such lower mutation rates would be advantageous in groups that evolved toward increased parental investment in their offspring—that is, reduced birth rate and greater postnatal care.

Wilson and coworkers (1987), in contrast, propose that the differences in evolutionary rates among taxonomic groups that Britten cites are probably exceptional. Instead, they suggest that an examination of many different genes in both bacteria and mammals shows fairly similar rates of nucleotide substitutions at synonymous

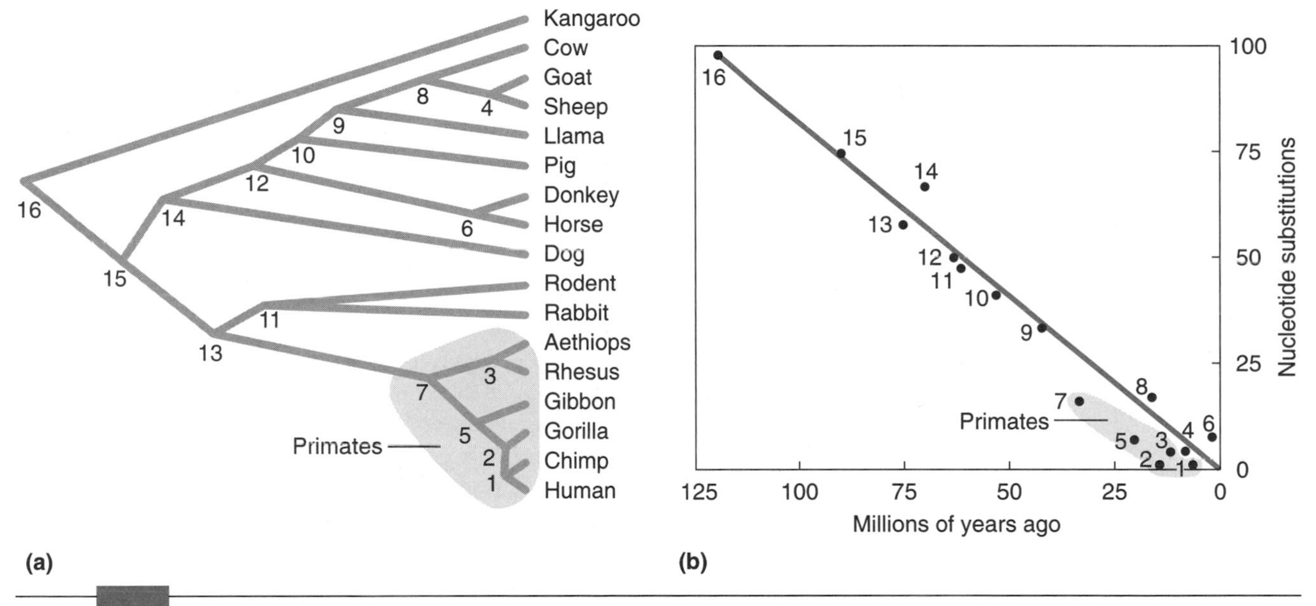

FIGURE 12–20 (*a*) A phylogeny of 17 vertebrate species determined from amino acid sequences examined in seven different proteins. Each numbered nodal point represents a common ancestor for its two diverging branches. (*b*) The relationship between the average number of nucleotide substitutions at each nodal point to the estimated time, based on fossil evidence, at which its branches diverged. This "calibrates" the evolutionary clock for each interval, relating genetic change to time. A *straight line* runs from the origin to the farthest nodal point (number 16), which represents the placental-marsupial divergence approximately 120 million years ago used to calibrate the evolutionary clock. The slope of this line indicates a substitution rate of 0.47 nucleotides per billion years, and, with the exception of primate rates of evolution (numbers 1, 2, 3, 5, 7), most nodal points fall fairly close to this value. The source for the decreased nucleotide substitution rate in primates is not known, although researchers have suggested that the number of nucleotide substitutions is more variable in this group, or estimated dates for primate divergences are less than those commonly accepted, or some primate mutation rates have decreased, or a combination of these and possibly other factors applies. Britten discusses this pattern of "primate slowdown" (see also Maeda et al.) and Easteal and Collert dispute it. (*Adapted from Fitch and Langley.*)

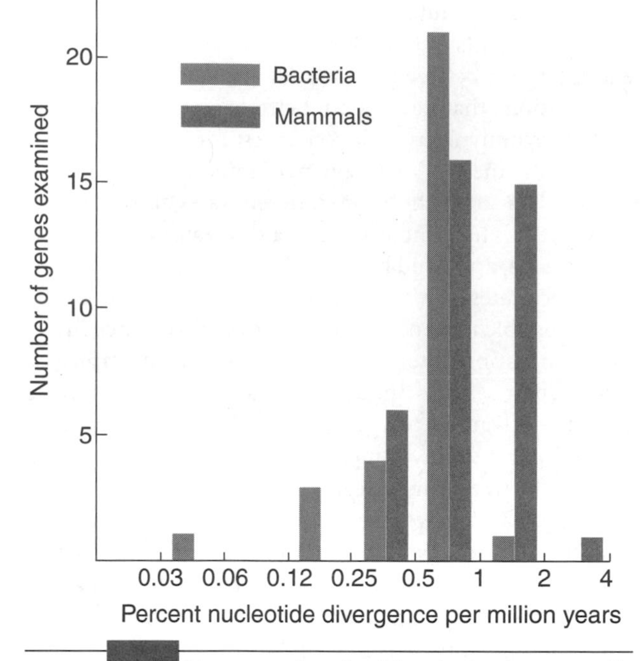

FIGURE 12–21 The rates of nucleotide substitutions per million years at synonymous codon sites in 30 bacterial and 38 mammalian genes, according to Wilson et al.

codon sites (Fig. 12–21). Riley has disputed this view for such sites in *Drosophila* species, so all one can say at present is that although evidence may exist for an evolutionary clock in various lineages, it apparently does not tick at the same rate in all taxonomic groups (Li 1993, Gibbs and Dugaiczyk, Ayala). Reasons offered for such variations include differences in selection intensity among genes and among groups, differences in DNA repair efficiency, different mutagenic experiences, different metabolic rates, and different nucleotide generation times.

Regulatory Genes and Some Evolutionary Consequences

One frequent observation that has emerged from comparing different organisms on the molecular level is that so many share the same kinds of proteins. For example, whether organisms are prokaryotes or eukaryotes, they share similar enzymes involved in basic biochemical processes such as glycolysis, amino acid synthesis, DNA replication, and protein synthesis. These dehydrogenases, kinases, polymerases, proteases, ferredoxins, cytochromes, nucleases, and numerous other gene products are common enzymes that are universally distributed. When we examine them closely, the distinctive structural features of different organisms within any group, such as vertebrates, seem less dependent on differences among the kinds of proteins organisms have than on how they organize and regulate various common proteins such as actin, myosin, collagen, and albumin.

In comparing the anatomy of cat with dog or cat with mouse, differences primarily seem to depend on the extent and location of their common tissues (for example, bone, muscle, nerve). This control over the quantity and placement of tissues arises from regulatory events during development that easily can be modified by mutational changes in the eukaryotic counterparts of prokaryotic regulator genes discussed in Chapter 10 and later more fully in Chapter 15. As Figure 12–22 shows, simple gene rearrangements such as deficiencies, duplications, inversions, and transpositions (translocations) can markedly change regulatory control over gene function. Because regulation is so important, a number of evolutionists have emphasized changes in regulation as being responsible for major changes in evolution—"new bottles for old wine."

Wilson and coworkers (1974) based one study on comparisons between frogs and placental mammals. Early frogs appear in Triassic deposits of about 200 million years ago, whereas placental mammals do not appear in the fossil record until some time during the Cretaceous, about 90 million years ago. Despite their more ancient fossil history and their numerous array of species, frogs have undergone few phenotypic changes in evolution compared to the enormous adaptive radiation of placental mammals.

The more than 3,000 species of frogs look very much alike (Fig. 12–23*a*) and taxonomists have consigned them to a single order (Anura), whereas the 4,300 species of placental mammals (of which about 2,000 are rodents) diverged widely and usually have been classified into about 18 orders ranging from bats to primates to whales (Fig. 12–23*b*). Since the amino acid sequences in proteins of both groups seem to have evolved at approximately the same rate, Wilson suggested that the phenotypic similarity among frogs indicates that relatively few regulatory mutations have established themselves in frogs compared to mammals.

The importance of **regulatory changes** seems obvious in the sharp phenotypic contrast among related species such as humans and African apes. These two groups differ enough (brain size, facial structure, bipedal locomotion, and so on) for many anthropologists to place them in different taxonomic families (Hominidae, Pongidae), yet their composition of structural proteins is strikingly similar: they have almost identical myoglobin and hemoglobin chains, cytochrome *c* proteins, and even fibrinopeptides (Table 20–3). In fact, comparisons for any given protein between these two groups show an average of more than 98 percent identity in amino acid sequence (King and Wilson).

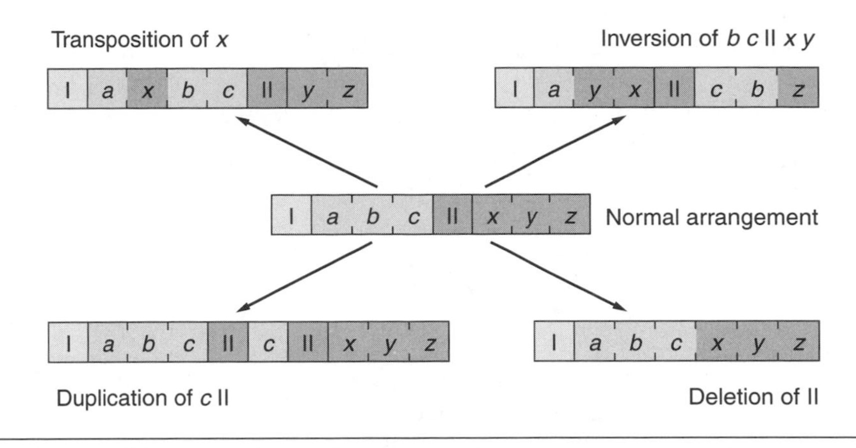

FIGURE 12-22 Some of the possible rearrangements that can occur in a gene sequence containing two sets of structural genes (*a, b, c* and *x, y, z*), each set controlled by separate regulator genes (*I, II*) and each change capable of causing striking mutant effects. For example, the deletion of regulator gene *II* places control of structural genes *x, y,* and *z* under regulator gene *I,* and the indicated inversion reverses the previous regulatory controls over genes *b, c* and *x, y.* (*After Strickberger, adapted from Wilson.*)

Clearly, these and other examples show that regulatory mutations can play a larger role in the morphological and functional differentiation of species than many structural gene changes (see also p. 360). Some evolutionists have used this observation to support the notion that new species, new genera, or new orders can arise from regulatory changes that have a large phenotypic effect producing major taxonomic divisions over relatively short periods of time (**macroevolution**). That is, a population undergoing only small gradual changes (**microevolution**) may persist that way for long periods until pronounced regulatory changes are incorporated that allow one or more of its small isolated groups to evolve rapidly into a higher taxonomic category.

Since some of the fossil data for vertebrates, mollusks, and other animals show such periodic bursts of evolutionary activity, paleontologists such as Eldredge and Gould have given the name **punctuated equilibria** to what they see as long-term fossil uniformity of populations punctuated by geologically short periods of rapid speciation. Others argue that because the fossil record has gaps, punctuated equilibria are only apparent, not real: what seems rapid in geological time may actually involve many thousands of generations. However, although this dispute has generated many arguments and counterarguments (see the discussion in Chapter 24), all evolutionists agree that both gradual and rapid changes occur during evolution. What we have not yet resolved is the relative importance of these changes in explaining speciation and the evolution of higher taxonomic categories. On the phenotypic level, some speciation events certainly seem to be gradual, so that sibling species of *Drosophila,* for example, look very much alike, whereas other related species that are probably no older than some *Drosophila* sibling species (for example, chimpanzees and humans) look very different.

Among questions these observations raise are:

- How many genetic differences does it take to make a species?
- Do differences in morphology between species correlate with the number of genetic differences?

Although these questions are still difficult to answer (see also Chapter 24), molecular studies are now providing important information. For example, DNA mapping of male sterility factors in *Drosophila simulans-mauritania* hybrids indicates there may be more than 100 genes involved in reproductive isolation between these morphological and molecularly similar sibling species (Davis and Wu). We don't yet know whether such high numbers of genes explain reproductive isolation between other closely related species, but such findings certainly indicate that some speciation events depend more on genetic interaction between many loci (epistasis) expressed in hybrid sterility, than on morphological or developmental novelty in one or few loci. In support are findings of Omland of a coupling between the amounts of morphological and molecular evolution in some groups.

The prospect for constructing phylogenies for groups whose relationships long seemed obscure are now attainable because of improved nucleotide sequencing techniques and a rapidly increasing fund of molecular data. Thus, ribosomal RNA sequence analysis has defined a new major group among prokaryotes, the Archaebacteria (Fig. 9–16), and is also being used to delineate relationships among eukaryotic phyla (see, for example, Fig. 14–5). Investigators have even extended the use of sequencing comparisons to studying molecular evolution on the laboratory level (**Box 12–3**, on pp. 290–291).

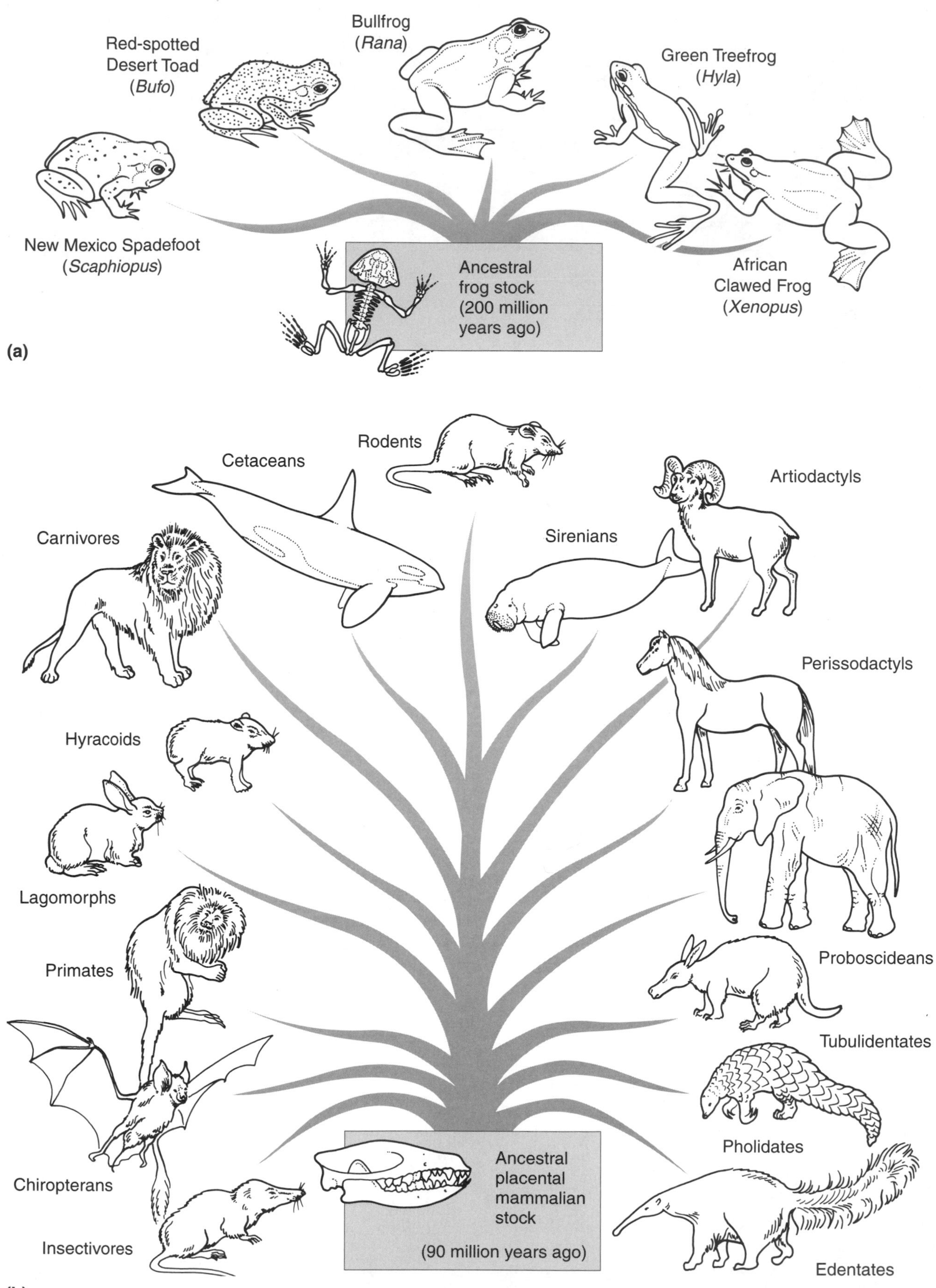
Red-spotted
Desert Toad
(Bufo)
Bullfrog
(Rana)
Green Treefrog
(Hyla)
New Mexico Spadefoot
(Scaphiopus)
Ancestral
frog stock
(200 million
years ago)
African
Clawed Frog
(Xenopus)
(a)
Rodents
Cetaceans
Artiodactyls
Carnivores
Sirenians
Perissodactyls
Hyracoids
Lagomorphs
Primates
Proboscideans
Tubulidentates
Pholidates
Chiropterans
Ancestral
placental
mammalian
stock
(90 million years ago)
Insectivores
Edentates
(b)

FIGURE 12-23 (*a*) Some of the major forms of present-day frogs along with a Triassic froglike fossil dating back to about 200 million years ago. (*b*) Representatives of 15 present-day orders of placental mammals and the fossil skull of a Cretaceous placental-like mammal dating back to about 90 million years ago. Note that although frogs are similar in adult phenotype, their development can differ considerably among species, producing immature forms ranging from tadpoles to froglets, in media ranging from water to brood pouches (pp. 414–415). (*Adapted from Strickberger.*)

SUMMARY

In searching for methods to determine phylogenetic relationships, researchers have used many new techniques that allow molecular comparisons among organisms, even among those without common morphological features. Among these techniques are immunological methods that test proteins from different organisms for antigenic similarities.

Another useful technique surveys amino acid sequences in individual proteins and can be informative about the evolution of individual genes. For example, amino acid sequencing indicates that myoglobin and the many different forms of hemoglobin arose from one ancestral globin gene that subsequently underwent a series of duplications. Such events were apparently common, with many of the duplicated genes evolving along different lines and carrying out disparate functions.

Researchers can also determine phylogenetic distances by comparing amino acid changes in the same protein in different species and calculating the number of mutations necessary to convert one amino acid to another. To construct a realistic phylogenetic tree, however, we need information from many proteins. Techniques involving measurement of DNA levels generally show that the quantity of DNA is not always proportional to the taxonomic status of the group. Analyses of the types of DNA found in organisms indicate that much DNA is repetitive, sometimes highly so. A method some investigators use is to measure the degree to which DNAs from different organisms hybridize, to detect homologies that can help determine phylogenetic lineages and date divergence points of various taxa.

Restriction enzymes, which cleave DNA into fragments at particular sites, also allow comparisons among DNAs of different species. The most difficult, and most informative, of these techniques is a comparison of exact nucleotide sequences of DNAs from different sources. For example, researchers have used data from 5*S* ribosomal RNA to compose a phylogenetic tree in which nucleotide changes serve as a measure of evolutionary distance. Researchers may construct phylogenies using both DNA and protein sequencing. Attempts are also being made to analyze sequences in fragments extracted from dead or fossil organisms ("ancient DNA").

Some workers have assumed that mutations are incorporated into genotypes at a fairly regular rate (the evolutionary clock), and the rate of molecular change seems constant over time for some genes. However, no single clock applies universally; different parts of the genome have different clocks, as do different taxonomic groups.

Major evolutionary change, even speciation, may occur because of mutations in regulatory genes, rather than in genes coding for protein structure. Therefore amino acid sequences for many proteins may be almost identical in different groups, but the phenotypes may vary considerably. Some workers have accredited sudden changes in the fossil record to such mutations.

Recently, scientists have been studying molecular evolution by inducing mutations in microorganisms and subjecting them to selection pressures in the laboratory. Under these conditions, evolutionary changes, such as enzyme adaptation to a new substrate or the conversion of a protein to a new function, have resulted. Researchers have also demonstrated the evolution of viral genomes and even smaller nucleotide sequences under molecular selection pressures.

BOX 12-3 MOLECULAR EVOLUTION IN THE TEST TUBE

The phenomenal growth of molecular information in biology has sparked various attempts to demonstrate evolutionary processes in the laboratory, where detailed analysis of successive changes is more possible than in nature. Since this approach usually demands considerable biochemical analyses and rapid generation times, as well as rigid control over genetic and environmental conditions, researchers have performed most of these studies with microbial organisms.

A common technique of **test tube evolution** is to subject a strain of bacteria to a new carbon source (for example, xylitol) or nitrogen source (for example, butyramide), which the cells cannot metabolize properly. When researchers simultaneously expose such cells to a mutagenic agent, mutations increase in frequency, and some adaptive mutations may then arise. Natural selection then proceeds "directionally" (p. 542) toward a new goal by allowing survival of those bacterial strains with improved metabolic efficiency for the new demanding environment. Evolutionary changes of this kind often occur through the adaptation of enzymes that were initially inefficient on the new substrate, since they were used primarily for other purposes and begin their new role as a partial "preadaptation" (p. 33). Various experimenters have noted a variety of such adaptational changes (Clarke):

1. Synthesis of the inefficient enzyme may become constitutive through a regulatory mutation (Fig. 10-25*d*), thereby increasing the amount of this enzyme in the presence of the new substrate.
2. A regulatory mutation may enable the new substrate to induce synthesis of the inefficient enzyme.
3. Mutations may occur that enable the substrate to enter the cell more easily.
4. A gene duplication may occur that enables increased production of the inefficient enzyme.
5. Mutations may occur in the enzyme's structural gene enabling the former inefficient enzyme to metabolize the new substrate more efficiently.

In an experiment on *E. coli* protein, Campbell and coworkers showed a striking demonstration of enzymatic evolution. Instead of trying to adapt bacteria to an unusual artificial medium, they exposed a strain carrying a deletion of the β-galactosidase *Z* gene (Fig. 10-25*a*) to lactose medium. In the absence of β-galactosidase, lactose does not hydrolyze, and researchers can recognize such bacterial colonies by their red color on a special indicator medium, in contrast to the white color of lactose-using colonies.

Campbell and coworkers found that within one month of growth on a lactose-containing medium, the *Z*-deficient strain gave rise to white colonies that could use lactose, although inefficiently. Further growth and selection among these new lactose-using cells then gave rise to a more efficient strain. On lactose medium unsupplemented with other sugars, the final selected strain of bacterial cells, called *evolved β-galactosidase* (*ebg*), could form colonies as rapidly as could wild-type *E. coli.*

Various tests then showed that the *ebg* strain had evolved a lactose-hydrolyzing enzyme (EBG) completely different from β-galactosidase. This new enzyme had a greater molecular weight, different immunological properties, and different ionic sensitivities. Interestingly, as Hall and Hartl showed, lactose could regulate the enzyme's appearance employing regulatory mutations similar to those that control β-galactosidase. After nucleotide sequencing of the genes involved, Stokes and Hall concluded that the EBG system is a remnant of an ancient duplication of the *E. coli lac* enzyme system. In sum, these experiments demonstrate that a protein with only vague affinities for a particular function can assume that function with remarkable efficiency by a stepwise evolutionary process of mutation and selection (Hall, Hartl).

On the nucleotide level, Mills and coworkers narrowed down test-tube evolutionary experiments to some very small self-replicating molecules. They began with the RNA nucleic acid of a Qβ virus that was about 4,220 nucleotides long, a molecule that they could replicate in test tubes on adding a replicase enzyme and various chemical components. Successive transfers of only the earliest replicating molecules to new cultures then caused selection for rapid replication.

Under these conditions successful Qβ molecules only need retain those sequences that let them be recognized by the replicase enzyme in the culture. That is, they no longer need genes that formerly coded for what are now unnecessary proteins—the coat protein and the replicase enzyme. The researchers then further intensified selection by placing fitness advantages on RNA molecules that could replicate when only a single such molecule is present in an entire culture. This single-stranded molecule (or plus strand) must rapidly attract a replicase enzyme to form a complement (or minus strand) which then forms a new plus strand, and so on. These techniques selected short, independently replicating RNA molecules, including one type called *midivariant* which was only about 220 nucleotides long (Fig. 12-24)!

Eigen and his group then showed that when these experiments are reversed—that is, when mixtures begin without any Qβ RNA sequences at all—the Qβ replicase splices together nucleotides on its own without a template. Moreover, evolution in such mixtures can produce a variety of de novo RNA sequences capable of adapting to different environmental conditions, including sequences that by accretion converge evolutionarily to reach that optimal self-replicating midivariant length of 220 nucleotides! Natural selection operates as it does elsewhere; establishing reproductively successful genotypes through a series of successively adaptive stages.

These and other experiments have led Eigen and coworkers to conclude

FIGURE 12-24 Nucleotide sequence and secondary structure of the smallest RNA molecule that can replicate independently in a mixture containing the replicase enzyme of Qβ virus. (*Adapted from Miele et al.*)

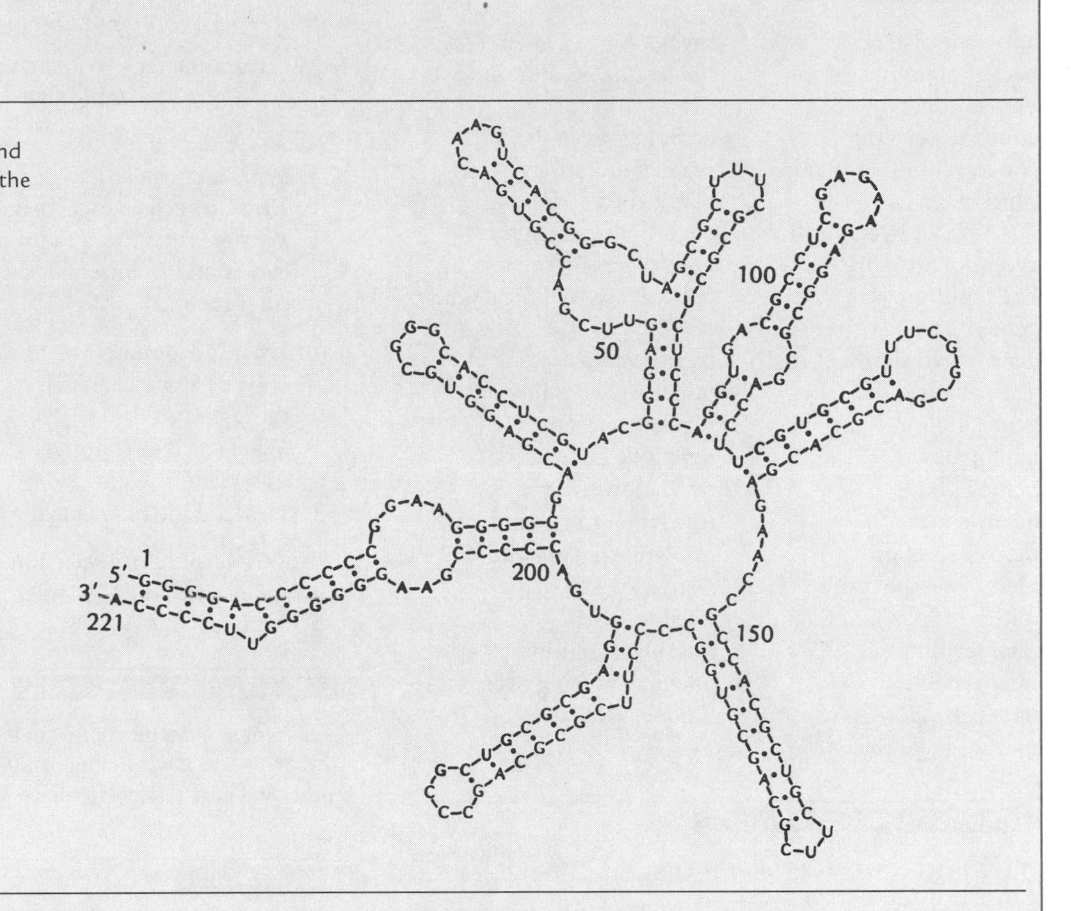

that the number of nucleotides (information content) in an RNA strand determines the frequency at which mutant sequences arise, as well as the number of reproductive cycles necessary for the selection of an optimal mutant with a high reproductive rate. The prevailing (wild-type) genotype achieves stability when its selective advantage is great enough to overcome the error rate of replication for that nucleotide sequence. If the error rate is too high ("mutational meltdown"), the genotype loses adaptive information. If the error rate is too low, the capacity for further adaptation declines, and a lineage may more easily become extinct compared to lineages with more optimal mutation rates. Such factors govern the nucleotide length of a molecule in these experiments. For example, Eigen points out that since RNA polymerases do not replicate nucleotides as accurately as DNA polymerases, the RNA molecules in single-stranded RNA viruses are usually no longer than 10^4 nucleotides, a value that can be calculated theoretically from error rates and selective advantages.[16]

The advantages of test tube evolution in rapidly creating entirely new molecules that can perform new biological roles attract many experimenters. For example, Joyce and coworkers have demonstrated that RNA molecules half the size of the Qβ sequence in Figure 12-24 evolved that have their own signals for test-tube replication (Breaker and Joyce). Subseuqently, Wright and Joyce improved the efficiency of an RNA catalyst ("ribozyme") to 14,000 times its initial value in only 52 hours, by allowing selection to occur in a continuous self-evolving system.

That researchers can also introduce "sex" into test-tube evolution has been shown by Stemmer, who broke genetic sequences into subsidiary fragments and then let them recombine in various ways using the polymerase chain reaction technique (Fig. 12-15). These new recombined molecular sequences were thousands of times more effective in their selected function (antibiotic resistance) than those sequences produced in the absence of recombination—testifying to an important advantage of sex (see also Chapter 23), albeit confined to the test tube. Experimental molecular evolution is becoming a highly promising field for understanding how changes on the molecular level can occur, for helping us grasp their biochemical and evolutionary significance, and for letting us use their effects in producing new functional molecules.

[16] Of course, genomes need to achieve longer lengths in coding for enzymes that need longer sequences. As explained on p. 142, one method of overcoming the separation and competitiveness of small nucleotide sequences is a *hypercycle* in which these subunits join into a mutual symbiotic group, each subunit enhancing the replication of the next.

KEY TERMS

antigenic distances
back mutations
bootstrapping
concerted evolution
convergent molecular evolution
DNA–DNA hybridization
evolutionary clock
fixed mutations
gene cluster
gene conversion
gene duplications
gene family
haplotypes
hemoglobin
heterochromatin
macroevolution
melting temperature
microcomplement fixation
microevolution
microsatellites
molecular cloning
multigene family
myoglobin
nucleotide substitution rates
orthologous
parallel mutations
paralogous
parsimony principle
phylogenetic tree
polymerase chain reaction (PCR)
pseudogenes
punctuated equilibria
regulatory changes
repetitive DNA
restriction enzymes
restriction fragment length polymorphisms (RFLPs)
restriction site maps
satellite DNA
test tube evolution
unequal crossing over
unique DNA

DISCUSSION QUESTIONS

1. What are the advantages in using proteins and nucleic acids to determine phylogenies?
2. How do researchers use antigen–antibody reactions in making phylogenetic determinations?
3. What accounts for the presence of paralogous and orthologous genes? Provide examples of their evolution.
4. How do geneticists use the parsimony method of determining phylogenetic distance?
5. Does the relative amounts of DNA among organisms reflect their phylogenetic positions? Explain.
6. What arguments, pro and con, have evolutionists made on the validity of phylogenies based on DNA–DNA hybridizations?
7. In a choice between DNA sequences that change slowly during evolution and sequences that change rapidly, which provide a better estimate for establishing a phylogeny among closely related species?
8. What arguments can you offer to support the principle of parsimony (p. 263) in determining phylogenies?
9. Why have researchers used 5*S* RNAs for nucleic acid phylogenies?
10. How do the findings of Barnabas and coworkers (Fig. 12–19) support the concept that eukaryotic mitochondria and chloroplasts originated from symbiosis with prokaryotes?
11. What arguments, pro and con, have been offered on the validity of the molecular evolutionary clock?
12. Gene regulation
    a. What are regulatory gene changes?
    b. How do such changes arise?
    c. Why are such changes presumed to have greater evolutionary consequences than changes in structural genes?
13. Enzymatic evolution
    a. How do researchers select enzymatic changes in the laboratory?
    b. What kinds of adaptive enzymatic changes are observed?
    c. How did the EBG enzyme strain of *E. coli* evolve?
14. How did laboratory selection reduce the genome length of the Qβ RNA virus?

EVOLUTION ON THE WEB

Explore evolution on the web! Visit the accompanying web site for *Evolution,* 3/e at www.jbpub.com/evolution for web exercises and links relating to topics covered in this chapter.

REFERENCES

Andrews, P., 1987. Aspects of hominoid phylogeny. In *Molecules and Morphology in Evolution: Conflict or Compromise?* C. Patterson (ed.). Cambridge University Press, Cambridge, England, pp. 23–53.

Austin, J. J., A. J. Ross, A. B. Smith, R. A. Fortey, and R. H. Thomas, 1997. Problems of reproducibility—Does geologically ancient DNA survive in amber-preserved insects? *Proc. Roy. Soc. Lond.* (B), **264,** 467–474.

Avise, J. C., 1994. *Molecular Markers, Natural History and Evolution.* Chapman & Hall, London.

Ayala, F. J., 1999. Molecular clock mirages. *BioEssays,* **21,** 71–75.

Barnabas, J., R. M. Schwartz, and M. O. Dayhoff, 1982. Evolution of major metabolic innovations in the Precambrian. *Origins of Life,* **12,** 81–91.

Benner, S. A., M. A. Cohen, G. H. Gonnet, D. B. Berkowitz, and K. P. Johnsson, 1993. In *The RNA World,* R. F. Gesteland and J. F. Atkins (eds.). Cold Spring Harbor Laboratory Press, Cold Spring Harbor, NY, pp. 27–70.

Bird, A. P., 1995. Gene number, noise reduction and biological complexity. *Trends in Genet.,* **11,** 94–100.

Boguski, M. S., 1998. Bioinformatics—A new era. In *Trends Guide in Bioinformatics.* Elsevier Trends Journals, Haywards Heath, West Sussex, England, pp. 1–3.

Breaker, R. R., and G. F. Joyce, 1994. Emergence of a replicating species from an *in vitro* RNA evolution reaction. *Proc. Nat. Acad. Sci.,* **91,** 6093–6097.

Britten, R. J., 1986. Rates of DNA sequence evolution differ between taxonomic groups. *Science,* **231,** 1393–1398.

Britten, R. J., and E. H. Davidson, 1971. Repetitive and nonrepetitive DNA sequences and a speculation on the origins of evolutionary novelty. *Q. Rev. Biol.,* **46,** 111–138.

Britten, R. J., and D. E. Kohne, 1968. Repeated sequences in DNA. *Science,* **161,** 259–540.

Brown, T. A., and K. A. Brown, 1994. Ancient DNA: Using molecular biology to explore the past. *BioEssays,* **16,** 719–726.

Campbell, J. H., J. A. Lengyel, and J. Langridge, 1973. Evolution of a second gene for β-galactosidase in *Escherichia coli. Proc. Nat. Acad. Sci.,* **70,** 1841–1845.

Charlesworth, B., P. Sniegowski, and W. D. Stephan, 1994. The evolutionary dynamics of repetitive DNA in eukaryotes. *Nature,* **371,** 215–220.

Clarke, P. H., 1978. Experiments in microbial evolution. In *The Bacteria, Vol. 6: Bacterial Diversity,* L. N. Ornston and J. R. Sokatch (eds.). Academic Press, New York, pp. 137–218.

Cooper, A., C. Mourer-Chauviré, G. K. Chambers, A. von Haesler, A. C. Wilson, and S. Pääbo, 1992. Independent origins of New Zealand moas and kiwis. *Proc. Nat. Acad. Sci.,* **89,** 8741–8744.

Davidson, E. H., and R. J. Britten, 1973. Organization, transcription, and regulation in the animal genome. *Q. Rev. Biol.,* **48,** 565–613.

Davis, A. W., and C-I. Wu, 1996. The broom of the sorcerer's apprentice: The fine structure of a chromosomal region causing reproductive isolation between two sibling species of *Drosophila. Genetics,* **143,** 1287–1298.

Dayhoff, M. O. (ed.), 1978. *Atlas of Protein Sequence and Structure,* vol. 5, supp. 3. National Biomedical Research Foundation, Washington, DC.

Dayhoff, M. O., R. M. Schwartz, and B. C. Orcutt, 1978. A model of evolutionary change in proteins. In *Atlas of Protein Sequence and Structure.* vol. 5, supp. 3, M. O. Dayhoff (ed.). National Biomedical Research Foundation, Washington, DC, pp. 345–352.

Dene, H. T., M. Goodman, and W. Prychodko, 1976. Immunodiffusion evidence on the phylogeny of primates. In *Molecular Anthropology,* M. Goodman and R. E. Tashian (eds.). Plenum Press, New York, pp. 171–195.

Doolittle, R. F., 1998. Microbial genomes opened up. *Nature,* **392,** 339–342.

Doolittle, W. F., 1997. Fun with genealogy. *Proc. Nat. Acad. Sci.,* **94,** 12751–12753.

Easteal, S., and C. Collert, 1994. Consistent variation in amino-acid substitution rate, despite uniformity of mutation rate: Protein evolution in mammals is not neutral. *Mol. Biol. and Evol.,* **11,** 643–647.

Edwards, A. W. F., 1996. The origin and early development of the method of minimum evolution for the reconstruction of phylogenetic trees. *Syst. Biol.,* **45,** 79–91.

Eigen, M. 1983. Self-replication and molecular evolution. In *Evolution from Molecules to Man,* D. S. Bendall (ed.). Cambridge University Press, Cambridge, England, pp. 105–130.

Eldredge, N., and S. J. Gould, 1972. Punctuated equilibria: An alternative to phyletic gradualism. In *Models in Paleobiology,* T. J. M. Schopf (ed.). Freeman, Cooper, San Francisco, pp. 82–115.

Felsenstein, J., 1988. Phylogenies from molecular sequences: Inference and reliability. *Ann. Rev. Genet.,* **22,** 521–565.

Feng, D.-F., G. Cho, and R. F. Doolittle, 1997. Determining divergence times with a protein clock: Update and reevaluation. *Proc. Nat. Acad. Sci.,* **94,** 13028–13033.

Ferris, S. D., A. C. Wilson, and W. M. Brown, 1981. Evolutionary tree for apes and humans based on cleavage maps of mitochondrial DNA. *Proc. Nat. Acad. Sci.,* **78,** 2432–2436.

Fitch, W. M., 1976. Molecular evolutionary clocks. In *Molecular Evolution,* F. J. Ayala (ed.). Sinauer Associates, Sunderland, MA, pp. 160–178.

Fitch, W. M., and C. J. Langley, 1976. Protein evolution and the molecular clock. *Fed. Proc.,* **35,** 2092–2097.

Fitch W. M., and E. Margoliash, 1967. Construction of phylogenetic trees. *Science,* **155,** 279–284.

———, 1970. The usefulness of amino acid and nucleotide sequences in evolutionary studies. *Evol. Biol.,* **4,** 67–109.

Frazier, W. A., R. H. Angeletti, and R. A. Bradshaw, 1972. Nerve growth factor and insulin. *Science,* **176,** 482–488.

Galau, G. A., M. E. Chamberlin, B. R. Hough, R. J. Britten, and E. H. Davidson, 1976. Evolution of repetitive and nonrepetitive DNA. In *Molecular Evolution,* F. J. Ayala (ed.). Sinauer Associates, Sunderland, MA, pp. 200–224.

Galis, F., and J. A. J. Metz, 1998. Why are there so many cichlid species? *Trends in Ecol. and Evol.,* **13,** 1–2.

Gibbs, P. M., and A. Dugaiczyk, 1994. Reading the molecular clock from the decay of internal symmetry of a gene. *Proc. Nat. Acad. Sci.,* **91,** 3413–3417.

Golding, G. B., and A. M. Dean, 1998. The structural basis of molecular adaptation. *Mol. Biol. and Evol.,* **15,** 355–369.

Goldman, N., 1998. Effects of sequence alignment procedures on estimates of phylogeny. *BioEssays,* **20,** 287–290.

Goodman, M., 1975. Protein sequence and immunological specificity. In *Phylogeny of the Primates,* W. P. Luckett and F. S. Szalay (eds.). Plenum Press, New York, pp. 219–248.

———, 1976. Toward a genealogical description of the primates. In *Molecular Anthropology,* M. Goodman and R. E. Tashian (eds.). Plenum Press, New York, pp. 321–353.

Goodman, M., A. E. Romero-Herrera, H. Dene, J. Czelusniak, and R. E. Tashian, 1982. Amino acid sequence evidence on the phylogeny of primates and other eutherians. In *Macromolecular Sequences in Systematic and Evolutionary Biology,* M. Goodman (ed.). Plenum Press, New York, pp. 115–191.

Gould, S. J., 1982. The meaning of punctuated equilibrium and its role in validating a hierarchical approach to macroevolution. In *Perspectives on Evolution,* R. Milkman (ed.). Sinauer Associates, Sunderland, MA, pp. 83–104.

Gouy, M., and W-H. Li, 1989. Molecular phylogeny of the kingdoms animalia, plantae and fungi. *Mol. Biol. and Evol.,* **6,** 109–122.

Gray, M. W., D. Sankoff, and R. J. Cedergren, 1984. On the evolutionary descent of organisms and organelles: A global phylogeny based on a highly conserved structural core in small subunit ribosomal RNA. *Nuc. Acids Res.,* **12,** 5837–5852.

———, 1989. On the evolutionary origin of the plant mitochondrion and its genome. *Proc. Nat. Acad. Sci.,* **86,** 2267–2271.

Hagelberg, E., 1994. Ancient DNA studies. *Evol. Anthropol.,* **2,** 199–207.

Hall, B. G., 1983. Evolution of new metabolic functions in laboratory organisms. In *Evolution of Genes and Proteins,* M.

Nei and R. K. Koehne (eds.). Sinauer Associates, Sunderland, MA, pp. 234–257.
Hall, B. G., and D. L. Hartl, 1974. Regulation of newly evolved enzymes. I. Selection of a novel lactase regulated by lactose in *Escherichia coli. Genetics,* **76,** 391–400.
Hardison, R. C., 1996. A brief history of hemoglobins: Plant, animal, protist, and bacteria. *Proc. Nat. Acad. Sci.,* **93,** 5675–5679.
Hartl, D. L., 1989. Evolving theories of enzyme evolution. *Genetics,* **122,** 1–6.
Hasegawa, M., Y. Iida, T. Yano, F. Takaiwa, and M. Iwabuchi, 1985. Phylogenetic relationships among eukaryotic kingdoms inferred from ribosomal RNA sequences. *J. Mol. Biol.,* **22,** 32–38.
Herrmann, B., and S. Hummel (eds.), 1994. *Ancient DNA.* Springer-Verlag, New York.
Higuchi, R., B. Bowman, M. Freiberger, O. A. Ryder, and A. C. Wilson, 1984. DNA sequences from the quagga, an extinct member of the horse family. *Nature,* **312,** 282–284.
Hill, R. L., K. Brew, T. C. Vanaman, I. P. Trayer, and J. P. Mattock, 1969. The structure, function, and evolution of α-lactalbumin. *Brookhaven Symp. Biol.,* **21,** 139–152.
Hillis, D. M., C. Moritz, and B. K. Mable, 1996. *Molecular Systematics,* 2d ed. Sinauer Associates, Sunderland, MA.
Hinegardner, R., 1976. Evolution of genome size. In *Molecular Evolution,* F. J. Ayala, (ed.). Sinauer Associates, Sunderland, MA, pp. 179–199.
Holmquist, R., M. M. Miyamoto, and M. Goodman, 1988. Analysis of higher-primate phylogeny from transversion differences in nuclear and mitochondrial DNA by Lake's method of evolutionary parsimony and operator metrics. *Mol. Biol. and Evol.,* **5,** 217–236.
Hori, H., and S. Osawa, 1979. Evolutionary change in RNA secondary structure and a phylogenetic tree of 54 5*S* RNA species. *Proc. Nat. Acad. Sci.,* **76,** 381–385.
Hummel, S., and B. Herrmann, 1994. General aspects of sample preparation. In *Ancient DNA,* B. Herrmann and S. Hummel (eds.). Springer-Verlag, New York, pp. 59–68.
Ingram, V. M., 1963. *The Hemoglobins in Genetics and Evolution.* Columbia University Press, New York.
Jeffreys, A. J., S. Harris, P. A. Barrie, D. Wood, A. Blanchetot, and S. M. Adams, 1983. Evolution of gene families: The globin genes. In *Evolution from Molecules to Men,* D. S. Bendall (ed.). Cambridge University Press, Cambridge, England, pp. 175–195.
Jukes, T. H., 1996. How did the molecular revolution start? What makes evolution happen? In *Evolution and the Molecular Revolution,* C. R. Marshall and J. W. Schopf (eds.). Jones and Bartlett, Sudbury, MA, pp. 31–52.
Kimura, M., 1979. The neutral theory of molecular evolution. *Sci. Amer.,* **241**(5), 94–104.
King, M. C., and A. C. Wilson, 1975. Evolution at two levels: Molecular similarities and biological differences between humans and chimpanzees. *Science,* **188,** 107–116.
Kocher, T. D., J. A. Conroy, K. R. McKaye, and J. R. Stauffer, 1993. Similar morphologies of cichlid fish in Lake Tanganyika and Lake Malawi are due to convergence. *Mol. Phylog. and Evol.,* **2,** 158–165.
Kornegay, J. R., J. W. Schilling, and A. C. Wilson, 1994. Molecular adaptation of a leaf-eating bird: Stomach lysozyme of the hoatzin. *Mol. Biol. and Evol.,* **11,** 921–928.
Krings, M., A. Stone, R. W. Schmitz, H. Kainitzki, M. Stoneking, and S. Pääbo, 1997. Neanderthal DNA sequences and the origin of modern humans. *Cell,* **90,** 19–30.
Lawn, R. M., K. Schwartz, and L. Patthy, 1997. Convergent evolution of apolipoprotein (a) in primates and hedgehog. *Proc. Nat. Acad. Sci.,* **94,** 11992–11997.
Lee, M. S. Y., 1999. Molecular phylogenies become functional. *Trends in Ecol. and Evol.,* **14,** 177–178.
Li, W.-H., 1983. Evolution of duplicate genes and pseudogenes. In *Evolution of Genes and Protein,* M. Nei and R. K. Koehne (eds.). Sinauer Associates, Sunderland, MA, pp. 14–37.
———, 1993. So what about the molecular clock hypothesis? *Current Opinion Genet. Devel.,* **3,** 896–901.
———, 1997. *Molecular Evolution.* Sinauer Associates, Sunderland, MA.
Li, W-H., C-C. Luo, and C-I. Wu, 1985. Evolution of DNA sequences. In *Molecular Evolutionary Genetics,* R. J. MacIntyre (ed.). Plenum Press, New York, pp. 1–94.
Maeda, N., C. Wu, J. Bliska, and J. Reneke, 1988. Molecular evolution of intergenic DNA in higher primates: Pattern of DNA changes, molecular clock, and evolution of repetitive sequences. *Mol. Biol. and Evol.,* **5,** 1–20.
Maley, L. E., and C. R. Marshall, 1998. The coming age of molecular systematics. *Science,* **279,** 505–506.
Martin, W., H. Brinkmann, C. Savonna, and R. Cerff, 1993. Evidence for a chimeric nature of nuclear genomes: Eubacterial origin of eukaryotic glyceraldehyde-3-phosphate dehydrogenase genes. *Proc. Nat. Acad. Sci.,* **90,** 8692–8696.
McCarthy, B. J., and M. N. Farquhar, 1972. The rate of change of DNA in evolution. *Brookhaven Symp. Biol.,* **23,** 1–41.
McClure, M. A., T. K. Vasi, and W. M. Fitch, 1994. Comparative analysis of multiple protein-sequence alignment methods. *Mol. Biol. and Evol.,* **11,** 571–592.
Meyer, A., T. D. Kocher, P. Basasibwaki, and A. C. Wilson, 1990. Monophyletic origin of Lake Victoria cichlid fishes suggested by mitochondrial DNA sequences. *Nature,* **347,** 550–553.
Miele, E. A., D. R. Mills, and F. R. Kramer, 1983. Autocatalytic replication of a recombinant RNA. *J. Mol. Biol.,* **171,** 281–295.
Miklos, G. L. G., 1985. Localized highly repetitive DNA sequences in vertebrate and invertebrate genomes. In *Molecular Evolutionary Genetics,* R. J. MacIntyre (ed.). Plenum Press, New York, pp. 241–321.
Mills, D. R., F. R. Kramer, and S. Spiegelman, 1973. Complete nucleotide sequence of a replicating RNA molecule. *Science,* **180,** 916–927.
Mirsky, A. E., and H. Ris, 1951. The deoxyribonucleic acid content of animal cells and its evolutionary significance. *Jour. Gen. Physiol.,* **34,** 451–462.
Miyamoto, M. M., B. F. Koop, J. L. Slightom, M. Goodman, and M. R. Tennant, 1988. Molecular systematics of higher primates: Genealogical relations and classification. *Proc. Nat. Acad. Sci.,* **85,** 7627–7631.
Mullis, K. B., F. Ferré, and R. A. Gibbs (eds.), 1994. *PCR: The Polymerase Chain Reaction.* Birkhäuser, Boston.
Nadeau, J. H., and D. Sankoff, 1997. Comparable rates of gene loss and functional divergence after genome duplications early in vertebrate evolution. *Genetics,* **147,** 1259–1266.
Nei, M., 1987. *Molecular Evolutionary Genetics.* Columbia University Press, New York.

Nowak, R., 1994. Mining treasures from junk DNA. *Science,* **263,** 608–610.

Ohno, S., 1970. *Evolution by Gene Duplication.* Springer-Verlag, New York.

Omland, K. E., 1997. Correlated rates of molecular and morphological evolution. *Evolution,* **51,** 1381–1393.

Pääbo, S., 1993. Ancient DNA. *Sci. Amer.,* **269**(5), 86–92.

Palumbi, S. R., 1996. Nucleic acids II: The polymerase chain reaction. In *Molecular Systematics,* 2d ed., D. M. Hillis, C. Moritz, and B. K. Mable (eds.). Sinauer Associates, Sunderland, MA, pp. 205–247.

Piatagorsky, J., and G. J. Wistow, 1989. Enzyme/crystallins: Gene sharing as an evolutionary strategy. *Cell,* **57,** 197–199.

Queller, D. C., J. E. Strassmann, and C. R. Bridges, 1993. Microsatellites and kinship. *Trends in Ecol. and Evol.,* **8,** 285–288.

Rey, M., S. Ohno, J. A. Pinter-Toro, A. Llobell, and T. Bentez, 1998. Unexpected homology between inducible cell wall protein QID74 of filamentous fungi and BR3 salivary protein of the insect *Chironomus. Proc. Nat. Acad. Sci.,* **95,** 6212–6216.

Riley, M. A., 1989. Nucleotide sequence of the *Xdh* region in *Drosophila pseudoobscura* and an analysis of the evolution of synonymous codons. *Mol. Biol. and Evol.,* **6,** 33–52.

Russo, C. A. M., N. Takezaki, and M. Nei, 1996. Efficiencies of different genes and different tree-building methods in recovering a known vertebrate phylogeny. *Mol. Biol. and Evol.,* **13,** 525–536.

Sarich, V. M., and J. E. Cronin, 1976. Molecular systematics of the primates. In *Molecular Anthropology,* M. Goodman and R. E. Tashian (eds.). Plenum Press, New York, pp. 141–170.

Sarich, V. M., and A. C. Wilson, 1966. Quantitative immunochemistry and the evolution of primate albumins: Microcomplement fixation. *Science,* **154,** 1563–1566.

Sibley, C. G., and J. E. Ahlquist, 1984. The phylogeny of primates as indicated by DNA-DNA hybridization. *J. Mol. Evol.,* **20,** 2–15.

———, 1987. Avian phylogeny reconstructed from comparisons of the genetic material, DNA. In *Molecules and Morphology in Evolution: Conflict or Compromise?* C. Patterson (ed.). Cambridge University Press, Cambridge, England, pp. 95–121.

Smouse, P. E., and W.-H. Li, 1987. Likelihood analysis of mitochondrial restriction-cleavage patterns for the human-chimpanzee-gorilla trichotomy. *Evolution,* **41,** 1162–1176.

Sparrow, A. H., H. J. Price, and A. G. Underbrink, 1972. A survey of DNA content per cell and per chromosome of prokaryotic and eukaryotic organisms: Some evolutionary considerations. *Brookhaven Symp. Biol.,* **23,** 451–493.

Stemmer, W. P. C., 1994. Rapid evolution of a protein by DNA shuffling. *Nature,* **370,** 389–391.

Stewart, C. B., and A. C. Wilson, 1987. Sequence conversion and functional adaptation of stomach lysozymes from foregut fermenters. *Cold Sp. Harbor Symp. Quant. Biol.,* **52,** 891–899.

Stokes, H. W., and B. G. Hall, 1985. Sequence of the *ebgR* gene of *Escherichia coli:* Evidence that the EBG and LAC operons are descended from a common ancestor. *Mol. Biol. and Evol.,* **2,** 478–483.

Straus, N. A., 1976. Repeated DNA in eukaryotes. In *Handbook of Genetics,* vol. 5, R. C. King (ed.). Plenum Press, New York, pp. 3–29.

Strickberger, M. W., 1985. *Genetics,* 3d ed. Macmillan, New York.

Swofford, D. L., G. J. Olsen, P. J. Waddell, and D. M. Hillis, 1996. Phylogenetic inference. In *Molecular Systematics,* 2d ed., D. M. Hillis, C. Moritz, and B. K. Mable (eds.). Sinauer Associates, Sunderland, MA, pp. 407–514.

Templeton, A. R., 1986. Relations of humans to African apes: A statistical appraisal of diverse types of data. In *Evolutionary Processes and Theory,* S. Karlin and E. Nevo (eds.). Academic Press, Orlando, FL, pp. 365–388.

Watson, J. D., M. Gilman, J. Witkowski, and M. Zoller, 1992. *Recombinant DNA,* 2d ed. Scientific American Books, New York.

Wilks, H. M., et al., 1988. A specific, highly active malate dehydrogenase by redesign of a lactate dehydrogenase framework. *Science,* **242,** 1541–1544.

Wilson, A. C., 1975. Evolutionary importance of gene regulation. *Stadler Symp.,* **7,** 117–133.

Wilson, A. C., L. R. Maxson, and V. M. Sarich, 1974. Two types of molecular evolution: Evidence from studies of interspecific hybridization. *Proc. Nat. Acad. Sci.,* **71,** 2843–2847.

Wilson, A. C., H. Ochman, and E. M. Prager, 1987. Molecular time scale for evolution. *Trends in Genet.,* **3,** 241–247.

Winnepenninckx, B., T. Backeljau, L. Y. Mackey, J. M. Brooks, R. De Wachter, S. Kumer, and J. R. Garey, 1995. 18*Sr* RNA data indicate that Aschelminthes are polyphyletic in origin and consist of at least three distinct clades. *Mol. Biol. and Evol.,* **12,** 1132–1137.

Wistow, G. J., 1993. Identification of lens crystallin: A model system for gene recruitment. *Methods in Enzymol.,* **224,** 563–575.

Wright, M. C., and G. F. Joyce, 1997. Continuous in vitro evolution of catalytic function. *Science,* **276,** 614–617.

13 Evolution in Plants and Fungi

Most evolutionists believe that the photosynthetic eukaryotic organisms that were the ancestors of vascular land plants were **algae** similar to some members of the present Chlorophyta (green algae) and Charophyceae (stoneworts). These algae are presumed to have originated from single-celled flagellated organisms somewhat like *Chlamydomonas,* which in turn evolved from eukaryotic cells that had been invaded by prokaryotic chloroplast-like symbionts (Chapter 9).

In time, multicellular photosynthetic colonial organisms appeared, probably aided by the ease of association between their cell division products and the advantages that accrue to larger structures whose component cells can undertake a division of labor. Although probably not land plant ancestors, some steps in this evolutionary sequence may echo in the presently observed series that extends from *Chlamydomonas* to *Volvox* (Fig. 13–1): that is, a progression from organisms in which most or all cells can reproduce the entire body, to organisms in which most cells are somatic and only a few are reproductive. At some unknown historical point, one or more groups of green algal organisms became sessile by losing flagellar motility, and began to follow an evolutionary direction toward higher plants.

Terrestrial Algae

Present sessile forms of algae are not necessarily direct relics of the ancient progenitors of land plants. Many grow terrestrially on soil or as epiphytes on trees, and like motile forms, may also follow unicellular or multicellular organization. Some land-dwelling algae such as *Coleochaete* bear morphological similarities to *Parka,* a fossil plant more than 400 million years old, dating from the Upper Silurian to the Lower Devonian (Fig. 13–2). The possibility that *Coleochaete* may represent a prototype of present land plants is discussed by Graham, and includes biochemical similarities (Delwiche et al.) as well as its use of a cell plate during cell division rather than cytoplasmic constriction or cell furrowing.

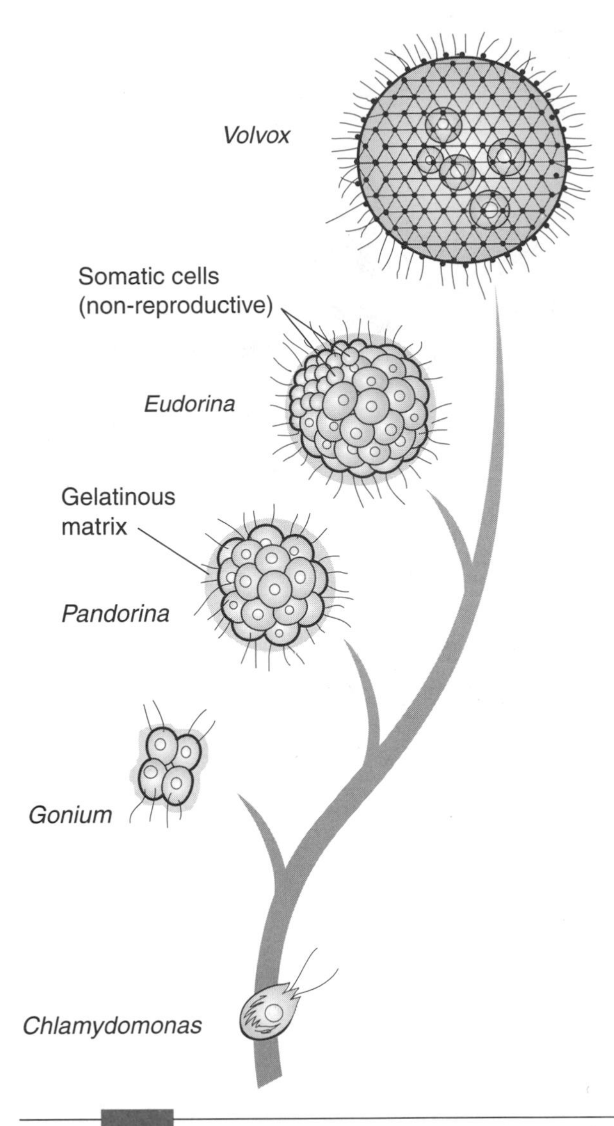

FIGURE 13-1 Possible phylogenetic sequence showing the origin of some multicellular algal aggregates such as *Volvox.* Land plants may have had a similar origin although the intermediary types were probably different.

Another green algae, perhaps farther from land plant ancestry than *Coleochaete* but bearing other terrestrial adaptations, is *Fritschiella tuberosa* (Stewart and Rothwell), a species whose rhizoids penetrate the ground and also maintains branched, multicellular filaments that are both prostrate and erect (Fig. 13–3). Some researchers report that the *Fritschiella* life cycle alternates between the haploid **gametophyte** (n) and diploid **sporophyte** (2n) phases common to some other green algae and so-called higher plants (Fig. 13–4). That is, although both phases are multicellular and grow through regular mitotic cell division, the algae change from diploid to haploid through a meiotic reductional division in the sporophyte. Haploid **spores** produced by the sporophyte develop into the gametophyte phase, which then produces sexual gametes mitotically. These unite, in turn, to form again the diploid zygote and subsequent sporophyte.

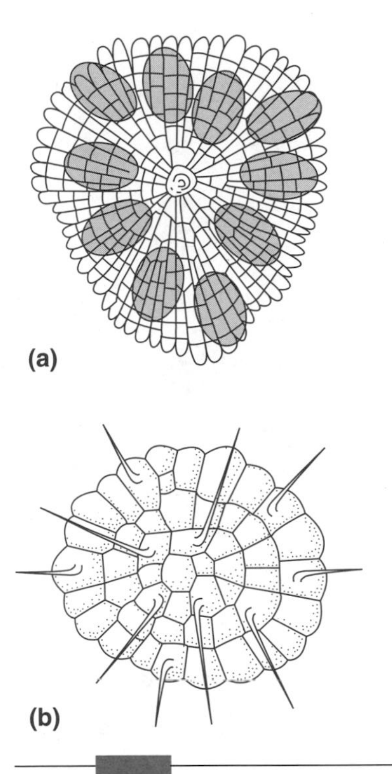

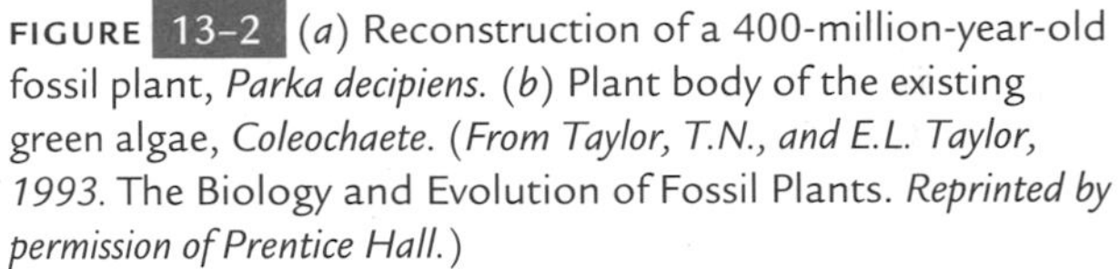

FIGURE 13-2 (*a*) Reconstruction of a 400-million-year-old fossil plant, *Parka decipiens.* (*b*) Plant body of the existing green algae, *Coleochaete.* (*From Taylor, T.N., and E.L. Taylor, 1993.* The Biology and Evolution of Fossil Plants. *Reprinted by permission of Prentice Hall.*)

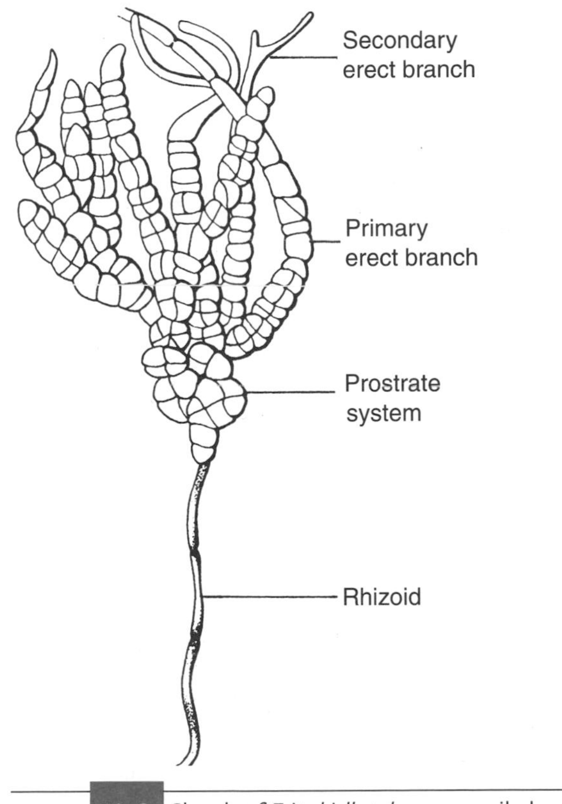

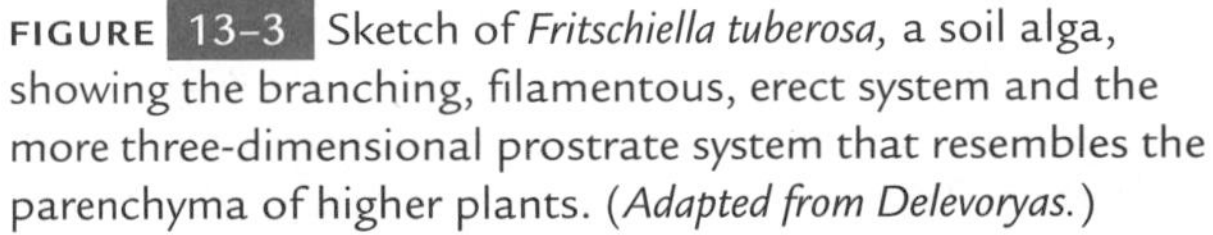

FIGURE 13-3 Sketch of *Fritschiella tuberosa,* a soil alga, showing the branching, filamentous, erect system and the more three-dimensional prostrate system that resembles the parenchyma of higher plants. (*Adapted from Delevoryas.*)

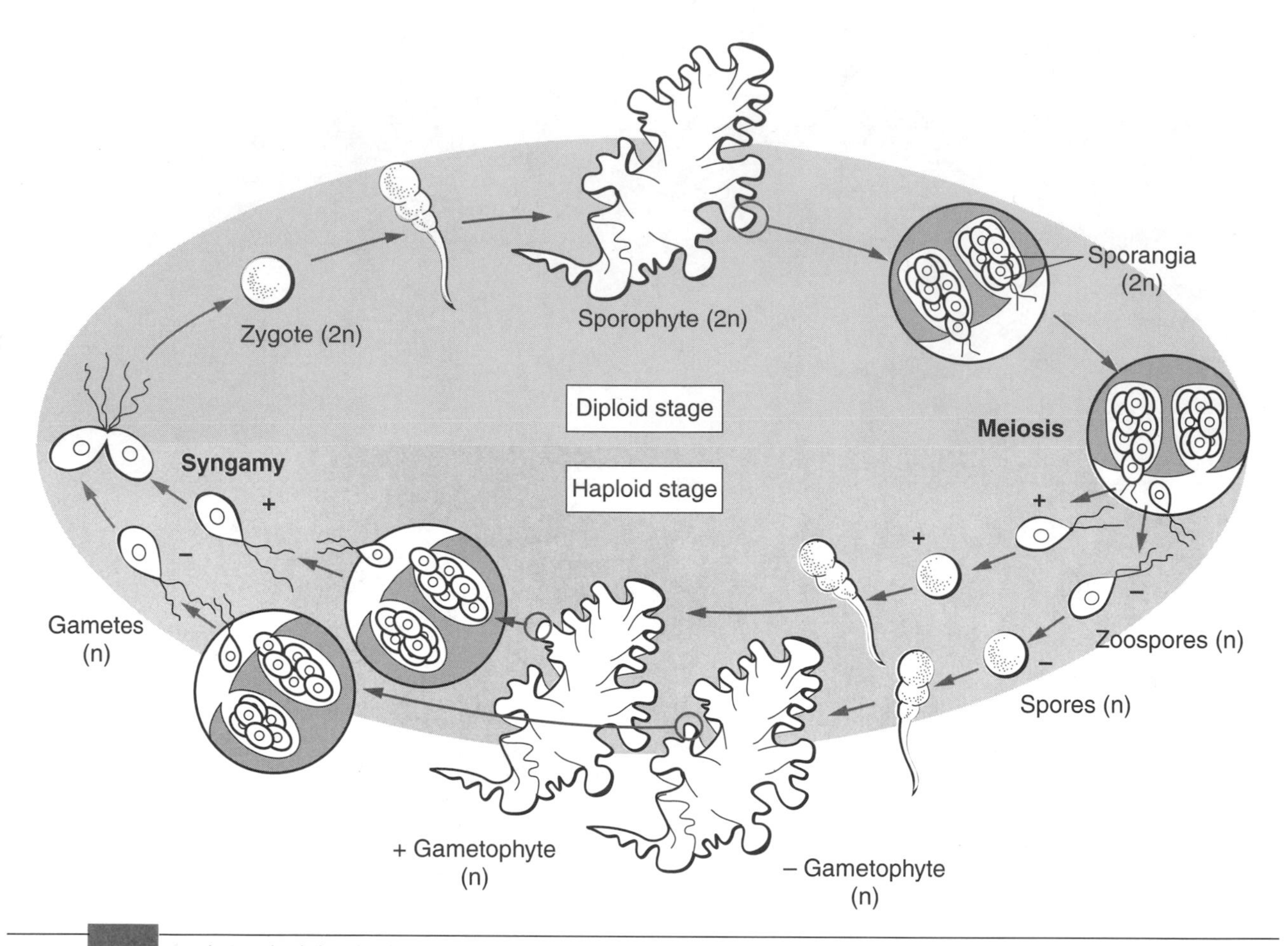

FIGURE 13-4 Mode in which haploid and diploid generations alternate in some green aquatic algae (for example, *Ulva*), showing similar appearing (isomorphic) gametophytes (n) and sporophytes (2n). In land plants the sporophyte embryo is retained and nourished within the gametophyte tissue and matures into a different form from the gametophyte (heteromorphic alternation of generations). In bryophytes, as explained in the text, the sporophyte remains dependent on the gametophyte, whereas the sporophyte is independent in higher vascular plants (Tracheophyta).

Among further similarities between green algae and higher plants is evidence that green algae store their carbohydrate reserves in the form of starch and many have rigid, cellulose-reinforced cell walls. In addition, both green algae and higher plants use similar types of chlorophyll (*a* and *b*) and carotenoids (α and β). We now know a number of green algae, such as *Ulva* and *Caulerpa* (Fig. 13–5), whose membranous forms simulate the appearance of some higher vascular plants yet show their evolutionary ancestry by passing through an algalike, filamentous stage.

Perhaps the most significant aspect of adaptation to land was the prevention of water loss because of cell surface evaporation, a problem that does not exist in most aquatic algae. In those instances where dehydration (**desiccation**) can occur in algae, two major mechanisms of coping with this difficulty evolved. One mechanism used in algae such as *Trentepohlia* has been simply to confine cellular growth to aquatic conditions and to become dormant under dry conditions. The absence of large, watery vacuoles in *Trentepohlia* cells, as in air-dispersed spores of other plants, enables such cells to suffer relatively little change in shape and volume during dehydration compared to those with vacuoles.

In contrast, some species with water-filled vacuolated cells, such as *Cladophorella* and *Fritschiella,* seem to maintain a waxy **cuticle** on their airborne parts, which retards water loss. The vacuoles, in turn, enable cells to continue their metabolic activity as though under a constant marine environment and also provide mechanical rigidity, or turgor, that prevents cellular collapse. The large volume that vacuoles occupy also forces the cytoplasm into a relatively thin sheet along the perimeter of the cell, maximizing the available photosynthetic surface.

In a number of important qualities, some shallow-water or mud-dwelling green algae seem eminently preadapted to begin the journey to land. Which group made this transition is still speculative, although most botanists now incline toward a land plant ancestry from a lineage that also gave rise to *Coleochaete.* Nevertheless, as Graham suggests, terrestrial algal land invasion was not a singular event: "there have probably been multiple colo-

FIGURE 13-5 *Caulerpa,* a green alga with leaflike forms. (*Adapted from Delevoryas, based on other sources.*)

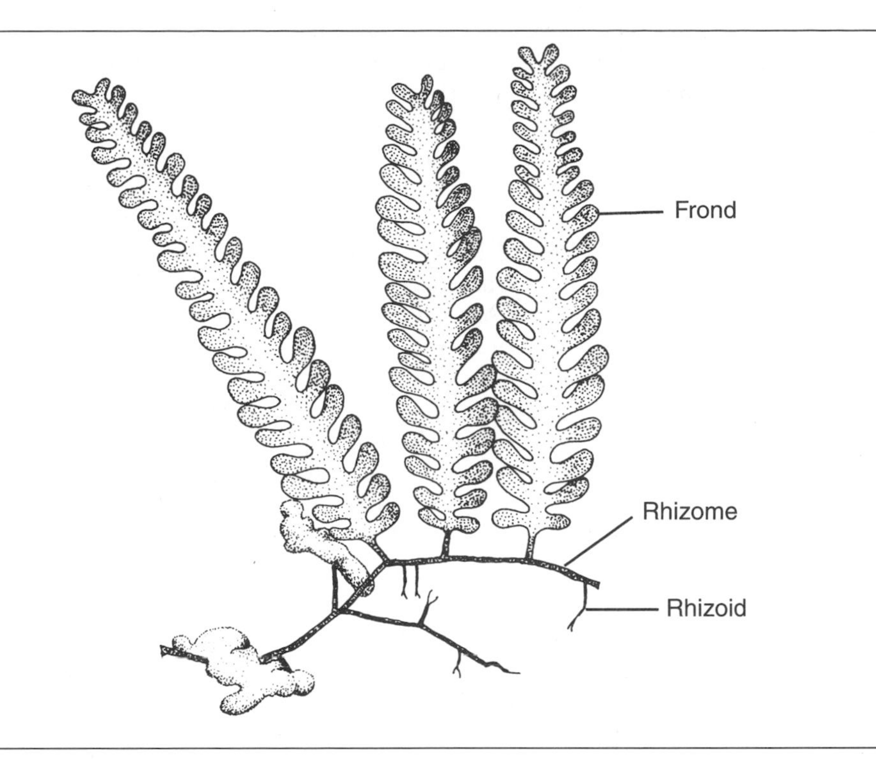

nizations of the terrestrial environment by green algae." Perhaps only one or two of these gave rise to higher plants.

We do not know when the algal journey to land began, but botanists suggest it probably occurred in a post-Cambrian period. Tiffney and others point out that a fall in sea level during Ordovician glaciations would have caused aquatic plants in shoreline communities to undergo selection for resistance to desiccation. Chapman has reviewed other environmental conditions contributing to land plant evolution. These include an increase in atmospheric oxygen that helps in forming highly oxygenated polymers, such as cutin (waxy cuticle material used in waterproofing) and lignin (a stiffening polymer used for mechanical support and water-conducting tissues), and produces an ozone screen against harmful ultraviolet rays (Chapter 9).

Among the algae themselves, evolutionary relationships are not entirely clear, although most botanists recognize that the golden-brown algae (Chrysophyta) and brown algae (Phaeophyta) show relatively advanced features, especially in respect to differentiated structures. Both these types of algae have **planktonic** (motile) and **benthic** (nonmotile) forms, the latter attaching to the sea floor in shallow areas. In some brown algae such as *Fucus* (Fig. 13–6), cell division appears localized, as in higher plants, to a specific meristematic growth area below the elongating tip of the plant, and to differentiated organs that produce sperm and eggs. Also, the relatively complex body tissues include specialized conducting cells that function similarly to the sieve tubes in the phloem of higher plants—probably an example of convergent

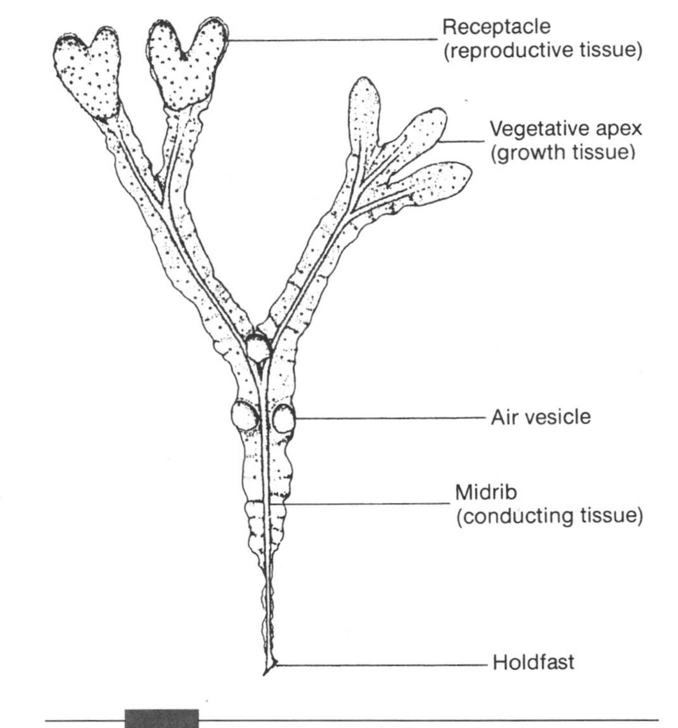

FIGURE 13-6 The plant body of *Fucus vesiculosis,* a brown alga that commonly grows in the intertidal zone. (*Adapted from Bold et al.*)

evolution. Nevertheless, botanists generally do not consider these algae ancestral to land plants since their pigmentation (chlorophyll c, fucoxanthins, and so on) and storage products are so different. Also, among factors they share with other algae, Chrysophyta and Phaeophyta lack the waterproof cuticles that would prevent desiccation on land.

Reproductive organs also distinguish algae from land plants. "True" land plants—photosynthetic eukaryotes marked by the ability to survive and sexually reproduce on land (Niklas)—are called **embryophytes.** Ranging from simple bryophytes to complex angiosperms, embryophytes are characterized by reproductive structures consisting of one or more multicellular layers that help protect and develop gametes. For the egg, these surrounding "sterile" cells also provide embryonic nutrients. By contrast, algal reproductive structures are less complex, and gamete development lacks such multicellular enclosures.

Bryophytes

Botanists traditionally classify the simplest land plants—liverworts, hornworts, and mosses—into a single group, **bryophytes.**[1] These plants have features common to land plants: multicellular reproductive structures, a cuticle in their aerial parts, and many epidermal pores (stomata) that permit the transfer of carbon dioxide, water vapor, and oxygen between their tissues and the atmosphere. Some of the bryophytes have food and water transport tissues, although these do not seem as efficient as the phloem and xylem of the more advanced vascular plants. Limitations in food and water transport apparently restrict bryophytes to small stature, and they live mostly in moist environments where they can transport water along their surfaces. In arctic or arid environments, they usually suspend growth until the warm, moist season begins.

Among characteristics bryophytes bear in common is an **alternation of generations** in which the haploid gametophyte generation is free-living, and the diploid sporophyte generation remains parasitically attached. In liverworts such as *Marchantia* (Fig. 13–7) and *Sphaerocarpos,* the sporophyte is relatively undifferentiated, whereas it is considerably more complex in the hornwort *Anthoceros.*

Although we have little clue as to the direct ancestry of bryophytes or even the phylogenetic relationships between their major groups, many botanists believe these plants have an algal origin. They point to the presence of both spore-forming and gamete-forming tissues in various algae and to the similarity between the filamentous growth pattern of some green algae and the branching filamentous protonema stage observed in many mosses. The aquatic environment of most bryophytes and their dependence on water for fertilization also points to an aquatic origin.

One evolutionary sequence, often popular in the past, proposes that algae evolved into bryophytes, which then evolved into vascular plants (for example, ferns). Today most botanists believe bryophytes and higher plants differ notably, and both may have had an independent algal origin. In fact, one molecular phylogeny suggests that bryophytes themselves are probably polyphyletic, different groups arising independently (Lewis et al.). Distinct algal origins for bryophytes and vascular plants seem even more likely, since the sporophyte generation of bryophytes depends for nutrients and support on the gametophyte, whereas the sporophyte of higher plants is completely independent. Furthermore, the earliest unequivocal appearance of bryophytes in the fossil record is in the Devonian period for liverworts and in the Carboniferous for mosses, whereas recognizable fossils of vascular land plants appear in earlier Silurian strata. According to this hypothesis, not homology but parallel evolution caused many of the similarities between bryophytes and vascular plants.

Whatever their relationship, it is interesting to note that so-called "simpler" organisms with seemingly only marginal or intermediate adaptations for a terrestrial existence persist in spite of the presence of so-called complex organisms with more advanced adaptations. Nonvascular plants did not become extinct because of the evolution of vascular plants, and nonseed plants such as ferns still survive in the presence of seed plants. Botanists know approximately 22,000 species of bryophytes and 10,000 species of ferns. Apparently, some environmental conditions confer no overwhelming advantage on later evolutionary inventions; that is, evolution among these plants did not go in only one direction leading to a single "higher" and "complex" form.

Sex, Meiosis, and Alternation of Generations

As indicated in Chapter 10, an important consequence of sex is to help provide the variability that enables a population of organisms to produce a wide array of genotypes. This variability arises because when two parents contribute chromosomes to an offspring, these chromosomes reshuffle in the offspring's sexual meiotic tissues to produce chromosomal and genetic combinations different from those either parent originally donated. The gametes

[1]Some authors restrict this name to the mosses and call the liverworts and hornworts *Hepatophytes.*

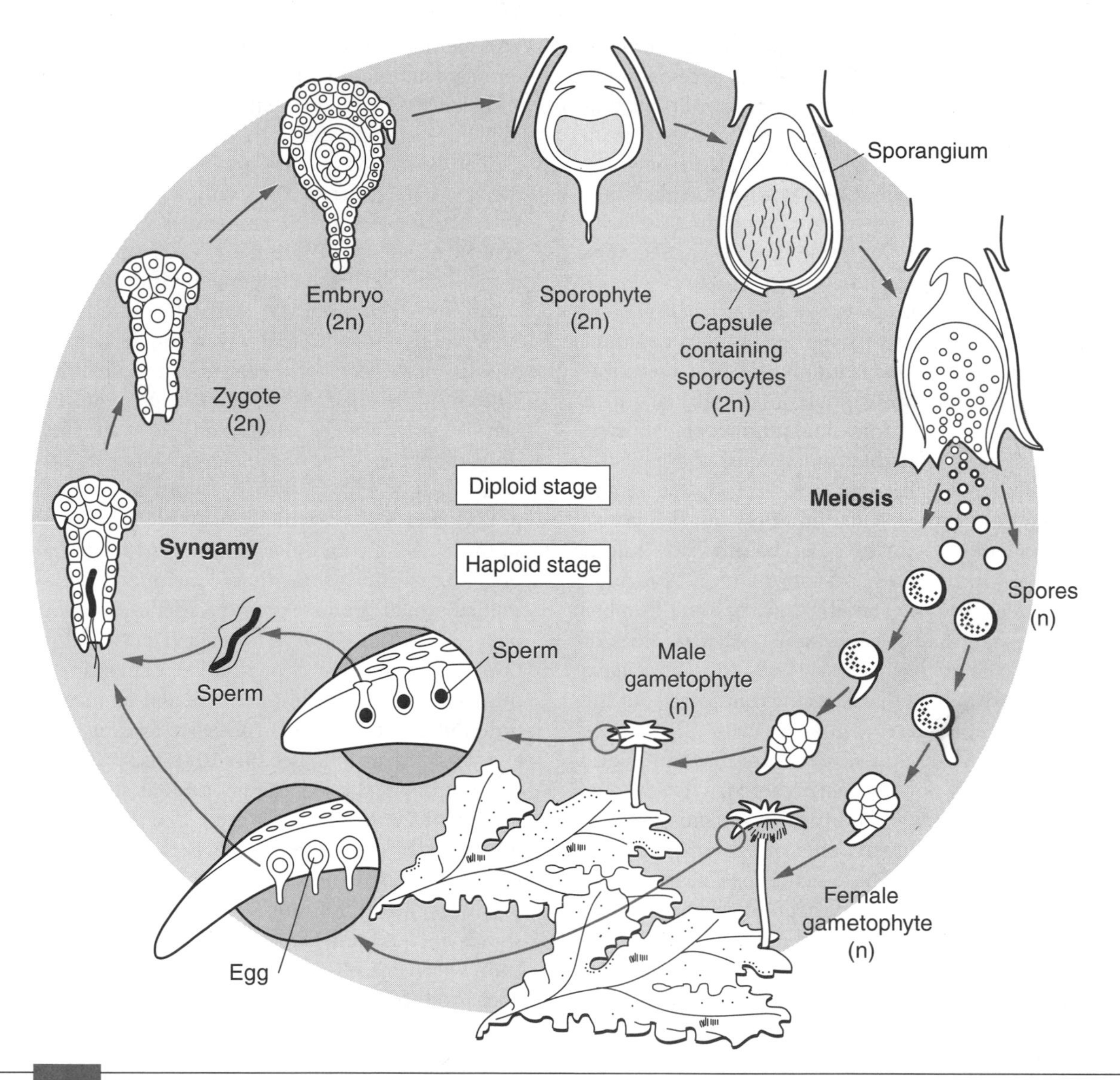

FIGURE 13-7 Life cycle of a liverwort species in the genus *Marchantia*. The gametophyte is a flattened, chlorophyll-bearing thallus with rootlike rhizoids on its undersurface. The spherical sporophyte grows on the tissue of the gametophyte, and each of its interior sporocytes divides meiotically to form a tetrad composed of four haploid spores. Germination of the spore leads to the gametophyte, and the cycle then continues as shown. In gametophytes, the *antheridium* is the sperm-producing tissue, and the *archegonium* is the egg-producing tissue.

containing these new combinations can combine to form the zygotes of the next generation, which then have new genetic combinations; these reshuffle in the following generation, and the process continues (Chapter 23).

For populations continually encountering different environments, sex is an obvious advantage to a lineage in producing new combinations of genes, some of which may be adaptive and allow the lineage to survive. Lineages without the variability introduced by sex can more easily become extinct under changing circumstances.[2] However, in a long-standing population that continually endures the same environment, genotypes probably will evolve that are eminently adapted to that environment and most, if not all, new genetic combinations will have lower adaptive value than the parentals. Under such constant circumstances, the advantages of sex are not apparent.

In fact, many plants have abandoned sexual reproduction and replaced it with asexual methods such as the spread of vegetative somatic tissues or by parthenogenesis (reproduction through unfertilized eggs—found also

[2]Data from the incidence of asexuality point to the likelihood that asexual species generally have higher extinction rates than sexual species. That is, asexuality only rarely appears as a prevailing character of large taxonomic groups, but is more common among smaller taxa such as occasional asexual species who are members of larger, more inclusive sexual taxa. The scarcity of asexual families and asexual higher taxa indicates that asexuality rarely survives long enough to become the predominant character of a large taxonomic group.

in some animals). Although some asexual plants may appear over wide geographical ranges, their success is often restricted to specific environments or to conditions that severely limit cross-fertilization because only very small inbred populations can survive. However, since environmental conditions are not often constant, eukaryotes generally use meiotic forms of sexual reproduction for at least some part of their life cycle, and asexual groups rarely survive over long evolutionary periods.

As yet, we have no exact knowledge of when meiosis originated, although it must have appeared in conjunction with, or soon after, the beginning of sexual fertilization.[3] The reason is simply that, in the absence of a mechanism to reduce chromosome numbers in gametes, sexual union leads to doubled nuclei and consequently doubled chromosome numbers. With each succeeding sexual generation, chromosome numbers would increase almost exponentially, forming large, unwieldy nuclei with difficulties in function and coordination. A meiotic mechanism reducing the gametic chromosome number to half would have had selective value. At what stage of the life cycle of primitive organisms would meiosis have occurred? The answer to this is again conjectural, but the following argument seems reasonable.

Since a doubling of chromosome number in somatic cells would probably not have been immediately advantageous to the primitive haploid organisms, meiosis probably took place immediately after fertilization in the diploid zygote cell itself. These sexual organisms would have immediately regained their haploid condition without the intervention of an extended diploid state, a situation similar to that found in algae such as *Chlamydomonas* and fungi such as *Neurospora.*

There are, however, advantages to lengthening the diploid stage, not the least being that such cells may have two kinds of genetic information, one from each parental sex, enabling a single organism to use different developmental pathways in responding to different environmental conditions. In addition, the two alleles of a gene in a diploid may each produce unique products that can then buffer each other to ensure developmental uniformity in any particular environment ("heterozygote advantage," Chapter 22). Diploidy also provides the opportunity for dominant genetic relationships that mask the effect of deleterious recessive alleles yet, at the same time, let a population evolve further by helping retain recessive alleles that may be advantageous under future conditions. Bearing pairs of homologous chromosomes may also let one member of a pair act, during recombination, as a template in repairing damages in the other (p. 225).

It can also be argued that when meiosis takes place immediately after fertilization the gametes have relatively limited genetic variability because only one reductional division has occurred. For example, a single diploid cell with three pairs of chromosomes (or three pairs of genes), A^1, A^2, B^1, B^2, and C^1, C^2, might produce four haploid gametes from a meiotic division that are of constitutions $A^1B^1C^2$, $A^1B^1C^2$, $A^2B^2C^1$, and $A^2B^2C^1$.

A multicellular diploid organism, in contrast, could produce a greater variety of gametes, since numerous kinds of reduction divisions can take place in a large number of parental cells undergoing meiosis (meiocytes). Thus some meiocytes of such an organism could produce $A^1B^1C^2$ and $A^2B^2C^1$ gametes, others could produce $A^1B^2C^1$ and $A^2B^1C^2$ gametes or $A^2B^1C^1$ and $A^1B^2C^2$, and so forth. Along with other advantages (p. 336), a population of organisms whose diploid meiotic tissues are multicellular would produce greater genetic variability among offspring and therefore have greater potential for evolutionary change than a population containing a similar number of organisms in which the diploid meiotic stage is unicellular. (For further hypotheses to explain the origin and persistence of sex, see Box 10–1 and pp. 563–564.)[4]

In animals, the lengthened diploid stage became the dominant feature of the life cycle, and the haploid stage is now mostly restricted to the gametes themselves. In plants, the lengthened diploid stage, or sporophyte, also produces meiotic products as in animals, but these are spores rather than gametes. The meiotically produced spores develop into haploid gametophytes, which only later produce gametes by mitosis.

A further animal–plant difference lies in how germ line cells separate from other tissues. In animals, extensive cell motility during development enables germ plasm to localize in specific reproductive organs—ovaries and testes. Plant cells, by contrast, mostly maintain their relative positions during development, and form germinal ovules and pollen only by transforming somatic cells in different vegetative regions. Animal reproduction attains improved genetic stability by restricting germ plasm to a single tissue that has selective value in avoiding somatically derived mutations common to plant germ cells.

---

[3]Some researchers favor the proposal that the haploid–diploid cycle preceded the origin of sex and survived in some asexual protists because it helped eliminate mutations during the haploid stage while providing the advantages of increased functional genetic material or other benefits during the diploid stage. According to this hypothesis, only after the meiotic reduction mechanism evolved to allow a regularized transition from diploid to haploid stages could sexual union occur (Kondrashov).

[4] Persistence of haploidy as a major life cycle stage in some organisms may be related to haploid rapidity in eliminating deleterious alleles unprotected by the diploid stage. According to Mable and Otto, such advantages favor haploids "if (1) sex is rare, (2) recombination is rare, (3) selfing is common, or (4) assortative mating [p. 529] is common." However, they also point out that:

> Once a certain ploidy level has become dominant within a taxonomic group, it may be difficult to expand the alternate ploidy phase, either because the necessary mutations simply do not arise or because individuals with atypical ploidy levels are unable to develop normally. . . . [A]n organism may evolve developmental pathways that depend on having the appropriate ploidy level.

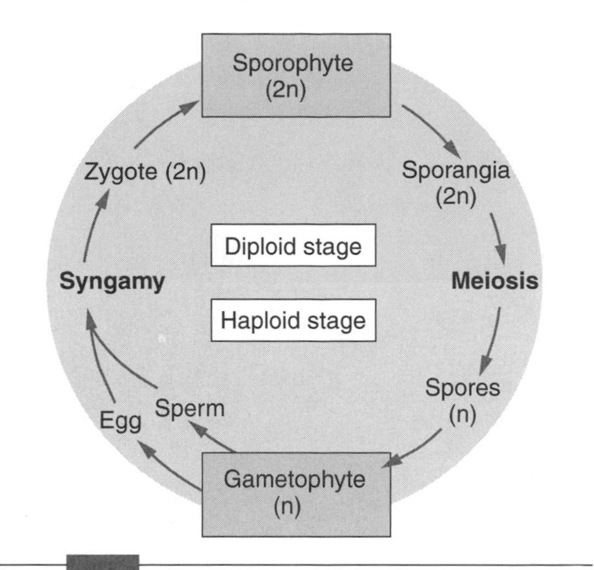

FIGURE 13-8 Alternation of gametophyte and sporophyte generations in the plant life cycle (see also Fig. 13–4). In some plants, the gametophyte is unisexual, either male or female. In others, the gametophyte is bisexual or hermaphroditic, producing both male and female tissues. (*Reprinted by permission of John Wiley & Sons, Inc.*)

The sporophyte–gametophyte alternation of generations in plants (Fig. 13–8) has long been puzzling, and we can offer various explanations. For example, aside from the advantages of maintaining a diploid state, the sporophyte produces dispersible, encapsulated spores that resist desiccation. By contrast, plant as well as animal gametes are relatively unprotected and generally depend on an aqueous environment for dispersion and fertilization. In fact, the vulnerability of sexual gametes to terrestrial conditions probably accounts for the persistence of the gametophyte stage in plants, a stage that easily disappears in animals. That is, animals are sufficiently mobile to allow the transfer of gametes by direct contact between organisms in the diploid stage, whereas plants are sessile, and the transfer of gametes among them is restricted to moist environmental conditions. Thus, the features of plant sporophytes that enable resistance to desiccation and conquest of the land are apparently quite different from gametophyte features necessary for the aqueous transfer of plant gametes.

The evolutionary development of the plant sporophyte from dependence on the gametophyte stage to independence reflects the advantages of diploidy as well as spore production, and we can consider the persistence of the gametophyte stage a reflection of the advantage for immobile gamete-producing individuals to grow in aqueous proximity to each other.

Traditionally, botanists have concentrated their disputes about the alternation of generations on whether the two generations were initially similar or different. According to the **antithetic,** or interpolation, theory of sporophyte origin, early plants were all gametophytes that produced diploid zygotes, some of which underwent a period of delayed meiosis yet kept dividing mitotically. These parasitic diploid tissues were the initial rudimentary sporophytes, and thus differed both functionally and morphologically from the parental gametophyte. With the colonization of land, an increasing proportion of sporophyte tissues converted to vegetative purposes such as photosynthesis, and it was then that the sporophyte became independent.

A contrasting **homologous,** or transformation, theory suggests that the sporophyte showed little initial difference from the gametophyte because they both share the same genetic constitution derived from the same organism and should therefore have shared the same patterns of growth. Supporting the homologous theory is the similarity between sporophyte and gametophyte in many algae, as well as structural similarities between their stems in the fernlike *Psilotum* and in some primitive ferns such as *Stromatopteris.* There also appear to be some basic developmental similarities: sporophyte tissue can arise in some gametophytes without the intervention of gametic fertilization (apogamy), and some sporophytes may produce gametophytes without spore formation (apospory). Some homologous proponents suggest that land plant evolution eventually led to morphological divergence between sporophyte and gametophyte, the former becoming erect, and the latter more prostrate.

We have not yet resolved the issue of sporophyte origin, although evidence suggests that some gametophytes may have been among the early land plants (Remy). On the whole, increased sporophyte importance best characterizes land plant evolution. In fact, the vascular tissues of land plants are generally restricted to sporophytes.

Early Vascular Plants

hatever the origin of the land plants, the fossil record shows their rapid evolutionary radiation from the time of their first appearance in the Silurian period more than 400 million years ago. By the end of the Devonian, 75 to 100 million years later, forests containing woody trees of relatively great variety had established themselves well. These successful land plants were **vascular,** bearing conductive tissue (xylem) that enables water to reach the erect parts of the plant, associated with tissue (phloem) that enables food to be distributed. Botanists have often given them the name **Tracheophyta** because they have tracheids, fluid-conducting tubes impregnated with an organic substance (for example, lignin) that also provides mechanical support for erect growth.

The earliest of the Silurian vascular fossils include a number of simple plants with leafless stems classified in

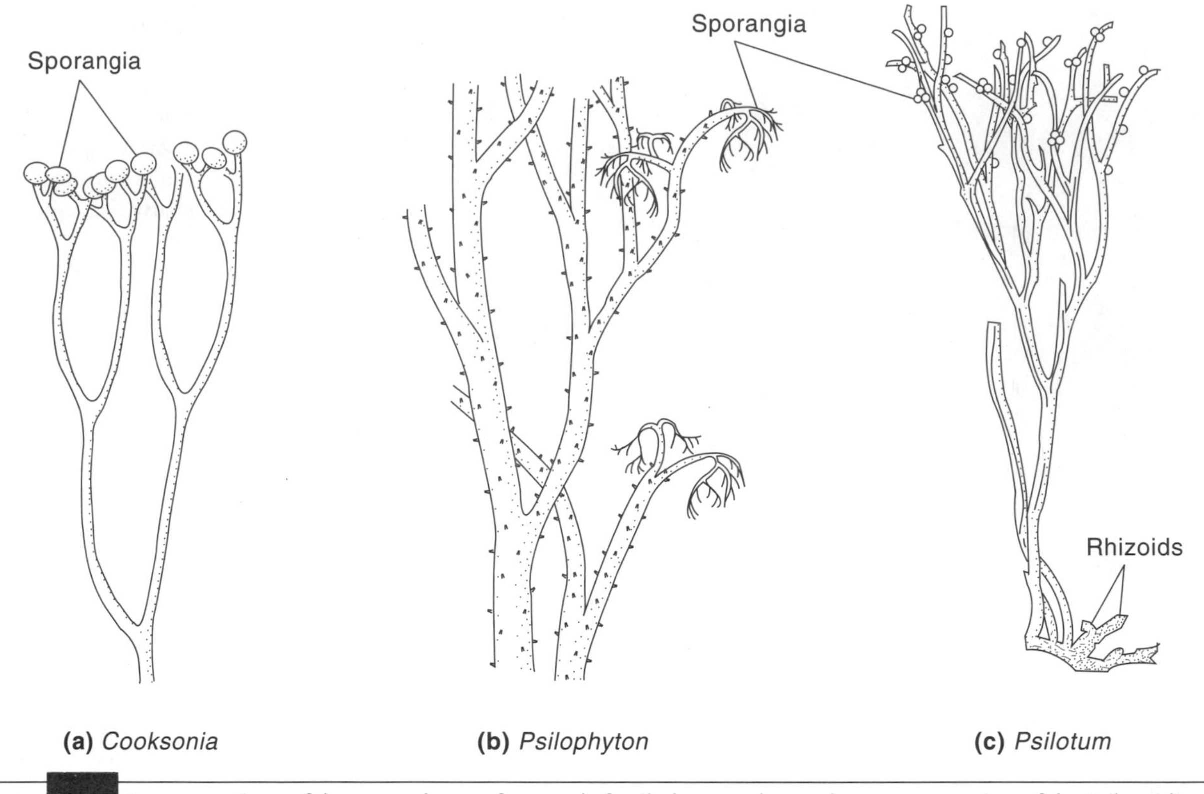

FIGURE 13-9 Reconstructions of the sporophytes of two early fossil plants and a modern representation of the Psilopsida. (*a*) An Upper Silurian plant, *Cooksonia caledonica,* about an inch or so high, which had no distinctive leaves or roots but showed naked, dichotomously branched axes with terminal sporangia. Similar fossils were long known from Devonian rocks in Rhynie, Scotland, and have been classified together in a group called **Rhyniophyta,** or Rhynia-type plants. (*b*) *Psilophyton princeps,* section of a spiny, leafless fossil plant that first appears less than 10 million years after *Cooksonia.* It had a main stem axis with lateral branches terminating in sporangia and a vascular structure that seems to have been larger than *Cooksonia,* so *Psilophyton* may have grown taller. (*c*) The modern plant *Psilotum nudum* has a number of features that resemble the fossil forms: simple stems, nondiscernible leaves, and absence of a modern root system. It is not clear whether this plant is a "fossil" or a secondary descendant of a more advanced form (Stewart and Rothwell). (a *and* b *adapted from Taylor and Taylor,* c *adapted from Bold et al.)*

the genus *Cooksonia* (Fig. 13–9*a*), some of which may have had terminal spore-bearing organs (**sporangia**), and even some Chinese forms with tracheids (Cai et al.). Together with other leafless and rootless fossil plants, these somewhat resemble plants in the modern genus *Psilotum* (**Fig.** 13–9*c*). According to Banks and others, such early plants were ancestors of multibranched plants that rapidly evolved into taxa such as *Psilophyton* (Fig. 13–9*b*).

Also in the Devonian period primitive leafless plants appear that differ from *Cooksonia* types in carrying their sporangia laterally along branches rather than terminally. Botanists have presumed these plants, of which *Zosterophyllum* is an example (Fig. 13–10*a*), to be the ancestors of the early **club mosses,** or **lycopods,** such as *Asteroxylon* (**Fig.** 13–10*b*), which in turn led to arborescent lycopods such as *Lepidodendron* (Fig. 13–10*c*), so abundant during the Carboniferous age. Herbaceous lycopods appear even today.

The Devonian period also saw the origin of **sphenopsids,** horsetail plants with segmented stems and whorled leaves and branches. These plants were common until the Mesozoic period, contributing huge trees to the Carboniferous coal forests (Fig. 13–11*a*), but now only the genus *Equisetum* (Fig. 13–11*b*) remains, consisting of a group of about 25 herbaceous species.

In terms of evolutionary persistence, an enduring group among these early spore-bearing plants was the **ferns,** Pterophyta, now numbering about 10,000 species in which the sporangia are carried directly on the leaves (Fig. 13–12). These include, then and now, small forms and large tree ferns (Fig. 13–13). Ferns were apparently the first plants to exploit the use of large, prominent leaves, megaphylls, in contrast to the smaller leaves, microphylls, used by the lycopods and sphenopsids. We do not know the origin of either leaf type, although botanists have offered various theories.

The **telome theory,** as Zimmermann developed it, suggests that primitive thin branches, called *telomes,* evolved in two major alternative directions: the first toward greater complexity and vascularization, leading

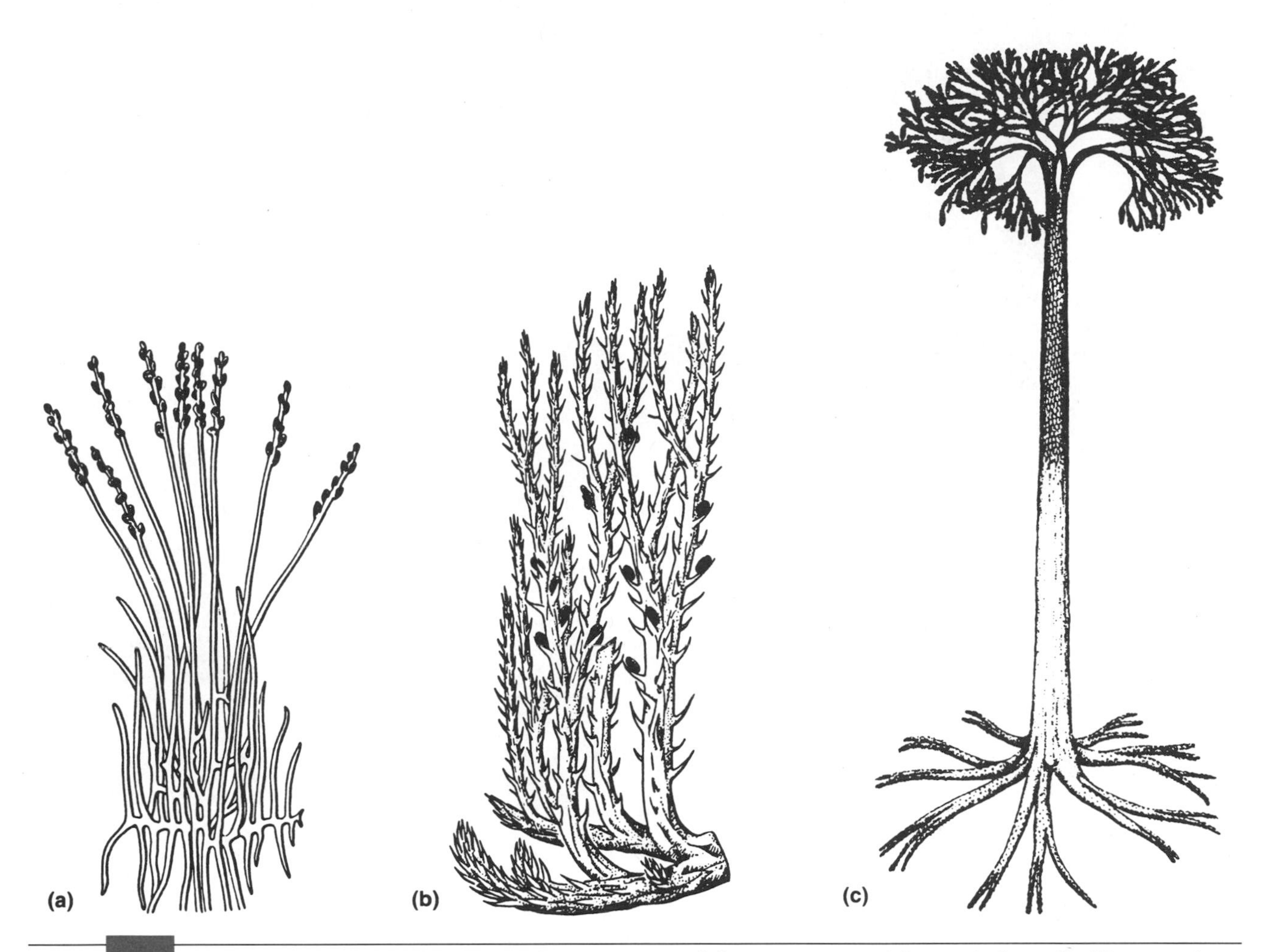

FIGURE 13–10 Reconstructions of several Devonian plants. (*a*) *Zosterophyllum myretonianum* (about 7 inches tall). (*b*) *Asteroxylon mackiei* (about 2 feet tall). (*c*) *Lepidodendron* species (about 150 feet tall). The scars along the upper stem of *Lepidodendron* are leaf cushions, where the long, filamentous leaves had attached during earlier growth. The heavy, pendulous cones carry the sporangia. (a *and* b *adapted from Foster and Gifford,* c *adapted from Stewart and Rothwell.)*

eventually to the leaves and branches of ferns and higher plants; the second in a retrogressive direction toward a single unbranched form, leading eventually to bryophytes such as the hornwort *Anthoceros*. According to this hypothesis, leaves originated from small branches that lay in the same plane. As shown in Figure 13–14*a*, webs formed between such planated branches could have produced leaflike structures.

A different proposal, the **enation theory** (Fig. 13–14*b*), suggests primitive leaves arose from small flaps of tissue along the stem, somewhat like microphylls in the fossil lycopod *Asteroxylon*. Only later did these leaves vascularize.

As yet, we do not know which theory is correct, although evolutionary changes clearly affected almost every aspect of these early plants. For example, beginning with the first spore-bearing plants, there is a progressive change from **homospory,** in which all spores are alike, to **heterospory,** in which sporophytes produce both large-diameter **megaspores** (200 μm) and smaller-diameter **microspores.** By the Upper Devonian period the heterosporous lines had evolved megaspores more than 2,000 μm in diameter. This increase in megaspore size apparently resulted from a reduced number of cells in the megasporangium, so that it produces only a single tetrad of spores, of which three spores abort and one enlarges (Fig. 13–15). As botanists have often pointed out, we can view such a huge megaspore as the bearer of a female gametophyte with enhanced nutritional resources that, on fertilization, will provide stored food for the developing sporophyte embryo—in a sense, the prototype of a **seed** (Fig. 13–20)—the beginning of the most successful sexual reproductive method in vascular plants.

Complexities of stem structures involved in conduction, support, and storage also originated from a primitive form, the **protostele,** in which these vascular tissues took, at first, a simple arrangement. Later in evolution they divided into various lobed and concentric arrangements, as **Figure 13–16** shows.

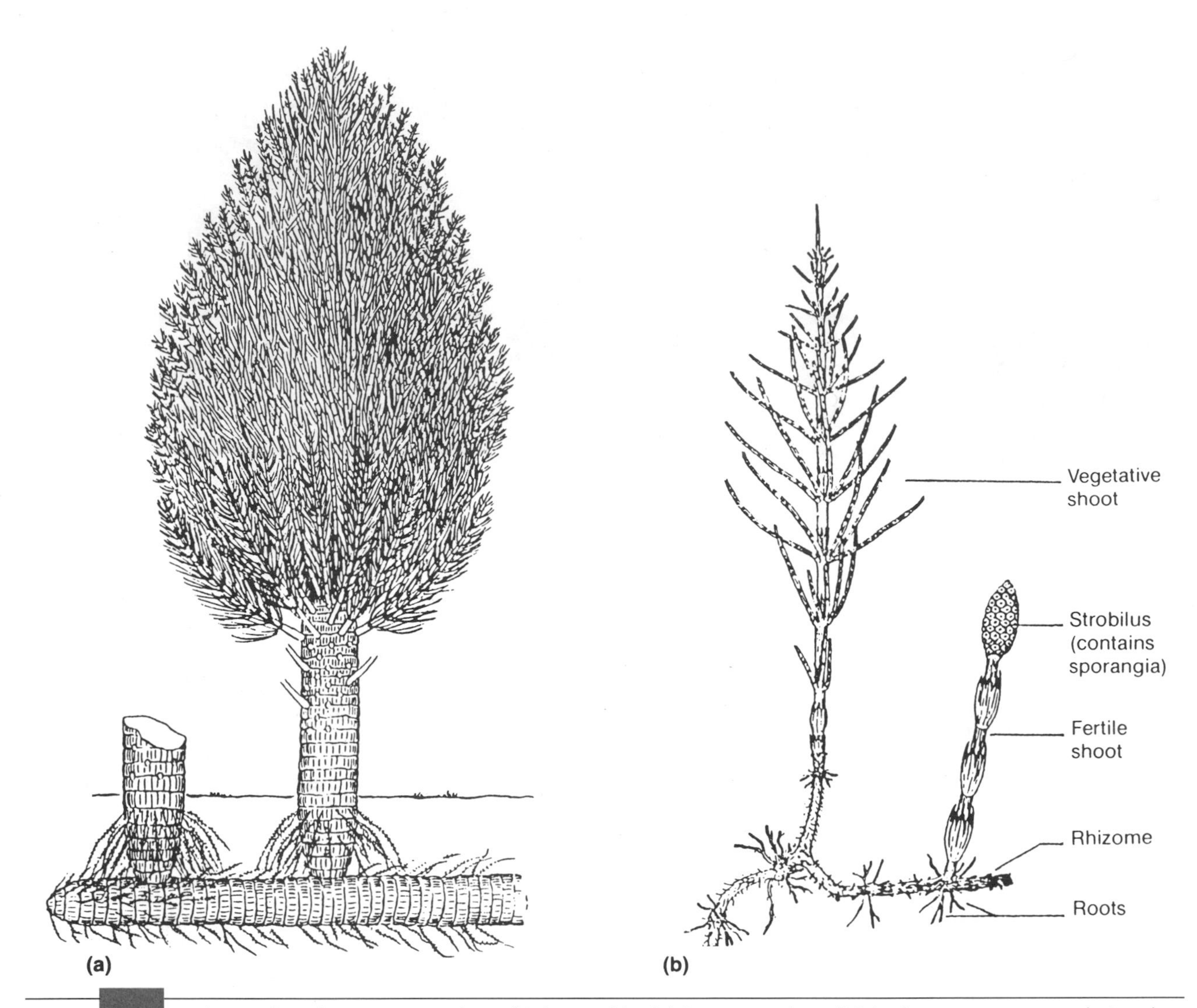

FIGURE 13–11 Ancient and modern representatives of Sphenopsida. (*a*) Reconstruction of *Calamites,* a common tree during the Carboniferous period, that reached heights of 90 to 100 feet with trunks 2 feet thick. (*Adapted from Foster and Gifford.*) (*b*) Modern *Equisetum arvense,* showing both vegetative and fertile shoots. Its reproductive system is homosporous: the strobili bear sporangia that produce spores alike in size. (*Adapted from Bold et al.*)

The structural and functional advantages many of these innovations provided bolstered the vertical development of plants and let large trees and shrubs of all types evolve. By the Carboniferous period, lush and extensive forests had developed in vast swamps along the eastern coast of North America and similar coastal regions of Europe and North Africa. The absence of annual growth rings in many tree trunks of this period indicates that the climate was mostly tropical and growth was rapid. In this environment, rapid submergence below the watery swamp surface of many fallen trees and shrubs inhibited their decay, making them immune to attack by all but anaerobic bacteria. As the sea level fluctuated in these areas, successive generations of swamps formed and submerged, and thick layers of organic strata compressed into peat. Further sedimentation and compression led to the escape of volatile hydrocarbons, allowing the enormously thick and extensive coal seams to form.

From Swamps to the Uplands

Successful as they were, Carboniferous spore-bearing plants were limited to a moisture-laden environment because the motile male gamete depended on aqueous transmission to the female gametophyte. To extend their range onto dry land, plants

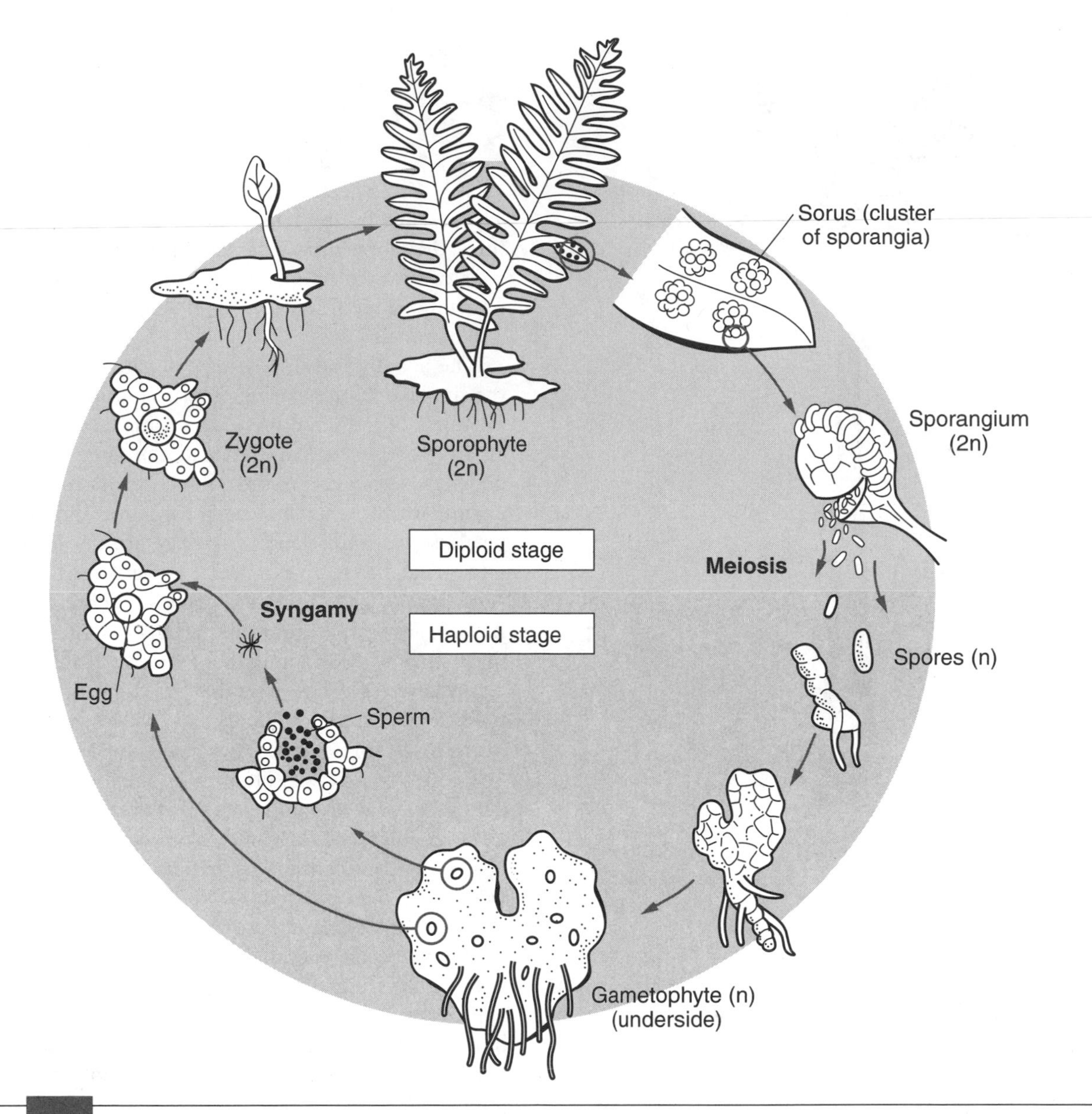

FIGURE 13–12 Life cycle of a common fern *Polypodium vulgare.*

had to await the evolution of enclosed desiccation-protected gametophytes in which cross-fertilization could occur by nonaqueous devices such as wind dispersal. Similarly, a considerable advantage attended the evolution of protected sporophyte embryos whose distribution could be independent of their parental gametophytes. In essence, the size reduction of the gametophyte and the evolution of easily dispersible **pollen** (male gametophytes) and seeds (sporophyte embryos) helped further the conquest of dry land.

We don't yet know how early vascular plants evolved into pollen-producing, seed-bearing plants. According to the fossil record, **gymnosperms** (naked seeds) appear in notable frequency during the Carboniferous period, eventually giving rise to modern representatives that include ginkgos and cycads as well as conifers such as pines, cedars, and sequoias, about 750 different species. The **angiosperms** (covered seeds), now the dominant land plants, accounting for about 220,000 (or more than 80 percent) of all plant species, have no identifiable fossils earlier than the Cretaceous period of the Mesozoic. We have not yet discovered transitional fossil forms leading directly to either gymnosperms or angiosperms, although some fossils such as *Archaeopteris,* dating to the Devonian period, may represent a group ancestral to all seed-bearing plants.

As Figure 13–17 shows, *Archaeopteris* was a tall tree resembling a modern conifer with a crown of leafy branches. Botanists propose that its stem bore a number of features in common with gymnosperms, although it

FIGURE 13-13 Reconstruction of the Carboniferous tree fern *Psaronius,* about 25 feet tall. Leaf scars left by earlier fronds that have fallen away are visible near the top of the trunk, and surrounding adventitious roots that increase in thickness toward the base cause the trunk's long pyramidal shape. The root structure suggests that these trees grew in swampy habitats. (*Adapted from Foster and Gifford, based on other sources.*)

produced free spores rather than protected seeds. Beck, Banks, and others believe that Paleozoic plants of this kind, called **progymnosperms,** combining pteridophytic, sporulating reproductive modes with more advanced anatomical structures such as large trunks, gave rise to the gymnosperms that became so successful during the relatively dry Mesozoic (Fig. 13–18).

Other fossil groups, perhaps also arising from the progymnosperms, were fernlike plants that bore seeds rather than spores (Fig. 13–19). These **seed ferns** (Pteridospermales) show a variety of seed forms that suggest a progression in the method by which the seed integument encloses the female gametophyte (Fig. 13–20). However, from the available phylogeny (Fig. 13–21), we don't know whether seeds originated in only one group of progymnosperms (monophyletic origin) or more than one group (polyphyletic origin). In any event, such evolution may have occurred early: Pettitt and Beck, for example, have described a fossil seed that dates back to the Upper Devonian, 350 million years ago.

Angiosperms

Angiosperms were the last major plant group to evolve, appearing first in the early Cretaceous and considerably abundant and variable by the late Cretaceous. Among other features, they share unique **flower** structures that enable insects or birds to pollinate many of them, and they also bear unique seeds that are often adapted to dispersal by other animals.

The adaptive advantage of pollination by animals is the simple one of ensuring cross-fertilization with other members of the same species by using a relatively small amount of pollen, compared to the large amounts of pollen necessary in random wind pollination. As a result, angiosperm flowers, derived from leaves modified into petals, sepals, and related structures, are among the most intricate and attractive organs that plants ever developed (Fig. 13–22). They have size, color, and odor differences that can attract specific animal pollinators—an advantage that can spread so rapidly that even some closely related plants have evolved flowers that can discriminate among pollinators, while pollinators have "coevolved" mechanisms to feed on specific flowers (Fig. 13–23).

Coevolution between flowers and pollinators was evident to Darwin who postulated that even a most unusual flower among orchids would have a matching pollinator. His example was a Madagascar star orchid whose nectary was at the base of a corolla tube ten inches long, yet whose pollinator was unknown at the time. Darwin's prediction that such a pollinator existed was borne out years later with the discovery of a giant hawkmoth bearing an appropriately long tongue (Nilsson). Apparently, some hawkmoths have been selected for longer tongues to pollinate star orchids selected for longer corolla tubes to be pollinated by specific hawkmoths.

Animal pollinators undoubtedly affected the sexual organization of the angiosperm flower, since it would also be to the advantage of a plant to contribute its pollen to a mobile animal pollinator at the same time that pollen from another plant fertilized its own ovules. That is, flowers would be selected in which pollen transfer and fertilization occurred in a single visit. The flowers of early angiosperms relying on insect pollination were probably bisexual, in contrast to their wind-pollinated ancestors, which would mostly have used unisexual flowers to help prevent self-fertilization between pollen and ovules of the same flower.[5]

[5]The fact that some groups of flowering plants have species that are also wind pollinated indicates that evolutionary reversals can occur from one form of pollination to the other. In the case of the evolution of figs, such reversals may well have occurred more than once: the first reversal being in the order Urticules from insect pollination to wind pollination, followed by a second subsequent reversal to insect pollination in the family Moraceae, in which the genus *Ficus* is almost exclusively pollinated by species of chalcid wasps.

FIGURE 13-14 Diagrammatic representations of two theories explaining the origin of leaves. (*a*) Flattening (planation) of a branch system according to the telome theory, followed by webbing between the branches to form a flat, veined megaphyll. (*b*) Evolution of microphylls according to the enation theory. (*Adapted from Bold et al., from other sources.*)

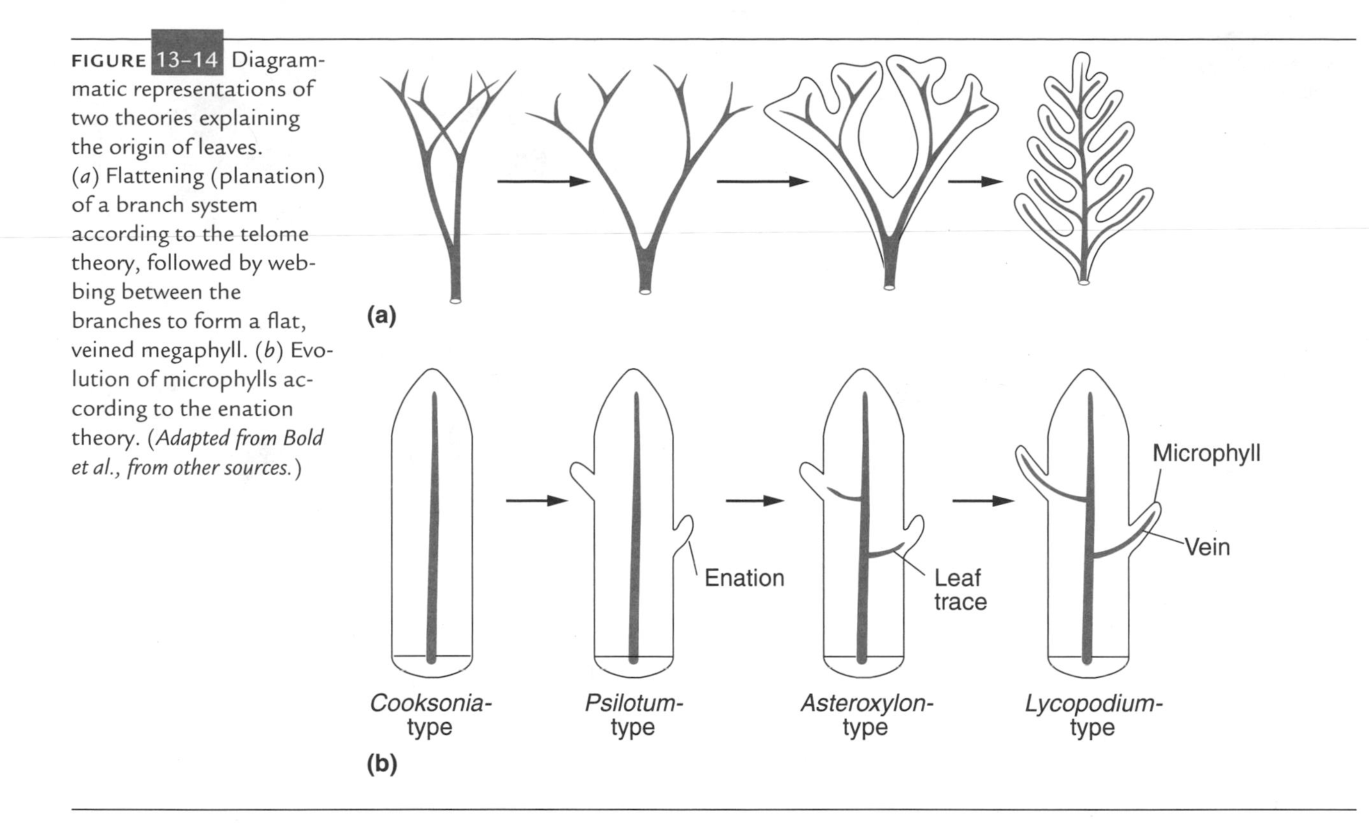

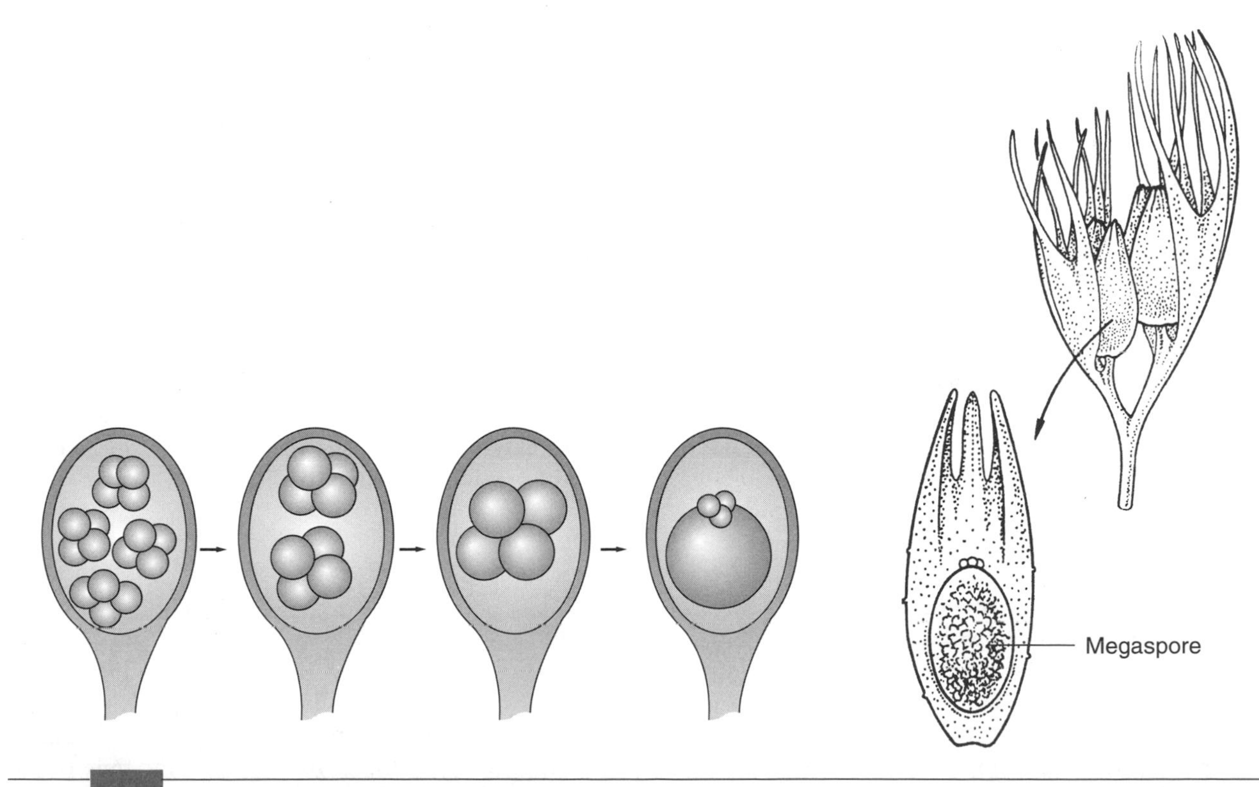

FIGURE 13-15 Proposed stages in the evolution of large megaspores that develop into egg-bearing gametophytes. As shown in a reconstruction of a Devonian gymnosperm seed (*Archaeosperma arnoldii*) on the right, the selective advantage of this trend was the development of large eggs that provide increased embryonic nutrition. (*Adapted from Niklas with additions.*)

FIGURE 13-16 Proposed evolutionary relationships among some of the vascular cylinders (steles) found in plants. The various protosteles (*a*) are considered primitive and occurred in Rhynia-type plants; siphonosteles and dictyosteles (*b*) characterize many ferns; various seed plants have eusteles; and some of the complex atactosteles appear in flowering plants (*c*).

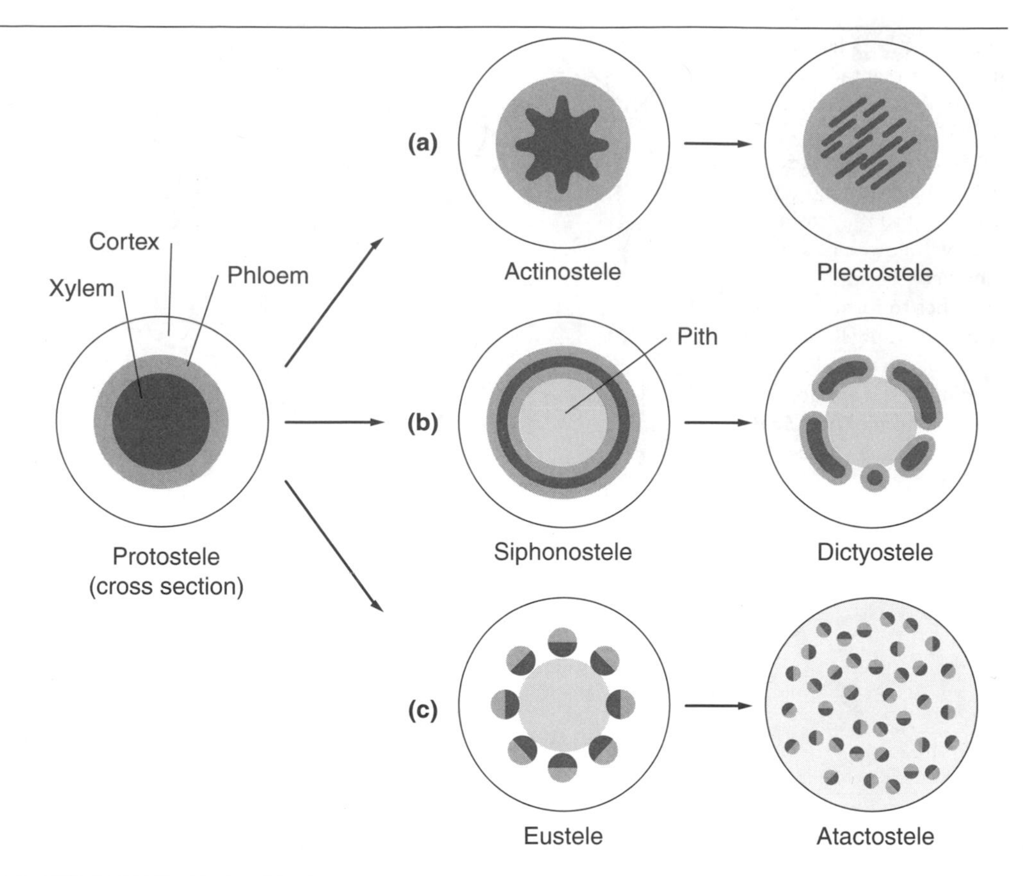

On a basic genetic level, an extremely important mechanism for preventing self-fertilization was the development of **self-incompatibility** (or **self-sterility**) alleles, so that pollen bearing the same allele as an ovule could not grow on the female style. For example, a haploid pollen grain carrying self-sterility allele S^1 will not grow well on a female style carrying S^1S^3 but can successfully fertilize a plant carrying S^2S^3 or S^3S^4, and so on. The consequences of this system were not only to protect the genetic variability that sexual reproduction produced, but also to establish new differences based on the self-sterility alleles themselves (p. 541).

Also distinctive in angiosperms is **double fertilization**: two gametic nuclei of the pollen tube fertilize the female gametophyte, one producing the diploid (2n) embryonic nucleus, and the other often producing a polyploid [commonly triploid (3n)] **endosperm** nucleus used for embryonic nutrition. Maturation of the fertilized ovule leads to an angiosperm seed that has two integuments rather than the single integument found in gymnosperms. As the name angiosperm ("seed vessel") implies, these seeds often have covers, either fleshy, fruity tissues, adhesive burs, feathery parachutes, or devices that let either animals or the elements disperse them.

Dispersal ability is only one of the selective forces acting on seeds. Among others are the ability of the seed coats to protect the seed against predators and the elements, the necessity of adequate food storage for embryonic development, and the programming of seed germination to coincide with the developmental period available. All these factors lead to specific anatomical and physiological adaptations, although noticeable quantitative adaptations have occurred as well. For example, some of these forces, such as selection for wide dispersal, put a premium on small size and large numbers, whereas others, such as selection for vigorous competitive embryos, emphasize large seeds and smaller numbers (discussed in Chapter 23 as "*r*" and "*K*" selection). On the whole, the size and number of seeds a particular species produces is a compromise between these various factors and the physical limitations of the plant.

Evolution of Angiosperms

A century ago, Darwin called angiosperm origin "an abominable mystery"—a mystery still unsolved today. What botanists dispute is not only the source but also the time of angiosperm origin, with estimates ranging from Early Permian to Late Carboniferous periods in the Paleozoic

FIGURE 13–17 Reconstruction of the progymnosperm *Archaeopteris,* about 75 feet high. (*Adapted from Foster and Gifford, from Beck.*)

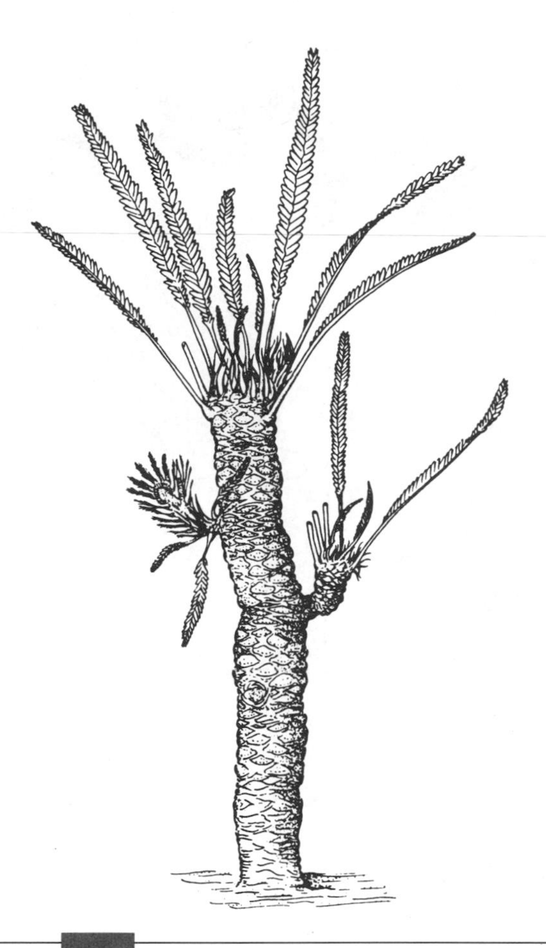

FIGURE 13–18 Reconstruction of a cycad gymnosperm, *Williamsonia sewardiana,* from Jurassic period rocks in India. The cycads were very abundant contemporaries of the dinosaurs, and this period is also known as the Age of the Cycads. Like the dinosaurs, most of this group became extinct, although 100 species of cycads exist, most in the tropics. Another "living fossil" is the *Ginkgo biloba* tree, sole remnant of a gymnosperm class Ginkgopsida, also common during the Mesozoic era. (*From Andrews, from other sources.*)

(Savard et al.) to a date corresponding with the earliest angiosperm fossils in the later Mesozoic/Early Cretaceous period. Since most discovered plant fossils appear associated with wet lowland areas where organic decomposition could be inhibited by silt and mud, those researchers who propose an earlier Mesozoic or Paleozoic origin for angiosperms suggest that they first arose in upland mountainous areas where fossil deposits rarely persist because of active erosion. Justification for this view lies in the finding that the first angiosperm fossil leaves already show considerable differentiation, as though preceded by a lengthy evolutionary period (Fig. 13–24). Discovery of Late Jurassic fossil insects with mouthparts adapted to flower pollination (Ren) lends further support to pre-Cretaceous angiosperm evolution.

On the other hand, the earliest fossil pollen that we can confidently ascribe to angiosperms is in the early Cretaceous period in the form of single-furrowed (monocolpate) grains, followed soon after by new pollen types. Doyle and Hickey propose that angiosperm evolution was rapid during the early Cretaceous, and it is this rapid diversification that accounts for the variety of fossil forms found in this period, rather than a much earlier unobserved origin.

Based on summaries of a large amount of information, Stebbins hypothesized that ancestral angiosperms were shrubs with spirally arranged, simple leaves and woody tissues formed from a single vascular cylinder. These progenitors bore bisexual flowers at the ends of branches, with the short male stamens lying in peripheral bundles and producing monocolpate pollen. The infolded female carpels had terminal stigmas and bore their ovules near the folded margins. After fertilization, development of the embryo and endosperm proceeded rapidly by nuclear division without cell wall formation (coenocytic), producing a two-leafed (dicotyledonous) embryo surrounded by considerable endosperm.

FIGURE 13-19 Reconstruction of a seed fern, *Medullosa,* about 12 to 15 feet tall. (*Adapted from Andrews, from other sources.*)

According to Stebbins, the ecological impact that led to the evolution of angiosperms was the alternation of dry and wet seasons, with its emphasis on rapid gametophyte and embryonic development. Rainy periods followed by calms after storms also would provide an opportune time for flowering and insect pollination, as well as promoting selection for protective seed structures such as closed carpels. It is not clear where such conditions might have appeared, but Stebbins suggests that semiarid mountainous regions with annual droughts would have offered early evolutionary opportunities for angiosperms, similar perhaps to those inferred from the rapid evolutionary rates observed among angiosperms that now inhabit mountainous regions in South Africa, Ethiopia, Ecuador, and Mexico.[6] Doyle and Hickey and others support this view.

Raven, however, points out that tropical lowland conditions with their large insect populations are more favorable for plants that depend on insect pollination than on wind pollination. Since tropical climates expanded significantly during the Cretaceous period, he suggests that this environment allowed the early angiosperms to disperse and become dominant.

Their place of origin is only one aspect of the "abominable mystery" of angiosperms; the other is, of course, their ancestry. One approach to angiosperm phylogeny has been to search for groups that have structures like those now carried by "primitive" angiosperm orders such as Magnoliales. For example, botanists once hypothesized that the magnolia flower was strikingly similar to the axial grouping of sporangia-bearing structures (strobili) of gymnosperms, cycads, and an extinct taxon called Bennetitales. Now botanists believe this similarity is only superficial since these groups differ fundamentally in respect to sexual organization, vascular anatomy, and general morphology.

In contrast to the large-flowered magnolia, small-flowered diminutive herbaceous plants have been proposed as the ancestral angiosperm type, based on a fossil Taylor and Hickey describe. They suggest that:

> . . .the lack of pre-Albian [pre-Early Cretaceous] fossil angiosperm wood is due to their diminutive habit and that the failure to recognize protoangiosperm fossils results from their diminutive size and an incorrect search image.

So far, excluding other possible progenitors, botanists generally believe that the most likely candidates for angiosperm ancestors come from among the pteridosperms (Thomas and Spicer). Some botanists suggest that the unique characteristics of angiosperms indicate a monophyletic origin, since it is doubtful that such characteristics arose independently in different groups or even that they arose more than once in the same group. Among these characteristics, reproductive mechanisms stand out. Mitosis reduces to only two cell divisions between formation of the haploid microspore and production of the male gamete, and to only three cell divisions between the megaspore and the multinucleate embryo sac. Also, only angiosperms use double fertilization to produce si-

[6]Data that Stebbins collected along the U.S. Pacific Coast for more than 8,000 plant species belonging to more than 800 genera show that some ecologically specialized regions such as alpine areas, deserts, lakes, swamps, and bogs have fewer plant species per genus than appear in more ecologically variable habitats such as fields, meadows, and open woods. Numerically, fewer than 4 or 5 species exist in each genus in ecologically specialized regions, compared to about 10 species per genus in ecologically variable regions. Apparently the latter provides greater evolutionary opportunities than the former. Stebbins points out that, by these criteria, tropical flora with their large numbers of different species seem to exist in evolutionarily restricted areas since the number of species per tropical genus is probably not much more than 5. It is rather the large number of genera in the tropics, along with the large number of families, that accounts for tropical floral diversity. Presumably much of this diversity reflects the continued persistence of species that are relics of genera and families that underwent evolutionary radiation into the tropics in the past, rather than examples of new rapid speciation. Stebbins calls tropical plant communities "museums"—"plant communities that have suffered the least disturbance during the past 50 to 100 million years and so have preserved the highest proportion of archaic forms in an essentially unchanged condition." Tropical communities mostly would represent geographical depositories for ancient plant groups rather than sites of origin. That is, this argument suggests that angiosperm diversity in tropical flora reflects dispersion into the tropics rather than origin from the tropics.

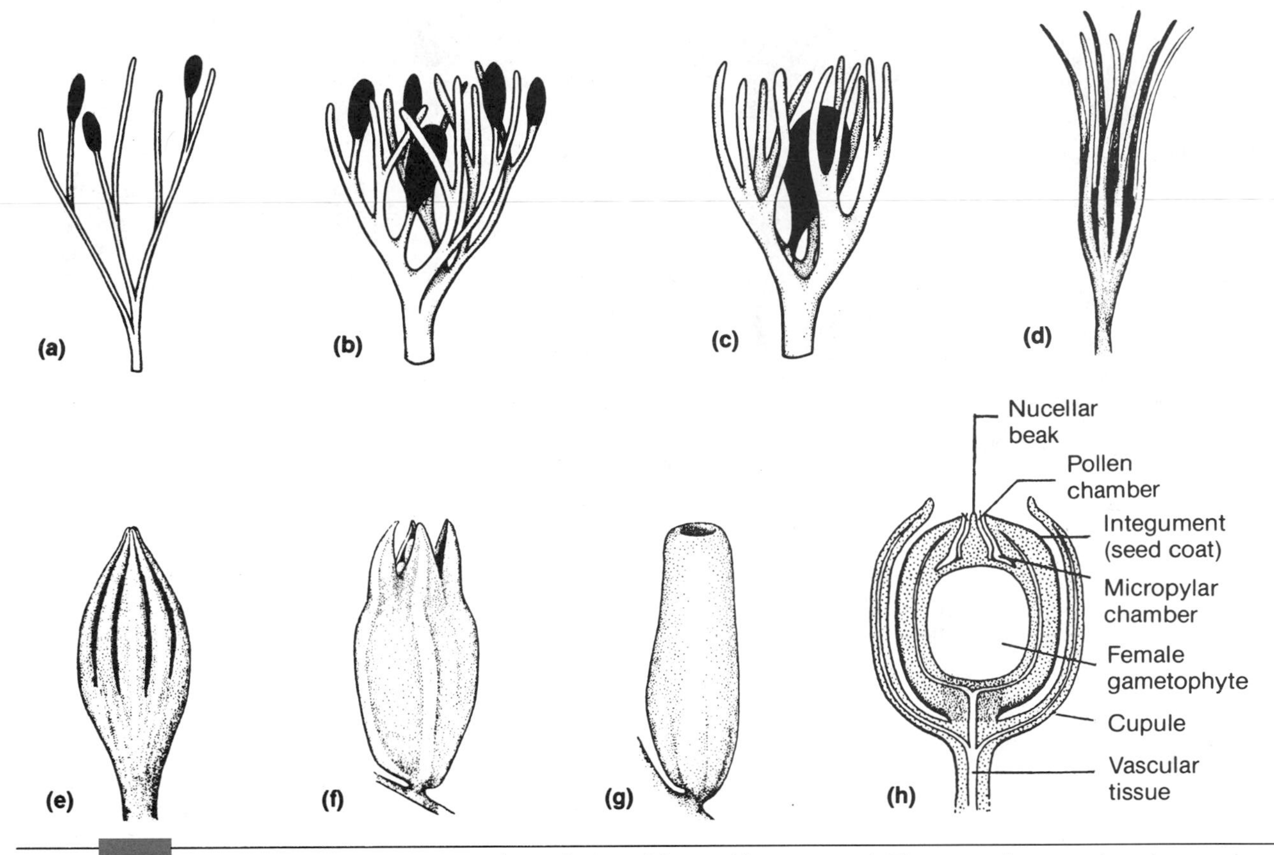

FIGURE 13–20 (*a–g*) One possible sequence in the evolution of the pteridosperm seed. The exposed sporangium that produces megasporocytes is gradually enclosed, enabling the female gametophyte (produced by the megaspore) to be completely protected within sporophyte tissue (nucellus). Fertilization takes place when a pollen tube (part of the male gametophyte) grows through the micropyle, thus enabling sperm to reach the female gametophytic egg. The complete seed that envelops the zygote and developing embryo is coated with an integument produced by the parental sporophyte. (*d–g*) Seeds of fossil pteridosperms. (*h*) Section of a pteridosperm ovule. (*Adapted from Foster and Gifford.*)

multaneously both the diploid sporophyte zygote and the commonly triploid nutritionally supporting endosperm.

Yet characteristics such as the sepals and petals of flowers, xylem vessels, and other traits (Stewart and Rothwell) are not universally found in all angiosperms, and may be considered similar to structures in gymnosperms and other vascular plants. Therefore some botanists propose that the combinations of characteristics that place plants in the angiosperm taxon may have arisen in more than one ancestral group, and the angiosperm taxon may have had a polyphyletic origin.

However this matter will be resolved, angiosperm advantages in rapid gametogenesis, biparental contributions to the endosperm, improved pollination, and fruity seed coverings clearly enabled this group to radiate into widely different ecological habitats and become the dominant group in many of them. Angiosperms, with their protected and nutritionally endowed seeds, like mammals with their fetuses, developed forms adapted to dry climates, wet climates, and various types of terrain. Some reinvaded the sea, others became parasitic, and some such as sundews and Venus flytraps are even carnivorous.

Not surprisingly, these bountiful adaptations also produced convergences. That is, selection under similar environmental conditions produced similar plant phenotypes even in different lineages residing in different geographical localities. A prominent example are some New World cacti and African euphorbs, both occupying desert environments and both highly similar in appearance, possessing sharp spines or thorns to dissipate heat and to guard their succulent water-laden stems (Fig. 13–25). Such evolutionary convergences, like those of animals (Fig. 3–7), derive some similarities by modifying different genetic pathways: the cactus spine is a modified leaf and the euphorb thorn is a modified branch.

Figure 13–26 shows a cross section of a hypothetical evolutionary tree with the various orders of angiosperms arranged as branches around an ancestral complex that served as the primeval trunk. The figure is drawn so that orders close to the ancestral complex are more primitive in respect to early angiosperm characteristics than those farther away. Although much is conjectural, such a tree gives us some idea of the successful radiation of angiosperms, their remarkable evolutionary plasticity, and

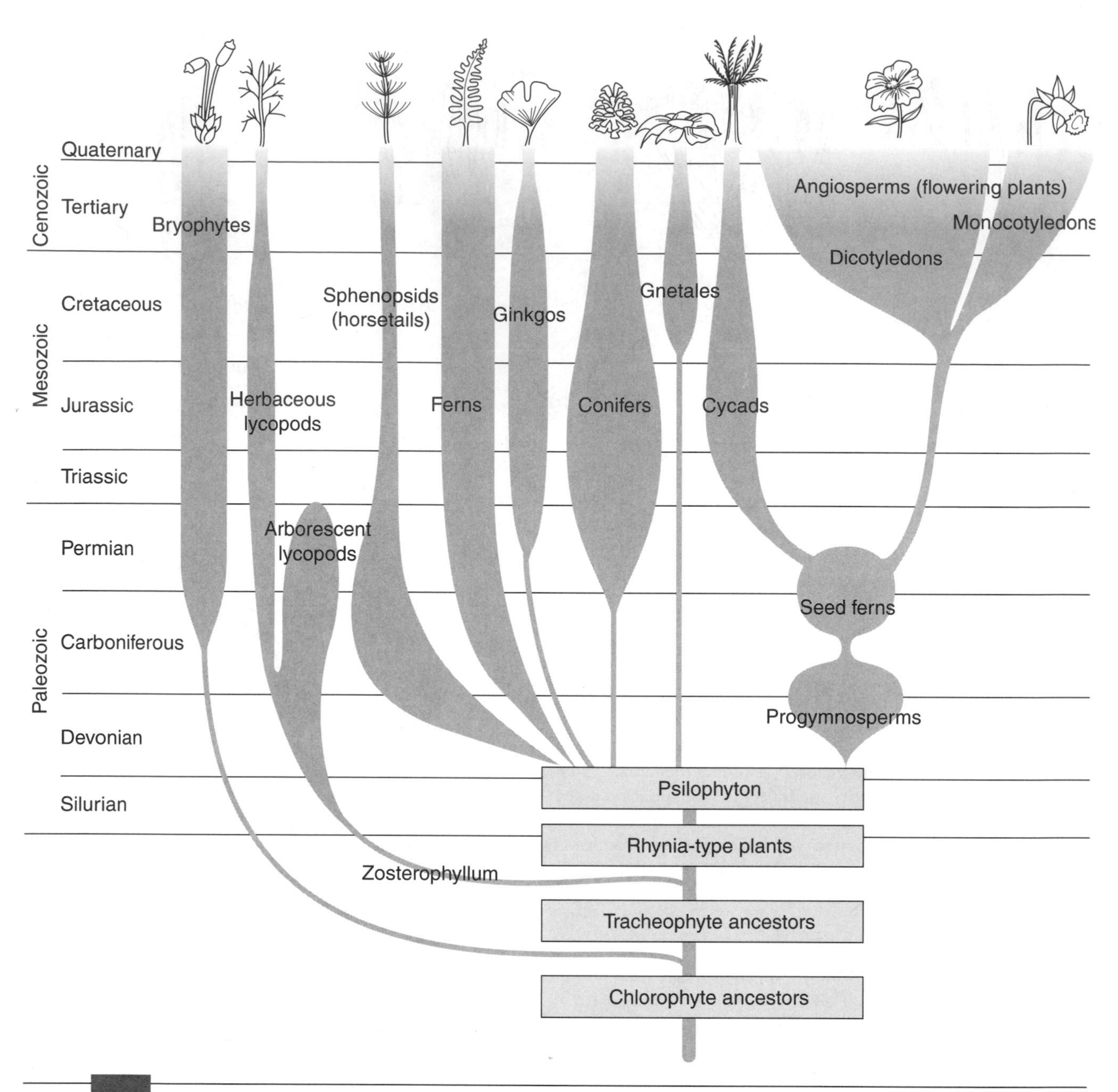

FIGURE 13-21 Possible evolutionary phylogeny stemming from some of the early plant forms. Widths of shaded sections provide very general estimates of the relative abundance of particular plant groups.

some of the phylogenetic relationships among them. A recent review article by Crane and coworkers and the books of Beck, Cronquist, Hughes, Hutchinson, Stebbins, and Takhtajan provide further views and information.

Fungi

In the past, biologists included fungi within the plant kingdom because they have cell walls and produce spores, and often defined them as "simple plants without chlorophyll." Because of their many unique attributes, this classification has changed in recent years, and biologists now generally place them within their own kingdom.[7] Some of the 120,000 fungal species are unicellular, such as yeasts, whereas others have vegetative stages that are mostly in the form of branched multicellular or multinuclear filaments called **hyphae** which aggregate into a mass called the **mycelium.** Restricted by their growth form and absence of chlorophyll, they are, as a result, heterotrophic,

[7]In fact, according to some molecular phylogenies (see, for example, Wainright et al.; Baldauf and Palmer; also, Fig. 9–16) they are the "sister group" of multicellular animals.

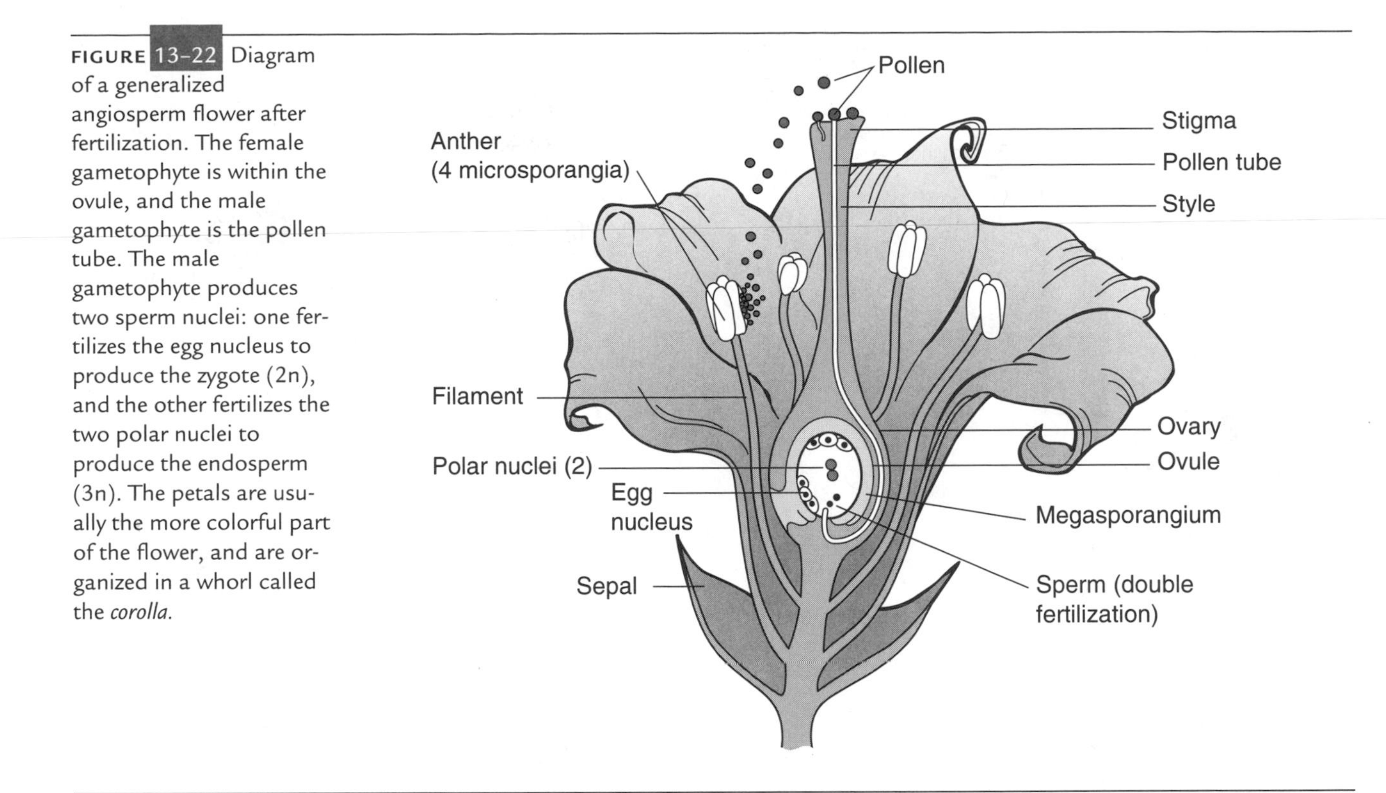

FIGURE 13-22 Diagram of a generalized angiosperm flower after fertilization. The female gametophyte is within the ovule, and the male gametophyte is the pollen tube. The male gametophyte produces two sperm nuclei: one fertilizes the egg nucleus to produce the zygote (2n), and the other fertilizes the two polar nuclei to produce the endosperm (3n). The petals are usually the more colorful part of the flower, and are organized in a whorl called the *corolla*.

deriving their nutrition either **parasitically** (live hosts) or **saprophytically** (dead organic material). Most classifications separate fungi from the various slime molds that have amoeboid stages, such as Myxomycetes, which form plasmodial, acellular aggregates, and the cellular Acrasiales.

Prior to Darwin, botanists often suggested that fungi were a form of algae, and grouped them with algae into a single division, Thallophyta, that had branched, thread-like filaments and produced motile, algalike zoospores. With the appearance of *On the Origin of Species,* botanists sought the ancestry of fungi among the algae, especially the red algae. They believed that the fungal lineage lost the algal chloroplasts responsible for its former photosynthetic mode of nutrition and consequently became exclusively parasitic or saprophytic. Because of the possibility that resemblances between fungi and plant algae (presence of cell walls, nonmotile habit) arise from convergence rather than ancestry, researchers no longer universally hold this view, and have put forth other hypotheses.

One suggestion is that early fungi shared a common ancestry with chemotrophic flagellated cells using inorganic sulfur or nitrogen for energy. Those chemotrophs that gave rise to fungi then evolved saprophytic forms that depended on organic materials synthesized by previous organisms. The fact that fungal cells are eukaryotic has led some workers to propose that some of the intermediary steps occurred through protozoan-like forms. Adopting a parasitic existence on live hosts would then have offered the opportunity for early aquatic fungi (perhaps related to the present forms, Oomycetes and Chytridiomycetes) to resist desiccation in the host tissues of land plants and evolve subsequently into more modern forms that use aerially dispersed spores (Zygomycetes, Ascomycetes, Basidiomycetes, and the asexual Deuteromycetes; see Fig. 13–27).

Another view suggests that evolution proceeded in the direction from obligate parasitism on host-provided nutrients to a more independent saprophytic habit capable of synthesizing simple protoplasmic products into more complex useful compounds. According to Raper:

> The eventual competence of an escaping parasite to continue to flourish on the remains of its dead host inevitably increased its reproductive and dispersal potential and conferred enhanced fitness in competition with related forms that remained totally dependent upon living host tissue.

Whatever their origin, the parasitic–saprophytic transition probably occurred a number of times in fungal evolution and in both directions. According to some interpretations of the fossil record, their success as a group stems from the Precambrian era, and they seem to have evolved into most of their presently observed forms by the end of the Paleozoic (Fig. 13–28). Although their outward appearances do not seem to have changed much, the parasitic fungi are still actively evolving on the gene level

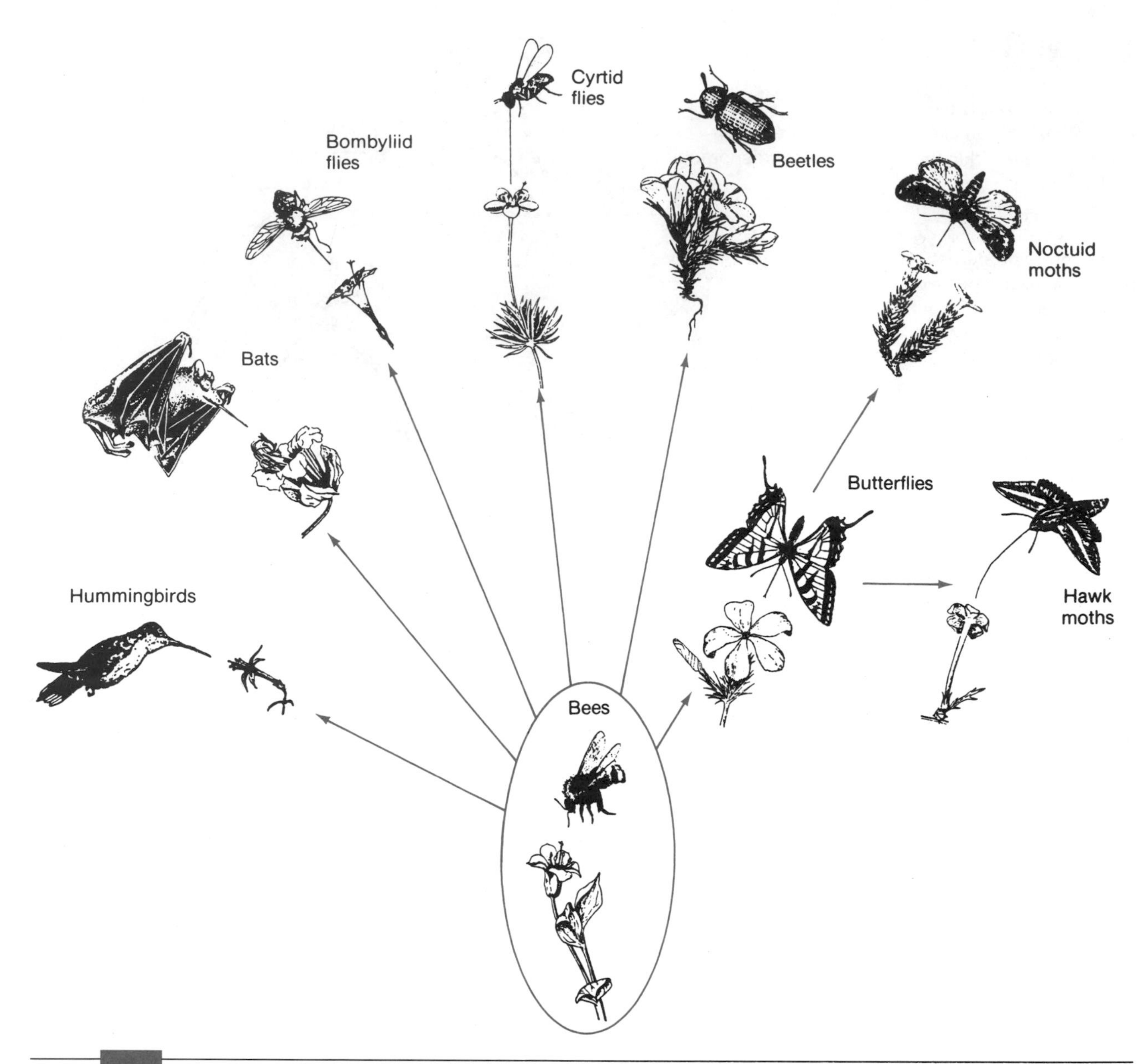

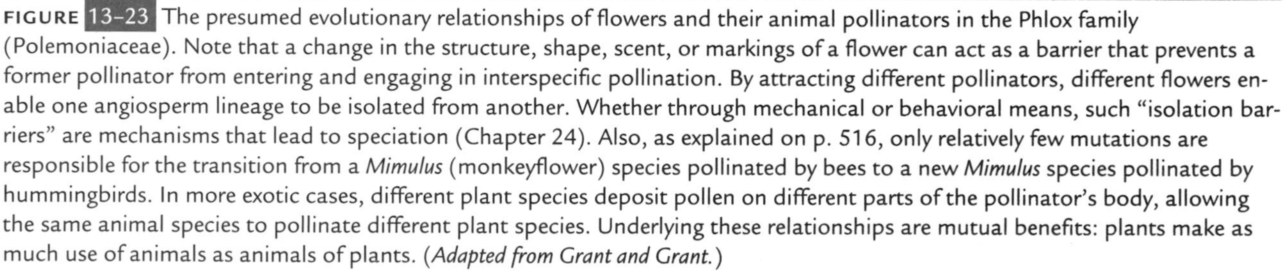

FIGURE 13-23 The presumed evolutionary relationships of flowers and their animal pollinators in the Phlox family (Polemoniaceae). Note that a change in the structure, shape, scent, or markings of a flower can act as a barrier that prevents a former pollinator from entering and engaging in interspecific pollination. By attracting different pollinators, different flowers enable one angiosperm lineage to be isolated from another. Whether through mechanical or behavioral means, such "isolation barriers" are mechanisms that lead to speciation (Chapter 24). Also, as explained on p. 516, only relatively few mutations are responsible for the transition from a *Mimulus* (monkeyflower) species pollinated by bees to a new *Mimulus* species pollinated by hummingbirds. In more exotic cases, different plant species deposit pollen on different parts of the pollinator's body, allowing the same animal species to pollinate different plant species. Underlying these relationships are mutual benefits: plants make as much use of animals as animals of plants. (*Adapted from Grant and Grant.*)

because of an "arms race" between host and parasite: each new genetic variant of a host that confers resistance against a fungal parasite is often overcome by selection for increased frequency of a fungal genetic variant that permits host susceptibility (p. 575).

As to phylogenetic relationships among the various major groups of fungi, molecular studies are producing much new information (see, for example, Bruns et al.), and mycologists are still debating and speculating on conclusions.

FIGURE 13-24 Appearance of angiosperm pollen and leaves at different times in the mideastern United States Cretaceous outcroppings (Potomac Group). The earliest pollens possess single furrows (monocolpate), but these become more sculpted as time goes on. The leaves generally show a hierarchy of vein complexities: primary veins, secondary veins, intercostal veins, and so on. (*Adapted from Doyle and Hickey.*)

FIGURE 13-25 Convergent evolution between representative desert species of American Cactaceae (*left*) and African Euphorbiaceae (*right*). (*From Niklas,* The Evolutionary Biology of Plants, *1997. Reprinted by permission of the University of Chicago Press.*)

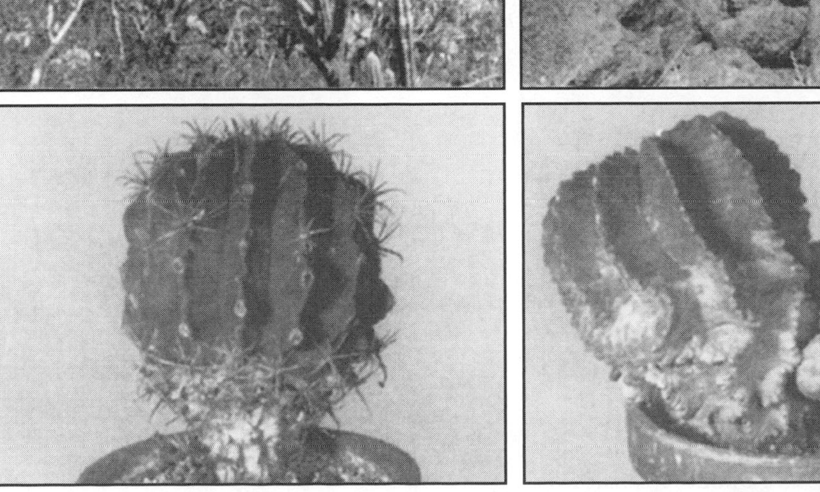

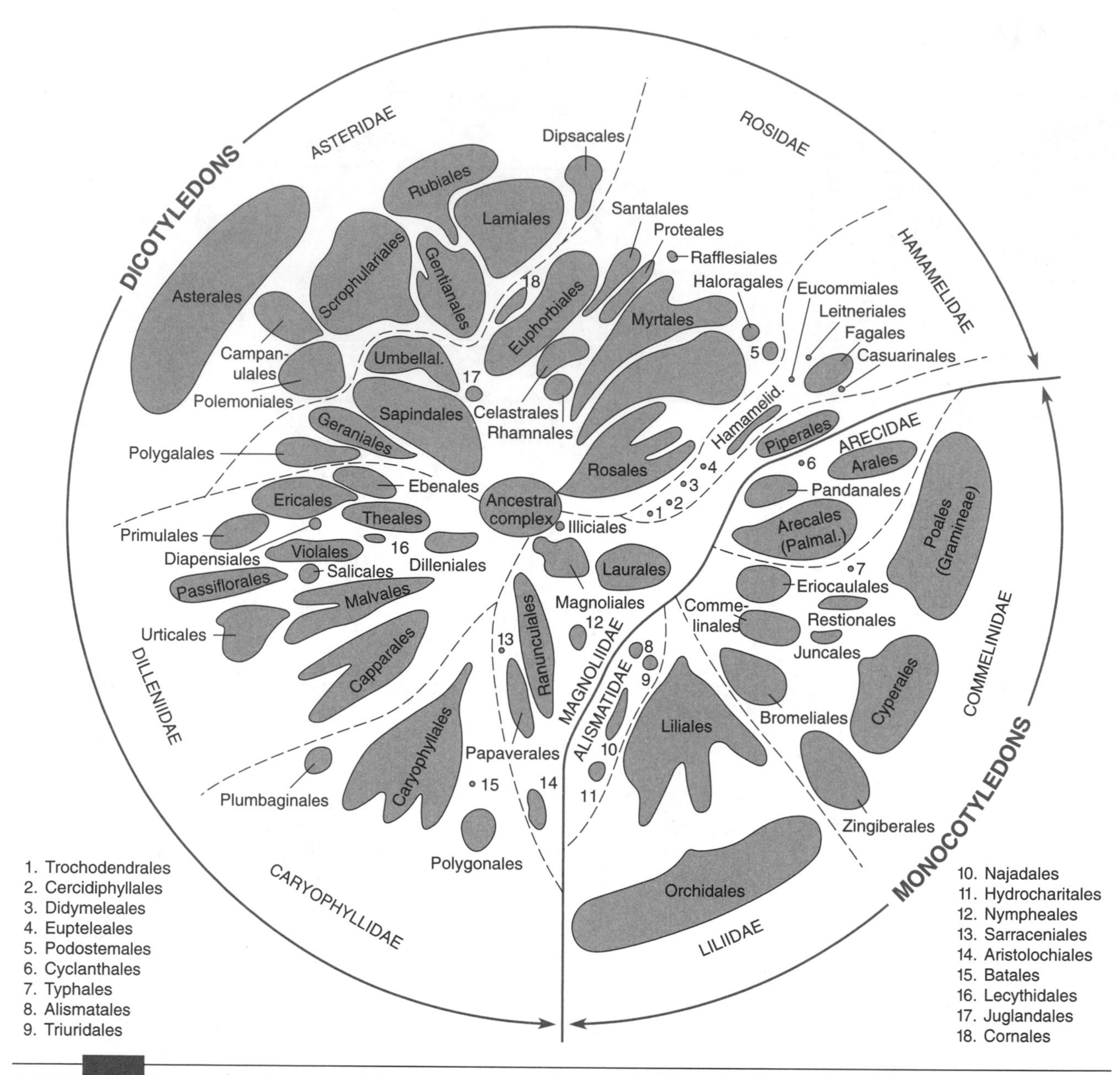

FIGURE 13-26 Proposed evolutionary relationships among major groups of angiosperms, according to Stebbins. Many botanists presume that the class of **dicotyledons** (two embryonic leaves) evolved before the **monocotyledons** (one embryonic leaf). The phylogenetic relationships between subclasses and between orders are shown by their physical proximities in the illustration (for example, the subclass Hamamelidae may derive from the subclass Magnoliidae). Within each subclass, each order occupies an area indicative of its relative population size. (Similar diagrams, but with different placements of subclasses and orders, can be found in Sporne and also Thorne.) (*Adapted from Stebbins with modifications.*)

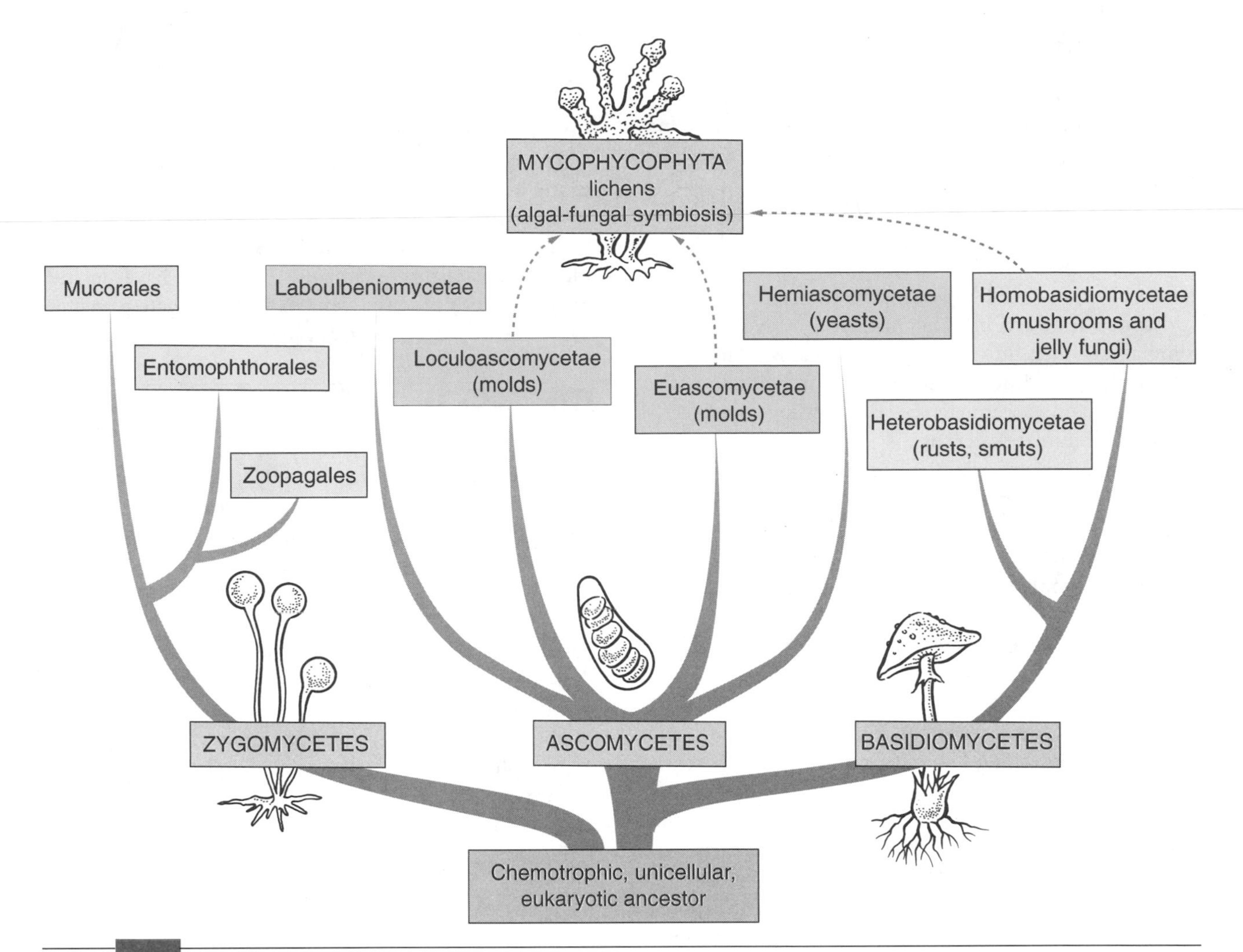

FIGURE 13-27 Conjectural phylogenetic relationships among the nonflagellated fungi, showing three subdivisions. Zygomycetes, Ascomycetes, and Basidiomycetes. A fourth subdivision, Deuteromycetes (fungi imperfecti), consists of about 25,000 species and includes some common fungi such as *Penicillium.* Researchers believe some species among this latter group evolved from Ascomycetes, others from Basidiomycetes. In addition to these nonflagellated fungi (Amastigomycota), are flagellated forms that include the classes Oomycetes and Chytridomycetes (Mastigomycota), and the separately classified slime molds (Gymnomycota). As the text mentions, molecular studies are now helping resolve fungal phylogenies. (*Adapted from Margulis and Schwartz.*)

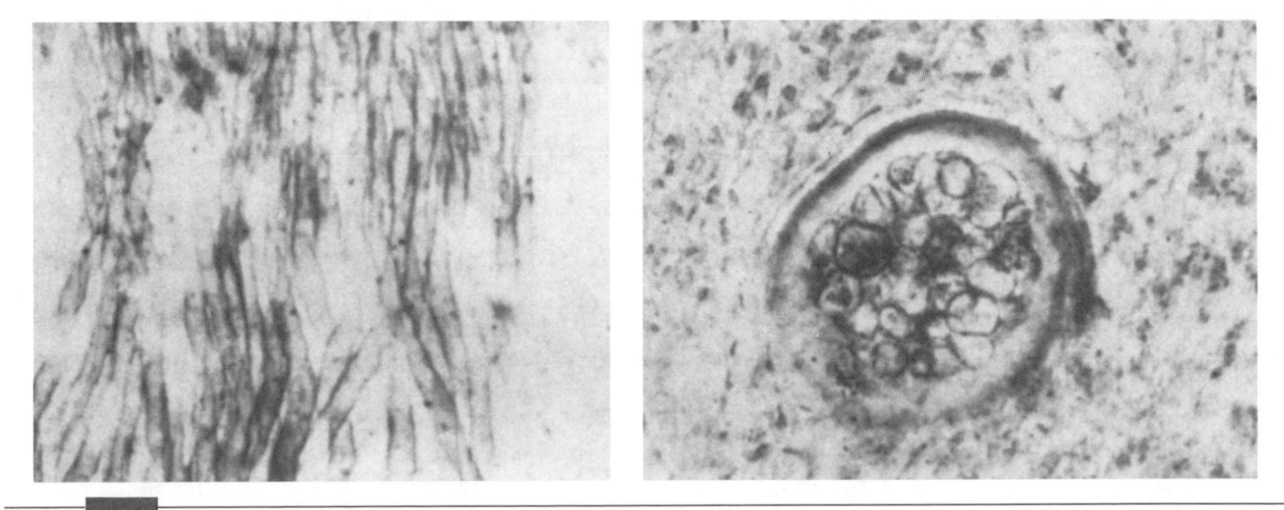

FIGURE 13-28 Fossils of different parts of a Devonian fungus, *Palaeomyces gordonii.* (*left*) Hyphae. (*right*) Spore. (*From* Paleobotany, *1981 by T.N. Taylor. Reprinted by permission of The McGraw-Hill Book Company.*)

SUMMARY

Vascular land plants share certain features with green algae, which botanists believe are similar to ancestral forms from which land plants arose. Among these characteristics are starch storage, cell plate formation during cytokinesis, and the use of chlorophylls a and b in photosynthesis. Other land plant characteristics that may have been present in their algal ancestors were (1) alternation of diploid and haploid generations and (2) meiotic spore production. The progenitors evolved into colonial and then into multicellular organisms in which the cells became differentiated but lost individual motility. When plants became terrestrial, they developed techniques to minimize water loss.

The evolutionary origins of bryophytes, the simplest land plants, are obscure. They live in moist areas because of their simple water- and food-distributing tissues and feature a distinctive life cycle in which the diploid generation, the sporophyte, parasitizes the haploid plant, the gametophyte.

According to many biologists, the increased variability among offspring meiosis produces has caused it to become entrenched in the reproductive processes of most eukaryotes. Whatever its origin, it may have served to reestablish the prevailing haploid condition in early sexually reproducing organisms. Probably because of the advantages in having two sets of genetic information, the initially brief diploid stage has considerably lengthened in many forms.

In plants, the products of meiosis are resistant spores that germinate into a haploid structure, the gametophyte. Fertilization of gametes that these structures produce yields the diploid sporophyte. Because of the need to provide the vulnerable gametes with an aqueous environment, the gametophyte generation has persisted in all land plants, but the sporophyte has become dominant and independent of the gametophyte.

The earliest vascular plants were leafless and rootless, and they are the ancestors of the club mosses, horsetails, and ferns. The evolutionary trends in these plants included the development of two types of spores (heterospory), the formation of large leaves (megaphylls), and the differentiation of increasingly elaborate vascular tissues.

When embryos gained independence from water and became enclosed in seeds, the gametophyte was greatly reduced. Plants, in the form of the gymnosperms, which bear naked seeds, and the angiosperms, with covered seeds, became truly terrestrial. The angiosperms feature elaborate flower structures, which attract pollinators, and double fertilization, in which they produce a triploid nutritional endosperm as well as a diploid embryo.

We know little of angiosperm origins, although fossil pollen appears in early Cretaceous deposits. Angiosperms may have first developed as woody shrubs in climatically unstable highland areas, or they may have been lowland tropical plants. Because of their unique characteristics, particularly reproductive features such as extreme truncation of gametogenesis, efficient pollination, and protected seeds, the angiosperms have become the dominant group in many environments.

The fungi—an ancient group, and one that is still actively evolving—are unicellular or multicellular saprophytic or parasitic organisms. They may have derived from chemotropic protistans or from obligate parasites, some of which became saprophytic.

KEY TERMS

algae
alternation of generations
angiosperms
antithetic theory
benthic
bryophytes
club mosses
cuticle
desiccation
dicotyledons
double fertilization
embryophytes
enation theory
endosperm
ferns
flowers
fungi
gametophyte
gymnosperms
heterospory
homologous theory
homospory
hyphae
lycopods
megaspores
microspores
monocotyledons
mycelium
parasitic
planktonic
pollen
progymnosperms
protostele
Rhyniophyta
saprophytic
seed ferns
seeds
self-incompatibility alleles
self-sterility alleles
sphenopsids
sporangia
spores
sporophyte
telome theory
Tracheophyta
vascular plants

DISCUSSION QUESTIONS

1. On what bases have biologists proposed the origin of land plants from algae?
2. Sex and meiosis
   a. What advantages does meiosis offer to sexual organisms?
   b. How does multicellularity increase the genetic variability produced by meiosis?
3. Alternation of generations
   a. What explanations have botanists offered for the gametophyte–sporophyte alternation of generations in plants?
   b. What can account for the persistence of the multicellular gametophyte stage in plants and its absence in animals?

4. What were the major evolutionary changes among the earliest land plants in respect to their leaves, spores, and vascular structures?
5. How do botanists believe seeds originated?
6. Angiosperms
   a. What are the unique characteristics of angiosperms, and what adaptive advantages do they offer?
   b. What proposals have botanists made as to angiosperm origin?
7. What proposals have botanists and mycologists offered to explain the origin and evolution of fungi?

EVOLUTION ON THE WEB

Explore evolution on the web! Visit the accompanying web site for *Evolution*, 3/e at www.jbpub.com/evolution for web exercises and links relating to topics covered in this chapter.

REFERENCES

Andrews, H. N., Jr., 1961. *Studies in Paleobotany.* Wiley, New York.

Baldauf, S. L., and J. D. Palmer, 1993. Animals and fungi are each other's closest relative: Congruent evidence from multiple proteins. *Proc. Nat. Acad. Sci.,* **90,** 11558–11562.

Banks, H. P., 1970. *Evolution and Plants of the Past.* Macmillan, London.

Beck, C. B. (ed.), 1976. *Origin and Early Evolution of Angiosperms.* Columbia University Press, New York.

Bold, H. C., C. J. Alexopoulos, and T. Delevoryas, 1980. *Morphology of Plants and Fungi,* 4th ed. Harper & Row, New York.

Bruns, T. D., T. J. White, and J. W. Taylor, 1991. Fungal molecular systematics. *Ann. Rev. Ecol. Syst.,* **22,** 525–564.

Cai, C., S. Ouyang, Y. Wang, Z. Fang, J. Rong, L. Geng, and X. Li, 1996. An Early Silurian vascular plant. *Nature,* **379,** 592.

Chapman, D. J., 1985. Geological factors and biochemical aspects of the origin of land plants. In *Geological Factors and the Evolution of Plants,* B. H. Tiffney (ed.). Yale University Press, New Haven, CT, pp. 23–45.

Crane, P. R., E. M. Friis, and K. R. Pedersen, 1995. The origin and early diversification of angiosperms. *Nature,* **374,** 27–33.

Cronquist, A., 1968. *The Evolution and Classification of Flowering Plants.* Nelson, London.

Darwin, C., 1862. *On the Various Contrivances by Which British and Foreign Orchids Are Fertilized by Insects, and on the Good Effects of Intercrossing.* Murray, London.

Delevoryas, T., 1977. *Plant Diversification,* 2d ed. Holt, Rinehart and Winston, New York.

Delwiche, C. F., L. E. Graham, and N. Thomson, 1989. Lignin-like compounds and sporopollenin in *Coleochaete,* an algal model for plant ancestry. *Science,* **245,** 399–401.

Doyle, J. A., and L. J. Hickey, 1976. Pollen and leaves from the mid-Cretaceous Potomac Group and their bearing on early angiosperm evolution. In *Origin and Early Evolution of Angiosperms,* C. B. Beck (ed.). Columbia University Press, New York, pp. 139–206.

Foster, A. S., and E. M. Gifford, Jr., 1974. *Comparative Morphology of Vascular Plants,* 2d ed. Freeman, San Francisco.

Gensel, P. G., and H. N. Andrews, 1984. *Plant Life in the Devonian.* Praeger, New York.

Graham, L. E., 1993. *Origin of Land Plants.* Wiley, New York.

Grant, V., and K. A. Grant, 1965. *Flower Pollination in the Phlox Family.* Columbia University Press, New York.

Hughes, N. F., 1976. *Palaeobiology of Angiosperm Origins.* Cambridge University Press, Cambridge, England.

Hutchinson, J., 1969. *Evolution and Phylogeny of Flowering Plants: Dicotyledons; Facts and Theory.* Academic Press, London.

Kondrashov, A. S., 1994. The asexual ploidy cycle and the origin of sex. *Nature,* **370,** 213–216.

Lewis, L. A., B. D. Mishler, and R. Vilgalys, 1997. Phylogenetic relationship of the liverworts (Hepaticae), a basal embryophyte lineage, inferred from nucleotide sequence data of the chloroplast gene *rbcL. Mol. Phylog. and Evol.,* **7,** 377–393.

Mable, B. K., and S. P. Otto, 1998. The evolution of life cycles with haploid and diploid phases. *BioEssays,* **20,** 453–462.

Margulis, L., and K. V. Schwartz, 1998. *Five Kingdoms,* 3d ed. Freeman, New York.

Mauseth, J. D., 1998. *Botany: An Introduction to Plant Biology,* 2d ed. Jones and Bartlett, Sudbury, MA.

Niklas, K. J., 1997. *The Evolutionary Biology of Plants.* University of Chicago Press, Chicago.

Nilsson, L. A., 1998. Deep flowers for long tongues. *Trends in Ecol. and Evol.,* **13,** 259–260.

Pettitt, J. M., and C. B. Beck, 1968. *Archaeosperma arnoldii*—A cupulate seed from the Upper Devonian of North America. *Contr. Mus. Paleontol. Univ. Michigan,* **22,** 139–154.

Pickett-Heaps, J. D., 1975. *Green Algae: Structure, Reproduction, and Evolution of Selected Genera.* Sinauer Associates, Sunderland, MA.

Raper, J. R., 1968. On the evolution of fungi. In *The Fungi: An Advanced Treatise,* Vol. III, G. C. Ainsworth and A. S. Sussman (eds.). Academic Press, New York, pp. 677–693.

Raven, P. H., 1977. A suggestion concerning the Cretaceous rise to dominance of the angiosperms. *Evolution,* **31,** 451–452.

Remy, W., 1982. Lower Devonian gametophytes: Relation to the phylogeny of land plants. *Science,* **215,** 1625–1627.

Ren, D., 1998. Flower-associated Brachycera flies as fossil evidence for Jurassic angiosperm origins. *Science,* **280,** 85–88.

Savard, L., P. Li, S. H. Strauss, M. W. Chase, M. Michaud, and J. Bousquet, 1994. Chloroplast and nuclear gene sequences indicate Late Pennsylvanian time for the last common ancestor of extant seed plants. *Proc. Nat. Acad. Sci.,* **91,** 5163–5167.

Sporne, K. R., 1976. Character correlations among angiosperms and the importance of fossil evidence in assessing their significance. In *Origin and Early Evolution of Angiosperms,* C. B. Beck (ed.). Columbia University Press, New York, pp. 312–329.

Stebbins, G. L., 1974. *Flowering Plants: Evolution Above the Species Level.* Harvard University Press, Cambridge, MA.

Stewart, W. N., and G. Rothwell, 1993. *Paleobotany and the Evolution of Plants,* 2d ed. Cambridge University Press, Cambridge, England.

Takhtajan, A., 1969. *Flowering Plants: Origin and Dispersal.* Oliver & Boyd, Edinburgh.

Taylor, D. W., and L. J. Hickey, 1990. An Aptian plant with attached leaves and flowers: Implications for angiosperm origin. *Science,* **247,** 702–704.

Taylor, T. N., and E. L. Taylor, 1993. *The Biology and Evolution of Fossil Plants.* Prentice Hall, Englewood Cliffs, NJ.

Thomas, B. A., and R. A. Spicer, 1987. *The Evolution and Palaeobiology of Land Plants.* Croom Helm, London.

Thorne, R. F., 1976. A phylogenetic classification of the Angiospermae. *Evol. Biol.,* **9,** 35–106.

Tiffney, B. H., 1985. Geological factors and the evolution of plants. In *Geological Factors and the Evolution of Plants,* B. H. Tiffney (ed.). Yale University Press, New Haven, CT, pp. 1–21.

Wainright, P. O., G. Hinkle, M. L. Sogin, and S. K. Stickel, 1993. Monophyletic origins of the metazoa: An evolutionary link with fungi. *Science,* **260,** 340–342.

Zimmermann, W., 1952. Main results of the "telome theory." *Paleobotanist,* **1,** 456–470.

14 From Protozoa to Metazoa

Although we can trace unicellular eukaryotic fossils back about 1.6–1.8 billion years or so (Mendelson), and there are claims of eukaryotic algae 2.1 billion years old (Han and Runnegar), diverse and more complex unicellular forms appeared later, about 1 billion years ago (Knoll). We still don't know how these changes led to the appearance of multicellular animal eukaryotes (**metazoans**), but by the beginning of the Cambrian period (about 545 million years ago), many differently skeletonized groups materialize. Within a relatively short geological time span, an explosive radiation of multicellular eukaryotes marks the emergence of the largest number of uniquely distinctive animal body plans that ever appeared in the fossil record (Fig. 14–1).

This chapter summarizes some of the possible explanations for these events, but we have to keep in mind that many uncertainties persist because of the obscurities caused by the large gap between the present and that far-distant past, such as the paucity of intermediary fossils and the lack of environmental and climatological information.

The Cambrian "Explosion"

Although the ancestral connections are unclear, some or many of the skeletonized forms that appeared in the Cambrian period represent entirely new adaptive radiations of Precambrian forms. Evolutionary molecular clocks (p. 284), based on a considerable amount of nucleotide sequence data, indicate that divergences among major Cambrian lineages had already begun about 700 million years ago (Ayala et al.) or even more than one billion years ago (Wray et al.). To more ancient soft-bodied organisms, Cambrian modifications added hard parts and skeletons that undoubtedly initiated changes in the location and attachment of tissues, leading to entirely novel animals. Solid mineralized structures provided leverage for muscles, support for body organs, enclosures for gills and filtering systems, and more dramatically, protective shells and spines for prey species, and teeth for predators. Such prey–predator adaptations would have quickly escalated into a persistent co-evolving

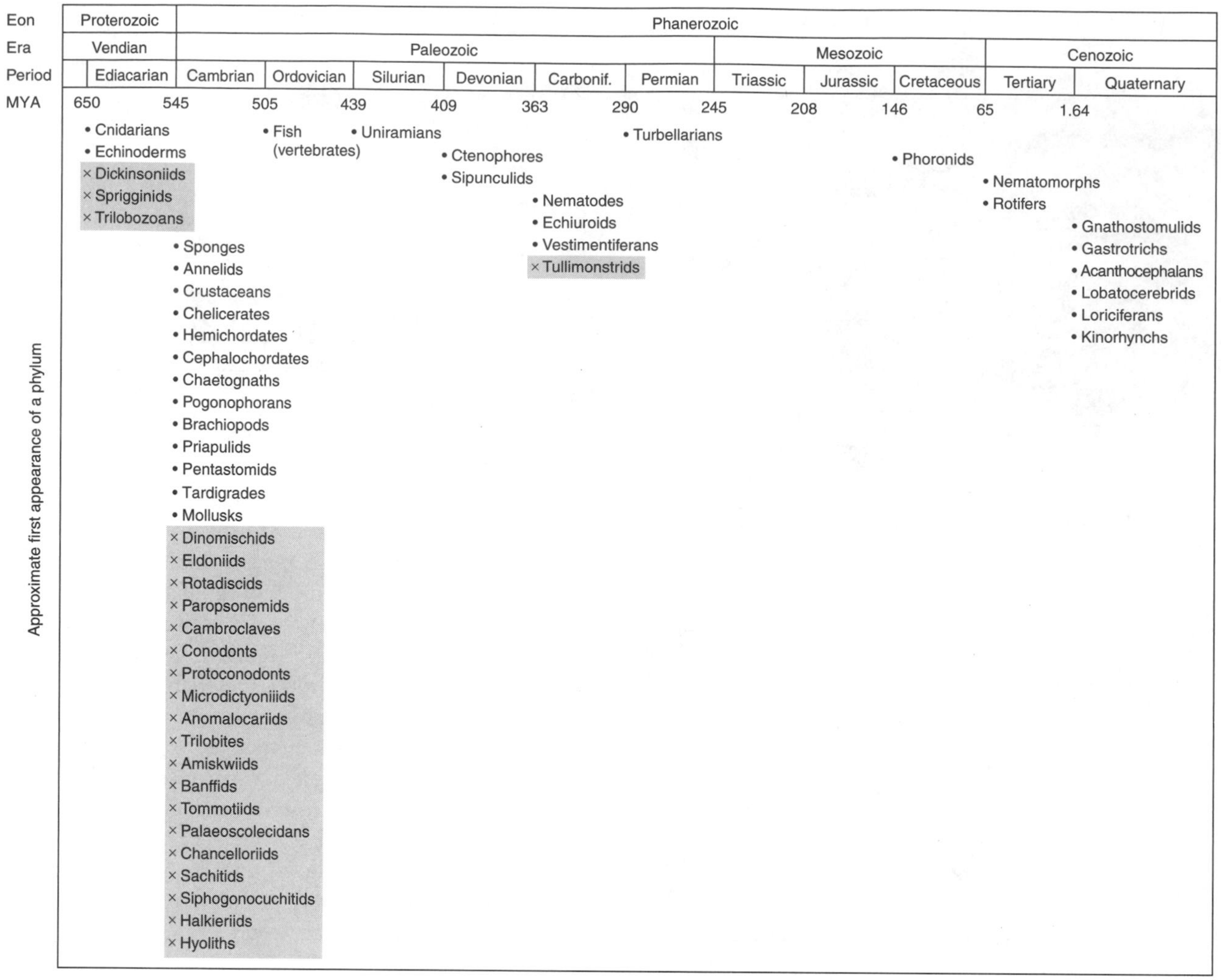

Eon	Proterozoic	Phanerozoic										
Era	Vendian	Paleozoic						Mesozoic			Cenozoic	
Period	Ediacarian	Cambrian	Ordovician	Silurian	Devonian	Carbonif.	Permian	Triassic	Jurassic	Cretaceous	Tertiary	Quaternary
MYA	650	545	505	439	409	363	290	245	208	146	65	1.64

Approximate first appearance of a phylum

- Cnidarians
- Echinoderms
- × Dickinsoniids
- × Sprigginids
- × Trilobozoans
- Sponges
- Annelids
- Crustaceans
- Chelicerates
- Hemichordates
- Cephalochordates
- Chaetognaths
- Pogonophorans
- Brachiopods
- Priapulids
- Pentastomids
- Tardigrades
- Mollusks
- × Dinomischids
- × Eldoniids
- × Rotadiscids
- × Paropsonemids
- × Cambroclaves
- × Conodonts
- × Protoconodonts
- × Microdictyoniids
- × Anomalocariids
- × Trilobites
- × Amiskwiids
- × Banffids
- × Tommotiids
- × Palaeoscolecidans
- × Chancelloriids
- × Sachitids
- × Siphogonocuchitids
- × Halkieriids
- × Hyoliths
- Fish (vertebrates)
- Uniramians
- Ctenophores
- Sipunculids
- Nematodes
- Echiuroids
- Vestimentiferans
- × Tullimonstrids
- Turbellarians
- Phoronids
- Nematomorphs
- Rotifers
- Gnathostomulids
- Gastrotrichs
- Acanthocephalans
- Lobatocerebrids
- Loriciferans
- Kinorhynchs

FIGURE 14-1 Approximate times in which various major metazoan groups first appear in the fossil record. Unshaded taxa marked with filled-in circles still have existing descendants, although the original species representing these groups are long extinct. Shaded taxa marked with **x**'s represent extinct orders or phyla that have no known surviving descendants. As this illustration shows, the Cambrian was a period that produced many different metazoan body plans, yet which, through extinction, also restricted their number. The rapid increase in Cambrian phyla suggests an adaptive radiation in which many new ecological habitats were made available for organisms that could evolve in diverse directions (see, for example, Figure 3-4). Extinctions among these phyla in such quickly saturated environments must have been caused, as in other periods, by both selective and accidental factors. That is, increasing competition for available resources among so many groups with disparate yet overlapping characters led to selection among them, and any environmental calamities that followed had serious impact on groups whose numbers, habitats, or lifestyles made them vulnerable to extinction—both "survival of the fittest" and "survival of the luckiest." Many biologists suggest that the multicellular phyla that survived the Cambrian radiated in so many directions, using such a wide range of body plans and habitats, that only few later-emerging phyla could successfully compete with them. In general, body plans that came first limited opportunities for those that came last. (The Cambrian absence of some expected fossils, such as platyhelminths—a primitive and basic multicellular form—is most probably because they lack a fossilizable epidermal cuticle.) (*Based on data derived from Conway Morris.*)

"arms race"—successive rounds of selection for adaptive predator responses to their preys' protective devices, and of adaptations by prey to their predators' aggressive devices. Each new environmental interaction would have allowed selection for new adaptive features and survival strategies, thus promoting diversity.

As true for many other Cambrian conditions, factors that initiated mineralization of tissues are not yet known. Perhaps the Cambrian marks a warming trend allowing mineralization. Or perhaps the atmospheric accumulation of oxygen through photosynthesis reached sufficient levels to permit mineralization and energize oxygen-dependent synthesis of metazoan connective tissue proteins (collagen). Certainly, the ability of oxygen to form a protective blanket of ozone (p. 165) facilitated the rapid expansion and radiation of multicellular animals in shallow waters and on varied surfaces, providing the opportunity for further skeletal and other adaptations.

Fenchel and Finlay suggest that a basic reason why the evolution of large organisms had to await an aerobic environment is that anaerobic metabolism has a low energy yield (p. 170) and therefore low biomass production (only 10 percent growth efficiency). This limits the food chain to perhaps only two steps, from bacteria who consume organic matter to single-celled eukaryotes who consume bacteria. Oxygen environments, in contrast, are about four times as efficient in energy yield, producing sufficient biomass to allow additional levels in the food chain, each with increasing morphological size (although numbering fewer individuals at each higher level).[1]

A change in oxygen concentration is only one of many possible causes for the Cambrian explosion. A formerly popular concept was that most signs of life disappeared in strata earlier than the Phanerozoic because of the heat and pressure involved in geological processes such as mountain building. A further notion was that living forms evolved mostly in freshwater areas, and their fossils are absent in Precambrian sediments, which are primarily of marine origin. However, since considerable evidence now shows that we can clearly identify both prokaryotic and eukaryotic Precambrian organisms, the Cambrian discontinuity seems real and not merely the result of geological metamorphism or imperfect fossilization. If the Precambrian ancestors of later metazoan phyla seem absent, that may be because they were quite small, no larger than a few millimeters, and therefore did not fossilize well, or at all.

Among other physical causes that researchers have offered for the Cambrian explosion are changes in the shape and extent of shorelines because of continental drift, profoundly transforming both climate and environment (see Chapter 6). The sea level changes that accompany glaciation undoubtedly would have had similar effects, and some proposals even suggest new tidal effects caused by the moon. Although we can exclude none of these ideas with certainty, and each may have had some influence on metazoan evolution, we cannot support any of them with convincing evidence (Valentine et al.).

Biologists commonly suggest that the evolution of eukaryotic sexual genetic exchange and/or regulatory genes that control multicellular development (see Chapter 15) could have sparked the diversity and anatomical complexity that fueled the Cambrian explosion. To these proposals we can add the hypothesis that since there were few Precambrian multicellular species, their exponential increase could only become notable after reaching a threshold number at the start of the Cambrian. Again, no demonstrable support exists for these proposals—in fact, we can claim that both sex and developmental regulatory elements existed long before the Cambrian—nor is there any way of ascertaining a "threshold number" of species.

A further biological cause, suggested by Stanley, depends on the principle of **cropping.** In cropping, predators feed on the most abundant prey species, thereby reducing their numbers and letting other species use resources formerly monopolized by the dominant prey. For example, cropping of a field once restricted to a single dominant plant species soon opens many niches for other plant species, which now can grow in an area where they were formerly excluded. The evolutionary value of cropping also extends to predators through a **feedback cycle,** since the diversification of prey species leads in turn to the diversification of predator species.

Once multicellular heterotrophic **herbivores** appeared, only a few evolutionary steps would have led them from being a herbivore to becoming a **carnivore** that feeds on herbivores, and then to carnivores that feed on other carnivores, initiating the explosive adaptive radiation of the Cambrian period. The appearance of hard exoskeletons may therefore have provided a common function to a wide variety of Cambrian organisms in offering varying degrees of protection from predation. The fact that eukaryotes by this time had evolved sexual reproductive methods also undoubtedly enhanced their ability to diversify into new functions and new available habitats.[2]

Although attractive, much of Stanley's hypothesis depends on the novel appearance of effective heterotrophic croppers immediately before the Cambrian period, a view that other workers in this field do not share. According to Signor and Lipps, heterotrophic nonphotosynthetic organisms were no novelty in the Precambrian, and most probably originated far back in the origin-of-life period. Yet, recent findings of Lower Cambrian filter-bearing crustaceans, enabling them to feed on even the smallest planktonic autotrophs, may be indicative of the

[1]Berkner and Marshall suggested that metazoan evolution, dependent on aerobic metabolism, had to await an oxygen atmosphere sufficient to sustain it at a level that was perhaps 1 percent of present atmospheric oxygen. Unfortunately, data on Cambrian and Precambrian oxygen pressures are not available, and it is even possible that plentiful supplies of free oxygen produced by algae may have been present during a billion-year interval before the Cambrian (for example, see Cloud). According to Knoll, even if Proterozoic oxygen pressure was generally low, some evidence exists for major increases in oxygen during brief Precambrian periods that, in conjunction with new sophisticated developmental controls, may have led to the evolutionary burst of large multicellular metazoans. Ohno claims that animals capable of exploiting Early Cambrian oxygen possessed a battery of common genes including those for lysyloxidase, which uses oxygen to crosslink collagen in ligaments and tendons; hemoglobin, a conveyor of molecular oxygen; and homeobox proteins that help orient body plans along a directional axis (see Chapter 15).

[2]The Precambrian fossil record indicates the widespread existence of algal stromatolites (p. 171) growing under conditions unhampered by limitations other than those of available light and nutrients. As Stanley says, "We can envision an all-producer Precambrian world that was generally saturated with [autotrophic] producers and biologically monotonous." In the absence of cropping, new algal prokaryotic species and even new algal eukaryotes must have been preempted from occupying these areas. In fact, Brock finds that prokaryotic algae do not exist in acid conditions (perhaps because their chlorophyll molecules are relatively unprotected in their sites on the plasma membrane), suggesting that such environmentally restricted areas may have provided the first opportunity for colonization by eukaryotic algae.

JAMES W. VALENTINE

Birthday:
November 10, 1926

Birthplace:
Los Angeles, CA

Undergraduate degree:
Phillips University, Enid, Oklahoma, 1951

Graduate degree:
M.A. University of California at Los Angeles, 1954
Ph.D. University of California at Los Angeles, 1958

Present position:
Professor of Integrative Biology Emeritus and Curator, Museum of Paleontology
University of California at Berkeley

WHAT PROMPTED YOUR INITIAL INTEREST IN EVOLUTION?

I was a World War II G.I. who had not planned on going to college but who took advantage of the G.I. Bill for Education, choosing geology rather blindly out of a long list of majors available at first enrollment. Paleontology was the most interesting course to me, and on reading "Tempo and Mode in Evolution" by George Gaylord Simpson I realized that the fossil record could be used as evidence in formulating hypotheses about evolutionary processes. Combined with data from genetics, zoology, botany and other life sciences, it was clear that paleontology could make important contributions to evolutionary theory.

WHAT DO YOU THINK HAS BEEN MOST VALUABLE OR INTERESTING AMONG THE DISCOVERIES YOU HAVE MADE IN SCIENCE?

The most fun has been not so much to make discoveries as to pursue an approach: to attempt to treat the fossil record as an ecological theater, to use G. Evelyn Hutchinson's metaphor. There has been lots of pleasure in looking for and finding clues that reveal the processes that produced the evolutionary play.

WHAT AREAS OF RESEARCH ARE YOU (OR YOUR LABORATORY) PRESENTLY ENGAGED IN?

I am now most interested in the "Cambrian explosion," when the remains of animals with the body plans of many living phyla first appeared during a 9- to 10-million year period beginning about 530 million years ago. This is one of the most spectacular events recorded by fossils, and must have involved extensive early evolution of metazoan developmental systems, but is so unique and so remote in time that it has been difficult to interpret. Dating of those ancient fossil assemblages by geophysical laboratories with new, extremely accurate techniques, and an integration of the fossil evidence with findings from molecular biology laboratories, has begun to clarify the fine structure of the evolutionary events leading to the explosion. In my laboratory we combine observations of living organisms with molecular studies to model the sorts of evolutionary pathways that were required to produce such a broad array of body plans.

IN WHICH DIRECTIONS DO YOU THINK FUTURE WORK IN YOUR FIELD NEEDS TO BE DONE?

Evolutionary hypotheses generated by the fossil record need to be tested by the tools of molecular biology; these seemingly disparate fields have much to offer each other.

WHAT ADVICE WOULD YOU OFFER TO STUDENTS WHO ARE INTERESTED IN A CAREER IN YOUR FIELD OF EVOLUTION?

Aside from the obvious need to master the basics of paleobiology and appropriate ancillary subjects, there are three bits of advice that seem particularly valuable. One is to question authority, a generally good idea in any area. A second is to contrast notions prevalent in different fields, because transfer of approaches from one field to another can lead to most creative results, and because contradictions between fields can lead to new insights as well. And finally, as you become interested in research, do as Peter Medawar has suggested and attack the most important problem you think you can solve, or at least advance.

sudden appearance of efficient "cropping" organisms that could have fueled the Cambrian expansion (Butterfield). A single source for this expansion is not yet generally accepted, and the Cambrian events may well have arisen from a combination of conditions.

Whatever the cause for the proliferation of Cambrian shelled animals, the soft-bodied fossils found in the Precambrian **Ediacaran strata** of South Australia as well as in tidal deposits at many other places around the globe are, if not ancestral, at least indirectly related to the later skeletonized metazoans. As Figure 14–2 shows, some of these Ediacaran (or Vendian) fossils resemble modern coelenterates such as jellyfish and sea pens, whereas others bear likenesses to segmented annelids and, more distantly, to mollusks and echinoderms. Because their organization seems relatively advanced, with intricate surface structures and some obvious tissue complexities, these fossil organisms clearly must have undergone considerable prior evolution. By the Middle Cambrian, the basic body plans to which they or other Precam-

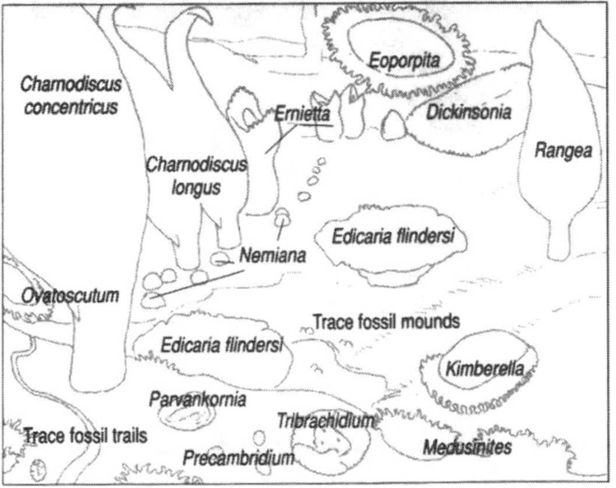

FIGURE 14–2 A panorama of soft-bodied animals found in Ediacaran tidal flat deposits of South Australia and 10 or so other places throughout the globe, most occurring approximately 545 to 650 million years ago, just prior to the Cambrian period. Some of these fossils (*Edicaria, Charniodiscus*) resemble modern cnidarians such as jellyfish and sea pens, whereas others bear possible likenesses to segmented annelid-like animals (*Dickinsonia*) and perhaps to echinoderms (*Tribachidium*) or mollusks (*Kimberella*). However, since researchers have not clearly detected any basic anatomical features of later animals such as eyes, mouths, anuses, intestinal tracts, or locomotory appendages in any of the Ediacaran fossils, biologists have debated their relationship to Cambrian metazoans. Glaessner proposed that these organisms were early representatives of modern phyla, whereas Seilacher placed them in a distinctive taxon, "Vendozoa," unrelated to any of the later phyla. The absence of features usually associated with prey capture and a digestive tract has led to suggestions that many of these Ediacaran animals may have depended on photosynthetic or other types of symbionts. Retallack suggests most were really lichens—symbiotic associations between fungi and algae—flattened by compaction in quartzite deposits. Others dispute this view (Fedonkin). Whatever their affinities, their morphological differences indicate that a variety of Ediacaran animal groups with already diverse evolutionary histories existed right up to the Cambrian boundary. Although some of these forms survived into the Cambrian, they became extinct probably because their lack of armor made them easy prey for new mobile Cambrian predators. (*From Erwin, D., J. Valentine, and D. Jablonski, 1997. "The origin of animal body plans."* The American Scientist, *85, 126–137. Reprinted with permission of D.W. Miller.*)

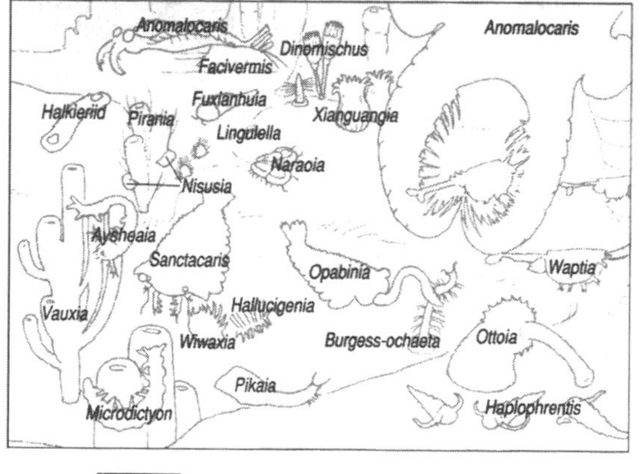

FIGURE 14-3 A variety of Cambrian animals including some exotic forms from the Middle Cambrian Burgess Shale of Canadian British Columbia. Many are arthropods, including some trilobite-like animals (*Naraoia*). *Aysheaia,* seen feeding on a sponge, is considered in the ancestral arthropod lineage, related to onychophora (velvet worms). *Pikaia,* with its notochord and fishlike zig-zag muscle arrangement, is the first fossil found of a chordatelike animal—a possible vertebrate ancestor. The possible lifestyles of these and other Burgess Shale animals are described by Conway Morris. (*From Erwin, D., J. Valentine, and D. Jablonski, 1997. "The origin of animal body plans."* The American Scientist, *85, 126–137. Reprinted with permission of D.W. Miller.*)

brian lineages gave rise were established, including some exotic forms that did not survive much further (Fig. 14–3). What were the first steps in multicellular animal evolution?

Protistan Ancestry

ost recent taxonomic schemes classify all unicellular eukaryotes into an exclusive kingdom, Protista. These organisms include (1) protozoans, which ingest their food directly, (2) photosynthetic algae, and (3) some saprophytic fungi. It is a large group, covering more than 100,000 species, and Margulis and coworkers have classified it (under the name "Protoctista") into 36 phyla.

Evolutionarily, protistan ancestry is old, dating back about 1.6–1.8 billion years. Their relationship to metazoans is commonly accepted,[3] and has received recent molecular support from nucleotide analysis of 16*S* RNA sequences (Wainright et al.) showing that the closest living metazoan relative is a single-celled choanoflagellate (collared spongelike zooflagellate; see Chapter 16). Thus, aside from their diverse forms and the many ways in which they affect other forms of life, biologists believe protistans are the essential link between the early progenote cells (p. 177) and all multicellular eukaryotes.

As discussed in Chapter 9, biologists now generally presume that endosymbiotic events provided protistans with mitochondria, chloroplasts, and perhaps other constituents. In addition to these organelles, protistans share many features with other eukaryotes, including a nuclear membrane as well as **cilia** and **flagella** whose substructures are organized into nine pairs of microtubules circling two microtubules in the axial center ("9 + 2" arrangement).

Because not all photosynthetic protistans contain the same light-gathering pigments, Sleigh and others suggest that photosynthetic prokaryotes invaded eukaryotic cells more than once.[4] Some of these endosymbiotic events resulted in protistan algae, completely reliant on autotrophic nutrition, whereas other events produced protistans that alternated from autotrophic to heterotrophic nutrition, such as *Euglena.* The lack of chloroplasts, whether caused by their initial absence or later loss, gave rise to the large diversity of protistan heterotrophs, or protozoans.

Until molecular phylogenies advance further, Figure 14–4 offers one proposed scheme for protistan radiation (see also Cavalier-Smith). In this case, protistan evolution begins with a **protoflagellate,** which had flagella and had already established endosymbiotic relations with a mitochondrion. Its nutrition was heterotrophic, based on phagocytosis through its naked cell membrane. Cell division among these early protistans evolved in two main directions: (1) retention of the nuclear membrane during mitosis somewhat similar to chromosomal division in prokaryotes (closed division), and (2) mitotic division accompanied by breakdown of the nuclear envelope (open division).

Events that followed according to this scheme include the appearance of a cytostome (mouth) in some forms, the differentiation of flagella into uneven lengths (heterokonts), or their loss, as well as the development of tests (shells), pseudopodia (cytoplasmic extensions), and internal spindles (mitotic microtubules within the nuclear membrane). Exact phylogenetic proposals are still conjectural, especially the matter of how the protozoan–metazoan transition took place developmentally and morphologically, an issue that has long raised considerable discussion and debate.

Hypotheses of Metazoan Origin

The change from unicellularity to multicellularity has apparently occurred a number of times and in different groups. For example, although they are not metazoans, we find forms of multicellular organization in such widely unrelated groups as filamentous cyanobacteria, slime bacteria (myxobacteria), aggregating amoebae (for example, dictyostelids), algae (brown, red, and green), and colonial ciliated protozoans (Zoothamnium).

Although some biologists used these examples to indicate the likelihood of multiple (polyphyletic) origins for animal (metazoan) multicellularity, recent molecular evidence has gone in the opposite direction, pointing to a close relationship between metazoans and protistan choanoflagellates (Fig. 14–5). For the most part, biologists now accept the view that most, if not all, existing metazoan phyla had a common ancestry. Whether we agree to monophyletic or polyphyletic metazoan multicellularity, however, one of the most difficult questions in biology still remains—how did unicellular ancestry transform into metazoan descendants? Debates on this process have gone on for more than a century, and have led to various proposals, some receiving more attention than others (see Willmer). Four concepts are briefly described next.

EVOLUTION FROM PLANTS

A hypothesis Hardy proposed bypasses the unicellular–multicellular transition, and suggests that metazoans arose from multicellular plants (**metaphyta**) that were forced into heterotrophic modes of nutrition because they were deprived of phosphates and nitrates. As these carnivorous plants presumably became successful in absorbing and capturing other organisms, they lost their chloroplasts and turned into metazoans.[5] Objections to this concept generally point to the difficulty of adapting the rigid cellulose cell wall

[3]According to the five-kingdom classification system first proposed by Whittaker, the Protista stand between the prokaryotic Monera (Archaebacteria and Eubacteria) and the three multicellular kingdoms of Plantae, Fungi, and Animalia (see Fig. 9–18).

[4]Although the number of endosymbiotic events accounting for chloroplast invasion into eukaryotes is still in question, most biologoists agree that the chloroplast source was probably cyanobacterial. Delwiche and coworkers offer data for monophyletic cyanobacterial origin based on sequence analysis of a gene (*tufA*) coding for an elongation factor used in protein synthesis.

[5]Shades of a musical comedy called "The Little Shop of Horrors"!

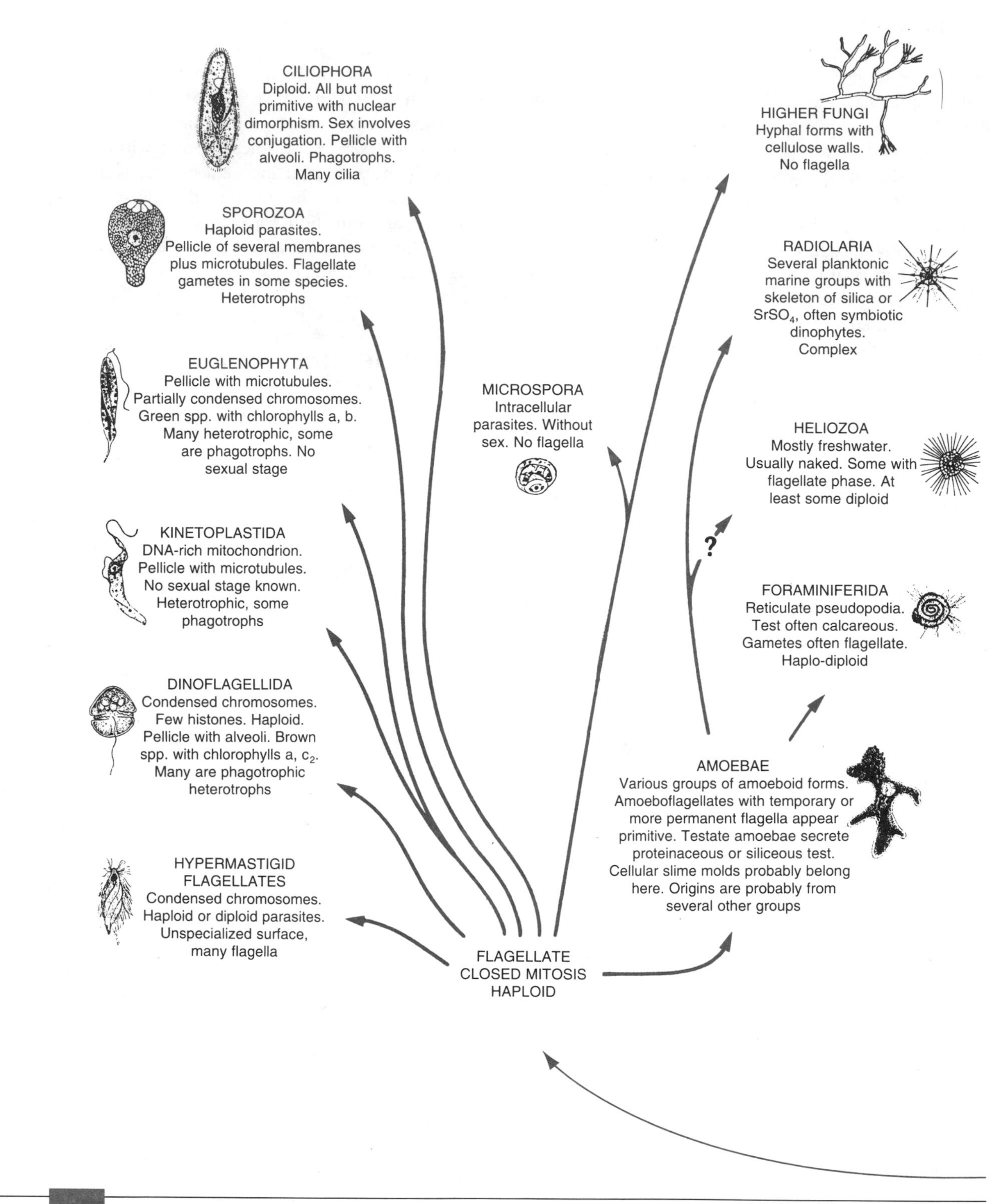

FIGURE 14-4 One proposed phylogeny for various protistan groups leading to metazoans, higher fungi, and land plants. How to decide between both the many different classifications and the many possible phylogenetic relationships offered for those protistans has long been a subject of debate. Fortunately, molecular data derived from nucleotide sequencing of ribosomal RNA and amino acid sequencing of various proteins are now helping make such decisions as well as determining more precise relationships between protistans, metazoans, and fungi (see, for example, Fig. 9–16). Schlegel has reviewed much of this information and proposes that "we will achieve a virtually complete picture of eukaryote phylogeny in the not too distant future." (*Adapted from Sleigh.*)

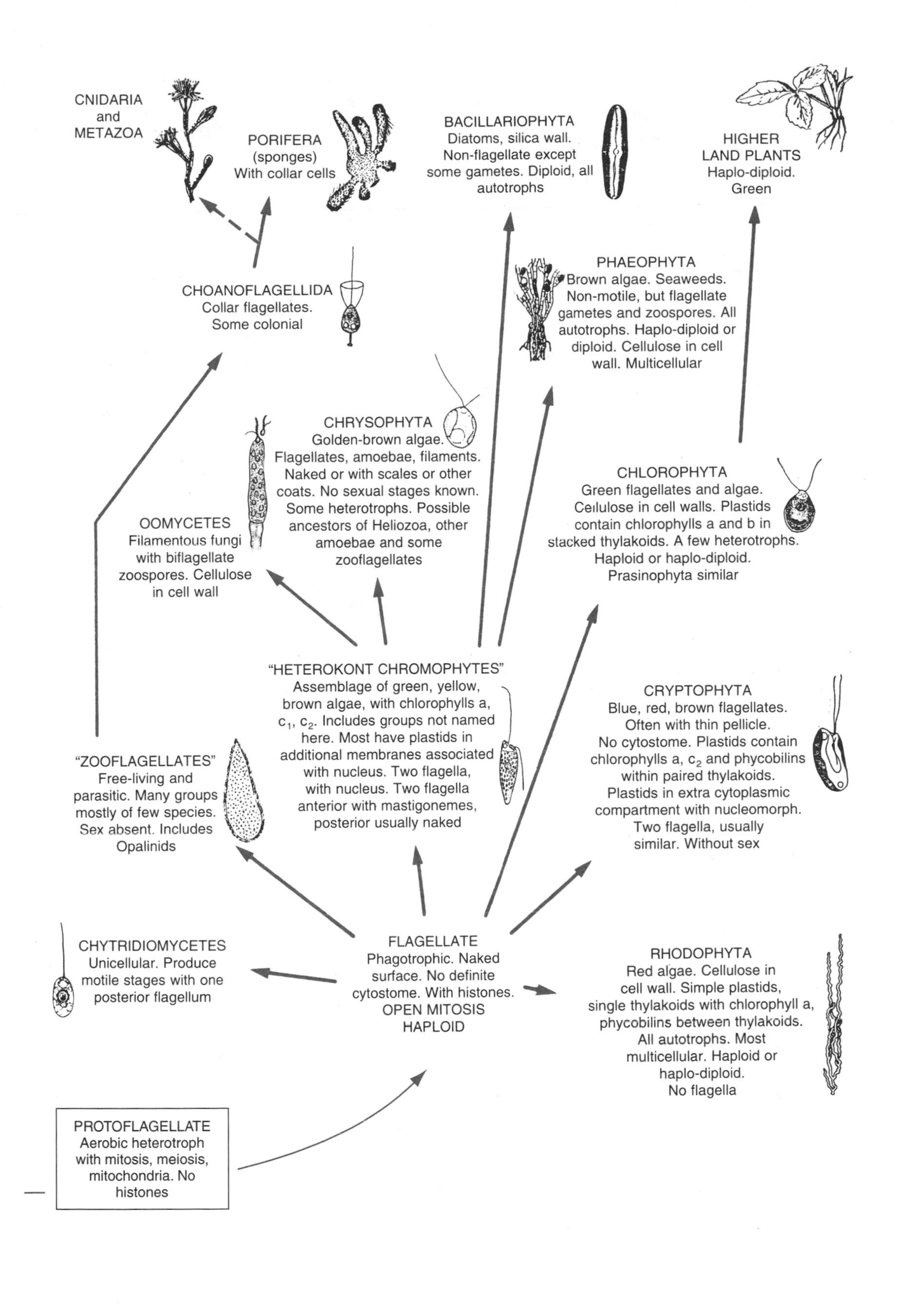
CNIDARIA
and
METAZOA
PORIFERA
(sponges)
With collar cells
BACILLARIOPHYTA
Diatoms, silica wall.
Non-flagellate except
some gametes. Diploid, all
autotrophs
HIGHER
LAND PLANTS
Haplo-diploid.
Green
PHAEOPHYTA
Brown algae. Seaweeds.
Non-motile, but flagellate
gametes and zoospores. All
autotrophs. Haplo-diploid or
diploid. Cellulose in cell
wall. Multicellular
CHOANOFLAGELLIDA
Collar flagellates.
Some colonial
CHRYSOPHYTA
Golden-brown algae.
Flagellates, amoebae, filaments.
Naked or with scales or other
coats. No sexual stages known.
Some heterotrophs. Possible
ancestors of Heliozoa, other
amoebae and some
zooflagellates
CHLOROPHYTA
Green flagellates and algae.
Cellulose in cell walls. Plastids
contain chlorophylls a and b in
stacked thylakoids. A few heterotrophs.
Haploid or haplo-diploid.
Prasinophyta similar
OOMYCETES
Filamentous fungi
with biflagellate
zoospores. Cellulose
in cell wall
"HETEROKONT CHROMOPHYTES"
Assemblage of green, yellow,
brown algae, with chlorophylls a,
c_1, c_2. Includes groups not named
here. Most have plastids in
additional membranes associated
with nucleus. Two flagella,
with nucleus. Two flagella
anterior with mastigonemes,
posterior usually naked
CRYPTOPHYTA
Blue, red, brown flagellates.
Often with thin pellicle.
No cytostome. Plastids contain
chlorophylls a, c_2 and phycobilins
within paired thylakoids.
Plastids in extra cytoplasmic
compartment with nucleomorph.
Two flagella, usually
similar. Without sex
"ZOOFLAGELLATES"
Free-living and
parasitic. Many groups
mostly of few species.
Sex absent. Includes
Opalinids
CHYTRIDIOMYCETES
Unicellular. Produce
motile stages with one
posterior flagellum
FLAGELLATE
Phagotrophic. Naked
surface. No definite
cytostome. With histones.
OPEN MITOSIS
HAPLOID
RHODOPHYTA
Red algae. Cellulose in
cell wall. Simple plastids,
single thylakoids with chlorophyll a,
phycobilins between thylakoids.
All autotrophs. Most
multicellular. Haploid or
haplo-diploid.
No flagella
PROTOFLAGELLATE
Aerobic heterotroph
with mitosis, meiosis,
mitochondria. No
histones

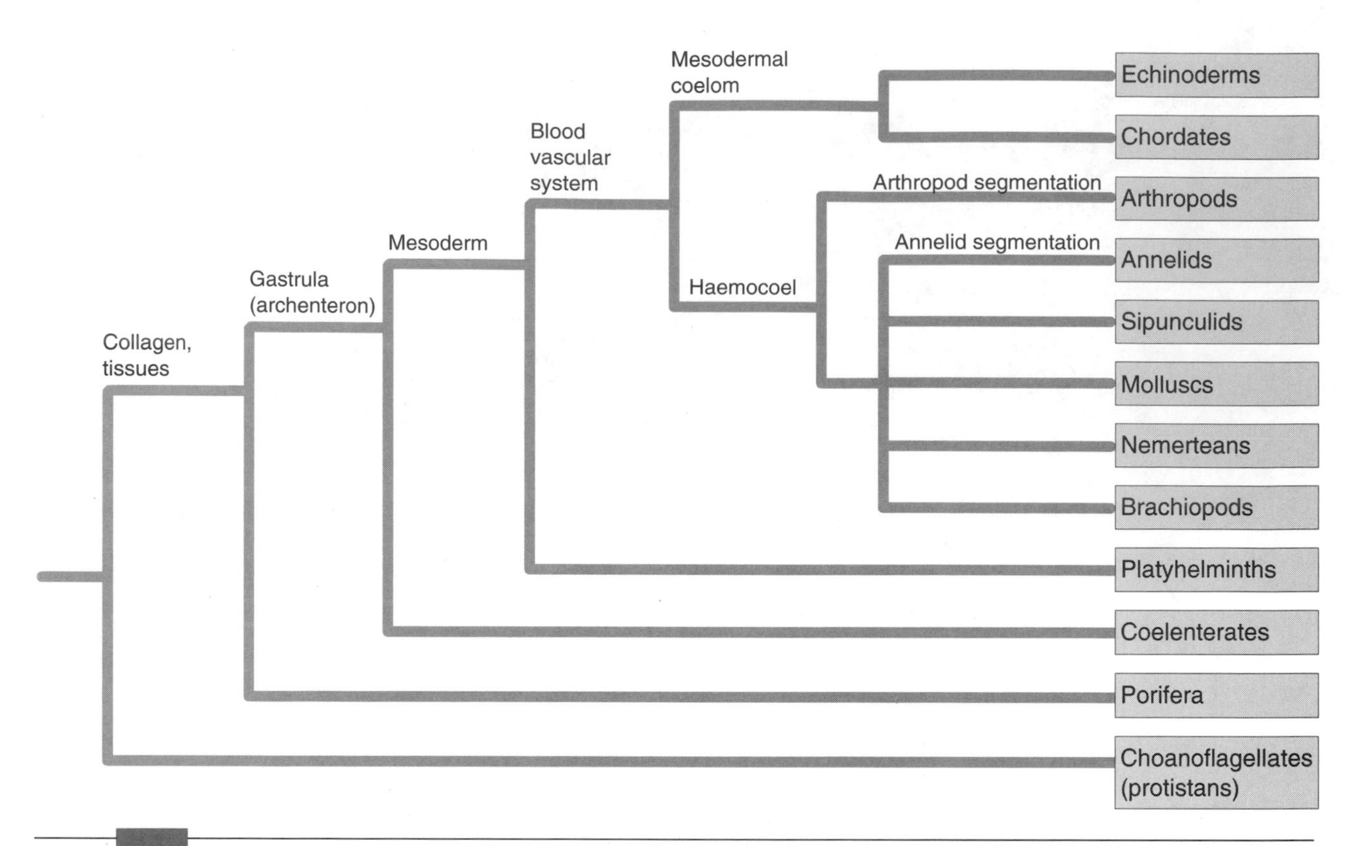

FIGURE 14–5 A metazoan phylogeny derived from a number of studies comparing ribosomal RNA sequences (16*S* as well as 18*S* components) among different phyla. These RNA molecules are essential for ribosomal protein synthesis, and, like 5*S* RNA (p. 276), portions of their sequences are conserved across a very wide range of organisms. Features believed to have been introduced at particular stages in metazoan evolution are placed in the order in which they presumably appeared. As described by Valentine, some of these innovative features appear to be tied to organismic complexity. Briefly: *collagen* = a distinctive fibrous protein used in animal connective tissue; *gastrula* = an embryonic stage in which a cup formed of two tissue layers (ectoderm and endoderm) encloses an inner primitively digestive cavity called the archenteron; *mesoderm* = a third tissue layer formed after gastrulation; *blood vascular system* = vessels used in the transmission of blood fluids; *mesodermal coelom* = a fluid-filled cavity formed in mesoderm tissue; *haemocoel* = a fluid-filled cavity derived from the hollow interior (blastocoele) of the early embryonic blastula; *annelid segmentation* = segments arising in a sequence along the anterior–posterior axis of the embryo, primarily in mesoderm; *arthropod segmentation* = segments arising by successive splitting or doubling of primarily ectodermal units. We do not know the geological dates when these features appeared, but many, or even most, may have originated in the Precambrian. Although not fossil evidence, Bromham and coworkers support this view with molecular data showing that metazoan phyla most probably diversified during an extended Precambrian period (see also Fig. 16–1). (*Based on Valentine.*)

of plants to the task of extruding flexible pseudopods that would enable motion, absorption, and predation. In present-day insectivorous plants, the cellulose cell walls remain intact, no intestinal cavity forms, and digestion takes place entirely on the external surface.

If there is a plant origin for animals, says Hanson, it lies at the unicellular level, in the origin of protozoans from eukaryotic algae such as *Euglena* and Chrysomonads that lost their chloroplasts.

CELLULARIZATION OF A MULTINUCLEATE PROTOZOAN

A hypothesis first suggested in the middle of the nineteenth century and that Hadzi later developed proposes that some early **multinucleate protozoans,** bilaterally organized along the anterior–posterior axis, gave rise to primitive flatworms similar to those in the phylum Platyhelminthes (Fig. 14–6). This evolutionary step occurred through formation of partial or complete plasma membranes around some of the protozoan nuclei, leading to the opportunity for tissue specialization and further enlargement by increase in cellular size and numbers.

This event presumably produced various forms of platyhelminth-like animals, and Hadzi suggested that it is the **acoelan turbellarians** (Fig. 14–6*c* and Fig. 16–6*a*) that most resemble the earliest metazoan ancestor since these bilaterally organized wormlike animals show neither complete cellularization of their digestive tissues, nor gut, nor other body cavity. Further primitive features that seem to link them to protozoans are small size (1 to 2 mm long), ciliated epidermis, ventral mouth, absence of excretory or-

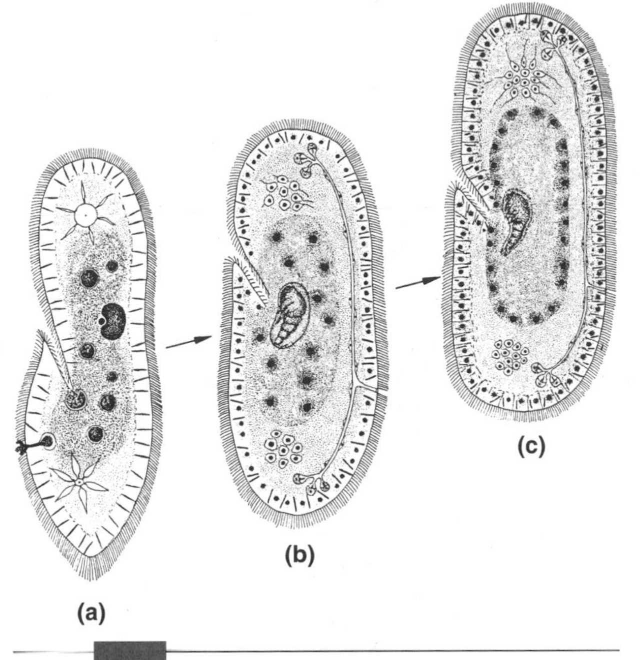

FIGURE 14–6 A hypothetical transformation of a ciliated paramecium-like protozoan (*a*) into a primitive turbellarian metazoan (*c*). (*Adapted from Hadzi.*)

gans, intracellular digestion, and relatively little differentiation (for example, cords of eggs and sperm lying side by side rather than organized into ovaries or testes).

Since one fairly common mode of acoelan nutrition appears highly specialized, being dependent on internal symbiotic algae, zoologists dispute whether these animals were necessarily the most primitive bilaterians. They also dispute Hadzi's proposal that the coelenterates derive from platyhelminths and that coelenterate anthozoans showing traces of bilateral symmetry, such as sea anemones, are ancestral to the radially organized hydrozoans, such as *Hydra*. Instead, most invertebrate zoologists consider **diploblastic** coelenterates (two tissue layers) more primitive than **triploblastic** flatworms (three tissue layers) and, in contrast to Hadzi, derive the relatively complex anthozoa from the simpler hydrozoa.

An attempt to salvage at least part of the hypothesis of protozoan-acoelan evolution was the proposal that coelenterates may have originated from protozoans independently and that coelenterates and flatworms are therefore not directly related. The study by Field and coworkers comparing various regions of 18*S* RNA molecules in a wide variety of eukaryotes (Fig. 14–7) seemed to support

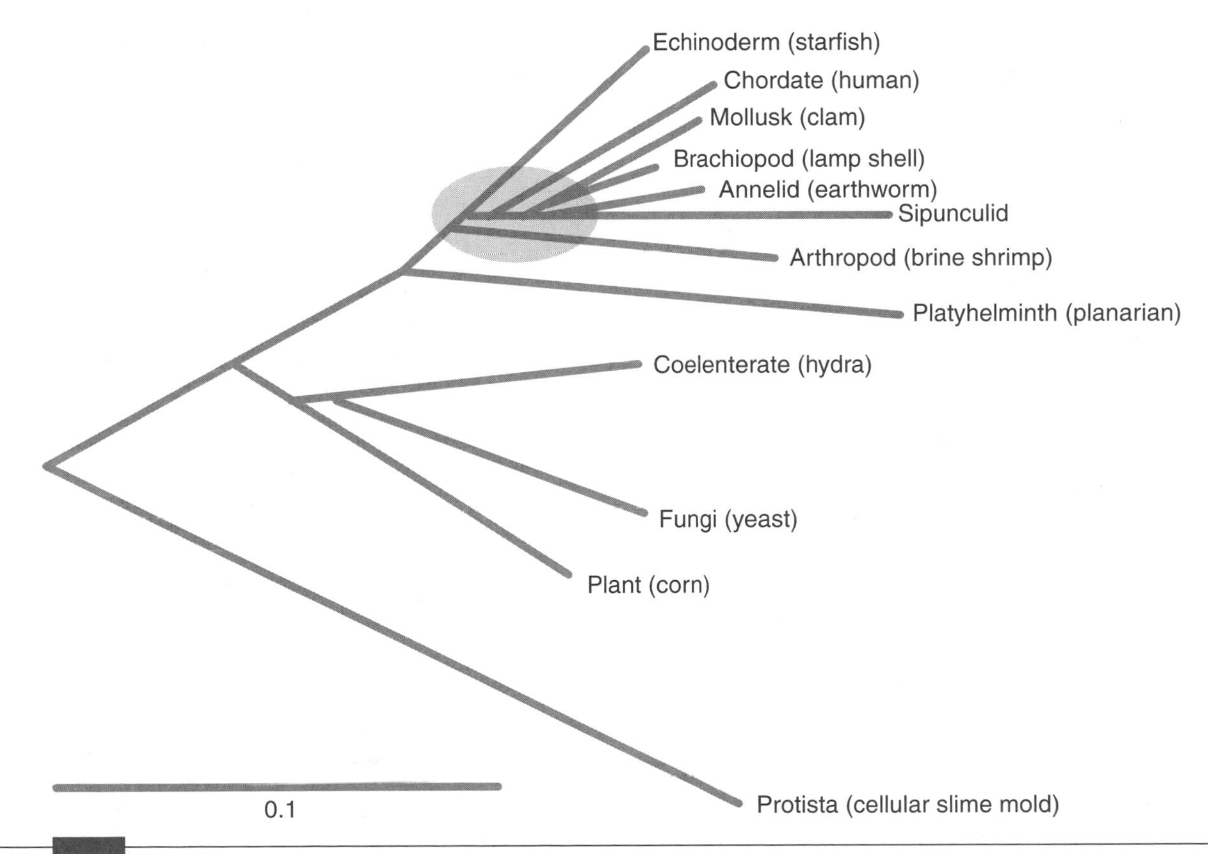

FIGURE 14–7 An evolutionary tree based on comparisons between 18*S* RNA sequences from the different groups shown. As indicated, the platyhelminths and later eucoelomate groups appear to have diverged from a protistan ancestry separate from the protistan ancestry of coelenterates, fungi, and plants. Unfortunately, the data cannot resolve the exact branching order within the *shaded ellipse,* although it seems clear that these different groups all radiated relatively rapidly from what may well have been a common coelomate ancestor that was also metameric. The *scale bar at the lower left* represents an evolutionary distance calculated as 0.1 substitutions for each nucleotide position in the sequences compared (a total of more than 800 nucleotides). (*Adapted from Field et al., Figs. 2 and 5.*)

such a polyphyletic origin of metazoans. According to this study, sequence comparisons show that coelenterates, fungi, and plants derive from a distinctly separate protistan origin from that of other metazoan groups. However, even if we consider that all metazoans have a monophyletic origin (Wainright et al., Borchiellini et al., see also Fig. 14–5), acoelan flatworms may well represent the earliest of bilateral triploblasts. According to molecular studies by Ruiz-Trillo and coworkers, acoelans, more than any other group, stand at the base of all existing triploblastic metazoans. Present difficulty in accepting Hadzi's proposal therefore lies in accepting the origin of a complex flatworm directly from a unicellular protozoan. A less abrupt, more orderly transition from an aggregation of protozoans, as described in the following hypotheses, evinced more interest.

GASTRULATION OF A COLONIAL PROTOZOAN

In the 1870s Haeckel proposed that hollow-balled colonies of flagellated protozoans, not unlike the modern alga *Volvox* (see Fig. 13–1) but lacking chloroplasts, developed an anterior–posterior axis as they swam through primitive waters. Ciliary action swept food particles in this primitive **blastula** (presumably recapitulated in the blastula embryonic stage of many metazoa) toward its posterior pole, and cells at that end specialized for digestive functions.

Haeckel and his followers claimed these digestive cells invaginated through a circular **blastopore** into the hollow interior of the organism to form an internal digestive tract or **archenteron** (Fig. 14–8). They believed this new bilayered, cuplike organism with **ectoderm** on the outside and **endoderm** on the inside was similar to one of the developmental stages in some present-day metazoa, the **gastrula.** The **gastraea hypothesis** suggests that the primitive nature of sponges and coelenterates lies in their persistence at this diploblastic gastrula level.

An extension of the gastraea hypothesis is that an important body cavity of most metazoans, the coelom, originated from lateral pockets formed in the archenteron. Zoologists hold different views on the number of pockets involved, but supporters of this hypothesis agree that coelomic formation allowed the development of a third tissue layer, the **mesoderm,** lying between the ectoderm and endoderm. Further evolution, either gastraeal or from a later stage, the trochaea (see Nielsen), then proceeded to form the various triploblastic phyla.

A major objection to the gastraea hypothesis is that gastrulation by invagination is not common in the embryological development of many metazoans. Even in hydrozoan coelenterates, which Haeckel presumed to exemplify the gastrula stage of evolution, endodermal tissues are formed by ectodermal cells that appear to wander in from an intact epithelial surface, rather than by a cuplike folding process. To the extent that developmental patterns are conserved in evolution (p. 40, also Chapter 15), gastrulation by invagination does not appear to be a primitive pattern among such coelenterates (unless one considers anthozoans ancestral to hydrozoans).

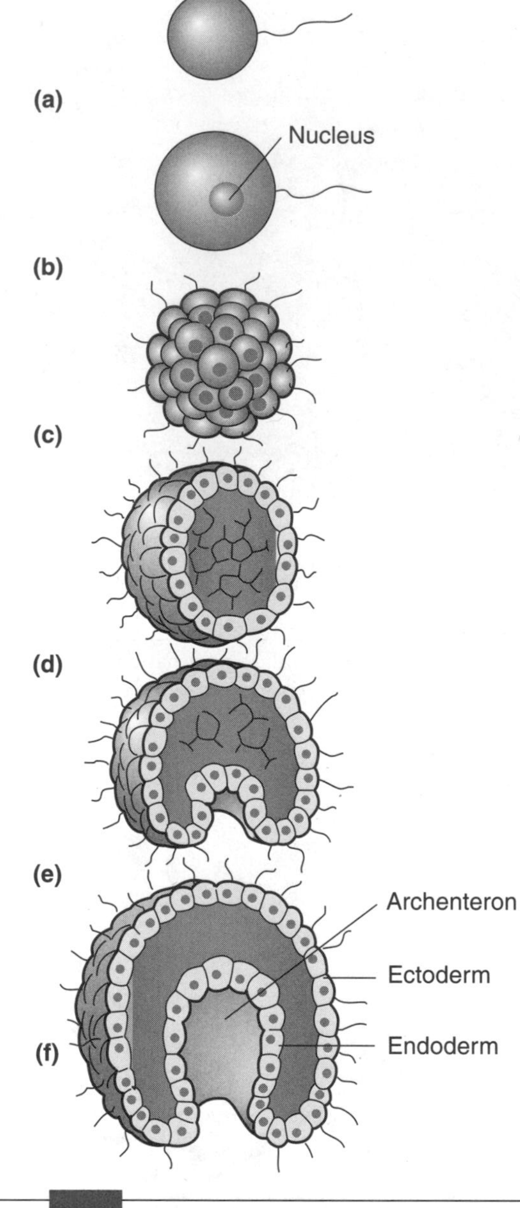

FIGURE 14–8 Stages in the evolution of a multicellular organism according to Haeckel. The monerula (*a*) has no nucleus; the cytula (*b*) is nucleated; the morula, in Haeckel's gastraea hypothesis (*c*) is a compacted solid ball of cells; the blastula (*d*) is a hollow, single-layered cellular sphere; and the gastrula (*e, f*) is a bilayered organism with an exterior opening. (*Adapted from Kerkut, based on Haeckel.*)

Many workers in this field are also unwilling to go along with the implication that the coelomic sacs formed in early gastrula-like organisms gave rise directly to the segmented coelomic cavities found in invertebrate animals such as phoronids and pterobranchs. These segmented coelomates would presumably have been ancestral to acoelomates such as platyhelminths and nemerteans as well as to nonsegmented coelomates such as nematodes and sipunculids. As most zoologists have come to believe that the platyhelminths are more primitive than segmented coelomic animals and that the coelom is probably a feature that arose in fairly large, nonciliated animals to aid in burrowing and swimming (see later discussion), the phylogeny derived from the gastrulation hypothesis is in serious dispute.

PLANULA HYPOTHESIS

A fourth, more popular hypothesis at present is that Haeckel's blastula was followed not by gastrulation but by the formation of a solid ball of cells (**planula**) in which the ectodermal cells specialized for locomotion and the endodermal cells for digestion (Fig. 14–9). As Metschnikoff and others in the nineteenth century showed, many lower metazoans do not use a mouth and

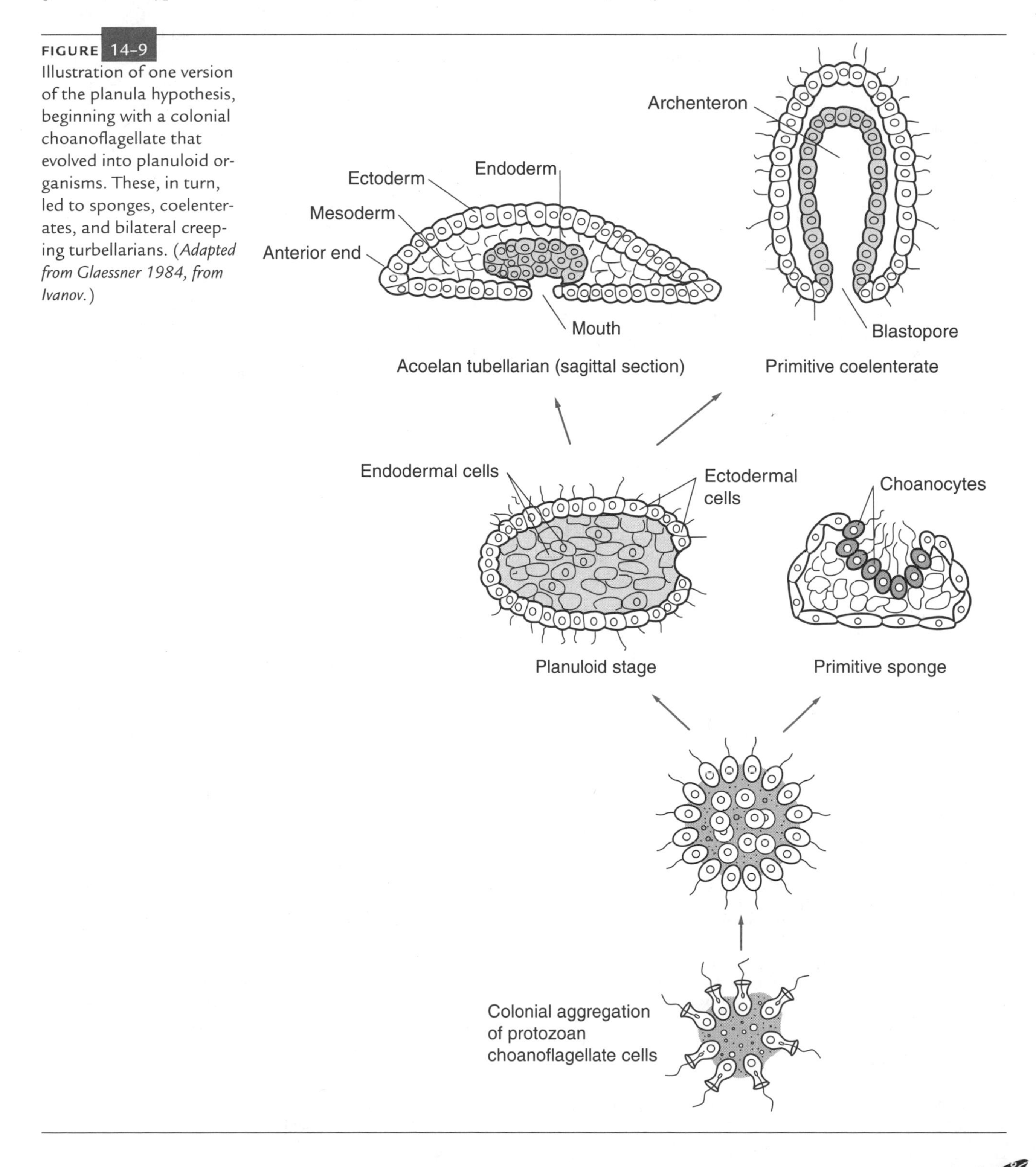

FIGURE 14–9 Illustration of one version of the planula hypothesis, beginning with a colonial choanoflagellate that evolved into planuloid organisms. These, in turn, led to sponges, coelenterates, and bilateral creeping turbellarians. (*Adapted from Glaessner 1984, from Ivanov.*)

digestive tube, since digestion is phagocytic and intracellular. The finding of planula-type larvae in primitive metazoan phyla such as sponges and coelenterates and the observation that various groups among the platyhelminths and pogonophorans have a solid gut filled with endodermal cells indicate that primitive planula-type organisms would have been viable.

According to the planula hypothesis, the formation of a hollow archenteron and open blastopore would have occurred during later evolutionary stages (Fig. 14–9, *upper right*). "Contrary to Haeckel's opinion it is probable that entoderm formation by invagination is a derived rather than the original method, and represents one of those short cuts common in embryology" (Hyman 1940).[6]

Forward from Multicellularity

In spite of their differences, all these hypotheses assume the advantages of multicellularity. A multicellular organism's food-gathering surface increases by extending its cells to places that it could not have reached were it small and unicellular. This increase ensures a more stable food supply to all its cells even where food distribution is uneven, and also allows multicellular organisms to attack and digest larger particles of food by secreting greater quantities of digestive enzymes than single cells can secrete.

Aiding such development are signaling systems (Chapter 15) that direct cells to move, aggregate, divide, and specialize into different tissues ("division of labor"). Multicellularity, based on increased gene numbers and regulatory pathways, provides morphological and functional innovations that broaden the scope of protection, dispersion, food gathering, reproduction, excretion, and other functions.[7]

According to some of these views, early metazoans were **pelagic** animals swimming above the sea bottom mainly by ciliary motion. At a subsequent evolutionary stage some became **benthic,** crawling along the ocean floor and feeding on accumulated detritus. Writers such as Clark have proposed that a number of further evolutionary steps would inevitably accompany a benthic existence. The scattering of food sources would give a selective advantage to organisms that could eat more food more rapidly, leading to an increase in size and to evolution of a mouth and gut that would permit selective digestion. Ciliary motion, by its nature slow and cumbersome for a large animal, would (as discussed later) give way to leechlike and pedal locomotion using circular and longitudinal muscles.

The increased success and proliferation of bottom feeders would open a niche for carnivorous animals that would emphasize speed of locomotion and development of a grasping mouth or other prehensile organs. However, just as improvements are selected in predators, means of defense and escape would be selected in prey, causing an arms race. Competition among all the different varieties of prey and predators would lead to an explosive evolutionary radiation, generating a large variety of morphological forms and adaptive strategies.

The Coelom

The **coelom,** an internal cavity between the ectodermal and endodermal tissues (as Haeckel defined it in 1872), was one of the most successful of the early metazoan adaptations. It is lined with an epithelium that often contains testes or ovaries and has ducts to the exterior used to transmit gametes or waste products. As Figure 14–10 shows, the "coelomate" term is used for two types of phyla: (1) **pseudocoelomates** (false coelomates), in which the body cavity (also called haemocoele) derives from a persistent blastocoele and is only partially lined with mesoderm, and (2) **eucoelomates** (true coelomates), in which the coelom arises as a cavity within mesodermal tissue and is completely lined with mesoderm.

In both pseudocoelomate and eucoelomate organisms, the body cavity is filled with fluid, enabling it, among other functions, to act as a **hydrostatic skeleton** that can transmit pressure from one part to another. Thus, the efficiency of peristaltic motions in coelomate animals is considerably better than in noncoelomates, because waves of circular and longitudinal muscle contraction can be transmitted more easily through the hydrostatic skeleton (see p. 339). This dynamic flow enables undulatory swimming movements as well as improved burrowing activity—that is, improvement in both speed of capture and speed of escape.

---

[6] A recent proposal by Collins based on extensive analysis of 18*S* ribosomal RNA sequence suggests that planula-type larvae produced by ancestral cnidarians became reproductive before reaching the adult stage. This developmental change, called "paedomorphosis" (the incorporation of adult features into immature stages), enabled these larvae to depart from cnidarian radial development, elaborate mesodermal tissue, and become bilateral (see Fig. 16–1).

[7] Given multicellularity's many advantages, we can question why vulnerable unicellular stages, such as the zygote, persist in sexual organisms. Grosberg and Strathmann discuss two explanations:

1. Deleterious mutations carried in a single cell are more easily eliminated by selection than when spread out among many different cells, thereby efficiently reducing "mutational load."
2. Parasitic elements, such as pathogens and cancer cells, which replicate cellularly within a host organism are eliminated in the unicellular zygotic stage.

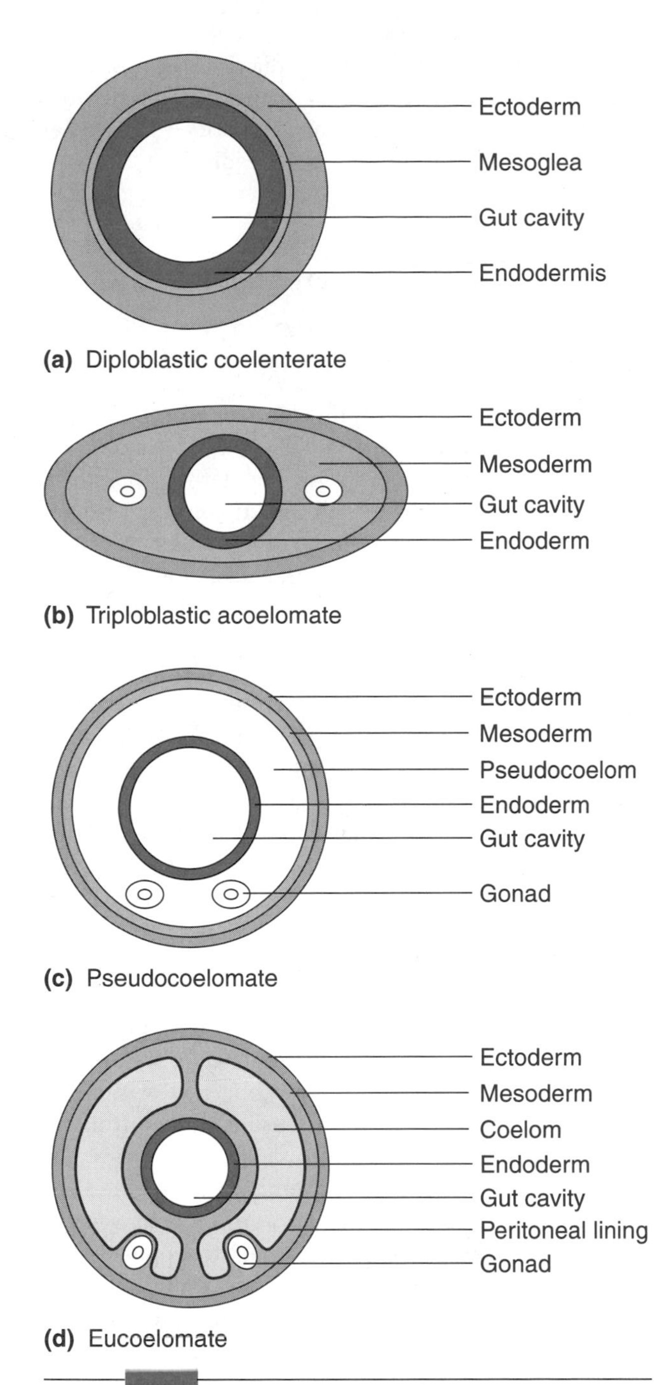

FIGURE 14–10 Diagrammatic illustrations of general kinds of metazoan body cavities. (*a*) Diploblastic body plan in coelenterates such as *Hydra*. (*b*) Triploblastic plan in which the coelom is absent (acoelomate), as it appears in platyhelminths. (*c*) Pseudocoelomate plan in various aschelminth phyla in which the body cavity is only partially lined with mesoderm. (*d*) Eucoelomate body plan in phyla such as arthropods, annelids, chordates, and echinoderms.

For the eucoelomates, biologists have not solved the problem of how the coelom originated and have debated it since Haeckel's time. Hypotheses range from coelomic-type pouches in the gonads of the coelomic ancestor (gonocoel hypothesis) to sacs within nephridial excretory organs (nephrocoel hypothesis) to outpocketings from the gastric cavity (enterocoel hypothesis) to intercellular fluid-filled cavities within mesodermal structures (schizocoel hypothesis). Clark discusses arguments for and against each of these hypotheses, and evidence exists that at least the last two modes of development (**enterocoely, schizocoely**) occur in coelomate phyla, supporting belief in a polyphyletic coelomate origin (Valentine).

In whatever manner it originated, the coelom conferred an important advantage in some lineages by providing a mechanical hydrostatic function. As these lineages evolved and diverged, some made use of the coelom in unsegmented form (for example, priapulids), whereas others adopted various segmental organizations (see Table 16–1) that had profound effects on their future evolution.

Metamerism

The serial **segmentation** of the body along an anterior–posterior axis, also called **metamerism,** appears in a variety of metazoan phyla, including some coelenterates, platyhelminths, annelids, arthropods, chordates, and others. Organ systems such as nephridia, gonads, and nerve ganglia often repeat within each segment, or metamere, and the segments are commonly marked by constrictions of the body wall musculature and by the repetition of coelomic cavities. Although no existing animals have identical segments throughout—the head and anal segments differ from other metameres in all known cases—some animals such as the polychaete annelids show remarkable identity among many of their segments.

Depending on the tissues involved, biologists usually consider segmentation development to be of two general kinds: (1) *mesodermal,* beginning in the mesoderm and proceeding from the interior of the animal outward, such as in annelids, arthropods, and chordates; and (2) *superficial,* which begins externally from the cuticular surface and then proceeds inwardly, often involving only the body wall musculature, as found in the Acanthocephala and other Aschelminthes phyla.

As with the coelom, biologists have offered a variety of hypotheses to explain the adaptiveness of metamerism. For example, one hypothesis points to the fact that many internal organs such as nerve ganglia, gonads, and excretory organs are serially repeated in some pseudometameric animals in which body wall segmentation is absent (platyhelminth turbellarians and nemerteans). The hypothesis suggests that the development of metameric organization in such animals would allow simple organ replacement if the animal is injured; that is, a nearby intact organ could replicate itself and thereby replace an adjacent injured or missing organ. However, the advantage of pseudometamerism might be to provide multiple excretory organs to an animal with an inefficient circulatory system.

Another hypothesis proposes that the uniformity of mesodermal growth along the longitudinal axis may be broken for various embryological reasons, such as the introduction of a pulsating pattern, and this could lead to metamerism. Still another hypothesis suggests that the segmentation of muscular tissue originated from improved undulatory swimming motions conferred on a flexible animal.

So far, there are arguments against each of these hypotheses (Clark), and they all suffer from a fossil record that has provided little information on the evolution of either coelom or metamerism. We don't know whether these structures evolved separately or together, and biologists debate whether these were monophyletic or polyphyletic events. If we follow the molecular phylogeny presented by Valentine (Fig. 14–5), initiation of annelid and arthropod segmentation were separate events. However, we cannot always clearly interpret the embryological and anatomical evidence, and considerable ambiguity remains.

Among the more promising modern studies are those which reveal development of segmentation patterns by tracing individual cell lineages and their relationships with other cells. These studies, which primarily began with *Drosophila,* use genetic techniques that label early embryonic cells as well as molecular techniques that analyze DNA and its transcriptional products, and evaluate developmental changes caused by mutation. Thus, as discussed in Chapter 15, we know that specific genes exercise control over segmental patterns, segmental borders, and the ability of segmental tissues to differentiate into particular structures such as wings and legs (see also Lawrence). Moreover, as Chapter 15 also discusses, genes that govern these and other developmental patterns in a wide range of animals, segmental as well as nonsegmental, share homologous nucleotide sequences, similar linkage orders, and even similar developmental targets. Successful genetic analysis of development thus shows promise of providing a deep understanding of what developmental *changes* occur, how such *changes* function, and which *changes* may be selected and transmitted during evolution.

Postponing for the present the discussion of development and evolution, a somewhat different approach toward gaining an understanding of at least some aspects of the evolution of various body structures is to study their selective advantages in phenotypic terms of locomotion and function. Such studies, although theoretical, are applicable to a wide variety of organisms that face similar environments and occupy similar roles within these environments. Using this approach, we can consider a body form that was probably at the base of all major forms of metazoan evolution—a worm. Recent findings of Precambrian wormlike burrows in India raise the possibility that such triploblastic animals originated long before their diversification in the Cambrian period (Seilacher et al.). What is a worm, and how did it evolve?

Evolutionary Solutions to Problems of Locomotion

The task of obtaining food for animals is inextricably bound with a variety of adaptations: sensory, locomotory, ingestatory, and others that support and enhance fulfillment of this primary need. Among these adaptations, differences in locomotory behavior have provided workers such as Clark the opportunity to examine some basic concepts of adaptive change.

On the unicellular level, small size enables locomotion through relatively simple ciliary, flagellar, pseudopodial, or even Brownian motion. Once the metazoan grade of organization is reached, locomotory cells face the problem of moving relatively large masses in concerted activity. The earliest of metazoan animals, perhaps planula-like organisms, probably moved by ciliary activity not unlike the motion of the small acoelan turbellarians. The acoelans, many of a size no larger than a millimeter, use ciliary cells on their ventral surface for creeping, and some degree of swimming can also be attained this way. Motion in a directional fashion quickly confers an anterior–posterior orientation on the animal, making the anterior portion more concerned with those adaptations necessary for both sensing and confronting the environment being entered.

Bilaterality, or the distinction between right and left sides, is an immediate consequence of a dorso–ventral, anterior–posterior anatomy and leads to opportunities for organizational complexity. However, as animals grow larger, the use of cilia alone limits more rapid locomotion because of the relatively small forces cilia can generate (Fig. 14–11), and a range of other methods are employed, all dependent on organized muscular tissues.

In its simplest form, tissue organization in triploblastic animals takes the shape of a **worm,** which can be defined as a long, flexible tube of constant volume enclosed by a muscular body wall. To allow coordinated activity, the muscle tissue is organized into two major groups: circular muscles whose contraction reduces the diameter of the animal and increases its length, and longitudinal muscles that contract with opposite effect by reducing length and increasing diameter. The opposed activity of these muscle tissues means essentially that for one type of muscle to extend, the other must contract.

Also, the impact of a localized muscle contraction depends strongly on what other muscles do: if a circular mus-

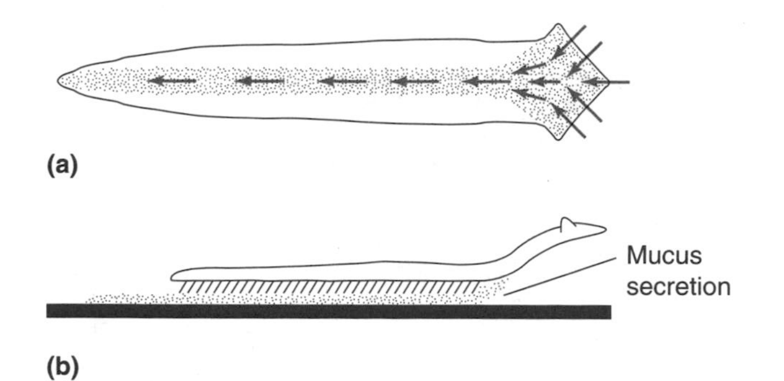

FIGURE 14–11 Ciliary distribution and locomotion in a platyhelminth turbellarian flatworm, *Planaria.* (*a*) Ventral surface of *Planaria* showing the direction of ciliary beats. (*b*) Mode of creeping by means of ciliary beating on a mucus secretion deposited on the substratum. The efficiency of such ciliary creeping generally depends on the relatively small size and flattening of the turbellarian body to present as large a ventral surface as possible. In platyhelminths with larger and more circular dimensions, ciliary creeping is mostly, if not entirely, abandoned. (*Adapted from Clark, from other sources.*)

cle contracts, then a neighboring area will expand unless its circular muscles also contract. If a longitudinal muscle contracts, then the animal will flex in that direction if the longitudinal muscles on the other side of the body relax. Limitations on the extent of movement depend on the size of the muscles, their locations, attachments, and the degree to which the body wall can distort (for example, the deformability of the cuticular basement membrane).

In many platyhelminth turbellarians that have reached sizes much larger than the acoelans, locomotory movements are almost entirely transmitted through a pedal longitudinal "foot." Pedal locomotory waves arise by contraction and relaxation of those ventral longitudinal muscles that contact the surface. This somewhat inefficient creeping mechanism allows the locomotion of animals whose bodies are mostly solidly filled with cells, since muscular effects on the body wall are restricted to relatively short distances.

As fluid accumulates in either cells or sinuses within the wormlike body, effects of muscular changes can be transmitted through greater distances and improve locomotion. For example, although a coelom is lacking in the ribbon worms (phylum Nemertea), some of these animals have a gelatinous parenchyma letting the effects of muscular contractions transfer more easily than through solid tissue. Undulatory swimming movements can then occur, produced by contraction of longitudinal muscles on opposite sides of the body. Furthermore, in some nemerteans we see the early signs of **peristaltic movements** that become an important feature of animals possessing a true fluid-filled coelom.

Essentially, the coelomate condition of a continuous body cavity represents a significant evolutionary advance in providing a fluid skeleton that eliminates cellular bar-

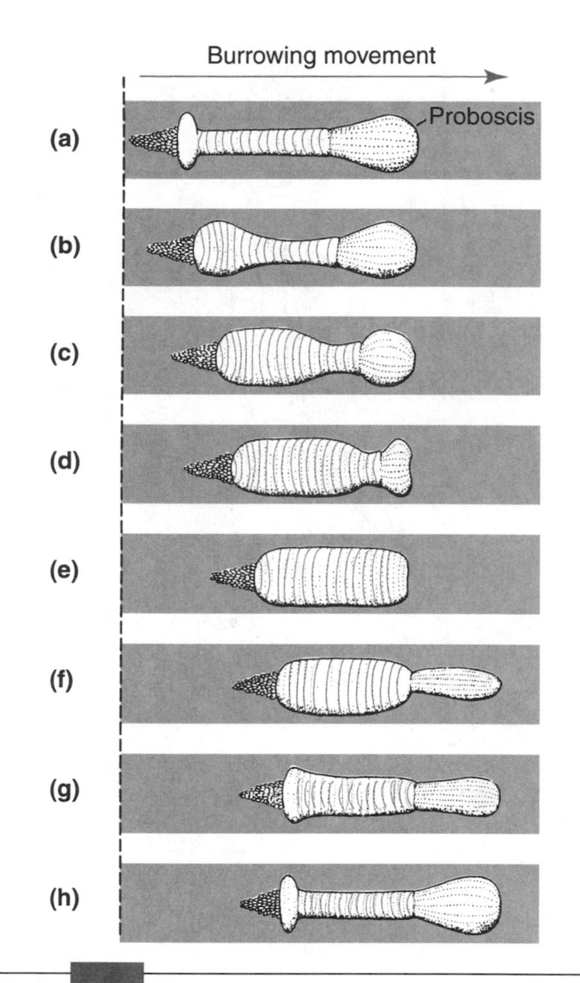

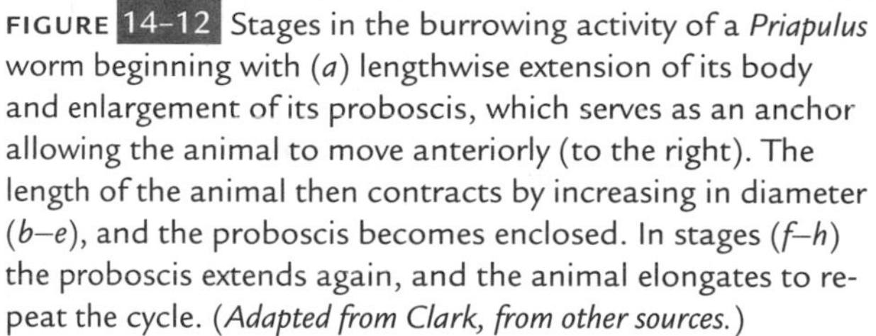

FIGURE 14–12 Stages in the burrowing activity of a *Priapulus* worm beginning with (*a*) lengthwise extension of its body and enlargement of its proboscis, which serves as an anchor allowing the animal to move anteriorly (to the right). The length of the animal then contracts by increasing in diameter (*b–e*), and the proboscis becomes enclosed. In stages (*f–h*) the proboscis extends again, and the animal elongates to repeat the cycle. (*Adapted from Clark, from other sources.*)

riers to hydrostatic pressure. This feature lets the effects of contractions in one part of the body immediately transfer to other parts. Peristaltic motion—a wave of circular muscle contraction followed by longitudinal muscle contraction—can generate much larger forces than in acoelomates because the entire coelomic hydrostatic skeleton and all the body wall musculature is involved.

For example, peristalsis adds adaptive value to a burrowing animal by enabling it to use the entire circumference of the body in thrusting through the substrate. Furthermore, a fluid-filled coelom also allows the rapid eversion of a proboscis or lophophore by simple hydrostatic pressure, as you can see in the rapid burrowing movements of priapulids (Fig. 14–12) and the extension and withdrawal of the tentacular polypide in animals such as ectoprocts (Fig. 14–13).

Despite its advantages, a large coelom has the disadvantage that sustained peristaltic movement is not localized

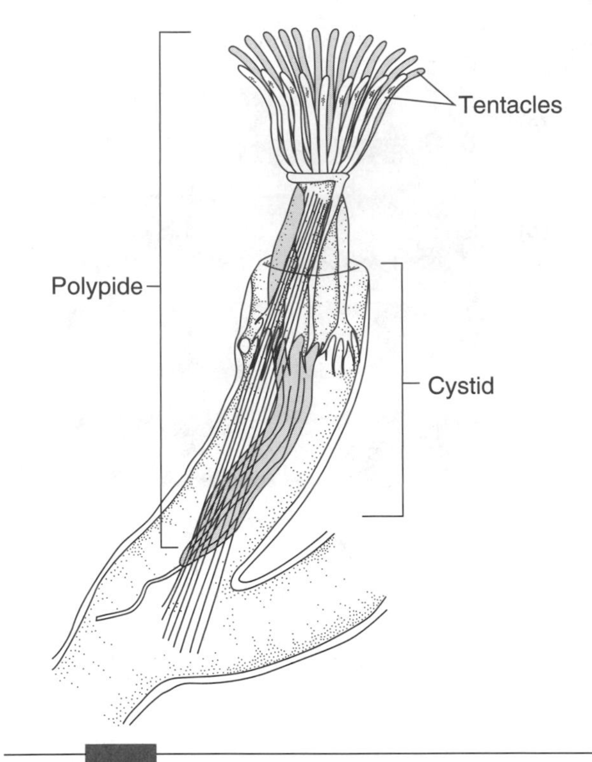

FIGURE 14–13 The ectoproct, *Fredericella sultana*, with everted polypide. (*Adapted from Clark, from other sources.*)

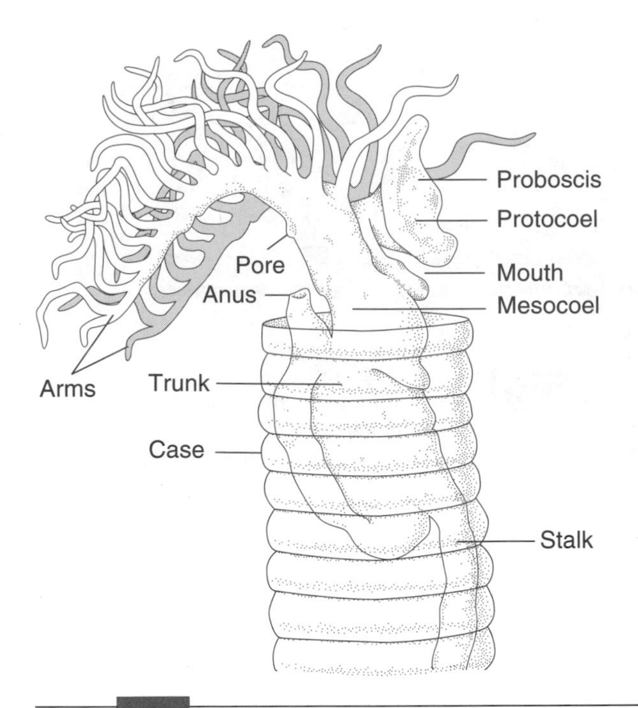

FIGURE 14–14 *Rhabdopleura,* a species of colony formers (pterobranchs) that zoologists believe are related evolutionarily to chordates. They are members of the phylum Hemichordata, a probable polyphyletic group of animals. These individuals have three coelomic regions: the proboscis, or cephalic shield contains the protocoel; the collar has a pair of mesocoels that extend into the tentacle-bearing arms; and the trunk contains the paired metacoel cavities. (*Adapted from Borradaile et al.*)

but involves the entire musculature, and pressures transmit even to those parts that are changing in a different direction. This causes the body wall muscles to operate at either relatively low coelomic pressure to prevent fatigue or at high coelomic pressure for only short periods of time. Unsegmented coelomic worms such as sipunculids and echiuroids that cannot localize their hydrostatic pressures are generally slow-moving and relatively sedentary. Segmentation of the body by metamerism is a mechanism that allows the localized establishment of pressure gradients and overcomes the generalized coelomic pressures that affect the entire musculature simultaneously.

Selection for segmentation of the coelom seems to have taken at least two major directions. One was toward the **oligomerous** animals, which have three main coelomic areas, the most anterior being an unpaired pocket, the protocoel, followed by a pair of mesocoels and a posterior pair of metacoels. Hemichordates, such as *Balanoglossus* (the acorn worm) and the pterobranchs (Fig. 14–14), as well as echinoderms, have all three coelomic sacs to various degrees, whereas the protocoels mostly disappear in phoronids, ectoprocts, and brachiopods. In another evolutionary direction are metameric animals such as annelids, which have multiple body wall divisions that provide a smooth transition of the peristaltic wave. The sustained and efficient burrowing by oligochaete annelids such as earthworms is a direct consequence of their numerous segments (Fig. 14–15).

Once segmentation appeared, further locomotory adaptations rapidly evolved. In many of the polychaete class of annelids, motion occurs primarily by means of oarlike **parapodia** (Fig. 14–16), with consequent reduction in the circular muscles. The polychaete septa provide rigid attachment points for the parapodial muscles, enabling turgor in the parapodium so the animal can move it as a single unit.

Nevertheless, despite its locomotory advantages, segmentation has drawbacks, since each segment, separated from its neighbor by a septum, must have its own set of organs, such as nerve ganglia, nephridia, gonads, and musculature. So the numbers and kinds of segments, as well as the coelomic cavities, inevitably become modified or reduced with changes in habit or function. For example, although arthropods originated from wormlike, segmented ancestors, their partitioned skeleton has become rigid, and localized muscular movement can now occur in the absence of either coelom or septa.[8]

[8]In vertebrates, a group that also has a metameric body plan (Chapter 17), the selective value of segmentation appears to derive from their increased swimming efficiency, given an axial skeleton. That is, vertebrates with an axial notochord can best achieve undulatory swimming movements by applying longitudinal muscular contractions to small sections of the axial skeleton. In early chordates whose spinal columns were most likely similar to the notochords of tunicate larvae (see Figure

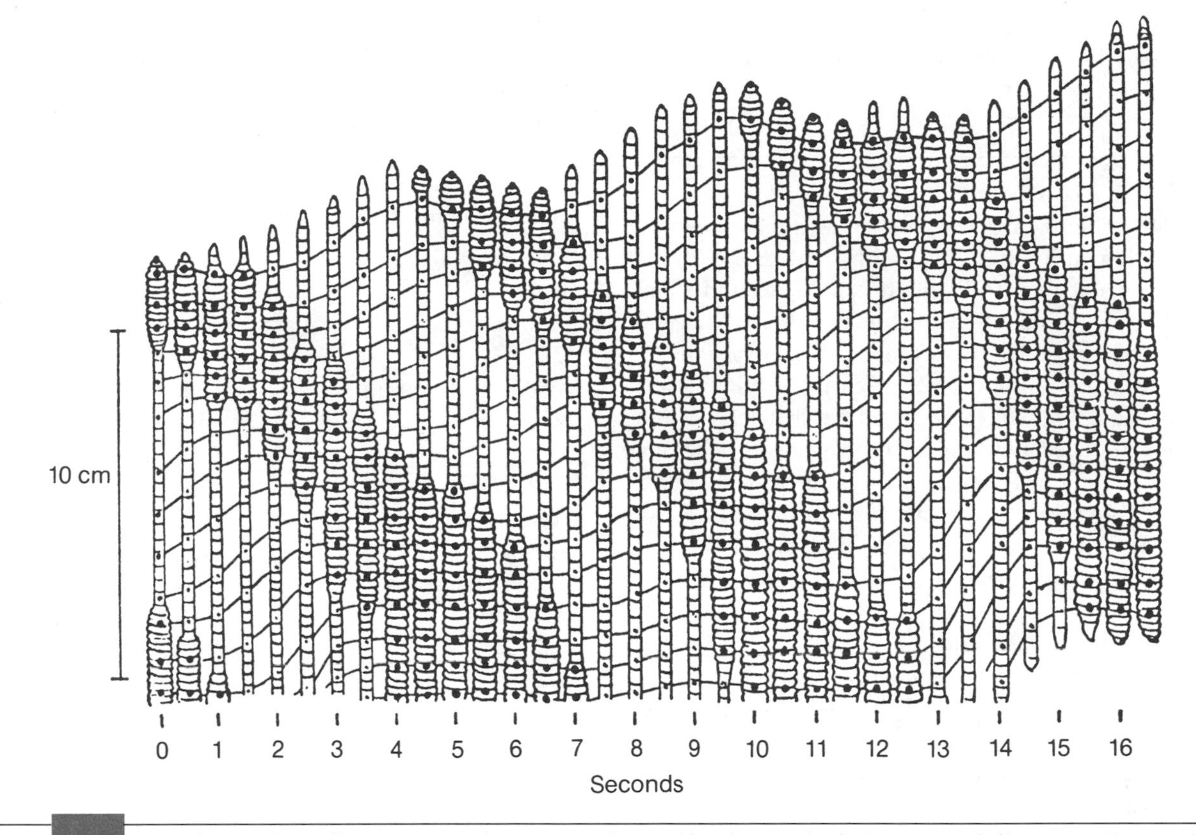

FIGURE 14–15 Peristaltic motion of burrowing earthworms as observed by changes in their segmental diameters. Contracting longitudinal muscles in a group of segments widens these segments and enables the animal to wedge against the sides of the burrow. Segments behind this group are then pulled up by further longitudinal muscle contraction, shortening the body and increasing the number of segments in the "anchor." Some of the widened segments in the anchor undergo circular muscle contraction, elongating the body and extending these segments in a forward direction. These elongated segments contract and widen in turn, and the peristaltic cycle repeats. Connecting lines indicate the relative motion of particular segments. (*Adapted from Clark, from Gray and Lissmann.*)

17–7), these muscles probably inserted directly on the notochord itself. This was then followed by segmentation of the spine through the occurrence of vertical septa at repeated intervals (as seen in the ammocoete larva of the lamprey *Petromyzon*), since such organization would provide increased mechanical advantage for longitudinal muscles. At some point in the evolution of fish, these vertical septa were replaced or transformed into inclined septa (myocommata), which allow transmission of both longitudinal and lateral forces to the long axis of the animal. Because of this evolutionary history it is important to note that the segmentation the myocommata produced does not exactly correspond to the segmentation of the axial skeleton (the vertebrae); to enable smooth locomotory movement each myocomma has insertions on a number of adjacent vertebrae, and each vertebra bears insertions from two or more myocommata.

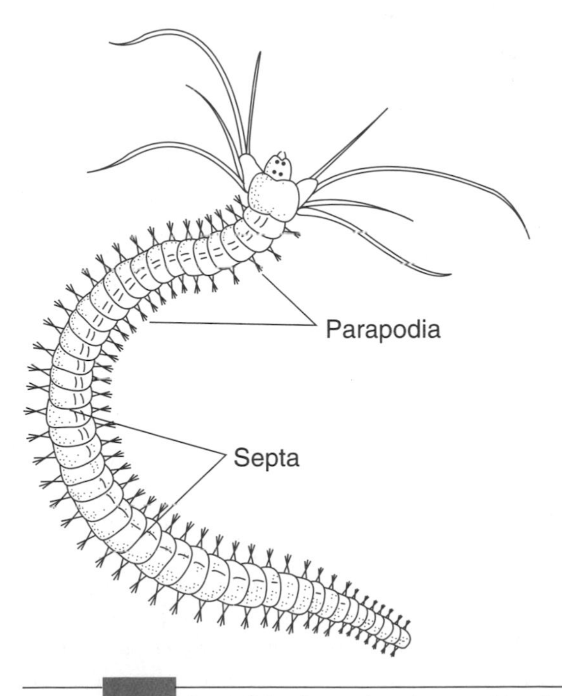

FIGURE 14–16 One of the free-swimming nereid polychaete annelids, *Platynereis*. (*Adapted from Smith.*)

Evolution among a variety of invertebrates, ranging from acoelomates to coelomates, will be discussed in Chapter 16.

SUMMARY

About 545 million years ago there was an apparent explosive radiation of animal phyla with hard exoskeletons, but why it occurred is still a matter of debate, and many proposals have been offered. One hypothesis suggests that prior to this time a homogeneous assemblage of autotrophs may have dominated the earth, but significant evolution could not have occurred until heterotrophic "croppers" opened new niches by preying on the autotroph population. From herbivores to carnivores was then only a small succession of steps. For protection, organisms developed the exoskeletons now found as Cambrian fossils, but evidence exists that their soft-bodied precursors arose much earlier.

All unicellular eukaryotes, including mainly photosynthetic algae, the heterotrophic protozoans, and a few fungi, are classified as Protista. They link prokaryotes and the kingdoms of multicellular organisms. Among several possible pathways by which heterotrophic protistans became multicellular animals, the two most seriously considered are:

1. A proposal by Haeckel that flagellated colonial protistans became bilaterally symmetrical and developed a gut by means of invagination of "digestive" cells, thus producing a gastrula-like structure, the gastraea.
2. That a solid ball of cells, the planula, differentiated into interior digestive cells and exterior locomotory cells, as occurs in some lower metazoan larvae.

However it arose, multicellularity was highly advantageous and permitted cell specialization and more efficient food gathering.

The coelom, the body cavity, is of uncertain origin and acts in many invertebrates as a hydraulic skeleton. Some animals have a body divided into segments (metameres) in which many organ systems serially repeat. The evolutionary source of segmentation is still unknown, and hypotheses offered to explain these events range from ease of organ replacement if segments are lost to improved flexibility in swimming.

Originally small ciliated creatures, metazoans had to make many adaptations for movement when they enlarged and became bilaterally symmetrical. The early, three-layered metazoans had a wormlike shape, whose movement depended on coordinated layers of circular and longitudinal muscles. With the development of a fluid-filled body cavity, these muscles could rapidly transmit peristaltic waves over the entire length of the body, enabling effective burrowing activity. Segmented animals can apportion hydrostatic pressure more efficiently to localized areas. Few animals are completely metameric, however, because it becomes necessary to provide a set of organs for each segment. Further adaptations, especially in hard-bodied forms, led to changes in body plan, and both coelom and segments have become modified in most animal phyla.

KEY TERMS

acoelan turbellarians
archenteron
benthic
bilaterality
blastopore
blastula
Cambrian explosion
carnivores
cilia
coelom
cropping
diploblastic
ectoderm
Ediacaran strata
endoderm
enterocoely
eucoelomates
feedback cycle
flagella
gastraea hypothesis
gastrula
herbivores
hydrostatic skeleton
mesoderm
metamerism
metaphyta
metazoans
multinucleate protozoans
oligomerous
parapodia
pelagic
peristaltic movements
planula hypothesis
Protista
protoflagellate
pseudocoelomates
segmentation
schizocoely
triploblastic
worm

DISCUSSION QUESTIONS

1. How do paleobiologists use the principle of "cropping" to explain the proliferation of multicellular animals during the time around the Cambrian period?
2. Would you classify all Precambrian multicellular organisms into the same phyla used for classifying organisms from the Cambrian onward?
3. Metazoan origins
   a. What are the advantages of animal multicellularity?
   b. What are the arguments, pro and con, for the various hypotheses of metazoan origin?
4. How would the development of a coelomic cavity have influenced the evolution of locomotion in wormlike organisms?
5. What advantages does metamerism (segmentation) offer to metazoans?

EVOLUTION ON THE WEB

Explore evolution on the web! Visit the accompanying web site for *Evolution,* 3/e at www.jbpub.com/evolution for web exercises and links relating to topics covered in this chapter.

REFERENCES

Ayala, F. J., A. Rzhetsky, and F. J. Ayala, 1998. Origin of the metazoan phyla: Molecular clocks confirm paleontological estimates. *Proc. Nat. Acad. Sci.,* **95,** 606–611.

Barnes, R. D., 1985. Current perspectives on the origins and relationships of lower invertebrates. In *The Origins and Relationships of Lower Invertebrates,* S. Conway Morris, J. D. George, R. Gibson, and H. M. Platt (eds.). Clarendon Press, Oxford, England, pp. 360–367.

Berkner, L. V., and L. C. Marshall, 1965. On the origin of oxygen concentration in the earth's atmosphere. *J. Atmosph. Sci.,* **22,** 225–261.

Borchiellini, C., N. Boury-Esnault, J. Vacelet, and Y. Le Parco, 1998. Phylogenetic analysis of the Hsp70 sequences reveals the monophyly of metazoa and specific phylogenetic relationships between animals and fungi. *Mol. Biol. and Evol.,* **15,** 647–655.

Borradaile, L. A., F. A. Potts, L. E. S. Eastham, and J. T. Saunders, 1959. *The Invertebrata,* 3d ed., revised by G. A. Kerkut. Cambridge University Press, Cambridge, England.

Brasier, M. D., 1979. The Cambrian radiation event. In *The Origin of Major Invertebrate Groups,* M. R. House (ed.). Academic Press, London, pp. 103–159.

Brock, T. D., 1973. Lower pH limit for the existence of blue-green algae: Evolutionary and ecological implications. *Science,* **179,** 480–483.

Bromham, L., A. Rambaut, R. Fortey, A. Cooper, and D. Penny, 1998. Testing the Cambrian explosion hypothesis by using a molecular dating technique. *Proc. Nat. Acad. Sci.,* **95,** 12386–12389.

Butterfield, N. J., 1994. Burgess Shale-type fossils from a Lower Cambrian shallow-shelf sequence in northwestern Canada. *Nature,* **369,** 477–479.

Cavalier-Smith, T., 1993. Kingdom protozoa and its 18 phyla. *Microbiol. Rev.,* **57,** 953–994.

Clark, R. B., 1964. *Dynamics in Metazoan Evolution.* Clarendon Press, Oxford, England.

Cloud, P., 1976. Beginnings of biospheric evolution and their biogeochemical consequences. *Paleobiology,* **2,** 351–357.

Collins, A. G., 1998. Evaluating multiple alternative hypotheses for the origin of Bilateria: An analysis of 18*S* rRNA molecular evidence. *Proc. Nat. Acad. Sci.,* **95,** 15458–15463.

Conway Morris, S., 1993. The fossil record and the early evolution of the metazoa. *Nature,* **361,** 219–225.

———, 1998. *The Crucible of Creation: The Burgess Shale and the Rise of Animals.* Oxford University Press, Oxford, England.

Delwiche, C. F., M. Kuhsel, and J. D. Palmer, 1995. Phylogenetic analysis of *tufA* sequences indicates a cyanobacterial origin of all plastids. *Mol. Phylogenet. and Evol.,* **4,** 110–128.

Dobzhansky, Th., F. J. Ayala, G. L. Stebbins, and J. W. Valentine, 1977. *Evolution.* Freeman, San Francisco.

Dougherty, E. C. (ed.), 1963. *The Lower Metazoa.* University of California Press, Berkeley.

Erwin, D., J. Valentine, and D. Jablonski, 1997. The origin of animal body plans. *Amer. Sci.,* **85,** 126–137.

Fedonkin, M. A., 1994. Vendian body fossils and trace fossils. In *Early Life on Earth,* S. Bengtson (ed.). Columbia University Press, New York, pp. 370–388.

Fenchel, T., and B. J. Finlay, 1994. The evolution of life without oxygen. *Amer. Sci.,* **82,** 22–29.

Field, K. G., G. J. Olsen, D. J. Lane, S. J. Giovannoni, M. T. Ghiselin, E. C. Raff, N. R. Pace, and R. A. Raff, 1988. Molecular phylogeny of the animal kingdom. *Science,* **239,** 748–753.

Glaessner, M. F., 1983. The emergence of metazoa in the early history of life. *Precambrian Res.,* **20,** 427–441.

———, 1984. *The Dawn of Animal Life.* Cambridge University Press, Cambridge, England.

Gray, J., and H. W. Lissmann, 1938. Studies in animal locomotion. VIII. The earthworm. *J. Exp. Biol.,* **15,** 506–517.

Grosberg, R. K., and R. R. Strathmann, 1998. One cell, two cell, red cell, blue cell: The persistence of a unicellular stage in multicellular life histories. *Trends in Ecol. and Evol.,* **13,** 112–116.

Hadzi, J., 1963. *The Evolution of the Metazoa.* Macmillan, New York.

Haeckel, E., 1874. The gastraea-theory, the phylogenetic classification of the animal kingdom and the homology of the germ-lamellae. *Q.J. Micr. Sci.* **14,** 142–165, 223–247.

Han, T. M., and B. Runnegar, 1992. Megascopic algae from the 2.1 billion-year-old Negaunee Iron Formation, Michigan. *Science,* **257,** 232–235.

Hanson, E. D., 1977. *The Origin and Early Evolution of Animals.* Wesleyan University Press, Middletown, CT.

Hardy, A. C., 1953. On the origin of the metazoa. *J. Microscop. Sci.,* **94,** 441–443.

House, M. R. (ed.), 1979. *The Origin of Major Invertebrate Groups.* Academic Press, London.

Hyman, L., 1940. *The Invertebrates: Protozoa Through Ctenophora.* McGraw-Hill, New York.

———, 1951. *The Invertebrates: Platyhelminthes and Rhynchocoela, the Acoelomate Bilateria.* McGraw-Hill, New York.

Ivanov, A. V., 1968. *The Origin of Multicellular Animals* (in Russian). Nauka, Leningrad.

Jagersten, G., 1972. *Evolution of the Metazoan Life Cycle.* Academic Press, London.

Kerkut, G. A., 1960. *Implications of Evolution.* Pergamon Press, Oxford, England.

Kershaw, D. R., 1983. *Animal Diversity.* University Tutorial Press, Slough, Great Britain.

Knoll, A. H., 1992. The early evolution of eukaryotes: A geological perspective. *Science,* **256,** 622–627.

Lawrence, P. A., 1992. *The Making of a Fly: The Genetics of Animal Design.* Blackwell Scientific, Oxford, England.

Margulis, L., J. O. Corliss, M. Melkonian, and D. J. Chapman (eds.), 1990. *Handbook of Protoctista.* Jones and Bartlett, Boston.

McMenamin, M. A. S., and D. L. S. McMenamin, 1990. *The Emergence of Animals: The Cambrian Breakthrough.* Columbia University Press, New York.

Mendelson, C. V., 1993. Acritarchs and prasinophytes. In *Fossil Prokaryotes and Protists,* J. H. Lipps (ed.). Blackwell Scientific, Boston, pp. 77–104.

Metschnikoff, E., 1884. Researches on the intracellular digestion of invertebrates. *Q.J. Micr. Sci.* **24,** 89–111.

Nielsen, C., 1995. *Animal Evolution: Interrelationships of the Living Phyla.* Oxford University Press, Oxford, England.

Ohno, S., 1996. The notion of the Cambrian pananimalia genome. *Proc. Nat. Acad. Sci.,* **93,** 8475–8478.

Retallack, G. J., 1994. Were the Ediacaran fossils lichens? *Paleobiology,* **20,** 523–544.

Ruiz-Trillo, I., M. Riutort, D. T. J. Littlewood, E. A. Herniou, and J. Baguñà, 1999. Acoel flatworms: Earliest extant bilaterian metazoans, not members of Platyhelminthes. *Science,* **283,** 1919–1923.

Schlegel, M., 1994. Molecular phylogeny of eukaryotes. *Trends in Ecol. and Evol.,* **9,** 330–335.

Seilacher, A., 1989. Vendozoa: Organismic construction in the Proterozoic biosphere. *Lethaia,* **22,** 229–239.

Seilacher, A., P. K. Bose, and F. Pflüger, 1998.Triploblastic animals more than 1 billion years ago: Trace fossils from India. *Science,* **282,** 80–83.

Signor, P. W., and J. H. Lipps, 1992. Origin and early radiation of the metazoa. In *Origin and Early Evolution of the Metazoa,* J. H. Lipps and P. W. Signor (eds.). Plenum Press, New York, pp. 3–23.

Sleigh, M. A., 1979. Radiation of the eukaryote Protista. In *The Origin of Major Invertebrate Groups,* M. R. House (ed.). Academic Press, London, pp. 23–54.

Smith, J. E. (ed.), 1971. *The Invertebrate Panorama.* Universe Books, New York.

Stanley, S. M., 1973. An ecological theory for the sudden origin of multicellular life in the late Precambrian. *Proc. Nat. Acad. Sci.,* **70,** 1486–1489.

Trueman, E. R., 1975. *The Locomotion of Soft-Bodied Animals.* Arnold, London.

Valentine, J. W., 1994. Late Precambrian bilaterians: Grades and clades. *Proc. Nat. Acad. Sci.,* **91,** 6751–6757.

Valentine, J. W., S. M. Awramik, P. W. Signor, and P. M. Sadler, 1991. The biological explosion at the Precambrian–Cambrian boundary. *Evol. Biol.,* **25,** 279–356.

Wainright, P. O., G. Hinkle, M. L. Sogin, and S. K. Stickel, 1993. Monophyletic origins of the metazoa: An evolutionary link with fungi. *Science,* **260,** 340–342.

Whittaker, R. H., 1969. New concepts of kingdoms of organisms. *Science,* **163,** 150–160.

Willmer, P., 1990. *Invertebrate Relationships: Patterns in Animal Evolution.* Cambridge University Press, Cambridge, England.

Wray, G. A., J. S. Levinton, and L. H. Shapiro, 1996. Molecular evidence for deep pre-Cambrian divergences among metazoan phyla. *Science,* **274,** 568–573.

15 Differentiation and the Evolution of Development

The term **differentiation** is often used to describe developmental processes that cause structural or functional distinction among parts of an organism. In single-celled organisms such developmental processes affect, by definition, only one cell or parts of a cell. In multicellular organisms large groups of cells such as tissues and organs come to differ from each other. Differentiation in both types of organism has in common the fact that changes generally arise as a result of the appearance of gene products differing in respect to quality or quantity from those in other parts of the organism, or because of differences in the efficiency with which a gene product functions in different parts of the organism.

Viral and Bacterial Development

An elaborate viral developmental system in which gene activity is responsible for each change in morphology (**morphogenesis**), from start to finish, is that of T4 bacteriophage production and assembly shown in Figure 15–1. The phage DNA first uses bacterial host DNA-dependent RNA polymerase to synthesize phage mRNA. This phage-induced mRNA then translates on host ribosomes to produce a number of "early" proteins, many of which are necessary for the subsequent synthesis of phage DNA. Within 5 to 7 minutes of infection, these early enzymes lead to formation of a pool of "vegetative" phage DNA fibrils.

After early viral protein synthesis has ceased, a number of "late" proteins appear, including an inducing protein that acts as a scaffold to position the *head protein* molecules that form the viral head capsule. As T4 phage DNA enters this viral shell, the scaffold protein is destroyed, and DNA packaging is completed when a "headful" of DNA has been enclosed. Other late proteins include those involved in the various tail structures as well as the lysozyme used to rupture the host cell wall. Altogether, a completed phage particle is composed of about 30 to 40 different components, each genetically

FIGURE 15-1 Sequence of T4 phage development and the genes involved in the various morphological steps, according to the studies of Edgar, Wood, and others. Although many of the genes (indicated by numbers) produce proteins or polypeptides directly incorporated into the T4 assembly, the products of other genes (for example, *38, 57, 63*) are not themselves incorporated but are necessary for the assembly. (*After Wood, modified.*)

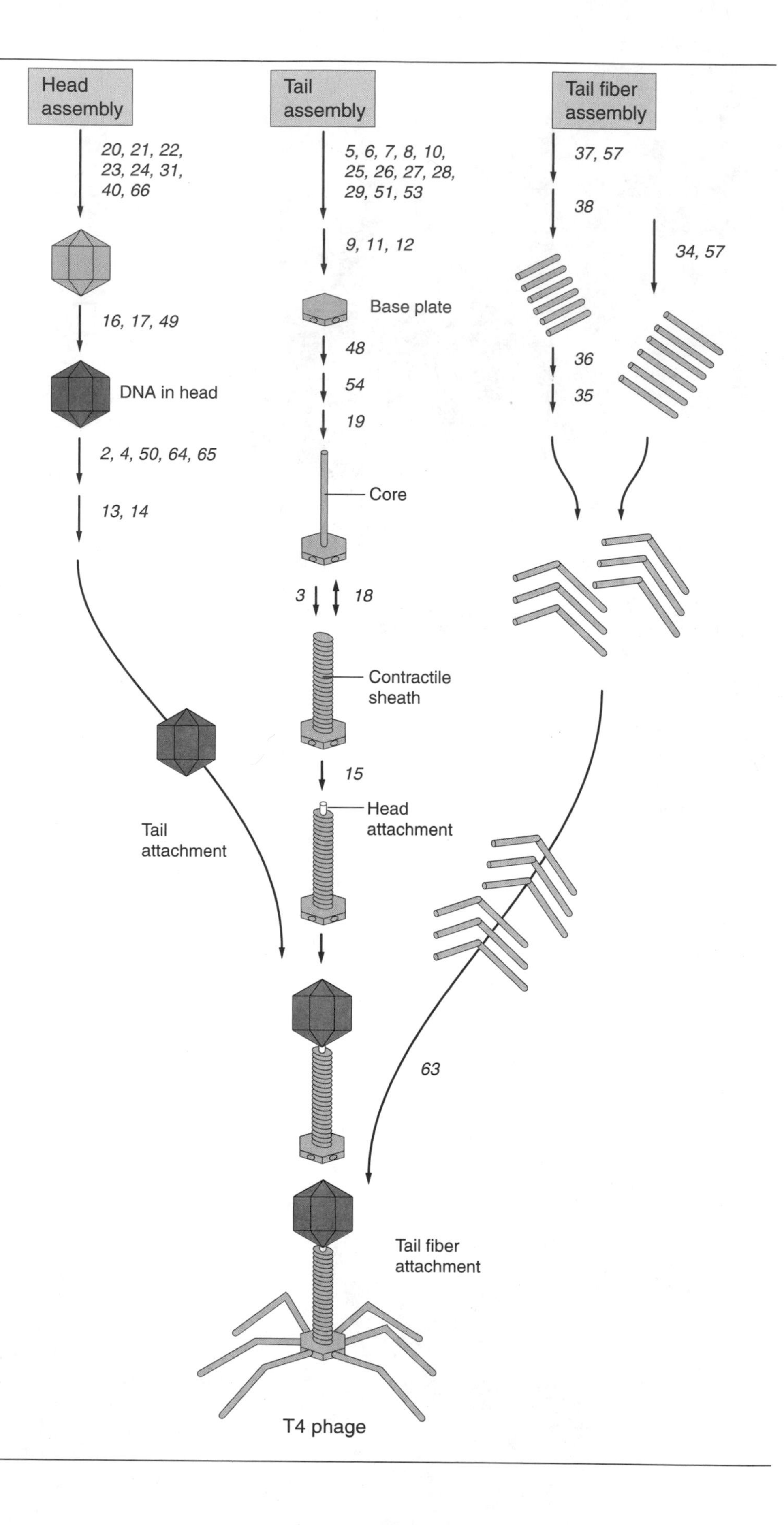

produced in a sequence that enables coordinated morphogenic interactions.

Although viruses may seem passive parasitical replicators, some possess genes that have been evolutionarily selected to allow them to choose whether to replicate or not. One such example occurs in the Lambda (λ) bacteriophage, which can remain quiescently incorporated within its host in "temperate" form, or can replicate and destroy its host by "lysis." This choice depends on the state of its bacterial host, and λ's response is precisely governed by a sequence of gene actions that produce host-sensitive proteins. Thus, the *cI* gene protein normally prevents viral replication by acting as a **repressor** that inhibits viral genes necessary for replication, whereas host conditions such as stress or ultraviolet radiation cause cleavage of the *cI* protein, leading to lysis (Ptashne).

A classic molecular example of how the activity of a bacterial cell can be regulated to perform differently under one condition (for example, presence of lactose sugar) than another (for example, absence of lactose) is illustrated in Figure 10–25. In that instance, production of the β galactosidase enzyme that metabolizes lactose is regulated by an **allosteric protein** that has two sites (p. 219):

1. A site that binds to a particular DNA sequence that prevents (represses) transcription of the messenger RNA, which would ordinarily be used to synthesize β galactosidase.
2. A site that binds to the inducer (produced by lactose) that changes the stereochemical conformation of the regulatory protein, causing it to vacate its normally repressive position on DNA.

Thus, for β galactosidase synthesis, the allosteric regulatory protein acts as a repressor to prevent synthesis unless lactose is present. **Activator** regulatory proteins, on the other hand, act oppositely. An inducer or specific condition empowers the activator to bind to a DNA sequence (*promoter*, Fig. 10–26) enabling or enhancing messenger RNA transcription. Genetic control, negative or positive, is precisely tuned to allow specific functional shifts.

Because microbial organisms offer advantages in dissecting and analyzing biochemical details of various developmental stages and their causative genetic elements, even some fairly complex morphological changes such as bacterial **sporulation** can be molecularly understood. For example, the decision to form a spore in some bacteria, such as *Bacillus subtilis,* is usually caused by a limitation in available nutrients, especially carbon and nitrogen. As shown in Figure 15–2, this new pathway is followed by a succession of enzymatic steps that enable the cell to change morphologically. Some bacterial enzymes increase in amount, others decrease, and still others appear that are unique to this type of development—an exact pattern that appears to be under precise genetic transcriptional control. As reviewed by Errington, new forms of RNA polymerase are produced through a change in its component proteins, which allow new sets of genes to be transcribed into messenger RNA. The spore that results neither replicates nor shows metabolic activity, yet may survive for hundreds of years under adverse conditions. When exposed to a proper growth environment it can germinate and resume normal morphology and function for vegetative growth and replication.[1]

Eukaryotic Development

In eukaryotes, interest in differentiation had its origin in observations as far back as Aristotle, who noted that embryos with few or no observable structural differences, such as a chick egg, give rise to complex differentiated organisms. The fact that development produces an exact replica of the parents was long believed to be caused by the transmission of adult structures in miniature form. As discussed in Chapter 1, this **preformationist** doctrine was replaced in the eighteenth and nineteenth centuries by the **epigenetic** view that adult structures were absent from the early embryo but appeared *de novo* during embryonic development.

With the onset of Darwinism, some embryologists began to explore evolutionary areas, emphasizing development as a unique evolving entity that can be studied by comparing observable embryological relationships. They provided large literatures on developmental changes in vertebrate embryo morphology, suture relationships in molluskan shells, and other comparative data. Out of these works came emendations of Haeckel's "biogenetic law" (Chapter 3), leading to the concept that ontogeny (development of an individual) can reflect an ancestral sequence of developmental processes. Embryological stages can be used as morphological traits to evaluate relationships between some organisms.

Although these investigations offered considerable descriptive information, they left unanswered questions of what factors are responsible for *changes* in the egg-to-adult process, how such changes are transmitted between generations, and what forces enable different developmental changes to become established in different generations and different groups. Based on research begun by

[1]Eukaryotic counterparts to sporulation, such as cellular changes leading to gametogenesis, are expected to involve more gene products operating in more complex pathways. In the simple eukaryotic example of budding yeast (*Saccharomyces cerevisiae*), a complex sporulation cascade occurs when the diploid stage is nutritionally depleted, producing four thick-walled haploid spores meiotically. According to Chu and coworkers, the major stages in this process are genetically controlled, involving the increased expression of about 500 genes accompanied by diminished expression of more than 600.

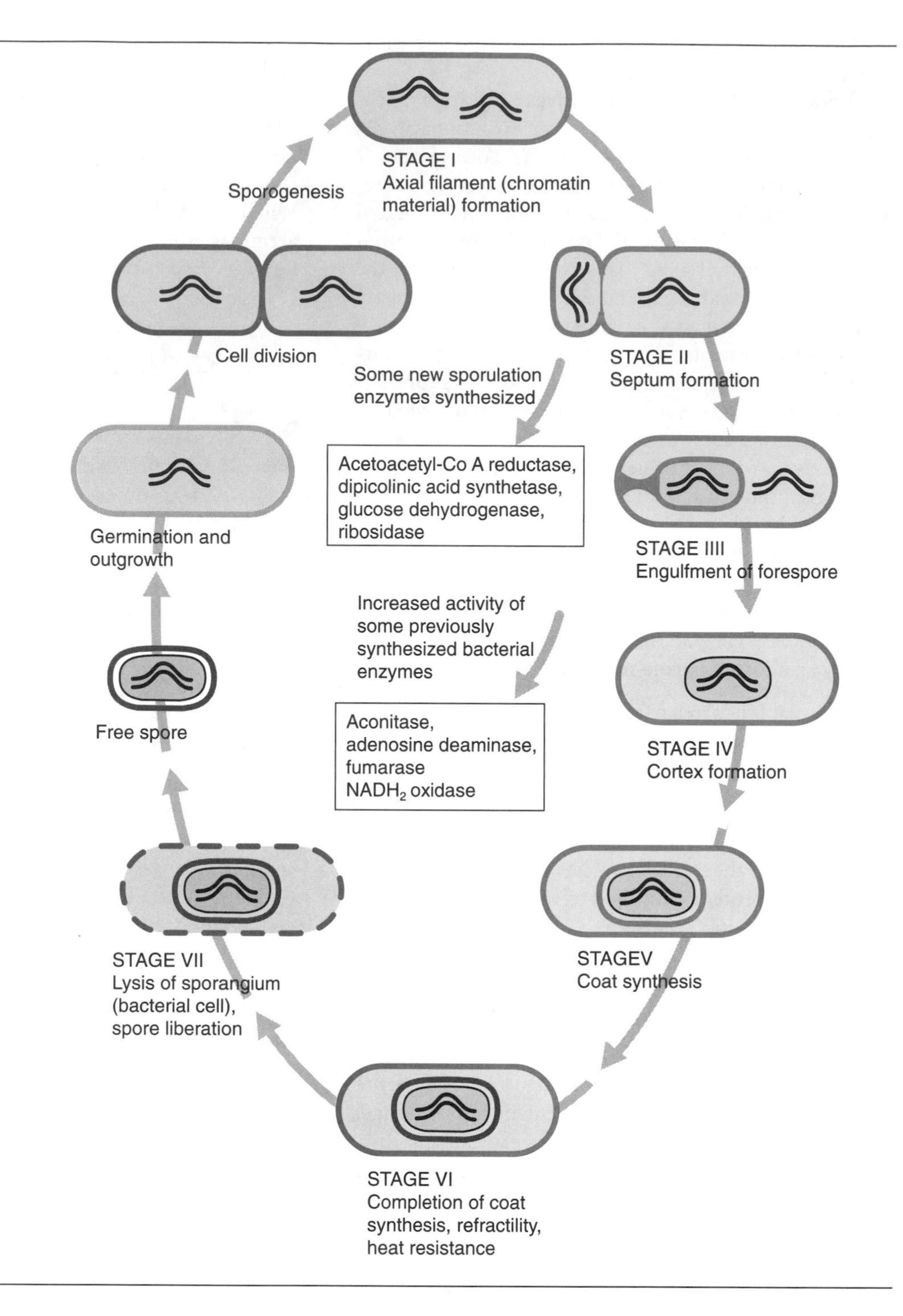

FIGURE 15-2 Sequence of morphological steps in sporogenesis in *Bacillus subtilis* and some of the enzymatic events involved. From the onset of sporulation to the production of a finished spore is about 6 to 8 hours. Coated by layers of special proteinaceous materials absent in vegetative cells, the final spore is liberated by cell lysis. Spores are resistant to heat, ultraviolet radiation, organic solvents, enzymes, and desiccation. (*Based on Halvorson and Szulmajster.*)

Goldschmidt, Dunn, Beadle and Ephrussi, and many others, the science for answering these questions shifted from embryology to developmental genetics.[2] Within the

[2]By 1950, even before molecular genetics blossomed, various geneticists investigated the impact of genes on developmental processes, including: rates of phenotypic reactions (R. B. Goldschmidt); kinds of mutant phenotypes (H. Muller); human metabolic changes (A. E. Garrod); pigment development in flowers (H. Onslow, R. Scott-Moncrieff); eye pigments in *Drosophila* (G. W. Beadle and B. Ephrussi); spinal vertebral development in mice (L. C. Dunn, S. Gluecksohn-Schoenheimer); pleiotropic effects in *Drosophila* (E. Hadorn); developmental changes in mice (H. Grüneberg); and metabolic sequences in *Neurospora* (G. W. Beadle and E. L. Tatum).

Some other developmental genetic studies of the time are reviewed in articles by Curt Stern and Hermann Muller in the 1949 classic *Genetics, Paleontology, and Evolution* (G. L. Jepsen, E. Mayr, and G. G. Simpson, editors), a basic document in the "Neo-Darwinian Synthesis" (Chapter 21).

Despite this literature and the broad research that followed, many morphologists insisted that the gap between genotype and phenotype was great enough to be "unbridgeable," and genetics could offer little information to explain development. For example, de Beer, a prominent comparative embryologist, maintained through editions of his classic *Embryos and Ancestors,* "It may be definitely stated that the internal factors which were inherited from the parents are *not* [his emphasis] sufficient to account for the development of the animal." As discussed later (p. 356), some biologists still share such views.

last decade or so, such genetic investigations have accelerated rapidly with the phenomenal expansion of biomolecular techniques.

Following the lead developed in microbial organisms, it became clear to molecular geneticists that regulatory processes also control eukaryotic gene expression, both through negative (repressor) and positive (activator) mechanisms. That is, how a cell differentiates in form and function depends on which of its genes are available for transcription. Transcription, in turn, depends on stimuli—"signals"—that may come from other cells by diffusion, or direct contact, or from the environment. Such signals, which can be organic or inorganic, may enter the cell directly through the plasma membrane or, commonly in development, attach to membrane receptors that then amplify the signal within the cytoplasm—**signal transduction.**

In general, as Figure 15–3 illustrates, signal transduction begins with an extracellular molecule that binds (ligates) to a cell-surface receptor specific for that "ligand." A cascade of cytoplasmic reactions follows which can then activate specific cytoplasmic transcriptional precursors, either by removing their inhibitors or by allowing them to associate functionally. These proteins, in turn, can then form multiprotein nuclear complexes that stimulate or repress transcription of specific genes that possess appropriate binding sites, such as promoters and enhancers.[3]

Furthermore, because development is a dynamic process, a cell's responsiveness to signals changes with its history; the same signal pathway can express or inhibit different genes depending on a cell's position in time and space (Freeman). This flexibility enables an economy of signal pathways for a multitude of purposes. Signaling pathways, such as that illustrated in Fig. 15–3, and many others (see Gerhart and Kirschner), are conserved in a wide variety of metazoans and applied to a wide variety of functions. For example, depending on the extracellular signal, phosphorylating enzymes such as mitogen-activated protein kinases (MAPKs) can activate different transcriptional proteins by following different routes from cell membrane to nucleus (Elion). In plants, common signals such as ethylene act in transduction pathways affecting a variety of functions, such as seed germination, fruit ripening, and cell development (Theologis).

As can be expected, choices among binding positions on the DNA molecule itself also allow choices among possible gene expressions. For example, Davidson and coworkers identified more than 20 regulating DNA sites for the actin protein gene (used for cytoskeletal and muscle filaments) in the sea urchin, *Strongylocentrotus purpuratus;* some were used to activate transcription, and others for repression. Interactions between regulatory proteins can also occur, converting activators to repressors, and *vice versa* (Ptashne).

Regulatory processes affect not only transcription but can act at many other basic molecular levels:

- Changes in chromatin can affect DNA segments or even entire chromosomes, such as those that cause *Drosophila* X-chromosome inactivation (Chapter 10).
- Post-transcriptional modification of messenger RNA can occur through different splicing patterns that produce different mature mRNAs from the same precursor molecule, or directly modify mRNA nucleotides ("RNA editing") by transitions (C → U), deletions, or insertions. For example, modified intron splicing during the evolution of domesticated rice caused a single base change in the *Waxy* gene leading to less waxy protein (Hirano et al.).
- Translation of messenger RNA into protein can be regulated by determining mRNA degradation rate (Ross), or by binding proteins or complementary RNA sequences to the mRNA molecule to prevent translation ("antisense control").
- Post-translational modification of proteins can occur through different splicing patterns that remove amino acid sequences (Cooper and Stevens), or by chemically modifying amino acid residues by adding acetyl, sugar, or phosphate side-chains (acetylation, glycosylation, phosphorylation).

Clearly, it is the variety of these many **regulatory pathways** and their specific activity in time and place that accounts for the variety of different kinds of cells in a multicellular organism.[4] The emerging picture is that all levels of gene activity, from DNA replication onwards, are exquisitely and sensitively genetically controlled and regulated. Selective processes acting on these genes and their alleles over time guide the form, function, and direction of development.

[3]The intricate relationship among these gene products, such as between a ligand and its cell surface receptor, indicates the result of **coevolution,** in which changes in one element (for example, ligand) select for changes in the other (for example, receptor).

[4]The sequential pattern of development—that cells and organs go through successive stages of differentiation—is reminiscent of the sequential pattern of organismic evolution itself. That is, organisms and their organs do not appear *de novo* out of a "lucky" and improbable combination of events, but result from a sequential process in which each successive change produces an additional and incremental feature (Fig. 3–1). In multicellular development, organs begin with only a few types of cells that undergo a series of successive differentiations as they divide and increase, rather than collectively undergoing a single spontaneous differentiation into their final individual forms. Apparently, to get from cell type A to a more complex cell type Z is more easily achievable by transforming intermediate step-wise stages A → B, B → C, C → D, etc. (Britten's "precursor groups"), rather than requiring a large and improbable number of simultaneous signaling and transcriptional events, A → Z.

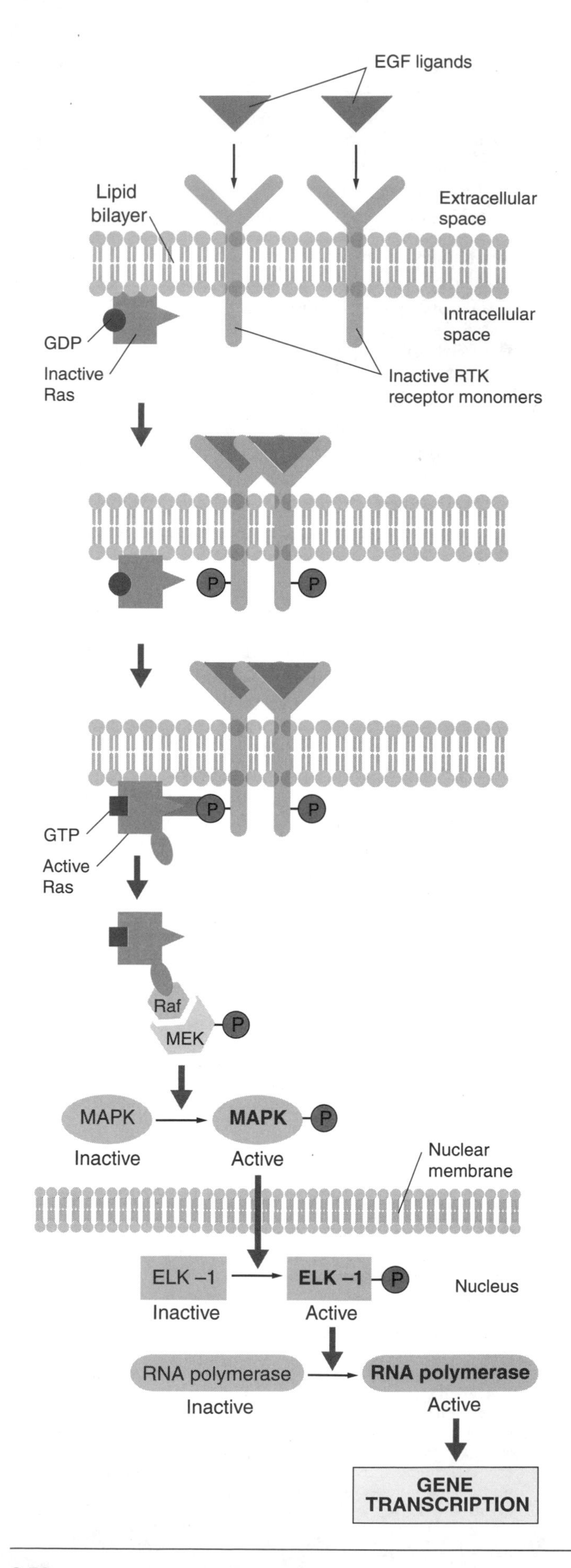

Extracellular EGF ligands move toward RTK receptor binding sites, extending into intracellular space from the lipid bilayer of the cell plasma membrane. Within the membrane are also "Ras" proteins, inactive because of binding to guanosine diphosphate (GDP).

Ligand binding causes the RTK receptor to dimerize, followed by kinase activity causing phosphorylation of several tyrosine amino acids —(P).

RTK phosphorylation recruits intermediaries that activate the "Ras" protein by exchanging GTP (guanosine triphosphate) for GDP.

Activated "Ras" binds to the kinase protein "Raf" which phosphorylates and activates MEK protein. MEK, in turn, phosphorylates and activates MAPK (mitogen-activated protein kinase).

Activated MAPK phosphorylates and activates latent transcription factors such as ELK-1 within the nucleus.

The activated transcription factor activates RNA polymerase to transcribe a specific gene sequence into messenger RNA.

FIGURE 15-3 Diagrammatic illustration of signal transduction in which an extracellular epidermal growth factor (EGF) binds to a transmembrane receptor (RTK) with tyrosine kinase activity that activates reactions in a chain of cellular proteins, leading to messenger RNA transcription. Such signaling systems show extensive homology across various phyla. Ras protein controls, for example, are involved in determining genital structures (vulva) in the *Caenorhabditis* nematode, in eye development in *Drosophila,* and in cellular proliferation and differentiation in mammals. When inactive, "G proteins" such as Ras are commonly complexed with the guanine nucleotide GDP, and are activated when GDP is phosphorylated to GTP. (After their activation and subsequent function, G proteins normally return to inactivity by phosphotase hydrolysis of GTP to GDP.) In some cases, targeted proteins in such cascades produce "second messengers," such as cyclic adenosine monophosphate (cAMP) or cyclic guanosine monophosphate (cGMP), that can affect glucose metabolism, fat storage, and cellular responses such as aggregation and secretion.

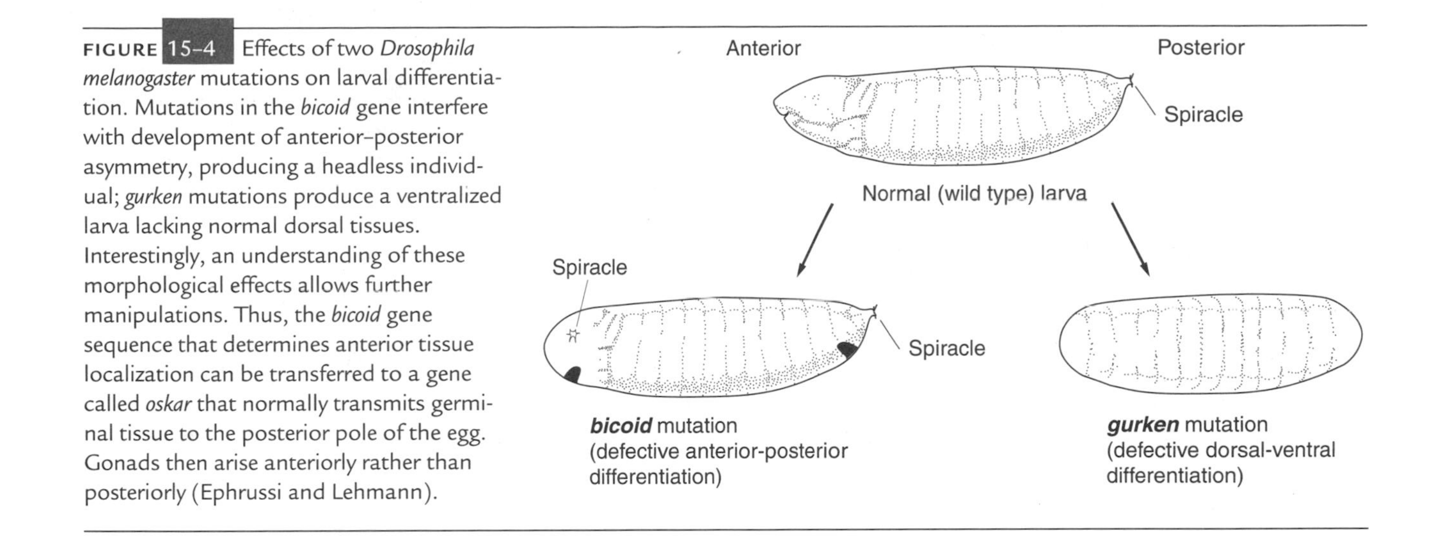

FIGURE 15-4 Effects of two *Drosophila melanogaster* mutations on larval differentiation. Mutations in the *bicoid* gene interfere with development of anterior–posterior asymmetry, producing a headless individual; *gurken* mutations produce a ventralized larva lacking normal dorsal tissues. Interestingly, an understanding of these morphological effects allows further manipulations. Thus, the *bicoid* gene sequence that determines anterior tissue localization can be transferred to a gene called *oskar* that normally transmits germinal tissue to the posterior pole of the egg. Gonads then arise anteriorly rather than posteriorly (Ephrussi and Lehmann).

Genetic Control of Embryonic Space

In multicellular eukaryotes, an obvious and distinctive feature is, of course, the spatial arrangement of their tissues and organs. That is, not only do these body parts differ from each other, but they also occupy localized positions that are generally uniform in all individuals of a species. For such precise and repeatable structures to occur, they must be preceded by a precisely ordered pattern of development: a process that creates and maintains a cell's special relationships to other cells must direct its fate.

In a general way the commencement of pattern begins with the egg. Differences, such as those imposed by the uneven deposition of maternal substances in the egg cytoplasm or by external factors such as gravity, undoubtedly help initiate localized differences in development. The developmental differences make use of genetically produced molecules that act as morphogenetic agents ("morphogens") enabling different cellular responses. The *bicoid* gene in *Drosophila,* for example, helps determine the anterior–posterior pattern of the egg by producing a protein whose concentration follows a gradually decreasing gradient along its length. Should *bicoid* be defective, a headless embryo forms with tail structures at both ends. The gene *gurken* acts similarly by producing messenger RNA that helps determine dorso–ventral differentiation (Fig. 15–4).

Figure 15–5 diagrams how some cells or tissues can obtain information as to their positions because they are affected by such gradients. This **positional information,** in turn, can influence subsequent activity so that different groups of cells follow different developmental patterns, such as moving to new locations, and/or causing the production of substances (**inducers**) that set up developmental gradients of their own. Researchers have suggested that each new set of positional coordinates activates special **selector genes** that cause cell lineages to commit themselves to a particular developmental direction in particular compartments. Subsequent exposure to further positional influence can activate a different battery of selector genes leading to a further commitment, and so on.

The uniformity of the developmental pattern in different individuals of the same species lies in the fact that the succession of regulatory events, or pattern history, is identical or similar. When the history of the pattern changes, such as a change in the type of inducer or gradient produced by a tissue, or a change in the ability of a tissue to recognize positional information, the phenotype

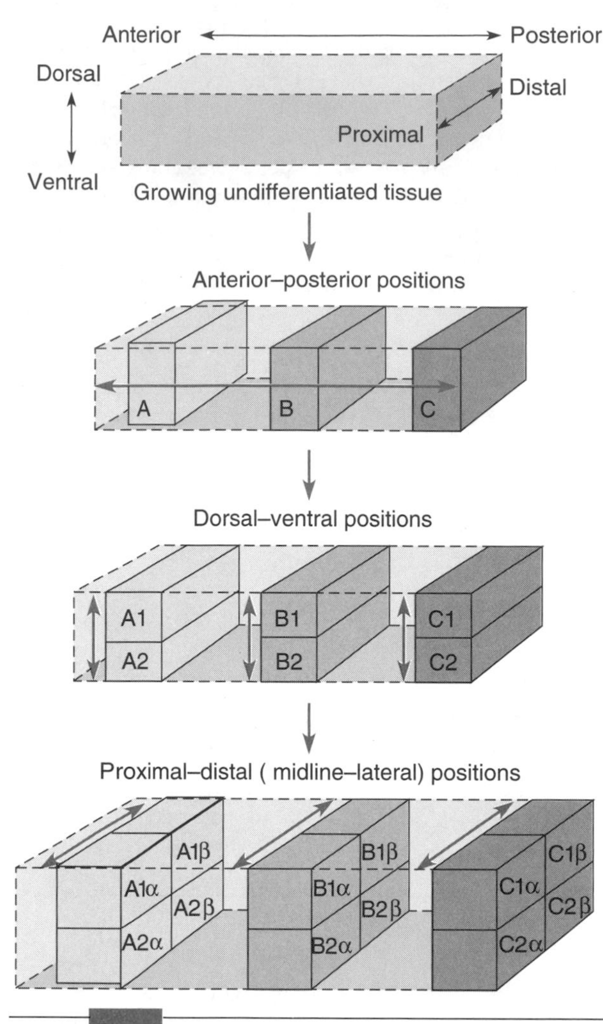

FIGURE 15-5 Diagrammatic representation of how some cells or tissues located along the major body axes assume positional information during development. A gradient along one axis provides cells or tissues A, B, and C with information as to their relative anterior-posterior positions. (For simplicity, only three anterior-posterior tissue blocks are shown. In normal development each of these blocks may be further differentiated "horizontally" into three or more anterior-posterior subdivisions, so there may be a total of nine or more blocks of differentiated tissue extending along the anterior-posterior axis.) Further growth and differentiation confers dorsal-ventral information upon clones of each A, B, and C tissue labeled 1 and 2. Subsequent cell divisions of these tissues provide their α and β subclones with information as to their proximal-distal positions. As a result of their geographical position and of the activity of special "selector" genes, some or many cells of tissue A, for example, may come to possess a unique set of developmental responses that let them develop differently from tissues B or C, or even from other subclones within A. A simple gradient model in which different morphogen concentrations produce different cellular colors is called the "French flag": high levels "blue," intermediate levels "white," and low levels "red." (*After Strickberger.*)

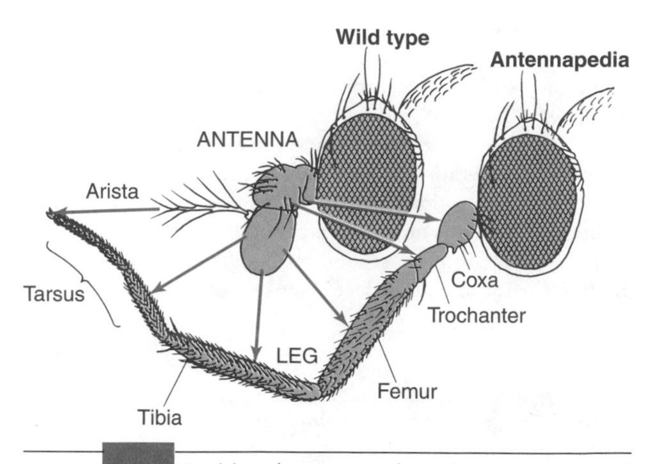

FIGURE 15-6 Positional correspondence between antennal sections and leg sections produced in *Drosophila melanogaster* flies carrying the homeotic *Antennapedia* mutation. The mutational substitution of leg tissues for antennae tissues follows the arrows (*color*) so that distal cells, for example, interpret their position as tarsal segments, and proximal cells as coxa or trochanter.

can also change. For example, the *Drosophila* mutation *Antennapedia* transforms the antennal structure of the head into a leg. Such mutations are called **homeotic** because they change a particular organ in a segment to resemble an organ normally found in a different segment along the body axis (**homoeosis**). The correspondence between antennal parts and leg parts indicates that both structures bear the same positional information, but that cells carrying the *Antennapedia* mutation interpret these positions to produce leg tissues rather than antennal tissues (Fig. 15–6).

The function of the normal *Antennapedia* gene product, *Antp*$^{+}$, seems to be that of a selector gene whose activity is required for normal leg development in thoracic segments. According to this view, the *Antennapedia* mutation causes this gene to malfunction so that it is also active ("gain of function") in the anterior head segment repressing genes which would normally allow antennae to be formed (Casares and Mann). A cluster of *Antennapedia*-linked loci, called the *antennapedia complex,* seems to possess somewhat similar regulatory functions. Like the antennapedia complex, the *bithorax complex* investigated by Lewis also causes homeotic effects, and consists of at least a dozen genes that control the fates of various structures from the posterior part of the second thoracic segment to the tip of the abdomen. Thus, should expression of the normal *Ultrabithorax* gene fail in the anterior abdomen, these segments develop as though they were thoracic rather than abdominal (see also Fig. 10–28).

Since both the bithorax and antennapedia complexes confer unique identities on *Drosophila* segments and on substructures within these segments, some genes in both complexes probably arose by duplication from a common

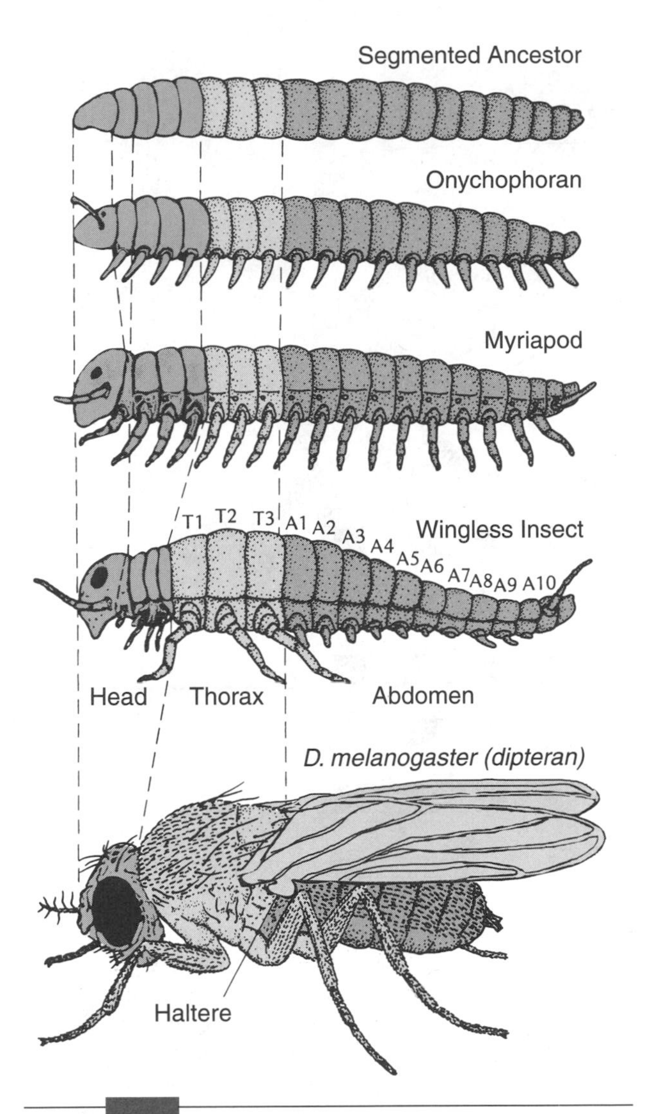

FIGURE 15–7 Proposed evolutionary progression to account for the segmental organization found in insects. Body segments were, at first, relatively uniform, and then became more complex in different evolved groups. In dipteran insects (*bottom figure*) the segments not only bear complex structures but, except for some abdominal segments, also differ from each other quite a bit. For example, the three thoracic segments of *Drosophila* each have a pair of legs, but only the second thoracic segment has wings, and only the third thoracic segment has the haltere balancing organs. The three thoracic segments are numbered consecutively from the head boundary as T1, T2, and T3; and the eight recognized abdominal segments in *Drosophila* are numbered consecutively from the thoracic boundary. (We don't know the exact number of segments involved in dipteran head formation.) (*After Strickberger.*)

ancestral gene which initially specified cell differentiation of only simple uniform segments. Duplications of this primitive gene or gene complex, and subsequent divergent mutation of each duplicate, allowed different segments to undergo different developmental pathways, yet each duplicate gene complex still maintained sequences necessary to produce or interact with the basic underlying segmentation process. As evolution proceeded from simple uniform segments in the insect ancestor to more complex segments found in dipteran flies such as *Drosophila* (Fig. 15–7), the number of regulatory genes successively increased, enabling unique structures in each segment to be controlled. For example, because of selection in dipterans, halteres (balancing organs) appear only in the third thoracic segment but not in others: genetic changes in duplicate genes allowed new regulatory functions.

From *Hox* Genes to the Zootype

In support of evolutionary duplication and differentiation of developmental genes is the finding by McGinnis and coworkers that a family of related DNA sequences called **homeoboxes** can be found in various locations in the *Drosophila* genome, including loci within the bithorax and antennapedia complexes. Each homeobox in these "*Hox*" gene complexes includes coding for a polypeptide sequence about 60 amino acids long called a **homeodomain.** The homeodomain, in turn, is part of a transcription factor that binds to DNA, thereby regulating messenger RNA production. The genes regulated by these homeoboxes are apparently those that affect cell positioning and differentiation in a wide variety of organisms (Kappen et al.). Interestingly, the linkage order of these homeobox-containing genes in the *Drosophila* antennapedia–bithorax clusters accords with their phenotypic expression along the anterior–posterior axis of the animal. Whatever the cause, the observation that very similar homeodomains as well as linkage orders of homeobox genes now appear in many other animals (Fig. 15–8) indicates that certain types of positional information in different phyla most probably have a common evolutionary origin.

Furthermore, a particular homeobox protein may at times perform a similar function in different organisms; for example, much or all of the function of the *Drosophila* homeobox gene *Antennapedia* can be assumed by a protein produced by a homologous *Hox-b* gene in mice (Malicki et al.). As Patel points out in his review, effects of homeobox genes are not the only developmental genetic homologies; segmentation stripes produced by the nonhomeobox *engrailed* gene in *Drosophila melanogaster* are also produced by *engrailed* homologues in other insects, as well as in crustaceans, annelids, and vertebrates. If we consider that anterior–posterior segmentation, although expressed in different structures, is a common *Hox* gene feature relating arthropods and vertebrates,

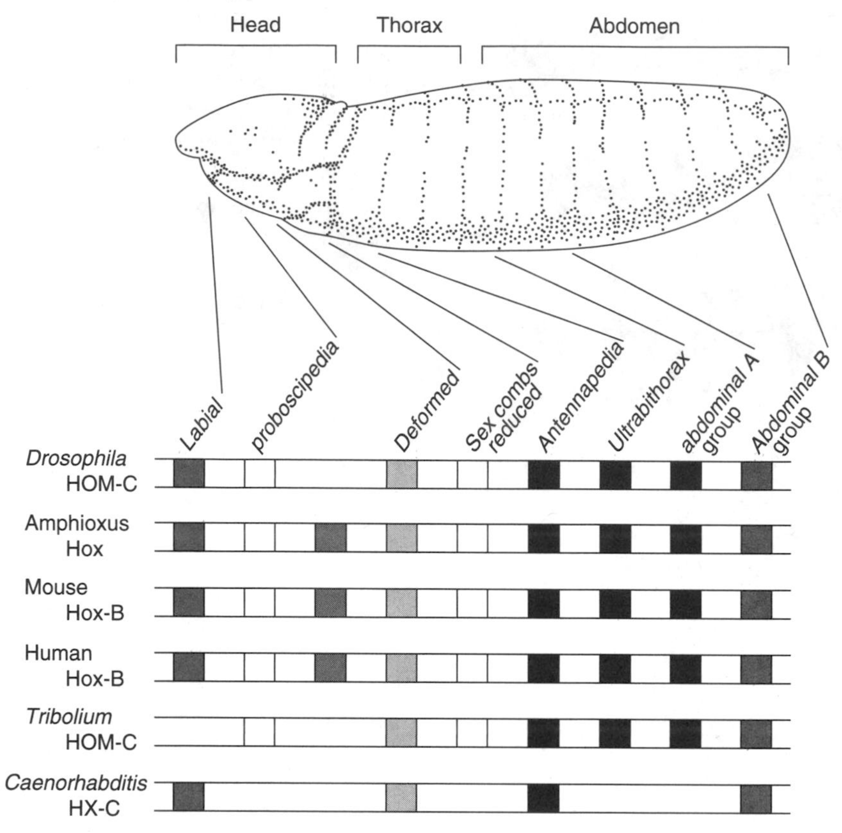

FIGURE 15-8 Linkage relationships between *Drosophila melanogaster* homeobox genes showing their positional effects on tissues along the anterior-posterior axis of the embryo. Shown also are clusters of homeobox genes from a cephalochordate (amphioxus), mammals (mouse, chromosome 17; humans, chromosome 11), beetles (*Tribolium castaneum,* chromosome 2), and nematodes (*Caenorhabditis elegans,* chromosome III) with homologous genes aligned vertically. These regulatory genes are being discovered in a wide array of metazoans, from cnidarians to echinoderms (Chapter 16). Almost all cases illustrated here show a highly conserved relationship between a gene's position along the chromosome and its developmental function along the anterior-posterior axis, whether the axis is segmented epidermis (for example, insects) or central nervous system polarity (vertebrates). This does not mean that gene expression is always restricted to a single developmental stage: some genes expressed early in development are also expressed later (Salser and Kenyon). That is, although "ontogeny" (development) is sequential, genes that produce it can act intermittently. The homeobox cluster shown for mammals is only one of four (*Hox*-A, B, C, D), each organized along similar lines, with a total of 39 homeobox genes. These four clusters are believed to have arisen in vertebrates by successive duplications of an ancestral chordate "D"-type cluster, correlated perhaps with successive increases in body complexity (Bailey et al.). (*Adapted from Kappen and Ruddle, and from Garcia-Fernàndez and Holland, with additions and modifications.*)

Geoffroy Saint-Hilaire's speculative judgment, made more than 150 years ago, seems quite perceptive: "every animal lives within [arthropods] or without [vertebrates] its vertebral column [linear segmentation]."

Homologous gene performance of similar functions is also involved in the growth of neuronal axons in nematodes and vertebrates, as well as muscle development in nematodes, *Drosophila,* and mice. Even the development of anterior sense organs (for example, eyes) and the central nervous system in organisms as different as humans, fish, tunicates, mollusks, nematodes, insects, and nemerteans shows evolutionary relationship through a common regulatory *pax–6* gene sequence, indicating such homologies most probably originated in the Precambrian triploblastic ancestor of invertebrates and vertebrates (Nilsson). Also, homeobox proteins expressed in *Drosophila* can function even in forming *Caenorhabditis* nonsegmental patterns (Hunter and Kenyon). Appendages in animal phyla, from protostomes to deuterostomes, also make use of homologous genes that regulate body wall outgrowths along a proximal-distal axis (Panganiban et al.).

Such homologies not only indicate a common genetic evolutionary origin, but also the conservation of some basic developmental pathways such as those that establish or implement a positional axis. Slack and coworkers, for example, propose that all animals use a common genetic developmental system, the **zootype,** that governs the ba-

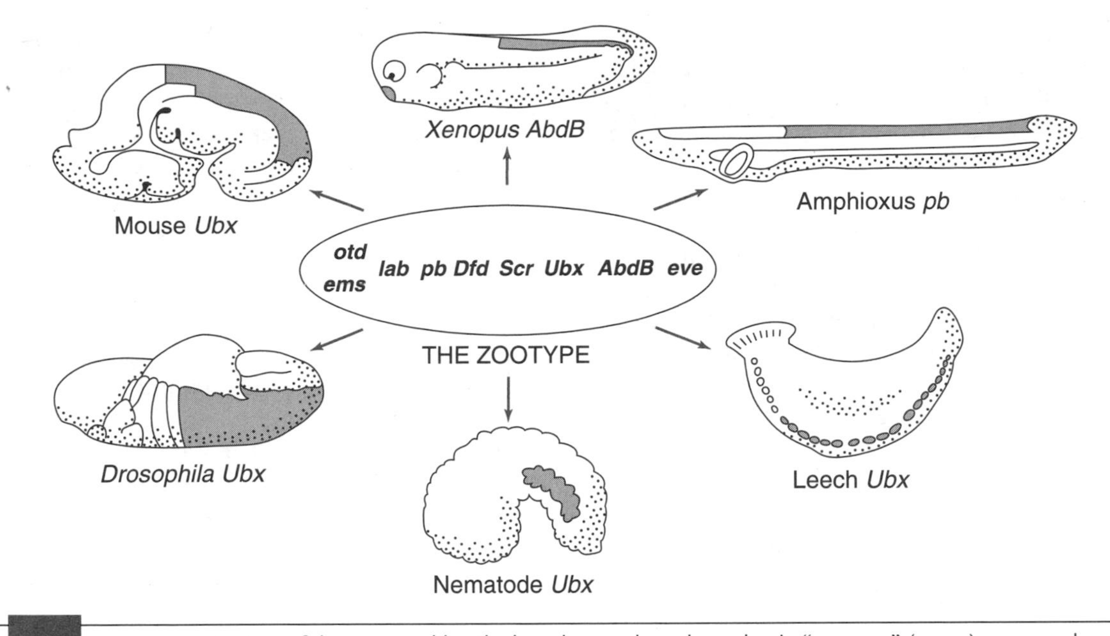

FIGURE 15–9 Diagrammatic view of the proposal by Slack and coworkers that a basic "zootype" (*center*) governs the spatial development of different multicellular animals in their presumed unique "phylotypic" stage (phylotype). The genes in the zootype are designated by standard *Drosophila* abbreviations, and the expression of particular zootype genes in different animals is shown in *colored shading.* (*otd* = *orthodenticle*, *ems* = *empty spiracles*, *lab* = *labial*, *pb* = *proboscipedia*, *Dfd* = *Deformed*, *Scr* = *Sex combs reduced*, *Ubx* = *Ultrabithorax–Antennapedia–Abdominal A* group, *AbdB* = *Abdominal B*, *eve* = *even-skipped.*) The phylotypic stage characterizes each phylum's distinctive development, and is called the pharyngula in vertebrates (Fig. 3–10). Homeobox genes can be used in many different body plans. However, in addition to homeobox genes, it probably takes many genetic interactions to produce such phylotypes. (*After Slack et al.*)

sic spatial arrangement of their tissues, and this feature—established by their common genetic ancestry and retained because of common selective developmental value—characterizes the kingdom Animalia. According to them, "the zootype is expressed most clearly at a particular stage of embryonic development." This stage, called the **phylotypic stage** or **phylotype,** is illustrated in Figure 15–9 for each metazoan group in which the zootype's developmental genes are conserved and expressed. This stage may owe its persistence in each group to its role as a necessary focal point upon which further development depends.

This concept of conserved developmental genes applies also to angiosperms and fungi, now shown to produce homeodomain proteins much like those of metazoans. Such proteins must have functioned early in eukaryotic history, and probably arose from duplication events before plant, fungal, and metazoan groups diverged (Bharathan et al.). In metazoan lineages, Zhang and Nei suggest that homeobox genes began as two linked loci, each locus conferring a distinctive role in anterior or posterior development. These genes then duplicated further, eventually generating the linked clusters of homeobox genes in triploblastic animals (Finnerty and Martindale).

Despite these findings, some biologists suggest that at least some developmental traits are chiefly caused by factors other than genetic influence. That is, such traits are really determined by developmental constraints rather than by genetic instruction, a point discussed below.

Developmental Interactions and Constraints

The presence of a developmental system with its regulated pathways raises the question whether such a system acts as a constraint that biases or limits structural and functional variation, and thereby molds and directs future evolution. For example, one can ask why insects have six legs and land vertebrates four, when almost any number of legs, or even none at all, can be used for locomotion. Do **constraints** in insect and vertebrate developmental systems dictate the number of legs—a number that is not the result of selection?

The answer is not entirely simple: limb abnormalities of all kinds do arise in both these taxa, indicating limb number can change, but since such mutations generate major bodily disruptions, they often produce an abortive developmental "monster" unable to survive or reproduce. Thus, although many developmental changes are possible, certain of these changes, especially those of large degree (for example, "hopeful monsters," p. 599) are mostly lethal ("hopeless") because their effects cannot be successfully integrated with other developmental stages: if such integration is deficient, development of the organism is often fatally distorted.

The need for integration dictates developmental constraints whose limits, according to Raff and coworkers, depend on the extent of interaction among the metabolic pathways, cells, tissues, organs, and other components of a developing system. That is, development of those embryonic stages with greater numbers of interactions is more crucial, and therefore more subject to constraint, than development of other stages with fewer interactions. The common gill-arched phylotypic stage ("pharyngula") that different vertebrates experience (Fig. 3–10) can be considered to incorporate many basic controls that influence further vertebrate development.

An anatomical example of developmental constraint is the position of the recurrent laryngeal nerve in mammalian vertebrates. As was shown in Figure 3–11, because the sixth arterial arch in vertebrates is first formed anteriorly during a "phylotypically" crucial embryonic stage, it governs the distance traveled by the later-appearing recurrent laryngeal nerve which extends from the cranium and loops around the sixth arch to reach its laryngeal site. This distance is greatly increased in mammals in which the arch has become displaced to the thorax and the nerve must double back a distance that may be longer than six feet in giraffes! That is, arterial arch development (rather than adaptive selection) appears responsible for the nerve's length and positioning. Also relevant is a constant number of seven cervical vertebrae in practically all mammals, whether short-necked (whales) or long-necked (giraffes). Why such constancy if not developmental constraint?

Using similar examples, some biologists propose that genetic influences are either absent or, at best, secondary to many developmental factors that have assumed the role of prime morphological determinants for both variations and novelties: "The causality for the origin of novel structures lies not within the genome but in epigenesis [developmental interaction]" (Müller). In contrast to the common view that ontogeny involves coordinated interactions between genotype and phenotype, such morphologists and developmentalists view ontogeny primarily as a succession of phenotypic interactions in which the genotype may perform an initial role of "prime mover," but then disappears. Phenotypic interaction is deemed separate from genetic interaction, and the phenotype can then transmit itself and change evolutionarily by nongenetic devices such as direct environmental influence.

One proposed example of phenotypic transmission is Ho's view of how the predator-prey "arms race" evolves (pp. 426, 453, 574). Instead of selection affecting the reproductive success of different predator and prey genotypes, Ho suggests such experiences cause neural sensory changes and muscular contractile changes in both types of animal, that are then passed on to following generations—an essentially Lamarckian notion (p. 24).[5] This view maintains that development ("phenotypic interaction") and the limits of structural design act independently of genes, and are the primary agents responsible for both the principal rules and prevailing forms found in both ontogeny *and* phylogeny (see also p. 583 footnote 1).

Nevertheless, to practically all other biologists, genes provide the necessary historical information that allow cells to differentiate so they can interact with each other and the environment to assume their ultimate phenotypic shape and relationship (see also Wolpert). Such interactions may include development of complex patterns from limited instructions. For example, each of the millions of different possible antibodies that can be produced by the mammalian immune system is not individually coded by the genome. Instead, they are produced by a developmental system that can produce almost any antibody to interact with almost any environmental antigen by selecting among an array of newly generated nucleotide sequences (Golub and Green). Inherited mutations in the immune system, and not environmental antigens, are responsible for producing the variety of potential antibody types and numbers. What is *evolutionarily* adaptive, therefore, is not the appearance of one or more specific antibodies, but a genotype selected to confer the potentiality to make them.

As discussed by Frank, other examples, such as neuronal connections in animals and root connections in plants, indicate the presence of similar genetic systems,

[5]Mechanisms that support such Lamarckian concepts have yet to identify biological coding agents other than DNA or RNA that can pass on information directing simple undifferentiated cells to produce highly differentiated tissues, such as nerves and muscles (epigenetic development). Perhaps, because phenotypes are descriptive objects rather than instructions, Lamarckian proponents believe they are caused by mechanisms other than genes, since genes seem only to provide instructions but not descriptions. To these biologists, who vary in their emphasis on nongenetic developmental mechanisms (for example, Goodwin; Ho and Saunders; Jablonka and Lamb), phenotypic change, whether ontogenetic or phylogenetic, generally comes from interaction between phenotypes or between phenotypes and environment, and selection is either absent or minimal.

However, we should recognize that entities which remain unaffected by selection and genetic change would be malleable only by physical forces, and would be no more lifelike in sustaining metabolic, reproductive, and adaptive attributes—essentials of life—than petrified fossils. That is, although evolution is marked by changes in phenotypic characters, such changes depend on, derive from, and transmit through genotypic change. Although Klar and others insist that "genotype" includes non-DNA factors influencing transcription, such as DNA methylation (p. 270) and heterochromatinization (p. 200), their effects are on DNA and their transmission depends on attachment to DNA.

Evolutionary innovations, as described later, come from genetic changes that allow the exploitation of new environmental resources, rather than from a Lamarckian mystical "drive" that induces a nongenetic specific phenotypic change to exploit a specific resource. So far, the only generally accepted form of nongenetically based biological (Lamarckian) evolution comes from the transmission and modification of learning and culture by intelligent agents such as humans (Chapter 25).

producing arrays of potential phenotypic variants. These systems generate complex physiological, morphological, and behavioral patterns that allow organisms to tune their responses to particular environmental stimuli by enabling a choice of gene action among different inherited possibilities. That is, such stimuli "trigger" genetic systems that respond to environmental signals, whether organic or inorganic, as in signal transduction (see, for example, Fig. 15–3). In sum, phenotypic plasticity is as much a selected genetic trait as phenotypic uniformity: genotypes that produce the former offer more responses to environmental differences than the latter. We can therefore say that the "musical notes" in the developmental "symphony" are painstakingly written in genetic form, and are expressed in organisms through "melodies" that involve interaction between genes and their environment.[6]

Heredity and Developmental Constraint

To repeat their functional roles each generation, developmental factors and their interactions must have an underlying hereditary basis. As for any other trait, heredity ties organismic characters to evolutionary processes with selection as a primary influence affecting transition and survival (Chapter 22). Thus, the superfluous length of the mammalian recurrent laryngeal nerve—seemingly embedded in mammals like a useless vestige—is a by-product of a more basic genetically determined developmental system that was subject to selection in the past, and whose conservative features are still selected in the present. Laryngeal nerve function and its circuitous positioning seem determined by separate genetic pathways, each maintained by selection—one for neurological activity and another for skeletal structure.

The unusual location of the sixth arterial arch in the mammalian thorax is caused by selection for neck-lengthening in mammals, rather than from some inherent or self-directed Lamarckian process isolated from selection. This means that biological evolution expressed through *changes* in factors such as morphology, physiology, and behavior arises from developmental *changes* caused by genes, rather than from nongenetic vitalistic causes such as mysteriously appearing "archetypes" (p. 6), "inheritance of acquired characters" (p. 24), and "orthogenetic" drives (p. 429). For example, Shubin and coworkers point out that "morphological laws," seemingly independent of genetic influence, such as the stability in vertebrate tetrapods of their inside digits (III, IV) and the lability of their outside ones (I, II, V), are probably based on the sequential effects of regulatory genes.

Even when observed, developmental constraints are tied to past evolutionary events and contingencies which impose constraints, some more deep-seated than others. What appears as a "lethal" mutation is the inability of a genetic variant to interact successfully with previously evolved systems that enable birth and survival (metabolic, physiological, anatomical, behavioral, and so on). Thus, in broad perspective, a mutation constrained in one organism need not be constrained in others that occupy other habitats, endure other circumstances, and whose organismic processes differ because of different selective histories.

It seems clear, for example, that the developmental constraints that prevent whales from evolving into horses, and moles into birds emanate from their long selective histories of swimming and burrowing. It is these antecedent selective histories and their intricately stabilized networks of developmental genetic interactions that channel their subsequent evolution and keep whale limbs from selection for running and mole limbs from selection for flying. Compatibility of new features with established genetic functions serves as a powerful selective mechanism.[7]

Although absence of an appropriate adaptive mutation can be declared a constraint, even more so is selection, which by leading adaptation in one direction constrains it in others by making "adaptive" mutations unadaptive for entirely different functions. Using previous examples, even if selection for excellence in swimming is successful, it will impact on traits that enable excellence in running. Similarly, selection for excellence in running sacrifices excellence in flying, and so forth. Organismic (biological) adaptation that can successfully face all possible eventualities is unattainable in a single lineage, simply because organisms do not possess unlimited internal resources to

[6]According to terminology introduced early in the development of mendelian genetics and used later by Goldschmidt and others, environmental comparisons would measure the **norm of reaction** of individual genotypes or the extent to which they are phenotypically affected by environmental change. For some genotypes this norm of reaction is relatively constant, and for others it may be highly variable. Human blood types (for example, *ABO* and *MN*) seem to be relatively unaffected by environmental changes and, once genotypically determined, persist unchanged throughout life. Other genotypes, such as those causing diabetes, can produce phenotypes quite sensitive to environmental changes (for example, diet or insulin). Schlichting and Pigliucci claim that the norm of reaction also applies to phenotypes encountering different *developmental* environments. That is, selection among genotypes occurs for developmental plasticity to produce adaptive phenotypes that can appear in the face of genetic variability and interaction—a concept strongly related to Waddington's notion of "canalization," described later.

[7]Some authors (for example, Schwenk) make a point of separating "internal" genetic and developmental constraints that limit character expression from "external" selective constraints that affect lineages. However, this distinction is often blurred since these constraints can overlap—selection that affects a phylogenetic trend certainly operates on constituent characters, and character constraints can certainly affect evolutionary direction.

continually enlarge one feature without affecting others (Nijhout and Emlen), nor can genes wholly adaptive to every possible eventuality interact successfully in a single organism. Selection toward becoming a "master of one trade" prevents an organism from becoming "jack of all trades."

"Adaptive Constraint" and Its Modifications

Because adaptive limitations arise from selective history, we call such channeling **adaptive constraint,** meaning that it is affected by genes that evolved and were selected previously. Selection can thus act as a phenotypic constraint on two levels: current selection may not favor appearance of a trait and/or past selection may not permit it. Nevertheless, adaptive history also indicates that constraints are not irrevocable: organismic components restricted in structure by a particular condition may evolve in new directions under different conditions. For example, given sufficient time and new opportunities, vertebrate lineages that evolved from fish into crawling terrestrial forms reevolved into still-different swimming marine forms (for example, ichthyosaurs) and even into entirely novel winged aerial forms (for example, bats). Similarly, some mammals did escape the constancy of seven cervical vertebrae, such as three-toed sloths (eight–nine) and manatees (six), and some reptiles had as many as 76 (*Elasmosaurus*). What appears as developmental constraint need not constrain evolutionary change.

Nevertheless, as we might expect, environmental conditions can be restrictive as well as opportunistic, limiting adaptation in only certain directions. Organisms must cope with physical laws that govern their selective environment, such as hydrodynamic laws involved in selection for aquatic speed, and aerodynamic laws involved in aerial flight and in pollen/seed dispersal. Environments can even cause extinction when the requirements for survival exceed the developmental limits of the organism (p. 452). For example, despite advantages in withstanding mutagenic radiation and skin-piercing predation, terrestrial organisms have never evolved epidermal lead or metal armor.

From such broader perspective, adaptive constraints may also be imposed by other groups with which organisms interact. For example, the absence of new metazoan body plans after the Cambrian radiation could well have been caused by the restricted access of new phyla to "saturated" ecological niches already occupied by highly adapted preexisting phyla (Fig. 14–1). Also, competition with successful resident phyla even in new "unsaturated" environments could have limited the development of body plans to variations of older forms rather than make room for distinctively new forms. However, where species can enter relatively unoccupied environments, opportunistic adaptations and morphological diversity are much more likely, as in cases of adaptive radiation (for example, Fig. 3–4).

Although not always obvious,[8] genetic dependency of adaptive constraints is also evidenced by the many regulatory genes that affect development. To these, we can add a genetic mechanism, called **canalization** by Waddington (1962) and **developmental homeostasis** by others, that accounts for uniform expression of traits or patterns of development despite slight changes in environment or genotype.

How Are Phenotypes "Canalized" (Genetically Constrained)?

Among Waddington's canalization experiments were demonstrations in *Drosophila* showing that the phenotypically uniform expression of the normal *Ultrabithorax* gene in the face of environmental stress depended on other "background" genes.[9] The genetic basis for such constant expression became apparent when Gibson and Hogness identified specific genetic loci that support *Ultrabithorax* transcriptional stability.

Figure 15–10 diagrams a scheme devised by Rendel showing how canalization is selected in a population so

[8]The genetic basis for developmental constraint may not be apparent because a phenotypic effect may be at the end of a long chain of interactions. For example, although a gene provides a primary effect in terms of a structural or regulatory product (for example, Fig. 10–25), it may also have a second order, third order, or even more distant effect because its product undergoes a series of interactions with different genetic and environmental factors. Perhaps uncommon, but of serious consequence, are genes with pleiotropic effects influencing many different developmental aspects of an organism, such as the sickle cell mutation and its normal counterpart (Fig. 10–24). Thus, the claim that organismic phenotypes can develop outside of genetic influence because particular phenotypes are not ascribed to particular genes does not really challenge the evidence that organismic development has ultimately an underlying genetic basis built on networks of gene interactions. It is important to realize that although genes are not regarded as traits, they are the instructions that enable traits to develop.

[9]Waddington (1957) also pointed out that traits that depend on environmental stimuli, such as crossveinless wings in *Drosophila* and callosities in vertebrates, can become genetically incorporated ("genetic assimilation") so they appear developmentally without the stimulus. However, this is not a Lamarckian process of direct instruction by the environment, but occurs because of selection for genotypes capable of such response. What seems a close fit between organismic flexibility and environmental change may be a product of underlying genetic components—"the genetic background."

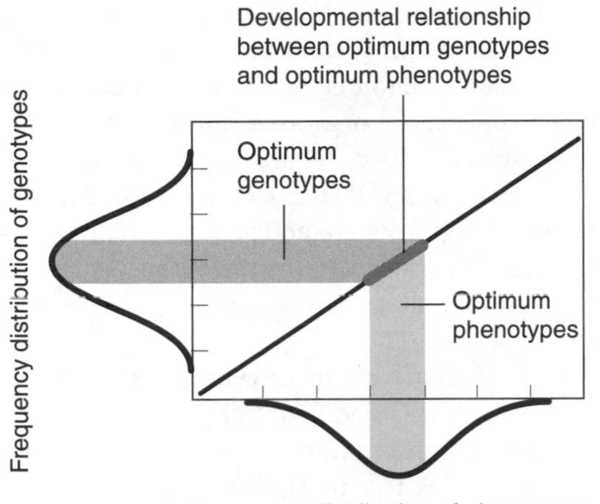

(a) Before canalizing selection

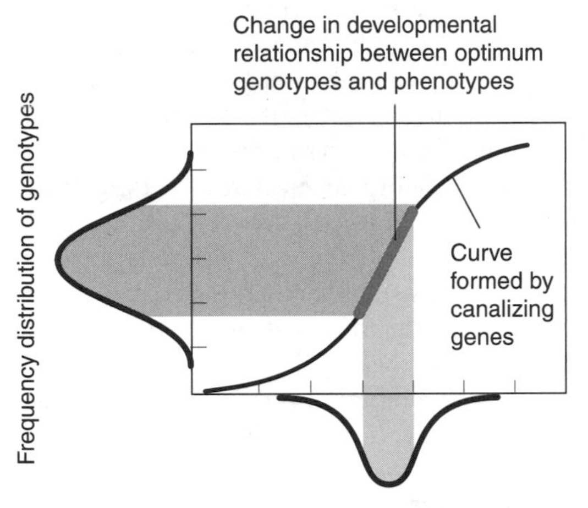

(b) Selection for canalization

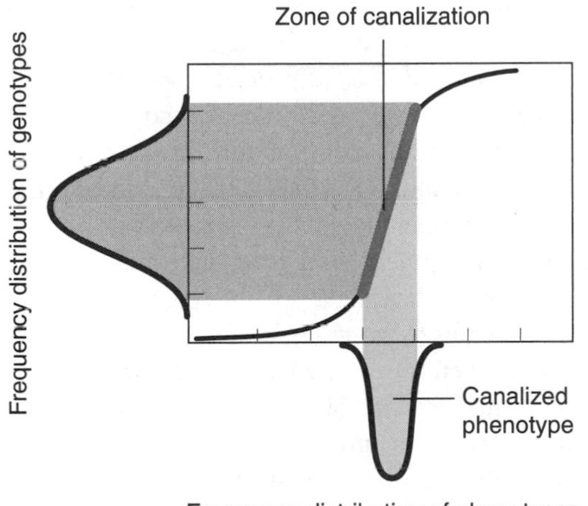

(c) After canalizing selection is completed

FIGURE 15-10 Sequence of selection for a canalized phenotype, or "zone of canalization." The *diagonal,* running from lower left to upper right, represents the developmental relationship between genotype (*vertical axis*) and phenotype (*horizontal axis*); the steeper this developmental curve, the greater the number of different genotypes that produce the same phenotype. (*a*) At this stage the developmental curve is relatively flat, and only a small section of the genotypic distribution produces the optimum phenotype. (*b* and *c*) As selection for canalization proceeds, the developmental curve assumes more of an "S" shape, and larger portions of genotypic variants produce the optimum phenotype. Thus, selection occurs for genotypes that can produce the same phenotype in spite of their variability. The same canalization process explains selection for genes that can preserve phenotypic constancy in the face of environmental change. (*Adapted from Strickberger.*)

that varied genotypes produce a particular optimum phenotype. Thus, in a *Drosophila* population selected for a change in bristle number caused by the *scute* gene mutation, selection can occur on two levels:

1. Selection for change in the primary expression of the major bristle-determining genotype in a mutant scute stock—from an average of two bristles to almost normal four bristles.
2. Selection for change in the zone of canalization so that even varied genotypes produce the same desired number of bristles.[10]

A developmental innovation that produces a new adaptive feature or new pattern can be channeled ("constrained") to produce that adaptation under varied circumstances by the selection of modifying genes that affect the trait's **zone of canalization.** Selection not only affects a trait's variability, but also significantly can change its pattern of variability, evolving an entirely new range of phenotypes.

However, even under selective change, constraints persist. Because organismic features neither arise nor exist independently, and any phenotype is a product of interaction both with other characters and the environment, constraint of some sort affects all characters. For example, we know that genetically correlated, mutually interdependent traits cannot change easily without causing a disruption in other traits (pp. 355–356), thus dampening their rapid response to selection. Although these effects can be difficult to trace, particular constraints must be identified developmentally, and placed in both historical and phylogenetic contexts.

[10]These two types of selection are later expressed as *directional* selection for the first form, and *stabilizing* selection for the second (p. 542 and Fig. 22–8).

From Constraints to Innovations

Despite their prevalence, some constraints are bypassed and developmental innovations do occur when opportune mutations can successfully integrate into a genetic network. Among such events are genetic changes in positional information, such as the homeotic mutations discussed above, which cause structures to be placed, repeated, or omitted in body segments. Other innovations can be caused by genetic changes in timing, when a developmental stage or event shifts in sequence relative to other stages or events **(heterochrony).** As discussed later (p. 397), the precocious onset of an adult stage such as sexual maturity in immature tunicate larvae is one such instance, as is the presence of tadpole features in the adult Mexican axolotl —the classic example of neoteny. Also, a genetic change that affects a seemingly simple but fundamental trait such as an increase in egg size, can lead to subsequent changes that radically modify larval development in sea urchins (Raff, 1992). Among other novelties are developmental growth rate differences between organs—**allometry**—that change between species. For example, as shown in Fig. 20–25, head and leg growth in humans follows a unique allometric path among primates.

In general, developmental patterns that seem deeply imbedded, such as segmentation, may shift toward new directions under changed selective conditions. As discussed in Chapter 16, arthropod segments differ extensively in number and function (see also Fig. 15–7), and segmentation was significantly reduced or even disappeared in mollusks. The presence of homeobox-containing genes in metazoans as diverse as nematodes, arthropods, and vertebrates indicates that even homologous regulatory molecules can be recruited for use in entirely different developmental pathways leading to unique morphologies nonhomologous to those in other phyla.[11]

Some homeobox genes are expressed in entirely novel vertebrate body structures such as the neural crest, limb buds, and gill arches (Benoit et al.). Among insects, even when homeobox genes are organized into identical clusters, they can still lead to different segmentation pathways: "short-term" insects (for example, beetles) in which abdominal segments are added sequentially from one end, and "long-term" insects (for example, *Drosophila*) in which all segments form simultaneously (Raff 1996).

New functions for such regulator genes most probably involve interaction with modified receptors and cofactors. For example, a transcriptional protein that regulates function A by attaching to a particular RNA polymerase promoter (Fig. 10–26) may be recruited to regulate function B by combining with a new transcriptional cofactor that attaches to a different promoter. Or a promoter binding site that stimulates transcription in one species can change, repressing gene function in another species (Singh et al.). Similarly, an enhancer DNA region that influences availability of promoter sites to the polymerase enzyme, and thus affects transcription, can mutate causing a change in regulatory gene activity (Belting et al.) Developmental innovations can therefore appear simply and economically by employing old regulatory genes in new roles (see also p. 349), rather than waiting for appropriate regulators to form entirely *de novo*. At least some (if not many) major phenotypic changes derive from large effects in small numbers of genes rather than an accumulation of small effects from large numbers of genes.

Averof and Patel provide a prime example of how a change in regulatory gene function can produce a major evolutionary impact. The *Hox* genes, *Ubx* and *abdA* (see Fig. 15–8), whose expressions normally produce legs in crustacean thoracic segments, are expressed more posteriorly in advanced groups of these arthropods. This delayed expression allows their anterior thoracic segments to become more headlike—that is, to produce headlike maxillary appendages with larger and more muscular thoracic features. These new limbs can function for both feeding and walking ("maxillipeds"). Among other such findings are homologies between regulatory genes that pattern crustacean gill development with genes involved in insect wing development, indicating a possible regulatory pathway leading from gills to wings (Averof and Cohen).

Proposals for even more extensive regulatory changes are those presumed responsible for dorsal–ventral inversion between arthropods and vertebrates. In arthropods, as in annelids, the circulatory system is dorsal and the nerve cord ventral, whereas these positions are exactly re-

[11]Because different organismic features can be caused by homologous genes and similar "convergent" features can be caused by nonhomologous genes (Koonin et al., Wray and Abouheif), "homology" can be an elusive concept when its genetic basis is ignored and its usage is restricted to morphology and function. Distinctions between homology and convergence should be made genetically and developmentally. In the words of Abouheif and coworkers:

> Homology is a powerful concept. In order to use it consistently when making comparisons across taxa, features should be termed homologous if, and only if, they share a common evolutionary origin. Other criteria, particularly those based on functional similarity, can be misleading. Homology is a hypothesis about the evolutionary origins of a trait, and gene expression data can be an extremely valuable source of evidence supporting homology of a morphological feature, although they cannot be the sole criteria. Any hypothesis of morphological homology based on gene expression data should include: (1) a robust phylogeny of the taxa; (2) a reconstructed evolutionary history of the genes whose expression is being compared; (3) extensive taxonomic sampling, including a broad range of evolutionary informative species; and (4) a detailed understanding of comparative anatomy and embryology. Further, we should regard proposed homologies as falsifiable, and test the possibility that overtly similar gene expression patterns might be due to convergence or recruitment, rather than common ancestry.

versed in vertebrates (Fig. 17–2). Among agents responsible for this inversion are two pairs of homologous genes, one pair (*dpp*, *bmp*) produces products that dorsalize development in arthropods (*Drosophila*) and ventralize development in vertebrates, while the other pair (*sog*, *chordin*) acts oppositely (Ferguson).

On the broadest scale, we can conclude that regulatory mutations of all kinds have been among the key agents of organismal evolution. Like a child's Lego set that produces differently shaped structures by rearranging modular blocks, developmental evolution uses regulatory mutations to produce a variety of new functions by rearranging constituent activity, such as changing the signals, pathways, and targets of signal transduction (p. 349). Such possible changes are inherent in the way development works. Akam, for example, points out that the regulatory effect of the *Hox* gene *Ultrabithorax*, in differentiating the normal *Drosophila* hind wing structure (the haltere), impacts on signals and pathways involving at least thirty target genes. Variation in such genes has been documented in both animals and plants (Moriyama and Powell, Purugganan and Suddith). As regulatory pathways extend, change, and interact, their increasing complexity enables increasing developmental novelty.

The particular effects and histories of these modular changes, however, are only now unraveling, and a full understanding of development and its manifold evolutionary relationships is yet to come. Although development emphasizes ontogenetic changes in "design" or structure, and evolution emphasizes phylogenetic changes in "fitness" based on hereditary variation, these subjects, long kept separate, are being united through an understanding of genetics, which, through its analysis of gene expression, selection, and mutation (for example, Chapters 10, 22, 23), provides a foundation for the relationship among form, variation, and fitness—the underlying process of evolution.

SUMMARY

Development—the recurrent sequence of organismic changes in structure and function from inception to maturity—is primarily a consequence of differential gene activity. In viruses and prokaryotes, molecular analysis of development has provided many such details. For example, we know that genes interact with their environment to exercise both negative and positive controls on viral growth, bacterial metabolism, and sporulation.

In eukaryotes, understanding of development that initially focused entirely on morphology eventually extended to uncovering genetic causes for developmental change. New emphasis was placed on identifying genes that produce differences among embryological stages and how such genetic effects are brought about. In combination with discoveries in molecular genetics, at least ten common themes emerge:

1. Changes in cell structure and function during development derive from changes in the presence or activity of cellular proteins produced by gene transcription.
2. Which genes are transcribed into messenger RNA, which transcriptions are translated into proteins, and which proteins are activated are the result of interactions that often begin with "signals" that can initiate different genetically produced regulatory pathways.
3. As the developmental positions of cells and tissues change in time and space they become subject to new sets of "signals" that lead to further developmental consequences.
4. Cell lineages that share common cellular histories (genes formerly turned on or off) are thus transformed into specific tissues that perform specific functions.
5. Genetic changes can modify each developmental stage by affecting various regulatory agents and processes, from signal reception to transcription and translation.
6. Because of genetic variability, many kinds of developmental genetic changes occur, including those that shift a gene's expression from its normal anatomical position and cause misplacement of organs from expected body segments (homeotic mutations). Developmental novelties also arise from changes in timing of gene expression (heterochrony), changes in growth rate (allometry), and changes in other normally regulated events. Increasing regulatory complexity enables increasing developmental novelty.
7. Whether new mutations can incorporate successfully into an organism's development depends on how they interact with existing developmental processes. Such interactions distinguish between "hopeful" and "hopeless" mutations, and thus "constrain" development in specific directions.
8. Among evolutionary effects that act upon development when organisms face a variety of different stimuli are selection of genotypes that allow either production of an array of phenotypic variants (for example, a variety of antibodies) or maintenance of a single adaptive phenotype (for example, canalization).
9. As for any other trait in which genes and their alleles have been selected for their functional (adaptive) value, it is now clear that an organism's development is the outcome of a historical evolutionary process.
10. Because of many possible complex interactions, detecting developmental evolutionary relationships

can still be difficult, since homologous genes do not always cause developmental similarities, nor do nonhomologous genes necessarily produce developmental differences. (See also p. 251.)

In sum, we can broadly characterize development as a "symphony" of interactive effects during growth, extending from molecules to cells to tissues to organs, whose "melodies" are the various developmental pathways, but whose transmitted "notes" are written in genes, and whose evolutionary relationships must be genetically deciphered.

KEY TERMS

activator
adaptive constraint
allometry
allosteric protein
canalization
coevolution
constraints
developmental constraints
developmental homeostasis
differentiation
epigenetic
genetic constraints
heterochrony
homeoboxes
homeodomain
homeotic
homoeosis
hox genes
inducers
morphogenesis
norm of reaction
phylotypic stage (phylotype)
positional information
preformationist
regulatory pathways
repressor
selector genes
signal transduction
sporulation
zone of canalization
zootype

DISCUSSION QUESTIONS

1. Explain your opinion of the following statements:
   a. "Heritable developmental effects are produced by environmental changes."
   b. "Heritable developmental effects respond to environmental changes."
2. Would you expect changes at early or late stages of development to have a greater effect on development? To be more "constrained"? Explain.
3. Can you suggest ways that a signal transduction pathway can change from transcribing one gene to transcribing another, using the same extracellular signal?
4. Would you expect to find the same kinds of constraint on development before and after a mass extinction? Explain.
5. How does an *Antennapedia* mutant act to modify a head segment? *Ultrabithorax* to modify a thoracic segment?
6. Von Baer (p. 40) proposed that a taxonomic group's general features appear earlier in development than unique features of individual taxa. What support can you offer for this rule?
7. Explain the kinds of information needed to designate a mutation as heterochronic. allometric.
8. Explain whether heterochronic mutations support or violate Haeckel's "biogenetic law" (Chapter 3) that developmental stages recapitulate the phylogenetic sequence.

EVOLUTION ON THE WEB

Explore evolution on the web! Visit the accompanying web site for *Evolution*, 3/e at www.jbpub.com/evolution for web exercises and links relating to topics covered in this chapter.

REFERENCES

Abouheif, E., M. Akam, W. J. Dickinson, P. W. H. Holland, A. Meyer, N. H. Patel, R. A. Raff, V. L. Roth, and G. A. Wray, 1997. Homology and developmental genes. *Trends in Genet.*, **13,** 432–433.

Akam, M., 1998. Hox genes: From master genes to micromanagers. *Current Biol.*, **8,** R676–R678.

Averof, M., and S. M. Cohen, 1997. Evolutionary origin of insect wings from ancestral gills. *Nature*, **385,** 627–630.

Averof, M., and N. H. Patel, 1997. Crustacean appendage evolution associated with changes in *Hox* gene expression. *Nature*, **388,** 682–686.

Bailey, W. J., J. Kim, G. P. Wagner, and F. H. Ruddle, 1997. Phylogenetic reconstruction of vertebrate *Hox* cluster duplications. *Mol. Biol. and Evol.*, **14,** 843–853.

Beadle, G. W., and B. Ephrussi, 1937. Development of eye colors in *Drosophila:* Diffusable substances and their interrelations. *Genetics*, **22,** 76–86.

Belting, H.-G., C. S. Shashikant, and F. H. Ruddle, 1998. Modification of expression and *cis*-regulation of *Hoxc8* in the evolution of diverged axial morphology. *Proc. Nat. Acad. Sci.*, **95,** 2355–2360.

Benoit, R., D. Sassoon, B. Jacq, W. Gehring, and M. Buckingham, 1989. *Hox-7*, a mouse homeobox gene with a novel pattern of expression during embryogenesis. *Eur. Mol. Biol. Org. J.*, **8,** 91–100.

Bharathan, G., B.-J. Janssen, E. A. Kellogg, and N. Sinha, 1997. Did homeodomain proteins duplicate before the origin of angiosperms, fungi, and metazoa? *Proc. Nat. Acad. Sci.*, **94,** 13749–13753.

Britten, R. J., 1998. Underlying assumptions of developmental models. *Proc. Nat. Acad. Sci.*, **95,** 9372–9377.

Casares, F., and R. S. Mann, 1998. Control of antennal versus leg development in *Drosophila. Nature*, **392,** 723–726.

Chu, S., J. DeRisi, M. Eisen, J. Mulholland, D. Botstein, P. O. Brown, and I. Herskowitz, 1998. The transcriptional program of sporulation in budding yeast. *Science*, **282,** 699–705.

Cooper, A. A., and T. H. Stevens, 1995. Protein splicing—Self-splicing of genetically mobile elements at the protein level. *Trends Biochem. Sci.*, **20,** 351–356.

Davidson, E. H., 1990. How embryos work: A comparative view of diverse modes of cell fate specification. *Development*, **108,** 365–389.

de Beer, G. R., 1951, 1958. *Embryos and Ancestors,* 2d and 3d eds. Oxford University Press, Oxford, England.

Dunn, L. C., 1964. Abnormalities associated with a chromosome region in the mouse. *Science,* **144,** 260–263.

Elion, E. A., 1998. Routing MAP kinase cascades. *Science,* **281,** 1625–1626.

Ephrussi, A., and R. Lehmann, 1992. Induction of germ cell formation by *oskar. Nature,* **358,** 387–392.

Errington, J., 1993. *B. subtilis* sporulation: Regulation of gene expression and control of morphogenesis. *Microbiol. Rev.,* **57,** 1–33.

Ferguson, E. L., 1996. Conservation of dorsal-ventral patterning in arthropods and vertebrates. *Current Opinion Genet. Devel.,* **6,** 424–431.

Finnerty, J. R., and M. Q. Martindale, 1998. The evolution of the *Hox* cluster: Insights from outgroups. *Current Opinion Genet. Devel.,* **8,** 681–687.

Frank, S. A., 1996. The design of natural and artificial adaptive systems. In *Adaptation,* M. R. Rose and G. V. Lauder (eds.). Academic Press, San Diego, pp. 451–505.

Freeman, M., 1998. Complexity of EGF receptor signaling revealed in Drosophila. *Current Opinion Genet. Devel.,* **8,** 407–411.

Garcia-Fernàndez, J., and P. W. H. Holland, 1994. Archetypal organization of the amphioxus *Hox* gene cluster. *Nature,* **370,** 563–566.

Gerhart, J., and M. Kirschner, 1997. *Cells, Embryos, and Evolution: Toward a Cellular and Developmental Understanding of Phenotypic Variation and Evolutionary Adaptability.* Blackwell Science, Malden, MA.

Gibson, G., and D. S. Hogness, 1996. Effect of polymorphism in the *Drosophila* regulatory gene *Ultrabithorax* on homeotic stability. *Science,* **271,** 200–203.

Goldschmidt, R. B., 1940. *The Material Basis of Evolution.* Yale University Press, New Haven, CT.

Golub, E. S., and D. R. Green, 1991. *Immunology: A Synthesis,* 2d ed. Sinauer Associates, Sunderland, MA.

Goodwin, B., 1994. *How the Leopard Changed Its Spots: The Evolution of Complexity.* Scribner, New York.

Halvorson, H., and J. Szulmajster, 1973. Differentiation: Sporogenesis and germination. In *Biochemistry of Bacterial Growth,* J. Mandelstam and K. McQuillen (eds.). John Wiley, New York, pp. 494–516.

Hirano, H.-Y., M. Eiguchi, and Y. Sano, 1998. A single base change altered the regulation of the *Waxy* gene at the post-transcriptional level during the domestication of rice. *Mol. Biol. and Evol.,* **15,** 978–987.

Ho, M.-W., 1988. On not holding nature still: Evolution by process, not by consequence. In *Evolutionary Processes and Metaphors,* M.-W. Ho and S. W. Fox (eds.). Wiley, Chichester, England, pp. 117–144.

Ho, M.-W., and P. T. Saunders (eds.), 1984. *Beyond Neo-Darwinism: An Introduction to the New Evolutionary Paradigm.* Academic Press, London.

Hunter, C. P., and C. Kenyon, 1995. Specification of anteroposterior cell fates in *Caenorhabditis elegans* by *Drosophila Hox* proteins. *Nature,* **377,** 229–232.

Jablonka, E., and M. J. Lamb, 1995. *Epigenetic Inheritance and Evolution: The Lamarckian Dimension.* Oxford University Press, Oxford, England.

Kappen, C., and F. H. Ruddle, 1993. Evolution of a regulatory gene family: HOM/HOX genes. *Current Opinion in Genet. Devel.,* **3,** 931–938.

Kappen, C., K. Schughart, and F. H. Ruddle, 1993. Early evolutionary origin of major homeodomain sequence classes. *Genomics,* **18,** 54–70.

Klar, A. J. S., 1998. Propagating epigenetic states through meiosis: Where Mendel's gene is more than a DNA moiety. *Trends in Genet.,* **14,** 299–301.

Koonin, E. V., A. R. Mushegian, and P. Bork, 1996. Non-orthologous gene displacement. *Trends in Genet.,* **12,** 334–336.

Lewin, B., 1997. *Genes VI.* Oxford University Press, Oxford, England.

Lewis, E. B., 1982. Control of body segment differentiation in *Drosophila* by the bithorax gene complex. In *Embryonic Development, Part A: Genetic Aspects,* M. M. Burger and R. Weber (eds.) Alan Liss, New York, pp. 269–288.

Malicki, J., K. Schughart, and W. McGinnis, 1990. Mouse *Hox 2.2* specifies thoracic segmental identity in *Drosophila* embryos and larvae. *Cell,* **63,** 961–967.

McGinnis, W., M. S. Levine, E. Hafen, A. Kuroiwa, and W. J. Gehring, 1984. A conserved DNA sequence in homeotic genes of the *Drosophila* antennapedia and bithorax complexes. *Nature,* **308,** 428–433.

McGinnis, W., R. L. Garber, J. Wirz, A. Kuroiwa, and W. J. Gehring, 1984. A homologous protein-coding sequence in *Drosophila* homeotic genes and its conservation in other metazoans. *Cell,* **37,** 403–408.

Moriyama, E. N., and J. R. Powell, 1996. Intraspecific nuclear DNA variation in *Drosophila. Mol. Biol. and Evol.,* **13,** 261–277.

Müller, G. B., 1990. Developmental mechanisms at the origin of morphological novelty: A side-effect hypothesis. In *Evolutionary Innovations,* M. H. Nitecki (ed.). University of Chicago Press, Chicago, pp. 99–130.

Nijhout, H. F., and D. J. Emlen, 1998. Competition among body parts in the development and evolution of insect morphology. *Proc. Nat. Acad. Sci.,* **95,** 3685–3689.

Nilsson, D-E. , 1996. Eye ancestry: Old genes for new eyes. *Current Biol.,* **6,** 39–42.

Panganiban, G., et al., 1997. The origin and evolution of animal appendages. *Proc. Nat. Acad. Sci.,* **94,** 5162–5166.

Patel, N. H., 1994. Developmental evolution: Insights from studies of insect segmentation. *Science,* **266,** 581–590.

Ptashne, M., 1992. *A Genetic Switch: Phage λ and Higher Organisms,* 2d ed. Blackwell Scientific, Cambridge, MA.

Purugganan, M. D., and J. I. Suddith, 1998. Molecular population genetics of the *Arabidopsis CAULIFLOWER* regulatory gene: Nonneutral evolution and naturally occurring variation in floral homeotic function. *Proc. Nat. Acad. Sci.,* **95,** 8130–8134.

Raff, R. A., 1992. Direct-developing sea urchins and the evolutionary reorganization of early development. *BioEssays,* **14,** 211–218.

———, 1996. *The Shape of Life: Genes, Development, and the Evolution of Animal Form.* University of Chicago Press, Chicago.

Raff, R. A., G. A. Wray, and J. J. Henry, 1991. Implications of radical evolutionary changes in early development for concepts of developmental constraint. In *New Perspectives on Evolution,* L. Warren and H. Koprowski (eds.). Wiley, New York, pp. 189–207.

Rendel, J. M., 1967. *Canalization and Gene Control.* Logos Press, London.

Ross, J., 1996. Control of messenger RNA stability in higher eukaryotes. *Trends in Genet.,* **12,** 171–176.

Salser, S. J., and C. Kenyon, 1996. A *C. elegans Hox* gene switches on, off, on and off again to regulate proliferation, differentiation, and morphogenesis. *Development,* **122,** 1651–1661.

Schlichting, C. D., and M. Pigliucci, 1998. *Phenotypic Evolution: A Reaction Norm Perspective.* Sinauer Associates, Sunderland, MA.

Schwenk, K., 1995. A utilitarian approach to evolutionary constraint. *ZACS,* **98,** 251–262.

Shubin, N., C. Tabin, and S. Carroll, 1997. Fossils, genes, and the evolution of animal limbs. *Nature,* **388,** 639–648.

Singh, N., K. W. Barbour, and F. G. Berger, 1998. Evolution of transcriptional regulatory elements within the promoter of a mammalian gene. *Mol. Biol. and Evol.,* **15,** 312–325.

Slack, J. M. W., P. W. H. Holland, and C. E. Graham, 1993. The zootype and the phylotypic stage. *Nature,* **361,** 490–492.

Strickberger, M. W., 1986. *Genetics,* 3d ed. Macmillan, New York.

Theologis, A., 1998. Redundant receptors all have their say. *Current Biol.,* **8,** R875–R878.

Waddington, C. H., 1957. *The Strategy of the Genes: A Discussion of Some Aspects of Theoretical Biology.* Allen & Unwin, London.

———, 1962. *New Patterns in Genetics* and *Development.* Columbia University Press, New York.

Wolpert, L., 1995. Development: Is the egg computable or could we generate an angel or a dinosaur? In *What is Life? The Next Fifty Years,* M. P. Murphy and L. A. J. O'Neill (eds.). Cambridge University Press, Cambridge, England, pp. 57–66.

Wood, W. B., 1980. Bacteriophage T4 morphogenesis as a model for assembly of subcellular structure. *Q. Rev. Biol.,* **55,** 353–367.

Wray, G. A., and E. Abouheif, 1998. When is homology not homology? *Current Opinion Genet. Devel.,* **8,** 675–680.

Zhang, J., and M. Nei, 1996. Evolution of *Antennapedia*-class homeobox genes. *Genetics,* **142,** 295–303.

16 Evolution Among Invertebrates

Ninety-nine percent or more of all present metazoan species do not have vertebral axial skeletons. These **invertebrates** include jellyfish, worms, squids, starfish, shrimp, flies, and myriad other forms and adaptations. Aside from parasitic invertebrates that have tended to become smaller and less differentiated during evolution, free-living invertebrates have become more complicated, with each innovation often helping the animal gain further control over its particular environment by enabling it to become a better burrower, crawler, swimmer, or food selector. Such adaptations often involve many organs (integument, muscles, nerves, and so on) that act in concert and form part of the architectural framework, or **body plan,** of the animal in dealing with its specialized problems of survival and reproduction.

Differences in body plans provide the basis for separating invertebrates into more than 30 different present phyla, some of which share developmental as well as morphological features. Thus, as discussed in Chapter 14, lacking a coelom separates the Platyhelminthes from many other phyla. Similarly, the radial organization of the Cnidaria (to which some researchers add the Ctenophora) separates them from bilaterally organized phyla. On the embryological level, as described later, the fate of the blastopore (whether it becomes mouth or anus) separates the **protostome superphylum** from the **deuterostome superphylum.** Using such criteria, Table 16–1 shows one approach towards classifying metazoan phyla.

From an evolutionary point of view, these and other morphological and developmental differences reflect some progressive stages in the ancestry of these organisms. Figure 16–1 provides only one of a sampling of proposals offered to relate various phyla evolutionarily. As shown in the lower half of this figure, ancestral lineages which gave rise to the different phyla date to Precambrian times, perhaps as early as 700 million to 1 billion years ago. However, sharp morphological distinctions among these lineages did not appear, or may not have even developed, until the Cambrian period or the immediately prior Ediacaran (Chapter 14). Further molecular studies, based on comparing sequences from the billion or more

TABLE 16–1 One scheme for classifying metazoan phyla according to morphological and developmental criteria

Criteria	Phylum
I. Differentiated tissues and organs poorly defined or absent	Porifera
	Placozoa
	Mesozoa[a]
II. Differentiated tissues and organs	
A. Radially symmetrical	Cnidaria
	Ctenophora
B. Bilaterally symmetrical[b]	
1. Acoelomates	Platyhelminthes
	Gnathostomulida
2. Pseudocoelomates (some authors group these together as the phylum or superphylum Aschelminthes)	Gastrotricha
	Rotifera
	Acanthocephala
	Nematoda
	Nematomorpha
	Kinorhyncha
	Priapulida
	Loricifera
3. Uncertain affinity	Chaetognatha
4. Coelomates	
a. Protostomes	
i. With lophophore (tentacled food-gathering crown)	Bryozoa
	Entoprocta
	Phoronida
	Brachiopoda
ii. Without lophophore	
(a) Nonmetameric or pseudometameric organization	Mollusca
	Sipunculida
	Nemertea
	Tardigrada
	Pentastomida
(b) Metameric organization	Annelida
	Pogonophora
	Echiura
	Onychophora
	Arthropoda
b. Deuterostomes	
	Echinodermata
	Hemichordata (Pterobranchia, Enteropneusta)
	Chordata

[a]Although this classification of phyla is common, some zoologists disagree. Also, some minor groups considered "enigmatic" are not included, and the status of some others are not fully resolved (see Nielsen). One proposal suggests that mesozoans be divided into two separate phyla (Fig. 16–3).

[b]Based on 18*S* ribosomal RNA sequence analysis and the discovery of triploblastic homeobox-containing (*Hox*) genes, the phylum Myxozoa, formerly considered protistan, has now been placed among the metazoans, as possible parasitic hydrozoans.

Source: Adapted from Lutz.

base pairs in each metazoan genome, will undoubtedly help clarify and change many of these relationships, and new paleontological studies may help date their origins more precisely (Conway Morris). Since so much is still conjectural, and a detailed review of possible phylogenetic schemes for all 30-odd invertebrate phyla would be beyond the scope of this book, only some aspects are discussed here.

Porifera (Sponges), Placozoa, and Mesozoa

Sponges are among the most simply constructed metazoan phyla. Their body plan (Fig. 16–2) enables them to extract food particles from water currents and digest them intracellularly. Within sponges, currents are generated by collared, flagellated cells called **choanocytes,** whose flagella move water out through a large exhalant body opening (osculum), thereby drawing water in through small inhalant pores. To filter sufficient water for food and expel the effluent far enough so it does not flow back, the many tiny choanocyte flagella combine to produce a forceful exhalant current that exceeds more than 6 inches a second. For this system to function efficiently, the spongiform body must stand erect, which is accomplished by networks of collagenous fibers called *spongin* and skeletons made of small spicules consisting of calcium or silicon compounds.

Various groups of sponges differ in body organization and kinds of spicules. In all cases, sponges seem to consist of perhaps eight or ten different cell types which do not organize into tissued organs of the types found in more advanced metazoans. For example, sponges have no organized digestive organ, muscular tissue, or nervous network. Function is localized mostly in specific cells, with coordinated movements based on direct cellular contact that usually extends no further than a small area. Moreover, cell determination is often quite flexible. The most generalized sponge cell, the **archaeocyte,** is a large amoeboid cell that can differentiate into all the various other cell types, many of which can dedifferentiate into archaeocytes. A classic experiment showed that a sponge strained through a sieve could redifferentiate into a complete organism.

Since sponge tissue organization seems so easily modified, zoologists have generally considered it inappropriate to characterize the tissue layers by the traditional terms *ectoderm, endoderm,* and *mesoderm.* Nevertheless, there are morphological differences between its external (pinacoderm) and internal (choanoderm) layers; also, a mostly gelatinous intermediary layer (mesohyl) carries archaeocytes and a variety of other cells. Archaeocytes are widely used during asexual reproduction, being incorporated into buds and fragments or into small, hardy, sporelike spheres coated with spongin, called *gemmules.* Sexual reproduction based on meiosis also takes place, producing radially symmetrical, free-swimming larvae that provide the principal means of dispersal.

Taxonomists have used the simplicity of sponges, compared to other metazoans, to place them in either a primitive metazoan phylum (Porifera) or a primitive metazoan subkingdom (Parazoa). Their archaic features include the absence of various structures found in higher organisms: they lack a distinctive mouth, tissued organs, tightly bound cellular sheets (epithelia), and distinguishable anterior and posterior ends. Unusual also are their choanocyte cells, which strongly resemble choanoflagellate protozoans (Figs. 14–4 and 14–9) and other cell types normally absent in animal phyla. Although some workers have suggested that sponges bear some affinity to Archaeocyathids, an extinct phylum, no other animal phyla seem to have evolved from sponges or are obviously related to them. At least from the Cambrian period onward, their simple body plan has successfully generated water currents for feeding, and they are still successful today, with about 5,000 species distributed widely from freshwater to marine areas and from shallow regions to great depths.

Morphologically allied to other metazoa are two phyla of very simple multicellular animals. One, called **Placozoa,** consists of only a single known species, *Trichoplax adhaerens,* a flattened, free-living marine organism only a few millimeters in diameter (Fig. 16–3*a*). It has an upper and lower layer of flagellated epithelial cells enclosing a sheet of loose, fibrous mesenchymal cells that some consider to be mesodermal tissue. Its body can assume irregular shapes as it creeps along the substratum like an oversized amoeba, enveloping food particles and digesting them through its lower surface. Asexual reproduction occurs by fission and budding, and researchers have also found sexually produced eggs in the "mesenchyme" inner layer. Although we know little about this primitive metazoan, it is undoubtedly a remnant of a very early metazoan offshoot.

Also of simple body plan is a group generally called **Mesozoans** (Fig. 16–3*b, c*), of which we know about 50 species, all parasites of marine invertebrates. These too are quite small but apparently have more complex life cycles than Placozoa, involving male and female differentiation and various larval stages. Because of their parasitism, zoologists have proposed that mesozoans are really degenerate platyhelminths that abandoned a free-living lifestyle, reducing and simplifying their tissues in the process. Most biologists, however, still classify them into either one or two separate phyla that may have become parasitic very early in their evolution. As yet, we have not determined relationships among Placozoa, Mesozoa, and other metazoans.

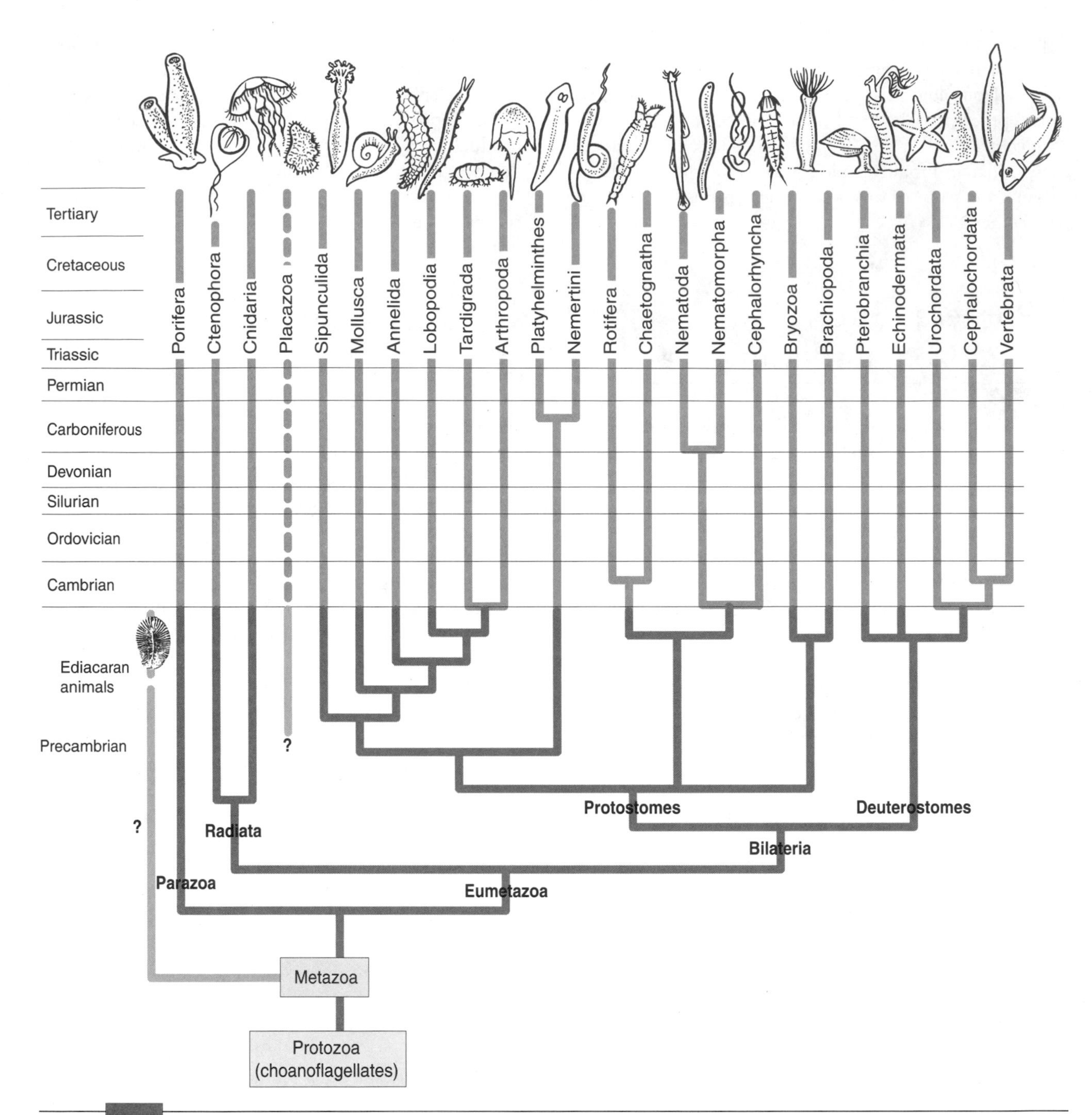

FIGURE 16-1 Major metazoan phyla arranged according to one possible phylogenetic scheme, along with an illustrated sample species for each phylum. Colored lines represent fossil lineages believed to extend, in most cases, to Early Cambrian. In older strata are found Ediacaran fossils and some others of questionable association (Fig. 14-2). The striking divergence among taxa, already obvious in the Cambrian, indicates considerable Precambrian history with possible relationships diagrammed in the lower part of the figure. (Linear distances connecting Precambrian lineages are not scaled chronologically.) Although early metazoan fossils have not yet been found, perhaps because they were unshelled and microscopically sized, some workers have made claims of wormlike burrows dating 1.1 billion years ago (p. 338). Molecular studies based on differently calibrated evolutionary "clocks," provide estimates of protostome-deuterostome divergence ranging from about 700 million to one billion years ago (p. 323). Because of such uncertainties, very early branchings are still in a state of flux, although molecular information challenges former morphologically based proposals. For example, Aguinaldo and coworkers divide protostomes into *Ecdysoza* which moult (for example, arthropods, priapulids, nematodes) and *Lophotrochozoa* with tentacles (for example, mollusks, annelids, brachiopods). Other metazoan phylogenies derive deuterostomes from protostomes (Sidow and Thomas), or place biradial ctenophores as the sister group to deuterostomes (Nielsen). By contrast, using an extensive 18*S* ribosomal RNA sequence analysis, Collins proposes that the extremely simple placozoans (Fig. 16-3) are the closest relatives of bilaterians (protostomes and deuterostomes). In Collins's scheme, a cnidarian-type planula larva (Chapter 14) bearing a layer of mesodermal tissue became

Radiata

Two phyla, Cnidaria and Ctenophora—often called **Radiata** or **Coelenterates**—show a major step forward in metazoan organization by developing a mouth and a specialized gastrovascular digestive cavity (coelenteron). These expandable organs allow coelenterates to ingest much larger food particles than sponges can filter, even permitting them to break down entire prey organisms extracellularly before absorbing them intracellularly.

Tissue organization is also more advanced in radiates than in sponges, with a distinctively organized outer epidermis and inner gastrodermis, which biologists believe to be homologous with the ectodermal and endodermal layers, respectively, of more advanced metazoans. The middle tissue layer of radiates, the mesoglea, is primarily gelatinous, and the other two layers carry on most body functions. Radiates have, for example, both epidermal and gastrodermal muscular tissue whose activity they coordinate with simple nerve nets that enable various body movements for locomotion and food capture. However, researchers have found no specific cells or tissues devoted exclusively to circulatory, respiratory, or excretory purposes.

Reproductively, radiates may use both sexual and asexual modes. Gamete formation, when it occurs, leads to a fertilized egg that develops into a solid, externally ciliated ball of cells, the **planula** (p. 335). Further development of the planula varies in different groups, but the planula itself is a universal feature of sexual reproduction in Cnidaria and also appears in one Ctenophora genus.

As the name Radiata indicates, both phyla are radially organized so that almost any plane through the central oral axis of the animal cuts it into two approximate mirror-image halves. Also, both phyla are soft-bodied and use flexible tentacles to bring food to their extendable oral cavity. Although this soft-bodied structure allows varied changes in shape—in some stages by using the gastric cavity as a hydrostatic organ (p. 336)—the radiate has no hard parts on which antagonistic muscles can operate; that is, it has no levers or fulcra that can amplify movements by using flexor and extensor muscles.

Coelenterates move quite slowly, and their muscles must function over a broader and less efficient range of contraction and expansion than in organisms with hard skeletal structures. Moreover, as the animals grow larger, their mesogleal tissues increase, and they must transmit body wall contractions through bulkier layers. In addition, the gastric cavity has an external oral opening that prevents its use as a hydrostatic organ when the animal is feeding.

Nevertheless, the Radiata have succeeded throughout metazoan history, and now include about 9,000 described species. Present radiates are carnivorous, although the method of food capture varies between the two phyla. In the more common of the two, the Cnidaria, the tentacular epidermis (and often sections of the gastrodermis) is armed with specialized cells, cnidocytes, that contain miniature stinging, harpoonlike organelles called **nematocysts** that immobilize prey and let them be brought to the gastric cavity. This cnidarian feature, used for both offense and defense, apparently dates back to the phylum's early history and may have originated from glandular secretory cells because there are no obviously homologous protozoan nematocysts (Robson). However it arose, this unusually effective mode of food capture certainly helps account for cnidarian evolutionary persistence.[1]

Many cnidarians also undergo developmental changes that produce one of two body forms, the **polyp** and the **medusa** (Fig. 16–4). The polyp is mostly a stationary (sessile) form with a tubular body. Tentacles surround the polypoid mouth at its oral end, and it often attaches to the substratum by a basal disk at its aboral end. The medusa, in contrast, is usually a free-swimming form resembling an inverted umbrella-shaped polyp. Its concave undersurface bears a centrally located mouth surrounded by tentacles that hang down from the umbrella margin.

You can see the close relationship between these two body forms in some cnidarians in the transformation from medusa to polyp when it attaches to a solid substrate. Generally, the polyp functions for stationary food gathering and the medusa for dispersion, although both forms may assume different importance in the various cnidarian classes. Thus, medusae are entirely absent in the Anthozoa (sea anemones and corals), polyps are absent in some of the Scyphozoa (jellyfish) and inconspicuous in

[1]Various shallow-water cnidarians, including coral-reef-builders, harbor symbiotic algae, called zooxanthellae, that provide nutrition through photosynthesis.

FIGURE 16–1 (CONTINUED) prematurely adult ("paedomorphosis," p. 397), giving rise to new triploblastic groups. Supporting this view are findings by Ruiz-Trillo and coworkers, also using 18S ribosomal RNA sequences, who suggest that acoelan flatworms, usually classified in a platyhelminth order, are even more ancient than platyhelminths, and represent the earliest of existing triploblastic metazoans. Valentine (1997) offers a phylogeny based on sequence analysis of the small subunits of ribosomal RNA, and on minimizing mutational changes in embryonic cleavage patterns (Fig. 16–7). He proposes that radial cleavage preceded spiral cleavage, making the deuterostome lineage ancestral to protostomes. From these and other studies (see also Adoutte et al.), we can see that clarifying metazoan relationships is an ongoing process as more molecular, genetic, and developmental information becomes available. (*Adapted from Fortey et al., with modifications.*)

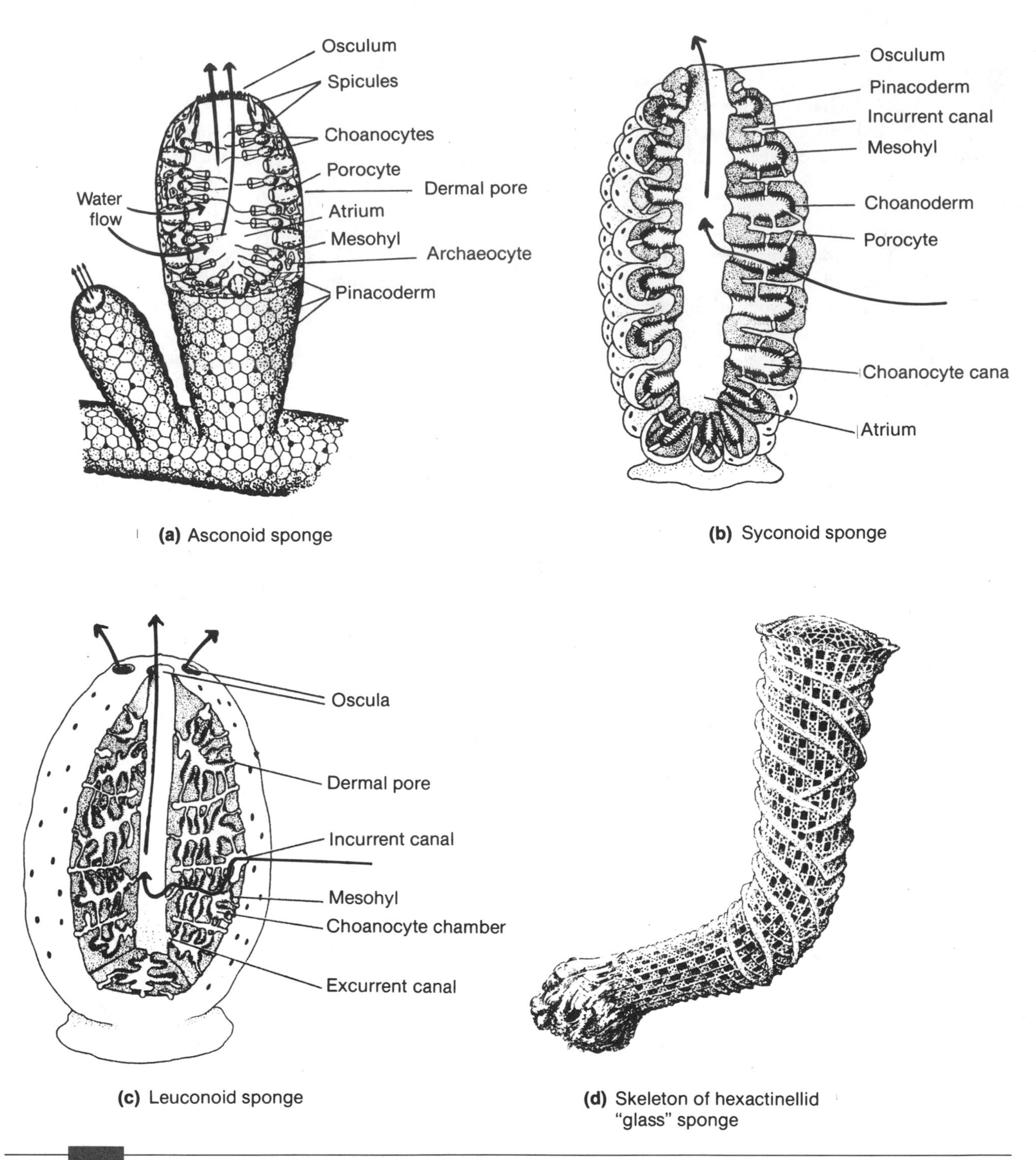

FIGURE 16-2 Models of the three major types of sponge morphology: (*a*) asconoid, (*b*) syconoid, and (*c*) leuconoid. All three forms appear in the Calcarea class, whereas the classes Demospongiae and Sclerospongiae include only the leuconoid type. Because the fourth class, the Hexactinellida, or glass sponges, have a distinctive skeletal framework (*d*), do not have the same type of pinacoderm found in other groups, and lack cell wall separation between many of their cells (syncytium), some researchers have proposed they be placed in a separate phylum (Bergquist).

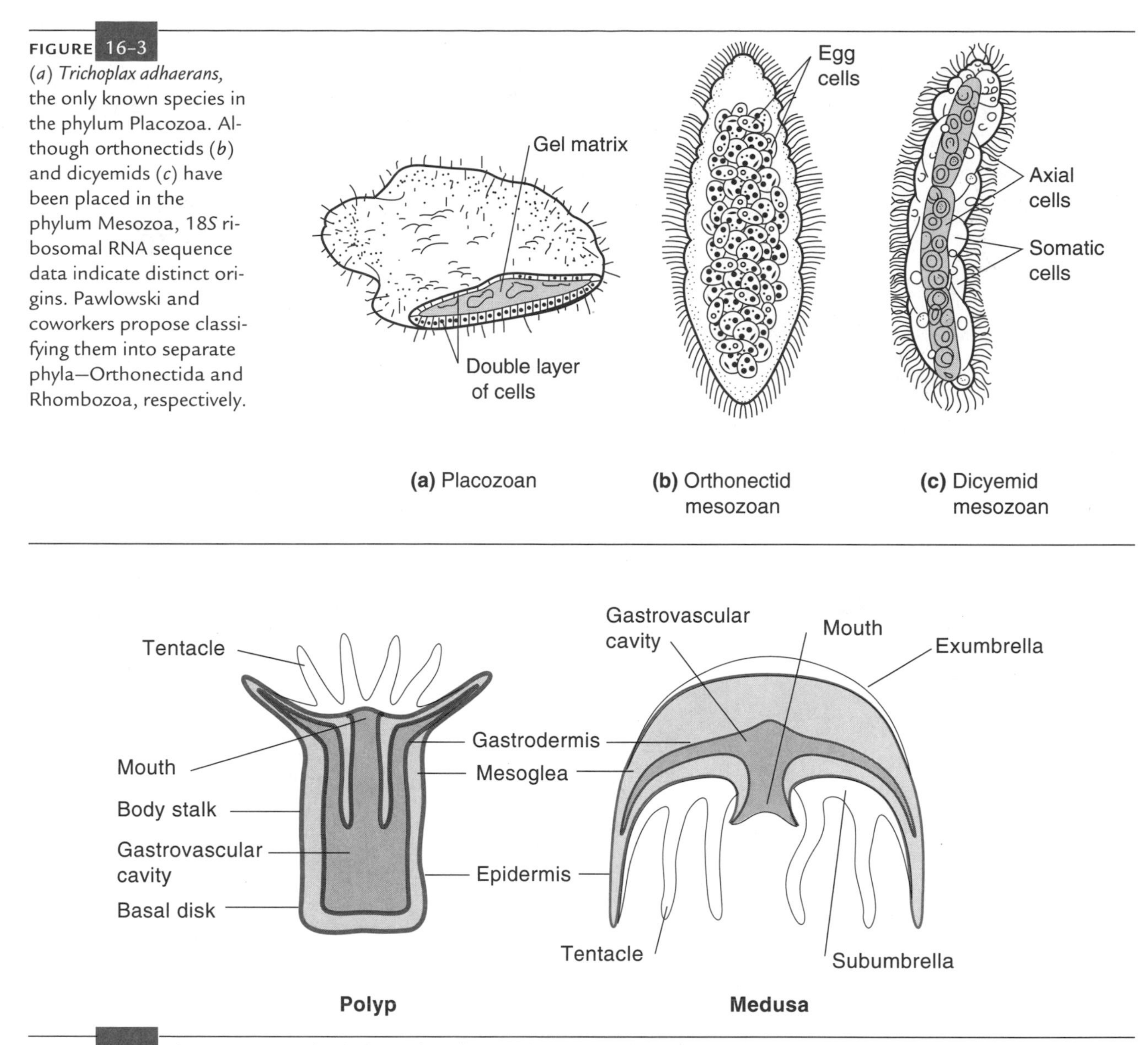

FIGURE 16-3 (*a*) *Trichoplax adhaerans,* the only known species in the phylum Placozoa. Although orthonectids (*b*) and dicyemids (*c*) have been placed in the phylum Mesozoa, 18*S* ribosomal RNA sequence data indicate distinct origins. Pawlowski and coworkers propose classifying them into separate phyla—Orthonectida and Rhombozoa, respectively.

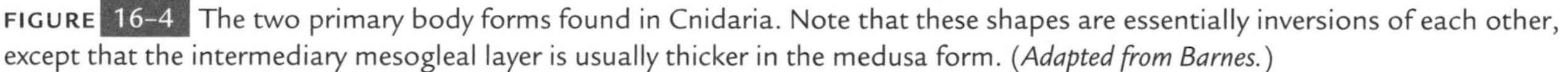

FIGURE 16-4 The two primary body forms found in Cnidaria. Note that these shapes are essentially inversions of each other, except that the intermediary mesogleal layer is usually thicker in the medusa form. (*Adapted from Barnes.*)

others, but both forms can occur in some species of Hydrozoa (hydra and its various solitary and colonial derivatives) and Cubozoa (sea wasps).

Zoologists have offered different views on the evolutionary relationships between these classes, but many researchers agree that a Precambrian group capable of producing larval polyps and adult medusa gave rise to the various Cnidaria and also probably to the Ctenophora (Fig. 16–5). The latter phylum, whose most typical forms are known as *comb jellies,* uses rows of ciliated plates (combs) for locomotion and mostly uses special adhesive cells (collocytes) for food capture. Although ctenophorans differ from cnidarians in tentacle attachment and other traits, many zoologists (but not all) believe they are closely related phylogenetically. (Nielsen argues that ctenophore mesoglea is mesodermal tissue, whose blastopore origin removes Ctenophora from cnidarian affinity and places it as a sister group to deuterostomes.)

Platyhelminthes and Other Acoelomates

Crossing the boundary from **diploblastic** animals with only two embryonic cell layers (ectoderm and endoderm) to **triploblastic** animals possessing also mesoderm, correlated with an increase in number and organizational complexity of

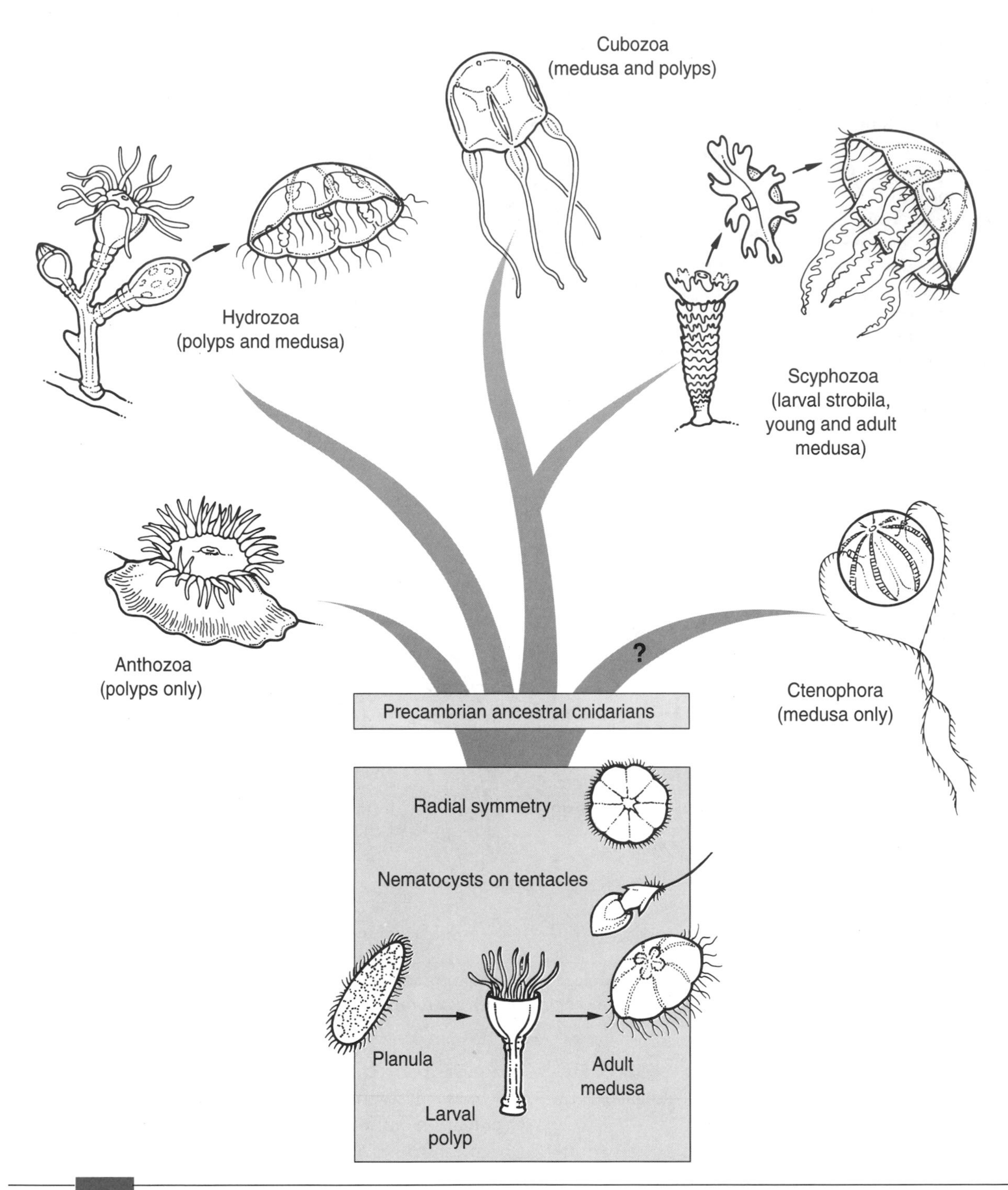

FIGURE 16-5 A possible phylogeny of existing radiates. Zoologists believe the present classes of Cnidaria arose early in evolutionary history, probably in the Precambrian and Cambrian periods, and almost in parallel with the origin of the phylum Ctenophora.

mesodermal cells.[2] These advances enabled a wide range of novel tissues and organs to be generated, such as bundles of circular and longitudinal muscles, excretory organs, circulatory channels and tissues, and complex reproductive systems. Organism size also increased and, together with active locomotion, an anterior–posterior

[2]Although homeobox-like genes have been observed in diploblastic animals (cnidarians, placozoans), their linkage relationships are still unknown, and comparisons cannot be made with organized triploblastic clusters shown in Fig. 15–8 (Schierwater and Kuhn). How far back the "zootype" concept (Fig. 15–9) extends in metazoan history is therefore uncertain. Nevertheless, the cnidarian *Hox* gene is known to influence anterior–posterior axial patterning in *Hydra,* and may perform similarly in other diploblasts.

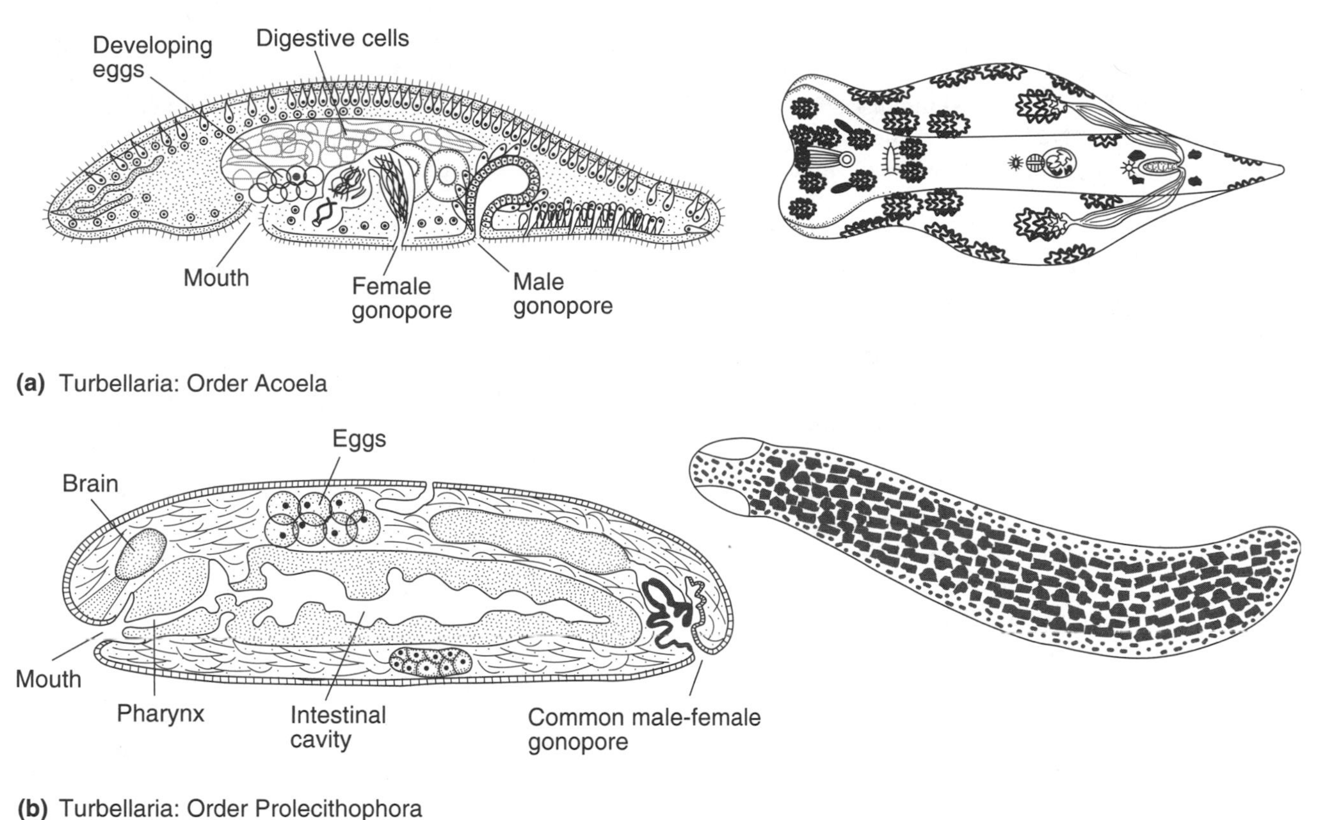

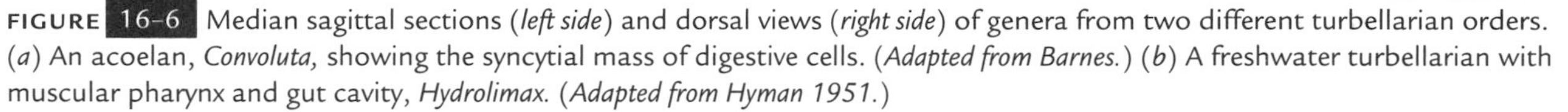

FIGURE 16-6 Median sagittal sections (*left side*) and dorsal views (*right side*) of genera from two different turbellarian orders. (*a*) An acoelan, *Convoluta,* showing the syncytial mass of digestive cells. (*Adapted from Barnes.*) (*b*) A freshwater turbellarian with muscular pharynx and gut cavity, *Hydrolimax.* (*Adapted from Hyman 1951.*)

orientation, or polarity, emerged that aided food gathering and provided various animal groups with bilateral ("left-right") symmetry.

The phylum **Platyhelminthes** represents one of the early, successful stages in the triploblastic progression, comprising at present more than 12,000 species. These *flatworms* have a permanent mesodermal layer from which they derive muscular tissue, an extensive hermaphroditic reproductive system, and relatively simple osmoregulatory organs (protonephridia). A circulatory system for gas exchange and excretion is absent and is apparently not essential in these dorsoventrally flattened animals, whose interior cells generally lie near either the external surface or an internal gut surface. Morphologically significant is the platyhelminth anterior–posterior organization with nervous and sensory structures concentrated at the cephalic end.

In the class Turbellaria, which includes the free-living flatworms, the mouth serves as both entrance and exit for the digestive organ. In the acoelan turbellarians, this organ is a communal cellular mass (Fig. 16–6*a*), whereas in other turbellarian orders it consists of one or more blind sacs similar to the coelenterate digestive cavity (Fig. 16–6*b*). In all cases tissue fills spaces between the internal organs and body wall, and there is no coelomic cavity. Generally, turbellarian locomotion is restricted to cilial movement and/or ventral (pedal) muscular creeping (Fig. 14–11).

The other two classes of platyhelminths are entirely parasitic: the Trematoda (flukes), like turbellarians, have a mouth and digestive cavity, whereas the Cestoda (tapeworms) depend entirely on absorbing host nutrients through the body wall. Once relatively large potential host organisms evolved, **parasitism** became a successful way of life for many platyhelminths because their small flattened bodies do not seriously or immediately hinder host functions. Flatworm parasitic adaptations involve devices that fasten onto host tissues, such as hooks and suckers, as well as the reduction or loss of sensory and digestive organs that are no longer needed for a dependent existence. Most important, in the continuous "arms race" between parasite and host, parasites have evolved integuments that protect them against host enzymes and antibodies, which evolved as protection against parasites.

Although parasitism may have simplified various organs, in many cases it also increased the complexity of the parasite's life cycle. Some tapeworms, for example, may pass through a few intermediate hosts ranging from arthropods to fish before the adult stage develops in the primary mammalian host. Such developmental networks must often have followed opportunities that other evolutionary events provided. Hyman (1951) has suggested that some members of one group of trematodes (order Digenea) originally only infected mollusks. After fish and other vertebrates evolved, many digenetic trematodes invaded these newer groups and retained mollusks as the intermediate host.

The adaptive advantage of such life cycle complexity lets the parasite build up population numbers in intermediate hosts to improve its chances for infecting a primary host. Also, spreading its early stages among intermediate hosts does not exhaust primary host resources and allows the adult parasite to remain productive for relatively long periods. For example, some adult *Schistosoma* trematodes, the source for the widespread tropical disease schistosomiasis, may live for 30 years, and some human tapeworms are active for 20 years or more.

Characteristic of these parasites are their enormous reproductive powers. Asexual reproductive stages often supplement sexual reproduction so that a single adult in some species can potentially produce hundreds of thousands, if not millions, of offspring. Various trematodes and cestodes seem to devote almost their entire anatomy and physiology to reproduction. Since their offspring have extremely low survival rates because of the many chance factors and hazards in parasite distribution and infection, such features must have been long selected. Limited survival opportunities explain why usually only a minority of an appropriate host species is infected by a particular parasite at any one time.

Evolutionarily, zoologists generally believe that the parasitic platyhelminth classes derived from a turbellarian ancestor, since the turbellarians appear to be the most primitive group in the phyla. We do not yet know whether the ancestral turbellarian was an acoelan or had a simple gut of the type shown in Figure 16–6*b*. Molecular studies by Ruiz-Trillo and coworkers suggest that acoelans stand close to the evolutionary base of all bilateral triploblastic metazoans (see also p. 334) and should even be placed in their own distinctive group, different from Platyhelminthes.

Pseudocoelomate Aschelminthes Phyla

Although their form and structure may vary considerably, the simplest metazoans to show a distinctive, fluid-filled body cavity include phyla that biologists often group together under the name **Aschelminthes** (Table 16–2). The body cavity of these phyla characteristically encloses a thin-walled digestive cavity that lacks peritoneal linings, muscles, and supporting mesenteries (Fig. 14–10). Since animals with a true coelom have such structures, biologists generally call the aschelminth phyla **pseudocoelomates.** Many aschelminths also have an epidermal cuticle, adhesive organs, constant cell numbers (eutely), and a digestive tract with mouth, anus, and muscular pharynx that pumps food into the flaccid gut cavity.

The most common and perhaps most representative of these phyla, **nematodes,** have a tubular shape maintained by high internal pressures, which distend the animal to the extent permitted by its thick cuticle (almost like an overstuffed sausage). Overall changes in length are slight since it is almost always fully extended. Nematodes are not highly adapted for burrowing, which demands peristaltic activity. Instead, they function as undulatory swimmers and coilers by means of antagonistic longitudinal muscle contractions. The well-known *Caenorhabditis elegans* is a common experimentally used nematode whose complete genome has recently been sequenced (Chapter 12).

Some authors have proposed that the aschelminths share enough features to unite these groups into one phylum. However, most zoologists find phylogenetic relationships among the aschelminths still difficult to discern, and their distinctions seem great enough to justify classifying them separately. Among them, gastrotrichs are generally considered the most primitive phylum because they are aquatic and ventrally ciliated.

Researchers have also proposed that although some of these phyla are probably related, the pseudocoelomate features of the others may be the result of convergence, and the aschelminth assemblage is most probably polyphyletic (Winnepenninckx et al.). For example, many zoologists suggest that the gastrotrichs, nematodes, and nematomorphs share a common heritage in their derivation from a single group of acoelan turbellarians, whereas the four remaining phyla derived independently from other acoelan groups. There are also suggestions that some aschelminths originally had true coelomic structures, but lost them because of severe size reduction.

Nevertheless, judging from the large numbers of species in these phyla, the pseudocoelomate condition and its various adaptations have endowed many of its bearers with continued evolutionary persistence. Clearly a fluid-filled tube of whatever nature, provided with circular and longitudinal muscles, offered significant advantages both as a hydrostatic organ for locomotion and for carrying metabolites, wastes, and gases throughout the body.

Coelomates

In terms of known numbers of species, distribution of habitats, and total mass, the so-called higher or true coelomates are the most successful metazoans. We classify them into 16 to

TABLE 16-2 Some characteristics of the Pseudocoelomate Aschelminthes phyla

Phylum[a]	Approximate Number of Described Species	Adult of a Sample Species	Lifestyles	Habitat	Features
Nematoda (roundworms)	12,000		Both free-living and parasitic	Marine, fresh-water, and soil	Complex flexible cuticle; lack flagella or cilia; tubular excretory system; mostly dioecious sexual reproduction
Nematomorpha (horsehair worms)	230		Adults free-living, but larvae parasitic in arthropods	Mostly fresh-water and damp soil	Thick, flexible cuticle; digestive tract absent in adults; dioecious
Gastrotricha (gastrotrichs)	400		All free-living	Marine and freshwater	Ciliated ventral surface; pseudocoel is diminished or absent; mostly hermaphroditic, but some parthenogenetic reproduction
Rotifera (wheel animals)	1,800		Mostly free-living; some sessile and colonial forms; a few are parasitic	Mostly fresh-water; others in marine habitats and in bryophytes	Ciliated crown; grinding pharynx (mastax); sexually dioecious, but some parthenogenesis
Acanthocephala (spiny-headed worms)	500		Parasitic	Larval stages in arthropods; vertebrates are final hosts	Both the retractable proboscis and body wall covered with short spines; digestive tract absent; dioecious
Kinorhyncha (kinorhynchs)	100		All free-living	Burrow in marine sediments	Segmented cuticle; lack external cilia; have large movable spines on trunk; dioecious
Loricifera	1		Free-living	Marine sediments	Spiny head; telescoping mouth; abdomen enveloped by plats (lorica); two oarlike tail appendages in larvae for swimming and climbing; dioecious

[a]Some authors include an eighth phylum, Priapulida in this group, related to kinorhynchs and loriciferans.

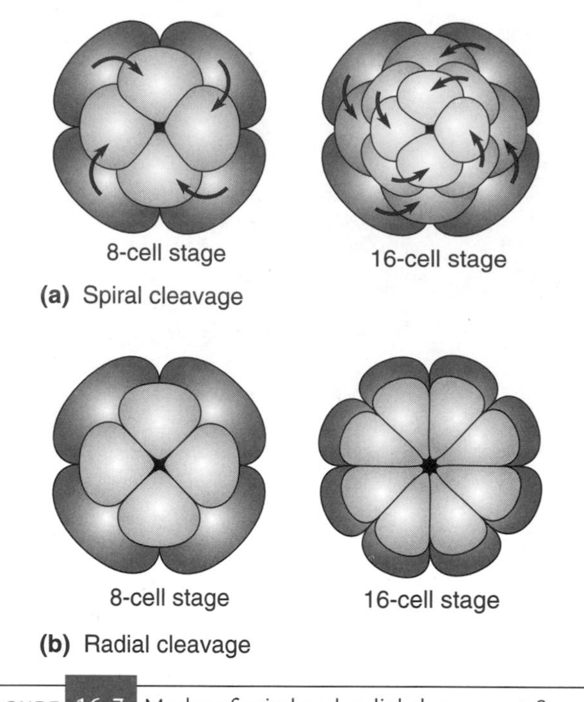

FIGURE 16-7 Modes of spiral and radial cleavages at 8- and 16-cell stages. According to Valentine (1997), cleavage patterns affect traits such as mesodermal tissue origin and coelomic development.

18 phyla, each characterized to varying degrees by a coelom surrounded by mesodermal tissues (Fig. 14–10). Further divisions commonly separate these phyla into two major groups, mostly distinguished by the embryonic location of the mouth. In protostomes the mouth develops at or near the blastopore, which is the blastula groove (Fig. 14–9) that invaginates to form the primitive gut. In deuterostomes, the blastopore develops into the anus, and the mouth develops elsewhere. In addition, coelomic development is mostly schizocoelous in protostomes and enterocoelous in deuterostomes.

Early embryonic cleavage patterns often differ between the two groups, with **spiral cleavage** common in protostomes and **radial cleavage** in deuterostomes (Fig. 16–7). Also, in protostomes, development is mostly **determinate** because regions of the egg differ; embryonic cells descended from differentiated egg regions are committed to their fate at very early stages and cannot develop into a complete animal when separated from other cells. In most deuterostomes, eggs are more homogeneous, and development is **indeterminate**: individual early-cleavage cells can generally develop into complete organisms.[3]

[3]Determinate and indeterminate development commonly have been called "mosaic" and "regulative," respectively. Whatever names used, their developmental distinctions seem based more on gene expression timing than on fixed cellular programs: in mosaic development, regulator genes that determine cell function may merely act earlier than those in regulative development.

We can also make further distinctions among protostome phyla in respect to metamerism: some are segmented, whereas others show no or little segmentation (*ametameric* or *pseudometameric*). All deuterostome phyla show some degree of segmentation. In both superphyla embryonic segmentation traces to a common ancestry, since both are affected by homologous developmental genes, such as *engrailed,* which specifies compartmental distinctions within segments (Chapter 15). Also, both protostomes and deuterostomes have phyla that are primarily sessile and feed mainly by capturing food particles with ciliated tentacular arms called **lophophores.**

Although the fossil record is incomplete for small soft-bodied organisms, most coelomate phyla appear quite ancient, probably of Precambrian origin. Many of the large soft-bodied specimens that appear in the Ediacaran fossil strata resemble coelomates (Fig. 14–2) and may well have had a history that extends 100 million or more years before the Ediacaran period (Fig. 16–1). Certainly by the Cambrian period the protostome–deuterostome divergence was completed, and the Paleozoic era marked the firm establishment of most metazoan body plans.

Since a discussion of each of these phyla is beyond the scope of this book, only some general evolutionary trends in some common coelomate invertebrates are discussed. For further information, refer to textbooks by Brusca and Brusca, Lutz, Nielsen, and Willmer, as well as various collections of articles, such as those edited by Conway Morris et al. and House.

Mollusca

Mollusca, a phylum that now numbers over 50,000 species of **mollusks,** were perhaps among the earliest of metazoan herbivores and had a body plan based on creeping over shallow marine substrates (Fig. 16–8). One of their distinctive features is a **radula,** a rasplike organ bearing chitinous teeth that unrolls from the mouth and scrapes algae from rocks. Some members of the phylum abandoned this herbivorous tool because of new feeding habits, but its vestiges remain.

Because of their early herbivorous habits, many mollusks have long intestinal tracts, dorsally located so that the ventral creeping surface, the **foot,** can remain free. Correlated with protecting and providing respiration for this visceral bulk is a hard shell and a set of active gills enclosed in a body fold called the **mantle.** Various groups of mollusks using this initial architecture have evolved subsidiary changes in structures that are often recognizably molluskan but differ considerably from one another (Fig. 16–9). A few are described below.

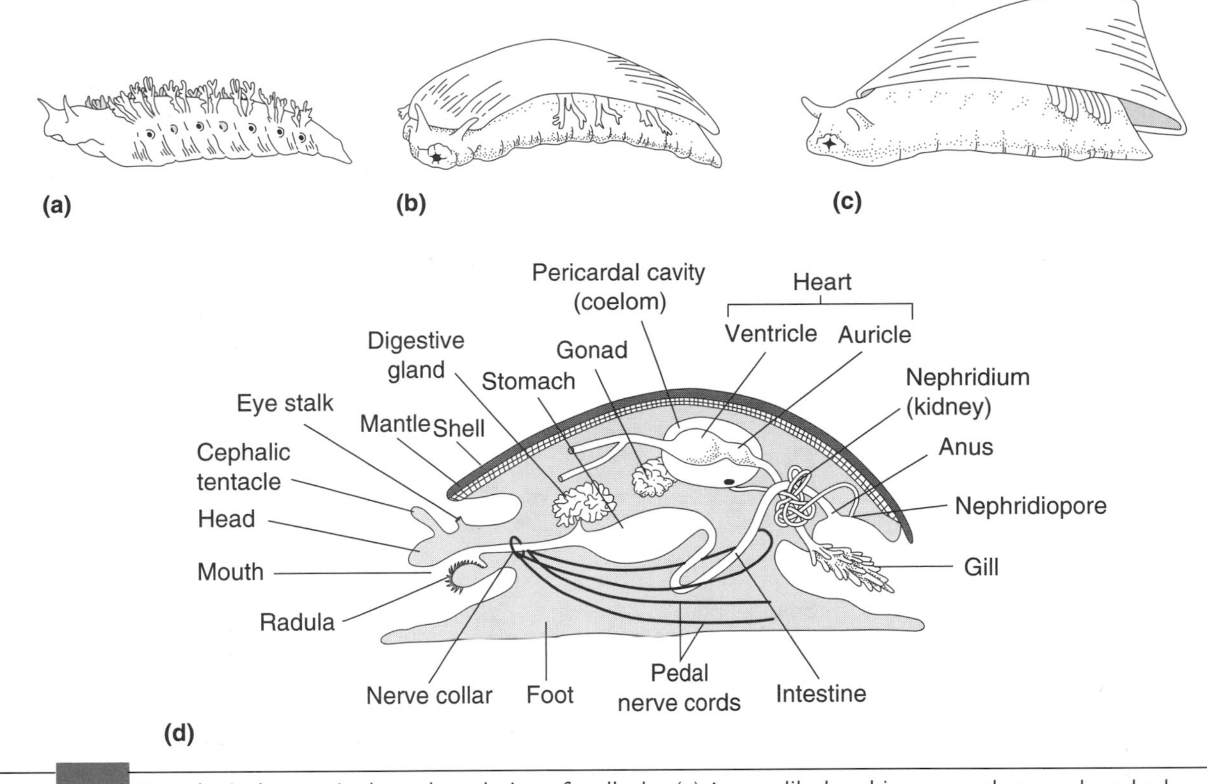

FIGURE 16-8 Hypothetical stages in the early evolution of mollusks. (*a*) A wormlike benthic creeper that may have had repeated sets of external gills and excretory pores. (*b*) Evolution of a protective calcified dorsal shield. (*c*) Development of the shield into a shell that moves forward to cover the head. (*d*) A lateral view of a later evolutionary stage showing the basic body plan and organs of a hypothetical ancient mollusk. (*Adapted from Lutz and from Solem.*)

Gastropods ("stomach-feet"), which include snails, whelks, and limpets, often have a cone-shaped shell, which was originally straight and served not merely as a dorsal shield but as a protected retreat for enclosure of the entire animal. As these animals evolved into larger forms, longer-length shells were apparently difficult to balance, and asymmetric, spirally coiled shells were selected in various lines. At some point, this evolution reached the gastropod stage when the mantle bearing the gills rotated from facing posteriorly to facing anteriorly (**torsion**) with an accompanying rotation of the viscera. Among the explanations zoologists offer for torsion is that it provided room for the head to withdraw and for water to enter the mantle cavity frontally rather than posteriorly.

With torsion, the problem of sanitation became important in gastropods, since the mantle enclosed excretory organs whose products could foul the respiratory gills. In keyhole limpets, one solution was to have a small hole in the mantle and shell immediately above the anus, positioned so the animal can excrete its feces externally instead of passing them down the gills. A more common solution was to eliminate one member of the original pair of gills so that incoming water circulates through the remaining gill and then propels outward along the other side, which now contains the anal and kidney duct orifices. Interestingly, species in one gastropod subclass (opistobranchs) show "detorsion," with a tendency towards reducing the remaining gill and forming new bilateral gills and bilateral symmetry.

Bivalvia (also called Pelecypods or "hatchet feet") are the hinge-shelled **bivalves** such as clams and mussels flattened from side to side. In these animals the gills are much larger than in gastropods, since most bivalves also use gill ciliary tracts to carry captured food particles to the mouth. Because these particles are relatively small and dispersed, the animals no longer need the radula and can collect food in a sedentary fashion. The head greatly reduces, since the animal no longer needs sensory orientation, and the foot either reduces or converts into a burrowing tool. In some bivalves evolution has led to completely separate inhalant and exhalant water currents.

Cephalopods ("head-feet") include the most mobile of mollusks such as squids and octopi. The head, bearing the largest and most complex invertebrate brain, now occupies the main locomotory position formerly occupied by the foot, and the foot has transformed into a ring of tentacles around the head. This style of life represents a transition from relative passivity to rapid mobile aggression and was apparently very successful through the Mesozoic era, as the large numbers of fossil ammonites

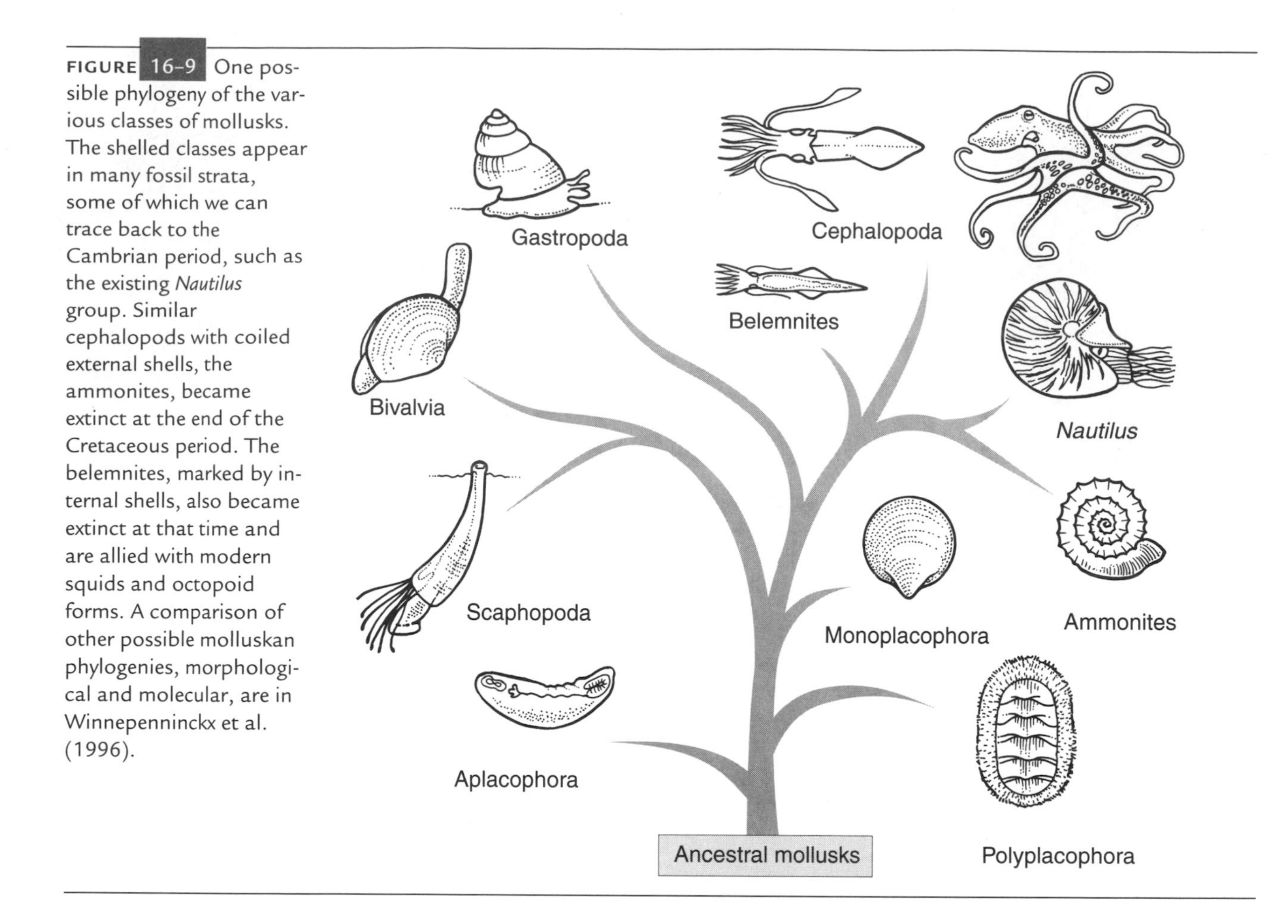

FIGURE 16–9 One possible phylogeny of the various classes of mollusks. The shelled classes appear in many fossil strata, some of which we can trace back to the Cambrian period, such as the existing *Nautilus* group. Similar cephalopods with coiled external shells, the ammonites, became extinct at the end of the Cretaceous period. The belemnites, marked by internal shells, also became extinct at that time and are allied with modern squids and octopoid forms. A comparison of other possible molluskan phylogenies, morphological and molecular, are in Winnepenninckx et al. (1996).

and belemnites bear witness. The mantle is heavily muscularized in this group, and the animal draws water around the sides and squirts it out like a syringe through a tubelike opening called the **siphon.** The cephalopod can aim the siphon in any direction, moving by jet propulsion in the opposite direction.

As a concomitant to hunting, both the cephalopod nervous system and vision develop extensively, and the eyes form images that are probably as clear as those of the similar and evolutionarily convergent eyes of vertebrates (Fig. 3–1). Modern cephalopods, with the exception of *Nautilus* and *Spirula,* show reduced shell size and complexity; in squids the shell, reduced to a chitinous plate, has now adopted the function of an internal skeleton. In the octopus, only tiny shell remnants remain.

Other mollusk groups include two that show some metameric organization, the chitons, or polyplacophorans (multiplated shell), and the monoplacophorans (single flat shell). The latter group was believed to have died out at the end of the Devonian period until a Danish expedition discovered a modern member, *Neopilina,* in the 1950s. Many of its organs, such as gills, muscles, and nervous system, show serial repetition, indicating that mollusks and annelids are probably closely related. Supporting this view is the occurrence of a **trochophore larva** in species of both phyla (Fig. 16–10).

Other zoologists point out that serial repetition of parts is not necessarily the same as annelid segmentation, although both phyla may have shared a common coelomate ancestor. Clark (1979) and others carry the argument further and suggest that the presumed molluskan coelom, its pericardial cavity, probably originated independently of the annelid-type coelom. Since no other evidence of a mollusk coelom exists, they propose that mollusks arose from benthic acoelomate animals that may have been similar to turbellarians and nemerteans. How this question will finally be resolved is unclear, although molecular studies (Fig. 14–5) suggest that the ancestral group from which coelomic metameric arthropods are derived probably also gave rise to coelomic metameric annelids and mollusks.

As in many other instances, the difficulty in discovering intermediate forms between one phylum and another probably stems from the fact that many transitional events occurred in Precambrian times among very small soft-bodied organisms that fossilized poorly, if at all (Fig. 16–1). Moreover, different phyla represent different major types of organization adapted to entirely different ways of life, each phylum often using a new mechanism

FIGURE 16–10
Trochophore larval stages in an annelid (*a*) and a mollusk (*b*). (*Adapted from Lutz and from Barnes.*)

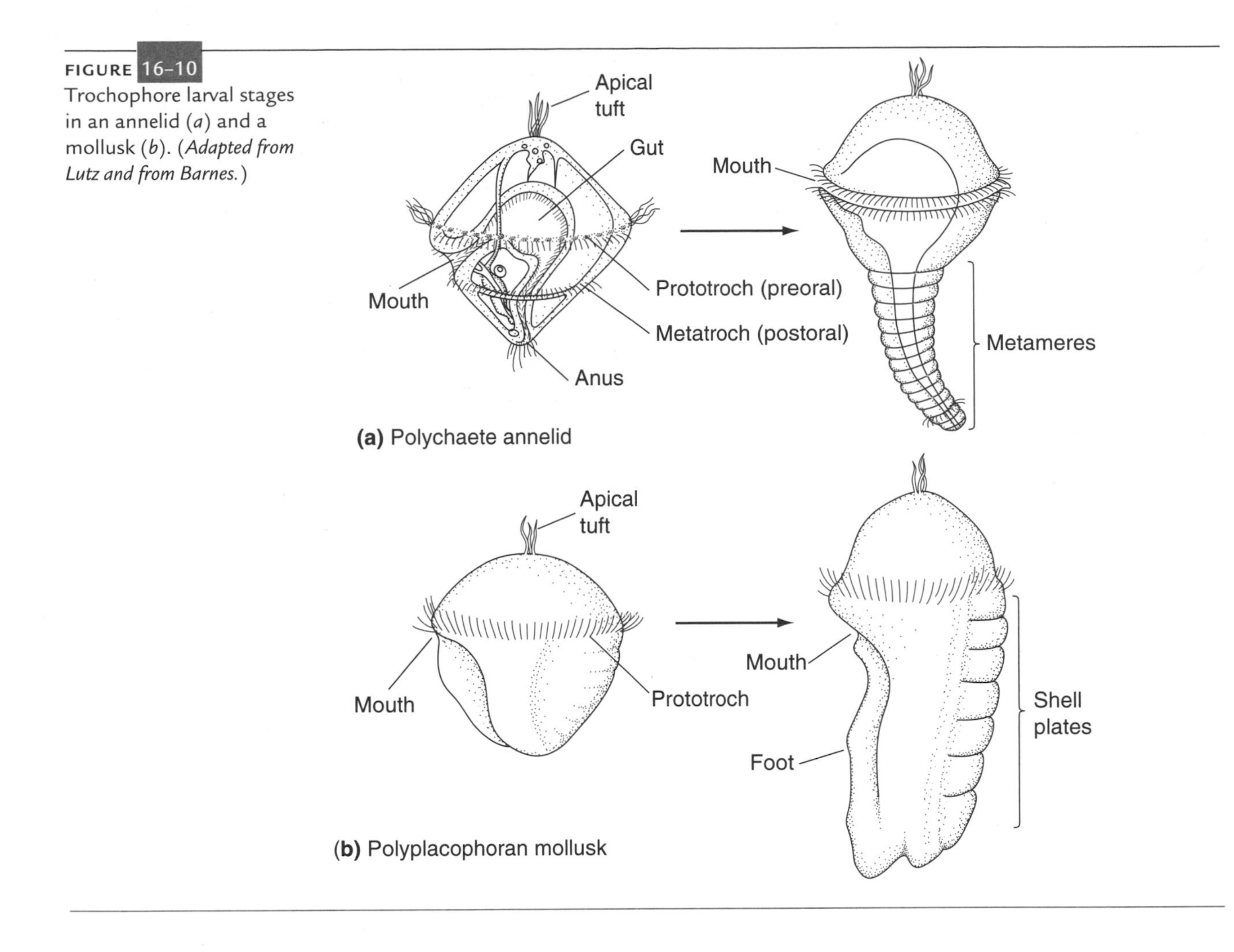

of obtaining food along with distinct metabolic needs, reproductive modes, and so on. Their origin would have been confined to a small number of individuals undergoing very rapid genotypic and morphological changes, with few, if any, opportunities for extensive fossilization at each stage.

Furthermore, these few newly emerging forms would most likely have suffered considerable competitive disadvantages, once better adapted forms arose. The scarcity of most transitional forms, either among living species or in the fossil record, reflects their typical short-term survival and helps explain the relatively wide evolutionary gaps between major groups that occupy different adaptive zones.

Annelida

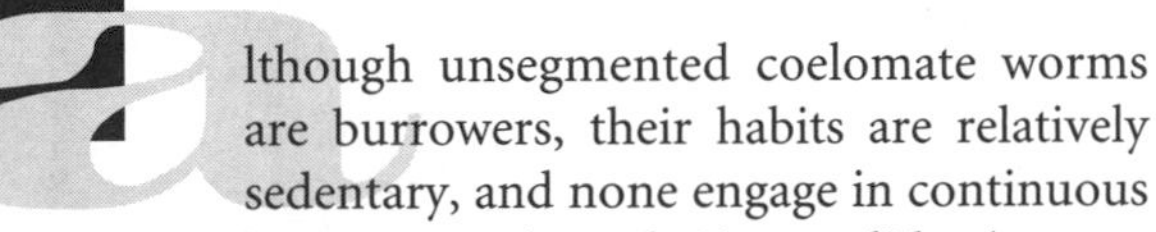

lthough unsegmented coelomate worms are burrowers, their habits are relatively sedentary, and none engage in continuous burrowing. The sausage-shaped Sipunculida (peanut worms) and the similarly shaped Echiura (proboscis worms) are two coelomate phyla that use peristalsis to move slowly through marine substrata. As discussed in Chapter 14, only segmented coelomates such as **annelids** localize hydrostatic pressures to specific segments, primarily because of selection to sustain active burrowing.

Annelids are a group of about 15,000 species that have a soft, wormlike body with various numbers of segmented coelomic compartments separated by transverse septa. Capping their anterior and posterior ends are unique structures, the prostomium and pygidium, different from other segments. Of the three existing annelid classes, Oligochaeta, Hirudinea, and Polychaeta (Fig. 16–11), the first represents an offshoot of what was probably a very early annelid benthic stock. **Oligochaetes** include the familiar earthworms that use spinelike chaetae (or setae) for traction during the continuous burrowing that so enhances soil fertility.

Allied to Oligochaeta are the Hirudinea (**leeches**). Both these groups are hermaphroditic and have a glandular organ called the **clitellum** that usually covers five to ten segments near the anterior end. The clitellum secretes a mucous coat to help bind two copulating animals together and also produces a "cocoon" for depositing fertilized eggs. Because this organ seems homologous in both

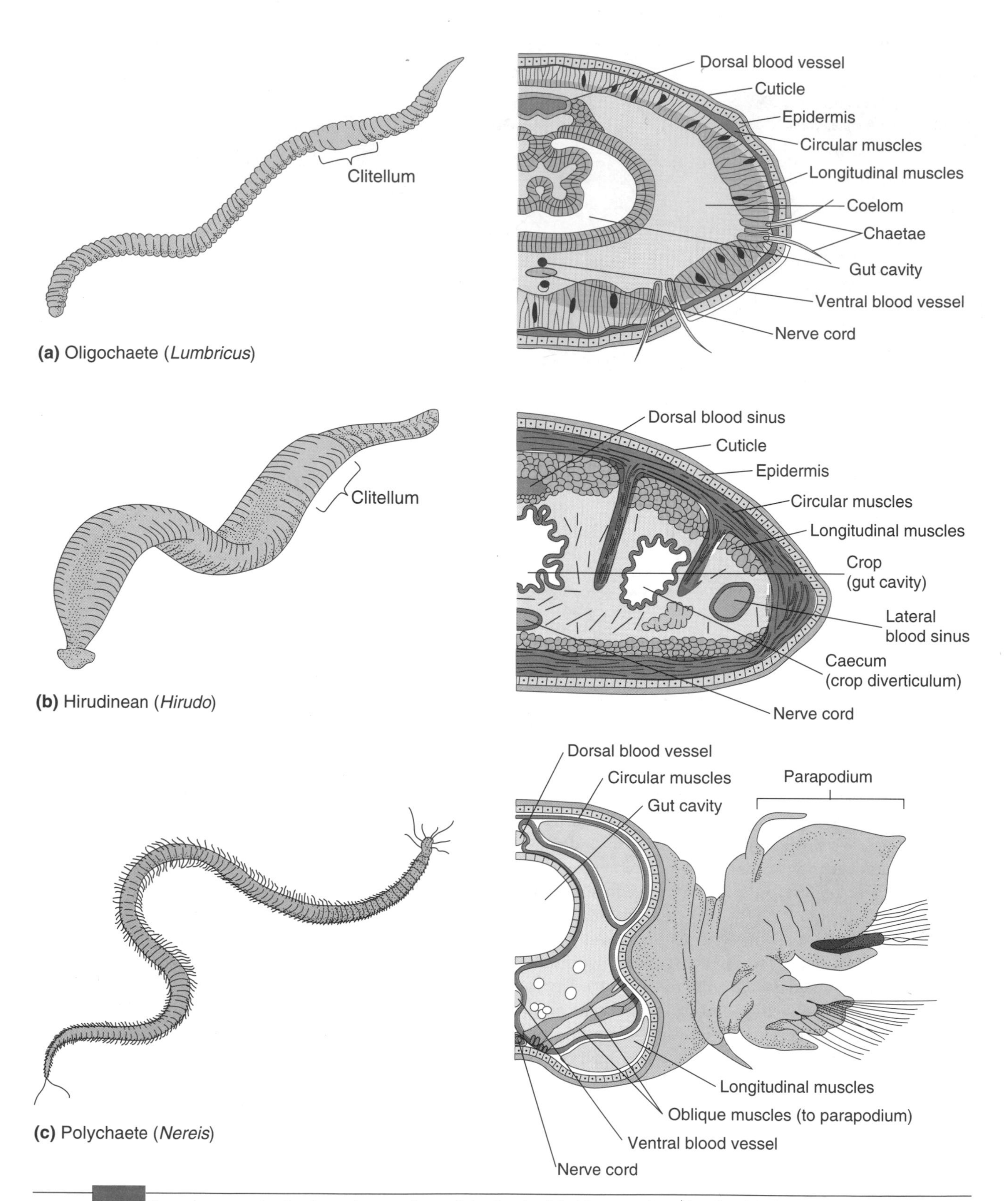

FIGURE 16–11 Sample genera in the three classes of annelids. On the *left side* are entire animals with their anterior ends toward upper right. On the *right side* are diagrammed transverse cross sections of a segment in each animal.

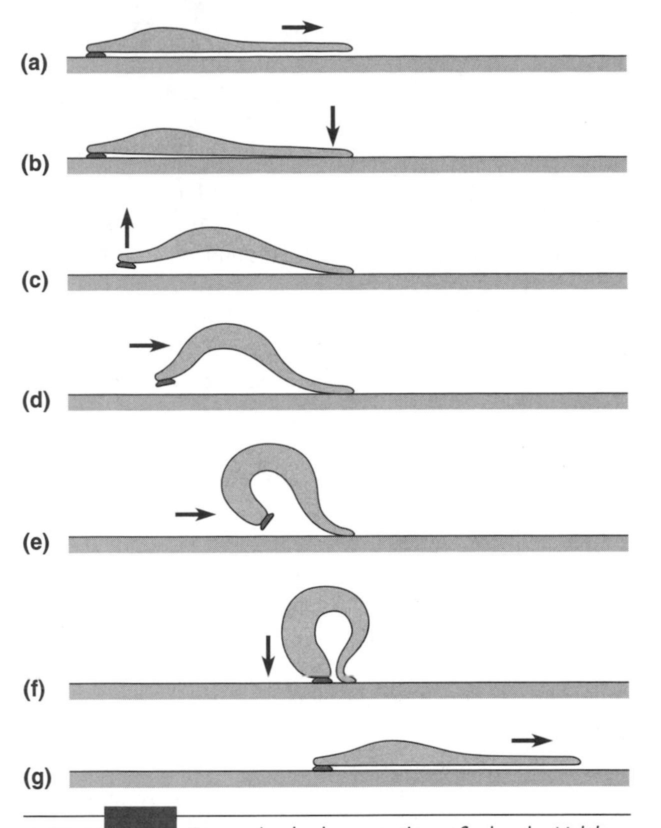

FIGURE 16–12 Stages in the locomotion of a leech, *Helobdella stagnalis*. (*a*) Posterior sucker (left side) fixes to the substrate, and anterior end (right side) extends in direction of arrow. (*b*) Anterior sucker is fixed. (*c–e*) Posterior sucker releases, and animal contracts. (*f*) Posterior sucker fixes immediately behind the anterior sucker, and the cycle begins again (*g*). (*Adapted from Clark 1964.*)

groups, some zoologists consider oligochaetes and hirudineans as subgroups within a "Clitellata" class or subphylum. Leeches, however, have abandoned burrowing in favor of predation or blood-sucking parasitism, and their locomotory habits are mostly based on "looping", in which they contract the entire circular and longitudinal musculature as a unit and don't need peristalsis at all (Fig. 16–12). Since peristaltic locomotion is absent, selection occurs in a reverse direction: they no longer need septa, and even the coelom reduces, since a hydrostatic organ for maintaining circularity and transmitting muscular pressure need not be as great as in a burrowing animal.

By contrast to oligochaetes and leeches, many free-living **polychaetes** such as the marine worm *Nereis* evolved lateral appendages or **parapodia** (see also Fig. 14–16). These structures let them mostly abandon burrowing and provided an adaptation for mobility in loose dispersed material rather than compacted substrates. Parapodial animals would have had increased opportunities to feed directly on the surface bottom rather than hide below it. Once parapodia appeared, many groups of animals must have further developed parapodial musculature for swimming above the substrate.

Using appendages, however, breaks up the body's circular and longitudinal muscles since the parapodial muscles necessary to move them must insert across the internal coelom (in polychaetes, into the midventral line). Thus, the longitudinal muscles do not cover the parapodial region, and the circular muscles are confined to regions between parapodia. These limitations in circular muscles make peristalsis difficult or impossible to perform, and the parapodial animal is now an effective crawler or swimmer, but no longer a good burrower.

Parapodia in a soft-bodied organism such as the annelid polychaete is only a partial answer to the development of appendages, since the "hinge" of the appendage cannot fix firmly in a soft body, and the animal cannot gain maximum leverage for its limbs. This lack of fixity also produces variable lengths of the parapodial muscles, and so they are not as efficient as they would be in a hard-bodied organism. The next great evolutionary advance in rapid locomotion came with the development of hard-bodied skeletons, as in arthropods.

Arthropoda

In **arthropods** a tough, chitinous cuticle provides an exoskeleton with fixed hinges for the jointed appendages. Except for some arthropod larvae that are burrowers and soft-bodied, this hardened skeleton eliminates the need for a hydrostatic scaffold, and the coelom contains only the excretory organs. Although internal septa disappear, various degrees of segmentation remain, indicating an ancestry that probably progressed through animals similar to onychophoran "velvet worms" (Fig. 16–13). Molecular studies cited previously (Fig. 14–5) suggest that segmentation in arthropods and annelids are convergences rather than shared derived characters, but the relationship between these two phyla is nevertheless strong, as indicated by a similarly structured central nervous system with ventral nerve cords and a contractile, dorsal, tubular "heart."

Hardened, chitinous exoskeletons with their inner projections (apodemes, or endophragma) for muscle attachments, jointed appendages, and the development of new kinds of cephalic structures provided many opportunities for arthropod evolutionary radiation: 80 percent of all known animal species today are arthropods, and these appear in almost every conceivable ecological habitat. Because phylogenetic relationships among major groups are not yet clear, zoologists classify the diversity of arthropod species in various ways. One common classification system, shown in Table 16–3, divides the phylum into four subphyla, a view that the molecular studies of Ballard and coworkers (who also include the onychophorans) support. Homologous genes for leg development point to a

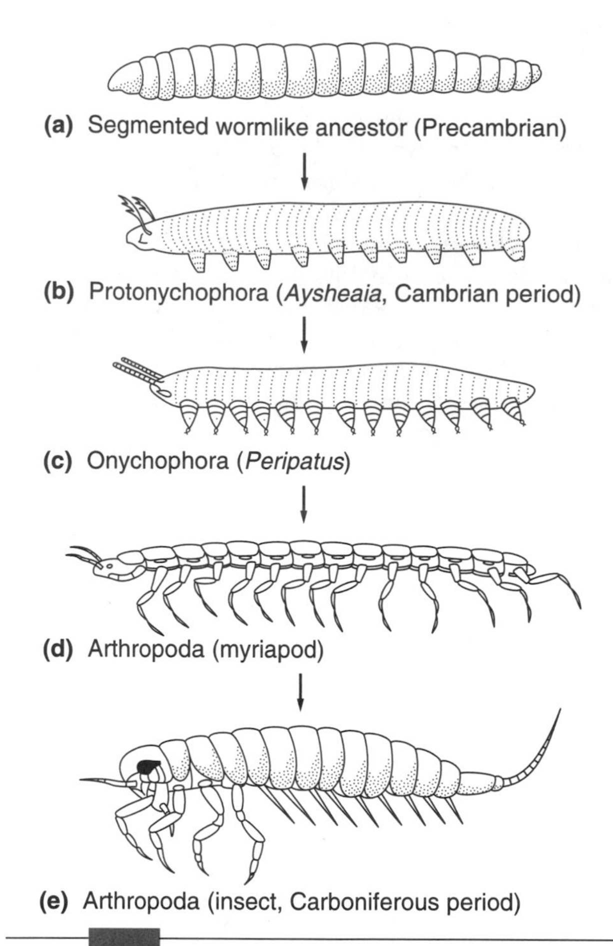

FIGURE 16–13 Evolutionary progression from a segmented wormlike ancestor to arthropods through hypothetical intermediary forms that may have been similar to those found among the Onychophora (see also Fig. 15-7). (*Adapted from Clarke.*)

close relationship between at least two of the arthropod groups, crustaceans and insects (Panganiban et al.).[4]

Whatever their origins, the arthropod armored and articulated body plan clearly gave them almost immediate success. During the Cambrian period, trilobite and merostome arthropods became dominant marine animals in terms of size, mobility, and predatory powers. (Some fossil eurypterids reached almost 10 feet long!) In the later Paleozoic era, these particular groups declined because of competition with more advanced arthropods and other invertebrates, as well as with vertebrates (Chapter 17).

After their marine origin, perhaps the most significant arthropod advance was the invasion of land, which gave rise to terrestrial arachnids and uniramians in addition to some terrestrial crustaceans. Preadaptations for a terrestrial existence would have included a hardened cuticle that, with waxy waterproofing, could act as a barrier to desiccation; ability to burrow into shoreline sand and soil enabling a terrestrial foothold; and protected gills that, with minor changes, could act as lungs. Among arthropods, **insects** (class Hexapoda) underwent what is probably the most explosive radiation of any metazoan group, occupying practically all of the many varied terrestrial habitats. Entomologists have described approximately three-quarters of a million insect species, and probably more than twice that number remain undescribed—perhaps millions in the tropics alone (Erwin)!

At present we find it hard to trace the causes for such relatively rapid evolutionary changes and widespread radiation, but they are undoubtedly tied to the small size of insects, their rapid generation times, their evolution of winged forms,[5] and their apparently endlessly malleable structures. Practically all sections of insect anatomy—such as legs, mouth parts, wings, and eyes—show a capacity for evolving new structures and functions that superficially seems as easy as inserting interchangeable parts in a child's modular building toy (Fig. 16–14). As pointed out by Shubin and coworkers:

> [T]he "arms race" of the Cambrian explosion may well have been a "limbs race" among arthropods to evolve better sensory, locomotory, feeding, grasping, and defensive appendages.

Such modular novelties seem to derive from the facility with which regulatory changes, such as homeotic mutations (Chapter 15), can modify individual segments in a serially repetitive structure, and provide a wide spectrum of adaptations (see, for example, Whiting and Wheeler). Used in this fashion, different modular replacements and substitutions lead to "mosaic" evolution or "tinkering"—a much more rapid tempo of evolution than attempting to fabricate a completely novel architecture from undifferen-

[4]A different view, which Manton and others promoted, is to classify each of the four groups as a separate phylum because each probably originated from a different annelid-like ancestor, probably in Precambrian times. That is, arthropods evolved polyphyletically, and we should consider them a grade or **superphylum** rather than a single monophyletic phylum. This view, based on development and morphology, points to embryonic distinctions, differences in leg structure, and differences in metameric segments. Opposed to these arguments are claims that such differences represent new features derived from a monophyletic origin (Raff). In support of monophyly are molecular studies showing that *Hox* (homeobox-containing) gene expressions affecting head segmentation are the same in chelicerates, myriapods, crustaceans, and insects (Damen et al.). Some relationships seem uncertain, and firmer conclusions await further sequencing (Regier and Shultz).

[5]Marden and Kramer suggest that insect wings evolved in aquatic forms who used gill plates for rowing and skimming across water. As evidence for a water-skimming origin, they point to stoneflies that flap their wings along the water surface to gain aerodynamic thrust without aerodynamic "lift." Once evolved for such purpose, stronger and larger water-skimming "wings" could then undergo selection for aerial flight. According to Averof and Cohen, the genetic basis for transforming gill appendages into wings comes from changes in regulatory genes used in developing gill structures in biramous (branched) crustacean legs. Some genes, such as crustacean homologues of *Drosophila's engrailed,* help develop anterior–posterior compartments, while others such as homologues of *apterous* help develop a dorsal–ventral pattern.

TABLE 16-3 Some characteristics of the four subphyla of Arthropoda

Subphylum	Approximate Number of Existing Species	Adult of a Sample Species	Habitat	Pairs of Antennae	Pairs of Legs	Characteristics and Features
Trilobitomorpha			Marine	1	Many	Body with variable numbers of segments organized into three longitudinal lobes; biramous (two-lobed) legs
Trilobites	Extinct					
Chelicerata			Marine	0	5	Anterior appendages are chelicerae (pincers or fangs); uniramous (one-lobed) legs; book gills, waxy cuticle, and muscular pumping pharynx (arachnids)
Merostomes (horse-shoe crabs; eurypterids—extinct	4					
Arachnids (scorpions, spiders, ticks, mites, and others)	73,400		Mostly terrestrial	0	4	
Crustacea			Mostly marine, some freshwater and terrestrial	2	Many	Biramous legs; carapace (dorsal cover); paired compound eyes; larvae when present are of nauplius type
Crustaceans (crabs, lobsters, shrimp, copepods, isopods, barnacles, and others)	40,000					
Uniramia			Terrestrial	1	Many	Uniramous legs; 2 pairs of maxillae; malphigian tubule excretory system; paired compound eyes, and complete metamorphosis through pupal stages (many insects), respiration via air tubules (trachea)
Myriapods (centipedes, millipedes)	10,500					
Insects (flies, beetles, bugs, bees, locusts, and so on)	750,000		Terrestrial, aerial, and aquatic	1	3	

tiated elements. Nevertheless, there are limits to the number and kinds of adaptations an organism can incorporate successfully because of both restricted internal resources and conflicting developmental interactions (p. 357). The unrelenting impact of selection drives biology toward available and immediate pragmatic solutions rather than toward ultimate (and unattainable!) perfection.

Other advantages accrue to some insect orders that possess specialized larval stages living and feeding differently from adult forms. This divided lifestyle allows these groups, especially those with winged adults, to finely partition their resources and exploit a wide range of environments and diets. Beetles, with more than 300,000 described species, have an additional adaptation in the

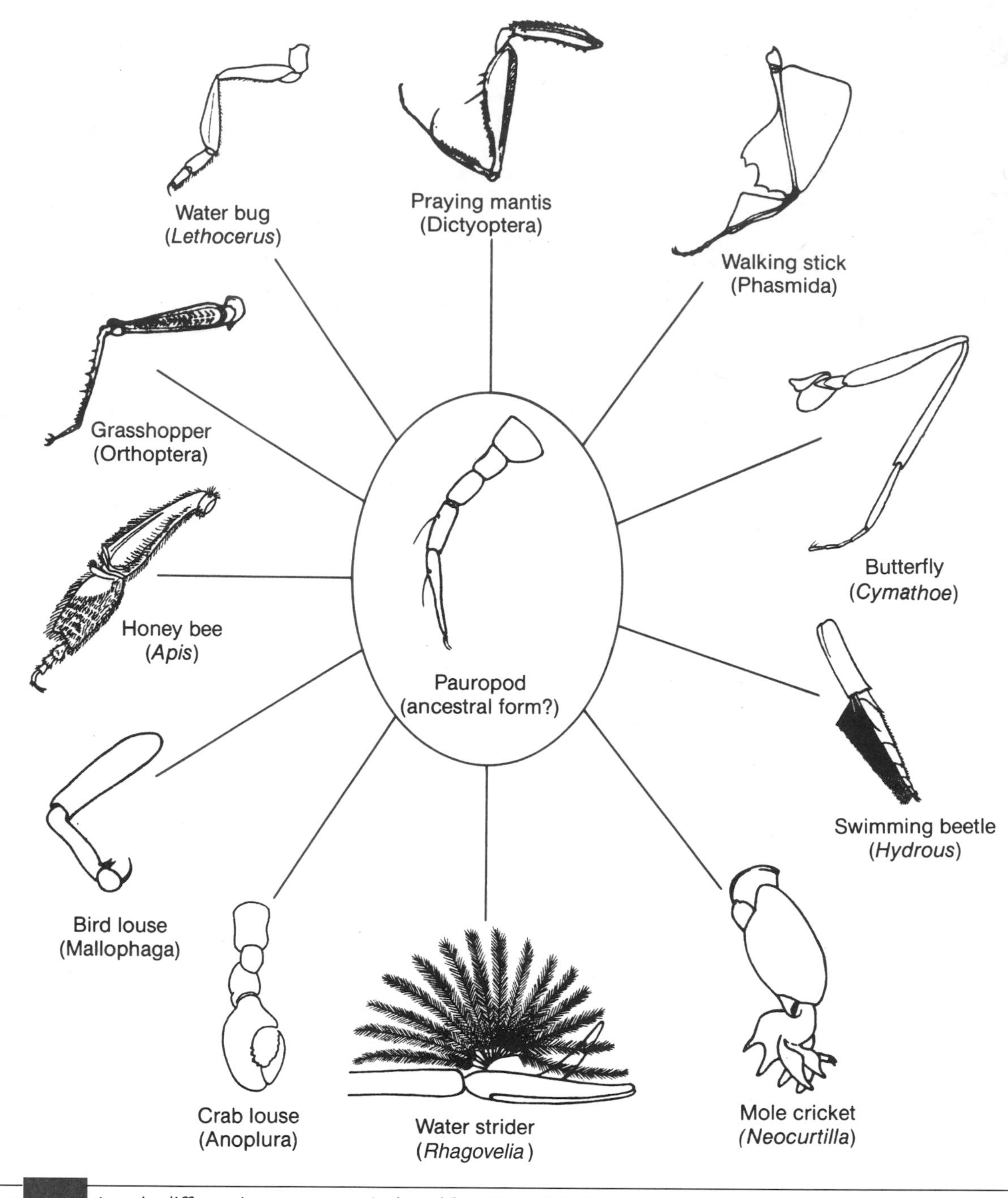

FIGURE 16–14 Legs in different insect groups. (*Adapted from Fox and Fox.*)

extra-thick chitinous armor that protects even their vulnerable wings against predators and parasites.

It seems reasonable that once such highly adaptive traits appeared among these arthropods, enabling them to enter and thrive in almost any small vacant terrestrial niche, the chances for new phyla to achieve body forms that could successfully compete with these arthropods diminished. Thus, among metazoans occupying terrestrial insect-sized niches, insects would have triumphed by arriving first and adapting first.

It is also among insects that the only social animal organizations appear until we reach the vertebrates. **Social organization** entails a division of labor among different members of a group, a phenomenon far beyond the kinds of simple aggregation in swarms of migrating locusts or in the cooperative "tents" that some caterpillars construct. The truly social insect societies include the order Isoptera (termites) and various members of the order Hymenoptera (ants, bees, and wasps).

In these socialized insects, different morphological types (castes) or age groups assume different social functions. Each colony, for example, usually has only one fertile queen engaged in egg production, one or more fertile males for egg fertilization, and one or more classes of ster-

FIGURE 16–15 Schematic illustration of the effect of hymenopteran haplodiploidy on the relationship between sisters ("workers") derived from the same mother ("queen") and father, using three alleles of a hypothetical gene, *R*. Fifty percent of the genes shared by such workers come from their haploid father who gives the same set of genes to all his daughters. All these female offspring also share half their remaining genes (.50 × .50 or 1/4 of their total complement) because genes in different haploid eggs produced by their diploid mother have the probability of being 50 percent alike. The total frequency of shared genes among workers is therefore .50 + (.50 × .50) = .75. In contrast, these workers (R^1R^3 or R^2R^3) share only 50 percent of their genes with their queen mother (R^1R^2), since queen and workers have different fathers (for example, R^2 and R^3 males), who contribute 50 percent of all female genes. This helps account for workers' reliance on their mother to produce their more closely related sisters then engage in producing their own more distantly related daughters or allow their sisters to produce even more distantly related female offspring. Genetic relationship also leads to a conflict of interest about a colony's sex ratio. Being more related to sisters, it is to the workers' genetic benefit that the colony support more females and fewer distantly related males. Queens, by contrast, being related equally to sons and daughters, benefit genetically with an equal male:female offspring ratio. Since workers control egg-to-adult development, fratricide can be expected at various stages (Seger). When the queen mates more than once, genetic relationships between female workers decline, but a worker is still more closely related to her mother's female offspring than to her sister's female offspring, again accounting for the destruction of workers' eggs by other workers. Genetic relationship between workers and male offspring follows a somewhat different pattern, depending on whether queens mate once or more than once (Ratnieks and Visscher).

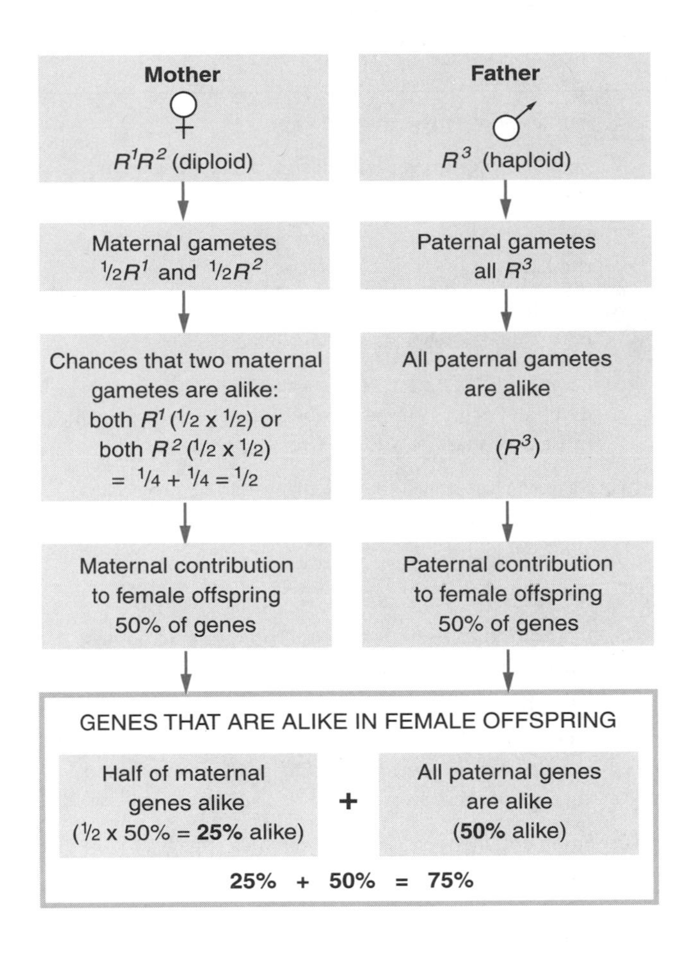

ile workers that are exclusively engaged in food gathering, cooperative brood care, nest maintenance, and defense of the colony. In worker honeybees, age-related divisions of labor primarily tie younger bees to caretaking and maintenance, and older bees to food foraging and defense.

These behavioral differences correlate with differences in juvenile hormone, younger bees possessing lesser amounts than older bees. Circadian rhythm also varies with age because of different expression of a gene homologous to the *Drosophila* gene, *period*: it is turned on in foragers, imparting diurnal activity, and turned off in younger bees who perform brood care in an irregular daily pattern (Robinson). Chemical influence on development and behavior also differentiates fertile queens and sterile workers. The queen produces "pheromones" that suppress fertility in workers, but that various environmental influences such as special foods (for example, royal jelly in honeybees) can overcome.

In termites both sexes derive from fertilized eggs, but in hymenopterans the distinction between sexes derives from development of males from unfertilized eggs and females from fertilized eggs (haploid egg males and diploid egg females).[6] Researchers have pointed out that this hymenopteran system of **haplodiploidy** especially is prone to evolve sociality since, as shown in Figure 16–15, the diploid female offspring of a queen share more genes with their sisters (75 percent) than they share with their own daughters (50 percent).

According to Hamilton, haplodiploidy encourages **kin selection** (a name John Maynard Smith invented) in

[6]Although the haplodiploid:haplodiploidy hymenopteran system may produce different sex ratios based on various environmental factors, in one case the sex ratio decision seems clearly associated with maximizing the fecundity of offspring. In this instance, *Lariophagus distinguendus* wasps that parasitize the larvae of common granary weevils, the female lays differently sexed eggs in response to the size of the wheat grain containing the larval host (Charnov). If the wheat grain is relatively large, the wasp inserts a single fertilized egg (female) into the weevil larva; if the grain is relatively small, an unfertilized egg (male) is injected. This difference appears caused by the difference in resources needed for the fertility of male and female offspring: a larval host in a larger grain enables female offspring to be more fecund because it supplies more resources, whereas a larval host in a smaller grain can still supply enough resources to enable a male offspring to produce a large number of viable sperm.

TABLE 16-4	One proposed sequence of behavioral changes leading to the evolution of sociability and caste divisions in wasps
1.	Female stings prey, then lays egg.
2.	Female stings prey, places it in a convenient niche, then lays egg.
3.	Female stings prey, constructs a nest on the spot, then lays egg.
4.	Female builds a nest, stings prey, transports it to nest, then lays egg.
5.	Female builds a nest, stings and transports a prey item, lays egg, then mass provisions egg with several more prey that are added before egg hatches.
6.	As in (5) but prey items are progressively provided as the larva grows.
7.	As in (6) but progressive provisioning occurs from the start.
8.	In addition to provisioning in a preconstructed nest, female macerates prey items and feeds the pieces directly to the larva.
9.	The founding female is long-lived, so that offspring remain with her in the nest. Offspring add cells and lay eggs of their own.
10.	Small colony of cooperating females engages in tropholaxis (liquid food exchange).
11.	Behavioral differences between a dominant queen caste and a subordinate worker caste appear; unfertilized workers may still lay male (haploid) eggs.
12.	Larvae are fed differentially; queen and workers that result are physically distinct, but intermediates remain common.
13.	Worker caste is physically strongly differentiated, and intermediates are rare or absent.

Source: From Hölldobler and Wilson, after Evans.

which it is to the genetic advantage of females to invest their energy in raising sisters, who are more closely related to them, than in producing daughters, who are more distantly related. In evolutionary terms, this means that the **altruistic behavior** of sterile female workers in helping raise more sterile worker progeny for their mother (the queen) is a phenomenon that we would expect to arise frequently in Hymenoptera. That such social behavior has arisen independently in three different hymenopteran groups—at least once in ants, eight times in bees, and twice in wasps (Table 16-4)—supports Hamilton's proposal.

Although the benefits of sociality seem obvious, not all haplodiploid hymenopteran species are social, and evolutionary biologists are still not sure of its specific causes. They do agree that among likely socializing factors are opportunities for communal nesting, joint protection against predators and parasites, and shared foraging for food. Cooperative behaviors implementing such traits would have had high selective value among genetically related individuals, conferring special advantages on the group as a whole ("group selection," p. 569ff), even overcoming social problems caused by increased infective disease. Although human sociality may also have had a selective history, proposals to apply genetic determinism to human social behaviors have led to serious disputes (Chapter 25).

In summary, insect success extends to social and nonsocial forms and has prevailed in almost every living terrestrial environment probably dating back at least to the Carboniferous period. Apart from insects and vertebrates, relatively few invertebrate phyla have been able to colonize the land as successfully, and these—nematodes, earthworms, and gastropods, for example—are often confined to special humid or soil-like conditions.

Nevertheless, it is interesting to note that although some early insects reached relatively large sizes (some Carboniferous dragonflies measured 2 feet between wing tips), they have tended to remain small, especially by comparison with most species in our own vertebrate phylum. Probably there are limits on the volume of tissue in which insect tracheal tubules can effectively exchange gases. An even more limiting factor may be that the insect exoskeleton would have had to become much heavier and more unwieldy if insects became larger—which did not happen in competition with predatory terrestrial vertebrates, whose endoskeletons could more efficiently support larger organisms.

Echinodermata

Aside from chordates, **echinoderms** are the largest of the deuterostome phyla, containing at present about 7,000 species. These "spiny-skinned" invertebrates are mostly marine bottom-dwellers that have three major distinguishing features:

- A mesodermal skeleton of small calcite plates or spicules (often fused in sea urchins) lying just below the outer epidermis
- Bilateral symmetry in larvae, followed mostly by a five-rayed (pentameral) symmetry in adults
- A water vascular system derived from the coelom in which small surface tube feet used for locomotion and feeding connect to internal canals of circulating seawater

All the present classes of echinoderms (as well as the extinct classes) can be found in various Paleozoic strata, indicating that this group had probably evolved considerably before they appeared in the fossil record. Given their larval forms, which are pelagic and bilateral, zoologists generally believe that echinoderms arose from a free-living, deuterostomate, bilaterally organized ancestor whose coelomic pouches eventually subdivided to include the water vascular system, the perivisceral coelom surrounding the gut, and various sinuses used for circulatory purposes. Presumably some echinoderm features date back to this early stage, such as the osmotic similarity between their coelomic fluids and seawater (they have no excretory or osmoregulatory organs) and their simple nervous system (they have no brain).

According to Arenas-Mena and coworkers echinoderm larval development may hold a clue to the evolution of some or many bilateral metazoans. In their investigation of echinoderm homeobox (*Hox*) gene expression, very few homeobox genes are active in larvae, the remainder functioning in adult development. They propose this pattern is basic for bilateral animals, which evolved from simple larval-type organisms, needing only few regulatory genes, to more complex adult body-plan stages, utilizing batteries of homeobox regulators.

Following an early free-living existence, an ancestral echinoderm group apparently adopted a sessile mode of life in which the animals attached to the marine substrate and their oral surfaces faced upward, surrounded by food-gathering lophophore-like tentacles. Many Paleozoic echinoderms, such as the eocrinoids, blastoids, and others that Figure 16–16 illustrates, show the radial organization that accompanies such sessile habits. Interestingly, although they lack head structures, they retain the same *Hox* genes used for developing anterior regions in other deuterostomes (Martinez et al.). Again, regulatory genes can apparently be channeled in different groups to perform different functions.

As for the question of why echinoderm radial symmetry assumed a pentamerous form, zoologists have offered at least two hypotheses. One hypothesis suggests that a five-tentacled lophophore (the pentactula) was the echinoderm ancestor, and its **pentameral radial symmetry**

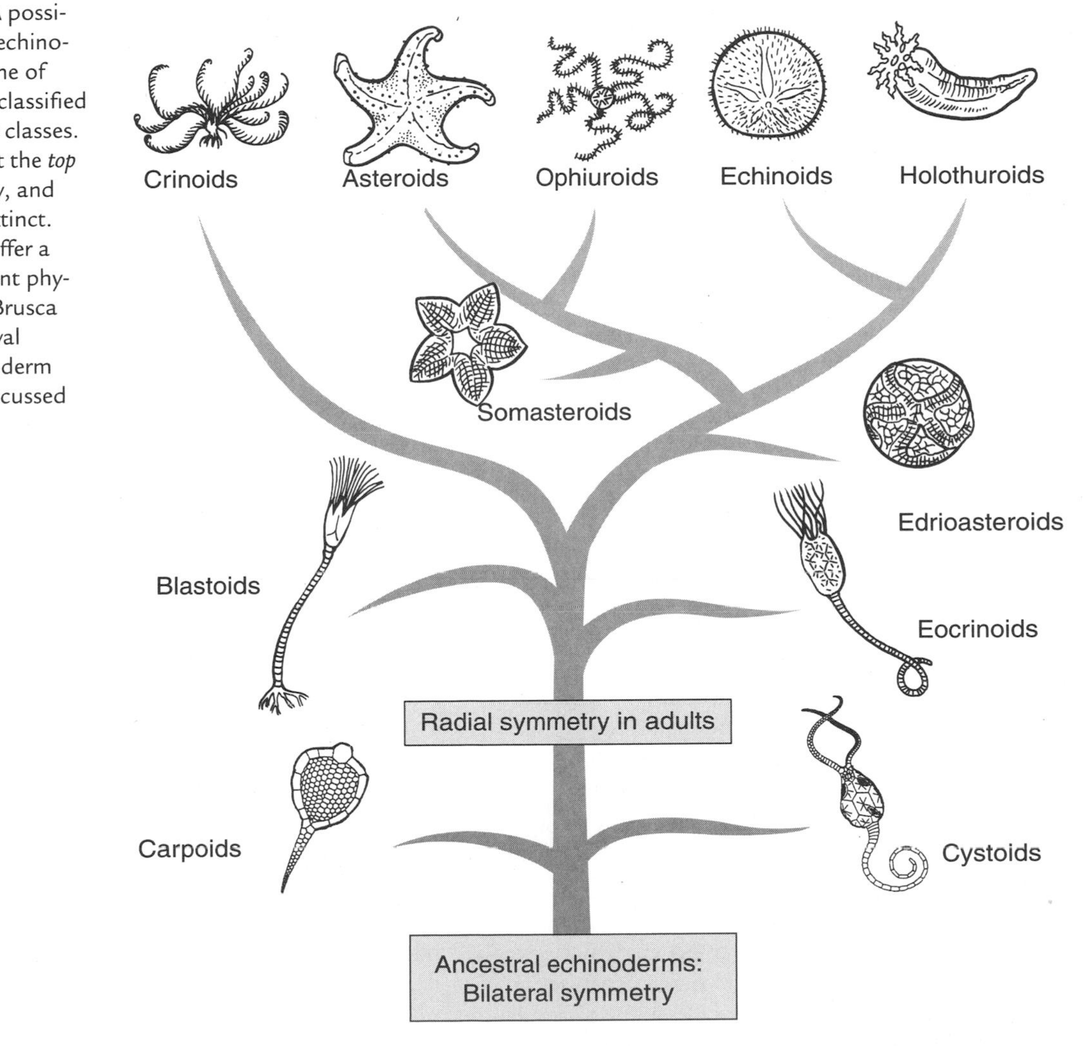

FIGURE 16–16 A possible phylogeny of echinoderm groups, some of which have been classified as subphyla or as classes. The five groups at the *top* are contemporary, and those *below* are extinct. Paul and Smith offer a somewhat different phylogeny (see also Brusca and Brusca). Larval aspects of echinoderm phylogeny are discussed by Smith.

then persisted in later forms. Another hypothesis proposes that pentameral organization ensured that torsional strains that would tend to cleave a suture between plates surrounding an essential central area of the animal would not easily transmit to a suture that was immediately opposite. That is, in an animal with peripheral plates, a structure such as

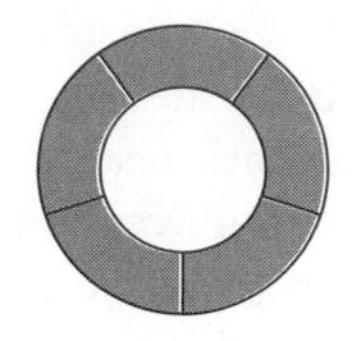

is less likely to break apart than

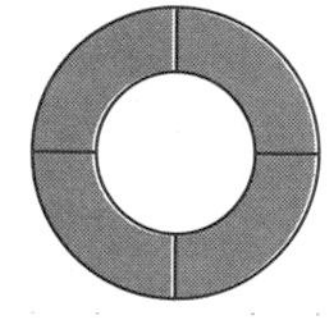

In the absence of fossilized echinoderm ancestors, it is difficult to decide which of these hypotheses is correct.

A further difficulty concerns the transition from a sessile habit with oral surface upward to a mobile habit with oral surface downward, as in groups such as starfish. These changes meant not merely that the animal inverted but that many of its organs drastically repositioned. Unfortunately, we have not yet identified any transitional forms that bridge this gap. Nevertheless, enough fossils exist to provide tentative relationships among most major echinoderm groups, and Figure 16–16 illustrates one possible phylogeny.

SUMMARY

The metazoa have evolved many complex body plans which, combined with embryological features, form the basis for invertebrate taxonomy. Some of the criteria for separating groups into phyla are the presence or absence of tissues or organs, body symmetry, metameric organization, and presence of a body cavity and its mode of origin.

Of the phyla, Porifera (sponges), Mesozoa, and Placozoa are the least complex, with a variety of cell types but no true tissues or organs. The sponge generally is organized into an outer layer (pinacoderm) and an inner layer (choanoderm) on either side of a gelatinous mesohyl.

Next in order of complexity are the organisms with radial symmetry, the Cnidaria and Ctenophora, which have a digestive cavity and more advanced tissue organization than sponges. Characteristic of cnidarians are the planula larva, specialized stinging cells or nematocysts, and two body forms, the polyp and the medusa, both of which may appear in the life cycle of a species. In other invertebrates a third tissue layer, the mesoderm, permits the development of complex organ systems and parallels an increase in size and the appearance of bilateral symmetry. Platyhelminths (flatworms) are filled with mesodermal tissue and have complex reproductive systems, as well as simple osmoregulatory, nervous, and digestive systems.

Although the many groups classified in the phylum Aschelminthes may not be closely related, they share characteristics such as a partially mesodermal body cavity (the pseudocoel), an exterior cuticle, and a digestive tract with both mouth and anus. Layers of circular and longitudinal muscle pressing on the fluid-filled cavity provide a hydraulic skeleton for these animals.

All other invertebrate phyla have a mesoderm-lined body cavity, the coelom, and zoologists divide coelomates into two groups mainly on embryological criteria. The deuterostomes are characterized by development of the blastopore into the anus, a coelom formed by outpocketing of the mesoderm (that is, they are enterocoelous), and radial, indeterminate cleavage. In contrast, in the protostomes, the blastopore becomes the mouth, the coelom forms from splitting the mesoderm (schizocoelous), and cleavage is spiral and determinate. Certain coelomate groups show some degree of metamerism, and some groups capture food by means of ciliated arms called *lophophores.*

Mollusks are a diverse group of mostly shelled coelomates specialized for creeping on a muscular foot and covered by a fold of the body wall, the mantle. These structures have been much modified in the different classes, and the cephalopods have lost the foot and most of the shell to become specialized for propulsive swimming. Some molecular evidence exists that mollusks, annelids, and arthropods may have arisen from the same ancestral group.

In the annelid worms, septa divide the coelom into compartments, and this segmentation allows peristalsis and an active burrowing lifestyle. The polychaetes have developed lateral appendages, or parapodia, for swimming, but their soft bodies are not effective foils for limb muscles.

In the arthropods, the exoskeleton provides a firm attachment for muscles. Since the coelom is no longer essential for this purpose, it has been considerably reduced, as has internal segmentation. Because of their advantageous body plan, arthropods have evolved into many diversified and enormously successful groups, the insects in particular. They have invaded most terrestrial habitats, although their exoskeleton and dependence on tracheal gaseous exchange has limited their size.

While the preceding phyla are protostomes, the echinoderms are deuterostomes with a mesodermal skeleton and a water vascular system for movement and circulation. They have probably retained many of their characteristics—rudimentary nervous system, isosmolarity with seawater, bilaterally symmetrical larvae—from a free-living bilateral ancestor, although they have subsequently become radially symmetrical with a pentamerous organization. Alternatively, they may have derived from a pentamerous lophophore.

KEY TERMS

altruistic behavior
annelids
archaeocyte
arthropods
Aschelminthes
bivalves
body plan
cephalopods
choanocytes
clitellum
coelenterates
determinate development
deuterostome superphylum
diploblastic
echinoderms
foot (mollusk)
gastropods
haplodiploidy
indeterminate development
insects
invertebrates
kin selection
leeches
lophophores
mantle (mollusk)
medusa
mesozoans
mollusks
nematocysts
nematodes
oligochaetes
parapodia
parasitism
pentameral radial symmetry
Placozoa
planula
Platyhelminthes
polychaetes
polyp
protostome superphylum
pseudocoelomates
radial cleavage
Radiata
radula
siphon
social organization
spiral cleavage
sponges
superphylum
torsion
triploblastic
trochophore larva

DISCUSSION QUESTIONS

1. Should we use the criteria for classifying metazoan phyla (Table 16–1) also for determining their phylogenetic relationships? Why or why not?
2. Would you consider sponges more related to early metazoans than coelenterates? Explain.
3. How could the complex life cycle of various platyhelminth parasites have evolved?
4. Would you unite the various pseudocoelomate groups (Table 16–2) into a single phylum, Aschelminthes? Why or why not?
5. What major features do zoologists use for separating protostome from deuterostome coelomate phyla?
6. What are some changes in foot and shell that occurred in the evolution of major molluskan groups?
7. Would you consider mollusks to have originated from a metameric coelomate? Why or why not?
8. What explanations can you offer to account for the general absence of transitional forms between the major animal phyla?
9. What prevented annelids with appendages (for example, polychaetes) from radiating into as many evolutionary niches as arthropods?
10. Why is haplodiploid reproduction (haploid egg males, diploid egg females) important in the evolution of socially organized insects?
11. Considering their widespread evolutionary radiation, why are insects generally small in size compared to vertebrates?
12. What factors can account for the pentameral symmetry of echinoderms?
13. To which, if any, invertebrate phyla can we ascribe polyphyletic origins? Explain.

EVOLUTION ON THE WEB

Explore evolution on the web! Visit the accompanying web site for *Evolution,* 3/e at **www.jbpub.com/evolution** for web exercises and links relating to topics covered in this chapter.

REFERENCES

Adoutte, A., G. Belavoine, N. Lartillot, and R. de Rosa, 1999. Animal evolution: The end of the intermediate taxa? *Trends in Genet.,* **15,** 104–108.

Aguinaldo, A. M. A., J. M. Turbeville, L. S. Linford, M. D. Rivera, J. R. Garey, R. A. Raff, and J. A. Lake, 1997. Evidence for a clade of nematodes, arthropods and other moulting animals. *Nature,* **387,** 489–493.

Arenas-Mena, C., P. Martinez, R. A. Cameron, and E. H. Davidson, 1998. Expression of the *Hox* gene complex in the indirect development of a sea urchin. *Proc. Nat. Acad. Sci.,* **95,** 13062–13067.

Averof, M., and S. M. Cohen, 1997. Evolutionary origin of insect wings from ancestral gills. *Nature,* **385,** 627–630.

Ballard, J. W., G. J. Olsen, D. P. Faith, W. A. Odgers, D. M. Rowell, and P. W. Atkinson, 1992. Evidence from 12*S* ribosomal RNA sequences that onychophorans are modified arthropods. *Science,* **258,** 1345–1348.

Barnes, R. D., 1980. *Invertebrate Zoology,* 4th ed. Saunders, Philadelphia.

Bergquist, P. R., 1985. Poriferan relationships. In *The Origin and Relationships of Lower Invertebrates,* S. Conway Morris, J. D. George, R. Gibson, and H. M. Platt (eds.). Clarendon Press, Oxford, England, pp. 14–27.

Boudreaux, H. B., 1979. *Arthropod Phylogeny with Special Reference to Insects.* Wiley-Interscience, New York.

Brusca, R. C., and G. J. Brusca, 1990. *Invertebrates.* Sinauer Associates, Sunderland, MA.

Charnov, E. L., 1982. *The Theory of Sex Allocation.* Princeton University Press, Princeton, NJ.

Clark, R. B., 1964. *Dynamics in Metazoan Evolution.* Clarendon Press, Oxford, England.

———, 1979. Radiation of the metazoa. In *The Origin of the Major Invertebrate Groups,* M. R. House (ed.). Academic Press, London, pp. 55–102.

Clarke, K. U., 1973. *The Biology of Arthropods.* Arnold, London.

Clarkson, E. N. K., 1993. *Invertebrate Palaeontology and Evolution,* 3d ed. Chapman & Hall, London.

Collins, A. G., 1998. Evaluating multiple alternative hypotheses for the origin of Bilateria: An analysis of 18*S* rRNA molecular evidence. *Proc. Nat. Acad. Sci.,* **95,** 15458–15463.

Conway Morris, S., 1998. Metazoan phylogenies: Falling into place or falling to pieces? A paleontological perspective. *Current Opinion Genet. Devel.,* **8,** 662–667.

Conway Morris, S., J. D. George, R. Gibson, and H. M. Platt (eds.), 1985. *The Origins and Relationships of Lower Invertebrates.* Clarendon Press, Oxford, England.

Damen, W. G. M., M. Hausdorf, E.-A. Seyfarth, and D. Tautz, 1998. A conserved mode of head segmentation in arthropods revealed by the expression pattern of *Hox* genes in a spider. *Proc. Nat. Acad. Sci.,* **95,** 10665–10670.

Erwin, T. L., 1982. An evolutionary basis for conservation strategies. *The Coleopterist's Bull.,* **36,** 74–75.

Evans, H. E., 1958. The evolution of social life in wasps. *Proc. 10th Int. Congr. Entomol.,* **2,** 449–457.

Fortey, R. A., D. E. G. Briggs, and M. A. Wills, 1997. The Cambrian evolutionary 'explosion' recalibrated. *BioEssays,* **19,** 429–434.

Fox, R. M., and J. W. Fox, 1964. *Introduction to Comparative Entomology.* Reinhold, New York.

Hamilton, W. D., 1964. The evolution of social behavior. *J. Theoret. Biol.,* **1,** 1–52.

Hölldobler, B., and E. O. Wilson, 1990. *The Ants.* Harvard University Press, Cambridge, MA.

House, M. R. (ed.), 1979. *The Origin of Major Invertebrate Groups.* Academic Press, London.

Hyman, L., 1940. *The Invertebrates: Protozoa Through Ctenophora.* McGraw-Hill, New York.

———, 1951. *The Invertebrates: Platyhelminthes and Rhynchocoela, the Acoelomate Bilateria.* McGraw-Hill, New York.

Kershaw, D. R., 1983. *Animal Diversity.* University Tutorial Press, Slough, Great Britain.

Lutz, P. E., 1986. *Invertebrate Zoology.* Addison-Wesley, Reading, MA.

Manton, S. M., 1977. *The Arthropoda: Habits, Functional Morphology, and Evolution.* Clarendon Press, Oxford, England.

Marden, J. H., and M. G. Kramer, 1994. Surface-skimming stoneflies: A possible intermediate stage in insect flight evolution. *Science,* **266,** 427–430.

Martinez, P., J. P. Rast, C. Arena-Mena, and E. H. Davidson, 1999. Organization of an echinoderm *Hox* gene cluster. *Proc. Nat. Acad. Sci.,* **96,** 1469–1474.

Nielsen, C., 1995. *Animal Evolution: Interrelationships of the Living Phyla.* Oxford University Press, Oxford, England.

Panganiban, G., A. Sebring, L. Nagy, and S. Carroll, 1995. The development of crustacean limbs and the evolution of arthropods. *Science,* **270,** 1363–1366.

Paul, C. R. C., and A. B. Smith, 1984. The early radiation and phylogeny of echinoderms. *Biol. Rev.,* **59,** 443–481.

Pawlowski, J., J.-I. Montoya-Burgos, J. F. Fahrni, J. Wüest, and L. Zaninetti, 1996. Origin of the Mesozoa inferred from 18*S* rRNA gene sequences. *Mol. Biol. and Evol.,* **13,** 1128–1132.

Raff, R. A., 1996. *The Shape of Life: Genes, Development, and the Evolution of Animal Form.* University of Chicago Press, Chicago.

Ratnieks, F. L. W., and P. K. Visscher, 1989. Worker policing in the honeybee. *Nature,* **342,** 796–797.

Regier, J. C., and J. W. Shultz, 1997. Molecular phylogeny of the major arthropod groups indicates polyphyly of crustaceans and a new hypothesis for the origin of hexapods. *Mol. Biol. and Evol.,* **14,** 902–913.

Robinson, G. E., 1998. From society to genes with the honey bee. *Amer. Sci.,* **86,** 456–462.

Robson, E. A., 1985. Speculations on coelenterates. In *The Origin and Relationships of Lower Invertebrates,* S. Conway Morris, J. D. George, R. Gibson, and H. M. Platt (eds.). Clarendon Press, Oxford, England, pp. 60–77.

Ruiz-Trullo, I., M. Riutort, D. T. J. Littlewood, E. A. Herniou, and J. Baguñà, 1999. Acoel flatworms: Earliest extant bilaterian metazoans, not members of Platyhelminthes. *Science,* **283,** 1919–1923.

Schierwater, B., and K. Kuhn, 1998. Homology of *Hox* genes and the zootype concept in early metazoan evolution. *Mol. Phylogenet. and Evol.,* **9,** 375–381.

Seger, J., 1996. Exoskeletons out of the closet. *Science,* **274,** 941.

Shubin, N., C. Tabin, and S. Carroll, 1997. Fossils, genes and the evolution of animal limbs. *Nature,* **388,** 639–648.

Sidow, A., and W. K. Thomas, 1994. A molecular evolutionary framework for eukaryotic model organisms. *Current Biol.,* **4,** 596–603.

Smith, A. B., 1997. Echinoderm larvae and phylogeny. *Ann. Rev. Ecol. Syst.,* **28,** 219–241.

Solem, G. A., 1974. *The Shell Makers.* Wiley, New York.

Trueman, E. R., and M. R. Clarke (eds.), 1985. *The Mollusca: Vol. 10, Evolution.* Academic Press, Orlando, FL.

Valentine, J. W., 1989. Bilaterians of the Precambrian–Cambrian transition and the annelid-arthropod relationship. *Proc. Nat. Acad. Sci.,* **86,** 2272–2275.

———, 1997. Cleavage patterns and the topology of the metazoan tree of life. *Proc. Nat. Acad. Sci.,* **94,** 8001–8005.

Whiting, M. F., and W. C. Wheeler, 1994. Insect homeotic transformation. *Nature,* **368,** 696.

Willmer, P., 1990. *Invertebrate Relationships: Patterns in Animal Evolution.* Cambridge University Press, Cambridge, England.

Winnepenninckx, B., T. Backeljau, L. Y. Mackey, J. M. Brooks, R. De Wachter, S. Kumer, and J. R. Garey, 1995. 18*S* rRNA data indicate that Aschelminthes are polyphyletic in origin and consist of at least three distinct clades. *Mol. Biol. and Evol.,* **12,** 1132–1137.

Winnepenninckx, B., T. Backeljau, and R. De Wachter, 1996. Investigation of molluscan phylogeny on the basis of 18*S* rRNA sequences. *Mol. Biol. and Evol.,* **13,** 1306–1317.

17 The Origin of Vertebrates

Separating vertebrates as a group from invertebrates as a group reflects an exclusive quality we humans confer on organisms that resemble us compared to organisms that do not. We also use other factors in making this distinction, of course, since vertebrates are members of a unique phylum, **Chordata,** and among the most varied and successful of all animals. They number about 42,000 species, ranging in size from minuscule fish to giant whales, and have invaded a wide variety of habitats from oceanic depths to soaring heights above the earth. In what characters are vertebrates unusual?

Although Figure 17–1 shows a composite of two different groups, fish and human, widely separated along the vertebrate spectrum, it portrays some special morphological features of this phylum:

- A paired series of clefts, or **gills,** that lead outward from the pharynx and commonly present in an embryonic stage, called the **pharyngula,** as well as in adults of some vertebrate groups (Fig. 3–10).
- An internal skeletal structure oriented along the anterior–posterior axis that derives from the embryonic presence of a flexible rod, the **notochord.** Although a column of cartilaginous or bony **vertebrae** replaces the notochord in the adults of many groups (subphylum Vertebrata), it is this structure that gives this phylum the name Chordata.
- A **tail** that may be quite prominent and extends beyond the anus in embryos of all groups, although not in all adults.
- A single **hollow nerve cord** that runs **dorsally** above the notochord.

These characters seem unusual compared to those of other phyla, and the question arises whether we can find the basis for such organization elsewhere, or perhaps find a direct fossil connection to another phylum. The answers are not encouraging.

FIGURE 17-1 Composite drawing of man and fish, showing various vertebrate characteristics. (*Adapted from Ohno.*)

Hypotheses on the Origin of Vertebrates

By the time vertebrates appear in the fossil record, they are quite distinctive, showing all the major attributes that characterize them as chordates. Since reliable fossil connections between vertebrates and ancestral forms are absent, researchers have investigated vertebrate origins using morphological and, more lately, molecular comparative methods to detect homologies with other phyla.

Among early proposals were suggestions for an annelid- or arthropod-type origin based on some general similarities between the groups (Table 17–1*a*). However, as shown in Fig. 17–2, this transition meant repositioning the ventral annelid/arthropod nerve cord dorsally and reversing the direction of blood flow to attain a dorsal vertebrate nerve cord and circulatory system—the animal had to be turned upside down, and its mouth relocated from top to bottom. Because of the need for such radical morphological manipulation and other major differences between the groups (Table 17–1*b*), researchers generally abandoned that hypothesis until it was recently revived on molecular evidence (p. 360–361).

TABLE 17-1 Summary of the main morphological arguments for and against an annelid/arthropod origin for vertebrates

	Annelids/ Arthropods	Vertebrates
(a) Arguments for an annelid origin		
Bilateral symmetry	Yes	Yes
Presence of coelom	Yes	Yes
Metameric organization	Yes	Yes
Terminal growth	Yes	Yes
Dorsal and ventral longitudinal blood vessels	Yes	Yes
Anterior "brain"	Yes	Yes
(b) Arguments against an annelid origin		
Complete segmentation through the body wall	Yes	No
Coelom embryology	Schizocoelous	Enterocoelous
Position of nerve cord	Ventral	Dorsal
Skeleton	External	Internal
Gill slits	No	Yes
Flow of dorsal blood vessel	Anteriorly	Posteriorly
Flow of ventral blood vessel	Posteriorly	Anteriorly
Fate of blastopore	Mouth (protostome)	Anus (deuterostome)

Source: Adapted from Neal and Rand.

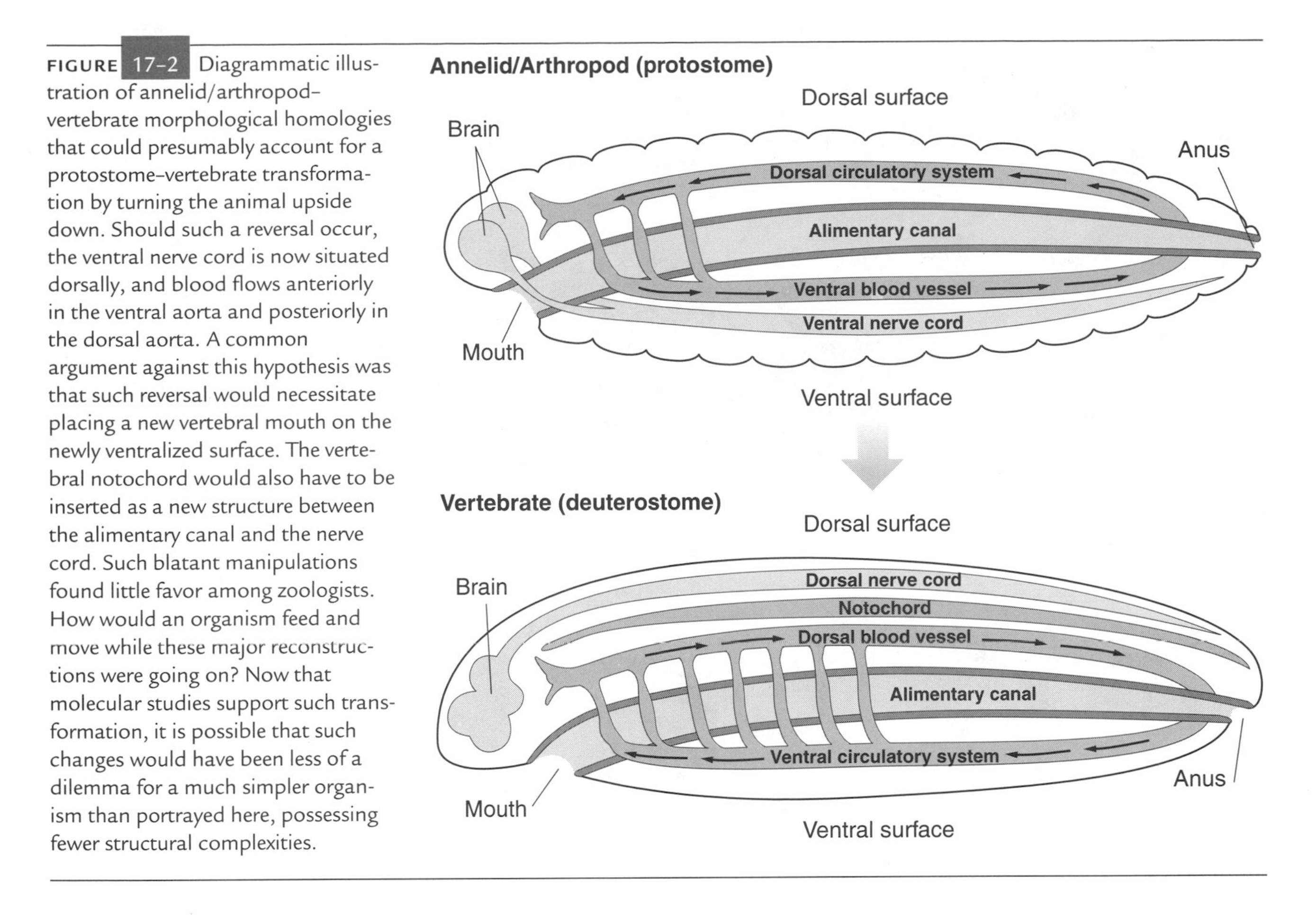

FIGURE 17-2 Diagrammatic illustration of annelid/arthropod–vertebrate morphological homologies that could presumably account for a protostome–vertebrate transformation by turning the animal upside down. Should such a reversal occur, the ventral nerve cord is now situated dorsally, and blood flows anteriorly in the ventral aorta and posteriorly in the dorsal aorta. A common argument against this hypothesis was that such reversal would necessitate placing a new vertebral mouth on the newly ventralized surface. The vertebral notochord would also have to be inserted as a new structure between the alimentary canal and the nerve cord. Such blatant manipulations found little favor among zoologists. How would an organism feed and move while these major reconstructions were going on? Now that molecular studies support such transformation, it is possible that such changes would have been less of a dilemma for a much simpler organism than portrayed here, possessing fewer structural complexities.

We now know that among genes affecting dorsal–ventral development in arthropods and vertebrates are two homologous pairs (DeRobertis and Sasai; Ferguson). One pair (*dpp, bmp*) produces proteins that dorsalize arthropods and ventralize vertebrates, and the other pair (*sog, chordin*) acts oppositely. The finding that the genes in each pair share common ancestry, and that they elicit complementary dorsal–ventral developmental activity, indicates that a common arthropod–vertebrate ancestor gave rise to both body plans, and that one is the inversion of the other. This most probably occurred in a small soft-bodied Precambrian lineage with little opportunity for fossilization, making such transitions difficult to visualize. Questions remain: What was the sequence of changes, and how were mechanical problems overcome? Until more details of a protostome relationship are uncovered, many workers concentrate on searching for a more conceivable vertebrate ancestry among deuterostomes themselves. What connects deuterostomes?

Deuterostome Affinities

Probably the most popular hypothesis today is the concept of a common ancestry shared by echinoderms and vertebrates. This concept rests, for the most part, on a variety of traits that biologists presume to be strongly conservative and have generally used to distinguish the echinoderm superphylum (**deuterostomes**) from the annelid superphylum (**protostomes**):

1. In echinoderms and vertebrates the blastopore produces the adult anus (*deuterostome*), whereas in the annelid group the blastopore becomes the mouth (*protostome*).
2. Both echinoderms and vertebrates show *radial cleavage* of early zygotic cells (blastomeres) rather than the *spiral cleavage* that seems the rule in the annelid group (Fig. 16–7). Furthermore, because isolated blastomeres of echinoderms and amphibian vertebrates develop into normal embryos, zoologists consider their developmental process *indeterminate,* in contrast to the abnormal embryos produced by isolated blastomeres undergoing spiral cleavage (*determinate* development).
3. The origin of the coelom in echinoderms and vertebrates is enterocoelous, whereas it is schizocoelous in the annelid superphylum.
4. The skeleton of vertebrates and echinoderms originates embryonically from mesodermal tissue, and originates from ectodermal tissue in the annelid superphylum.

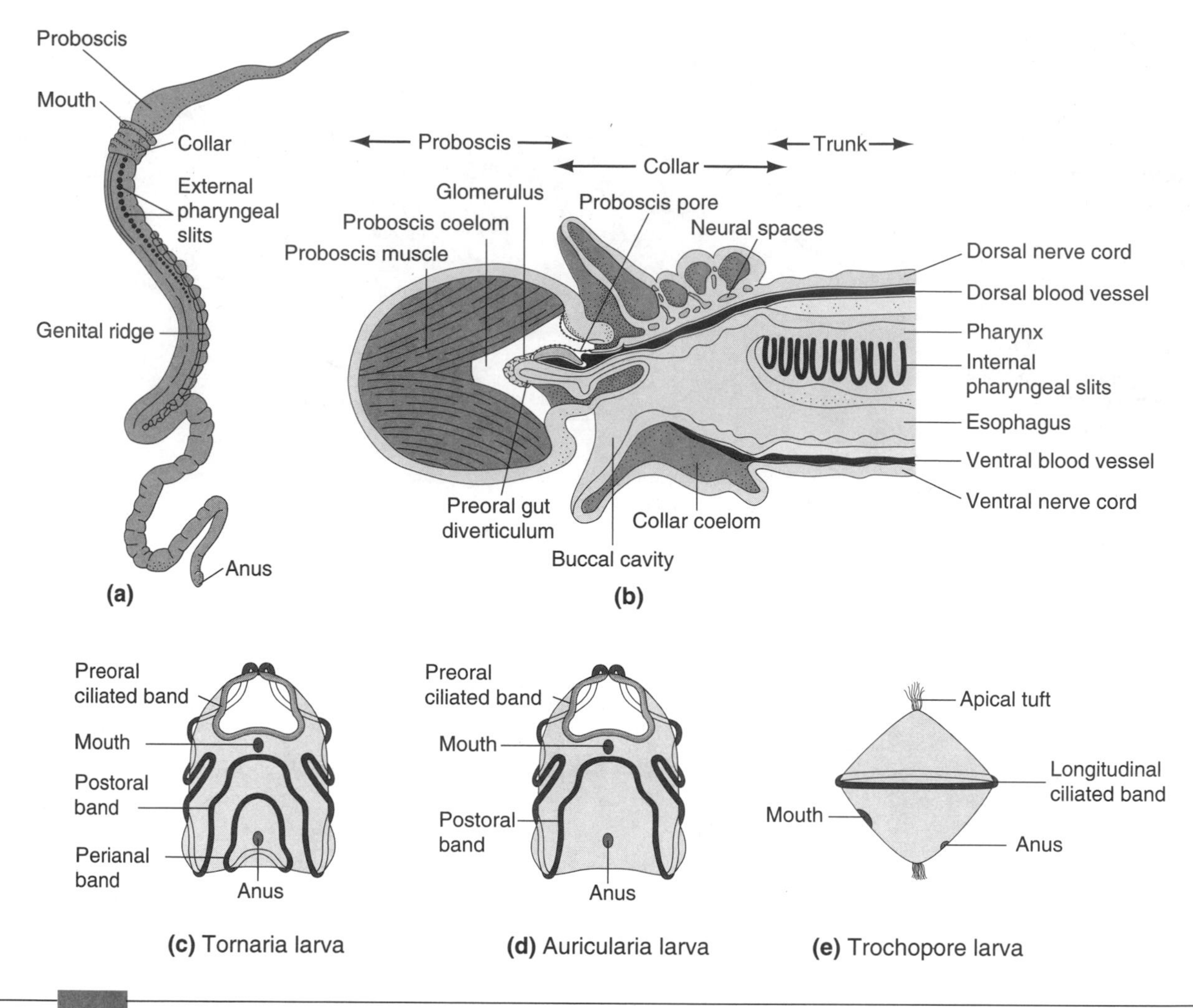

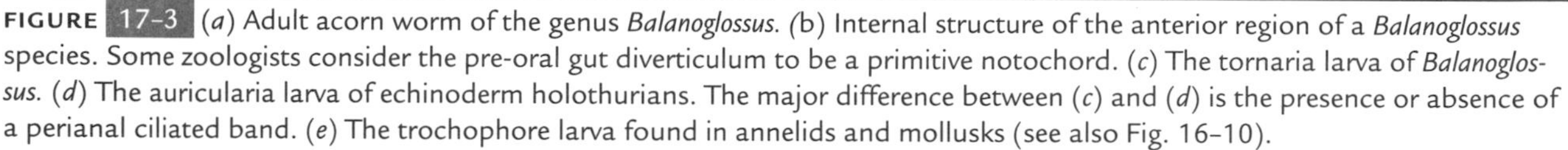

FIGURE 17-3 (*a*) Adult acorn worm of the genus *Balanoglossus*. (b) Internal structure of the anterior region of a *Balanoglossus* species. Some zoologists consider the pre-oral gut diverticulum to be a primitive notochord. (*c*) The tornaria larva of *Balanoglossus*. (*d*) The auricularia larva of echinoderm holothurians. The major difference between (*c*) and (*d*) is the presence or absence of a perianal ciliated band. (*e*) The trochophore larva found in annelids and mollusks (see also Fig. 16–10).

5. Jefferies has proposed that some groups among the echinoderm carpoids (Fig. 16–16) have lateral openings that resemble gill slits. These and other suggested vertebrate-like features, such as tails, may directly link echinoderms and chordates.

6. The larvae in echinoderms are of the pluteus-type or a variation of it, such as the auricularia larvae in holothurians (sea cucumbers). By contrast, the annelid superphylum produces trochophore larvae. Since the acorn worm, *Balanoglossus* (Fig. 17–3*a*), has an auricularia-type larva (called *tornaria*) and biologists generally believe this animal to be a chordate, although primitive (**hemichordate**; subphylum Hemichordata), it seems to follow that echinoderms and vertebrates are closely allied (Fig. 17–3*c*, *d*).

Although many biologists accept some of these arguments for echinoderm–vertebrate affinity, others have raised enough criticism against almost every point to moderate enthusiasm. For example, the blastopore, at the base of the protostome–deuterostome division, does not always develop consistently in a particular phylum: in some so-called protostome annelids, arthropods, brachiopods, and mollusks, the blastopore closes completely and the mouth forms elsewhere. Cleavage patterns are also not well defined: other than in some crustaceans, arthropods do not show spiral cleavage, nor is cleavage consistently determinate in mollusks.

Zoologists also dispute the embryological source of the coelom for both echinoderms and vertebrates, with some observations indicating a schizocoelous rather than enterocoelous origin. Even the presumed common mesodermal origin of the skeleton in echinoderms and vertebrates has been difficult to accept unequivocally since this structure differs so greatly in composition and pattern between the two phyla.

The prominent involvement of ectodermal tissues in forming enamel in vertebrate teeth and in the outer layer of shark scales further challenges the presumed restriction of vertebrate skeletal formation to mesodermal tissue. Jefferies's proposal of a carpoid origin for chordates is also difficult to accept, considering that the carpoid tail is much too thin to contain a notochord and nerve cord, and in contrast to carpoids, early vertebrates had already evolved calcium phosphate skeletons. Moreover, the 18*S* RNA investigations by Field and coworkers cited previously (Fig. 14–7) indicate that the divergence between echinoderms and vertebrates most likely occurred before echinoderm groups such as carpoids had evolved.

Larval similarity between *Balanoglossus* and some echinoderms may be the least questionable of vertebrate–echinoderm affinities, but we can dispute even that idea because it is based on accepting *Balanoglossus* as a chordatelike invertebrate—a point that has been argued from the time Bateson first proposed it in 1886. Bateson believed the *Balanoglossus* preoral gut diverticulum was a notochord of sorts (see Fig. 17–3*b*), although no one has found anything like a rod or notochord in it. Nevertheless, many zoologists still consider *Balanoglossus* to have chordate similarities, basing this association primarily on the presence of gill slits and on molecular similarities for homeobox-containing (*Hox*) genes used in spatial differentiation (see Pendleton et al.).[1]

These considerations only slightly seem to support the hypothesis of a vertebrate–echinoderm relationship. As Stahl pointed out,

> All extant echinoderms develop a radial symmetry and a semisessile habit that are far distant from the vertebrate condition. A connection between echinoderms and vertebrates is not impossible on this account, but to maintain it, one must assume that a protovertebrate group diverged from the echinoderm line in Precambrian times before the development of the specialized radially symmetrical forms.

On the morphological level, we have not resolved the question of which phylum provided or was most related to the first vertebrate ancestor, although researchers are making proposals that rely on histology and physiology, including Løvtrup's scheme resuscitating the tie between arthropods and chordates. However, since molecular comparisons can transcend morphological and embryological difficulties, many biologists are now placing more emphasis on nucleotide sequence comparisons. As noted previously in Figure 14–5, such studies show a close relationship between echinoderms and chordates. Keeping in mind their large differences in body plan—one bilaterally linear (chordates) and the other pentamerally radial (echinoderms)—we may still accept a common ancestry and recognize that it must have been very far in the distant past, certainly Precambrian. Once their lineages diverged, changes may have proceeded rapidly. According to Wada and Satoh, the deuterostomes are monophyletic and "may have evolved during a very short period of time."

Whichever phylogeny actually occurred, many steps in the origin of chordates are still hidden. Since we know so little, perhaps we should simply accept the existence of a primitive chordatelike animal and ask general questions about how it may have lived and under what conditions it may have evolved further. Following this approach, zoologists turned to small marine animals that bear some relationship to vertebrates: cephalochordates and urochordates.

Cephalochordates and Urochordates

In addition to vertebrates, biologists usually classify several other groups as subphyla of Chordata. Although none have vertebrae or certain other vertebrate characteristics, they share features that have prompted zoologists to suggest what early chordates were like. Among these groups, the most vertebrate-like are **cephalochordates** such as the marine lancelet, *Branchiostoma,* known also by the common name of **amphioxus.** These 1- to 2-inch-long fishlike marine animals swim by contracting metamerically organized muscles (myotomes) placed alongside a semirigid notochord that runs from tip to tip (Fig. 17–4).

As an adult, the animal is mostly sedentary, burrowing in the sea bottom, then extending its anterior end into the waters above to filter feed on passing food particles. They have a large pharyngeal "branchial basket," penetrated by up to 200 gill slits, that filters water drawn through the mouth by ciliary currents and then passes extracted food (primarily algae) along a mucosal strand into the digestive tract.

Evolutionists have often pointed out that cephalochordates are probably not in the direct line of vertebrate ancestry because their notochord does not end at a "brain" but extends to the very anterior tip of the animal, and they have no sense organs related to those of vertebrates. However, recent findings point to head structures, including an eye spot, that are homologous to vertebrate head structures, as well as to homologous body segments recognized by homologous homeobox (*Hox*) genes (Fig. 17–5).

The cephalochordate notochord, gill slits, dorsal nerve cord, metamerically organized myotomes, posterior direction of blood flow in the dorsal vessels and anterior direction in the ventral vessels, as well as vertebrate-like

[1]Nübler-Jung and Arendt claim that hemichordates represent a transitional stage between annelid/arthropods and vertebrates in which the proposed inversion from one group to the other, described previously, was not completed.

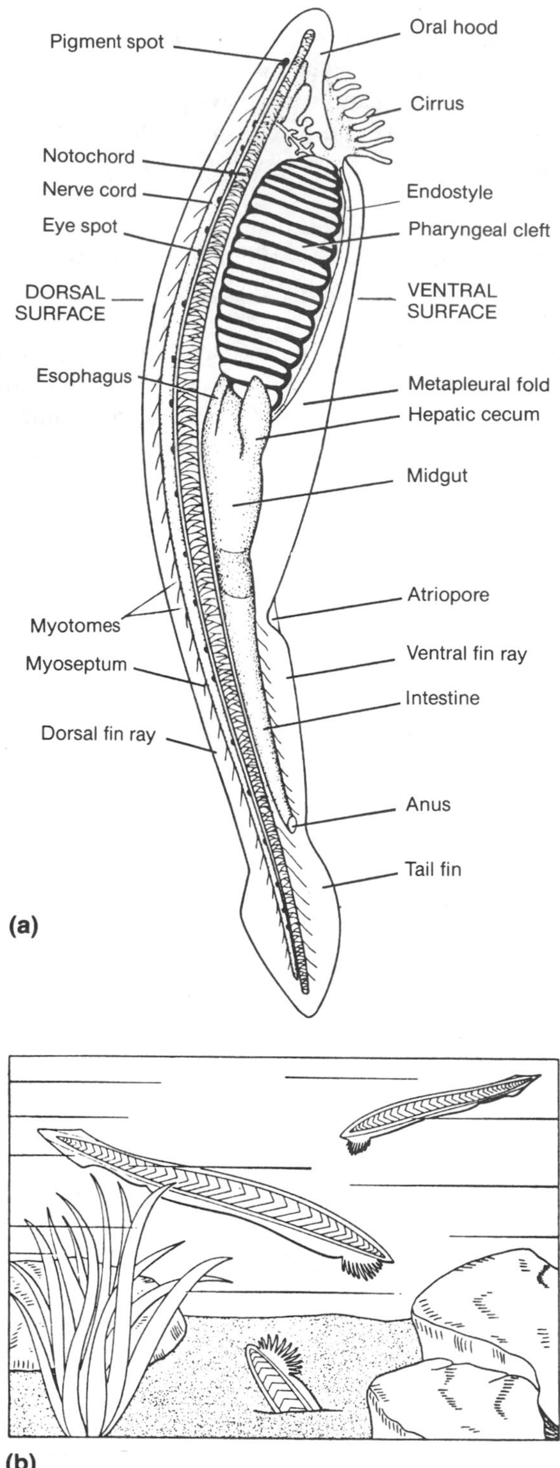

FIGURE 17-4 Anatomical features (*a*) and habitat (*b*) of the cephalochordate *Branchiostoma.*

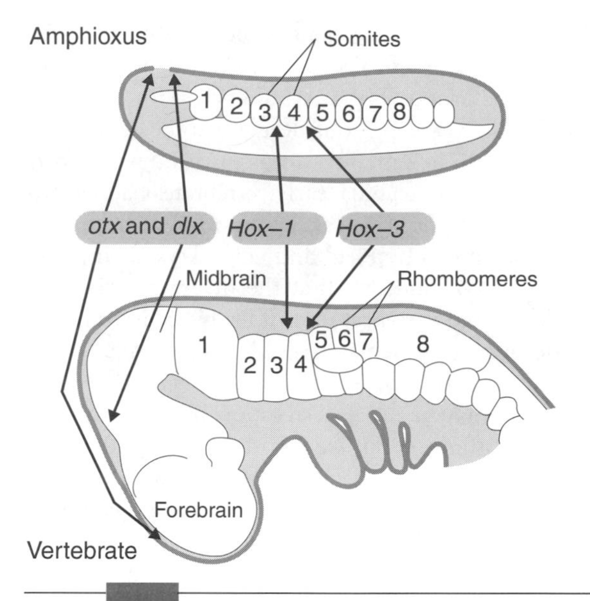

FIGURE 17-5 Expression pattern of *Hox* genes in an *Amphioxus* and generalized vertebrate embryo. Numbered segments represent somites (trunk segments) in *Amphioxus* embryos corresponding to rhombomeres (nervous system segments) in vertebrate embryos. The *Hox* genes are expressed in similar locations in both embryos indicating these are homologous structures and that *Amphioxus* has a forebrain. Homology between vertebrate and *Amphioxus Hox* gene clusters is shown in Fig. 15-8. (*After Stokes and Holland.*)

organs such as the thyroid also indicate they are closely related to vertebrates. We would find it hard to accept an accidental convergence of so many basic characters in two unrelated groups. It seems reasonable to propose that cephalochordates, although perhaps not directly ancestral to modern vertebrates, represent a mode of life that early chordatelike animals probably shared, that is, swimming and **filter feeding.**[2]

At least two points of evidence support this view. One is the observation that lampreys, a primitive group of jawless and boneless vertebrates but vertebrates nevertheless, have an **ammocoete** larval form that is also a swimming filter feeder remarkably similar to cephalochordates. Another point, discussed later, is that the earliest fossil chordates were also undoubtedly filter feeders, using a muscular pharynx to suck water into the gill chamber.

According to many zoologists, a group of small marine animals, **urochordates,** also filter feeders, provide the connection showing how filter feeding may have progressed to active notochordial swimming. Among the

[2]A song popular among summer students who attended or worked at the Marine Biological Laboratory at Woods Hole on Cape Cod, followed the World War I tune "It's a Long Way to Tipperary":

It's a long way from Amphioxus
It's a long way to us;
It's a long way from Amphioxus
To the meanest human cuss;
Good-bye fins and gill slits,
Welcome teeth and hair;
It's a long, long way from Amphioxus
But we came from there!

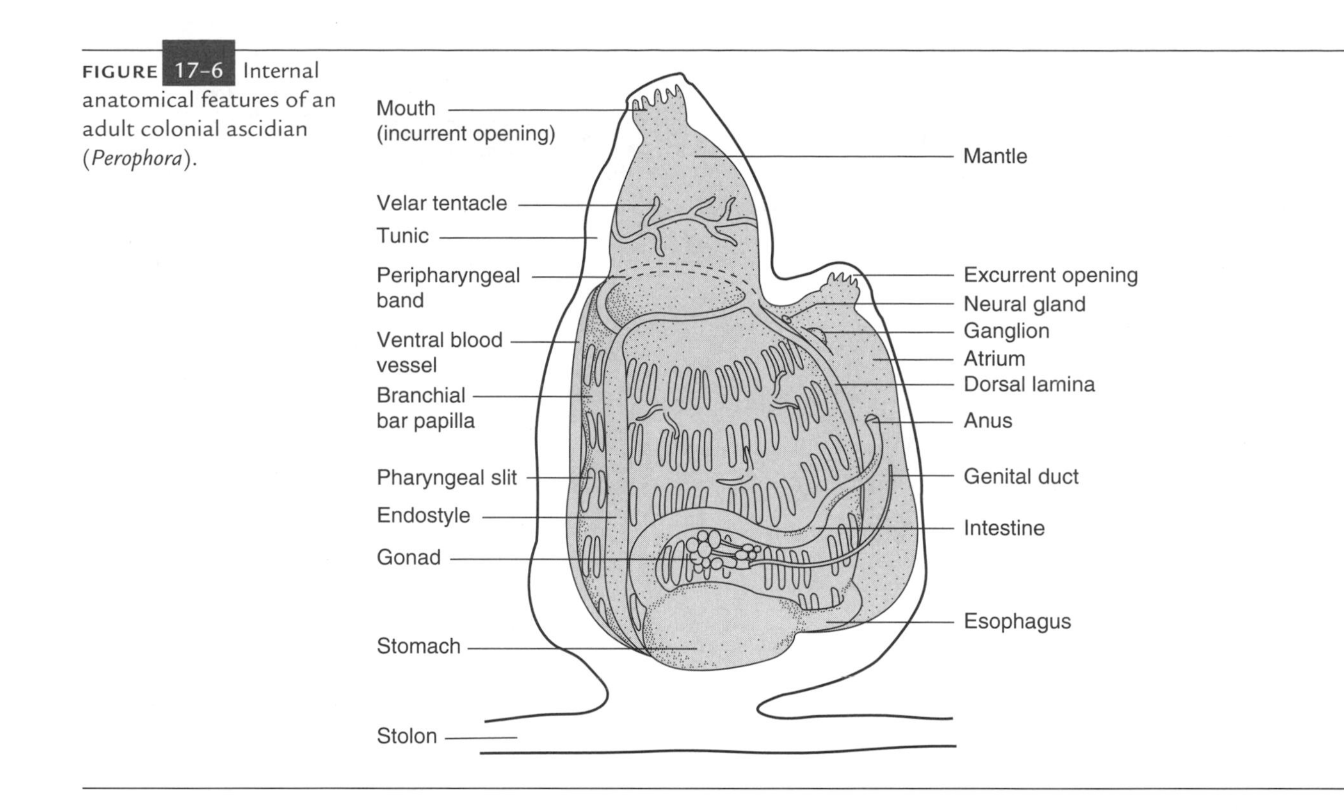

FIGURE 17-6 Internal anatomical features of an adult colonial ascidian (*Perophora*).

urochordates is a large subgroup called ***ascidians*** (also called **tunicates** or sea squirts) that are sessile in the adult, feeding by means of a large pharyngeal basket through which they draw and filter water and then expel it through many gill slits (Fig. 17–6). Although aside from gill slits no immediate evidence exists of chordate structure, the organization of the ascidian larvae differs remarkably from the adult. As Figure 17–7*a* shows, this larva has a notochord and dorsal nerve cord as well as gill slits. Its function is to actively seek out suitable habitats and distribute itself as widely as possible before metamorphosing into the sessile adult form (Fig. 17–7*b*, *c*).

Garstang developed the notion, quickly accepted by many biologists, that this metamorphosis could have been increasingly delayed over successive generations until the swimming larval form itself became sexually mature. This process, termed **paedomorphosis** ("shaping like a child"), involved the incorporation of adult sexual features into earlier immature stages. Garstang hypothesized that some urochordate groups such as the Larvacea arose by paedomorphosis and can now be considered as sexually mature larvae.

Extending Garstang's argument, biologists suggested that early free-swimming chordates passed through this same process of paedomorphosis in descending from ancestors that had chordatelike larval stages; that is, the swimming, filter-feeding larvae of some prechordate animals replaced their adult sessile forms, perhaps because they could better follow and search out new food supplies as well as escape predators. Selection for preserving larval characteristics led to precocious sexuality that bypassed the sessile adult, yielding mobile mature "chordates."[3]

In general, the relationship of vertebrates to cephalochordates and urochordates makes it likely that the earliest of chordate groups were actively swimming filter feeders, and paedomorphosis may well explain their origin. This process and its auxiliary, **neoteny** (retaining some immature morphological traits into adult stages), are not unusual events; for example, some salamanders such as the mudpuppy and axolotl reproduce in gilled, immature forms.[4]

[3]For those who believed that the similarity between the tornaria larva of the hemichordate *Balanoglossus* and the auricularia larva of echinoderms stems from a phylogenetic relationship, Garstang's proposal had the added attraction of explaining that the transition between the two phyla occurred through a larval form rather than through changes in the considerably different adult forms. He visualized echinoderms evolving into chordates by developing dorsal and neural folds from ciliated bands of the auricularia and adding gill slits and a notochord. The hypothesis did not explain the origin of the latter structures, nor did it make clear how this unusual larva reached the adult sexual stage.

[4]Authors use these terms and others referring to evolutionary changes in developmental rates (heterochrony) in various ways. For example, some researchers define paedomorphosis as the *result* of neoteny. That is, when the developmental rate of nonreproductive compared to reproductive tissues slows down (neoteny), the result is a full-sized adult organism with juvenile features that is sexually mature (paedomorphosis). Garstang provided a poetic description of paedomorphosis in the Mexican axolotl (*Ambystoma*):

> Ambystoma's a giant newt who rears in swampy waters,
> As other newts are wont to do, a lot of fishy daughters:
> These Axolotls, having gills, pursue a life aquatic,
> But, when they should transform to newts, are naughty and erratic.

FIGURE 17-7 (*a*) Free-swimming ascidian larva before metamorphosis. (*b*) As the larva attaches to a substrate by its anterior suckers, the notochord degenerates and its internal structures rotate. (*c*) Stage at which the tail is almost completely resorbed and the body parts are assuming their adult positions.

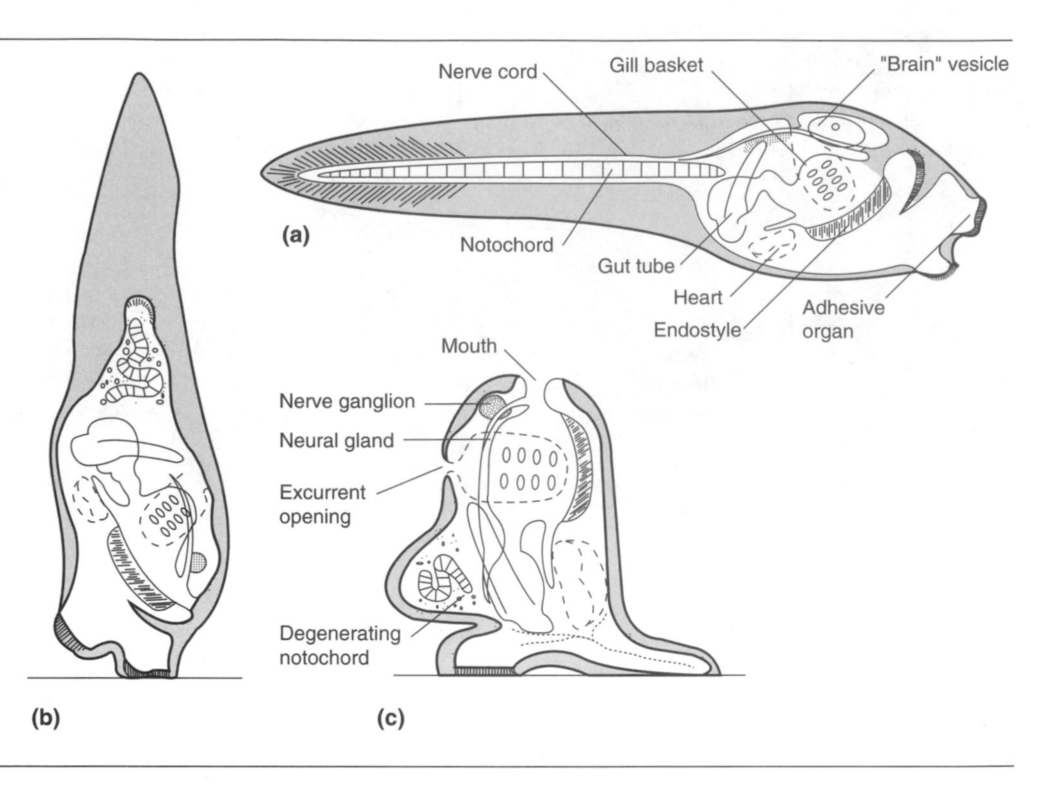

Since hemichordates such as the pterobranch *Rhabdopleura* (see Fig. 14–14) use a lophophore (group of ciliated tentacles) for food capture, biologists have suggested that links in the chordate chain extend back to a lophophorate-type ancestry that chordates may have shared with other lophophorates such as phoronids, brachiopods, and ectoprocts, as well as with the crinoidlike ancestors of the echinoderms. A proposed phylogeny of this type, extending from lophophore ancestor to vertebrates, appears in Figure 17–8.

However neat these solutions may seem, keep in mind that ancestral branching patterns extending back to Precambrian times are not easily settled. For example, clearly related as urochordates and chordates may seem morphologically, they are not so on all molecular levels analyzed so far: sequence analysis using one gene (that for 18*S* ribosomal RNA) indicates a very distant relationship, whereas another gene (for the actin muscle protein) indicates a close relationship. Biologists offer other hypotheses, such as chordate origin from free-living animals rather than from neotenous forms of sessile urochordates (Wada and Satoh), as well as a polyphyletic origin of chordates (Turbeville et al.).

Nevertheless, it seems clear to many researchers that tadpole larvae of some urochordates broadly reflect a very likely chordate body plan—a slotted pharyngeal basket swimming by attachment to a muscularly endowed undulating flexible notochord. Among the developmental homologies that support this view is a *Brachyury* gene active in differentiating notochords in both mammals and ascidian larvae (DiGregorio and Levine). Although the actual "hard" fossil evidence of a Precambrian chordate lineage remains elusive, molecular information can fill important gaps.

However, if we pass from the deep unknowns of the Precambrian, fossils do begin to accumulate from the Cambrian onward—enough to offer a more detailed vertebrate history. What do we know paleontologically?

They change upon compulsion, if the water grows too foul,
For then they have to use their lungs, and go ashore to prowl:
But when a lake's attractive, nicely aired, and full of food,
They cling to youth perpetual, and rear a tadpole brood.
(From Garstang, W., 1951. *Larval Forms and other Zoological Verses.* Oxford: Blackwell Publishers.)

Among other forms of paedomorphosis and heterochrony (see McKinney and McNamara) is *progenesis,* used to describe an increased rate of sexual development leading to early sexual maturity in an adult that remains quite small in size, such as the tiny parasitic *Bonellia* males (p. 198).

Fossil Jawless Fish (Agnatha)

The earliest known vertebrates are a group of **jawless fishes,** whose teeth and armored skin plates (dermal bones) appear in marine deposits of the Late Cambrian period and whose fossils extend into the Late Devonian. Some, called conodonts, were soft-bodied eel-like creatures, a few inches long, that captured prey and ground them into

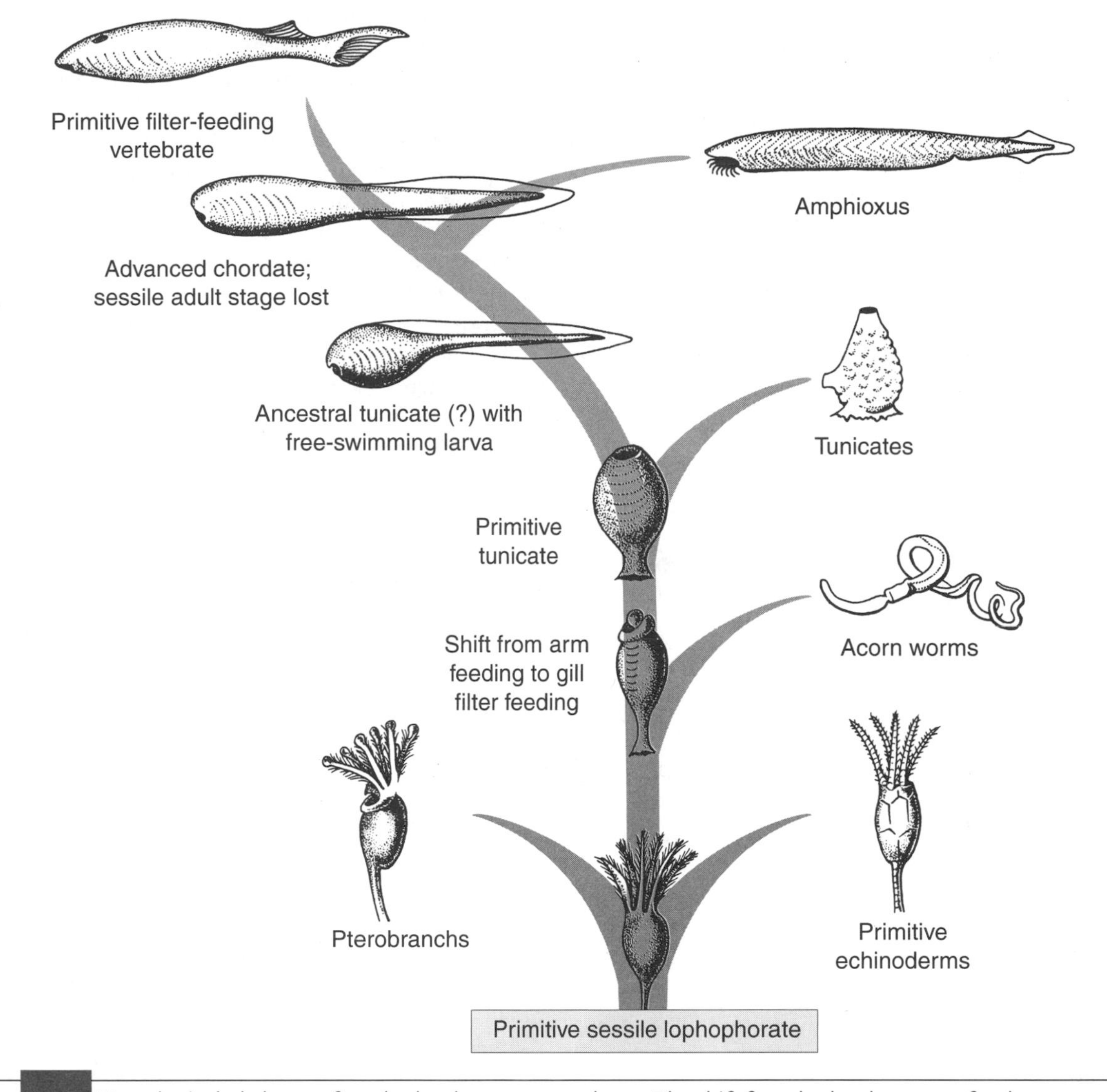

FIGURE 17-8 Hypothetical phylogeny from lophophorate to vertebrate. The shift from lophophore-type food capture to sessile filter feeding occurred with the introduction of gill slits, and the subsequent shift to mobile, vertebrate-type food capture occurred when paedomorphosis in a tunicate-like larva introduced a notochord. Among the earliest of these filter-feeding vertebrates is a fossil, *Pikaia,* found in the Burgess Shale of Canada dating back to the Middle Cambrian period (Fig. 14-3). It is shaped somewhat like amphioxus, with repeating myotomes along its length and a notochord that extends from the tip of the tail to the anterior third of the animal. Conway Morris suggests that "it may not be far removed from the ancestral fish." That both adult *Pikaia* and embryonic gill-slitted stages of modern vertebrates (Fig. 3-10) share similarity to free-swimming filter feeders indicates a common chordate ancestor that probably appeared and functioned like a tunicate larvae. (*Adapted from Romer and Parsons.*)

food with arrays of sharp bony pharyngeal elements. These sawlike structures, commonly found in various Paleozoic strata, have recently been associated with the conodont animal, and may represent the earliest bony vertebrate tissue.

Other early fish, called heterostracans, were usually small, 8 to 12 inches long, encased in large dermal plates anteriorly and smaller plates or scales posteriorly (Fig. 17–9*a*). In addition to two laterally placed eyes, they also had a mid-dorsal opening for a median eye or pineal organ. Various forms of these fishes appear, some bottom feeders, mostly with ventrally placed mouths, and others with mouths on the dorsal surface that may have fed on plankton.

Both the heterostracans and an associated group that were about half their size, the coelolepids (hollow-scaled, Fig. 17–9*b*), had two nasal openings, distinguishing them from other jawless fishes with single nostrils. The latter include the heavily armored osteostracans (bony shells, Fig. 17–9*c*) and the lightly armored but apparently more maneuverable "shieldless" anaspids (Fig. 17–9*d*), both forms appearing in the mid-Silurian and becoming extinct during the Devonian.

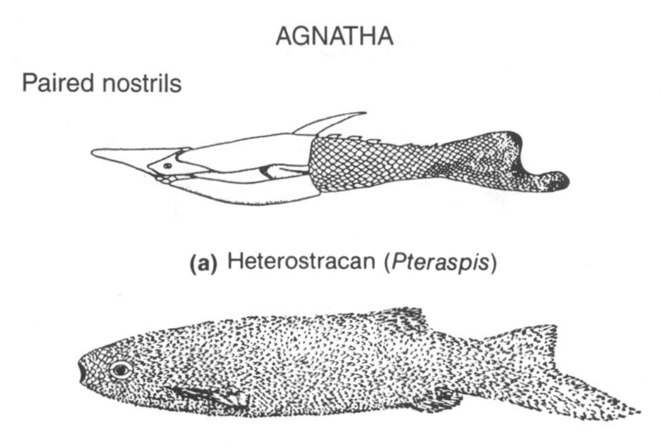

(a) Heterostracan (*Pteraspis*)

(b) Coelolepid (*Phlebolepis*)

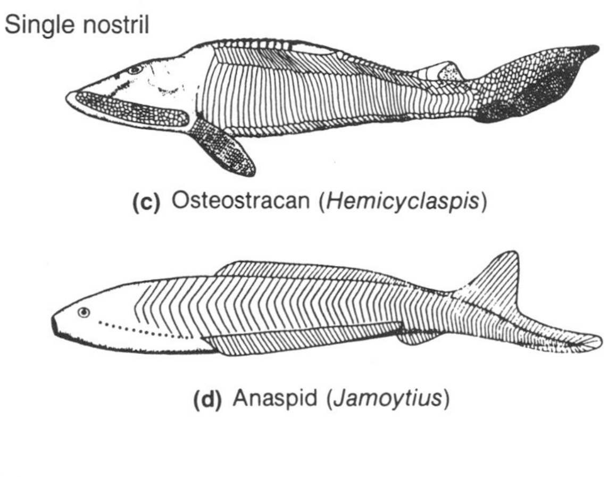

(c) Osteostracan (*Hemicyclaspis*)

(d) Anaspid (*Jamoytius*)

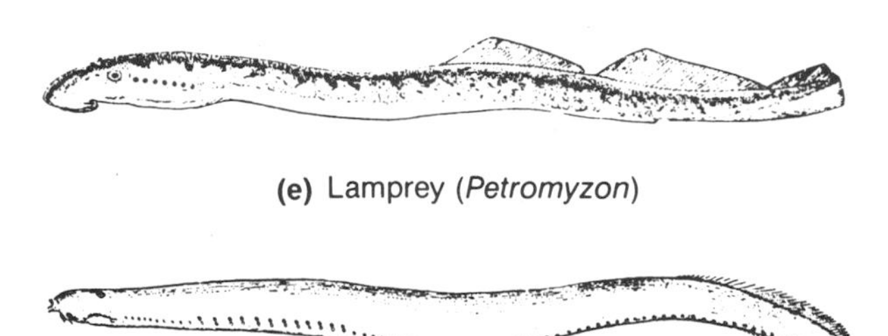

(e) Lamprey (*Petromyzon*)

(f) Hagfish (*Bdellostoma*)

FIGURE 17-9 Fossil (*a–d*) and present-day (*e, f*) jawless fish (Agnatha). Note that tail structures differ: the axial supporting element may be in the upper lobe (heterocercal) or in the lower lobe (hypocercal). The living agnathans, called *cyclostomes,* have round, sucking mouths and lack paired fin structures and dermal bones. Because of other primitive features that include a lensless eye and simplified hindbrain, Forey and Janvier suggest that hagfishes represent the most ancient of all vertebrate lineages.

Moy-Thomas and Miles have suggested that all four of these jawless groups, often called **ostracoderms** (bony skins), used a large, **muscular pharynx** that let them suck up food-laden water more rapidly and in much larger amounts than the ciliary activity of invertebrate lophophorates and filter feeders could achieve. In contrast, Northcutt and Gans propose that although the early vertebrate pharynx may have originally been used for pumping water and filter feeding, it evolved into an active predatory organ in adult ostracoderms. According to them, jawless fish used the pharynx for scooping up small soft-bodied or lightly armored bottom-dwelling animals.

Whether through improved filter feeding or active benthic predation, most paleontologists agree that pharyngeal adaptations were important in accounting for ostracoderm early success. They established themselves throughout Late Cambrian and Early Ordovician marine environments (Repetski) and attained widespread Devonian distribution in fresh or mixed fresh- and saltwater areas in North America and Europe.[5]

After the Devonian period, ostracoderms no longer appear in the fossil record, but there are good indications that one or more agnathan lineages persisted. As **Figures 17–9*e* and *f*** show, there are two modern groups of round-mouthed jawless fishes called **cyclostomes.** These are the lampreys (order Petromyzontiformes) and hagfishes (order Myxiniformes), both now occupying restricted ecological niches, either as ectoparasites (lampreys) or as burrowing detritus feeders and scavengers (hagfish). Although their precise ancestral pattern is unclear, they are probably related to ostracoderms. Paleontologists have proposed, for example, that the fossil anaspid *Jamoytius* may be in the direct line of descent of ostracoderms, modern lampreys, and perhaps hagfishes as well.

In any case, new jawed and finned vertebrates arose during the mid-Devonian using hard bony tissue for structure and defense. Although cartilage is somewhat softer than bone, it also appears as the major structural tissue in some later lineages such as sharks. Bone—an organic protein matrix mineralized with the calcium phosphate salt, hydroxyapatite—turned out to be an exemplary tissue with substantial evolutionary advantages for vertebrates:

[5]Smith, from physiological evidence, suggested that early vertebrates originated in freshwater streams and lakes and only later entered the saltwater marine environment. He proposed that the vertebrate kidney arose as an organ regulating osmosis in a freshwater environment in which the concentration of ions is much lower than in cellular tissues. To prevent "swamping" of body tissues by incoming water, the kidney glomeruli pump out excess water, while the kidney tubules resorb necessary ions and small molecules back into the circulatory system. In contrast to these views, researchers have pointed out that the presence of vertebrate kidney glomeruli does not necessitate a freshwater origin but may have arisen in areas of dilute seawater, such as the brackish coastal estuaries where continental rivers empty their contents. The osmoregulatory function of the kidney Smith emphasized may also be secondary to its excretory function; that is, the adaptive value of the kidney was primarily to get rid of waste products accumulated by an animal with a high rate of metabolism. The likelihood of a marine and estuarine origin for early vertebrates is also strengthened by the observation that presumed ancestral or closely related forms (cephalochordates, urochordates, and hemichordates) are all marine fauna, found mostly in shallow waters.

1. Excess calcium ions, diffusing through the skin and gills, can be deposited in the skin as an osmotically inert substance—bone—thus conserving energy that the organism would otherwise expend in excreting these ions.
2. Tissue deposits of calcium and phosphate provide a metabolic reserve that the animal can mobilize, when needed, by partial bone decalcification. (Animals can also calcify and use cartilage for this purpose.)
3. Calcium phosphate tissues such as dentin and enamel can crystallize near electrosensory organs (for example, "lateral line" systems that detect electrical currents emitted by prey), insulating them from internal electrical body currents, thereby improving their directional resolution. Underlying dermal bone structures would mechanically stabilize the position of these organs (Northcutt and Gans).
4. The elaboration of such bone structures would become important in providing defensive dermal armor.
5. Hardened tooth surfaces can evolve for grasping and masticating food.
6. The evolution of ossified internal skeletons would offer rigid supporting structures for muscle and organ attachment, far stronger than the notochord itself.

Paleontologists have suggested that the development of defensive bony plates was an essential element enabling early vertebrates to withstand predation by the voracious and widely distributed scorpion-like eurypterids—that is, until the vertebrates themselves evolved into important predators.

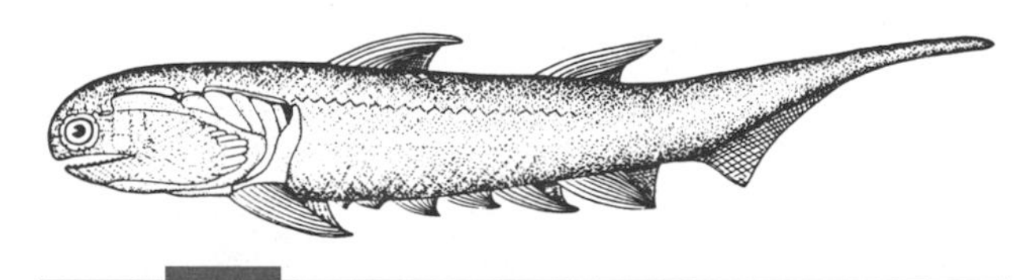

FIGURE 17–10 The acanthodian *Climatius* showing the broad-based spiny fins running mid-dorsally and (in two rows) ventrally. Small armored scales completely covered these fins. *(Adapted from Colbert and Morales.)*

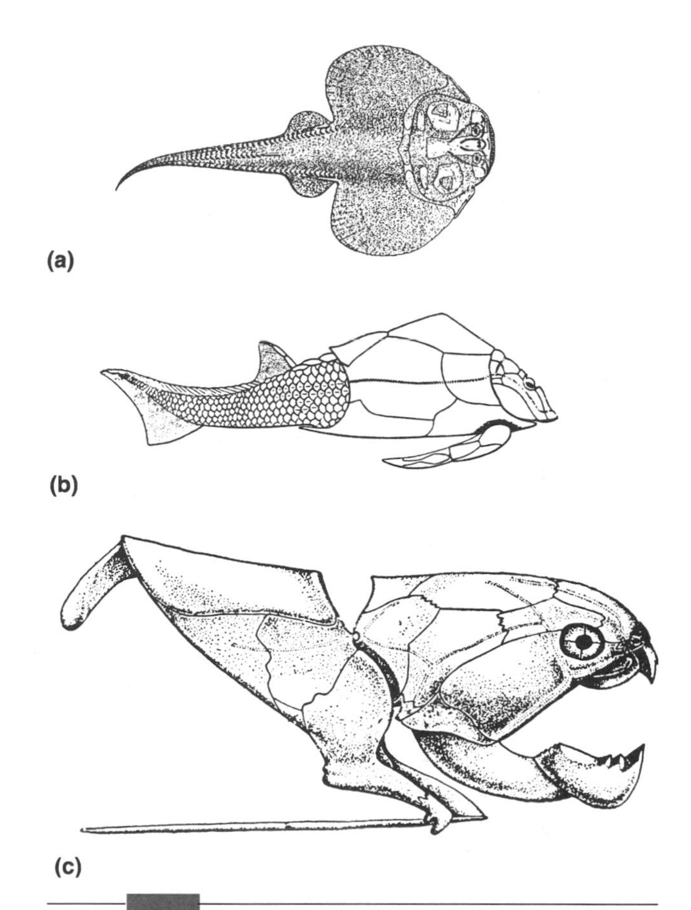

FIGURE 17–11 Placoderms. (*a*) Skatelike rhenanid (*Gemuendina*) with large lateral fins. (*b*) An antiarch (*Pterichthyodes*) showing the scaled posterior portion and the heavily armored anterior trunk and head regions. The pectoral "fins" of antiarchs were encased in bony plates that may have enabled them to crawl along the sea bottom. (*c*) The 10-foot-long anterior bony shield of the arthrodire *Dunkleosteus.* The animal was about 30 feet long. (*Adapted from Romer.*)

Evolution of Jawed Fishes (Gnathostomata)

The first jawed fossil vertebrates appeared during the Silurian period and are divided into two groups. The earlier of these, the **acanthodians** (spiny sharks), are generally represented by *Climatius,* which was only a few inches long and characterized by both paired and unpaired spiny fins (Fig. 17–10). Although the earliest known forms are in marine sediments, they were, for the most part, freshwater animals found in river, lake, and swamp deposits, many surviving up to Late Paleozoic times.

Some time after the first appearance of these spiny fishes, toward the end of the Silurian period, another group of jawed fishes evolved called **placoderms** (plate-skinned), which flourished during the Devonian, then rapidly became extinct. Some placoderms were bottom dwellers, such as the skatelike rhenanids (Fig. 17–11*a*) and the antiarchs with their stiltlike jointed pectoral appendages (Fig. 17–11*b*). Others were predators of gigantic proportions such as the arthrodire *Dunkleosteus,* with a length of more than 30 feet (Fig. 17–11*c*).

The presence of **jaws** and the further development of **paired fins** were significant features in both acanthodians and placoderms. Jaws revolutionized the way of life for these early vertebrates by offering them new food

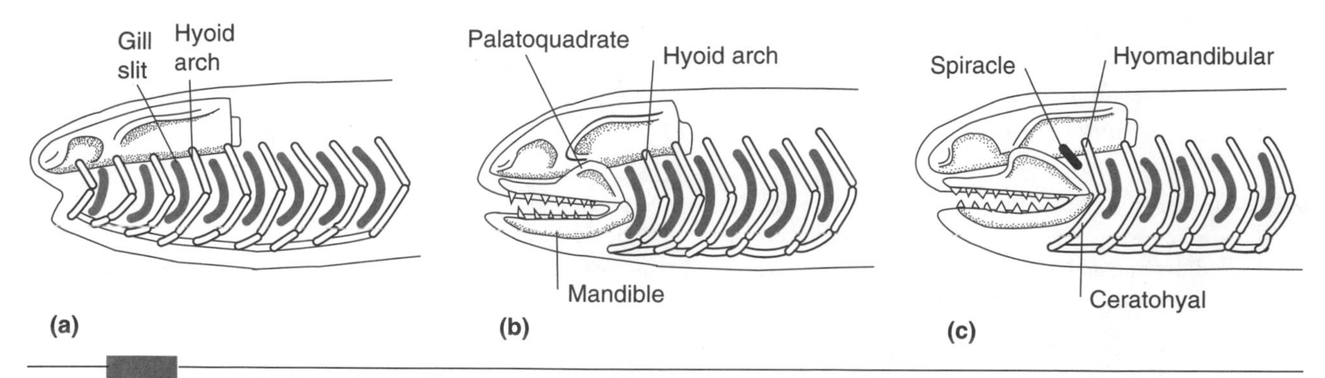

FIGURE 17–12 Stages in the evolution of jaws according to the hypothesis that the vertebrate jaw derives from one of the anterior gill arches. (*a*) The jawless condition. (*b*) The conversion of an anterior gill arch into jaws. (*c*) Incorporation of bones from the hyoid arch to support the hinge of the jaw. In this progression, the gill slit anterior to the hyoid arch reduced to the spiracle. In further evolutionary steps, the ceratohyal and hyomandibular become, respectively, the articular and quadrate bones in the jaw joint of amphibians and reptiles. In mammals, this joint is replaced by a squamosal-dentary hinge, and selection for improved hearing converts the articular/quadrate connection into malleus/incus ossicles (Fig. 19-3). (*Adapted from Romer and Parsons.*)

resources previously excluded because of limitations in filter feeding and sucking. Fishes could now extended carnivorous behavior to all sizes of prey by grabbing, tearing, and chewing. Flattened, opposed teeth could now ground and mill hard or armored food materials (such as mollusks) that were formerly inaccessible.

In addition, jaws allowed defensive and aggressive behaviors that these fishes could use both intra- and interspecifically and offered them greater opportunities to manipulate the environment in building nests or grasping mates. The teeth that provided primitive jaws with their cutting function evolved either from skin "denticles" that initially served as armor plating in these early vertebrates or from cutting edges on the jaws themselves, as in some of the placoderms.

We do not know the intermediary steps between jawless and jawed fish from the fossil record, although some biologists have suggested that jaws evolved from the transformation of pharyngeal **gill arches** previously used in filter feeding and perhaps respiration as well (Romer and Parsons). These gill arches are paired on each side and supported by V-shaped hinged structures whose apices point backward. However, the fate of the anterior pairs of gill arches remains unclear; the arches may have disappeared, incorporated into the base of the cranium, or formed one or more of the mouth structures. Whatever happened, the first gill arch posterior to these anterior pairs changed so that the upper part of the hinge became the upper jaw, or palatoquadrate bone, and the lower part became the mandible (Fig. 17–12). Behind these structures, an arch called the *hyoid* incorporated into the complex by contributing its dorsal portion, the hyomandibular, to anchor the hinge of the jaw to the braincase.

Among the evidence to support this view is the archlike appearance of palatoquadrate and mandible in acanthodians as well as in later sharks and bony fish (Fig. 17–13). Further supporting a jaw–gill slit connection, the trigeminal cranial nerve in the shark (and some of the other cranial nerves) runs a branch down to the lower jaw and another anteriorly to the upper jaw as though a gill slit had been enclosed at one time (Fig. 17–14). Researchers have used the presence of cranial nerves anterior to the trigeminal to indicate there once were gill slits anterior to those involved in jaw formation.

A different hypothesis for the evolution of jaws proposes that no clear sign of a gill arch remnant anterior to the mouth appears in either ancient or modern vertebrates. According to this view, the palatoquadrate and mandibular bones may never have served as gill supports, and the mouth, as well as its supporting structures, was always distinctly separate from the pharynx (Carroll). Although we do not know which of these hypotheses is correct, vertebrate paleontologists generally agree that, because their jaw structures are so similar, the major modern forms of fish must have derived their jaw pattern from a common ancestral group.

Chondrichthyes and Osteichthyes

Evolving from a group of jawed fish that was most probably separate from placoderms, entirely new forms began to appear during the Devonian period. Improved swimming efficiency resulting from increased nervous and muscular coordination and progressively streamlined body forms probably influenced their success. Among these groups taxonomists classify **cartilaginous fishes** as members of the class **Chondrichthyes** and **bony fishes** in the class **Osteichthyes.**

Although a variety of subgroups exists in each of these classes, sufficient common features within each class in-

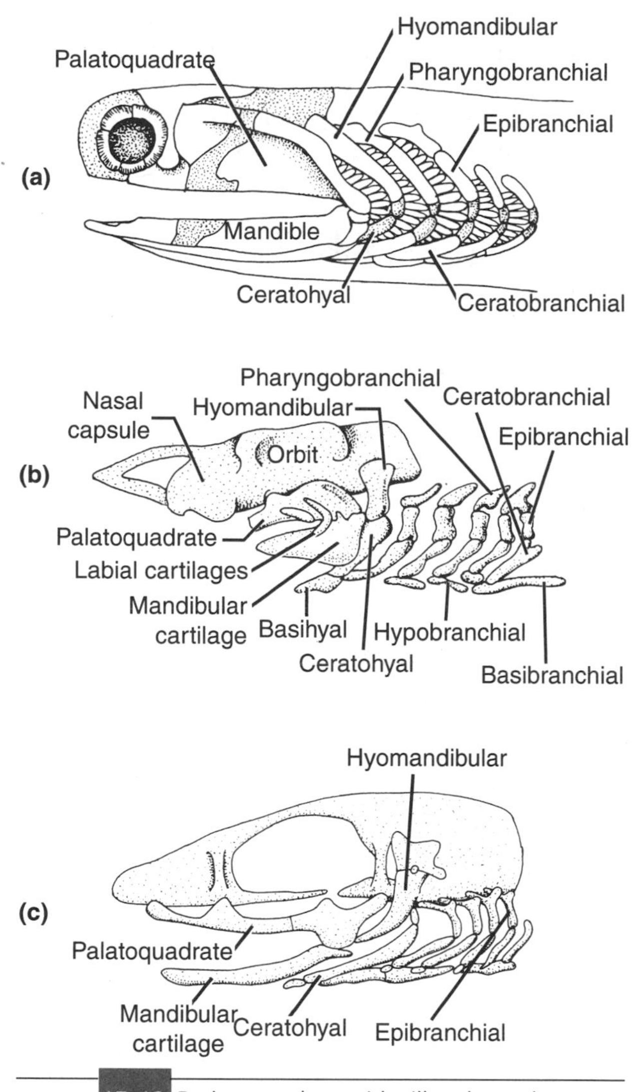

FIGURE 17-13 Braincases along with gill and mouth structures (visceral arches or skeletons), in a fossil acanthodian (*a*), a shark (*b*), and a bony teleost embryo (*c*). (*Adapted from Romer and Parsons.*)

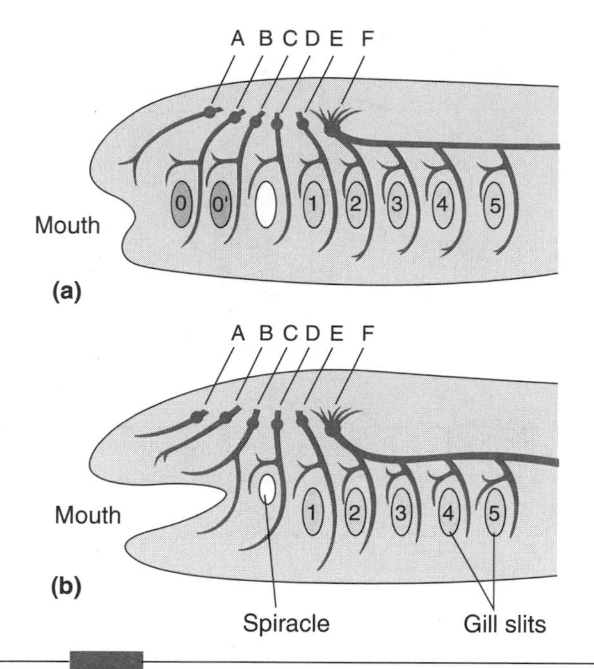

FIGURE 17-14 Diagrams of the dorsal root cranial nerves that innervated the gill arches (*a*) in a hypothetical primitive fish, and (*b*) in a later jaw-bearing fish such as the shark. The letters *o* and *o′* represent gill slits lost with the acquisition of jaws, *s* represents the gill slit later used as the spiracle in sharks, and *1* to *5* are gill slits posterior to the spiracle. *A* to *F* are cranial nerves. *A* represents the terminal cranial nerve found in numerous vertebrates that may have innervated the anteriormost member of the gill series. *B* represents a nerve that is independent in lower vertebrates but combines with the trigeminal nerve in mammals. *C* is the trigeminal nerve that innervates the upper and lower jaws in all present vertebrates. *D, E,* and *F* represent, respectively, the facial, glossopharyngeal, and vagus nerves. (See Fig. 3-11 for a comparison of innervations of the vagus nerve in fish and in mammals.) (*Adapted from Romer and Parsons.*)

dicate its members share some degree of homology. The Chondrichthyes are cartilaginous, with no identifiable bone tissue, although some parts can become calcified. Almost all these cartilaginous fish have heterocercal tails, a feature also common in early bony fish. Common ancestry of modern chondrichthyians—sharks, skates, and rays—can be traced to the Jurassic period, but a strange group called *chimaeras* traces to more ancient Carboniferous forms. The Osteichthyes are characterized by the bony composition of skull, jaws, gill cover (operculum), scales, vertebrae, and ribs. Also, in contrast to cartilaginous fish, most bony fish use a hydrostatic organ, the **swim bladder,** for buoyancy control.

Compared to the older acanthodians and placoderms, both groups show new uses and arrangements of fins including (1) a caudal (tail) fin used for propulsory motion, (2) usually stationary dorsal and ventral fins that act as keels to prevent rolling and side-slipping, and (3) mobile, paired pectoral and pelvic fins that provide vertical controls, "brakes," and "bilge rudders" (Fig. 17–15).

Under the name "fin-fold" theory, some workers have suggested that the dorsal, caudal, and anal fins are derived from a continuous fold of skin originally present along the dorsal midline of the body, and the paired pectoral and pelvic fins from similar folds along the flanks. No paleontological evidence yet supports this view, although clearly fin structures have evolved considerably in the placement of fins, in their supporting structures, and in their mobility and flexibility. For example, the bases of the paired fins in Devonian sharklike fishes, such as *Cladoselache,* are quite wide (Fig. 17–16), but these bases become narrower and more mobile in later forms. Among other fishes, the dorsal fin in some forms (*Dorypterus*) is almost as long as the fish itself.

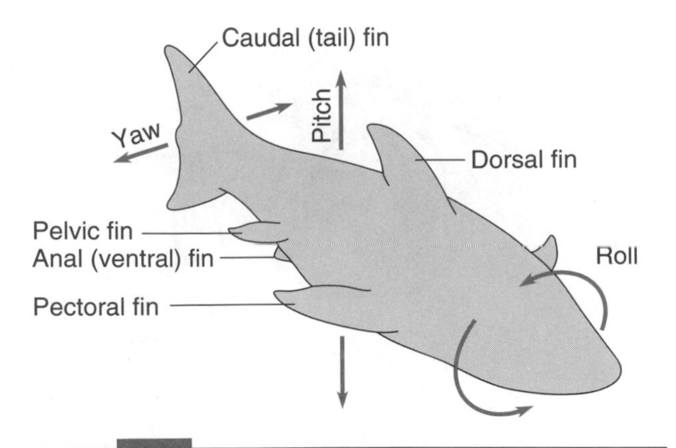

FIGURE 17–15 Generalized design of modern fish that enables it to cleave through the water rapidly with considerable control and cope efficiently with disturbing forces such as yaw, pitch, and roll. (*Adapted from Waterman et al.*)

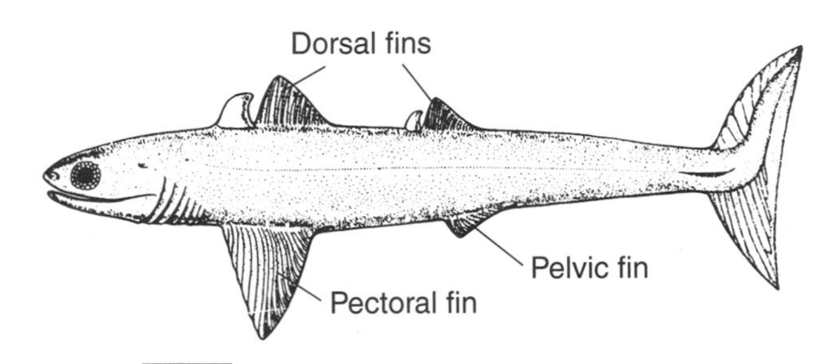

FIGURE 17–16 *Cladoselache,* a late Devonian sharklike fish ranging from 1.5 to 4 feet long. It had paired pectoral and pelvic fins as well as fins on the mid-dorsal line, each supported by a broadened row of unconnected rods of cartilage. Some studies question whether *Cladoselache* was ancestral to modern lineages of sharks and skates, and suggest instead their more recent Mesozoic origin (Rasmussen and Arnason). (*Adapted from Romer and Parsons.*)

Since land vertebrates evolved through lineages of Osteichthyes, paleontologists have paid much attention to them. The Middle Devonian strata in which they first appear indicate their presence and major evolution in shallow waters. Such environments would have included stagnating lakes and slow-moving rivers poor in oxygen (hypoxia) because of high temperatures or the growth of algae and microorganisms (eutrophication). This would have placed selective advantages on developing specialized respiratory tissues and structures (**lungs**) to help oxygenate blood in oxygen-depleted waters. Other views suggest that early marine fish entering shallow waters would have carried saclike air or swim bladders previously used for buoyancy that could evolve into lungs (see Long).

However lungs arose, whether for buoyancy or respiration, Farmer suggests their primary value was to increase oxygen supply to the heart, enabling air-breathing fish to increase muscular activity even in nonhypoxic environments. Graham proposes that pulmonary air breathing evolved more than once ("as many as 67 times"). Lungs and air bladders may therefore have reversed functions because of selection in different conditions, each structure serving as a preadaptation for the other (see also p. 33).

The Bony Fish

The first appearances of bony fish in Devonian sediments already show their division into two subclasses based on fin structure and other features: the **actinopterygians** and **sarcopterygians.**

ACTINOPTERYGII

The actinopterygians are ray-finned fish in which the fins are supported by parallel bony rays whose movements are controlled almost entirely by muscles within the body wall. They all seem to derive from a basic ancestral form that had paired pectoral and pelvic fins as well as a single dorsal fin balanced by a single anal fin on the ventral surface. Various groups of ray-finned fish evolved, differing in the degree of ossification of the skeleton, types of scales, tail structure, jaw structure, and position of the fins.

The most primitive actinopterygians (Chondrostei) still have some surviving species today, such as the sturgeon and the Mississippi paddlefish. More complex forms, the Neopterygei, gave rise to a number of groups including the modern bowfin and garpike, as well as to the most recent of all bony fish, the **teleosts,** characterized by a highly ossified skeleton, very thin scales, and many fin specializations. Teleosts have expanded in both abundance and diversity from the Cretaceous period onward until they now number 20,000 living species classified into 40 orders, almost all the bony fish found today. Their advantages probably lie in their capacity to exploit a wide variety of environments because of flexibility in feeding apparatus (see, for example, Fig. 12–18) as well as new fin and scale structures associated with locomotory adaptations.

Note that the replacement of each type of fish, chondrosteans by neopterygians and primitive neopterygians by later teleosts, was not necessarily accompanied by highly significant changes in shape or size. Most innovations appeared relatively minor, probably indicating that within the competitive struggle for existence, even minor changes that improve locomotion and feeding adaptations (Carroll) can significantly increase survival ability.

SARCOPTERYGII

Within the sarcopterygians, biologists group the flesh-finned ("lobe-finned") fish that supported the fin with small individual bones, arranged either along the fin axis or in rows parallel to the body. Muscles within the fin itself apparently mostly controlled its movements. Early flesh-finned fish had two dorsal fins and a pineal opening at the top of the skull, which bore a median eye. Some also had internal nostrils that functioned in air breathing.

Among sarcopterygian descendants, one group still persists in the form of African, Australian, and South American **lungfishes,** called Dipnoi (**dipnoans**). During dry seasons, when their pools stagnate, the African and South American fish encyst themselves in mud, breathing air into a pair of vascularized lungs through openings in their burrows. Australian lungfish, considered more primitive, behave differently since they cannot survive out of the water: they come to the surface during the dry season and use their single lung to breathe air. Some have claimed that the present distribution of lungfish in Africa, South America, and Australia derives from the proximity between these Gondwana continents during the Mesozoic era (Chapter 6).

Because one or more sarcopterygian groups made the transition to land vertebrates (**tetrapods,** Chapter 18), their classification and phylogenetic relationships are of considerable interest. Benton, for example, offers six possible cladograms for sarcopterygian–tetrapod relationships (see also Schultze). Until there is common agreement, we can use a traditional classification that divides sarcopterygians into two subgroups, the dipnoans described above, and a polyphyletic **crossopterygian** group marked by distinctive modes of skeletal ossification and other features. Crossopterygians are then further divided into **rhipidistians** and **coelacanths** that trace back to the Devonian, the former being primarily freshwater fish while the latter became primarily marine. The coelacanths had no internal nostrils, and their lung often evolved into a calcified swim bladder. As mentioned earlier (p. 49), coelacanth fossils appear as far back as the middle Devonian, and biologists presumed they became extinct sometime during the Cretaceous. This view prevailed until a living coelacanth, *Latimeria chalumnae,* was found in 1938 (Fig. 3–15). Since then, many fish of this species have been caught in the Indian Ocean between Africa and Madagascar, and some as far east as Indonesia.

Among the rhipidistians, two of the lineages that evolved—**panderichthyids** and **osteolepiforms** (Fig. 17–17)—show strong similarities to later amphibian tetrapods in their internal nostril openings and limb and skull structures (Chapter 18). Although such comparative paleontological judgments seem simple, molecular sequence analyses of living groups has added complexity. According to Zardoya and coworkers, coelacanths, generally considered crossopterygian descendants, are probably further from the tetrapod lineage than are lungfishes. Which sarcopterygian lineage is closest to tetrapods, crossopterygians or dipnoans, is in question, although it seems obvious that both groups possess features that would lend themselves to evolution on land.

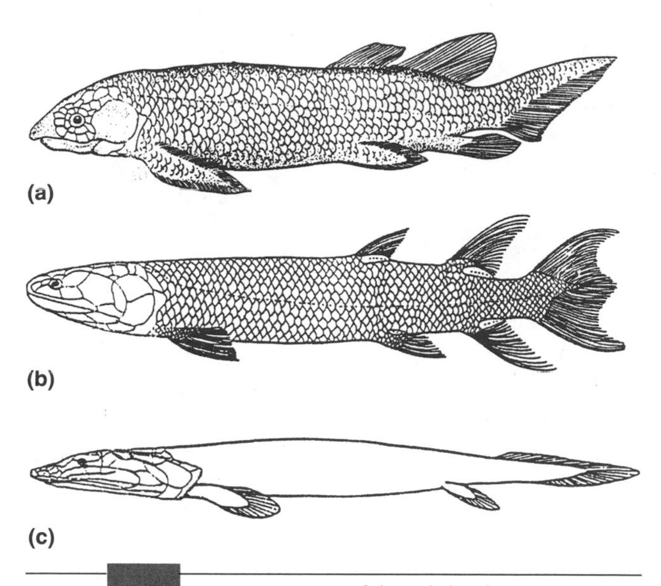

FIGURE 17–17 Reconstruction of three lobe-finned Devonian fishes of lineages that may be ancestral to early land-living amphibians. (*a*) The dipnoan lungfish *Dipterus.* (*b*) and (*c*) Rhipidistians, *Eusthenopteron* (an osteolepiform), and *Panderichthys.* (*From various sources, see Benton.*)

Figure 17–18 offers a possible phylogeny of major groups of fish mentioned in this chapter.

SUMMARY

The vertebrates have several singular features such as pharyngeal gill clefts, an embryological notochord, an internal skeleton mostly derived from mesodermal tissue, a tail, and a hollow dorsal nerve cord. Although we have not located an ancestral fossil shared with other phyla, zoologists have tried to draw lines of descent from annelids, arthropods, and, more plausibly, from a common ancestor with the echinoderms. Embryological similarities between echinoderms and vertebrates include deuterostomy, radial indeterminate cleavage, enterocoelous origin of the coelom, and mesodermally derived skeletons. However, many inconsistencies appear in the data, and the relationship between the two groups is quite distant. Molecular studies may provide more definitive answers.

Some living chordate subgroups may resemble the primitive vertebrate ancestor. Although probably not ancestral to vertebrates, cephalochordates are related and have a lifestyle—swimming and filter feeding—characteristic of the putative protovertebrates. Many larval urochordates have gill slits, notochords, and dorsal nerve cords, structures that are lost in the sessile adults. If such larval types underwent paedomorphosis and reproduced, they might have evolved into free-swimming chordates and, eventually, vertebrates.

The jawless fishes are the most ancient fossil vertebrates known, and many were covered with dermal bony

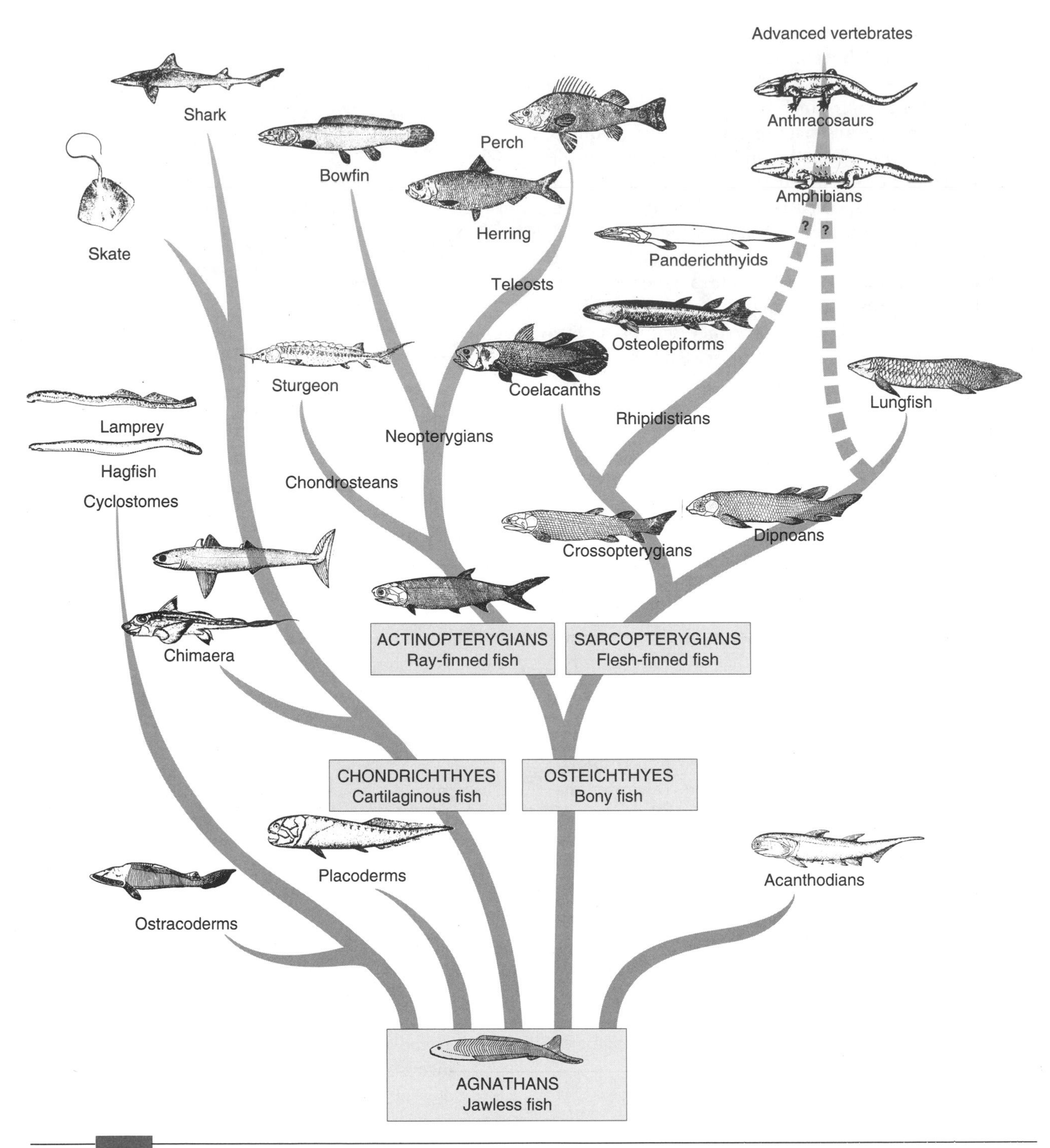

FIGURE 17–18 Phylogenetic relationships among different lines of fish. The ancestor of each given lineage is not necessarily the species illustrated, but paleontologists believe them to have been similar. Nevertheless, there are still disputes on various transitions and lineages; for example, whether the cyclostomes are more closely related to each other than lampreys are to jawed vertebrates, whether the amphibian tetrapods derive from a crossopterygian or lungfish ancestor, and whether some groups such as osteolepiforms are monophyletic or not (Ahlberg and Johanson). On the molecular level, Rasmussen and Arnason question whether sharks and skates originated earlier than the Mesozoic era, and whether "advanced vertebrate" ancestry is osteichthyan as portrayed by paleontologists rather than derived from an even more ancient group of jawed fishes. Deciding such issues depends on obtaining further data.

plates and scales. They had a muscular pharynx, which they at first probably used for filter-feeding and pumping water but which later converted into a predatory organ. The lampreys and hagfishes are modern representatives of these fishes, which jawed and bony fishes mostly replaced. Interestingly, the cartilaginous fishes *appear* later in the fossil record than the bony fishes, although we can debate which one actually *arose* first. In many lineages, bone apparently offers many advantages over cartilage, particularly as a supporting tissue.

The jaws of later fishes probably evolved from bony pharyngeal gill arches used for filter feeding, and this adaptation made fish into very effective carnivores. The earliest jawed groups, acanthodians and placoderms, died out, and more efficient swimmers, the Chondrichthyes (cartilaginous fishes) and Osteichthyes (bony fishes with swim bladders), replaced them. Both groups have elaborate and highly evolved fin patterns including a caudal fin, dorsal and ventral fins for stability, and paired pectoral and pelvic fins for control. Beginning with marine swim bladders, lungs evolved to deal with problems of low oxygen content in freshwater environments. In fish re-entering a marine environment, the lung reconverted to a swim bladder, but in other cases the lung allowed vertebrates to colonize the land.

The early bony fishes are divided into Actinopterygii (ray-finned fish) and Sarcopterygii (flesh-finned fish). Living derivatives of the former group are the sturgeon, garpike, and teleosts. Among these, teleosts are the most successful of bony fishes because of changes in skull, fin, and scale structures associated with new feeding and locomotory adaptations. The sarcopterygians had a pineal eye and fins supported by small bones, and often had internal nostrils and lungs. The lungfishes and the recently discovered coelacanths, which paleontologists previously thought extinct, are relics of the ancient sarcopterygians. Also among sarcopterygian fossils are osteolepiforms and panderichthyids, that, along with dipnoans, are lineages close to the ancestry of early amphibians.

KEY TERMS

acanthodians
actinopterygians
ammocoete larva
amphioxus
bony fish
cartilaginous fish
cephalochordate
Chondrichthyes
Chordata
coelacanths
crossopterygians
cyclostomes
deuterostomes
dipnoans
dorsal hollow nerve cord
filter feeding
gills
gill arches
hemichordate
jawed fish
jawless fishes
lungfishes
lungs
muscular pharynx
neoteny
notochord
Osteichthyes
osteolepiforms
ostracoderms
paedomorphosis
paired fins
panderichthyids
pharyngula
placoderms
protostomes
rhipidistians
sarcopterygians
swim bladder
tail
teleosts
tetrapods
tunicates
urochordates
vertebrae

DISCUSSION QUESTIONS

1. Which features distinguish chordates from other phyla?
2. What arguments, pro and con, have paleontologists offered for the origin of chordates from other known phyla?
3. How do paleontologists use paedomorphosis to explain the origin of early vertebrates?
4. In the evolution of fish, what advantages can we ascribe to
   a. A muscular pharynx
   b. Bone tissue
   c. Jaws
   d. Fins
   e. Lungs
5. What hypotheses have paleontologists offered to explain the morphological evolution of jaws?
6. What are the major subgroups and proposed phylogenetic relationships among the flesh-finned fish (sarcopterygians)?

EVOLUTION ON THE WEB

Explore evolution on the web! Visit the accompanying web site for *Evolution,* 3/e at www.jbpub.com/evolution for web exercises and links relating to topics covered in this chapter.

REFERENCES

Ahlberg, P., and Z. Johanson, 1998. Osteolepiforms and the ancestry of tetrapods. *Nature,* **395,** 792–794.

Bateson, W., 1886. The ancestry of the Chordata. *Q. J. Micr. Sci.,* **26,** 535–571.

Benton, M. J., 1997. *Vertebrate Paleontology,* 2d ed. Chapman & Hall, London.

Berrill, N. J., 1955. *The Origin of Vertebrates.* Clarendon Press, Oxford, England.

Carroll, R. L., 1988. *Vertebrate Paleontology and Evolution.* Freeman, New York.

Colbert, E. H., and M. Morales, 1991. *Evolution of the Vertebrates: A History of the Backboned Animals Through Time,* 3d ed. Wiley, New York.

Conway Morris, S., 1979. The Burgess Shale (Middle Cambrian) fauna. *Ann. Rev. Ecol. Syst.,* **10,** 327–349.

DeRobertis, E. M., and Y. Sasai, 1996. A common plan for dorsoventral patterning in Bilateria. *Nature*, **380**, 37–40.

DiGregorio, A., and M. Levine, 1998. Ascidian embryogenesis and the origins of the chordate body plan. *Current Opinion Genet. Devel.*, **8**, 457–463.

Farmer, C., 1997. Did lungs and the intracardiac shunt evolve to oxygenate the heart in vertebrates? *Paleobiology*, **23**, 358–372.

Ferguson, E. L., 1996. Conservation of dorsal-ventral patterning in arthropods and chordates. *Current Opinion Genet. Devel.*, **6**, 424–431.

Forey, P., and P. Janvier, 1994. Evolution of the early vertebrates. *Amer. Sci.*, **82**, 554–565.

Garstang, W., 1928. The morphology of the Tunicata, and its bearings on the phylogeny of the Chordata. *Q. J. Micr. Sci.*, **72**, 51–187.

Gee, H., 1996. *Before the Backbone: Views on the Origin of the Vertebrates.* Chapman & Hall, London.

Graham, J. B., 1994. An evolutionary perspective for bimodal respiration: A biological synthesis of fish air breathing. *Amer. Zool.*, **34**, 229–237.

Jarvik, E., 1977. The systematic position of acanthodian fishes. In *Problems of Vertebrate Evolution*, S. M. Andrews, R. S. Miles, and A. D. Walker, eds. Academic Press, London, pp. 199–225.

———, 1980. *Basic Structure and Evolution of Vertebrates*, vols. 1 and 2. Academic Press, New York.

Jefferies, R. P. S., 1986. *The Ancestry of Vertebrates.* British Museum (Natural History), London.

Long, J. A., 1995. *The Rise of Fishes.* Johns Hopkins University Press, Baltimore, MD.

Løvtrup, S., 1977. *The Phylogeny of Vertebrata.* Wiley, London.

McKinney, M. L. and McNamara, K. J., 1991. *Heterochrony: The Evolution of Ontongeny.* Plenum Press, New York.

Moy-Thomas, J. A., and R. S. Miles, 1971. *Palaeozoic Fishes*, 2d ed. Saunders, Philadelphia.

Neal, H. V., and H. W. Rand, 1939. *Comparative Anatomy.* Blakiston, Philadelphia.

Northcutt, R. G., and C. Gans, 1983. The genesis of neural crest and epidermal placodes: A reinterpretation of vertebrate origins. *Q. Rev. Biol.*, **58**, 1–28.

Nübler-Jung, K., and D. Arendt, 1996. Enteropneusts and chordate evolution. *Current Biol.*, **6**, 352–353.

Ohno, S., 1970. *Evolution by Gene Duplication.* Springer-Verlag, New York.

Olson, E. C., 1971. *Vertebrate Paleozoology.* Wiley-Interscience, New York.

Pendleton, J. W., B. K. Nagai, M. T. Murtha, and F. H. Ruddle, 1993. Expansion of the *Hox* gene family and the evolution of chordates. *Proc. Nat. Acad. Sci.*, **90**, 6300–6304.

Rasmussen, A.-S., and U. Arnason, 1999. Molecular studies suggest that cartilaginous fishes have a terminal position in the piscine tree. *Proc. Nat. Acad. Sci.*, **96**, 2177–2182.

Repetski, J. E., 1978. A fish from the Upper Cambrian of North America. *Science*, **200**, 529–531.

Romer, A. S., 1966. *Vertebrate Paleontology*, 3d ed. University of Chicago Press, Chicago.

Romer, A. S., and T. S. Parsons, 1977. *The Vertebrate Body*, 5th ed. Saunders, Philadelphia.

Schultze, H.-P., 1994. Comparison of hypotheses on the relationships of sarcopterygians. *Systematic Biol.*, **43**, 155–173.

Smith, H. W., 1961. *From Fish to Philosopher.* Doubleday, New York.

Stahl, B. J., 1974. *Vertebrate History: Problems in Evolution.* McGraw-Hill, New York.

Stokes, M. D., and N. D. Holland, 1998. The lancelet. *Amer. Sci.*, **86**, 552–560.

Turbeville, J. M., J. R. Schulz, and R. A. Raff, 1994. Deuterostome phylogeny and the sister groups of the chordates. *Mol. Biol. and Evol.*, **11**, 648–655.

Wada, H., and N. Satoh, 1994. Details of the evolutionary history from invertebrates to vertebrates, as deduced from the sequences of 18S rDNA. *Proc. Nat. Acad. Sci.*, **91**, 1801–1804.

Waterman, A. J., B. E Frye, K. Johansen, A. G. Kluge, M. L. Ross, C. R. Noback, I. D. Olsen, and G. R. Zug, 1971. *Chordate Structure and Function.* Macmillan, New York.

Zardoya, R., Y. Cao, M. Hasegawa, and A. Meyer, 1998. Searching for the closest living relative(s) of tetrapods through evolutionary analysis of mitochondrial and nuclear data. *Mol. Biol. and Evol.*, **15**, 506–517.

18 From Water to Air: Amphibians, Reptiles, and Birds

Various sarcopterygian fish, such as some rhipidistians (panderichthyids, osteolepiforms) or dipnoan lungfish, were apparently preadapted for moving out of water onto land. They had functioning lungs and two pairs of bone-strengthened muscular fins on which they could move their bodies and support themselves terrestrially without depending on the buoyancy of water. Although it would seem that many of these fish needed relatively few further changes to attain a primitive terrestrial existence (Fig. 18–1), the question of why some of them abandoned their shallow-water habitats and went onto land is difficult to answer with certainty.

A classic hypothesis that Romer (1968) popularized suggests that when the shallow, hypoxic habitats of ancient crossopterygians dried up or stagnated further, some varieties that were preadapted to breathing atmospheric oxygen would have searched for new pools of water and probably survived on land for short periods of time. According to Romer, seasonal droughts were common in the Devonian, and selection during such periods would lead to increased intervals of terrestrial exploration until some groups could eventually maintain themselves out of water for significant parts of their life cycle.

Another hypothesis, more commonly accepted now (McFarland et al.), suggests that these aquatic forms escaped to land because of population pressures resulting from predation (probably other fish) as well as from competition for space, food, and breeding sites in these warm, swampy habitats. The transition to land in a moist tropical climate might have produced relatively little stress in such terrestrially preadapted sarcopterygians, and some invertebrate food sources on land may not have been much different from in the swamps themselves. Certainly, throughout the Devonian, land plants were establishing themselves in increasing number and variety (p. 304), and arthropods, among other invertebrates, had already made a successful terrestrial transition (Little).

Whether because of drought or expansion or both, once existence on land was established as an important stage in survival, further selection would operate on many levels to improve air breathing, eliminate carbon

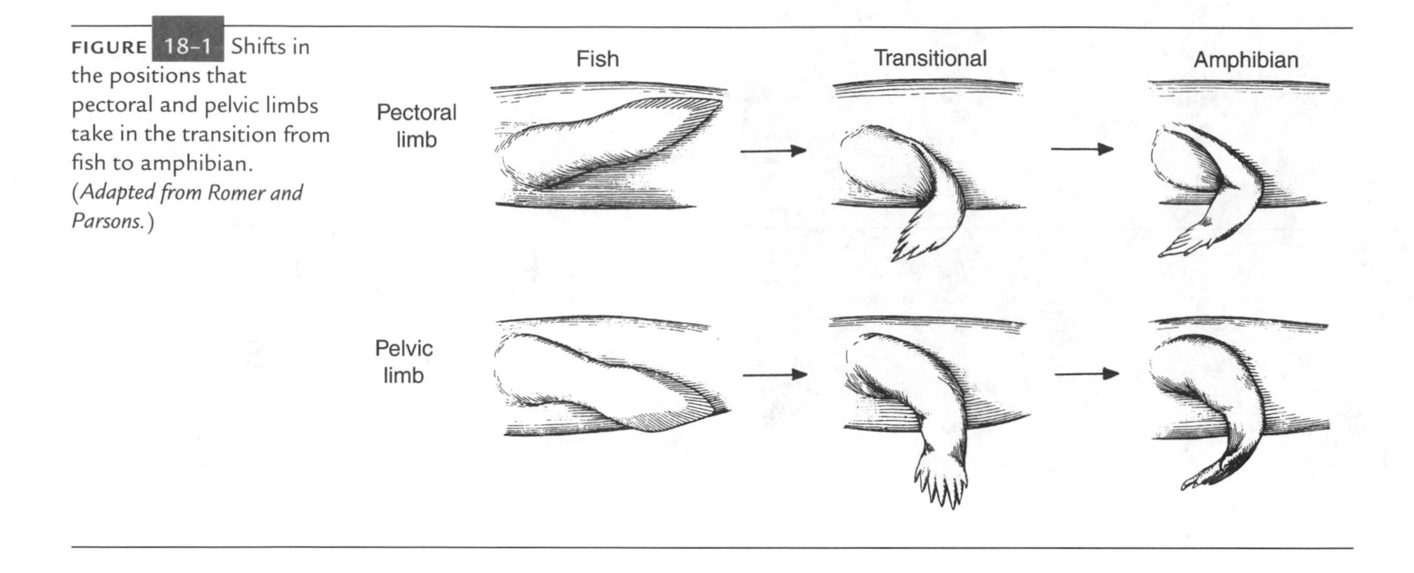

FIGURE 18–1 Shifts in the positions that pectoral and pelvic limbs take in the transition from fish to amphibian. (*Adapted from Romer and Parsons.*)

(a)

(b)

FIGURE 18–2 Two Late Devonian tetrapod skeletons (not to scale). (*a*) *Acanthostega,* a 24-inch primitive tetrapod, which retained fishlike internal gills and led an aquatic life, probably using its limbs in water rather than on land (see Coates and Clack 1991). (*b*) The stout-limbed 3-foot long *Ichthyostega,* which maintained many fishlike features including scales, dorsal tail fin, and a hydrodynamic shape, probably indicating a shallow-water environment. Their mixture of aquatic and terrestrial features indicates these animals were truly transitional forms, "neither fish nor frog."

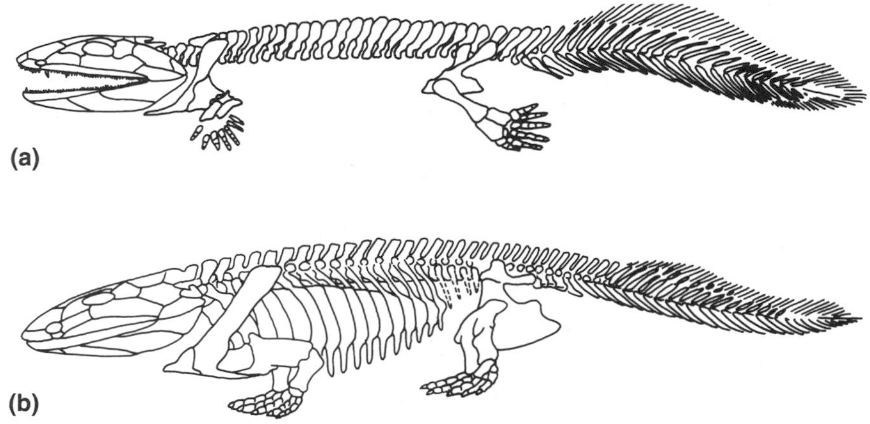

dioxide, increase resistance to desiccation, increase head mobility, and enhance further transformations. You can see that such changes were possible in the observation that some present-day fish such as mudskippers, climbing perch, and walking catfish have developed various terrestrial adaptations, even to the point of climbing trees and capturing food.

Early Amphibians

Although the details are not yet fully known (Chapter 17), many paleontologists agree that land vertebrates, however they first evolved, were related to sarcopterygian lobe-finned fishes. The transition from fish to crawling four-legged **tetrapod** occurred by the end of the Devonian period, about 360 million years ago during a relatively short geological interval—no more than probably 15 or 20 million years—and encompassed perhaps three or more separate lineages (Carroll 1995). The earliest of such identified **amphibians** in the fossil record, called *Acanthostega* and *Ichthyostega* (Fig. 18–2), show their relationship to rhipidistian forms in a number of features:

1. Many dermal bones in the skulls of panderichthyids and Devonian tetrapods appear similar, occupying relatively similar positions (Fig. 18–3). Even the remnant of a preopercular bone is present in these primitive amphibia although they possessed no operculum (gill cover).
2. The fins of osteolepiforms and their supporting girdles have bones that we can easily consider homologous to those of early tetrapods (Fig. 18–4).
3. The tooth structure of both osteolepiforms and Devonian tetrapods shows similar complex labyrinthine foldings of the pulp cavity (Fig. 18–5). In fact, because of the prevalence of these unusual teeth,

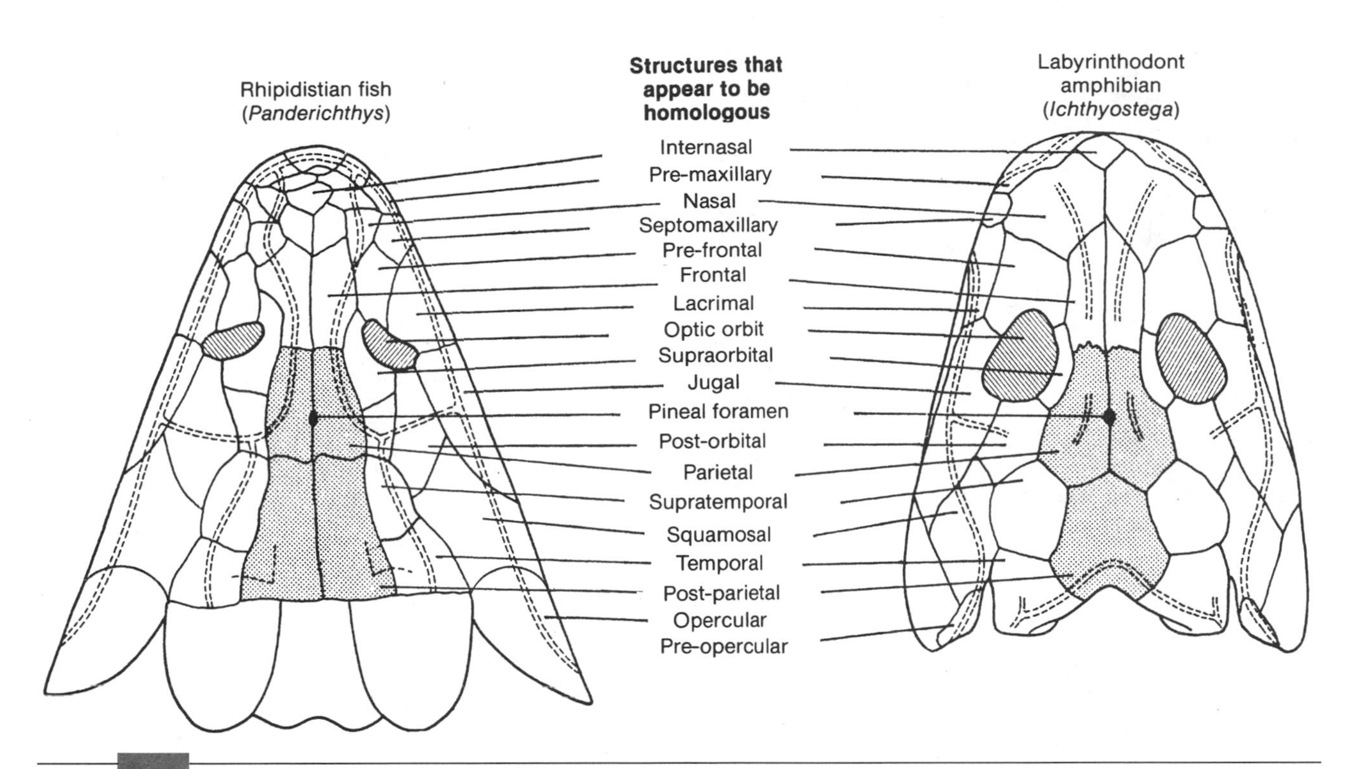

FIGURE 18-3 Dorsal views of the tabular bones in skulls of an osteolepiform and early amphibian compared in terms of likely homologous structures. *Dashed lines* indicate sensory canals. (*Adapted from Duellman and Trueb.*)

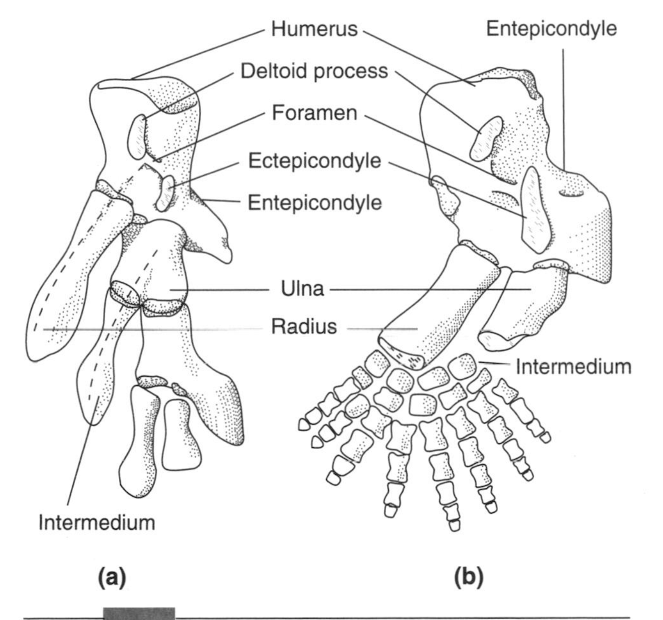

FIGURE 18-4 Comparison between bones in the pectoral fin of an osteolepiform, *Eusthenopteron* (*a*) and those in the forelimb of the early fossil amphibian, *Acanthostega* (*b*). As shown, the early tetrapods were polydactylous. *Acanthostega* had eight digits on its forelimbs, and *Ichthyostega* had seven on its hindlimbs. Later pentadactyl limb may have evolved from the loss of supernumerary digits. (*Adapted from Coates and Clack 1990.*)

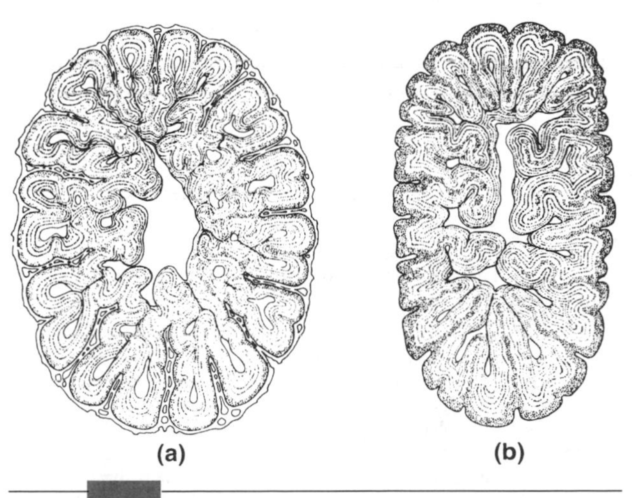

FIGURE 18-5 Cross sections of teeth from an osteolepiform, *Polyplocodus* (*a*), and a labyrinthodont amphibian, *Benthosuchus* (*b*). (*Adapted from Stahl, from Bystrow.*)

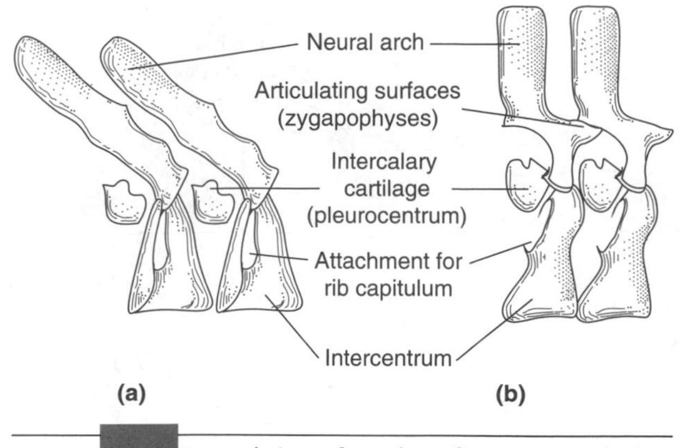

FIGURE 18–6 Lateral view of vertebrae from an osteolepiform, *Eusthenopteron* (*a*), and from the fossil amphibian, *Ichthyostega* (*b*). (*Adapted from Romer 1966.*)

paleontologists have given the name Labyrinthodontia to these tetrapods and also to two other orders of fossil amphibians, the anthracosaurs and temnospondyls. (Researchers believe reptiles derive from early **anthracosaurs.**)

4. The sensory lateral line system of osteolepiforms that extended into the skull appears homologous to a similar pattern of sensory canals embedded in the tetrapod skull (Fig. 18–3).
5. The early amphibians possessed a fin-rayed caudal tail that showed obvious fishlike ancestry.
6. The structure of early tetrapod vertebrae had changed relatively little from the vertebral structure of rhipidistian osteolepiforms (Fig. 18–6).

Unfortunately, we know almost nothing of the soft-body structures of these early amphibians, and we have no obvious clue as to how they prevented desiccation. Very probably, like many modern amphibians, they spent a considerable part of their development in water, and the adult form never wandered too far from moist surroundings. It seems reasonable to suppose that, like present fish and other vertebrates, they possessed a developmental stage, the *pharyngula,* characterized by pharyngeal arches and other embryonic features derived from an aquatic experience.[1] Nevertheless, their many anatomical post-aquatic innovations indicate that they and their successors were evolving adaptive solutions to at least some mechanical difficulties that land-dwelling vertebrates face.

One important early terrestrial problem was the need to prevent compression of the internal organs by pressure transmitted from the limbs and limb girdles. A second problem, pertaining specifically to the forelimbs, was that of keeping the impact of terrestrial locomotion from transmitting to the braincase since the osteolepiform pectoral fins connect to the skull. As Figure 18–7 shows, these difficulties were solved by supporting the spine at two major points along its axis with pectoral and pelvic girdles supported by limbs and feet. The spine thus became a "suspension bridge" that (1) absorbed the impact of terrestrial motion, (2) freed internal organs from pressure, and (3) enabled the head to turn and lift independently of the body by placing it on a forward cantilever of cervical vertebrae.

Selection for terrestrial skeletal rigidity had various consequences, emphasizing strengthened vertebral elements and increasing contact between adjacent vertebrae. Originally, osteolepiform vertebrae consisted of a neural arch, an intercentrum, and intercalary cartilages (Fig. 18–8). In the later *rhachitomous* condition, the intercentrum and pleurocentrum (which paleontologists believe homologous to the intercalary elements) remained but became increasingly ossified in amphibian groups called *temnospondyls.*

These amphibians (also distinguished by features such as the size and position of their cranial tabular bones) flourished from the Carboniferous through the Triassic periods and produced a diversity of forms that ranged from the alligator-like *Eryops* to the smaller, dorsally armored *Cacops.* Out of the temnospondyls evolved Triassic organisms in which the intercentrum alone became the main vertebral central element. Among this new group, called *stereospondyls,* are found various forms with flattened heads, including one short-faced, armored type, *Gerrothorax,* which had obviously returned to an aquatic existence by maintaining gills in the adult stage.

In the anthracosaur line of amphibians vertebral structure evolved in a direction opposite to that of temnospondyls; that is, the pleurocentrum gradually increased in size rather than the intercentrum. Fewer in number and variety than temnospondyls, the anthracosaurs had a shorter fossil history, first appearing during the early Carboniferous period and becoming extinct by the end of the Permian. During this interval, one anthracosaur lineage led back to predominantly aquatic forms called *embolomeres* who had very flexible vertebral

[1]According to Langeland and Kimmel, the pharyngula embryonic stage, common to vertebrates (Fig. 3–10), can be described as follows:

> At this stage, the embryo is supported along its axis by a differentiated notochord. Brain morphogenesis is advanced, with the hindbrain fully segmented into rhombomeres. The primordia of the pharyngeal arches are just recognizable, partitioned by a series of grooves in the pharyngeal wall. In the trunk, somite-derived segmental muscle blocks, or myotomes, are innervated by the axons of primary motoneurons, and spontaneous activity in these neurons mediates vigorous muscular contractions. Primary sensory neurons in both the body trunk and head have grown axons to the skin and have connected centrally in the brain and spinal cord. Primary interneurons have made long central axonal pathways, some of these cells projecting between the sensory and motoneurons so as to form sensory motor circuits. Indeed, shortly after the segmentation period is over, a light touch to the embryo will elicit the first reflexive contractile responses.

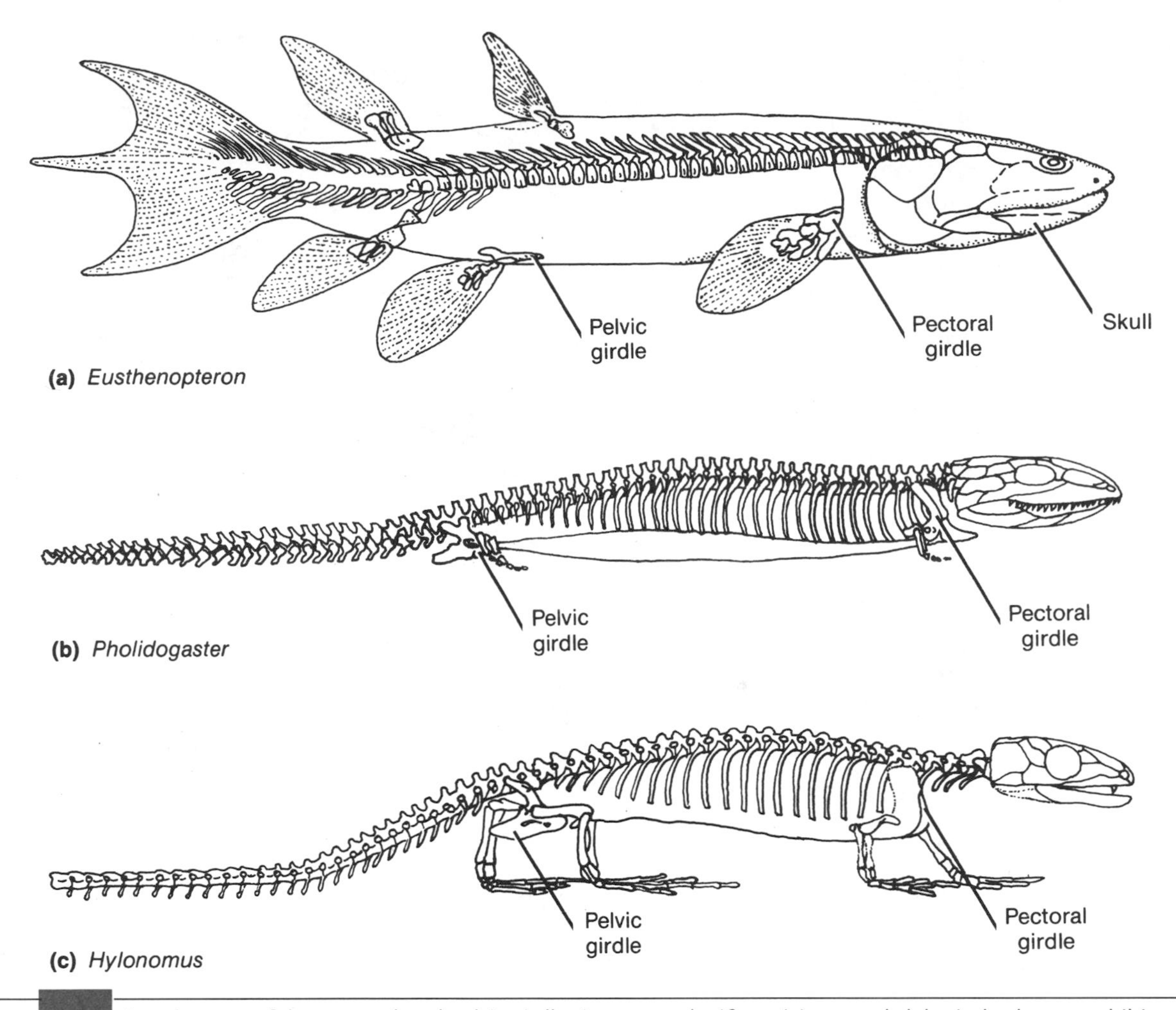

FIGURE 18–7 Attachments of the pectoral and pelvic girdles in an osteolepiform (*a*), an early labyrinthodont amphibian (*b*), and a very early reptile (*c*). In the change from aqueous swimming to terrestrial walking, the pectoral girdle separated from direct attachment to the skull by intermediary cervical vertebrae. The pelvic girdle enlarged to allow leg mobility and became firmly attached to sacral ribs of the vertebral column. Benton (1997) points out, "A tetrapod is rather like a wheelbarrow, since the main driving forces in walking come from the hindlimbs, and the sacrum and pelvis had to become rigid to allow more effective transmission of thrust." (*Parts (a) and (b) adapted from Romer 1967; (c) adapted from Carroll 1991.*)

columns and greatly reduced limbs. Another anthracosaur lineage led to stout-legged terrestrial animals such as *Seymouria* (Fig. 18–9*a*), which researchers believed to have been reptilian until they discovered gilled amphibian larvae among some of their group.[2]

Although *Seymouria* and most of these early amphibians were carnivorous, probably depending on a diet of either invertebrates, fish, or other amphibians, one Late Carboniferous–Early Permian group, represented by *Diadectes* (Fig. 18–9*b*), was probably one of the earliest, if not the first, of the tetrapod herbivores. It had evolved a massive bone structure with heavy vertebrae, probably to support considerable weight. However, in many features *Diadectes* and other diadectomorphs seem intermediate between amphibians and reptiles, suggesting to some paleontologists that, among amphibian groups, this group is probably most closely related to the reptilian ancestor.

Other fossil amphibians also possessed unique vertebral and skeletal structures. One order, often assembled under the name *microsaurs,* had extremely variable body proportions and also vertebral structures that seem to parallel those found in anthracosaurs. Although they, too, have been proposed as possible reptilian ancestors, Carroll and others have pointed to their many peculiar specializations as evidence against this hypothesis. The group to which microsaurs belong, lepospondyls, are characterized by vertebrae much like those of modern amphibia in which a single cylindrical, bony spool surrounds the notochord. Other lepospondyls were either

[2]Panchen (1977) separates the seymouriamorphs from the anthracosaurs and considers them as the two suborders of the order Batrachosauria.

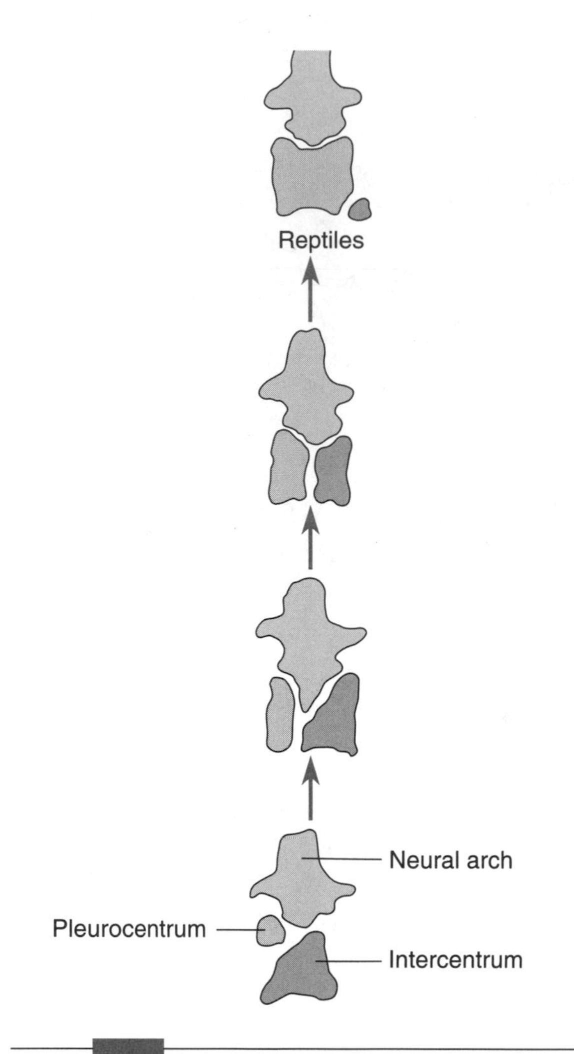

FIGURE 18–8 The evolution of vertebral central elements (lateral views) beginning with an early amphibian labyrinthodont and proceeding through stages found in amphibian anthracosaurs up to the reptilian grade. In advanced reptiles, birds, and mammals, the entire vertebral central element consists of the expanded pleurocentrum. (*Adapted from Romer 1966.*)

limbless and snakelike (aistopods) or had very reduced limbs; and some had peculiar hornlike projections of the rear skull bones (nectridians). These small aquatic forms flourished during the Carboniferous period, but disappeared by the close of the Permian.

Unfortunately, we have not discovered transitional forms between many of these early amphibian groups, nor do fossils yet permit us to supply many details between the earliest amphibians known and their lobe-finned ancestors. In the absence of a clear phylogeny, some paleontologists suggest that at least some of the amphibian diversity stemmed from a polyphyletic origin in which different rhipidistian groups served as ancestors for different groups of amphibians (Jarvik). Others argue that evolution of the characteristic bone arrangements in amphibian limbs and the five-toed foot would probably

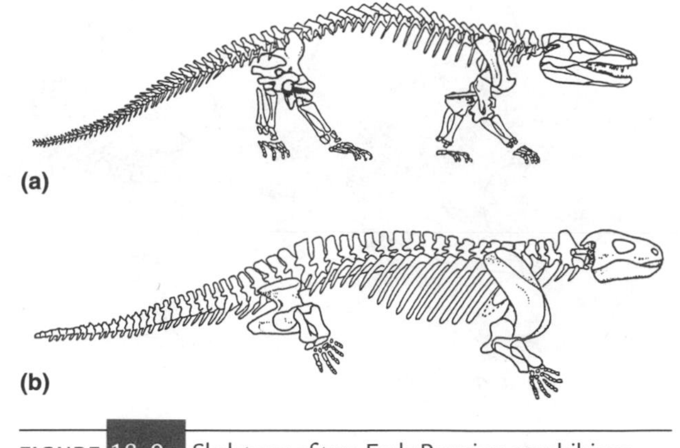

FIGURE 18–9 Skeletons of two Early Permian amphibians with reptile-like features. *Seymouria* (*a*), about 20 inches long. *Diadectes* (*b*), about 8–9 feet long. As noted in the text, ascribing gilled larvae to the seymouriomorphs apparently excludes this group from the reptile classification. (*From Benton, M.J., 1997.* Vertebrate Paleontology, 2nd Edition. *HarperCollins, London.*)

not have arisen more than once.[3] A monophyletic amphibian origin is now more commonly accepted, and Figure 18–10 represents one possible phylogenetic scheme.

Modern Amphibians

Among the intriguing unanswered questions is the origin of modern Amphibia. Zoologists usually classify these together under the subclass **Lissamphibia** and separate them into three orders: caecilians (legless, wormlike burrowers), anurans (frogs and toads), and urodeles (newts and salamanders). In contrast to the scaled skins of early amphibians, modern lissamphibians have a permeable, glandular skin that allows for considerable water and gaseous exchange. (Caecilian skins have, in some species, small embedded scales.) Although this feature limits important segments of their activity to moist or aqueous surroundings, lissamphibians have achieved a variety of remarkable adaptations.

Some toads, for example, are desert inhabitants capable of surviving long periods of drought in underground burrows by drawing moisture from the soil (*Scaphiopus*) or by encapsulating their water-soaked bodies in relatively impermeable membranes (*Cyclorana*). Lissamphibians also show a wide range of reproductive patterns, from traditional egg laying in water and gilled larvae, to the viviparous production of well-developed offspring. The eggs of some frog species, for example, develop conventionally into free-swimming tadpoles, whereas other species raise

[3]We should nevertheless note that the earliest fossil tetrapods known, *Acanthostega* and *Ichthyostega,* had limbs with eight and seven digits, respectively (Fig. 17–4), and there may also have been other numerical variations.

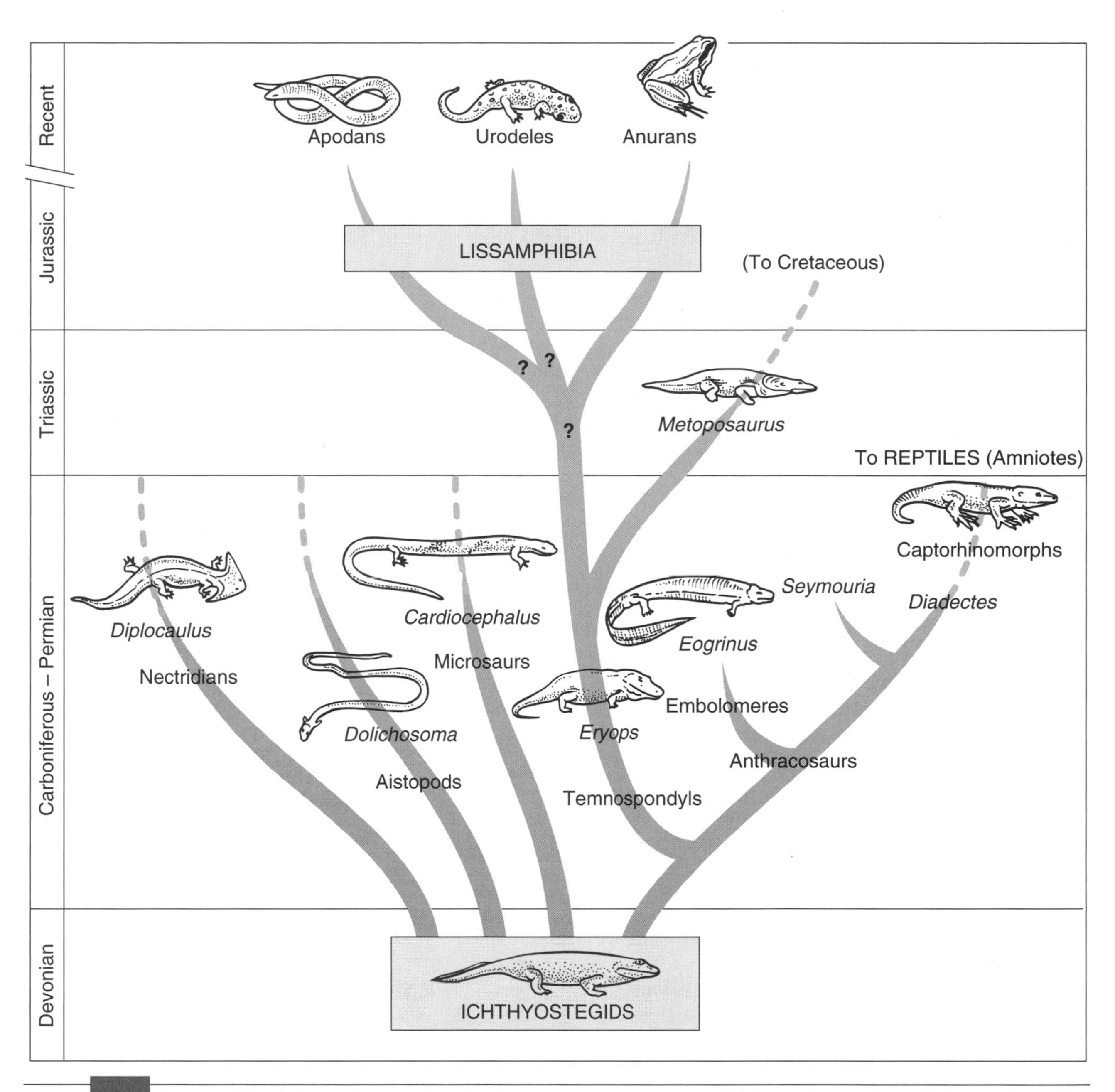

FIGURE 18–10 One possible phylogeny for some of the primary lineages among amphibians. Most major early forms of amphibians seem to have become extinct by the end of the Permian period, with the exception of the stereospondyls, which survived into the Early Cretaceous. We cannot yet trace any present-day amphibia (Lissamphibia) directly to any specific Paleozoic (pre-Triassic) form. (For more detailed phylogenies, see Ahlberg and Milner, and Benton 1997.)

their tadpoles or froglets in parental brood pouches. In some caecilians (*Typhlonectes*), offspring at birth can be almost half the maternal length, having fed on a thick, milky substance supplied by the maternal oviduct.

The many distinctions and specializations among lissamphibians make them difficult, as yet, to connect phylogenetically to any of the Paleozoic forms for the following reasons:

- Most lissamphibian species have unique pedicellate teeth in which a zone of fibrous tissue separates the calcified base and crown.
- They generally show a marked reduction of bone.
- Compared to fossil amphibians other than temnospondyls, the hands of modern anurans and urodeles have four digits rather than five.
- The anurans and urodeles have two auditory ossicles, the stapes and operculum, rather than only the stapes of earlier tetrapods.

Although we have found some Mesozoic frog and urodele fossils (Fig. 18–11), these are already so differentiated from

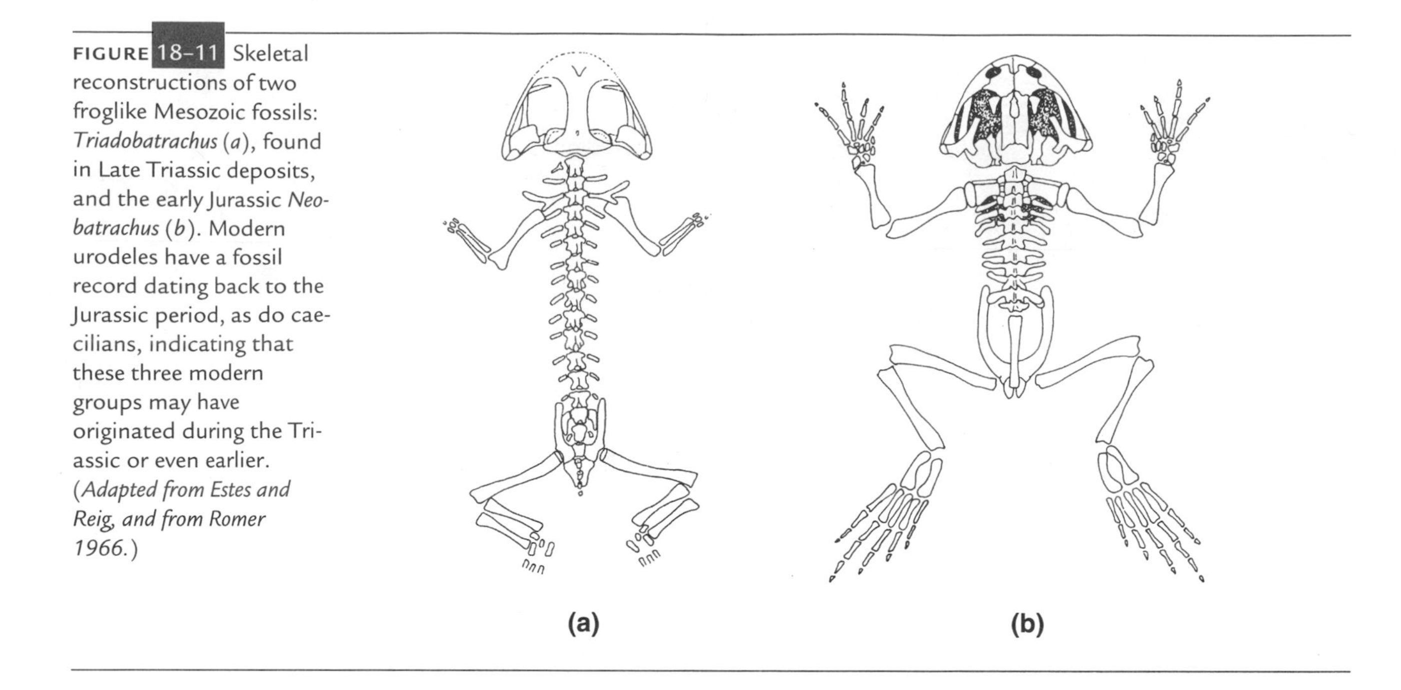

FIGURE 18–11 Skeletal reconstructions of two froglike Mesozoic fossils: *Triadobatrachus* (*a*), found in Late Triassic deposits, and the early Jurassic *Neobatrachus* (*b*). Modern urodeles have a fossil record dating back to the Jurassic period, as do caecilians, indicating that these three modern groups may have originated during the Triassic or even earlier. (*Adapted from Estes and Reig, and from Romer 1966.*)

earlier amphibians that we cannot easily trace their ancestries. Like many such phylogenetic gaps, it raises the question of monophyly or polyphyly for the origin of this group. According to some workers, the lissamphibian taxon has a polyphyletic origin: frogs appear derived from a temnospondyl group, and urodeles and caecilians may each have had separate ancestries among the lepospondyl microsaurs. Carroll (1997) points out,

> There seems to have been a succession of radiations among [amphibian] assemblages with common anatomical patterns, including the Paleozoic 'labyrinthodonts' and 'lepospondyls,' the early Mesozoic 'stereospondyls,' and the Mesozoic and Cenozoic 'lissamphibians,' but none of these groups can be established as being monophyletic.

This issue, however, can be debated, and there is some support for lissamphibian monophyly in a recent discovery of a group of Early Jurassic caecilians showing some similarities to frogs and urodeles (Jenkins and Walsh).

Perhaps, as Schmalhausen has suggested, some, if not most, lissamphibia evolved in isolated, poorly fossilized mountainous ponds and streams that protected these land vertebrates because the environment was cold and relatively inhospitable to other early tetrapods. Modern amphibia are even now more frequent in some cooler areas than are reptiles. The special jumping adaptations anurans developed, enabling them to escape predators with a few giant leaps into or out of water and to swim rapidly by "frog kicking," were probably among the advantageous features that let them re-enter the tropics. In any case, when considered in terms of their continued persistence for more than 200 million years and the significant numbers of existing species (about 4,000), modern amphibians are a successful group.

From Amphibian Tetrapods to Amniotes

A major evolutionary innovation occurred during the Carboniferous period characterized by the hard-shelled **amniotic egg** with its protected embryo (Fig. 18–14). We call the vertebrates in which this innovation first appeared *reptiles,* and their evolution has been particularly interesting for many reasons:

- By evolving a shelled amniotic egg, reptiles were freed from reproductive dependence on an aqueous environment and allowed to enter a full terrestrial existence.
- Large numbers of reptile fossils appear from the late Paleozoic era onward.
- Reptilian phylogenetic relationships seem clearly delineated in a number of groups.
- Reptiles represent, in terms of numbers, size, and mass, the ruling land vertebrates throughout the long Mesozoic era.
- Reptiles clearly gave rise to birds, and also to mammals, our own vertebrate class. In fact, many systematists classify all the amniote derivatives—reptiles, birds, and mammals—into a single presumably monophyletic group, **Amniota.**

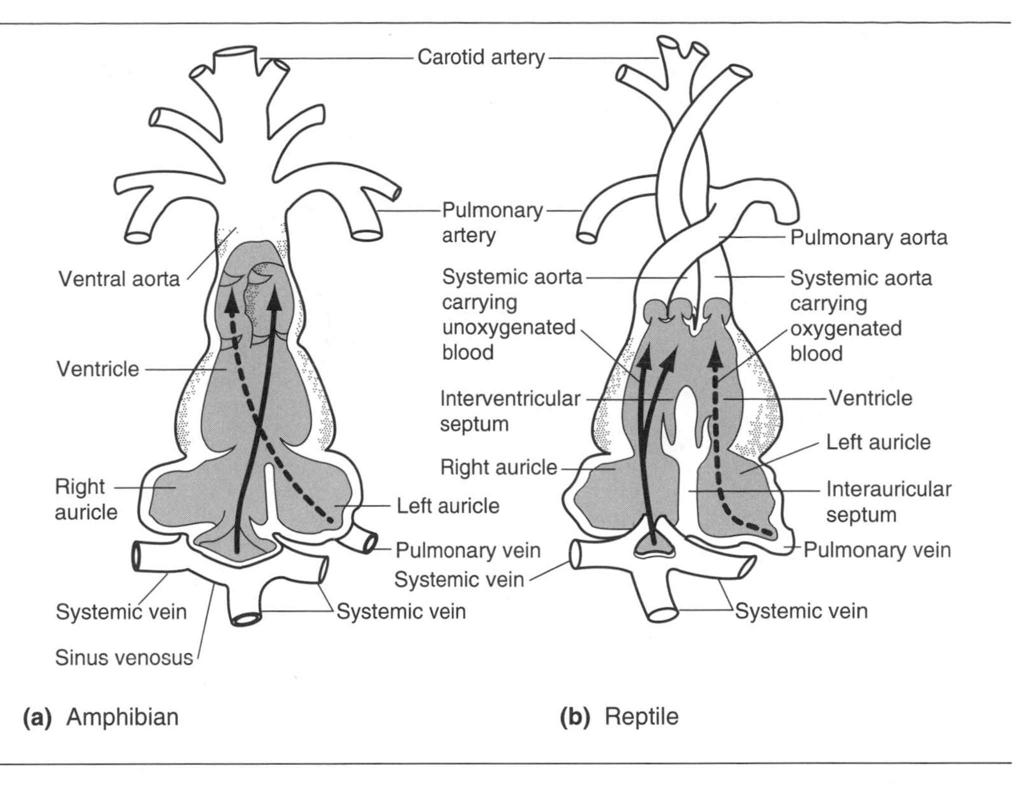

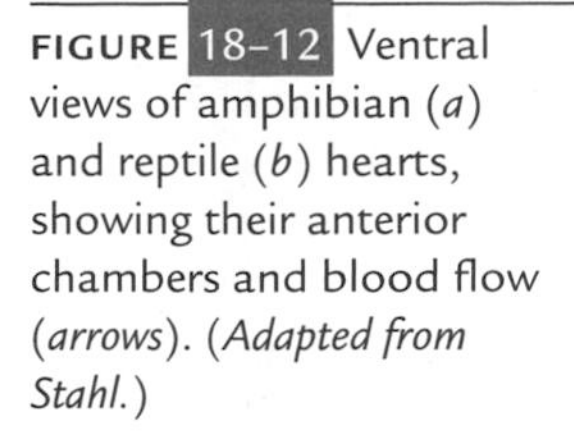

FIGURE 18–12 Ventral views of amphibian (*a*) and reptile (*b*) hearts, showing their anterior chambers and blood flow (*arrows*). (*Adapted from Stahl.*)

The features that distinguish reptiles from present-day amphibians are fairly easy to note and include the following:

1. SKULL AND SKELETAL DIFFERENCES Modern reptilian skulls have one occipital condyle compared to two such condyles in modern amphibians, and the reptilian sacrum incorporates at least two vertebrae, compared to only one in amphibians.
2. HEART Amphibians have a single ventricle, whereas the reptilian ventricle is at least partially divided—the left side sending oxygenated blood to the carotid artery, and the right side sending venous blood to the pulmonary artery (Fig. 18–12).
3. EPIDERMIS The amphibian epidermis is generally soft and moist, allowing some degree of gaseous and aqueous exchange, whereas the more heavily cornified reptilian epidermis acts as a barrier to such exchange.
4. GONADIC DUCTS AND EXCRETION In many amphibians a single excretory duct system services both the gonads and the kidneys, whereas reptiles have separate ducts for each of these systems (Fig. 18–13). Also, reptiles can concentrate the nitrogenous products of excretion, urea and uric acid, and do not need a large flow of water to remove them. In many amphibians, the urine is quite dilute and may contain considerable ammonia.
5. EGGS AND EMBRYONIC MEMBRANES Reptiles produce protected, shelled eggs consisting of membranes that have no counterpart in the gel-coated eggs of amphibians. Although all vertebrate embryos have a yolk sac membrane continuous with the wall of the gut, reptilian embryos (as well as those of birds and mammals) produce an additional membrane continuous with the embryonic body wall that folds around it to yield an outer chorion and inner amnion (Fig. 18–14). The amniotic cavity prevents adhesions by isolating the embryo from direct contact with the shell and helps provide the embryo with protection against temperature fluctuations. In addition, amniotic eggs have a sac called the **allantois** that grows out of the embryonic hindgut and rapidly covers the inner surface of the chorion. The allantochorion membrane complex is well supplied with blood vessels and acts as a respiratory organ that allows inward diffusion of oxygen through the permeable shell as well as the outward passage of carbon dioxide. Nitrogenous wastes of the reptilian embryo are deposited into the allantoic cavity as relatively insoluble, nontoxic precipitates such as uric acid that the animal need not immediately eliminate.

Of all traits that characterize the reptilian advance, the amniotic egg appears most significant. With the exception of some viviparous forms, the moisture-dependent amphibian egg is an important element in maintaining amphibian aquatic ties. Reptiles, in contrast, can lay their eggs in a large variety of terrestrial environments, and in some lizards and snakes, embryonic development can proceed even with a loss of water. To most evolutionists, the complexity of the amniotic egg suggests that the transition

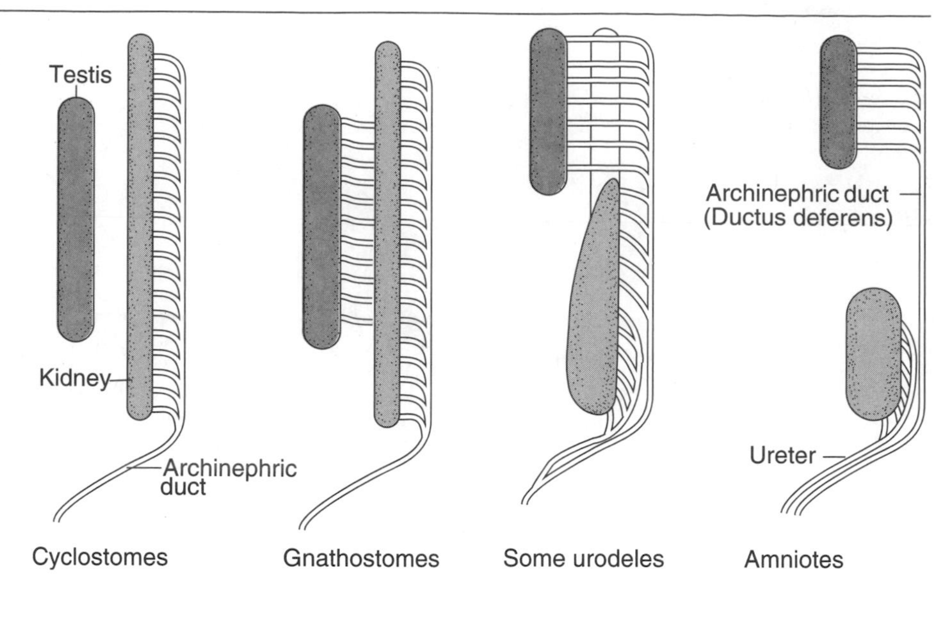

FIGURE 18–13 Male urinary and genital duct systems in various vertebrate groups. In primitive vertebrates, such as cyclostomes, the gonad is not connected to the urinary system. In primitive jawed fishes (gnathostomes), such as the sturgeon and garpike, the testis has multiple connections to the kidney, which drains into the archinephric duct. Many sharks and amphibian urodeles show replacement of the anterior portion of the kidney by testicular ducts that drain directly into the urinary duct system. In more advanced amniotes, the archinephric duct serves as the gonadic duct, and a single ureter is used for kidney drainage in both sexes. (*Adapted from Romer and Parsons.*)

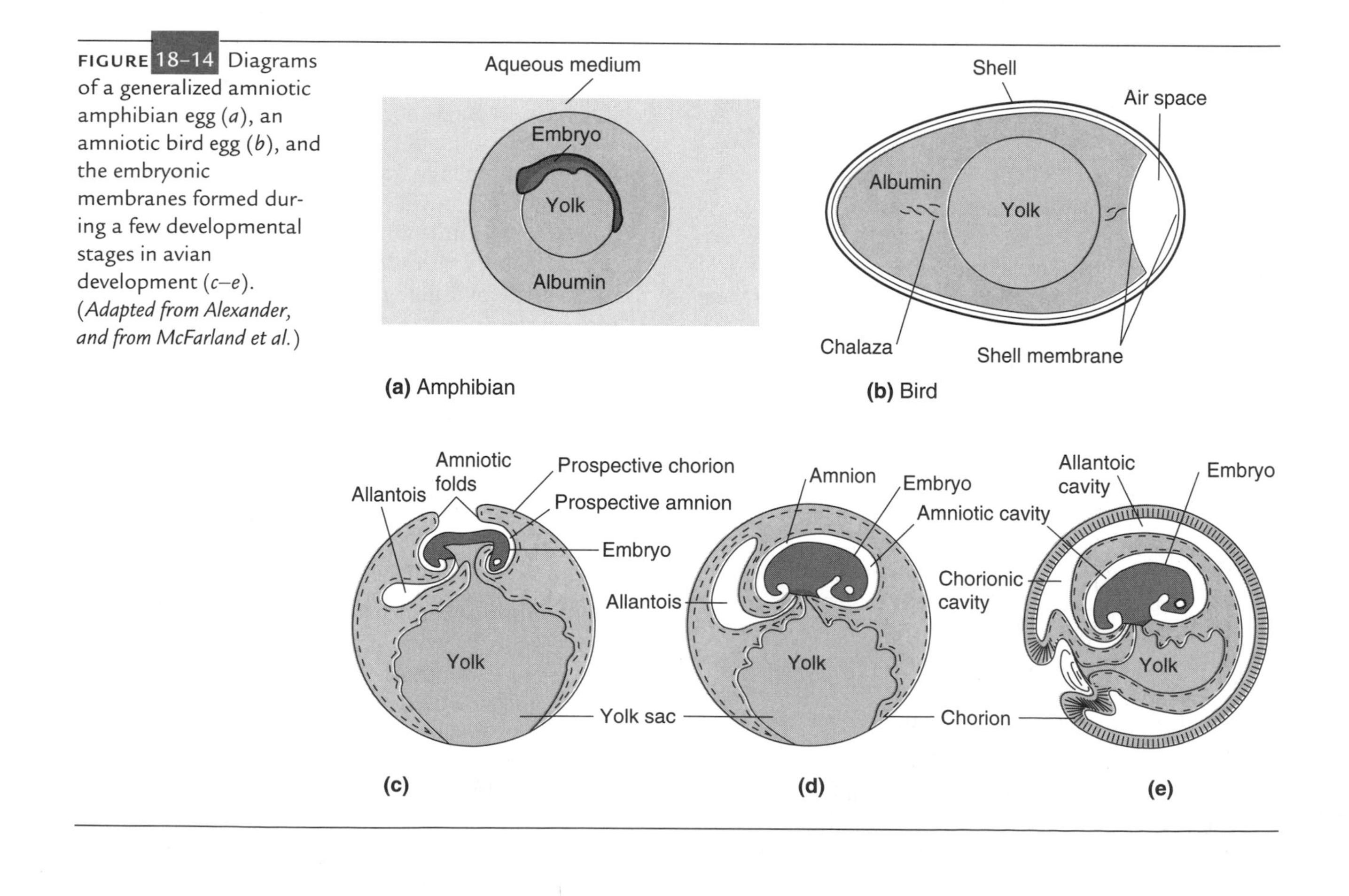

FIGURE 18–14 Diagrams of a generalized amniotic amphibian egg (*a*), an amniotic bird egg (*b*), and the embryonic membranes formed during a few developmental stages in avian development (*c–e*). (*Adapted from Alexander, and from McFarland et al.*)

from amphibians to reptiles did not occur more than once, and the reptilian grade of evolution is therefore most probably monophyletic. Unfortunately, since eggs rarely fossilize, we don't know when this transition occurred,[4] although clearly the following steps must have preceded it:

- Since a shelled amniotic egg must be fertilized before egg laying, internal fertilization must have appeared first in this evolutionary line.
- The habit of laying eggs on land must also have arisen, because the shelled reptilian embryo depends on gas exchange and could not have obtained enough oxygen while immersed in water.
- For the amniotic embryo to be born on land, the stage of an aquatic gilled larva was probably absent.
- To these attributes, Carroll (1991) adds that the early eggs laid on land must have been fairly small if they were to maintain a satisfactory rate of gas exchange before the amniotic membranes evolved.

Paleontologists believe reptilian organization was achieved during the Carboniferous period by a group called the **captorhinomorphs,** considered the earliest of the stem reptiles. According to a number of authors, these animals evolved from a line of small, lizardlike amphibians, a stock that probably diverged very early from its labyrinthodont ancestors.

Since many reptilian traits involve soft tissues that are rarely, if ever, preserved, paleontologists seeking criteria for crossing the amphibian–reptilian boundary concentrate on skeletal characteristics such as structure of the palate: in early reptiles the pterygoid bone in the skull begins to show a transverse flange associated with what becomes the largest jaw-closing muscle, the pterygoideus. The large palatal fangs of labyrinthodont amphibians disappear, and the postparietal, tabular, and supratemporal skull bones also reduce. Other changes include increased heterogeneity of reptilian teeth compared to the uniformly shaped amphibian teeth, as well as changes in the proportions and degree of ossification in the pectoral and pelvic limb girdles, probably associated with selection for improved support and locomotion in less aquatic habitats.

By the end of the Permian period a large variety of different reptilian lines appear, and, because connections between some of them are not yet clear, paleontologists have grouped them in various ways. The common classification of amniotes focuses on openings, called **fenestrae,** in the temporal region of the skull behind the optic orbits. In the earliest reptiles the skull has a solid roof, and the temporalis muscles used to close the jaw run between the inside (medial) surface of the lower jaw and the braincase, within the outer bony layer of the skull.

[4]The earliest fossil purported to be a reptilian egg dates to the Early Permian period.

This unfenestrated condition defines members of the subclass **Anapsida** ("an" = without, "apse" = arch), which have a relatively rigid skull structure (Fig. 18–15*a*). Nevertheless, this anapsid structure may have had disadvantages because the outer bony covering restricts expansion of the temporalis muscle. According to this view, fenestral openings in the cheek region of the skull would have enabled jaw muscles to increase in size and allow a stronger bite and more efficient mastication. Frazzetta has also suggested that the bony edges of fenestral openings serve as a much stronger anchorage for jaw muscles than do the internal surfaces of the skull bones.

Reptiles with single temporal openings, the **Synapsida** (Fig. 18–15*b*), are the first group to diverge from the ancestral anapsid stocks and include the mammal-like reptiles. The presence of fenestra above and below a bar formed by joining the postorbital and squamosal bones defines the subclass **Diapsida** (Fig. 18–15*c*), now separated into the infraclasses Lepidosauria and Archosauromorpha. Following this approach, we can apply the classification system in Table 18–1 to most reptilian orders, although some placements, such as the mesosaurs, placodonts, and ichthyosaurs, are still unclear.

In addition to these taxonomic qualifications, note that cladistic systematists, as discussed in Chapter 11, do not accept this classification of Reptilia because it is paraphyletic: it does not include birds and mammals descended from dinosaurs and synapsid reptiles, respectively. Traditional classifications systems, however, find no difficulty with the paraphyletic taxon, recognizing that the classes Aves and Mammalia can arise from an older class (Reptilia). Perhaps as somewhat of a compromise, classifiers are now generally using the taxon Amniota to include all reptiles, birds and mammals, although the class Reptilia as defined here is still paraphyletic.

Reptilian Evolution

Figure 18–16 shows a general scheme for some major reptilian phylogenetic relationships, and also indicates approximate times during which various groups became extinct. In brief, the first reptiles appear in Pennsylvanian deposits of the Carboniferous period,[5] although paleontologists suggest they

[5]Carroll (1988) has pointed out the unusual nature of their fossilization, which illustrates the importance of accidental factors in such processes:

> These fossils are not found in normal coal-swamp deposits, such as those from which the majority of Carboniferous tetrapods have been found, but rather within the upright stumps of the giant lycopod Sigillaria. These trees grew in areas that were subject to periodic flooding, which resulted in the burial of the trees in several meters of sediments. The trees died and the central portion rotted out, but the bark was stronger and retained the cylindrical shape of the stump. After the withdrawal of the water, animals living on the newly developed land surface would occasionally fall into the hollow stumps. Eventually they died and were covered with sediments and fossilized.

FIGURE 18-15
Schematic diagrams illustrating various kinds of reptilian postorbital temporal openings are on *the left* and illustrative fossil skulls with their fenestra are on *the right.* The euryapsid pattern (*d*), also called *parapsid,* is found in groups such as ichthyosaurs, nothosaurs, and plesiosaurs. According to Carroll (1988), the euryapsids derive from early diapsids, and their fenestral pattern evolved by loss of the temporal bar beneath the lower fenestrum, accompanied by thickening of the postorbital and squamosal bones. (*Adapted from Colbert and Morales.*)

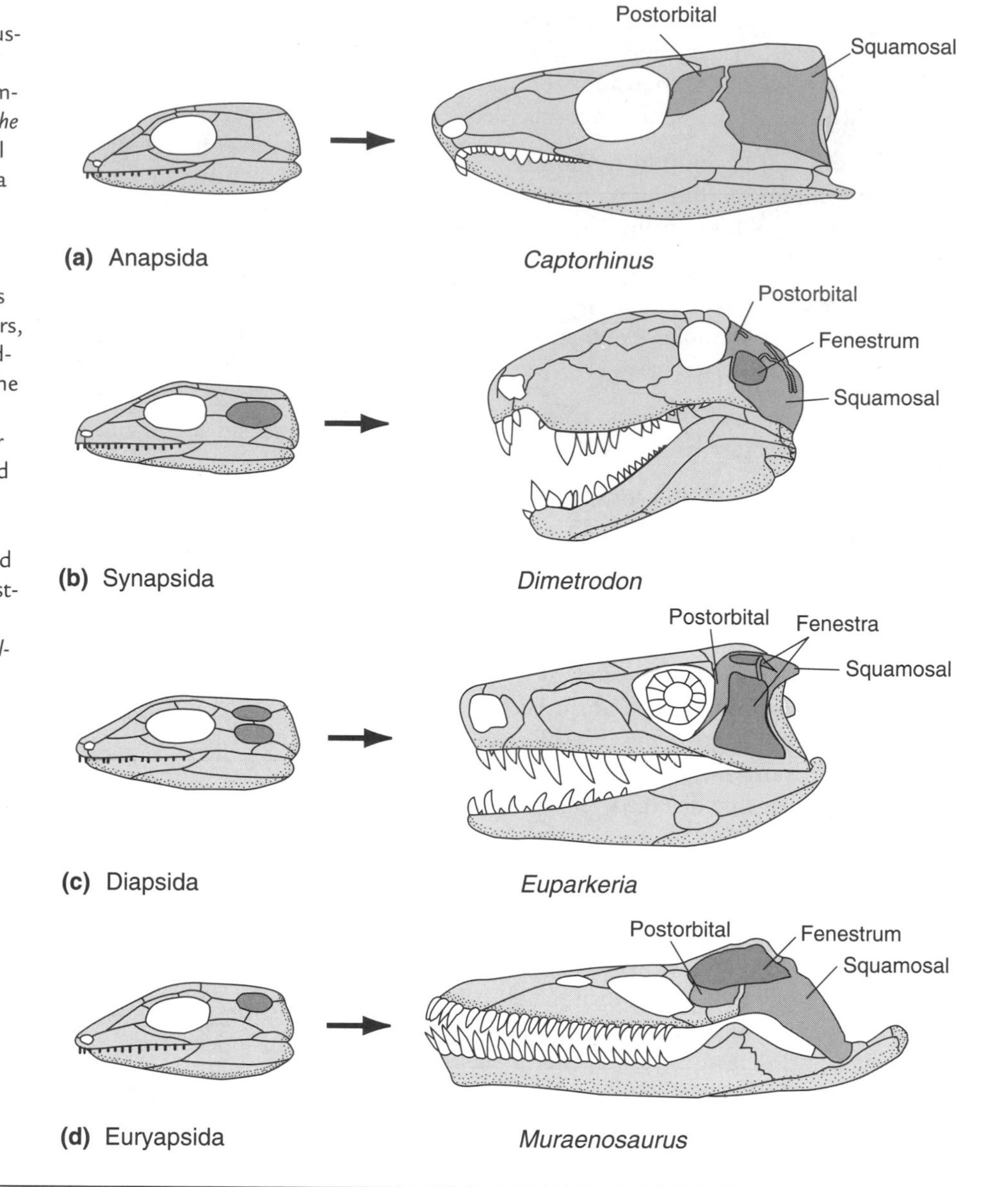

may well have evolved from anthracosaurs earlier, perhaps during the late Mississippian. By the beginning of the Permian, many new reptilian groups appear side by side with many varieties of amphibians. The end of the Permian and beginning of the Triassic record a decline in amphibian fossils (with the exception of stereospondyls) and also mark a striking expansion of the mammal-like reptiles, the **therapsids.** By the time of the Jurassic, almost all the major reptilian groups have emerged, accompanied by a drastic fall in numbers of therapsids.

From the Jurassic onward, reptilian adaptations enabled widespread dispersion to many habitats, including the aquatic, and the next 100 million years or so was a veritable age of dinosaurs, pterosaurs, and marine reptiles. This reptilian dominance lasted until the end of the Cretaceous, when almost all reptilian groups disappeared, except for lizards, snakes, turtles, crocodiles, and the New Zealand tuatara (*Sphenodon*). Although theorists offer many explanations for this remarkable drama of reptilian radiation and decline, only a sample of these hypotheses can be discussed here.

The early reptilian captorhinomorphs appear first as small, slender animals, about 1 to 2 feet long, at a time in the Carboniferous during which many insects evolved terrestrial forms. The exact relationship between insects and reptiles is obscure, but paleontologists have suggested that captorhinomorphs functioned primarily as insectivores in the terrestrial food chain. Adaptation for a terrestrial existence seems also to have been shared by other captorhinomorph-derived reptiles, such as small

TABLE 18–1 One classification system for the class Reptilia

Subclass Anapsida
Order Captorhinida (Cotylosauria): stem reptiles
Order Mesosauria: mesosaurs (aquatic fresh-water reptiles)
Order Testudinata: turtles
Subclass Synapsida
Order Pelycosauria: pelycosaurs (includes "sail-backed" reptiles)
Order Therapsida: mammal-like reptiles
Subclass Diapsida
Infraclass Lepidosauria
Order Eosuchia: early diapsids
Superorder Lepidosauria
Order Sphenodontida: sphenodontids
Order Squamata: lizards and snakes
Superorder Sauropterygia: marine Mesozoic reptiles
Order Nothosauria: nothosaurs
Order Plesiosauria: plesiosaurs
Order Placodontia: placodonts
Superorder Ichthyopterygia: ichthyosaurs
Infraclass Archosauromorpha
Order Prolacertiformes: protorosaurs
Order Trilophosauria: trilophosaurids
Order Rhynchosauria: rhynchosaurs
Superorder Archosauria
Order Thecodontia: early (Triassic) archosaurs
Order Crocodylia: crocodiles and alligators
Order Pterosauria: flying reptiles
Superorder Dinosauria
Order Saurischia: lizard-hipped dinosaurs
Order Ornithischia: bird-hipped dinosaurs

Note: For other reptile classification systems, see Benton (1997), and Colbert and Morales. Disputes as to the origin of some groups, such as turtles, still persist.
Source: Adapted from Carroll (1988).

synapsids called **pelycosaurs** that also appear during the Pennsylvanian epoch.

The rapid radiation of these early forms led, by the end of the Carboniferous period, to exploitation of various environments. In aquatic habitats lived mesosaurs with long, toothy jaws (Fig. 6–9). Among terrestrial forms were probably some of the larger pelycosaurs, whereas others of this group were more aquatic and also preyed on fish and amphibious vertebrates. Early reptiles who specialized on nonaquatic food sources were either herbivores, such as the pareiasaurs and caseids, or predators on various upland insectivorous and herbivorous forms, such as the carnivorous therapsids.

Strong evidence indicates that pelycosaurs such as *Dimetrodon* (Fig. 18–17) are close to the line that gave rise to the therapsids, and these, in turn, later gave rise to the mammals discussed in Chapter 19. *Dimetrodon,* however, was a specialized animal with extremely long neural spines that paleontologists believe supported a dorsal "sail" used in regulating body temperature. A sail of this kind, well supplied with blood vessels, would have enabled an animal that had cooled off at night to resume an optimum metabolic temperature soon after daybreak by placing its body perpendicular to the sun's rays. The animal would have accomplished further heating and cooling by increasing or decreasing blood flow into this large, heat-exchanging dorsal surface.

Dimetrodon may also have cooled off during very warm periods by moving into the shade or orienting itself parallel to the sun's rays. Since continuous enzymatic activity in muscle and other tissues depends on maintaining optimum body temperatures, selection for such mechanisms would be important in helping pelycosaurs and their therapsid cousins engage in longer periods of active predation or escape.

Paleontologists think the therapsids, however, used temperature-regulating mechanisms other than dorsal sails, and quite possibly some of these new forms were the first of the **endothermic** vertebrates; that is, more constant internal metabolic heat-producing reactions rather than more variable external **ectothermic** influences, now supported optimum body temperatures. As Bennett and Ruben point out, higher metabolic rates in endothermic animals not only raise body temperatures, but (together with greater numbers of mitochondria and increased aeration) also allow higher levels of oxygen use. Increased aerobic metabolism, in turn, supports more sustained activity and greater stamina than ectotherms can achieve, who become rapidly exhausted because they rely mostly on anaerobic metabolism.

Given the advantage of a high body temperature, selection for insulating mechanisms to help maintain it would have led to thicker layers of subcutaneous fat as well as to modifying scales into hair and feathers. By contrast, such insulation would disadvantage ectotherms because they achieve their optimum body temperature through heat exchange with the environment.

Because an animal must eat much more food to provide energy for high metabolic rates, the cost of endothermy is relatively high. For a given body weight, an endothermic mammal or bird needs five to ten times more energy to maintain the same body temperature than does an ectothermic reptile or amphibian. Nevertheless, despite their stamina limitations, ectotherms are capable of short bursts of activity through anaerobic metabolism. Thus ectotherms can survive well under conditions that stress low energy expenditure and may even compete successfully with endotherms when predatory pursuit or escape requires only short distances. Lizards, for example, have

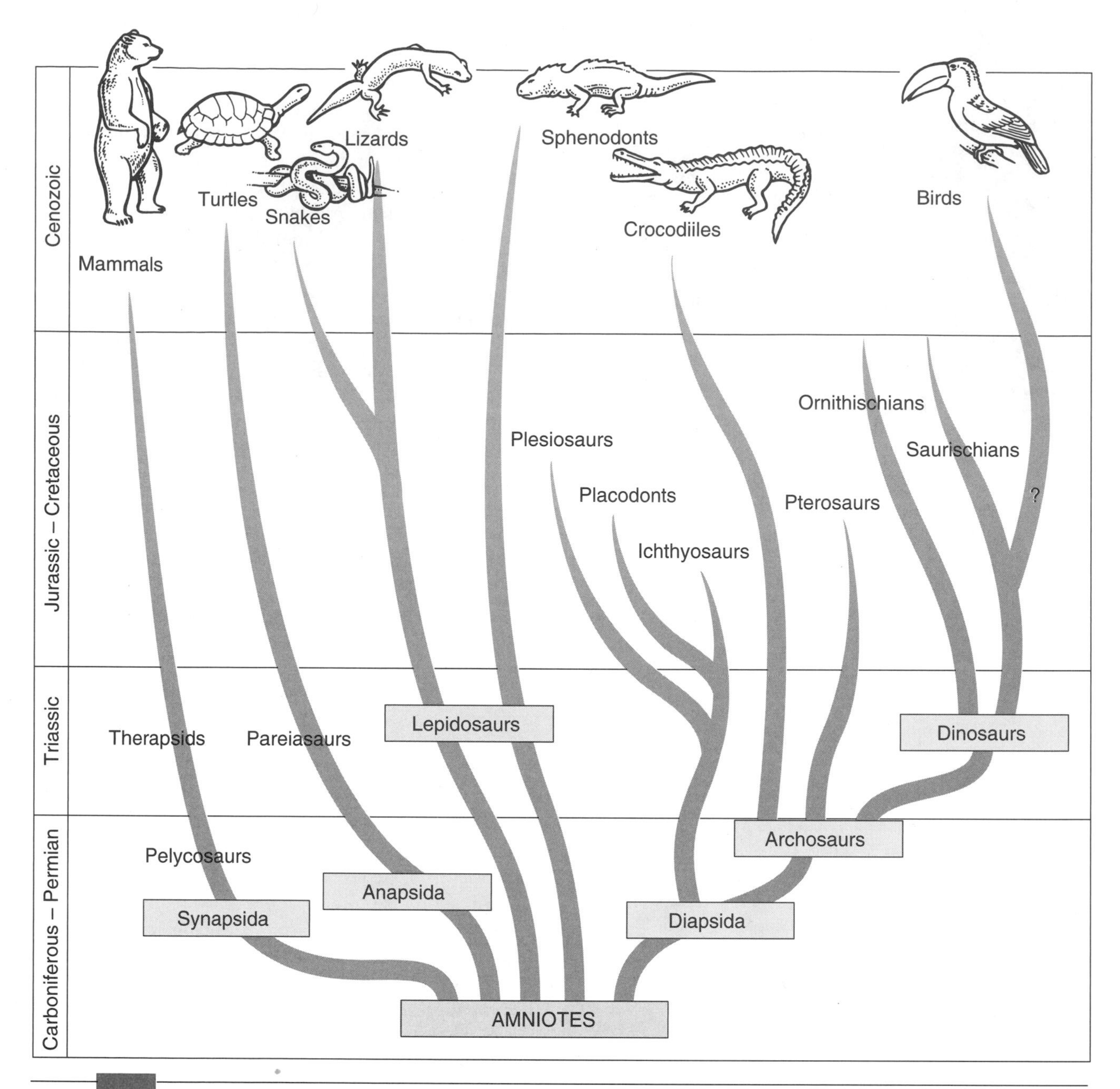

FIGURE 18–16 General evolutionary scheme showing relationships among the major reptilian groups, beginning with their origin in the Paleozoic era. As indicated, further major evolutionary events include the origin of mammals from synapsids and the origin of birds from a lineage that may also have given rise to dinosaurs. Conflicting with traditional phylogeny, mitochondrial DNA analysis by Zardoya and Meyer places turtles as a diapsid lineage rather than anapsid, suggesting that diapsid skull fenestration was lost during turtle evolution.

prospered in many environments where they can move quickly in and out of protected surroundings, and now number about 3,800 species. Similarly, snakes (3,000 species), many of which are highly poisonous and therefore among the most dreaded vertebrate predators, can use energy bursts for both offense and defense.

In environments where animals must sustain activity, the race has gone to endotherms, of which therapsids may have been the earliest forms. Along with improved locomotory adaptations that moved the legs beneath the body close to the median plane,[6] these forms succeeded well both as herbivores and carnivores (Fig. 18–18), some reaching sizes 10 to 12 feet long. By the end of the Permian period about six out of seven reptilian fossils were therapsids.

[6]According to Alexander (1991), this adaptation diminishes reptilian side-to-side bending while running, thus improving rib movement necessary for breathing.

MICHAEL J. BENTON

Birthday:
April 8, 1956

Birthplace:
Aberdeen, Scotland

Undergraduate degree:
University of Aberdeen (Zoology)

Graduate degree:
Ph.D. University of Newcastle-Upon-Tyne, 1981

Postdoctoral training:
University of Oxford 1982–1983

Present position:
Professor of Vertebrate Paleontology
Department of Earth Sciences
University of Bristol
Bristol, England

WHAT PROMPTED YOUR INITIAL INTEREST IN EVOLUTION?

I first got into palaeontology when I was seven or eight. I was given a small colour book, *The Golden Guide to Dinosaurs* by Zim and Shaffer, and I was hooked. Then, I read books about Darwin, and I became fascinated by the interdisciplinary nature of the study of evolution. It's no different now from Darwin's day: zoologists, botanists, palaeontologists, ecologists, experimental biologists, and philosophers all have important contributions to make.

WHAT DO YOU THINK HAS BEEN MOST VALUABLE OR INTERESTING AMONG THE DISCOVERIES YOU HAVE MADE IN SCIENCE?

- One of the first efforts at a cladogram of basal diapsid reptiles, and the discovery that the split between lepidosauromorphs (the lizard group) and archosauromorphs (the bird-crocodile group) goes deep in time
- The demonstration that the fossil record is not as bad as some people have suggested, assessed by quantitative comparisons of phylogenies and stratigraphies

WHAT AREAS OF RESEARCH ARE YOU (OR YOUR LABORATORY) PRESENTLY ENGAGED IN?

- Determination of long-term patterns of the diversification of life, and assessment of whether they follow equilibrium or non-equilibrium patterns
- Tests of the quality of cladograms: how well do they reconstruct phylogeny?
- Excavations at a huge dinosaur bonebed in the Mid Cretaceous of Tunisia in North Africa

IN WHICH DIRECTIONS DO YOU THINK FUTURE WORK IN YOUR FIELD NEEDS TO BE DONE?

We are living through an exciting time in the study of evolutionary patterns, which began about 1970. We now have two pretty well independent methods for reconstructing phylogeny (patterns of evolution), cladistics and molecular phylogenies. Biologists and palaeontologists working in this area are real pioneers. Old ideas can be tested, and some dramatic new discoveries have been made about patterns of the one great evolutionary tree of life. This is original "one-off" enterprise, and in centuries to come, people will look back to the time from 1970 to 2010, when the outlines of the evolution of life were pinned down in a testable way.

WHAT ADVICE WOULD YOU OFFER TO STUDENTS WHO ARE INTERESTED IN A CAREER IN YOUR FIELD OF EVOLUTION?

Students who wish to make original contributions to the growing field of phylogeny reconstruction and macroevolution must master a broad field of knowledge in biology and geology. Luckily, much of the work is reported in excellent, readable, popular books. Students must then read really widely in the current professional journals to be really up-to-date. They need enthusiasm and excitement, and there's no harm in dreaming about dinosaurs, huge asteroid impacts, and the vastness of geological time. But, it's important to master the necessary quantitative approaches, and to adopt a rigorous questioning approach.

Surprisingly, although many therapsids crossed the Permian boundary into the Mesozoic era, their numbers significantly diminished before the end of the Triassic. Perhaps the warm, constant climate of the Mesozoic reduced the importance of therapsid temperature-regulating advantages by letting many ectothermal reptiles maintain stable high body temperatures on less food intake. That is, the same amount of ingested energy needed to maintain the high metabolic rate of relatively few endotherms could now produce increased numbers of active reptilian ectotherms.

Also interesting is the fact that ruling reptiles from an entirely different subclass, the diapsid **archosaurs** characterized the remainder of the Mesozoic, from the Jurassic to the close of the Cretaceous. Many other reptilian groups also persisted and evolved such as turtles, lizards, and snakes, but it was among the archosaurs that the ruling **dinosaurs** appear, some of which still dwarf any other land vertebrate yet evolved.

FIGURE 18–17 Reconstruction of the early Permian pelycosaur, *Dimetrodon.* (*Adapted from Romer 1968.*)

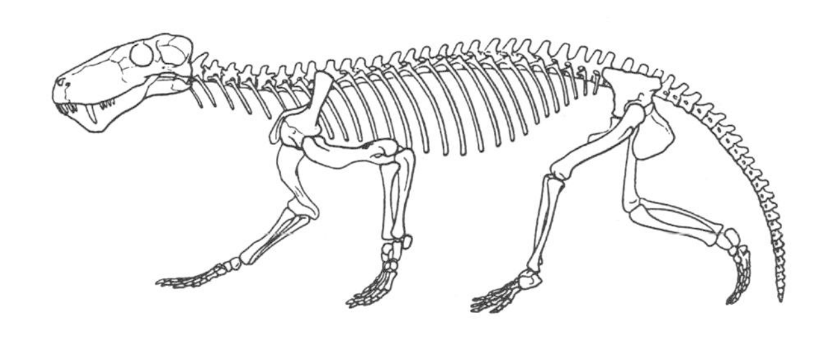

FIGURE 18–18 Skeleton and reconstruction of a carnivorous therapsid (*Lycaenops*) from the late Permian, showing a number of mammal-like features. It was about the size of a wolf, with large upper canines functioning as "saber teeth." (*Adapted from Colbert and Morales, and from Romer and Parsons.*)

Early Archosaurs

We know the earliest diapsids from Upper Pennsylvanian deposits, and they probably had a captorhinomorph ancestry (Reisz). These animals bore two fenestrae behind the optic orbit and an additional opening near the tip of the snout. During the radiation of these diapsids, two main infraclasses arose, lepidosauromorphs and archosauromorphs, differentiated by many traits, but most importantly by a unique ankle-and-foot structure in archosauromorphs that facilitates an upright posture. By the end of the Permian and beginning of the Triassic period, archosauromorph evolution had proceeded far enough to produce a variety of groups, including **bipedal** forms in which the forelegs were shorter than the hind legs. Also, these animals had teeth set in sockets (**thecodonts**) rather than fused to the jaw margins.

We still don't know what environmental pressures account for the innovation of archosaur bipedalism, but its development and persistence seem associated with selection for improved running speed as well as selection for large size. Even small early Triassic archosaurs begin to show the effects of selection toward bipedalism.

For example, *Euparkeria,* an Early Triassic thecodont, had hind legs about 1.5 times the length of the forelegs and bore two fenestrae (one in the lower jaw and the other anterior to the orbit) that anticipate those found in the later dinosaurs. It was a carnivorous form, about 2 feet long, lightly built with hollow bones, and as Figure 18–19 shows, may have been partially bipedal. Some paleontologists place *Euparkeria* in the lineage leading to bipedal dinosaurs, and others do not (Benton and Clark).

By the middle of the Triassic, bipedal innovations seem further developed in a number of thecodont lines, whereas other lines such as phytosaurs and crocodiles preserved an obligate four-footed gait. In some bipedal forms, selection for increased length of stride by the hind legs had led to a tibia about as long as the femur—animals that must have been quite speedy. (In the racehorse, the tibia–femur ratio is about 0.9.) Despite various thecodont adaptations, dinosaur radiation had already begun during the Triassic, and many tetrapod groups, including thecodonts and therapsids, became extinct before the close of that period. According to Benton (1997), such Triassic ex-

FIGURE 18–19 Reconstruction of *Euparkeria,* an early Triassic thecodont. It was about 2 to 3 feet long, with a short trunk counterbalanced by a heavy, muscular tail that might have enabled it to run bipedally.

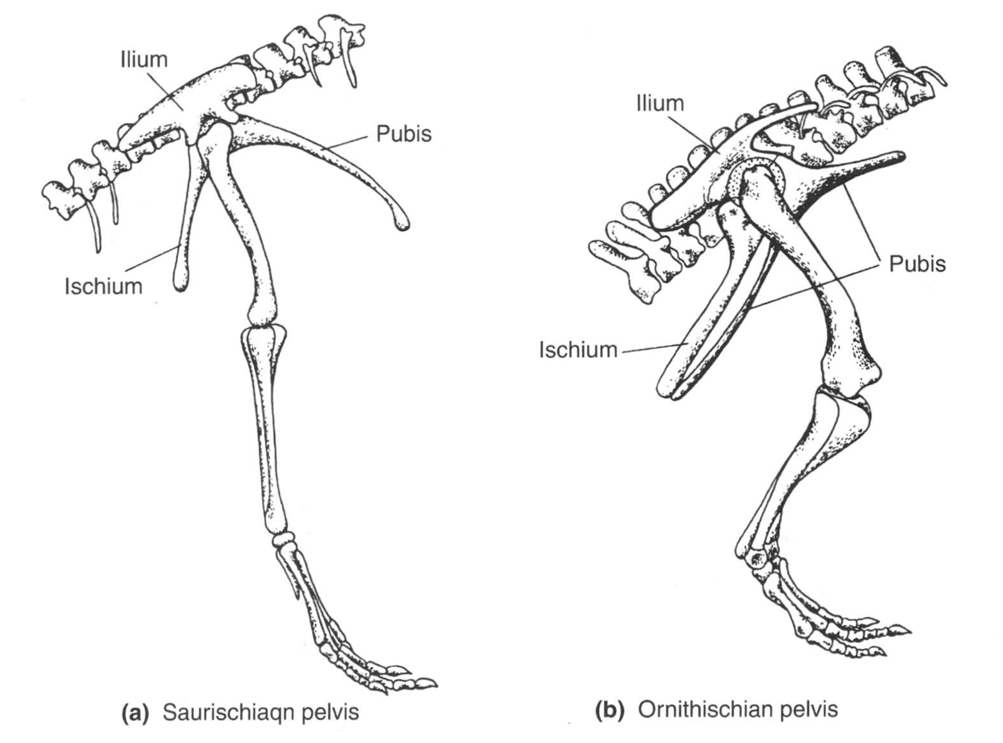

FIGURE 18–20 The two general types of pelves found in dinosaurs. In primitive reptiles, the pelvis is a solid, platelike structure from which the femur projects horizontally (parallel to the ground). With the evolution of bipedalism and a vertical femur, greater leverage for moving the hind legs arose by attaching the limb muscles to fore and aft extensions of the pelvis. In birds, the posterior extension of the pubis helps support the ischium, but the anterior pubis is not well developed. (*Adapted from Romer 1968.*)

tinctions, caused or accompanied by major climatic changes, enabled dinosaurs to enter a variety of vacant ecological niches, accounting for their initial radiation.

The Dinosaurs

People think of the dinosaurs as a single group, but they actually include two major orders, **Saurischia** and **Ornithischia,** both perhaps sharing a common thecodont ancestry during the Triassic. (Charig questions such common ancestry, and Carroll 1988 supports it.) A primary difference in pelvic structure between the two orders appears early in their evolutionary history: the Saurischia show the original thecodont triradiate structure in which the pubis extends anteriorly and the ischium posteriorly (Fig. 18–20*a*), while the Ornithischia have a tetraradiate pelvis with the pubis usually parallel to the ischium (Fig. 18–20*b*). The two groups show their bipedal ancestry in shorter forelimbs than hind limbs, although some dinosaurs returned to a quadrupedal stance with lengthened forelegs.

As Figure 18–21 shows, phylogenetic branching among the dinosaurs proceeded throughout the Mesozoic, each of the two orders producing a variety of suborders as well as infraorders. Among the saurischians, paleontologists have traditionally classified the carnivorous **theropods** (four or fewer toes on the hind feet) into two groups based on presumably distinct body proportions, the smaller coelurosaurs and larger carnosaurs. Since some theropods appear to share traits from both groups, these distinctions no longer hold fast. However classified, by the Cretaceous period bipedal theropods had radiated widely, evolving into forms such as the small, ostrichlike *Ornithomimus*, small to medium-sized deinonychosaurs such as *Deinonychus*, and large theropods such as *Tyrannosaurus* that may have weighed 6 to 8 tons and stood 20 feet above the ground.

The increased body size in some lines of theropods seems to have paralleled the increase in body size in some lines of their herbivorous prey, probably because an **arms race** develops in which protection that larger size offers to prey selects in turn for larger size in their predators. This cycle of selection for body size elicited on one hand the large theropods, and on the other large herbivorous ornithischians and even larger herbivorous sauropods (five toes on the hind legs) such as *Apatosaurus* (or *Brontosaurus*) that were 70 to 80 feet long, and weighed 50

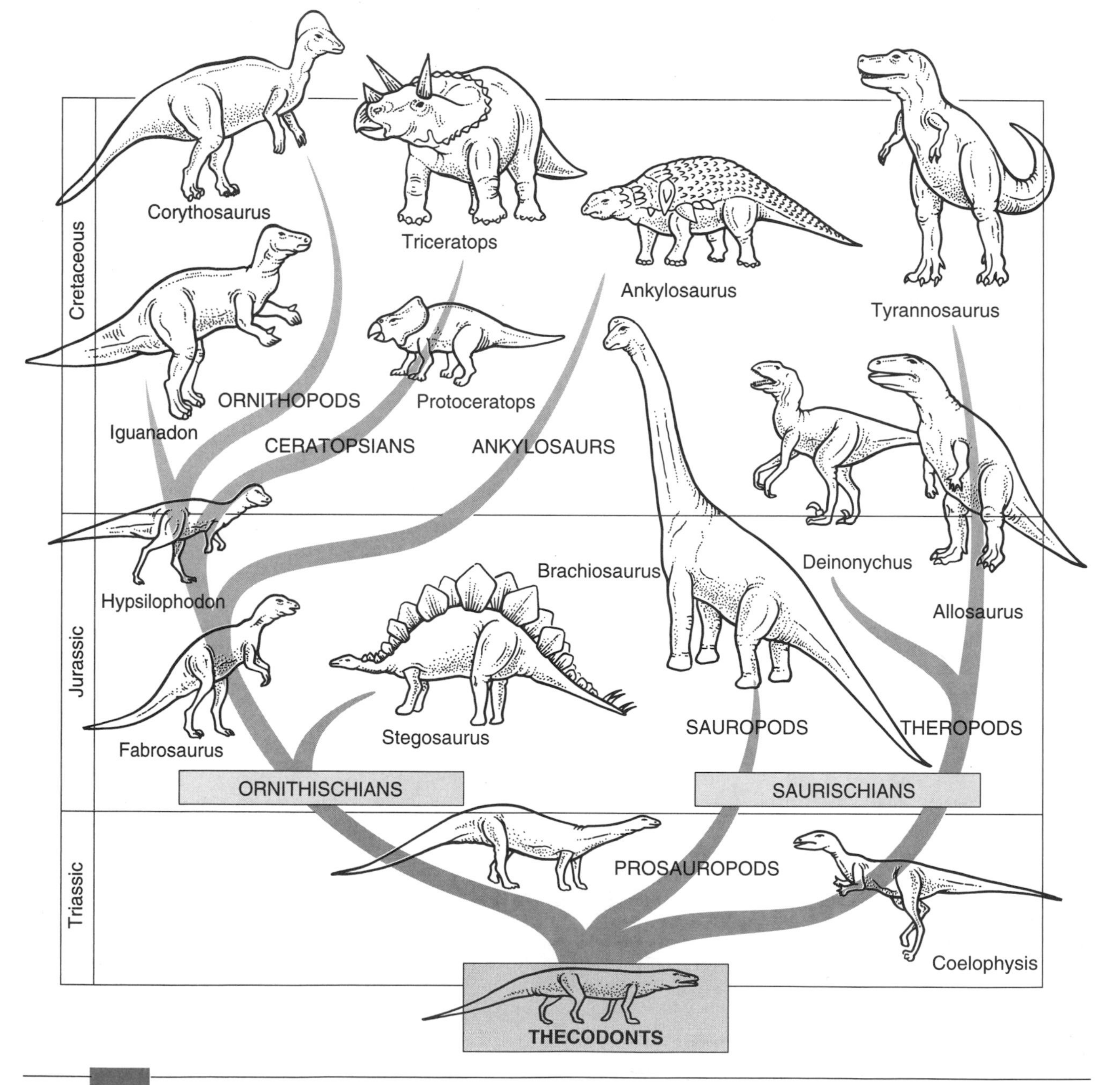

FIGURE 18–21 Dinosaur phylogeny from the Triassic to the end of the Cretaceous. (*Adapted from Colbert and Morales.*) The standard review of all major dinosaur groups is in the volume edited by Weishampel et al.

tons or more. The existence of very large fossil bones indicates that some forms probably exceeded even these impressive dimensions, and terrestrial vertebrate body size may not have reached its upper limit when the dinosaurs became extinct at the end of the Cretaceous.

A classic hypothesis suggested that many giant sauropods were aquatic because only water could have buoyed up their immense weights. Opposed to this view are findings of a terrestrial existence: the fossilized stomach contents in one animal indicates a diet of woody, leafy material, and an early Cretaceous trackway, found in Texas, indicates they traveled in socially organized herds. At least some of the long-necked sauropods were ground feeders, gathering food from low-lying plants or from shallow lakes and ponds (Stevens and Parrish). Their large size, powerful tails, and social organization probably gave them considerable protection against all but the largest predators.

Various lines of the exclusively herbivorous ornithischians did not have the same degree of bipedalism and size as some lines of saurischians, but many were nevertheless quite large, reaching lengths of 30 feet and weights of 5 tons or more. Among the ornithischian groups are the duck-billed hadrosaurs, the armored stegosaurs and ankylosaurs, and the horned ceratopsians. Some of these groups seem to have developed considerable variation and specialization; hadrosaurs, for example, were the most diverse of the dinosaurs in terms of skull morphology, with a large variety of crested forms (Fig. 18–22). Their relative success accompanied this diversity, since hadrosaurs may have comprised as much as 75 percent of terrestrial vertebrate biomass in many places by the end of the Cretaceous.

Endothermy Versus Ectothermy

Clearly, dinosaurs were successful creatures throughout much of the Mesozoic, and no other group of reptiles or mammals approached their size or importance in terrestrial deposits until after the Cretaceous period. Some paleontologists such as Bakker and Ostrom (1974) have suggested that dinosaur success stemmed at least partially from the high metabolic rates that endothermy provided, allowing continuous high levels of activity. A large, sluggish, ectothermal reptile, whether herbivore or carnivore, could hardly have competed successfully with endothermal therapsids and their mammalian descendants. That is, if dinosaurs were only typical ectothermal reptiles, why did endothermal mammals remain small and insignificant throughout the Mesozoic?

Marshaled in further support of dinosaur endothermy are a number of arguments:

1. The fully erect posture of many dinosaurs, with their limbs extending vertically downward beneath the body, appears only among present-day endotherms (mammals and birds), whereas all modern ectotherms (reptiles and amphibians) have a

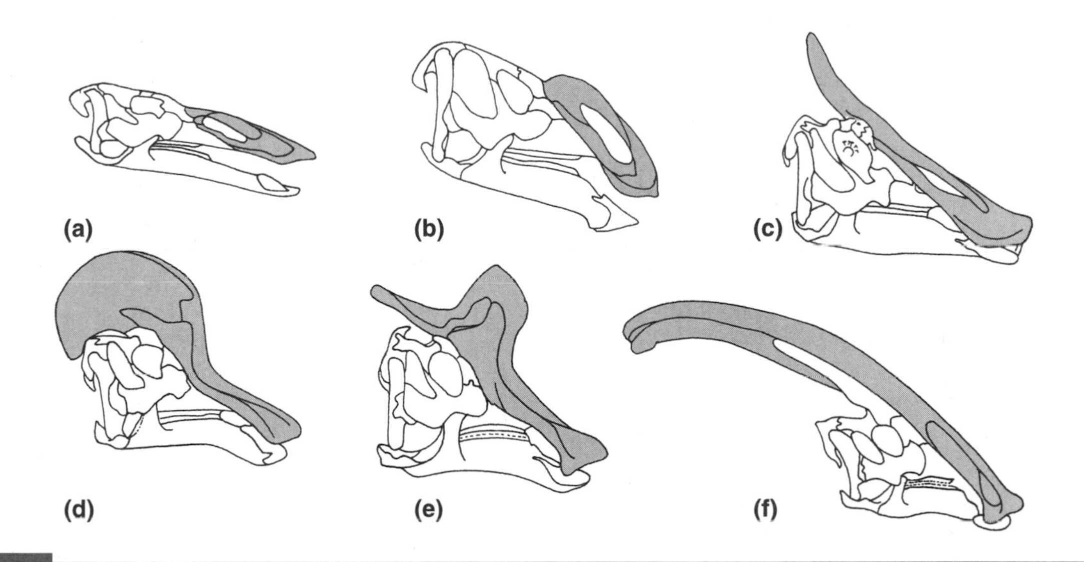

FIGURE 18–22 Skulls of various duck-billed hadrosaurs of the Upper Cretaceous: (*a*) *Anatosaurus;* (*b*) *Kritosaurus;* (*c*) *Saurolophus;* (*d*) *Corythosaurus;* (*e*) *Lambeosaurus;* (*f*) *Parasaurolophus.* Behind a flattened, toothless beak lay rows of teeth (as many as 700 per jaw ramus!) apparently used to grind tough vegetable matter. As shown, the nasal and premaxillary bones (*shaded*) assumed unusual shapes in different groups, producing long loops of nasal passages for purposes that are still undetermined. Among suggestions for their use have been improvement in sense of smell, visual signals for species recognition, male weapons or shields during mating competition, and resonators for amplifying sound. (*From Romer,* Vertebrate Paleontology, 3rd ed. *1966. Reprinted by permission of the University of Chicago.*)

sprawling gait. The speed and agility that accompanies an erect posture would presumably have had its source and sustenance in high metabolic activity.

2. The microscopic bone structure of dinosaurs shows a high density of blood-carrying haversian canals similar to those in mammals, whereas modern ectotherms show few such canals.
3. The distribution of dinosaur fossils during the Cretaceous extends to areas in Canada that must have been close to the Arctic Circle. Assuming a cool climate in these localities, this indicates that dinosaurs, like endotherms, were animals that could supply their own body heat.
4. The likelihood that birds such as *Archaeopteryx* originated from small carnivorous dinosaurs (Carroll 1988) carries with it the corollary that primitive featherlike structures may have insulated their ancestral dinosaur groups against loss of body heat.
5. The predator-to-prey ratio of carnivorous-to-herbivorous dinosaurs in some fossil deposits is on the order of 3 to 100. This ratio is similar to that for modern endothermal predators who need a large prey population to support their high-energy metabolism. By contrast, ectothermal predators with lower metabolic requirements can exist on a prey population 10 times less numerous (for example, a predator-to-prey ratio of 3 to 10).

Since these and other arguments suggest that the distinction between dinosaurs and reptiles may be as profound as the distinction between mammals and reptiles or birds and reptiles, Bakker and other endotherm proponents propose removing dinosaurs from the class Reptilia and putting them in a separate vertebrate class, Dinosauria.

However interesting these arguments may be, other paleontologists have disputed them. Erect posture, for example, may have little to do with endothermy but may be instead the only stance a large, heavy, terrestrial animal can assume without unduly bending its supporting limbs. Similarly, haversian bone structures may be more related to growth rate, body size, and other factors than to endothermy, especially since some ectothermal reptiles such as turtles show these structures and some small mammals and birds do not.

Dinosaur radiation to northern latitudes is not sufficient evidence for endothermy since Cretaceous climates were warmer than they are now, continental drift may have moved their ancient habitats northward, and such northern deposits contain fossils of ectothermal crocodiles and turtles. Also, researchers dispute whether feathers were originally used for insulation rather than for flight (Feduccia 1985), and as yet, no firm evidence of feathers among dinosaurs exists. According to Ruben and coworkers, dinosaur respiratory anatomy also differs markedly from endothermic mammals in lacking nasal turbinate bones, a feature highly correlated with endothermal physiology (Fig. 19–2).

Some biologists also dispute the predator-to-prey ratios Bakker cites for some localities. Others point out that it is difficult to discriminate between large ectotherms and endotherms on predator-to-prey ratios alone, since both types may require similar amounts of food. In general, incomplete fossilization makes it difficult, if not impossible, to determine the relative abundance of different species in a community and adds to the uncertainty of specifying predator-to-prey ratios; for example, one fossil deposit in Texas contains only carnivorous coelurosaurs.

Although we have not resolved the issue of endothermy (Thomas and Olson, Farlow), the dominance of dinosaurs over other terrestrial vertebrates throughout most of the Mesozoic is an undisputed fact whose source must lie in at least one or more special adaptations. For those long-necked sauropods believed to have fed on the upper branches of tall Mesozoic conifers, there must have been mechanisms, such as the four-chambered heart, that would efficiently pump blood to the sauropod head many feet above heart level. The possibility that dinosaurs were also endothermal, as a recent oxygen isotope study of *Tyrannosaurus rex* may indicate (Barrick and Showers), would explain many additional elements for their success, yet serious questions remain:

- Why have no adult dinosaurs been found smaller than about 20 pounds in weight, whereas practically all endothermal Mesozoic mammals (and many ectothermal reptiles) were below this size?
- What accounts for the extinction of all dinosaurs at the end of the Mesozoic, yet the survival of various other vertebrate groups, including mammalian endotherms?
- Why couldn't presumed endothermal dinosaurs adapt to ecological and climatological conditions to which endothermal mammals adapted?

If dinosaurs were ectothermal, large size alone may have affected dinosaur success as well as their limitations. As discussed previously, large ectotherms in a warm climate would have been able to preserve body heat and perhaps attain fairly high rates of metabolism and activity without paying the high cost of endothermy. Given such dependence on high body temperatures, smaller dinosaurs, subject to greater temperature fluctuations because of their size and lack of insulation, would not have been as successful.[7]

[7]Temperature stability in modern ectothermal reptiles is a function of size: the larger the animal, the more stable its body temperature (Spotila).

Also, if dinosaurs were primarily ectothermal, even large size would not have protected them against the more variable climate that probably inaugurated the Cenozoic era. For example, a large ectotherm with its reduced ability to change body temperature rapidly would have had considerable difficulty losing heat during a hot summer as well as in gaining enough heat during winter's prolonged cold periods. By contrast, endothermal mammals were able to survive the end of the Cretaceous and increase during the Cenozoic because their activity did not depend as much on external temperatures.

The Late Cretaceous Extinctions

Although we know of no dinosaurs that survived the Cretaceous, other groups also suffered, and paleontologists estimate that more than half of all animal species, classified into the various groups given in Table 18–2, became extinct during a relatively short geological period and produced no further lineages. Since the extinctions covered so many different kinds of organisms, the extent to which dinosaur ectothermy or endothermy affected their survival is unclear: we know no land vertebrate larger than 50 pounds to have survived the Cretaceous, and, with the exclusion of crocodiles and turtles, the extinctions embraced numerous marine organisms of varying sizes and metabolic features.

To account for such a wide spectrum of extinction, workers have offered many possible causes, often with considerable debate (see discussions in Russell, Kerr, Hallam, and McGhee). Among these hypotheses are intense volcanic activity, epidemics of disease, changes in plant composition, shifting continental profiles, elevated carbon dioxide level (greenhouse effect), changes in sea level or ocean salinity, high doses of ultraviolet radiation, dust clouds caused by collisions with comets or asteroids, and ionizing radiation from supernova explosions or other sources.[8]

The most popular of these theories, the hypothesis of collision with a comet or large meteorite, gathered considerable support in the 1980s from the discovery of iridium deposits in strata marking the Cretaceous–Tertiary boundary (Alvarez). Iridium is a rare earth element, often found in meteorites, and its worldwide presence in these strata along with high-impact particles (glasslike spherules and shocked, fractured quartz) strongly indicated collision with an extraterrestrial body (Box 18–1).

In spite of this evidence, many paleontologists have objected, noting that dinosaurs and other animal groups had already declined in numbers or disappeared before these impact layers were deposited (Sloan et al.). Contrary to the immediate effects of an extraterrestrial impact, paleontologists point out that the dinosaur extinction process may have taken a million years or more, and according to some claims, may even have extended into the Paleocene epoch of the Cenozoic era (Rigby et al., Van Valen).

Some paleontologists compromise by suggesting a combination of stressful environments and an extraterrestrial impact: "a literally earth-shaking event magnified the differences between species doing well and species not so well" (Archibald). For others, the question of how to decide whether extinction was gradual or catastrophic for various vertebrate groups still remains (Dingus and Rowe).

Whatever the cause for the Late Cretaceous mass extinction, an important issue is whether it negates the effects of selection in evolution. To paraphrase Raup (1991), to what extent must we abandon "bad genes" as an explanation for species loss and instead emphasize "bad luck"? The answer so far seems equivocal: the fact that entire genera and families can become extinct in a single event indicates little if any discrimination among their subsets of populations and species—obviously "bad luck." Thus, Jablonski found that the Late Cretaceous extinction destroyed many species of gastropod mollusks on a fairly random basis; for example, it made no difference whether these were from genera with many species or from genera with only one or two species.

However, some species or their subpopulations do survive in the midst of such extinction, indicating that certain traits may well have been beneficial. That is, traits of benefit during extinction may well differ from advantageous traits before extinction. The extinction of all dinosaurs would be difficult to ascribe only to "bad luck" since many mammalian species managed to endure the Cretaceous extinctions, as did many birds. Small size, and perhaps efficient endothermy, may have been crucial for terrestrial survival in the Late Cretaceous. In a sense, both "good

[8]A once popular hypothesis sought to explain the extinction of dinosaurs by internal rather than external causes. Just as individuals are born, grow old, and die, this hypothesis suggested that races, species, and other taxonomic categories follow a similar life history driven by internal **orthogenetic** factors that cause evolution to proceed in a direction unrelated to selection and adaptation. Many observed evolutionary successions seemed to support this concept, called **racial senescence,** and followers believed its cause lay in the gradual decline of some unknown vital force in each group, leading ultimately to the appearance of bizarre and nonadaptive characters. Schindewolf designated such traits as "abnormal" and "phenomena of decadence." Thus, some people long considered the large, seemingly clumsy 11-foot-wide antlers of the "Irish Elk" (Fig. 24–7, p. 589) to be a cause or corollary of its senescence and extinction. This notion is certainly not true. In the case of dinosaurs, far from being senescent, various groups were progressively adaptive for more than 100 million years, and even toward the end of the Cretaceous period new groups, such as the ceratopsians, seemed to be continually evolving in adaptive directions. No evidence indicates that mechanisms other than the failure of their adaptations to cope with new environmental or competitive challenges caused the extinction and replacement of dinosaurs or any organism.

TABLE 18-2 Number of genera that lived during the interval that began 20 million years before the extinctions at the end of the Cretaceous period and ended with the beginning of the Cenozoic era, compared with the number that existed 10 million years afterward

	Before Extinctions	After Extinctions	Percentage of Genera After Extinctions
Freshwater organisms			
Cartilaginous fishes	4	2	
Bony fishes	11	7	
Amphibians	9	10	
Reptiles	12	16	
	36	35	97
Terrestrial organisms (including freshwater organisms)			
Higher plants	100	90	
Snails	16	18	
Bivalves	0	7	
Cartilaginous fishes	4	2	
Bony fishes	11	7	
Amphibians	9	10	
Reptiles	54	24	
Mammals	22	25	
	226	183	81
Floating marine microorganisms			
Acritarchs	28	10	
Coccoliths	43	4	
Dinoflagellates	57	43	
Diatoms	10	10	
Radiolarians	63	63	
Foraminifers	18	3	
Ostracods	79	40	
	298	173	58
Bottom-dwelling marine organisms			
Calcareous algae	41	35	
Sponges	261	81	
Foraminifers	95	93	
Corals	87	31	
Bryozoans	337	204	
Brachiopods	28	22	
Snails	300	150	
Bivalves	399	193	
Barnacles	32	24	
Malacostracans	69	52	
Sea lilies	100	30	
Echinoids	190	69	
Asteroids	37	28	
	1,976	1,012	51
Swimming marine organisms			
Ammonites	34	0	
Nautiloids	10	7	
Belemnites	4	0	
Cartilaginous fishes	70	50	
Bony fishes	185	39	
Reptiles	29	3	
	332	99	30
Overall Totals	2,868	1,502	52

Note: The record for terrestrial organisms is limited to North America but is global for marine organisms.

Source: From Russell. Reproduced with permission from the *Annual Review of Earth and Planetary Sciences*, Volume 7, © 1979 by Annual Reviews Inc.

genes" and "good luck" enabled the explosive Cenozoic radiation of mammalian species (Chapter 19), a conclusion that serves to remind us that evolution is opportunistic.[9]

Extinction is, of course, not confined to the past. Wilson estimates that about 17,500 tropical terrestrial species now become extinct annually, many because humans have destroyed their habitats. Other researchers suggest that extinctions in even widely diverse groups and environments reach 100 to 1,000 times that of pre-human levels (Pimm et al.). Although some contest these numbers, such appraisals, even if reduced by half, portend a bleak future for many species living in or near human-occupied areas. We humans not only usurp natural resources used by other species for our own use, but we also destroy natural resources by erosion and pollution. Many writers warn that humans have become agents of extinction, perhaps as powerful as global climatic change and the impact of extraterrestrial bodies.

What appears necessary is a change from passive destructive activity to mindful, accountable behavior. As discussed in Chapter 25, because of our new-found ability to consciously direct biological change and affect the environment, we now play the role of architects of evolutionary change—a highly responsible role that demands concerned awareness of other creatures, rational goals, and control of our own behavior.

Reptilian Flight: Pterosaurs

Escape from extinction, through "good luck" and "good genes," is the earmark of biological survival, and may at times depend on the ability to move rapidly from threatened environments. Airborne flight is, therefore, a highly adaptive form of locomotion providing a number of advantages:

- Rapid escape from terrestrial predators and menacing conditions
- Access to feeding and breeding grounds that would otherwise be difficult or impossible to reach
- Relatively swift transit between localities

[9]According to Benton (1995), a general increase in the number of families of all life forms has been continuing throughout the Phanerozoic. Starting with about 280 families in the Early Cambrian period, their number diversified to about 600 at the end of the Paleozoic era, 1,260 at the end of the Mesozoic era, and more than 2,000 at present. He points out that "High measures of origination indicate bursts of diversification into new habitats, and many of the high frequencies follow after mass extinction events, when empty ecospace was filled." Such instances of "macroevolutionary" events, defined as major increases in organismic diversity (pp. 287, 598) are obviously the result of improved adaptive opportunities for genetic variants—adaptive radiation (pp. 37, 452). That is, surviving populations can enter, proliferate, and subdivide in ecological and geographical zones formerly closed to them, each zone subject to novel selection environments that enable populations to embark in new evolutionary directions.

Although gliding forms capable of parachuting for short distances arose in various vertebrate groups including fish, known adaptations for sustained ascending flights have appeared only three times in two vertebrate classes: twice in reptiles (**pterosaurs** and **birds**) and once in mammals (bats). The two reptilian-derived flying forms differ in respect to mechanisms and accompanying adaptations: in pterosaurs a flight membrane was formed by a thin fold of skin stretched between the trunk and elongated fourth finger of each hand, while in birds the flying surface consists of many stiff wing feathers that project posteriorly from the front limb.

Although primitive forms of pterosaurs appear in the Upper Triassic and birds in the Jurassic, earlier evolution of these two types has left no obvious record, perhaps because they started out as small arboreal creatures in poorly fossilized highland habitats. Nevertheless, their skeletal features provide a number of fairly clear homologies with earlier reptiles. For pterosaurs, one phylogeny suggests they originated from early Triassic bipedal thecodonts (Fig. 18–24, *left side*). The ancestry of birds includes an additional dinosaur intermediate (Fig. 18–24, *right side*), since the similarity between the first fossil bird (*Archaeopteryx*) and dinosaurs is so striking that paleontologists would most probably have classified *Archaeopteryx* as a dinosaur if it had no feathers. In general, by the time of their first fossil appearance, both types were well adapted to active gliding and probably to sustained flying.

A well-described Jurassic pterosaur, *Rhamphorhynchus*, (Fig. 18–25*a*), was about 2 feet long with a typical diapsid archosaurian skull and an additional preorbital fenestra that helped lighten the head. The **rhamphorhynchoid** tail was long with a small, rudderlike flap of skin at the end, the bones were light and hollow, and the elongated jaws were armed with strong, pointed teeth. According to Padian, the sternum and its accessory bones provided sufficient surface to have allowed attachment of large flight muscles like those of modern flying birds. He also claimed that the wing membrane did not cover the hind limb, and the legs were therefore free for bipedal locomotion somewhat like existing large birds. In contrast, Wellnhofer and others support a more traditional notion that pterosaur motion on land was quite awkward, based on a sprawled quadrupedal stance. Whichever concept is correct, terrestrial locomotion may not have been a serious requirement: their fossil locations generally indicate that these pterosaurs roosted near large lakes and coastal areas, perhaps on offshore islands in trees or protected cliffs, and hunted for fish swimming at the surface.

By the late Jurassic, rhamphorhynchoid success had given rise to another group, the **pterodactyloids,** which continued well into the Cretaceous. In the pterodactyloids (Fig. 18–25*b*), the tail was almost completely absent, a special bony element anchored the shoulder girdle to a number of spinal vertebrae, and the teeth tended to be

BOX 18-1 EXTINCTIONS AND EXTRATERRESTRIAL IMPACTS

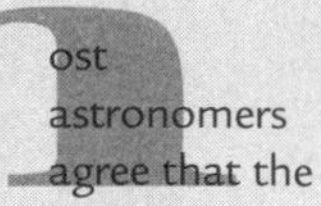

Most astronomers agree that the universe is a violent place. It had a violent birth, its stars and galaxies were born in the midst of violent interactions, its elements were created from the debris of many violent episodes, and violent impacts still occur—even in our small solar system. Whether from impacts, volcanism, or plate tectonics (Chapter 6), we recognize occasional catastrophes in earth's history that may have had major effects on geology and life, and we no longer follow a philosophy restricted to gradual change, such as Lyell expounded (pp. 15–16), as the principal explanation for all earthly events.

As a result of the *Apollo* space program, we know that a series of heavy extraterrestrial impacts battered the moon about 4 billion years ago, indicating that very similar events must have occurred on the earth's surface. The sterilizing heat such impacts generated, whether caused by asteroids, comets, or collisions with even larger planetlike bodies, may indeed mark the period after which present forms of life arose (Fig. 9-13).

Although their frequency has greatly diminished from the Hadean age, impacts have persisted, and geologists have identified more than 100 craters on earth. Table A relates the size of the crater to a very general estimate of the size of the celestial body causing it, and estimates the time between such events (Raup 1991).

Such impacts, depending on their size, may have enormous environmental effects for periods ranging from months to years. The impact crater throws large amounts of particles up into the atmosphere, producing dust clouds that interfere with photosynthesis, causing collapse of the food chain in various localities. Depending on whether the impact hits land or sea, it can also cause large climatic temperature changes ranging from an immediate "winter" (dust) followed by a high-temperature "greenhouse effect" (water vapor). The heat generated by the object entering the atmosphere and the heated material it ejects on impact can spark raging forest fires as well as nitrous oxides that seed acid rains that destroy vegetation and marine organisms. Alvarez and Asaro point out that the impact of a 10-kilometer diameter body would produce an explosion equal to that of 100 million atomic warheads (10^8 megatons). Given such colossal effects, the question arises whether we can discern a relationship between extraterrestrial impacts and mass extinctions.

As far as we can tell from fossil data, there have been five major mass extinctions dating from the Cambrian period onward (Table B), each extinction marked by the relatively abrupt disappearance of at least 75 percent of marine animal species. (Because marine environments are the major sites of sediment deposit, aquatic organisms more commonly fossilize than do terrestrial ones, and paleontologists consider estimates of their species frequency more reliable.)

According to Raup and Sepkoski, these large-scale extinctions are not isolated events but seem allied to other extinctions, perhaps caused by a series of impacts from extraterrestrial bodies that occurred throughout the Phanerozoic eon. They point out that mass extinctions of families and genera occur with a periodicity of about 26 million years, and Fox's analysis of their data supports this view (Fig. 18-23). However, since no clear cause appears for this periodicity, researchers have disputed the Raup-Sepkoski hypothesis, although some offer extraterrestrial factors such as the effects of a possible companion star to our sun (dubbed the "Death Star" or "Nemesis") as an as yet unproved explanation.

Table A

Meteorite Diameter	Crater Diameter (km)	Average Time (millions of years) Between Impacts
10 km	> 150 km	100
·	> 100	50
·	> 50	12.5
·	> 30	1.2
·	> 20	0.4
1 km	> 10	0.11

Table B

Extinction Period	Approximate Date (millions of years ago)	Estimated Percentages of Marine Animal Extinctions	
		Genera	Species
Late Ordovician	440	61	85
Late Devonian	365	55	82
Late Permian	245	84	96
Late Triassic	208	50	76
Late Cretaceous	65	50	76

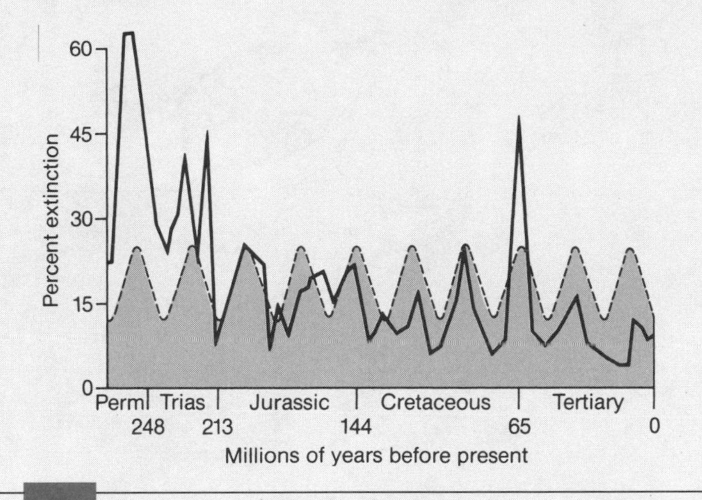

FIGURE 18–23 *Heavy solid lines* show the percentage of marine animal extinctions during the approximate 260-million-year interval from the Permian period to the present. *Dashed lines* and *shaded areas* are based on a periodicity of 26 million years (calculated as the tenth Fourier harmonic). Although Fox finds the relationship between the extinctions and 26-million-year periods statistically significant, other workers question the statistical as well as taxonomic basis for the periodicity. McGhee as well as Benton (1995), for example, point out that some peaks are missing, and others occur that are not predicted. *(Adapted from Fox.)*

Periodical or not, it is still quite clear that mass extinctions have recurred, and Raup (1991) estimates that extinction events on the order of the type just tabulated occur on average about every 100 million years. As we might expect, extinctions with less effect are more common, but the average waiting time between such events is still long enough so that most species rarely face extinction (for example, an event that eliminates 5 percent of species occurs on average only once every million years).

Despite extensive environmental effects of such impacts, good evidence for a close association with a foreign body impact exists only in the case of the Late Cretaceous extinction. As the text indicates, an anomalous iridium-rich layer at the Mesozoic–Cenozoic boundary now found in more than 100 different localities on earth, lends credence to a large meteorite impact at the end of the Cretaceous period (Alvarez and Asaro). Such an explosion would have carried the observed iridium and high-velocity particles—spherules, and shocked quartz—to the top of the atmosphere and then spread them worldwide.

Among the best candidates for a crater large enough to produce such Late Cretaceous global effects is the Chicxulub crater off the coast of the Yucatán peninsula in Mexico. According to recent measurements, it has an outer diameter of about 195 kilometers (120 miles) and "records one of the largest collisions in the inner solar system since the end of early period of heavy bombardment almost 4 billion years ago. . . . Earth probably has not experienced another impact of this magnitude since the development of multicellular life approximately a billion years ago." (Sharpton et al.)

However, the issue is whether this impact accounts for all the Late Cretaceous extinctions. Paleogeologists have proposed volcanic eruptions, which, when explosive, may deposit iridium globally in atmospheric dust and ash (and even when nonexplosive may affect world climate) as a primary or auxiliary cause for these extinctions. An enormous outpouring of nearly one million cubic miles of volcanic lava covering one-third of India (the "Deccan Traps") during the Late Cretaceous extinctions supports this hypothesis.

As to proposals that extinctions other than the Late Cretaceous relate to impacts, no firm supporting evidence has yet appeared. Iridium deposits in strata associated with other extinctions are not great enough to assume extraterrestrial impact, and strata in the Late Permian extinction, the greatest of all extinctions (Erwin), show almost no iridium. Nevertheless, the likelihood that the Permian extinction occurred over a geological period between only 10,000 and 165,000 years (Bowring et al.), makes it as catastrophic as a conspicuous impact.

Knowledge of the cause(s) for mass extinctions and the precise manner in which they manifested their effects still awaits more exact information. Extinctions certainly had serious evolutionary consequences by producing a "major restructuring of the biosphere wherein some successful groups are eliminated, allowing previously minor groups to expand and diversify" (Raup 1994). In our unpredictable solar system, such events may occur again: of the many asteroids that cross the Earth's orbit at various times, perhaps a thousand or more can have an impact equal to the asteroid that caused the extinction of the dinosaurs.

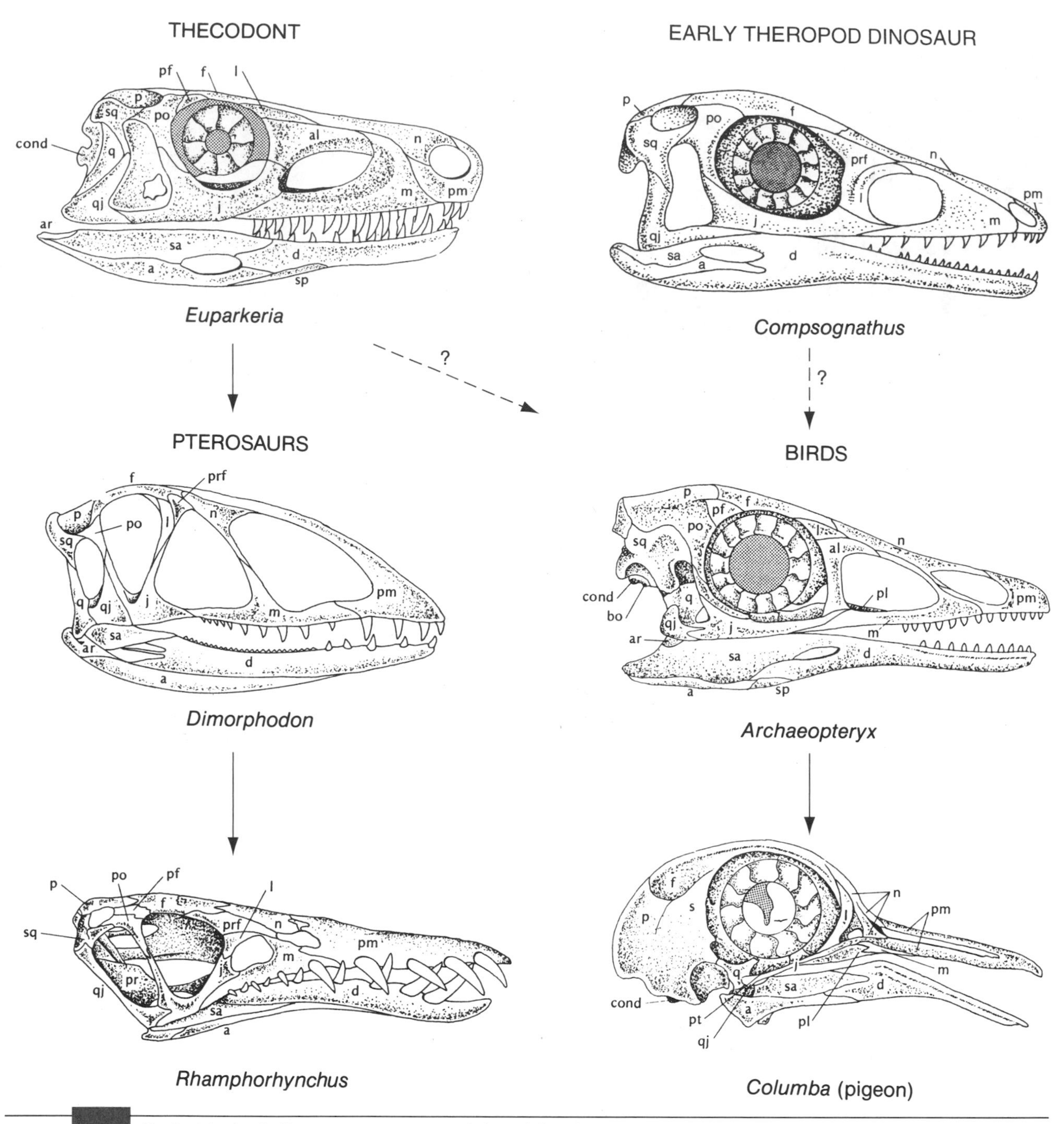

FIGURE 18-24 Similarities in skull structures among early bipedal archosaurs such as *Euparkeria* and the pterosaur and bird lineages that researchers believe may be derived from them. Abbreviations: *a,* angular; *al,* adlacrimal; *ar,* articular; *bo,* basioccipital; *cond,* occipital condyle; *d,* dentary; *f,* frontal; *j,* jugal; *l,* lacrimal; *m,* maxilla; *n,* nasal; *p,* parietal; *pf,* postfrontal; *pl,* palatine; *pm,* premaxilla; *po,* postorbital; *pr,* prootic; *prf,* prefrontal; *pt,* pterygoid; *q,* quadrate; *qj,* quadratojugal; *sa,* surangular; *sp,* splenial; *sq,* squamosal. (*Adapted from Stahl, from Romer, and from Heilmann.*)

reduced, leading eventually to a long, toothless beak. In *Pteranodon,* a long skull crest extending behind the optic orbit doubled the length of the head. Size differences were pronounced, the largest fossil pterydactyloid being a Late Cretaceous form found in Texas, *Quetzalcoatlus,* with a wingspread that may have reached 40 feet! These animals must have been successful gliders, probably using sea or land thermals for lift, and may have been capable of powered flapping for short distances.

Since pterosaurs could not have used the various environmental devices for temperature regulation that were available to terrestrial ectotherms, such as movement in and out of shady areas, paleobiologists have suggested that they must have been endothermal, and some

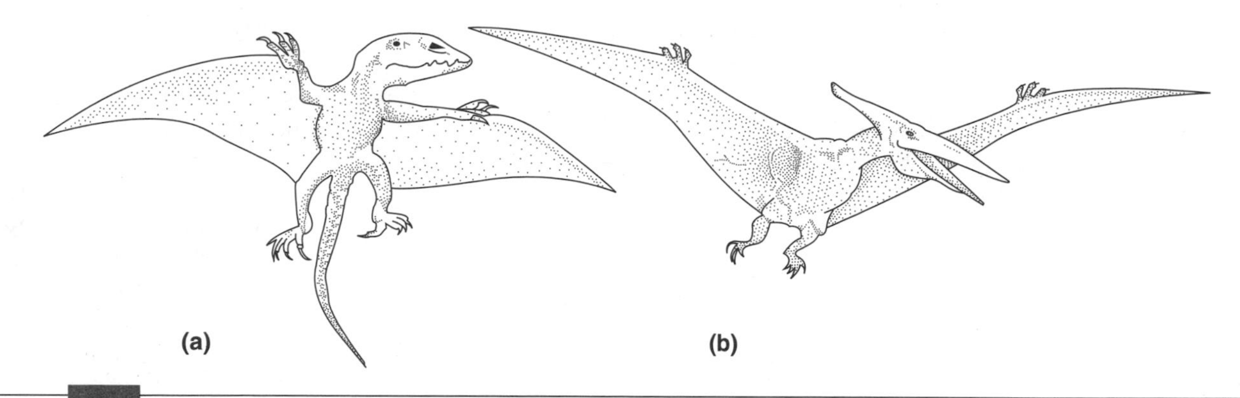

FIGURE 18-25 Comparison between rhamphorhynchoids (*a*) and pterodactyloids (*b*). The rhamphorhynchoids were tailed with wing spans ranging between 1 foot and 7 feet. The pterodactyloids were relatively untailed, with wing spans from 6 inches to as much as 40 feet. Other interpretations of pterosaur fossils suggest that, although narrow at the tips, the wings broadened at the trunk to attach from neck to ankles (Unwin and Bakhurina.) This would have made pterosaur mobility on land hardly more than bat-like, a view that is supported by their sprawling quadrupedal posture (Clark et al.).

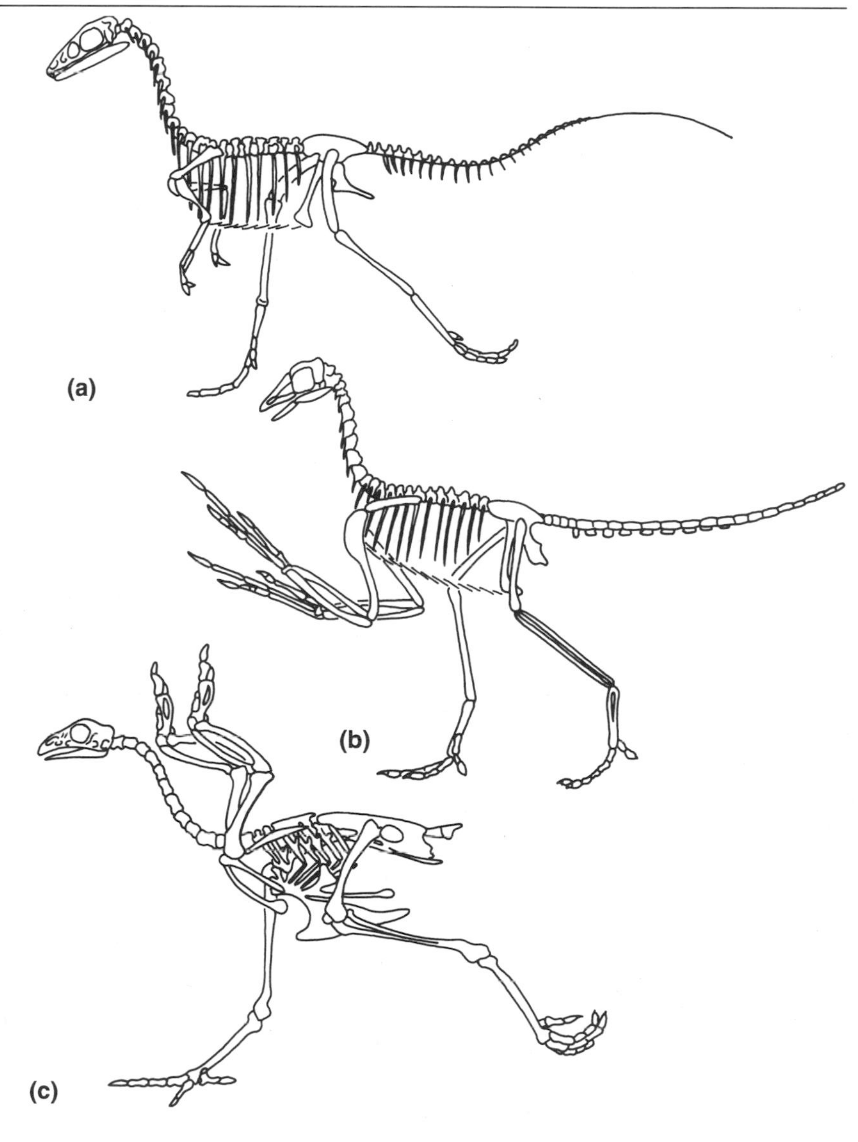

FIGURE 18-26 Skeletons of (*a*) *Compsognathus*, a dinosaur, (*b*) *Archaeopteryx*, a fossil bird, and (*c*) *Gallus*, a modern chicken. Although *Archaeopteryx* had proportionately longer arms and hands than *Compsognathus*, it is easy to see that birds and dinosaurs share some common skeletal features. Whether such shared features indicate that birds were derived from dinosaurs, or only that they both share a common reptilian origin, is in dispute. Feduccia (1999) argues that there are major anatomical differences between birds and dinosaurs, such as the embryological origin of their forelimb digits: the dinosaur three-fingered hand is composed of digits 1, 2, and 3, whereas the bird forelimb uses digits 2, 3, and 4. Feduccia also questions why birdlike dinosaur theropods such as *Velociraptor* first occur 80 million years later than *Archaeopteryx*, whereas Triassic theropods that are contemporary with *Archaeopteryx* have few, if any, specifically birdlike features. Padian and Chiappe oppose these claims, and present evidence arguing that birds must have descended from an early dinosaur lineage. However such disputes will be resolved, it is clear that birds underwent marked structural changes from fossil to modern forms: the pelvis and sacrum coalesced into a single structure, the sternum enlarged for the attachment of flight muscles, the hand bones fused, and the long, bony tail diminished. [(*a*) *and* (*b*) *Reprinted with permission of Cambridge University Press from* Patterns and Processes of Vertebrate Evolution, *Robert Lynn Carroll, © 1997.* (*c*) *From Dingus, L., and T. Rowe, 1998.* The Mistaken Extinction: Dinosaur Evolution and the Origin of Birds. *New York: W.H. Freeman.*]

pterosaur fossils show evidence of hairlike scales that may have served as insulation. Perhaps, along with dinosaurs, pterosaurs merit nonreptilian class status. In any case, flight did not protect them from extinction, and they too, like the dinosaurs, did not survive the Cretaceous; instead, their archosaurian cousins, the birds, became the most widely distributed of Cenozoic flyers.

Birds

The first feathered, birdlike fossils, all classified as **Archaeopteryx,** were found in Upper Jurassic limestone deposits in Bavaria. Among seven skeletons we know about, four are nearly complete, and some show the flight feathers of wing and tail in their natural position (Fig. 3–13*a*). Most striking are their similarities to dinosaurs (Fig. 18–26), marked by teeth, separate clawed fingers, long bony tail, and dozens of other features that indicate their status as a true transitional "missing link" between theropods and birds (Ostrom 1991).

Unfortunately, fossilization in such fine-grained silts was rare, and no clearly feathered intermediates have been found between *Archaeopteryx* and its Jurassic dinosaur ancestors. However, within a short geological period—by the Early Cretaceous—a range of aquatic birds and shorebirds began to appear, marking a transition to more modern forms. A few such fossils represent groups such as flamingos, loons, cormorants, and sandpipers, although some, such as *Hesperornis,* still retained reptilelike teeth.

Among its various consequences, the absence of a pre-*Archaeopteryx* soft tissue record preserves the mystery of when and how feathers first evolved from reptilian scales and makes the origin of avian flight a matter of dispute.

- Were feathers primarily an adaptation for insulating the presumed endothermic reptilian ancestors of birds, or were they primarily associated with flight and only secondarily with insulating qualities?
- Were primitive ancestral birds originally **arboreal** reptiles that used their developing wings to glide from branch to branch, or were they **cursorial,** ground-dwelling creatures whose primitive feathers formed planing surfaces enabling them to increase running speed?

Different opinions extend also to *Archaeopteryx* itself, although researchers generally agree that *Archaeopteryx* represents a fairly advanced stage in a long history of bird evolution that may have begun much earlier. The primary feathers of *Archaeopteryx* are remarkably similar in vane structure to the primary (flight) feathers of modern flying birds, whereas nonflying birds have feathers with different structures (Feduccia and Tordoff): *Archaeopteryx* could fly.

Ostrom (1974) presents the view that feathers evolved primarily as means to control heat loss in some endothermal dinosaurs, and these feathers, especially on the forelimbs, could then help capture prey such as insects. A feathered "insect net" of this type, along with accompanying muscular adaptations such as enlarged pectorals, would then serve as an incipient **wing,** preadaptive for powered flight. The report by Chen and coworkers of a "downy feathered" mane on a *Compsognathus*-like fossil would, if substantiated, add support to Ostrom's notion of a feathered-dinosaur origin of birds. Other paleontologists argue that wings in a cursorial animal would have hindered rather than aided rapid movement, since wing lift would have interfered with ground traction necessary for the hind legs. On the other hand, if *Archaeopteryx* had an exclusively gliding arboreal history, what would explain its bipedal stance, which it derived from ground-dwelling forms?

We have no clear answers to these questions as yet, although there are some suggestions that bird evolution may not have been a straightforward process but went through a number of stages, beginning with a bipedal, cursorial reptile that later became arboreal. Burgers and Chiappe provide calculations showing that "[A] running *Archaeopteryx* . . . could have achieved the velocity necessary to become airborne by flapping feathered wings." In contrast, Feduccia (1999) claims support for the notion that a cursorial origin of avian flight is "a near biophysical impossibility," and long-feathered gliding forms must have developed in trees. Even if *Archaeopteryx* was primarily cursorial, its presence might mean that, out of an extensive avian adaptive radiation, only one transitional ground-dwelling form fossilized at that time.

Once past its initial stages, bird evolution seems to have advanced rapidly from the Cretaceous period through the beginning of the Tertiary. The Cretaceous birds show such modern features as an enlarged brain, fused skull bones, and reduced temporal fenestrae. The sternum greatly enlarged in some forms, indicating the attachment of powerful flight muscles, and there were also skeletal changes such as fused pelvis and sacrum. Although bird fossils are never too plentiful, there are sufficient numbers and kinds of Eocene and Oligocene deposits to indicate that almost all modern orders of birds had evolved by then (Feduccia 1995), with some lineages perhaps tracing back to the Late Cretaceous (Dingus and Rowe).

One dramatic adaptive opportunity that the extinction of the dinosaurs caused at the end of the Cretaceous was to open terrestrial niches into which various large, flightless, ground birds evolved, such as the 7-foot-tall *Diatryma* (Fig. 18–27) and others (Phorusrhacidae) that may have reached a height of 10 feet or more. Some of

FIGURE 18–27 Reconstructions of some extinct large flightless birds showing their relative sizes. *Diatryma* was an early Cenozoic bird, the others date from the much later Pleistocene epoch. (*Adapted from Feduccia 1980.*)

these giant forms were widely distributed until they became extinct later in the Cenozoic because of competition with advanced mammalian carnivores and predation by humans. Relatively few flightless ground birds now survive, such as the ostriches of Africa and the rheas of South America; the smaller flightless species such as kiwis and island rails are mostly confined to gradually diminishing habitats.

The evolution of flightlessness seems to have involved changes in the direction of selection, caused either by:

- ABSENCE OF PREDATION In some protected or island habitats where major carnivorous forms were absent, local birds could evolve to dominate the terrestrial food chain. Once assuming such roles, selection among mutations that vestigialized former flight structures would have reduced their energy expenses.

- MARINE HABITATS Birds becoming adapted to productive marine habitats would have been subject to selection for mutant wing modifications that improve underwater propulsion but reduce flight ability, such as penguins and steamer ducks.

At present, systematists classify birds into a total of about 35 orders subdivided into about 200 families. The distinctions among these groups are sometimes subtle, mostly based on external traits rather than on pronounced anatomical and skeletal differences used in separating groups among mammals and other vertebrate classes. Among the existing 8,800 species of birds are a variety of feeding and locomotor adaptations, ranging from flesh eating to nectar feeding, and from rapid running to gliding, swooping, diving, and swimming (Fig. 18–28). Although some workers have proposed evolutionary changes that may have given rise to some of these adaptations, the absence of fossil intermediates makes relationships among avian groups uncertain, especially among those in arboreal habitats. The molecular approach to this problem based on DNA studies (Chapter 12), is providing considerable information and shows promise of clarifying many disputed issues (Sibley and Ahlquist, Sheldon and Bledsoe, see also Box 12–2).

SUMMARY

Sarcopterygian fish were preadapted to terrestrial life because they had lungs and fleshy fins. They may have invaded the land when their swampy habitats desiccated, or when competition and predation forced them to seek new habitats. Similarities of teeth, vertebrae, and other bones indicate that the earliest amphibia arose from these

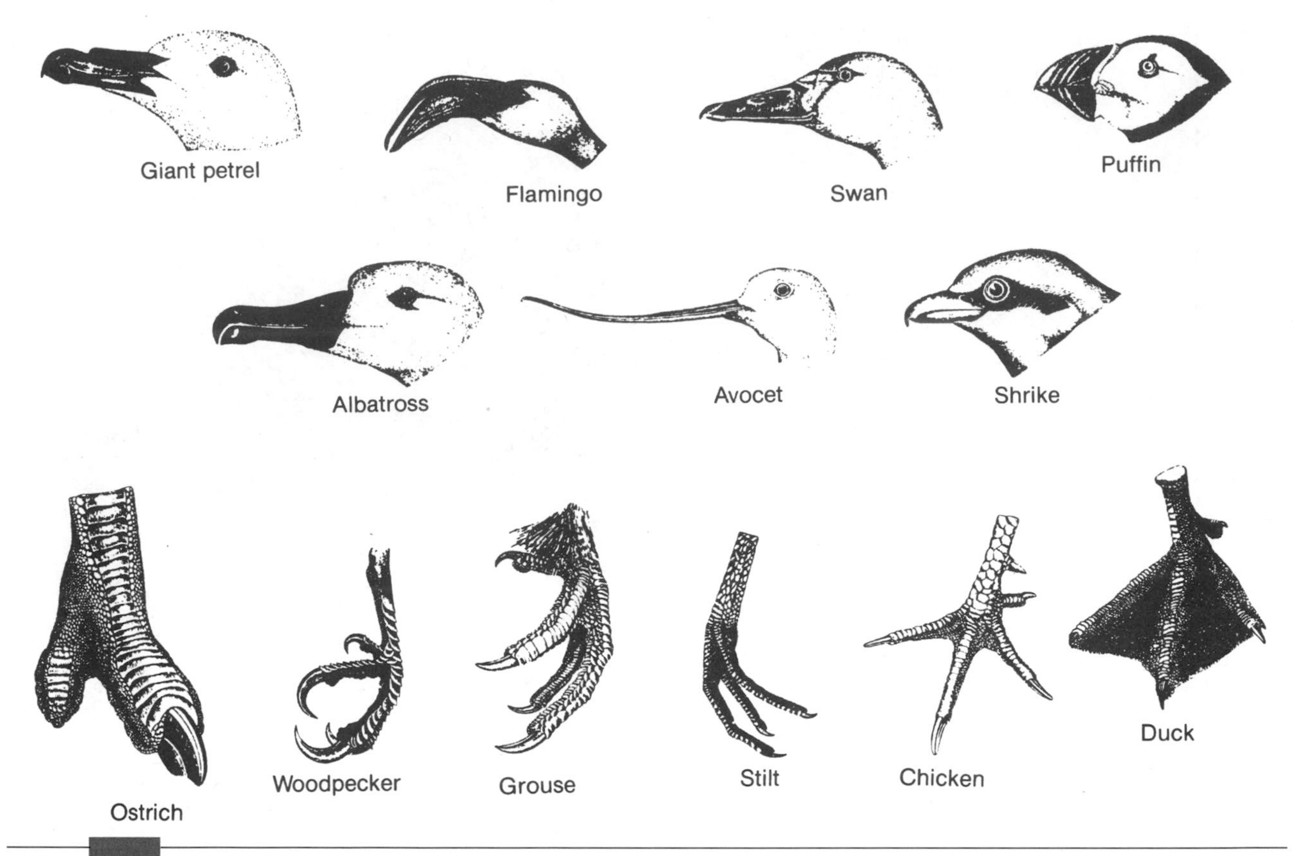

FIGURE 18–28 Some of the many adaptations of bird bills (*above*) and feet (*below*). (*Adapted from Feduccia 1980.*)

fish. A terrestrial environment favored certain structural modifications. A rigid spine provided suspension for limb girdles, freed internal organs from pressure, and acted as a cantilever for the head. In certain amphibia, the pleurocentrum rather than the intercentrum became the primary vertebral element, a trait retained by reptiles and mammals.

Modern amphibia live in many different habitats and are quite specialized. Their unique teeth, reduced skeletal bone, four-digit hands and unusual ear structure make it hard for us to trace their phylogenetic relationships as yet.

The reptiles, arising from small, lizardlike amphibia, have many novel attributes such as a partially divided ventricle, a cornified epidermis, separate reproductive and excretory ducts, and a shelled egg containing a membrane-enclosed embryo. The development of the amniotic egg liberated reptiles from dependency on water and was preceded by terrestrial egg laying and internal fertilization. Fenestrae in the temporal bone—a trait paleontologists often use to classify fossil reptiles—provided better anchorage for jaw muscles.

Reptiles diversified into many terrestrial habitats and became the dominant vertebrate group throughout the Mesozoic era. Therapsid reptiles, ancestors of the mammals, appear to have had temperature-regulating mechanisms and may have been the first endotherms. Special limb bones allowed the archosaur reptiles to become bipedal, increasing their speed and size. Their descendants, the dinosaurs, radiated into almost every terrestrial habitat, and some became so enormous as to approach terrestrial size limits. Some paleontologists claim that dinosaurs were endothermic, based on bone structure, posture, and biogeographical distribution. According to these views, endothermy, in combination with their size and other favorable attributes, may have enabled dinosaurs to dominate all other land vertebrates. At the end of the Cretaceous period, dinosaurs along with other large marine reptiles and various other groups became extinct. This extinction may have been caused by a unique single event or a succession of events or may have been only one in a cycle of periodic extinctions.

Adaptations for sustained flight appeared twice in reptiles, giving rise to pterosaurs and birds. The pterosaurs evolved from dinosaurs by developing hollow bones and flight membranes between trunk and forelimbs and in some cases by losing the reptilian tail and teeth, leaving the jaw as a beak. Birds may have originated from bipedal, ground-dwelling endothermic reptiles, their feathers derived from scales used as insulating mechanisms and/or as structures that aided gliding. Birds evolved rapidly, altering bone structures and enlarging sternum and brain. On the paleontological level, we know little of evolutionary relationships among birds since bird fossils are relatively rare. However, molecular techniques are now beginning to offer new phylogenetic information.

KEY TERMS

allantois
Amniota
amniotic egg
amphibians
Anapsida
anthracosaurs
arboreal
Archaeopteryx
archosaurs
arms race
bipedalism
birds
captorhinomorphs
cursorial
Diapsida
dinosaurs
ectothermic
endothermic
fenestrae
Late Cretaceous extinctions
Lissamphibia
Ornithischia
orthogenetic
pelycosaurs
pterosaurs
pterydactyloids
racial senescence
reptilian flight
rhamphorhyncoids
Saurischia
Synapsida
tetrapod
thecodonts
therapsids
theropods
wing

DISCUSSION QUESTIONS

1. What explanations have paleontologists offered to account for the transformation of crossopterygian fishes to a land-based existence?
2. What evidence indicates that early amphibians, such as *Ichthyostega,* were related to crossopterygian (rhipidistian) fishes?
3. How did changes in the spinal column and limb girdle attachments improve locomotion on land for early tetrapods?
4. What proposals have researchers made for the origins of modern amphibia (lissamphibia)?
5. What advantages did the amniotic egg offer to reptiles, and what preceding stages were necessary?
6. What factors can account for the radiation and extinction of therapsid reptiles?
7. What are the advantages and disadvantages of both endothermy and ectothermy?
8. What selective factors can account for
   a. The very large size of many dinosaurs
   b. Dinosaur bipedalism, especially in theropods
9. What are the arguments, pro and con, for dinosaur endothermy?
10. What proposals have paleontologists offered to explain the mass extinctions at the close of the Cretaceous period?
11. What is the relationship between reptilian pterosaurs and birds?
12. What arguments do paleobiologists use to support the proposal that primitive birds were arboreal? cursorial?

EVOLUTION ON THE WEB

Explore evolution on the web! Visit the accompanying web site for *Evolution,* 3/e at www.jbpub.com/evolution for web exercises and links relating to topics covered in this chapter.

REFERENCES

Ahlberg, P. E., and A. R. Milner, 1994. The origin and early diversification of tetrapods. *Nature,* **368,** 507–514.

Alexander, R. M., 1975. *The Chordates.* Cambridge University Press, Cambridge, England.

———, 1991. Apparent adaptation and actual performance. *Evol. Biol.,* **25,** 357–373.

Alvarez, L. W., 1983. Experimental evidence that an asteroid impact led to the extinction of many species 65 million years ago. *Proc. Nat. Acad. Sci.,* **80,** 627–642.

Alvarez, W., and F. Asaro, 1992. The extinction of the dinosaurs. In *Understanding Catastrophe,* J. Bourriau (ed.). Cambridge University Press, Cambridge, England, pp. 28–56.

Archibald, J. D., 1996. *Dinosaur Extinction and the End of an Era.* Columbia University Press, New York.

Bakker, R. 1986. *The Dinosaur Heresies.* Longman, Harlow, England.

Barrick, R. E., and W. J. Showers, 1994. Thermophysiology of *Tyrannosaurus rex*: Evidence from oxygen isotopes. *Science,* **265,** 222–224.

Bennett, A. F., and J. A. Ruben, 1979. Endothermy and activity in vertebrates. *Science,* **201,** 649–654.

Benton, M. J., 1995. Diversification and extinction in the history of life. *Science,* **268,** 52–58.

———, 1997. *Vertebrate Paleontology,* 2d ed. Harper Collins, London.

Benton, M. J., and J. M. Clark, 1988. Archosaur phylogeny and the relationships of the Crocodylia. In *The Phylogeny and Classification of the Tetrapods: vol. 1. Amphibians, Reptiles, Birds,* M. J. Benton (ed.). Oxford University Press, Oxford, England, pp. 235–338.

Bowring, S. A., D. H. Erwin, Y. G. Jin, M. W. Martin, K. Davidek, and W. Wang, 1998. U/Pb zircon geochronology and tempo of the End-Permian mass extinction. *Science,* **280,** 1039–1045.

Burgers, P., and L. M. Chiappe, 1999. The wing of *Archaeopteryx* as a primary thrust generator. *Nature,* **399,** 60–62.

Carroll, R. L., 1977. Patterns of amphibian evolution: An extended example of the incompleteness of the fossil record. In *Patterns of Evolution, as Illustrated by the Fossil Record,* A. Hallam (ed.). Elsevier, Amsterdam, pp. 405–437.

———, 1988. *Vertebrate Paleontology and Evolution.* Freeman, New York.
———, 1991. The origin of reptiles. In *Origins of the Higher Groups of Tetrapods: Controversy and Consensus,* H.-P. Schultze and L. Trueb (eds.). Cornell University Press, Ithaca, New York, pp. 331–353.
———, 1995. Between fish and amphibian. *Nature,* **373,** 389–390.
———, 1997. *Patterns and Processes of Vertebrate Evolution.* Cambridge University Press, Cambridge, England.
Charig, A., 1983. *A New Look at the Dinosaurs.* British Museum, London.
Chen, P.-J., Z.-M. Dong, and S.-N. Zhen, 1998. An exceptionally well-preserved theropod dinosaur from the Yixian Formation of China. *Nature,* **391,** 147–152.
Clark, J. M., J. A. Hopson, R. Hernández, D. E. Fastovsky, and M. Montellano, 1998. Foot posture in a primitive pterosaur. *Nature,* **391,** 886–889.
Coates, M. I., and J. A. Clack, 1990. Polydactyly in the earliest known tetrapod limbs. *Nature,* **347,** 66–69.
———, 1991. Fish-like gills and breathing in the earliest known tetrapod. *Nature,* **352,** 234–236.
Colbert, E. H., and M. Morales, 1991. *Evolution of the Vertebrates: A History of the Backboned Animals Through Time,* 4th ed. Wiley, New York.
Desmond, A. J., 1976. *The Hot-Blooded Dinosaurs: A Revolution in Paleontology.* Dial Press/Wade, New York.
Dingus, L., and T. Rowe, 1998. *The Mistaken Extinction: Dinosaur Evolution and the Origin of Birds.* Freeman, New York.
Duellman, W. E., and L. Trueb, 1986. *Biology of Amphibians.* McGraw-Hill, New York.
Erwin, D. H., 1993. *The Great Paleozoic Crisis: Life and Death in the Permian.* Columbia University Press, New York.
Estes, R., and O. A. Reig, 1973. The early fossil record of frogs: A review of the evidence. In *Evolutionary Biology of the Anurans,* J. L. Vial (ed.). University of Missouri Press, Columbia, pp. 11–63.
Farlow, J. O., 1990. Dinosaur energetics and thermal biology. In *The Dinosauria,* D. B. Weishampel, P. Dodson, and H. Osmólska (eds.). University of California Press at Berkeley, pp. 43–55.
Feduccia, A., 1980. *The Age of Birds.* Harvard University Press, Cambridge, MA.
———, 1985. On why dinosaurs lacked feathers. In *The Beginnings of Birds,* M. K. Hecht, J. H. Ostrom, G. Viohl, and P. Wellnhofer (eds.). Freunde des Jura-Museums Eichstätt, Willibaldsburg, Eichstätt, Germany, pp. 75–79.
———, 1995. Explosive evolution in Tertiary birds and mammals. *Science,* **267,** 637–638.
———, 1996. *The Origin and Evolution of Birds.* Yale University Press, New Haven, CT.
———, 1999. 1,2,3 = 2,3,4: Accommodating the cladogram. *Proc. Nat. Acad. Sci.,* **96,** 4740–4742.
Feduccia, A., and H. B. Tordoff, 1979. Feathers of *Archaeopteryx*: Asymmetric vanes indicate aerodynamic function. *Science,* **203,** 1021–1022.
Fox, W. T., 1987. Harmonic analysis of periodic extinctions. *Paleobiology,* **13,** 257–271.
Frazzetta, T. H., 1969. Adaptive problems and possibilities in the temporal fenestration of tetrapod skulls. *J. Morphol.,* **125,** 145–158.
Hallam, A., 1987. End-Cretaceous mass extinction event: Argument for terrestrial causation. *Science,* **238,** 1237–1242.
Heilmann, G., 1927. *The Origin of Birds.* Appleton, New York. (Reprinted 1972, Dover Books, New York.)
Jablonski, D., 1986. Background and mass extinctions: The alternation of macroevolutionary regimes. *Science,* **231,** 129–133.
Jarvik, E., 1980. *Basic Structure and Evolution of Vertebrates,* vols. 1 and 2. Academic Press, New York.
Jenkins, F. A. Jr., and D. M. Walsh, 1993. An Early Jurassic caecilian with limbs. *Nature,* **365,** 246–250.
Kerr, R. A., 1987. Asteroid impact gets more support. *Science,* **236,** 666–668.
Langeland, J. A., and C. B. Kimmel, 1997. Fishes. In *Embryology: Constructing the Organism,* S. F. Gilbert and A. F. Raunio (eds.). Sinauer Associates, Sunderland, MA, pp. 383–407.
Little, C., 1990. *The Terrestrial Invasion: An Ecophysiological Approach to the Origins of Land Animals.* Cambridge University Press, Cambridge, England.
McFarland, W. N., F. H. Pough, T. J. Cade, and J. B. Heiser, 1985. *Vertebrate Life,* 2d ed. Macmillan, New York.
McGhee, G. R. Jr., 1990. Catastrophes in the history of life. In *Evolution and the Fossil Record,* K. C. Allen and D. E. G. Briggs (eds.). Smithsonian Institution Press, Washington, DC, pp. 26–50.
Olson, E. C., 1971. *Vertebrate Paleozoology.* Wiley-Interscience, New York.
Ostrom, J. H., 1974. *Archaeopteryx* and the origin of flight. *Q. Rev. Biol.,* **49,** 27–47.
———, 1991. The question of the origin of birds. In *Origins of the Higher Groups of Tetrapods: Controversy and Consensus.* H.-P. Schultze and L. Trueb (eds.). Cornell University Press, Ithaca, New York, pp. 467–484.
Padian, K., 1985. The origins and aerodynamics of flight in extinct vertebrates. *Paleontology,* **28,** 413–433.
Padian, K., and L. M. Chiappe, 1998. The origin of birds and their flight. *Sci. Amer.,* **278,** (2), 38–47.
Panchen, A. L., 1977. The origin and early evolution of tetrapod vertebrae. In *Problems in Vertebrate Evolution* (Linnaean Society Symposia Series, vol. 4), S. M. Andrews et al. (eds.). Academic Press, London, pp. 289–318.
——— (ed.), 1980. *The Terrestrial Environment and the Origin of Land Vertebrates.* (Systematics Association Special Volume No. 15.) Academic Press, New York.
Pimm, S. L., G. J. Russell, J. L. Gittleman, and T. M. Brooks, 1995. The future of biodiversity. *Science,* **269,** 347–350.
Randall, D. J., W. W. Burggren, A. P. Farrell, and M. S. Haswell, 1981. *The Evolution of Air Breathing in Vertebrates.* Cambridge University Press, Cambridge, England.
Raup, D. M., 1991. *Extinction: Bad Genes or Bad Luck?* Norton, New York.
———, 1994. The role of extinction in evolution. *Proc. Nat. Acad. Sci.,* **91,** 6758–6763.
Raup, D. M., and J. J. Sepkoski, Jr., 1986. Periodic extinction of families and genera. *Science,* **231,** 833–835.
Reisz, R. R., 1977. *Petrolacosaurus,* the oldest known diapsid reptile. *Science,* **196,** 1091–1093.
Rigby, J. K. Jr., K. R. Newman, J. Smit, S. Van der Kars, R. E. Sloan, and J. K. Rigby, 1987. Dinosaurs from the Paleocene part of the Hell Creek Formation, McCone County, Montana. *Palaios,* 2, 296–302.

Romer, A. S., 1966. *Vertebrate Paleontology,* 3d ed. University of Chicago Press, Chicago.
———, 1967. Major steps in vertebrate evolution. *Science,* **158,** 1629–1637.
———, 1968. *The Procession of Life.* World Publishing, Cleveland.
Romer, A. S., and T. S. Parsons, 1977. *The Vertebrate Body,* 5th ed. Saunders, Philadelphia.
Ruben, J. A., T. D. Jones, and N. R. Geist, 1998. Respiratory physiology of the dinosaurs. *BioEssays,* **20,** 852–859.
Russell, D. A., 1979. The enigma of the extinction of the dinosaurs. *Ann. Rev. Earth Planet Sci.,* **7,** 163–182.
Schindewolf, O. H., 1993. *Basic Questions in Paleontology: Geologic Time, Organic Evolution, and Biological Systematics.* (English translation of the 1950 German edition.) University of Chicago Press, Chicago.
Schmalhausen, I. I., 1968. *The Origin of Terrestrial Vertebrates.* Academic Press, New York.
Sharpton, V. L., et al., 1993. Chicxulub multiring impact basin: Size and other characteristics derived from gravity analysis. *Science,* **261,** 1564–1567.
Sheldon, F. H., and A. H. Bledsoe, 1993. Avian molecular systematics. *Ann. Rev. Ecol. Syst.,* **24,** 243–278.
Sibley, C. G., and J. E. Ahlquist, 1990. *Phylogeny and Classification of Birds.* Yale University Press, New Haven.
Sloan, R. E., J. K. Rigby, Jr., L. M. Van Valen, and D. Gabriel, 1986. Gradual dinosaur extinction and simultaneous ungulate radiation in the Hell Creek Formation. *Science,* **232,** 629–633.
Spotila, J. R., 1980. Constraints of body size and environment on the temperature regulation of dinosaurs. In *A Cold Look at the Warm-Blooded Dinosaurs,* R. D. K. Thomas and E. C. Olson (eds.). AAAS Selected Symposium **28.** Westview Press, Boulder, CO, pp. 233–252.
Stahl, B. J., 1974. *Vertebrate History: Problems in Evolution.* McGraw-Hill, New York.
Stevens, K. A., and J. M. Parrish, 1999. Neck posture and feeding habits of two Jurassic sauropod dinosaurs. *Science,* **284,** 798–800.
Sumida, S. S., and K. L. M. Martin (eds.), 1997. *Amniote Origins: Completing the Transition to Land.* Academic Press, San Diego, CA.
Thomas, R. D. K., and E. C. Olson (eds.), 1980. *A Cold Look at the Warm-Blooded Dinosaurs.* AAAS Selected Symposium **28.** Westview Press, Boulder, CO.
Unwin, D. M., and N. N. Bakhurina, 1994. *Sordes pilosus* and the nature of the pterosaur flight apparatus. *Nature,* **371,** 62–64.
Van Valen, L., 1988. Paleocene dinosaurs or Cretaceous ungulates in South America. *Evol. Monogr.,* **10,** 1–79.
Weishampel, D. B., P. Dodson, and H. Osmólska (eds.), 1990. *The Dinosauria.* University of California Press, Berkeley.
Wellnhofer, P., 1988. Terrestrial locomotion in pterosaurs. *Hist. Biol.,* **1,** 3–16.
Wilson, E. O., 1992. *The Diversity of Life.* Harvard University Press, Cambridge, MA.
Zardoya, R., and A. Meyer, 1998. Complete mitochondrial genome suggests diapsid affinities of turtles. *Proc. Nat. Acad. Sci.,* **95,** 14226–14231.

19 Evolution of Mammals

Mammals derive their name from the maternal **mammary glands** used to suckle their young after birth. Mammalian uniqueness, however, extends to many other anatomical, physiological, and behavioral traits that evolved throughout much of the Mesozoic era.

In addition to mammaries, the soft-body features of mammals are:

- **Live birth** (except for monotremes)
- Body temperature control (**endothermy**) augmented with adaptations such as hairy coverings to control heat loss and sweat glands to enhance evaporation and cooling
- A **diaphragm** to increase the inspiration of oxygen and expiration of carbon dioxide, both necessary for high metabolic activity
- A **four-chambered heart** that completely separates oxygenated arterial blood from venous blood; and
- Greater intelligence, derived from expansion of the **neocortex** of the brain

Among the hard-body, or skeletal, differences that set living mammals apart from other vertebrates are:

- A **double occipital condyle** at the rear of the skull that articulates with the first cervical vertebra
- A **mandible** consisting of a single bone (the dentary) with a condyle that articulates with the squamosal bone of the skull
- The transformation of the reptilian quadrate and articular bones, formerly used for jaw articulation, into the incus and malleus **ear ossicles** used for sound transmission
- A **bony secondary palate** separating the nasal passages from the mouth
- A **single nasal opening** in the skull
- A relatively **large braincase**

FIGURE 19–1
Reconstruction of *Cynognathus,* a carnivorous cynodont therapsid of the early Triassic, about 4 feet long. (*Adapted from Colbert and Morales.*)

- Greater differentiation among teeth (**heterodont** dentition), characterized largely by **multirooted cheek teeth** (molars and premolars) with **multicusped crowns**

Since we find mammalian soft-body features as yet impossible to trace, our hypotheses on the origin of mammals from egg-laying amniotes mainly derive from studies of fossilized skeletal materials. The most prominent candidates for mammalian ancestors lie among extinct groups of **synapsids** that initially appeared in the Carboniferous period.[1] Although at first these were ungainly looking pelycosaurs (Fig. 18–17), by the Permian they had evolved into a variety of therapsid forms adapted primarily to a terrestrial existence (Fig. 18–18). A number of paleontologists consider it likely that by the Early Triassic some advanced therapsids had become the first of the vertebrate endotherms and may have had other mammalian soft-body features.

Skeletally, the therapsids were distinct from other reptiles and certainly evolving in a mammalian direction. For example, the sprawled reptilian stance had changed in some of the doglike **cynodont therapsids** to a more vertical placement of limbs, with the knees pointing forward and elbows backward (Fig. 19–1). This elevated the body from the ground and enhanced mobility by enabling direct fore and aft leg motion. Although the jaw articulation was still reptilian, the later therapsid mandible was almost entirely composed of the **dentary bone,** as in the therapsid-derived mammals.

[1]Paleontologists disagree on how to classify synapsids. Some workers find this problem important because if we include synapsids (or therapsids, for that matter) among reptiles and exclude their descendants (mammals), the class Reptilia becomes a paraphyletic taxon that does not include all descendants of the original reptilian ancestor (see also p. 421). These researchers find it more acceptable to establish a single taxon that includes all synapsid lineages and their descendants, extending this group to therapsids and mammals. According to Kemp (1988), this taxon, called Synapsida, can be distinguished from the taxon, called Sauropsida, that includes all reptilian relatives. Many workers, however, find separating synapsids and therapsids from reptiles quite arbitrary, and most classifications continue to place them in the class Reptilia. To indicate their reptilian but transitional status, therapsids are generally called "mammal-like reptiles."

Teeth

Among other highly significant mammal-like features in therapsids was the development of differently specialized (heterodont) teeth—incisors, canines, premolars, and molars. The animal apparently used the various cusplike surfaces on the crowns of the molar and premolar "cheek" teeth to cut and break food into small particles rather than to gulp large chunks or swallow whole prey in reptilian fashion. Since continued breathing is essential for the metabolic needs of mammals, the consequences of retaining food orally while chewing led to a variety of innovative changes. In most amphibians and terrestrial reptiles the nasal openings are in the anterior portion of the mouth, and the animal can temporarily interrupt breathing without ill effect while the mouth is full of food. Mammals, in contrast, depending more on constant aerobic respiration, would asphyxiate if the food bolus blocked inspired air long enough to chew a mouthful. Selection for an extended secondary palate therefore occurred in mammal-like lines, allowing air to be carried to and from a point beyond the mouth near the trachea (Fig. 19–2).

Tooth replacement was another area in which considerable evolution had to occur before the mammalian grade was reached. In general, the addition and replacement of teeth is closely associated with relationships between the size of the teeth and the size of the skull during growth. There is considerable value for a growing animal whose head and mouth are enlarging to keep pace with the increased size of its food by increasing the number of large teeth.

In newborn reptiles, teeth along the jaw margins are small in accord with the size of the animal, and these are shed and replaced by larger teeth in an alternating pattern. That is, as each tooth matures and becomes more firmly fixed to the jaw, adjacent older teeth weaken and are replaced. Since a fully mature reptile may have a skull ten times longer than when it begins tooth replacement, this alternating cycle ensures that firm teeth of appropriate size are always present.

In mammals such continuous replacement of alternate teeth would interfere with the precise fit between the upper and lower teeth necessary for effective chewing.

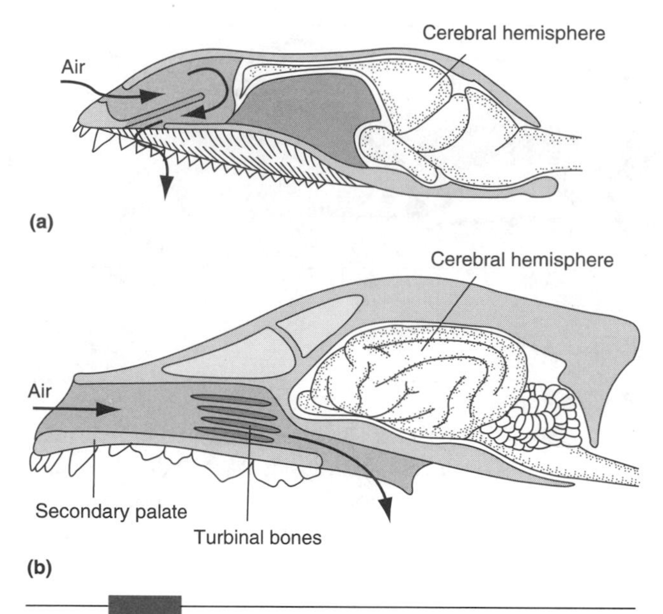

FIGURE 19-2 Air pathways in reptiles (*a*) and mammals (*b*), showing the long mammalian secondary palate that separates air entering the pharynx from food being retained in the mouth. Mucous membranes that cover the mammalian turbinal bones warm and moisten the entering air. According to Hillenius, two therapsid suborders from the Late Permian, including the cynodont mammalian ancestors, had nasal turbinate bony ridges to which respiratory membranes attached. He points out that such features, used to reduce "desiccation associated with rapid and continuous pulmonary ventilation," are unique to endothermic mammals and birds, and are absent in reptiles and other ectotherms. (*Adapted from Romer 1968.*)

Therefore, only one set of teeth is replaced in mammals, the **deciduous** or milk teeth, which include incisors and canines as well as the postcanine deciduous molars (cheek teeth). In immature mammals, the deciduous molars perform the chewing function, and permanent premolars replace them when the more posterior adult molars have emerged.

The need for only a single replacement of mammalian teeth probably derives from the relatively large size of mammalian newborns, who can survive without teeth by suckling at maternal mammaries until their heads are even larger than at birth. By the time the deciduous teeth have erupted and the weaned young mammal is subsisting on adult food, their skulls have reached about 80 percent of postnatal growth. Thus they need only relatively little more skull growth before permanent teeth replace their deciduous set.

Jaws and Hearing

The precise fit between upper and lower mammalian cheek teeth results from selection for improved chewing activity and correlates with changes in both jaw muscles and tooth shape. In addition to the relatively limited grasping and puncturing functions of the reptilian jaw, mammalian evolution has emphasized shearing, grinding, and crushing activity in both premolar and molar regions.

There are now muscles (masseter and temporalis) that let the animal bite down with considerable force in these regions, muscles (internal pterygoid) that move the jaw from side to side during chewing, and muscles (buccinator and tongue) that move the food within the mouth. Vertebrate zoologists believe that the rearrangement and change in mammalian jaw muscles, along with lengthening of the coronoid and angular processes of the dentary, reduced strain at the jaw articulation and correlated with the transition of former posterior bony elements of the jaw, the **articular** and **quadrate,** to assume auditory functions within the middle ear.

A major stimulus for these changes—improved hearing to capture prey and escape predators not in the direct line of vision—must have been one of the primary selective forces acting on early land vertebrates. This was especially true for early mammals, who survived the Mesozoic "reptilian tyranny" by entering a **nocturnal** environment that emphasized auditory and olfactory perception. A possible evolutionary sequence for these events, diagrammed in Figure 19–3, unfolds as follows.

The **tympanic membrane,** which functions as a taut, drumlike receptor for airborne sound, may well have been lacking in early land vertebrates and even perhaps in early synapsid reptiles, who relied mainly on ground-transmitted vibrations (Fig. 19–3*a*). In these therapsid ancestors, sound was mostly transmitted from ground to inner ear through bone, via the relatively thick stapes that maintained contact with both the quadrate in the skull and the articular in the lower jaw (Fig. 19–3*b*).

Kermack and Mussett suggest that the evolution of a tympanum in mammal-like reptiles would have allowed hearing for airborne sounds but would nevertheless have been inefficient in detecting a wide range of frequencies because of the relatively large mass and immobility of the bones between the tympanic membrane and inner ear (Fig. 19–3*c*). Since the difficulty in reducing the size of the articular and quadrate bones in these lineages derives from their use as the jaw hinge, one solution was to use other bones for this purpose and let the articular and quadrate diminish in size and confine themselves to sound conduction.[2]

This process is already apparent in some Triassic therapsids such as *Diarthrognatus,* who show a new mammal-like jaw articulation involving the dentary bone in the lower jaw and the **squamosal** bone in the upper jaw, in ad-

[2]Lineages leading to modern reptiles (sauropsids) apparently did not encounter this difficulty, since the tympanic membrane was more posterior, allowing the stapes to serve as a single bone connecting the tympanum and inner ear without the intervention of the quadrate and articular (which remained as the reptilian jaw joint).

FIGURE 19-3 Proposed stages in the evolution of the ear apparatus, beginning with a land tetrapod that picks up ground vibrations through bone conduction (*a, b*). In reptilian synapsid lineages that lead to the mammal-like therapsids (*c*), a tympanic membrane picks up airborne sound and transmits it to the articular and quadrate bones of the jaw hinge and into the stapes that connects to the inner ear. As therapsid evolution proceeds, a new mammalian jaw joint evolves (squamosal-dentary) because of selection for improved molar chewing abilities. In *Morganucodon,* an early Triassic mammal (*d*), both jaw joints are present, although the size of the articular–quadrate– stapes bones have diminished. In late Triassic mammals (*e*), the squamosal-dentary joint has become the only jaw hinge, and the articular–quadrate–stapes bones are now entirely involved in hearing. The diagram in (*f*) presents a more anatomical view of the shape and positioning of these bones in the ear of a modern mammal. (*Adapted from Kermack and Mussett.*)

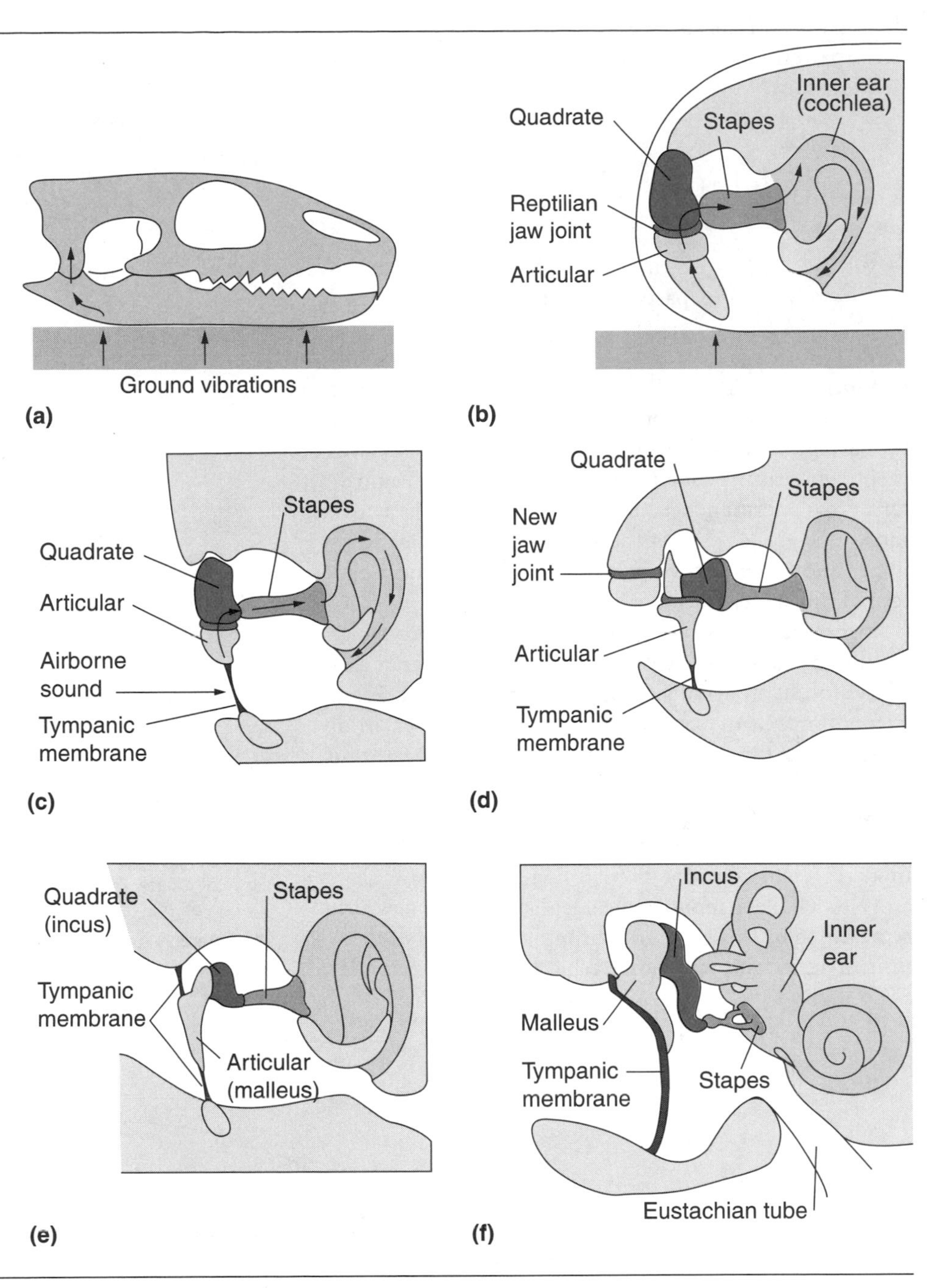

dition to the old reptilian articular–quadrate joint. In early mammals such as *Morganucodon,* the articular and quadrate had reduced even further, although they still formed part of the jaw hinge (Fig. 19–3*d*). Apparently only in the Jurassic were the quadrate and articular freed from their function in jaw articulation and incorporated into the mammalian middle ear as small ossicles, the **incus** (anvil) and **malleus** (hammer), respectively (Fig. 19–3*e, f*). Supporting this view are embryological studies showing that the quadrate and articular in the mammalian fetus first occupy a reptilian position on the side of the jaw and later transform into the mammalian ear ossicles.

Very early mammalian fossils are scarce, so we still don't know when and where therapsids evolved into mammals. Early mammals seem to have been small, often about the size of mice or rats, and their bodies usually rapidly disarticulated even in well-fossilized areas. With some exceptions, the complete mammalian skeletons that have been discovered date no earlier than the Late Cretaceous. As a result, researchers have derived possible evolutionary lineages among earlier Mesozoic mammals almost entirely from fossil teeth and jaws. Hard, enameled dentition is the most easily preserved part of the vertebrate body, and the cusps, ridges, and depressions on the surfaces of teeth follow heritable genetic patterns that can point the way to phylogenetic relationships.

Early Mammals

We find the earliest mammalian departure from the therapsid line in a geographically widespread group of Late Triassic–Early Jurassic fossils called **morganucodontids** (Fig. 19–4). For the first time the postcanine teeth differentiate into premolars and molars, with only the last premolars showing evidence of tooth replacement, whereas the molars followed the mammalian pattern of permanence. Among other distinctive morganucodont traits is the precise occlusion between upper and lower jaws, which produced a consistent pattern of molar wear facets (Fig. 19–5).

In general, the structure of a morganucodont molar was three cusps aligned along the anterior–posterior axis of the tooth, an arrangement called **triconodont.** Fossil triconodonts with patterns similar to or derived from these appear throughout the remainder of the Mesozoic era, and are among the variety of groups traditionally classified in the subclass **Prototheria.**

As mentioned previously (p. 102 and Fig. 6–16), the modern remaining prototherian lines are the Australian and New Guinean **monotremes,** egg-laying mammals now represented by the grub- and shrimp-eating platypus (*Ornithorhyncus*) and ant-eating echidna (*Tachyglossus* and

FIGURE 19–4 Proposed skeletal and full-body reconstructions of a Late Triassic–Early Jurassic mammal, the morganucodontid *Megazostrodon.* It was about 4 inches (10 centimeters) long and weighed approximately 1 ounce. (*Adapted from Crompton et al.*)

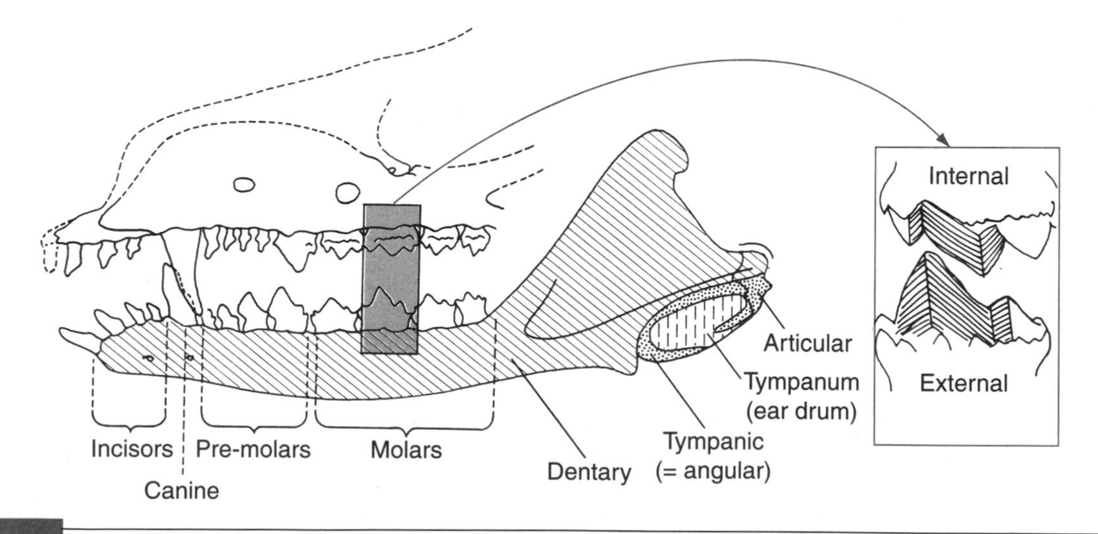

FIGURE 19–5 Lateral view of the inch-long jaws of a late Triassic morganucodontid, *Morganucodon.* The molar teeth occlude more precisely than in any reptile, and, as can be seen in the inset, show matching wear facets between the internal surface of the upper molars and the external surface of the lower molars. (To illustrate the internal surface of the upper molar, the tooth is drawn as though it were transparent.) (*Adapted from Crompton and Jenkins.*)

Zaglossus).[3] Although these animals are now quite rare and their fossil record is meager, some of their features may reflect those of their more numerous Mesozoic prototherian ancestors.

For example, the monotremes still use the reptilian cloaca (in addition to its egg laying function) as a common chamber for both the rectal and urogenital openings and show a number of reptilian skull characters. Also, in contrast to the more advanced marsupials and placentals, monotreme pectoral girdles lack a scapular spine and contain an interclavicle, much like the Triassic–Jurassic mammalian docodonts, another primitive mammalian group. The absence of teeth in monotremes is a rare specialization that helps make their phylogeny difficult to determine but that can be explained by an evolutionary history confined either to mud burrowing or ant eating. It is probably the persistence of such unique specializations in the relatively isolated Australian continent that enables these Mesozoic relics to continue surviving for such a long period.

Discoveries of a variety of fossil teeth indicate that by the end of the Triassic period an important mammalian division occurred, separating the morganucodontids and their subsequent prototherian lineages from a new group called the **therians,** marked by a more sophisticated molar structure. Therian molars, called **tribosphenic** or trituber-cular, differ from prototherian types in having a triangular arrangement of cusps in the upper molars, one of which (protocone) fitted closely into a lower molar basin (talonid), much like a pestle into a mortar (Fig. 19–6*a*, *b*).

The crushing action of cusp-to-basin was supplemented by shearing and cutting surfaces that progressed in number from three to six as therian lineages evolved (Fig. 19–6*c*–*g*). These activities were further enhanced by evolution of a narrower lower jaw suspended in a sling of muscle that enabled side-to-side grinding action, a feature whose beginnings already appear in earlier mammals. These improved oral-pulverizing mechanisms probably accompanied new dietary opportunities as well as increased digestive efficiency.

A possible forerunner of the tribosphenic molar is found among contemporaries of the morganucodonts, called *kuehneotheriids* (Fig. 19–6*c*), and some paleontologists have used this evidence to suggest a diphyletic or polyphyletic origin of mammals from different therapsid stocks. That is, more than one line of mammal-like therapsids, distinguished from other groups by cusped molar teeth, presumably gave rise separately to morganucodonts, kuehnotheriids, and perhaps also to some other early mammalian lineages (Kermack and Kermack, Maio).

[3]Using an analysis of an early Cretaceous jaw fragment identified as monotreme, Kielan-Jaworowska and coworkers suggest that rather than being prototherians, monotremes originated from an ancestral therian mammal (see text later). However, other monotreme skeletal features are so primitive that some workers do not readily accept this view (Carroll).

Important braincase similarities between therian and atherian mammals, however, have prompted other paleontologists to adopt a monophyletic position (Kemp 1988). A firm decision between these views will obviously depend on overcoming the skimpiness of the fossil record and tracing the complexity of early mammalian evolution among the various groups that fall under the broad category of cynodont therapsids.

Early Mammalian Habitats

Even with many gaps in the fossil evidence, we can reconstruct some aspects of early mammalian lives and habitats. Certainly, improved dentition and mastication in the small Mesozoic mammals would have helped maintain constant body temperatures by promoting rapid food absorption. Some researchers have suggested that these primitive mammals, because of their endothermy, were primarily nocturnal and insectivorous, functioning in the cool of the evening when their ectothermal reptilian predators were inactive. In support of this nocturnal role, Jerison and others point out that early mammalian brains were three or four times larger than those of even advanced therapsids, and that we can attribute a significant portion of this increase to selection for additional neural connections that provided enhanced auditory (and perhaps also olfactory and visual) acuity associated with adaptation to a nocturnal habitat.

The increased specialization of the mammalian auditory apparatus, partly accomplished by freeing the articular and quadrate bones from the jaw and transforming them into more effective middle-ear ossicles, is perhaps further evidence of selection for improved sensory ability in a nocturnal, light-diminished habitat. A nocturnal mammalian ancestry, according to some researchers, is also supported by the finding that the retinas of many present-day primitive mammals, such as insectivores, are extremely rich in rod photoreceptors sensitive to dim light, in contrast to the daylight-adapted retinae of reptiles such as lizards (almost entirely composed of cone photoreceptors).

Dental changes in early mammalians, which a rodent-like, nocturnal way of life probably prompted, seem to have continued throughout the Jurassic; Figure 19–7 diagrams such changes in fossil molars. In the Early Cretaceous, the first truly tribosphenic molar appears (*Aegialodon*), and toward the end of that period the two major modern therian groups emerge in relative paleontological abundance: **marsupials** in North America and **placentals** in both North America and Asia.[4]

[4]In contrast to this paleontological grouping, Janke and coworkers, using both nucleotide and amino acid sequencing, argue against combining

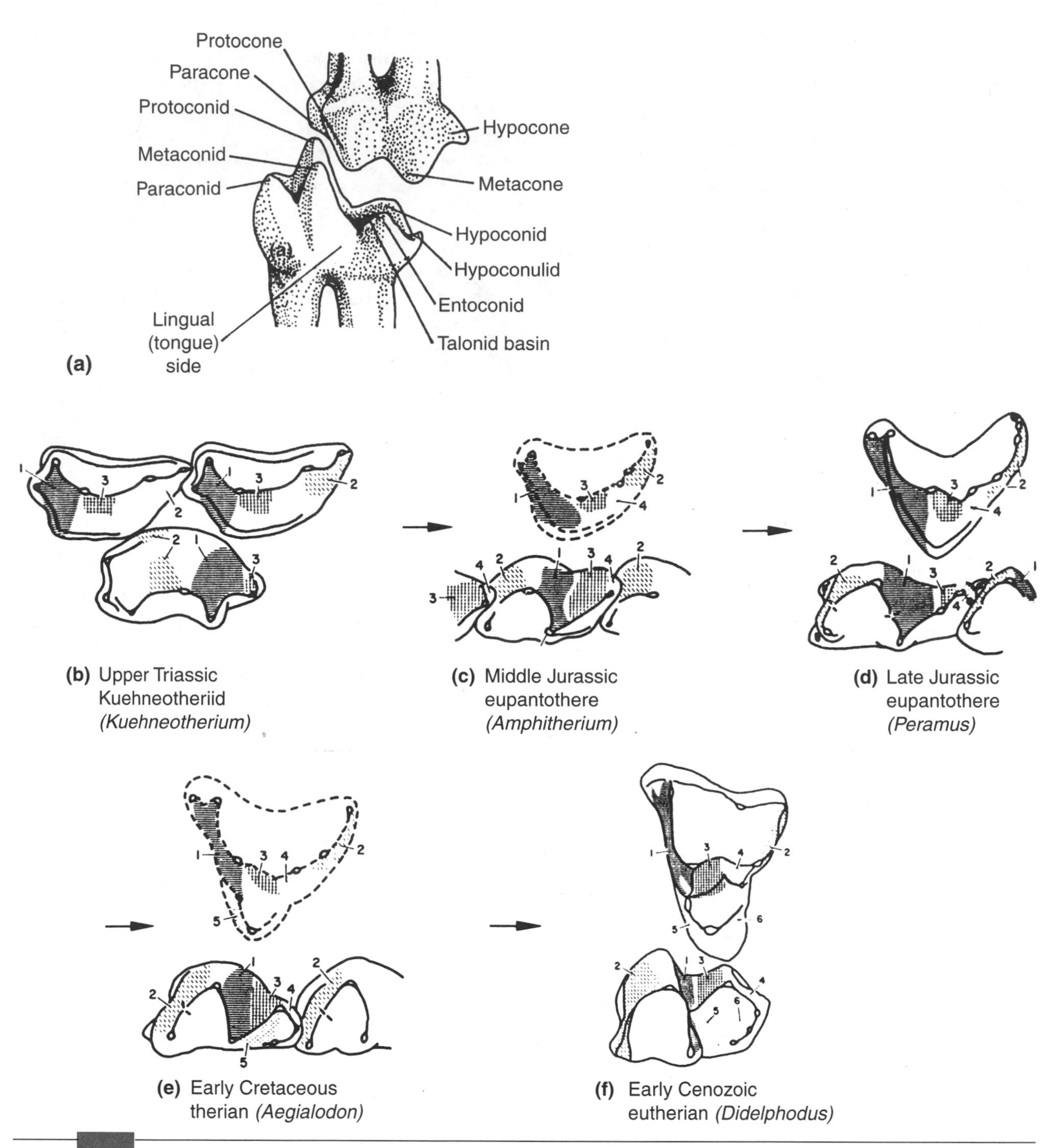

FIGURE 19-6 (*a*) Generalized upper and lower tribosphenic molars, based on those of a modern therian, the opossum *Didelphis* (oriented with the anterior of the animal to the left). (*b–f*) Crown (occlusal) views of upper molars (*above*) and lower molars (*below*) with matching wear facets shaded alike, representing general stages in the evolution of the therian tribosphenic molar (only lower molars are actually known for fossils *c* and *e*). (*Adapted from McFarland et al., and from Bown and Kraus.*)

marsupials and placentals into a therian subclass. They suggest instead a monotreme/marsupial sister group that separated from placentals in the Early to Mid Cretaceous, about 130 million years ago. According to their "molecular clock," only 15 million years later did the monotreme–marsupial divergence take place. However, other molecular time scales place the initial marsupial divergence even further into the Mesozoic era, about 170 million years ago (Kumar and Hedges). Resolution of these differences depends on further studies.

Marsupials and Placentals

Fossils of these two major therian groups differ in both skull and tooth structure, with marsupials showing the following:

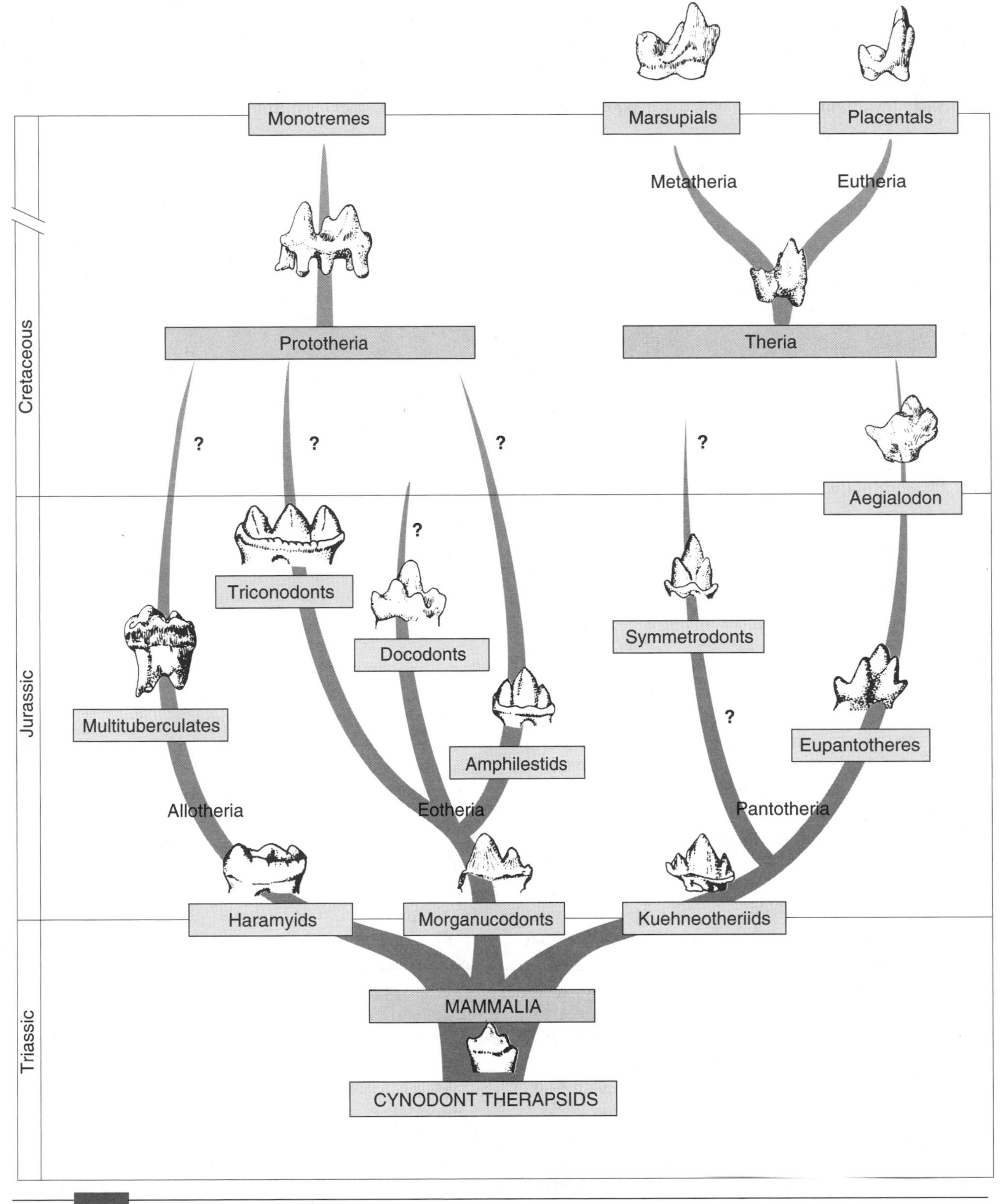

FIGURE 19–7 A phylogeny of mammals from the Triassic to the Cenozoic, shown in terms of changes in molar teeth. A complete triconodont skeleton found in China by Ji and coworkers indicates that triconodont tooth structure (and features such as the pectoral girdle) were not unique events but the result of convergence in different mammalian lines. (*The Prototheria–Theria division is based on Carroll.*)

- A relatively small braincase
- Unique bony composition of the auditory bulb
- Deciduous teeth reduced to the posterior premolars, which, in turn, often markedly differ from the anterior molars
- A relatively large number of incisors (eight or more in the complete upper jaw)
- Distinctive arrangements of molar cusps and ridges
- Some postcranial Cretaceous skeletons also show the presence of pelvic epipubic bones associated with supporting a marsupial-type pouch.

Since a major difference between marsupials and placentals lies in reproductive modes, the absence of soft tissue fossilization makes it difficult to recapitulate exactly how they evolved. Among various prevailing hypotheses, we can offer the following progression (see Lillegraven). Very early mammals, distinguished by small size, endothermy, and heterodont dentition, most likely laid eggs, which were probably small. Under such circumstances, a significant selective advantage would have accrued to animals that could raise their young past the immature stages of hatching. **Lactation** offered by maternal mammary glands was apparently one successful solution to the problem, and present-day monotremes are presumably a relic of this stage of evolution.

Given a system that provided maternal care, protection, and nourishment, selection could then have continued in the direction of smaller eggs and more rapid development of the fetus before hatching. Endothermy would have facilitated such evolution, since the animal could keep hatched offspring close to the maternal body at optimum enzymatic temperatures. At some point, **viviparous** reproduction could replace **oviparity,** because it would probably take only a few additional mutational steps for hatching to occur within the maternal oviduct itself. The embryo could then be nourished on maternal fluids within a portion of the oviduct that would eventually become the **uterus.** Although we have no way of proving this hypothesis yet, researchers believe that mammalian viviparity was probably restricted to therian lines, that is, to marsupials and placentals.

In marsupials, a thin, permeable eggshell surrounds the embryo for the major part of the pregnancy period, which ranges between 11 and 38 days for different species. During this time, the marsupial embryo receives nourishment from both egg nutrients and maternal uterus, but because of the short pregnancy, emerges from the vagina in highly immature form.[5] It may take two to three months after birth before marsupial offspring are capable of terrestrial locomotion.

In placentals, a shelled embryonic stage is no longer discernible, pregnancy extends considerably, and emerging offspring are larger than those of marsupials and also far more advanced—often capable of independent locomotion shortly after birth. The difference in marsupial–placental gestational periods may derive, according to some authors, from differences in maternal immunity response to the fetus: the marsupial fetus is not protected against the maternal immune system, and must therefore abandon the uterus soon after egg hatching, before maternal leukocyte invasion can damage it. The trophoblastic membranes surrounding the placental fetus, in contrast, ordinarily prevent exchange between maternal and fetal tissues and thus act as barriers that help keep the fetus from being immunologically rejected.

Once immunologically protected, uterine retention of the fetus could be prolonged by incorporating both maternal and fetal membranes into the eutherian placenta. Such a placenta, sustained by various endocrine secretions, nourishes the fetus, provides the oxygen requirements for rapid developmental growth, and acts as a waste removal system. Compared to any other mode of reproduction, uterine development probably confers greater protection to the embryonic organism during its most vulnerable stages. Also, because mother–child attachments continue past birth through mammary feeding, the stage is set for prolonged family relationships that emphasize learning and intelligence.

It is nevertheless significant that however profound placental advantages may have been, they did not eliminate marsupials: marsupials have at least as long a history as placentals, are still prevalent in Australia, include forms such as the opossum that compete successfully with placentals in various placental-dominated localities, and number about 270 living species. One basic reason for the persistence of marsupials derives from their relatively minor reproductive investment: their birth size is so small that they can abandon offspring soon after birth without great maternal loss. Placentals, in contrast, commit much greater resources to early reproductive stages, and a pregnancy often continues in the face of serious maternal sacrifice.

Thus marsupial reproduction can more easily adjust to appropriate environmental conditions; that is, reproduction and nursing continue when conditions are advantageous, and marsupials incur little expense in discarding their minuscule newborn offspring when conditions turn

[5]This immaturity significantly restricts the directions that marsupial evolution can take. For example, the absence of marsupial forms with front flippers, as placental seals have, and the absence of marsupial hoofed forms stem from the need for marsupial offspring to have forelegs with claws in order to crawl from the vagina to the maternal mammary teat. A fully aquatic life, such as found among placental cetaceans, is also closed to marsupials since they could not provide their very immature young with air during that considerable period of weeks or months in which the young remain continually fastened to the teat, nor could tiny marsupial offspring survive the temperature stress caused by complete immersion in water.

poor. Placentals take greater reproductive risks, because their commitment to their offspring is greater (pregnancy being often difficult to interrupt) and involves considerable cost.

The Mesozoic Experience

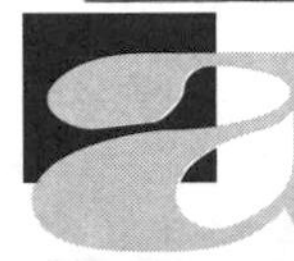s from all evolutionary history, we can learn a number of lessons from the Mesozoic experience. First, dominance of a particular group at a particular time is not necessarily a measure of its long-term evolutionary success. This lesson has repeatedly proven true: various therapsid groups replaced each other, dinosaur groups replaced therapsids, and later dinosaur groups replaced earlier ones. At the end of the Cretaceous period, even dinosaurs, which had dominated in various forms for more than 100 million years and were the largest land vertebrates that ever existed, ceded dominion to a different vertebrate class, mammals.

The transition from one group to the other offers a second lesson, the importance of preadaptation. Not all reptilian lines evolved into dinosaurs or into therapsids, nor did all therapsid lines evolve into mammals. Rather, the reptilian groups that made these significant evolutionary advances had incorporated important preadaptive characters by the fortunes of their own evolutionary histories. For example, dinosaur bipedalism and mammalian endothermy trace, respectively, to the beginning of a bipedal stance in some early archosaurs and to the beginning of endothermy among some therapsids. Although most if not all characters are, or have been, adaptive, the inability of organisms to anticipate future evolutionary needs often makes it a matter of rare chance as to which of these characters are preadaptive. The limitation in what will become preadaptive for a particular future environment may help explain why polyphyletic evolution is not common.

The third lesson offered is the unpredictability of long-term evolutionary succession. All the many kinds of biological and environmental changes involved in evolution, although individually understandable in terms of cause and effect, are seen to act largely at random when we view them together over long (or even relatively short) periods of evolutionary time. Evolution is tied to historical contingencies (uncertainties). For example, during the Triassic period could one have predicted which lines of primitive mammals would provide descendants that would survive into the Cenozoic 150 million years distant, or even last for another 70 million years through the Jurassic?

Thus, out of the hazards that await any particular lineage, a fourth lesson emerges: new modes of biological organization can enhance the opportunity for survival. True, not all Triassic lines of mammals survived into the Cenozoic, but some did, and many carried with them improvements in temperature regulation, reproductive mode, nursing care, sensory perception, brain development, blood circulation, oxygen use, locomotion, dentition, and so forth. It was undoubtedly because of many or all of these biological innovations, along with their small size, that some mammals were among the groups that withstood the Mesozoic "reptilian tyranny" and made safe passage through the Late Cretaceous extinctions that destroyed the dinosaurs.[6]

A fifth lesson derives from the complexity of major biological adaptations: the evolution of a new level of organization, like that of mammals, is marked by coordinated changes in many different traits often occurring over a considerable period of time. For example, although we can characterize a mammal by one or another of its unique traits, the trait itself, such as dentition or reproductive mode, actually results from a number of successive mutations, each of which must coordinate with its entire genetic architecture. That is, most if not all characters cannot evolve independently to maximize only a single adaptive function, but must coevolve with other characters so that the many possible developmental interactions between them do not decrease fitness. "Hopeful monsters" (p. 356) are rare, if at all viable. The transition from reptile to placental mammal may well have taken 75 to 100 million years because a wide range of **coadaptive mutations** had to integrate into evolving organisms.

Also, in support of monophyletic evolution, the complexity of this integrative process makes it unlikely that many different lines would have continued to undergo the same succession of identical genetic changes. If different lines evolve similarly, as observed in therapsid–early mammalian transitions, such apparent polyphyletic events may

[6]Although one can claim that time provides the arrow that orients the direction of evolutionary change and that an evolutionary trend represents "progress," there is more than one direction and many lines of progress. Some may consider the criterion for progress to be increasing morphological complexity exemplified by the increasing number of cell types (Fig. 22–10), but many evolutionary lineages show no such tendency. Parasites which lose organs that consume needless energy can certainly replace those who retain such structures: rather than complexity, simplification and reduction often direct parasitic "progress." Even if we restrict the term *progress* to increased complexity, its measurement is still unresolved (McShea), although most biologists would agree that more complex organisms are alive today than were alive 3.5 billion years ago. If there is a common evolutionary thread that runs through organisms and their many different lifestyles, it is the opportunism that became embedded in the earliest of their ancestors. Like many authors in the past, we can call the results of this opportunism "progress" (see Nitecki), but in view of its semantic ambiguities and contradictions, and socially judgmental overtones, it is questionable whether this term or other value judgments help us understand evolution. In Darwin's words, ". . . natural selection, or the survival of the fittest, does not necessarily include progressive development—it only takes advantage of such variations as are beneficial to each creature under its complex relations of life." "Progress" is therefore, at best, not a cause but a description given to an evolutionary outcome. Biological evolution only tracks opportunistic pathways, and is blind to destinations other than survival.

really spring from parallel or convergent evolution; that is, similar characters evolved through *different* genetic events, and therefore their genetic lineages differ. Furthermore, the "sweepstakes" nature of evolution makes unlikely that even such parallel evolution could have continued in a variety of therapsid lines throughout the Mesozoic, with all such different lines attaining all the same placental mammalian features in the Cretaceous.

A sixth lesson is that once a new adaptive innovation appears, or adaptive organization reaches a new grade, opportunities for widespread radiation can follow. Mammalian endothermy must have opened a nocturnal niche into which many lines entered, just as improvements in mammalian dentition opened a dietary niche that few, if any, reptiles had ever fully exploited. The diversity produced by such radiations are not "passive," but arise from selective acts on the mutational differences with which organisms confront environmental differences.

A seventh lesson of the Mesozoic is that the survival or extinction of any group may be closely connected to the survival or extinction of other groups. That is, individuals of a particular group depend on the existence of entire constellations of associated organisms. Replacement of therapsids by dinosaurs as the new dominant land vertebrates was, for example, a mass phenomenon, involving many genera and families. Also, as already mentioned and to be discussed later, a significant cause for mammalian radiation during the Cenozoic was the many new functional roles and habitats that the extinction of the dinosaurs made available to mammals.

Perhaps a final lesson, like many others not confined to the Mesozoic, is the very high prospect of extinction, of the loss or replacement of lineages because of *genetic constraint*— a group's genomes cannot adapt to every environmental change it may possibly encounter. That is, although selection among genotypes can lead to adaptation in one or more directions, it constrains adaptation in others. Even if selection for excellence in climbing is successful, it will impact on traits that enable excellence in swimming. Similarly, selection for excellence in swimming sacrifices excellence in running, and so forth. Organismic adaptation that can successfully face all possible eventualities appears unattainable in a single lineage (pp. 357–358).

Even when adaptations appear to be ingeniously provident, such as those found in some desert animals and plants that can delay development during long unfavorable periods until opportune circumstances arise, extinction still occurs because some stressful physical or biological environmental impacts can only be circumvented by adaptations that would exceed the limit of species tolerance and developmental ability.

To put this last lesson somewhat differently: in the game of life engaged by a species, extinction (death) can win because environmental impacts that threaten organisms can happen more quickly than the adaptational changes necessary for a species to respond successfully. Short-term adaptations do not necessarily confer long-term advantage. Extinction is caused by both "bad luck" and "bad genes" (see also pp. 429–431).

The Cenozoic Era: The Age of Mammals and the Northern Continents

However important the Mesozoic was to early mammalian evolution, the full flowering of mammalian radiation burst forth in the Cenozoic. The extinction of the dinosaurs seemed to have removed many mammalian Mesozoic constraints: they could invade herbivorous and carnivorous niches formerly closed to them and could become active **diurnally** as well as nocturnally. New mammalian lifestyles began to appear during the Paleocene epoch, but observable morphological differences accumulated slowly.

In addition to dinosaur extinction opening new adaptive niches, a significant stimulus for mammalian radiation must have been the breakup of the large Pangaea landmass that began in the Mesozoic and the movements of tectonic plates that continued throughout the Cenozoic. As shown previously, these geological changes not only established new continents with their varying connections and separations (Fig. 6–12) but also dispersed and isolated major mammalian groups (Fig. 6–17). To these land movements with their marked effects on climate, environment, and regionalization, we can also add the uplifts of mountain systems that took place from the Cretaceous onward leading to chains such as the Rockies, Andes, Alps, and Himalayans; the submersions and regressions of shallow seas; and the delineations of new shorelines. Changes in vegetation, especially the emergence of angiosperms, also took place during these periods leading to new landscapes of grasslands, savannas, and forests (Fig. 19–8). These novelties and modifications shaped new and different habitats, affecting mammalian adaptation, variation, and distribution.

By the Middle and Late Paleocene, evidence for radical evolutionary changes appears in a few major centers, especially in North America, but also in Europe and Asia. In quantitative terms mammalian radiation was extraordinary: from about 21 diverse mammalian families found at the end of the Cretaceous, to 37 by the Early Paleocene, and more than quadrupled (86) by the Late Paleocene (Benton, 1997). A greater partitioning of the environment also took place: North American fossil faunas that usually contained about 20–30 mammalian

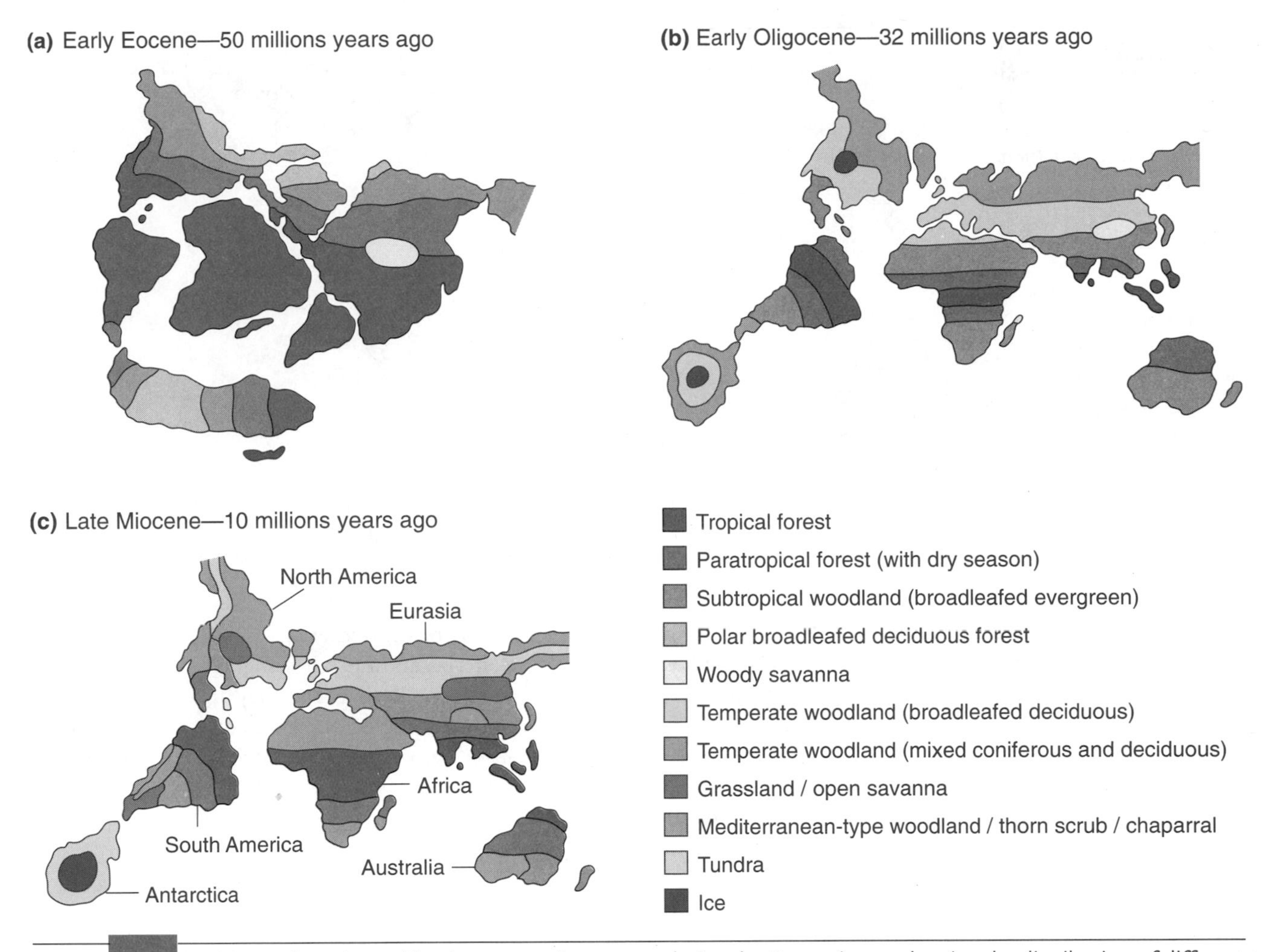

FIGURE 19–8 Positions of continental landmasses at three stages during the Cenozoic era, showing the distribution of different kinds of vegetation. (*Adapted from Janis.*)

species in the Late Cretaceous had 50–60 species by the Middle Paleocene.

If we continue onward to the Middle and Late Eocene, 20 to 30 million years after the Cretaceous, mammalian skeletal adaptations for creeping, running, digging, swimming, flying, and climbing had morphologically differentiated many major fossil groups, from primitive whales to bats (see Fig. 19–11). These and further changes led, for example, to modifications in the lower limbs of some terrestrially mobile animals from a flat-footed stance (**plantigrade**) to running on the digits (**digitigrade**) or on the tips of the toes (**unguligrade**). A reduction in the number of toes and lengthening of the limbs and foot bones (Fig. 19–9) accompanied increased speed in groups such as horses and other hoofed animals. Cerebral brain size, a mark of the ability to integrate sensory and motor information, increased relative to body size in both mammalian prey and predators as part of a continuous "arms race" in which predators are selected for greater skill in capturing prey and prey are selected for greater skill in avoiding predators (Fig. 19–10).[7]

Such evolutionary changes, as well as many others, led to the replacement of most of the primitive Mesozoic and Early Cenozoic mammalian forms, a process that continued throughout the Tertiary period. By the end of the Pliocene, about 2 million years ago, many mammalian groups had evolved, such as horses, cattle, deer, pigs, elephants, rodents, carnivores, and even primates (Fig. 19–11).

Notable as are these morphological distinctions, discerning their evolutionary ties and divergences has been

[7]Success in the arms race is crucial to both prey and predator but especially to prey, who often invest more resources in defense that do predators in offense. As Dawkins points out, the struggle involves different needs, "the rabbit runs faster than the fox, because the rabbit is running for his life, while the fox is only running for his dinner." It is to the advantage of rabbits to concentrate major resources on speed and evasion, whereas it is to the advantage of foxes to also look for different prey and use strategies other than running.

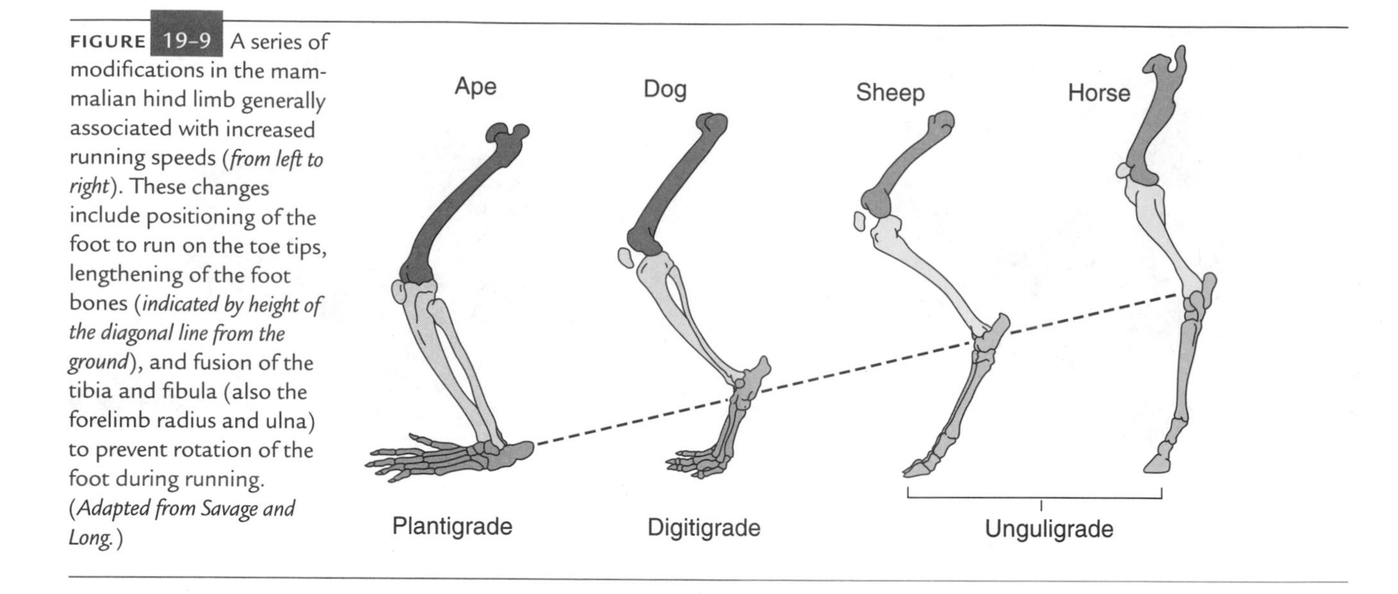

FIGURE 19–9 A series of modifications in the mammalian hind limb generally associated with increased running speeds (*from left to right*). These changes include positioning of the foot to run on the toe tips, lengthening of the foot bones (*indicated by height of the diagonal line from the ground*), and fusion of the tibia and fibula (also the forelimb radius and ulna) to prevent rotation of the foot during running. (*Adapted from Savage and Long.*)

difficult, although more accessible now with increasing molecular information. Among various molecular studies (see De Jong), a recent report by Stanhope and coworkers analyzed mitochondrial and nuclear gene sequences in a large number of mammalian taxa (Fig. 19–12). Their findings support some previously mentioned relationships such as between whales and artiodactyls (p. 252), but also propose a polyphyletic origin of Insectivora, separating golden moles and tenerecs from moles and shrews. Some common insectivoran morphologies are thus ascribed to convergence (homoplasy) and not to homology.

However, fossils still remain our primary source for data on ancient morphological organismic change. Whether we begin in the Tertiary or even earlier, we know that mammalian diversity continued through the Pleistocene, an epoch that marked the appearance of many mammals in what we consider their modern forms.

Climatically, the Pleistocene also marks a period of at least seven glaciations, called the Ice Ages, which at times covered one-third of the Earth's surface. Woolly mammoths and woolly rhinoceroses made their appearances in the northern continents during this interval, along with giant deer, giant cattle, and large cave bears. Interestingly, these large mammals all became extinct in North America about 11,000 years ago, in addition to horses, camels, and various other groups. Among approximately 79 mammalian species weighing more than 100 pounds, 57 (more than 70 percent!) became extinct at that time (Martin). Similar extinctions, although not as far-ranging, occurred in Europe.

Among possible explanations for these Late Pleistocene events is the hypothesis that climatic advantages for large animals deteriorated rapidly as the ice sheets retreated, and thus caused their extinction (Webb). Another explanation is the predatory role of humans: stone-age hunters who entered formerly glaciated areas of North America and Europe slaughtered ("overkilled") these large mammals because they made easy targets. Whether or not some of our ancestors played this role, our present role as agents of extinction has unfortunately grown from incidental to flagrant (p. 431).

Two Island Continents: Australia and South America

Most of the new adaptive radiations occurred among placentals, although marsupials on two continents, Australia and South America, also experienced significant evolutionary changes. Some reasons for the marsupial radiation in the two southern continents derive from the isolation of these landmasses because of continental drift during the Late Cretaceous and Early Cenozoic (Chapter 6). Although paleontologists have offered various scenarios, most workers generally believe that marsupials originated in North America during the mid-Cretaceous, and, along with a couple of very primitive placental groups, migrated down an arc of Central American islands into South America before the end of that period.

From South America, marsupials dispersed into an Antarctican continent that was considerably warmer than at present and, unaccompanied by placentals, reached Australia during the Early Eocene, about 50 million years ago. By mid-Eocene, perhaps 5 million years later, Australia separated from Antarctica and began its northern journey toward Asia, carrying along its isolated marsupial

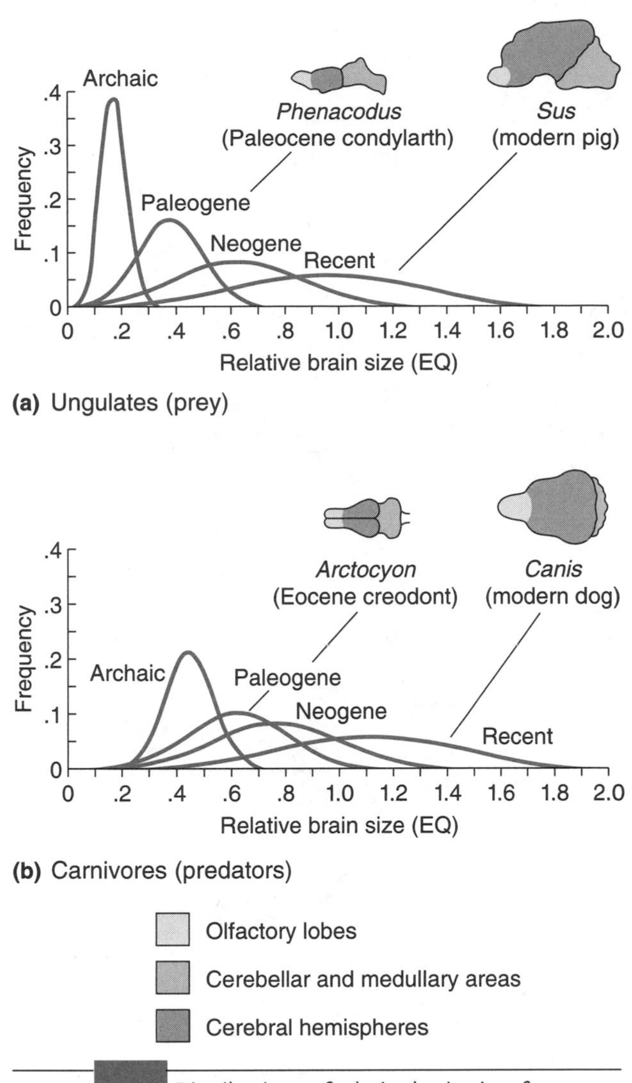

FIGURE 19-10 Distributions of relative brain sizes for mammalian ungulates (*a*) and carnivores (*b*) during different Cenozoic periods showing brain outlines of some similar body-sized species. (*Cerebral hemispheres are indicated by dark shading, olfactory lobes are light, and cerebellar and medullary areas are medium.*) The relative brain size estimate is given as the **encephalization quotient (EQ),** calculated as the ratio of actual brain weight to the brain weight expected for an animal of the same body size. [The expected brain weight for a broad sample of mammals, according to Jerison, is $0.12 \times (\text{body weight in grams})^{.67}$]. Although there is no exact correlation between brain size and intelligence, a large difference in EQ probably denotes a significant difference in mental capacity. An animal with an EQ of 0.5 possesses a brain that is half the size of an "average" modern mammal of that body weight, most likely indicating fewer intellectual powers and less complex behavior. By contrast, an EQ of 2.0 would signify twice the expected brain size and probably greater than expected mental ability. Encephalization quotients for hominids are given in the table on p. 500. (*Brain diagrams after Lull, from Osborn; EQ distributions from Jerison.*)

population.[8] New Zealand apparently separated from the Gondwana continents even earlier, probably during the Cretaceous, and neither native nor fossil terrestrial mammals have been found there.

Paleontologists ascribe the success and diversity of herbivorous and carnivorous marsupials produced during their Australian radiation (diagrammed in Fig. 3–4) to the absence of placental rivals. The only other mammalian subclass present in Australia, monotremes, were probably too primitive or too specialized to offer much competition.[9] As a result, by mid-Miocene times, at least 15 families of marsupials existed. Many of these were browsers that probably fed on temperate rain forest vegetation, along with at least two groups that were carnivorous.

By the Late Miocene, drier conditions led to an expansion of grasslands, followed by the evolution of many different kinds of grazing kangaroos, including one Pleistocene species whose adults were 10 feet high. The invasion of Australia by humans, both during the Pleistocene and more recently, has been accompanied by other placental groups, including dogs, rabbits, sheep, and rodents. Given their vulnerability to placental competition, many present Australian marsupial groups will probably not survive without protection.

In South America, marsupial radiation followed a different pattern, because the presence of placental herbivores and edentates channeled the marsupials into carnivorous and insectivorous niches. These animals ranged from many species of opossum-like didelphids to jumping, gnawing, and doglike forms. Paleobiologists believe that some of the latter, members of the borhyaenid family, are ancestral to *Thylacosmilus,* the marsupial saber-toothed "tiger" of the South American Pliocene (Fig. 19–13*a*), a carnivore strikingly similar to the large placental saber-toothed cat, *Smilodon,* of the North American Pleistocene (Fig. 19–13*b*). The borhyaenids also show marked similarities to the Australian marsupial family of thylacines that included the Tasmanian wolf (Fig. 19–14). Evolutionary convergence or parallelism, caused by selection for a similar way of life, produced similar structures in the genetically different placentals and marsupials even on different continents.

[8]A colony of North American marsupials also reached Europe during the Early Eocene and short-lived lineages made their way to Asia and Africa, probably through a North Atlantic–Greenland–Europe connection. During the Miocene, marsupial populations of both northern continents became extinct, and only when Pliocene events re-established a North American–South American land bridge did some marsupials reinvade North America from the south.

[9]Fossil evidence for monotremes is extremely poor, although Pleistocene deposits show monotremes in Australia at that time in the forms of both platypus and echidna genera, and researchers have given a middle Miocene date to teeth that may have belonged to a platypus-type animal. A few Cretaceous platypus-like fragments have also been reported, indicating monotremes probably reached or arose in Australia by the Late Jurassic–Early Cretaceous (see also p. 102).

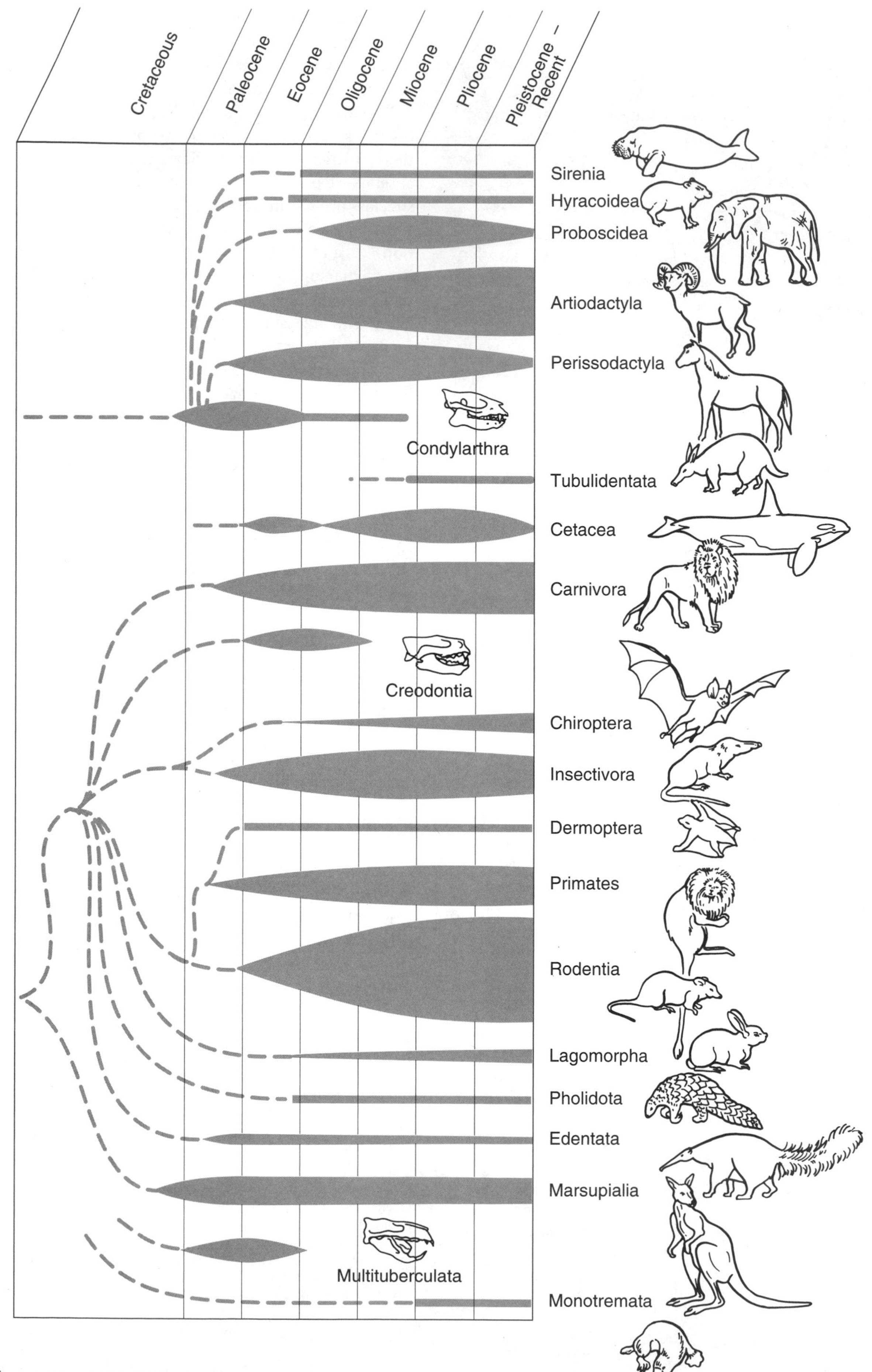
Cretaceous
Paleocene
Eocene
Oligocene
Miocene
Pliocene
Pleistocene – Recent
Sirenia
Hyracoidea
Proboscidea
Artiodactyla
Perissodactyla
Condylarthra
Tubulidentata
Cetacea
Carnivora
Creodontia
Chiroptera
Insectivora
Dermoptera
Primates
Rodentia
Lagomorpha
Pholidota
Edentata
Marsupialia
Multituberculata
Monotremata

FIGURE 19–11 Radiation pattern of mammalian orders beginning with the Cretaceous period, including three extinct groups (multituberculates, condylarths, and creodonts). Widths of *shaded bars* indicate rough estimates of relative fossil abundances at various times. Exact phylogenetic relationships among many of these mammalian orders (*dotted lines*) are still disputed by paleontologists (Benton 1988; Novacek et al.), and molecular phylogenies, such as illustrated in Fig. 19–12, show novel relationships, such as between elephants and some insectivores. Estimates of dates of divergence also differ between the fossil record and molecular studies: paleontological findings point to major mammalian radiations at or near the 65-million-year-old Cretaceous–Tertiary boundary, whereas some molecular divergence times are more than double (Bromham et al., Foote et al.). Such differences are yet to be resolved. (*Data from Gingerich.*)

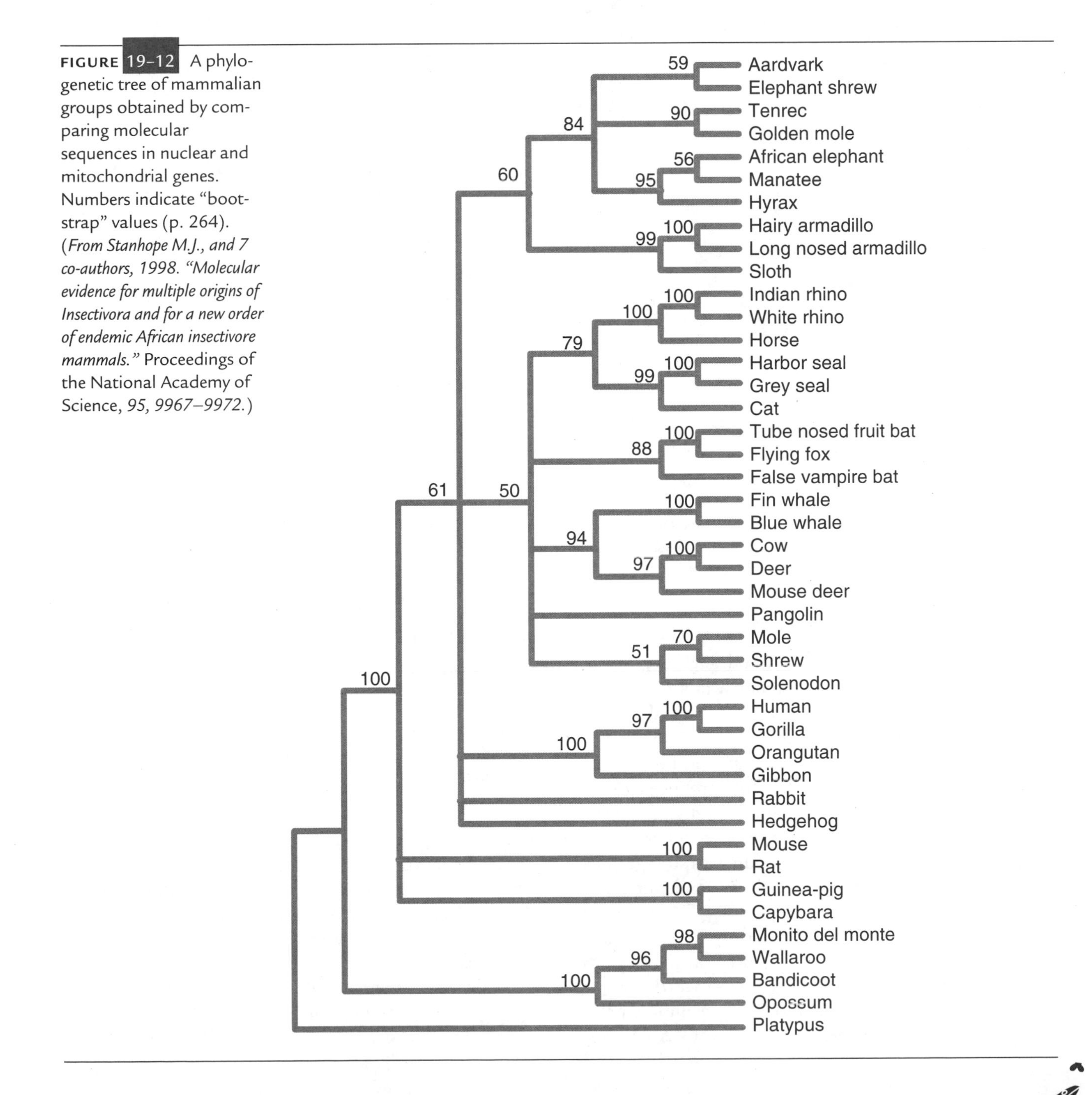

FIGURE 19–12 A phylogenetic tree of mammalian groups obtained by comparing molecular sequences in nuclear and mitochondrial genes. Numbers indicate "bootstrap" values (p. 264). (*From Stanhope M.J., and 7 co-authors, 1998. "Molecular evidence for multiple origins of Insectivora and for a new order of endemic African insectivore mammals."* Proceedings of the National Academy of Science, *95, 9967–9972.*)

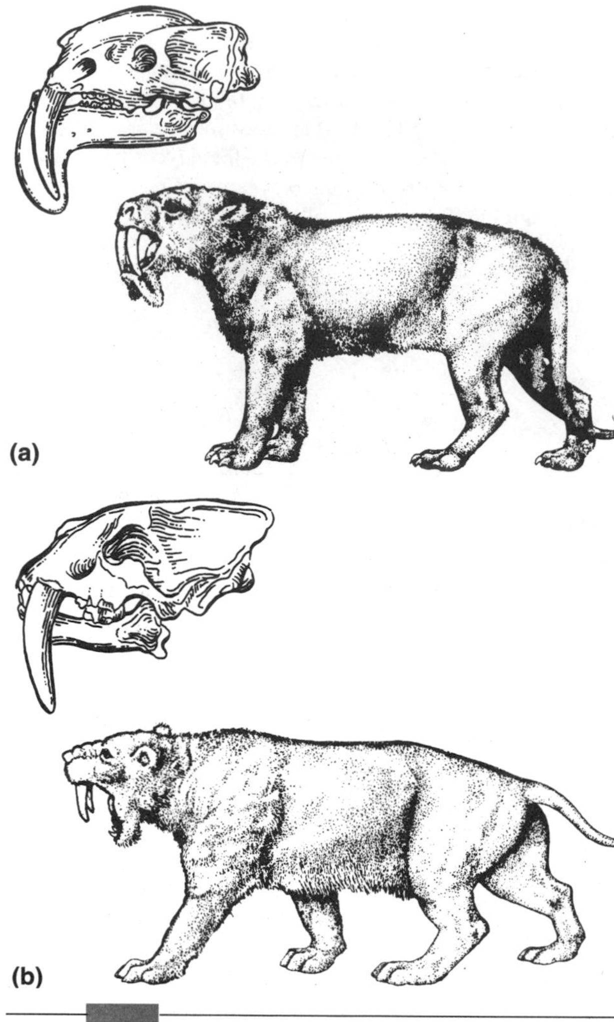

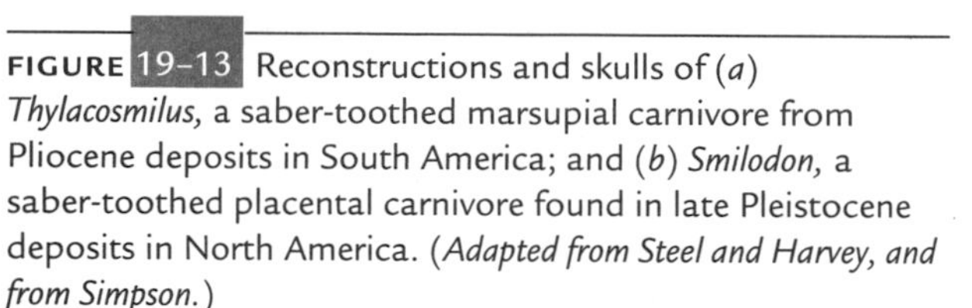

FIGURE 19–13 Reconstructions and skulls of (*a*) *Thylacosmilus,* a saber-toothed marsupial carnivore from Pliocene deposits in South America; and (*b*) *Smilodon,* a saber-toothed placental carnivore found in late Pleistocene deposits in North America. (*Adapted from Steel and Harvey, and from Simpson.*)

(a) Borhyaenid marsupial (Miocene, Argentina)

(b) Marsupial Tasmanian wolf (Tasmania, Australia)

(c) Placental wolf (North America)

FIGURE 19–14 (*a*) *Prothylacynus patagonicus,* a borhyaenid marsupial from the early Miocene period in southern Argentina. (*b*) *Thylacinus cynocephalus,* the recently extinct marsupial Tasmanian wolf. (*c*) *Canis lupus,* the modern placental North American wolf. (*From* Aspects of Vertebrate History, *1980, p. 345–386 by L.Q. Marshall, L.L. Jacob (ed.) Reprinted by permission.*)

The South American placentals, although beginning only with some ungulates and **xenarthrans** ("strange-jointed"), radiated perhaps even more rapidly than did marsupials on that isolated continent. By the Early Eocene, within 15 to 20 million years of their initial Late Cretaceous colonization, placentals had produced 75 to 100 new genera, which we divide into about 15 families. The xenarthrans (also called **edentates** because of their reduced or suppressed dentition) produced a strange bestiary of armadillos, glyptodonts, sloths, and anteaters (Fig. 19–15*a*).

Also radiating widely were the mostly hoofed ungulates, which paleontologists believe originated from an ancestral herbivorous stock called **condylarths** (Fig. 19–15*b*). Again, convergent or parallel evolution produced striking similarities: some South American litopterns, apparently selected for grazing and rapid running, had become, by the Early Miocene, remarkably similar to the one-toed horses that first developed about 20 million years later in North America.

In the Early or Late Oligocene, a similar rapid radiation began among the rodents and primates that had reached South America at that time from Africa, probably by "island hopping" along the island chains on the oceanic ridges cast up in the South Atlantic ocean. The

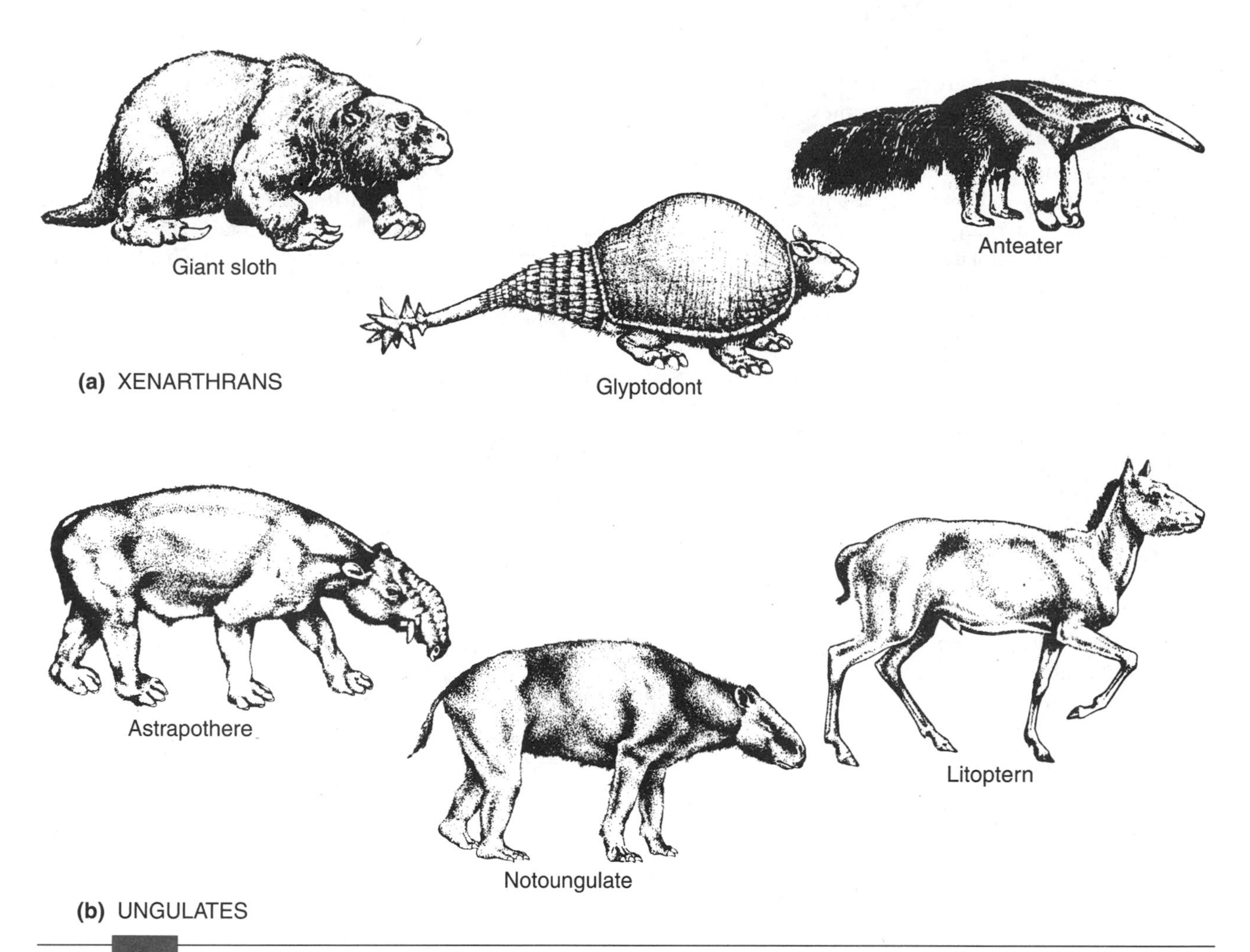

FIGURE 19–15 Reconstructions of some of the (*a*) xenarthrans (edentates) and (*b*) ungulates produced by the South American placental mammalian radiation. Flynn and Wyss point out that the marsupial immigration from South America to Australia would also lead us to expect an Australian presence of other South American groups such as xenarthrans and ungulates. Their absence in Australia is, so far, unexplained. (*Adapted from Steel and Harvey.*)

rodents produced a great diversity of caviomorphs (cavies) distinguished by special jaw muscle attachments. The primates, confined mostly to tropical areas, produced the wide array of New World monkeys collected in the superfamily Ceboidea.

When South America was next united with North America, 30 to 35 million years later in the Pliocene, an extensive interchange between the mammals of these two continents followed. Many South American groups became extinct, including marsupial carnivores and many placental ungulates, at least partially because of competition with more advanced North American placentals. Some of the successful invading North American groups also diversified quite rapidly. For example, cricetid rodents (field mice) evolved into about 60 genera and 300 species within the 5- or 8-million-year period of their South American immigration (Engel et al.). As elsewhere, extinction and radiation in South America seemed to go hand in hand, testifying again to the basic opportunism of evolutionary change.

SUMMARY

Many traits distinguish mammals from their reptilian predecessors. Among these are mammary glands, live birth, a four-chambered heart, a diaphragm, skeletal changes in the skull and head regions, and adaptations for homeothermy. The therapsid reptiles, which paleontologists think are ancestral to mammals, were probably endothermic and had legs placed directly below the trunk. In addition they had cusped teeth, allowing them to grind food material, and a secondary palate to keep food from blocking the nasal openings.

Mammals shed their teeth only once, whereas their reptilian ancestors continuously replaced their teeth. As the structure and function of teeth changed, so did the jaw and its muscles. The articular and quadrate bones articulating the jaw with skull became the ossicles of the inner ear, accommodating the need for keen hearing in these small animals.

The earliest known mammals are a fossil group dating from the Triassic period, the morganucodontids, which developed tricuspid molars and a precise occlusion of upper and lower cheek teeth. From them probably came the Prototheria, egg-laying mammals, the modern representatives of which are the platypus and the echidna. A later branching gave rise to therians with more elaborately cusped molars and jaws capable of grinding motions enabling them to use new food sources. Because of reptilian predation during the Mesozoic era, most mammals were probably nocturnal, which favored selection for excellent sensory organs.

Selection pressures on the therians, which had small eggs, favored viviparity, maternal protection, and rapid fetal development. In the therian marsupial branch, the egg remains shelled for most of the brief embryonic period, and the fetus emerges from the oviduct at a very immature stage to be nourished by maternal mammary structures. Placental mammals have lost the eggshell, and the fetus develops for a much longer time inside the uterus, where the placenta provides nourishment, oxygen, and waste removal. Most probably marsupials have persisted along with placental mammals because marsupials expend relatively little energy on their undeveloped newborn offspring.

A number of principles emerge from an examination of mammalian evolution:

- We cannot predict continued evolutionary success of a particular group because of its dominance at a particular time.
- Crucial evolutionary advantages may fall to groups that already have characteristics adaptable to new circumstances (preadaptation).
- Long-term evolutionary replacement among groups is not predictable.
- New modes of biological organization can enhance group survival.
- New levels of organization occur because of complex coordinated changes in many traits over long intervals of time.
- Once such new levels have been attained, widespread radiation can often begin.
- Evolving groups are often interdependent.
- Extinction is common, if not inevitable, because of genomic constraints on the ability of a species to adapt to large and rapid environmental changes.

After the extinction of the dinosaurs at the end of the Cretaceous period, mammals diversified into many habitats because of new adaptations such as specialized limbs, and mammals replaced reptiles as the dominant land vertebrates. Marsupials isolated in South America and Australia by continental drift radiated widely. In South America, competition from placental mammals forced marsupials into specialized niches, and when South and North America united in the Pliocene period, many of the marsupials became extinct. Invading North American placental mammals then had the opportunity to diversify rapidly.

KEY TERMS

articular
bony secondary palate
coadaptive mutations
condylarths
cynodont therapsids
deciduous teeth
dentary bone
diaphragm
digitigrade
diurnal
double occipital condyle
ear ossicles
edentates
encephalization quotient (EQ)
endothermy
four-chambered heart
heterodont
incus
lactation
large braincase
live birth
malleus
mammary glands
mandible
marsupials
monotremes
morganucodontids
multicusped crowns
multirooted cheek teeth
neocortex
nocturnal
oviparity
placentals
plantigrade
Prototheria
quadrate
single nasal opening
squamosal
synapsid reptiles
therians
tribosphenic molar
triconodont
tympanic membrane
unguligrade
uterus
viviparity
xenarthrans

DISCUSSION QUESTIONS

1. What major features distinguish mammals from other vertebrates?
2. Why do paleontologists consider mammals to have had an ancestry among the therapsid reptiles?
3. How did selection for endothermy and continuous metabolic activity affect the mammalian palate? mammalian dentition?
4. What evolutionary stages can account for the transformation of the posterior elements of the reptilian jaw (articular and quadrate bones) into the mammalian ear ossicles?

5. How have biologists used changes in fossil teeth in constructing hypotheses about early mammalian evolution?
6. What lifestyle have paleontologists proposed for Mesozoic mammals, and what evidence supports this view?
7. How can we explain the evolution of major differences in reproductive modes among monotremes, marsupials, and placentals?
8. In terms of the directions taken by evolution (such as radiation patterns, long-term predictability, new levels of organization, and group interactions and replacements), what are some lessons that we can learn from reptilian and mammalian evolution during the Mesozoic era?
9. How can we explain the evolution and distribution of major groups of both fossil and present mammals in South America and Australia?

EVOLUTION ON THE WEB

Explore evolution on the web! Visit the accompanying web site for *Evolution*, 3/e at www.jbpub.com/evolution for web exercises and links relating to topics covered in this chapter.

REFERENCES

Archer, M., and G. Clayton (eds.), 1984. *Vertebrate Zoogeography and Evolution in Australia.* Hesperian Press, Carlisle, Australia.

Benton, M. J., 1988. The relationships of the major group of mammals: New approaches. *Trends in Ecol. and Evol.,* **3,** 40–45.

———, 1997. *Vertebrate Palaeontology,* 2d ed. Chapman & Hall, London.

Bown, T. M., and M. J. Kraus, 1979. Origin of the tribosphenic molar and metatherian and eutherian dental formulae. In *Mesozoic Mammals: The First Two-Thirds of Mammalian History,* J. A. Lillegraven, Z. Kielan-Jaworowska, and W. A. Clemens (eds.). University of California Press, Berkeley, pp. 172–181.

Bromham, L., M. J. Phillips, and D. Penny, 1999. Growing up with dinosaurs: Molecular dates and the mammalian radiation. *Trends in Ecol. and Evol.,* **14,** 113–118.

Carroll, R. L., 1988. *Vertebrate Paleontology and Evolution.* Freeman, New York.

Colbert, E. H., and M. Morales, 1991. *Evolution of the Vertebrates: A History of the Backboned Animals Through Time,* 4th ed. Wiley, New York.

Crompton, A. W., and F. A. Jenkins, Jr., 1979. Origin of mammals. In *Mesozoic Mammals: The First Two-Thirds of Mammalian History,* J. A. Lillegraven, Z. Kielan-Jaworowska, and W. A. Clemens (eds.). University of California Press, Berkeley, pp. 59–73.

Crompton, A. W., and P. Parker, 1978. Evolution of the mammalian masticatory apparatus. *Amer. Sci.,* **66,** 192–201.

Crompton, A. W., C. R. Taylor, and J. A. Jagger, 1978. Evolution of homeothermy in mammals. *Nature,* **272,** 333–336.

Dawkins, R., 1986. *The Blind Watchmaker.* Longmans, Harlow, Essex, England.

De Jong, W. W., 1998. Molecules remodel the mammalian tree. *Trends in Ecol. and Evol.,* **13,** 270–275.

Eisenberg, J. F., 1981. *The Mammalian Radiations: Evolution, Adaptation, and Behavior.* University of Chicago Press, Chicago.

Engel, S. R., K. M. Hogan, J. F. Taylor, and S. K. Davis, 1998. Molecular systematics and paleobiogeography of the South American sigmodontine rodents. *Mol. Biol. and Evol.,* **15,** 35–49.

Flynn, J. J., and A. R. Wyss, 1998. Recent advances in South American mammalian paleontology. *Trends in Ecol. and Evol.,* **13,** 449–454.

Foote, M., J. P. Hunter, C. M. Janis, and J. J. Sepkoski, Jr., 1999. Evolutionary and preservational constraints on origins of biologic groups: Divergence times of eutherian mammals. *Science,* **283,** 1310–1314.

Gingerich, P. D., 1977. Patterns of evolution in the mammalian fossil record. In *Patterns of Evolution as Illustrated by the Fossil Record,* A. Hallam (ed.). Elsevier, Amsterdam, pp. 469–500.

Hillenius, W. J., 1994. Turbinates in therapsids: Evidence for Late Permian origins of mammalian endothermy. *Evolution,* **48,** 207–229.

Janis, C. M., 1993. Tertiary mammal evolution in the context of changing climates, vegetation, and tectonic events. *Ann. Rev. Ecol. Syst.,* **24,** 467–500.

Janke, A., X. Xu, and U. Arnason, 1997. The complete mitochondrial genome of the wallaroo (*Macropis robustus*) and the phylogenetic relationship among Monotremata, Marsupialia, and Eutheria. *Proc. Nat. Acad. Sci.,* **94,** 1276–1281.

Jerison, H. J., 1973. *Evolution of the Brain and Intelligence.* Academic Press, New York.

Ji, Q., Z. Luo, and S.-A. Ji, 1999. A Chinese triconodont mammal and mosaic evolution of the mammalian skeleton. *Nature,* **398,** 326–330.

Kemp, T. S., 1982. *Mammal-like Reptiles and the Origin of Mammals.* Academic Press, London.

———, 1988. Interrelationships of the Synapsida. In *The Phylogeny and Classification of the Tetrapods,* vol. 2, M. J. Benton (ed.). Oxford University Press, Oxford, England, pp. 1–22.

Kermack, D. R., and K. A. Kermack, 1984. *The Evolution of Mammalian Characters.* Croom Helm, London.

Kermack, K. A., and F. Mussett, 1983. The ear in mammal-like reptiles and early mammals. *Acta Palaeontolgica Polonica,* **28,** 147–158.

Kielan-Jaworowska, Z., A. W. Crompton, and F. A. Jenkins, 1987. The origin of egg-laying mammals. *Nature,* **326,** 871–873.

Kumar, S., and B. Hedges, 1998. A molecular timescale for vertebrate evolution. *Nature,* **392,** 917–920.

Lillegraven, J. A., 1979. Reproduction in Mesozoic mammals. In *Mesozoic Mammals: The First Two-Thirds of Mammalian History,* J. A. Lillegraven, Z. Kielan-Jaworowska, and W. A. Clemens (eds.). University of California Press, Berkeley, pp. 259–276.

Lillegraven, J. A., Z. Kielan-Jaworowska, and W. A. Clemens (eds.), 1979. *Mesozoic Mammals: The First Two-Thirds of Mammalian History.* University of California Press, Berkeley.

Lull, R. S., 1940. *Organic Evolution.* Macmillan, New York.

Maio, D., 1991. On the origin of mammals. In *Origins of the Higher Groups of Tetrapods: Controversy and Consensus,* H.-P. Schultze and L. Trueb (eds.). Cornell University Press, Ithaca, NY, pp. 579–597.

Marshall, L. G., 1980. Marsupial paleobiogeography. In *Aspects of Vertebrate History,* L. L. Jacobs (ed.). Museum of Northern Arizona Press, Flagstaff, AZ, pp. 345–386.

Marshall, L. G., S. D. Webb, J. J. Sepkoski, Jr., and D. M. Raup, 1982. Mammalian evolution and the great American interchange. *Science,* **215,** 1351–1357.

Martin, P. S., 1984. Catastrophic extinctions and Late Pleistocene blitzkrieg: Two radiocarbon tests. In *Extinctions,* M. H. Nitecki (ed.). University of Chicago Press, Chicago, pp. 153–189.

McFarland, W. N., F. H. Pough, T. J. Cade, and J. B. Heiser, 1985. *Vertebrate Life,* 2d ed. Macmillan, New York.

McShea, D. W., 1996. Metazoan complexity and evolution: Is there a trend? *Evolution,* **50,** 477–492.

Nitecki, M. H. (ed.), 1988. *Evolutionary Progress.* University of Chicago Press, Chicago.

Novacek, M. J., A. R. Wyss, and M. C. McKenna, 1988. The major groups of eutherian mammals. In *The Phylogeny and Classification of the Tetrapods,* vol. 2, M. J. Benton (ed.). Oxford University Press, Oxford, England, pp. 31–71.

Olson, E. C., 1971. *Vertebrate Paleozoology.* Wiley-Interscience, New York.

Rich, P. V., and E. M. Thompson (eds.), 1982. *The Fossil Vertebrate Record of Australia.* Monash University Press, Clayton, Australia.

Romer, A. S., 1966. *Vertebrate Paleontology,* 3d ed. University of Chicago Press, Chicago.

———, 1968. *The Procession of Life.* World Publishing, Cleveland.

Romer, A. S., and T. S. Parsons, 1977. *The Vertebrate Body,* 5th ed. Saunders, Philadelphia.

Savage, R. J. G., and M. R. Long, 1986. *Mammal Evolution: An Illustrated Guide.* British Museum (Natural History), London.

Simpson, G. G., 1980. *Splendid Isolation: The Curious History of South American Mammals.* Yale University Press, New Haven, CT.

Stahl, B. J., 1974. *Vertebrate History: Problems in Evolution.* McGraw-Hill, New York.

Stanhope, M. J., et al., 1998. Molecular evidence for multiple origins of Insectivora and for a new order of endemic African insectivore mammals. *Proc. Nat. Acad. Sci.,* **95,** 9967–9972.

Steel, R., and A. P. Harvey, 1979. *The Encyclopaedia of Prehistoric Life.* Mitchell-Beazley, London.

Webb, S. D., 1984. Ten million years of mammalian extinctions in North America. In *Quaternary Extinctions: A Prehistoric Revolution,* P. S. Martin and R. G. Klein (eds.). University of Arizona Press, Tucson, pp. 189–210.

Young, J. Z., 1981. *The Life of Vertebrates,* 3d ed. Clarendon Press, Oxford, England.

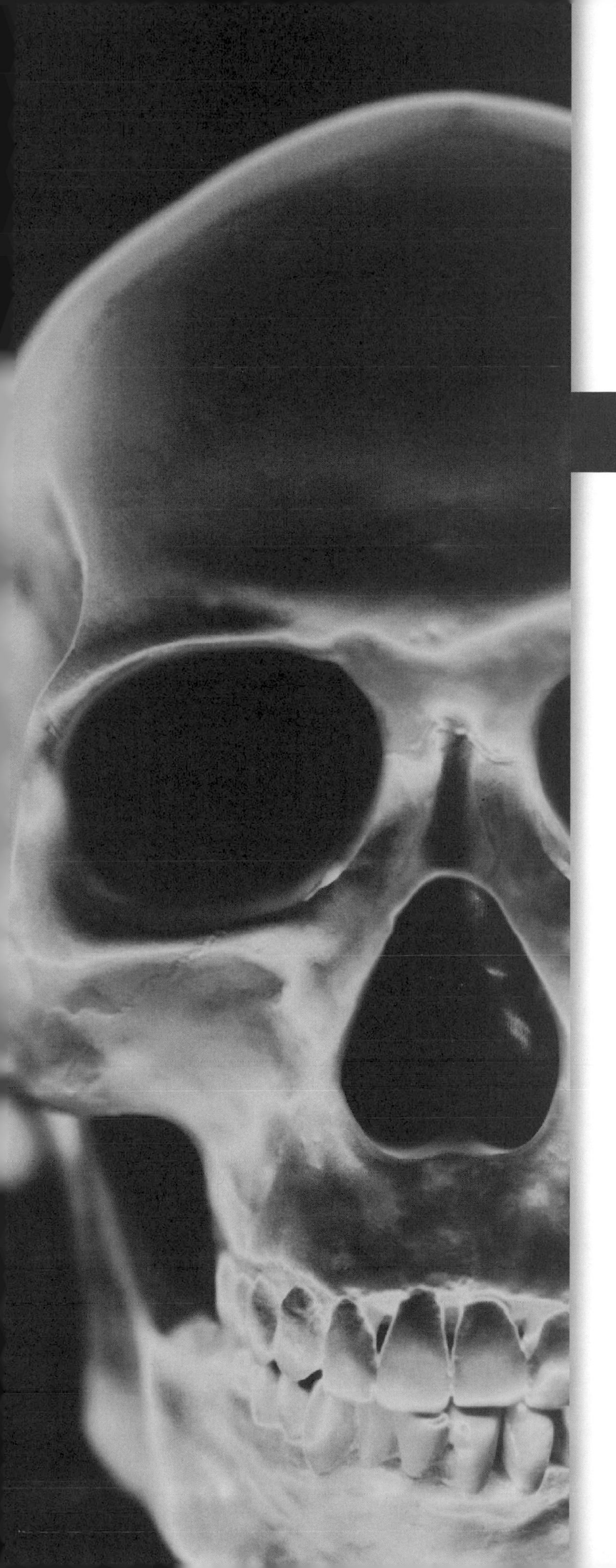

20 Primate Evolution and Human Origins

Primates, the mammalian order that includes humans, are species that have a number of adaptations indicating an arboreal (tree-living) ancestry. Among others that Table 20–1 lists, these adaptations include

1. Ability to move the four limbs in various directions
2. Grasping power of the hands and feet
3. Slip-resistant cutaneous ridges (dermatoglyphs) on the ventral pads of these extremities, which also contain specialized tactile-sensitive organs (Meissner's corpuscles)
4. Retention of the clavicle (collar bone) to support the pectoral girdle in positioning the forelimb
5. Flexibility of the spine to allow twisting and turning

In addition to having their highly developed brain, anthropoid primates (monkeys, apes, and humans) also undergo a relatively long postnatal growth period accompanied by considerable parental care for a relatively small number of offspring. The selective value of this trait probably arises from the limited number of offspring that can be successfully born and carried by highly mobile primates, along with the long-dependent learning period needed to cope with many complex environmental and social variables. Although not every feature Table 20–1 mentions characterizes every primate, all existing primates have enough of these features to distinguish them from other mammalian arboreal groups such as shrews, squirrels, and raccoons.

Primate Classification

There are presently about 230 primate species, which primatologists usually classify into two suborders, **prosimians** and **anthropoids** (Table 20–2). The prosimians, or lower primates, generally retain earlier mammalian features (for example, claws, long snout, lateral-facing eyes) than do the anthropoid primates. With the exception of Madagascar,

TABLE 20–1 Traits and tendencies found in primate groups

- Independent mobility of the digits
- An opposable first digit in both hands and feet (thumb, big toe)
- Replacement of claws by nails to support the digital pads on the last phalanx of each finger and toe
- Teeth and digestive tract adapted to an omnivorous diet
- A semierect posture that enables hand manipulation and provides a favorable position preparatory to leaping
- Center of gravity positioned close to the hind legs
- Well-developed hand–eye motor coordination
- Optical adaptations that include overlap of the visual fields to gain precise three-dimensional information on the location of food objects and tree branches
- An eye completely (anthropoids) or fractionally (prosimians) encased by bone (bony orbits)
- Shortening of the face accompanied by reduction of the snout
- Diminution of the olfactory apparatus in diurnal forms
- Compared with practically all other mammals, a very large and complex brain in relation to body size

TABLE 20–2 Classification of existing subgroups in the order Primates, with common names for some members in each group

Suborder Prosimii
- Superfamily Lemuroidea: lemurs
- Superfamily Lorisoidea: lorises, galagos (bush babies)
- Superfamily Tarsioidea: tarsiers

Suborder Anthropoidea
- Infraorder Platyrrhini (New World)
  - Superfamily Ceboidea
    - Family Callitrichidae: marmosets, tamarins
    - Family Cebidae: capuchins, howler monkeys, spider monkeys
- Infraorder Catarrhini (Old World)
  - Superfamily Cercopithecoidea
    - Family Cercopithecidae: macaques, baboons, guenons, vervet monkeys
    - Family Colobidae: langurs, colobines
  - Superfamily Hominoidea
    - Family Hylobatidae: gibbons, siamangs
    - Family Pongidae:[a] orangutans, gorillas, chimpanzees
    - Family Hominidae: humans

Note: Primate taxonomy generates considerable debate (see Aiello). Some primatologists rename the prosimians as the suborder Strepsirhini (moist, doglike muzzle between nose and lip) and the anthropoids as the suborder Haplorhini (dry skin or fur between nose and lip) and include the tarsiers as a haplorhine infraorder (Tarsiiformes).

[a]In some molecular classifications based on immunological studies (Fig. 12–2) and DNA sequencing (see Miyamoto et al.), researchers put all the great apes together with humans in a single family, Hominidae, which they then subdivide into two subfamilies: one containing the orangutans (Ponginae), and the other containing humans, chimpanzees, and gorillas (Homininae). (See also Tattersall et al. and Goodman et al.)

an island that separated from Africa before anthropoids had evolved, prosimians have small bodies and are nocturnal.

The anthropoids, which include monkeys, apes, and humans, are mostly larger than prosimians and are generally diurnal rather than nocturnal. Compared to prosimians, anthropoids have more of the primate features enumerated previously, such as a shortened face, forward-directed eyes, and a larger, more complex brain. Figure 20–1 illustrates species members from the two primate suborders, which we can briefly describe as follows.

LEMURS

The lemurs appear exclusively in Madagascar, which probably separated from the African continent sometime during the late Cretaceous period. In this relatively protected area, lemurs have produced a range of small and large species that often parallel the role of forest monkeys on the mainland. But lemurs are more primitive than monkeys, having a longer snout and moist philtrum between nose and upper lip that accentuates their sense of smell. They also have a special toilet claw on the second toe, thick fur, sensitive facial hairs (vibrissae), and a dental comb formed by the nearly horizontal (procumbent) orientation of the lower incisors and canines that is used for both grooming and feeding. Because lemurs have a "tapetum lucidum" (retinal layer that reflects incoming light back through the retina), some remain exclusively nocturnal (mouse and dwarf lemurs) while others are active during the dim crepuscular light of late dusk and early dawn as well as diurnally (true lemurs).

LORISES

The lorises are found in forests of both Africa (pottos and galagos, or bush babies) and Southeast Asia (slender and slow lorises). The snout is shorter than in lemurs, and the relatively large eyes face forward, indicating adaptation to a larger forebrain and perhaps also to increased emphasis on visual predation. These and other adaptations, including a retinal tapetum, permit either nocturnal or crepuscular activity. Like lemurs, lorises have dental tooth combs, a toilet claw, and a moist philtrum.

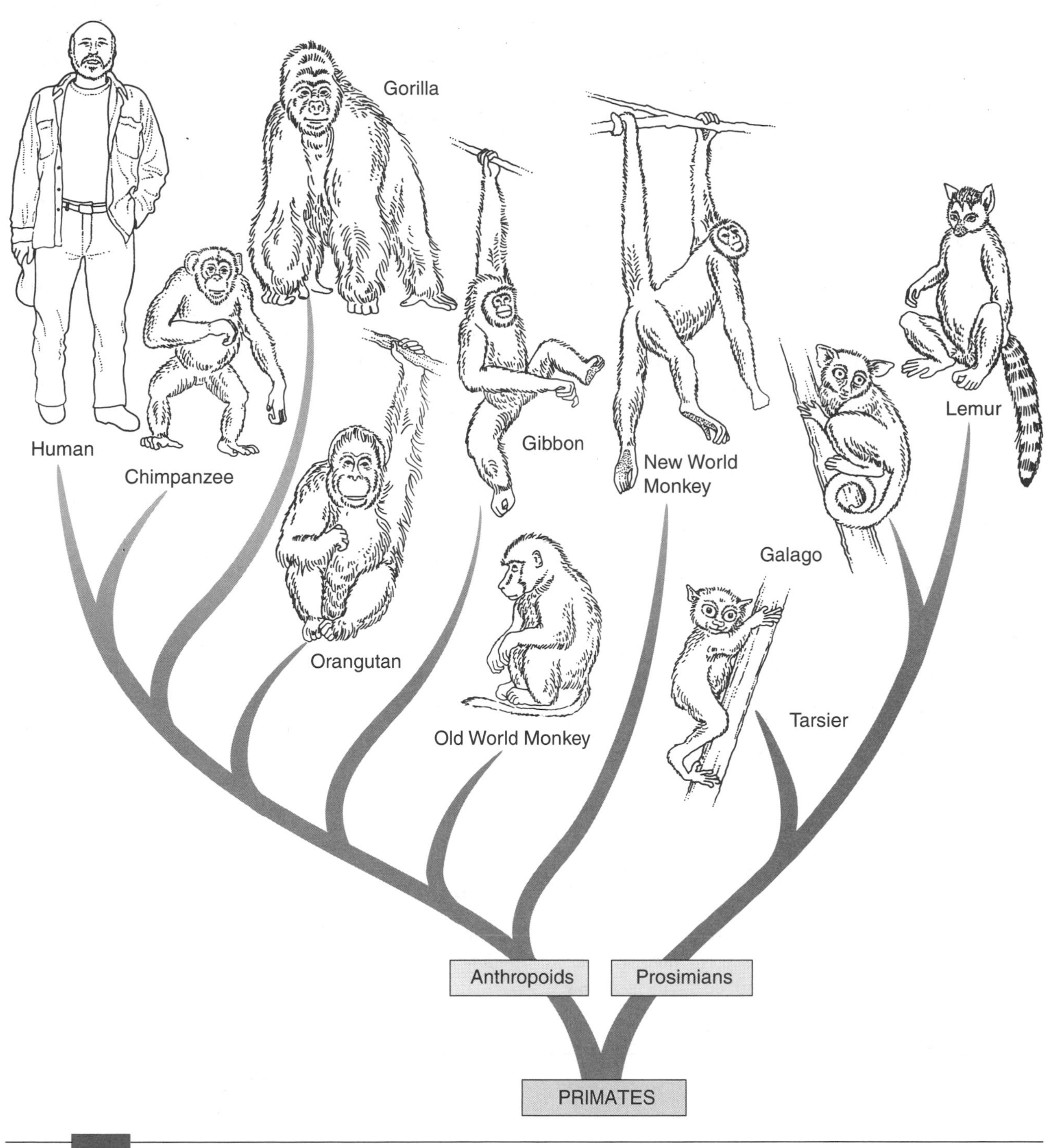

FIGURE 20–1 Various living representatives of the order Primates.

TARSIERS

The tarsiers, nocturnal Southeast Asian primates, seem to stand between prosimians and anthropoids. Although they have two toilet claws on each foot and enormous eyes relative to head size, like anthropoids, they lack the retinal tapetum that characterizes lemurs and lorises. Tarsiers also show anthropoid characteristics in the replacement of the moist philtrum by a dry, furry space between nose and lip, as well as upright lower incisors, and a partially closed bony orbit around the eyes. Unique to tarsiers is the fusion of tibia and fibula in the lower leg, an adaptation that apparently helps them make single leaps as long as 6 or 7 feet.

PLATYRRHINES

The platyrrhine infraorder designates the New World monkeys found in Central and South America, all of which are arboreal. They are characterized by broad noses with widely spaced nostrils facing laterally and three premolars on each side of the jaw. In one family (Callitrichidae) are marmosets and tamarins, small animals that have claws on all digits except the big toe. Species in the other platyrrhine family (Cebidae) have nails instead of claws, and some also have prehensile tails.

CATARRHINES

This infraorder includes two superfamilies, the Cercopithecoidea (Old World monkeys) and Hominoidea (apes and humans). They share narrowly spaced nostrils facing downward and a dental formula of 2.1.2.3 (two incisors, one canine, two premolars, and three molars on each side of the centerline in both upper and lower jaws). The catarrhine monkeys are mostly larger than the New World monkeys, lack prehensile tails, and have produced terrestrial (baboons, mandrills, vervets, patas monkeys, and some macaques) as well as arboreal forms.

HOMINOIDS

In this catarrhine superfamily of apes and humans appear a number of adaptations to brachiating (arm-hanging and -swinging) arboreal locomotion along with different degrees of adaptation to a ground-dwelling existence. Perhaps because arboreal hominoids had to adapt to holding on to overhead tree branches, their posture is more erect than that of monkeys. Also, as an aid in brachiation, the arms and shoulders are more flexible, the wrists and elbows are more limber, and the spine is shorter and stiffer. Other hominoid attributes that distinguish them from monkeys are a broader and larger pelvis to support more vertical weight; visceral attachments and arrangements that provide more vertical support for stomach, intestines, and liver; loss of the tail; five-cusped lower molars rather than the four cusps in monkeys; a broad but shallow thorax because of the change to a more vertical posture; and scapulae placed dorsally on the thorax to position the shoulder joint so the arms can be extended laterally. Other hominoid features include a larger body size compared to Old World monkeys, and a life history with greater emphasis on extended postnatal development and complex social interactions.

GIBBONS

Perhaps the most primitive existing hominoids are the gibbons and siamangs, an almost entirely arboreal group confined to Southeast Asia. They share with many Old World monkeys a relatively small size (none are more than 25 pounds) and ischial callosities (cornified sitting pads fused to the ischial bones). Compared to other hominoids, they are superb acrobatic brachiators who swing with elongated arms through the trees, their legs often folded beneath them.

ORANGUTANS

The orangutans are large apes (some males may weigh more than 200 pounds) restricted to Borneo and Sumatra. With the exception of adult males, they are mainly arboreal and on the ground move mostly quadrupedally with clenched fists to support the upper torso. However, like chimpanzees and gorillas, they lack ischial callosities. As discussed previously (Chapter 12), molecular data indicate that this group separated at an early stage in hominoid evolution from the group that includes chimpanzees, gorillas, and humans.

CHIMPANZEES

The chimpanzees are found in equatorial Africa where they live in groups of about 40 to 50, socially organized in a dominance hierarchy. Although they sleep and do most of their feeding arboreally, they are less specialized for arboreal pursuits than Asiatic apes and spend more time on the ground. Like the gorillas, they travel terrestrially by "knuckle-walking," using friction pads on the middle phalanges of nonthumb digits as forelimb support. Their diet is mainly frugivorous, but observers have seen them eat termites as well as capture and eat young baboons, monkeys, and occasionally even young chimpanzees. Compared to all other primates except humans, chimpanzees show a remarkably wide array of expressions, postures, and gestures.

GORILLAS

The gorillas are the largest apes (some males may weigh 500 pounds or more) and inhabit equatorial Africa in two main distributions: lowland gorillas west of the Congo basin, and mountain gorillas eastward. Adult males make their sleeping nests on the ground and rarely climb in trees. Their social groups, usually fewer than 10 individuals, organize around a single dominant male ("silverback"), with other adult males occasionally present. They do not seem as active as chimpanzees and have a diet that seems almost entirely herbivorous.

Human-Ape Comparisons

Compared to other hominoids, humans (**hominids**) present the greatest number of adaptations to bipedal terrestrial locomotion. Their hind limbs are longer relative to their forelimbs than in any of the apes, and their hands, freed from

TABLE 20-3 Amino acid differences between chimpanzees and humans for nine proteins

Protein	Amino Acids: Number in Protein	Amino Acids: Human–Chimpanzee Differences
Hemoglobins		
α chain	141	0
β chain	146	0
$^{G}\gamma$ chain	146	0
$^{A}\gamma$ chain	146	0
δ chain	146	1
Myoglobin	153	1
Cytochrome *c*	104	0
Fibrinopeptides A and B	30	0
Carbonic anhydrase I	259	3
Totals	1,271	5

Source: From Diamond, J.M., 1995. "The evolution of human creativity." In *Creative Evolution?!,* J.H. Campbell and J.W. Schopf (eds.), 1995, Sudbury, MA: Jones and Bartlett Publishers, www.jbpub.com. Reprinted with permission.

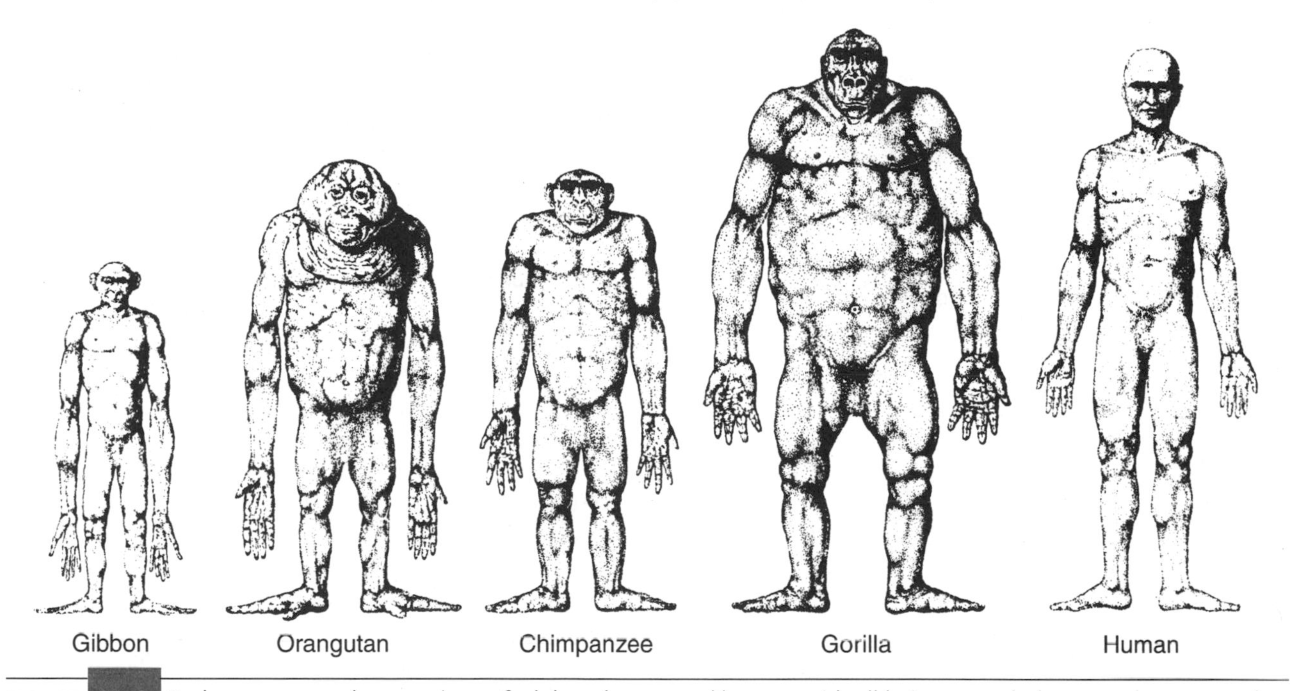

FIGURE 20-2 Body contours and proportions of adult male apes and humans with all hair removed, drawn to the same scale. (*Adapted from Schultz.*)

supporting the body, provide the most refined of manipulatory controls. Additional uniquely human anatomical traits include a relatively large brain and small face, shorter canines, less body hair, and many cranial, dental, skeletal, and other features. Despite these differences, researchers observe a large number of molecular similarities. For example, comparisons of protein and DNA sequences for chimpanzees and humans show more than 98 percent identity, indicating that molecular differences are probably no more than those between other related species (Table 20–3).

Anatomically, if we compare bone for bone, muscle for muscle, organ for organ, humans also strikingly resemble apes, although differing in proportions (Fig. 20–2). You can see this similarity between apes and humans in some of their motions and postures:

- Because their arms extend laterally in brachiator fashion and their elbows and shoulders are remarkably mobile, both scratch the back of their head from the side, rather than from the front as do monkeys.

FIGURE 20–3 A chimpanzee and movie actress (Dorothy Lamour) resting on a 1938 movie set of the motion picture *Jungle Love*. (*From a photograph in Mann.*)

- As you can see from Figure 20–3, positioning of the limbs can be remarkably alike—both support the chin with their hands and cross their legs. Even some facial expressions seem similar (Fig. 20–20).
- In walking, the heel touches the ground first, whereas in monkeys the metatarsals touch first.
- The knuckle-supporting stance of crouching American football players is the conventional stance for ground movement in chimpanzees and gorillas.

The Fossil Record

According to the fossil record, many Mesozoic mammals were very much like extant tree shrews (Fig. 20–4*a*; see also Fig. 19–4), probably adapted to an insectivorous lifestyle that encompassed both the forest floor and trees and shrubs. Although the primates that evolved from these early forms enhanced many of these basic arboreal adaptations, the point at which primate origin took place is still obscure. Part of the difficulty lies in the relative absence of fossilized forest animals in sedimentary rocks. Generally, arboreal animals disarticulate completely soon after they reach the forest floor, and only rare chance events wash their skeletons down into rivers, lakes, or marine sediments where they can more easily fossilize.

Nevertheless, Paleocene deposits show the presence of an early archaic primate group, classified in the suborder Plesiadapiformes (Fig. 20–4*b*). These animals have uniquely structured auditory regions and dentitions that differentiate them from related insectivorous forms (Fig. 20–4*a*). Along with some other groups, such as tree shrews (Scandentia) and bats (Chiroptera), they show a change to an arboreal and omnivorous lifestyle, and may be closely allied to the original primate lineage. By the Eocene, some such groups had evolved further changes such as a bony ring around their optical orbits and digital nails rather than claws. One Eocene family (adapids, Fig. 20–4*c*) may have given rise to modern lemurs and lorises and another (omomyids) to tarsiers.

The next evolutionary stage, which led to the platyrrhine (New World) and catarrhine (Old World) anthropoids, has been the subject of considerable debate. Although most primatologists consider both anthropoid groups to have had a monophyletic origin, questions remain as to whether they evolved from the adapids or omomyids and whether the platyrrhine monkeys arrived in the New World via dispersal from North America or by island hopping from Africa. At present, although there seems a preference for an omomyid anthropoid ancestry (Kay et al.), there is no general agreement. According to some researchers, it is even possible that neither adapids or omomyids were anthropoid ancestors but a different primate stock entirely (Martin 1993, Beard et al.). Platyrrhine geographic origin seems more soluble: since there are no Cenozoic prosimian fossils in South America or Cenozoic anthropoid fossils in North America, the platyrrhines probably did not originate in either continent. Rather, the New World monkeys may well have come from Africa, which, because of continental drift, lay closer to South America during the early and middle Cenozoic than it does today (Fig. 6–12). During this time, the two continents were probably spanned by one or more chains of islands that later submerged, some of which might have served as intermediate stations for the platyrrhine journey.

In any case, both paleontological and molecular dating suggest that an anthropoid lineage may be quite ancient, diverging from other primates by the Paleocene epoch more than 50 million years ago (Martin 1993, Takahata and Satta). This ancestral group then separated into platyrrhine and catarrhine forms: the former eventually colonizing South America as New World monkeys (ceboids); the latter evolving into Old World monkeys (cercopithecoids) and apelike forms (hominoids). Along the catarrhine trajectory is *Aegyptopithecus*, (Fig. 20–4*d*), a fairly primitive anthropoid fossil in the Fayum Province of Egypt, dating to the 30 million-year-old Oligocene epoch. In the Early Miocene epoch (about 20 or so million years ago) are found perhaps the earliest of hominoid-like fossils, the African *Proconsul* (Fig. 20–4*e*). More obvious hominoids, such as a Eurasian group called the *dryopithecines*, appear later in the Miocene.

Compared to their relatively fewer numbers today, apelike forms were more common during the Early and

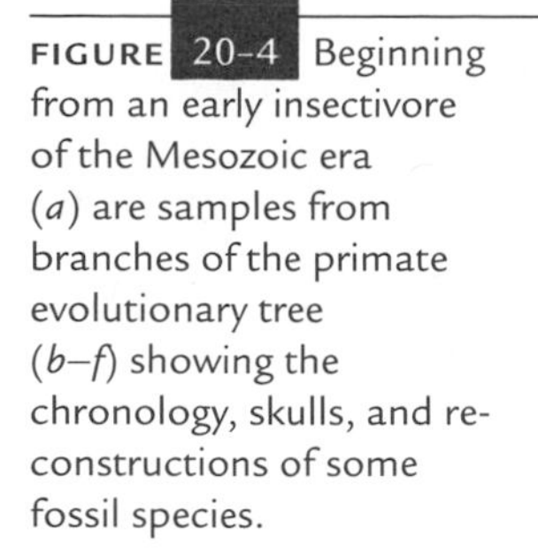

FIGURE 20-4 Beginning from an early insectivore of the Mesozoic era (*a*) are samples from branches of the primate evolutionary tree (*b–f*) showing the chronology, skulls, and reconstructions of some fossil species.

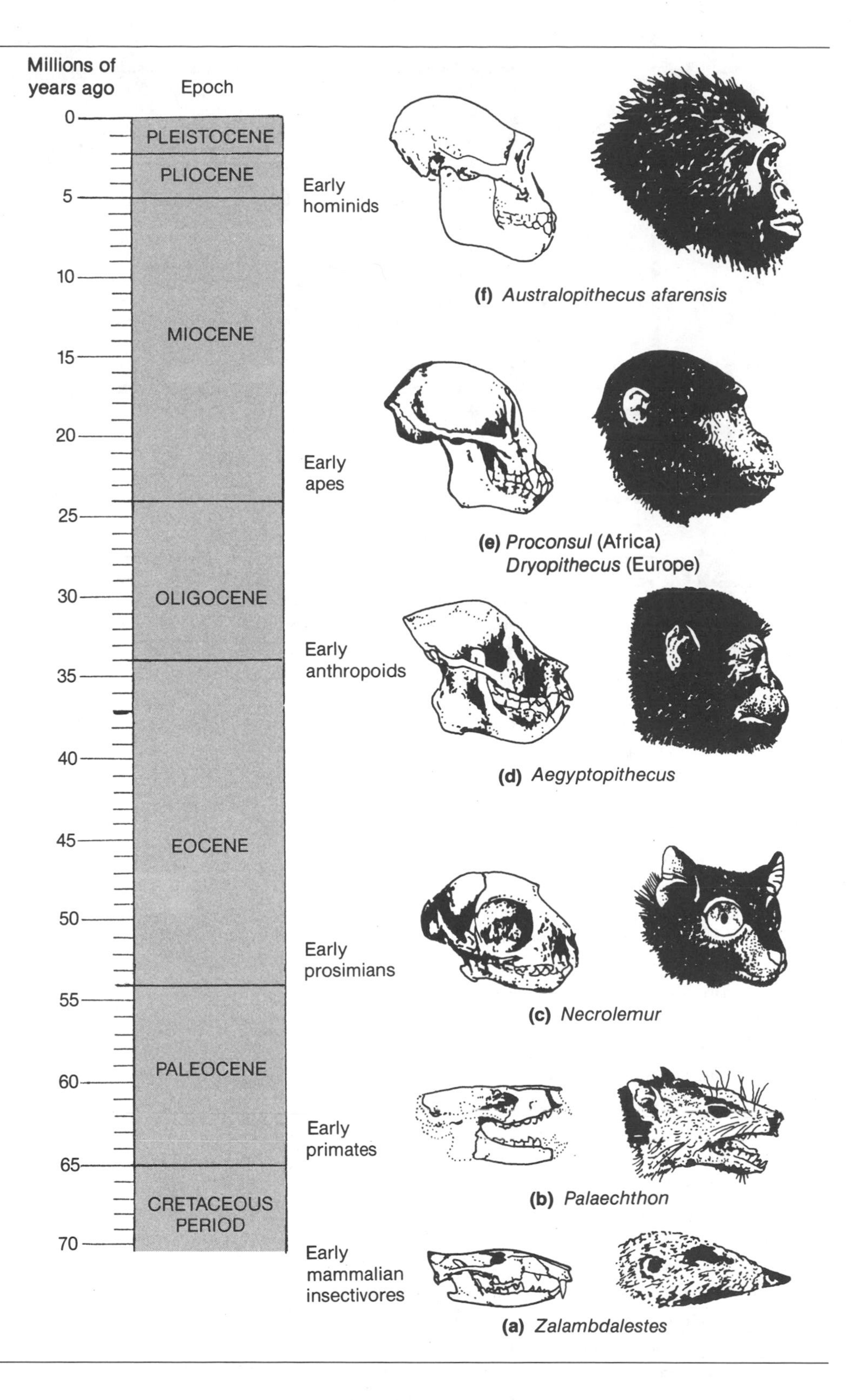

Middle Miocene than monkeys. This situation was reversed during the Late Miocene when monkeys became much more numerous and widespread than apes. Andrews (1981) suggests that dietary changes in Old World monkeys during the Miocene enabled them to compete successfully against many of the arboreal apes, perhaps by developing the ability to eat and digest fruits before they ripened enough for the hominoids. The more rapid reproductive rate of monkeys may also have allowed them to compete successfully with apes as well as radiate into

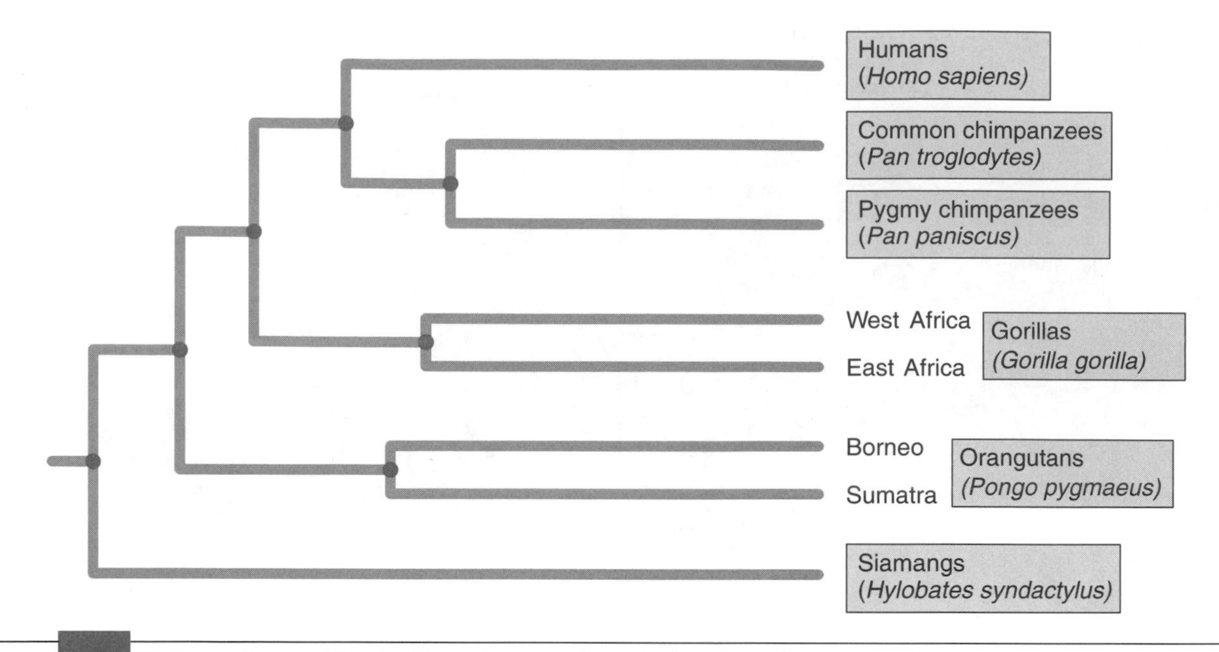

FIGURE 20–5 Phylogenetic tree of hominoid species and subspecies based on comparing nucleotide DNA sequences in the mitochondrial gene that produces the protein cytochrome oxidase subunit II. Connecting lines between taxa reflect relationship: longer lengths denote greater nucleotide differences and therefore greater evolutionary distance. Each node (*small circle*) represents the ancestral population that produced the derived taxa that follow to its right. The similarity between these results and data gathered from DNA-DNA hybridization (Fig. 12–11) indicates again that humans are more related to chimpanzees than to other hominoids. Moreover, these data point to the possibility that the genetic separation between gorillas in East and West Africa, and between orangutans in Borneo and Sumatra, may be sufficient to support species distinctions. (*Abridged and modified from Ruvolo et al. 1994.*)

new habitats such as savannas, whereas most apes (with the exception of protohominid lineages) became restricted to wet forest habitats.

One important consequence of monkeys replacing apes in various arboreal habitats was probably an increase in selective pressure among a few ape (and some monkey) species for ground-dwelling adaptations. The ancestral humans probably evolved from a group of these terrestrial Miocene apes, although we don't know the exact timing of this event. One view, originally based on the presumed humanlike lower jaws of a fossil ape called *Ramapithecus,* was that the ape–human split occurred about 12 to 14 million years ago or earlier. However, with the recent discovery of more complete ramapithecine fossils, researchers have discarded this view and now consider this group linked to a much earlier apelike lineage that may have been ancestral to orangutans (Wolpoff 1982; Andrews 1983; Pilbeam 1984). African apes and humans may therefore have diverged a long time after the ramapithecine radiations.

Supporting these lineages and divergences are molecular data of the types discussed in Chapter 12. Delarbre and coworkers, for example, show that the Homininae (chimpanzees–gorillas–humans) form a monophyletic group that split from other apes (orangutans and gibbons) about 8 million years ago. According to these researchers, one gene used in the immune system for recognizing antigens (CD8 β-chain gene) duplicated after the split, and its rate of mutation indicates an age of 8 to 9.5 million years.

For the human–chimpanzee divergence, the molecular dates reduce further, ranging from 3.5 to 5.5 million years ago (Sarich and Cronin) to 5.5 to 7.7 million years ago (Sibley and Ahlquist, see Fig. 12–11). Horai and coworkers, who performed one of the most intensive molecular studies of this problem by comparing all nucleotide base pairs in mitochondrial DNA, confirm this view and propose a chimpanzee–human separation of 4.9 million years ago. As shown in Figure 20–5, this separation took place within a chimpanzee–human group that had branched off earlier from a group that became ancestral gorillas. That no unequivocal hominidlike fossils (such as the australopithecines described below) appear before the Pliocene, about 4.5 million years ago, indicates that the hominid lineage is probably not much older than that.

The Australopithecines

In 1925, anthropologist Raymond Dart reported an early hominid fossil from a lime quarry at Taung in the Cape Province of South Africa. Ascribed to a new genus, *Australopithecus* (southern ape), the fossil consisted of the front part of the skull and most of the lower jaw (Fig. 20–6*b*) of a 6-year-

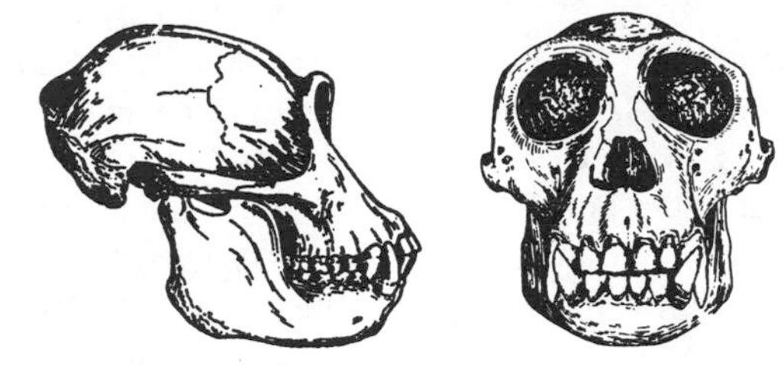

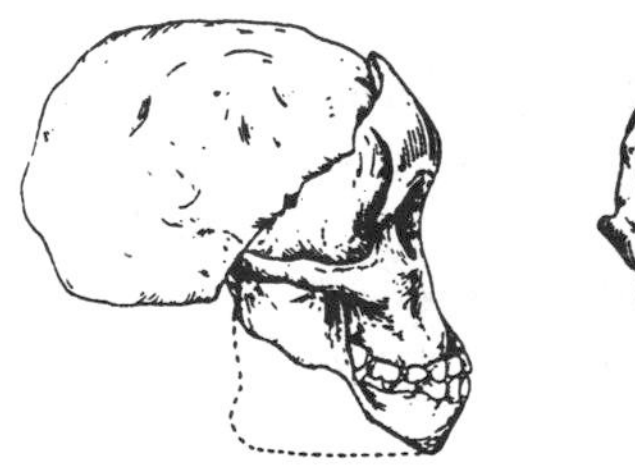

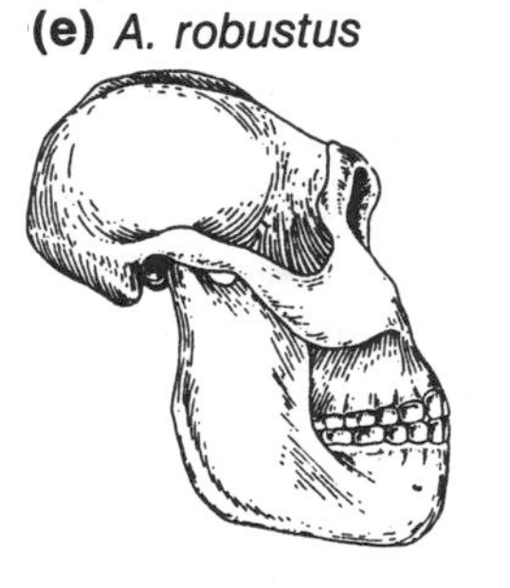
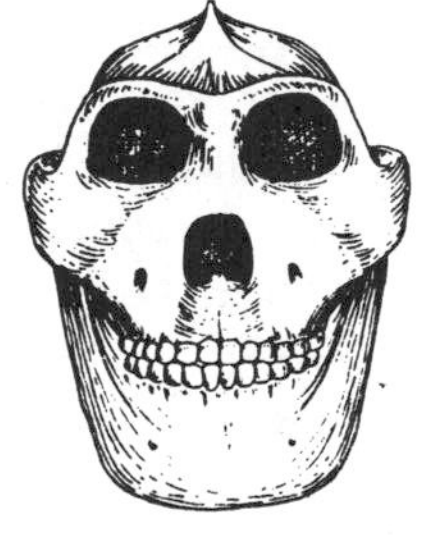

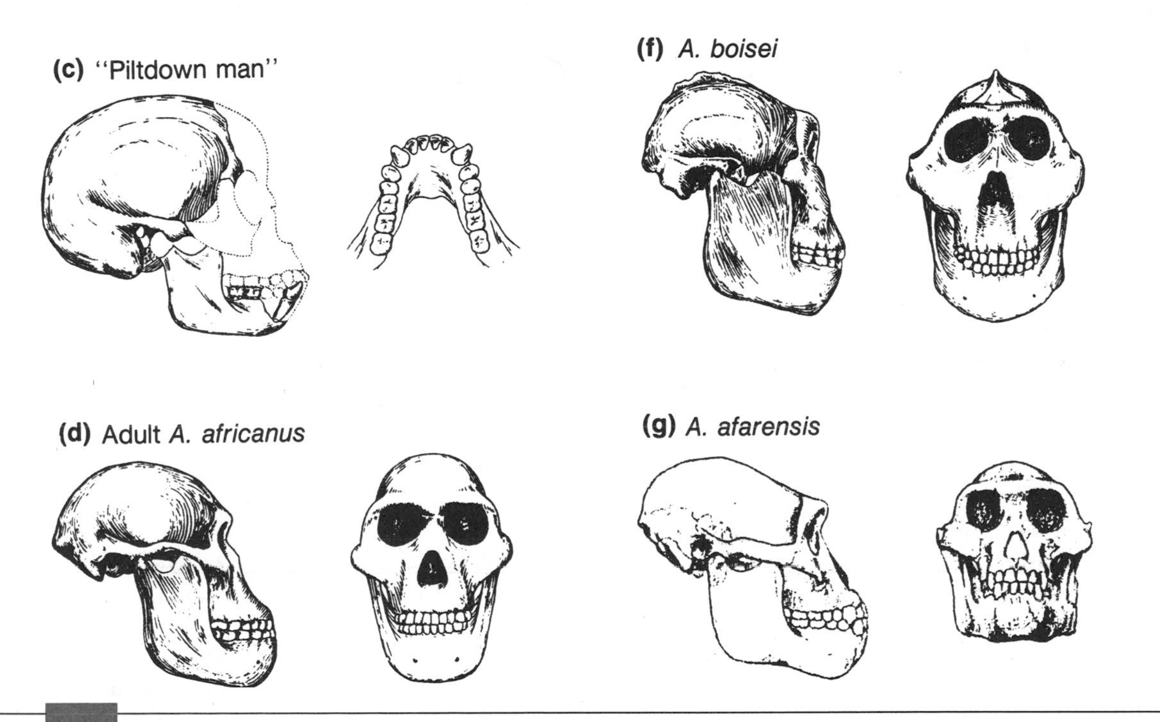

FIGURE 20-6 Skulls of a chimpanzee (*a*) and fossil hominids described in the text, including cranial and lower jaw fragments of the Piltdown forgery (*c*). (*Adapted from Johanson and Edey.*)

old (the "Taung child"), which he named *A. africanus.* All the deciduous teeth were present as well as the first of the permanent replacement molars. Although these teeth were generally larger than those of humans, they showed humanlike features in the multicusped nature of the anterior milk molar, which is single-cusped in apes. Also, the lunate sulcus (the anterior border of the visual area in the brain) in the endocranial cast was further back than its usual position in apes, more like that of humans.[1] Some interpretations of the Taung skull indicated that the adult brain volume was about 450 cc (midway between chimpanzee and gorilla) and that the adult body was probably

[1] In humans this sulcus is pushed back largely because the parietal cerebral areas responsible for symbolic associations connected with language and sequential reasoning have expanded.

smaller than that of chimpanzees, with weights between 40 and 70 pounds.

At the time of the Taung discovery, however, most anthropologists believed that primitive humans had large braincases and apelike jaws, with large canines. Evidence for this belief came from a so-called fossil cranium and lower jaw found in 1912 at Piltdown, England, that showed such features (**Fig. 20–6*c***). Unfortunately, many anthropologists accepted the Piltdown fossil as valid for about 40 years until a number of anthropologists showed that the entire skull was a hoax: the teeth had been artificially ground down, the cranium was of a different age than the jaw, artificial pigmentation had colored the bones, and the molar teeth had long roots like those of apes. Moreover, the associated animal fossils at the Piltdown site had a large accumulation of radioactive salts whose origin could be traced to a site in Tunisia. The **Piltdown man** therefore turned out to be a combination of a human cranium and the lower jaw of a female orangutan, a hoax perpetrated by someone who knew enough to destroy all obvious signs of the pseudofossil's true origin by removing the jaw joint and modifying other features.[2]

Because of the Piltdown forgery, it was more than 20 years after the Taung discovery before most anthropologists began to accept the humanlike nature of the australopithecine fossils. By then, anthropologists had found many such fossils in other sites in Africa, such as the adult australopithecine skulls at Sterkfontein not far from Taung (**Fig. 20–6*d***). The thickly enameled teeth of these fossils indicated heavy tooth usage that enabled the teeth to wear flat before the dentin was exposed, also perhaps indicating a longer life span than apes. The Sterkfontein fossils also showed that, although the australopithecine premolars and molars were larger than those of humans, the canines were smaller than those of apes and no longer projected above the tooth row. Paleontologists set the date for these fossils at about 2.5 to 3 million years ago.

Robinson and others showed post-cranial australopithecine skeletal material, such as the pelvis and vertebrae, to be humanlike, with a distinct lumbar curvature of the spine indicating erect posture. As in humans, vertical weight was transmitted through the outer condyle of the knee, and *Australopithecus* could walk bipedally, although this may not have been its exclusive mode of locomotion. Recent studies on australopithecine balancing organs (bony semicircular canals in the ear) suggest that bipedalism may have accompanied aerial climbing (Spoor et al.).

Uncovered in other South African sites is a somewhat larger hominid having a mature weight of about 80 or more pounds, called *A. robustus* (Fig. 20–6*e*). In addition to larger size, this group—often represented as an offshoot of *A. africanus*—had significantly larger teeth and jaws, which undoubtedly reflected a different, perhaps more herbivorous, diet (for example, seeds, nuts, and tubers) with emphasis on more powerful grinding. The brain, too, was larger than that of *A. africanus*, with a volume of about 550 cc, but this may have been primarily associated with increased relative body size rather than increased intelligence.

In East Africa a Pliocene australopithecine, *A. boisei* (Fig. 20–6*f*), apparently underwent selection for even larger molars than *A. robustus*. It is the largest of the australopithecines and shows various cranial adaptations for powerful masticating jaw muscles that indicate a diet of tough plant food such as seeds and fruits with hard husks and pods. Interestingly, an early robust australopithecine (*A. aethiopicus*) found in 2.5-million-year old deposits at Lake Turkana, Kenya, indicates that these massively built australopithecines may have been evolving separately but parallel to the *africanus–robustus* lineage. If we follow both these lineages forward from the Pliocene into the Pleistocene less than 2 million years ago, it seems that the molars were becoming larger while the face was shortening and becoming more vertical and humanlike. Their common skeletal similarities, according to McCollum, are the result of convergent evolution caused by "developmental by-products of dental size and proportion," rather than by shared cladistic synapomorphies. The cause for these convergent/parallel changes in two different lineages is difficult to understand, since we don't yet know the selective forces acting upon these groups.

Of an earlier age than the above groups are a series of fossils found in East African sites at Laetoli, Tanzania, and the Afar (Hadar) region of Ethiopia. Anthropologists have included these fossils, spanning an interval from about 3 to 3.9 million years ago, among the australopithecines under the species name *A. afarensis* (Fig. 20–6*g*). Figure 20–4*f* gives a reconstruction of the face of this very early hominid species, showing the heavy brow ridges, low forehead, and projecting (prognathous) mouth. In spite of some such similarities, *A. afarensis* displays a large number of important cranial, dental, and skeletal differences from apes. For example, although the canines are larger in *afarensis* males than in females, this sexual dimorphism is much less pronounced than in apes or early Miocene pongidlike fossils. Selection also apparently modified tooth positioning, enamel thicknesses, and the resulting wear facets to allow greater transverse jaw movements that retain cutting functions and improve side-to-side grinding.

[2] Spencer suggests that the Piltdown forgery was perpetrated by Charles Dawson, principal "discoverer" of the Piltdown fossils, in conspiracy with Arthur Keith, a leading British anatomist and physical anthropologist. Their motivations were presumably self-centered: Dawson's to become a Fellow in the Royal Society, and Keith's to promote his view of the antiquity of a large human brain. Although other authors suggest other suspects, no conclusive evidence has yet appeared identifying the perpetrator(s). Whoever caused this hoax, the result for a time was the preservation of false views with false facts. Fortunately, the events that followed showed that false facts can be challenged in science and false views replaced.

Perhaps the most primitive hominidlike fossils known to date are relics of an even earlier Ethiopian species named *Ardipithecus ramidus,* dating between 4.3 and 4.5 million years ago (White et al.). The species is believed to show hominidlike reduced sexual dimorphism for canine teeth as well as indications of an upright stance based on forward positioning of the foramen magnum (the aperture through which spinal nerves enter the base of the skull). Most *A. ramidus* fossils so far are teeth, which indicate relationships with both *A. afarensis* and the present great apes, especially chimpanzees. As its discoverers point out, "*A. ramidus* is the most apelike hominid ancestor known." Probably more than any other hominidlike fossil, *A. ramidus* deserves to be called the "missing link" between hominids and apes.

Unfortunately, because post-cranial fossils of *A. ramidus* are still to appear, questions about size and type of locomotion remain unanswered. It is therefore *A. afarensis,* among very early hominids, that has provided some relevant information. For example, the fairly complete skeleton of an *afarensis* female ("Lucy") found at Afar in Ethiopia shows a small, muscularly powerful body, perhaps only 3.5 to 4 feet in height, with relatively longer arms than modern humans but presenting a habitual bipedal stance and some form of bipedal locomotion. In support of such early bipedalism are footprints dated to about 3.7 million years ago, preserved under a layer of volcanic ash at Laetoli. These prints of two individuals who walked along the same path for a distance of more than 70 feet are of distinctly bipedal hominids, demonstrating that bipedalism must have preceded many other hominid adaptations such as increased brain size.

Since all australopithecine skeletal reconstructions starting with *afarensis* show the bipedal stance (Fig. 20–7), the later human genus, *Homo,* must have evolved from a population within this group. What is not yet clear are the exact lineages within these early hominid species: Was *ramidus* a bipedal australopithecine, or a tree-climbing

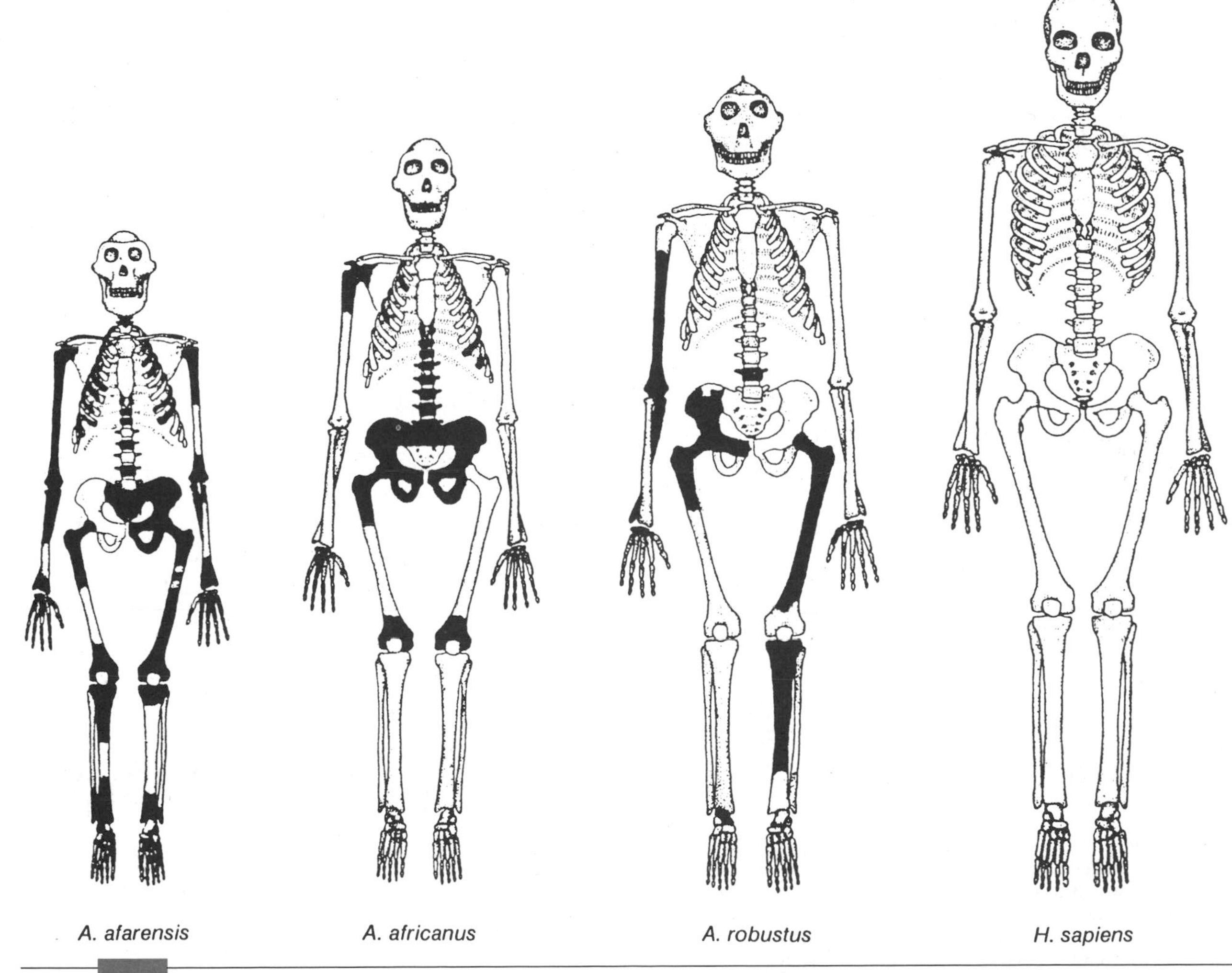

FIGURE 20–7 Skeletons of three australopithecines and a modern human. Black portions indicate the actual fossils found in the australopithecine skeletons. In all these homonids the pelvis is relatively shallow and rounded, and the femurs tilt toward the midline. This structure indicates that the pelvis is supporting the trunk, and body weight is transmitting directly downward through the hip and knees when the individual stands erect. Among other australopithecine humanlike traits are their forelimb proportions, relatively short toes, and presence of a large heel bone (calcaneous). *From* Lucy: The Beginnings of Humankind *by Donald Johanson & Maitland Edey. New York: Simon & Schuster, Inc., 1981. Copyright © 1981 by Donald C. Johanson and Maitland A. Edey.*

member of some other genus? Was *A. afarensis* actually an early form of *A. africanus?* Was *A. afarensis* so specialized in the direction of increased food grinding that its descendants could only have been heavy-jawed australopithecines such as *A. robustus* or *A. boisei*? Or do the fossils we ascribe to *A. africanus* represent two species, one allied to the "robust" australopithecines and the other ancestral to the *Homo* lineage? In any case, there is little question that *A. afarensis* stands at or very near the base of hominid phylogeny and indicates that we should regard the australopithecines as a group in which considerable evolutionary change was occurring, exemplifying rapid adaptive radiation of bipedal tropical apes.

Bipedalism

Although paleoanthropologists have established bipedalism as a long-standing feature in hominid lineages, they have disputed why and how it originated for over a century, and still find it a matter of controversy. Its importance is obvious, since it either accompanied or led to many adaptations that helped cast the future evolution of bipedal primates into a human framework. In a brief review of these controversies, Day (1986b) offers three arenas in which selective pressures might have enhanced bipedalism, each of which may have been influenced by the others:

- IMPROVED FOOD ACQUISITION Early hominids lived in a patchy environment of mixed woodland and savanna (relatively dry grassland and bushland with occasional trees) that provided seasonal food supplies.[3] This emphasized an omnivorous diet, demanding relatively more time spent searching for food over longer distances than in a more localized, homogeneous environment. An upright stance and bipedal striding would have enhanced **long-distance foraging** by enabling the manual transport of food gathered in different places. Tanner proposes that food gathering was originally a female function, prompted primarily by food sharing with their offspring, which led to the invention and use of food-gathering tools. Unfortunately, we don't know what kinds of food, plant or animal, were acquired and carried in these early journeys. A primary diet of dispersed plant foods would accord with the apelike teeth of early hominids. Yet some workers suggest that early hominids relied heavily on scavenging carcasses from migratory herds of ungulates, and bipedalism became important both for terrestrial locomotion and for manual transport of immature offspring.

  In a sense, bipedalism may have arisen as a byproduct of adaptations that reduced forelimb involvement in quadrupedal support and movement. As hands became increasingly specialized for grasping; manipulating; and carrying food, tools, and offspring, selection occurred for an upright stance and for transferring locomotion to hindlimbs.

- IMPROVED PREDATOR AVOIDANCE Since bipedalism enhances height, it improves a hominid's ability to see over tall grasses and obstructions and to wade in deeper water to pursue game or seek protection from predators. Day points out that the ability to climb trees would have helped in escaping predators and increased the field of view in detecting danger and surveying surroundings. The curved hand and foot bones and relatively long arms of early australopithecines and early *Homo* (*H. habilis*) point to persistent tree-climbing abilities.

- IMPROVED REPRODUCTIVE SUCCESS Lovejoy (1981) proposed that bipedalism enabled adult males to carry food manually to their females and offspring, who could then remain sequestered in a single locality, the **home base.** This mode of provisioning reduced the need for females to be continuously mobile in foraging both for themselves and their attached offspring as in other competing hominids, thereby offering three important advantages: (1) a relatively stable home base that provided more constant social relationships and perhaps closer mother–infant relationships that improved infant survival, (2) reduced infant injuries because infants no longer were attached to a continuously mobile mother, and (3) a reduction in the spacing between births by allowing parents to care for more offspring successfully.

Although people still debate the extent to which these proposals represent historical events, some with considerable vehemence (for example, see the collection edited by Kinzey), Foley (1987) points out that "evolution is as much about reproductive strategy as foraging behavior." It would certainly seem that a survival strategy dependent on bipedalism and a home base also uses other adaptations and preadaptations. For example, **sexual bonding**

[3] Evidence exists that beginning with the Late Miocene and extending through Pliocene–Pleistocene times, periodic decreases in global temperature took place, marked by the onset of ice sheet formation in Antarctica and glaciation in the northern hemisphere. As ice locked up water, various terrestrial areas became relatively dry, and open environments such as woodlands and grasslands replaced many rain forests in tropical regions. Tectonic changes may also have had important effects. For example, Coppens suggests that because of the Rift Valley, a depression that runs from Ethiopia in the Red Sea to Mozambique in the Indian Ocean, eastern Africa underwent environmental changes different from its western counterpart. According to Coppens, the striking absence of any trace of chimpanzee or gorilla stocks among the 200,000 East African vertebrate fossils dating around the time of early hominid evolution, points to a climatological barrier between East and West Africa that arose 7 million years ago.

between males and females can motivate male foragers to continue to provision their family group because of their ties to particular females and can also extend male involvement into helping parent their offspring, assuming they have good or reasonable assurance of paternity.

Certainly, one important element that encourages human sexual bonding and year-round copulation is the absence of seasonal estrus cycles marked by specific, externally recognizable signals ("concealed ovulation"). Human secondary sexual characteristics that persist from puberty onward stimulate continued interest of both sexes in sexual bonding. These traits include the relatively large penis in males, the enlarged mammaries and increased subcutaneous fat deposits in females (for example, buttocks),[4] as well as hair adornments and apocrine scent glands displayed in both sexes. Some of these sexual features are associated with the bipedal stance, and point to an evolutionary process that may have begun among australopithecines but developed more fully in *Homo.*

The lifestyle introduced by bipedalism, long-distance foraging, continuous sexual activity, and their many corollaries probably improved survivorship by

- Intensifying the involvement and "investment" of parents in their offspring
- Extending children's learning period
- Promoting a supportive and familial relationship among siblings whose births could be placed closer together

In an important sense, this evolutionary stage led to strengthening of ties between related individuals, on the level of both the hominid nuclear family and its more extended kinships. In effect, it enlarged the size and reach of the social group and increased interactions among its members.

To these significant advantages we can add that bipedalism fosters the use of manual weapons such as stick wielding and stone throwing, which extends the reach of hominids beyond the teeth, claws, and other defenses of animal competitors, predators, and prey. Bipedalism also allows hominids to carry even primitive tools and weapons from place to place, as well as to move offspring from one camp to another or from one food resource to another.

[4] Cant points out that there is (or was) a relationship between reproductive success and the size of female breasts and buttocks in the sense that these anatomical parts provide an easily visible signal to males of the degree of feminine fat reserves. Minimal levels of fat reserves are essential for continued ovulation and lactation, and fat breasts and buttocks probably interfere less with bipedal locomotion than fat deposits elsewhere. He suggests that the evolution of such localized fat deposits began as an expression of the feminine nutritional state and was probably reinforced by its use as an attractant in sexual bonding, helping reproductively successful choices to be made and maintained between sexual partners.

Anatomically, bipedalism is based on broadening of the pelvis, changes in hind-limb muscle origins and insertions, and convergence of the femora toward the knee, conferring a knock-kneed stance compared to apes. In quadrupedal vertebrates, the hip bone, or innominate, of the pelvis has three components (ilium, ischium, and pubis) that link the hind limbs to the spine and help provide the propulsive force for quadrupedal motion. In animals such as the tree shrew, the ilium is a long, narrow blade that lies alongside the sacral vertebrae. It lengthens further in primates as more hind-limb power is used to leap from place to place. The ischial tuberosities also widen in many primates, in association with selection for stability in sitting. The more erect anthropoids also used their pelvis for visceral support and, as a consequence of such selection, the ilium has become wider and more of the sacral vertebrae are fused into it (Fig. 20–8).

In bipedal hominids, the pelvis assumes a broad, shallow, bowl-like shape with a widened but shortened ilium, bringing the sacrum closer to the acetabulum, the femoral socket. These changes, together with forward curvature of the lower spine and further flattening of the thoracic cage, help transmit the weight of the trunk directly to the legs, producing a balanced center of gravity along a vertical axis (Fig. 20–9).

Normal hind-limb musculature in mammals involves flexors and extensors that move the femur forward or backward in relation to the pelvis, as well as abductors and adductors that control the lateral positioning of the trunk and pelvis on the legs. Since humans move terrestrially while standing erect, they have mostly attained the necessary leverages for these muscles through attachments provided by the broadened pelvis and its various bony projections, such as the quadriceps group (rectus femoris) that swings the leg forward (Fig. 20–10).

On the dorsal surface, the rearward development of the iliac spine enables the gluteus maximus muscle, formerly an abductor of the thigh, to become an extensor that provides the important power stroke in running and climbing. Other changes from apelike bipedalism include the transformation of the gluteus medius and minimus, formerly extensors, into abductors that balance the body laterally during walking. Moreover, since the hominid leg is aligned in a straight vertical line—from hip joint to knee joint to ankle joint to foot surface—body weight efficiently transfers downward directly through the bones without tension rather than through the muscles. Convergence of the femora toward the knees ensures that the axis of weight remains close to the center of gravity. The pelvis rotates around this axis during walking, and the body can be kept in a forward plane by swinging the arms.

Below the knee, the tibia bears all the weight of walking, and little, if any, transmits through the fibula. Whereas in tree-dwelling primates the fibula is useful in revolving the foot, such rotation has been reduced in humans since they walk and run on the ground, so the fibula is also reduced.

FIGURE 20-8 Comparison between the pelves of a chimpanzee (*a*), an australopithecine (*b*), and a modern human (*c*). Frontal views are on the *left* and lateral views on the *right*. Note that the distance between left and right acetabuli in *A. africanus* is less than in *H. sapiens*. This increase in interacetabular distance in humans apparently resulted from selection for a relatively large birth canal to permit the passage of newborn infants with larger crania than australopithecine newborns had. However, widening of the human pelvis is limited to the upper part since a further increase in the interacetabular distance of the lower part would splay the legs outward and make bipedal walking more difficult. As a result, during childbirth human babies enter the upper part of the pelvis facing sideways and then rotate 90 degrees to emerge from the lower narrower part with their head facing downward. It is this tight "corkscrew" squeeze that often causes obstetrical problems in human births and necessitates assistance. Some anthropologists point to these common obstetrical difficulties as evidence for an early selective impetus to improve communication and cooperation among humans that, among other factors, would have led to larger brain size (Rosenberg and Trevathan, see also Box 20-3). (*Adapted from LeGros Clark.*)

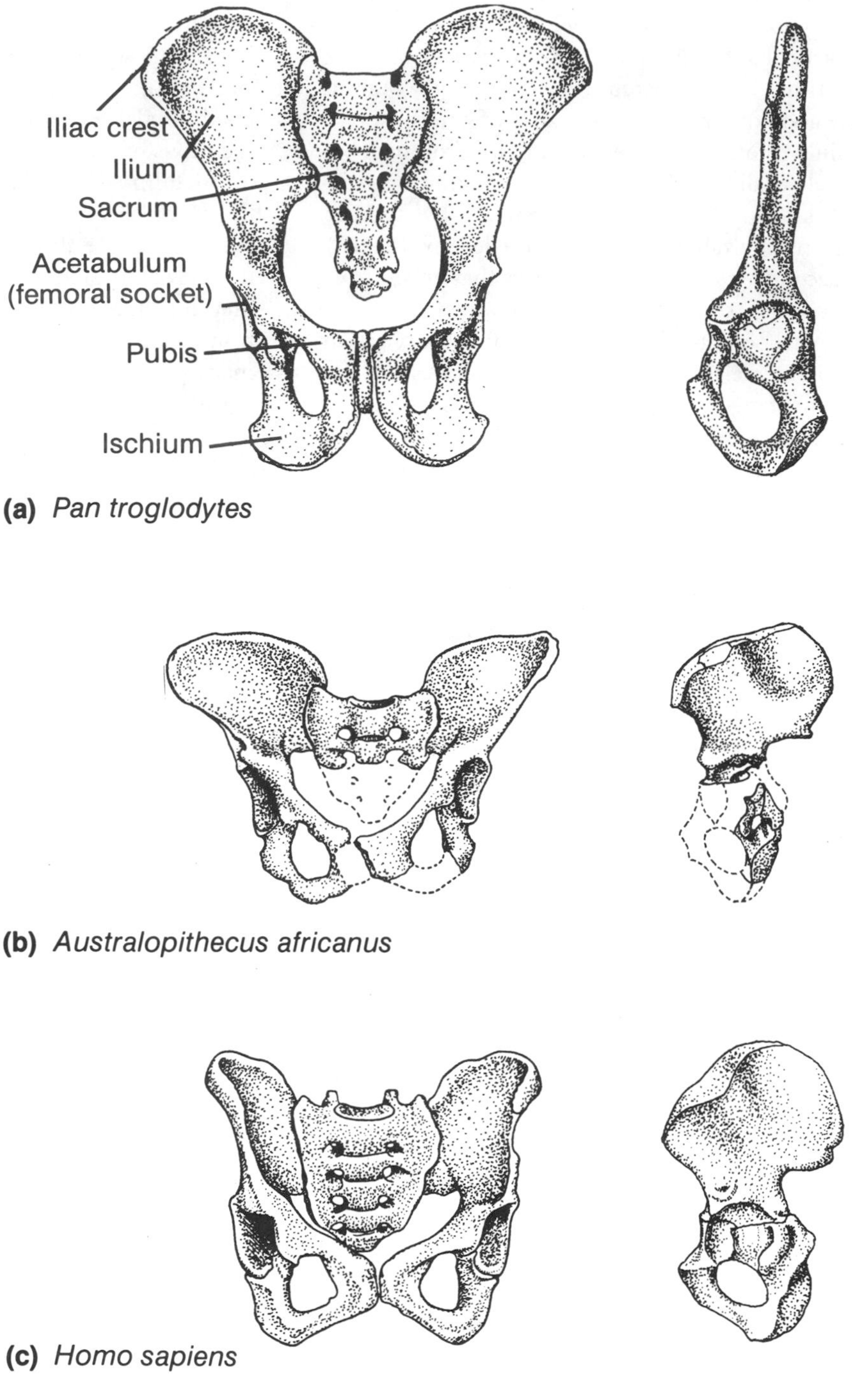

Iliac crest
Ilium
Sacrum
Acetabulum (femoral socket)
Pubis
Ischium

(a) *Pan troglodytes*

(b) *Australopithecus africanus*

(c) *Homo sapiens*

Walking also involves changes in the foot to provide both upright balance and striding power. Hominids achieved balance by a three-point weight distribution between the heel and two points on the ball of the foot, the inner (first) and outer (fifth) metatarsals. They achieved power by sequentially transferring weight from the heel to the fifth metatarsal to the first metatarsal to the big toe that serves as the "pushoff" (Fig. 20–11). In fact, the shape of the big toe alone is extremely helpful in determining whether an individual is capable of humanlike striding.

As Figure 20–9 implies, the transition to bipedalism did not come directly from a baboonlike quadrupedal form but from a knuckle-walking apelike form living mostly or entirely on the ground. Bipedal modifications allowed hominids to accomplish sustained walking without apelike bipedal shifting of the trunk from side to side and therefore with relatively minor expenditures of energy. On the physiological level, Wheeler has pointed out that bipedalism, along with hairlessness and increased sweat gland density, offers thermoregulatory advantages

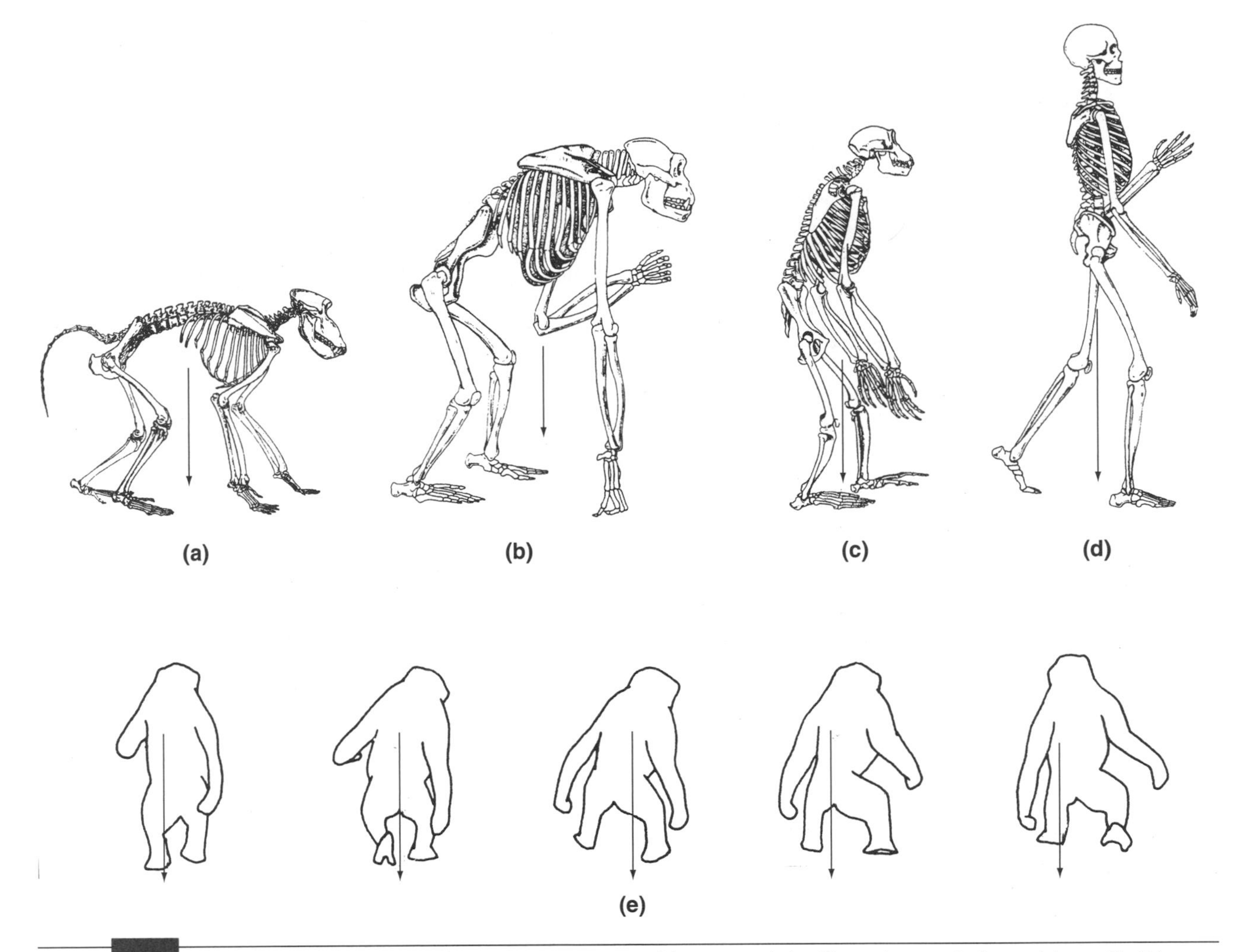

FIGURE 20–9 Centers of gravity (*arrows*) in the common stances of four ground-dwelling primates: baboon (*a*), gorilla (*b*), chimpanzee (*c*), and human (*d*). When humans walk bipedally the center of gravity remains in approximately the same vertical plane as the skeletal axis, the spinal lumbar curve helping to center upper body weight in line with hip and knee. Since their center of gravity lies in front of the hip, apes walking bipedally perform strenuous muscular activity to keep from falling forward—like humans trying to walk in a forwardly tilted or crouched position. Also, the abductor muscles between femur and hip used for lateral stability in human walking are used as extensors in apes, so a standing ape lifting its left leg, for example, will tilt toward the left unless it bends its trunk over the right leg to regain balance. A walking bipedal ape (*e*) thus tilts from side to side, causing shifts in the center of gravity that must be corrected by muscular exertion.

to a "naked ape" who must deal with heat stress while searching for food in a hot patchy savanna environment.

Homo

The earliest fossils ascribed to the genus *Homo* have been found in both East and South Africa and date between 2.2 and 1.8 million years ago at the Pliocene–Pleistocene boundary. These early forms, first named *Homo habilis* (Fig. 20–12*a*), were about as short as *A. afarensis,* with males at 4.5 feet tall; a recent fossil individual with a height of about 3 feet has also been found. Their cranial capacities, however, were at about 600 to 700 cc, pointing to a significant departure from the australopithecines. Associated with their appearance are artifacts, clearly indicating that these new hominids were engaged in making regularly patterned **stone tools,** products of the stone industry named Oldowan (**Fig. 20–13*a***).

These tools were possibly used in hunting and butchering animals (including small reptiles, rodents, pigs, and antelopes) and probably used also in scavenging carcasses of animals as large as elephants. Some tools date back 2.5 million years, indicating that even groups of more primitive australopithecines also may have engaged in scavenging and hunting, and had probably already embarked on using simple stone tools, bones, and sticks.

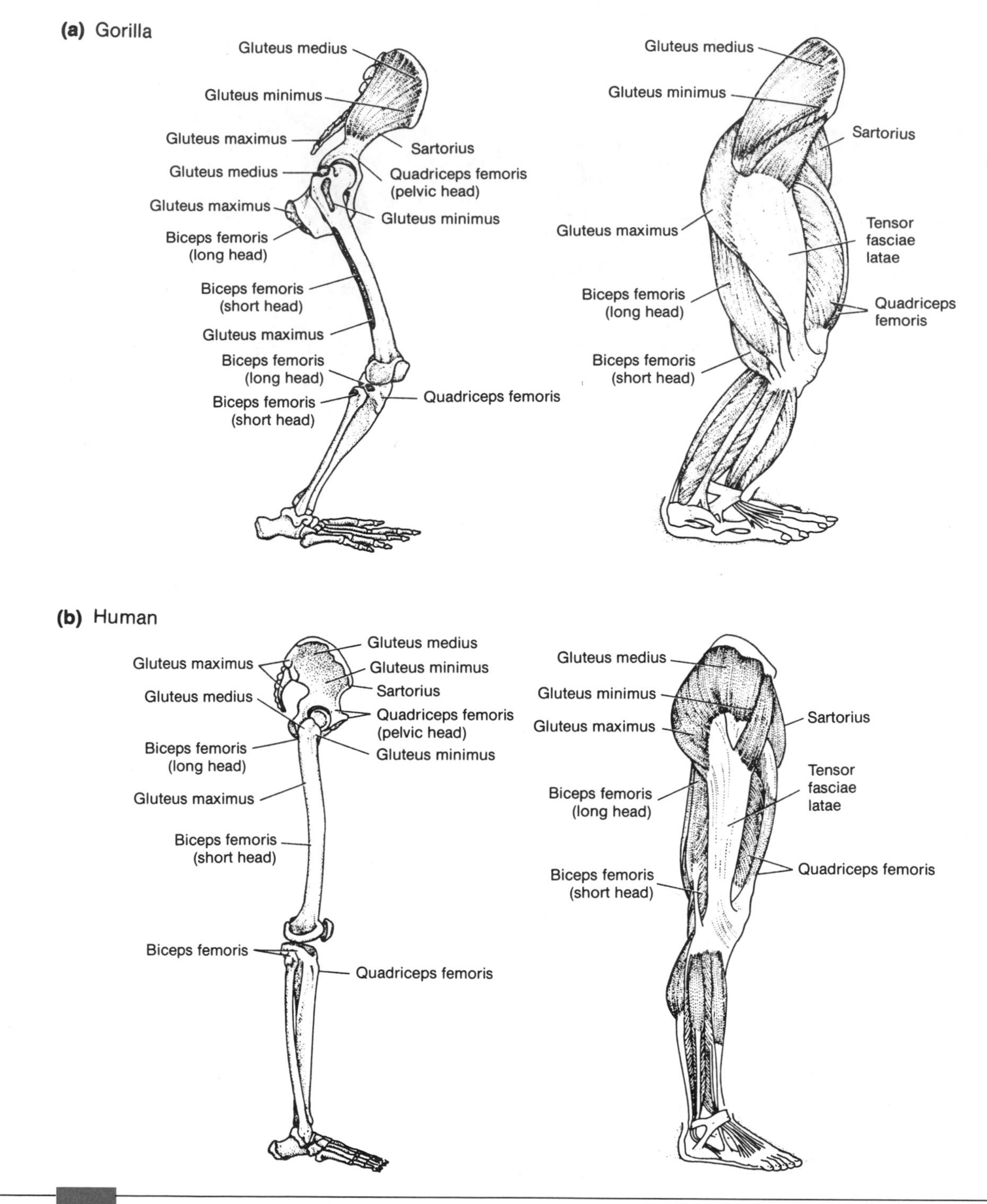

FIGURE 20-10 Bones of the hind limbs (*left side*) showing origins and insertions of the major muscles (*right side*) of a gorilla (*a*) and human (*b*). In both primates the muscles that cause the femur to swing forward in relation to the pelvis (flexion) are the sartorius and quadriceps. However, the broadened human pelvis provides leverage (pulling power) for these muscles when the individual is standing erect, whereas the gorilla can only gain such leverage in the bent position shown. Similarly, the human ability to swing the femur backward (extension) can be accomplished in the erect position because the gluteus maximus muscle that extends the femur attaches to a rearward projection of the pelvis. Extension of the gorilla femur, in contrast, again depends on its bent position, using the biceps femoris as an extensor. For side-to-side positioning of pelvis and femur during abduction (swinging the femur laterally outward in respect to the pelvis), the gorilla uses mainly the gluteus maximus muscle (converted to an extensor in humans), whereas humans rely on the gluteus medius and minimus. (*From* Paleoanthropology, *1980 by M.H. Wolpoff. Reprinted by permission of The McGraw-Hill Book Company. See also Lovejoy 1988.*)

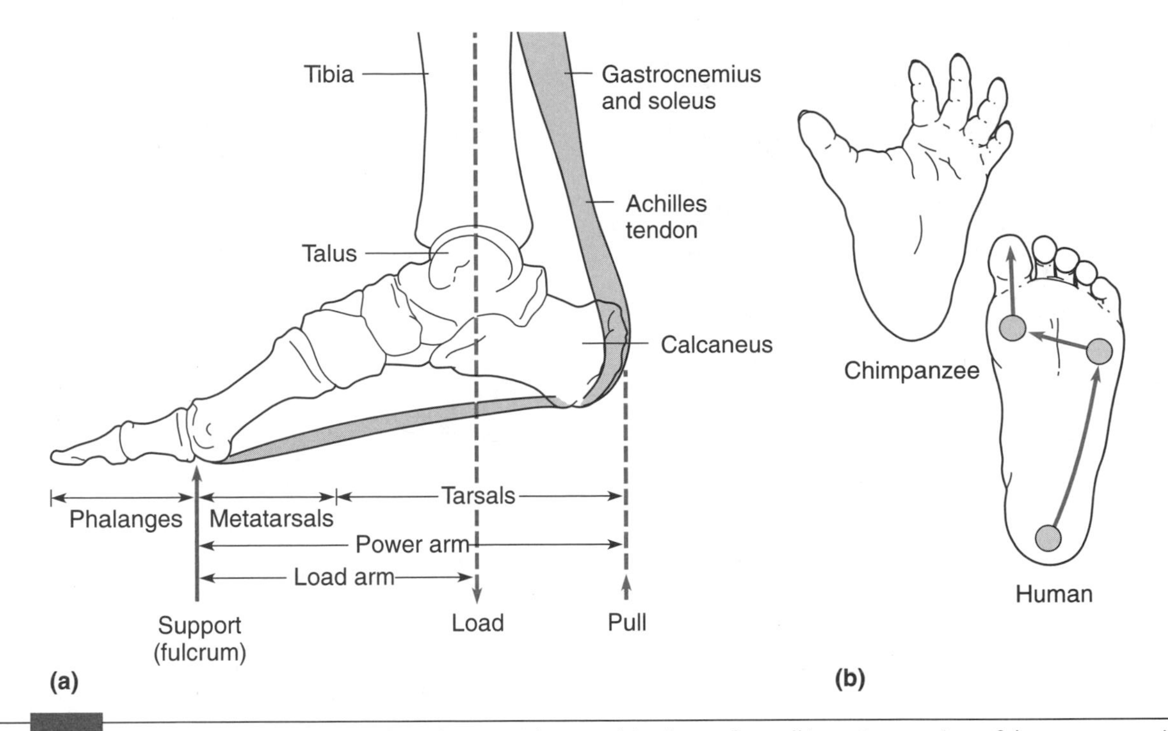

FIGURE 20–11 (*a*) Leverages in the human foot that provide propulsive forces for walking. Contraction of the gastrocnemius and soleus muscles in the calf of the leg pulls the Achilles tendon attached to the heel (calcaneus). This produces the power (power arm) that transfers the weight load (load arm) to the metatarsals. For light-bodied leaping primates the load arm is almost as long as the power arm, thus producing the springing jump, whereas in humans and heavier primates the load arm is relatively shortened, and power is increased at the expense of leaping. (*b*) Plantar view of chimpanzee and human feet. In the chimpanzee, the first metatarsal and its accompanying phalanx (big toe) are at a marked angle to the other metatarsals, the phalanges are relatively long and curved, and the foot can be used for grasping. In humans the phalanges are reduced, and the more robust first metatarsal and big toe are set parallel to the others, thus enhancing the ability to walk or run directly forwards. Note that although the human foot is narrow, it acts as a tripod on which weight is stably distributed (*indicated by the three circles*), and the *arrows* show how weight transfers between these three centers to the "pushoff" on the big toe. (*Adapted from Campbell.*)

About 1.8 million years ago, somewhat after the *H. habilis* fossil period, new groups of hominid fossils appear that are taller than their predecessors, reaching 5.5 feet or more. These individuals are also distinguished by thicker skulls, heavier brow ridges, smaller teeth, and larger brain volumes (750 to more than 1,000 cc). In 1891 in Trinil, Java, Dubois discovered the first of these fossils, now named *Homo erectus* (Fig. 20–12*b*), and others have since been found in Africa, China, and Europe. Associated with *H. erectus* are occasional signs of the use of fire and considerable use of stone tools including large hand axes of the Acheulean type (Fig. 20–13*b*). Much of the evidence indicates that most, if not all, *H. erectus* groups had entered into full-scale hunting, with large animals such as deer, elephant, and wild boar among their prey.

If we knew them all, the various distinctions among different *H. erectus* groups over time would probably be enough to mark off new evolutionary levels and perhaps even new species. These distinctions, however, encompass anatomical and behavioral traits that we cannot always see on the fossil level, and therefore we don't know at which points separations or transitions occurred. Were there parallelisms? Were there convergences? Obviously so. Common selective factors may well have caused adaptational similarities in different groups, clouding distinctions between *Homo erectus* and *Homo sapiens,* our own species.[5]

For example, hominid fossils found near the Solo River in Java, dated to less than 250,000 years ago, show brain volumes averaging 1,100 to 1,200 cc, significantly larger than those of middle Pleistocene *H. erectus* fossils from the same area; yet they are like older fossils in respect to prominence of brow ridges and some other features. European fossils (from Swanscombe in England, and Steinheim in Germany) dated to about 200,000 years ago also show such increased brain volumes as well as anatomical traits intermediate between *H. erectus* and *H. sapiens.*

If it is true that *H. sapiens* evolved from an *H. erectus* group, this change is not clear-cut and seems to be marked by gradual changes. Whether such findings indicate that

[5] Some writers insist there is too much variation among *H. erectus* fossils and too little distinction from later forms to define it as a species. "*H. erectus* is but an early version of *H. sapiens*" (Wolpoff and Caspari).

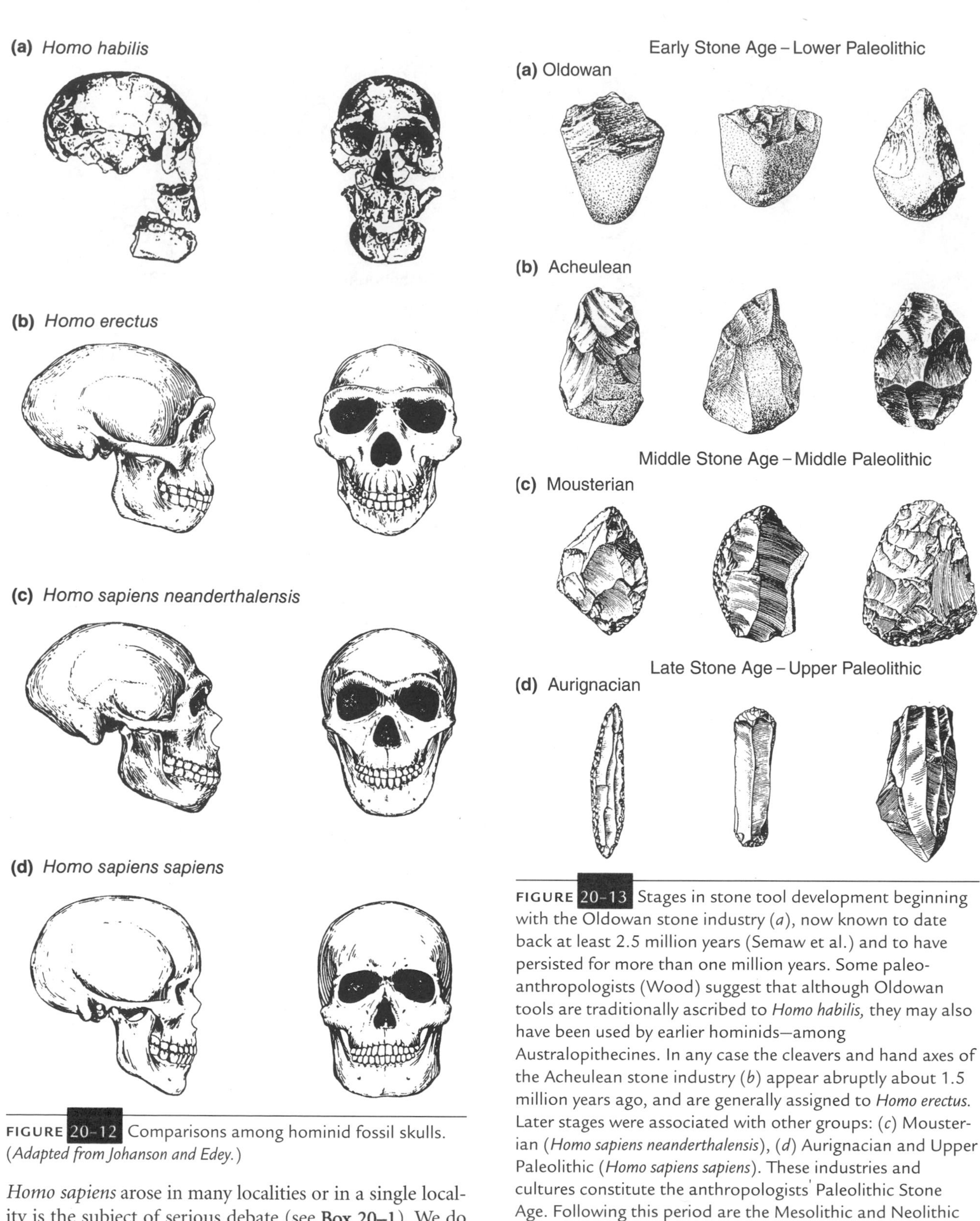

FIGURE 20-12 Comparisons among hominid fossil skulls. (*Adapted from Johanson and Edey.*)

FIGURE 20-13 Stages in stone tool development beginning with the Oldowan stone industry (*a*), now known to date back at least 2.5 million years (Semaw et al.) and to have persisted for more than one million years. Some paleoanthropologists (Wood) suggest that although Oldowan tools are traditionally ascribed to *Homo habilis,* they may also have been used by earlier hominids—among Australopithecines. In any case the cleavers and hand axes of the Acheulean stone industry (*b*) appear abruptly about 1.5 million years ago, and are generally assigned to *Homo erectus.* Later stages were associated with other groups: (*c*) Mousterian (*Homo sapiens neanderthalensis*), (*d*) Aurignacian and Upper Paleolithic (*Homo sapiens sapiens*). These industries and cultures constitute the anthropologists' Paleolithic Stone Age. Following this period are the Mesolithic and Neolithic ages, the latter beginning about 10,000 years ago and marked by polished stone tools, pottery, domesticated animals, cultivated plants, and woven cloth.

Homo sapiens arose in many localities or in a single locality is the subject of serious debate (see **Box 20–1**). We do know that by the time the European Neanderthals appear (*H. sapiens neanderthalensis*), along with similar types found in Shanidar (Iraq)—that is, between 50,000 to 100,000 years ago—humans had reached their present brain volume averages of 1,300 to 1,500 cc.

Although the Neanderthals are somewhat shorter than modern humans and show distinctive brow ridges, large

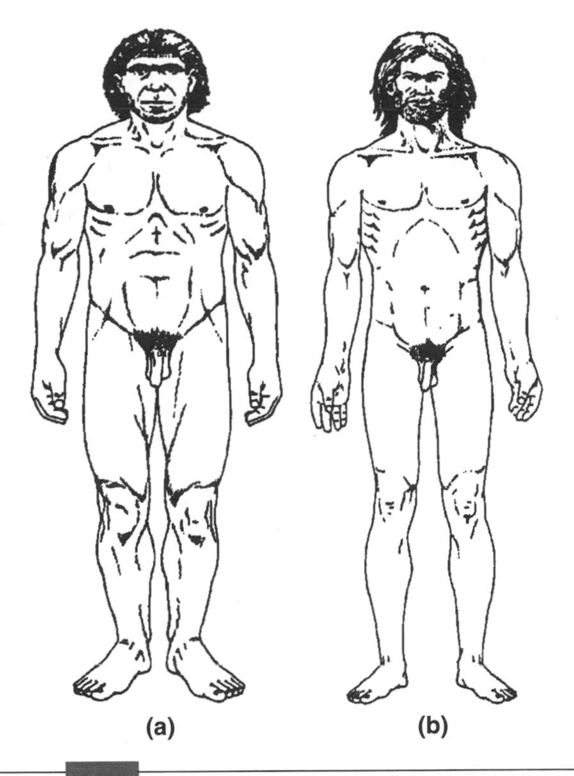

FIGURE 20–14 Reconstructions of the generalized Neanderthal male body form (*a*) and the anatomically modern Cro-Magnon (*b*). According to some anthropologists, the stockier body form of the Neanderthal came from selection for adaptation to colder climates (see also Fig. 24–4). Increased muscularity went along with their stockier build, giving Neanderthals a body weight about 30 percent greater than average modern humans (Ruff et al.). *Reprinted with permission from* Pour La Science: Pour La Sciencen, *64, 2/1983.*

jaws, small chins, and other anatomical relics (Fig. 20–14 and Fig. 20–12*c*), they were socially and behaviorally quite advanced in many respects: they were apparently skillful hunters of large animals such as the cave bear and mammoth; they produced many complex stone tools; and they apparently performed ritualistic social ceremonies, including placing flowers in graves of their dead. Many anthropologists, although not all (refer to footnote 6), feel that the Neanderthals probably deserve as full a membership in *H. sapiens* as the higher-skulled races, that is, *H. sapiens sapiens* (Fig. 20–12*d*), which began to replace the Neanderthals in various parts of the world about 40,000 years ago.

Among the earliest fossils of the more modern humans are those found in Mount Carmel in Israel, dated to about 90,000 years ago (Stringer et al.). Other transitional forms are all associated with the Mousterian stone age industry (Fig. 20–13*c*). Since the Mousterian culture is also associated with the Neanderthals, there must have been some important cultural overlap between these various *H. sapiens* groups.

About 35,000 years ago, an era named the Upper Paleolithic (the last part of the Old Stone Age) began in Europe, characterized by new methods of flaking flint to form stone tools. A marked change in human fossils accompanied these cultures, among which the earliest was the Aurignacian (Fig. 20–13*d*). The Neanderthals were apparently then replaced by anatomically distinct types (often called Cro-Magnon) with smaller brow ridges, higher skull vault, and a smaller and less prognathous face.

Behaviorally, as Table 20–4 shows, the evolving hominid lifestyle followed a pattern of increasing technological sophistication and expanded use and control of the environment. Many examples of representational art appeared, painted on cave walls and sculpted in clay or bone. Some anthropologists propose that Cro-Magnon success and enhanced artistic expression may have been the result of improved social organization and improved language ability (p. 504). In essence, "anatomically modern" humans had arrived in Europe and elsewhere, and we can consider these new forms to be among the present human races.[6]

Figure 20–15 shows one possible phylogenetic scheme that broadly traces relationships among various known hominid fossil groups beginning with forms dating back more than 4 million years. Because hominid fossil finds are spotty, we don't know specific evolutionary events that took place among these groups, although many of the fossils are transitional. Were there, for example, two separate australopithecine lineages, as the figure shows, or should we combine *A. robustus* and *A. boisei* in a single lineage? Also, because of marked pelvic distinctions and/or presumed limited language abilities (p. 495), some researchers suggest considering Neanderthals as a separate species, *Homo neanderthalensis*, rather than as a subspecies of *Homo sapiens*. Taking all the many hominid variations into account, it is clear that, "Instead of a ladder with humans at the pinnacle, there is a bush with humans as one little twig" (Foley 1995).

A major dispute concerns the last twig of this phylogeny—the origin of modern human races. This topic, discussed further in Box 20–1 (on pp. 486–488) gained considerable attention with a 1987 publication by

[6] Little is known about the abrupt disappearance of the Neanderthals; some of their populations may have died out, whereas others may have merged into the new dominant forms. One hypothesis suggests that the Neanderthals represent a separate offshoot of the human line, differing from both the *Homo sapiens* groups that preceded it and those that followed (Stringer and Gamble). In support of this view are findings that recovered sequences of Neanderthal mitochondrial DNA are outside the limits of normal *Homo sapiens* variability. According to Krings and coworkers (see also p. 279), this evidence indicates that "Neanderthal mtDNA and the human ancestral mtDNA gene pool have evolved as separate entities for a substantial period of time and gives no support to the notion that Neanderthals should have contributed mtDNA to the modern human gene pool." Rak points out that the extraordinarily large Neanderthal face resulted from biological innovations that allowed strong biting forces to be exerted on their front teeth, which were much larger and had deeper roots than in other *Homo sapiens* groups. The fact that their incisors and canines often show heavy wear indicates that the Neanderthals may have used these teeth for processing tough foods or hides, or both. Presumably, the expanded nasal chamber in their large face may have served as a radiator, warming and humidifying inspired air in the dry, cold, glacial climates that many European Neanderthals inhabited.

TABLE 20-4 Brief inferences about adaptions, behavior, and ecological factors in hominid evolution

Lineage[a]	Approximate Time (Years ago)	Adaptations, Behavior, and Habitats	Fossil and Archaeological Evidence
Hominid ancestors	8–5 million years	Relatively large-bodied apes distributed in Central and Eastern Africa across forest-woodland mosaics	No fossil evidence yet, but when found, expected to be a group or groups ancestral to humans and chimpanzees
Australopithecines	4–2 million years	Bipedal on the ground, occasionally arboreal Open savanna, and mosaic grassland and woodland habitats Fibrous plant diet that may also have included meat[b]	Extensive fossils in Eastern and Southern Africa Large teeth and jaws
Homo habilis	Pliocene–Pleistocene: 2–1.5 million years	Improved bipedalism Tools to procure and process food Habitats in drier areas indicating larger home ranges Scavenging and active animal hunting	Skeletal changes and increase in brain size Early stone tools
Homo erectus	Early-Mid Pleistocene 1.5–.5 million years	Entry into new habitats and geographical zones Definite preconception of tool form Manipulation of fire Increased level of activity and skeletal stress	Fossils found in formerly unoccupied areas of Africa, and outside Africa Development of a stone tool industry Archaeological hearths Increased cranial and postcranial development
"Archaic" *Homo sapiens*	Mid Pleistocene 500–150 thousand years	Geographical divergence and ecological adaptations More complex tools	Old World distribution with some distinct regional morphologies Bifacial axes: Acheulean–Mousterian stone tool industries
H. sapiens neanderthalensis	Late Pleistocene 150–35 thousand years	Large and robust individuals More social complexity and development of ritual Increasingly sophisticated tools	Massive cranial and postcranial development Intentional burial of the dead Increased number of stone-tool types
H. sapiens sapiens	Late Pleistocene to Present	Decreased levels of activity and skeletal stress Expansion of technology Development of complex cultures Increase in population size	Appearance of "anatomically modern" humans From Upper Paleolithic (Aurignacian) stone tools to satellite communication Beginnings and expansion of agriculture

Abridged from tables in Foley (1996) and Potts, with modifications.

[a] "Lineage" designates the name commonly given to a major group found in the specified period. As the text indicates, other names have been used for fossil groups in these periods (for example, *H. ergaster, H. rhodesiensis, H. heidelbergensis*). Various groups also overlapped.

[b] New findings of 2.5 million-year-old hominids in Ethiopia suggest that behavioral changes associated with lithic (stone tool) technology and enhanced carnivory (butchered antelopes, horses, and other animals) may have been coincident with the emergence of the *Homo* clade that arose from *Australopithecus afarensis* in East Africa (de Heinzelin et al.).

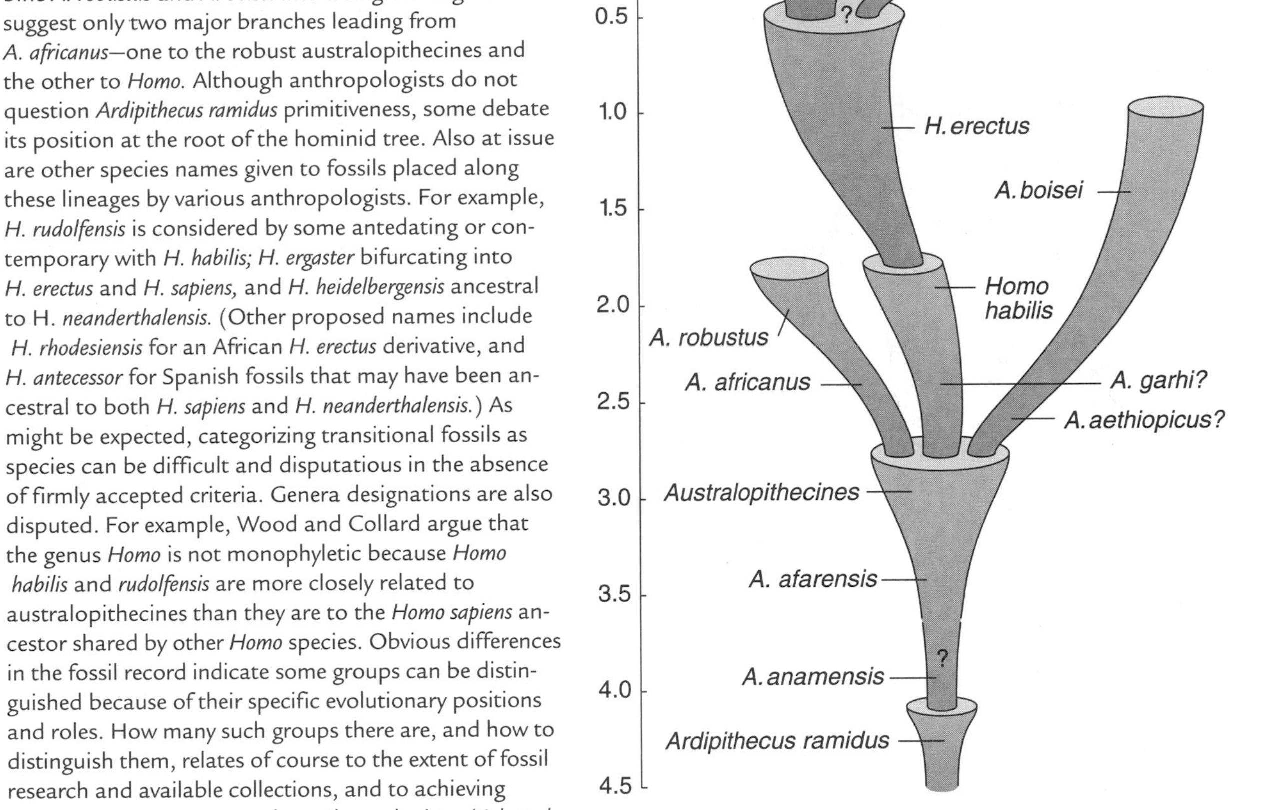

FIGURE 20–15 One of the possible phylogenetic schemes depicting evolutionary relationships among presently known hominid groups. This three-pronged tree is based on findings of an *A. boisei* form (*A. aethiopicus?*) that indicate the *boisei* line was evolving independently and parallel with the *A. africanus–A. robustus* line (McCollom), and that another australopithecine lineage, the Ethiopian *A. garhi,* is the "candidate ancestor for early *Homo*" (Asfaw et al.). Some schemes combine *A. robustus* and *A. boisei* into a single lineage and suggest only two major branches leading from *A. africanus*—one to the robust australopithecines and the other to *Homo.* Although anthropologists do not question *Ardipithecus ramidus* primitiveness, some debate its position at the root of the hominid tree. Also at issue are other species names given to fossils placed along these lineages by various anthropologists. For example, *H. rudolfensis* is considered by some antedating or contemporary with *H. habilis; H. ergaster* bifurcating into *H. erectus* and *H. sapiens,* and *H. heidelbergensis* ancestral to H. *neanderthalensis.* (Other proposed names include *H. rhodesiensis* for an African *H. erectus* derivative, and *H. antecessor* for Spanish fossils that may have been ancestral to both *H. sapiens* and *H. neanderthalensis.*) As might be expected, categorizing transitional fossils as species can be difficult and disputatious in the absence of firmly accepted criteria. Genera designations are also disputed. For example, Wood and Collard argue that the genus *Homo* is not monophyletic because *Homo habilis* and *rudolfensis* are more closely related to australopithecines than they are to the *Homo sapiens* ancestor shared by other *Homo* species. Obvious differences in the fossil record indicate some groups can be distinguished because of their specific evolutionary positions and roles. How many such groups there are, and how to distinguish them, relates of course to the extent of fossil research and available collections, and to achieving some agreement among paleoanthropologists. (*Adapted from Day 1986a, with additions.*)

Cann and coworkers indicating that all modern human mitochondrial DNA sequences probably originated in Africa between 140,000 and 290,000 years ago.

Among various possible molecular techniques (see Chapter 12), the use of mitochondrial DNA (mtDNA) for evolutionary studies had a number of important advantages:

- It is a circular molecule, 16,569 base pairs long, whose complete nucleotide sequence we know (Fig. 20–16).
- It is inherited primarily, if not entirely, through the maternal lineage as a sequestered extranuclear haploid unit (male sperm do not ordinarily transmit their cytoplasm to the egg during fertilization, and whatever few male mitochondria enter are soon diluted out in successive cell divisions by the large numbers of oocyte mitochondria) and does not ordinarily recombine either with nuclear DNA or with other mtDNA. [Although some researchers propose that mitochondrial recombination occurs, other researchers question such findings and consider such instances as, at most, rare events (Wallis).]
- Therefore, unless modified by mutation (usually by single nucleotide substitutions), an mtDNA molecule remains unchanged from one generation to the next, and is homogeneous within an individual.
- Since there are about 10^{16} molecules of mtDNA per individual and up to thousands of copies per cell, researchers can more easily isolate mtDNA from

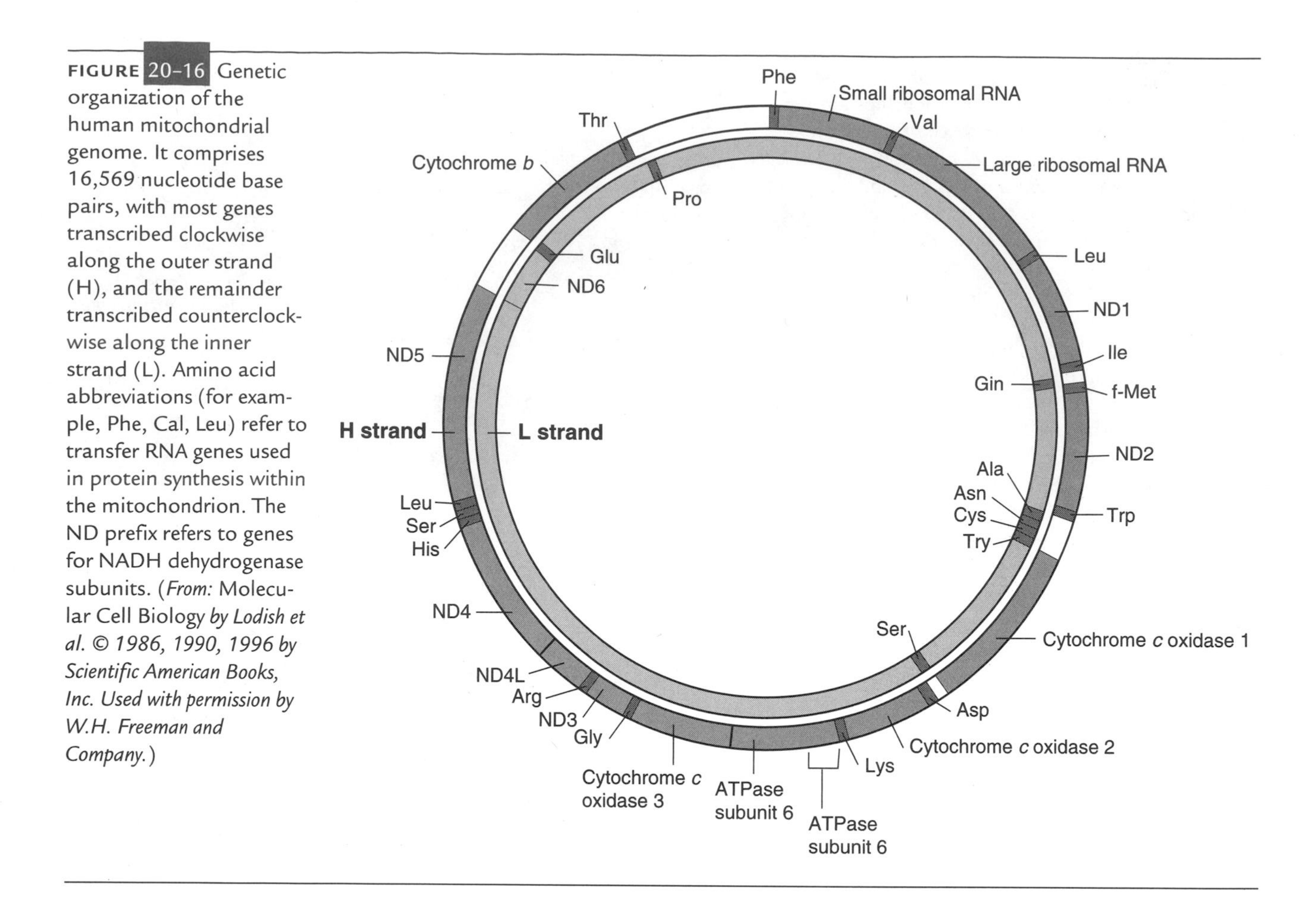

FIGURE 20-16 Genetic organization of the human mitochondrial genome. It comprises 16,569 nucleotide base pairs, with most genes transcribed clockwise along the outer strand (H), and the remainder transcribed counterclockwise along the inner strand (L). Amino acid abbreviations (for example, Phe, Cal, Leu) refer to transfer RNA genes used in protein synthesis within the mitochondrion. The ND prefix refers to genes for NADH dehydrogenase subunits. (*From:* Molecular Cell Biology *by Lodish et al. © 1986, 1990, 1996 by Scientific American Books, Inc. Used with permission by W.H. Freeman and Company.*)

human tissues than they can nuclear DNA genes, which have only two copies per cell.

- In contrast to nuclei, mitochondria lack repair enzymes, and mutations can accumulate up to 10 times faster than they can nuclear DNA mutations. Such rapid evolution enables comparisons between groups that would be more difficult to differentiate if researchers used slower evolving and more complex nuclear DNA sequences.
- Assuming that most mitochondrial DNA changes have little effect on viability, and that mutations accumulate at a fairly constant rate, differences between mitochondrial DNA sequences can act as a molecular clock, marking the time taken for these DNAs to diverge. A mitochondrial gene "tree" can be used as a chronological "tree," depending on how accurately the evolutionary clock is calibrated (Chapter 12), taking into account variability between taxa (Strauss).
- Most important, DNA sequencing techniques (discussed in Chapter 12) let researchers trace mtDNA differences among individuals, establishing branching pathways that help determine their evolutionary relationships.

As Figure 20–17 shows, Cann and coworkers proposed an evolutionary tree of modern human mtDNA that began with a single ancestral sequence in Africa (*a*).[7] This led to nine major descendant sequences (*b–j*) that were subsequently dispersed to populations in Africa and other geographical regions in which further branching occurred. Since we can trace any geographical race outside Africa to more than one unique mtDNA branch, apparently females carrying their particular mtDNA

[7] Although Cann et al. proposed that all our mtDNA can be traced to a single ancestral "Mitochondrial Eve," we should keep in mind that the sexual process greatly diluted her contribution of nuclear DNA (the major component of heredity): in the absence of inbreeding (Chapter 21) an individual inherits only half its nuclear DNA from each parent, one-quarter from each grandparent, one-sixteenth from each great grandparent, and so on. Although other mtDNAs in "Mitochondrial Eve's" generation became extinct, the difference in transmission between mitochondria (cytoplasmic inheritance) and nuclear DNA (chromosomal inheritance) may well have allowed her contemporaries to contribute the major portion of the 3 billion nucleotides in our nuclear genome. That is, since the breeding human population was probably large, "Mitochondrial Eve" was not alone in her population, nor were her female mtDNA-transmitting descendants alone in theirs. Estimates suggest average hominid population sizes of about 10,000 breeding individuals for the past one million years (Takahata), which may have been reduced to no less than several thousand at different times (Harpending et al.)! If we were to trace the ancestry of any *Homo sapiens* nuclear gene or gene sequence back to a single individual, that person may have been male as easily as female.

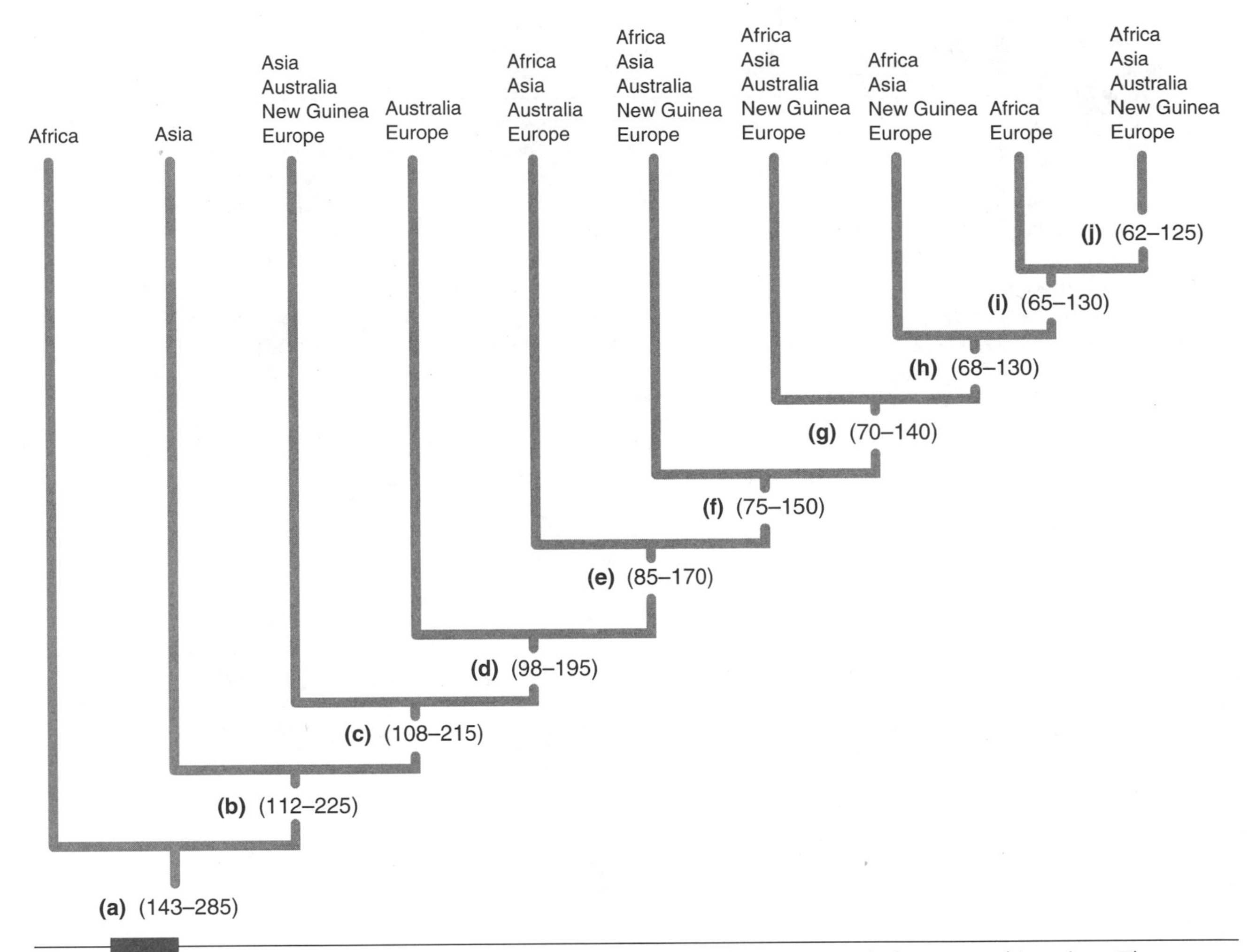

FIGURE 20-17 Phylogeny of mitochondrial DNA from 147 members of indigenous races in five geographic regions. The ancestral sequence is designated as *a,* and each node in this phylogeny (*b–j*) indicates a major descendant sequence that can be traced to *a*. Estimated dates for each node (in thousands of years ago) are given in parentheses and the areas colonized by each sequence are indicated at the bottom of the figure (for example, sequence *i* colonized Africa and Europe, which was also colonized by various other sequences). Not shown are subsidiary branches of these sequences also found among individuals in the various designated regions. (*Based on data that Cann et al. obtained from surveying 370 restriction enzyme sites per individual, covering about 1,500 bases of the mitochondrial genome.*)

sequences made many colonizations of each area. For example, Cann and coworkers suggested that mitochondrial genomes in their sample of New Guinea Highlanders had seven different maternal origins, most from Asia and the remainder probably from Australia.

Researchers base the timing of these migratory events on estimating the rate at which mtDNA sequences diverge, using measurements of differences between mtDNA sequences whose common ancestry can be approximately dated. According to Cann and coworkers, this divergence rate was most likely between 2 and 4 percent nucleotide change per million years in vertebrates, thus giving the dates shown in Figure 20–17. These results supported the view that early forms of *Homo sapiens* were present in Africa between 100,000 and 200,000 years ago (Clarke; Stringer and Andrews) and then radiated outward to different localities and differentiated into the various races, apparently replacing the indigenous races of *H. erectus*.

Novel and interesting as these data were, it did not take long for researchers to raise statistical objections. Various disputants pointed out that the phylogenetic tree Cann and coworkers offered was only one of many possible trees, some of which could better explain the data. They also challenged the date Cann and coworkers proposed for an African mtDNA origin, and later work provided new dates, some of which Box 20–1 gives. Although the dispute continues, Cann and coworkers' general conclusion for a single African origin of *Homo sapiens* still seems generally favored.

BOX 20-1 DID *HOMO SAPIENS* ARISE IN ONE PLACE ONLY OR IN MANY PLACES?

As Figure 20-18 diagrams, there are two main views of the origin of modern humans from a *Homo erectus* ancestor:

- The *single-origin hypothesis,* also called the "Out of Africa" or "Noah's Ark" model, proposes the origin of *Homo sapiens* in a single locality (Africa) followed by subsequent dispersal to other continents (Fig. 20-18*a*).
- The *multiple-origin hypothesis,* also called the "Candelabra" model because of its shape, proposes the parallel origin of *Homo sapiens* in different unconnected localities (Fig. 20-18*b*).

Among the information considered important in deciding between these alternatives is the date of the last common ancestor to modern humans. If this individual existed one million or more years ago, such a date might well coincide with one of the dispersals of *Homo erectus* from Africa. It would indicate that modern human races found in different continents are the present end points of evolutionary lineages that each began with *H. erectus* in these geographically separated localities—support for the multiple origin hypothesis.

However, if the last common *Homo sapiens* ancestor was much more recent—for example only 100,000 to 500,000 years old—then the dispersal of *Homo sapiens* occurred *after* a 1- or 2-million-year-old dispersal of *Homo erectus.* This would indicate that populations of *Homo sapiens* entered localities where *Homo erectus* had already been established, and eventually replaced these earlier hominids—supporting the single-origin hypothesis.

Because Cann and coworkers' original proposal of an approximate 200,000-year-old common mitochondrial DNA ancestor to modern humans was widely challenged, researchers undertook many new studies. Some of these studies calibrated the rate of mitochondrial nucleotide substitution using a 4- to 6-million-year-old date for sequence divergence between humans and chimpanzees (Chapter 12), while others used a 60,000-year-old date for sequence divergence among Papuans, based on the time they first colonized Papua New Guinea (Stoneking). As the table on p. 487 shows, almost all these studies support the relatively young 200,000-year-old single-origin date. Even high upper confidence limits of about 500,000 years are still too recent to fit the 1-million-year-old or more dispersal age expected in the multiple-origin hypothesis or in its multiregional variation.

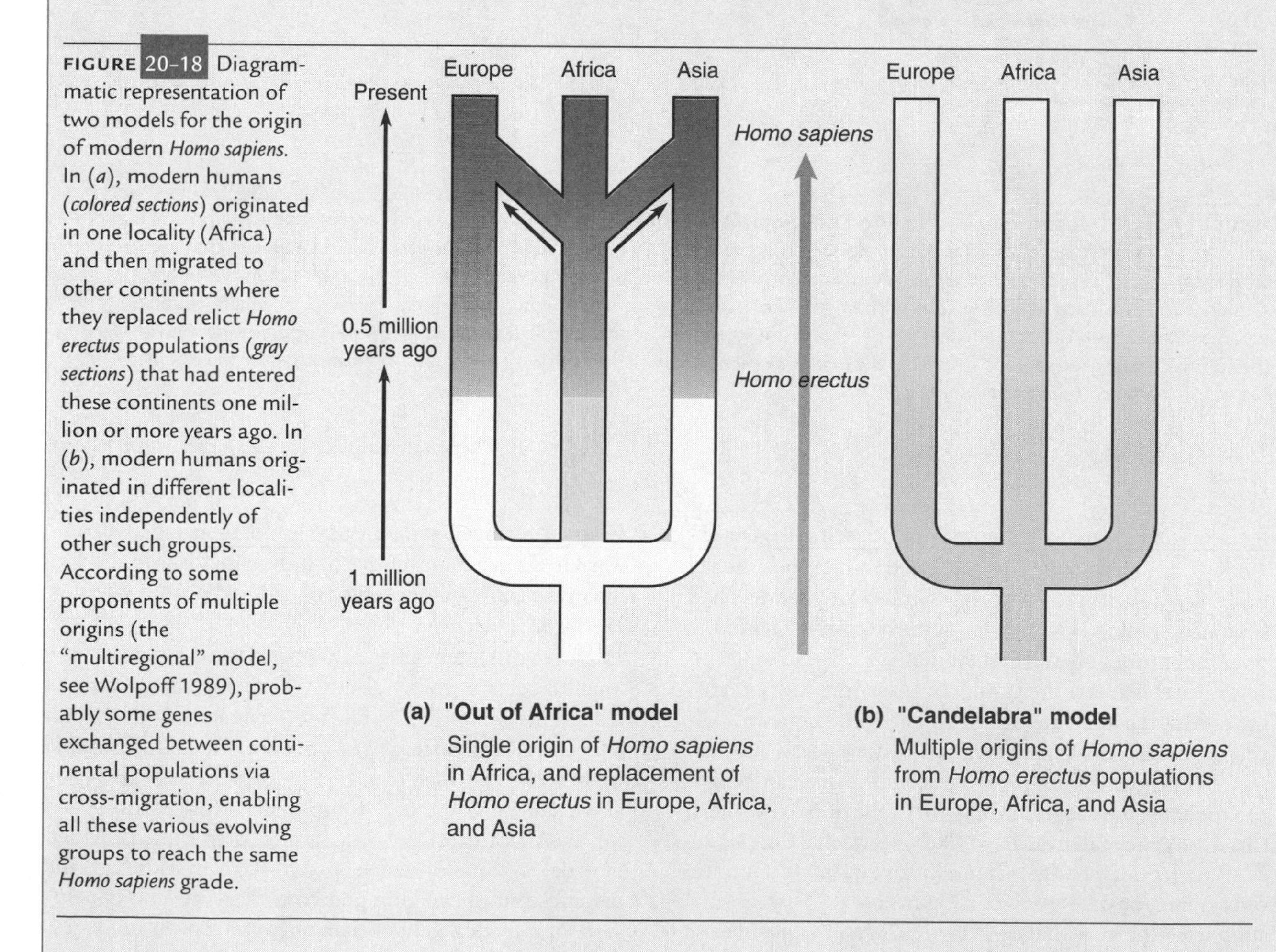

FIGURE 20-18 Diagrammatic representation of two models for the origin of modern *Homo sapiens.* In (*a*), modern humans (*colored sections*) originated in one locality (Africa) and then migrated to other continents where they replaced relict *Homo erectus* populations (*gray sections*) that had entered these continents one million or more years ago. In (*b*), modern humans originated in different localities independently of other such groups. According to some proponents of multiple origins (the "multiregional" model, see Wolpoff 1989), probably some genes exchanged between continental populations via cross-migration, enabling all these various evolving groups to reach the same *Homo sapiens* grade.

Estimates of the age of the common human mtDNA ancestor with 95 percent confidence intervals for these estimates

Study	Age of Common Ancestor	95 Percent Confidence Interval
M. Hasegawa and S. Horai 1991	280,000 years	180,000–380,000 years
G. Pesole et al. 1992	400,000 years	200,000–600,000 years
M. Nei 1992	207,000 years	110,000–504,000 years
K. Tamura and M. Nei 1993	160,000 years	80,000–480,000 years
M. Hasegawa et al. 1993 (control region sequences)	211,000 years	0–433,000 years
M. Hasegawa et al. 1993 (coding sequences)	101,000 years	0–205,000 years
A. R. Templeton 1993	213,000 years	102,000–389,000 years
M. Stoneking et al. 1992 (control region sequences): 2 methods	133,000 years 137,000 years	63,000–356,000 years 63,000–416,000 years
M. Ruvolo et al. 1993 (mtDNA cytochrome oxidase II)	195,000 years	85,000–349,000 years
S. Horai et al. 1995 (complete mtDNA sequences)	143,000 years	125,000–161,000 years

Source: From Table 1 in Stoneking, with additions. Stoneking provides full references for the first eight of these studies.

Among other support for the single-origin hypothesis are the following:

- All the non-African mtDNA sequences are variants of the African sequence. If the non-African mtDNA had been derived from resident non-African populations, much more non-African mtDNA variability would be expected. But there are apparently no non-African mtDNA types.
- Researchers find most mtDNA sequence variability among African populations, again suggesting that these are the oldest mtDNA populations among modern humans and the non-African populations are their derivatives.
- Because the multiple-origin hypothesis proposes that all populations evolved in parallel over long periods, we would expect them to have similar amounts of variability, a conclusion that the data contradict.
- Stoneking also points out that the age of the most common mtDNA ancestor is most likely older than the age at which the population bearing this ancestor diverged. For example, a 200,000-year-old date for an mtDNA ancestor means an even more recent date for the populations derived from this ancestor.

In spite of its advantages, mitochondrial DNA acts as only a single genetic unit whose genealogy may not necessarily coincide with other genetic units. The question then arises whether non-mitochondrial genes trace back to continents other than Africa. Other studies have therefore sought information from nuclear genes. A prominent example is the small human Y chromosome, inherited exclusively through the male line, a counterpart to maternally transmitted mitochondria. The Y has a nonrecombining portion in which mutations, like those in mitochondria, can be used to establish phylogenetic and chronological trees. One such study by Hammer and coworkers surveyed more than 1,500 individuals from all continents and many racial groups, and traced all Y chromosomes to a common ancestor living about 150,000 years ago. Interestingly, like "Mitochondrial Eve," this "Y-chromosome Adam" was of African origin, but African populations also received Y chromosomes returning from Asia. (Like mitochondria, we should keep in mind that our Y chromosome ancestor may have provided only a very small part of our genome, which contains 22 autosomal chromosome pairs plus the X.)

Broadening the data even further are studies that cover autosomal genes. Nei and Roychoudhury, for example, calculated genetic distances between 26 human populations for 29 different nuclear genes (Fig. 24-3). They also suggest a single African origin for *Homo sapiens* with subsequent widespread geographic divergence (Fig. 20-19). Mountain and coworkers present similar findings. Polymorphism for *Alu* chromosomal DNA sequences (p. 226) that are specific to humans, also support a recent African ancestry (Batzer et al.).

Whether mitochondrial, Y-chromosomal, or autosomal, the genetic and molecular data do not convince all paleontologists. A major argument some offer to support multiple origins of *Homo sapiens* is the continuity of anatomical features in Chinese and Australian humanoid fossils. Some fossil characters in these localities, such as brow ridges and cranial size, seem to have progressed from *Homo erectus*-like to modern *Homo sapiens*-like in fairly continuous sequence, pointing to their independent origin.

However, if each population evolved independently, how did they become so similar? Modern humans may differ in color and other minor attributes, but they all share basic *Homo sapiens* traits. Where and how did they get these similar traits? In answer, paleontologists such as Wolpoff (1989) proposed that *Homo sapiens* populations did not evolve completely isolated from each other but did exchange some genes. That is, interpopulational gene flow enabled the evolving groups to reach the same modern *Homo sapiens* grade—a new multiple-origin model given the name **multiregional evolution.**

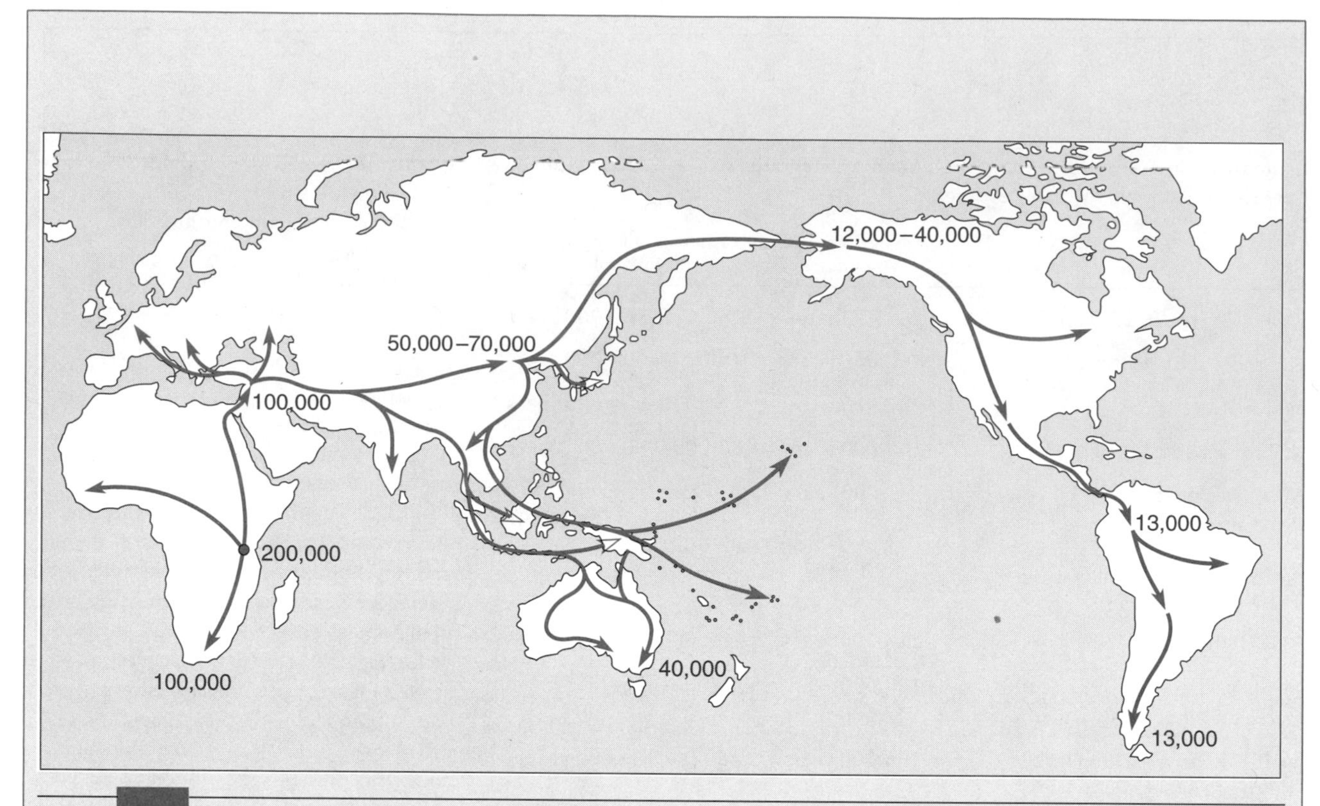

FIGURE 20-19 One proposed scenario for the geographic distribution of *Homo sapiens* from their African origin. The numbers derive from a study of genetic distances between 26 human populations and represent estimated dates at which these populations reached their various destinations years ago. According to Cavalli-Sforza and coworkers, these migrations correlate highly with many major patterns of linguistic evolution. Surprisingly, some Native American genes may owe their origin to a European/Asia Minor population, indicating a long migration occurred from Europe to North America that picked up Central or East Asian genes along the way (Brown et al.). (*Adapted from Nei and Roychoudhury.*)

The multiregional proposal, an apparent compromise between Out of Africa and Candelabra models, is also contested. Rouhani says that gene flow alone does not explain the necessary transitions, and Waddle says that quantitative tests on fossil morphology also contradict multiple-origins. On the other hand, Templeton makes a strong genetic argument supporting multiregionalism. He calls it a "trellis" model in which geographically mobile ancestral humans exchanged genes between populations to form a genetic trellis as they dispersed from their African origin. There was movement both in and out of Africa, but since gene exchange depends on proximity, genetic differences between distant populations increased ("isolation by distance").

Although much present molecular evidence seems to support the single-origin hypothesis, there are exceptions (see, for example, Ayala et al.), and the issue still generates debate as to how and when this occurred. According to Harris and Hey, sequence analysis of an X-chromosome gene indicates a major genetic division between African and non-African populations about 200,000 years ago. Since this date is earlier than the presumed 100,000–130,000-year-old origin of "anatomically modern" humans, such data appear to support the hypothesis that modern humans originated in different geographic localities later than their initial African/non-African separation.

To summarize: researchers are continually obtaining and evaluating further molecular information that may eventually prove one or the other of these hypotheses. Such procedures, however, depend more on intricate statistical analyses and population genetics (Mountain) that, for some paleontologists, lack the realism of actual fossils. For the future, both strategic hominid discoveries and increasing nuclear gene information will help decide the matter. At present, some workers are willing to accept a somewhat intermediate position: "An African origin, with some mixing of populations, appears to be the most likely possibility" (Jorde et al.).

Hunting Hominids

Apart from finding fossil and biochemical evidence of hominid origins, many anthropologists also want to understand how past environments affected and selected among various human traits, and how humans, in turn, affected and selected their environments. As discussed, the introduction of bipedalism and the home base, however they arose, would have profoundly effected the kinds of environments these early hominids could exploit. They could now move from forest to savanna with greater ease than ever before and cover much larger areas in their search for food. Food resources in the savanna, however, differ from those in the forest, and the impact of this new environment emphasized a host of new behaviors and adaptations.[8]

The relatively low rainfall in the savanna provided fewer high-quality plant foods than in the forests and made the distribution of such resources patchy, that is, present in some places and not in others. These resource irregularities, combined with plant seasonality, would have engendered further selection for increased hominid mobility, broad dietary habits, and flexible strategies in searching for food. Since the savanna grasslands also supported various herbivores, including migratory herds of large mammals, a selective advantage for **meat eating** may have appeared fairly early, including strategies for both avoiding and successfully competing with the large predators that preyed on these mammals.

An important change in the lifestyle of these hominids would have been an increase in the relative amount of meat in their diet. **Animal hunting** is not a novel trait in primates, and many investigators have recorded instances of baboon and chimpanzee groups engaged in purposeful hunting of animals smaller than themselves (Harding and Teleki; Goodall). Various baboon troops hunt small ungulates, other primates, and hares. Practically all chimpanzee groups studied to date engage in hunting (mostly other primates), and researchers estimate that about 3 percent of their caloric intake is animal food (Hill). Among primates, humans are the greatest meat eaters of all—a dietary habit that undoubtedly varied in degree at different times and places but probably became established early in human history. It therefore seems likely that even in their forest habitats, early hominids had become meat eaters to at least some minor degree.

An increase in meat consumption would have offered such early hominids many advantages:

[8] Boesch-Achermann and Boesch suggest that forest chimpanzees are much more versatile in behavior, tool use, hunting, and cooperative food-sharing than chimpanzees in savanna–woodlands. Because the chimpanzee–human phylogenetic relationship is so close, and the forest environment apparently so demanding and selective, they propose that early hominid evolution was primarily associated with forest experiences.

- Meat is a rich source of essential amino acids used in proteins such as lysine, tryptophan, and histidine.
- Meat provides more calories per unit weight than most plant foods.
- Meat is either packaged (small animals) or can be modified by cutting and tearing (large animals) into units easy to transport to a home base.
- Killing only one large animal feeds a group of individuals, often for more than one day.
- Meat remains available in dry seasons when plant food diminishes.
- As an added food source in a marginal environment, meat would have helped provide the additional energy to develop and sustain a larger brain (p. 502).

Although there are differences, we can get some idea of early hominid lifestyles from **hunter-gatherer societies** that continue today in places such as Central Africa (Mbuti Pygmies), South Africa (Kalahari Bushmen), and Australia (Aborigines). These groups consist of social communes or bands where males are usually the hunters and females the plant gatherers. Since their omnivorous diet depends on highly variable plant and animal food sources that are often seasonal, each band moves about several times a year over fairly wide ranges to different home bases or settlements.

Among some groups of the Kalahari Bushmen, researchers estimate that females gather about 60 percent of the diet in the form of vegetables and fruit, and male hunters bring in about 40 percent in the form of animal game. Since they usually share food, Bushmen waste little, and maximize the chances that all band members will get some. As Silberbauer points out, whatever the proportion of meat in the diet, hunting is a "prestigious activity. . . . The [Bushmen] are hungry for meat, and any description of the 'good life' always includes mention of a plentiful supply of it."

We still have no information on the proportions of plant and animal foods in the diets of ancient hominids, especially since many plants that such hominids may have used have not fossilized or are poorly preserved. Nevertheless, bone accumulations that appear with traces of hominid activities at East African sites indicate that hominid scavenging and probably hunting were most likely an important part of the lifestyle of various groups by the early Pleistocene, about 2 million years ago, if not earlier.

So, although we may not know exactly when hunting began in human history, it was a significant industry for a long enough period—in many societies, up to the agricultural (Neolithic) revolution about 15,000 to 10,000 years ago—to have seriously influenced human behavior. Even if we grant that plant food was as important or more important a food source than meat in early human history (Tanner, Shipman), it seems reasonable to ask: What selective forces and effects did hunting generate? As Tooby and DeVore stress, hunting "would elegantly and economically explain a number of the unusual aspects of hominid evolution."

First of all, successful medium and large game hunting requires active cooperation among hunters. We see this even in groups of foraging chimpanzees (usually two to five males) who will tree a monkey and then cut off its escape by assuming strategic positions around it.[9] Hominid hunters, empowered with simple weapons such as wooden spears, clubs, and hand axes, probably used such techniques and others, including tracking; stalking; and chasing game into cul-de-sacs and swamps, over cliffs, into ambushes, or by continuing the chase until the animal tired. With **cooperative hunting,** humans could bring down larger animals than could single hunters alone.

Second, cooperative hunting and the killing of large animals emphasized increased social cohesion both during hunting and the food sharing that followed. Transfer of information in successful hunts became vital and performed a necessary function in many later social interactions of the entire group. That is, improved communication became especially advantageous as individuals took more complex roles in planning, hunting, helping, food gathering, food sharing, infant care, child training, and other vital activities.

Third, successful hunting emphasized perceiving and retaining information on migratory pathways, watering sites, and home base settlements, whose geographical positions extended over home ranges (regions habitually occupied by a group) probably greater than those occupied by most other primates or carnivores (Foley 1987). Hominid hunters had to mentally dissect their experiences and observations into component geographical and ecological features, prey behaviors, weather effects, and seasonal changes, then store and synthesize this information into communicable mental maps that enable prediction, planning, and modification (Box 20–2). A genetic basis for the selection and evolution of visual-spatial reasoning can be observed in humans with Williams syndrome who lack such abilities because of a defective gene (Frangiskakis et al.)

Fourth, hominid hunting involved stresses that would have fostered increased locomotory adaptations such as persistence in the chase (humans can continue jogging for distances that are generally longer than many large animals can continue running), maneuverability in the kill, and long-distance traveling to or between home bases while carrying heavy burdens. Various writers have also pointed out that the need for increased diffusion of metabolic heat during these pursuits would have selected for the loss of body hair and increased numbers of sweat glands, features that among primates are unique to humans.

Fifth, the technological skills necessary for a clawless, canineless hominid to capture and butcher large prey promoted the making of a variety of snares, weapons, and tools, including the stone implements that date back at least to *H. habilis* (Fig. 20–13). Such technologies, especially evident in the fossil tools, involve shaping material according to some preconceived notion of what it should look like after the process is completed. These toolmaking skills involve not only manual dexterity, hand–eye coordination, and considerable concentration, but also the ability to plan and visualize an object that is not apparent in the raw material from which it is created. The artisan must conceptualize the final form of a tool in its three dimensions, and implement such concepts by mastering a series of techniques. These included finding and recognizing appropriate, workable stones in outcrops that were often widely dispersed, carrying these stones back to a base, and shaping them into tools by a sequence of precise strokes. The toolmakers also had to supplement the considerable mental abilities they used in toolmaking with social and communicatory abilities, to transmit such skills to other individuals who could continue the industry.

Finally, hunting placed further social emphasis on the home base to allow food exchange among foraging subgroups, particularly when the food supply was irregular, as it often is in hunting. A home base has value for nursing and pregnant females who could not always or easily cover the long distances necessary for large-scale hunting. The home base would have become a center for food sharing, shelter, hunting preparation, sexual bondings, child care, and other social exercises, in which communication skills tied all members together.

However, like the origin of bipedalism, researchers have considerably disputed the role of hunting among human ancestors, especially since we don't know the extent to which early human groups hunted (Harding and Teleki). Nevertheless, from what we can surmise from present hunting-gathering groups, and even from individuals in more modern societies who engage in hunting as a sport, the practice of hunting, whatever its role, was probably reinforced in various emotional ways: by the pleasures of seeking out and subduing prey; by the satisfactions of mastering the physical skills necessary for efficient aiming, throwing, and grappling; and by mastering the intellectual skills used in devising cunning offensive and defensive strategies. These behaviors arise early in human development, especially in play among juveniles and adolescents, and their perfection in adults has been socially approved and rewarded in every known historical culture. That most modern societies no longer need hunting for food has not lessened interest in these behav-

[9] Goodall lists more than 200 observed incidents where Gombe chimpanzees caught and/or ate colobus monkeys near Lake Tanganyika, in addition to cannibalism and the capture and consumption of many other mammals. According to Teleki, the chimpanzee kill rate in Gombe is about 225 to 300 hundred mammals a year, and agrees "with the kill rates of some large carnivores."

BOX 20-2 HUNTING, TECHNOLOGY, AND THE KALAHARI !KUNG BUSHMEN

The following is Carl Sagan's view of the technology and science involved in hunting by Kalahari Bushmen:

> It is very important to note that they [Bushmen] are highly technological. The technology is wood and stone and domestication-of-fire technology, but it's unambiguously technology. They are technological because their lives depend upon it. Chipping and flaking stone tools back before the external civilization sent a little trickle of metal into their economy is key. They did it superbly well. The archaeological and anthropological record is clear that we were technologists all the way back to the beginning. So the idea that science and technology is something new, unusual, and inaccessible to most people is completely backwards. Technology is, if anything, the most characteristically human activity, although, as I'll mention later, it is not exclusively a human activity.
>
> Now, hunter-gatherer game tracking techniques: A small group, with their bows and poison arrows and digging tools and a few other lightweight technological contrivances, is following the game. They come near a stand of trees. They take one close look at the ground. Immediately, they know how many animals went by, what their ages and sexes were, how long ago they passed; this one is lame in the back left foot; at the pace they're going we should be able to overtake them in another 2 hours if we hurry. Now, how do they know all this? In fact, what do they notice in order to follow the game on which their lives pretty well depend? One thing is the hoofprint. Different animals have different characteristic shapes of their hooves; different sized animals leave different sized hoofprints; but the decay of the hoof crater, the falling of pebbles in, the collapse of the raised rims, debris blown into it, tells you age. In fact, it reminds me of nothing so much as determining the ages of planetary surfaces by looking at how fresh the impact craters are. Maybe the reason that studying cratering physics seems so natural to us planetary scientists is because we've been doing it for a million years. The !Kung also know that animals in the hot Sun like to avoid sunlight. If there is a shadow on the ground, they will deviate from their path to run through the cool shadow. But where the shadow is depends on where the Sun is, and therefore, when you see the deviation of the trail from a straight line, you know that there had to be a shadow at that spot when they passed. Well, where in the sky did the Sun have to be in order to cast that shadow? Oh, it was eleven o'clock this morning.
>
> Now, I don't claim that every hunter-gatherer made such a scientific calculation, did the trigonometry of the angle of the Sun, and so on. This was tradition; each generation taught the next. But someone had to have figured it out, and that someone had to be a scientist. This is another reminder that we've been scientists and technologists from the beginning.

iors, and athletes (and "warriors") who develop such skills are often greatly esteemed. Who are our heroes?[10]

Communication

Communication is the means through which a stimulus from one individual can trigger a response in others. Communication methods may include signals transmitted through any of the sensory channels: scent, touch, vision, and sound. Practically all animals that interact with each other use one or more communicatory methods, but they are especially well developed in social animals where information is essential in providing cues to other individuals about factors such as food sources, predator encounters, territorial boundaries, sexual readiness, social ranking (dominance), and emotional states. We can find examples of these throughout the primate order (Jolly).

For instance, various prosimians and monkeys use urine or scents that special glands emit to mark trails and territorial boundaries. They also commonly use **olfactory cues** to attract sexual partners and signal the onset of ovulation. Such communication has its counterpart in humans, who emit odors from their axillary and genital regions. Although in Western culture people now generally wash off or disguise these odors by deodorants and perfumes, tests have shown that many can use such body scents to distinguish between the two sexes as well as among individuals.

Tactile communication assumes its most common primate form as grooming, or fur cleaning with fingers, lips, and teeth. It is one of the most obvious and frequent kinds of interaction we see in many mammalian groups

[10] It is often pointed out that the risks of the hunter and warrior "heroic" lifestyles were compensated by improved access to females in groups where the division of labor between the sexes made child-raising females vulnerable and dependent on males for security and for added nutrition in the form of meat. Such females would have been attracted to males showing protective and resourceful behavior, presumably among the traits exemplified by heroes (see also pp. 588 and 611).

and seems to serve as the main social cement that binds pairs of individuals or group members together. Chimpanzees supplement grooming with other tactile behaviors such as holding hands, patting, embracing, and kissing (Goodall). Since humans have relatively little fur, they don't engage in the traditional form of primate grooming, and not surprisingly they mostly confine tactile social reassurances to other tactile behaviors.

Primate **visual signals** include physical gestures or anatomical displays such as the postures, genital swellings, and colorations used to signal sexual receptivity. As Figure 20–20 shows, facial expressions may be quite varied and are easily visible in hairless faces. Some of these expressions, such as the glare and scream call (Fig. 20–20*a* and *c*), probably signal threat messages throughout the primate order and mark an aggressive attitude even in human cultures (Fig. 20–20*b* and *d*).

Compared to visual displays, **vocalizations** have the advantage that they go from mouth to ear and leave the hands and body free for other activities. Oral sounds also have the advantage of providing feedback by letting the vocalizer hear his or her own vocalizations and thereby evaluate and control them while (or perhaps even before) uttering them. Moreover, although sound fades rapidly, it can be transmitted over long or short distances, in all directions, even around obstacles that would interfere with visual communication.

In general, because sound leaves no record, most primates usually send short, simple messages, denoting, for example, predator alarms or territorial calls. Yet some primate vocalizations have a variety of gradations, each providing a subtle meaning. Thus, Japanese macaques use particular variations of the "coo" sound in specific situations, such as a male separated from the group, females contacting their young, dominants contacting subordinates, and females in estrus (Green and Marler).

Other primate vocalizations not only reflect the emotional state of the vocalizer but also direct attention to specific external events. A prominent example is that of three different alarm calls vervet monkeys give, each designating a specific kind of predator:

- The monkey emits a "rraup" upon detecting a hawk, prompting the troop to look up and then seek cover in lower branches.
- The vervet "chirps" on seeing a mammalian predator, prompting the troop to ascend to the forest canopy;
- The monkey uses a "chutter" when detecting a snake, and the troop may then adopt aggressive positions on the ground.

Each of these calls is **symbolic** in the sense that it denotes an object that has no direct relationship to the call

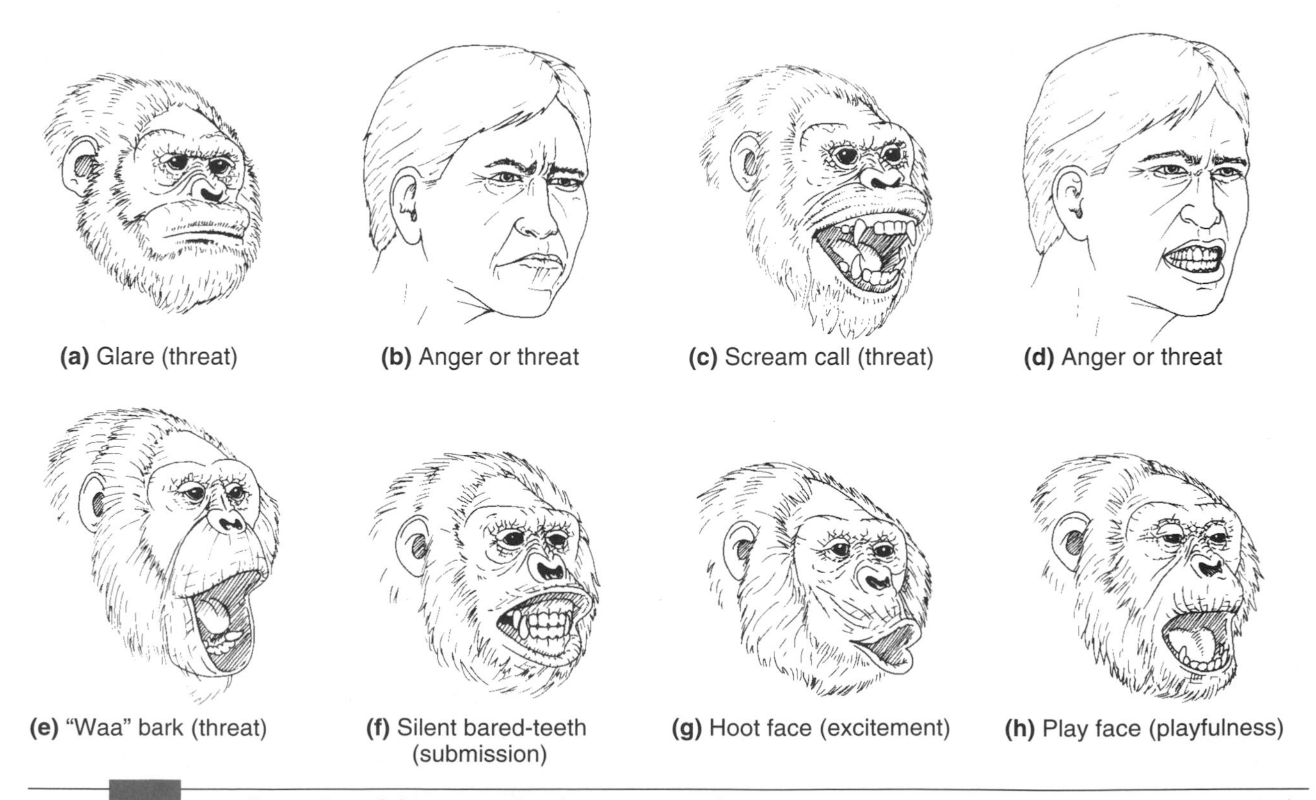

FIGURE 20-20 A small sampling of chimpanzee facial expressions indicating various emotional states, along with two seemingly related human expressions in (*b*) and (*d*). (*Adapted from Chevalier-Skolnikoff.*)

itself (for example, a chirp is not a leopard) and we may consider the calls a primate preadaptation for human communication in which the sounds of words do not correspond to their meanings.

Speech

The most symbolic of primate vocalizations is human **speech** and **language.** Here, we have introduced new characteristics through a wide range of different-sounding syllables that we can string together in various ways to provide a vocabulary of different meanings (words). Compared to a sequence of sounds limited to single tones, human speech provides a rapid means of communication. For example, we can interpret a sequence of dots and dashes, as in the Morse code, at a rate often much less than 50 words a minute, whereas we can often easily understand a sequence of spoken syllables delivered at 150 words a minute.

As in other mammalian vocalizations, the **larynx,** in the upper part of the tracheal tube, provides the basis for speech. Its origin is not connected with sound but stems from early air-breathing fish. These ancestral vertebrates, like the lungfish of today, opened a valve in the floor of their pharynx to help swallow air into their lungs when they were out of the water, and closed the valve when in the water. As selection for air breathing continued in terrestrial vertebrates, this laryngeal valve developed fibers and cartilages that more precisely controlled laryngeal dilation and closure and let the animal breathe more air when necessary. Like so many other evolutionary features, the ability of the larynx to generate oral sounds was a preadaptation of an organ originally used for a different purpose.

In producing sound, the larynx acts like a woodwind reed that controls vocal pitch by opening and closing rapidly so that expired air from the lungs is interrupted to form puffs: the greater the frequency of puff formation, the higher the pitch. However, to produce the vowels of humanlike speech, laryngeal puffs must pass through a tubelike airway (the pharynx) whose length and shape determine the eventual frequency-patterns emitted and thus the quality of the different vowel sounds (Fig. 20–21).

All terrestrial, air-breathing animals that produce oral sounds, from frogs to mammals, use this basic mode of vocalization—a laryngeal-like output and a supralaryngeal "filter." In addition, neural auditory units in these animals seem to react with maximum sensitivity to specific ranges of frequencies. These specific neural sensitivities allow bullfrogs, for example, to respond to the mating calls of their own species and not to those of others. Humans seem to have special neural brain circuits that can perceive and identify various categories of sound combinations and thus distinguish between different kinds of spoken syllables—an ability that has apparently evolved from primates with more limited powers of distinction (Lieberman).

Figure 20–22*a* shows a diagram of the adult human upper respiratory tract, with its sound-producing airway that begins at the larynx and proceeds through the pharynx and mouth. Note that compared to that of the chimpanzee (Fig. 20–22*b*), the adult human mandible extends forward for a relatively shorter distance. Among the consequences of reduced mandibular size[11] and lower positioning of the human larynx in the vocal airway are the thickening and rounding of the tongue to form the anterior wall of the pharynx. In the chimpanzee (and newborn human infant), the pharyngeal section of the oral tract is shorter, and the epiglottis (used to cover the trachea during food intake) overlaps the soft palate. As a result, the

[11] Excessive tooth crowding is apparently one of the biological costs of reducing the mandibular body.

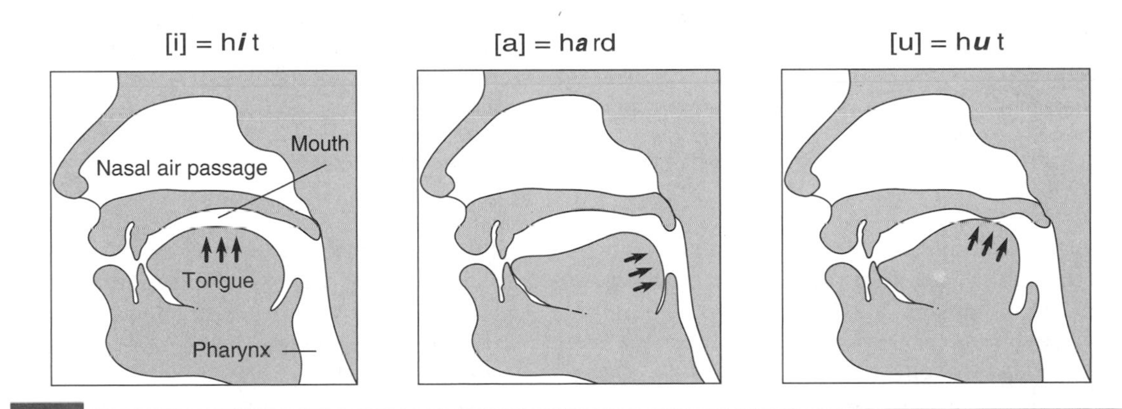

FIGURE 20–21 Diagrammatic views of how adult humans produce three vowel sounds by positioning the tongue (*arrows*) in different parts of the oral airway. Note that a sharp bend (formed by the hard palate above the mouth and rear wall of the pharynx) partitions this airway into right-angled mouth and pharyngeal sections that are essential for these vowel sounds. By contrast, the vocal airways of chimpanzees and newborn human infants are shorter and primarily in the shape of a slightly curved tube, making these vowels much less distinct. (*After Aiello and Dean.*)

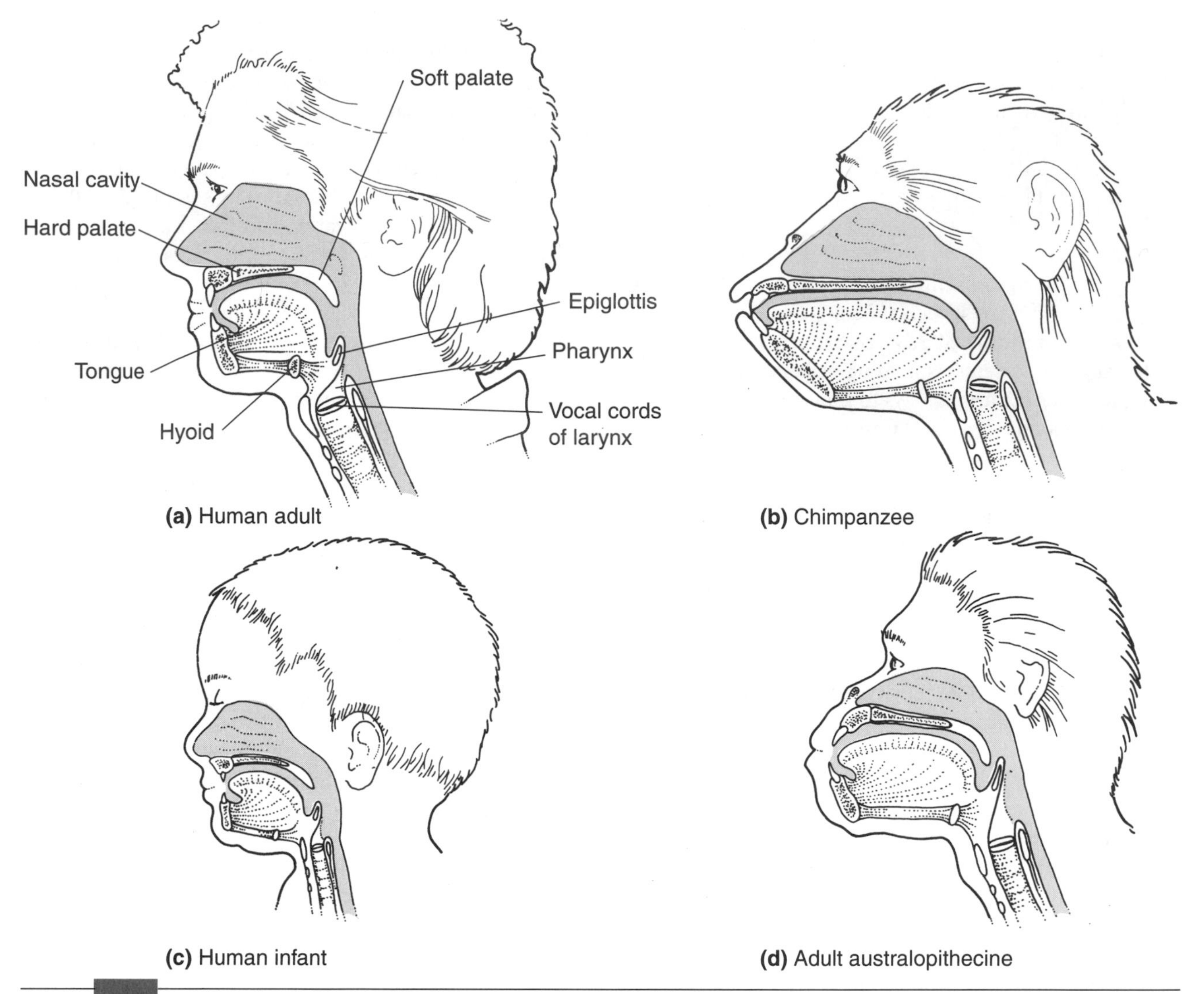

FIGURE 20-22 Upper respiratory systems of an adult human (*a*), chimpanzee (*b*), human infant (*c*), and adult australopithecine (*d*) showing important structures associated with vocalization. The pharynx is much longer in human adults than in chimpanzees because the larynx is displaced downwards in the neck, and the bulging tongue formed by shortening of the mandible (note also the lower position of the hyoid bone and epiglottis) now forms the anterior wall of the pharynx. As a result, humans can enunciate vowels and syllables more clearly by positioning the tongue in both mouth and pharynx. (Newborn human infants (*c*) show the same overlap between epiglottis and soft palate as nonhuman primates (*b*) but the pharynx lengthens considerably during infancy and childhood, transforming humans from obligate nose breathers as infants to the adult condition of voluntary mouth breathers.) In (*d*) is a reconstruction of the presumed upper respiratory system of an adult australopithecine. As in nonhuman primates, the epiglottis overlaps the soft palate, the back of the tongue does not reach the pharynx, and the larynx is relatively high in the vocal tract. (*Based on Conroy after Lieberman.*)

tongue is isolated from the pharynx and respiration normally proceeds through the nasal cavity. Nonhuman primates and newborn humans (Fig. 20–22*c*) can drink and breathe simultaneously because the nasal respiratory pathway to the lungs can remain open while liquid passes around it into the esophagus. Thus, although there is a better separation between breathing and swallowing in nonhuman primates and newborn human infants, they cannot produce adult human speech sounds by manipulating the tongue in the pharyngeal cavity.

Because speaking depends so much on soft, nonfossilizable tissues, we find it hard to uncover the phylogenetic history of speech. We can, however, make important correlations between skull structure and the positioning of the larynx, size of the tongue, and length of the pharynx. According to such studies (Lieberman 1984, 1991), the vocal tract of the australopithecines (Fig. 20–22*d*) probably was no different from that of nonhuman primates, and this was probably also true for most, if not all, of the lineages classified as *H. erectus.*

In fact, Lieberman and others suggest that *H. sapiens neanderthalensis* had a chimpanzee-like vocal tract that would have considerably limited its speech patterns. Sounds dependent on pharyngeal shape and control, such as [i], [a], [u], [k], and [g], would probably have been considerably distorted. Like chimpanzees and newborn human infants, Neanderthals would have nasalized speech because they could not close off the nasal cavity from the pharynx.[12] If this view is correct, a primary reason for the divergence between Neanderthals and modern humans may have been differences in phonetic ability and consequent differences in the kinds of languages that each could employ and the level of social and technological organization that could be achieved.

A contrasting view is based on the discovery of a 60,000-year-old Neanderthal skeleton with an intact hyoid bone used for laryngeal muscle control. According to Arensburg and coworkers, this crucial bone shows little difference from that of present-day humans, and they suggest that it signifies Neanderthal capability for human speech. Since we do not know the actual position of the larynx and other soft tissues in this fossil or in earlier ones, we have no evidence for the extent of Neanderthal vocalization. And even if Neanderthals could vocalize, their language abilities may have been quite primitive, no more than the "pidgin" level, because of limited cerebral associative neural capacities. Possibly hominids experienced various anatomical and neurological evolutionary trials among different vocal systems over the last 200,000 or so years, only one of which led to that of modern humans.

Whether these hypotheses are true or not, the sophistication of modern human language is a crucial difference in separating humans from all other animals, and some of the features and possible evolutionary characteristics of this language are worth considering.

Language and Self-Awareness

In contrast to Darwin, who stated in *The Descent of Man* that it would be "impossible to fix on any definite point when the term 'man' ought to be used," Max Müller, a linguist, soon laid down the challenge that "language is our Rubicon, and no brute [ape] will dare cross it." Müller's barrier could only be overcome by teaching apes a human language. Various trials made throughout the early twentieth century were unsuccessful, although clearly apes often understood far more than they could communicate.

Among the first serious, long-term attempts to bridge the language gap was the 1947 undertaking by Keith and Catherine Hayes, who raised an infant chimpanzee (Viki) in a normal human environment and tried to teach her to articulate human speech. The attempt was a failure, and after 6 years, Viki could not utter more than four distinguishable words: "Papa," "Mama," "cup," and "up." As already pointed out, a chimpanzee can only produce a quite limited range of sounds, because its vocal tract is relatively short and apes seem unable to control its shape. This is not surprising, since chimpanzees in their natural habitats are usually silent except when aroused.

A different and more successful approach was begun in 1965 by the Gardners, who raised a chimpanzee (Washoe) from the age of 10 months in an environment where its human caretakers used American Sign Language (ASL), which does not involve speech. By the time Washoe was 5 years old, she had learned to use at least 132 different signs covering a variety of names, actions, modifiers, and functions. Her sentences, however, rarely extended to more than one or two words or their repetitions, indicating that her language abilities stopped at about the level of a 2- to 3-year-old human child. This has generally also been true for other ASL-taught chimpanzees since Washoe.

Despite their limitations, chimpanzees have linguistic abilities, although on a more elementary level than humans. For example, as Table 20–5 shows, chimpanzees trained in ASL by the Gardners, Roger Fouts, and others, as well as Koko the gorilla, trained by Francine Patterson,

TABLE 20–5 Sign sequences (word combinations) created by chimpanzees and gorillas for items not in their ordinary vocabulary

Item	Sign Sequence
Onion, radish	"Cry fruit", "cry hurt fruit"
Watermelon	"Candy fruit", "drink fruit"
Alka Seltzer	"Listen Drink"
Cigarette lighter	"Metal Hot"
Ring	"Finger bracelet"
Swan	"Water bird"
Ostrich	"Giraffe bird"
Brazil nut	"Rock berry"
Hateful objects	"Dirty . . ." (probably signifying fecal), for example, "dirty leash", "dirty monkey", "dirty Roger."

[12] Adult modern humans can open or close at will access from the pharynx to the nasal cavity by lowering or raising a flap (velum) of the soft palate. However, although some languages use nasalized vowels (for example, Portuguese), we have more difficulty in distinguishing small differences between such vowels compared to distinguishing small differences between nonnasalized vowels. In general, most modern human languages tend to avoid nasalized vowels.

can use various appropriate and imaginative combinations of signs to signify items not in their vocabulary.

Such symbols can also refer to events that are distant in time and place (**displacement**)—one of the important features of language. Washoe demonstrated this ability in ASL conversation with a human companion in which Washoe repeatedly asked for an orange, then signed, "You go car gimme orange hurry," indicating that she could communicate an event that was to occur in the future at some other place. (Washoe had been in a car more than two years earlier and knew that one could get oranges in a store.)

Even some simple word orders can be grasped by chimpanzees—Roger Fouts observed that Lucy could distinguish between "Roger tickle Lucy" and "Lucy tickle Roger." Another chimpanzee, Ally, showed that ape discourse can extend beyond their own persons to comment on the environment when she signed "George smell Roger" to her trainer George after Roger Fouts lit a pipe. Chimpanzee communications also include deceptions, as when chimpanzee Booee asked for a tickle from his chimpanzee companion, Bruno, while obviously trying to get to the raisins that Bruno possessed.[13]

A somewhat more abstract two-dimensional language that David Premack taught to chimpanzees involves metal-backed plastic tokens of arbitrary shape (lexigrams) to represent words that are arranged in vertical sequence on a magnetized board. Duane Rumbaugh further developed this approach through the use of computer-connected keys embossed with lexigrams that, when pressed, light up sequentially on a screen above the keyboard console.

Lana, one of the computer-trained chimpanzees, demonstrated that she could use this system to ask "intelligent" questions. For example, having been taught the symbol "name of," she used it to elicit the unknown name of the object ("box") that contained a desirable item, candy. First she asked, "Tim give Lana name of this," and, when provided with the name, called for the "box." Perhaps a greater linguistic feat was that of Sarah, trained in Premack's plastic token language, who used it to understand "if–then" causal relationships: for example, "If Sarah take apple—then Mary give chocolate Sarah; if Sarah take banana—then Mary no give chocolate Sarah." Also impressive is Kanzi, a bonobo ("pygmy") chimpanzee, who could respond appropriately to sentences in which different verbs were used with the same nouns placed in reverse order: "Take the potato to the bedroom;" "Go to the bedroom and get the potato" (Savage-Rumbaugh and Rumbaugh).

These experiments show that although apes do not use and understand language in its adult human sophisticated forms (for example, see the descriptions of language by Jolly and by Vauclair), some can use and understand simple elements of language, such as simple sentences, and can even make use of some mathematical forms (Boysen and Berntson); they can:

- Use words (or signs) as arbitrary symbols for real objects and actions
- Create combinations of words for objects that are not in their vocabulary
- Refer to activities or objects distant in time or place
- Use words to deceive
- Use words to gain information
- Use words to comment on the world about them
- Understand word sequences that use the logic of cause and effect
- Classify newly presented objects into groups (for example, fruit, tool)
- Use Arabic numerals to count
- Learn signs from other chimpanzees

According to many observers (but not all), apes show they have gone beyond merely perceiving events and objects to conceiving how they occur and how they are related (Savage-Rumbaugh and Rumbaugh). In their natural habitat, these accomplishments are obvious in chimpanzee toolmaking and tool use, such as the way they employ twigs and vines in fishing for termites. Termite fishing involves being aware of its most favorable months (October and November), locating the sealed termite tunnels, often importing the necessary supply of tools from distances as far away as half a mile, shaping some of the tools by removing leaves, biting off the ends of the tools to achieve an optimum length, inserting the tool with a proper twisting motion that can follow the curves of the termite tunnel, vibrating it gently to bait the termite soldiers, and retracting it carefully to avoid tearing off the termites. Learning these tasks takes years, and even an anthropologist (Geza Teleki) who studied the technique for months was no better at it than a chimpanzee novice.

[13] Perhaps even more clever deceptions by apes are those based on tactical planning. De Waal recounts an episode at the Arnhem Zoo in Holland in which a male chimpanzee, Nikkie, was pursuing a female, Spin, and caught her by a strategic ruse when she ran behind a tree trunk. As De Waal puts it:

> Nikkie started to turn to the left, and Spin responded by moving to the right. At the moment Spin appeared around the corner [tree], Nikkie threw a brick, but almost without losing speed, so that he was able to catch his victim when she jumped back to the left in order to avoid the projectile. . . . That Nikkie anticipated her jump into his arms was evident from the fact that he did not wait to see her reaction to the projectile. The smoothness and speed with which the whole maneuver was executed even suggests that Nikkie may have planned five steps ahead. The five hypothetical steps are:
>
> (1) I move to the left,
> (2) Target will move to the right,
> (3) I throw stone,
> (4) Target will jump back,
> (5) I reach target.

Similar demanding techniques used by nut-cracking chimpanzees involve finding a properly sized stone or hardwood club to be used as a "hammer," choosing a well-shaped tree root as an "anvil," and precisely positioning each nut on the anvil, keeping in mind that nuts from different species must be positioned differently. The hammer must then be gripped in its most effective position, swung with the proper force, and aimed so that it hits the nut in exact locations to extract the maximum amount of nutmeat.[14]

Lieberman and others suggest that the "rules" involved in sequencing such motor-controlled operations are preadaptations for language, which also must follow **sequencing rules,** since a sentence is a sequence of phrases made up of components such as noun, verb, adjective, and subject. The ability to devise and follow such rules apparently lies in the association centers of the brain, and glimmerings of this can even be found in simple vertebrates such as frogs who can perform a sequential set of actions such as fly catching.[15]

Clearly, to enable a word (for example, an auditory stimulus) to stand as a symbol for an object or action (for example, a visual stimulus), a neural associative center must allow such connections to be made. Such **cross-modal associations,** as they are called, are already present in many mammals, because they understand and can act on some of the words spoken to them. In the case of apes, and even monkeys, cross modality extends to the ability to correlate objects that they cannot see but can feel to objects that they can see. Tests of apes, for example, show that even a glimpse of the photograph of an object lets them choose correctly by feeling for the object among different objects hidden from view.

We do not yet know which neurological areas of the brain allow cross-modal associations and matching in apes, but neurologists believe one such human center to be the **angular gyrus** shown in Figure 20–23. Near this region, on the parietal lobe of the left cerebral hemisphere, is **Wernicke's area,** concerned with formulating and comprehending intelligent speech; people who have lesions in this area emit informationless, wordy babble. Patients with lesions in **Broca's area** on the left frontal lobe have difficulty speaking, since this region apparently serves to coordinate vocal muscular movements.

Although these speech and language areas most often lie on the left cerebral hemisphere, there is some variability, and left-handed individuals may have these areas localized on the right hemisphere. The existence of such functional laterality, with one cerebral hemisphere dominant over the other, seems present in other primates, even in rhesus monkeys, and may be a preadaptation for developing the associative centers used in human speech and language. That is, the cross-modal associations used in connecting various faculties, such as being able to visualize a predator on hearing a particular alarm call, may have led to the dominance of one cerebral hemisphere, and these areas were then further enhanced in human lineages during selection for the acquisition of language (see also Box 20–3).[16]

Interesting as these findings are, we could claim, as many already have, that language can only develop in organisms that have a concept of "I," or **self-awareness,** so they can intellectually separate themselves from the rest of the world. That is, to be capable of language one must analyze external events by taking them apart and putting them back together in a symbolic, thoughtful, manipulative context (the intellectual "I") different from one's own immediate, nonthoughtful, reflexive reactions. However, even from this challenging point of view, apes can demonstrate the kernel of self-awareness as psychologist Gordon Gallup has shown. Gallup noted that self-awareness could be tested by a relatively simple procedure—the reaction of an individual to its image in a mirror.

In human children, the realization that the mirror reflects themselves rather than another child appears at the age of about 20 months. This achievement indicates that the child has already attained a concept of itself as distinct from that of other children, a concept whose imaged reflection can be identified in the mirror. The child then uses the reflection to observe and examine itself rather

---

[14] Although young chimpanzees generally learn this technique from their mothers by imitation, Boesch also recounts incidents in which mothers actively intervened in their offspring's unsuccessful nut-cracking attempts by taking the hammer, positioning the nut, and demonstrating the proper technique—active teaching!

[15] According to Chomsky (1972, 1976), inherited neural structures determine the ability to structure words into meaningful sentences (syntax, p. 503), no matter what the language. That is, the conceptual basis of language—the ability to designate and distinguish its fundamental sentence components such as *agent, subject, goal*—appears universal. Verbs are always surrounded by modifiers in predictable patterns (Cinque), and questions of ownership begin with the same phrase (for example, "whose") in all languages. This does not mean that any human child necessarily communicates by language without hearing or speaking one. In fact, children who by misfortune have grown up in isolation without the chance to communicate with other humans do not acquire language and remain seriously deficient in social and intellectual development. The neuronal components of language and speech apparently must be exercised at a crucial early stage to become functional, just as kittens temporarily deprived of vision in one eye between 4 and 12 weeks of age do not possess binocular vision even when full vision is restored. Although the ability to acquire language appears genetically innate in humans, they must learn by experience and imitation the particular language used to communicate, because the morphology of language symbols (word sounds) and patterns for their linear arrangement in sentences are culturally determined. That is, basic language ability is genetic, but its usage and morphology are environmental. The need for learning seems especially true for producing understandable vocalizations, since children take much longer to speak understandably than to comprehend what is spoken.

[16] On the intellectual plane, preadaptations are as important as they are on the physical plane: chimpanzees were not selected in the past to learn American Sign Language nor humans to practice higher mathematics. The abilities that let these primates perform these intellectual feats are neurological preadaptations that evolved for other selected purposes, such as evaluating and responding to intricate social and environmental factors and relationships.

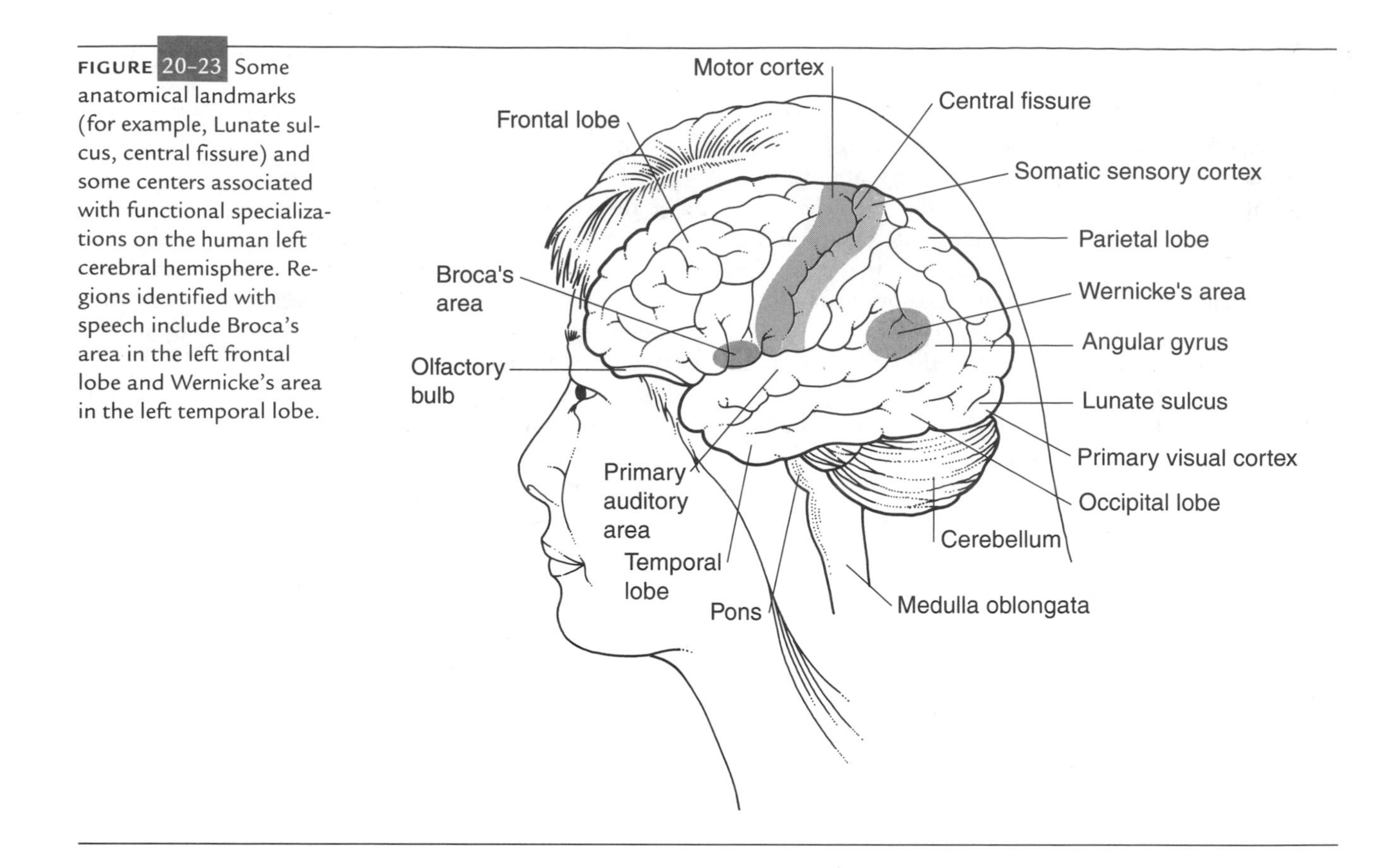

FIGURE 20–23 Some anatomical landmarks (for example, Lunate sulcus, central fissure) and some centers associated with functional specializations on the human left cerebral hemisphere. Regions identified with speech include Broca's area in the left frontal lobe and Wernicke's area in the left temporal lobe.

than examine the image on the mirrored surface, signifying an awareness of its own individuality. Birds, dogs, and most other animals, including monkeys, cannot recognize themselves in the mirror but only a different member of the same species, no matter how long they are exposed to the mirror image. The "I" seems to be less developed or missing in these animals.

Most interestingly, when Gallup exposed wild-born chimpanzees to full-length mirrors he found the development of "self-directed" behavior within a few days: the chimpanzees used the mirrors to examine and manipulate different parts of themselves. Moreover, when such chimpanzees were anesthetized and painted with bright red spots that they could see only in the mirror, they used their reflections to detect these spots on themselves, to touch the spots, and then smell their fingers. Since self recognition of this kind is certainly a sign of self-awareness, we can reasonably conclude that if self-awareness is necessary for the development of language, its presence in apes helps explain their language abilities, elementary as these may be.

Although obvious differences appear in the degree to which these traits develop in the two groups, all these findings on language and self-awareness (and perhaps even the concept of death; see Desmond) point to extensive similarities between apes and humans. We can reasonably conclude that some of the unique intellectual attributes of humans can probably be ascribed to evolutionary events that occurred early enough in history for the two groups to share them. Both groups are composed of individuals who continually interact in complex and changing social patterns, and this was most likely true for their common ancestors as well. Since self-awareness and language depend almost entirely on social interactions (isolated young chimpanzees do not develop self-awareness, and isolated human children do not develop language), an environment of complex interpersonal relations, as well as one that presented continually varying challenges in food gathering and capture, probably played an essential role in the evolution of these traits.

We can reasonably say that many flexible and complex human behavioral traits are adaptations to conditions in which early groups lived, and in turn, contributed to these conditions. Once started, such a reciprocal process could have accelerated rapidly, leading to increased behavioral complexities that emphasized selection for improved communication (language), reasoning, and creative attributes, all contributing to the evolution of the brain (Box 20–3).

Altruism and Morality

The importance of social interactions in developing behavioral and communication skills can be seen throughout primate groups in the panoply of calls, grimaces, gestures, and activities they use to indicate social positions (for example, dominance, subordination, group affiliation); needs (for example, food, sex, reassurance); and changes in any of these areas (for example, new social positions, alliances, sexual states, or dietary interests). Such behaviors range from transmitting only information on themselves as individuals to actions that may immediately affect the survival of other group members.

Information affecting the survival of other group members is most obvious, for example, when a monkey encounters a leopard and reacts with a loud scream, signaling nearby listeners to take refuge. Thus, although this warning signal may call the predator's attention to the screamer and diminish its own chances for survival, the effect can help preserve its relatives or compatriots.

Population geneticists beginning with Haldane and Wright (Chapter 23) have suggested that there were genetic advantages in such **altruistic behavior** in which individuals may even go so far as to endanger their own genetic future for those who carry closely related genotypes. In 1964, Hamilton popularized this cooperative process under the name **kin selection,** and provided formulas by which some of its benefits could be evaluated (p. 385). As Maynard Smith has pointed out, "the main reason for thinking that kin selection has been an important mechanism in the evolution of cooperation is that most animal societies are in fact composed of relatives."

Some years after Hamilton's proposals, Trivers introduced a concept of altruism that seemed to have special applicability to human social behavior. Trivers's theory of **reciprocal altruism** suggested that altruism can become established in a group where the frequency of interaction among individuals is high and the life span sufficiently long to enable recipients of altruistic acts to return favors to the altruists. The benefits to individuals who partake in such reciprocal altruism can far outweigh the costs, since even slight expenditures of altruistic energy (such as throwing a life preserver to a drowning individual) may have significant benefits to the altruist when it is reciprocated by the previous beneficiary or other group members. Frequent interaction and exchange of roles ("sometimes an altruist, sometimes a beneficiary") are necessary in order to recognize "cheaters" early on—individuals who would otherwise continually try to act as beneficiaries and exploit the altruists (see also p. 587).[21] Emphasis is placed on precise accounting and balancing of exchanges among individuals.

By *expecting* altruistic behavior from other community members and by rejecting cheaters, through either punishment or exile, **moral sentiments** of approval and disapproval are developed and enhanced in such cooperative groups. As pointed out by Trivers, the maintenance of such systems is supported by introducing or reinforcing a variety of **emotional traits:**

- *Friendship* The emotional bonds established among individuals who behave altruistically toward one another
- *Moral indignation and resentment* The feelings of injustice and hostility toward cheaters and oppressors that can lead to retribution against them
- *Gratitude* The emotional responses of recipients to what they perceive as altruistic acts
- *Sympathy, kindness, and generosity* Emotional motivations that help individuals perform altruistic acts
- *Guilt and repentance* Emotions engendered as a result of active cheating or its contemplation that either prevent cheating from happening or lead to reparation and thereby help prevent the rupture of social bonds.

Although these emotions are not uniformly felt or expressed by all individuals under all circumstances, they, like other feelings such as love and security, help preserve social groupings based on intricate reciprocal relationships.[22] It seems very likely that human morality and its accompanying sentiments (such as fairness and justice) did not emerge full blown from human thought but derive from an evolutionary history that probably traces back to early hominid groups or even earlier. The emotions just listed (as well as others such as empathy and remorse) led to abilities we call "moral judgment" and "moral conscience," enabling humans to conceive and incorporate "ethics" of right and wrong.

As Goodall and others show, there are hints of the existence of some of these emotions in other primates, in addition to well-expressed emotions such as fear that all

[21] The other side of this social coin can be seen where **dominance relations** are allowed to hold sway, and cheating by one or more dominant individuals replaces cooperative relationships. Under such circumstances, reciprocal altruism would be absent or tend to diminish except in those areas in which social alliances are forged ("You help me dominate X, and I'll help you dominate Y"), or where dominance is excluded (chimpanzees who capture game are considered its proprietors whether they are dominant or subordinate, and the entire troop may then congregate for handouts). However, even in human societies opposed to overt dominance relations, cheating may be practiced where it can remain undetected, or where local conditions may allow it to occur (for example, smuggling). In Triver's words, humans differ "in the degree of altrüism they show and in the conditions under which they will cheat." Thus, even in societies where cheaters are considered to be "criminals," social and economic exploitation can be practiced and even institutionalized by those who use and manipulate a community's social resources for purposes other than social needs. Apparently, *without firm reciprocal social controls,* appeals made only to "conscience" have little power in sustaining altruistic behavior and preventing exploitation by conscienceless cheaters.

[22] In *On the Origin of Species,* Darwin emphasized the value of social interactions on the evolution of instinctive "moral faculties," pointing out that these led to "love" of praise and "dread" of blame.

BOX 20-3 EVOLUTION OF THE HUMAN BRAIN

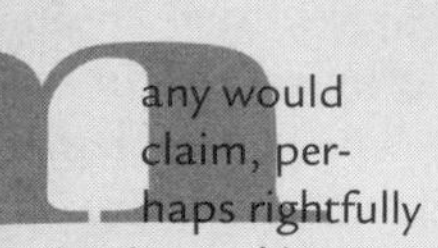

Many would claim, perhaps rightfully so, that the most distinctive and interesting character in human evolution is our brain. This organ separates humans from other primates in both behavior and communication, and also gives humans unique insight into their own thought processes, a feature commonly called **consciousness.**

Because of the brain's value and importance, scientists of all kinds, from anthropologists to psychologists and zoologists, have offered explanations to account for its evolution. For the most part, these hypotheses break down into two general types:

- The evolution of a large brain accommodated the increased number of neural cells necessary for increased processing capacity.
- The evolution of more complex substructures and intricate neuronal circuits allowed refinement of neural function.

These two broad concepts are not mutually exclusive, since a brain can be both large and complex, and neurobiologists well understand that the human brain must have achieved its present state because of both types of evolution. In respect to brain size, we have already pointed to the marked increase in brain size among mammalian ungulates and carnivores during the Cenozoic era (Fig. 19-10), and the table below carries this forward to primates by comparing average brain volumes in fossil hominids, humans, chimpanzees, and gorillas.

As the data in the last column show, relative brain size in present hominids is just about triple that of early australopithecines, and is also triple that of the present great apes. Most of this increase, as Figure 20-24 indicates, is largely associated with increased area and thickness of the cerebral cortex. Like some other traits that also increased disproportionately relative to body size (for example, shorter arms and longer legs relative to apes), brain size increase in hominids as an example of **allometry:** a difference in rate at which a particular feature grows during development or evolution relative to the growth rate of other structures (Fig. 20-25).

To achieve this relatively large brain size without accompanying large body size, human brain development follows a unique pattern compared to other mammals or even other primates. In most mammals and primates, brain growth is rapid relative to body growth during the fetal stages, but this rate diminishes after birth. In humans, prenatal brain growth is also quite rapid relative to body growth, but this rate does not significantly diminish until infants are past 1 year of age (Martin 1990). In that first year of postnatal growth, human brain weight almost triples from about 300 grams to about 900 grams, and then follows the usual primate brain-body growth rate, reaching its full size of about 1,350 grams in adulthood (≈ 15-20 years). (Were brain growth to consistently follow the slower primate pattern from birth onward, adults with 1,350-gram brains would weight about 1,000 pounds.)

This emphasis on rapid early postnatal brain growth essentially extends the human gestation period from 9 months to 21 months by adding 12 months of extrauterine development

Relative brain size calculations in some hominoid species

Species	Dates (millions of years before present)	Estimated Average Body Weight (kilograms)[a]	Average Brain Volume (cubic centimeters)	Relative Brain Size (EQ)[b]
Homo sapiens	.4–present	54	1350	5.8
Late *Homo erectus*	.5–.3	58	980	4.0
Early *Homo erectus*	1.8–1.5	55	804	3.3
Homo habilis	2.4–1.6	42	597	3.1
Australopithecus robustus	1.8–1.0	36	502	2.9
Australopithecus boisei	2.1–1.3	42	488	2.6
Australopithecus aethiopicus	2.7–2.3	?	399	?
Australopithecus africanus[c]	3–2.3	36	420	2.5
Australopithecus afarensis	4–2.8	37	384	2.2
Chimpanzee (*Pan troglodytes*)	Present	45	395	2.0
Gorilla (*Gorilla gorilla*)	Present	105	505	1.7

[a] An average based on combining male and female body weights given by McHenry.

[b] Relative brain size is presented as the encephalization quotient (EQ) calculated as the ratio of actual brain weight or volume to the weight or volume expected for a mammal of that body size (Fig. 19-10). McHenry calculated expected brain volume in these data as $0.0589 \times (\text{species body weight in grams})^{0.76}$.

[c] Conroy and coworkers recently report an endocranial capacity of 515 cc in a particular *A. africanus* skull. They suggest that other measurements may need to be reevaluated.

Source: From McHenry, with some modifications.

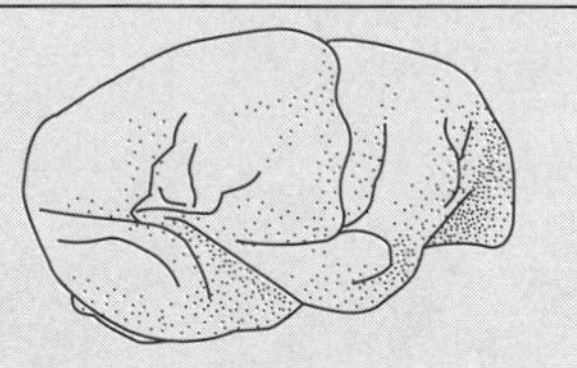

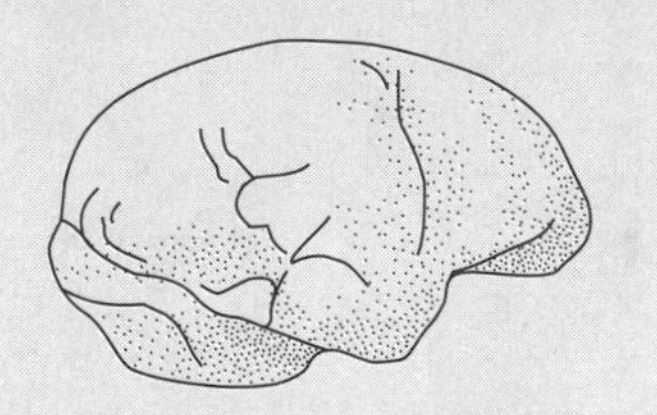

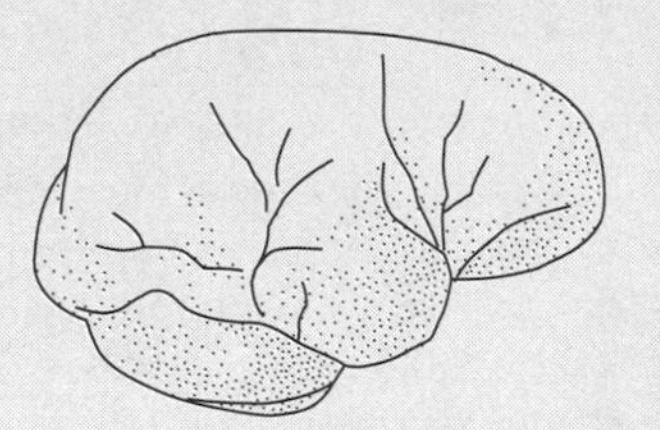

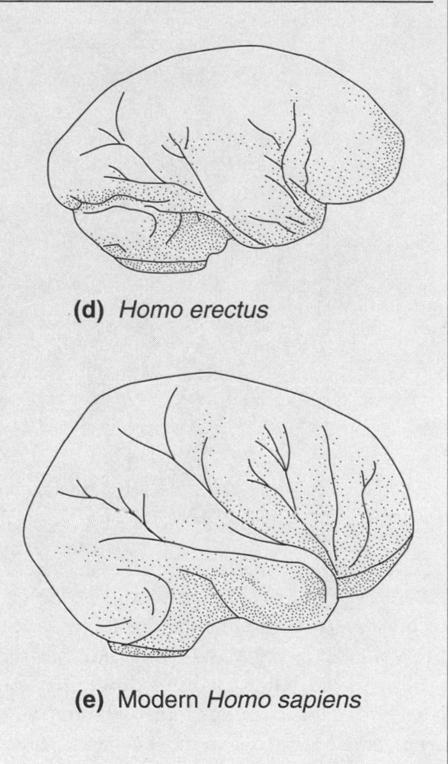

FIGURE 20-24 Casts taken of the inner cranial surfaces (**endocranial casts**) of a chimpanzee (*a*), three fossil hominids (*b–d*), and modern *Homo sapiens* (*e*), showing a lateral view (right side) of each brain, with frontal lobes on the right. Most of the increase in brain size compared to apes is caused by an increase in height, which continues onward to *Homo sapiens,* mostly associated with an expansion of the cerebral cortex. (*From Aiello and Dean, after Holloway.*)

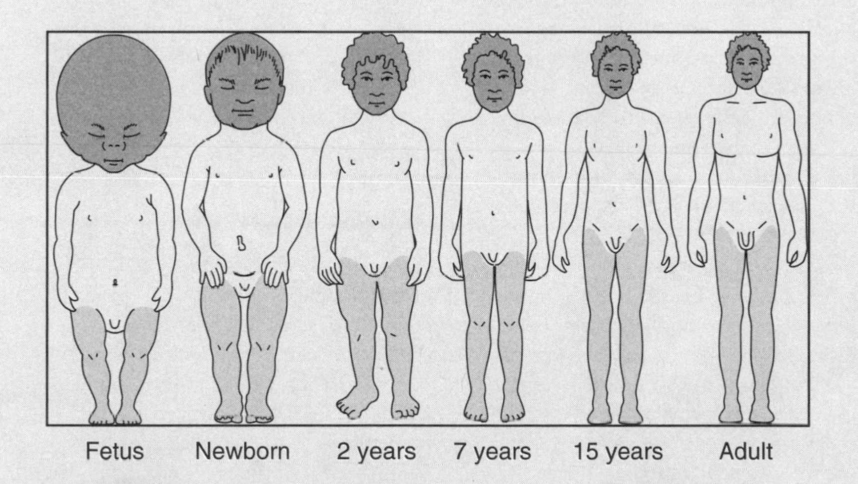

FIGURE 20-25 Different relative growth rates of human body parts—allometry. In proportion to body size increase that occurs during normal growth, the head (30% shaded) grows less rapidly and the legs more rapidly (10% shaded). Head and leg growth in other primates follow a less pronounced allometric pattern. (*Based on Beck et al.*)

(Portmann). Infant dependency, marked by helplessness and vulnerability, is a corollary of the advantages of increased brain size[17] and must have had considerable selective power enhancing those social interactions required to support such dependency, interactions that also provide the learning experiences that help stimulate postnatal brain development. According to

[17] Instead of explaining extended human childhood caused by a larger brain and need for a longer learning period, Hawkes and coworkers explain lengthened immaturity resulting from increased female longevity—the appearance of long-lived postmenopausal (nonfertile) grandmothers. They suggest these older females could supply food to the offspring of their childbearing daughters, enabling children to be weaned earlier, thus extending the juvenile learning period between infancy and sexual maturity. By contrast, others propose that the interval between menopause and death—about 10 to 20 years in humans—primarily lets a female help her last-born child reach independence (Packer et al., see also p. 614). Different views for causes of the lengthened human childhood are not yet resolved.

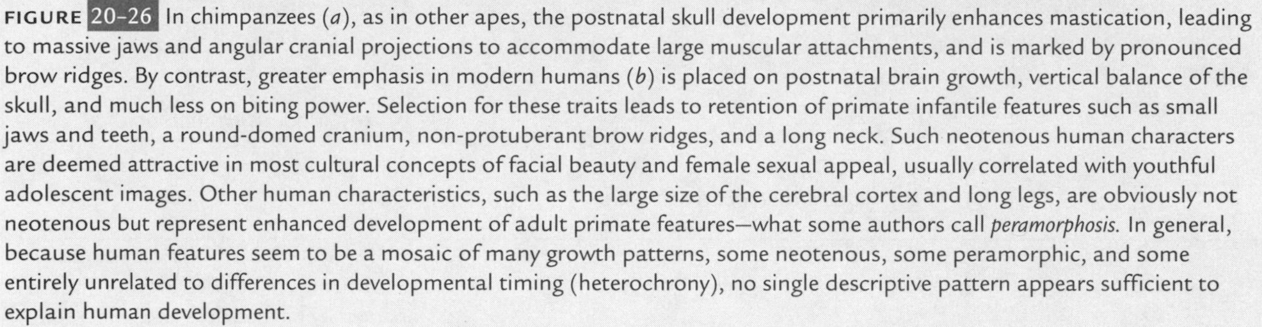

FIGURE 20-26 In chimpanzees (*a*), as in other apes, the postnatal skull development primarily enhances mastication, leading to massive jaws and angular cranial projections to accommodate large muscular attachments, and is marked by pronounced brow ridges. By contrast, greater emphasis in modern humans (*b*) is placed on postnatal brain growth, vertical balance of the skull, and much less on biting power. Selection for these traits leads to retention of primate infantile features such as small jaws and teeth, a round-domed cranium, non-protuberant brow ridges, and a long neck. Such neotenous human characters are deemed attractive in most cultural concepts of facial beauty and female sexual appeal, usually correlated with youthful adolescent images. Other human characteristics, such as the large size of the cerebral cortex and long legs, are obviously not neotenous but represent enhanced development of adult primate features—what some authors call *peramorphosis.* In general, because human features seem to be a mosaic of many growth patterns, some neotenous, some peramorphic, and some entirely unrelated to differences in developmental timing (heterochrony), no single descriptive pattern appears sufficient to explain human development.

some authors, prolonged human infancy with its increased brain development has led to cranial "neoteny"—the retention of juvenile facial features in sexually mature adults (Fig. 20-26).

A further cost of brain expansion comes from its anatomical requirement—large cranial volume in newborns—helping to explain the difficulties faced by human females in giving birth to large-skulled offspring (Fig. 20-8) and the common necessity for social support and obstetrical assistance. Moreover, the energy requirements needed to sustain such a large mass of metabolically active neural tissue are also significant (Parker, Martin 1990). That is, although the adult human brain represents only 2 percent of total body weight, it can consume as much as 20 percent of the energy budget.

How then can we explain the beneficial effect of increased brain size that would counter its obvious costs? Although commonly assumed, the present data have generally not supported the proposal that increased brain size always correlates with increased intelligence (however measured). Including extreme examples, brain volumes can reach more than 1,500 cubic centimeters in some "idiots," and less than 1,000 cubic centimeters in some "geniuses." Nor do we know of any correlation between overall brain volume among humans and particular behaviors and skills.

The obvious progression in hominid cranial size over the last 4 million years must signify some crucial changes in mental capacity, especially since it is so expensive anatomically and metabolically. Among the questions raised are: What advantageous mental functions could have been selected to explain this increase in brain size? Can we identify their cerebral locations? How did these functions evolve? Because paleoneurology is a very young science, many answers are yet to come, but we can still list a few primary mental functions that most changed during the last few million years:

LINGUISTIC ABILITIES

As previously discussed (p. 497), the cerebral location of some functions such as speech comprehension (Wernicke's area) and speech motor control (Broca's area) are anatomically localized, usually to the left cerebral hemisphere (Fig. 20-23). According to Falk, this development is not restricted to *Homo sapiens:* endocranial casts show that expansion of Broca's area in the left hemisphere had already begun in *Homo habilis* and extended further in *Homo erectus.* Other aspects of language, such as grammatical structure (syntax) and vocabulary, seem associated with neural circuits in the prefrontal cerebral cortex that lie somewhat forward of Broca's area. The disproportionate enlargement of the prefrontal cortex is an obvious feature of modern *Homo sapiens* (see Fig. 20–27).

Language acquisition is not based exclusively on speech and hearing but, as many deaf people can attest, can also be acquired through sign language using manual gestures. Deaf children exposed to sign language from birth (that is, born to deaf parents) learn this language as easily and rapidly as hearing children learn spoken language. The capacity for language does not seem restricted to a particular mode of acquisition but seems to be a property of neural networks in the cerebral cortex that can be used both vocally and manually (Calvin).

TECHNICAL APTITUDES

The suggestion that tool use and oral language may share a common neurological root is not far distant from the preceding proposal that auditory and gestural languages are neurologically related. Both manual tool manipulation and language are sequential processes, and, since the appearance of stone tools coincides with the appearance of Broca's area in hominoids, both may have begun their evolutionary maturation together. Linkage between the two functions is also indicated by their common localization in the left "dominant" hemisphere of right-handed individuals. (About 93 percent of the human population is right-handed.) For those left-handed people whose manual control center lies in the right hemisphere, language control is often found there as well.[18]

This "lateralization" of the brain into left or right dominant hemispheres has evolutionary antecedents in many organisms in which an important behavioral pattern locates in one hemisphere rather than the other. Song production in many birds is predominantly associated with the left hemisphere, as is the auditory perception of "coo" signals in Japanese macaque monkeys (p. 492). Right-hemisphere specializations in humans also appear, especially spatial perceptions and other nonverbal traits such as rhythm and musical abilities. Similarly, spatial mapping that rodents use for maze running is localized to the right hemisphere. Hemisphere dominance may therefore be a feature that enables specialization by allowing close interconnection of neural circuits devoted to a particular function.

CAPACITIES FOR SOCIAL INTERACTIONS

Basic elements of social interaction involve awareness of the behavior of others, recognition of the various stages in behavioral sequences, and both knowledge and ability to choose among the possible responses to different behaviors. To individuals, the advantages that appropriate social behaviors confer can include access to food, access to mates, and benefits to offspring. To a large degree, an individual's behavior is also coordinated with complementary behaviors by others in the group, often to reach goals that may or may not be immediately evident or favored by all members. These factors emphasize mastery of highly sophisticated behavioral skills. As Ingold points out,

> The potential for intragroup conflict is as great as the need to maintain community solidarity. The management of relationships with other individuals in the group therefore entails considerable skill, for at every moment the animal has to anticipate not only the immediate effects of its own actions, but also the ways these actions might be perceived by others on the basis of their own perceptions.

For an individual to function socially in human culture might therefore by somewhat like a chesspiece that must develop its own strategies and tactics in confronting aggressive, defensive, and deceptive maneuvers and coalitions, or making alliances with other pieces on the chess board. These behaviors involve stepwise processes of thinking, which some have called "Machiavellian intelligence," that entail a rapid succession of thoughts evaluating perceptions, combining past information, visualizing alternatives, making predictions, and reaching decisions.[19] Social behavior may therefore share some or many neural circuits with other stepwise processes such as language and tool use.

Clearly, a human who must exercise considerable mental flexibility in making appropriate social choices needs much more brainpower than when performing solitary functions. For example, Ridley points out that social complexities in vampire bats are probably responsible for the largest cerebral neocortex of any bat species. For survival, these bats need to recognize and keep track of those neighbors willing to provide them with blood meals (by regurgitation) when their own hunt has been unsuccessful: that is, to discriminate between individuals, and help those who helped them and reject those who did not ("reciprocal altriusm," p. 499).

[18] Wynn, in contrast, suggests that "evolution of a more human-like tool behavior could have long preceded the appearance of language. Long-term memory capacity and problem solving ability, the core abilities in tool behavior, could well have evolved without the appearance of the domain specific features necessary for language."

[19] Such social dynamics have been explored in primate societies by many anthropologists. Cummins writes,

> The struggle for survival in chimpanzee societies is best characterized as a struggle between dominance and the outwitting of dominance, between recognizing your opponent's intentions and hiding your own. The evolution of mind emerges from this scene as a strategic arms race in which the weaponry is ever-increasing mental capacity to represent and manipulate internal representations of the minds of others. If you are big enough to take what you want by force, you are sure to dominate available resources—unless your subordinates are smart enough to deceive you. If you are subordinate, you must use other strategies—deception, guile, appeasement, bartering, coalition formation, friendship, kinship—to get what you need to survive.

LANGUAGE EVOLUTION

Once it could be vocalized, even primitively, further development of language would have been stimulated by its mutual ties with developing social interactions. Given the verbal means for symbolically representing and comprehending experience, language facility could embark on an evolutionary trajectory of its own, allowing an ever-increasing list of subjects to be communicated: food sources, health, safety and security, duration, feelings, possessions, obligations, measurements, comparisons, events, techniques, and tactics.

Vocabularies that include word symbols for these subjects and others, when connected with prepositions, qualifiers, conditional clauses, and nested phrases in meaningful structured sentences—**syntax,** can lead to more complex terms and concepts signifying truth, order, belief, hypothesis, evidence, value, justice, change, and probability. An example of linguistic evolution would be a transition from the simple phrase "Joe hit dog" to the sentence, "When accused by James, Joe admitted to hitting the dog because he believed that it failed to perform its function in guarding the sheep from predators." However, although the means by which such changes were acquired are still unknown, it would seem that, like other evolved complex traits (for example, vision; Fig. 3-1), the evolution of syntax involved a succession of selective stages. Probably any basic language improvement would have had selective value in groups or lineages competitive with others in developing an increasing list of sophisticated technologies, relationships, and social issues. Again, one does not need a full complement of language to use and understand some simple elements (p. 496).

Bickerton suggests that an early pre-syntax beginning may have been a **protolanguage** not much different from "pidgin" jargons used for communicating between people of different languages. These dialects omit articles, prepositions, and tense markers ("You no good man," "Me no have got," "Watee namee you?") reminiscent of expressions used by two-year-old children ("Want mommy read," "Where birdie?" "Get grape juice cup") and apes ("You tickle me," "Hurry gimme," "Choose hide you," see also p. 496). According to Bickerton:

> Brains got bigger because protolanguage created bigger brains and yielded an immediate adaptive payoff in terms of survival: it became possible to warn of more remote dangers, to pool useful information, to plan foraging activities, and the like. But that payoff did not include cultural acceleration, because protolanguage—a slow, clumsy, ad hoc stringing together of symbols—would not support the kind of thinking that led to radical innovation. With no reliable syntax, there was no way of constructing complex propositions. If complex propositions (cause-and-effect, if . . . then, not . . . unless) could not be constructed, there was no escape out of "doing things the way we always did, but a bit better" into "doing things in ways no one thought of before." So for all their ever-growing brains, brains that grew in the end to a size greater than our own [Neanderthals], hominids found no escape from cultural stagnation.

Although not all agree, Bickerton proposes that it was the transition from protolanguage to the longer more complex sentences of full syntactical language that enabled anatomically modern humans to make greater use of their large brains, and accounted for their relatively rapid geographic expansion (Fig. 20-19) and replacement of Neanderthals and more ancient *Homo erectus:*

> If you have anything approximating human language, you have a tool for thinking that has enormous power: a tool not for just any old thinking, but specifically for the kind of thinking that changes your technology and your culture, that gives you fish-hooks, coracles, harpoons, game traps, domestic animals, houses, bridges, wheels.

It certainly seems possible, if not likely, that neural faculties employed in developing complex linguistic patterns and symbols would also allow mental manipulation of events and relationships: to evaluate the past, to consider the future, and to create ideas and imaginative scenarios. However related to language, such conceptualizations transcend immediate experience and mean that individuals can distance themselves from their own ideas, and delay or negate an instantaneous response to their thoughts and sentiments—to consider these separately from their physical selves and thus *contemplate their contemplations.* The awareness of one's own thoughts and feelings is what some would call **consciousness,** and provides humans with an important adaptive social feature that helps understand, predict, and relate to the thoughts and feelings of others.[20]

* * * * *

We see that these functions may be interrelated, involving neural processes and circuitry, which are also probably associated. Former "mosaic" concepts of the cerebrum, in which each section acts independently of others, have given way to an "interconnective" concept in which impulses in one section can stimulate or inhibit cerebral function in others. Moreover, some functions such as visual memory and linguistic syntax cannot localize to specific neurons and seem to exist as "fields"—that is, as diffuse sets of neural synapses that can vary in the frequency and intensity in which their synaptic connections are maintained. In sum, perhaps only minor portions of the human cerebral cortex, no greater than about 20 percent, are sensory and motor areas linked to distinct body parts, and the remainder is associative.

From all accounts, the primary functions of the cerebrum appears to be to

[20]Weiskrantz points out that such conscious awareness has costs: "The fact that we have the capacity to think about what others might be thinking produces a condition that is probably species-specific to humans: namely paranoia."

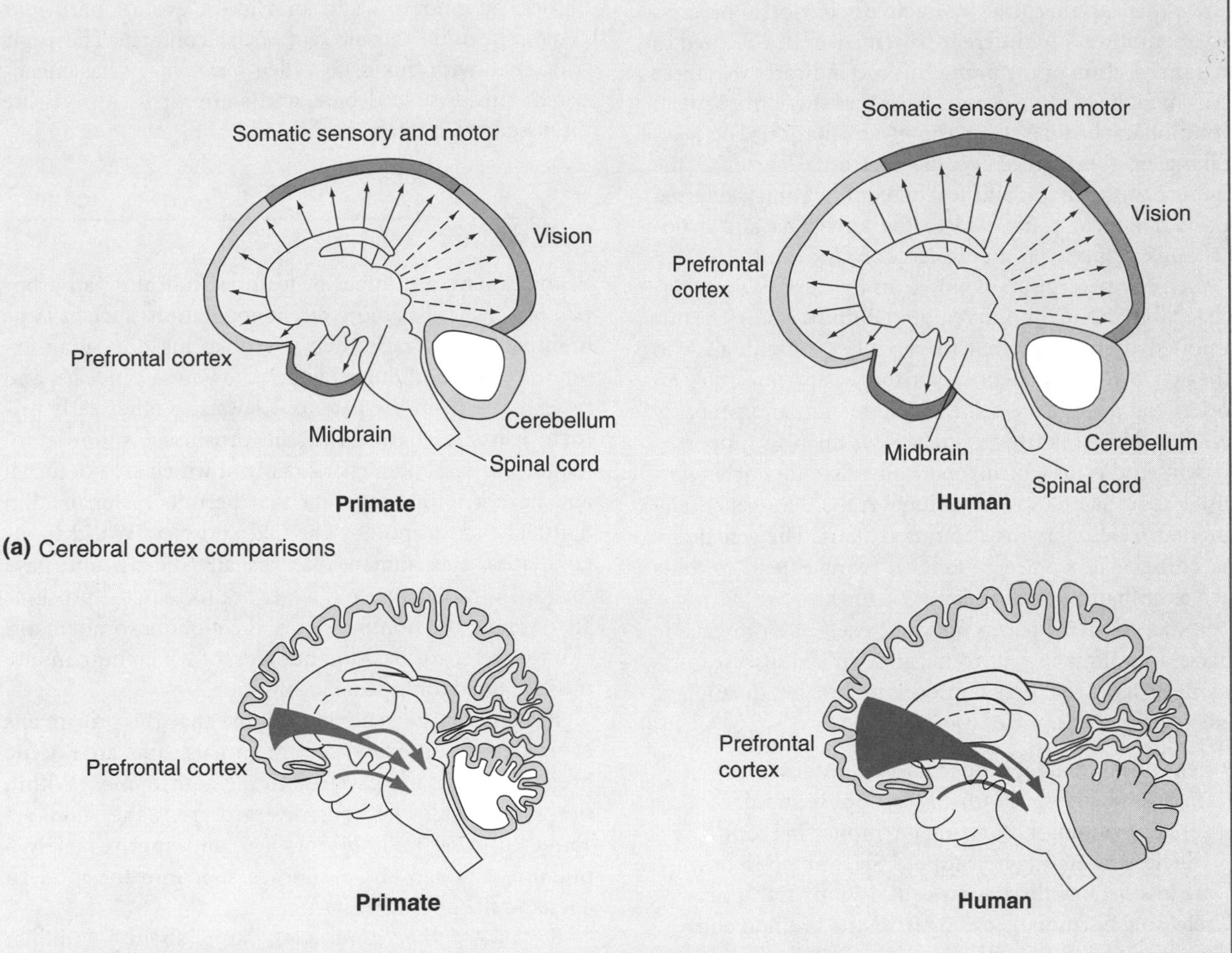

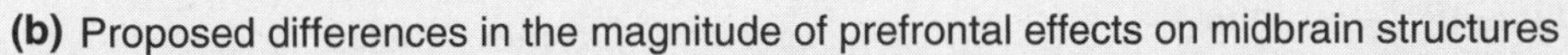

FIGURE 20-27 Comparisons between cortical areas in a typical primate (*left side*) and those in a human with an enlarged cerebrum (*right side*). Most of the disproportionate cortical enlargement in humans is of the prefrontal cortex (*a*), which produces, among other consequences, a greater effect on the midbrain limbic system that governs many emotional responses (*b*). According to Deacon, the prefrontal cortex is more than twice as large as one would predict for a comparably sized ape brain. This prefrontal effect may therefore have accounted for the "reduced repertoire of stereotypic [instinctive] call types in humans" by replacing them with more complex and learned vocalizations. (*After Deacon, modified.*)

analyze and select information received from sensory nerves and coordinate it with stored information (that is, memory), providing connections that can allow the resulting impulses to be further stored or transmitted to other parts of the brain such as motor nerves. Since cerebral analysis depends on selecting among the choices offered by sensory, memory, and motor circuits, the greater the array of choices available the more precise and coordinated can be the responses.

The increased neural connections led to expansion of this elaborately structured cerebral organ and enabled development of those uniquely human traits that appear so much more complex and so much higher in order than other primates. As Gibson states, "expansion of the association areas yielded increased capacities for hierarchical construction in linguistic, manipulative and social domains." To these elements, other workers (for example, Calvin) add sequential traits such as elaborate hunting procedures and aimed missile throwing that may also have provided the selective drive for brain expansion. Taking these many factors together, Deacon has emphasized that "the correlated reorganization of underlying neural circuitry" may well account for the disproportionate enlargement of the human prefrontal lobes, since these cortical areas are involved in a variety of processes that include organizing sequential activity (Fig. 20-27).

mammals probably share.[23] Since neurologists have shown that certain social-interactive behaviors ("processing of emotion") in different patients can be localized to the same section of the brain, this too indicates that these traits must have some genetic basis and therefore have an evolutionary history (Damasio et al.). Rational decision making has been traced to the prefrontal cortex in the forepart of the brain, and fearful conduct and depressed behavior appear connected to the amygdala and hypothalamus, respectively, in the center of the brain.

Even our perception of emotions in others is found localized to particular brain regions (Phillips et al.) The uninhibited friendliness that marks most individuals who inherit Down or Williams syndromes also indicates an underlying genetic basis for an important, and probably complex, human attitude. Such views are also supported by twin studies in which genetically alike identical twins raised apart share more emotional traits than genetically different fraternal twins also raised apart. The genetically shared behaviors among identical twins extend to traits such as empathy and altruism (Loehlin).

Sexual behavior is certainly not exempt from genetic effects. In Kallmann syndrome, an X-linked disease causing males to lose all sexual interest in females, the mutant gene triggers a cascade of events:

> Behavioral libido is reduced *because* of low testosterone levels, in turn *because* of reduced gonadotrophins, luteinizing hormone (LH) and follicle-stimulating hormone (FSH), which in turn are low *because* there is no GnRH [gonadotrophin-releasing hormone] coming from the brain to enter the pituitary, *because* there are no GnRH cells in the brain, *because* GnRH neuronal migration has failed, *because* of the absence of the protein produced by a gene at Xp-22.3 (Pfaff).

From what we can see of their effects, the origin of emotions is certainly no mystery: these devices help individuals perform tasks necessary for their own preservation or that of the group. The evolution of emotions is also not a mystery: an emotion, like a preadaptation, can be used for one purpose and then later serve or be modified for another purpose. Thus love, an emotion originally tied to reproduction, has extended into a device that helps bond individuals into familial and social groups. The strength of emotional ties is reflected in emotional responses when ties are broken: closer ties often cause greater grief. Such significant behavioral effects raise the question of the manner and extent to which evolutionarily derived emotions and strategies ("evolutionary psychology") guide our present social conduct. The topic concerned with this issue, called *sociobiology,* has engendered considerable debate, and some of its aspects are discussed in Chapters 24 and 25.

[23] In an analysis seeking to understand how chimpanzees think, Povinelli and Godfrey point out:

> Chimpanzees may share with humans certain attributional processes, including some form of self-awareness, the attribution of attention, role-taking, and perhaps the attribution of knowledge. The evidence is weaker, but chimpanzees may also be capable of some degree of empathic concern, pretense, and intentional deception.

SUMMARY

Primates have a number of features indicating an arboreal past, as well as more recent adaptations such as large brain size. They raise their few offspring during an extended childhood during which the young can learn and receive care from the parents. Claws and other early primate features characterize the prosimian suborder of lemurs, lorises, and tarsiers, many of which are nocturnal and have a retinal tapetum that permits vision in dim light. The anthropoids—the Old and New World monkeys, apes, and humans—are usually diurnal and have forward-directed eyes and larger brains than the prosimians. Among the hominoids are the gibbons, orangutans, chimpanzees, gorillas, and humans, of which humans are most adapted for bipedal locomotion.

Paleontologists generally assume that the prosimians evolved from insectivorous mammals, and an Eocene prosimian line later gave rise to the anthropoids. Within the anthropoids the apes diverged from the monkeys about 20 to 30 million years ago, and approximately 5 million years ago an ape lineage split into the modern apes and the hominid line.

So far, *Ardipithecus ramidus,* dating about 4.4 million years ago, represents the most ancient and apelike hominid. It is followed by a succession of more humanlike species in the genus *Australopithecus* whose postcranial fossils show good evidence that they were bipedal.

Bipedalism may have been favored because of advantages in foraging, evading predators, carrying provisions, and using tools and weapons. Among the eventual consequences of hominid bipedalism in a patchy environment were probably establishment of a home base, sexual bonding, and parental involvement in offspring. Changes in the pelvis, legs, and feet accompanied the new body alignment.

The earliest fossils of the genus *Homo, H. habilis,* are about 2 million years old and have brains much larger than those of the australopithecines. *Homo erectus,* a widespread species that used distinguishable stone tools, appeared somewhat later and gave rise to *Homo sapiens.* The Neanderthals, early representatives of this new species, had a modern brain size, heavy jaws and brow ridges, and were socially advanced. They yielded about 35,000 years ago to anatomically modern humans, who had an advanced culture and were modern in appearance. Although some molecular studies indicate a single origin (Africa) for modern humans, there has been considerable dispute, and some paleoanthropologists point to multiple origins.

At some point hominids began supplementing their foraging activities with hunting and its increasing importance enhanced social cohesion, cooperation, and communication, as well as favoring improvements in technology.

Primates can transmit signals to other individuals through any of the sensory mechanisms: scent, grooming behavior, gestures and facial expressions, displays, and vocalization. The development of the tongue, soft palate, and larynx let humans produce a variety of sounds not available to other primates.

Although attempts to teach chimpanzees how to speak have failed, a number of investigators have shown that chimpanzees have conceptual linguistic skills that do not depend on the structure of the vocal tract. Perhaps language can arise only in organisms, humans and apes, that have awareness of their "self" or separateness from their environment, traits that arise because of the complexity of social interactions. Other traits, such as altruism, morality, friendship, and guilt, may be based on frequent interactions among individuals in a group whcrc such bchaviors and emotions are adaptations that promote survival and social cohesion.

In summary, what long appeared to be unbridgeable gaps between apes and humans are now being bridged by discoveries that language, toolmaking, planning capacity, self-awareness, complex social interaction, and emotional development—traits that many formerly considered exclusively human—are also present in apes, but generally to much lesser degrees. If we feel compelled to define our unique human behavior, we can probably best present it as the attainment of many new hierarchical levels at which these traits are now developed and expressed.

KEY TERMS

allometry
altruistic behavior
angular gyrus
animal hunting
anthropoids
australopithecines
bipedalism
Broca's area
catarrhines
chimpanzees
communication
consciousness
coopcrativc hunting
cross-modal associations
displacement (language)
dominance relations (behavior)
emotional traits
endocranial casts
gibbons
gorillas
home base
hominids
hominoids
Homo
hunter-gatherer societies
kin selection
language
larynx
lemurs
long-distance foraging
lorises
meat eating
moral sentiments
multiregional evolution
olfactory cues
orangutans
Piltdown man
platyrrhines
prosimians
protolanguage
reciprocal altruism
self-awareness
sequencing rules
sexual bonding
speech
stone tools
symbolic calls
syntax
tactile communication
tarsiers
visual signals
vocalizations
Wernicke's area

DISCUSSION QUESTIONS

1. What major features distinguish primates from other mammals?
2. What features characterize the major subgroups in the order Primates?
3. What features do humans share with anthropoid apes?
4. What proposals have researchers made to account for the increase in ground-dwelling apes during the Miocene epoch?
5. What time periods have primatologists suggested for the ape–human divergence, and how do they justify these suggestions?
6. What are the major fossil lineages among early ominids?
7. How did the Piltdown forgery affect interpretation of the pattern of human evolution?
8. Bipedalism
   a. What advantages could bipedalism have offered to early hominids?
   b. What anatomical changes have accompanied hominid bipedalism?
9. What effects on human evolution have anthropologists proposed for the establishment of a home base and sexual bonding?
10. The genus *Homo*
    a. What major fossil groups of *Homo* have paleontologists found, and in what time periods?
    b. What is the connection between these fossil groups and types of stone tools?
    c. What does the fossil evidence suggest about the origins of modern humans?
    d. How have geneticists used mitochondrial DNA to trace the origin and migration of *Homo sapiens*?
    e. Assuming that a common ancestral Y-chromosome sequence was found, would you expect the individual carrying that original sequence (the "Y-chromosome Adam") to be a contemporary of the "Mitochondrial Evc" (p. 487)? Why or why not?
11. Hunting
    a. What advantages does meat eating offer a primate?
    b. What social and behavioral characteristics are of selective value in hunting societies?

12. Communication
    a. What modes of communication do primates use?
    b. What advantages does vocalization confer over other modes of communication?
    c. How do symbolic and nonsymbolic communication differ?
13. Speech
    a. What anatomical features distinguish the human vocal tract from that of other primates?
    b. How do these features affect vocalization?
14. Language
    a. What techniques have people used to communicate with apes?
    b. What kinds of linguistic abilities have apes demonstrated?
15. How can primates demonstrate self-recognition and self-awareness in nonverbal form?
16. What selective factors lead to altruistic behavior? to reciprocal altruism?
17. What types of emotions do societies dependent on complex and reciprocal social relationships develop or enhance?
18. Is morality a sentiment that has no evolutionary antecedents but is exclusive to *Homo sapiens*? Explain.
19. What primate and human characteristics, physical and mental, would you consider to have an origin that can be ascribed to preadaptation? Explain.
20. Since humans had an ancestry among ancient apelike primates, would you agree that we can briefly describe humans as intelligent, naked, bipedal apes? Why or why not?

EVOLUTION ON THE WEB

Explore evolution on the web! Visit the accompanying web site for *Evolution*, 3/e at www.jbpub.com/evolution for web exercises and links relating to topics covered in this chapter.

REFERENCES

Aiello, L. C., 1986. The relationships of the Tarsiioformes: A review of the case for the Haplorhini. In *Major Topics in Primate and Human Evolution*, B. Wood, L. Martin, and P. Andrews (eds.). Cambridge University Press, Cambridge, England, pp. 47–65.

Aiello, L., and C. Dean, 1990. *An Introduction to Human Evolutionary Anatomy.* Academic Press, London.

Andrews, P., 1981. Species diversity and diet in monkeys and apes during the Miocene. In *Aspects of Human Evolution*, C. B. Stringer (ed.). Taylor & Francis, London, pp. 25–61.

————, 1983. The natural history of *Sivapithecus*. In *New Interpretations of Ape and Human Ancestry*, R. Ciochon and R. Corruccini (eds.). Plenum Press, New York, pp. 441–464.

Arensburg, B., A. M. Tillier, B. Vandermeersch, H. Duday, L. A. Schepartz, and Y. Rak, 1989. A middle Paleolithic human hyoid bone. *Nature,* **338,** 758–760.

Asfaw, B., T. White, O. Lovejoy, B. Latimer, S. Simpson, and G. Suwa, 1999. *Australopitchecus garhi:* A new species of early hominid from Ethiopia. *Science,* **284,** 629–635.

Ayala, F. J., A. Escalante, C. O'hUigin, and J. Klein, 1994. Molecular genetics of speciation and human origins. *Proc. Nat. Acad. Sci.,* **91,** 6787–6794.

Batzer, M. A., et. al., 1994. African origin of human-specific polymorphic *Alu* insertions. *Proc. Nat. Acad. Sci.,* **91,** 12288–12292.

Beard, K. C., T. Qi, M. R. Dawson, B. Wang, and C. Li, 1994. A diverse new primate fauna from middle Eocene fissure-fillings in southeastern China. *Nature,* **368,** 604–609.

Beck, F., D. B. Moffatt, and J. B. Lloyd, 1973. *Human Embryology and Genetics.* Blackwell, Oxford, England.

Bickerton, D., 1995. *Language and Human Behavior.* University of Washington Press, Seattle.

Boesch, C., 1993. Aspects of transmission of tool-use in wild chimpanzees. In *Tools, Language and Cognition in Human Evolution,* K. R. Gibson and T. Ingold (eds.). Cambridge University Press, Cambridge, England, pp. 171–183.

Boesch-Achermann, H., and C. Boesch, 1994. Hominization in the rainforest: The chimpanzee's piece of the puzzle. *Evol. Anthropol.,* **3,** 9–16.

Boysen, S. T., and G. G. Berntson, 1990. The development of numerical skills in the chimpanzee (*Pan troglodytes*). In *"Language" and Intelligence in Monkeys and Apes: Comparative Developmental Perspectives,* S. T. Parker and K. R. Gibson (eds.). Cambridge University Press, Cambridge, England, pp. 435–450.

Brown, M. D., et al., 1998. MtDNA haplogroup X: An ancient link between Europe/Western Asia and North America? *Amer. J. Hum. Genet.,* **63,** 1852–1861.

Calvin, W. H., 1993. The unitary hypothesis: A common neural circuitry for novel manipulations, language, plan-ahead, and throwing? In *Tools, Language and Cognition in Human Evolution,* K. R. Gibson and T. Ingold (eds.). Cambridge University Press, Cambridge, England, pp. 230–250.

Campbell, B., 1985. *Human Evolution,* 3d ed. Aldine, New York.

Cann, R. L., M. Stoneking, and A. C. Wilson, 1987. Mitochondrial DNA and human evolution. *Nature,* **325,** 31–36.

Cant, J. G. H., 1981. Hypothesis for the evolution of human breasts and buttocks. *Amer. Nat.,* **117,** 199–204.

Cavalli-Sforza, L. L., E. Minch, and J. L. Mountain, 1992. Coevolution of genes and languages revisited. *Proc. Nat. Acad. Sci.,* **89,** 5620–5624.

Chevalier-Skolnikoff, S., 1973. Facial expressions of emotion in nonhuman primates. In *Darwin and Facial Expression,* P. Ekman (ed.). Academic Press, New York, pp. 11–89.

Chomsky, N., 1972. *Language and Mind.* Harcourt, Brace, and Jovanovich, New York.

————, 1976. On the nature of language. In *Origins and Evolution of Language and Speech,* S. R. Harnad, H. D. Steklis, and J. Lancaster (eds.). New York Academy of Sciences, vol. 280, pp. 46–57.

Cinque, G., 1999. *Adverbs and Functional Heads: A Cross-Linguistic Approach.* Oxford University Press, Oxford, England.

Clarke, R. J., 1985. A new reconstruction of the Florisbad cranium, with notes on the site. In *Ancestors: The Hard Evidence,* E. Delson (ed.). Liss, New York, pp. 301–305.

Conroy, G. C., 1997. *Reconstructing Human Evolution: A Modern Synthesis.* Norton, New York.

Conroy, G. C., et al., 1998. Endocranial capacity in an early hominid cranium from Sterkfontein, South Africa. *Science,* **280,** 1730–1731.

Coppens, Y., 1994. East side story: The origin of humankind. *Sci. Amer.,* **270,** 88–95.

Cummins, D. D., 1998. Social norms and other minds: The evolutionary roots of higher cognition. In *The Evolution of Mind,* D. D. Cummins and C. Allen (eds.). Oxford University Press, New York, pp. 30–50.

Damasio, H., T. Grabowski, R. Frank, A. M. Galaburda, and A. R. Damasio, 1994. The return of Phineas Gage: Clues about the brain from the skull of a famous patient. *Science,* **264,** 1102–1105.

Dart, R., 1925. *Australopithecus africanus:* The man-ape of South Africa. *Nature,* **115,** 195–199.

Day, M. H., 1986a. *Guide to Fossil Man,* 4th ed. Cassell, London.

———, 1986b. Bipedalism: Pressures, origins and modes. In *Major Topics in Primate and Human Evolution,* B. Wood, L. Martin, and P. Andrews (eds.). Cambridge University Press, Cambridge, England, pp. 188–202.

Deacon, T. W., 1990. Rethinking mammalian brain evolution. *Amer. Zool.,* **30,** 629–705.

de Heinzelin, J., et al., 1999. Environment and behavior of 2.5 million-year-old Bouri hominids. *Science,* **284,** 625–629.

Delarbre, C., H. Nakauchi, R. Bontrop, P. Kourilsky, and G. Gauchelin, 1993. Duplication of the CD8 β-chain gene as a marker of the man-gorilla-chimpanzee clade. *Proc. Nat. Acad. Sci.,* **90,** 7049–7053.

Delson, E. (ed.), 1985. *Ancestors: The Hard Evidence.* Liss, New York.

Desmond, A. J., 1979. *The Ape's Reflexion.* Dial Press, New York.

De Waal, F. B. M., 1986. Deception in the natural communication of chimpanzees. In *Deception: Perspectives on Human and Nonhuman Deceit,* R. W. Mitchell and N. S. Thompson (eds.). SUNY Press, Stony Brook, NY, pp. 271–292.

Diamond, J. M., 1995. The evolution of human creativity. In *Creative Evolution?!,* J. H. Campbell and J. W. Schopf (eds.). Jones and Bartlett, Boston, pp. 75–84.

Falk, D., 1993. Sex differences in visuospatial skills: Implications for human evolution. In *Tools, Language and Cognition in Human Evolution,* K. R. Gibson and T. Ingold (eds.). Cambridge University Press, Cambridge, England, pp. 216–229.

Fleagle, J. G., 1988. *Primate Adaptation and Evolution.* Academic Press, San Diego, CA.

Fleagle, J. G., T. M. Bown, J. D. Obradovitch, and E. L. Simons, 1986. Age of the earliest African anthropoids. *Science,* **234,** 1247–1249.

Foley, R., 1987. *Another Unique Species: Patterns in Human Evolutionary Ecology.* Longman, Harlow, Great Britain.

———, 1995. *Humans Before Humanity.* Blackwell, Oxford, England.

———, 1996. The adaptive legacy of human evolution: A search for the environment of evolutionary adaptedness. *Evol. Anthropol.,* **4,** 194–203.

Frangiskakis, J. M., A. K. Ewart, C. A. Morris, C. B. Mervis, et al., 1996. LIM-kinase 1 hemizygosity implicated in impaired visiospatial constructive cognition. *Cell,* **86,** 59–69.

Gallup, G. G., Jr., 1977. Self-recognition in primates: A comparative approach to the bidirectional properties of consciousness. *Amer. Psychol.,* **32,** 329–338.

Gardner, R. A., and B. T. Gardner, 1969. Teaching sign language to a chimpanzee. *Science,* **165,** 664–672.

Gibson, K. R., 1993. Overlapping neural control of language, gesture and tool use. In *Tools, Language and Cognition in Human Evolution,* K. R. Gibson and T. Ingold (eds.). Cambridge University Press, Cambridge, England, pp. 187–192.

Goodall, J., 1986. *The Chimpanzees of Gombe: Patterns of Behavior.* Harvard University Press, Cambridge, MA.

Goodman, M., et al., 1998. Toward a phylogenetic classification of primates based on DNA evidence complemented by fossil evidence. *Mol. Phylogenet. and Evol.,* **9,** 585–598.

Green, S., and P. Marler, 1979. The analysis of animal communication. In *Handbook of Behavioral Neurology,* vol. 3, P. Marler and J. G. Vandenbergh (eds.). Plenum Press, New York, pp. 73–158.

Haldane, J. B. S., 1932. *The Causes of Evolution.* Harper, London.

Hamilton, W. D., 1964. The evolution of social behavior. *J. Theoret. Biol.,* **1,** 1–52.

Hammer, M. F., et al., 1998. Out of Africa and back again: Nested cladistic analysis of human Y chromosome variation. *Mol. Biol. and Evol.,* **15,** 427–441.

Harding, R. S. O., and G. Teleki (eds.), 1981. *Omnivorous Primates: Gathering and Hunting in Human Evolution.* Columbia University Press, New York.

Harpending, H. C., M. A. Batzer, M. Gurven, L. B. Jorde, A. R. Rogers, and S. T. Sherry, 1998. Genetic traces of ancient demography. *Proc. Nat. Acad. Sci.,* **95,** 1961–1967.

Harris, E. E., and J. Hey, 1999. X chromosome evidence for ancient human histories. *Proc. Nat. Acad. Sci.,* **96,** 3320–3324.

Hawkes, K., J. F. O'Connell, N. G. Blurton Jones, H. Alvarez, and E. L. Charnov, 1998. Grandmothering, menopause, and the evolution of human life histories. *Proc. Nat. Acad. Sci.,* **95,** 1336–1339.

Hayes, K. J., and C. Hayes, 1951. The intellectual development of a home-raised chimpanzee. *Proc. Amer. Philos. Soc.,* **95,** 105–109.

Hill, K., 1982. Hunting and human evolution. *J. Hum. Evol.,* **11,** 521–544.

Horai, S., K. Hayasaka, R. Kondo, K. Tsugane, and N. Takahata, 1995. Recent African origin of modern humans revealed by complete sequences of hominoid mitochondrial DNAs. *Proc. Nat. Acad. Sci.,* **92,** 532–536.

Howells, W., 1993. *Getting Here: The Story of Human Evolution.* Compass Press, Washington, DC.

Ingold, T., 1993. Tool-use, sociality and intelligence. In *Tools, Language and Cognition in Human Evolution,* K. R. Gibson and T. Ingold (eds.). Cambridge University Press, Cambridge, England, pp. 429–445.

Johanson, D., and M. Edey, 1981. *Lucy: The Beginnings of Mankind.* Simon and Schuster, New York.

Jolly, A., 1985. *The Evolution of Primate Behavior.* Macmillan, New York.

Jones, S., R. Martin, and D. Pilbeam (eds.), 1992. *The Cambridge Encyclopedia of Human Evolution.* Cambridge University Press, Cambridge, England.

Jorde, L. B., M. Bamshad, and A. R. Rogers, 1998. Using mitochondrial and nuclear DNA markers to reconstruct human evolution. *BioEssays,* **20,** 126–136.

Kay, R. F., C. Ross, and B. A. Williams, 1997. Anthropoid origins. *Science,* **275,** 797–804.
Kinzey, W. G. (ed.), 1987. *The Evolution of Human Behavior: Primate Models.* SUNY Press, Albany, NY.
Krings, M., H. Geisert, R. W. Schmitz, H. Krainitzki, and S. Pääbo, 1999. DNA sequence of the mitochondrial hypervariable region II from the Neanderthal type specimen. *Proc. Nat. Acad. Sci.,* **96,** 5581–5585.
LeGros Clark, W. E., 1978. *The Fossil Evidence for Human Evolution,* 3d ed. University of Chicago Press, Chicago.
Lieberman, P. 1984. *The Biology and Evolution of Language.* Harvard University Press, Cambridge, MA.
———, 1991. *Uniquely Human: The Evolution of Speech, Thought, and Selfless Behavior.* Harvard University Press, Cambridge, MA.
Lodish, H., D. Baltimore, A. Berk, S. L. Zipursky, P. Matsudaira, and J. Darnell, 1995. *Molecular Cell Biology,* 3d ed. Scientific American Books, New York.
Loehlin, J. C., 1992. *Genes and Environment in Personality Development.* Sage, Newbury Park, CA.
Lovejoy, C. O., 1981. The origin of man. *Science,* **211,** 341–350.
———, 1988. The evolution of human walking. *Sci. Amer.,* **259** (5), 118–125.
Mann, W. M., 1938. Monkey folk. *Nat. Geograp.,* **73,** 615–655.
Martin, R. D., 1990. *Primate Origins and Evolution: A Phylogenetic Reconstruction.* Chapman & Hall, London.
———, 1993. Primate origins: Plugging the gaps. *Nature,* **363,** 223–234.
Maynard Smith, J., 1983. Game theory and the evolution of cooperation. In *Evolution from Molecules to Man,* D. S. Bendall (ed.). Cambridge University Press, Cambridge, England, pp. 445–456.
McCollum, M. A., 1999. The robust australopithecine face: A morphogenetic perspective. *Science,* **284,** 301–305.
McHenry, H. M., 1994. Tempo and mode in human evolution. *Proc. Nat. Acad. Sci.,* **91,** 6780–6786.
Miyamoto, M. M., B. F. Koop, J. L. Slightom, M. Goodman, and M. R. Tennant, 1988. Molecular systematics of higher primates: Genealogical relations and classification. *Proc. Nat. Acad. Sci.,* **85,** 7627–7631.
Mountain, J. L., 1998. Molecular evolution and modern human origins. *Evol. Anthropol.,* **7,** 21–37.
Mountain, J. L., and L. L. Cavalli-Sforza, 1994. Inference of human evolution through cladistic analysis of nuclear DNA restriction polymorphisms. *Proc. Nat. Acad. Sci.,* **91,** 6515–6519.
Mountain, J. L., A. A. Lin, A. M. Bowcock, and L. L. Cavalli-Sforza, 1992. Evolution of modern humans: Evidence from nuclear polymorphisms. *Phil. Trans. Roy. Soc. London (Biol.*), **337,** 159–165.
Napier, J. R., and P. H. Napier, 1985. *The Natural History of Primates.* British Museum, London.
Nei, M., and A. K. Roychoudhury, 1993. Evolutionary relationships of human populations on a global scale. *Mol. Biol. and Evol.,* **10,** 927–943.
Packer, C., M. Tater, and A. Collins, 1998. Reproductive cessation in female animals. *Nature,* **392,** 807–811.
Parker, S. T., 1990. Why big brains are so rare: Energy costs of intelligence and brain size in anthropoid primates. In *"Language" and Intelligence in Monkeys and Apes,* S. T. Parker and K. R. Gibson (eds.). Cambridge University Press, Cambridge, England, pp. 129–154.
Pfaff, D. W., 1997. Hormones, genes, and behavior. *Proc. Nat. Acad. Sci.,* **94,** 14213–14216.
Phillips, M. L., et al., 1997. A specific neural substrate for perceiving facial expressions of disgust. *Nature,* **389,** 495–498.
Pilbeam, D., 1984. The descent of hominoids and hominids. *Sci. Amer.,* **250,** 84–96.
———, 1986. Hominoid evolution and hominoid origins. *Amer. Anthropol.,* **88,** 295–312.
Portmann, A., 1990. *A Biologist Looks at Humankind.* (Translated from an earlier German edition by J. Schaefer.) Columbia University Press, New York.
Potts, R., 1992. The hominid way of life. In *The Cambridge Encyclopedia of Human Evolution,* S. Jones, R. Martin, and D. Pilbeam (eds.). Cambridge University Press, Cambridge, England, pp. 325–334.
Povinelli, D. J., and L. R. Godfrey, 1993. The chimpanzee's mind: How noble in reason? How absent of ethics? In *Evolutionary Ethics,* M. H. Nitecki and D. V. Nitecki (eds.). SUNY Press, Albany, NY, pp. 277–324.
Premack, D., 1971. Language in the chimpanzee? *Science,* **172,** 808–822.
Rak, Y., 1986. The Neanderthal: A new look at an old face. *J. Hum. Evol.,* **15,** 151–164.
Rasmussen, D. T. (ed.), 1993. *The Origin and Evolution of Humans and Humanness.* Jones and Bartlett, Boston.
Ridley, Matt, 1996. *The Evolution of Virtue: Human Instincts and the Evolution of Cooperation.* Viking, New York.
Robinson, J. T., 1972. *Early Hominid Posture and Locomotion.* University of Chicago Press, Chicago.
Rosenberg, K., and W. Trevathan, 1996. Bipedalism and human birth dilemma revisited. *Evol. Anthropol.,* **4,** 161–168.
Rouhani, S., 1989. Molecular genetics and the pattern of human evolution: Plausible and implausible models. In *The Human Revolution: Behavioural and Biological Perspectives on the Origin of Modern Humans,* P. Mellars and C. Stringer (eds.). Edinburgh University Press, Edinburgh, pp. 47–61.
Ruff, C. B., E. Trinkaus, and T. W. Holliday, 1997. Body mass and encephalization in Pleistocene *Homo. Nature,* **387,** 173–176.
Rumbaugh, D. M., 1977. *Language Learning by a Chimpanzee: The Lana Project.* Academic Press, New York.
Ruvolo, M., D. Pan, S. Zehr, T. Goldberg, T. R. Disotell, and M. von Dornum, 1994. Gene trees and hominoid phylogeny. *Proc. Nat. Acad. Sci.,* **91,** 8900–8904.
Ruvolo, M., S. Zehr, M. von Dornum, D. Pan, B. Chang, and J. Lin, 1993. Mitochondrial COII sequences and modern human origins. *Mol. Biol. and Evol.,* **10,** 1115–1135.
Sagan, C., 1997. The age of exploration. In *Carl Sagan's Universe,* Y. Terzian and E. Bilson (eds.). Cambridge University Press, Cambridge, England, pp. 141–160.
Sarich, V. M., and J. E. Cronin, 1976. Molecular systematics of the primates. In *Molecular Anthropology,* M. Goodman and R. E. Tashian (eds.). Plenum Press, New York, pp. 141–170.
Savage-Rumbaugh, E. S., and D. M. Rumbaugh, 1993. The emergence of language. In *Tools, Language and Cognition in Human Evolution,* K. R. Gibson and T. Ingold (eds.). Cambridge University Press, Cambridge, England, pp. 86–108.
Schultz, A. H., 1933. Die körperproportionen der erwachsenen catarrhinen Primaten, mit spezieller Berücksichtigung der Menschenaffen. *Anthropol. Anz.,* **10,** 154–185.
Semaw, S., et al., 1997. 2.5-million-year-old stone tools from Gona, Ethiopia. *Nature,* **385,** 333–336.

Shipman, P., 1985. The ancestor that wasn't. *The Sciences,* **25** (2), 43–48.

Silberbauer, G., 1981. Hunter/gatherers of the Central Kalahari. In *Omnivorous Primates: Gathering and Hunting in Human Evolution,* R. S. O. Harding and G. Teleki (eds.). Columbia University Press, New York, pp. 455–498.

Spencer, F., 1990. *Piltdown: A Scientific Forgery.* Oxford University Press, Oxford, England.

Spoor, F., B. Wood, and F. Zonneveld, 1994. Implications of early hominid labyrinthine morphology for evolution of human bipedal locomotion. *Nature,* **369,** 645–648.

Stoneking, M., 1993. DNA and recent human evolution. *Evol. Anthropol.,* **2,** 60–71.

Strauss, E., 1999. Can mitochondrial clocks keep time? *Science,* **283,** 1435–1438.

Stringer, C. B., and P. Andrews, 1988. Genetic and fossil evidence for the origin of modern humans. *Science,* **239,** 1263–1268.

Stringer, C. B., and C. Gamble, 1993. *In Search of the Neanderthals: Solving the Puzzle of Human Origins.* Thames and Hudson, London.

Stringer, C. B., and R. Grün, H. P. Schwarcz, and P. Goldberg, 1989. ESR dates for the hominid burial site of Es Skhul in Israel. *Nature,* **338,** 756–758.

Takahata, N., 1993. Allelic genealogy and human evolution. *Mol. Biol. and Evol.,* **10,** 2–22.

Takahata, N., and Y. Satta, 1997. Evolution of the primate lineage leading to modern humans: Phylogenetic and demographic inferences from DNA sequences. *Proc. Nat. Acad. Sci.,* **94,** 4811–4815.

Tanner, N. M., 1987. The chimpanzee model revisited and the gathering hypothesis. In *The Evolution of Human Behavior: Primate Models,* W. G. Kinzey (ed.). SUNY Press, Albany, NY, pp. 3–27.

Tattersall, I., E. Delson, and J. Van Couvering (eds.), 1988. *Encyclopedia of Human Evolution and Prehistory.* Garland, New York.

Teleki, G., 1974. Chimpanzee subsistence technology: Materials and skills. *J. Hum. Evol.,* **3,** 575–594.

Templeton, A. R., 1997. Out of Africa? What do genes tell us? *Curr. Opinion Genet. Devel.,* **7,** 841–847.

Tooby, J., and I. DeVore, 1987. The reconstruction of hominid behavioral evolution through strategic modeling. In *The Evolution of Human Behavior: Primate Models,* W. G. Kinzey (ed.). SUNY Press, Albany, NY, pp. 183–237.

Trivers, R., 1985. *Social Evolution.* Benjamin/Cummings, Menlo Park, CA.

Vauclair, J., 1990. Primate cognition: From representation to language. In *"Language" and Intelligence in Monkeys and Apes: Comparative Developmental Perspectives,* S. T. Parker and K. R. Gibson (eds.). Cambridge University Press, Cambridge, England, pp. 312–329.

Waddle, D. M., 1994. Matrix correlation tests support a single origin for modern humans. *Nature,* **368,** 452–456.

Wallis, G. P., 1999. Do animal mitochondrial genes recombine? *Trends in Genet.,* **14,** 209–210.

Weiskrantz, L., 1995. The origins of consciousness. In *Origins of the Human Brain,* J. P. Changeux and J. Chavaillon (eds.). Clarendon Press, Oxford, England, pp. 239–248.

Wheeler, P. E., 1991. The thermoregulatory advantages of hominid bipedalism in open equatorial environments: The contribution of increased convective heat loss and cutaneous evaporative cooling. J. *Hum. Evol.,* **21,** 107–116.

White, T. D., G. Suwa, and B. Asfaw, 1994. *Australopithecus ramidus,* a new species of early hominid from Aramis, Ethiopia. *Nature,* **371,** 306–312.

Wolpoff, M. H., 1980. *Paleoanthropology.* Knopf, New York.

———, 1982. *Ramapithecus* and hominid origins. *Curr. Anthropol.,* **23,** 501–510.

———, 1989. Multiregional evolution: The fossil alternative to Eden. In *The Human Revolution: Behavioural and Biological Perspectives on the Origin of Modern Humans,* P. Mellars and C. Stringer (eds.). Edinburgh University Press, Edinburgh, pp. 62–108.

Wolpoff, M. H., and R. Caspari, 1997. *Race and Human Evolution.* Simon and Schuster, New York.

Wood, B., 1997. The oldest whodunnit in the world. *Nature,* **385,** 292–293.

Wood, B., and M. Collard, 1999. The human genus. *Science,* **284,** 65–71.

Wright, S., 1949. Adaptation and selection. In *Genetics, Paleontology, and Evolution,* G. L. Jepson, G. G. Simpson, and E. Mayr (eds.). Princeton University Press, Princeton, NJ, pp. 365–389.

Wynn, T., 1993. Layers of thinking in tool behavior. In *Tools, Language and Cognition in Human Evolution,* K. R. Gibson and T. Ingold (eds.). Cambridge University Press, Cambridge, England, pp. 389–406.

IX The Mechanisms

21 Populations, Gene Frequencies, and Equilibrium

At the center of Darwin's evolution theory was the concept that small inherited changes provided the **continuous variation** on which natural selection acted, and each species represented a unique accumulation of such small changes: "Species are only strongly marked varieties with the intermediate gradations lost." Soon after publication of *On the Origin of Species,* Francis Galton (Darwin's cousin) became convinced that a mathematical approach to heredity showed that evolution must have proceeded in sharp, discontinuous steps. By 1871, Galton had already disproved Darwin's pangenesis hypothesis (p. 28) to his own satisfaction by showing that transfusing blood between rabbit strains had no effect on heredity.

Similar to August Weismann's later germ plasm theory, in 1875 Galton suggested that instead of somatically acquired "gemmules," the hereditary material was passed on between generations with little or no change. It seemed apparent to Galton that parents who deviate significantly from the average for some continuous quantitative trait (such as height) tend to produce offspring that are closer to the average than themselves (Galton's law of regression). He surmised that continuous variation was not the agency that leads to the origin of new species but that nonblending, **discontinuous variation** ("sports") provided the abrupt changes between species.

Mutationists and Selectionists

By the end of the nineteenth century, two schools of thought had established themselves in England based on their preference for continuous or for discontinuous variation. Upholding the importance of continuous variation in evolution were the mathematically oriented **biometricians,** Weldon and Pearson, and in opposition to them were Bateson and his supporters. When Mendel's 1865 paper on the genetics of peas was rediscovered in 1900, these two camps polarized further. Bateson and the **Mendelians**

proposed that most hereditary characteristics were discontinuous and could be explained by the segregation of mendelian factors, whereas the biometricians insisted that most characteristics were continuous and mendelian factors were only involved in exceptional traits. To Mendelians such as De Vries and others, evolution could only be effective if selection operated on large mutations of the kind that produced races and species in the evening primrose, *Oenothera,* whereas the biometricians allied themselves with Darwin's original concept, that selection acting on small differences was the primary mechanism for evolutionary change.

Apparently supporting the mutationist position were Johannsen's experimental observations that selection was ineffective in quantitatively changing the size of beans descended from homozygous **pure lines.** Furthermore, even when selection was practiced on beans descended from crosses between different pure lines, size differences among their descendants seemed to show relatively little change from the range of values initially observed in the F_2 generation of the cross. It seemed that marked changes in the size of beans could only come from mutations with large effect (**macromutations**), rather than from selection among the small differences observed in Johannsen's experiments.

Various biologists extended these views to propose that new species can arise in only one or a few mutational steps driven perhaps by mutation pressure in a particular and even nonadaptive direction, a view known as **saltation.** According to the saltationists, the slow, plodding process of Darwinian selection was no longer necessary to explain evolution.[1]

Castle and Phillips, in contrast, demonstrated that selection could lead to entirely new coat color patterns in hooded rats: some selected lines had "less pigment than any known type other than albino," whereas others were "so extensively pigmented that they would readily pass for the 'Irish type' which has white on the belly only." Although these studies identified no specific genes with quantitative effect, similar results with other organisms (Fig. 10–31) did point to the likelihood that selection could act on small continuous characters to produce marked changes in phenotype.

The rift between **mutationists** and **selectionists** in explaining the basic mechanisms of evolution remained until the 1920s and 1930s, but the gap had already been bridged by further experimental work on **quantitative characters.** For example, Nilsson-Ehle and East both showed experimentally that a number of different gene pairs (multiple factors) may affect a single quantitative character so that a wide array of possible genotypes can occur, each with a different phenotype. Genes that segregate in typical mendelian patterns are responsible for many observed distributions of continuous traits (Fig. 10–32). No real difficulty arose in providing a mendelian interpretation of Darwinian selection for quantitative traits, and dependence on the introduction of mutations with very large effect no longer seemed necessary to explain most basic evolutionary changes.

The emphasis on evolution through small continuous characters does not mean that alleles with large effects on phenotype are always unadaptive. As Orr and Coyne point out, "mutations of large effect clearly play a substantial role in animal and plant breeding." Natural populations also show such effects. For example, insecticide resistance is often caused by only a few favorable mutations, and polymorphism for large and small beak size in an African seed-eating bird is the result of segregation for only two alleles at a single locus (Smith).

That genes with large effect can be selected causing considerable morphological difference, is supported by studies on the evolution of cultivated maize from its wild ancestor, teosinte. Although these plants are quite different in appearance (Fig. 21–1), changes in about five genes are probably responsible (Doebley et al.). Similarly, in *Mimulus* (monkeyflower), Bradshaw and coworkers show that several mutational changes in flower structure (red flower color, long beak-shaped corolla tube, protruding anthers and stigma) can account for a shift from pollination by bees to hummingbirds, and thereby help give rise to a new form, *Mimulus cardinalis.* Along with other examples, some adaptations involve genes with major phenotypic effects, but these effects, taken singly, are not really those envisioned by macromutationists—that is, they do not create a complex organ or new species in a single stroke.

The Neo-Darwinian Synthesis

Concurrent with the disputes between selectionists and mutationists, Yule, Castle, Hardy, and Weinberg were proposing important new concepts. These workers em-

[1]Variations on this theme have been proposed a number of times. For example, many people who consider selection merely a passive "sieve" that acts only to remove the "unfit" but does little to create the "fit," generally substituted mutation as the primary creative force in evolution. This creativity was often presumed to be caused either by a single mutation of major effect that led directly to a new species (macromutation) or by a succession of somewhat smaller changes that influenced development in a particular evolutionary direction (orthogenesis). As discussed later (p. 599), some paleontologists used macromutation, in conjunction with saltation concepts, to explain the unevenness of the fossil record and the presumed origin of new taxa. Orthogenesis via mutation also had its adherents (see, for example, Berg), although concepts of major evolutionary trends completely free of selection seem as mystical as Lamarck's *"feu éthéré"* that supposedly prompts organisms to evolve only in adaptive directions (p. 24). Mutagenic orthogenesis seems to ascribe to nucleotides the supernatural ability to choose only those mutational changes that cause a singular phenotypic trend when alternative changes are possible (see also pp. 45, 227, 429).

FIGURE 21-1 (*a*) Teosinte (*Zea mays parviglumis*), the wild ancestor of cultivated maize, showing the mature plant and a kernel-bearing ear, and (*b*) a mature plant and ear of its descendant, modern corn (*Zea mays mays*). Although strikingly different in plant and ear architecture, these two forms differ in relatively few genes.

(a) Teosinite

(b) Maize

TABLE 21-1 Characteristics of individuals compared to those of populations

Characteristic	Individual	Population
Life span	One generation	Many generations
Spatial continuity	Limited	Extensive
Genetic characteristics	Genotype	Gene frequencies
Genetic variability	None	Considerable
Evolutionary characteristics	No changes, since an individual has only one genotype and is limited to only a single generation	Can evolve (change in gene frequency), since evolution occurs between generations

phasized that populations rather than individuals were an important evolutionary focus, and researchers had to pay attention to population gene frequencies, rather than only to whether a gene was present or absent. It became clear that the collection of gametes a population contributes to the next generation can be considered as a giant gene pool from which offspring draw their various genotypic combinations at random. In the absence of selection and other factors that could change gene frequencies, these frequencies tend to be conserved, as the **Hardy–Weinberg equilibrium** demonstrated (p. 520ff).

During the 1920s, Fisher, Wright, and Haldane developed in considerable detail the approach toward considering evolution as a change in gene frequencies. They demonstrated this approach in various papers that dealt with the effects of inbreeding, the evolution of dominance, and the effect of selection on gene frequencies, as well as the effects of mutation, migration, and genetic drift. In the early 1930s these studies culminated in a variety of papers and books that laid the foundations for **population genetics,** the study of gene frequencies and their changes.

Along with these mathematical models, Chetverikoff in the Soviet Union and others elsewhere observed considerable genetic variation in natural populations on which selection could act. All these studies helped establish the concept that the population has the variability necessary to explain evolutionary genetic change through space and time, whereas an individual is extremely limited in these dimensions (Table 21–1). Differences among a population's genotypes enable different reproductive rates among them, whereas an individual's genotype is constant from birth to death. That is, evolutionary changes depend on differentiation among genotypes: *populations evolve, not individuals.*

This emphasis on the genetics of populations helped transform evolutionary thinking into its more modern form, often called the **Neo-Darwinian** (or **modern**)

synthesis. At the base of this synthesis is the concept that mutations occur randomly and furnish the fuel for evolution by introducing genetic variability. We can then define evolution as an ongoing process in which random mutation introduces genes whose frequencies change through time, with natural selection usually considered as the most important, although not the only, cause for such change. (Among other factors are migration and random genetic drift, discussed in Chapter 22.) In contrast to other biological disciplines that emphasized static typology (p. 10), genetics offered the advantage of understanding and accentuating the transmission, persistence, and modification of inherited variation—the elements that enable evolution to occur.

The accumulation of gene frequency differences, by whatever means, eventually leads to more pronounced (racial) differences among populations in geographically different localities. When gene exchange between racial groups can no longer occur because of reproductive barriers, separate species become established. Essentially, this approach gave a genetic slant to the biological species concept (Chapter 11) by conceiving of a species as a population of individuals bearing distinctive genes and gene frequencies, separated from other species by biological mechanisms that prevent gene exchange. Furthermore, mechanisms such as mutation and selection that lead to the origin of races and species are generally no different from mechanisms that lead to the origin of higher taxa such as genera, families, and orders, although the formation of higher taxa usually takes place over longer periods of time. This approach led to the realization that whichever groups occupy a particular environment, whether designated as genotypes or higher taxa, *selective* interactions between them becomes an essential factor in their evolution.

In other words, the Neo-Darwinian synthesis helped explain how mutation led to variation and how selection led to adaptation and "design." The synthesis proposed that what begins quantitatively in populations as gene frequencies and changes in gene frequencies, becomes through time and its environmental vicissitudes and interactions, qualitative changes that transform some groups into races, and races into species, whose similarities and differences may then be organized into even higher taxonomic categories. To briefly summarize this process—evolution cycles continuously from mutation to variation to selection to adaptation to organismic diversity. The modern synthesis provides a conceptual sequence that helps explain the chain of events going from genes to organisms to their communities, and back again.[2]

By providing the general ideological framework in which to understand this continuum, the Neo-Darwinian synthesis helped motivate many evolutionary biologists to ask and answer more detailed questions as to how and why particular evolutionary events and adaptations actually occurred. Such questions began at basic levels: What are the gene frequencies in populations? How do gene frequencies change? Why do they change? What genetic differences separate races? species? At what rates do these differences arise? What historical phenomena can account for these differences?

Perhaps the most prominent of such studies was a series on *Drosophila* species by Dobzhansky and coworkers, called the *Genetics of Natural Populations.* By the 1950s, these papers were widely influential in:

> Comprising a model of how genetical variation in natural populations could be studied [and] included observations of temporal variation and stability in polymorphism, estimates of migration and effective population size, evidence for the existence of selective differences in nature, and the creation of laboratory model populations in which selection could be demonstrated and estimated. (Lewontin)

Because of this fundamental materialistic approach, which both Dobzhansky and Huxley popularized in books, the Neo-Darwinian synthesis made itself widely felt in biology by effectively eliminating Lamarckian concepts and other mystical or semimystical theories such as saltation and orthogenesis described earlier. The influence of population genetics, the essential component of this synthesis, also extends to many other fields such as demography, ecology, epidemiology, plant and animal breeding, and other areas in which gene variation and distribution affect the relationships and life patterns of organisms. Although population genetics has emphasized mathematical models that cannot reflect reality in all its myriad details, it has provided logically precise concepts that help us understand many common populational features, such as the frequently observed conservation of gene and genotype frequencies and the general effects of forces that change these frequencies.

[2] One can claim that the Neo-Darwinian synthesis, although influential in helping provide a common evolutionary genetic theme for fields as diverse as embryology, systematics, and paleonotology (Jepsen et al.), did little more than introduce a genetic basis for Darwin's fundamental concepts of variation and divergence. That is, although Darwin was unaware of the source, measurement, and extent of variation, he understood that a species maintains sufficient variation letting it evolve varieties and subspecies that diverge from each other because of selection for new and different environmental conditions (Ospovat). It was this gradual divergence—the confluence of variation and selection—that produced in time the striking abundance of evolutionary hierarchies, from species to phyla, and their many complex interactions. The genetics of Neo-Darwinism made selection and variation scientifically understandable and helped reinforce a search for the genetics of evolutionary form and adaptation—developmental genetics (pp. 347–348). In recent years, molecular techniques introduced into such studies demonstrate many of the intricate connections among genes and morphological patterns and how these can change (Chapter 15).

TABLE 21-2 Techniques for obtaining gene frequencies for a diploid population of 200 individuals

A. Using numerical gene counts (there are 400 genes in 200 diploid individuals):

$$T = 180 \text{ (in } TT) + 60 \text{ (in } Tt) = \frac{240}{400} = .60$$

$$t = 100 \text{ (in } tt) + 60 \text{ (in } Tt) = \frac{160}{400} = .40$$

Total 1.00

B. Using genotype frequencies:

$$T = .45\ TT + 1/2\ (.30\ Tt) = .45 + .15 = .60$$

$$t = .25\ tt + 1/2\ (.30\ Tt) = .25 + .15 = .40$$

Total 1.00

Note: These individuals are of the following types:

90 *TT* + 60 *Tt* + 50 *tt* = 200 individuals

[.45 *TT* + .30 *Tt* + .25 *tt* = 1.00 (genotypes)]

Populations and Gene Frequencies

Geneticists usually define a population as a group of sexually interbreeding or potentially interbreeding individuals. Since mendelian laws apply to the transmission of genes among these individuals, Wright has called such a group a **mendelian population.** The size of the population may vary, but it is usually considered to be a local group (also called **deme**), each member of which has an equal chance of mating with any other member of the opposite sex. Most theory and experiments have so far emphasized populations of diploid organisms, and the discussions that follow deal mostly with such cases. However, whether diploid or haploid, populations have two important attributes: **gene frequencies** (also called *allele frequencies*)[3] and a **gene pool.**

Gene frequencies are simply the proportion of the different alleles of a gene in a population. To get these proportions, we count the total number of organisms with various genotypes in the population and estimate the relative frequencies of the alleles involved. Except for gametes and occasional mutation, the genetic complements of all cells in a multicellular organism are the same. We may therefore adopt the convention that a haploid organism has only one gene at any one locus, a diploid has two, a triploid three, and so on.

For example, we can presume that the difference between humans who can and cannot taste the chemical phenylthiocarbamide resides in a single gene difference between two alleles, *T* and *t*. Since the allele for tasting, *T*, is dominant over *t*, two genotypes (homozygous *TT* and heterozygous *Tt*) represent tasters and the nontasters are *tt*. A population of 200 individuals composed of 90 *TT*, 60 *Tt*, and 50 *tt* will therefore have a total of 400 alleles at this locus. As Table 21–2*a* shows, 240 of these are *T* (a frequency of .60), and 160 are *t* (a frequency of .40). We can also calculate the same gene frequencies from the frequencies of the three genotypes, according to the formula: frequency of a gene = frequency of homozygotes for that gene + 1/2 frequency of heterozygotes, who each contain one such gene out of two (Table 21–2*b*).

The gene pool is the sum total of genes in the reproductive gametes of a population. It can be considered as a gametic pool from which samples are drawn at random to form the zygotes of the next generation. The genetic relationship between an entire generation and the subsequent generation is very similar to the genetic relationship between a parent and its offspring. Since the frequencies of genes in the new generation will depend, to some degree at least, on their frequencies in the old, we might say that gene frequencies rather than genes are inherited in populations. In what form can we express and analyze these gene frequency relationships between generations?

One of the first attempts at using the concept of gene frequencies occurred in the dispute mentioned earlier between the biometricians and Mendelians. Some argued that dominant alleles, no matter what their initial frequency, would be expected to reach a stable equilibrium frequency of three dominant individuals to one recessive, since this was the mendelian segregation pattern for these genes. That such ratios were not observed for very low-frequency dominant alleles such as brachydactyly (short fingers) was offered as evidence that populations did not follow mendelian rules, and gene frequencies could be ignored in evolutionary studies.

Although such arguments were widely accepted at first, in 1908 both Hardy in England and Weinberg in

[3] The precise, appropriate term describing or comparing frequencies of different alleles of a gene in a population should be *allele frequency* rather than *gene frequency*. However, the term *gene frequency* was introduced early in the history of population genetics and has remained too common to enable an easy change.

Germany disproved them. They demonstrated that gene frequencies do not depend upon dominance or recessiveness but remain essentially unchanged from one generation to the next under certain conditions. Such **conservation of gene frequencies** is briefly discussed in this chapter, while Chapter 22 deals with the forces that can change gene frequencies. Various books, including those of Crow, Crow and Kimura, Falconer, Gillespie, Hartl and Clark, Hedrick, Spiess, and Wallace more fully treat the basic theoretical principles of population genetics and provide more complete formula derivations and extensive examples.

Conservation of Gene Frequencies

The principle Hardy and Weinberg discovered may be illustrated using the tasting example previously mentioned. For example, let us place on an island a group of children of the genotypes .45 *TT*/.30 *Tt*/.25 *tt*, where gene frequencies are .60 *T* and .40 *t*. Let us assume that the number of individuals in this newly formed population is large and that tasting or nontasting has no effect on survival (viability), fertility, or attraction between the sexes.

As these children mature, they will choose their mates at random from those of the opposite sex regardless of their tasting abilities. We can then predict matings between any two genotypes solely on the basis of the genotypic frequencies in the population. As Table 21–3 shows, nine different types of matings can occur, of which three matings are reciprocals of others (for example, *TT* × *tt* = *tt* × *TT*). In all, these six different mating combinations will produce offspring in the ratios shown.

Note that although **random mating** has altered the frequencies of genotypes, the gene frequencies among the offspring have not changed. For *T* the offspring gene frequency is equal to .36 + 1/2 (.48) = .60, and the frequency of *t* is .16 + 1/2 (.48) = .40, exactly the same as before. Under these conditions, no matter what the initial frequencies of the three genotypes, the gene frequencies of the next generation will be the same as those of the parental generation. For example, if the founding popu-

TABLE 21-3 Gene frequencies produced by random mating among individuals in a population having the frequencies given in Table 21-2

		Males		
		TT = .45	*Tt* = .30	*tt* = .25
Females	*TT* = .45	.2025 ①	.1350 ②	.1125 ③
	Tt = .30	.1350 ④	.0900 ⑤	.0750 ⑥
	tt = .25	.1125 ⑦	.0750 ⑧	.0625 ⑨

Parents				Offspring		
Matings of Genotypes	Box Number(s) from Above		Mating Frequency	*TT*	*Tt*	*tt*
TT × *TT*	①	=	.2025	.2025		
TT × *Tt*	② + ④	=	.2700	.1350	.1350	
TT × *tt*	③ + ⑦	=	.2250		.2250	
Tt × *Tt*	⑤	=	.0900	.0225	.0450	.0225
Tt × *tt*	⑥ + ⑧	=	.1500		.0750	.0750
tt × *tt*	⑨	=	.0625			.0625
			1.0000	.3600	.4800	.1600

Gene frequencies among offspring:

T = .36 *TT* + 1/2 (.48 *Tt*) = .45 + .15 = .60

t = .16 *tt* + 1/2 (.48 *Tt*) = .25 + .15 = .40

Total 1.00

lation of this island contained .40 *TT*, .40 *Tt*, and .20 *tt*, the gene frequency for *T* would be .40 + 1/2 (.40) and .20 + 1/2 (.40) for *t*, the same as before. However, as Table 21–4 shows, despite the different initial genotype frequencies, offspring are again produced in the ratio .36 *TT*/.48 *TT*/.16 *tt*, or a gene frequency of .60 *T*/.40 *t*.

Two important conclusions follow:

1. Under conditions of random mating (**panmixia**) in a large population where all genotypes are equally viable, gene frequencies of a particular generation depend on the gene frequencies of the previous generation and not on the genotype frequencies.
2. The frequencies of different genotypes produced through random mating depend only on the gene frequencies.

Both these points mean that by confining our attention to genes rather than to genotypes, we can predict both gene and genotype frequencies in future generations, providing outside forces are not acting to change their frequency and random mating occurs between all genotypes. To continue our previous illustration, we may predict that under these conditions the initial gene frequencies in taster–nontaster populations will not change in the next or succeeding generations. Also, after the first generation, the genotype frequencies will remain stable, that is, at **equilibrium.**

The theory describing this genotypic equilibrium, based on stable gene frequencies and random mating, is known as the *Hardy–Weinberg principle* (or *law*) and has served as the founding theorem of population genetics. Perhaps its main contribution to evolutionary thought lies in demonstrating that genetic differences in a randomly breeding population tend to remain constant unless acted on by external forces: a point contrary to the pre-mendelian concept that heredity involves a blending of traits that become more dilute with each generation of interbreeding. Figure 21–2 outlines major assumptions and steps in the Hardy–Weinberg principle.

The general relationship between gene frequencies and genotype frequencies can be described in algebraic terms by means of the Hardy–Weinberg principle as follows: if p is the frequency of a certain gene in a panmictic population (for example, *T*) and q the frequency of its allele (for example, *t*), so that $p + q = 1$ (that is, there are no other alleles), the equilibrium frequencies of the genotypes are

TABLE 21–4 Gene frequencies produced by random mating among individuals in a population that has genotypic frequencies of .40 *TT*, .40 *Tt*, and .20 *tt* (gene frequencies: .60 *T*, .40 *t*)

		Males		
		TT = .40	*Tt* = .40	*tt* = .20
Females	*TT* = .40	.1600 (1)	.1600 (2)	.0800 (3)
	Tt = .40	.1600 (4)	.1600 (5)	.0800 (6)
	tt = .20	.0800 (7)	.0800 (8)	.0400 (9)

Parents				Offspring		
Matings of Genotypes	Box Number(s) from Above		Mating Frequency	*TT*	*Tt*	*tt*
TT × *TT*	(1)	=	.1600	.1600		
TT × *Tt*	(2) + (4)	=	.3200	.1600	.1600	
TT × *tt*	(3) + (7)	=	.1600		.1600	
Tt × *Tt*	(5)	=	.1600	.0400	.0800	.0400
Tt × *tt*	(6) + (8)	=	.1600		.0800	.0800
tt × *tt*	(9)	=	.0400			.0400
			1.0000	.3600	.4800	.1600

Gene frequencies among offspring:

T	=	.36 *TT*	+	1/2 (.48 *Tt*)	– .45 + .15	=	.60
t	=	.16 *tt*	+	1/2 (.48 *Tt*)	= .25 + .15	=	.40
					Total		1.00

given by the terms $p^2(TT)$, $2pq(Tt)$, and $q^2(tt)$. If the gene frequencies of T and t are $p = .6$ and $q = .4$, respectively, the equilibrium frequencies will then be

$$(.6)^2(TT) + 2(.6)(.4)(Tt) + (.4)^2(tt) =$$
$$.36\ TT + .48\ Tt + .16\ tt$$

You can visualize this relationship by drawing a checkerboard in which the genotype frequencies stem from random union between alleles that are in the frequencies of p and q (Fig. 21–3). The same results also derive from the **binomial expansion** $(p + q)^2 = p^2 + 2pq + q^2$. Therefore with any given p and q and random mat-

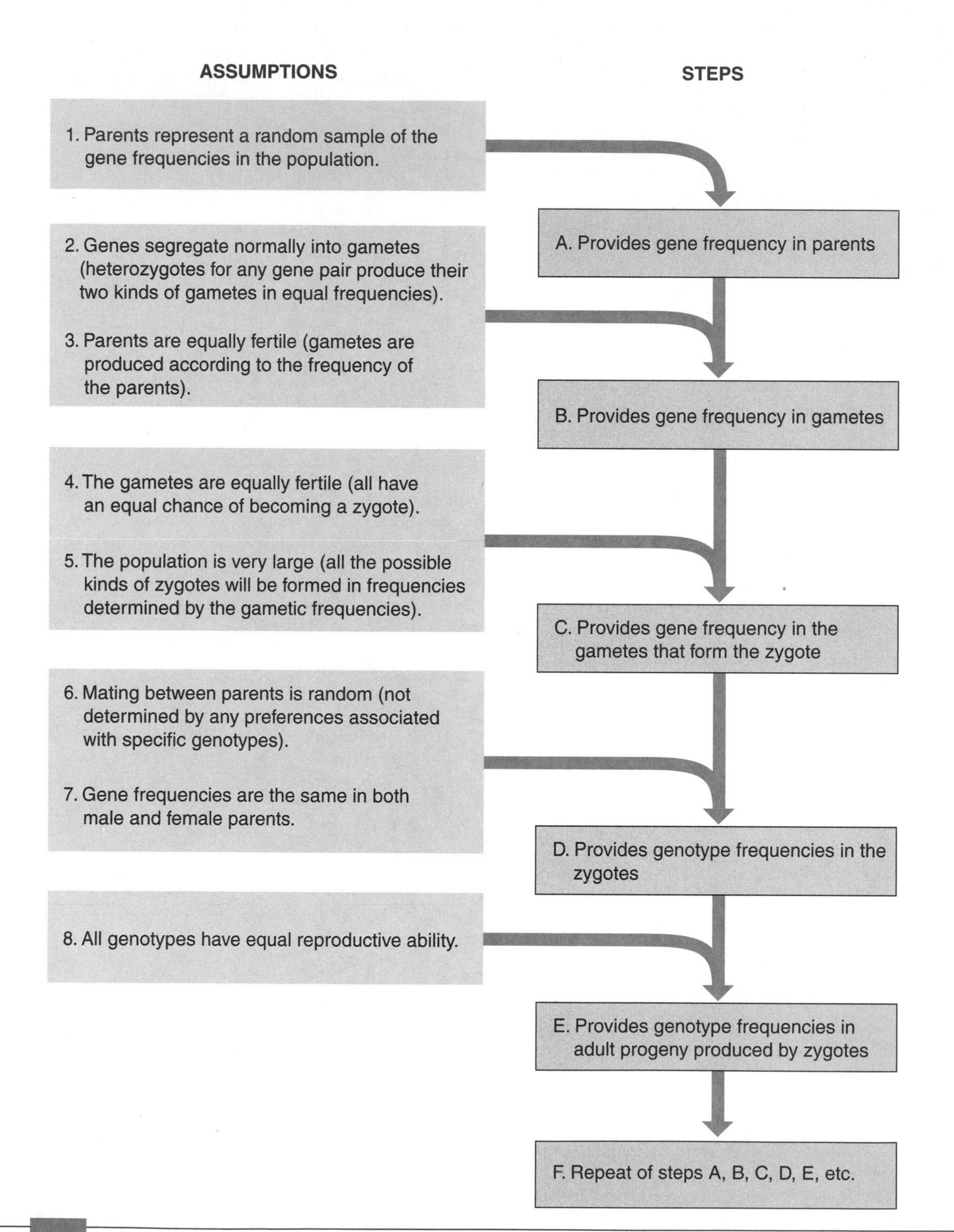

FIGURE 21–2 Assumptions and steps in the Hardy–Weinberg equilibrium. (*Based on Falconer and Mackay.*)

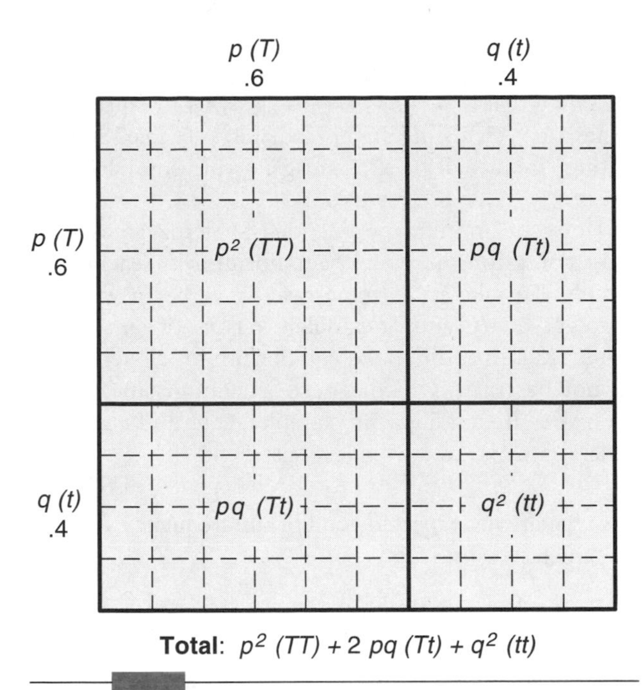

FIGURE 21-3 Genotypic frequencies generated under conditions of random mating for two alleles, T and t, at a locus when their respective frequencies are $p = .6$ and $q = .4$. Equilibrium genotypic frequencies are therefore .36 *TT*, .48 *Tt*, and .16 *tt*.

ing between genotypes, one generation of a population in which generations do not overlap is enough to establish equilibrium for the frequencies of genes and genotypes. Once established, the equilibrium will persist until the gene frequencies are changed.

Figure 21–4 shows the genotypic frequencies at Hardy–Weinberg equilibrium for a two-allele locus, where the frequency of each allele ranges from 0 to 1. Note that the frequency of heterozygotes never exceeds .50, but is significantly higher than the frequency of homozygotes for a rare allele (for example, when *a* is .1, *aa* is .01 but *Aa* is .18).

The presence of two alleles at a locus is only one example for which we may want information. There are often more than two alleles at a locus, and we must then consider each allelic frequency as an element in a multinomial expansion. For example, if there are only three possible alleles at a locus, A_1, A_2, and A_3, with respective frequencies p, q, and r, so that $p + q + r = 1$, the **trinomial expansion** $(p + q + r)^2$ determines the genotypic equilibrium frequencies. The six genotypic values are then

$$p^2\, A_1A_1 + 2pq\, A_1A_2 + 2pr\, A_1A_3 + q^2\, A_2A_2 + 2qr\, A_2A_3 + r^2\, A_3A_3$$

Since each haploid gamete contains only a single allele for any one gene locus, zygotic combinations will depend only on the frequency of each allele (Fig. 21–5), and, as when there are only two alleles, equilibrium is established in a single generation of random mating.

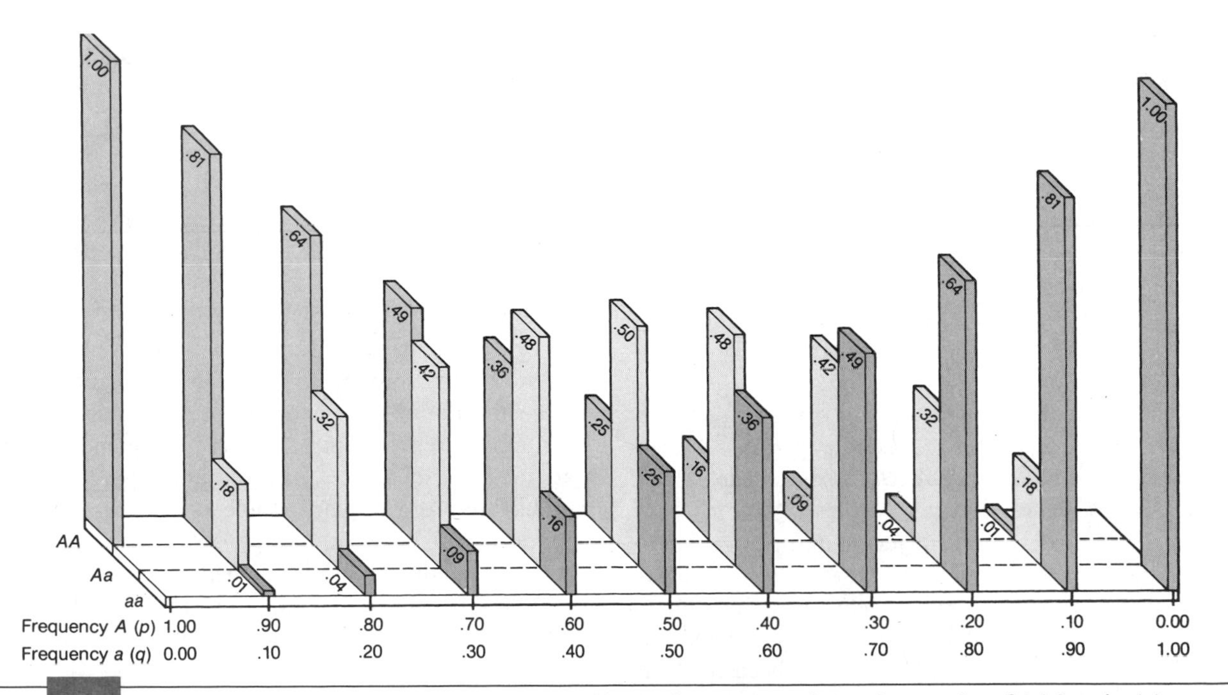

FIGURE 21-4 Genotypic frequencies at Hardy–Weinberg equilibrium for a variety of gene frequencies of *A* (*p*) and *a* (*q*). (*Adapted from Wallace.*)

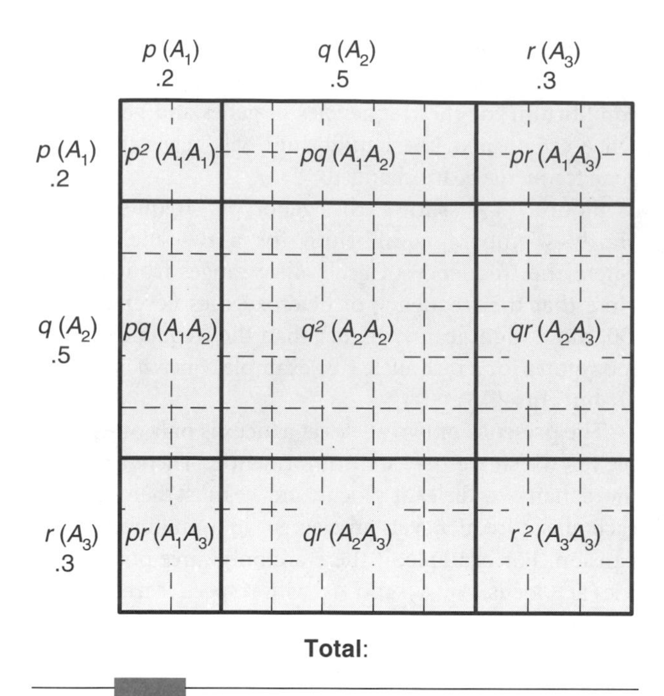

FIGURE 21-5 Genotypic frequencies generated under conditions of random mating when there are three alleles, A_1, A_2, and A_3, present at a locus. For purposes of illustration, the respective gene frequencies of these alleles have been given as $p = .2$, $q = .5$, and $r = .3$. Equilibrium genotypic frequencies are therefore .04 A_1A_1, .20 A_1A_2, .12 A_1A_3, .25 A_2A_2, .30 A_2A_3, and .09 A_3A_3.

Attainment of Equilibrium at Two or More Loci

Establishment of equilibrium in one generation holds true as long as we consider each single gene locus separately without being concerned about what is happening at other gene loci. If, however, we consider the products of two independently assorting gene pair differences simultaneously—for instance, *Aa* and *Bb*—the number of possible genotypes increases to 3^2 (*AABB, AABb, AaBB, AaBb,* and so on). As expected, more terms are now involved in the multinomial expansion, so that if we call *p, q, r,* and *s* the gene frequencies of *A, a, B,* and *b,* respectively, the equilibrium ratios of their genotypes are expressed as $(pr + ps + qr + qs)^2$, or p^2r^2 *AABB*, $2p^2rs$ *AABb*, $2p^2s^2$ *AAbb*, $2pqr^2$ *AaBB*, . . . , q^2s^2 *aabb*.

This equilibrium formula depends on the terms *pr, ps, qr, qs,* which are the equilibrium frequencies of the gametes *AB, Ab, aB,* and *ab,* respectively. Once the gametic frequencies reach these equilibrium values, the equilibrium genotypic frequencies will also have been reached. The problem of attainment of equilibrium resolves itself to the time that it takes for the gametic frequencies to reach these values. If we begin only with heterozygotes ($AaBb \times AaBb$) in which the frequencies of all genes are the same (that is, $p = q = r = s = .5$), all four types of gametes (*AB, Ab, aB, ab*) are immediately produced at equilibrium frequencies (.25), and genotypic equilibrium is reached within one generation.

However, an initial population of heterozygotes is the only condition in which equilibrium is reached so rapidly. To take an extreme case, if we begin with the genotypes *AABB* and *aabb,* only two types of gametes are produced (*AB* and *ab*) and equilibrium for all genotypes cannot be reached in the next generation since many genotypes are missing (for example, *AAbb* and *aaBB*). In general, we may ask two questions:

- What are the expected equilibrium frequencies of gametes?
- How rapidly are these frequencies achieved?

To deal with these questions, we can characterize *AB* and *ab* gametes as nonrecombinant or in **coupling,** and *Ab* and *aB* gametes as recombinant or in **repulsion.** However defined, both gametic types carry the same alleles (*A, a, B, b*), meaning that the frequency of each allele in one type (for example, repulsion) is equal to its frequency in the other (for example, coupling). We would then expect the *products* of the frequencies of both types of gametes to be equal at equilibrium: $(AB) \times (ab) = (Ab) \times (aB)$.

For example, if the frequencies of *A* and *B* are each .6, and the frequencies of *a* and *b* are each .4, then at equilibrium (.36)(.16) = (.24)(.24), or both products equal .0576. If the coupling and repulsion products in the initial population differ, this difference represents the change in gametic frequencies that must occur for equilibrium values to be reached. If we call this difference **disequilibrium,** or *d,* and it is positive so that coupling − repulsion > 0, for example, $(AB)(ab) - (Ab)(aB) = +d$, then at equilibrium this fraction will have been added to each of the coupling gametes and subtracted from each of the repulsion gametes. If *d* is negative, the reverse operation will occur. In both cases, disequilibrium will have diminished to zero, and equilibrium will have been established.

Until the final gametic ratios are reached, half the difference from equilibrium reduces each generation, so that within four to five generations more than 90 percent of this difference from equilibrium frequency has been attained by all gametes, or less than 10 percent of disequilibrium value remains. Table 21–5 shows how to calculate *d* and how the changes in gametic frequencies occur until equilibrium is attained. For three gene pairs, the speed of approach to equilibrium is even further diminished, and it becomes slower still as more gene pairs are involved.

As you might expect, linkage between two loci complicates reaching equilibrium since the chances that all the

TABLE 21-5 Calculation of d and the equilibrium frequencies of gametes for a population in which the frequencies of two unlinked gene pairs *Aa* and *Bb* are $A = b = .6$. and $a = B = .4$, and the initial genotypic frequencies are $AABB = AAbb = aabb = .30$ and $aaBB = .10$

EQUILIBRIUM FREQUENCY OF GAMETES

Initial Population	Gametes: Type	Gametes: Initial Frequency	Gametes: Equilibrium Frequency	
30% *AABB*	*AB*	.3	$.3 - d$	$(.6 \times .4 = .24)$
30% *AAbb*	*Ab*	.3	$.3 + d$	$(.6 \times .6 = .36)$
30% *aaBB*	*aB*	.1	$.1 + d$	$(.4 \times .4 = .16)$
10% *aabb*	*ab*	.3	$.3 - d$	$(.4 \times .6 = .24)$

$$d = (AB)(ab) - (Ab)(aB) = (.3)(.3) - (.3)(.1) = .06$$

ATTAINMENT OF EQUILIBRIUM
WHEN RECOMBINATION IS UNHINDERED BY LINKAGE

Generation	Amount Added (*AB*, *ab*) or Subtracted (*Ab*, *aB*)	Proportion of Disequilibrium Remaining	Gametes: *AB*	Gametes: *Ab*	Gametes: *aB*	Gametes: *ab*
1		1.0d	.3	.3	.1	.3
2	.5d	.5d	.27	.33	.13	.27
3	.75d	.25d	.255	.345	.145	.255
4	.875d	.125d	.2475	.3525	.1525	.2475
5	.9375d	.0625d	.24375	.35625	.15625	.24375
·	·	·	·	·	·	·
·	·	·	·	·	·	·
·	·	·	·	·	·	·
Equilibrium	d	0d	.24	.36	.16	.24

Source: Adapted from Strickberger, modified.

different types of dihybrid gametes will be found depend on crossover frequencies between the two loci. The closer the linkage, the longer it will take for the frequency of coupling gametes to equal the frequency of repulsion gametes. In other words, such **linkage disequilibrium** depends on recombination frequency, and lower recombination frequencies between linked loci delay the attainment of equilibrium accordingly (Fig. 21–6). This does not mean that the eventual equilibrium values for linked genes will differ from those attained in the absence of linkage; d depends on gametic frequencies and not on linkage. Once equilibrium is attained we have no way of distinguishing linked or unlinked genes except through tests for departures from independent assortment (recognizing, however, that even genes on different chromosomes may show linkage disequilibrium if the chromosomes do not assort independently).

Despite these theoretical considerations, not all gametes of linked loci in natural populations reach equilibrium frequencies, and researchers ascribe this phenomenon to various causes. For example, some linkage disequilibrium appears for genes between which recombination is extremely rare. Researchers have also discovered cases in which linkage disequilibrium persists because certain linked allelic combinations seem beneficial so that reproductive success of alleles at one locus depends upon the presence of particular alleles at another locus (epistasis, p. 197). Thus, the third chromosome gene arrangements common in *Drosophila pseudoobscura* probably represent linked groups of genes that are advantageous under particular environmental conditions (p. 228). Because these genes are included within inversions that restrict recombination, their linkage can be preserved for relatively long periods of time, thereby forming **coadapted gene complexes** (Wallace).[4]

[4] Other linkage effects are possible when a "neutral" allele at one locus, that itself has little or no discernible influence on reproductive success, is closely linked to an allele that has an adaptive effect. Because of linkage disequilibrium, the neutral allele will then increase its frequency in conjunction with the linked advantageous allele through this type of "hitchhiking" (see also p. 561).

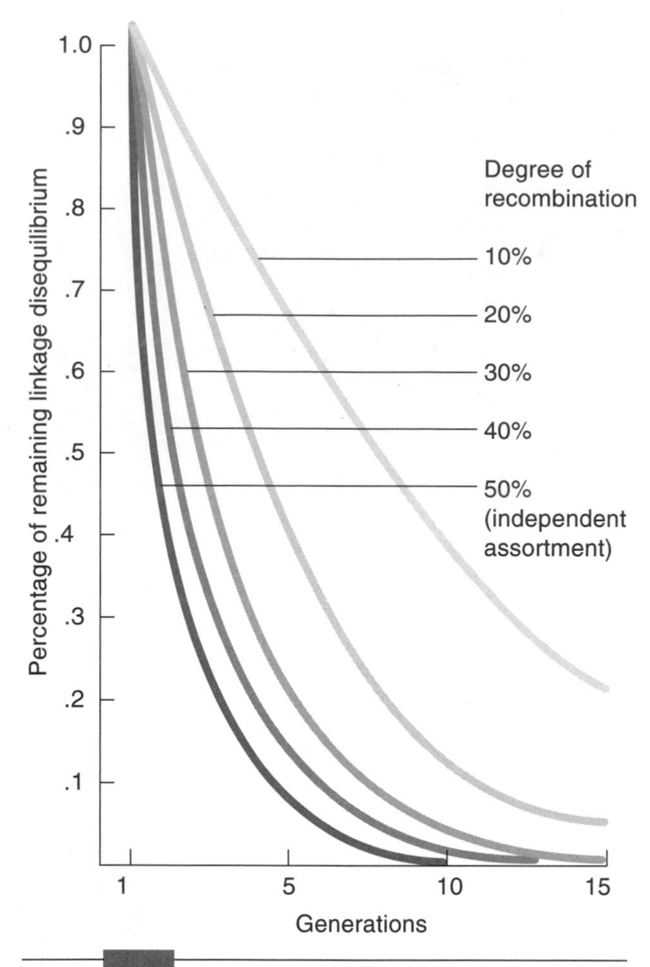

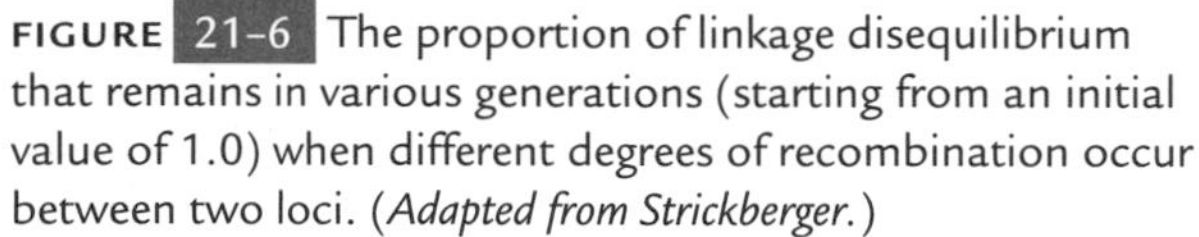
FIGURE 21-6 The proportion of linkage disequilibrium that remains in various generations (starting from an initial value of 1.0) when different degrees of recombination occur between two loci. (*Adapted from Strickberger.*)

Sex Linkage

For sex-linked genes, the number of possible genotypes is increased because of the difference in number of sex chromosomes between the homogametic and heterogametic sexes. If females are chromosomally XX and males XY, five genotypes can occur for a sex-linked pair of alleles *A* and *a:* three in females (*AA, Aa, aa*) and two in males (*A* and *a*). If we assign the frequencies *p* and *q* to *A* and *a*, respectively, the equilibrium genotypic values in females are the same as for an autosomal gene, p^2 *AA*, $2pq$ *Aa*, and q^2 *aa*, but are expressed directly in hemizygous males as *p A* and *q a* genotypes. Thus, at equilibrium the sex-linked gene frequencies are the same in both sexes, although the genotypes differ.

Assuming all genotypes are equally viable for the sex-linked gene, a difference in gene frequencies between males and females indicates that the population is not at equilibrium. For example, in a population with the proportions .20 *A*/.80 *a* in males and .20 *AA*/.60 *Aa*/.20 *aa* in

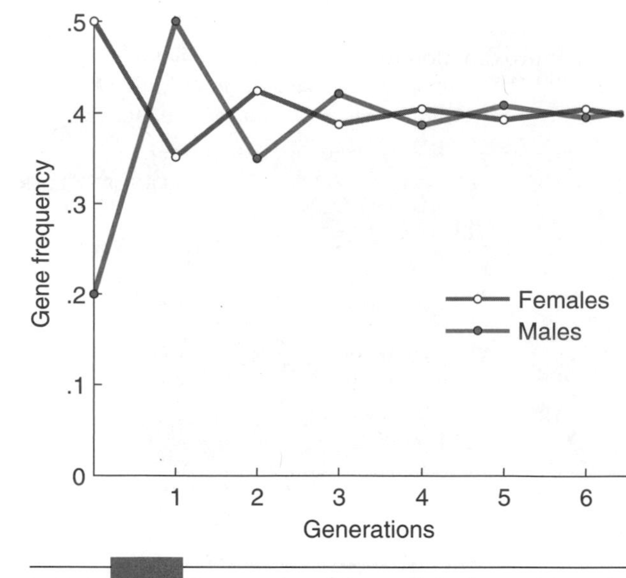

FIGURE 21-7 Frequencies of the sex-linked gene *A* in males and females in successive generations under conditions of random mating when the initial frequency of *A* is .2 in males and .5 in females.

females, the frequency of *A* is .2 in males and .5 in females. We can then calculate equilibrium frequencies of all five genotypes by considering that since there is only one X chromosome in males and two in females, the average frequency of a sex-linked gene in a breeding population with equal numbers of males and females is the sum of one-third of its frequency in males plus two-thirds of its frequency in females, or $p = 1/3\ (p_{\text{males}}) + 2/3\ (p_{\text{females}}) = (p_{\text{males}} + 2p_{\text{females}})/3$. In the present example, this translates into an *A* (*p*) frequency of $[.2 + 2(.5)]/3 = (1.2)/3 = .4$, and an *a* (*q*) frequency of .6. The equilibrium genotypic values expected are .16 *AA*/.48 *Aa*/.36 *aa* in females, and .4 *A*/.6 *a* in males.

In contrast to single autosomal loci with two alleles, however, equilibrium values for these genotypes will not be reached in a single generation. Because males inherit their X chromosomes only from their mothers, the frequency of a sex-linked gene among them is the same as its maternal frequency, whereas the frequency of the gene among daughters is an average of paternal and maternal frequencies, because they each inherit one paternal and one maternal X chromosome. Therefore, if the females in a founding population had a frequency of *A* equal to .5, but the males had an *A* frequency of only .2, the daughters would have an *A* frequency of $(.2 + .5)/2 = .35$, while their brothers would have the .5 frequency of their mothers.

Thus, in the first generation of random mating the daughters will not reach the *A* equilibrium value of .4, and the sons will exceed it. In the second generation, the difference from equilibrium values will diminish, but this time the sons will not achieve equilibrium ($A = .35$) and the daughters will exceed it [$A = (.35 + .5)/2 = .425$]. As Figure 21–7 shows, each succeeding generation will show a similar reversal, nevertheless achieving a successively closer approximation to the final equilibrium values.

Equilibria in Natural Populations

In natural populations we can reliably estimate gene frequencies where we can score all segregants of a gene at a single locus. Using these values, we can then easily compare observed genotype frequencies to their expected equilibrium values. To take a simple example, codominance at the *MN* blood group locus (using only two alleles, *M* and *N*) enabled Boyd to classify 104 American Ute Indians into genotype frequencies of .59 *MN*, .34 *MN*, and .07 *NN*. Since the gene frequencies are .59 + .17 = .76 for *M*, and .07 + .17 = .24 for *N*, the expected genotype frequencies are $(.76)^2$ = .58 for *MM*, 2(.76)(.24) = .36 for *MN*, and $(.24)^2$ = .06 for *NN*. The close correlation between observed and expected genotypic values indicates that this population has reached Hardy–Weinberg equilibrium.

When researchers know the genotypes of mated couples in the population, they can also test the assumption of random mating. Under random mating the frequencies of the different mating combinations should depend only on the frequencies of their genotypes. An actual set of data that Matsunaga and Itoh collected provided the blood types of 741 couples (or 1,482 individuals) in a Japanese town that showed genotypic frequencies of .274 *MM*, .502 *MN*, and .224 *NN*. (Since the gene frequencies are .525 *M* and .475 *N*, the expected equilibrium genotypic frequencies are .276 *MM*, .499 *MN*, and .225 *NN*, again indicating equilibrium for this locus.)

Table 21–6 demonstrates random mating among these individuals, with the number of observed matings of different combinations given in the last column. To the left of this column, the expected mating combination frequencies are calculated on the basis of the gene frequencies, p for *M* and q for *N*. For example, since the frequency of the genotype *MM* is p^2 at equilibrium, the random mating combination *MM* × *MM* should be p^4. When we calculate the frequencies of all expected mating combinations this way, comparing observed and expected agrees very well with the assumption of random mating.

A more common case in natural populations is when the effect of one allele at a locus is completely dominant over another so that we cannot phenotypically distinguish the heterozygous genotype (for example, *Aa*) from the homozygous dominant (for example, *AA*). Under such circumstances we cannot obtain gene frequencies directly, as in codominance, because we don't know two of the genotypic frequencies (*AA*, *Aa*), and instead must rely on the distinctive recessive homozygote (*aa*), whose genotypic frequency coincides with its phenotypic frequency. That is, if we assume that such a population has reached Hardy–Weinberg equilibrium (p^2 *AA*, $2pq$ *Aa*, q^2 *aa*), the recessive homozygotes are present in a frequency q^2 equal to the square of the recessive gene frequency, q. If, let us say, q^2 is .49, then q is $\sqrt{.49} = .70$, and the frequency of the dominant allele p is $1 - q$, or .30. The homozygous dominants, therefore, have the frequency $p^2 = (.30)^2 = .09$, and the heterozygotes have the frequency $2pq = 2(.30)(.70) = .42$.

One consequence of this analysis is that when recessive phenotypes are rare, it is surprisingly common to find that the **carrier heterozygotes,** phenotypically disguised as dominants, are present in relatively high frequency. Albinism, for example, affects only about 1 in 20,000 humans in some populations, or $q^2 = 1/20{,}000 = .00005$. The gene frequency, q, of the albino gene is therefore .007, and the frequency, p, of the nonalbino allele is .993. The frequency of heterozygous albino carriers is therefore 2(.993)(.007) = .014, or approximately 1 in 70 individuals. Thus, there are .014/.00005 = 280 times as many heterozygotes for this trait as there are homozygotes. Similarly high proportions of carriers of other recessive traits (Table 21–7) point to the difficulty of

TABLE 21–6 Comparison of mating combinations expected according to random mating and those observed in 741 couples by Matsunaga and Itoh (p = .525, q = .475)

Mating Combination	Expected Frequency					Expected Number Observed (frequency × 741)	Observed Number
MM × *MM*	$(p^2)(p^2)$	=	p^4	=	.0760	56.3	58
MM × *MN*	$2 \times (p^2)(2pq)$	=	$4p^3q$	=	.2749	203.7	202
MM × *NN*	$2 \times (p^2)(q^2)$	=	$2p^2q^2$	=	.1244	92.2	88
MN × *MN*	$(2pq)(2pq)$	=	$4p^2q^2$	=	.2487	184.3	190
MN × *NN*	$2 \times (2pq)(q^2)$	=	$4pq^3$	=	.2251	166.8	162
NN × *NM*	$(q^2)(q^2)$	=	q^4	=	.0509	37.7	41
					1.0000	741	741

Source: Adapted from Strickberger.

TABLE 21-7 Genotype frequencies for some human diseases caused by recessive genes

Disease	Population	Gene Frequency (q)	Frequency of Homozygotes (q^2)	Frequency of Heterozygous Carriers ($2pq$)	Ratio of Heterozygous Carriers to Homozygotes ($2pq/q^2 = 2p/q$)
Achromatopsia	Pingelap (Caroline Islands)	.22	1 in 20	1 in 2.8	7/1
Sickle cell anemia	Africa (some areas)	.2	1 in 25	1 in 3	8/1
Albinism	Panama (San Blas Indians)	.09	1 in 132	1 in 6	21/1
Ellis-van Creveld syndrome	Old Order Amish	.07	1 in 200	1 in 8	26/1
Sickle cell anemia	U.S. blacks	.04	1 in 625	1 in 13	48/1
Cystic fibrosis	U.S. whites	.032	1 in 1000	1 in 16	60/1
Tay-Sachs disease	Ashkenazi Jews	.018	1 in 3000	1 in 28	108/1
Albinism	Norway	.010	1 in 10,000	1 in 50	198/1
Phenylketonuria	United States	.0063	1 in 25,000	1 in 80	314/1
Cystinuria	England	.005	1 in 40,000	1 in 100	400/1
Galactosemia	United States	.0032	1 in 100,000	1 in 159	630/1
Alkaptonuria	England	.001	1 in 1,000,000	1 in 500	2,000/1

Source: From *Genetics Third Edition* by Monroe W. Strickberger. Copyright © 1985 by Monroe W. Strickberger. Reprinted by permission of Prentice Hall, Inc., Upper Saddle River, NJ.

eliminating rare harmful recessives, since such deleterious alleles are carried mostly in the unexpressed heterozygous condition.

When more than two alleles are present at a locus, the Hardy–Weinberg equilibrium is based on a multinomial expansion such as that described on page 523. For that example of three alleles (A_1, A_2, A_3), we expect six genotypes, and we can calculate the gene frequency of each allele (p, q, r) from the following equations:

$$p = \frac{2(A_1A_1) + (A_1A_2) + (A_1A_3)}{2N}$$

$$q = \frac{2(A_2A_2) + (A_1A_2) + (A_2A_3)}{2N}$$

$$r = \frac{2(A_3A_3) + (A_1A_3) + (A_2A_3)}{2N}$$

where A_1A_1, A_1A_2, A_1A_3, and so on refer to the numbers of genotypes in each category, and N refers to the total number of individuals scored.

A system of this type appears in human populations bearing different forms of the red blood cell enzyme acid phosphatase, which researchers can score into six different phenotypes, AA, BB, CC, AB, BC, or AC, as determined by all possible combinations of the alleles *A*, *B*, and *C* at a single locus. As Table 21–8 shows, investigations of a Brazilian population indicate that the observed phenotypic frequencies of the acid phosphatase combinations conform closely to those the Hardy–Weinberg equilibrium predicts. Although exceptions arise (Spiess), conformities to the Hardy–Weinberg equilibrium seem quite common for both autosomal and sex-linked genes and appear in a variety of sexually outbreeding organisms. In general, as noted earlier, these studies emphasize that populational gene and genotype frequencies do not change without cause.

TABLE 21-8 Comparison of observed acid phosphatase phenotypes and those expected according to Hardy–Weinberg equilibrium in a sample of 369 Brazilian individuals

Phenotypes	AA	BB	CC	AB	AC	BC
Observed	15	220	0	111	4	19
Expected	14.4	219.9	0.4	112.2	4.4	17.7

Source: From Lai et al.

Inbreeding

One set of conditions that interferes with the Hardy–Weinberg equilibrium is **nonrandom mating.** An important example occurs when related individuals of similar genotype mate preferentially with each other in a phenomenon called **inbreeding.** (An extreme form of inbreeding is when two gametes of a single individual unite to form a fertile zygote, **self-fertilization.**) Although the effect of inbreeding will not change the overall gene frequency, it will lead to an excess of homozygous genotypes. Inbreeding will thus cause a rare recessive allele to appear in greater homozygous frequency than under random mating, offering increased opportunity for selection to act on rare recessives.

We usually quantify inbreeding by an **inbreeding coefficient,** *F,* which measures the probability that the two alleles of a gene in a diploid zygote are identical—descended from a single ancestral allele. For example, if we begin with a heterozygous diploid, A^1A^2, normal mendelian segregation will confer a ½ probability that each F_1 offspring will receive the same A^1 allele. The allele transmitted by an F_1 individual to its offspring, in turn, has a ½ chance of being the same as the ancestral allele. This means that if two such F_1 offspring mate, the chances that the two alleles in one of *their* offspring (an F_2) are identical by descent (A^1A^1) is ½ × ½ × ½, or the inbreeding coefficient is 1/8.

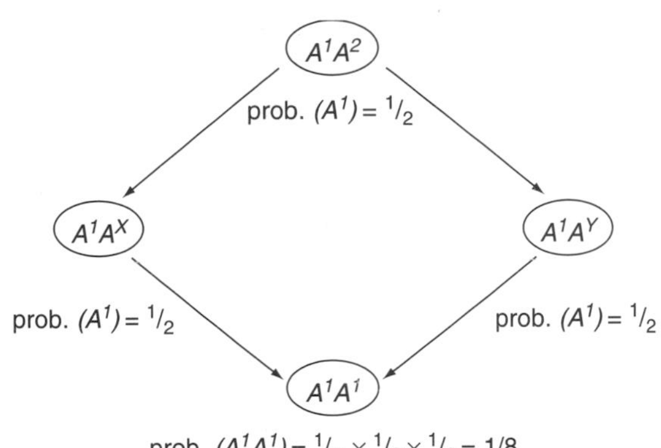

Having identical alleles, of course, means homozygosity, and *F* can range from one (complete homozygosity) to zero (complete heterozygosity). Of the inbred proportion measured by *F,* some will be *AA* and some *aa,* the frequencies of each depending on their respective population gene frequencies *p* and *q*. Thus inbreeding will produce *pF AA* and *qF aa* genotypes. In addition to these, however, the remaining individuals in this population $(1 - F)$ will bear genotypes whose frequencies are determined according to the Hardy–Weinberg equilibrium of p^2 *AA*, $2pq$ *Aa*, and q^2 *aa*. The three genotypes will have the following frequencies:

$$AA = p^2(1 - F) + pF = p^2 - p^2F + pF - p^2 + pF(1 - p) = p^2 + pqF$$

$$Aa = 2pq(1 - F) = 2pq - 2pqF$$

$$aa = q^2(1 - F) + qF = q^2 - q^2F + qF = q^2 + qF(1 - q) = q^2 + pqF$$

It is now easy to see that the increase in the frequency of each homozygote type by a factor of pqF flows from an equivalent fall in the heterozygote frequency ($-2pqF$). Note also that this reduction in heterozygotes affects the gene frequencies *p* and *q* equally, so that only the genotypic frequencies change. When inbreeding is absent, $F = 0$, and the preceding equations reduce to the Hardy–Weinberg frequencies p^2 *AA*, $2pq$ *Aa*, and q^2 *aa*. When inbreeding is complete, $F = 1$, $2pq - 2pqF = 0$, and the only remaining genotypes are *pAA* and *qaa*.

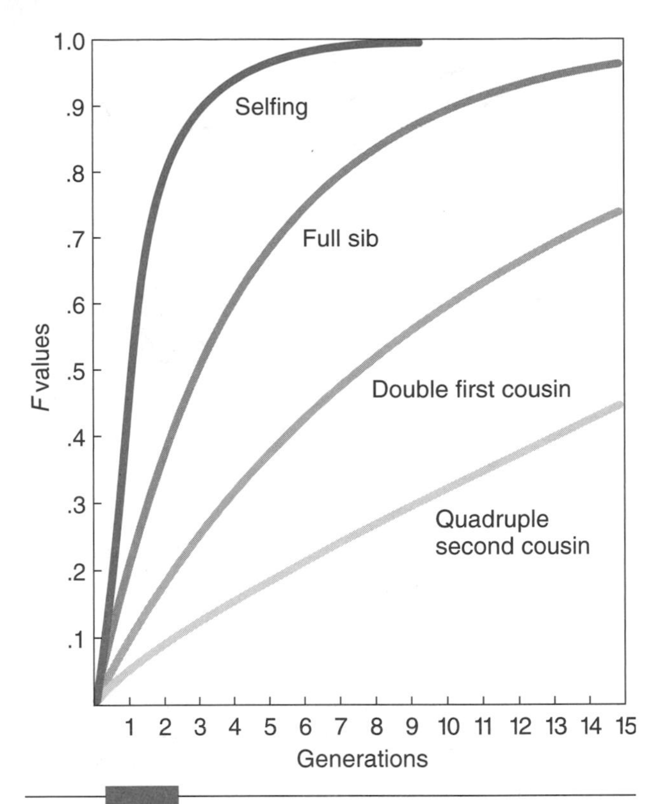

FIGURE 21-8 Inbreeding coefficients at generations 1 to 15 for four different systems of inbreeding (pedigrees given in Strickberger). You can obtain or derive formulas for calculating F in other inbreeding systems from Wright's (1921) basic work in this field.

The mating system in which inbreeding is greatest is under self-fertilization (for example, hermaphrodites), where *F* is equal to .5 in the first generation and approaches 1 within four or five generations. As Figure 21–8 shows, any other mating scheme slows the rate of inbreeding.[5] Similarly, as Figure 21–9 shows, population size can also affect inbreeding, since the smaller the size, the greater the opportunity for related individuals to mate.

[5] We can call systems such as brother–sister mating and first-cousin mating, in which individuals mate on the basis of their genetic relationship, *genetic assortative mating.* Phenotypic similarity, however, may also cause preferential mating, and in many human societies mates are chosen that share characteristics such as height, color, facial form, muscular build, and intelligence. In such phenotypic assortative matings homozygosity can also increase, but only for those loci involved in the trait(s) on which the preferred matings are based. This is in contrast to the genetic assortative mating of inbreeding, which tends to increase homozygosity at all loci. A further type of mating practice is disassortative mating in which individuals of unlike genotype or phenotype form mating pairs, thereby preventing inbreeding and helping to maintain heterozygote frequency (**heterozygosity**). There are various examples of such systems, including alleles in plants that cause sterility of male gametes when they attempt to fertilize ova of the same genotype (self-sterility alleles; see p. 310).

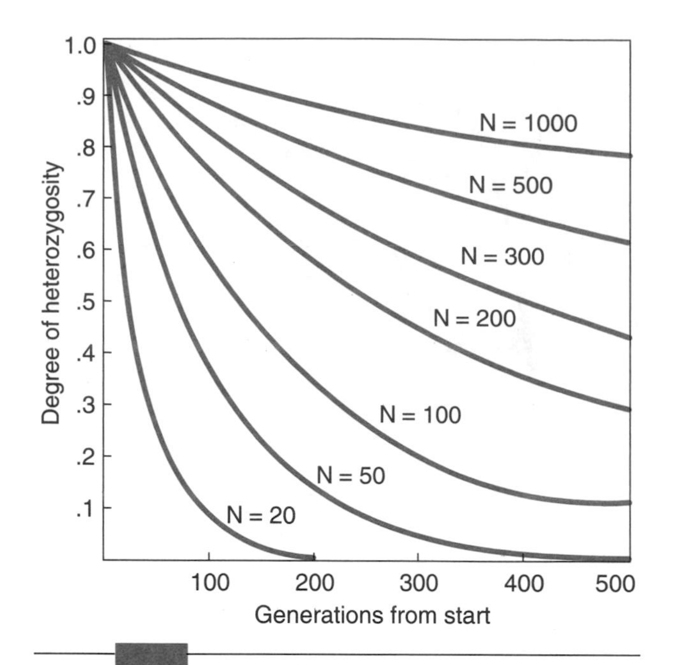

FIGURE 21-9 Degrees of heterozygosity (heterozygote frequency) remaining in populations of different sizes after given generations of random union between gametes. Calculations are based on 1.00 as the initial degree of heterozygosity (or when $F = 0$). *(Adapted from Strickberger.)*

Although some degree of inbreeding occurs in most outbreeding populations, significant amounts of inbreeding can cause **inbreeding depression** in which rare deleterious recessives may now appear with increased homozygous frequency. If a recessive disease with genotype *aa* occurs with frequency q^2 in a random outbred population, its frequency will increase by pqF in an inbred population as just derived. The ratio of inbred to outbred frequency for the homozygous recessive will therefore be

$$\frac{q^2 + pqF}{q^2} = \frac{q(q + pF)}{q^2} = \frac{q + pF}{q}$$

Obviously, if q is large and F is small, the inbreeding increment pF will be relatively small, and the increased frequency of homozygous recessives will hardly be noticeable. However, if q is very small (rare) and p is large, then pF provides a notable increase in recessives even when F is fairly small. For example, if q is .5, first-cousin mating ($F = .0625$) will produce an inbred-to-outbred ratio of homozygotes of

$$\frac{q + pF}{q} = \frac{.5 + (.5)(.0625)}{.5} = \frac{.53125}{.5} = 1.06$$

However, if q is .005, this ratio increases to $.067/.005 = 13.4$. When $q = .0005$, the increase of homozygotes because of first-cousin mating is .0630/.0005, or 126 times that of randomly bred populations.

Should homozygous recessives have a quantitative effect on one or more traits, inbreeding would cause the measured values of these traits to tend in the direction of recessive values. Thus, in various outbred populations, such as corn, inbreeding depression can reduce height, yield, and other characters. Inbreeding depression, however, is not a universal phenomenon in all species, certainly not in many species that are normally self-fertilized and have eliminated most or all of their deleterious recessives. On the whole, ample evidence shows that most normally cross-fertilizing species deteriorate on consistent inbreeding, leading even to extinction (Saccheri et al.), although some strains may escape because they carry relatively few deleterious recessive genes.

SUMMARY

Following the publication of *On the Origin of Species,* Darwin proposed that natural selection operated on small, continuous hereditary variations, while Galton and others maintained that variations were sharp and discontinuous. The controversy resolved when researchers showed that several genes, each with small effect, can nevertheless have a large effect when they mutually influence expression of a single phenotypic trait. By the 1930s it became clear through genetics that evolution is a population phenomenon that we can represent as a change in gene frequencies because of the action of various natural forces such as selection and genetic drift, and these changes can lead to differences among races, species, and higher taxa. Along with other concepts, this populational view of evolution became known as the Neo-Darwinian (modern) synthesis.

Gene frequencies and the gene pool are two major attributes of a population, which can be defined as a group of potentially interbreeding organisms. Gene frequency is the ratio of the different alleles of a gene in a population without regard to their homozygosity or heterozygosity. A gene pool consists of all alleles in the gametes of a population and therefore represents all the genes available for the next generation.

According to the Hardy–Weinberg principle, gene frequencies are conserved in a random mating population unless external forces act on it, and the equilibrium of genotype frequencies (for example, $p^2 + 2pq + q^2$) derive from the gene frequencies.

If there are two or more pairs of independently assorting genes, there are many more possible genotypes, and the more gene pairs, the longer it will take to achieve overall genotypic equilibrium. In the case of linkage, the higher the frequency of recombination between linked genes, the shorter the time needed to reach equilibrium. When genes are linked on the X chromosome, gene frequencies at equilibrium will be equal in both sexes, but this may take a number of generations if frequencies between the two sexes differ initially.

In natural populations we can determine genotype frequencies quite easily if no allele is dominant, and such observations generally show that Hardy–Weinberg equilibrium has been achieved. If one of the two alleles is dominant, we can compute gene frequencies by assuming Hardy–Weinberg equilibrium and using the frequency of homozygous recessive individuals as q^2 in the genotypic equilibrium formula. When we do this for various recessive conditions present in low frequency, the frequency of heterozygous "carriers" is surprisingly high. In the case of multiple alleles, we can calculate genotype frequencies using a multinomial expansion of the Hardy–Weinberg equation. Most studies indicate that gene pools are quite stable and generally remain at equilibrium unless selection or other conditions interfere.

Inbreeding does not affect gene frequencies but does increase homozygosity, allowing relatively rare recessive alleles to be expressed. If these alleles are harmful, inbreeding depression may result.

KEY TERMS

binomial expansion
biometricians
carrier heterozygotes
coadapted gene complexes
conservation of gene frequencies
continuous variation
coupling
deme
discontinuous variation
disequilibrium
equilibrium
gene frequencies
gene pool
Hardy–Weinberg equilibrium
heterozygosity
inbreeding
inbreeding coefficient
inbreeding depression
linkage disequilibrium
macromutations
mendelian population
Mendelians
modern synthesis
mutationists
Neo-Darwinian synthesis
nonrandom mating
panmixia
population genetics
pure lines
quantitative characters
random mating
repulsion
saltation
selectionists
self-fertilization
trinomial expansion

DISCUSSION QUESTIONS

1. Continuous versus discontinuous variation
   a. What controversy arose between Mendelians and biometricians?
   b. What controversy arose between mutationists and selectionists?
   c. What role did "pure lines," "macromutations," "saltations," and "quantitative characters" play in these arguments?
   d. How were these various issues resolved?
2. What are the elements of the Neo-Darwinian (modern) synthesis?
3. How do researchers determine gene frequencies in a diploid population when they can identify the frequencies of all genotypes?
4. Hardy–Weinberg principle
   a. Under what conditions are gene frequencies conserved?
   b. How do geneticists derive genotype frequencies according to the Hardy–Weinberg principle?
5. Equilibrium between genes at two or more loci
   a. Why does a population rarely, if ever, attain multilocus equilibrium in a single generation?
   b. How do linkage and recombination frequencies affect such attainment of equilibrium?
6. How do an autosomal locus and a sex-linked locus differ in reaching genotypic equilibrium?
7. How can we test the assumption of random mating in the Hardy–Weinberg principle for a particular gene segregating in a natural population?
8. How can we derive gene and genotype frequencies in a diploid population when we only know the frequency of recessive homozygotes?
9. What is the relationship between (a) the frequency of a recessive allele and (b) the ratio of heterozygous carriers to homozygotes for that allele?
10. How and why does inbreeding affect the frequency of homozygous genotypes?

EVOLUTION ON THE WEB

Explore evolution on the web! Visit the accompanying web site for *Evolution*, 3/e at www.jbpub.com/evolution for web exercises and links relating to topics covered in this chapter.

REFERENCES

Berg, L. S., 1969. *Nomogenesis, or Evolution Determined by Law.* MIT Press, Cambridge, MA. (Translated from the 1922 Russian edition.)

Boyd, W. C., 1950. *Genetics and the Races of Man.* Little, Brown, Boston.

Bradshaw, H. D., Jr., S. M. Wilbert, K. G. Otto, and D. W. Shemske, 1995. Genetic mapping of floral traits associated with reproductive isolation in monkeyflowers (*Mimulus*). *Nature,* **376,** 762–765.

Castle, W. E., and J. C. Phillips, 1914. *Piebald Rats and Selection.* Carnegie Insti. Wash., Publ. No. 195, Washington, DC.

Chetverikov, S. S., 1926. On certain aspects of the evolutionary process from the standpoint of modern genetics. (Translated from Russian to English, 1961, *Proc. Amer. Phil. Soc.,* **105,** 167–195.)

Crow, J. F., 1986. *Basic Concepts in Population, Quantitative, and Evolutionary Genetics.* Freeman, New York.

Crow, J. F., and M. Kimura, 1970. *An Introduction to Population Genetics Theory.* Harper & Row, New York.

Dobzhansky, Th., 1937, 1941, 1951. *Genetics and the Origin of Species* (three eds.). Columbia University Press, New York.

———, 1938-1976. *Dobzhansky's Genetics of Natural Populations I–XLIII,* R. C. Lewontin, J. A. Moore, W. B. Provine,

and B. Wallace (eds.). Columbia University Press, New York. This volume is a collection of the 43 papers in Dobzhansky's influential series on *Drosophila* population genetics, with various coworkers, along with introductory articles by Provine and Lewontin.

Doebley, J., A. Stec, and C. Gustus, 1995. *Teosinte branched 1* and the origin of maize: Evidence for epistasis and the evolution of dominance. *Genetics,* **141,** 333–346.

East, E. M., 1916. Studies on size inheritance in *Nicotiana. Genetics,* **1,** 164–176.

Falconer, D. S., and T. F. C. Mackay, 1996. *Introduction to Quantitative Genetics,* 4th ed. Longman, London.

Fisher, R. A., 1930. *The Genetical Theory of Natural Selection.* Clarendon Press, Oxford, England. (2d ed., 1958, Dover, New York.)

Gillespie, J. H., 1998. *Population Genetics: A Concise Guide.* Johns Hopkins University Press, Baltimore, MD.

Haldane, J. B. S., 1932. *The Causes of Evolution.* Harper & Row, New York. (Reprinted 1966, Cornell University Press, Ithaca, NY.)

Hardy, G. H., 1908. Mendelian proportions in a mixed population. *Science,* **28,** 49–50.

Hartl, D. L., and A. G. Clark, 1997. *Principles of Population Genetics,* 3d ed. Sinauer Associates, Sunderland, MA.

Hedrick, P. W., 1983. *Genetics of Populations.* Science Books International, Boston.

Huxley, J., 1942. *Evolution: The Modern Synthesis.* Allen & Unwin, London.

Jepsen, G. L., E. Mayr, and G. G. Simpson (eds.), 1949. *Genetics, Paleontology, and Evolution.* Princeton University Press, Princeton, NJ.

Johannsen, W., 1903. *Über Erblichkeit in Populationen und in reinen Linien.* Fischer, Jena.

Lai, L., S. Nevo, and A. G. Steinberg, 1964. Acid phosphatases of human red cells: Predicted phenotype conforms to a genetic hypothesis. *Science,* **145,** 1187–1188.

Lewontin, R. C., 1997. Dobzhansky's *Genetics and the Origin of Species:* Is it still relevant? *Genetics,* **147,** 351–355.

Matsunaga, E., and S. Itoh, 1958. Blood groups and fertility in a Japanese population, with special reference to intrauterine selection due to maternal-fetal incompatibility. *Ann. Hum. Genet.,* **22,** 111–131.

Nilsson-Ehle, H., 1909. Kreuzungsuntersuchengen an Hafer und Weisen. *Lunds Univ. Aarskr. N. F. Afd.,* ser. 2, vol. 5, no. 2, pp. 1–122.

Orr, H. A., and J. A. Coyne, 1992. The genetics of adaptation: A reassessment. *Amer. Nat.,* **140,** 725–742.

Ospovat, D., 1981. *The Development of Darwin's Theory: Natural History, Natural Theology, and Natural Selection, 1838–1859.* Cambridge University Press, Cambridge, England.

Provine, W. B., 1971. *The Origins of Theoretical Population Genetics.* University of Chicago Press, Chicago.

Saccheri, I., M. Kuussaari, M. Kankare, P. Vikman, W. Fortelius, and I. Hanski, 1998. Inbreeding and extinction in a butterfly metapopulation. *Nature,* **392,** 491–494.

Smith, T. B., 1993. Disruptive selection and the genetic basis of bill size polymorphism in the African finch *Pyrenestes. Nature,* **363,** 618–620.

Spiess, E. B., 1977. *Genes in Populations.* Wiley, New York.

Strickberger, M. W., 1985. *Genetics,* 3d ed. Macmillan, New York.

Wallace, B., 1981. *Basic Population Genetics.* Columbia University Press, New York.

Weinberg, W., 1908. Über den Nachweis der Vererbung beim Menschen. *Jahreshefte des Vereins für Vaterändlische Naturkunde in Württemburg,* **64,** 368–382.

Wright, S., 1921. Systems of mating. *Genetics,* **6,** 111–178.

———, 1931. Evolution in Mendelian populations. *Genetics,* **16,** 97–159.

Yule, G. U., 1902. Mendel's laws and their probable relations to intra-racial heredity. *New Phytol.,* **1,** 193–207, 222–238.

Changes in Gene Frequencies

For populations to evolve—that is, to change their gene frequencies—mutation must first introduce the nucleotide differences from which such changes arise. The mere appearance of new genes (alleles), however, is no guarantee that they will persist or prevail over others. For example, no certainty exists that a newly mutated gene such as *a* (for example, $A \rightarrow a$) will transmit to the next generation, since its carrier (for example, *Aa*) may or may not survive and may or may not mate. Even if the *Aa* mutant carrier does mate ($Aa \times AA$), the chances of *a* transmission declines since a significant proportion (40 percent) of matings in most stable populations produce families with zero surviving offspring (*a* is lost) or only one offspring (*a* has a 50 percent chance of being lost during meiosis). Larger families may also lose the gene since, for example, even when the $Aa \times AA$ family produces two offspring the mutant gene has a 25 percent chance ($.5 \times .5$) of not being transmitted to either of them.

Fisher has calculated that the chance that a newly mutated gene may be eliminated within one generation is more than 33 percent because of its possible random loss in families of such different sizes. By the time 30 generations have passed after its introduction, the probability of elimination has risen to almost 95 percent. To explain the persistence of many mutations and their increase in frequency, we must look elsewhere than the original mutational event.

Mutation Rates

One factor that can be expected to affect gene frequency is the frequency of mutation. If gene *A* continually mutates to *a* and the reverse mutation never occurs, the chances improve that *a* will increase in frequency with each generation. Given a long enough period of time and a persistent mutation rate in a population of constant size, *a* can eventually replace *A*. Of course, the mutation rate does not always occur in only one direction. For example, if *u* is the mutation rate of *A* to *a*, the allele *a* may mutate back to *A*

with frequency v. We can estimate these effects quantitatively by calling the initial frequencies of alleles A and a, p_0 and q_0, respectively, and noting that a single generation of mutation will produce a frequency of A equal to $p_0 + vq_0$ and a frequency of a equal to $q_0 + up_0$.

If we now confine our attention to only one of the alleles, a, clearly it has gained the fraction up_0 (new a alleles) but lost the fraction vq_0 (new A alleles). In other words, the change in the frequency of a, which we call delta q (Δq), can be expressed as $\Delta q = up_0 - vq_0$. Thus, if p were relatively large and q small, Δq would be large and q would increase rapidly; when q became larger and p became smaller, Δq would diminish. The point at which Δq is zero—that is, the point where there is no further change and p and q are balanced in relation to their mutation frequencies—we call the **mutational equilibrium** (frequency $a = \hat{q}$, or "q hat"): $\Delta q = 0 = up - vq$, or $up = vq$ at $\hat{q}$. However, since there are only two alleles, A and a, $p = 1 - q$, which leads to

$$up = vq$$

$$u(1 - q) = vq$$

$$u = uq + vq = q(u + v)$$

$$\hat{q} = \frac{u}{u + v}$$

The same procedure applied to the frequency of A gives $\hat{p} = v/(u + v)$, so that $\hat{p}/\hat{q} = [v/(u + v)]/[u/(u + v)] = v/u$. Thus when the mutation rates are equal ($u = v$) the **equilibrium gene frequencies** $\hat{p}$ and $\hat{q}$ will be equal. If the mutation rates differ, so will the equilibrium frequencies. For example, if $u = .00005$ and $v = .00003$, the equilibrium frequency $\hat{q}$ equals $5/8 = .625$ and $\hat{p} = 3/8 = .375$. However, the rate at which mutation reaches this equilibrium frequency is usually quite slow and we can derive it by calculus methods from Δq as

$$(u + v)n = \ln [(q_0 - \hat{q})/(q_n - \hat{q})]$$

where n is the number of generations required to reach a frequency q_n when starting with a frequency q_0. For the example just considered, the number of generations necessary for q to increase from a frequency of one-eighth to three-eighths is

$$(.00008)n = \ln \frac{.125 - .625}{.375 - .625}$$

$$= \ln 2.00 = .69315$$

$$n = \frac{.69315}{.00008} = 8{,}664 \text{ generations}$$

Thus the approach to equilibrium based on the usually observed mutation rates of 5×10^{-5} or less (Table 10–4) is very slow, and mutational equilibrium is probably rarely if ever reached, especially since mutation rates are probably not constant. As a rule, the attainment of mutational equilibrium does not appear to be the sole cause for existing gene frequencies. A more efficient mechanism that can help explain how gene frequencies change is the effect of selection, the "scrutinizing process" that Darwin proposed.

Selection

The knowledge that phenotypes can differ in viability and fertility can evidently influence the frequencies of their genotypes[1]. If individuals carrying gene A are more successful in producing viable and fertile offspring than individuals carrying its allele a, and sufficient numbers of advantageous genotypes have arisen to overcome their loss by chance, then the A allele frequency will tend to increase relative to a. Survival and fertility mechanisms that affect the reproductive success of a genotype are known as **selection,** and the extent to which a genotype contributes to the offspring of the next generation relative to other genotypes in a given environment is commonly known as its **fitness, selective value,** or **adaptive value.** That is, we can describe selection as a composite of the forces that limit the reproductive success of a genotype and we can describe fitness as the comparative ability of a genotype to withstand selection. The genetic effect of selection on a particular trait in a population is confined to fitness differences among the different genotypes that affect that particular trait. When the selective process operates, gene frequencies tend to change among generations, unless the population has reached a genetic equilibrium, as described later in this chapter.[2]

[1]Some writers dispute whether selection should be considered primarily phenotypic or genotypic. The view followed here is that although evolution is marked by phenotypic changes, the transmission of such changes between generations (evolution) is entirely genotypic: phenotypic differences are "descriptive" and not "instructive," and must therefore be genetically based in order to be transmitted (see also p. 356).

[2]If we define evolution as hereditary changes over time, selection, although important, is not the only process that can cause such changes; other evolutionary mechanisms considered in this chapter (mutation, migration, and random genetic drift) can also affect gene frequencies. For example, the frequency of a harmful recessive gene may remain stable despite selection against homozygotes for that gene if there is heterozygote advantage (p. 538). In other words, the relative fitness of a genotype may not be the only reason for its survival, and the statement that evolutionary theory merely proposes the "survival of the fittest" is misleading and incorrect. In fact, Fisher begins his classic treatise *The Genetical Theory of Natural Selection* with the statement "Natural Selection is not Evolution."

Critics often claimed that "survival of the fittest" is a circular, tautological, or unprovable statement that cannot be challenged because it defines those who survive as fittest, and fittest as those who survive. Some philosophers, such as Waters, propose that this notion be abandoned, whereas others, such as Resnik, suggest it has practical value. In general, we can see that natural selection can be a *cause* for change in gene frequency (evolution) but is not the *same* as a change in gene frequency: different gene frequencies ("survival") may result from causes

In simplest form, fitness and selection are measured by the number of descendants produced by one genotype compared to those produced by another. For example, if individuals of genotype A produce an average of 100 offspring that reach full reproductive maturity while genotype B individuals produce only 90 in the same environment, the adaptive value of B relative to A is reduced by 10 offspring, or the fraction $10/100 = .1$. If we designate the adaptive value of a genotype as W and the selective force acting to reduce its adaptive value as s (the **selection coefficient**), then we can say that $W = 1$ and $s = 0$ for A in the preceding example, and $W = .9$ and $s = .1$ for B. The relationship between W and s for a particular genotype is $W = 1 - s$, or $s = 1 - W$.

Selection against a genotype may occur in either the haploid (gametic) or diploid (zygotic) stage or both, depending at which of these stages gene expression influences survival or fertility (Fig. 22–1). In any of these stages, selection may be obvious or subtle, its effects ranging from complete lethality or sterility ($s = 1$) to only slight reductions in adaptive value (for example, $s = .01$). When selection occurs among haploids, there is no difference between dominant and recessive genes, since their carriers phenotypically express both kinds of alleles.

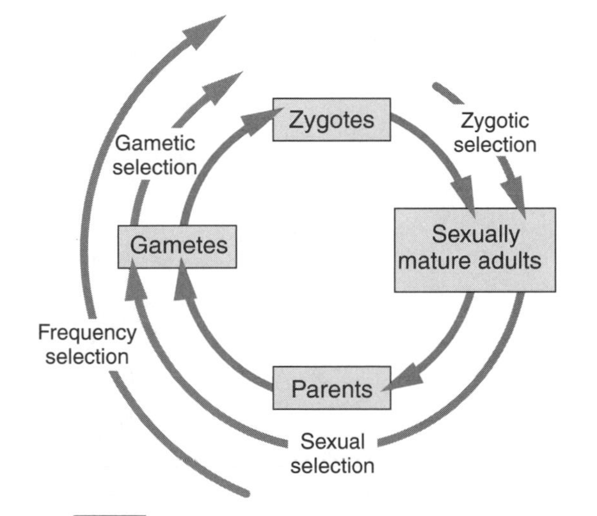

FIGURE 22-1 Simplified diagram of selection acting on the life stages of an organism during a single generation, from zygote to zygote ("egg to progeny"). In addition to sperm and pollen competition, sexual selection includes choice of sexual partners (pp. 588–589). Other complexities occur in organisms such as mammals, where fitnesses overlap between generations because a juvenile's fitness depends not only on its own attributes but also on receiving parental care—on the fitness of individuals in the preceding generation. (*Adapted from Christiansen.*)

other than selection ("fitness"), and selection itself is subject to change when the environment changes. Moreover, selection is a process, and survival is a result—they are not the same. Neither natural selection nor its colloquial expression, "survival of the fittest," is a tautology.

Thus, as we might expect, the effect of selection on haploids is much more rapid and direct than on diploids because **deleterious recessive** alleles cannot be hidden from selection among heterozygotes as they are in diploids. Table 22–1 provides estimates of the number of generations necessary to change the frequency of deleterious genes in haploids under a variety of selective conditions. Note that in contrast to diploids (see Table 22–3), haploids completely eliminate a **deleterious lethal** gene ($s = 1$) in one generation, and even lesser selection coefficients result in relatively rapid gene frequency changes.

In most higher animals and plants selection takes place primarily in the diploid or zygotic and post-zygotic stages. In diploids however, there are three possible genotypes for a single gene difference (for example, AA, Aa, aa), so that the effectiveness of selection depends, among other things, on the degree of dominance. Table 22–2 shows calculation of the change in gene frequency (Δq) of a for one generation when complete dominance exists and selection occurs only against the recessive aa.

Table 22–3 summarizes the number of generations necessary to change such deleterious recessive gene frequencies when we project selection over periods of time and for various selection coefficients. Note that the initial change in gene frequency from .99 to .10 is relatively rapid for selection coefficients from $s = 1$ to $s = .10$. Further reductions in gene frequency are considerably slower: to reduce the gene frequency below .01 may take thousands of generations, even when the selection coefficient is relatively high. As indicated in Chapter 21, the reason for the relative inefficiency of selection against rare recessives is simply that most recessive genes are present in heterozygotes where they are protected from selection: the more rarely a gene appears in a population, the more frequently it occurs in heterozygotes compared to homozygotes (see Table 21–7).

The selective situation can reverse so that the dominant allele is selected against and the recessive is favored. Selection will then be more effective, because the **deleterious dominant** gene is subject to selection in all genotypes in which it occurs. For example, should a dominant allele become lethal, its frequency falls to zero in a single generation. However, as the selection coefficient against the dominant allele decreases, replacement by the recessive is considerably slower. For the general case, selection against a dominant allele of gene frequency p results in a change of

$$-sp(1-p)^2/[1-sp(2-p)]$$

as Table 22–4 shows.

Note that if s is small, the denominator is close to 1, and Δp is effectively equal to $-sp(1-p)^2$. Since $1 - p$ is q and p is $1 - q$, this means that Δp is now $-sq^2(1-q)$, or Δp is identical to Δq for a deleterious recessive at low

TABLE 22-1 Number of generations required for given frequency changes (q_0 to q_n) of a deleterious gene under different selection coefficients in haploids

Change in Gene Frequency		Number of Generations for Different *s* Values						
From (q_0)	To (q_n)	$s = 1$	$s = .80$	$s = .50$	$s = .20$	$s = .10$	$s = .01$	$s = .001$
.99	.75		4	7	17	35	350	3496
.75	.50		1	2	5	11	110	1099
.50	.25		1	2	5	11	110	1099
.25	.10	1	1	2	5	11	110	1099
.10	.01		3	5	12	24	240	2398
.01	.001		3	5	12	23	231	2312
.001	.0001		3	5	12	23	230	2303

Source: From *Genetics Third Edition* by Monroe W. Strickberger. Copyright © 1985 by Monroe W. Strickberger. Reprinted by permission of Prentice Hall, Inc., Upper Saddle River, NJ.

TABLE 22-2 Calculation of Δq for a deleterious recessive gene (***a***)

	AA	*Aa*	*aa*	Total	Frequency of *a*
Initial frequency	p^2	$2pq$	q^2	1	q
Adaptive value	1	1	$1 - s$		
Frequency after selection	p^2	$2pq$	$q^2(1 - s)$	$p^2 + 2pq + q^2 - sq^2 = 1 - sq^2$	
Relative frequency after selection	$\frac{p^2}{1 - sq^2}$	$\frac{2pq}{1 - sq^2}$	$\frac{q^2(1 - s)}{1 - sq^2}$		$\frac{pq + q^2(1 - s)}{1 - sq^2} = \frac{pq + q^2 - sq^2}{1 - sq^2} = \frac{q(1 - sq)}{1 - sq^2}$

Δq = relative frequency of *a* after selection − initial frequency of *a*

$$\Delta q = \frac{q(1 - sq)}{1 - sq^2} - q = \frac{q(1 - sq)}{1 - sq^2} - \frac{q(1 - sq^2)}{1 - sq^2} = \frac{q - sq^2 - q + sq^3}{1 - sq^2} = \frac{-sq^2 + sq^3}{1 - sq^2} = \frac{-sq^2(1 - q)}{1 - sq^2}$$

Source: Adapted from Strickberger.

selection coefficients. Under these conditions, we may apply the values of Table 22–3 in reverse order. That is, for a selection coefficient of .10 against the dominant allele (in favor of the recessive), 90,023 generations are necessary to increase the frequency of the recessive from .0001 to .001, or to reduce the frequency of the dominant from .9999 to .9990. Subsequent changes in frequency are more rapid as the favored recessive homozygotes become more frequent.

When dominance of the advantageous allele is incomplete, heterozygotes will show the effect of a deleterious gene since the heterozygous phenotype is at least partially harmful. If dominance is completely absent and the heterozygote has a phenotype exactly intermediate between the two homozygotes, its selection coefficient will be exactly half that in the deleterious homozygotes $[(1) + (1 - 2s)]/2 = (½) + (½ - s) = 1 - s$. As Table 22–4 shows, the resultant change in gene frequency in one generation $[-sq(1 - q)]/[1 - 2sq]$ is almost identical to that for gametic selection $[-sq(1 - q)]/[1 - sq]$.

In other words, the absence of dominance uncovers deleterious alleles and makes all of them available for se-

TABLE 22–3 Number of generations required for a given change in frequency (q_0 to q_n) of a deleterious recessive allele in diploids under different selection coefficients

Change in Gene Frequency		Number of Generations to Attain Given Gene and Genotype Frequencies for Different *s* Values					
From (q_0)	To (q_n)	$s = 1$ (lethal)	$s = .80$	$s = .50$	$s = .20$	$s = .10$	$s = .01$
.99	.90	↓	3	5	13	25	250
.90	.75	1	2	3	7	13	132
.75	.50	↓	2	3	9	18	176
.50	.25	2	4	6	15	31	310
.25	.10	6	9	14	35	71	710
.10	.01	90	115	185	462	924	9,240
.01	.001	900	1,128	1,805	4,512	9,023	90,231
.001	.0001	9,000	11,515	18,005	45,011	90,023	900,230

Source: From *Genetics Third Edition* by Monroe W. Strickberger. Copyright © 1985 by Monroe W. Strickberger. Reprinted by permission of Prentice Hall, Inc., Upper Saddle River, NJ.

TABLE 22–4 Single-generation changes in gene frequency for diploid genotypes subject to given selection coefficients under different conditions of dominance

	Adaptive Values for Genotype Frequencies Initially in Hardy-Weinberg Equilibrium			
Dominance Relations for the Three Given Genotypes	*AA* p^2	*Aa* $2pq$	*aa* q^2	Change in Gene Frequency[a]
Complete dominance: selection against the recessive allele	1	1	$1 - s$	$\frac{-sq^2(1-q)}{1-sq^2}$
Complete dominance: selection against the dominant allele	$1 - s$	$1 - s$	1	$\frac{-sp(1-p)^2}{1-sp(2-p)}$
Absence of dominance: selection against the *a* allele also occurs in the heterozygote	1	$1 - s$	$1 - 2s$	$\frac{-sq(1-q)}{1-2sq}$
Overdominance: selection against both homozygotes	$1 - s$	1	$1 - t$	$\frac{pq(ps-qt)}{1-p^2s-q^2t}$

[a]As mentioned in Chapter 21, mathematical derivations for various formulas have been omitted for simplicity but can be found in many population genetics textbooks.

lection, allowing rapid changes in gene frequencies mostly on the order of those observed in Table 22–1. The effectiveness of selection therefore strongly depends on the degree to which the heterozygote expresses the deleterious gene. Since population geneticists believe most recessive genes have some heterozygous expression, selection efficiency for or against them probably falls between the extremes of slow progress for complete dominance and rapid progress for absence of dominance.

Heterozygous Advantage

The examples of selection just considered always go in one direction, toward **elimination** of the deleterious allele and establishment or **fixation** of the favored allele. As long as the selection coefficient does not change, equilibrium between favored and unfavored alleles is impossible without new mutations.

Various conditions, however, permit the establishment of an equilibrium through which both alleles may remain indefinitely within the population. One such condition, **overdominance,** occurs when the heterozygote has superior reproductive fitness to both homozygotes.[3]

In general, if the heterozygote *Aa* has an adaptive value of 1.00 while the fitnesses of the homozygotes *AA* and *aa* reduce by the selective coefficients *s* and *t*, respectively, the change in frequency of *a* in a single generation is that shown in Table 22–4. When Δq is zero, equilibrium has been reached and gene frequency will not change further. Note that three possible conditions will cause the numerator $[pq(ps - qt)]$ to be equal to zero and therefore Δq to equal zero. Under the first two conditions, when either p or q are zero, neither allele will be present in the population at the same time, and balance, or equilibrium, will be absent. The third condition occurs when $ps = qt$, so that the numerator of Δq is $pq(0) = 0$. When this happens, the following relationships can be derived:

$$ps = qt$$

Add qs to both sides of preceding equation	Add pt to both sides of preceding equation
$ps + qs = qt + qs$	$ps + pt = qt + pt$
$s(p + q) = q(s + t)$	$p(s + t) = t(p + q)$

(Now, since $p + q = 1$)

$$q = \frac{s}{s + t} \qquad p = \frac{t}{s + t}$$

It is easy to see that if *s* and *t* are constant values, both p and q will reach a stable equilibrium: if q departs from the equilibrium value, selection pressure will force it back. That is, if Δq is positive, the gene frequency q increases, but if Δq is negative, q decreases, the negative or positive sign of Δq depending on whether q is above or below its equilibrium value. For example, when $s = .2$ and $t = .3$, the equilibrium value for q is $s/(s + t) = .2/(.2 + .3) = .4$. Values of q below .4 cause Δq to be positive, which increases q, whereas values of q above .4 cause Δq to be negative, which decreases q. As you can see from Figure 22–2, the effect of such heterozygote superiority is to drive the frequencies of the two alleles in the population to a stable equilibrium at $q = .4$.[4]

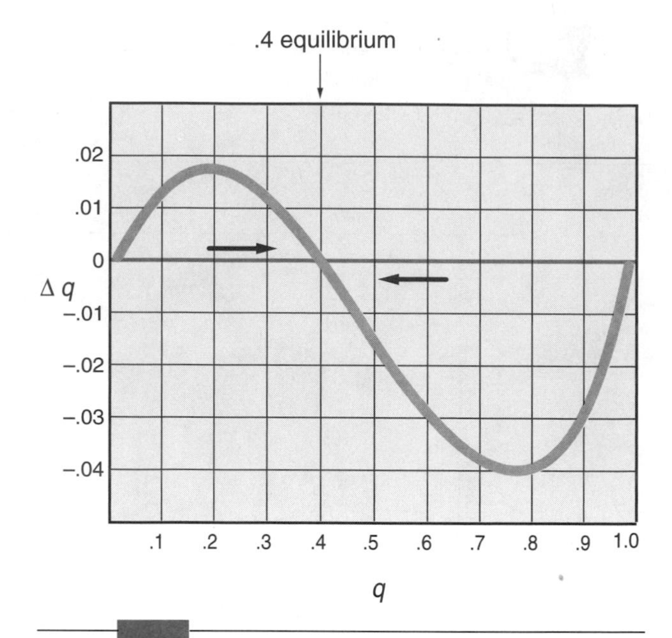

FIGURE 22–2 Change in the frequency (Δq) of allele *a* when the genotypic adaptive values are $AA = .8$, $Aa = 1.0$, $aa = .7$, and population size is infinite. These values provide a stable balanced polymorphism ($\Delta q = 0$) at $q = .4$. That is, Δq is positive (q increases) if q is less than .4 and negative (q decreases) if q is more than .4. Should one allele be accidentally eliminated—that is $q = 0$ or 1—then Δq is of course zero, but polymorphism is lost. (*Adapted from Li.*)

Selection and Polymorphism

The persistence of different genotypes through heterozygote superiority is an example of **balanced polymorphism,** a term Ford invented to

[3]The superiority of the heterozygote, often called **heterosis** or **hybrid vigor,** may show itself in improved fitness characters such as longevity, fecundity, and resistance to disease. An oft-cited example of heterosis is the dramatic increase in agricultural yield of hybrid corn, achieved by crossing selected inbred lines. Beginning with an average yield of about 25 bushels per acre in the 1920s, hybrid corn has enabled increases to as much as 140 bushels per acre (Crow). However, researchers still debate whether such hybrid vigor arises from the superiority of the heterozygote for particular gene differences (overdominance) or from other causes such as the introduction of favorable dominant alleles at particular loci that were formerly homozygous for deleterious recessives.

[4]Not all equilibria are permanent or stable. We consider them unstable if any disturbance of equilibrium frequencies causes the frequency of one of the alleles to continue moving away from equilibrium. One such unstable equilibrium is possible when selection acts against the heterozygote at a gene locus with two alleles. If both homozygotes have equal adaptive value and the heterozygote is inferior, equilibrium will arise only when the frequency of each of the two alleles exactly equals .5. At this value the alleles are perfectly balanced, since equal amounts of each of the two are being removed in the heterozygote, that is, when genotype frequencies are .25 *AA*, .50 *Aa*, and .25 *aa*. However, any slight departure from these frequencies will cause the less frequent allele to have proportionally more of its genes in heterozygotes than the more frequent allele does. For example, if the gametic frequency of *A* rose accidentally to .6 and that of *a* fell to .4, then the genotypic frequencies under random mating are .36 *AA*, .48 *Aa*, and .16 *aa*, and the heterozygotes now contain a greater proportion of the *a* alleles than they do of the *A* alleles ($.24/.16 > .24/.36$). Thus, if the heterozygotes were lethal, the *A* gene frequency would become $.36/(.36 + .16) = .69$, and *a* would become $.16/(.36 + .16) = .31$. In the next generation, continued lethality of the heterozygotes would lead to an increase of the *A* frequency to .83, and the *a* frequency would fall to .17. Within a relatively short time, the *A* allele would go to fixation and the *a* allele to elimination.

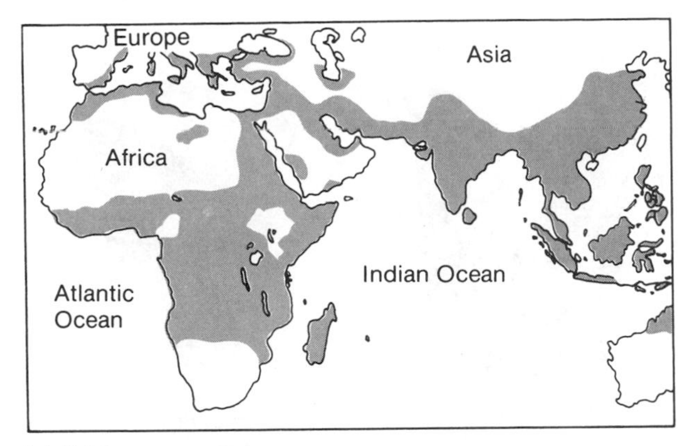

(a) Falciparum malaria

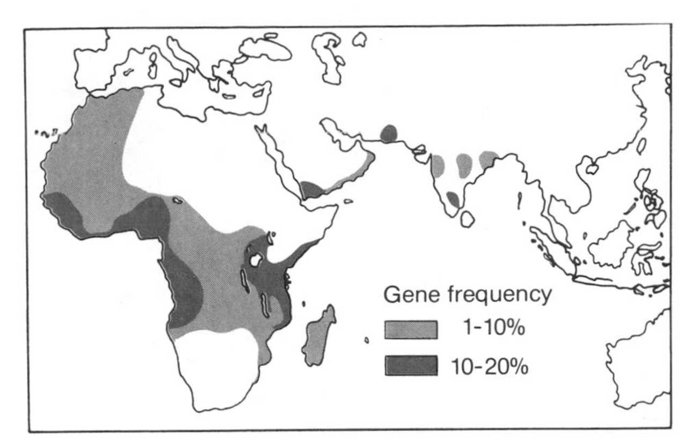

(b) Sickle cell anemia

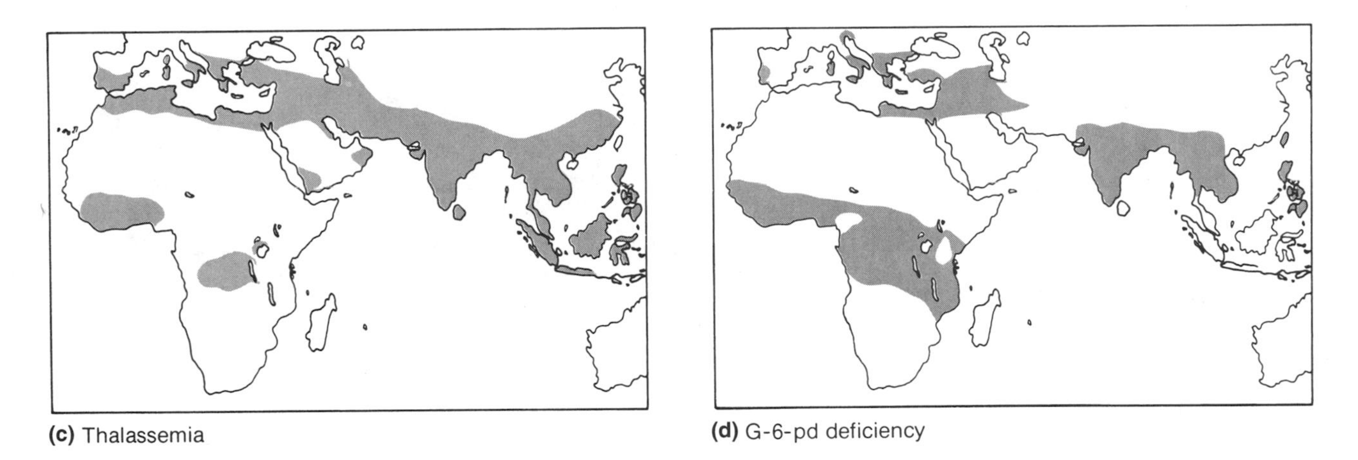

(c) Thalassemia

(d) G-6-pd deficiency

FIGURE 22–3 Relationship between the geographic distributions of malaria and genes that confer resistance against the disease. *(a)* Distribution of falciparum malaria in Eurasia and Africa before 1930. *(b)* Distribution of the gene for sickle cell anemia (Hb^S). *(c)* Distribution of the gene for β-thalassemia. *(d)* Distribution of the sex-linked gene for glucose-6 phosphate dehydrogenase deficiency in males in frequencies above 2 percent. *(From Strickberger, adapted from Allison.)*

describe the preservation of genetic variability through selection. In general, we consider a gene locus polymorphic if at least two alleles are present, with a frequency of at least 1 percent for the second most frequent allele. Although selection coefficients are difficult to measure in natural populations, such polymorphisms are certainly ubiquitous in practically all populations researchers have examined so far, both on the chromosomal level (Fig. 10–35) and on the genic level (Table 10–5). One prominent example of polymorphism that overdominance causes is the sickle cell gene in humans (p. 217), where heterozygotes (Hb^A/Hb^S) survive the malarial parasite more successfully than either normal (Hb^A/Hb^A) or sickle cell homozygotes (Hb^S/Hb^S). As Figure 22–3 shows, this gene (as well as others that appear to offer protection against malaria) persists in notable frequencies in geographical areas where malaria is endemic (see also Rotter and Diamond).

In laboratory populations, where researchers can more easily control and measure genetic variability, many experiments achieve balanced polymorphism, apparently by some sort of overdominance. In *Drosophila pseudoobscura*, for example, Dobzhansky and Pavlovsky have shown that the frequencies of the Standard (ST) and Chiricahua (CH) third-chromosome arrangements come to a stable equilibrium when flies carrying these arrangements are placed together in a population cage kept continuously for a year or longer (Fig. 22–4). The superiority of the heterozygote can be seen in the relative adaptive values calculated for the various third-chromosome combinations: ST/ST = 0.90, ST/CH = 1.00, CH/CH = .41.

In fact, we can calculate that even lethal recessive genes may remain in a population if they confer only a small heterozygous advantage. For example, gene *a*, lethal in *aa* homozygous condition but providing a 1 percent advantage to the *Aa* heterozygote compared to the *AA* homozygote, would reach a frequency of approximately 1 percent at equilibrium:

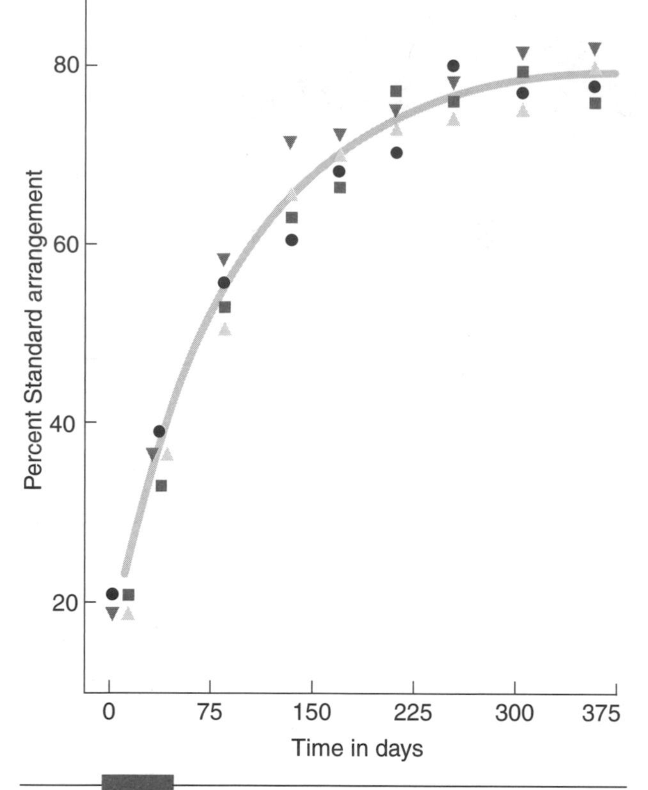

FIGURE 22-4 Results of four *Drosophila pseudoobscura* population cage experiments in which two third-chromosome arrangements are competing, Standard (ST) and Chiricahua (CH). Each population is denoted as a *circle, square,* or *triangle,* and was begun with 20 percent ST and 80 percent CH, reaching equilibrium values of 80 to 85 percent ST after approximately 1 year. The solid curve represents the frequencies of the ST arrangement expected according to the adaptive values ST/ST = .90, ST/CH = 1.00, CH/CH = .41. (*Adapted from Dobzhansky and Pavlovsky.*)

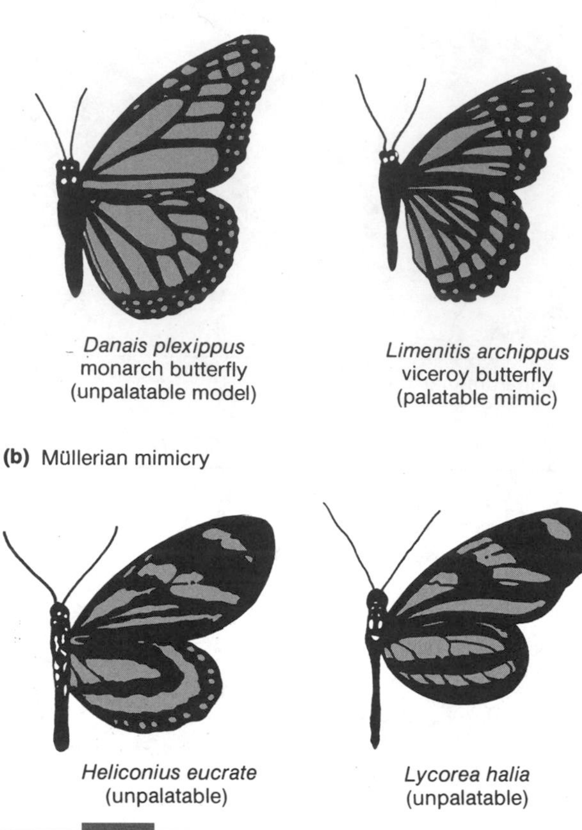

FIGURE 22-5 Mimicry in different species of butterflies. (*a*) Batesian mimicry by a North American species, in which the more palatable viceroy butterfly (*right*) mimics the more unpalatable monarch (*left*). Resemblance between two South American unpalatable species in (*b*) provides a common warning pattern to predators and helps protect both prey species (Müllerian mimicry). Whether it is Batesian or Müllerian, mimicry is one of the most obvious examples of convergent evolution.

	Genotypes		
	AA	Aa	aa
Adaptive value	.99	1.00	0
Selection coefficient	$s = .01$	0	$t = 1$

$$\hat{q}\text{ (frequency of } a) = \frac{s}{s+t} = \frac{.01}{1.01} = .0099$$

Other conditions responsible for polymorphism may include a change in selection coefficients so that genes detrimental at one time are advantageous at another. Also, selection against a gene may depend on its frequency and may be reversed when it is at low frequency, before it can be eliminated. An example of such **frequency-dependent selection** is **Batesian mimicry** in which palatable species that mimic distasteful models are protected against predators. In general, the more frequent the mimic and the less frequent its model, the greater the chances that predators will attack the mimic; conversely, the less frequent the mimic compared to the model, the greater the chances that the mimic will be protected.

As Figure 22–5 shows, mimicry also occurs when a palatable mimic imitates a conspicuous warning (aposematic) coloration or pattern shared by two or more different unpalatable species. Mimicry between different unpalatable species (**Müllerian mimicry**) benefits all such species by enabling predators to learn a single warning pattern that applies to all these potential but distasteful prey. Selection that favors such warning patterns is probably again a matter of frequency dependence. When very rare, conspicuous warning patterns on unpalatable individuals probably offer little protection since predators have few chances to learn their distastefulness. Distinctive patterns, however, offer greater protection to unpalatables when they are at higher densities, as Sword has shown in grasshopper experiments.

In plants, **self-sterility genes** that prevent fertilization between closely related individuals are also frequency-dependent. For example, a haploid pollen grain carrying a self-sterility allele, S^1, will not grow well on a diploid female style carrying the same allele, such as S^1S^2, but can successfully fertilize a plant carrying S^2S^3 or S^3S^4. Once an allele becomes common (for example, S^1), its frequency is reduced by the many sterile mating combinations to which it is now exposed. Rare alleles, in contrast, will successfully fertilize almost every female plant they meet, until they, too, become common. Thus, because of frequency dependence, self-sterility systems of considerable numbers of alleles can become established, reaching, for example, as high as 200 alleles or more in red clover.

Polymorphism may also become established when selection coefficients are not constant but vary from one environment to another. A population sufficiently widespread to occupy many environments may therefore maintain a variety of genotypes, each of which is superior in a particular habitat. A prominent example is the polymorphism associated with the phenomenon known as **industrial melanism.** Certain moths and butterflies show increased proportions of dark-colored, or melanic, forms, usually caused by the increased frequency of a dominant gene in industrial areas where air pollution darkens vegetation because of coal smoke deposits.

In the English industrial city of Birmingham, Kettlewell and others sought to explain the selective advantage of such melanic genes by releasing known numbers of both light and melanic forms of the British peppered moth, *Biston betularia*, and recapturing a significantly greater proportion of melanic forms. Their data suggested that sooty areas offer greater protection to melanic forms than to light-colored forms, since more of the former survived to be recaptured. The adaptive value of the melanic types may lie, at least partly, in their ability to remain concealed on darkened twigs or tree trunks from bird predators (Fig. 22–6). In nonindustrial areas, in contrast, trees covered with normal gray lichens offer decided advantages to the light-colored moths. Whether because of environmental camouflage, or other as yet unknown factors, English *B. betularia* populations show various degrees of polymorphism, ranging from high frequencies of the melanic gene in industrial areas to almost zero in many rural areas.[5]

FIGURE 22–6 Light-colored and dark-colored tree trunks, each with a melanic and nonmelanic *Biston betularia* moth. The light-colored trunks derive their appearance from lichens, a symbiotic association between fungi and algae in which fungi receive products of algal metabolism and algae are protected from desiccation by fungal tissue. Although trees are commonly shown as resting sites for these moths, their actual resting habits are unknown. Some authors suggest "caution" in crediting color camouflage as the full explanation for *Biston betularia* polymorphism (Sargent et al.). (*Science VU/Visuals Unlimited.*)

Interestingly, passage of clean air legislation in Britain in 1956 has reduced industrial smoke and sulfur dioxide in many formerly polluted areas. This reduction in pollution is now correlated with "reverse evolution": the frequency of melanic forms of *B. betularia* and other insects has declined dramatically (Brakefield).

Levins has pointed out that both the spatial and temporal organization of the environment may significantly affect the extent to which a population will rely on genetic polymorphism as an adaptive strategy. **Coarse-grained environments,** in which different individuals in a population endure different experiences, promote greater genetic polymorphism than **fine-grained environments,** in which all individuals experience the environmental differences. Hartl and Clark discuss other mechanisms that can maintain polymorphism.

[5]Based on dates of British amateur and museum collections, one can estimate that it took about 40 generations (one generation per year) during the nineteenth century for the frequency of nonmelanic phenotypes of *B. betularia* to decrease in some industrial areas from about 98 percent to about 5 or 6 percent. Using such data we can arrive at an approximate selection coefficient for industrial melanism in this moth by noting that the nonmelanic phenotypes are homozygotes (frequency q^2) for the recessive nonmelanic allele (frequency q), and therefore q was reduced during this 40-generation interval from $\sqrt{.98} = .99$ to $\sqrt{.06} = .25$. From Table 22–3, it would take about 44 generations (13 + 7 + 9 + 15) to reduce q from .99 to .25 when the selection coefficient is .20. In other words, the selection coefficient against the nonmelanic gene in some of these industrial areas was fairly intense, at about .20 or somewhat greater.

The Kinds of Selection

When selection has occurred for particular conditions over long periods of time, we can consider most populations to have achieved phenotypes that are optimally adapted to their surroundings. That is to say, many phenotypes will tend to cluster around some value at which fitness is highest. We can expect individuals that depart from these **optimum phenotypes** to show less fitness than those closer to the optimal values. In a classic 1899 study on sparrows that survived a storm, Bumpus showed that measurements taken on eight of nine different characteristics tended to cluster around intermediate phenotypic values, while sparrows killed by the storm showed much greater variability. In Bumpus's terms, "it is quite as dangerous to be conspicuously above a certain standard of organic excellence as it is to be conspicuously below the standard." Many studies on a variety of organisms, including snails, lizards, ducks, and chickens have since supported this view (Lerner).

In humans, measurements of birth weights of newborn babies, among other characteristics, show selection for optimum values. As you can see from Figure 22–7, most survivors cluster around a birth weight of 8 pounds, and those who depart from this value have fewer chances for survival. This reduction in frequency of extreme phenotypes has been called **stabilizing,** or **centripetal, selection,** because it signifies selection for an intermediate stable value (Fig. 22–8*a*). Since mutation continually introduces departures from optimum character values, this mode of selection acts genetically to inhibit or reduce variation.

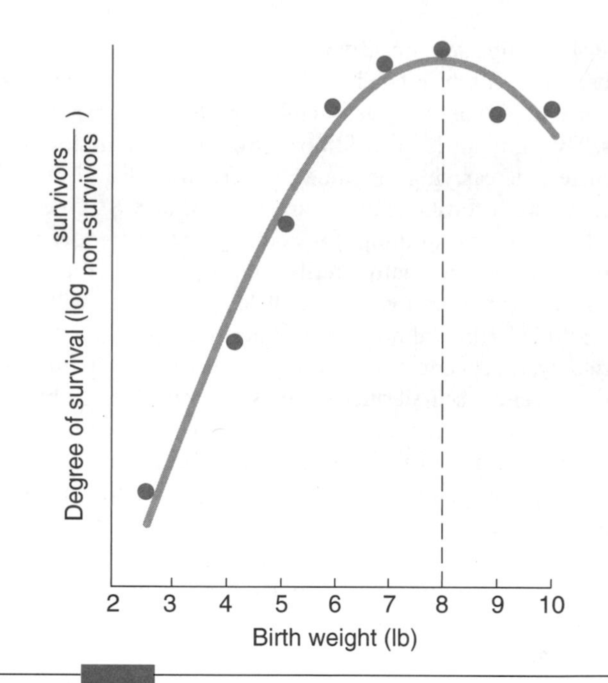

FIGURE 22–7 Relationship between birth weight and the degree of survival in female births in a London obstetric hospital. Of 6,693 births, there were 6,419 survivors one month later, or a mortality rate of 274/6,693 = 4.1 percent. Since mortality in the "optimum" 8-pound class was only 1.2 percent, this means that 4.1 − 1.2 = 2.9 percent of deaths occurred among the nonoptimal classes, indicating that selection against nonoptimal phenotypes causes a fairly high proportion (2.9/4.1 = 70.8 percent) of the deaths between birth and 1 month of age. (*Adapted from Karn and Penrose.*)

However, not all character selection is stabilizing, because selection may well favor an extreme phenotype by proceeding in one or the other direction of a phenotypic distribution (Fig. 22–8*b*). Animal and plant breeders, who select for extremes of yield, productivity, resistance to disease, and so forth, commonly practice such **directional selection** (Fig. 22–9). Its role in evolution is especially important when the environment of a population is changing and only extreme phenotypes happen to be adapted for new conditions.

Selection, whether stabilizing or directional, may act in a constant fashion if the selective environment is uniform. However, when conditions are changeable, a population may be subjected to divergent or cyclically changing (oscillating) environments to which different genotypes among its members are most suited (Gibbs and Grant). Such selection is **disruptive, diversifying,** or **centrifugal,** because it establishes different optima within a population (Fig. 22–8*c*).

Because environmental conditions can be quite changeable, these different types of selection do not remain separate, but may combine in different ways. For example, disruptive selection may be followed by directional selection, which may then yield to stabilizing selection. The genetic means through which these forms of selection are expressed may also vary, from genes with large effect (p. 516) to polygenes with smaller effect, some causing simple developmental changes and others more complex canalizing processes (Fig. 15–10).

To ultimately affect evolution, selection (however it occurs) must change the frequencies of genes or genotypes involved in fitness. This means that genetic variability must be present, and **pure lines** that are homozygously uniform for such fitness genes offer no opportunity for selection to produce any noticeable evolutionary change. Fisher formulated this principle mathematically as a fundamental theorem that essentially states, "*The greater the genetic variability upon which selection for fitness may act, the greater the expected improvement in fitness.*"

One consequence of Fisher's theorem is that we would expect populations long subjected to selection—and this includes all populations—to have little remaining variability for genes affecting fitness, because selection would have diminished such variability. The continued existence of selection therefore implies that variability itself is favorably selected: continuous changes in the environment af-

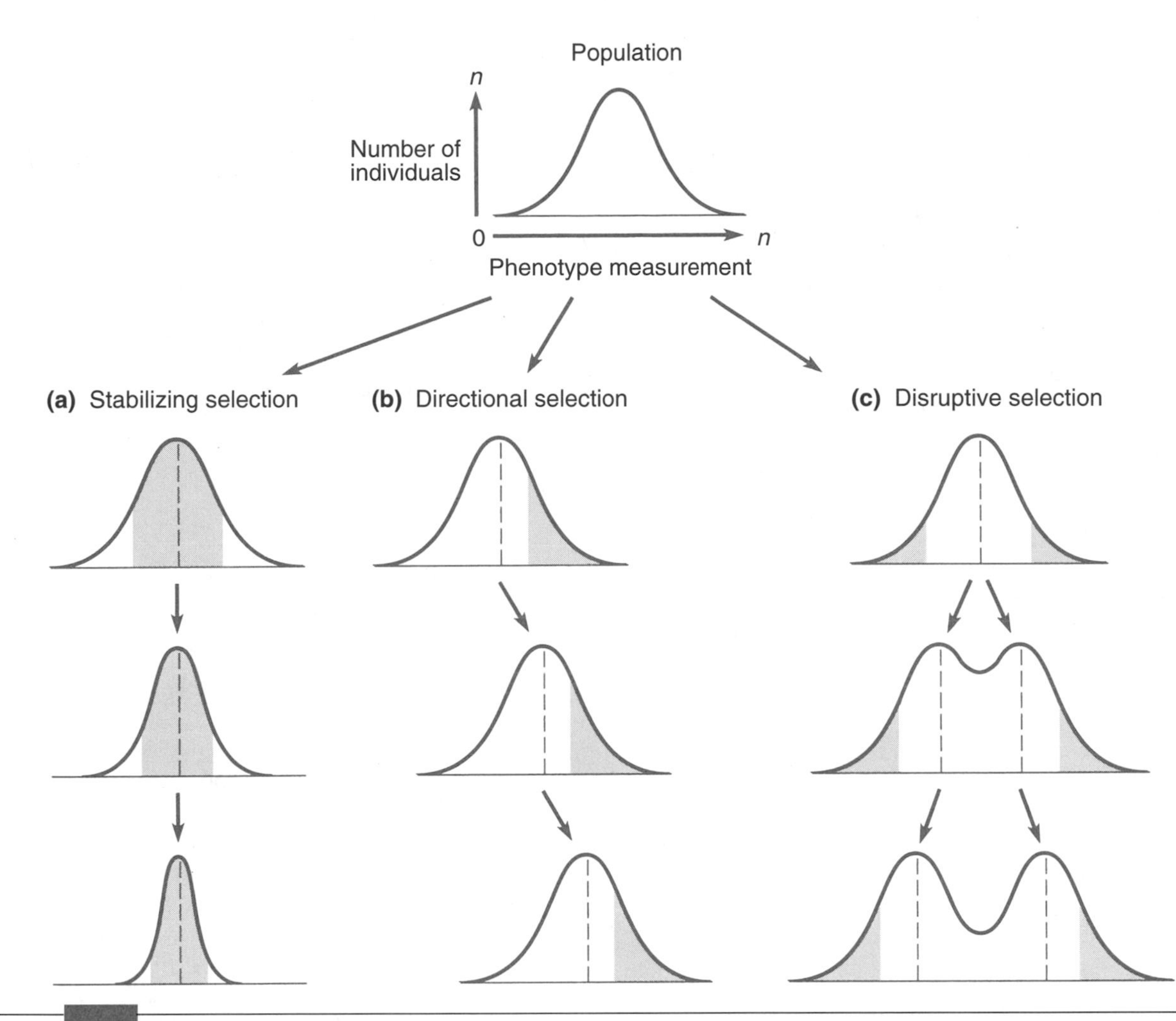

FIGURE 22-8 Three basic modes of selection and their effects on the mean *(dashed lines)* and variation of a normally distributed quantitative character. The horizontal axis of each bell-shaped curve represents measurements of a quantitative character (for example, from low on the left end to high on the right end), and the vertical axis represents the number of individuals found at each measurement. Shaded areas represent the individuals selected as parents of the next generation.

fect formerly unselected genes whose presence now offers new opportunities for improving fitness. Such environmental changes include changes in resources, supplies, waste products, and predator and parasite populations.

We do know that variability for fitness genes can persist despite continued selection, such as when allelic differences are retained through devices discussed in the previous section (for example, frequency dependence). In general, however, populations tend to change genetically in directions that improve fitness for their environment, and Endler lists more than 160 cases in natural populations where selection has been demonstrated. Such findings indicate that genetic variability for fitness must have provided the baseline on which selection acts. Given persistent and recurrent genetic variability in factors such as differential mortality, differential fecundity, and differential mating success, gene frequency change caused by selection must be a constant feature of most or all populations. This seems especially so since populations cannot long remain immune to the repeated onslaught of environmental changes that affect these components of fitness.

Interaction with the environment and other species is continuous. Van Valen has proposed that species generally compete with each other for resources, so that an advantage, or improvement in fitness, for one species represents a deterioration in the environment of others. He points out that species survival is very much in accord with the remark made by the Red Queen whom Alice meets in Lewis Carroll's *Through the Looking Glass*: "Here, you see, it takes all the running you can do to keep in the same place." According to this **Red Queen hypothesis,** each species continually faces new selective challenges because of environmental changes often associated with variations in the fitness of its interacting populations. Species must constantly confront and overcome recurring threats to fitness in order to survive. Nature perpetuates a cyclical process where adaptations in any one organism

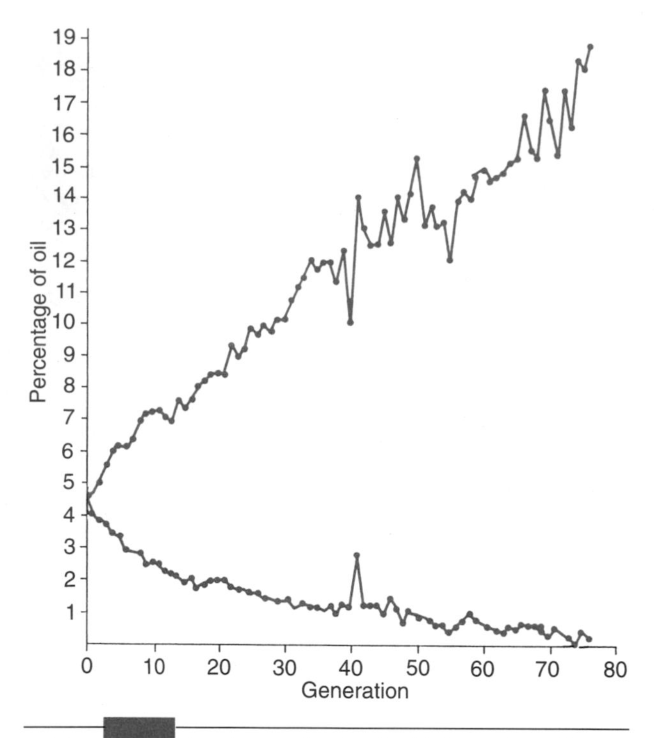

FIGURE 22-9 Results of selection for high and low oil content in corn kernels in an experiment begun in 1896 at the University of Illinois and continuing to the present. Selection for high oil content still continues to yield increases, whereas the effect of selection for low oil content has tapered off on reaching the 0 percent lower limit. (*Adapted from Dudley.*)

continually elicit selection for adaptations in others: sooner or later, species face an "arms race" with a changing biological environment. Or as Darwin stated in *On the Origin of Species,* "If some of these many species become modified and improved, others will have to be improved in a corresponding degree or they will be exterminated."

The long-term consequence of the Red Queen's reign is to increase the competitive fitness of each interacting population, a view that has gained support from experiments with RNA viruses (Clarke et al.). The adaptations resulting from this process will, of course, vary between organisms, but researchers have suggested that we can discern a pattern of increasing complexity over time. These changes include a steady increase in genome size from prokaryotes to eukaryotes, from about 10^6 to 10^9 or more nucleotide base pairs (Fig. 12–9); and among vertebrates, a marked increase in relative brain mass from fish to reptiles to mammals to humans (Fig. 3–5).

According to Valentine and coworkers, an additional manifestation of such pattern is the increasing morphological complexity among metazoans as measured by their estimated number of different somatic cell types. As Figure 22–10 shows, this number has increased at an average rate of about one cell type per 3 million years, starting with the Precambrian–Cambrian period, with no evidence of any downward trend. Note, however, that the

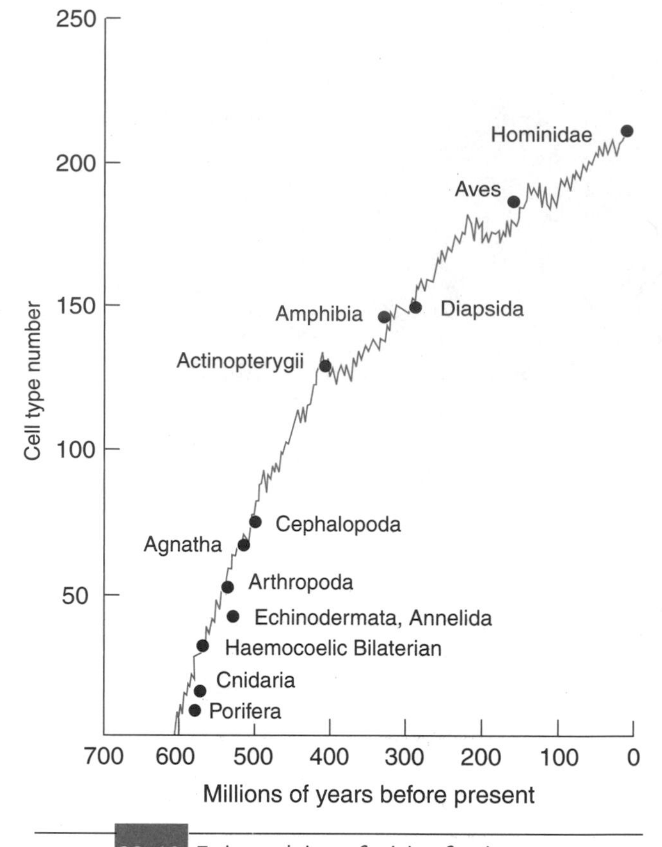

FIGURE 22-10 Estimated time of origin of various metazoans with their estimated somatic cell-type numbers. The marked increase from low to high is considered by many authors to reflect an increase in complexity. (*Valentine, J. W., A. G. Collins, and C. P. Meyer, 1994. Morphological complexity increases in metazoans.* Paleobiology, *20, 131–142. Reprinted by permission.*)

increased cell-type numbers that provide new adaptational opportunities do not necessarily imply replacement of all organisms that have fewer cell types (see also pp. 300 and 450). In a sense the race for survival goes on at many levels—at lower as well as higher morphological complexities—with survivors at each level, yet with seeming pressure to generate new levels of interaction.[6]

Equilibrium Between Mutation and Selection

For convenience we have considered changes in gene frequency to be caused by either mutation or selection acting separately. In nature, however, mutation and selection are simultaneous

[6]Some authors would consider increased morphological complexity as "progress" (p. 451), yet in various lineages such change is only one evolutionary trend among many others. Aside from obvious "evolutionary opportunism," no singular evolutionary direction applies to every lineage. As McShea points out, "Something may be increasing [in evolution]. But is it *complexity?*"

processes, and both factors influence gene frequency values. Predictions on the basis of one factor alone may be misleading.

For example, even though a recessive gene is detrimental in homozygous condition, it may nevertheless persist in a population because of its mutation frequency. That is, a population reaches a certain equilibrium point at which the number of genes being removed by loss of homozygotes through selection is replaced by the same number of genes introduced into heterozygotes through mutation. We may determine this **mutation–selection equilibrium** frequency by the following argument.

We have seen that the change in gene frequency per generation for a deleterious recessive *a* with frequency *q* is equal to a loss of $sq^2(1 - q)/(1 - sq^2)$. If *s* is small, we can consider the denominator 1, and the loss in frequency is then $sq^2(1 - q)$. The frequency of newly mutated *a* genes, however, is equal to the mutation rate (u) of $A \rightarrow a$ multiplied by the *A* frequency, which is $1 - q$. Thus the loss of *a* genes through selection is exactly balanced by the gain of newly mutated *a* genes when

$$sq^2(1 - q) = u(1 - q)$$

$$sq^2 = u$$

$$q^2 = \frac{u}{s}$$

$$q = \sqrt{\frac{u}{s}}$$

The equilibrium frequency of a mutant gene in a population is thus a function of both the mutation frequency and the selection coefficient. As you can see in the hydraulic model of this relationship in Figure 22–11, when the mutation rate increases the equilibrium gene frequency also increases, but the equilibrium frequency decreases when the selection coefficient increases.

For a deleterious dominant allele, similar algebraic manipulations point to an equilibrium frequency of about u/s, a value almost identical to the equilibrium frequency of genes that lack dominance. Since such dominant or partially dominant genes are of considerable disadvantage to heterozygotes, Fisher proposed that their deleterious effect probably diminishes in most organisms by selection of **modifier genes** at other loci that change the degree of dominance. For example, mutant alleles at a particular locus *A* (for example, $A^1, A^2, A^3 \ldots$) may act as partial dominants in the presence of the wild-type allele A^+. Since these mutant alleles are mostly deleterious, modifier genes at other loci (for example, $B^1, B^2 \ldots$, or $C^1, C^2 \ldots$, etc.) that increase the dominance of A^+ will be selected until the effects of mutations at the *A* locus are relatively recessive.

The successful selection for dominant and recessive modifiers that Ford demonstrated in the currant moth *Abraxas grossulariata* provides evidence for Fisher's view.

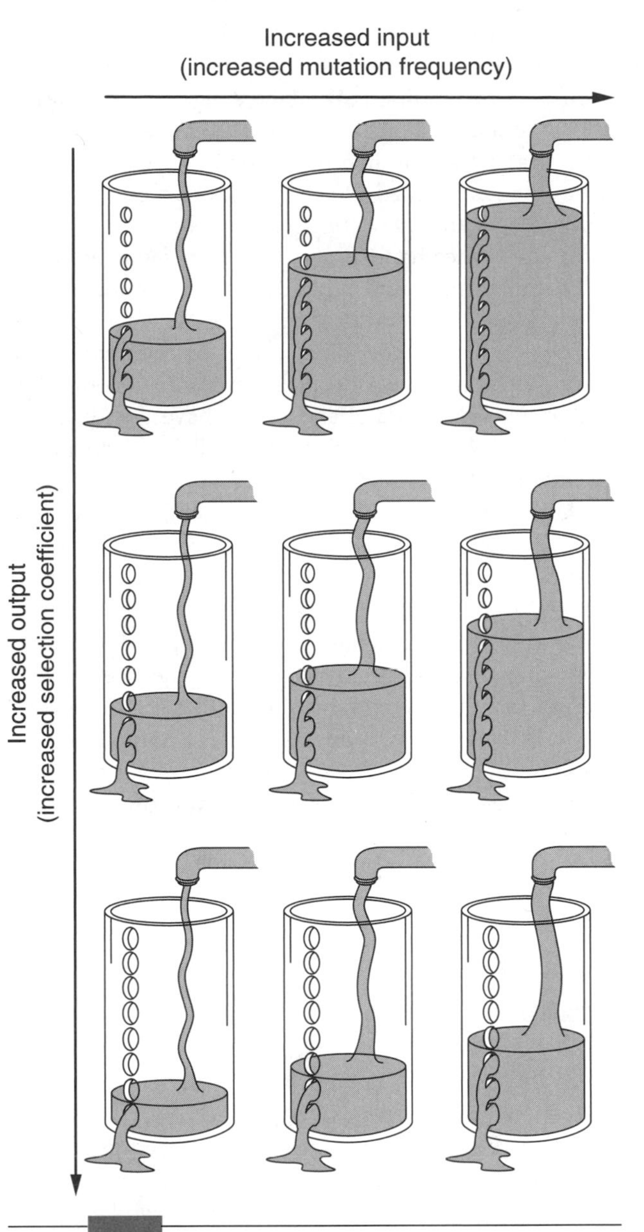

FIGURE 22–11 Hydraulic model of mutation–selection equilibrium. Each container is analogous to a population in which the water level represents the equilibrium frequency of a gene. As the water input (mutation frequency) increases, the standing water level (equilibrium gene frequency) increases. When the overflow holes are small (small selection coefficient) the water levels are higher for the same input (mutation frequency) than when the overflow holes are larger (large selection coefficient). (*Adapted from Stern with additions.*)

In this moth a single gene, *lutea*, in homozygous condition, produces yellow instead of the normal white ground color but has an intermediate effect as a heterozygote. After four generations of selecting moths for greater and lesser expression of the *lutea* phenotype in the heterozygote, Ford obtained two distinct strains: in one case *lutea* acted almost as a complete dominant and in the other case almost as a complete recessive. In each strain special modifiers had been chosen, some enhancing and some detracting from the dominance of this particular gene.

Instead of modifiers, Haldane suggested that special wild-type alleles are selected (for example, A^{X+}, A^{Y+}, A^{Z+}) that act as dominants in the presence of a mutant allele (for example, A^1, A^2, A^3, . . .). On the other hand, Orr presents the view that selection has little to with dominance since haploid algae (*Chlamydomonas*) artificially transformed into diploids display dominance despite the lack of modifiers or alleles selected for this purpose. Whatever the initial cause for dominance (see also p. 197), modifiers or alleles may affect dominance once it appears. Harland and others have demonstrated that both dominance modifiers and alleles occur in two species of cotton, *Gossypium barbadense* and *G. hirsutum.* In these plants certain alleles show simple dominance when variants of the same species cross. Interspecific crosses, in contrast, show the effect of many modifying genes on these traits, as well as differences in the degree of dominance of particular alleles. Despite these dominance-producing mechanisms, many harmful genes are probably still not completely recessive and seem to have some effect in heterozygous condition. Thus, in natural populations the equilibrium frequencies of harmful genes are probably higher than for dominants but lower than for pure recessives.

Migration

Mutation is not the only mechanism by which new genes enter a population. A population may receive alleles by **migration** (also called **gene flow**) from a nearby population that maintains an entirely different gene frequency. When this occurs, two factors are important to the recipient population: (1) the difference in frequencies between the two populations and (2) the proportion of migrant genes that are incorporated each generation. If we designate q_0 as the initial gene frequency in the recipient, or hybrid, population, Q as the frequency of the same allele in the migrant population, and m as the proportion of newly introduced genes each generation, then the gene frequency in the hybrid population will suffer a loss of q_0 equal to mq_0 and a gain of Q equal to mQ. Over n generations of migration, when the gene frequency of the hybrid population becomes q_n, one can calculate that the relationship between these factors will reach

$$q_n - Q = (1 - m)^n(q_0 - Q)$$

$$\text{or } (1 - m)^n = \frac{q_n - Q}{q_0 - Q}$$

For populations where this equation can be applied, we must know four of these factors to calculate the fifth. One such example can be found in human populations where blood group gene frequencies are known for both American blacks and American whites, two populations between which gene exchange has occurred. In general, although some black genes undoubtedly enter the white population, the white population is so large that this introduction probably makes little difference in white gene frequencies. In contrast, the black population is much smaller and has remained isolated from its African origin for two or more centuries. (Twelve percent of the present U.S. population is black, according to the 1990 census.) On this basis the white population can be considered as the gene donor or migrant population (Q) and the present black population as the hybrid (q_n).

To obtain the original gene frequency of one of the Rh blood group alleles, R^0, in the black population (q_0), researchers used data of present East African blacks on the assumption that these data may reflect the original gene frequencies of 200 to 300 years ago. Among the East Africans, R^0 showed a frequency of .630, indicating that the frequency of this gene had fallen in American blacks to its present frequency of .446. This fall in frequency could be ascribed to interbreeding with the American white population, where the frequency of R^0 is about .028, much lower than among blacks. According to Glass and Li, this reduction had begun at the time of the initial introduction of blacks into the American colonies 300 years ago and probably continued throughout the 10 generations since. Substituting these values into the preceding formula, we obtain

$$(1 - m)^{10} = \frac{q_{10} - Q}{q_0 - Q} = \frac{.446 - .028}{.630 - .028} = .694$$

$$1 - m = \sqrt[10]{.694} = .964$$

$$m = .036$$

This value of m means that, excluding all other causes such as mutation, 36 genes per 1,000, or 3.6 percent of genes in the black population, entered from the white population each generation. Since $1 - m$ represents the proportion of nonintroduced genes, $(1 - m)^{10} = .694$ is the proportion of genes that have remained of African origin over the 10-generation period. Supported by somewhat similar estimates in more recent studies, blood group gene frequencies generally indicate that the American black population is genetically about 70 to 80 percent African and 20 to 30 percent white, with some differences between Southern and Northern blacks (Adams and Ward).

Where we lack exact information on gene frequency exchanges between populations, and this includes most populations, considerable discussion and dispute have flourished about the importance of migration. According to Mayr and to Stanley, migration can hinder local evolutionary changes by infusing genes from populations that are not adapted to local conditions. For example, some populations of mammals who live on dark, formerly volcanic lava flows have dark fur when they are isolated from neighboring populations who live on lighter colored backgrounds, but do not have dark fur when they receive immigrants from the lighter-colored surroundings. In contrast, Ehrlich and coworkers describe populations of the butterfly *Euphydras editha* which show no phenotypic changes whether or not they are subject to migration from phenotypically different populations. In the absence of genetic information, these issues are difficult to resolve (Slatkin).

Random Genetic Drift

The three forces we have considered up to now—mutation, selection, and migration—share one important quality; they usually act directionally to change gene frequencies progressively from one value to another. Unopposed, these forces can fix one allele and eliminate all others; when balanced, they can lead to equilibrium between two or more alleles. However, in addition to these directional forces, there are also changes that have no predictable constancy from generation to generation. **Random genetic drift,** one of the most important of such **nondirectional forces,** arises from variable sampling of the gene pool each generation.

This is apparent if we consider that, in the absence of directional forces to change gene frequencies, there is always a strong likelihood of obtaining a good sample of the genes of the previous generation as long as the number of parents in a population is consistently large. However, since real populations are limited in size, genetic drift will cause gene frequency changes because of **sampling errors.** For example, if only a few parents are chosen to begin a new generation, such a small sample of genes may deviate widely from the gene frequency of the previous generation.

The extent of the deviation for all sizes of populations can be measured mathematically by the standard deviation of a proportion $\sigma = \sqrt{pq/N}$, where p is the frequency of one allele, q of the other, and N the number of genes sampled. For diploid parents, each carrying two alleles, $\sigma = \sqrt{pq/2N}$, where N is the number of actual parents. For example, if we begin with a large diploid population, where $p = q = .5$, and continue this population each generation by using 5,000 parents, then $\sigma = \sqrt{(.5)(.5)/10{,}000} = \sqrt{.000025} = .005$. The values of such populations will fluctuate mostly around $.5 \pm .005$, or between .495 and .505. A choice of only two parents as founders will produce a standard deviation of $\sqrt{(.5)(.5)/4} = \sqrt{.0625} = .25$, or values of $.50 \pm .25$ (from .25 to .75).

In other words, sampling accidents because of smaller population size can easily yield gene frequencies that depart considerably from the initial .5 values in a single generation. If the population remained small and the next generation began with either of these extremes—that is, a gene frequency of .25 or .75 for a particular allele—in the following generation the frequency of that allele may fall to almost zero ($.25 \pm \sqrt{(.25)(.75)/4} = .25 \pm .22$: a range of .03 to .47) or increase almost to 1 ($.75 \pm \sqrt{(.75)(.25)/4} = .75 \pm .22$: a range of .53 to .97). Should such small populations continue each generation, the likelihood increases that one or more will eventually reach fixation for one of the alleles. The proportion of such populations that attain fixation—that is, the **rate of fixation**—will eventually reach $1/2N$. Obviously, if N is large, fixation proceeds slowly, but even large populations can show some degree of drift, as diagrammed in Figure 22–12.

This reliance of drift on population number emphasizes the importance of what is called **effective population size** (N_e). It differs from the observed population size because not all members of a population are necessarily parents and because parentage can also be limited by a reduced number of one of the sexes. For example, if out of a total population of 1,000, 3 males mated to 300 females produced the next generation, the effective population size is more than 6 but still less than 303. Wright

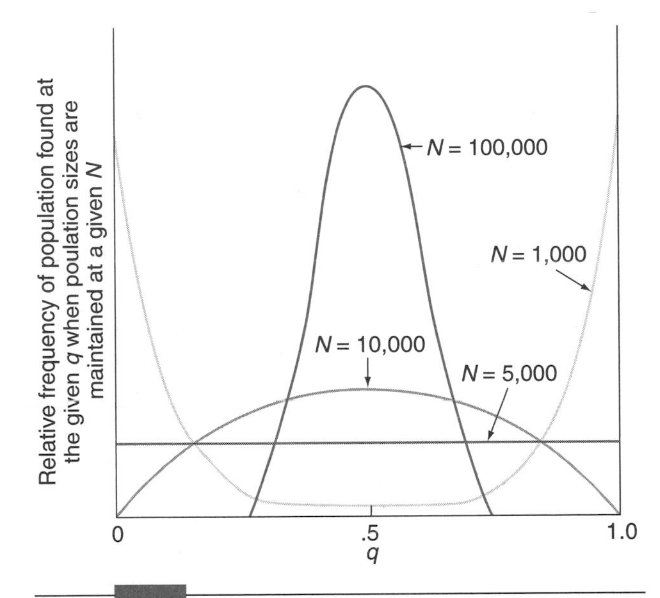

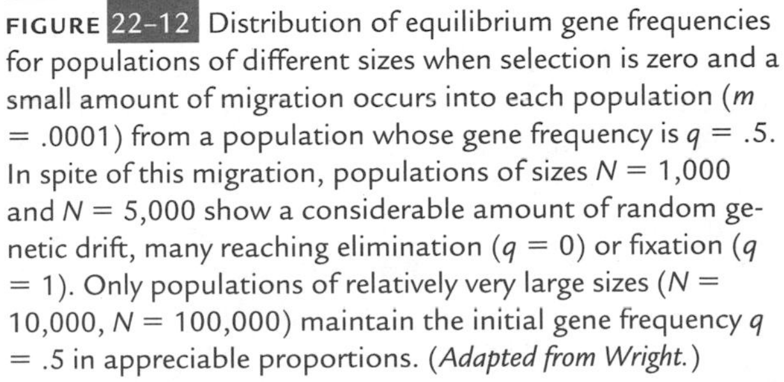
FIGURE 22–12 Distribution of equilibrium gene frequencies for populations of different sizes when selection is zero and a small amount of migration occurs into each population ($m = .0001$) from a population whose gene frequency is $q = .5$. In spite of this migration, populations of sizes $N = 1{,}000$ and $N = 5{,}000$ show a considerable amount of random genetic drift, many reaching elimination ($q = 0$) or fixation ($q = 1$). Only populations of relatively very large sizes ($N = 10{,}000$, $N = 100{,}000$) maintain the initial gene frequency $q = .5$ in appreciable proportions. (*Adapted from Wright.*)

has expressed the relationship as $N_e = 4N_fN_m/(N_f + N_m)$, where N_f is the number of parental females and N_m the number of parental males. In the preceding case N_e would be 4(300)(3)/303 = 11. Inequalities in numbers of offspring among different parents will also reduce the effective population size.

Wright has therefore proposed that genetic drift may be quite important in changing gene frequencies among populations when their effective sizes are small. Among the observations illustrating this concept is that of Buri, who set up 107 separate lines of *D. melanogaster,* each line carrying two alleles at the *brown* locus (*bw* and bw^{75}) at initially equal frequencies of 50 percent. Buri then continued the lines for 19 generations by randomly selecting 8 males and 8 females as parents from each preceding generation (N = 16 = 32 *brown* alleles) and scoring the frequency of the two different *brown* alleles.

As Figure 22–13 shows, by the first generation of the experiment a number of Buri's populations already showed departures from the original 50 percent bw^{75} frequency, and genetic drift continued to increase successively so that by generation 19 more than half the 107 populations reached fixation for either the *bw* or bw^{75} alleles. Despite these genetic differences, note that the average frequencies of *brown* alleles when combining all populations remain at about .5. Genetic drift therefore increases variation between populations, but on the average, not in any particular direction. These results also indicate that, because of genetic drift, selection in very small populations, unless intense, may have little or no effect on a deleterious gene frequency such as bw^{75}.

Although the persistence of small population size over many generations is a cause of genetic drift, occasional size reductions for only one or a few generations may also have pronounced effects on gene frequencies and future evolution. At the extreme of such reductions, which Mayr called the **founder principle,** a population may occasionally send forth only a few founders to begin a new population. Whatever genes or chromosome arrangements these founders take with them, detrimental or beneficial, all stand a chance of becoming established in the new population because of this sudden sampling accident (Fig. 22–14).

Thus Carson, by carefully analyzing salivary chromosome banding patterns, has shown that the more than 100 native Hawaiian "picture-winged" *Drosophila* species can derive from founder events in which each island was settled by relatively few individuals whose descendants evolved into different species. For example, the 41 species unique to the Maui island complex (Fig. 22–15) derive from only 12 founders—10 from Oahu and 2 from Kauai—with each single founder providing unique chromosome arrangements that we can trace in the descendant species.

There are by now numerous examples in many organisms, including humans, of unique gene frequencies that

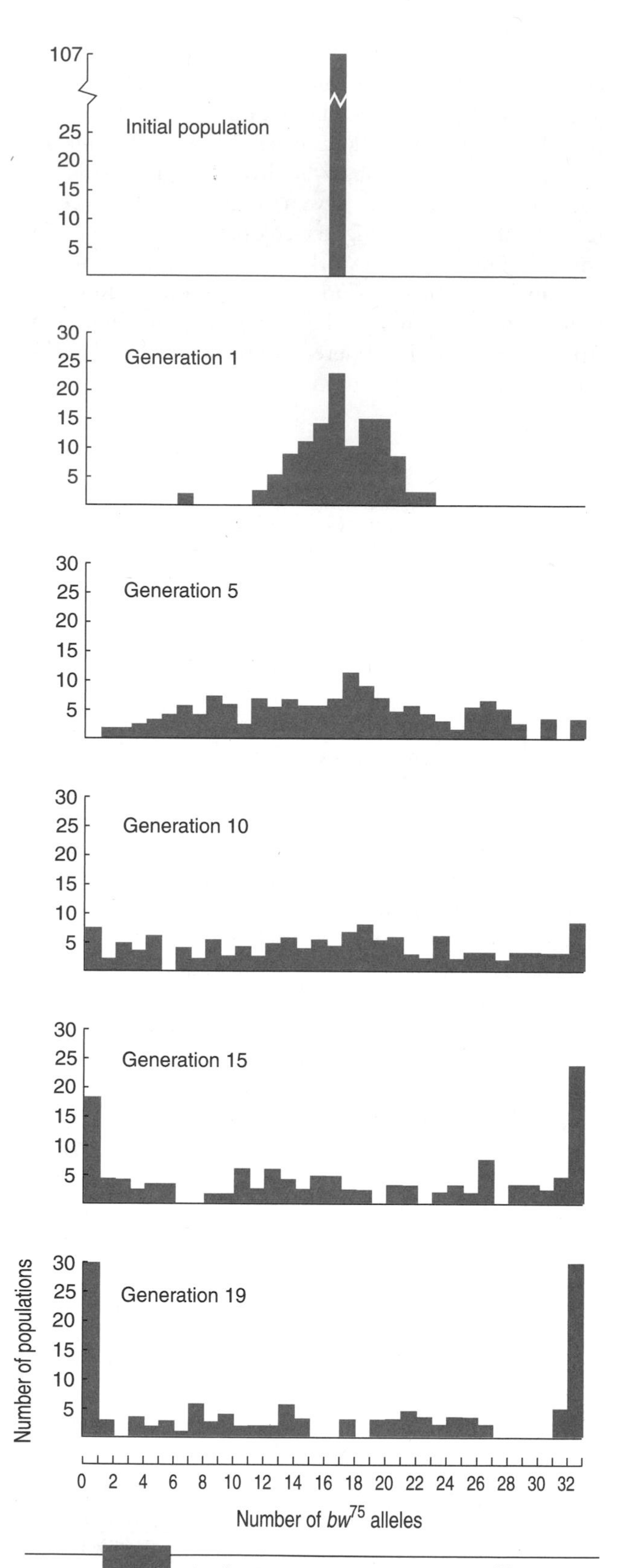

FIGURE 22–13 Distributions of the numbers of bw^{75} alleles in 107 lines of *D. melanogaster,* each with an initial frequency of .5 bw^{75}. Buri continued the lines for 19 generations, using 16 parents to start each generation (32 alleles at the *brown* locus), and the number of bw^{75} alleles found are given for the various lines. Note that by generation 19, the bw^{75} allele had been eliminated from 30 of these lines (0 alleles) and had been fixed in 28 of these lines (32 alleles). (*Data from Buri's series I cultures.*)

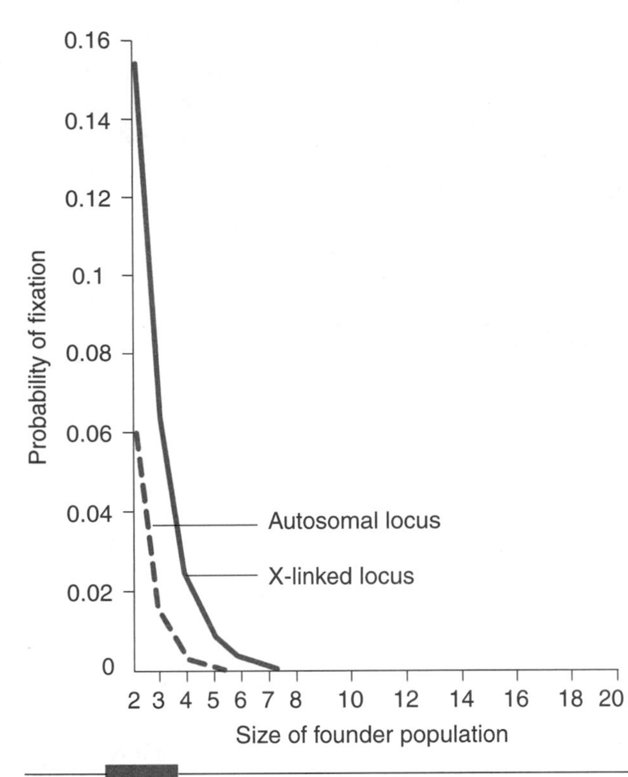

FIGURE 22-14 Theoretical probability of fixation (frequency = 1) of an allele that begins with a frequency of 0.5 in different founder population sizes. The dashed line is for an autosomal locus. The solid line is for an X-linked locus in a founder population consisting of twice as many males as females. In both cases, the smaller the founder population, the greater the chances that the allele will be fixed, but, even so, note that the chances for fixation are not especially great. As pointed out on p. 595, footnote 6, less common alleles may have a greater "founder effect" on developmental processes, but their low frequency further reduces their chances for incorporation into a founder population. (*Adapted from Templeton.*)

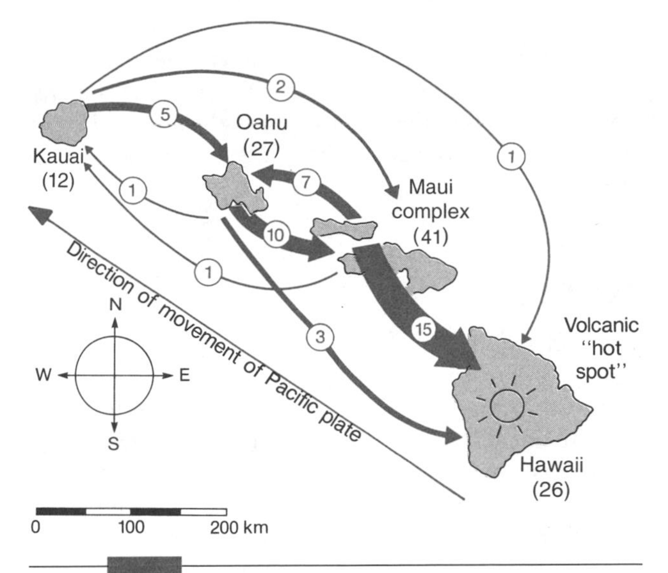

FIGURE 22-15 Colonization pattern showing the founder events that Carson proposed to explain the origin of native "picture-winged" *Drosophila* species found on the Hawaiian islands. The width of each *arrow* is proportional to the number of founders (*circled*), and the number of *Drosophila* species now present on each island is given in *parentheses.* Each successful founder is presumed to have been a fertilized female, usually from a geologically older island. The oldest Hawaiian island is Kauai (about 5.6 million years old), and 10 of the 12 species it has are believed to be the most ancient *Drosophila* elements in the islands. The youngest island, Hawaii, has been colonized entirely by founders from the older islands. Altogether the Drosophilidae family has more than 800 native Hawaiian species with an ancestry that probably dates back to founding episodes on islands even older than Kauai, some 30 or so million years ago (Beverly and Wilson). Note that tectonic events formed these islands: a localized "hot spot" in the Earth's mantle, lying under the large island of Hawaii, pierces the lithosphere and produces volcanic eruptions that form a succession of islands as the Pacific plate moves northwestward. In time various older islands erode; first becoming atolls, then seamounts (submerged volcanoes), which form a series that extends from Hawaii to Midway to a point near the far-western Aleutians. (*Adapted from Carson.*)

seem best explained by such founder events, or **bottleneck effects.** It therefore seems likely that at least some populations began with only a few "Adams" and "Eves" carrying genotypes that may have differed greatly in frequency from their parental populations. Certainly the relatively high incidences of some genes that Table 21–8 lists, such as achromatopsia among the Pingelapese and Ellis-van Creveld syndrome (polydactylous dwarfism) among the Lancaster County Amish, are difficult to explain except as founding accidents, since they seem to confer no advantage on either their homozygous or heterozygous carriers. The same conclusion is true for chromosomal translocations, which are usually selected against because they can cause sterility in heterozygotes (pp. 209–210), yet are nevertheless common features in the evolution of many mammalian lines.

Bottleneck effects may counter the effects of previous selection for a short period of time—an interval during which previously favorable mutations may be lost and deleterious mutations may be fixed. However, it is difficult to imagine that any genetic trait that affects the armament with which organisms face their environment can long continue to escape selective environmental pressures. Through nonselective genetic changes, a bottleneck may cast the evolution of a population in a new direction, but, from all we understand of evolution, this direction could not remain nonadaptive without extinction.

In general, the consequences of founding events on gene frequencies are largely unpredictable compared to variability estimates (that is, σ) that can be arrived at when population size is constant. Other factors that make the variability of gene frequencies unpredictable are unique historical events such as a change in the direction or intensity of selection because of a radical change in environment, an unusually favorable mutation, a rare hybridization event with another variety, or an unusual swamping of a population by mass immigration. The effects of these factors on evolutionary changes may be quite important, although genetic data for such events are still difficult to obtain.

SUMMARY

To effect evolutionary change new alleles that appear because of mutation must persist in populations. An allele *a* increases in frequency if the mutation rate (*A* to *a*) exceeds the frequency of the reverse mutation (*a* to *A*) until the alleles reach equilibrium. Since mutational equilibrium is rarely attained, other mechanisms must influence the frequency of alleles that mutation introduces.

One such factor is selection, because if certain alleles improve the reproductive success of the carrier (that is, fitness), they tend to become more frequent. Selection acts on the fitness differences among different alleles affecting a given trait, so gene frequencies tend to change from generation to generation.

Alleles whose effects are deleterious will decline in frequency, the rate of decline depending on the allelic frequency and on the gene's recessiveness or dominance. For example, deleterious recessive alleles decline rapidly from high frequencies but only very slowly from low frequencies. In any case, selection may eventually remove the deleterious allele from the population, and persistence of the allele (equilibrium) depends on factors such as new mutations.

Selection can preserve genetic variability (polymorphism) if the heterozygous genotype is more fit than either homozygote, as in sickle cell anemia, or when an allele's frequency affects its fitness. Also, allele fitness may vary in different environments, as in the case of the British peppered moth, where lighter forms are favored in unpolluted areas and melanic forms in industrial regions.

Selection can operate on a population in several ways: (1) if phenotypes far from the norm are less fit (stabilizing selection); (2) if an extreme phenotype has adaptive value (directional selection); and (3) if different environmental circumstances favor different phenotypes (disruptive selection). For selection to occur, genetic variability must be present: the greater the genetic variability, the greater the chance for improving fitness.

Since both mutation and selection act on gene frequencies, mutation may maintain even deleterious genes in a population, although selection will remove many individuals bearing them. In addition to mutation and selection, gene frequencies may also change because genes can migrate from one population into another, as in the case of blood group genes in the American black population.

Mutation, selection, and migration change gene frequencies directionally and can lead to (1) fixation of certain genes and loss of others or to (2) equilibrium. However, nondirectional forces such as genetic drift may also significantly change gene frequencies. This is especially true if a new population begins with a small sample ("founders") in which the gene frequency varies from that of the original population. Drift and other random factors may be important in establishing unpredictable populational variability.

KEY TERMS

adaptive value
balanced polymorphism
Batesian mimicry
bottleneck effects
centrifugal selection
centripetal selection
coarse-grained environments
deleterious dominant
deleterious lethal
deleterious recessive
directional selection
disruptive selection
diversifying selection
effective population size
elimination
equilibrium gene frequencies
fine-grained environments
fitness
fixation
founder principle
frequency-dependent selection
gene flow
heterosis
heterozygous advantage
hybrid vigor
industrial melanism
migration
modifier genes
Müllerian mimicry
mutation rates
mutational equilibrium
mutation–selection equilibrium
nondirectional forces
optimum phenotypes
overdominance
polymorphism
pure lines
random genetic drift
rate of fixation
Red Queen hypothesis
sampling errors
selection
selection coefficient
selective value
self-sterility genes
stabilizing selection

DISCUSSION QUESTIONS

1. Why does the origin of an allele by a single mutational event rarely lead to its persistence in a population?

2. How can we calculate mutational equilibrium, and why does a population rarely attain it?

3. Fitness
   a. How would you define fitness? Would you equate fitness with survival? Why or why not?
   b. How would you measure fitness?

4. Selection
   a. How are selection and fitness related?
   b. Why is selection generally more effective against an allele in haploids than in diploids? Would you say that this effectiveness confers an advantage on diploidy? Why or why not?

c. Why is selection generally less effective in diploids against rare deleterious recessive alleles than against common deleterious recessive alleles?
d. Does the action of artificial selection exclude natural selection? Explain.

5. Heterozygote superiority
   a. Why does heterozygote superiority (overdominance) lead to gene frequency equilibria?
   b. How can we calculate such equilibria?
   c. Can gene frequency reach equilibrium when an allele is lethal to homozygotes?
6. What selective conditions can explain balanced polymorphisms and the persistence of harmful genes in populations?
7. For selection on a particular quantitative character, what are the consequences if that selection is stabilizing, directional, or disruptive?
8. Why would you or would you not expect the effects of directional selection on a character to continue indefinitely?
9. Since selective success for increased fitness depends on genetic variability, could an increase in fitness occur in the absence of new mutation? Explain.
10. Is the population that always has the least remaining variability the most fit population in a stable environment? Explain.
11. Do you think a species could evolve to the point where it could escape selection? Why or why not?
12. How do researchers determine equilibrium gene frequency when both mutation rate and selection are acting simultaneously?
13. What hypotheses may explain the evolution of dominance at a particular locus?
14. How can we calculate the effect of migration on gene frequency?
15. How can we calculate the effect of random genetic drift on gene frequency?
16. Will the long-term effect of random genetic drift differ when population size is large compared to when it is small? Why or why not?
17. How would you support the argument that a very small number of founders ("founder effect," "bottleneck effect") can cause a radical change in genotype in a new population?
18. How, and under what conditions, would you rank (a) mutation, (b) selection, (c) migration, (d) random genetic drift, and (e) founder effect, as forces that cause rapid changes in gene frequencies?

EVOLUTION ON THE WEB

Explore evolution on the web! Visit the accompanying web site for *Evolution, 3/e* at www.jbpub.com/evolution for web exercises and links relating to topics covered in this chapter.

REFERENCES

Adams, J., and R. H. Ward, 1973. Admixture studies and the detection of selection. *Science,* **180,** 1137–1143.

Allison, A. C., 1961. Abnormal hemoglobin and erythrocyte enzyme-deficiency traits. In *Genetical Variation in Human Populations,* G. A. Harrison (ed.). Pergamon, New York, pp. 16–40.

Beverley, S. M., and A. C. Wilson, 1985. Ancient origin for Hawaiian Drosophilinae inferred from protein comparisons. *Proc. Nat. Acad. Sci.,* **82,** 4753–4757.

Brakefield, P. M., 1987. Industrial melanism: Do we have the answers? *Trends in Ecol. and Evol.,* **2,** 117–122.

Brower, L. P. (ed.), 1988. *Mimicry and the Evolutionary Process.* University of Chicago Press, Chicago.

Bumpus, H. C., 1899. The elimination of the unfit as illustrated by the introduced sparrow. *Biol. Lect. Woods Hole,* pp. 209–226.

Buri, P., 1956. Gene frequency in small populations of mutant *Drosophila. Evolution,* **10,** 367–402.

Carson, H. L., 1992. Inversions in Hawaiian *Drosophila.* In *Drosophila Inversion Polymorphism,* C. B. Krimbas and J. R. Powell (eds.). CRC Press, Boca Raton, FL, pp. 407–439.

Christiansen, F. B., 1984. The definition and measurement of fitness. In *Evolutionary Ecology,* B. Shorrocks (ed.). Blackwell, Oxford, England, pp. 65–79.

Clarke, D. K., E. A. Duarte, S. F. Elena, A. Moya, E. Domingo, and J. Holland, 1994. The Red Queen reigns in the kingdom of RNA viruses. *Proc. Nat. Acad. Sci.,* **91,** 4821–4824.

Crow, J. F., 1998. 90 years ago: The beginning of hybrid maize. *Genetics,* **148,** 923–928.

Dobzhansky, Th., and O. Pavlovsky, 1953. Indeterminate outcome of certain experiments on *Drosophila* populations. *Evolution,* **7,** 198–210.

Dudley, J. W., 1977. Seventy-six generations of selection for oil and protein percentages in maize. In *Proceedings of the International Conference on Quantitative Genetics,* E. Pollak, O. Kempthorne, and T. B. Bailey, Jr. (eds.). Iowa State University Press, Ames, pp. 459–473.

Ehrlich, P. R., R. White, M. C. Singer, W. W. McKechnie, and L. E. Gilbert, 1975. Checkerspot butterflies; A historical perspective. *Science,* **188,** 221–228.

Endler, J. A., 1986. *Natural Selection in the Wild.* Princeton University Press, Princeton, NJ.

Fisher, R. A., 1930. *The Genetical Theory of Natural Selection.* Clarendon Press, Oxford, England. (2d ed., 1958, Dover, New York.)

Ford, E. B., 1940. Genetic research in the *Lepidoptera. Ann. Eugenics,* **10,** 227–252.

Gibbs, H. L., and P. R. Grant, 1987. Oscillating selection on Darwin's finches. *Nature,* **327,** 511–513.

Glass, H. B., and C. C. Li, 1953. The dynamics of racial admixture: An analysis based on the American Negro. *Amer. J. Hum. Genet.,* **5,** 1–20.

Haldane, J. B. S., 1939. The theory of the evolution of dominance. *J. Genet.,* **37,** 365–374.

Harland, S. C., 1936. The genetic conception of species. *Biol. Rev.,* **11,** 83–112.

Hartl, D. L., and A. G. Clark, 1989. *Principles of Population Genetics,* 2d ed. Sinauer Associates, Sunderland, MA.

Karn, M. N., and L. S. Penrose, 1951. Birth weight and gestation time in relation to maternal age, parity, and infant survival. *Ann. Eugenics,* **161,** 147–164.

Kettlewell, H. B. D., 1973. *The Evolution of Melanism.* Clarendon Press, Oxford, England.

Lerner, I. M., 1954. *Genetic Homeostasis.* Wiley, New York.

Levins, R., 1968. *Evolution in Changing Environments.* Princeton University Press, Princeton, NJ.

Li, C. C., 1955. The stability of an equilibrium and the average fitness of a population. *Amer. Nat.,* **89,** 281–295.

Mayr, E., 1942. *Systematics and the Origin of Species.* Columbia University Press, New York.

McShea, D. W., 1996. Metazoan complexity and evolution: Is there a trend? *Evolution,* **50,** 477–492.

Orr, H. A., 1991. A test of Fisher's theory of dominance. *Proc. Nat. Acad. Sci.,* **88,** 11413–11415.

Resnik, D. B., 1988. Survival of the fittest: Law of evolution or law of probability? *Biol. and Phil.,* **3,** 349–362.

Rotter, J. I., and J. M. Diamond, 1987. What maintains the frequencies of human genetic diseases? *Nature,* **329,** 289–290.

Sargent, T. D., C. D. Millar, and D. M. Lambert, 1998. The "classical" explanation of industrial melanism. *Evol. Biol.,* **30,** 299–322.

Slatkin, M., 1985. Gene flow in natural populations. *Ann. Rev. Ecol. Syst.,* **16,** 393–430.

Stanley, S. M., 1979. *Macroevolution: Process and Product.* Freeman, San Francisco.

Stern, C., 1973. *Principles of Human Genetics,* 3d ed. Freeman, San Francisco.

Strickberger, M. W., 1985. *Genetics,* 3d ed. Macmillan, New York.

Sword, G. A., 1999. Density dependent warning coloration. *Nature,* **397,** 217.

Templeton, A. R., 1996. Experimental evidence for the genetic transilience model of speciation. *Evolution,* **50,** 909–915.

Valentine, J. W., A. G. Collins, and C. P. Meyer, 1994. Morphological complexity increase in metazoans. *Paleobiology,* **20,** 131–142.

Van Valen, L., 1973. A new evolutionary law. *Evol. Theory,* **1,** 1–30.

Waters, C. K., 1986. Natural selection without survival of the fittest. *Biol. and Phil.,* **1,** 207–225.

Wickler, W., 1968. *Mimicry in Plants and Animals.* Weidenfeld & Nicolson, London.

Wright, S., 1951. The genetic structure of populations. *Ann. Eugenics,* **15,** 323–354.

23 Structure and Interaction of Populations

The structure and relationships of natural populations depart from many of the ideal conditions that would make their evolutionary behavior simple to understand; populations are not of constant size; nor uniformly distributed in space; nor always of the same mating pattern; nor subject to invariable conditions of mutation, migration, and selection. Because of multiple alleles, there are usually more than three diploid genotypes for any one locus and, through developmental interactions, the fitness these genotypes confer must depend on genes at other loci.

Moreover, the environmental contexts in which populations evolve are usually changing. These changes include elements in their physical environment such as moisture, temperature, pressure, and sunlight–shade, as well as elements in their biological environment such as prey, predators, parasites, hosts, and competitors. The relationship of a population to various ecological factors is also more than that of a passive recipient, since a population often modifies its physical and biological environment in ways that can diminish or enhance both its own resources and those of other populations. Because of all these interactions, no wonder populations "must continue running in order to keep in the same place" (the "Red Queen" hypothesis, p. 543). It is also not surprising that the structure of populations is difficult if not impossible to predict mathematically in detail, even by the most elaborate techniques.

Nevertheless, mathematical models, like other kinds of generalities, let us apply information from populations under one set of conditions to others under similar conditions. Various populations share common features in their response to factors such as random breeding, selection, mutation, migration, and genetic drift. A popular approach has been to measure various characteristics present in natural and experimental populations and then, with mathematical analysis, to use these observations to derive some broad evolutionary concepts about the structure of populations. This chapter briefly surveys some such attempts at the ecological and genetic levels.

Some Ecological Aspects of Population Growth

One area of mathematical modeling that deals with the evolutionary potential of populations applies to their reproductive powers. After all, the capacity for reproduction is always counterpoised against selection in the evolutionary process. In its simplest form, as in rapidly growing, asexual, unicellular species, the early stages of population growth in an environment well supplied with resources can occur exponentially so that a single individual produces 2 offspring, who then produce 4 in the next generation, 8 in the following, then 16, 32, . . . and so on, until there are 2^t individuals at t generations. Assuming the persistence of uniform reproductive properties for each individual in each generation, this provides an exponential growth curve of the type Figure 23–1*a* shows, which would be limitless if space and resources were limitless.

Quantitatively, we can describe the rate of numerical change (ΔN) in populations by noting that this change equals the difference between birth (b) and death (d) rates multiplied by the number of individuals (N): $\Delta N = (b - d)N$. Thus, when the birth rate exceeds the death rate, ΔN is positive and population size increases; equal birth and death rates yield $\Delta N = 0$ and an unchanged N, while a death rate higher than the birth rate yields negative ΔN, decreasing population size. Ignoring other causes in this simple illustration, $b - d$ is a primary factor determining population size, and we can describe it as the **rate of increase** (r) so that $\Delta N = rN$. In environments where a population is free of those factors that limit its growth, it attains what is called an **intrinsic rate of natural increase** (r_m).

Environmental resources and space are not limitless, of course, nor are individuals in a dense population unaffected by the waste products and toxins neighbors produce. So population growth is not limitless; population size eventually stabilizes at some constant value or may even suddenly "crash" to some very low number. (As Chapter 2 pointed out, it was Malthus's popularization of the idea that war, famine, and disease held in check human population exponential growth that led both Darwin and Wallace to the concepts of the struggle for existence and natural selection.)

In accordance with such limitations, Figure 23–1*b* shows how population size of a yeast strain grown under a particular set of conditions levels off from its early exponential direction to a plateau of about 665 individuals. The smooth S-shaped curve that results can be described mathematically as the modification of ΔN by a factor $(K - N)/K$ in the formula $\Delta N = rN\text{H}\times (K - N)/K$, so that when N is very small this factor is essentially 1, and population growth is then almost exponential, or $\Delta N = rN$ as before. However, as N increases in value closer to K, the $(K - N)/K$ factor becomes a fractional quantity closer to zero.

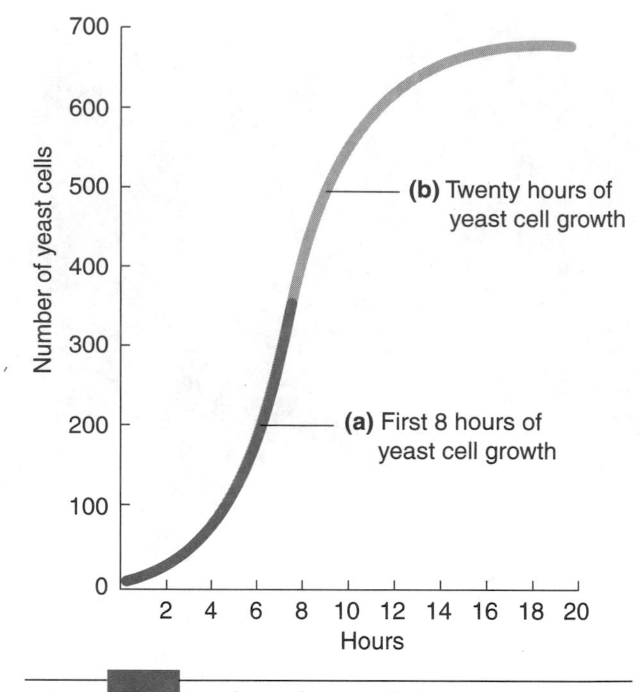

FIGURE 23–1 Numbers of yeast cells (*Saccharomyces cerevisiae*) in a defined volume of culture medium for two growth periods, beginning with approximately 10 cells per volume. (*a*) Exponential growth during the first 8 hours ($\Delta N = rN$, where $r = .5535$). (*b*) Sigmoidal growth curve approximating the logistic relationship [$\Delta N = rN \times (K - N)/K$, where $r = .5535$ and $K = 665$] for the 20-hour growth period. (*Data from Carlson.*)

Eventually N is large enough to equal K so that $\Delta N = rN \times 0 = 0$, and population size no longer changes, theoretically, but stabilizes at the value K. The growth model that provides the relationship $\Delta N = rN(K - N)/K$ is known as the **logistic growth model** and, for a particular environment, K is commonly called the **carrying capacity** of the population. P. F. Verhulst discovered the original equation for such logistic growth, and you can find discussions of its derivation, along with possible applications, in population ecology textbooks such as Emlen and Pianka and in Hutchinson's historical account.

In reality, populations rarely follow such smooth growth curves, and considerable fluctuations in numbers may occur (Fig. 23–2) that we can often ascribe to the impact of environmental agents. Some agents, such as climatic effects, may often be independent of population size and crowding and are therefore called **density independent;** whereas others, such as the effects of metabolically produced toxins, depletion of resources, and intrapopulational aggression, depend more on crowding, and are therefore called **density dependent.** Although obvious examples arise where one or both of these density factors can influence population size, their quantitative effects are often difficult to measure, and ecologists have considerably disputed their relative importance (for example, Strong).

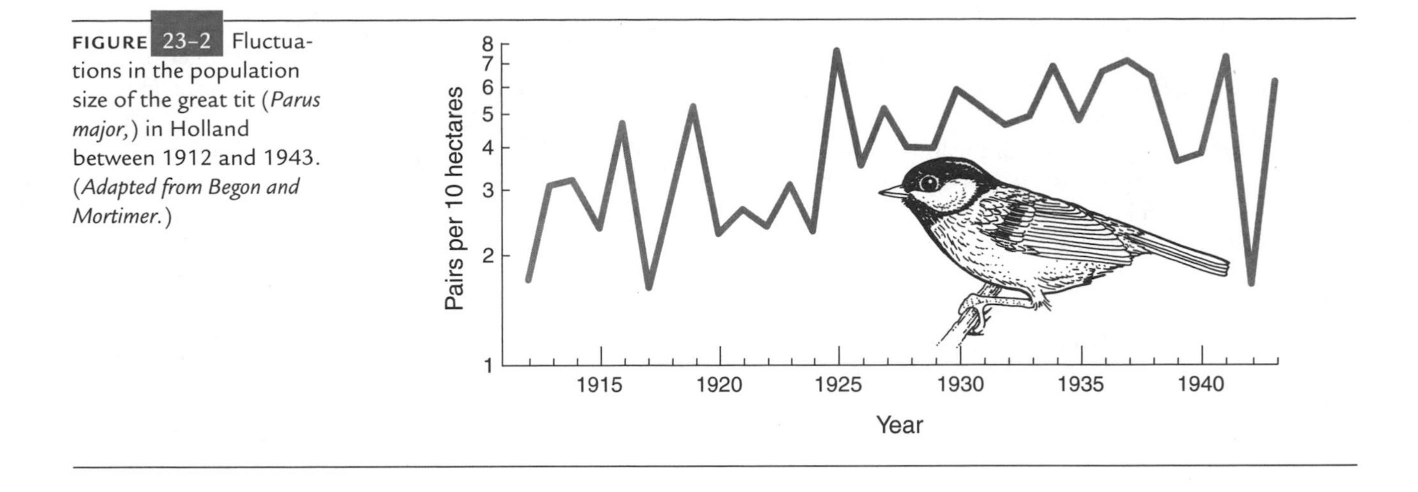

FIGURE 23–2 Fluctuations in the population size of the great tit (*Parus major,*) in Holland between 1912 and 1943. (*Adapted from Begon and Mortimer.*)

TABLE 23–1 Age structure of a hypothetically stable population with seven discrete age classes, ranging from nonreproductive juveniles (class 0, *lightly shaded*) to reproductive adults (classes 1–5, *darkly shaded*) to nonreproductive senescents (class 6, *lightly shaded*), illustrating the calculation of net reproductive rate (R_0)

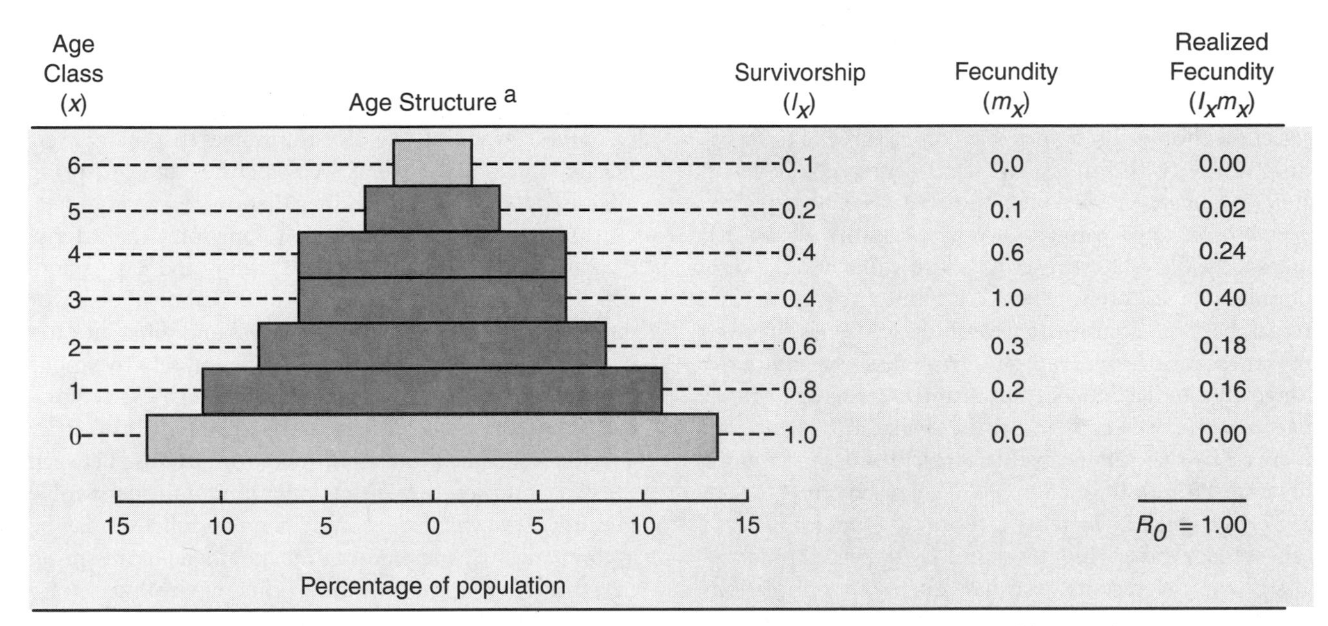

Age Class (x)	Age Structure [a]	Survivorship (l_x)	Fecundity (m_x)	Realized Fecundity ($l_x m_x$)
6		0.1	0.0	0.00
5		0.2	0.1	0.02
4		0.4	0.6	0.24
3		0.4	1.0	0.40
2		0.6	0.3	0.18
1		0.8	0.2	0.16
0		1.0	0.0	0.00
	15 10 5 0 5 10 15 Percentage of population			R_0 = 1.00

[a]Frequency of each age class in a stable population is related to l_x (Hutchinson) and can be noted here as the percentage graphed symmetrically on either side of the zero midline: for example, class 0 = 14.5 + 14.5 = 29.0 percent.

Source: Adapted from Pianka.

Aside from the effect of environmental agents, a most important aspect of a population's reproductive power is its age structure. If reproduction is associated with a particular age of individuals, the chances to survive to that age and the number of offspring that such individuals produce are essential attributes. The first of these, **survivorship,** we measure by a factor l_x, which represents that proportion of individuals who survive from age 0 to age x out of a group (cohort) who were all born during the same period. The second, **fecundity,** we represent by m_x, the average number of offspring produced by an individual of age x. The **net reproductive rate** of the entire population (R_0) is the sum of all its individuals multiplied by their fecundity at each age x:

$$R_0 = \Sigma\, l_x m_x$$

This relationship between survival and fecundity can be further elaborated in various ways (for example, Pianka) but, in general, a population that is stable in size will have a net reproductive rate (or replacement rate) of 1, as **Table 23–1** illustrates. Populations with R_0 less than 1 (for example, decreased survival of age class 3 in Table 23–1 and lacking compensating increases in the size and fecundity of other reproductive age classes) will decrease in size, and populations with R_0 greater than 1 will increase.

Older age classes such as class 6 in Table 23–1 are generally reproductively barren, with the lowest survival rate of all classes. In the past, researchers considered their aging (**senescence**) and death to have been selected as traits that benefit a population by removing individuals who might compete for its resources but who no longer contribute to its reproductive success. That is, selection occurred among populations (group selection, discussed later in this chapter) by conferring added fitness on populations in which nonreproductives died out because of the aging process. Although population benefits of this kind probably exist, such evolutionary explanations are presently in disfavor. Some workers have instead suggested that senescence evolves by mechanisms that let genes with deleterious effects on older, nonreproductive individuals spread through a population because they are either neutral or advantageous to younger, reproductive individuals (Partridge and Barton), or because somatic mutations that affect longevity accumulate as individuals grow older (Hughes).

Both these approaches reflect the concept that natural selection chooses genotypes that confer survival to reproductive age, with little or no selection for genotypes to survive longer. As Dawkins points out, "we inherit whatever it takes to be young, but not necessarily whatever it takes to be old." However, we do know that at least some populations maintain genetic variability for aging, and researchers are continually searching for genes involved in senescence. For example, *Drosophila melanogaster* longevity can decline when experimenters select individuals to reproduce early in the life cycle, or longevity can increase by selecting individuals to reproduce later in the life cycle (Rose). In one *Drosophila* experiment (Mueller, 1987) the decline in female fecundity (an estimate of senescence) was apparently caused by increased homozygosity for deleterious recessive alleles in populations whose individuals were selected for rapid growth rates (*r*-selection, see discussion, below). Finch and Tanzi provide a brief review of aging experiments in other organisms.

Whatever the causes for differences in age distribution, the strategies used in reproduction can differ among organisms. Many annual plants and insects, for example, breed only once during their lifetimes (**semelparous**), whereas many perennial plants and vertebrates breed repeatedly (**iteroparous**). The number of offspring a reproductive female produces at any one time also varies significantly, ranging, for example, from a single offspring in many larger mammals to the millions of eggs a female codfish lays. Furthermore, since the individual may die before reaching reproductive age, the sooner reproduction begins the greater the chances of producing offspring. Thus, in some organisms, reproductive stages follow soon after hatching, whereas other organisms—more dependent for survival on reaching larger size, obtaining greater experience, or needing more parental care—delay reproduction until a fairly late stage in the life cycle.

The alternatives of producing many offspring with little parental care or few offspring with greater parental care are among properties often ascribed to differences between what are called ***r***- and ***K*-selection** strategies. Organisms such as bacteria and plant weeds that tend to exhibit rapid populational growth in the face of wildly fluctuating environmental challenges and opportunities place reproductive emphasis on a rapid rate of increase (r) at low population densities, and researchers consider individuals in such populations to be *r*-selected. As stated by Darwin, "A large number of eggs is of some importance to those species which depend upon a fluctuating amount of food, for it allows them rapidly to increase in numbers."

Other organisms, such as large vertebrates, face more uniform or predictable environments with population sizes that are close to the environmental carrying capacity (K). In these *K*-selected organisms there is density-dependent competition for food, nesting space, and other resources, providing selective advantages that increase efficiency in resource use as well as ensure that offspring are raised to the stage when they themselves can compete.

Table 23–2 gives comparisons among some characteristics often deemed to be associated with *r*- and *K*-selection. Mueller (1988, also Mueller et al.) has demonstrated evolution of different competitive abilities and different rates of population growth under these different selective models in experimental *Drosophila* populations. Note, however, that these two types of selection are not strict alternatives; some populations may compromise between the two strategies. In Welsh populations of the periwinkle mollusk, *Littorina*, for example, larger adult size may be accompanied by increased numbers of smaller offspring in some environments, while other environments are populated by smaller adults who produce relatively fewer but larger offspring (Hart and Begon). Thus, although *r*- and *K*-selection schemes offer a pattern for understanding life histories in various populations, not all groups fit conveniently into such conventions. Boyce gives a fairly detailed discussion of some pros and cons of *r*- and *K*-selection theory.

In general, the ecological approach briefly reviewed in this section and in a later one ("Group Interaction," pp. 571–574) emphasizes how populations respond to their environment in terms of their numbers and distribution. Unfortunately, in the overwhelming majority of these studies, researchers have found it extremely difficult to discover and incorporate genetic information (for example, gene frequency estimates) that would explain how such changes correlate with evolutionary mechanisms (for example, mutation, selection, genetic drift) discussed

TABLE 23–2 Characteristics often associated with *r*- and *K*-selection

Characteristic	*r*-selection	*K*-selection
Climate	Variable or unpredictable	Fairly constant or predictable
Diversity of resources and habitats	Usually broad	Relatively narrow
Causes for mortality	Often catastrophic and density independent	Mostly density dependent
Survivorship	Very high mortality at younger stages, with high survivorship at later stages	Either constant rate of mortality at most stages, or little mortality until a certain stage is reached
Competitive interactions	Variable, mostly weak	Usually strong
Length of life	Relatively short, usually less than 1 year	Longer, usually more than 1 year
Selection pressure for	1. Rapid development 2. Rapid increase in numbers 3. Early reproduction 4. Small body size 5. Semelparity 6. Many small offspring 7. Increased productivity (quantity)	1. Slower development 2. Greater competitive ability 3. Delayed reproduction 4. Larger body size 5. Iteroparity 6. Fewer and larger offspring 7. Increase efficiency (quality)

Source: Adapted from Pianka.

in previous chapters. Attempts at relating population ecology to population genetics are proceeding, and they are beginning to receive considerable attention—but so far these efforts are mostly on a theoretical plane (for example, Roughgarden).

Genetic Load and Genetic Death

In contrast to the ecological approach, with its emphasis on population distributions, numbers, growth rates, and life histories, geneticists have placed more emphasis on the amounts and kinds of genetic variability present in populations and on uniting observations on natural populations with models of their genetic structure. These studies received special impetus from the 1920s and 1930s onward through the work of Chetverikoff, Dobzhansky, Fisher, Ford, Haldane, Wright, and many others. Most interesting in these studies was the demonstration that large amounts of genetic variability exist in practically all natural populations examined. As noted earlier (for example, Fig. 10–35 and Table 10–5), considerable polymorphism shows up on both the chromosomal and genic levels, and this variability allows genetic evolutionary changes to proceed.

However, despite the numerous advantages of genetic variability, many genes that natural populations maintain may handicap their carriers either in certain combinations or in homozygous condition. As Figure 22–7 showed, selection can account for a significant loss of nonoptimal individuals even in long-standing populations. Thus, if we consider genetic perfection as the elimination of all harmfully inferior gene combinations, then most, if not all, populations are genetically imperfect.

The extent to which a population departs from a perfect genetic constitution is called its **genetic load,** and is marked by the loss of some individuals through their **genetic death.** Genetic death is not necessarily an actual death before reproductive age but can be expressed through sterility, inability to find a mate, or by any means that reduces reproductive ability relative to the optimum genotype(s). We therefore phrase estimates of these values in terms of the proportion of individuals eliminated by selection. For example, if a gene is deleterious in homozygous condition, the homozygote frequency before and after selection will be as shown in the following table:

	Genotypes		
	AA	*Aa*	*aa*
Frequency at fertilization	p^2	$2pq$	q^2
Relative adaptive value	1	1	$1-s$
Frequency after selection	p^2	$2pq$	q^2-sq^2

The loss in frequency of individuals, or incurred genetic load, equals sq^2. *If* N individuals were in the population before selection, genetic load has now eliminated sq^2N.

This value of sq^2, however, also equals the mutation rate (u) at equilibrium for $A \rightarrow a$ (p. 545), which means that the genetic load that a deleterious homozygous recessive causes is equal to its mutation rate. An important feature of this relationship, which Haldane pointed out, is that if the mutation rate is constant, it will make little difference to the genetic load whether s is small or large. As you can see in each single column (constant mutation rate) of the hydraulic model in Figure 22–11, if s is small, q will be large at equilibrium, and if s is large, q will be small.

High selection coefficients eliminate the gene more rapidly (low q), and low selection coefficients let the gene stay longer in the population (high q). In either case, the genetic load is still $sq^2 = u$ and the total number of genetic deaths remains at sq^2N. Insofar as mutation produces deleterious recessives, any increase in their mutation rate causes a corresponding increase in genetic load and thus in genetic death.[1]

According to Crow (1992), any factor that produces differences in fitness among genotypes can create a genetic load, and he and others have devised techniques to evaluate their relative importance. The **mutational load,** just discussed, is only one essential factor researchers consider responsible for genetic load. Another is the **segregational** or **balanced load,** restricted to those instances in which a heterozygous genotype is superior to both types of homozygotes. For a gene with two alleles, the segregational load amounts to p^2s for the AA homozygotes plus q^2t for the aa homozygotes, since s and t are the selection coefficients against these homozygotes, respectively, when the heterozygote is overdominant (Table 22–4). If we substitute the equilibrium frequencies of p and q under those circumstances (p. 538) into $p^2s + q^2t$ we obtain

$$\left(\frac{t}{s+t}\right)^2 s + \left(\frac{s}{s+t}\right)^2 t = \frac{st^2 + ts^2}{(s+t)^2} = \frac{st(s+t)}{(s+t)^2} = \frac{st}{s+t}$$

Thus, if s and t are both about .1, the segregational load will be .01/.2, or .05. This value is considerably higher than most mutation rates and demonstrates the increased genetic load that we may expect segregation to cause in randomly breeding populations, compared with the load mutation causes ($sq^2 = u$).

Although most or all populations carry genetic loads of one kind or another—including even **recombinational loads** that can break up adaptive combinations of linked genes by crossovers (p. 525)—we have not yet determined the relative values of each type of load. For example, Crow (1993) points out that "The total deleterious mutation rate remains unknown in any animal except *Drosophila.*" From the viewpoint of evolution, probably no species ever reaches "genetic perfection" with its absence of genetic death, because environments usually change with time and the advantages of different genotypes change accordingly. On the one hand, it is conceivable that a population so perfectly adjusted to its environment—with little or no genetic load (no variability)—may become extinct within a short period because of rapid environmental change. On the other hand, a population with a relatively large genetic load may encounter a new environment in which formerly deleterious genes endow it with adaptations that help it survive.

Lacking a genetic load may therefore harm a population more than having one (C. C. Li, Brues)—a point made previously in relation to optimal mutation rates (p. 291). Therefore, although we can measure the genetic load in terms of departure from the optimum genotype, the evolutionary value of a particular optimum genotype may be very limited; the optimum genotype may change in time, or from place to place, or may even differ in the same place—if, for example, there is a division of labor between different genotypes (for example, male and female).

The Cost of Evolution and the Neutralist Argument

Whatever type of load a population bears, natural selection adds to the load by favoring some genes and discarding others. Since gene replacement is ongoing and pervasive, we can reasonably ask how many individuals a population must lose, in terms of genetic death, to replace a single gene by selection alone. If a dominant mutation arises that has greater selective value than the more frequent recessive, we have seen that the number of genetic deaths will be sq^2N for one generation of selection, where N is the number of individuals in the population.

For complete replacement of a deleterious allele, Haldane has calculated that the total number of genetic deaths is determined by a factor D ($D = \ln p_0$) that is based primarily on the initial frequency of the favored (p) allele, multiplied by the population number N. For a newly favored but rare dominant allele, D is about 10, so the cost of evolution (DN) for eliminating the deleterious recessive allele is about 10 times the average number of individuals in a single generation. This cost, of course, is spread over many generations, the rapidity of gene replacement depending on the selection coeffi-

[1]Holmquist and Filipski have pointed out that particular DNA sequences (for example, guanine–cytosine-rich sequences that replicate early during DNA synthesis) may, because of their composition, mutate at different rates and directions than other sequences. As a result, such sequences can bias the nucleotide genomic patterns yet still produce a minimal mutational load.

cient.[2] If the rare newly favored allele is recessive, DN increases to about $100N$, since the favored homozygotes are extremely rare.

According to Haldane, many favored alleles are probably intermediate in dominance, and he therefore proposed an average death value of $30N$ for the replacement of a single allele. He suggested that a population is probably capable of sacrificing about one-tenth of its reproductive powers for such selective purposes; that is, a gene substitution can occur at a rate $1/10 \times 30N$, or every 300 generations.

However, selection may not cause all gene substitutions, and Kimura has proposed that the rate of evolution at the molecular level is actually far more rapid than Haldane suggested. In vertebrates, for example, one can calculate that many hemoglobin protein amino acids are replaced at a rate of approximately one amino acid change per 10^7 years, yet the total amount of DNA is certainly more than necessary to code for 10^7 amino acids. So we would expect each vertebrate species to undergo at least one complete amino acid substitution per year if we assume that overall amino acid substitution rates are about the same for all proteins.

If we accept Haldane's value of about $30N$ for the cost of a single gene substitution by selection and estimate an average of about 3 years per vertebrate generation, then each vertebrate population must continually expend an enormous number of genetic deaths to maintain its size and to escape extinction. In the present example, the population would have to devote 90 times its number in each generation ($3 \times 30N$) if selection were the primary cause for gene frequency changes.

Because of this presumably high cost of selection, Kimura and coworkers (see Kimura 1983), as well as King and Jukes, have instead proposed that most amino acid changes are neutral in effect. Since selection does not act on **neutral mutations,** the fixation of such alleles incurs no genetic load and depends only on their mutation rates and on random genetic drift.[3] This neutralist hypothesis (also called by some **non-Darwinian evolution** because of its dependence on mutation and not on selection) seems supported by the widely observed and extensive degrees of enzyme and protein polymorphisms, indicating that allelic differences persist at many thousands of loci in many species. We could of course claim that allelic differences can persist in selectively advantageous heterozygotes without fixation and, since they will never be fixed, we can exclude the high cost of evolution that Haldane has shown necessary for gene substitution. However, maintaining polymorphism by selection may itself entail an enormous and intolerable genetic load.

If only the superiority of heterozygotes maintained polymorphism at a single locus, the segregational load would be $st/(s + t)$, as explained previously. Should the cause for this segregational load depend on the lethality of the two homozygotes (**balanced lethals**), the selection coefficients s and t are then both 1, the load is 1/2, and the remaining fitness of the population is $1 - 1/2 = 1/2$. For two pairs of balanced lethal genes acting independently of each other (for example, AA and aa are lethal, as are BB and bb, and only the $AaBb$ heterozygotes survive), the fitness of the population reduces to $(1 - 1/2)^2 = 1/4$. In general, no matter what the value of the selection coefficients, the average fitness of a population bearing a balanced or segregational load is $[1 - st/(s + t)]^n$ where n designates the number of gene pairs at which heterozygote superiority is being maintained. Kimura and Crow have calculated that the fitness of such a population approximately equals $e^{-\Sigma L}$ where e (2.718) is the base of natural logarithms and ΣL designates the sum of individual loads for each gene pair involved.

Not surprisingly, this load can be quite large, even if the selection coefficients acting against homozygotes are small, as long as many gene pairs are involved in maintaining superior heterozygotes. For example, if superior heterozygotes are being maintained at 100 loci, each bearing a genetic load of .01 (for example, $s = .02$, $t = .02$: $st/(s + t) = .0004/.04 = .01$), the average fitness of the population falls to about $e^{-100(.01)} = e^{-1} = .37$. For 500 gene pairs acting similarly, the fitness is approximately $e^{-5} = .007$, and it is $e^{-10} = .00005$ for 1,000 such gene pairs. Thus, to maintain polymorphism at 1,000 loci, even with only a very small selective advantage for the heterozygote at each gene, only one out of 20,000 offspring would survive. For every female to produce two surviving offspring (and thereby allow the population to survive), each such female would have to produce about 40,000 young for this selective purpose alone! Most polymorphic systems must therefore consist of neutral mutations, according to Kimura and his followers, who support this argument with observations of high

[2]When s is large, gene replacement is more rapid than when s is small, but the population now faces the danger of extinction because its numbers may be too few to ensure mating partners or survival in an accident. In the extreme case, when $s = 1$, replacement may occur in a single generation if the unfavored allele is dominant and only favored recessive homozygotes survive. Under such circumstances, however, extinction is fairly sure to occur if the favored gene is present in very low frequency, since very few favorable homozygous recessives will be available for survival.

[3]If the chance for any gene to mutate is u, a population of N diploid individuals bearing $2N$ genes will have $2Nu$ newly arisen mutations. Should all or most of these mutant genes have an equally neutral phenotypic effect, each will be present in about $1/2N$ frequency and will probably persist in this relative frequency because no one of them is any better than the other. Thus, there is a total of $2Nu$ neutral mutations, each with a $1/2N$ chance of fixation, or the fixation probability for a particular neutral allele is $2Nu \times 1/2N = u$. On an individual basis, the time needed to fix any particular neutral allele will depend on population size, since the random drift process in small populations greatly speeds up the fixation of these genes, although it does not change their probability of fixation. Kimura and Ohta calculate that the average number of generations necessary to fix a new neutral mutation is approximately four times the number of parents in each generation.

TABLE 23-3 Comparisons of evolutionary rates for two proteins

Comparisons Between Groups	Time of Divergence in Millions of Years Ago	Rate[a]	
		Superoxide Dismutase	Glycerol-3-Phosphate Dehydrogenase
Drosophila groups	45 ± 10	16.6	0.9
Drosophila subgenera	55 ± 10	16.2	1.1
Drosophila genera	60 ± 10	17.8	2.7
Mammals	70 ± 15	17.2	5.3
Drosophila families	100 ± 20	15.9	4.7
Animal phyla	650 ± 100	5.3	4.2
Multicellular kingdoms	1,100 ± 200	3.3	4.0

[a]Rate is measured as amino acid replacements per year $\times 10^{-10}$, after correction for multiple replacements. Both genes compare the same species in each group.

Source: From Ayala, F. J., 1999. Molecular clock mirages. *Bio Essays,* 21, 71–75.

frequencies of selectively neutral mutations in a rapidly mutating strain of *Escherichia coli* (Gibson et al.).

Among further evidence that neutralists use is the presumed constancy of amino acid substitution rates in particular proteins (the "evolutionary clock"; Chapter 12). Kimura and Ohta, for example, note that the same number of changes in the α-hemoglobin chain occur, relative to amino acids in β-hemoglobin, whether the α-chain comes from the same species or from a different species. Compared to the human β-chain, the human α-chain shows 75 differences, the horse α-chain shows 77 differences, and the carp α-chain shows 77 differences. Kimura and Ohta ask: Why should the α-hemoglobin chains of humans, horses, and fish, each with a different selective history, have diverged from the human β-chain at exactly the same rate? According to the neutralists, we can most easily explain this uniformity by a common rate of neutral mutation and drift rather than common selective conditions. As striking as this evidence is, some researchers dispute the regularity of the molecular clock. There are certainly exceptions to constant evolutionary rates in some proteins, and different rates of nucleotide substitution appear in comparisons among different proteins (Table 23–3, see also pp. 284–286). The erratic nature of the molecular clock and other "unsettled" issues of the neutral theory are reviewed by Takahata.

The Selectionist Argument

To counter the neutralist view, **selectionists** have proposed various selection schemes that would explain the persistence of many polymorphisms but would confer only minimal genetic loads. One such mechanism is frequency-dependent selection (p. 540) which entails genetic loads only when the frequency of a relatively rare selected allele is changing but produces no genetic load when the allele has reached equilibrium (Kojima and Yarbrough). However, since polymorphism seems to exist at thousands of loci, it is questionable that there are also thousands of individual frequency-dependent mechanisms in the environment.

Sved, King, and others have suggested that selection in a natural population probably lumps the effects of many individual genotypes into two main groups: the fit and the unfit. There is a "threshold" number of polymorphic loci in a population which separates these two groups: heterozygotes for more gene loci than the threshold number show no increased heterotic effect, and heterozygotes for fewer genes than the threshold are presumably all equally deleterious. The threshold thus acts as a form of **truncation selection** to eliminate or truncate an entire class of phenotypes, whatever their genotypes may be (Wills).

Although a threshold of this kind may shift, depending on the environmental stresses on the population (Milkman 1967; Wallace), the genetic loads such populations incur may be relatively small since, because of lumping, differences between each of the many genotypes above or below the threshold do not add to the load. The likelihood that genes are not individually replaced in evolution but can be selected in linkage blocks (Franklin and Lewontin), or perhaps as functional groups, supports the view that considerable interaction between genes may produce threshold effects. The possibility of such interactions also indicates that we may erroneously calculate theoretical genetic loads if we assume that each locus acts independently of all others. Although the threshold concept is attractive to some population geneticists, we have little direct evidence that thresholds exist.

At present, selectionists are emphasizing the following findings:

ASSOCIATION BETWEEN PROTEIN POLYMORPHISMS AND ECOLOGICAL CONDITIONS

Selectionists have presented the argument that a strong correlation between particular alleles and particular environmental conditions might indicate that selection is maintaining polymorphisms for protein variations—**allozymes.** One well-known example is the relationship between the gene for sickle cell anemia and malaria (Fig. 22–3), where the heterozygote is superior in fitness to both homozygotes. A further documented example of overdominance is that for an alcohol dehydrogenase gene in yeast (Hall and Wills). In such cases, however, maintaining polymorphisms by selection entails a significant expense in the loss of homozygotes.

A system that may have milder selective effects is the polymorphism Koehn discovered in the freshwater fish *Catostomus clarkii.* The distribution of two alleles of an esterase enzyme in this fish seems to follow a temperature cline along the Colorado River basin. The homozygote for the allele that is most frequent in the more southern (and warmer) latitudes produces an esterase enzyme that becomes more active as temperature increases, whereas the allele that is more frequent in the northern (and colder) latitudes forms an enzyme that is more active as temperature decreases. Not unexpectedly, the heterozygote for the two alleles forms an enzyme that is most active at intermediate temperatures.

In *Drosophila melanogaster,* allozymes produced by the alcohol dehydrogenase locus (*Adh*) also show correlations between their frequencies and environmental temperatures. According to experiments by Sampsell and Sims, the stability of different *Adh* allozymes under high temperatures relates to the fitnesses they confer on flies by enabling them to survive high alcohol concentrations in the food medium. Selection acting on such allozyme differences can probably help explain some of the gene frequency changes researchers observe along various north–south geographical gradients (for example, see Oakeshott et al.).

Although other such correlations between alleles and particular environments appear in plants (Allard and Kahler) as well as in animals (Gillespie, Koehn et al., Somero), in many cases no obvious correlation exists. Researchers find it hard to prove that such correlations are always necessarily causal and may not be accidental. Neutralists have also argued that some enzyme loci being scored for polymorphism may be strongly linked to a gene locus at which selection is operating, and the protein polymorphism researchers observe is only the effect of such linkage disequilibrium, or **hitchhiking.**

NONRANDOM ALLELIC FREQUENCIES IN ENZYME POLYMORPHISMS

Geneticists have pointed out a number of enzyme polymorphisms whose advantages we do not know but whose frequencies we find difficult to explain on a purely random basis. An early example of this kind was Prakash and coworkers' observation that populations of *D. pseudoobscura* ranging from California to Texas show remarkably similar allozyme frequencies for a number of proteins. Although some have suggested that such similarities arise because of migration among different populations rather than through common selective factors, migration could hardly explain similarities in gene frequencies among different species.

As Ayala and Tracey noted, genetically isolated species of the *D. willistoni* group share common gene frequencies for the alleles of many different enzymes. Furthermore, the pattern of similarity between them is not constant, and the data seem to show that different species share common selective factors for some enzymes but not for others. On a broad scale, coordinating both genetics and ecology, Nevo has surveyed 35 species from Israel, including insects, mollusks, vertebrates, and plants, and concludes that the amount of allozyme polymorphism in a species correlates to factors such as life habit and climate.

Perhaps even more striking is Milkman's (1973) finding that *E. coli* clones isolated from the intestinal tracts of animals as diverse as lizards and humans and from localities as widespread as New Guinea and Iowa seem to share common allozyme frequencies. For each of five different enzymes, Milkman found that one particular electrophoretic band was frequent in almost all samples. Since other allozymes of these proteins exist, the finding of such a narrow distribution of allozymes would seem difficult to explain on any basis other than selection.[4]

Selection must also operate to cause convergent evolution, in which the same functions arise in different taxonomic groups by different evolutionary pathways (for example, Fig. 3–7). On the molecular level, Yokoyama and Yokoyama have used DNA sequences to demonstrate such convergences in fish and humans, who independently evolved red visual pigments from green pigments. Also, highly conserved gene products, such as histone proteins (pp. 273 and 565) alike in both plants and animals, indicate that selection can actively reject any major change in an essential function.

ASSOCIATION BETWEEN ENZYME FUNCTION AND DEGREE OF POLYMORPHISM

Gillespie and Kojima first made the suggestion that the function of an enzyme influences the degree of polymorphism at its locus. They grouped enzymes into two classes, namely those involved in restricted pathways of energy metabolism such as glycolysis (for example, aldolase) and

[4]Selander and Levin nevertheless contest this view, suggesting that mutational and recombinational events in *E. coli* may have been rare enough that the genotypes of the initial founding populations persisted over long periods of time.

those that can use a variety of substrates (for example, esterases, acid phosphatases). Their findings and others show that enzymes with more restricted uses show significantly less polymorphism than enzymes whose substrates vary more. Johnson, extending this notion further, proposed that enzymes involved in regulating metabolic pathways (for example, glucose-6-phosphate dehydrogenase and phosphoglucomutase) are generally more polymorphic than enzymes whose functions are not primarily regulatory (for example, malate dehydrogenase and fumarase). Although we do not yet know the biochemical causes that sustain these differences in polymorphism, they are clearly not random, so many of the polymorphic alleles are not neutral.

POLYMORPHISMS FOR DNA CODING SEQUENCES

In general, nucleotide sequence analyses show that polymorphisms are significantly greater in those DNA sequences that do not determine amino acid sequences compared to those DNA sequences that transcribe and translate into amino acids. This argument suggests that selection reduces the variability in amino acid coding regions because such sequences have a greater effect on the phenotype than do noncoding regions. Or, put differently, those random "neutral" mutations mostly responsible for variability in noncoding regions do not determine the variability that remains in amino acid coding regions. Thus, we would expect selection to operate more strongly on the first two amino acid codon positions, because these are more involved in amino acid determination than the "wobbly" third-codon position, which is essentially responsible for degeneracy of the genetic code (Chapter 8).

Similarly, we would expect more polymorphism in introns (intervening sequences, Chapter 9), which do not code for amino acids, than in exons (expressed sequences), which do. According to Kreitman, both these expectations are fulfilled for the *D. melanogaster alcohol dehydrogenase* gene, in which he found 6 and 7 percent polymorphism among introns and third-codon positions but found polymorphism practically absent in exons and in the first two codon positions. Since random events can hardly explain such pronounced differences in polymorphism, these findings strongly suggest that selection must be the discriminating agent that determines which nucleotide base substitutions will become established in functionally different DNA sequences.

We can also see the active role of selection when we compare duplicate genes that are in the process of becoming functionally different. In such cases, we would expect greater differences when comparing amino acid coding nucleotide sequences between these duplicates for their functionally strategic sections than when comparing coding sequences for sections that are not becoming functionally differentiated. Ngai and coworkers studied duplicate genes involved in producing olfactory receptor proteins in catfish. They found greater differences among these duplicate genes for a particular amino acid coding region of the gene than for other regions, indicating very active selection for functional divergence rather than for neutral mutation.

Functional divergence also accounts for mutations in sex-related genes (sex determination, mating behavior, spermatogenesis, egg fertilization) that occur more frequently in amino acid determining codons than in synonymous codons which do not cause amino acid substitutions. Such findings in a wide variety of organism, including mollusks, arthropods, and mammals (for example, Whitfield et al., Civetta and Singh), indicate that changes in sex-gene amino acid sequences were actively selected as species diverged.

A further test, also derived from DNA sequencing, is based on the concept that, according to the neutral theory, we would expect most mutations in a specific gene to be neutral in whatever species they occur. That is, if we make both interspecific and intraspecific comparisons for a particular locus, the degree of nucleotide divergence leading to amino acid differences between two related species should correlate with presumably neutral variation in other traits, such as the degree of nucleotide divergence (polymorphism) within each species. Eanes and coworkers tested this neutral mutation model by comparing sequences in 44 copies of the glucose-6-phosphate dehydrogenase (*G6pd*) gene derived from two related *Drosophila* species, *D. melanogaster* and *D. simulans.*

In this test, Eanes and his collaborators found much more interspecific amino acid divergence (21 replacement differences) than we would expect from the degree of polymorphism within species and from the degree of interspecific synonymous mutational differences that do not cause amino acid changes. Evidently, their greater frequency indicates that these amino acid changes result from much more stringent selection than we would expect if they had experienced little selection—that is, if they had been neutral in effect.

* * * * *

In summary, selective forces unquestionably affect polymorphism and help maintain it under some circumstances. However, we do not know the number of genes on which selection operates at any one time, the linkage relationship between these genes, the kinds of selection that operate, and the size of the selection coefficients. Although researchers are beginning to explore some of these areas (Hey), we still cannot exclude neutral mutation and random genetic drift as important causes for some, or even many, polymorphisms. Some genetic variants may be neutral at certain times or under certain conditions but have selective value when the environment or genetic background changes (Hartl and Dykhuizen).

Some Genetic Attributes of Populations

As Table 21–1 showed, populations have unique evolutionary characteristics, and many genetic factors that affect the evolution of populations may act differently from our expectations if we consider individuals alone. Among these factors are the following:

SEX

Sex may be of little value to an individual (or even a disadvantage) but can be a distinct advantage to a population. For example, evolutionists have often noted that females who reproduce parthenogenetically (asexually) accomplish the work of the usual two sexes more efficiently since such females spare the expense of producing a superfluous male sex. Sex is also often costly to individuals in terms of the risks associated with finding the opposite sex, the risks of fertilizing gametes in exposed circumstances, and the extra resources expended during intrasexual competition (p. 588). Predation and parasitic infection are, for many species, common perils of courtship and copulation. Furthermore, the recombination events that accompany sex can rearrange and reassort genes whose previous combinations had already achieved high fitness, thereby lowering individual fitness. Various population geneticists point to populations rather than to individuals in seeking a cause for sexual reproduction.

Among the most popular of these proposals is that sexual crossing in a population allows single individuals to incorporate different beneficial mutations from other members through mating and recombination, whereas without sex, combinations of such beneficial mutations are more difficult to achieve. As Figure 23–3 shows,

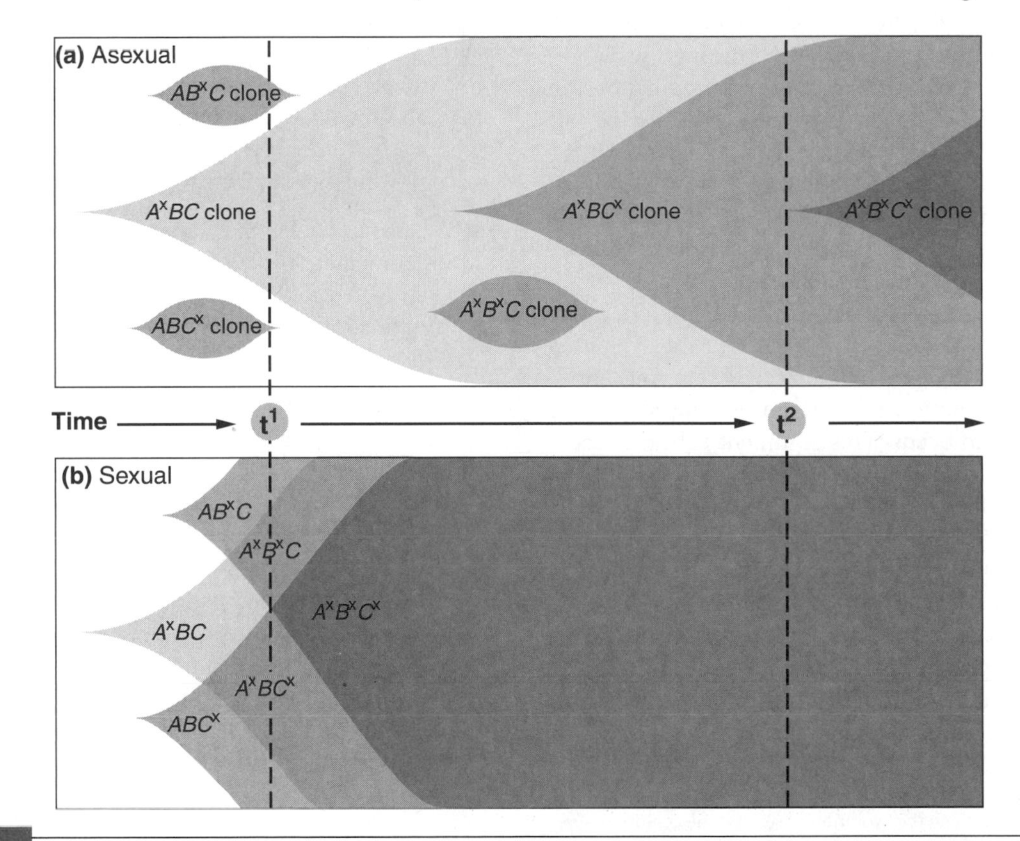

FIGURE 23–3 Muller's model of the difference between asexual and sexual populations in the speed at which they incorporate combinations of advantageous mutant alleles and lose deleterious alleles. In this illustration, we assume both kinds of populations begin with an initial state where three loci are fixed for the *A, B,* and *C* alleles. We call the advantageous alleles that arise at these loci A^x, B^x, and C^x, with individuals carrying A^x being more fit than those carrying B^x or C^x. Among the other combinations, we presume A^xBC^x to be more fit than A^xB^xC or AB^xC^x, while all individuals carrying all three mutant alleles, $A^xB^xC^x$, are the most fit of all. (*a*) In asexual clonal populations, each clone is independent of the others, so attaining the most fit genotype must await successive beneficial mutational events in a single clone ($A^xBC \rightarrow A^xBC^x \rightarrow A^xB^xC^x$, time t^2), while the least fit clones become extinct. (*b*) In large sexual populations, beneficial mutations need not occur successively in a single clone, and the $A^xB^xC^x$ genotype can be achieved relatively rapidly (time t^1) through recombination among individuals who carry the various advantageous alleles without the need of further mutational events. Also, as Peck pointed out, when deleterious mutations are common, beneficial mutations have a better chance of becoming established in sexual than in asexual populations because of the asexual requirement that beneficial mutations can only persist in clones that are already favorably selected, whereas such rigid clonal selection does not govern sexual populations. (*Adapted from Muller.*)

favorable mutations such as A^X, B^X, and C^X may arise in an asexual population but remain in separate individual lineages. Only if the asexual population becomes extremely large, which may take considerable time, is there a reasonable chance that a second favorable mutation will occur in a lineage that contains a previous favorable mutation. By contrast, individuals within a sexual population may rapidly incorporate favorable mutations soon after they occur, simply by mating between the separate lineages that carry the favorable mutations.

Furthermore, just as mating and recombination in sexual populations enable relatively rapid favorable combinations of newly arisen mutations, they also enable relatively rapid favorable combinations of already existing mutations. In both cases, the advantages achieved include the additional variability that lets a population persist in a changing environment (see also p. 301). For example, parasites usually evolve rapidly to counter resistance in their hosts and hosts usually evolve rapidly to counter infectivity of their parasites (p. 575). This "arms race" places a premium on rapid evolution through sexual recombination (Hamilton et al.).[5] Asexual populations, in contrast, must depend mostly on the variability they already have (Bell; Crow 1988; Maynard Smith 1978)—a static condition that can limit their opportunities for rapid change. It is no surprise that experimental competition between sexual and asexual strains of yeast can evolve in favor of the former (Birdsell and Wills).

In addition to an improved rate of adaptation, a further advantage of sexual populations, as Muller pointed out, is their relative ease in eliminating deleterious mutations via recombination rather than waiting for chance back mutations to occur. This argument is based on the concept that at equilibrium values of mutation–selection, individuals will be carrying varying numbers of deleterious mutations (*del*) at many loci (for example, $A^{del} B^+C^+D^+\ldots$, $A^+B^{del}C^+D^+\ldots$, $A^{del}B^+C^{del}D^+\ldots$, $A^{del} B^{del}C^+D^{del}\ldots$), but the class of individuals carrying a total of zero deleterious mutations (for example, $A^+B^+C^+D^+\ldots$) will be quite small. If sampling error (random genetic drift) should lose this zero class in an asexual population, only back mutation from a class carrying a single deleterious mutation could reconstitute it (for example, $A^{del}B^+C^+D^+\ldots \rightarrow A^+B^+C^+D^+\ldots$; and $A^+B^{del}C^+D^+\ldots \rightarrow A^+B^+C^+D^+\ldots$).

Similarly, if the single-mutation class is then lost through drift, its reconstitution depends on back mutation from a class carrying two deleterious mutations. Because of drift, therefore, classes with increasing numbers of deleterious mutations tend to replace those with fewer such numbers, a phenomenon called **Muller's ratchet.** As time goes on, it will become more difficult to eliminate deleterious mutations in asexual populations, and their genetic load will increase (for a confirming experiment, see Andersson and Hughes). On the other hand, Muller's ratchet will not operate in sexual populations because the zero mutation class can easily be reconstituted through recombination (for example, $A^{del}B^+C^+D^+\ldots \times A^+B^{del}C^+D^+\ldots \rightarrow A^+B^+C^+D^+\ldots$).

From such effect on fitness, we would expect recombination to be as subject to selection as any other character influencing reproductive success. Supporting this view, and demonstrating a range of recombinational variability, are experiments with *Drosophila* showing that both high and low crossover rates can be selected (Kidwell). Observed genomic recombination rates probably evolved as compromises between two contrasting effects:

- A beneficial effect that assembles specific alleles of different loci adapted for specific environments (Fig. 23–3), and also disrupts disadvantageous gene combinations by reducing linkage between them ("disequilibrium," see Fig. 21–6).
- A deleterious effect that can break up advantageous combinations, whether composed of genes that interact with each other (epistasis) or act independently. Only mechanisms that inhibit recombination in specific areas, such as inversion systems in *Drosophila* (Fig. 10–35 and p. 525), can prevent such loss. In this sense, lack of recombination provides an advantage to asexual populations that can persist long enough to accumulate strings of favorable mutations. (Note, however, that lack of recombination also enables an increase in linkage disequilibrium of deleterious genes and will cause a loss of such asexual clones.)

MUTATION

A further important attribute of populations is mutation. One might expect that since mutation is mostly random, as many new adaptive mutations arise as deleterious ones. It would therefore appear as though evolution merely awaited the occurrence of superior adaptive individuals before it proceeds. Although such views have been expressed (W. H. Li), it is likely that the role of new mutations is not often immediately significant.

A population that has long been established in a particular environment will have many genes adapted for pre-

[5] The relationship between the evolution of sex and the evolution of recombination is a matter of considerable interest. For example, one hypothesis suggests that sex (meiosis) originated as a way of overcoming DNA damage by using recombinational DNA repair mechanisms (p. 225), and the genetic variation that resulted was only an accidental by-product (Bernstein et al.). However, according to Maynard Smith (1988), even if recombination had a DNA repair origin, one can still argue that it was the genetic variability "by-product" that became the primary sexual mainstay. One can also contend that DNA repair mechanisms are not essential for sex since they are also quite efficient in asexual organisms including viruses, bacteria, and even long-standing asexual eukaryotes such as bdelloid rotifers. Various contributors in the collections of Stearns and of Michod and Levin discuss other aspects of this topic, including the proposal that parasitic elements, such as transposons and plasmids (pp. 225–226) initiated or promoted sexual fusion as a mechanism to infect other cells (Hickey and Rose).

vailing conditions. New mutations that arise, if not neutral in effect, will rarely be better and will likely be worse than the genes already present—a consequence not much different from the serious damage we can expect when a random change is introduced into any intricately organized and integrated system, such as computer wiring. Complex organisms are developmentally constrained by their evolutionary history (p. 357), so that advantageous mutant effects are generally confined to few of the many intricate and sensitive developmental processes. For phylogenetically crucial gene products, conserved molecular sequences are commonplace, and viable changes occur only rarely. For example, although plants and animals are separated by more than one billion years of evolution, there are only two differences between them in the 100 amino acids of histone 4—a eukaryotic protein that binds and folds DNA (see also p. 273).

Gross developmental monsters proposed to explain macroevolutionary events are therefore more "hopeless" than "hopeful" (p. 599), and even if some mutations are better, it is unlikely that they will much exceed the fitness of previously established genes. Assuming these beneficial mutations are not lost by chance (p. 533), their increase in frequency can still take many generations (p. 535).

In contrast, a change in environmental conditions may have a more important evolutionary effect, since many genes formerly in low frequency may suddenly have high adaptive value. We can see this in the rapid genetic changes in many insect populations exposed to pesticides such as Dieldrin and DDT (Fig. 23–4), where resistant alleles appear on all major chromosomes (see Fig. 10–37); in the *Biston betularia* populations that show large increases of melanic gene frequencies in industrialized regions (p. 541; see also Lees); in increased frequencies of resistant genes in some plant populations exposed to herbicides and metallic toxins (Bradshaw); and in the human populations that Figure 22–3 shows, in which genes that modify red blood cell physiology offer protection against malaria.

Nevertheless, new mutations can occasionally be adaptive and lead to divergence, such as those for new enzymatic functions derived from duplications (p. 562). Whether old or new, mutations supply the variation upon which selection acts, and which selection then incorporates into evolutionary novelty.

LINKAGE

Selection among individual mutant genes on separate chromosomes, however, is not the only method by which genetic progress is achieved. As Mather and others point out, recombination between linked genes may also markedly affect the response to selection. To illustrate this point, let us assume that each of four loci, *Aa, Bb, Cc, Dd,* influence a character quantitatively, and that all capital letter alleles have a positive ("plus") effect on the character and all small letter alleles have a negative ("minus") effect. If the phenotypic optimum is an intermediate one, as it is for many characters, the genotype would benefit if it were also intermediate; for example, *AaBbCcDd.*

One way of achieving such optimum genotypes is for tight linkage to be present between these four loci, in the fashion of *AbCd* on one homologue and *aBcD* on the other. Thus any combination of chromosomes will always have four plus genes and four minus genes, yet the population retains the variability of all the different alleles. Note, however, that such a linkage group must have three crossovers to form chromosomes containing all plus or all minus genes. If selection changes from an intermediate phenotype to an extreme phenotype (either all plus or all minus), some time may elapse before the appropriate crossovers can furnish the most adaptive combinations. Again, recombination rates may dictate the progress of selection by connecting as well as disrupting gene combinations affecting fitness.

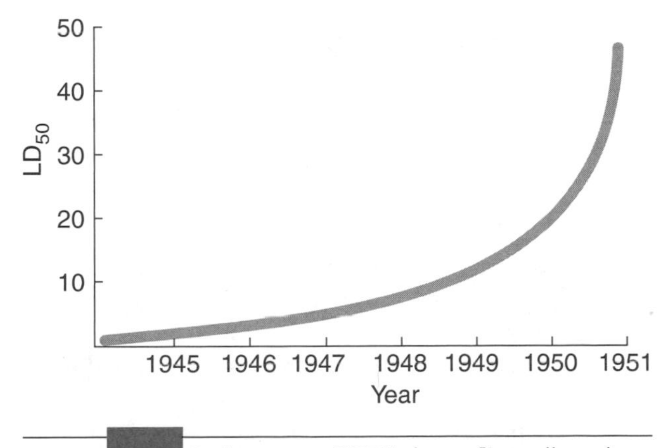

FIGURE 23–4 Resistance to DDT in houseflies collected from Illinois farms measured in terms of the lethal dose necessary to kill 50 percent of the flies (LD_{50}). *(Adapted from Strickberger, data from Decker and Bruce.)*

The Adaptive Landscape

In general, we can see that since more than one locus affects fitness in a population, increased fitness can evolve in many possible ways. For simplicity, we can consider a population containing only homozygous genotypes, where the same four loci just mentioned affect a character, so that equal numbers of capital letter and small letter alleles determine the optimum phenotype. A variety of six optimum genotype is then possible, for example, *AABBccdd, AAbbCCdd, aaBBccDD,* and so forth. According to Wright (1963, and earlier publications), we may consider each of these genotypes to occupy an **adaptive peak,** which means simply a position of high fitness associated with a specific environment. As long as no other factors change the fitness of these genotypes, each of these six peaks is of equal height, and a population consisting entirely of any one of these genotypes would therefore achieve maximum fitness for this phenotype.

When fitness involves more than four loci with more than two alleles at any locus, the number of possible adaptive peaks increases astronomically. A locus with only four alleles has 10 possible diploid gene combinations, and 100 loci with four alleles each have a total of 10^{100} possible gene combinations. Even limited to this relatively small number of loci, the number of possible combinations far exceeds the number of individuals in any species and even the estimated number of protons and neutrons in the universe (2.4×10^{70}). Thus, even if only a small portion of these gene combinations is adaptive, there are undoubtedly more possible adaptive peaks than a species can occupy at any one time.

Again, however, we must clearly differentiate between populations and individuals; that is, a potentially high adaptive peak for a population need not coincide with a high selective peak for a genotype within the population. This discrepancy arises because the selective values of genotypes are based on competition with other genotypes but may not indicate their effect on the population. For example, Haldane, Wright, and others have pointed out that **altruists** who sacrifice themselves for the benefit of shared genotypes may have low selective value as individuals, although a population bearing such altruistic genotypes may have higher reproductive values than one without them (pp. 386, 499, and 570).

Conversely, **social parasites** that increase their frequency at the expense of other genotypes in a population may have high individual selective value, although they depress the reproductive fitness of the population as a whole. Illustrations of the latter type are alleles that modify segregation ratios in their favor (**segregation distorters, meiotic drive**) so that the gametes produced by heterozygotes carrying these alleles consist mostly of such distorters rather than normal nondistorter alleles (p. 202). The frequency of such segregation distorters tends to increase in a population even though some distorters are associated with deleterious or even lethal phenotypic effects such as *tailless* alleles in mice (Silver).[6]

For simplicity, let us assume that the present example of four gene pairs does not involve such complications. Even so, the concept of many adaptive peaks with uniform height departs from real conditions. It is likely that the effects of each of these four pairs of genes may differ considerably, and we can, for example, assign adaptive values to the effects of the *A* and *B* genes on the basis that the greater the number of capital letter genes in these two pairs, the greater the fitness. When we combine them with the previous adaptive values, we can construct an **adaptive landscape** of peaks, as in Figure 23–5, showing the adaptive heights of different possible genotypes.

Note that now one peak superior to all others (*AABBccdd*) arises in a landscape of intermediate peaks, each surrounded by relatively inferior genotypes. A population may increase in fitness during evolution but nevertheless reach an intermediate peak that is not necessarily the most adaptive. To move from peak to peak until a population finds the highest one demands that it travel through inferior genotypes that occupy the lesser **adaptive valleys** of this landscape. Arrows in Figure 23–5 indicate such reductions in fitness, showing the general route a population at *aabbCCDD* might take to reach the highest peak at *AABBccdd.* At least two nonadaptive stages appear in this illustration, at which the population will suffer.

Once such an adaptive landscape has evolved, further evolution will depend on the origin of a new selective environment and the creation of new adaptive peaks. However, if conditions are not changing rapidly, the same set of adaptive peaks may remain for long periods of time. A population on one adaptive peak can then no longer reach a higher peak without going through a nonadaptive valley. Since constant selection can hardly occur for nonadaptation, it seems reasonable to ask: How can a population located on a relatively low adaptive peak evolve so that it occupies the highest or near-highest peaks on the adaptive landscape?

In answer to this problem, Wright proposed that many populations break into small groups of subpopulations. These local populations, or **demes,** are small enough to differ genetically through the nonselective process of random genetic drift but are not so widely separate as to completely prevent gene exchange and the introduction of new genetic variability. The adaptive landscape is therefore occupied by a network of demes, some at higher peaks than others. Thus selection takes place not only between genotypes competing within demes but also between demes competing within a general environment.

Wright called the kaleidoscopic pattern of evolutionary forces acting on these demes the **shifting balance process,** which we can simplify as follows:

1. Random genetic drift, acting on polymorphism and heterozygosity at various loci, allows a number of demes to change their gene frequencies and move across nonadaptive valleys to different parts of the adaptive landscape by developing new fitness values. (Wright believed that gene interaction—epistasis—was a major component of variation in fitness: "Genes favorable in one combination, are, for example, extremely likely to be unfavorable in another.")

[6] In contrast to common expectations that selection for adaptation improves survival, meiotic drive which increases a deleterious genotype's reproductive success can lead to its extinction. Such extremely "selfish" genes and devices can only exist as transient rarities.

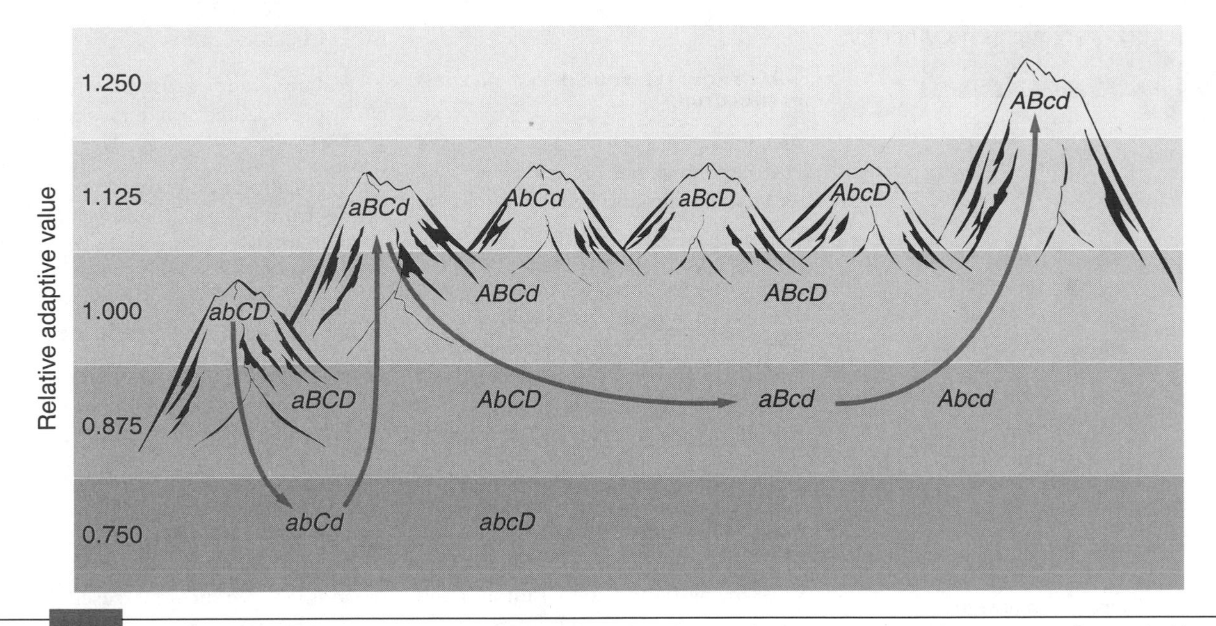

FIGURE 23-5 Adaptive landscape for homozygous genotypes at four loci in which six genotypes homozygous for capital letter alleles at two loci (*AAbbccDD, AAbbCCdd, aaBBccDD,* and so on) attain relatively high adaptive values (*peaks*). Further differences among them are caused by fitnesses of different alleles at the *Aa* and *Bb* loci so that genotypes bearing more capital letter alleles at these loci have higher adaptive values than those that do not (for example, *aaBBCCdd* > *aabbCCDD, AABBccdd* > *AAbbccDD).* If this landscape remains stable, *arrows* show one possible path in the progress of a population from a lower adaptive peak at *aabbCCDD* to the highest adaptive peak at *AABBccdd.* Movement of a population from peak to peak depends on various factors, especially on the size of the population: small populations are more subject to random genetic drift, so their frequencies vary more easily than in large populations. If selection is weak, and a population is not held firmly to a particular peak, movement across a "valley" becomes more favorable. We should also recognize that peaks can be transitory, making the adaptive landscape change like a rubber sheet with adaptive "bumps" arising at different places at different times (reminiscent of the "epigenetic developmental landscape" C. H. Waddington offered in his classic book, *The Strategy of the Genes*). Thus, as the environment changes, adaptive values of genotypes change, and as new alleles are introduced through mutation or migration, interaction with other alleles (epistasis) changes their adaptive values. What is a "valley" at one time need not be a valley at another, and a population's position on the landscape can fluctuate accordingly. (*Adapted from Wright 1963.*)

2. Selection pushes some of these demes up the nearest available adaptive peak by changing gene frequencies even further, that is, by making some loci homozygous or nearly so.
3. Polymorphism retained at other loci, or variability introduced through migration and mutation, provides further opportunity for genetic drift to trigger movement across the adaptive landscape, eventually enabling a population to occupy still higher adaptive peaks.
4. Also, because population subdivision impedes genetic recombination between demes, epistatic gene combinations producing novel advantageous interactions can persist (p. 564, and see also reviews by Fenster et al.). When we add the influence of genetic drift, demes can respond uniquely to selective pressure and evolve in distinctive and diverse directions.
5. A deme that has attained a high adaptive peak tends to displace other demes at lower peaks by expanding in size or dispersing outward and changing the genetic structure of other demes through migration.
6. Environmental change can act on populations like a stream of seismic earthquakes, continually producing new adaptive landscapes surfaced with new slopes and adaptive peaks. Channeling selection in new directions encourages populations to continually shift their genetic structures.[7]
7. Because selection, time, and genetic accident are needed to achieve the most optimum genotypes, while genetic loads and environmental contingencies can oppose such optimal achievement, highest adaptive peaks are potential and not necessarily realized. Populations can, at best, trail behind the summits of oncoming adaptive landscapes.

Fisher, by contrast, suggested that most populations are large and fairly homogeneous, so that selection tests each new allele independently in competition with all other alleles

[7] As Grant and Grant point out in a study of selective changes in Darwin's finches on the Galapagos Islands, "The population tracks a moving peak in an adaptive landscape under environmental fluctuations, and there is more than one individual fitness optimum within the range of phenotypes in the population."

B. ROSEMARY GRANT

Name:
B. Rosemary Grant

Birthday:
October 8, 1936

Undergraduate degree:
University of Edinburgh, Scotland, 1960

Graduate degrees:
Ph.D., University of Uppsala, Sweden, 1985

Present position:
Senior Research Scholar (Professor rank) Department of Ecology and Evolutionary Biology Princeton University, New Jersey

■

WHAT PROMPTED YOUR INITIAL INTEREST IN EVOLUTION?

As a youngster growing up in The Lake District in England, I roamed the woods and fells and became fascinated with the diversity of organisms and particularly the differences between individuals of the same species. Why are no two individuals alike, and to what extent are these differences inherited are questions I discussed first with my parents and much later with professors at the University of Edinburgh.

WHAT DO YOU THINK HAS BEEN MOST VALUABLE OR INTERESTING AMONG THE DISCOVERIES YOU HAVE MADE IN SCIENCE?

With my husband Peter Grant and colleagues I have carried out intensive studies on populations of finches in the Galapagos Islands for more than 25 years. The highlights of this study have been:

- Establishing the heritability of morphological traits and the evolutionary responses to natural selection over short periods of time in natural populations.
- Finding the extent to which genetic, ecological, and behavioral factors interact and the bearing this has on the formation of species.
- Determining the causes and consequences of low levels of hybridization for evolution and the speciation process.
- A recent exciting finding has been the role a culturally transmitted learned trait, imprinting on song, can play in evolution and the early stages of speciation.
- These long-term studies have enabled us to interpret the evolutionary dynamics of natural populations living in climatically variable environments.

WHAT AREAS OF RESEARCH ARE YOU (OR YOUR LABORATORY) PRESENTLY ENGAGED IN?

- Continuing our long-term study of individually marked birds in the Galapagos under the altered ecological conditions caused by the recent unprecedented severe El Nino of 1997–98.
- In collaboration with two postdoctoral fellows, Ken Petren and Lukas Keller, we are using molecular genetic techniques in a study of phylogeny, paternity, hybridization, and inbreeding in Darwin's finches.
- Investigating the consequences of inbreeding when populations fluctuate under extreme climatic conditions.
- Exploring implications of our work for conservation where human induced fragmentation of the environment produces increased incidences of inbreeding and hybridization.

IN WHICH DIRECTIONS DO YOU THINK FUTURE WORK IN YOUR FIELD NEEDS TO BE DONE?

- Understanding the developmental and genetic basis for the modification of traits.
- Investigating the interaction between culturally transmitted learned traits and genetic variation.
- Establishing the connection between patterns of evolution in the past and evolutionary dynamics in contemporary time.

WHAT ADVICE WOULD YOU OFFER TO STUDENTS WHO ARE INTERESTED IN A CAREER IN YOUR FIELD OF EVOLUTION?

It helps to know one organism or assemblage of organisms in depth, but at the same time to remain broadly interested and widely read. In this way intuitions arise and can be explored in a thoroughly understood system. It is important to appreciate that there are many routes to the same objective. A diversity of approaches by different people can work synergistically and lead to a fuller understanding of both the problem and its solution. Most discoveries are the work of many people, not one, and by fostering these individualistic roles we can extend our knowledge.

TABLE 23–4 Comparison of evolutionary processes in a single homogeneous population and in a subdivided population

	Homogeneous Population	Population Subdivided into Demes
What is selected	A gene, differing from its alleles in net selective value	Different gene frequencies
Source of variation	Gene mutation	Random drift between demes and selection toward new adaptive peaks
Process of selection	Selection among individuals	Selection among demes
Evolution under static conditions	Progress restricted to a single peak	Continued shifts as new adaptive peaks are encountered
Evolution under changing conditions	Progress up nearest adaptive peak	Selection between different demes for occupancy of all available peaks

Source: From *Methodology in Mammalion Genetics* by Wright and Burdette, editors. Reprinted by permission of Holden-Day, Inc.

in the population. This large population primarily improves its fitness by small, incremental ("additive") selective steps rather than major random genetic drift.

According to Wright's scheme (Table 23–4), subdivided populations have many evolutionary advantages over a single, large, homogeneous population, and this pattern helps explain how evolution really occurs in most sexually interbreeding species. Selection acting on the products of gene interaction causes the close fit between organismic adaptation and environment, and adaptational change is also influenced by population substructure in which forces such as genetic drift operate.

The dispute between the views of Fisher and Wright still continues. Coyne and coworkers present strong arguments against Wright's shifting balance theory, pointing to its dependence on unsupported assumptions of genetic drift, selection, and gene flow. In contrast, Wade and Goodnight present examples of deme structure and intergroup selection that supports Wright's theory. Until further evidence appears, perhaps both Fisherian and Wrightian populations exist, some large and homogeneous during one period and subdivided during another. Also selection may have multiple objects, acting both on genes with individual phenotypic effects and on genes with epistatic effect. Perhaps, as Wright (1988) stated, different aspects of populations demand different theoretical approaches:

> It is to be noted that the mathematical theories developed by Kimura, Fisher, Haldane, and myself dealt with four very different situations. Kimura's "neutral" theory dealt with the exceedingly slow accumulations of neutral biochemical changes from accidents of sampling in the species as a whole. Fisher's "fundamental theorem of natural selection" was concerned with the total combined effects of alleles at multiple loci under the assumption of panmixia in the species as a whole. He recognized that it was an exceedingly slow process. Haldane gave the most exhaustive mathematical treatment of the case in which the effects of a pair of alleles are independent of the rest of the genome. He included the important case of "altruistic" genes, ones contributing to the fitness of the group at the expense of the individual. I attempted to account for occasional exceedingly rapid evolution on the basis of intergroup selection (differential diffusion) among small local populations that have differentiated at random, mainly by accidents of sampling (i.e., by local inbreeding), exceptions to the panmixia postulated by Fisher.

Wright concludes, "All four are valid."

Group Selection

One important consequence of Wright's shifting balance theory has been to emphasize differences in survival or extinction among populations rather than only among individuals. Selection among individuals in a population is a conservative force that pushes the population up a single adaptive peak, whereas selection among populations (accompanied by random genetic drift) leads to occupation of higher adaptive peaks and replacement, extinction, or colonization of populations at lower adaptive peaks.

To some extent, we have already considered selection among groups in discussing the advantages of sexual reproduction. Because sex may involve hazards as well as significant expenditure of resources, individuals often incur considerable disadvantages. Benefits in the evolution of sex therefore seem most likely related to the variability conferred on the sexual population as a whole (Nunney). The truism of Table 21–1 repeats itself: "populations evolve, not individuals."

Perhaps another factor causing differences among populations that can lead to their differential reproductive success is mutation rate. As explained in Box 12–3, mutation rates that change nucleotide replication fidelity also change prospects for adaptation: high rates generate continual errors and break down adaptations, whereas absent or very low rates inhibit or prevent adaptation to new environments. Achievement of optimal mutation rates may be a matter of selecting among different lineages/populations for replication fidelity of their replicases and polymerases (see also p. 225).

A further example where population selection seems to operate on a level different from individual selection is the **kin selection** model that population biologists use to explain social behavior in some groups of hymenopteran insects—ants, bees, and wasps. As noted in Chapter 16, these groups have haploid males and diploid females ("haplodiploidy"), so that all females (sisters) derived from a single pair of parents are more closely related than are mothers to their own daughters. The genotype of their group, or kin, are therefore benefited by female workers who sacrifice their own reproductive ability and rely instead on their mother's reproductive ability by helping raise sisters rather than producing their own daughters.

Advantages of kin selection can also in be seen in the alarm calls of vervet monkeys (Chapter 20) and Belding's ground squirrels (Sherman). Although these cries enhance the caller's danger, they provide it with indirect benefits by helping its genetic relatives. Cooperative defense roles taken on by group members—lions, primates, cattle, birds, and so forth—involve shared genetic interests in survival. Even fish aggregate in schools that offer increased defense against predators compared to isolated individuals (Fig. 23–6).

A similar type of selection probably occurs in distasteful prey species where mutant individuals arise who possess an aposematic warning pattern (p. 540), but through their mortality help protect related genotypes that carry the same pattern. In the flour beetle, *Tribolium confusum,* Wade has shown experimentally that egg-eating cannibalism by larvae declines in groups in which larvae feed on genetically related eggs. This altruistic behavior of refraining from cannibalism is apparently selected because it enhances the survival of related individuals who would be considered prey in the absence of altruism.

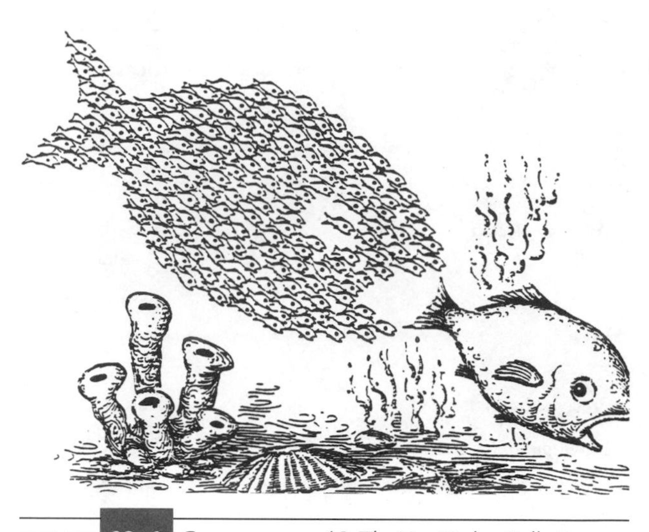

FIGURE 23–6 Group power. (© *The New Yorker Collection 1991 John O'Brien from cartoonbank.com. All Rights Reserved.*)

As the diagram below shows, the fitness effect of altruism on the altruist may be negative compared to selfishness, but it will have a positive effect on others.[8] On the genetic level, an altruist really benefits genes it shares with compatriots, although possibly sacrificing its own. Thus, individuals in a socially interacting group containing many altruists are better off (achieve higher fitness) because the "effect on other" is more positive than a group with fewer altruists.

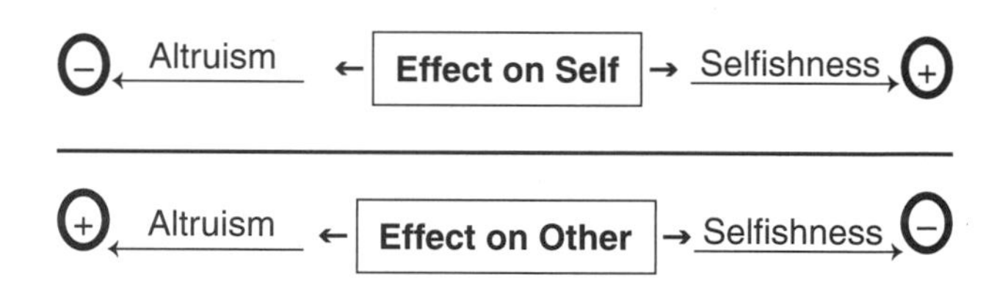

That is, the success of a group depends on a group property (the *frequency* of individuals expressing certain behaviors) rather than on characters confined to only single individuals. As Darwin (1871) put it for humans:

> It must not be forgotten that although a high standard of morality gives but a slight or no advantage to each individual man and his children over the other men of the same tribe, yet that an increase in the number of well-endowed men and an advancement in the standard of morality will certainly give an immense advantage to one tribe over another.

[8] A performer's social behavior may produce negative and positive effects, usually classified as follows:

		Effect on Self	
		Positive	Negative
Effect on Other	Positive	Mutual Benefit	Altruistic sacrifice
	Negative	Selfish advantage	Spiteful malevolence

TABLE 23-5 Interaction effects that can occur between two populations

Type	Effect on Species A	Effect on Species B	Nature of Interaction
Neutralism	0	0	Neither population affects the other.
Commensalism	0	+	Species A (for example, the host) is not affected but species B (the commensal) benefits from the relationship.
Amensalism	0	−	Species A is not affected but species B is inhibited.
Mutualism	+	+	Both species benefit (for example, Müllerian mimicry).
Predation or parasitism	+	−	Species A (predator or parasite) benefits at expense of species B (prey or host).
Competition	−	−	Each species inhibits the other.

Source: Adapted from Pianka.

Evolutionists have long debated the extent to which group selection occurs, or whether it occurs at all (Wilson). Among the arguments against group selection is some authors' insistence (Williams) that, looked at closely, we can explain many so-called group adaptations as arising from selection among individuals. Others (Maynard Smith 1989) show mathematically that selection among groups necessitates very high extinction rates and practically no gene flow.

Nevertheless, most researchers would agree that group selection is at least theoretically possible (Tanaka), and clearly, from the examples just given and others (for example, Stevens et al.), this idea is receiving much more attention and approval than in the past (Sober and Wilson). Some authors have even broadened the discussion to consider "hierarchies" of selection that include terms such as **species sorting** and **species selection;** that is, competition between species (species selection) or accidental factors such as catastrophes that caused survival differences among them (species sorting) (see Vrba).[9] Other mass interactions, such as susceptibility or resistance to parasitic infection, can also involve group survival or extinction. The manner in which natural selection can be evaluated—reproductive success or survival—need not be confined to individuals.

[9] The originator of the term *species selection* was Hugo De Vries, a macromutationist (p. 516) who claimed that selection between species differs from Darwinian natural selection that takes place within species. He presumed that selection between species arose from unique macromutational events responsible for speciation, whereas Darwinian selection could make slight changes but was incapable of speciation. Proponents of the punctuated equilibrium hypothesis discussed later have echoed a related view (pp. 599–600).

Group Interaction

The natural communities of organisms are assemblages of species or groups, each interacting with others in various ways. In terms of survival, growth, or fecundity, a slight increase in the numbers of one particular group may cause an increase (+) in numbers of another group, a decrease (−), or have no discernible effect (0). We generally classify group interactions between two groups according to the terminology of Table 23–5, with categories ranging from neutral interaction (0, 0) to mutual gain (+, +), predation (+, −), and competitive inhibition (−, −).

Ecologists have observed examples and variations of practically all such interactions, and have proposed an extensive variety of models to explain their mechanisms and effects, usually under the headings of population and community ecology (for example, the introductory texts by Begon and Mortimer and by Putman and Wratten). Although these ecological interactions have significant implications for any species, unraveling their exact evolutionary effects is still difficult. We can briefly review a few aspects of two of these interactions, **competition** and **predation.** In their various forms, competition and predation are interactions that often decide which species will be members of a localized community of organisms (Roughgarden and Diamond).

Competition arises when two groups depend on the same limited environmental resource(s) so that each group causes a demonstrable reduction in numbers of the other. Such reduced availability of common resources

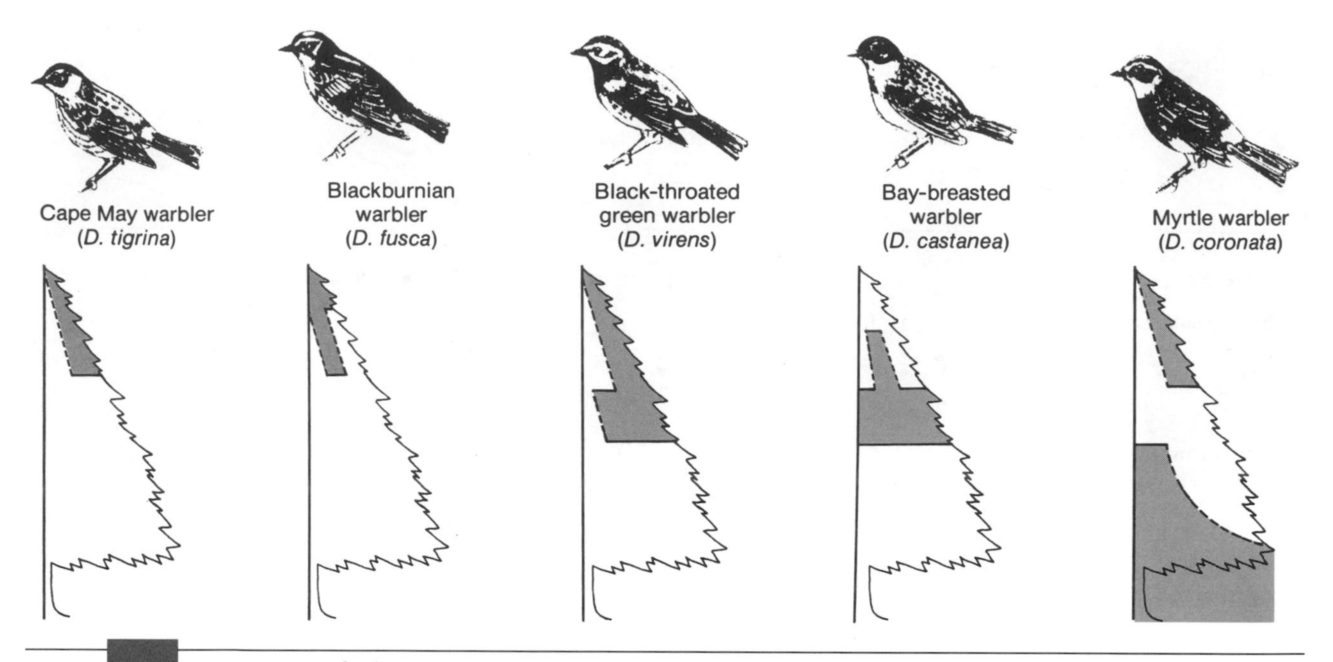

FIGURE 23–7 Most common feeding zones (*shaded*) in spruce trees for five species of northeastern U.S. warblers of the genus *Dendroica,* based on the number of birds observed. The Cape May warbler may be quite rare unless there is a large outbreak of insects. The myrtle warbler is also rare and less specialized than the other species. The generally more common warblers, Blackburnian, bay-breasted, and black-throated green, are different enough in feeding zone preferences to explain their coexistence. Should such resource partitioning be disrupted, and overlap occur, we would expect competition to lower the fitness of competing groups. A study by Martin supports this notion, showing that nest predation increases when nesting sites overlap between different species. (*Adapted from Krebs, based on MacArthur.*)

often has important ecological or behavioral consequences. For example, in some groups natural selection favors the evolution of **protective territorial mechanisms** that inhibit competitors' use of such resources. These devices may include growth-inhibiting chemicals (for example, toxins such as the creosote some plants produce) or aggressive encounters (for example, fighting in many vertebrates) often responsible for species dispersal.

Competition often leads to ecological diversity: it can be to the advantage of competing groups that they minimize the harmful effects of direct competition by using different aspects of their common environmental resources. Among the many examples of such **resource partitioning** is that illustrated in Figure 23–7 for five species of warblers, each using different parts of their spruce tree habitat. A further possible evolutionary response to competition is **character displacement,** where measurable phenotypic differences accompany resource partitioning among coexisting groups.

One prominent example of character displacement occurs among "Darwin's finches" in the Galapagos Islands where coexisting species show large differences in bill sizes, enabling each species to feed on differently sized seeds. In contrast, species isolated on different islands possess intermediate bill sizes enabling them to feed without partitioning seed resources. For example, one beak dimension measures about 8 mm for *Geospiza fuliginosa* and 12 mm for *G. fortis* on islands where both species exist together, whereas it measures about 10 mm for each species on islands where they exist separately (Fig. 23–8).

When competition is not checked by partitioning or fluctuation of resources and two competing species use exactly the same resources in the same environment—that is, they both occupy the same niche[10]—Gause and others have shown in laboratory experiments that one species commonly dies out (Fig. 23–9). This finding supports the **principle of competitive exclusion,** which states that two species cannot continue to coexist in the same environment if they use it in the same way. This principle, also called *Gause's axiom* or *law,* had been foreshadowed by Darwin's statement in *On the Origin of Species:*

> Owing to the high geometrical rate of increase of all organic beings, each area is already fully stocked with inhabitants; and it follows from this, that as the favored forms increase in number, so generally will the less favored decrease and become rare.

However, evolutionists have seriously questioned whether competitive exclusion alone accounts for the differences observed among coexisting species (for example,

[10] There are many niche definitions. Most essentially propose that a species's niche includes all the various environmental resources the species uses as well as the strategies it applies in exploiting these resources.

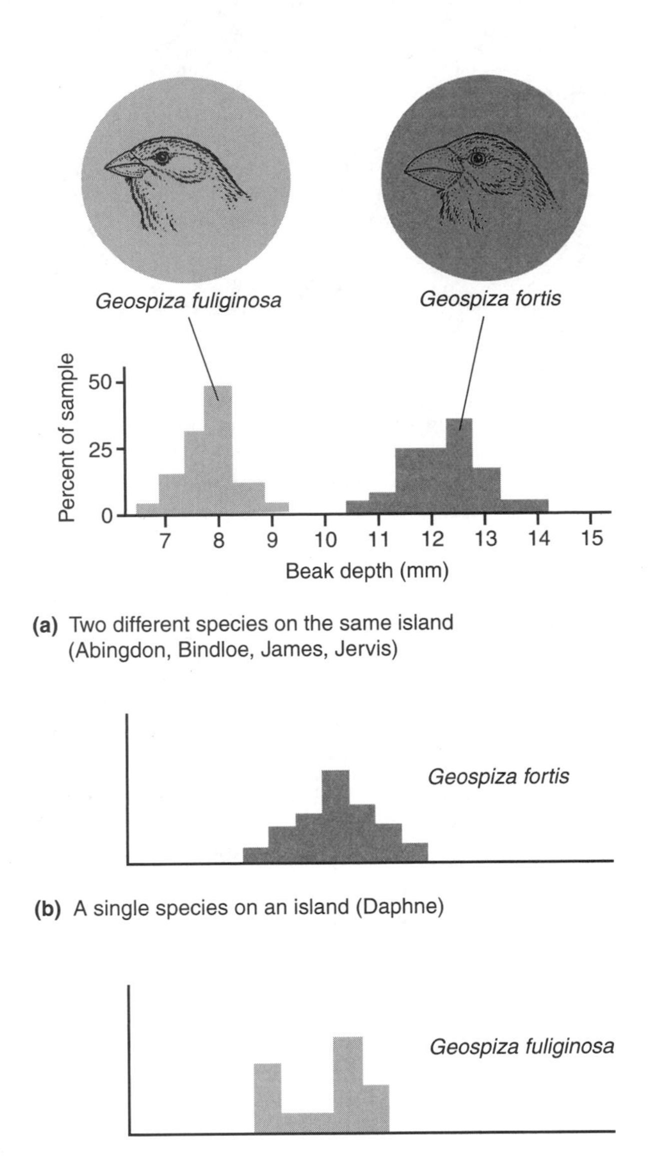

FIGURE 23-8 Character displacement among Darwin's finches in the Galapagos Islands. Coexisting species on four islands (*a*) show large differences in bill sizes, enabling each species to feed on different sized seeds. However, when either species exists alone on different islands (*b, c*), it posseses intermediate bill sizes (about 10 mm) enabling it to feed without partitioning seed resources. (*Adapted from Givnish, based on Grant.*)

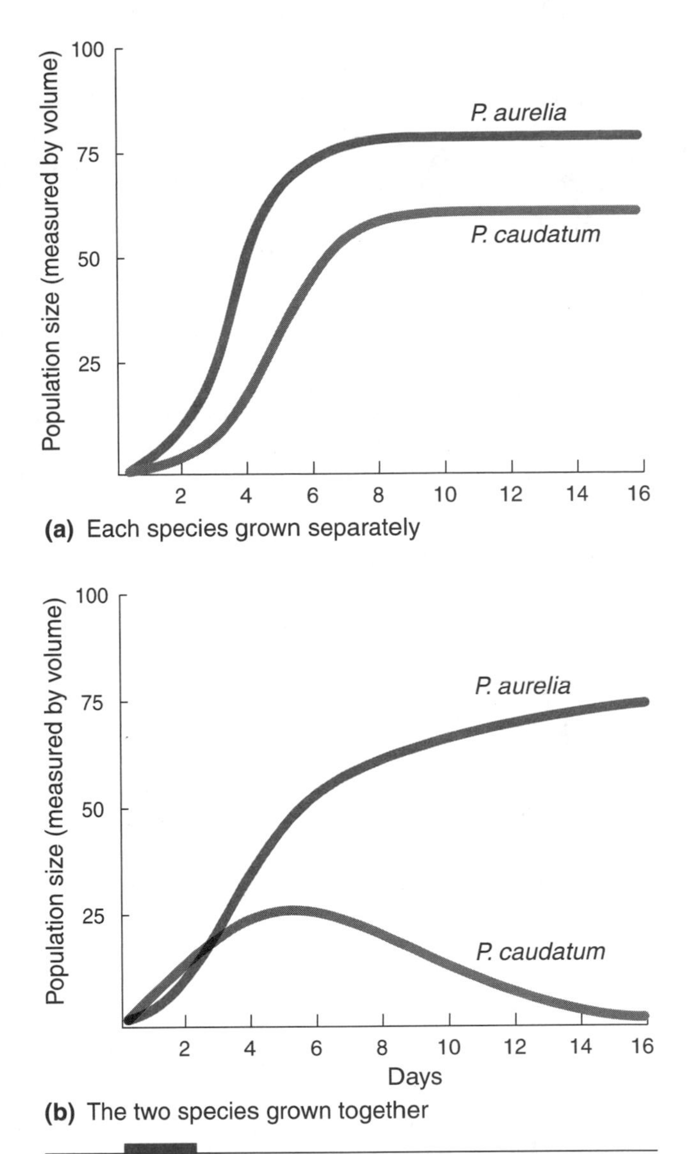

FIGURE 23-9 Growth of two species of *Paramecium* (*a*) in separate cultures, and (*b*) in mixed cultures. Although *P. aurelia* generally replaces *P. caudatum* as shown, in some mixed culture conditions *P. caudatum* multiplies faster than *P. aurelia.* Apparently *P. caudatum* is more sensitive to metabolic pollutants than is *P. aurelia,* so that removing such pollutants encourages *P. caudatum* population growth. The results of competitive interactions may be quite sensitive to external environmental factors. (*Adapted from Gause.*)

character displacement) or for the finding that closely related and potentially competitive species often occupy different habitats. Morphological differences among related coexisting species may have evolved in the past in places where these species did not compete, and related species that occupy different habitats may not have diverged because of competition but because of different food preferences, nesting sites, and so on (Den Boer).

In fact, one can argue that much more coexistence appears among related species (for example, species found in oceanic plankton) than we would expect if species were randomly distributed. Certainly, one important factor that probably diminishes exclusion among competing, coexisting species is predation: predators can reduce the ability of any single dominant species to reach its full potential carrying capacity, thus making room for other competitors.[11]

In predation, the predator entirely or partially consumes its prey, thus affecting the numbers of those organisms it feeds on. Predation may be exercised in various ways on both plants and animals, including overt

[11] This concept is similar to that described as "cropping" in Chapter 14.

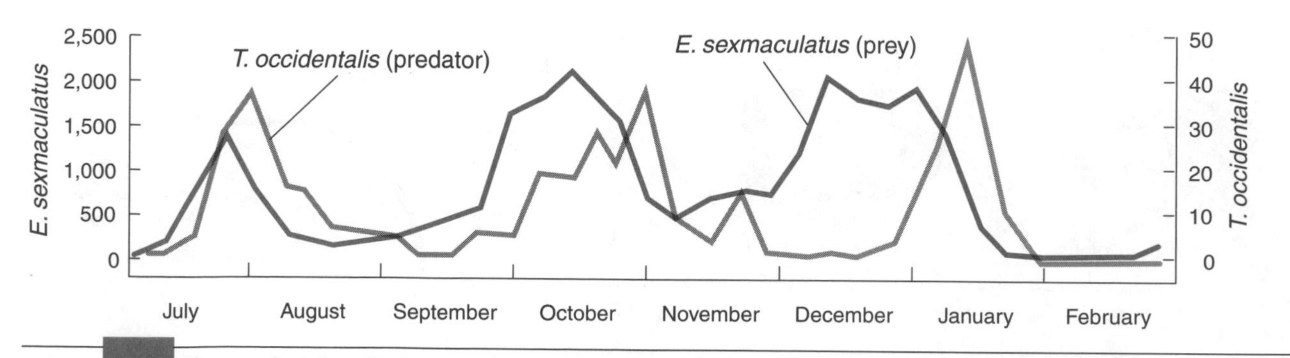

FIGURE 23–10 Three cycles of oscillating population numbers for two species of mites in a defined, controlled environment, one species (*Typhlodromus occidentalis*) being the predator and the herbivorous species (*Eotetranychus sexmaculatus*) being the prey. For each cycle, as numbers of prey increase, predator numbers follow, causing a "crash" in the prey population, followed by a crash in the predator population. (*Adapted from Pianka, from Huffaker.*)

attack and consumption of prey or **parasitism** (infestation and impairment of host tissues). The intimate dependence of predators on their prey often leads to a coupling of their relative abundances: an increase in numbers of prey allows an increase in predators that, in turn, can reduce prey that can then reduce predators. As Figure 23–10 shows, cyclic oscillations in population numbers may become a pattern, especially in the relationships of predators confined to a single species of prey.

Such simplicities are not the rule because some predators don't cause a crash in abundance of prey but, like the predation of wolves on caribou herds, consume mostly prey weakened by age or disease who have little reproductive value ("prudent predation"). Moreover, predator numbers may be buffered when more than one species of prey is being exploited, thereby reducing large oscillations in predator population size and spreading the effects of predation so that prey population sizes also remain fairly stable.

Although researchers have offered various mathematical models to generalize the intricacies of predator–prey relationships, there are no simple universal solutions. As Begon and Mortimer note:

> Predators and prey do not normally exist as simple, two-species systems. To understand the abundance patterns exhibited by two interacting species, these must be viewed in realistic multi-species context. . . . Before multi-species systems are even considered, we must abandon our expectation of universal prey–predator oscillations, and look instead, much more closely, at the ways in which predators and their prey interact in practice.

Such interactions, whether of competition or predation, can be unique—dependent, for example, on the past evolutionary history of the populations involved, spatial limitations, climatic conditions, soil nutrients, and the effects of other species in the community. Each interaction must be disentangled from others and explored separately—an extremely difficult task, but one that has engendered considerable interest and effort among population biologists (Diamond and Case).

Among such examples is evidence that different combinations of environmental factors can affect populations of Canadian snowshoe hares more differently than we would ordinarily expect (Krebs et al.). A choice of either reducing predation or increasing the food supply caused a two- or threefold increase in numbers whereas combining these factors caused an elevenfold increase. Obviously, different interactions occur (perhaps a large reduction in physiological stress) when predation is low and food is plentiful that do not occur when food is plentiful and predators are prevalent, or when predation is low and food is meager.

Coevolution

Group interaction has various consequences. Despite the lack of widely applicable theoretical models, evolutionary changes in one species can prompt evolutionary changes in a species with which it interacts ecologically. Paleontological evidence already points strongly to an evolutionary "arms race" between herbivorous and carnivorous mammals during which size, speed, and intelligence seemed to increase sequentially in various members of both groups (Chapter 19).[12] Similar evolutionary progressions in respect to size, speed, and protective devices must also have occurred among the reptilian dinosaurs

[12] To illustrate the selective mechanism that operates on individuals under predatory pressure, Greenwood recounts the following story of two men in Greenland trapped in a tent by a polar bear. "One, a native Greenlander, began to remove his boots. 'What are you doing?' whispered his Danish companion. 'You know that even without your boots you can't run faster than a polar bear.' 'Yes,' replied the Greenlander, 'But without my boots I can run faster that **you**.' " As Darwin wrote in a letter to Hooker, "Organic beings are not perfect, only perfect enough to struggle with their competitors."

(Chapter 18) and have probably been a consistent trend in many prey–predator groups.

Notable current examples of **coevolution** are those between parasites or pathogens and their hosts. Flor found 27 genes in the flax plant, *Linum usitatissium,* that confer resistance against a fungal rust pathogen, while the pathogen, in turn, had a similar number of genes allowing it to overcome resistance these host genes confer. In such cases and in others, one can reasonably claim that an increased frequency of a resistant mutation in the host will be followed by selection for an increased frequency of one or more mutant genes in the parasite or pathogen that overcomes resistance.

Not unexpectedly, coevolutionary events between host and parasite can be quite complex and may reduce the virulence of the parasite. One prominent example concerns a myxoma virus imported from South America to Australia to control the phenomenal populational growth of the European rabbit, *Oryctolagus cuniculus.* Although the virus caused only a mild disease in native South American cottontail rabbits (*Sylvilagus brasilensis*), it acted as a highly lethal pathogen among the Australian rabbits, being transmitted primarily by mosquitoes. Viral-caused lethality in 1950–1951 was as high as 99 percent among infected rabbits, and it seemed as though the Australian rabbit population would eventually be eradicated by the virus or persist only at very low numbers.

This expectation was not fulfilled, however, because although some Australian rabbits became increasingly resistant to the virus, surprisingly the virus itself became less virulent. Apparently, since mosquitoes feed only on live rabbits, the rate at which they can transmit the virus falls if the virus kills its immediate host too quickly. Highly virulent, rapidly replicating, strains of virus were therefore selected against, whereas strains with reduced virulence were favored because they let infected rabbits live long enough for the virus to spread more easily. (If the cause for these two different virulent modes arises partly from different kinds of interaction between viral particles, this may perhaps provide another example of group selection discussed earlier.)

Interestingly, this host–pathogen relationship is not static: as rabbit resistance to the virus increases, increased viral virulence can become a more favorable and selected trait. Thus, the relationship between virus and rabbit in Australia may eventually emulate the relationship in South America, in which a virulent virus has only a partially harmful effect on its native host. Such evolutionary outcomes are not unusual: some small DNA viruses are quite genetically stable compared to rapidly evolving RNA viruses, and can persist and coevolve with their hosts causing little disease (Shadan and Villerreal).[13]

Although competition, predation, and parasitism reflect the popular concept of "nature red in tooth and claw," instances of cooperation and mutualism certainly modify its impact. Among these examples are cooperative relationships discussed earlier under "Group selection," as well as many instances of symbiotic relationships. Such symbioses include those between cellular organelles and their eukaryotic hosts (Chapter 9), algae and fungi (Fig. 22–6), and cellulose digesting protozoans in termite intestinal tracts. Long periods of coevolution made many such associations obligatory, but others such as between some pollinators and plants may be facultative—neither species relying on the other for survival (Feinsinger).

Depending on how one defines the concept, many other examples of coevolution abound. These include mimics that evolve in step with the evolution of their models, ants that cospeciate with fungi they "farm" for food (Hinkle et al.), competing species that evolve changes between them to reduce competition (for example, character displacement), and many others (Futuyma and Slatkin).

Coevolution must therefore be a common phenomenon because the intimate ecological relationships among many species probably derive from coevolutionary events in which adaptive changes in one species follow adaptive changes in others. What begins as casual interactions between different species can develop into obligatory coevolving associations. There has been, and still is, a great deal of interdependence among many different life forms, and we cannot fully understand them except in an evolutionary framework (Loehle and Pechmann).

SUMMARY

The structure of populations is so complex that quantitative evaluation of their evolutionary potential is extremely difficult. Growth, however, is one characteristic that can be measured. Unlimited population growth is exponential, but as the environment imposes restrictions, the population will tend to stabilize at a size called the *carrying capacity,* or *K.* Strategies such as *r*-selection (increasing the numbers of offspring) or *K*-selection (increasing the selective advantages of offspring) enable populations to approach the carrying capacity.

[13] Reduced parasitic virulence, however, is not always a successful option for competing parasitic strains. As we have seen, parasitic infection and parasitic virulence seem positively correlated, indicating (1) that success among competing parasitic strains may depend on ability to replicate rapidly, and (2) rapid parasitic replication is probably a major factor in causing virulence. Viruses that incorporate into the host chromosome and are transmitted between generations ("vertical transmission") are called *temperate,* and generally have been selected for reduced virulence since they depend on host reproduction for survival. On the other hand, viruses that enter organisms exclusively through infection from other hosts ("horizontal transmission") are generally selected to increase infectivity by replicating rapidly, and thus almost always cause host destruction. Different infective opportunities come into play in selecting viral virulence. The sexually transmitted human immunodeficiency virus (HIV) has apparently increased its virulence in populations that have greater sexual promiscuity (Ewald).

Although some variability is advantageous, the immediate effect of many alleles or genetic combinations may not be, and the extent to which these detrimental genotypes affect a population is called the *genetic load.* Both mutation and heterozygote superiority may contribute to genetic load by perpetuating deleterious recessive alleles. Although this polymorphism reduces the frequency of optimum genotypes, it may increase the adaptiveness of the lineage at some future time.

Because we cannot explain all genetic polymorphisms as resulting from heterozygote superiority, some workers have proposed that most alleles are neutral in effect and incur no genetic load. Under these circumstances polymorphisms will remain in the gene pool, not because of selection but because of mutation and random genetic drift.

Selectionists, however, argue that relationships between ecological conditions and polymorphism, similarities in allozyme frequencies in different species, the association between polymorphism and enzyme function, and the higher frequency of polymorphism in noncoding DNA sequences cannot be entirely explained by random forces but must be due to selection pressures.

Sexual reproduction, recombination of linked genes, and mutation can all produce genetic combinations that increase fitness. When such advantageous genotypes occur, we say they occupy adaptive peaks of varying value according to their degree of fitness. However, to reach higher adaptive peaks, a population must often pass through less adaptive valleys, a process that, among other factors, involves random genetic drift, according to Wright.

Although the issue is still unresolved, in some instances (sexual reproduction and altruistic social behaviors) selection seems to occur for the benefit of the group even though it may harm the individual's own genetic future.

Population interactions such as competition and predation may favor the selection of particular traits. Competition favors niche and character distinctions between groups, in some cases eliminating one group entirely, while predation has complex effects on the size of the populations of both prey and predator. However intricate the interrelationships, clearly one species can influence the evolution of another with which it interacts.

KEY TERMS

adaptive landscape
adaptive peak
adaptive valleys
allozymes
altruists
balanced lethals
balanced load
carrying capacity
character displacement
coevolution
competition
demes
density dependence
density independence
enzyme polymorphisms
fecundity
genetic death
genetic load
group interaction
group selection
hitchhiking
intrinsic rate of increase
iteroparous
kin selection
logistic growth model
meiotic drive
Muller's ratchet
mutational load
net reproductive rate
neutral mutations
non-Darwinian evolution
parasitism
predation
principle of competitive exclusion
protective territorial mechanisms
r- and *K*-selection
rate of increase
recombinational load
resource partitioning
segregation distorters
segregational load
selectionists
semelparous
senescence
shifting balance process
social parasites
species selection
species sorting
survivorship
truncation selection

DISCUSSION QUESTIONS

1. What advantages does mathematical modeling offer evolutionary studies?
2. What is the relationship between the growth rate of a population and its carrying capacity? and its age structure?
3. How do *r*- and *K*-selection strategies differ?
4. Genetic loads
   a. How do geneticists calculate mutational and segregational (balanced) genetic loads?
   b. How do their effects differ?
   c. What changes in these genetic loads would you expect because of increased inbreeding?
5. In explaining the prevalence of genetic polymorphism, what arguments have workers used to support the importance of neutral mutation? of selection?
6. What are the comparative evolutionary advantages of sexual and asexual populations?
7. Does the survival of a population exposed to a new environment primarily depend on new adaptive mutations? Explain.
8. How can linkage and recombination affect adaptation?
9. Can genes that affect the survival of their individual carriers act differently for the survival of the population?
10. According to Wright, how can populations evolve to occupy high adaptive peaks when some genotypic combinations necessary to achieve these peaks are nonadaptive?
11. Why do populations not occupy all the adaptive peaks that are theoretically available?
12. Provide arguments, pro and con, for the concept of group selection.

13. Would you extend the notion of "hierarchies" of selection (p. 571) to include selection among genera, families, orders, classes, and phyla? Why or why not?

14. What are the various possible consequences of species competition? predation? What information would you deem necessary to declare these consequences predictable?

15. Provide examples in which interaction among different species affects their evolutionary direction.

EVOLUTION ON THE WEB

Explore evolution on the web! Visit the accompanying web site for *Evolution,* 3/e at www.jbpub.com/evolution for web exercises and links relating to topics covered in this chapter.

REFERENCES

Allard, R. W., and A. L. Kahler, 1972. Patterns of molecular variation in plant populations. In *Proceedings of the Sixth Berkeley Symposium on Mathematical Statistics and Probability,* vol. 5. University of California Press, Berkeley, pp. 237–254.

Andersson, D. I., and D. Hughes, 1996. Muller's ratchet decreases fitness of a DNA-based microbe. *Proc. Nat. Acad. Sci.,* **93,** 906–907.

Ayala, F. J., 1999. Molecular clock mirages. *BioEssays,* **21,** 71–75.

Ayala, F. J., and M. L. Tracey, 1974. Genetic differentiation within and between species of the *Drosophila willistoni* group. *Proc. Nat. Acad. Sci.,* **71,** 999–1003.

Begon, M., and M. Mortimer, 1986. *Population Ecology,* 2d ed. Blackwell, Oxford, England.

Bell, G., 1982. *The Masterpiece of Nature: The Evolution and Genetics of Sexuality.* University of California Press, Berkeley.

Bernstein, H., F. A. Hopf, and R. E. Michod, 1988. Is meiotic recombination an adaptation for repairing DNA, producing genetic variation, or both? In *The Evolution of Sex,* R. E. Michod and B. R. Levin (eds.). Sinauer Associates, Sunderland, MA., pp. 139–160.

Birdsell, J., and C. Wills, 1996. Significant competitive advantage by meiosis and syngamy in the yeast *Saccharomyces cerevisiae. Proc. Nat. Acad. Sci.,* **93,** 908–912.

Boyce, M. S., 1984. Restitution of *r-* and *K*-selection as a model of density-dependent natural selection. *Ann. Rev. Ecol. Syst.,* **15,** 427–447.

Bradshaw, A. D., 1984. The importance of evolutionary ideas in ecology—and vice versa. In *Evolutionary Ecology,* B. Shorrocks (ed.). Blackwell, Oxford, England, pp. 1–25.

Brues, A. M., 1964. The cost of evolution vs. the cost of not evolving. *Evolution,* **18,** 379–383.

Carlson, T., 1913. Über Geschwindigkeit und grösze der Hefevermehrung in Würze. *Biochemische Zeitschrift,* **57,** 313–334.

Chetverikoff, S. S., 1926. On certain aspects of the evolutionary process from the standpoint of modern genetics. (English translation: 1961, *Proc. Amer. Phil. Soc.,* **105,** 167–195.)

Civetta, A., and R. S. Singh, 1998. Sex-related genes, directional sexual selection, and speciation. *Mol. Biol. and Evol.,* **15,** 901–909.

Coyne, J. A., N. H. Barton, and M. Turelli, 1997. Perspective: A critique of Sewall Wright's shifting balance theory of evolution. *Evolution,* **51,** 643–671.

Crow, J. F., 1988. The importance of recombination. In *The Evolution of Sex,* R. E. Michod and B. R. Levin (eds.). Sinauer Associates, Sunderland, MA, pp. 56–73.

———, 1992. Genetic load. In *Keywords in Evolutionary Biology,* E. F. Keller and E. A. Lloyd (eds.). Harvard University Press, Cambridge, MA, pp. 132–136.

———, 1993. Mutation, mean fitness, and genetic load. *Oxford Surv. Evol. Biol.,* **9,** 3–42.

Darwin, C., 1871. *The Descent of Man, and Selection in Relation to Sex.* Murray, London.

Dawkins, R., 1995. *River Out of Eden: A Darwinian View of Life.* Basic Books, New York.

Decker, G. E., and W. N. Bruce, 1952. House fly resistance to chemicals. *Amer. J. Trop. Med. Hygiene,* **1,** 395–403.

Den Boer, P. J., 1986. The present status of the competitive exclusion principle. *Trends in Ecol. and Evol.,* **1,** 25–28.

DeVries, H., 1905. *Species and Varieties: Their Origin by Mutation.* Open Court, Chicago.

Diamond, J., and T. J. Case, 1986. *Community Ecology.* Harper & Row, New York.

Dobzhansky, Th., 1970. *Genetics of the Evolutionary Process.* Columbia University Press, New York.

Eanes, W. F., M. Kirchner, and J. Yoon, 1993. Evidence for adaptive evolution of the *G6pd* gene in *Drosophila melanogaster* and *Drosophila simulans* lineages. *Proc. Nat. Acad. Sci.,* **90,** 7475–7479.

Emlen, J. M., 1973. *Ecology: An Evolutionary Approach.* Addison-Wesley, Reading, MA.

Ewald, P. W., 1994. *Evolution of Infectious Disease.* Oxford University Press, Oxford, England.

Feinsinger, P., 1983. Coevolution and pollination. In *Coevolution,* D. J. Futuyma and M. Slatkin (eds.). Sinauer Associates, Sunderland, MA, pp. 282–310.

Fenster, C. B., L. F. Galloway, and L. Chao, 1997. Epistasis and its consequences for the evolution of natural populations. *Trends in Ecol. and Evol.,* **12,** 282–286.

Finch, C. E., and R. E. Tanzi, 1997. Genetics of aging. *Science,* **278,** 407–411.

Fisher, R. A., 1930. *The Genetical Theory of Natural Selection.* Clarendon, Oxford, England. (2d ed. 1958, Dover, New York.)

Flor, H. H., 1956. The complementary genic systems in flax and flax rust. *Adv. in Genet.,* **8,** 29–54.

Franklin, I., and R. C. Lewontin, 1970. Is the gene the unit of selection? *Genetics,* **65,** 707–734.

Futuyma, D. J., and M. Slatkin (eds.), 1983. *Coevolution.* Sinauer Associates, Sunderland, MA.

Gause, G. F., 1934. *The Struggle for Existence.* Williams & Wilkins, Baltimore.

Gibson, T. C., M. L. Schleppe, and E. C. Cox, 1970. On fitness of an *E. coli* mutation gene. *Science,* **169,** 686–690.

Gillespie, J. H., 1991. *The Causes of Molecular Evolution.* Oxford University Press, New York.

Gillespie, J. H., and K. Kojima, 1968. The degree of polymorphisms in enzymes involved in energy production compared to that in nonspecific enzymes in two *Drosophila ananassae* populations. *Proc. Nat. Acad. Sci.,* **61,** 582–585.

Givnish, T. J., 1997. Adaptive radiation and molecular systematics: Issues and approaches. In *Molecular Evolution and Adaptive Radiation,* T. J. Givnish and K. J. Sytsma (eds.). Cambridge University Press, Cambridge, England, pp. 1–54.

Grant, B. R., and P. R. Grant, 1989. Natural selection in a population of Darwin's finches. *Amer. Nat.,* **133,** 377–393.

Grant, P. R., 1986. *Ecology and Evolution of Darwin's Finches.* Princeton University Press, Princeton, NJ.

Greenwood, J. J. D., 1984. The evolutionary ecology of predation. In *Evolutionary Ecology,* B. Shorrocks (ed.). Blackwell, Oxford, England, pp. 233–273.

Haldane, J. B. S., 1960. More precise expressions for the cost of natural selection. *J. Genet.,* **57,** 351–360.

Hall, J. G., and C. Wills, 1987. Conditional overdominance at an alcohol dehydrogenase locus in yeast. *Genetics,* **117,** 421–427.

Hamilton, W. D., R. Axelrod, and R. Tanese, 1990. Sexual reproduction as an adaptation to resist parasites (a review). *Proc. Nat. Acad. Sci.,* **87,** 3566–3573.

Hart, A., and M. Begon, 1982. The status of general life-history strategy theories, illustrated in winkles. *Oecologia,* **52,** 37–42.

Hartl, D. L., and A. G. Clark, 1989. *Principles of Population Genetics,* 2d ed. Sinauer Associates, Sunderland, MA.

Hartl, D. L., and D. E. Dykhuizen, 1981. Potential for selection among nearly neutral allozymes of 6-phosphogluconate dehydrogenase in *Escherichia coli. Proc. Nat. Acad. Sci.,* **78,** 6344–6348.

Hey, J., 1999. The neutralist, the fly, and the selectionist. *Trends in Ecol. and Evol.,* **14,** 35–38.

Hickey, D. A., and M. R. Rose, 1988. The role of gene transfer in the evolution of eukaryotic sex. In *The Evolution of Sex,* R. E. Michod and B. R. Levin (eds.). Sinauer Associates, Sunderland, MA., pp. 161–175.

Hinkle, G., J. K. Wetterer, T. R. Schultz, and M. L. Sogin, 1994. Phylogeny of the attine ant fungi based on analysis of small subunit ribosomal RNA gene sequences. *Science,* **266,** 1695–1697.

Holmquist, G. P., and J. Filipski, 1994. Organization of mutations along the genome: A prime determinant of genome evolution. *Trends in Ecol. and Evol.,* **9,** 65–69.

Huffaker, C. B., 1958. Experimental studies on predation: Dispersion factors and predator-prey oscillations. *Hilgardia,* **27,** 343–383.

Hughes, K. A., 1995. The evolutionary genetics of male life-history characters in Drosophila melanogaster. *Evolution,* **49,** 521–537.

Hutchinson, G. E., 1978. *An Introduction to Population Ecology.* Yale University Press, New Haven, CT.

Johnson, G. B., 1974. Enzyme polymorphism and metabolism. *Science,* **184,** 28–37.

Kidwell, M. G., 1972. Genetic change of recombination value in *Drosophila melanogaster.* I. Artificial selection for high and low recombination and some properties of recombination-modifying genes. *Genetics,* **70,** 419–432.

Kimura, M., 1983. *The Neutral Theory of Molecular Evolution.* Cambridge University Press, Cambridge, England.

Kimura, M., and J. F. Crow, 1964. The number of alleles that can be maintained in a finite population. *Genetics,* **49,** 725–738.

Kimura, M., and T. Ohta, 1972. Population genetics, molecular biometry, and evolution. In *Proceedings of the Sixth Berkeley Symposium on Mathematical Statistics and Probability,* vol. 5. University of California Press, Berkeley, pp. 43–68.

King, J. L., 1967. Continuously distributed factors affecting fitness. *Genetics,* **55,** 483–492.

King, J. L., and T. H. Jukes, 1969. Non-Darwinian evolution: Random fixation of selective neutral mutations. *Science,* **164,** 788–798.

Koehn, R. K., 1969. Esterase heterogeneity: Dynamics of a polymorphism. *Science,* **163,** 943–944.

Koehn, R. K., A. J. Zera, and J. G. Hall, 1983. Enzyme polymorphism and natural selection. In *Evolution of Genes and Proteins,* M. Nei and R. Koehn (eds.). Sinauer Associates, Sunderland, MA., pp. 115–136.

Kojima, K., and K. M. Yarbrough, 1967. Frequency-dependent selection at the esterase 6 locus in a population of *Drosophila melanogaster. Proc. Nat. Acad. Sci.,* **57,** 645–649.

Krebs, C. J., 1985. *Ecology,* 3d ed. Harper & Row, New York.

Krebs, C. J., S. Boutin, R. Boonstra, A. R. E. Sinclair, J. N. M. Smith, M. R. T. Dale, K. Martin, and R. Turkington, 1995. Impact of food and predation on the snowshoe hare cycle. *Science,* **269,** 1112–1115.

Kreitman, M., 1983. Nucleotide polymorphism at the *alcohol dehydrogenase locus of Drosophila melanogaster. Nature,* **304,** 412–417.

Lees, D. R., 1981. Industrial melanism: Genetic adaptation of animals to air pollution. In *Genetic Consequences of Man Made Change,* J. A. Bishop and L. M. Cook (eds.). Academic Press, London, pp. 129–176.

Li, C. C., 1963. The way the load ratio works. *Amer. J. Hum. Genet.,* **15,** 316–321.

Li, W.-H., 1997. *Molecular Evolution.* Sinauer Associates, Sunderland, MA.

Loehle, C., and J. H. K. Pechmann, 1988. Evolution: The missing ingredient in systems ecology. *Amer. Nat.,* **132,** 884–899.

MacArthur, R. H., 1958. Population ecology of some warblers of northeastern coniferous forests. *Ecology,* **39,** 599–619.

Martin, T. E., 1996. Fitness costs of resource overlap among coexisting bird species. *Nature,* **380,** 338–340.

Mather, K., 1953. The genetical structure of populations. *Symp. Soc. Exp. Biol.,* **7,** 66–95.

Maynard Smith, J., 1978. *The Evolution of Sex.* Cambridge University Press, Cambridge, England.

———, 1988. The evolution of recombination. In *The Evolution of Sex,* R. E. Michod and B. R. Levin (eds.). Sinauer Associates, Sunderland, MA, pp. 106–125.

———, 1989. *Evolutionary Genetics.* Oxford University Press, Oxford, England.

Michod, R. E., 1993. Genetic error, sex, and diploidy. *J. Hered.,* **84,** 360–371.

Michod, R. E., and B. R. Levin (eds.), 1988. The *Evolution of Sex.* Sinauer Associates, Sunderland, MA.

Milkman, R. D., 1967. Heterosis as a major cause of heterozygosity in nature. *Genetics,* **55,** 493–495.

———, 1973. Electrophoretic variation in *Escherichia coli* from natural sources. *Science,* **182,** 1024–1026.

Mueller, L. D., 1987. Evolution of accelerated senescence in laboratory populations of *Drosophila. Proc. Nat. Acad. Sci.,* **84,** 1974–1977.

———, 1988. Evolution of competitive ability in *Drosophila* by density-dependent natural selection. *Proc. Nat. Acad. Sci.,* **85,** 4383–4386.

Mueller, L. D., P. Guo, and F. J. Ayala, 1991. Density-dependent natural selection and trade-offs in life history traits. *Science,* **253,** 433–435.

Muller, H. J., 1932. Some genetic aspects of sex. *Amer. Nat.,* **66,** 118–138.

Nevo, E., 1983. Population genetics and ecology. In *Evolution from Molecules to Men,* D. S. Bendall (ed.). Cambridge University Press, Cambridge, England, pp. 287–321.

Ngai, J., M. M. Dowling, L. Buck, R. Axel, and A. Chess, 1993. The family of genes encoding adorant receptors in channel catfish. *Cell,* **72,** 657–666.

Nunney, L., 1989. The maintenance of sex by group selection. *Evolution,* **43,** 245–257.

Oakeshott, J. G., J. B. Gibson, P. R. Anderson, W. R. Knibb, D. G. Anderson, and G. K. Chambers, 1982. Alcohol dehydrogenase and glycerol-3-phosphate dehydrogenase clines in *Drosophila melanogaster* on different continents. *Evolution,* **36,** 86–96.

Partridge, L., and N. H. Barton, 1993. Optimality, mutation and the evolution of ageing. *Nature,* **362,** 305–311.

Peck, J. R., 1994. A ruby in the rubbish: Beneficial mutations, deleterious mutations and the evolution of sex. *Genetics,* **137,** 597–606.

Pianka, E. R., 1988. *Evolutionary Ecology,* 4th ed. Harper & Row, New York.

Prakash, S., R. C. Lewontin, and J. L. Hubby, 1969. A molecular approach to the study of genic heterozygosity in natural populations. IV. Patterns of genic variation in central, marginal and isolated populations of *Drosophila pseudoobscura. Genetics,* **61,** 841–858.

Putman, R. J., and S. D. Wratten, 1984. *Principles of Ecology.* Croom Helm, London.

Rose, M. R., 1985. The evolution of senescence. In *Evolution: Essays in Honor of John Maynard Smith,* P. J. Greenwood, P. H. Harvey, and M. Slatkin (eds.). Cambridge University Press, Cambridge, England, pp. 117–128.

Roughgarden, J., 1979. *Theory of Population Genetics and Evolutionary Ecology: An Introduction.* Macmillan, New York.

Roughgarden, J., and J. Diamond, 1986. Overview: The role of species interactions in community ecology. In *Community Ecology,* J. M. Diamond and T. J. Case (eds.). Harper & Row, New York, pp. 333–343.

Sampsell, B., and S. Sims, 1982. Effect of *adh* genotype and heat stress on alcohol tolerance in *Drosophila melanogaster. Nature,* **296,** 853–855.

Selander, R. K., and B. R. Levin, 1980. Genetic diversity and structure in *Escherichia coli* populations. *Science,* **210,** 545–547.

Shadan, F. F., and L. P. Villarreal, 1993. Coevolution of persistently infecting small DNA viruses and their hosts linked to host-interactive regulatory domains. *Proc. Nat. Acad. Sci.,* **90,** 4117–4121.

Sherman, P. W., 1980. The limits of ground squirrel nepotism. In *Sociobiology: Beyond Nature/Nurture?* G. W. Barlow and J. Silverberg (eds.). Westview Press, Boulder, CO, pp. 505–544.

Silver, L. M., 1993. The peculiar journey of a selfish chromosome: Mouse *t* haplotypes and meiotic drive. *Trends in Genet.,* **9,** 250–254.

Sober, E., and D. S. Wilson, 1998. *Unto Others: The Evolution and Psychology of Unselfish Behavior.* Harvard University Press, Cambridge, MA.

Somero, G. N., 1986. Protein adaptation and biogeography: Threshold effects on molecular evolution. *Trends in Ecol. and Evol.,* **1,** 124–127.

Stearns, S. C., 1987. *The Evolution of Sex and Its Consequences.* Birkhaüser Verlag, Basel, Switzerland.

Stevens, L., C. J. Goodnight, and S. Kalisz, 1995. Multilevel selection in natural populations of *Impatiens capensis. Amer. Nat.,* **145,** 513–526.

Strickberger, M. W., 1985. *Genetics,* 3d ed. Macmillan, New York.

Strong, D. R., 1986. Density vagueness: Abiding the variance in demography of real populations. In *Community Ecology,* J. Diamond and T. J. Case (eds.). Harper & Row, New York, pp. 257–268.

Sved, J. A., T. E. Reed, and W. F. Bodmer, 1967. The number of balanced polymorphisms that can be maintained in a natural population. *Genetics,* **55,** 469–481.

Takahata, N., 1996. Neutral theory of molecular evolution. *Current Opinion Genet. Develop.,* **6,** 767–772.

Tanaka, Y., 1996. A quantitative genetic model of group selection. *Amer. Nat.,* **148,** 660–683.

Vrba, E. S., 1989. Levels of selection and sorting with special reference to the species level. *Oxford Surv. Evol. Biol.,* **6,** 111–168.

Wade, M. J., 1980. An experimental study of kin selection. *Evolution,* **34,** 844–855.

Wade, M. J., and C. J. Goodnight, 1998. Perspective: The theories of Fisher and Wright in the context of metapopulations: When nature does small experiments. *Evolution,* **52,** 1537–1553.

Wallace, B., 1970. *Genetic Load: Its Biological and Conceptual Aspects.* Prentice Hall, Englewood Cliffs, NJ.

Whitfield, L. S., R. Lovell-Badge, and P. N. Goodfellow, 1993. Rapid sequence evolution of the mammalian sex-determining gene *SRY. Nature,* **364,** 713–715.

Williams, G. C., 1966. *Adaptation and Natural Selection: A Critique of Some Current Evolutionary Thought.* Princeton University Press, Princeton, NJ.

Wills, C., 1981. *Genetic Variability.* Clarendon Press, Oxford, England.

Wilson, D. S., 1992. Group selection. In *Keywords in Evolution ary Biology,* E. Fox Keller and E. A. Lloyd (eds.). Harvard University Press, Cambridge, MA, pp. 145–148.

Wright, S., 1963. Genic interaction. In *Methodology in Mammalian Genetics,* W. J. Burdette (ed.). Holden-Day, San Francisco, pp. 159–192.

———, 1978. *Evolution and the Genetics of Populations: Vol. 4. Variability within and among Natural Populations.* University of Chicago Press, Chicago.

———, 1988. Surfaces of selective value revisited. *Amer. Nat.,* **131,** 115–123.

Yokoyama, R., and S. Yokoyama, 1990. Convergent evolution of the red- and green-like pigment in fish, *Astyanax fasciatus,* and human. *Proc. Nat. Acad. Sci.,* **87,** 9315–9318.

From Races to Species

We have learned that the interbreeding nature of a sexual species serves as an important cohesive force that holds it together and enables it to share a common gene pool. At the same time, we understand that such a species may consist of many individual populations with various degrees of interbreeding. For example, Epperson has shown that an entirely heterozygous model population (for example, all *Aa*) may produce irregularly dispersed clusters of genotypes (for example, *Aa, AA, aa*) if individuals persistently mate with their neighbors. We can then expect widely separated populations to have less opportunity to share gene pools than those closer together, thus separating a species into various genetically diverse geographical subunits. Because the forces acting on these subunits may change among localities, it will come as no surprise to find observable differences among populations.

In the yarrow plant *Achillea,* a transect across central California shows populations differing significantly in factors such as height and growing season (Fig. 24–1). We see the adaptive nature of most of these differences in the different responses of these populations when originating from different localities. Coastal plants are weak when grown at higher altitudes, and the high-altitude forms grow poorly at much lower altitudes (Fig. 24–2). Long ago, Turesson noted the adaptive features of many such plant populations as the genetic response of a population to a particular ecological habitat.

Where we can score gene frequencies, researchers have well documented many instances of changes between localities during various time intervals. As Figure 10–35 shows, the frequencies of third-chromosome arrangements in *Drosophila pseudoobscura* differ notably in a range of environments across the American Southwest and also undergo significant seasonal changes. Further genetic changes in this species extend over longer periods of time, such as the significant increase in the frequency of one arrangement (Pikes Peak) in many California populations from almost zero to as high as 10 percent over a 17-year period. Populations of the British peppered moth, *Biston betularia,* also show significant changes in melanism frequency for even longer periods, all associated with specific localities and environments (p. 541).

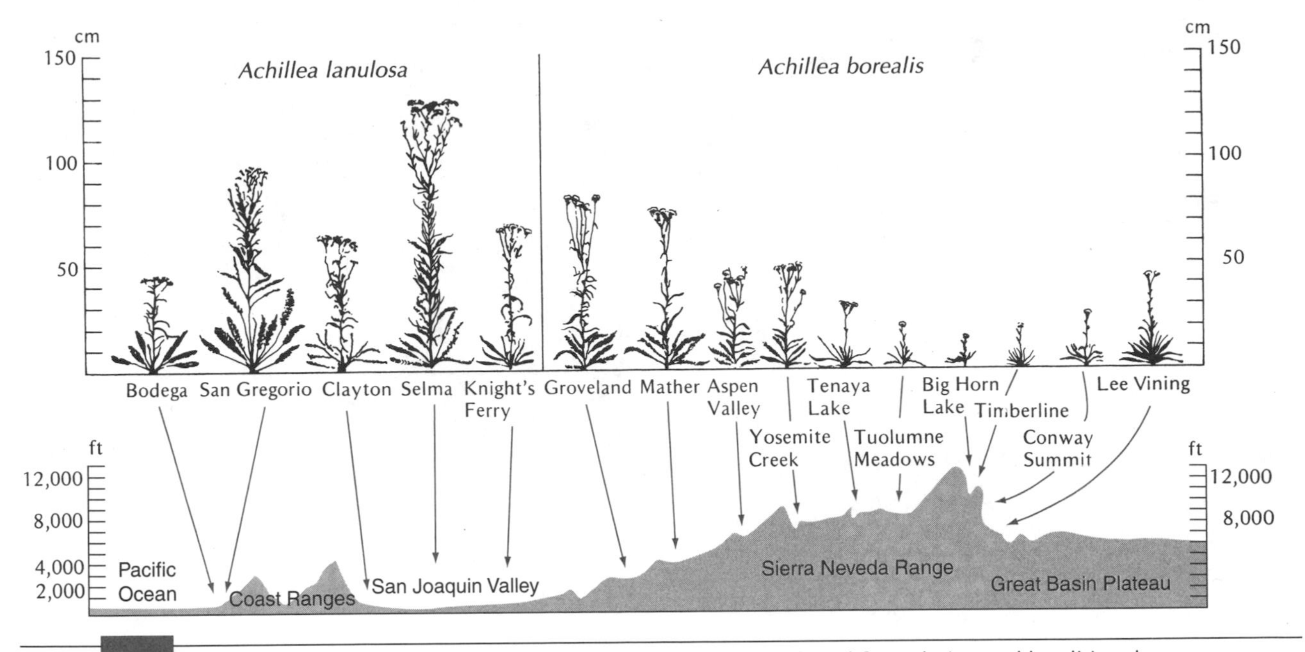

FIGURE 24–1 Representative plants from different populations of *Achillea* gathered from designated localities along a transect across central California and grown in a garden at Stanford, California. The fact that these populations, grown in a uniform environment, differ in terms of plant size, leaf shape, and other characteristics indicates that genetic differences have evolved among them. (*Adapted from Clausen et al.*)

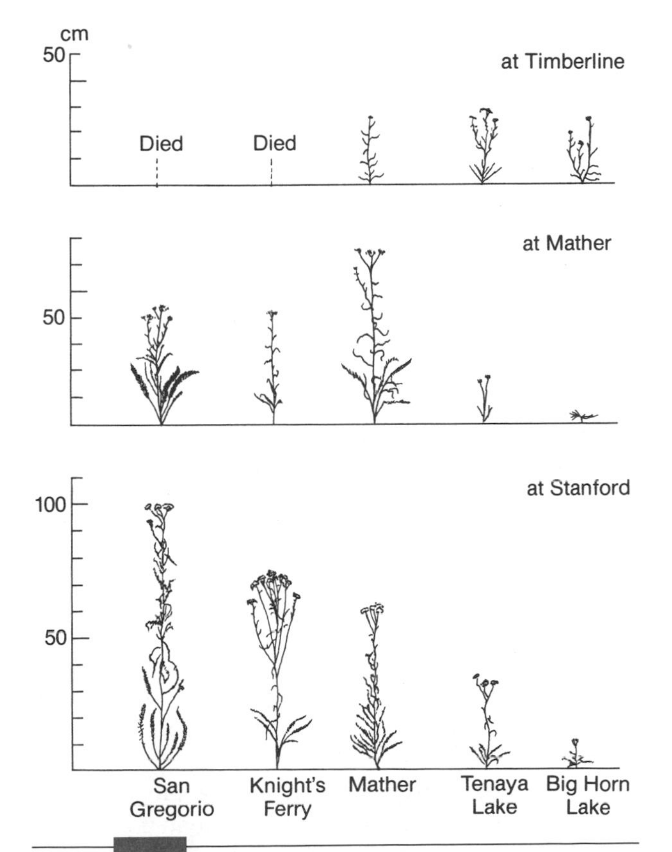

FIGURE 24–2 Responses of clones from representative *Achillea* plants originating from five localities in California and grown at three different altitudes: sea level (Stanford), 4,600 feet (Mather), and 10,000 feet (Timberline). (*Adapted from Clausen et al.*)

Races

In general, geneticists have characterized as **races** populations of the same species that differ markedly from each other. Races share the possibility of participating in the gene pool of the entire species, although they are sufficiently separated to exhibit individually unique gene frequencies. The distinction among races is not absolute: races may differ in the relative frequency of a particular gene, but these differences do not prohibit gene exchange.

For genes whose frequencies we can detect and score, racial distinctions are not simply discerned from the presence or absence of particular genes but are, in many instances, a matter of gene frequencies. Table 24–1 shows a comparison of frequencies for a variety of gene systems in three major human racial groups. In practically all these gene systems, knowledge of a particular genotype alone is not by itself enough to indicate to which race an individual belongs. An individual of O blood type who is also Rh positive may, for example, belong to any of the races listed if we look only at these genes.

It is interesting to note that differences among human populations have not reached the point where one population is fixed for one allele at a particular locus and another population is fixed for a different allele: when a population shows fixation for one allele, other populations are always polymorphic for it. To calculate genetic divergences under such circumstances, Nei proposed one

TABLE 24–1 Frequencies of alleles in samples of individuals taken from three broadly defined human races

Gene Locus	Allele	Caucasians (whites)	Africans (blacks)	Asians (Mongols)
PROTEINS				
Acid protein	Pa^1	.21	.14	.42
	Pa^0	.79	.86	.58
Adenylate cyclase	AK^1	.96	.99	1.00
	AK^2	.04	.01	—
Esterase D	ESD^1	.89	.97	.66
	ESD^2	.11	.03	.34
Glyoxylase I	GLO^1	.44	.26	.09
	GLO^2	.56	.74	.91
Haptoglobin-α	Hp^1	.43	.51	.24
	Hp^2	.57	.49	.76
BLOOD GROUPS				
ABO	A	.24	.19	.27
	B	.06	.16	.17
	O	.70	.65	.56
Duffy	Fy^a	.41	.06	.90
	Fy^b	.59	.94	.10
MN	M	.54	.58	.53
	N	.46	.42	.47
Rh (simplified to two alleles)	Rh^+	.62	.70	.95
	rh^-	.38	.30	.05

Source: Adapted from Strickberger, data from Nei and Roychoudhury.

procedure shown in Table 24–2. When we apply such an **index of genetic distance** (D) to human populations (Fig. 24–3), we can discern five major racial groups:

1. AFRICAN Includes various black tribes and groups that were indigenous to Africa.
2. CAUCASIAN Includes a variety of white European populations ranging from the Lapps of Scandinavia to the Mediterranean peoples of Southern Europe and North Africa.
3. GREATER ASIAN Includes Mongoloid peoples as well as Polynesians and Micronesians.
4. AMERINDIAN Includes North American Eskimos and Indians as well as South American Indians.
5. AUSTRALOID Groups native to Australia and Papua (Australopapuans).

Using the procedure in Table 24–2, Nei and Roychoudhury estimated that, when averaged over many loci, genetic distance in humans (D) accumulates at some constant rate that may be roughly 3.75 million years per unit of D. Thus for the 85 loci that allow comparisons among the three major races, the estimated times when these races initially diverged from each other are:

- Caucasian from Asian (D, .019) = 41,000 ± 15,000 years ago
- Caucasian from African (D, .032) = 113,000 ± 34,000 years ago
- African from Asian (D, .047) = 116,000 ± 34,000 years ago

The oldest divergences are apparently between the Africans and other races, a view that other findings, mentioned in Chapter 20, seem to support. Also striking in these data is the consistently high levels of variability for the many genes examined. According to Nei and Roychoudhury, the proportion of loci that were polymorphic in the three races ranged from 45 to 52 percent for proteins and from 34 to 56 percent for blood groups. The average frequency of heterozygotes per locus ranged from 13 to 16 percent for proteins and from 11 to 20 percent for blood groups.

The presence of so much genetic variability in human races indicates the fictional nature of concepts such as "pure" races. Members of a race are not genetically pure in the sense of sharing a uniform genetic identity, nor does genetic uniformity even apply to members of the same family. Templeton (1997) points out that about 84

TABLE 24–2 Example of the Nei procedure in calculating indices of genetic identity (I) and genetic distance (D) between Caucasians and Africans for two of the loci given in table 24–1

Locus	Allele	Caucasians	Africans
Acid protein	Pa^1	$p_1 = .21$	$p_2 = .14$
	Pa^2	$q_1 = .79$	$q_2 = .86$
ABO blood group	A	$p_1 = .24$	$p_2 = .19$
	B	$q_1 = .06$	$q_2 = .16$
	O	$r_1 = .70$	$r_2 = .65$

$$I = \frac{\text{arithmetic mean of the products of allele frequencies}}{\text{geometric mean of the homozygote frequencies}}$$

$$I = \frac{[(p_1 \times p_2) + (q_1 \times q_2) + (r_1 \times r_2)]/\text{number of loci}}{\sqrt{[(p_1)^2 + (q_1)^2 + (r_1)^2] \times [(p_2)^2 + (q_2)^2 + (r_2)^2]/(\text{number of loci})^2}}$$

$$I = \frac{[(.21 \times .14) + (.79 \times .86) + (.24 \times .19) + (.06 \times .16) + (.70 \times .65)]/2}{\sqrt{[.21^2 + .79^2 + .24^2 + .06^2 + .70^2] \times [.14^2 + .86^2 + .19^2 + .16^2 + .65^2]/(2)^2}}$$

$$I = \frac{1.2190/2}{\sqrt{(1.2194)(1.2434)/4}} = \frac{.6095}{\sqrt{1.5162/4}} = \frac{.6095}{.6157} = .9899$$

$$D \text{ (genetic distance)} = -\ln I = -\ln .9899 = .0101$$

Note: I can range from zero (no similar alleles between the two populations) to one (complete similarity of alleles and frequencies). When $I = 1$, genetic dissimilarity (D) is zero. When I approaches zero, D can increase to very large values, indicating that the alleles at some or many loci have been replaced one or more times.

Source: From Strickberger.

percent of the genetic variability among humans comes from differences among individuals and groups of the same race and only 16 percent comes from differences among races. From a genetic point of view we can only ascribe "purity" to asexual clones derived from a single individual. In clonal reproduction, however, the terms *race* and *species* may not be appropriate, and workers have devised other terms to describe populations among microorganisms (Sonneborn).

We essentially base our criterion for evaluating differences among populations of a species on gene frequency differences. When these differences are extensive, involving many genes, and it is advantageous to consider populations as separate entities, we may categorize them broadly as races. At times, observable morphological differences accompany racial differences, as among some human populations. At other times, observed racial differences extend only to gene or chromosomal differences such as those between Texas and California populations of *D. pseudoobscura* (Fig. 10–35).

We should keep in mind that terms specifying populational differences can vary: taxonomists often use *subspecies, variety,* or even *subvariety* rather than *race* to designate taxonomically distinct groups within species. Since it is quite difficult to distinguish biologically and evolutionarily among these various terms (as well as between other terms such as *geographical race, ecotype,* and *ecological race*), the commonly accepted term *race* is used here to designate any group that we can differentiate from another on the basis of its unique gene frequencies.

Adaptational Patterns

As we have seen, the forces producing racial differences are often adaptive; that is, at least some gene frequency changes are the response of a population to the selective forces operating within a particular environment.[1] Climate, terrain, prey,

[1]Evaluating what is or is not an adaptation can be controversial because a feature's underlying functional and selective values are not always obvious, especially since these often derive from unknown past events. Some authors even question the validity of searching for functional values, whether current or retrospective, since many characters may be nonfunctional by-products of processes unrelated to the feature itself. In a widely cited paper, Gould and Lewontin claim that the "spandrels" in the Venetian Church of San Marco (curved triangular spaces between arches

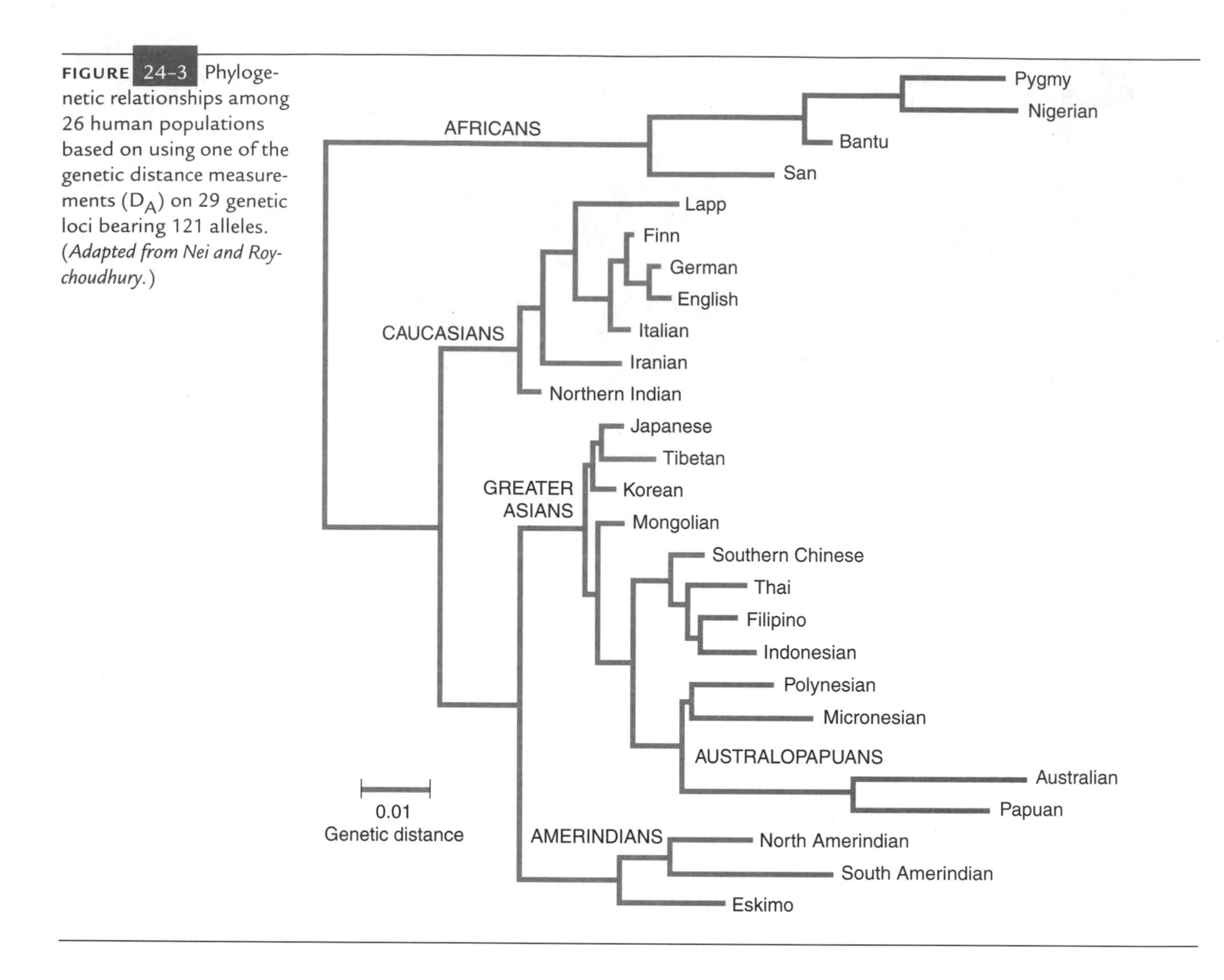

FIGURE 24-3 Phylogenetic relationships among 26 human populations based on using one of the genetic distance measurements (D_A) on 29 genetic loci bearing 121 alleles. (*Adapted from Nei and Roychoudhury.*)

and predators can evince specific adaptations to differentiate a population, as the many remarkable examples of camouflage and mimicry attest. In fact, a number of "rules" generalize the adaptive response of populations to certain ecological and geographical conditions. As with many other generalizations, exceptions to these **ecogeographical rules** exist, but the rules point to the importance of environmental selection in exacting parallel-convergent evolutionary changes in different species.

supporting the domed roof) are, like many organic structures, nonfunctional artifacts. That is, both spandrels and various organismic features are really nonadaptive, arising from the constraints of constructional design rather than from function and adaptation. For example, they suggest that we could explain the tiny forelegs of the dinosaur *Tyrannosaurus* as a "developmental correlate" that accompanies an increased size of the head and hind limbs, and they characterize adaptive explanations for this and other phenomena as "just-so stories." However, although the testing of adaptive explanations is often difficult (as is also true for nonadaptive explanations), the fact that every organismal lineage has been subjected to selection makes it extremely likely that selection must have affected most or all organismal characters, which therefore have or had some adaptive value. Even if the transmission of a character between generations can escape selection because nonselective forces influence its gene frequency (for example, genetic drift, mutation, migration), the character, if it is more than just transitory, will sooner or later confront selection because of some effect it has, either by itself or together with other characters, on the relationship between the organism and its environment. How many persistent phenotypic characters exist that have never affected, however subtly, organismic-environmental interactions? It is certainly reasonable to claim that nonselective causes that affect the gene frequencies involved in producing such a character do not continue to operate *ad infinitum* without selection intervening at some point. Thus, although we often may not know the adaptive explanation for a particular character, the search for such explanations in organisms whose survival depends on adapting to persistent selective environmental pressures is a reasonable enterprise in evolutionary biology. To extend Gould and Lewontin's dinosaur example, why did *Tyrannosaurus* have tiny forelegs but the 40-foot-long carnosaur, *Spinosaurus,* also of the late Cretaceous, have relatively large forelegs? Although Gould and Lewontin ridiculed it, the "adaptationist" position is basic to our understanding of biological evolution—that a primary feature of all organic life is its subjection to selection. Selection for foreleg reduction in bipedal carnivorous dinosaurs may well have been an adaptation that improves balance and speed. Even their claim that the spandrels of San Marco are "nonadaptive" has been disputed by the demonstration that spandrels help support the overlying dome and are functional improvements over less functional "squinches" (Mark). In fact, one can ask, why must a "spandrel" or "squinch" or any necessary structural concomitant of "mounting a dome on rounded arches," be considered of "nonadaptive origin" (Gould 1997)? Is the right-angled extension of legs from animal trunks a "nonadaptive" concomitant of locomotion? You can find further discussions of Gould and Lewontin's criticism of adaptationist explanations in Mayr (1983) and in the book Dupré edited.

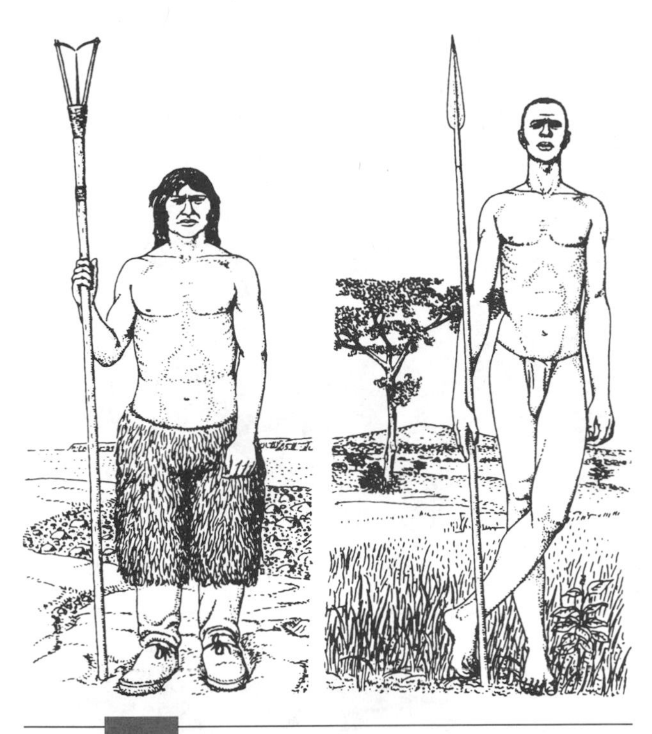

FIGURE 24-4 Difference in body proportions between a man from a group of arctic Eskimos (*left*) and a man from a black tribe in the Sudanese Nile (*right*), indicating differences in their adaptation to their prevailing climates. The proportionately greater bulk and smaller body surface of the Eskimo (approximately 39 kg/m^2) helps to conserve heat, and the proportionately greater body surface of the Sudanese (approximately 34 kg/m^2) helps to dissipate heat (Schreider). A similar explanation may account for differences in body form between Neanderthals and Cro-Magnon groups (Fig. 20-14), each originating in different climates; the former more cold adapted, and the latter more equatorial (Ruff). (*From "The Distribution of Man," by William W. Howells,* Scientific American, *September 1960. © 1960 Eric Mose. Reprinted with permission.*)

Among the best known of these climatic rules is **Bergmann's rule,** which relates body size in warm-blooded (endothermic) vertebrates to average environmental temperature. Bergmann's rule states that races of a species in cooler climates tend to be larger than those in warmer climates. This relationship derives primarily because bodies with larger volumes have proportionately less exposed surface areas than bodies with smaller volumes. Since heat loss relates to surface area, larger bodies can retain heat more efficiently in cooler climates, whereas smaller bodies can get rid of heat more efficiently in warmer climates.

Comparisons made among many North–South races of both terrestrial and marine birds and mammals corroborate the rule. For example, body size of American bushy-tailed woodrats (*Neotoma*) closely follows climatic fluctuations from the time of the last glacial period about 25,000 years ago: larger size in colder periods and smaller size in warmer periods (Smith et al.). In some human races, we can apply the rule by noting the ratio of body weight to body surface, comparing, for example, thick-chested, short-limbed Eskimo groups to slender, long-limbed Nilotic African tribes (Fig. 24–4).

An extension of Bergmann's rule is **Allen's rule,** which states that protruding body parts (for example, tail, ears) are generally shorter in cooler climates than they are in warmer climates. Other provisional ecogeographical regularities are given by **Gloger's rule** (races are more heavily pigmented in warm, humid areas than in cool, dry areas; Fig. 24–5) and **Rapoport's rule** (species adapted to cooler climates are distributed along a wider range of latitudes than species adapted to warmer climates), as well as rules that apply primarily to insects, reptiles, and amphibians (Mayr 1963).

Behavioral Adaptations and Strategies

In addition to morphological adaptations, organisms relate to their surroundings by assuming various motions and positions in escaping from predators, pursuing prey, interacting with conspecifics, and so on—all of which usually come under the name "behavior." Behaviors may be **innate,** needing no prior learning experience, or **learned,** improving over time through trial and error. Innate behaviors include **tropisms,** the directionally oriented growth patterns found in plants, fungi, and sessile animals, and **taxes,** directionally oriented locomotion among animals.

Examples of such relatively simple innate behaviors are movements toward or away from light, gravity, and environmental nutrients and chemicals. Some behaviors can also be prompted by chemicals produced within organisms themselves, such as **pheromones,** which are molecular substances that many animals use to attract mates, lay down trails, and warn off competitors. In the nematode, *Caenorhabditis elegans,* the distinction between solitary and social feeding behavior can change because of a single amino acid substitution in a cellular receptor sensitive to secreted neuropeptide signals (de Bono and Bargmann).

Relatively complex innate behaviors, often called **instincts,** may involve many behavioral components such as courtship patterns in most animals and "dancing" patterns used by honeybees to communicate the direction and distance of food sources. Whether simple or complex, innate behaviors are often quite uniform within a species, and therefore seem, like other species-specific traits, to be entirely or almost entirely genetically influenced. For example, laboratory experiments show that various genes and associated neuroanatomical locations are directly involved in *Drosophila* courtship behavior (**Fig. 24–6**), and that male–female sexual orientation can be genetically reversed (p. 198).

FIGURE 24-5 Distribution of skin color in human populations in Africa, Asia, and Europe before A.D. 1400. (*Adapted from Williams.*)

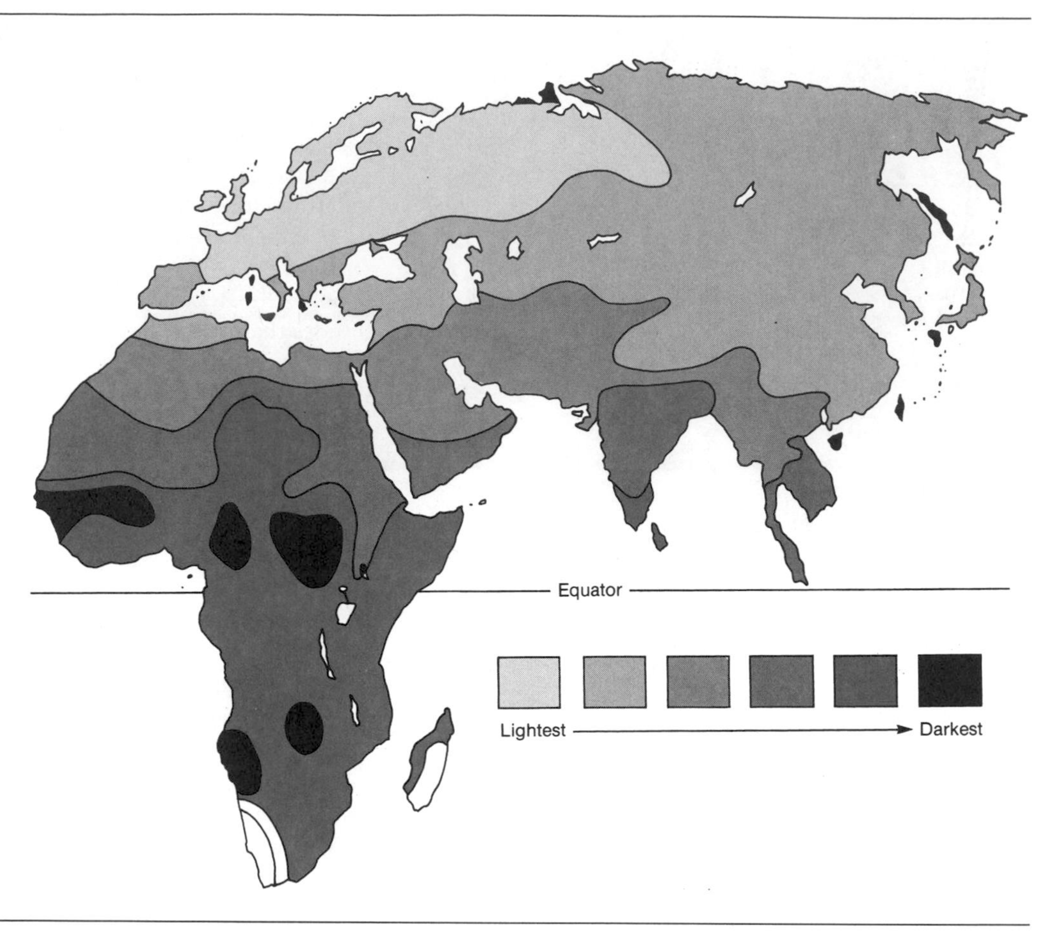

In *Drosophila melanogaster* natural populations, allelic differences in a *foraging* gene that produces a protein kinase used in signal transduction pathways (Chapter 15) affects feeding behavior: "rovers" move longer distances than "sitters" (Osborne et al.). These differences in foraging activity are apparently selected by differences in population density: rovers are adaptive at high density where larvae must travel longer distances to obtain food, and sitters are adaptive at lower densities where food is more available (Sokolowski et al.).

Learned behavior is more flexible and often more complex than innate behavior, since an individual can modify it to suit different environmental or social circumstances. Also, because these qualities are based on accommodating to a transitional and often unpredictable environment, learned behavior is apparently difficult or impossible to entirely program genetically. Instead, learned behavior derives largely from practice (for example, play and observation), which lets individuals modify their behavior on the basis of their own or others' past experiences. For example, naive mammalian juveniles without enough experience must often be protected until they can use learned behaviors to deal with environmental and social problems or hazards.

However, even learned behavior must have genetic components, because many neurological, muscular, and sensory structures involved in the ability to learn and practice must be programmed into the individual. Researchers have widely investigated such genetic influences (Ehrman and Parsons, Fuller and Thompson). For example, Scott and Fuller's experiments have shown that genes in different races of dogs influence tameness, playfulness, and aggressiveness. In mice, a mutation called *disheveled* has no observable effect other than on social behavior, causing reduced social interaction and deficient nest building (Lijam et al.). In humans we have considerable evidence for genes involved in various behavioral disorders such as schizophrenia, manic depression, and mental retardation (Wahlstrom). The presence of such genetic influence and variation indicates that the capacities for most behavioral traits, like so many other adaptations, probably result from evolutionary selective forces (see also p. 506).

An area of behavioral evolution that researchers have given much attention is that of social relationships in which members of a group or species interact for breeding, feeding, and defense. Chapter 20 discussed some behaviors involved in social interactions (**sociobiology**) as

FIGURE 24-6 Sequence of the normal male courtship pattern in *D. melanogaster* and localization of some of these stages to particular sections of the central nervous system, according to the findings of Hall and others. Some mutations that have a primary effect on certain stages in the courtship pattern are shown in the *left column*. (*From Strickberger.*)

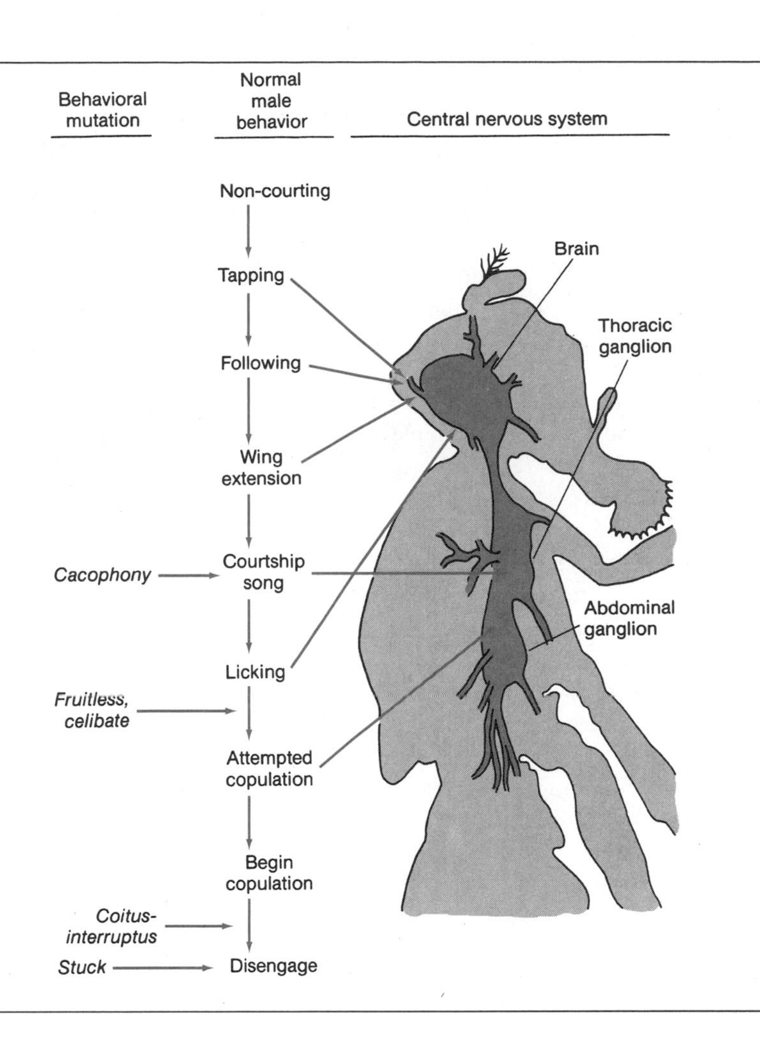

they relate to primates, and many behaviors, such as cooperation and dominance relations, also apply to other social groups (Wilson).

In evaluating such interactions between individuals, workers have proposed various models that deal with the advantages of cooperation as opposed to the advantages of cheating. For example, when two individuals confront each other in a social group, there may be conflicting interests between cooperation, in which each individual gains some advantage, and cheating, in which one individual gains greater immediate advantage than it could by cooperation.

Among the strategies that seem stable in the face of competing strategies (**evolutionarily stable strategies** or ESS) is one called "Tit for Tat," in which an individual behaves cooperatively in a game's first move or interaction, then repeats its opponent's previous move. Thus an opponent who acts selfishly is punished by a selfish response, and cooperative behavior is rewarded by a cooperative response. As Sigmund points out, "the advantage of *Tit for Tat* lies in it being quick to retaliate and quick to forgive."

May discusses some experiments that support the Tit for Tat model, and Maynard Smith and others have elaborated other aspects of applying game theory to social interactions.[2] When these concepts are extended to repetitive interactions among more than one pair of individuals, further strategies may develop, such as **reciprocal altruism** (p. 499), in which individuals cooperate only with those who also cooperate. "Cheaters" who do not cooperate can be identified early on, and may be banished, or suffer retaliation and punishment. Cooperation can become an especially effective strategy, even between strangers, if an individual's past record is commonly known (Nowak and Sigmund 1998). Theorists have hotly debated the relationship of sociobiological determinants to human behavior and culture, as discussed in Chapter 25.

[2]A danger in Tit for Tat strategy is that incorrectly evaluating an opponent's response as selfish can cause a cycle of retaliatory moves until a random correction occurs that reestablishes cooperation. One solution is called "Generous Tit for Tat," in which the probability of correcting such mistakes is greater than chance, because opponents occasionally overlook selfish responses. Nowak and Sigmund (1992) suggest that although Tit for Tat begins the process of successfully eliminating selfish exploiters, it will later be replaced by more cooperative strategies such as Generous Tit for Tat, or Tit for Two Tats, and so on.

Sexual Competition and Selection

Among the most important influences on social behavior is the mating system. Although both sexes benefit if their offspring survive, males and females often differ in the cost of reproduction, because females begin their reproductive careers by investing more resources in producing eggs than males do in producing sperm. Thus, a female's genes benefit if she discriminates in her choice of mates (**female choice**) to protect as much as possible her relatively expensive gametic output. For this same reason, there is an advantage for females to seek increased male **parental investment** in their offspring, a strategy especially notable among mammals and birds, where reproductive success can depend on relatively long-term commitment to their progeny.

Males, in contrast, can be more extravagant in disposing of their relatively inexpensive and more plentiful gametes. Genes carried by a male benefit when he fertilizes as many females as possible, often with relatively little discrimination. This sociobiological conflict of interest between the sexes leads to a variety of mating patterns that depend on various factors, including the degree of parental care necessary for egg or infant survival and which sex (or both) provides such care (see also p. 611). Even when males are normally involved in helping provide parental care, they may diminish such care if they have been "cuckolded," and the paternity of family offspring derives from another male (Dixon et al.).

In groups where females are primarily responsible for parental care—and these include many vertebrate species—males are likely to compete with each other for success in mating. As a result, selection can occur for traits that improve combative abilities of males (**intrasexual selection**) and/or traits that improve their attraction to females (**intersexual** or **epigamic selection**).[3] As Darwin put it:

> Sexual selection depends on the success of certain individuals over others of the same sex, in relation to the propagation of the species; whilst natural selection depends on the success of both sexes, at all ages, in relation to the general conditions of life. The sexual struggle is of two kinds; in the one it is between the individuals of the same sex, generally the males, in order to drive away or kill their rivals, the females remaining passive; whilst in the other, the struggle is likewise between the individuals of the same sex, in order to excite or charm those of the opposite sex, generally the females, which no longer remain passive, but select the more agreeable partners.

Certainly much evidence exists for intrasexual competition between males in **polygynous species,** where one male mates with many females. In such species, males may have special competitive armaments, such as horns and antlers, in order to gain access to females (Fig. 24–7*a*). Not surprisingly, these specialized masculine traits can lead to considerable sexual dimorphism. For example, males are generally larger than females in such groups, reaching a weight that is about eight times that of females in elephant seals (*Mirounga leonina*).

Because of female choice, male ornaments can also become quite conspicuous, such as the dramatic plumage of peacocks, birds of paradise, and even hummingbirds (Fig. 24–7*b*). In some cases we can reasonably claim that although male decorative traits may enhance their breeding success, traits of this kind can also cause increased susceptibility to predation. In fact, we can show mathematically that male decorative traits can become so exaggerated that their adaptive value in enhancing male fitness appears, at best, secondary to their value for sexual attraction.

As Fisher comments, this situation results because females continually choose mates whose attractiveness in such populations is passed on to their sons, and because the daughters of these females inherit their mother's preference for such phenotypically exaggerated males. What starts out as "fashion" in sexual attraction can escalate to extreme limits because of a selective cycle primarily devoted to mating success fed continually by genes that produce more exaggerated phenotypes and more exaggerated mating preferences (**runaway selection**).[4]

Nevertheless, female choice for nonadaptive traits in males is probably uncommon, and some reports now demonstrate that *Drosophila* fruit fly and *Colias* butterfly females choose sexual partners who confer greater fitness on their offspring (Taylor et al., Watt et al.). In houseflies, developmental stability as marked by symmetry in wing- and leg-length improves mating success and also confers resistance to predation and fungal infection (Møller). Among passerine birds, species with bright-colored males also appear to be associated with resistance to parasite infection (Read). The offspring of female tree frogs who choose males with long mating calls have clear advantages in growth and survival (Welch et al.). Zahavi and Zahavi propose that such exaggerated traits, like the peacock's tail, are adaptive and selected because they signal to others that the carrier is in sufficiently good condition to expend the extra energy to produce the trait in spite of its "handicap."

As for other adaptations, enough variations in ecological, genetic, and evolutionary factors arise to prevent

[3]The intra/inter terminology is somewhat arbitrary, since the competition for mates is commonly between members of one sex—males.

[4]At a behavioral extreme of sexual selection are redback spider males who place themselves within reach of females jaws so they can be cannibalized by their mating partners. Andrade shows that this suicidal trait has adaptive features in letting such males copulate longer and fertilize more eggs than their non-cannibalized male competitors.

FIGURE 24-7 Two examples of sexual dimorphism in species of mammals and birds, with *females on the left* and *males on the right.* (*a*) Mature males of the extinct giant deer of Europe, *Megaloceros,* (also called the giant Irish Elk) had antlers more than 11 feet wide and weighing about 100 pounds. (*From* Mammarion Evolution, *1985 by Savage and Long. Reprinted by permission.*) (*b*) A South American hummingbird, *Spathura underwoodi.* (*From Darwin.*)

strict adherence to any widely applicable rules of sexual behavior (see Clutton-Brock). For example, male features may be selected on both intra- and intersexual levels when females actively choose males that are more successful in intrasexual combat. Female choice can even extend beyond coitus into sperm utilization, where male competition can occur through differences in genitalia and hormonal secretions for direct access to eggs (Eberhard). For example, the *Drosophila* female reproductive tract is now known to be a coevolving battleground between the sexes. Males increase their sperm fertilization power by controlling female fertility, longevity, and behavior through "sexually antagonistic" genes (p. 199), and females attempt to respond with gene products that counter such effects (Clark et al.). On the whole, differences in sexual behavior between groups may evolve quite rapidly and lead easily to barriers in gene exchange, as briefly discussed later.

From Races to Species Barriers

However racial adaptations occur, whether through changes in morphology, physiology, or behavior, or through all three, race formation is potentially reversible because different races may interbreed and combine again into a single populational unit. Thus, a large extent of migratory activity (gene flow) among individuals of a species may impede race formation. Rensch, for example, has calculated that migratory species of birds average less than half the number of races of nonmigratory species: the greater the gene flow, the fewer the differences.

As a rule, therefore, barriers that reduce gene exchange between populations accelerate race formation. Initially such barriers are primarily geographical and occur when populations bud off from one another and occupy different areas or environmental habitats. The potential for gene exchange, however, lets us view all these different populations as members of a single species. Only when populations differ enough to inhibit any gene exchange at all do we commonly view them as separate species.

To biologists, the concept of a species as an interbreeding group distinct from other such groups arises in sexually reproducing organisms from the knowledge that such groups exist in nature and are mutually separated in many instances by "bridgeless gaps" across which interbreeding does not occur ("biological species concept," Chapter 11). The existence of species is also supported by evidence that both humans and other forms of life to whom such discrimination is essential recognize species as distinct groups. Predators of all kinds, for example, learn early to discriminate among varieties of prey and to select those that are palatable and can be used for food.

In groups that can verbalize the recognition of species, including primitive human societies, the distinctions made are, in many cases, strikingly similar to species classifications based on more sophisticated biological criteria. Thus a tribe of New Guinea islanders uses distinct names for 137 species of birds found in this region, almost equal to the exact number of 138 species recognized by ornithologists (Mayr 1969). Molecularly, Avise and Walker point out that taxonomic species designations in vertebrates also correlate, to a reasonable degree, with distinctions in mitochondrial DNA sequences. Since mitochondria are transmitted vertically without recombination, their DNA represents a lineage of historical events that can reflect species differences. However measured, morphologically or molecularly, a qualitative change accompanied by reproductive separation or isolation usually marks the transition of racial differences to species differences (**speciation**). What mechanisms prevent gene exchange between populations, and how do such mechanisms originate?

Isolating Mechanisms

Researchers have broadly termed factors that prevent gene exchange among populations **isolating mechanisms,**[5] and some authors include all barriers, even geographical and spatial isolation. Speciation events among such geographically separated—**allopatric**—populations (p. 239) have been described in various ways, related to the kind and degree of separation and the opportunity for gene exchange.

The most common allopatric concept (also called **vicariance**) is of a formerly unified population that splits because of a natural physical barrier or because intervening geographical populations became extinct. **Peripatric** speciation describes the budding off of a small completely isolated "founder" colony from its larger more widespread parental population. **Parapatric** is used for a population at the periphery of a species that adapts to different environments but remains contiguous with its parent so that gene flow is possible between them.

[5]Paterson insists the term *isolating mechanisms* implies selection specifically for the purpose of speciation. Since this is obviously not so—postmating isolating mechanisms such as hybrid sterility and inviability can result from selection for traits unconnected to direct selection for speciation—he suggests the term is inappropriate. Perhaps terms such as "isolating barriers" or "reproductive barriers" would be more fitting, but *isolating mechanisms* was the term Dobzhansky originally proposed and it is still the most common term for impediments to gene exchange. Paterson's main proposal is that selection for improved intraspecific fertilization fundamentally causes reproductive isolation [species are defined by sharing a common "mate recognition" or "fertilization" system (p. 239, footnote 4)], and we can lump both premating and postmating barriers between groups together as by-products ("effects") of such selection. He seems to claim that a trait such as mating behavior selected to serve one function (improved fertilization within a group) cannot also be selected, at times, to serve another function (isolation between groups). This view opposes the concept that selection may occur to reinforce sexual isolation between populations whose hybrids are deleterious (p. 591).

Where populations are sufficiently separated to prevent gene exchange, evolutionists have debated whether, given the opportunity, many such populations would remain reproductively isolated. Some authors propose that we restrict the term *isolating mechanisms* to those that prevent gene exchange among populations in the same geographic locality, that is, to mechanisms that isolate **sympatric** populations.

Mayr (1963) has classified sympatric isolating mechanisms into two broad categories: those that operate before fertilization can occur (**premating**), and those that operate afterward (**postmating**). Among the premating isolating mechanisms are:

- SEASONAL OR **HABITAT ISOLATION** Potential mates do not meet because they flourish in different seasons or in different habitats. For example, some plant species, such as the spiderworts *Tradescantia canaliculata* and *T. subaspera,* are sympatric throughout their geographical distribution, yet remain isolated because their flowers bloom at different seasons. Also, one species grows in sunlight and the other in deep shade.
- **BEHAVIORAL** OR **SEXUAL ISOLATION** The sexes of two species of animals may appear together in the same locality, but their courtship patterns are sufficiently different to prevent mating. The distinctive songs of many birds, the special mating calls of certain frogs, and the sexual displays of most animals are generally attractive only to mates of the same species. Many plants have floral displays that discriminate between insect and bird pollinators, or attract only certain insect pollinators (see Fig. 13–23 and p. 516). Even where the morphological differences between two species are minimal, behavioral differences may prevent cross-fertilization. Thus *D. melanogaster* and *D. simulans,* designated as sibling species because of their morphological similarity, normally do not mate with each other even when kept together in a single population cage. According to Coyne and coworkers (1994), male courtship in this group depends on their attraction to specific hydrocarbons in the female cuticle, and sexual isolation can be caused by only few genetic differences.
- **MECHANICAL ISOLATION** Individuals attempt to mate, but cannot achieve fertilization because of difficulty in fitting together male–female genitalia. This type of incompatibility, long thought to be a primary isolating mechanism in animals, we no longer consider important. There is little evidence that matings in which the genitalia differ markedly are ever seriously attempted, although some exceptions exist among damselfly species and some other groups (Paulson).

Among the postmating mechanisms that prevent a successful interpopulational cross, even though mating has taken place, are:

- **GAMETIC MORTALITY** In this mechanism, the interspecific cross destroys either sperm or egg. Pollen grains in plants, for example, may be unable to grow pollen tubes in the styles of foreign species. In some *Drosophila* crosses, Patterson and Stone and others have shown that an insemination reaction in the vagina of the female causes swelling and prevents successful fertilization of the egg.

- **ZYGOTIC MORTALITY** AND **HYBRID INVIABILITY** The egg is fertilized, but the zygote either does not develop, or develops into an organism with reduced viability. Researchers have found many such instances of incompatibility in both plants and animals. For example, Moore made crosses among 12 frog species of the genus *Rana* and found a wide range of inviability. In some crosses, no egg cleavage occurred; in others, the cleavage and blastula stages were normal but gastrulation failed; and in still others, early development was normal but later stages failed to develop.

- **HYBRID STERILITY** The hybrid has normal viability but is reproductively deficient or sterile. This is exemplified in the mule (progeny of a male donkey and female horse) and many other hybrids. Sterility in such cases may be caused by interaction between genes from the two different sources or by interaction between cytoplasm from one source and chromosomes from the other. For example, Dobzhansky long ago used hybrid sterility to provide a simple genetic answer to the question of how reproductive isolation can arise between two formerly interfertile populations. Assuming both populations begin with a two locus genotype *aabb*, different adaptive mutations can be selected in each now separated population: one becoming *AAbb*, and the other *aaBB*. Each population retains its fertility, but epistatic interaction between the *A* and *B* alleles in the *A–B–* hybrid causes sterility and/or inviability. As time goes on and more loci are differentially selected in each population, opportunity for such epistatic interactions increase. The observation that many genes are involved in *Drosophila simulans-mauritania* hybrid male sterility (p. 287) indicates that such epistatic interactions may be a leading cause for speciation.

In general, the barriers separating species are not confined to a single mechanism. The *Drosophila* sibling species *D. pseudoobscura* and *D. persimilis* are isolated from each other by habitat (*persimilis* usually lives in cooler regions and at higher elevations), courtship period (*persimilis* is usually more active in the morning, *pseudoobscura* in the evening), and mating behavior (the females prefer males of their own species). Although the distribution ranges of these two species overlap throughout large areas of the western United States, these isolating mechanisms are enough to keep the two species apart. To date, only a few cross-fertilized females have appeared in nature among many thousands of flies examined. Even when cross-fertilization occurs between these two species, however, gene exchange is still impeded, since the F_1 hybrid male is completely sterile and the progenies of fertile F_1 females backcrossed to males of either species show markedly lower viabilities than the parental stocks (**hybrid breakdown**).

Haldane's rule also calls for postzygotic isolation combining both lethality and sterility (p. 200). Haldane noted that the heterogametic sex is most commonly lethal or sterile in the F_1 of a cross between two races or species, and many observations now support this rule (Coyne and Orr 1989).

Modes of Speciation

In 1889 A. R. Wallace proposed that natural selection might favor the establishment of mating barriers among populations if the hybrids were adaptively inferior. That is, genotypes that did not mate to produce inferior hybrids would be selected over genotypes that did. According to this hypothesis, which Dobzhansky and others supported, selection for sexual isolation arises because most races and species are strongly adapted to specific environments. That is, speciation is the means by which populations preserve their adaptive advantages from the disruption of gene flow from nonadapted groups. Hybrids between two such highly adapted populations represent a genetic dilution of their parental gene complexes that can be of great disadvantage in the original environments. Genotypes that incorporate premating isolating mechanisms would have the advantage of not wasting their gametes in producing deleterious offspring.

Full use of this mode of speciation demands that the different populations producing deleterious hybrids be exposed to each other in the same locality; only then could the more sexually isolated genotypes be specifically selected. Speciation should, therefore, occur in the following sequence:

1. Genetic differentiation between allopatric populations
2. Overlap of these differentiated populations in a sympatric area
3. Subsequent selection (also called "reinforcement") for intensified sexual isolating mechanisms (**Fig. 24–8** left column)

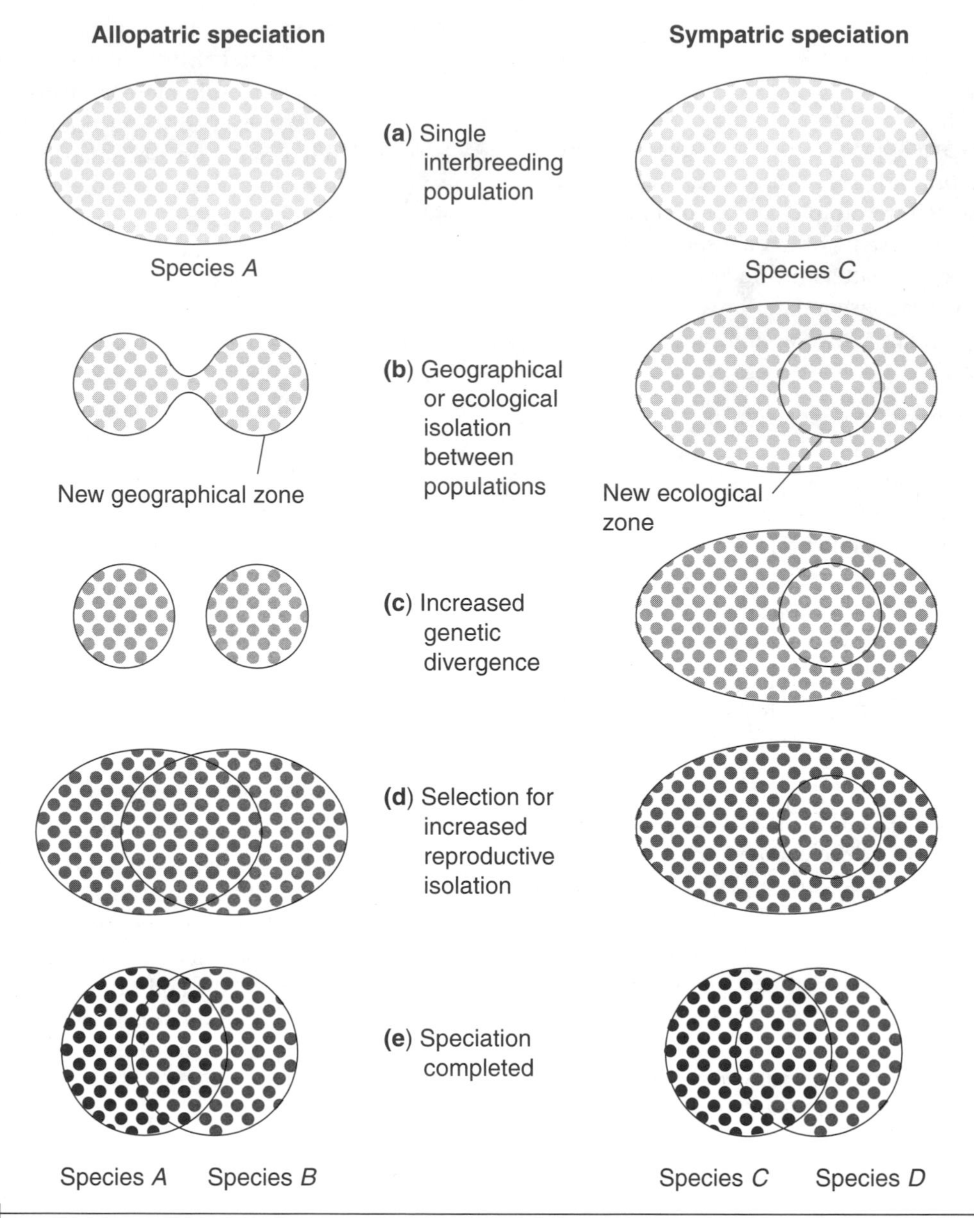

FIGURE 24-8 Simplified diagram of allopatric and sympatric speciation, showing different modes of divergence. In allopatric speciation (*left column*), a population (*a*) splits, or buds, into one or more new geographical zones (*b*) that allow genetic differentiation to occur among different geographical groups by means of both random genetic drift and selection (*c*). When each group has differentiated into a uniquely adapted genetic identity, geographical mixture of the groups (*d*) can result in selection for improved reproductive isolation mechanisms among them. When gene flow between the groups can no longer occur, even when they occupy the same locality, speciation is complete (*e*). In sympatric speciation (*right column*), a population (*a*) splits into one or more groups that occupy different ecological zones, such as special habitats or food sources, within a single geographical locality (*b*). Increased genetic differentiation between the groups (*c*) permits selection for reproductive isolation mechanisms (*d*) that eventually lead to complete speciation (*e*). The difference between these models is the extent of physical separation involved in the initial genetic divergence between the groups. White discusses many examples and variations of these models, and Table 1 of Barton and Charlesworth compares a variety of speciation models. (*From* Genetics Third Edition *by Monroe W. Strickberger. Copyright © 1985 by Monroe W. Strickberger. Reprinted by permission of Prentice Hall, Inc., Upper Saddle River, NJ.*)

Researchers have tried to demonstrate this sequence among natural populations by comparing the degree of sexual isolation among different sympatric and allopatric populations; sexual isolation should be strongest among sympatric populations of different related species, since they are close enough to produce deleterious hybrids, and weakest among allopatric populations of species that are too distantly separated to produce such hybrids.

In one such experiment, Wasserman and Koepfer tested the degree of sexual isolation between the sibling species *D. arizonensis* and *D. mojavensis* by attempting crosses in which the species strains derived from both allopatric and sympatric origins. Their findings showed that when the species strains came from sympatric origins the interspecific cross *arizonensis* × *mojavensis* occurred more rarely (14 out of 377 total matings) than when the strains came from allopatric origins (119 out of 473 total matings). In plants, V. Grant reported that of nine species in the annual herb *Gilia,* the most difficult to cross are the sympatric ones. The allopatric species, by contrast, show no barriers against intercrossing although all F_1 hybrids produced are sterile.

To these observations we can also add Phelan and Baker's examination of hundreds of moth species showing that male scent-emitting organs used to attract females are significantly more common among species associated with the same host plant than among species associated with different host plants. Since these organs produce species-specific courtship pheromones, we can view them as sexual-isolating mechanisms that are apparently more frequent in sympatric species (same host plants) than in allopatric species (different host plants).

In an experiment environmentally manipulating mating behavior, Koopman used the normally isolated sibling species *D. pseudoobscura* and *D. persimilis* to demonstrate that premating isolating mechanisms can actually be increased in sympatric populations. Although sexual isolation exists between these two species in nature and at normal temperatures in the laboratory, cold temperatures can apparently cause a significant increase in interspecific mating. By marking each of the two species with different homozygous recessive genes, Koopman was able to recognize hybrids formed under these low-temperature conditions and remove them from interspecific population cages.

When Koopman performed this operation each generation, he found that fewer and fewer hybrids appeared. For example, after five generations the frequency of hybrids in the mixed populations had generally fallen to 5 percent, from values that were initially as high as 50 percent. This was striking evidence that selection against hybrids had caused rapid selection for sexual isolation that reduced hybrid formation. Paterniani performed a somewhat similar experiment, planting a mixture of yellow sweet and white flint strains of corn; by eliminating plants that produced the greatest proportion of heterozygotes, he reduced intercrossing from about 40 percent to less than 5 percent in five generations.

According to Butlin and others, it is debatable whether selection for premating isolation mechanisms can occur between groups that have not already speciated. Butlin therefore suggests that since Koopman conducted his experiment between species that were already reproductively isolated, we cannot view increased sexual isolation between *D. pseudoobscura* and *D. persimilis* as a cause for new speciation; rather, Koopman only selected each species for increased mate recognition ("reproductive character displacement").

Nevertheless, the fact that we can experimentally increase premating isolation by selecting against hybrids indicates that this isolating mechanism may well function in other cases where hybrid fitness declines. Mating tests between *D. pseudoobscura* females and *D. persimilis* males taken from natural populations showed that sexual isolation is increased in areas where their populations overlap compared to areas where *D. persimilis* is absent (Noor). Similarly, when threespine stickleback fish are confronted with possible mates from different populations, they discriminate more between sympatric than between allopatric males (Rundle and Schluter). Also supporting this view is the extensive survey showing that sexual isolation between pairs of *Drosophila* species of similar age is greater for sympatric than for allopatric species (Fig. 24–9).

One may argue that, given sufficient time, even allopatric populations will accumulate enough genetic differences to show sexual isolation when they come together in the same locality (**allopatric speciation**). In the *virilis* group of *Drosophila* species, Patterson and Stone observed that the European *D. littoralis* is much more isolated from the American populations of *americana, texana,* and *novamexicana* than are American species in the same group.

Some experiments that separate a single population into two or more groups for a considerable period and then test these groups for reproductive isolation also support such allopatric differentiation. In one example, two replicate populations of *D. melanogaster,* raised in the laboratory under different conditions of temperature and humidity for six years, developed both sexual isolation and hybrid sterility (Kilias and Alahiotis).

Whatever the speciation process between geographically separated populations, a number of authors suggest that it may proceed quite rapidly under some circumstances (Mayr 1954; Carson and Templeton). They emphasize **founding accidents,** or **bottlenecks** (Chapter 22), in which a small, isolated peripatric population is subject to forces such as random genetic drift, increased homozygosity caused by inbreeding, and changes in the adaptive landscape (Chapter 23), followed by radical changes in selection pressure. The combined effect of such forces may produce novel, **coadapted gene combinations** affecting behavioral, morphological, and physiological traits that lead to reproductive isolation from neighboring and ancestral populations.

On the island of Hawaii, for example, Carson uses such concepts to help explain the origin of its 26 species of "picture-winged" Drosophilidae (Fig. 22–13) in what may have been less than half a million years. Some of the founding events that occurred in these Hawaiian Drosophilidae, according to Kaneshiro, caused radical changes in male courtship behavior so that less discriminating females in

FRANCISCO J. AYALA

Birthday:
March 12, 1934

Birthplace:
Madrid, Spain

Undergraduate degree:
B.S., University of Madrid, Spain

Graduate degrees:
M.A., Columbia University, New York, 1963
Ph.D., Columbia University, New York, 1963

Postdoctoral training:
Rockefeller University, New York, 1964–1965

Present position:
Donald Bren Professor of Biological Sciences
Department of Ecology and Evolutionary Biology
University of California at Irvine

WHAT PROMPTED YOUR INITIAL INTEREST IN EVOLUTION?

I was born in Spain and went to school there. However, my early interests in evolution developed primarily outside of school in trying to understand why the living world was so diverse and especially the origin of humans. These interests were fed by reading Spanish translations of various books including such twentieth-century classics as *Genetics and the Origin of Species* by Theodosius Dobzhansky, and *Evolution: The Modern Synthesis* by Julian Huxley. Once started, I also read books by Richard Goldschmidt and C. H. Waddington. The effect of this exposure was to emphasize the relevance of genetics to the study of evolution, and I undertook experimental work with *Drosophila* at the University of Salamanca. My teachers there encouraged me to develop my interests further, and in 1961 I came to the United States to study for a Ph.D. under Theodosius Dobzhansky at Columbia University.

WHAT DO YOU THINK HAS BEEN MOST VALUABLE OR INTERESTING AMONG THE DISCOVERIES YOU HAVE MADE IN SCIENCE?

I have been primarily involved in the study of genetic variation and the role it plays in evolution. Using techniques such as electrophoresis of enzymes, many of my experiments were oriented towards discovering the genetic differences that account for speciation in *Drosophila,* such as genetic distinctions between subspecies and between sibling species.

WHAT AREAS OF RESEARCH ARE YOU (OR YOUR LABORATORY) PRESENTLY ENGAGED IN?

More recently, I have also become interested in measuring genetic variation in parasitic protozoa such as trypanosomes, in attempting to understand the mechanisms by which these parasites rapidly adapt themselves to changes in host immune systems. In a sense, these genetic changes represent small, isolated, capsules of evolution. My publications also extend to various areas in the philosophy of biology such as teleology, reductionism, and the biological foundations of ethics. I have also written on the use of testimony by scientists in courts of law. A basic concern to which I have addressed considerable effort is the necessity for science education in schools, especially the teaching of evolution.

IN WHICH DIRECTIONS DO YOU THINK FUTURE WORK IN YOUR FIELD NEEDS TO BE DONE?

The tremendous power of molecular biology offers the opportunity to answer long-standing evolutionary questions such as how extensive are the genetic differences between species, of what kinds are they, and at what rates do they occur. I intend to continue my recent research on the evolutionary history and population structure of *Plasmodium,* which causes malaria, a disease affecting several hundred million people each year.

WHAT ADVICE WOULD YOU OFFER TO STUDENTS WHO ARE INTERESTED IN A CAREER IN YOUR FIELD OF EVOLUTION?

From my own experience, I believe it is extremely important for students to identify conceptual problems in their area of interest by reading the "masters," and by discussion and exchange with others in the field. Science is a community enterprise! Also, preparing for a career in science necessitates getting the best "tools" one can: in evolution these can include one or more disciplines such as biochemistry, systematics, mathematics, and statistics.

these small, isolated populations were selected to respond to such changes, whereas more discriminating females in ancestral or neighboring populations were unresponsive to such modified males. Courtship changes and ornamentation differences, however they arise, are powerful agents in initiating the speciation process. A high ratio of nonsynonymous (amino acid-changing) to synonymous ("silent") nucleotide substitutions in genes for mating behavior and other sex-related functions indicates that such genes play an important selective role in speciation (Civetta and Singh 1998a).

In opposition to these views on bottlenecks, Barton and Charlesworth suggest that the concept of speciation caused by single founder events has little theoretical support, since such events usually do not produce an immediately significant change in an isolated population. For

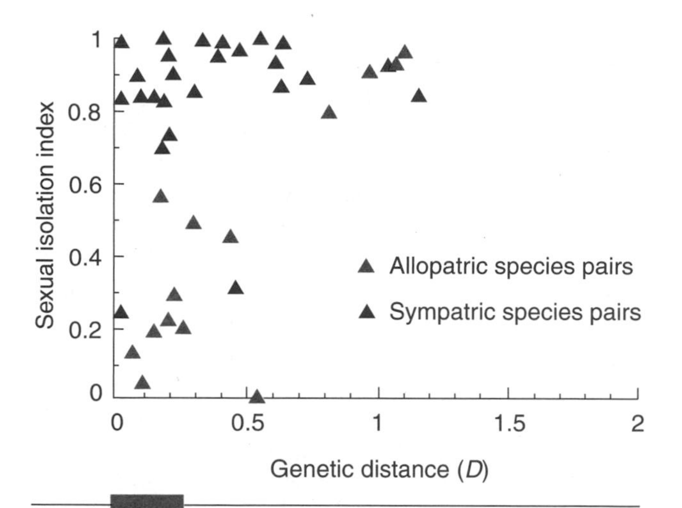

FIGURE 24-9 Measurements of the degree of sexual isolation for pairs of allopatric and sympatric *Drosophila* species where each pair has also been evaluated for Nei's *D* genetic distance (p. 583). The sexual isloation index is based on comparing the frequency of heterospecific matings (matings between individuals from different species in the pair) and homospecific matings (matings between individuals from the same species in the pair) according to the following formula: Isolation = 1 − (heterospecific mating frequency)/(homospecific mating frequency). As Coyne and Orr (1997) point out and which is obvious in this illustration, when the species in a pair are closely related (genetic distance between them is small, for example, 0.5), they are clearly more isolated from each other when they are sympatric than when they are allopatric. Also, for allopatric populations to achieve reproductive isolation requires a much greater genetic distance ($D = 0.54$) than reproductively isolated sympatric populations ($D = 0.04$). As a rough estimate, they propose that "speciation requires approximately 200,000 years among taxa that became sympatric and approximately 2.7 million years among taxa that remain allopatric." (*After Coyne and Orr 1997.*)

example, it may take many generations for random genetic drift to effectively modify gene frequencies. These authors state that "it is impossible to separate the effect of isolation, environmental differences, and continuous change by genetic drift [in moderately sized populations] from the impact of population bottlenecks [in small founder populations]." Even the impact of bottlenecks considered separately is in dispute, since some experiments do not support it (Moya et al.), and phenotypic variation may even increase rather than decrease in populations that pass through bottlenecks (Bryant et al.).[6]

As yet, the relative importance of each suggested mode of speciation remains unclear and open to different interpretations. Thus Rice and Hostert, in a broad survey of pertinent laboratory experiments, claim that reproductive barriers develop as secondary effects resulting from pleiotropy or hitchhiking when disruptive selection between populations occurs for other genetic differences. Male hybrid sterility, caused by epistatic interaction between genes that are otherwise adaptive or neutral within a population (pp. 200–201, 287, and 591; see also Turelli), may be an example of such speciation events. What then can we say about the genes responsible for speciation?

For many workers, the search for "speciation genes" leads to sexual traits, since these are often basic in erecting the barriers that isolate species. Following such approach, Civetta and Singh (1998b) measured variation between sexual traits such as testis length and nonsexual traits such as femur length for a group of *Drosophila* species. These comparisons showed that sexual traits exhibit greater variation between species and less variation within species, signifying that selection acts differently on sexual traits at different times. That is, sexual traits undergo greater selection for differences between populations during speciation, and greater selection for uniformity within species after attaining speciation. Civetta and Singh therefore propose that changes in such sexual traits probably correlate with early speciation events. Studies pointing to similar genetic roles for other sexual traits include investigations of pheromonal differences between *Drosophila* species (Buckley et al.), mating preferences in *Heloconius* passion-vine butterflies (McMillan et al.), evolution of mating type genes in *Chlamydomonas* (Ferris et al.), and sperm–egg fertilization interaction in animals (Vacquier).

From what we know at present, speciation events can occur in various ways and at various rates. In some groups such as the Hawaiian Drosophilidae, speciation has been dramatically rapid and may well have involved fewer genes with greater phenotypic effects than in the slower speciation events in some other *Drosophila* groups. In plants, relatively few mutations also seem to account for the rapid transition from teosinte to modern maize (Fig. 21–1), and from bee-pollinated to hummingbird-pollinated species in *Mimulus* (p. 516). In other groups allopatric speciation in the absence of bottlenecks

[6]Templeton counters such arguments by claiming that founder events can cause severe allelic frequency changes leading to homozygosity for some loci (Fig. 22–14). When such changes involve "major" genes that have epistatic and pleiotropic effects, genes and their modifiers may then be subject to new selective conditions. Templeton proposes these new circumstances "convert epistatic variance into additive genetic variance, thereby increasing—not diminishing—the overall levels of additive genetic variance and hence selective responsiveness immediately after the founder event." He calls this *the genetic transilience model of speciation.* Although such founder effect models seem attractive, they still arouse controversy. In addition to Barton and Charlesworth's contentions, it is difficult to accept that fixation of common alleles from a parental population would cause a significant developmental change and founder effect in a new population. By contrast, fixation of rare alleles might well cause greater developmental changes, but their chances for becoming founders and for their subsequent fixation are much less probable. Nevertheless, the founder effect model still gathers proponents: García-Ramos and Kirkpatrick point out that a peripheral population genetically isolated from its parent can evolve rapidly in new directions under strong selective conditions, sufficient to cause speciation.

may have been more common. Such allopatric speciation modes, however, do not exclude selection for sexual isolation between sympatric populations because of hybrid sterility or inviability, although as discussed earlier, that too has been disputed.

Hybridization

Where species barriers break down to produce viable and fertile hybrids—and such instances arise, especially in plants—**zones of hybridization** or **hybrid swarms** may develop whose genotypes and phenotypes differ from both parental species. If a unique and discrete habitat exists to which the hybrids are better adapted than the parents, the new population may eventually become isolated from its parental populations. This mode of speciation is supported by detailed demonstrations of changes in chromosome number (ploidy levels) in both plants (V. Grant) and animals (Bullini), although animal hybridization occurs more rarely because such chromosomal changes have greater impact on fitness (p. 206).

One well-investigated example of plant species hybridization occurs in sunflowers. In this genus, *Helianthus,* three western United States species studied by Ungerer and coworkers show the rapid evolution of a hybrid, *Helianthus anomalus,* that was initially formed from a cross between *H. annuus* and *H. petiolaris* probably less than 60 generations ago (Ungerer et al.). Interestingly, synthetic hybrids made experimentally by crossing the two parental sunflower species, and performing successive crosses and backcrosses, acquire genomes similar to the natural *H. anomalus* hybrid, incorporating similar parental genes while excluding others. Apparently, once the initial hybrid is formed, selection becomes an important factor in choosing genes that further develop its genetic architecture. Rieseberg and coworkers therefore conclude that "although the majority of interspecific gene interactions are indeed unfavorable or neutral, a small percentage of alien genes do appear to interact favorably in hybrids."

In some cases fertile hybrids can act as intermediaries introducing genes from one species into the other, thereby enhancing a species' ecological range and evolutionary flexibility: a phenomenon that Anderson has termed **introgressive hybridization** (see also Levin).[7] According to P. R. and B. R. Grant, hybridizations between Darwin's finches on one of the Galapagos islands have led to increased genetic variation in the interbreeding species. They claim this effect is "two to three orders of magnitude greater than that introduced by mutation," greatly facilitating new evolutionary change.

Of course, hybrid sterility is a barrier to further evolution, but even then, specifically in plants, polyploidy may arise in a vegetatively propagating hybrid, enabling it to produce fertile gametes (allopolyploids; see Fig. 10–14). Since these gametes are diploid relative to the haploid gametes of the parental species, a new species is born at one stroke, fertile with itself or other such polyploid hybrids but sterile in crosses with either parental species.

How often new hybrid species occur has been difficult to document (Rieseberg). According to Ellstrand and coworkers, the frequency of hybridization in vascular plants appears to vary between families: some with hardly any hybrid species, and some with 50 or more. Of the 250,000 described plant species, they estimate only about 10 percent are hybrids. Hybrid evolutionary impact may be more significant than their frequency, since some plant hybrids may be ancestral to entire lineages comprising many species that occupy many habitats and, through introgression, may also have affected parental populations. Similar claims are made by Dowling and Secor for the significance of species hybrids in animals. Arnold, discussing both plant and animal hybrids, argues that hybrid fitness is quite heterogeneous, and some hybrids may be superior even in parental environments. Although considered uncommon by evolutionists in the past, hybridization is beginning to receive more attention.

Can Species Differences Originate Sympatrically?

The sequence of evolutionary events in speciation seems, therefore, to begin with race formation and end with reproductive isolation. In this sequence evolutionary geneticists further dispute the degree to which geographical separation between populations is necessary to accumulate the initial genetic differences that lead to speciation. Many workers in this field believe that populations can only accumulate genetic differences when they are spatially separated enough to prevent the gene exchange that might eradicate these differences. They propose that the speciation process takes hold only after this important early period of geographical separation, either by the accidental origin of isolating mechanisms or by later selection of isolating mechanisms because of defective hybrids.

Other workers, especially Mather and Thoday, propose that a population in a single locality selected for adaptation to different habitats within that locality could produce an increase in genetic variability (see disruptive selection, p. 542) that would lead to polymorphism. One such example is the polymorphism that now appears in

[7]Researchers recognize that claims of introgression based on finding identical alleles in hybridizing species should exclude shared polymorphisms inherited from a common ancestor.

TABLE 24-3 Results of tests for mating preferences among *D. melanogaster* flies selected for high bristle numer (H) and low bristle number (L) and in which males and females are given a free choice of mates

Generation of Selection	H × H	H × L	L × H	L × L
7	12	3	4	12
8	14	2	6	10
9	10	4	6	7
10	8	4	3	13
19	27	2	8	20
	71	15	27	62

Source: From J.M. Thoday. *Disruptive Selection,* Proc. Royal Society of London (B) 182: 109–143, 1972. Reprinted by permission.

the British peppered moth *Biston betularia,* and a further important example is the polymorphism of mimicry in the butterfly *Papilio dardanus* (Sheppard).

Geneticists have also proposed that under some circumstances isolation between two or more selected groups might occur in the same locality, especially if the selected forms can exist independently of each other. Thoday and Gibson first presented evidence for this view in selection experiments on bristle number in *D. melanogaster.* They selected flies each generation for high (H) and low (L) bristle number and found that, although they permitted random mating, mating preferences of these flies went rapidly in the direction of positive assortative mating, H × H and L × L, with relatively few H × L and L × H matings, as Table 24–3 shows. Other experiments have also since achieved increased isolation by disruptive selection between populations in the same locality (Coyne and Grant, Soans et al.).

However, despite many attempts, researchers have not replicated some results of disruptive selection (Scharloo), and workers have asked whether any single locality in nature could consistently maintain divergent selective conditions long enough to produce **sympatric speciation** (Mayr 1963). The primary issue seems to revolve around selection's power and direction in an ecological isolate. Is selection strong enough to produce adaptive changes within a group causing hybrid inviability and/or sterility (reproductive isolation) while it continually faces gene flow from the surrounding population? Can reproductive isolation arise as a direct or indirect consequence of ecological adaptation?

However this matter will be resolved, Bush (1975) and others have used sympatric speciation (Fig. 24–9 right column) to explain the likelihood that various groups of insects speciated within a single geographical range by adapting to different kinds of host plants (apples and hawthorns) as food sources. Differences in the timing of fruit maturation between the plants cause these parasitic insects to emerge as adults at different times, thus providing a barrier preventing gene exchange between them. According to Feder and coworkers, this timing leads to selecting genotypes that help restrict insects to their host plants.

Cichlid fishes, a prominent example of very recent and rapid speciation (Fig. 12–18), may also have undergone sympatric speciation, especially in small crater lakes, where they can diversify ecologically but need not separate geographically (Schliewen et al.). Ecological heterogeneity can certainly account for genetic diversity, leading even toward reproductive isolation, as shown in a *Drosophila melanogaster* population selected for radically different experimental habitats (Rice and Salt). It seems likely that some or even many sympatric speciation events may have occurred (Barton et al., Bush 1994), and Seger and others have offered theoretical models to support these contentions.

Evolutionary Rates and Punctuated Equilibria

Although we are just beginning to discern the underlying mechanisms of speciation, the presence of so many fossil and existing species enables some estimates of **evolutionary rates** from geological and paleontological data (Chapters 6 and 13 through 20) or from biochemical changes and "molecular clocks" (Chapter 12). Nevertheless, rate determinations in both cases are beset with problems. Among the questions that arise are whether we should measure rates in geological duration and periods (chronological time) or in generations (biological time). Furthermore, what morphological or molecular features should we use to measure rates, and how do we determine the numbers and kinds of genes involved? We have reached no common agreement on solutions to these problems.

On the paleontological level, taxonomic difficulties also intrude, because "lumpers" and "splitters" may,

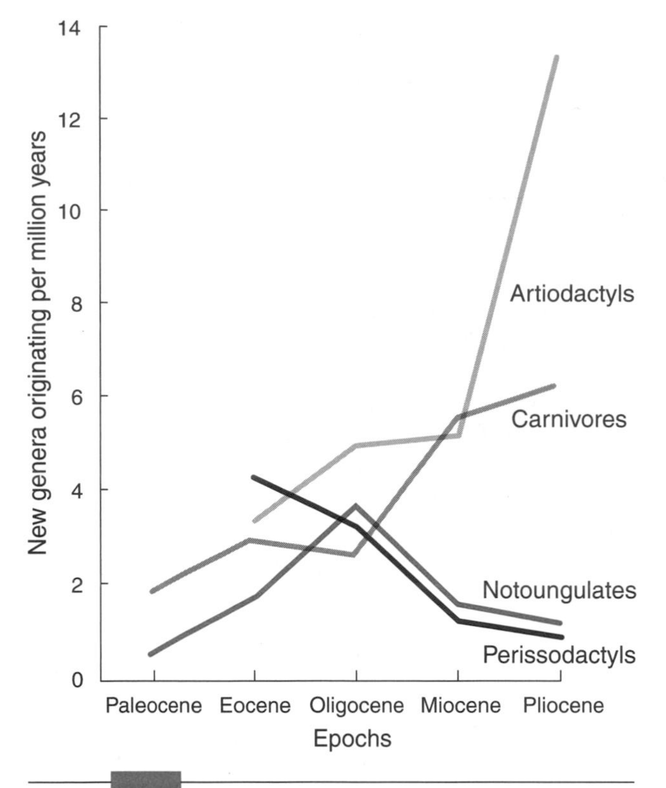

FIGURE 24–10 Evolutionary rates measured in terms of new genera originating per million years in four orders of mammals during the Tertiary period. (*Adapted from Simpson.*)

respectively, combine or split groups of organisms into different taxonomic categories (Chapter 11). This might lead, for example, to different numbers of genera for the same lineages and thus to different generic evolutionary rates. Incompleteness of the fossil record and the frequent absence of evolutionary intermediate groups also cloud many paleontological rate determinations. Nevertheless, as Figure 24–10 shows, evolutionary rates among known fossil taxa of various mammalian groups seem to differ significantly, and this also seems true of many other lineages for which we have fossil information.

Since rates often seem to vary over time within any particular group, one paleontological method has been to classify rates into those most commonly found (**horotely**) flanked by slower rates (**bradytely**) and faster rates (**tachytely**). "Living fossils" (for example, *Latimeria* and *Neopilina*, Chapter 3) lie at the more static bradytelic end of evolutionary rates in their respective groups, and artiodactyls at the tachytelic end of mammalian rates. To instances of very rapid evolutionary changes often marked by expansion into new adaptive zones and the origin of new taxa, Simpson gave the name **quantum evolution.** As noted in Chapter 12, the term **macroevolution** has been used to distinguish major taxonomic evolutionary changes from the presumed less radical changes that occur within a species, called **microevolution.**

Morphologically, a common proposal for evaluating evolutionary rate is the amount of change in a character divided by elapsed time, generally measured in *darwin* units proposed by Haldane. The value of such rates depends on completeness of the fossil record in providing valid phylogenetic information and accurate chronology. Given such information, a rate in darwins is calculated as the difference in a character's average dimension ($\bar{X}$) from time t_1 to time t_2, in natural logarithms (base $e = 2.718$):

$$\text{rate} = (\ln \bar{X}_{t_2} - \ln \bar{X}_{t_1})/(t_2 - t_1)$$

Such logarithmic calculations offer the advantage of enabling evolutionary rate comparisons among organisms of different sizes. For example, a small femur in a rodent lineage that increased by 0.1 darwin evolved twice as fast as a much larger elephant femur that changed by only .05 darwin, although the measured increase in size of the elephant's femur was larger.[8]

The question of what accounts for evolutionary rate differences, whether on micro- or macroevolutionary levels, fossil or current, is not yet satisfactorily answered, but many biologists agree that such differences probably rely on various factors:

1. THE STRUCTURE OF POPULATIONS—THEIR SIZE, GENETIC VARIABILITY, DISTRIBUTION, AND SO ON For example, Wright proposed that evolution can proceed more rapidly in a population subdivided into demes than in one more interconnected and homogeneous (pp. 566–569). Similarly, Mayr (p. 548) proposed the rapid evolution of new taxa in small "founder" groups bearing unique genotypes or gene frequencies that break off from the peripheries of large populations (also called peripatric speciation).
2. ADAPTIVE AND DEVELOPMENTAL CONSTRAINTS THAT LIMIT OR DICTATE THE STRUCTURES AND FUNCTIONS THAT ORGANISMS ARE ABLE TO ACHIEVE Humans and most extant mammals, for example, cannot develop wings to escape predators. Such constraints do not mean that selection is ineffective, only that evolution proceeds along one path rather than another (see also p. 357).

[8]Fenster and Sorhannus review these and other methods for obtaining numerical rates of morphological evolution. Interestingly, such rates generally show a contrast between slow changes in the fossil record measured in single darwins or less and more rapid evolutionary changes in present populations, such as color changes in British peppered moths (p. 541) measured in thousands of darwins (Kirkpatrick). A possible reason for such marked differences may reside in the numbers, kinds, and durations of selective and environmental interactions experienced by fossil and contemporary populations. In fossil populations, these interactions occur continuously over lengthy periods, and genetic and developmental processes affecting some characters appear to change slowly because of repeated compromises necessary to resolve conflicting selective/environmental pressures—probably through canalizing (p. 359) and stabilizing selection (p. 542). By contrast, contemporary organisms face shorter observation periods, experiencing fewer different selective and environmental interactions, and can therefore change some characters more easily and rapidly, although more transiently.

Constraints that seem apparent on the morphological level may not reflect constraints or changes at other developmental levels, thereby clouding the actual rate of evolution. Thus, frogs whose adult stages have remained morphologically similar for about 200 million years—showing an apparent low rate of evolution—have nevertheless undergone considerable preadult changes and evolved more than 3,000 thousand present species—showing an apparent high rate of evolution (p. 286).

3. CHANGES IN THE DIRECTION AND INTENSITY OF SELECTION AT DIFFERENT TIMES IN DIFFERENT GROUPS—THE HISTORICAL-ENVIRONMENTAL CONTINGENCIES THAT ALL POPULATIONS EXPERIENCE This factor only reflects the truism that changes in the biological and physical environments that define the ecological niche (for example, different habitats, food supplies, competitors, predators, or parasites) are the stimuli for adaptation.

Supporting the importance of selection among these factors is the finding that rate differences do not seem to correlate with length of generations or with available genetic variability measured as electrophoretic protein differences (Chapter 10). For example, mammals with short generation times, such as opossums, have evolved much more slowly than those with much longer generation times, such as elephants, and the apparently slow-evolving horseshoe crabs (*Limulus,* Fig. 11–3) show as much electrophoretic genetic variability as more rapidly evolving invertebrates.

An important point now in dispute is whether speciation and the origin of higher taxa involve macroevolutionary mechanisms uniquely different from the microevolutionary mechanisms that cause less noticeable changes within lineages. Most population geneticists propose that the speciation process may add new directions to evolution, but its mechanisms are similar to those used in nonspeciation changes.

In contrast, proponents of **punctuated equilibria,** such as Gould and Stanley, feel the mode of origin of new taxa is qualitatively unique, as evidenced by the rapidity of macroevolutionary events in the fossil record; that is, punctuationists see the fossil record as long intervals of microevolutionary **stasis,** or equilibrium, during which relatively little change occurs, punctuated by rapid macroevolutionary periods during which new taxa arise through entirely new causes and mechanisms.

In the 1930s and 1940s, Goldschmidt proposed that each macroevolutionary change derives from a single **macromutational** incident. This view had much in common with **saltationist** doctrines various paleontologists espoused, that species can arise suddenly because of unknown types of events (p. 516). According to Goldschmidt, mutations with large developmental effects presumably produced some "hopeful monsters" that could then enter into new adaptive zones. Because their effects must integrate with many other genetic changes (the genetic background), it now seems clear that single mutations large enough to cause instantaneous species differences would most probably not be viable. Such precipitous events are therefore regarded as more "hopeless" than "hopeful," and modern proponents of punctuated equilibria have therefore abandoned Goldschmidt's monsters.

However, as Chapter 12 discusses, punctuationists have taken comfort in the finding that regulatory mutations can have significant developmental effects and probably account for important differences among various groups. Similarly, punctuationists have endorsed founding accidents or bottlenecks (p. 593) as possible causes for or accompaniments of macroevolutionary events. For the most part, they search for causes other than natural selection to explain the diversity that accompanies macroevolutionary events such as speciation.

Despite punctuationist arguments, macroevolutionary changes are still quite compatible with what we know of population genetics. From a Neo-Darwinian as well as punctuationist view, the rate of evolution in a new populational offshoot may certainly be rapid compared to changes in its parental species, which remains tied to its more traditional ecological niche. The fact that paleontology can show evolutionary stasis in one or more branches of a group and rapid evolution in the group's other branches does not contradict the Neo-Darwinian concept of evolution. There is no novelty in the idea that stabilizing selection (p. 542) can reduce variability and conserve similar phenotypes even among different species (Spicer), or that new species originate through cladogenesis as well as through phyletic transformation (p. 241).

Nor does any Neo-Darwinian rule prescribe the rate of speciation: that rates must be uniform and cannot be variable or even change abruptly. Thus, in many plants, we recognize that allopolyploidy—the cause for reproductive isolation in chromosomally doubled hybrids (p. 206)—is a common source of rapid speciation. We have also long recognized that catastrophic events can account for considerable evolutionary changes (pp. 451–452), as can other unique historical phenomena (pp. 549–550), but their unpredictability makes them difficult to model mathematically. That organisms endured environmental interactions in the past different from the present does not detract from the modern Neo-Darwinian synthesis and provide cause to make an explainable process unexplainable. As we understand evolutionary change, it can be expressed in different patterns in different lineages, especially over different time scales (Gingrich).

Thus, although macroevolution may at times be associated with large environmental changes and microevolution with smaller ones, mechanisms used in one mode need not be excluded in the other, according to Neo-Darwinism. Regulatory mutations, bottlenecks, and new directions and intensities of selection, as well as other presumed macroevolutionary mechanisms, also help explain microevolutionary events. Lenski and Travisano's recent study of laboratory bacterial populations undergoing many thousands of generations of selection clearly shows that evolutionary changes that may seem "punctuational" are fully explainable as rapid increases in frequency of new favorable mutations followed by static periods until further such mutations appear. When the opportunity for diversity ("adaptive radiation") presents itself by exposure to novel environments, bacteria can adapt through simple mutations (Rainey and Travisano). Such "stop and go" events support the "streetcar theory of evolution" (Hammerstein), in which evolutionary advances pause until temporary genetic constraints are overcome by the entry of new genetic "passengers" (mutations) and replacement of old ones.

Fossil data used to support punctuationism (Jackson and Cheetham) can also be disputed by data supporting gradualism (Sheldon), as well as by data that indicate some speciation events are punctuationist and others are gradualist (Geary). Furthermore, what seems static when regarded at one level can be dynamic at another. Morphological differences among early mammalian Mesozoic fossils may seem relatively minor and inconspicuous, but we know that unperceived physiological changes, involving temperature regulation, lactation, and viviparous reproduction, were probably quite dynamic in effect and highly evolutionarily significant (Chapter 19). Even on a strict morphological level, stasis does not necessarily imply uniformity, since selection can at times alternate in opposite morphological directions in fluctuating environments to produce an ostensible average.

One can also question whether the fossil record offers enough information to show how rapidly macroevolutionary events have occurred. A very slow rate of change during each generation in a population may, in fact, lead to speciation in a period of geological time so short that paleontologists cannot detect it. For example, a duration of 100 years may produce a few thousand or more generations of flies in a *Drosophila* lineage, as could 1,000 years in a rodent lineage; yet both these periods could easily experience significant morphological changes through gradual microevolutionary mechanisms, neither necessarily leaving a fossil record. Illustrating the power of sustained natural selection, Reznick and coworkers report that guppy fish (*Poecilia reticulata*) transplanted from a high predator to low predator environment for a period of only several years evolved at a rate of thousands of darwins. Certainly the speed with which sexual isolation can develop between experimental populations shows how rapidly microevolutionary methods can achieve speciation while they appear macroevolutionary on the paleontological level. Rather than invoking a new hierarchical level requiring unprecedented novel explanations (Box 25–1), macroevolution may simply characterize cumulative evolutionary changes due to known processes that increase diversity among species (pp. 516–518).[9]

It is important to note that relatively long periods within lineages, such as 100,000 years, often go undetected in paleontology. Dawkins offers the hypothetical example of a mouse lineage that gradually reaches the size of an elephant in 60,000 years and points out that "evolutionary change too *slow* to be detected by microevolutionists [population geneticists] can nevertheless be too *fast* to be detected by macroevolutionists [paleontologists]." Because of all these and other arguments (for example, Ayala, Kellog, Levinton), it seems difficult to accept as yet that new evolutionary rules or mechanisms must be enlisted to explain paleontological observations of punctuated equilibria. In fact, Gould and Eldredge, the initiators of the punctuated equilibrium theory, seem to be moderating some of their early views, and they now declare that their dispute with Neo-Darwinism is more a matter of emphasis on "species sorting" (p. 571) and on "stasis" than on differences in speciation mechanisms.[10] One can only agree with Gould (1994) that "macro- and microevolution should not be viewed as opposed, but as truly complementary."

SUMMARY

All members of a species can share a common gene pool, although populations within it may vary genetically from each other. If the gene frequencies of these populations are sufficiently distinct, they are known as *races*. Humans,

[9]Other paleontologists have disputed even the claim (Gould 1989) that the Cambrian period represents the effect of a unique evolutionary mechanism different from those experienced by later organisms. Briggs and coworkers point out:

> We found no evidence for vastly greater disparity or variety of *Baupläne* in arthropods of the Cambrian than those of the Recent, either in terms of numbers of body plans we can recognize, or expressed phenetically by attribute space occupied. We therefore consider it likely that rates of evolution and levels of morphological "experimentation" in the Cambrian "explosion" can be explained in terms of traditionally modeled genetic mechanisms, rather than requiring processes unique to this period of history.

Perhaps even more apt is the argument made by Maynard Smith and Szathmáry:

> Much has been made of the fact that few, if any, wholly new body plans have emerged since the Cambrian, but it is hard to see how it could be otherwise. Organisms with an already evolved body plan could hardly give rise to descendants with a completely different one, so the only possibility is that new phyla should have evolved from single-celled ancestors. The prior existence of many highly elaborated multicellular animals has apparently prevented this.

[10]Vrba and Gould define *sorting* as differences among organisms and groups of organisms, distinguishing it from *evolution* by claiming that evolution is a product of sorting, and that sorting may have many causes, such as selection and genetic drift. Since most biologists define evolution as *any* genetic change over time—whatever the cause—distinctions between sorting and evolution seem confusing.

at least, show so much polymorphism that human races cannot be distinguished by the presence or absence of certain alleles but only by variations in a panoply of gene frequencies.

At least some racial differences, both morphological and behavioral, are adaptations to dissimilar environments. Both learned and innate behaviors have a genetic component, although it is more apparent in tropisms and instincts than in learned behaviors.

New species form when genetic exchange among races is impeded. Reproductive isolating mechanisms, which provide the barriers for genetic exchange, may be of various kinds. Behavioral, seasonal, and mechanical premating mechanisms obstruct zygote formation, while with postmating mechanisms offspring will be inviable or sterile. According to some proposals, more of these mechanisms should develop in sympatric populations than in allopatric ones, which are geographically isolated from each other.

New species may form in allopatric groups by the slow accumulation of genetic differences; or if they originated from only a few individuals (founder effect), that is, a group that has been substantially diminished in size (bottleneck). Speciation is presumed to occur in sympatric populations under diverse selection pressures, and these groups will become distinct because of persistent preferential mating.

The rate at which new taxa form is difficult to determine, as evolutionary rates differ even within phylogenetic groups. These inconsistencies probably relate to variations in selection pressures on the population at different times. Whether or not microevolutionary forces inducing change within species are identical to macroevolutionary forces generating new species is a matter of contention. Advocates of punctuated equilibrium believe that speciation is rapid and produced by unique forces; others feel that macroevolution is subject to the same forces (mainly natural selection) as is microevolution.

KEY TERMS

Allen's rule
allopatric populations
allopatric speciation
behavioral isolation
Bergmann's rule
bottlenecks
bradytely
coadapted gene combinations
ecogeographical rules
epigamic selection
evolutionarily stable strategies (ESS)
evolutionary rates
female choice
founding accidents
gametic mortality
Gloger's rule
habitat isolation
horotely
hybrid breakdown
hybrid inviability
hybrid sterility
hybrid swarms
index of genetic distance
innate behaviors
instincts
intersexual selection
intrasexual selection
introgressive hybridization
isolating mechanisms
learned behaviors
macroevolution
macromutation
mechanical isolation
microevolution
parapatric populations
parental investment
peripatric speciation
pheromones
polygynous species
postmating isolating mechanisms
premating isolating mechanisms
punctuated equilibria
quantum evolution
races
Rapoport's rule
reciprocal altruism
runaway selection
saltation
seasonal isolation
sexual isolation
sociobiology
speciation
stasis
sympatric populations
sympatric speciation
tachytely
taxes (behavior)
tropisms
vicariance
zones of hybridization
zygotic mortality

DISCUSSION QUESTIONS

1. Races
   a. How would you define a race?
   b. Do "pure" races exist?
   c. What factors are involved in increasing and decreasing the number of races in a species?
   d. How can we measure genetic distances among races?
   e. What ecogeographical rules can we apply to races?
2. Behavioral adaptations
   a. How do innate and learned behaviors differ?
   b. Do learned behaviors have a genetic basis?
   c. What model of behavioral strategy can evolve when both cooperative and selfish responses can occur in interactions between individuals?
3. What are the conflicting interests between the two sexes of polygynous species in how they choose mating partners? How can such conflicting interests lead to exaggerated male–female dimorphism?
4. However species are defined, would you support the concept that species represent natural groupings of organisms? Why or why not?
5. What response would you offer to the statement "All isolating mechanisms are equally efficient"?
6. What conditions would promote selection for premating isolating mechanisms? Would you say that such selection could occur between races of the same species, or that such selection could only occur between groups that have already speciated?
7. How would you support the concept that speciation can occur rapidly because of founding accidents or bottlenecks? What views oppose this concept?

8. How does sympatric speciation differ from allopatric speciation? What support is there for each of these modes of speciation?
9. How would you distinguish microevolution from macroevolution?
10. What arguments can you offer for and against the punctuated equilibrium hypothesis?

EVOLUTION ON THE WEB

Explore evolution on the web! Visit the accompanying web site for *Evolution,* 3/e at www.jbpub.com/evolution for web exercises and links relating to topics covered in this chapter.

REFERENCES

Anderson, E., 1949. *Introgressive Hybridization.* Wiley, New York.

Andrade, M. C. B., 1996. Sexual selection for male sacrifice in the Australian redback spider. *Science,* **271,** 70–72.

Arnold, M. L., 1997. *Natural Hybridization and Evolution.* Oxford University Press, Oxford, England.

Avise, J., and D. Walker, 1999. Species realities and numbers in sexual vertebrates: Perspectives from an asexually transmitted genome. *Proc. Nat. Acad. Sci.,* **96,** 992–995.

Ayala, F. J., 1983. Microevolution and macroevolution. In *Evolution from Molecules to Men,* D. S. Bendall (ed.). Cambridge University Press, Cambridge, England, pp. 387–402.

Barton, N. H., and B. Charlesworth, 1984. Genetic revolutions, founder effects, and speciation. *Ann. Rev. Ecol. Syst.,* **15,** 133–164.

Barton, N. H., J. S. Jones, and J. Mallet, 1988. No barriers to speciation. *Nature,* **336,** 13–14.

Briggs, D. E. G., R. A. Fortey, and M. A. Wills, 1993. How big was the Cambrian evolutionary explosion? A taxonomic and morphological comparison of Cambrian and recent arthropods. In *Evolutionary Patterns and Processes,* D. R. Lees and D. Edwards (eds.). Academic Press, London, pp. 33–44.

Bryant, E. H., S. A. McCommas, and L. M. Combs, 1986. The effect of an experimental bottleneck upon quantitative genetic variation in the housefly. *Genetics,* **114,** 1191–1211.

Buckley, S. H., T. Treganza, and R. K. Butlin, 1997. Speciation and signal trait genetics. *Trends in Ecol. and Evol.,* **12,** 299–301.

Bullini, L., 1994. Origin and evolution of animal hybrid species. *Trends in Ecol. and Evol.,* **9,** 422–426.

Bush, G. L., 1975. Sympatric speciation in phytophagous parasitic insects. In *Evolutionary Strategies of Parasitic Insects,* P. W. Price (ed.). Plenum Press, London, pp. 187–206.

———, 1994. Sympatric speciation in animals: New wine in old bottles. *Trends in Ecol. and Evol.,* **9,** 285–288.

Butlin, R., 1989. Reinforcement of premating isolation. In *Speciation and Its Consequences,* D. Otte and J. A. Endler (eds.). Sinauer Associates, Sunderland, MA, pp. 158–179.

Carson, H. L., 1986. Sexual selection and speciation. In *Evolutionary Processes and Theory,* S. Karlin and E. Nevo (eds.). Academic Press, Orlando, FL, pp. 391–409.

Carson, H. L., and A. R. Templeton, 1984. Genetic revolutions in relation to speciation phenomena: The founding of new populations. *Ann. Rev. Ecol. Syst.,* **15,** 97–131.

Civetta, A., and R. S. Singh, 1998a. Sex-related genes, directional sexual selection, and speciation. *Mol. Biol. and Evol.,* **15,** 901–909.

———, 1998b. Sex and speciation: Genetic architecture and evolutionary potential of sexual versus nonsexual traits in the sibling species of the *Drosophila melanogaster* complex. *Evolution,* **52,** 1080–1092.

Clark, A. G., D. J. Begun, and T. Prout, 1999. Female × male interactions in *Drosophila* sperm competition. *Science,* **283,** 217–220.

Clausen, J., D. D. Keck, and W. M. Hiesey, 1948. Experimental studies on the nature of species. III. Environmental responses of climatic races of *Achillea. Carnegie Inst. Wash. Publ. No. 581,* 1–129.

Clutton-Brock, T. H., 1983. Selection in relation to sex. In *Evolution from Molecules to Men,* D. S. Bendall (ed.). Cambridge University Press, Cambridge, England, pp. 457–481.

Coyne, J. A., 1992. Genetics and speciation. *Nature,* **355,** 511–515.

Coyne, J. A., A. P. Crittenden, and K. Mah, 1994. Genetics of pheromonal difference contributing to reproductive isolation in *Drosophila. Science,* **265,** 1461–1464.

Coyne, J. A., and B. Grant, 1972. Disruptive selection on I-maze activity in *Drosophila melanogaster. Genetics,* **71,** 185–188.

Coyne, J. A., and H. A. Orr, 1989. Two rules of speciation. In *Speciation and Its Consequences,* D. Otte and J. A. Endler (eds.). Sinauer Associates, Sunderland, MA, pp. 180–207.

———, 1997. "Patterns of speciation in *Drosophila*" revisited. *Evolution,* **51,** 295–303.

Darwin, C., 1871. *The Descent of Man and Selection in Relation to Sex.* Murray, London.

Dawkins, R., 1983. Universal Darwinism. In *Evolution from Molecules to Men,* D. S. Bendall (ed.). Cambridge University Press, Cambridge, England, pp. 403–425.

de Bono, M., and C. I. Bargmann, 1998. Natural variation in a neuropeptide Y receptor homolog modifies social behavior and food response in *C. elegans. Cell,* **94,** 679–689.

Dixon, A., D. Ross, S. L. C. O'Malley, and T. Burke, 1994. Paternal investment inversely related to degree of extra-pair paternity in the reed bunting. *Nature,* **371,** 698–700.

Dobzhansky, Th., 1970. *Genetics of the Evolutionary Process.* Columbia University Press, New York.

Dowling, T. E., and C. L. Secor, 1997. The role of hybridization and introgression in the diversification of animals. *Ann. Rev. Ecol. Syst.,* **28,** 593–619.

Dupré, J. (ed.), 1987. *The Latest on the Best: Essays on Evolution and Optimality.* MIT Press, Cambridge, MA.

Eberhard, W. C., 1996. *Female Control: Sexual Selection by Cryptic Female Choice.* Princeton University Press, Princeton, NJ.

Ehrman, L., and P. A. Parsons, 1981. *Behavior Genetics and Evolution.* McGraw-Hill, New York.

Eldredge, N., 1989. *Macroevolutionary Dynamics: Species, Niches, and Adaptive Peaks.* McGraw-Hill, New York.

Ellstrand, N. C., R. Whitkus, and L. H. Rieseberg, 1996. Distribution of spontaneous plant hybrids. *Proc. Nat. Acad. Sci.,* **93,** 5090–5093.

Epperson, B. K., 1995. Spotted distributions of genotypes under isolation by distance. *Genetics,* **140,** 1431–1440.

Feder, J. L., J. B. Roethele, B. Wlazlo, and S. H. Berlocher, 1997. Selective maintenance of allozyme differences among sympatric host races of the apple maggot fly. *Proc. Nat. Acad. Sci.,* **94,** 11417–11421.

Fenster, E. J., and U. Sorhannus, 1991. On the measurement of morphological rates of evolution. *Evol. Biol.,* **25,** 375–410.

Ferris, P. J., C. Pavlovic, S. Fabry, and U. W. Goodenough, 1997. Rapid evolution of sex-related genes in Chlamydomonas. *Proc. Nat. Acad. Sci.,* **94,** 8634–8639.

Fisher, R. A., 1930. *The Genetical Theory of Natural Selection.* Clarendon, Oxford, England. (2d ed. 1958, Dover, New York.)

Fuller, J. L., and W. R. Thompson, 1978. *Foundations of Behavior Genetics.* Mosby, St. Louis.

García-Ramos, G., and M. Kirkpatrick, 1997. Genetic models of adaptation and gene flow in peripheral populations. *Evolution,* **51,** 21–28.

Geary, D. H., 1990. Patterns of evolutionary tempo and mode in the radiation of Melanopsis (Gastropoda; Melanopsidae). *Paleobiology,* **16,** 492–511.

Gingrich, P. D., 1998. Vertebrates and evolution. *Evolution,* **52,** 289–291.

Goldschmidt, R. B., 1940. *The Material Basis of Evolution.* Yale University Press, New Haven, CT.

Gould, S. J., 1980. Is a new and general theory of evolution emerging? *Paleobiology,* **6,** 119–130.

———, 1989. *Wonderful Life: The Burgess Shale and the Nature of History.* Norton, New York.

———, 1994. Tempo and mode in the macroevolutionary reconstruction of Darwinism. *Proc. Nat. Acad. Sci.,* **91,** 9413–9417.

———, 1997. The exaptive excellence of spandrels as a term and prototype. *Proc. Nat. Acad. Sci.,* **94,** 10750–10755.

Gould, S. J., and N. Eldredge, 1993. Punctuated equilibrium comes of age. *Nature,* **366,** 223–227.

Gould, S. J., and R. C. Lewontin, 1979. The spandrels of San Marco and the panglossian paradigm: A critique of the adaptationist program. *Proc. Roy. Soc. Lond.,* **205,** 581–598.

Grant, P. R., and B. R. Grant, 1994. Phenotypic and genetic effects of hybridization in Darwin's finches. *Evolution,* **48,** 297–316.

Grant, V., 1985. *The Evolutionary Process.* Columbia University Press, New York.

Haldane, J. B. S., 1949. Suggestions as to quantitative measurement of rates of evolution. *Evolution,* **3,** 51–56.

Hall, J. C., 1979. Control of male reproductive behavior by the central nervous system of *Drosophila*: Dissection of a courtship pathway by genetic mosaics. *Genetics,* **92,** 437–457.

Hammerstein, P., 1996. Darwinian adaptation, population genetics and the streetcar theory of evolution. *J. Math. Biol.,* **34,** 511–532.

Howells, W. W., 1960. The distribution of man. *Sci. Amer.,* **203** (3), 114–127.

Jackson, J. B. C., and A. H. Cheetham, 1994. Phylogeny reconstruction and the tempo of speciation in cheilostome Bryozoa. *Paleobiology,* **20,** 407–423.

Kaneshiro, K. Y., 1983. Sexual selection and direction of evolution in the biosystematics of Hawaiian Drosophilidae. *Ann. Rev. Entomol.,* **28,** 161–178.

Kellog, D. E., 1988. "And then a miracle occurs"—Weak links in the chain of argument from punctuation to hierarchy. *Biol. and Phil.,* **3,** 3–28.

Kilias, G., and S. N. Alahiotis, 1982. Genetic studies on sexual isolation and hybrid sterility in long-term cage populations of *Drosophila melanogaster. Evolution,* **36,** 121–131.

Kirkpatrick, M., 1996. Genes and adaptation: A pocket guide to the theory. In *Adaptation,* M. R. Rose and G. V. Lauder (eds.). Academic Press, San Diego, pp. 125–146.

Koopman, K. F., 1950. Natural selection for reproductive isolation between *Drosophila pseudoobscura* and *D. persimilis. Evolution,* **4,** 135–148.

Lenski, R. E., and M. Travisano, 1994. Dynamics of adaptation and diversification: A 10,000-generation experiment with bacterial populations. *Proc. Nat. Acad. Sci.,* **91,** 6808–6814.

Levin, D. A. (ed.), 1979. *Hybridization: An Evolutionary Perspective.* Dowden, Hutchinson and Ross, Stroudsburg, PA.

Levinton, J., 1988. *Genetics, Paleontology, and Macroevolution.* Cambridge University Press, Cambridge, England.

Lijam, N., et al., 1997. Social interaction and sensorimotor gating abnormalities in mice lacking Dvl1. *Cell,* **90,** 895–905.

Mark, R., 1996. Architecture and evolution. *Amer. Sci.,* **84,** 383–389.

Mather, K., 1955. Polymorphism as an outcome of disruptive selection. *Evolution,* **9,** 52–61.

———, 1973. *Genetical Structure of Populations.* Chapman & Hall, London.

May, R. M., 1987. More evolution of cooperation. *Nature,* **327,** 15–17.

Maynard Smith, J., 1982. *Evolution and the Theory of Games.* Cambridge University Press, Cambridge, England.

Maynard Smith, J., and E. Szathmáry, 1995. *The Major Transitions in Evolution.* Freeman, New York.

Mayr, E., 1954. Change of genetic environment and evolution. In *Evolution as a Process,* J. S. Huxley, A. C. Hardy, and E. B. Ford (eds.). Allen & Unwin, London, pp. 156–180.

———, 1963. *Animal Species and Evolution.* Harvard University Press, Cambridge, MA.

———, 1969. The biological meaning of species. *Biol. J. Linn. Soc.,* **1,** 311–320.

———, 1983. How to carry out the adaptationist program. *Amer. Nat.,* **121,** 324–334.

McMillan, W. O., C. D. Jiggins, and J. Mallet, 1997. What initiates speciation in passion-vine butterflies? *Proc. Nat. Acad. Sci.,* **94,** 8628–8633.

Møller, A. P., 1996. Sexual selection, viability selection and developmental stability in the domestic fly *Musca domestica. Evolution,* **50,** 746–752.

Moore, J. A., 1949. Patterns of evolution in the genus *Rana.* In *Genetics, Paleontology, and Evolution,* G. L. Jepsen, E. Mayr, and G. G. Simpson (eds.). Princeton University Press, Princeton, NJ, pp. 315–355.

Moya, A., A. Galiana, and F. J. Ayala, 1995. Founder-effect speciation theory: Failure of experimental corroboration. *Proc. Nat. Acad. Sci.,* **92,** 3983–3986.

Nei, M., 1987. *Molecular Evolutionary Genetics.* Columbia University Press, New York.

Nei, M., and A. K. Roychoudhury, 1993. Evolutionary relationships of human populations on a global scale. *Mol. Biol. and Evol.,* **10,** 927–943.

Noor, M. A., 1995. Speciation driven by natural selection in *Drosophila. Nature,* **375,** 674–675.

Nowak, M. A., and K. Sigmund, 1992. Tit for tat in heterogeneous populations. *Nature,* **355,** 250–253.

———, 1998. Evolution of indirect reciprocity by image scoring. *Nature,* **393,** 573–577.

Osborne, K. A., et al., 1997. Natural behavior polymorphism due to a cGMP-dependent protein kinase of *Drosophila. Science,* **277,** 834–836.

Paterniani, E., 1969. Selection for reproductive isolation between two populations of maize, *Zea mays* L. *Evolution,* **23,** 534–547.

Paterson, H. E. H., 1993. *Evolution and the Recognition Concept of Species.* Johns Hopkins University Press, Baltimore.

Patterson, J. T., and W. S. Stone, 1952. *Evolution in the Genus Drosophila.* Macmillan, New York.

Paulson, D. R., 1974. Reproductive isolation in damselflies. *Systematic Zool.,* **23,** 40–49.

Phelan, P. L., and T. C. Baker, 1977. Evolution of male pheromones in moths: Reproductive isolation through sexual selection? *Science,* **235,** 205–207.

Rainey, P. B., and M. Travisano, 1998. Adaptive radiation in a heterogeneous environment. *Nature,* **394,** 69–72.

Read, A. F., 1987. Comparative evidence supports the Hamilton and Zuk hypothesis on parasites and sexual selection. *Nature,* **328,** 68–70.

Rensch, B., 1960. *Evolution Above the Species Level.* Columbia University Press, New York.

Reznick, D. N., F. H. Shaw, F. H. Rodd, and R. G. Shaw, 1997. Evaluation of the rate of evolution in natural populations of guppies (*Poecilia reticulata*). *Science,* **275,** 1934–1937.

Rice, W. R., and E. E. Hostert, 1993. Perspective: Laboratory experiments on speciation: What have we learned in forty years? *Evolution,* **47,** 1637–1653.

Rice, W. R., and G. W. Salt, 1988. Speciation via disruptive selection on habitat preference: Experimental evidence. *Amer. Nat.,* **131,** 911–917.

Rieseberg, L. H., 1997. Hybrid origins of plant species. *Ann. Rev. Ecol. Syst.,* **28,** 359–389.

Rieseberg, L. H., B. Sinervo, C. R. Linder, M. C. Ungerer, and D. M. Arias, 1996. Role of gene interactions in hybrid speciation: Evidence from ancient and experimental hybrids. *Science,* **272,** 741–745.

Ruff, C. B., 1993. Climatic adaptation and hominid evolution: The thermoregulatory imperative. *Evol. Anthropol.,* **2,** 53–60.

Rundle, H. D., and D. Schluter, 1998. Reinforcement of stickleback mate preferences: Sympatry breeds contempt. *Evolution,* **52,** 200–208.

Savage, R. J. G., and M. R. Long, 1986. *Mammal Evolution.* British Museum, London.

Scharloo, W., 1971. Reproductive isolation by disruptive selection: Did it occur? *Amer. Nat.,* **105,** 83–86.

Schliewen, U. K., D. Tautz, and S. Pääbo, 1994. Sympatric speciation supported by monophyly of crater lake cichlids. *Nature,* **368,** 629–632.

Schreider, E., 1964. Ecological rules, body-heat regulation and human evolution. *Evolution,* **18,** 1–9.

Scott, J. P., and J. Fuller, 1965. *Dog Behavior: The Genetic Basis.* University of Chicago Press, Chicago.

Seger, J., 1985. Intraspecific resource competition as a cause of sympatric speciation. In *Evolution: Essays in Honour of John Maynard Smith,* P. J. Greenwood, P. H. Harvey, and M. Slatkin (eds.). Cambridge University Press, Cambridge, England, pp. 43–53.

Sheldon, P. R., 1987. Parallel gradualistic evolution of Ordovician trilobites. *Nature,* **330,** 561–563.

Sheppard, P. M., 1961. Some contributions to population genetics resulting from the study of the *Lepidoptera. Adv. in Genet.,* **10,** 165–216.

Sigmund, K., 1993. *Games of Life: Explorations in Ecology, Evolution, and Behaviour.* Oxford University Press, Oxford, England.

Simpson, G. G., 1949. *The Meaning of Evolution.* Yale University Press, New Haven, CT.

Smith, F. A., J. L. Betancourt, and J. H. Brown, 1995. Evolution of body size in the woodrat over the past 25,000 years of climate change. *Science,* **270,** 2012–2014.

Soans, A. B., D. Pimentel, and J. S. Soans, 1974. Evolution of reproductive isolation in allopatric and sympatric populations. *Amer. Nat.,* **108,** 117–124.

Sokolowski, M. B., H. S. Pereira, and K. Hughes, 1997. Evolution of foraging behavior in *Drosophila* by density-dependent selection. *Proc. Nat. Acad. Sci.,* **94,** 7373–7377.

Sonneborn, T. M., 1957. Breeding systems, reproductive methods, and species problems in Protozoa. In *The Species Problem,* E. Mayr (ed.). American Association for the Advancement of Science, Washington, DC, pp. 155–324.

Spicer, G. S., 1993. Morphological evolution of the *Drosophila virilis* species group as assessed by rate tests for natural selection on quantitative characters. *Evolution,* **47,** 1240–1254.

Stanley, S. M., 1979. *Macroevolution: Process and Product.* Freeman, San Francisco.

Strickberger, M. W., 1985. *Genetics,* 3d ed. Macmillan, New York.

Taylor, C. E., A. D. Pereda, and J. A. Ferrari, 1987. On the correlation between mating success and offspring quality in *Drosophila melanogaster. Amer. Nat.,* **129,** 721–729.

Templeton, A. R., 1996. Experimental evidence for the genetic transilience model of speciation. *Evolution,* **50,** 909–915.

———, 1997. Out of Africa? What do genes tell us? *Current Opinion Genet. Devel.,* **7,** 841–847.

Thoday, J. M., 1972. Disruptive selection. *Proc. Roy. Soc. Lond.* (*B*), **182,** 109–143.

Thoday, J. M., and J. B. Gibson, 1962. Isolation by disruptive selection. *Nature,* **193,** 1164–1166.

Turelli, M., 1998. The causes of Haldane's rule. *Science,* **282,** 889–891.

Turesson, G., 1922. The genotypical response of the plant species to the habitat. *Hereditas,* **3,** 211–350.

Ungerer, M. C., S. J. Baird, J. Pan, and L. H. Rieseberg, 1998. Rapid hybrid speciation in wild sunflowers. *Proc. Nat. Acad. Sci.,* **95,** 11757–117562.

Vacquier, V. D., 1998. Evolution of gamete recognition proteins. *Science,* **281,** 1995–1998.

Vrba, E. S., and S. J. Gould, 1986. The hierarchical expansion of sorting and selection: Sorting and selection cannot be equated. *Paleobiology,* **12,** 217–228.

Wahlstrom, J., 1998. *Genetics and Psychiatric Disorders.* Elsevier Science, New York.

Wallace, A. R., 1889. *Darwinism: An Exposition of the Theory of Natural Selection with Some of Its Applications.* Macmillan, London.

Wasserman, M., and H. R. Koepfer, 1977. Character displacement for sexual isolation between *Drosophila mojavensis* and *Drosophila arizonensis*. *Evolution,* **31,** 812–823.

Watt, W. B., P. A. Carter, and K. Donohue, 1986. Females' choice of "good genotypes" as mates is promoted by an insect mating system. *Science,* **233,** 1187–1190.

Welch, A. M., R. D. Semlitsch, and H. C. Gerhart, 1998. Cell duration as an indicator of genetic quality in male gray tree frogs. *Science,* **280,** 1928–1930.

White, M. J. D., 1978. *Modes of Speciation.* Freeman, San Francisco.

Williams, B. J., 1979. *Evolution and Human Origins.* Harper & Row, New York.

Wilson, E. O., 1975. *Sociobiology: The New Synthesis.* Harvard University Press, Cambridge, MA.

Zahavi, Amotz, and Avishag Zahavi, 1997. *The Handicap Principle: A Missing Piece of Darwin's Puzzle.* Oxford University Press, New York.

25 Culture and the Control of Human Evolution

At the apex of our interest in evolution stands an interest in the state and future of our own species. How close are the ties between our culture and our biology? In which direction are humans evolving? Are human biological endowments satisfactory for human needs? What are the prospects for controlling human evolution? Our knowledge so far clearly offers us the chance to answer some aspects of these questions. However, before making this attempt, let us first consider some unique features of *Homo sapiens.*

Learning, Society, and Culture

The most distinctive feature of our species is probably our intelligence. However measured, this intelligence provides us with flexible adaptive behaviors that are far more complex than those attained by any other species. That is, humans can consistently **learn** from their environmental experiences by incorporating such experiences into their behavior and can create new environments over which they have considerable control.

Much human learning follows a Lamarckian pattern, in the conscious acquisition and transmission of those behavioral responses that answer the needs of specific situations. Although some learning occurs in other organisms (Chapter 24), they must primarily rely for survival on rigid and automatic responses genetically built into their nervous systems. Humans, in contrast, can grow up in different environments and learn to get food, defend themselves, find shelter, and perform various tasks in many specialized ways without depending on specialized genotypes.

Most important, more than any animal, humans can acquire and transmit such practices and behaviors, or **culture,** through social exchanges involving language, teaching, and imitation, both among individuals and among generations. Cultural transmission of learned behavior eliminates the hazards individuals encounter who must learn independently to cope with environmental variables, by trial and error. Instead, cultural transmission allows more successful imitative learning of adaptive practices that people have incorporated, often over more than a single lifetime, into the social and cultural heritage.

Because of such socially mediated transmission, cultural changes—unlike biological genetic changes—are not restricted to passage from one distinct generation to another, but may be proposed, accepted, and used during most stages in the human life cycle in interactions between both consanguineous and nonconsanguineous individuals. That is, the cultural "parents" of individuals need not be their biological parents, nor need cultural parents derive from the same geographical area as their cultural offspring. Thus, the kinds of isolation barriers that inhibit genetic exchange among biological species do not exist among human cultural groups: biological traits, with rare exceptions (pp. 225–226) transmit **vertically** within lineages, whereas cultural traits can transmit both vertically and **horizontally** within and among lineages.

In short, humans have two unique hereditary systems. One is the **genetic system** that transfers biological information from biological parent to offspring in the form of genes and chromosomes. The other is the **extragenetic system** that transfers cultural information from speaker to listener, from writer to reader, from performer to spectator, and forms our cultural heritage. Both systems are informational in that they produce their effects by instruction: the biological system through the information embodied in DNA via the coding properties of these cellular macromolecules, the cultural system through social interactions coded in language and custom and embodied in records and traditions.[1]

Relative Rates of Cultural and Biological Evolution

The changes cultural heredity has provided over the last 10,000 years have been most impressive. We know that somewhere during the Neolithic Age (Fig. 20–13) the long-prevailing lifestyle of hunting–gathering–fishing began to give way to the cultivation of food using domesticated plants and animals; that is, energies formerly expended in finding food were now directed into the more reliable and productive methods of farming–agriculture. Although originally developed in only a few localities—the Middle East, China, and Central America—such changes spread rapidly through migration. Within 1,000 years or so, many contiguous areas had begun some form of agriculture, and within 5,000 years agriculture and the technologies it stimulated extended widely (Table 25–1).

Perhaps the most immediate as well as far-reaching effect of agriculture was to increase the food supply manyfold, thus increasing both population size and population density in agricultural communities. Notwithstanding the uncertainties about whether sedentary communities preceded or followed agriculture, and whether agriculture was stimulated by climatic changes or by increased Neolithic social complexities, it apparently did not take many generations for villages to grow into towns, and towns into cities. These and other social effects of agriculture have been much discussed by archaeologists, historians, and others (see, for example, books by Diamond; Gowlett; and Maryanski and Turner), and have profoundly affected all areas of human interaction and creativity, from economics and politics to art, technology, and science.

What matters here was the change in emphasis on human function. From bands of food gatherers, hunters, and fishermen mainly concerned with satisfying hunger, we moved to complex urban societies in which such concerns occupy relatively little time for many of us. Skills, abilities, and behaviors that were often only modestly, if at all, emphasized in the past have become important adaptations for new technologies and lifestyles. Instead of hunting and primitive food gathering, an increased proportion of our efforts now concern cultural and technological tasks that we could not have foreseen 10,000 years ago or even a few generations ago. In fact, in some fields we can hardly predict from one year to the next what kinds of changes will appear.

According to one estimate (Holzmüller), the rate at which we gather new experience is now doubling at least every 15 years. Our present lifetime experience is therefore equivalent to about a 300-year life span of humans living just a few generations ago when the rate of gathering new experience was perhaps one-quarter or one-eighth what it is now. This remarkably rapid cultural and technological change, at least in the fields of science, promises even further increase if we consider the many scientists who now exist and are in training. Price has provided the widely quoted estimate that of all scientists who have ever lived, more than 90 percent are alive today!

In contrast to the rapid changes associated with cultural and technological heredity, changes in human biological heredity during this 10,000-year period seem

[1]Some writers claim that we can broaden the term *culture* to include any form of socially transmitted learned behavior whether transmitted by language *or* imitation. This definition extends culture to organisms lacking speech, such as chimpanzees, in whom variations in learned behaviors can be incorporated by imitation into different communities—behaviors based on different social histories ("traditions") rather than on genetic differences. Whiten and coworkers support this view by demonstrating that 39 "behaviour patterns, including tool usage, grooming and courtship behaviours, are customary or habitual in some [chimpanzee] communities," and differences in these patterns are unrelated to genetic differences between communities.

TABLE 25-1 Major human expansions from the neolithic age onward

Center of Origin	Area of Expansion	Time, Years Ago	Technologies
Middle East	Europe, North Africa, and Southwest Asia	10,000 to 5,000	Farming and domestication (wheat, barley, goats, sheep, and cattle)
North China	North China	9,000 to 2,000	Farming and domestication (millet and pigs)
South China	Southeast Asia	8,000 to 3,000	Farming and domestication (rice, pigs, and water buffalo)
Central America and North Andes	Americas	9,000 to 2,000	Farming (corn, squash, and beans)
West Africa	Sub-Saharan Africa	4,000 to 300	Farming (millet, sorghum, cowpea, and gourd)
Eurasian steppes	Eurasia	5,000 to 300	Pastoral nomadism (horses and warfare)
Southeast Asia or Philippines	Polynesia	5,000 to 1,000	Oceanic navigation
Greek colonization	Mediterranean	4,000 to 2,400	Navigation and trade

Source: From Cavalli-Sforza et al.

relatively small—if at all detectable. The most distinguished possession of *Homo sapiens,* the human brain, shows no change in size over the last 100,000 years, nor is there any clear indication that any qualitative change has occurred during this period. Our ancestors of many years ago, given our training, may well have shown the same range and distribution of mentality that we have today. Why this difference in speed between cultural and biological evolution?

By way of oversimplifying, although not too seriously, this contrast can be ascribed to differences between two distinct types of evolution: the mode of inheritance of acquired characters used by cultural evolution and the mode of inheritance through natural selection used by biological evolution. The **Lamarckian mode of cultural evolution** is an extension of the method by which humans learn. It depends on conscious agents—that is, humans with brains—who can modify inherited cultural information in a direction that offers them greater adaptiveness or utility. Furthermore, transmission occurs from mind to mind rather than through DNA. Thus, the information that humans receive from ancestors and contemporaries can be purposely changed to provide improved utility for themselves, their offspring, and others. This means that human minds have now become agents of a novel selection mechanism by consciously choosing among alternatives because of their consequences; that is, humans have brought teleology (p. 5) into evolution—the induction and adoption of changes for the sake of what we consider our own benefit.

The rate with which such purposeful modification takes place and the consequent rate of cultural change are limited by many factors, but—theoretically at least, and in the long run—primarily by human inventiveness. Furthermore, the generation time for cultural evolution may be as rapid as communication methods can make it. We can now move from place to place as fast as sound, and transfer ideas at electronic speed.[2]

In striking contrast to the high rate of cultural evolution is the slow progress of natural selection. The reasons for this are apparent from previous discussions. As far as we know, no cellular particles are sufficiently intelligent to detect or determine the direction of biological evolution and then change themselves accordingly. Organic evolution, as we have seen, can occur through a process of selection (among other forces) acting on random genetic changes. According to this view, chance differences that arise in genes or combinations of genes (mutation and recombination) produce a variety of effects on their carriers. These genetic differences furnish an array of genotypes among which the environment, in its different manifestations, selects only some for survival. Genetic evolution is slow because it must await fortuitous accidental genetic changes in DNA sequences and their organization before it can proceed, and each change may take many generations before it can be incorporated into the population.

[2]Because cultural evolution relies so heavily on communication among individuals and group interaction, the whole easily becomes more than the sum of its parts, in the sense that creations by a socially coordinated group of individuals—a city, a daily newspaper, an automobile factory, a cathedral, or a film—are quantitatively and qualitatively more than such individuals can create acting alone.

This relatively slow process of biological evolution is clearly quite different from the rapid, conscious selective process human minds use to choose among behavioral alternatives, indicating these processes evolve on separate methodological tracks. At the same time, the biological equipment needed to transmit and use cultural information (memory, perception, language ability, and so on) still connects them both.

Social Darwinism

The fact that human culture has at its source a biological foundation and that both culture and biology arise from informational systems that evolve over time have prompted various writers to suggest that general laws cover both society and nature, each sharing similar evolutionary mechanisms, especially that of natural selection.

During the nineteenth century, theorists developed these ideas into concepts, later called **Social Darwinism,** that we briefly describe as follows:

- Differences among human groups arose through natural selection.
- Natural selection was the mechanism that led to social class structures and to national differences in respect to economic, military, and social power.

Slogans such as "struggle for existence" and "survival of the fittest," when extended to social traits, enabled various English Social Darwinists, especially Herbert Spencer (1820–1903), to suggest that social evolution had progressed inevitably toward increased social and moral perfection and was approaching its culmination in Victorian society.[3] The religion and social customs of western Europe, especially England, could therefore be considered higher on the evolutionary scale than their counterparts elsewhere.

In its harsher forms, the Spencerian approach became popular in various circles in the United States, especially through the teachings of William Graham Sumner, the best known American Social Darwinist (Bannister, Haller, Hofstadter). Sumner concluded that:

> We cannot go outside of this alternative: liberty, inequality, survival of the fittest; not-liberty, equality, survival of the unfittest. The former carries society forward and favors all its best members; the latter carries society downwards and favors all its worst members.

As we might expect, many wealthy capitalists found such views to their liking. John D. Rockefeller, Jr., for example, whose father forged the gigantic Standard Oil trust by destroying many smaller enterprises, justified such behavior thus:

> The growth of a large business is merely a survival of the fittest. . . . The American Beauty rose can be produced in the splendor and fragrance which bring cheer to its beholder only by sacrificing the early buds which grow up around it. This is not an evil tendency in business. It is merely the working-out of a law of nature and a law of God.[4]

Sociologists such as Lester Ward reacted strongly against such blatant transposition of biological conduct into social conduct by pointing out that:

> If we call biologic processes natural, we must call social processes artificial. The fundamental principle of biology is natural selection, that of sociology is artificial selection. The survival of the fittest is simply the survival of the strong, which implies and would better be called the destruction of the weak. If nature progresses through the destruction of the weak, man progresses through the protection of the weak.

Along with criticisms made by others such as T. H. Huxley in his *Evolution and Ethics,* we can see that the difficulties in accepting Social Darwinism stem from its assumption that society (economics, politics, and so on) operates through the same laws as biology and for the same goals.[5] As people have repeatedly pointed out, this assumption is false because no evidence exists that what *is* in biology, *is* or *ought to be* in society.

For example, the laws of inheritance of wealth and power in society are legal and man-made, whereas the laws of biological inheritance do not result from human decision. It is also clear that, because we can consciously select them, we can direct social goals toward almost any objective that we humans choose for ourselves, such as wealth, poverty, chastity, obedience, and revolution.

[3]Spencer's writings ranged widely from biology to economics, philosophy, and sociology (Peel), with continued influence into the twentieth century (Hawkins). Interestingly, although Spencer invented the term "survival of the fittest," implying the action of selection, he remained a Lamarckian in respect to biological evolution until relatively late in life. Whatever its mechanisms, he conceived evolution as a powerful mystical force that governed all spheres of existence and therefore justified social and economic policies that supported those who were most "morally fit." For many Protestant intellectuals, Spencer's belief in such an evolutionary "cosmic" power helped reconcile science to their religion and made his writings extremely popular. His books sold more than 500,000 copies in the United States, and led the philosopher, William James, to criticize Spencer as "the philosopher whom those who have no other philosopher can appreciate."

[4]Quoted in Ghent.

[5]Some Social Darwinists, such as Wiggam, reacted quite strongly to proposals for social cooperation rather than for biological competition, echoing some still-prevailing attraction to the myth of "The Noble Savage" (p. 58 footnote 8):

> Evolution is a bloody business, but civilization tries to make it a pink tea. Barbarism is the only process by which man has ever organically progressed, and civilization is the only process by which he has ever organically declined.

"You needn't feel guilty. You earned the fortune you inherited by giving her great happiness while she was alive."

FIGURE 25-1 (© *The New Yorker Collection 1992 Henry Martin from cartoonbank.com. All Rights Reserved.*)

Biological goals, in contrast, are restricted to those of organismic evolution and follow opportunistic paths without conscious or moral direction. No moral or ethical qualities determine survival of the biologically fittest: the "bad" and "ugly" parasite can be even more biologically fit than its host, a "good" and "beautiful" human.

Distinctions between society and biology are also reflected in the fact that the social rewards bestowed on individuals or groups may be unrelated to biological merit or even to presumed social merit. One can be mentally or physically incapacitated, or dissolute, immoral, and criminal, yet exercise considerable social and economic power, and exploit social resources for asocial purposes. Certainly no genes guarantee social rewards—even dogs and cats have inherited wealth in our society (Fig. 25–1). The negative correlation between fertility and those classes with greater economic resources has also often been noted. Professional success and improved social status in the striving middle class are often achieved—through delayed marriage—at the expense of fertility. As Jones points out, wealth reproduces itself rather than people.

In general, we can see that Social Darwinism has no valid scientific basis—the socially "fit" are not necessarily the biologically "fit." Sumner's statement that "the law of the survival of the fittest was not made by man and cannot be abrogated by man," is simply untrue on the social–cultural level. Nevertheless, despite these contradictions, we can also see that by presuming a "scientific" basis for social stratification, Social Darwinism has been an attractive ideology to many individuals and groups who occupy or would like to occupy "superior" social positions.

This corruption of Darwinism has, in fact, often been used to justify or reinforce racism, genocide, and social and national oppression, and was incorporated into the views of many writers and educators who proposed socially or racially biased directions for the genetic improvement of humans (see Shipman). An extreme example is that of Germany during the 1930s and 1940s where the "racial health" movement was an important ideological element in the purposeful destruction of millions of people because they were considered members of "inferior" racial groups. Even in the United States, with its more democratic social heritage, laws were passed during the 1920s restricting immigration from eastern and southern Europe because of their "inferior" or "undesirable" races. (Except for some thousands recruited to work on the transcontinental railroad in the nineteenth century, further immigration of practically all Asians had been halted in 1882 by the Chinese Exclusion Acts—which were not repealed by U.S. Congress until 1943.)

Sociobiology

Despite the failings of Social Darwinism, we can hardly ignore underlying biological influences on society. At least some, if not many, biologically induced motivations are involved in friendship, sex, incest barriers, raising children, and attitudes toward strangers—behaviors common in all human groups, whatever cultural forms they take.

These and other observations have prompted modern **sociobiologists** such as Wilson and Alexander to argue quite logically that the capacity for culture among humans evolved from a noncultural state by means similar to the evolution of other biological traits. Carrying this idea further, they suggest that natural selection, by favoring genetic predispositions for cultural behaviors of the kinds just mentioned, must have been a major force acting to increase the frequency of such traits. In the case of culture, the sociobiologists point out that selection is not restricted to the fitness of an individual carrier of a favorable cultural "gene," but, like altruism (pp. 386 and 499), fitness is also evaluated by its effect on the genetic relatives of individuals carrying such favorable genes ("kin selection" or "inclusive fitness").

Sociobiologists presume that their approach provides the rationale to gauge how (and perhaps to what extent) biological causes account for social behaviors. Wilson, whose book *Sociobiology* furnished a major stimulus for modern interest in this field, defines sociobiology as "the systematic study of the biological basis of social behavior" and states that, because of interactions between genes and environment, "there is no reason to regard most forms of human social behavior as qualitatively different from physiological and non-social psychological traits." It has therefore seemed to some sociobiologists and their critics that sociobiology leads to the concept that most observed human social behaviors are biologically caused and that social inequities are justified—a concept very close to the views held by the Social Darwinists. Sahlins, for example, a critic of sociobiology, points out that:

> Darwinism, at first appropriated to society as "social Darwinism," has returned to biology as a genetic capitalism. . . . Natural selection is ultimately transformed from the appropriation of natural resources to the expropriation of others' resources . . . [or] social exploitation.

Some authors have seriously questioned the justice of such criticisms (for example, Ruse) since many sociobiologists certainly oppose the claim that biological exploitation justifies social exploitation. Nevertheless, the confidence with which sociobiologists reduce humans and their social behavior to the genetic level (to Wilson there is a "morality of the gene"—"the organism [social or nonsocial] is only DNA's way of making more DNA"), has made it easy for some partisans to "scientifically" vindicate those forms of social domination that suit their political views.

From a historical point of view, equating biological and social practices is clearly erroneous if we take into account the speed with which cultural changes occur and the resulting distance that has developed between human biology and culture. Many profound social and cultural changes, such as those involved in the transition from slavery to feudalism, or from feudalism to capitalism, or from "low" technology to "high" technology, are far too rapid to be caused by genetic changes in their human participants. Were there major differences in behavioral genes between the Tudor English, Victorian English, and modern English? Obviously, people can change their culture without changing their genes.

Even those behaviors that seem to have a sustained adaptive evolutionary basis can be socially confounded or manipulated toward ends that seem far beyond their original reproductive goals. For example, the altruism of preserving one's genes through kin selection can be socially transformed into the altruism of being a "team player" in a corporate firm in which none of the workers are biologically related to each other. Also, in some social circumstances, anger and rage can lead to behavior that is both self-destructive and offers no discernible biological benefit to related individuals.

Distinctions between socially and biologically influenced behaviors can be supported by further observations. Assault and murder, for example, are often subject to social punishment, whatever their biological–behavioral motivations. By law, traffic lights are to be obeyed whether or not drivers feel impatient, impetuous, aggressive, or weary. Drivers need not act altruistically in obeying traffic lights because of their altruistic genes, stockbrokers need not act selfishly in obeying specific stock exchange procedures because of their selfish genes, and military personnel need not act aggressively in obeying military orders because of their aggressive genes. The importance of sociobiology to humans has been considerably disputed; see, for example, the collections of articles that Barlow and Silverberg, Caplan, Fetzer, Gregory et al., and Montagu have edited.

It would seem that the most important lesson about humans to be learned from such disputes is that many forms of social behavior have two sides: behavioral patterns that were inherited biologically can be contained or modified within a cultural framework that must, at the same time, be accommodated to varying degrees within a biological framework. Both biology and culture are tied to each other by many strands—some apparently quite strong, while others are weak or imperceptible.

For example, the distorted behaviors caused by genetic factors involved in Down syndrome, Lesch-Nyhan syndrome, schizophrenia, manic-depressive psychosis, and various types of mental retardation indicate the importance of biological components in "normal" social behavior. Sexual behaviors in any society must also be influenced by biological factors that enable arousal and copulation, and these easily extend to courtship and mate selection. Supporting this view are data collected by Buss and his collaborators, who surveyed more than 10,000 men and women from a wide variety of geographical, racial, and cultural backgrounds to determine mating preferences. Not surprisingly (p. 588), these researchers found strategies differed between the two sexes, especially for short-term mating; with men generally seeking fertile (that is, young) women who are sexually accessible, and women seeking men who can provide resources and support beyond cursory sexual contacts. Various psychologists offer similar findings that point to congenital differences between genders in social behavior (see, for example, Freedman). To these, we can add other aspects of social behavior that appear to have evolutionary adaptive roots (pp. 499 and 506). Again, as true for other complex biological traits, we should keep in mind that social-behavioral responses are not fixed in any group or population, but are influenced by a wide spectrum of genetic variation.

Because some behaviors are affected by many levels of social interaction, they are more obviously tied to culture than to biology. The biology of political affiliations such as Democrat, Republican, or Socialist is indiscernible, although the emotions that may be associated with such affiliations (loyalty, fear, aggression, altruism) have a biological basis. When carried forward to activities that seem purely cultural, such as evaluating ideas in philosophy and science, the connecting strands between such evaluations and biology may be so thin and twisted as to be impossible to follow. Moreover, as human culture advances toward new social occupations and intricate behavioral interactions, social controls increase as well; what start out as simple rules and rituals prescribing

BOX 25–1 REDUCTIONISM AND HIERARCHIES

The difficulty of using biology to explain cultural changes echoes in some of the arguments against the philosophical concept of **reductionism.** Reductionists propose that explanations for events on one level of complexity can and should be reduced to (deduced from) explanations on a more basic level. Reductionists would claim that cultural events can be explained in terms of biology, biological events in terms of chemistry, chemical events in terms of physics, and to some proponents, physical events in terms of mathematics, and mathematical concepts in terms of symbolic logic. Thus, all events would have an ultimately singular level of explanation and reference.

Aside from the philosophical problems associated with reductionism (see, for example, Dupré, Hull, Rosenberg), it would cause considerable confusion by eliminating understandable generalizations used at each level of explanation, and substituting explanatory terms from a different level that would seem both incomprehensible and inadequate. For example, how a repressor molecule (p. 219) functions in a cell depends on its atomic structure, but the repressor function it shares with other such regulators is a specific molecular property, and would be difficult, if not impossible, to explain as an atomic property. This concept of uniquely shared properties can relate to many levels of complexity, such as:

- Glycolysis (p. 158) is a property shared by some metabolic pathways, but not of a molecule.
- Dominance (p. 197) is a property shared by some genes but not of a nucleotide.
- Sexual behavior (p. 588) is a property shared by some organisms, but not of a gene.
- Population density (p. 554) is a property of a group of organisms, but not of an organism.
- Exchange value (money) is a property of U.S. Treasury dollar bills, but not of a 2 1/2 × 6 inch sheet of paper.

We can therefore claim that a **hierarchy** extends across many levels, from atoms to molecules to cells to tissues to organs to individuals to populations to species to cultures, each with specific functional properties. Because of their multi-component subunit structure, hierarchies also have the advantage of being more stable and more easily constructed than a nonhierarchical structure in which all parts must be simultaneously assembled because the absence of any single component interferes with any kind of assembly at all. For example, a cell composed of multiple subsystems is less vulnerable to accident and more easily synthesized than a similar compartment which has no subsystems but can function only when all of its many chemical components match perfectly and aggregate simultaneously. In fact, it is difficult to visualize a nonhierarchical system that maintains complex functional properties.

It seems reasonable that hierarchical systems have aided and stabilized evolution, enabling organisms to incorporate new functional properties and to avoid cataclysmic fragmentation should any minor single component be defective because of change or substitution. A marked change in selection from competitiveness to cooperation

simply defined behaviors, become, with increased social complexity, immense legal compendia.[6]

The transformation from the evolution of biology to the evolution of culture generally marks a qualitative change from one level to another. New rules and laws come into play in understanding culture that are not apparent in biology. This is obvious from previous illustrations that social laws may restrict or limit biologically motivated behavior: to achieve human morality often means negating or circumventing biologically based impulses. In like manner, society and culture can route human biology into purposes other than biological reproductive success. It is also clear that cultural differences can easily arise from the quirks of social history rather than from biological differences. For example, although the ability to learn a language has an evolutionary and genetic basis (p. 497), the particular language learned by an individual depends not on biology but upon the history of the society in which he or she lives. The importance of social and cultural history also applies to individuals and groups in respect to their specific technologies, architectures, forms of artistic expression, and perhaps to even more biologically intimate matters such as modes of infant care and toilet training. We must therefore learn to distinguish capability from form: although we are capable of acting in socially acceptable fashions, particular social codes and conventions vary in different places and at different times.

[6]Wilson's (1978) oft-quoted statement that "genes hold culture on a leash" is a gross oversimplification that distorts both biology and culture. As Kaye incisively demonstrates, such sociobiological views attribute to biology mystical goal-seeking properties, yet sociobiologists are forced to admit that many forms of culture—such as those leading to self-destructive wars—have escaped their "leash" and become mysteriously nonadaptive and abiological. If anything, judging from its long toolmaking history, one can claim that society (culture), because of its many technological changes and social–economic–political innovations, has become at least a partial master of human biology (see also, for example, Kingdon).

among cells exemplifies such benefits in the transition from unicellular to multicellular organisms. Similar cooperative changes occur in the formation of hypercycles (p. 142) in which competing elements united into a mutually supportive entity, and in the transition from solitary individuals into social groups. Components of complex biological organizations can become modified by selection away from independence toward cooperation.

However, because hierarchical properties seem novel and "emergent," people have raised questions whether hierarchical levels are understandable and open to analysis based on cause and effect, or whether their complexity makes them opaque to common scientific methods. Some proponents of "holism" have suggested that a hierarchical entity such as life is so different from its chemical and physical components and antecedents that formerly mechanistic analyses no longer apply. Such views (p. 13) are essentially **vitalistic**—based on a belief in mystical states possessing unexplainable attributes.

In contrast to vitalism, scientists argue with considerable success that hierarchical levels are explainable. We can certainly analyze glycolysis biochemically (Fig. 9–1), dominance enzymatically (p. 197, footnote 4), sexual behavior genetically (Fig. 24–6), population density ecologically (p. 554), and so forth, even though the terms used are not easily, or at all, interchangeable between levels. For biology, vitalism is no longer an issue, although reductionism and anti-reductionism continue to be argued (Williams). Each side takes comfort in a distinctive view: the reductionist in life's ability to function through physical material, the anti-reductionist in life as a series of dynamic processes that can organize physical material.

Aspects of biological hierarchies, however, still pose problems. For example, because unique hierarchical properties confer distinct evolutionary values, some authors propose that each level is a special unit of selection—for example, "molecular selection," "tissue selection," "organ selection," and "species selection." Some writers take this to imply that some, if not all, levels possess uniquely separate hereditary systems which respond to selection differently from others (see, for example, p. 356). However, most biologists insist that only the nucleic acid genetic system can consistently transmit *biological* properties between generations at whatever hierarchical level these properties appear, from molecular to populational. This dependence on a single instructional system is not unlike the use of the same two-dimensional drafting system to plan different structures—mechanical, electrical, aeronautical, architectural, and so forth—each with its own shapes and terminology.

Nevertheless, despite genetic influence, we must recognize that hierarchical properties are not independent of environmental influence, since environmental interactions can modify phenotypic expression. Temperature, light, density, competition, and so forth (p. 553) are factors that can alter organismic properties. However, the distinction between biology and culture needs to be made: effects of environmental changes are not biologically transmitted unless genetic instructions are changed, but changes can be culturally transmitted by conscious agents through nonbiological instruction.

Understanding social change is a quite different task from understanding biological change: societies and cultures cannot be explained as biological behaviors any more than biological behaviors can be understood as atomic interactions. The changes that have taken place in biological evolution do not provide a sufficient understanding of mechanisms, sequence, or ethics in cultural changes. In disparaging attempts to reduce all biology to molecular terms, Wilson (1984) himself has pointed out that "molecular biology on its own is a helpless giant. It cannot specify the parameters of space, time, and history that are crucial to and define the higher levels of organization." Wilson's argument also applies to attempts to reduce an understanding of human culture to its biological components ("biological determinism") and shows the insufficiency of sociobiology when applied uncritically to humans. Clearly, different kinds of interactions and explanations are involved in different levels of complexity (Box 25–1).

Biological Limitations

The fact that cultural considerations can transcend biological considerations becomes apparent in dealing with the topic of human control over evolution. In which direction are we to guide evolution? What goals are we to set? These questions do not arise from unconscious biological laws but from the conscious cultural realization that we would like to improve ourselves and the world we live in, and from the social technology that allows us to achieve such goals. This quest for human improvement comes mostly from the disparity between our cultural needs and our biological limitations. Let us consider a few of these.

One area of biological inadequacy stems from our advanced technology: we are becoming to a large degree sedentary in occupation, but our intestines and appetites do not adapt accordingly. Many who live in surplus societies, such as the United States, tend to put on extra weight and suffer from the accompanying ills. The pains and problems of childbirth are probably a consequence of our erect posture and can be aggravated by the lack of physical conditioning. The stress of many aspects of social living, ambition, and competition finds much of the human species biologically unprepared, and we suffer from anxiety, ulcers, heart disease, and other socially aggravated illnesses (which Comfort calls "the ulcer belt syndrome"). Pollutions of various kinds caused by sewage, tobacco, automobiles, and industry lead to a variety of modern diseases ranging from induced cancer to emphysema and silicosis.

Perhaps one of the most important contrasts between what we are and what we would like to be lies in the difference between biological and cultural maturity. Biologically, our efficiency begins to fall soon after we reach the reproductive ages of 20 to 30 years. Our cultural efficiency, however, in the contributions we can make in various professions, often begins to increase during that period or even later. Our cultural development is thus limited by our biological decline; that is, our biological heritage stresses reproductive success and hardens the arteries afterward, while our cultural development asks for continued plasticity and **longevity.**

Postreproductive longevity, unfortunately, is a trait that tends to remain low in many organisms that reach reproductive maturity relatively early in their potential life span. Before civilization, only about half the human population passed the age of 20 and probably not more than 1 out of 10 lived beyond 40 years. These low longevity values extended into the period of the early Greeks and even into modern periods among primitive people. Life expectancy remained between 20 to 30 years until the Middle Ages, then rose somewhat and has risen sharply among Europeans and Americans in the last century, from about 40 years in 1850 to the present 75 years.

These statistics are important because they indicate that we now have among us an age group, those 40 to 50 years and older, on whose biological attributes natural selection has never directly operated. In other words, the adaptive traits of such older individuals are those they had in the years before their reproductive periods, and their postreproductive fitness is no longer reflected in their relative reproductive success.

For example, an individual who has produced three children and at the age of 50 develops cancer or other diseases with genetic components is by this fact no less reproductively successful than an individual of the same age who has produced three children but does not suffer from such diseases. One may, of course, argue that children with healthy grandparents are more fit than children with ill or absent grandparents since they get more attention and care. In present social situations, however, such caretaking functions can be easily assigned to other individuals, and it is unlikely that grandparental attention adds to reproductive fitness.[7]

Healthy old age is a trait that only relatively few genetic variants might be expected to attain. At present about 2 percent of the population reaches 90 years, and only about 1 per 1,000 individuals reaches age 100. There is little promise that average human life expectancy can be raised beyond 80 to 85 years, even with considerable medical progress.

A further contrast between past and present biological requirements is in fertility. **Clutch size,** a term used to describe the number of offspring born to a nesting pair of birds, is certainly as adaptive a factor in humans as in birds.[8] The ability of a mature human female under primitive conditions to produce eight or nine offspring during her reproductive period, of which an average of two survive, is of selective value in ensuring that her lineage will persist in the face of high infant mortality. However, in many modern countries, infant mortality rates have markedly decreased (Fig. 25–2). From 15 or more percent of all births in 1900, infant mortality rates have decreased to less than 1 percent in more than 20 countries. Much of this change is caused by improved prenatal care and medical control of infectious disease (Wegman). This overall imbalance between human fecundity and survival rate has led to an exponential growth of the human population, which is now doubling at the rate of about once every 40 years—almost 7,000 times as fast as in primitive Paleolithic societies!

This phenomenal growth rate has led to a **population explosion** that has produced the present global six billion people. Although this growth rate is diminishing, about

---

[7]Proposals that female menopause allows "grandmothers" to transfer maternal care to their grandchildren is in conflict with findings that menopause offers no such advantage to lions and baboons (Packer et al.). Instead, some suggest that the interval between menopause and death allows last-born offspring to be attended by their older mothers until they reach the age of independence. For lions and baboons this age is about 1 to 2 years, explaining why their menopause to death interval is usually no longer than twice that—about 4 to 5 years. If we apply this same ratio to humans, where children often remain dependent for 10 years or so, the menopause to death interval should be about 10 to 20 years. Since human menopause begins at about 40–50 years of age, this may account for female life spans reaching about 60 or so years in non-medically assisted hunter–gatherer societies.

The question whether menopause is adaptive or only a geriatric by-product of aging is related to the question of aging itself. Muller and others have pointed out that senescence and death may have evolutionary value to populations, since such means help ensure the turnover and replacement of older genotypes by new genotypes that may be better adapted. Other views, less based on group selection (pp. 569–570) suggest that senescence is caused by an accumulation of mutations as the organism ages or arises from postadult pleiotropic effects of genes whose earlier effects were beneficial to the organism (see Rose, and also p. 556).

[8]Lack pointed out that very small clutch sizes in birds may produce too few offspring for continued reproduction, and large clutch sizes may produce too many offspring for their parents to feed. "Optimal" clutch sizes, like optimal birth weights (Fig. 22–7), rest somewhere between extremes, probably dependent on factors such as the availability of food and parental longevity.

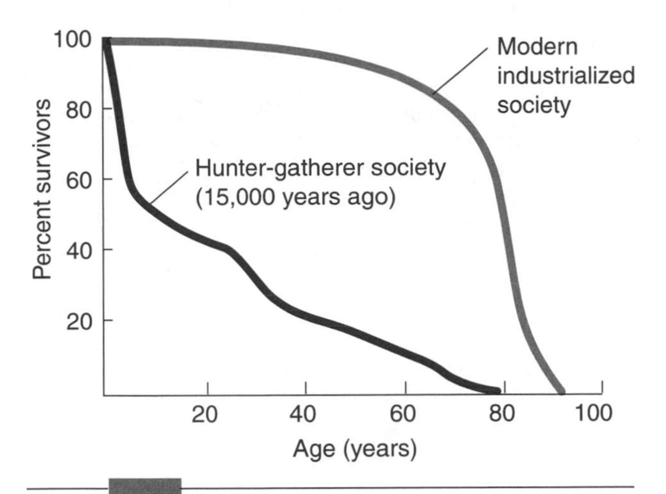

FIGURE 25-2 Survival curves for a population of hunter-gatherers who lived 15,000 years ago on the Mediterranean coast (based on skeletal remains) compared to a present-day population living in an industrialized society. (*Adapted from May.*)

200,000 people are being added every 24 hours, and the population may double to 12 billion people by the year 2050. The problem is serious if we consider that although food production has been increasing, it barely keeps up with present requirements. According to Fedoroff and Cohen, "an estimated three quarters of a billion people still suffer from malnutrition." Improving the quality of life for an increasing population is a critical issue if we assume that all people and those yet to come would like to live above the level of abject poverty. Famine remains a recurrent problem in some Third World countries where yield per acre is low, and small agricultural increases have come only from putting more land into cultivation.

The increase in agricultural acreage is a limited solution because of the limited amount of usable land. In the United States today, about 6 acres of agriculturally usable land per person supply domestic needs and provide surpluses sent to some of the agriculturally deficient countries. In England usable acreage falls to 1/10 this value and in Japan to 1/30. Considerable amounts of usable land are still available in some newly developing countries, such as in Africa, and some agricultural gains can be made by using new high-yield varieties and improved fertilizers, but even these increases will be inadequate if population growth continues. At present, most additions to arable croplands come from the worldwide destruction of forests, yet available agricultural land per capita is still declining (Pimentel and Giampetro). Other problems and uncertainties can also threaten the food supply locally as well as globally, such as soil erosion, fertilizer and pesticide contamination, water supply depletion, and extreme climatic events (Daily et al.).

Fortunately, a reduction in human fecundity can be instituted through presently known birth control methods without the need for evolving such a reduction biologically. The most serious challenge that lies ahead is educating people and institutions to use and promote such methods.

The same approach toward other biological inadequacies, such as insufficient longevity and the "ulcer belt syndrome," cannot be undertaken as easily. For these traits, we need some sort of controlled genetic change that would help put humans in biological harmony with their present or future cultural surroundings. As stated earlier, the question that arises is whether we can impose a direction on our biological evolution using conscious cultural means to approximate the speedier method of inheritance of acquired characters. If this were possible, what kinds of genetic change would be desirable?

Deleterious Genes

Let us briefly consider possible changes in the frequencies of genes that have obvious **deleterious effects.** The number of children classified as born markedly defective, either physically or mentally, is conservatively estimated at about 20 to 25 in 1,000 births, and the mortality rate ascribed to such congenital malformations in the United States is about 15 percent of all infant deaths. Many other defects, not immediately noted at birth, become apparent during childhood years and are more widely prevalent than may be imagined. In various studies, about 30 percent or more of hospital admissions for children and 50 percent of all childhood deaths are ascribed to birth defects or to complications that such defects may have caused. Although not all birth defects are genetic, the proportion of the genetically handicapped among them is undoubtedly high.

Table 25–2 lists estimates of the frequencies of some genetic disorders, which occur in approximately 3 percent of all births. If we include other genetic defects that appear later in life, such as muscular dystrophy and diabetes, the frequency probably doubles. If in addition we include less obvious defects that nevertheless have strong genetic components, such as impaired resistance to stress and infection and other physical and psychological weaknesses, the effects of harmful genes probably touch a majority of our population.

The ubiquity of deleterious genes with lethal effect has been dramatically demonstrated in studies made of the offspring of cousin marriages by Morton and coworkers and by others. These studies have used techniques of detecting and partitioning the genetic load (pp. 557–558) caused by inbreeding and have shown that outwardly normal individuals in our society carry a genetic load equivalent to that of approximately one to eight deleterious lethal genes (**lethal equivalents**) that, if homozygous, would cause early death. Two important questions then arise:

TABLE 25-2 Estimated frequencies (in percent) of some human genetic disorders

Disorder	Percent	Disorder	Percent
Single Gene Disorders		**Chromosomal Disorders**	
Autosomal recessive		Autosomal	
Mental retardation, severe	0.08	Trisomy 21	0.13
Cystic fibrosis	0.05	Trisomy 18	0.03
Deafness, severe (several forms)	0.05	Trisomy 13	0.02
Blindness, severe (several forms)	0.02	Other	0.02
Adrenogenital syndrome	0.01		0.20
Albinism	0.01	Sex chromosome	
Phenylketonuria	0.01	XO and X deletions	0.02
Other aminoacidurias	0.01	Other "severe" defects	0.01
Mucopolysaccharidoses (all forms)	0.005	XXY	0.1
Tay-Sachs disease	0.001	XXX	0.1
Galactosemia	0.0005	Others	0.015
	0.25		0.35
X-linked			
Duchenne muscular dystrophy	0.02	Total chromosomal disorders	0.55
Hemophilias A and B	0.01		
Others	0.02	**Multifactorial Disorders**	
	0.05	Congenital malformations	
Autosomal dominant		Spina bifida and anencephaly	0.45
Blindness (several forms)	0.01	Congenital heart defects	0.4
Deafness (several forms)	0.01	Pyloric stenosis	0.3
Marfan syndrome	0.005	Clubfoot	0.3
Achondroplasia	0.005	Cleft lip and palate	0.1
Neurofibromatosis	0.005	Dislocated hips	0.1
Myotonic dystrophy	0.005		1.65
Tuberous sclerosis	0.005		
All others	0.015	Total multifactorial disorders	>1.65
	0.06		
Total single-gene disorders	0.36	**Total frequency of listed genetic disorders**	>2.56

Source: From *American Scientist* 65: 703–711, 1977 by C.J. Epstein and M.S. Golbus. Reprinted by permission.

- What accounts for the prevalence of these harmful genes?
- What, if anything, can we do to get rid of them?

The reasons for their high frequency are not fully agreed on, although there is little question that they arise originally through mutation. One opinion, which the late Theodosius Dobzhansky and others held, is that such genes, although deleterious in homozygous condition, may offer considerable advantage to their heterozygous carriers by producing some sort of hybrid vigor. According to this theory, a gene will be maintained in the population although the homozygote produced by this gene is relatively inferior in fitness (pp. 538–540).

Another school, formerly headed by the late Hermann Muller, believes that such genes produce no advantage of any kind and that their frequency is now high because the usual effect of natural selection has been artificially reduced. According to Muller, genotypes that were formerly defective and would have been eliminated under more primitive conditions are now kept alive by medical techniques and enabled to pass on their defective genes to their offspring. As we know, a decrease in the selection coefficient against a particular gene causes an increase in the equilibrium frequency of the gene ($q = \sqrt{u/s}$ for a recessive gene, $p = u/s$ for a dominant gene; p. 545). Thus if deleterious genes are not eliminated by selection, they will gradually increase in frequency in accord with their mutation rate. Since the mutation rate is usually low, the frequency of any particular gene will increase rather slowly, but since there are many possible deleterious genes, the genetic load will increase significantly.

According to Muller's theory, humans cannot reach any biological harmony until most deleterious genes are removed. If they are not removed and continue to increase in frequency, Muller held out the prospect that the human species would end up with two types of individuals: one kind would be so genetically crippled that they could hardly move, and the other kind would be less crippled but spend all their time taking care of the first kind.

This specter is made even gloomier if we attempt to consider means by which such genes can be eliminated. Since we all probably carry at least a few deleterious recessive genes, most of which we do not know about, there is little prospect in eliminating them short of mass sterilization.

Serious as this argument may be for genes that produce severe handicaps, it is undoubtedly exaggerated for genes whose harmful effects can be treated relatively inexpensively. Nearsightedness, for example, is a trait whose frequency has most likely increased in recent periods, but can be corrected quite simply by an optometrist. Furthermore, the fact that natural selection no longer operates to eliminate many genotypes is not necessarily an undesirable feature of modern life. Few individuals would argue today that fire and clothes should be abolished because they are artificial devices that circumvent natural selection by permitting nonfurry genotypes to survive in cold climates. It would also be difficult for us to return to the "good old" prevaccination, presanitation days of smallpox, diphtheria, typhus, cholera, and plague.

However, despite medical and cultural progress, the effect of many deleterious genes cannot be easily treated, and although Muller may have been overly alarmed about their increase in frequency, we have become more aware of their widespread existence in recent years. Many geneticists have turned to exploring the possibility of controlling harmful effects by artificially changing gene frequencies.

Eugenics

In its modern form, suggestions for improving human genetic material have come under the name **eugenics,** a term proposed by Francis Galton before the turn of the century.[9] Galton, concerned with the heredity of quantitative characters such as intelligence, became aware, after reading Darwin, that the evolution of human traits through natural selection could be substituted by their evolution through social selection. ("What nature does blindly, slowly, and ruthlessly, man may do providently, quickly, and kindly.") However, like the Social Darwinists, many early eugenicists reflected their own personal and racial biases as to which characteristics were desirable and which undesirable.

C. B. Davenport, a leader of the eugenics movement in the United States, exemplified this racist approach by using New Englanders as the standard of comparison for all American nationalities. According to Davenport and other members of his Eugenics Record Office, most social characteristics had identifiable genetic components typical of particular groups (for example, Italian violence, Jewish mercantilism, Irish pauperism). These attitudes can be said to have reached their culmination in the "racial health" movement in Nazi Germany during the 1930s, when eugenic laws were promulgated establishing German Aryans as the "master race" and forbidding intermarriages with presumed racially inferior non-Aryans.

Although the particular social and political causes that fostered Nazi attitudes and their horrible consequences are not the subject of this book (see books by Müller-Hill, Proctor, and Weiss), it is important to recognize that ethnic stereotyping and fear of strangers (xenophobia) were essential elements used to foster racial prejudice, group violence, and ultimately, genocide. Unfortunately, xenophobic fears and sentiments are common to this day, perpetuated by myths about the degeneration of race and intelligence because of mixture with "inferior" types,[10] and by a world-view in which satanic evil forces and conspiracies are held responsible for economic, social, and personal difficulties (see also pp. 64–65).

In the United States, one of the true racial melting pots of the world, no evidence appears for the biological superiority of any particular race in respect to intelligence. Blacks, long at the bottom of the racial pecking order, show as wide a range of intelligence as do whites. According to Pettigrew and others, researchers have shown racial differences in intelligence (IQ) examinations to be remarkably plastic, influenced by such factors as prenatal

[9] Plato in his dialogue *The Republic* made one of the first proposals suggesting that humans could be improved through selective breeding. In his ideal philosopher-state only the most physically and mentally fit individuals were to be mated and their offspring raised by the state. Inferior types were to be prevented from mating or their offspring destroyed. Since family relations were absent in *The Republic,* superior and inferior types could be determined impartially, and the governing class was selected only "from the most superior." However, Plato's notion of superiority and inferiority was quite different from that of his contemporaries, who considered a conquering people superior and a subjugated people inferior. In ancient Sparta, for example, some measure of selective breeding seems to have been practiced with the purpose of raising a "superior" military ruling class to hold in subjugation the "inferior" servant classes. The fact that cultural disparities usually existed between conquerors and subjects reinforced such notions but did not seriously reflect whether any essential biological differences were responsible for these cultural differences. Were the Spartans biologically "superior" to the Helots, Corinthians, and Athenians?

[10] Women have not escaped pseudoscientific claims as being an "inferior" class. Gould, in his review of many biases about racial intelligence, quotes a nineteenth century French social psychologist, Gustave Le Bon, as an extreme example of such misogyny:

> In the most intelligent races, as among Parisians, there are a large number of women whose brains are closer in size to those of gorillas than the most developed male brains. This inferiority is so obvious that no one can contest it for a moment; only its degree is worth discussion. All psychologists who have studied the intelligence of women, as well as poets and novelists recognize that they represent the most inferior forms of human evolution and they are closer to children and savages than to an adult civilized man. They excel in fickleness, inconstancy, absence of thought and logic, and incapacity to reason. Without a doubt there exist some distinguished women, very superior to the average man, but they are as exceptional as the birth of any monstrosity, as, for example, of a gorilla with two heads; consequently, we may neglect them entirely.

Of course, the question arises why Le Bon did not oppose matings with such deprived creatures.

diet, early cultural surroundings, and even the color of the interviewer in the IQ examination. Analysis of IQ heritabilities by Devlin and coworkers suggests that maternal uterine environment plays a significant role, blurring class and racial differences in IQ proposed by Herrnstein and Murray.

Facts also contradict predictions that intelligence will steadily decline because of the higher reproductive rate of the lower, "unintelligent" social classes and their consequent increase in frequency. A Scottish survey that covered almost 90 percent of all 11-year-old children in 1932 and again in 1947, found no decrease in IQ. On the contrary, these studies showed a significant increase in average intelligence during this interval. (See also more recent studies in the book edited by Neisser.)

The fears expressed by various writers that the abolition of privileged classes in society will lead to "hybridization" and thus to the loss or dilution of superior genotypes are hardly scientific. The evidence at present is that high intelligence is not the exclusive genetic property of a particular social class, but rather that its expression can easily be masked in any group by deficiencies in diet, lack of cultural stimulation, and absence of opportunity. As environmental conditions improve, average intelligence scores may also be expected to improve, although genetic differences between individuals will remain.

We may predict that equality of economic and educational opportunity for all classes will enable each individual to more nearly achieve his or her true potential. Society will be the benefactor in producing more creative and inspired individuals such as Leonardo da Vinci, Voltaire, Newton, and Marie Curie, who might otherwise die anonymously among the dispossessed sections of our society. As Dobzhansky has stated, there is little to lament in "the passing of social organizations that used the many as a manured soil in which to grow a few graceful flowers of refined culture."

THE KINDS OF EUGENICS

Stripped of racism and provincial prejudice, eugenics may be considered a serious attempt to diminish human suffering and improve the human gene pool. It has been subdivided into two aspects:

- **Negative eugenics,** the attempt to decrease the frequency of harmful genes.
- **Positive eugenics,** the attempt to increase the frequency of beneficial genes.

Negative eugenics involves socially discouraging the reproduction of genotypes that are most obviously deleterious. For example, it would be foolish and self-destructive to encourage hemophiliacs, who are being preserved by blood transfusions, to reproduce. Similarly, where known, female carriers of the hemophilia gene should be made aware of their genetic problem and encouraged not to pass it on.

These eugenic programs will suffice to control suffering from a number of deleterious genes, although they will not eliminate them, and many of these educational measures are already in practice today. Many harmful genes present in high frequency, such as diabetes, and others where the carriers are not easily known, such as cystic fibrosis, cannot be controlled in this way because of the very inefficient elimination of recessives under selection (pp. 527–528 and 535).

It might be more encouraging to place emphasis on positive eugenics—increasing the frequency of beneficial traits rather than merely decreasing the frequency of deleterious genes. Unfortunately many characteristics we consider desirable, such as high intelligence, esthetic sensitivity, good physical health, and longevity, are not caused by single genes that are easily identified but by complexes of many genes acting together in appropriate environments.

In other organisms in which the development of beneficial gene complexes has been attempted, the methods involve complicated schemes based on selection of parents and families along with testing of progeny under controlled environmental conditions. As I. M. Lerner and others discussed, the results of these experiments have, in general, improved certain complex characters by some degree but have usually caused the deterioration of others. One characteristic that usually suffers most in such experiments is that overall quality called "fitness"; many highly selected lines end up physically debilitated and sterile.

Muller, Crow, and others have pointed to the likelihood that traits such as high intelligence and esthetic sensitivity have not been stringently selected for in the past, and considerable genetic variability for these traits probably exists. Thus, were selection to be instituted for these traits, the population might well respond rapidly without an accompanying fall in fitness. The means of selection themselves, however, assume paramount importance in humans. Eugenic measures dictating who is to mate with whom would be intolerable, even presuming that controls on human activity can be implemented to evaluate selection progress.

As a first approach toward a more acceptable method of positive eugenics than selective mating, Muller and others have proposed using sperm banks containing the preserved frozen sperm of outstanding creative men. According to this method, called **germinal choice,** or **eutelegenesis,** women volunteers would choose to be artificially inseminated by males that were long dead but had highly desirable characteristics.

Possible acceptance of this method has some precedence given that between 5,000 and 10,000 babies fathered by sperm donors are born annually in the United States. The cause for most of these donor fertilizations lies in the sterility of the husband, although in some cases genetic incompatibilities between husband and wife (for example, Rh factor) or genetic defects in the husband (for example, hemophilia) are responsible. Muller proposed to extend these donor fertilizations by educating couples to desire a highly superior genetic endowment for their children and demonstrating the increased proportion of genetically gifted children that will presumably be produced by this method.

Other proposed eugenic methods involve direct manipulation of human DNA by **genetic engineering,** techniques that are now being used to modify the DNA of bacteria, viruses, and higher organisms. The application of these techniques to humans holds considerable future promise for directly changing human genetic material. For example, various experiments now demonstrate that human gene sequences can be inserted into viruses, which can then serve as vectors to transfer these genes to mammalian cells (Box 25–2). This technology may eventually advance to the point where a large repertoire of genes isolated or produced in the laboratory can be incorporated into reproductive tissues and thus change the genetic constitution of entire lineages.

Another possible eugenic method is **parthenogenesis**—to induce females who have desirable genetic constitutions to lay diploid eggs that need not be fertilized. Such eggs would more truly reflect the constitutions of their mothers than fertilized eggs and thereby permit the replication of desirable maternal genotypes. New mammalian nuclear transfer technology described in Box 25–3 makes such "cloning" possible, even for humans. Entire genomes can now be replicated.

The Future

All we can say at present is that perhaps some form of eutelegenesis or genetic manipulation will be developed that will be acceptable and productive. It would seem that the need to improve the human gene pool will become more desirable the more we become aware of our biological limitations and the more our technology allows us to successfully perform genetic changes. We can most probably agree on the need for treatment and prevention of serious genetic defects and disease. Like other creatures, humans evolve, but unlike other creatures, humans *know* they evolve. The control of biological evolution lies in changing from reproductive success caused by natural selection and other forces to reproductive success caused by human choice. Bold as this sounds, it is no more daring than the methods by which many cultural advances have been and will be made.

Perhaps the most pertinent question we can ask of eugenics is: What is its goal? Even if we assign to eugenics the most moral of motives—the good of humankind—it still remains to be determined whether this "good" is known. Can we choose the direction of human evolution with the certainty that this direction leads to what is best for our descendants? Shall we populate the world with the weak or the strong? With the sensitive or the insensitive? Fortunately, the answers to such questions will probably not be limited to an unequivocal choice of one type or the other. There will probably be the opportunity, then as now, to choose many different genotypes, among which factors such as intelligence and longevity will undoubtedly rank high in value.

Such major genetic improvements are not ready for debate; they are tasks for a distant, more advanced, social and technological future. More relevant today is the need to channel and improve our cultural evolution: to overcome damaging human social behaviors. For example, to devise cultural solutions that deal successfully with our **provincialism**: our parochial social, racial, religious, economic, and political prejudices and structures that help make us into the most dangerous and destructive beasts that terrestrial evolution has ever known. Since provincialism seems to be a common feature of every human group, its basis seems built into our biology, probably derived from the millions of years during which small bands of hominids fought and protected themselves against others.

In the complex but fragile and threatening political world we now live in, these aggressive and destructive behaviors are anachronistic and no longer appropriate. The sentiments of group patriotism, chauvinism, and superiority—long supported by myths and rituals—were useful in the past in providing motivation for social defense and aggressive acquisitions in contacts among small local groups. Unfortunately, such attitudes are now enhanced technologically to permit military destruction on a broad national and international scale. The dividing line between group homicide (genocide) and suicide is being obliterated rapidly.

Certainly one need for continued human survival is to become consciously aware of these underlying behaviors and purposely control them by social and cultural means. Just as we see the need for traffic lights, we can also agree on the need for ethical concepts and social agencies that encourage behaviors necessary and appropriate for the kind of society in which we would like to live. For example, we can consciously devise mechanisms that increase respect for other humans as well as for our environment, and we have the power to correct the horrors and blind injustices of nature rather than perpetuate them. Our intellects and inventiveness, unique among all biological creatures, can provide us with the means to accomplish such goals.

From this point of view, the most important present human need is to accept differences among individuals and groups without indulging in provincial, exploitative, and chauvinistic practices: we must learn to assume our social identities without damaging others, and to assert a morality that respects all humans. Our lives should be sacred to us, not because evolution has

BOX 25-2 GENETIC ENGINEERING OR EVOLUTION BY INTERVENTION

Gene manipulation or genetic engineering began more than 20 years ago following the discovery that restriction enzymes can precisely cut sections of DNA at particular sequences (p. 274). It did not take long for experimenters to realize that they could then splice together DNA fragments from different sources cut by such enzymes to form entirely new **recombinant DNA** (Fig. 25-3).

Since those early days, an entire industry of genetic engineering has developed, allowing new kinds and combinations of genetic material to be artificially constructed in the laboratory by the controlled insertion and manipulation of nucleic acid sequences. A basic technique is to insert a new or modified nucleic acid sequence into a **vector,** such as a virus, plasmid (an extrachromosomal genetic element, p. 225), or yeast artificial chromosome ("YAC") that can carry these novel sequences into host cells where they can be propagated and amplified many times over. This process allows multiple copies of a piece of genetic material to be isolated and sequenced, and also, in various cases, to be transcribed into messenger RNA and translated into protein.

Briefly, we can divide gene manipulation into four major steps, diagrammed in Figure 25-4:

- Generating DNA fragment or fragments that are to be used or manipulated (Fig. 25-4*a, b*)
- Splicing the DNA fragments into a composite molecule, recombinant DNA, which can act as a vector, or be incorporated into one, for further transmission (Fig. 25-4*c*)
- Transferring the vector DNA into a cell in which it is to be replicated (Fig. 25-4*d*), a process called **transformation** of the host cell
- Selecting those cells that carry the desired recombinant DNA molecules, and replicating them as **clones** (Fig. 25-4*e, f*)

Other aspects of recombinant DNA have been widely reviewed, and various features are discussed in books such as that by Watson et al. What is of interest here is that this technology allows active intervention into genetic material of any organism; and genomes of many viruses, bacteria, protists, animals, and plants have already been manipulated—either by modifying their existing genes or introducing new genetic material. Accompanying these genetic alterations are, of course, physical and physiological changes that affect protein production, resistance to infective agents, agriculture yield, nutritional value, toxic susceptibility, environmental stamina, tumor resistance, and so forth (see, for example, Flavell).

In essence, genetic traits can be manipulated, and organismic evolution moved from the age-old province of random mutation and natural selection to human-directed evolution. Even humans are not exempt from such intervention, and medical technology is proceeding steadily toward directed gene therapy—the treatment of disease by transfer of corrective genes in vectors rather than by drug therapy (see the table on p. 623). As of 1998, Anderson reports that 300 clinical procedures have been approved, and genetically engineered cells have been taken up by more than 3,000 patients. Although the risks of adverse reactions appear low, inefficiency of transforming vectors remains the largest hurdle in gene therapy. Researchers are continually developing improved forms of retrovirus, adenovirus, herpes virus, and other agents that can be used for this purpose (Nabel, Palese). In addition, mapping of the entire human genome now underway (the "Human Genome Project," p. 276) will undoubtedly provide detailed nucleotide sequence information about many genes that cause defects and disease.

Because of these advances and many more to come, genetic engineering raises a fundamental issue: Are we justified in intruding in a process in which our intervention may cause unforeseen, detrimental consequences? The answer, as may be expected, is not simple. There is, for example, continual pressure, both humanitarian and commercial, to improve the yield of agricultural products by available means. There is also, as expected, considerable support for using gene therapy to treat serious human genetic diseases such as immune deficiencies, Huntington chorea, and others where drugs are totally or partly ineffective. At the same time, many scientists recognize that genetic engineering experience is still limited, and long-term effects are difficult to envisage, especially when genetically altered organisms are widely distributed and both present and future generations may be genetically or environmentally affected.

So far, some governmental agencies have assumed the role of supervising commercial use and release of genetically engineered organisms. There is also a consensus among scientists that human gene therapy be restricted to somatic cells and to correcting or ameliorating genetic diseases in individuals rather than attempting to modify human germ plasm. In Anderson's (1992) words, "The feeling of many observers is that germline gene therapy should not be considered until much more is learned from somatic cell gene therapy, until animal studies demonstrate the safety and reliability of any proposed procedure, and until the public has been educated as to the implications of the procedure" (see also Wivel and Walters). It is presently considered unethical to alter genes that would change or "enhance" nondisease characters such as an individual's appearance or height.

How these present considerations will affect future genetic practices is difficult to predict, since much will depend on whether society can develop consensual ethical principles to deal with genetic engineering, and with the effects of increasing genetic information and genetic screening (detection of genetic defects) on human privacy and other values (see Kevles and Hood, Murphy and Lappé, and Weir et al.). On one hand are justifiable fears that

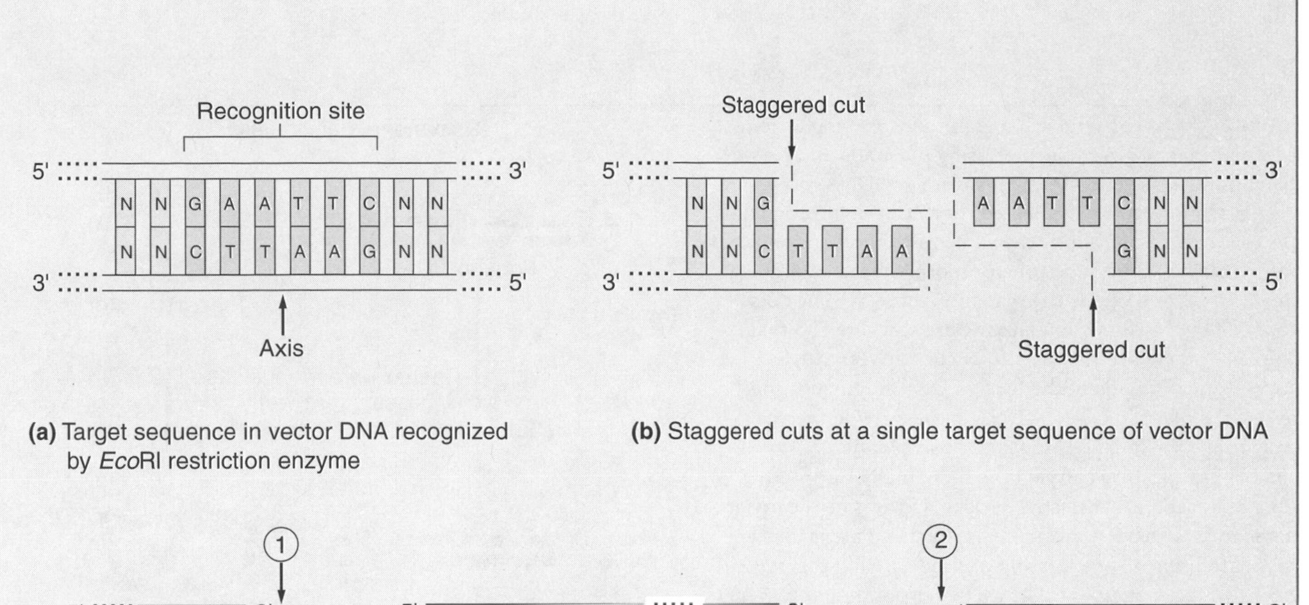

(a) Target sequence in vector DNA recognized by *Eco*RI restriction enzyme

(b) Staggered cuts at a single target sequence of vector DNA

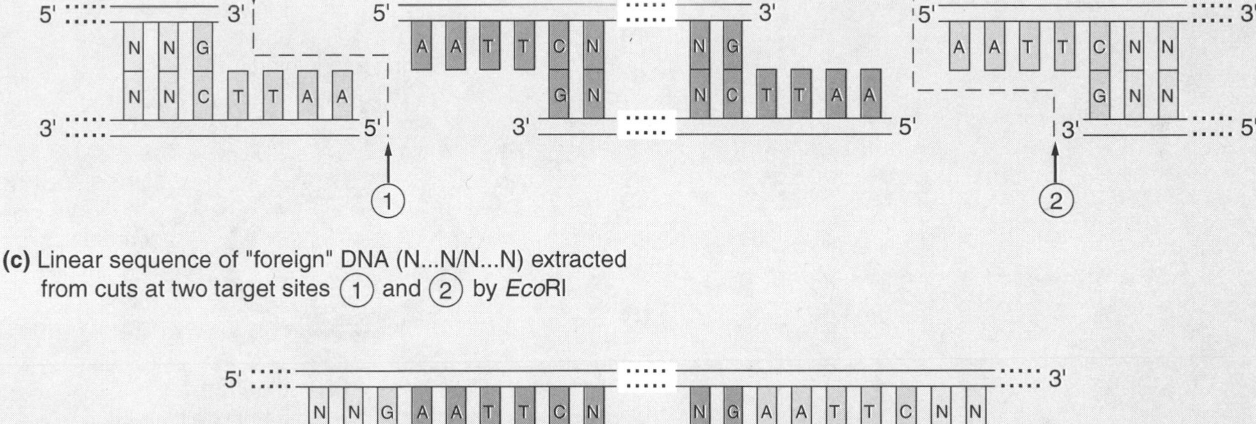

(c) Linear sequence of "foreign" DNA (N...N/N...N) extracted from cuts at two target sites (1) and (2) by *Eco*RI

(d) Insertion of extracted foreign DNA sequence into cut target sequence of vector DNA

FIGURE 25-3 As discussed in Chapter 12, restriction endonucleases are enzymes that recognize specific DNA nucleotide sequences, and cleave the DNA double helix at or near these specific sites. The kinds of restriction enzymes most used in gene manipulation are those which generally recognize short "palindromic" sequences that have an axis of symmetry. In this illustration, the palindrome recognized by the bacterial *Escherichia coli* enzyme *Eco*RI is the double-strand hexanucleotide sequence GAATTC/CTTAAG, which reads the same from each 5′ → 3′ or 3′ → 5′ direction toward the central axis (N represents unspecified nucleotides and . . . represents further extensions of the nucleotide sequence). In (*a*) this hexanucleotide target sequence is located in a vector, such as a virus or plasmid, which can later be inserted into a host cell. (*b*) Breakage of the target sequence is asymmetric, and the staggered cuts (*indicated by arrows*) produce two protruding but complementary single-strand ends, four nucleotides long, that can be considered as "sticky" or "cohesive." (*c*) DNA from a different ("foreign") source bearing the same restriction recognition sites can then be treated in the same way to generate one or more linear sequences, also with single-strand sticky ends. (*d*) Because of the cohesive nature of these ends, the foreign DNA (*shaded sequence*) can anneal by complementary base pairing with the vector DNA in which it is being inserted. The terminal nucleotides of the inserted foreign DNA can then be covalently joined to the adjacent nucleotides by a special DNA ligase enzyme, thus forming a recombinant DNA molecule derived from two (or more) sources. (*Adapted from Strickberger.*)

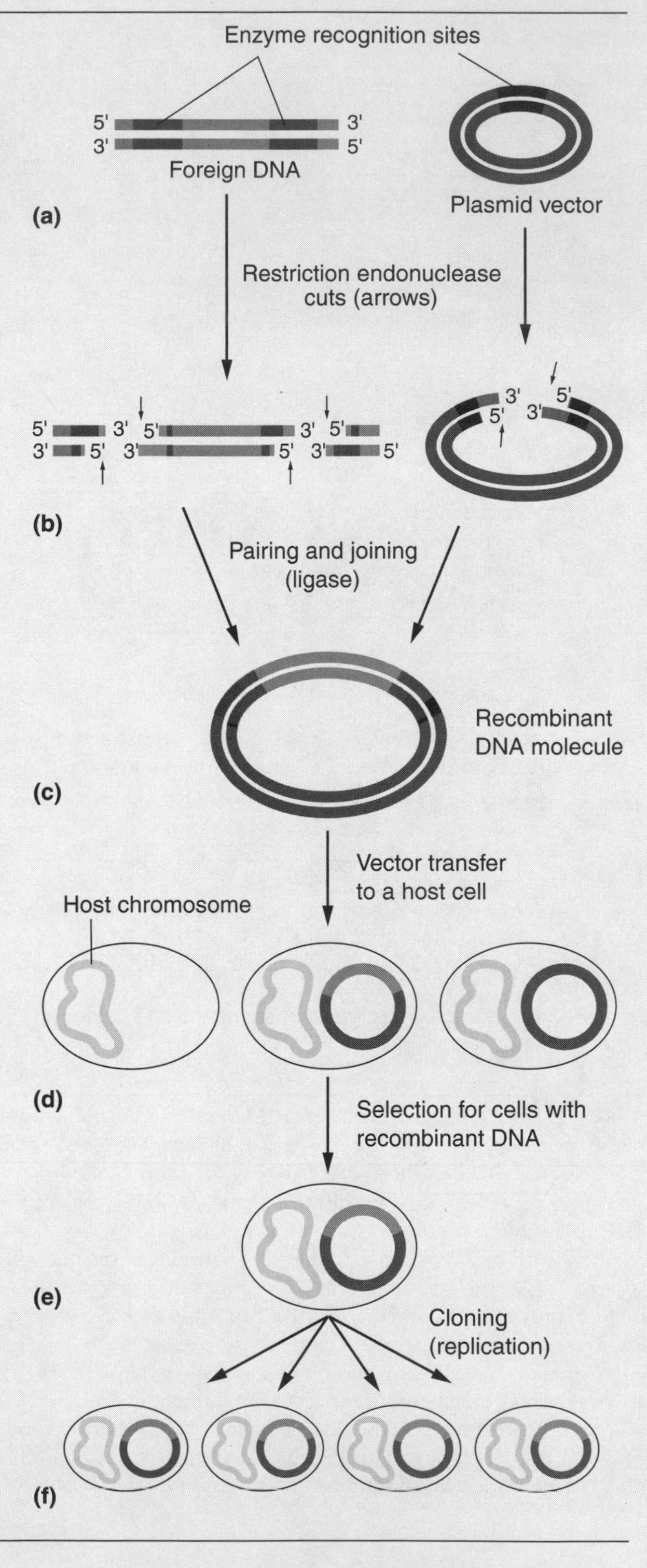

FIGURE 25-4 A general scheme for constructing a clone of recombinant DNA molecules using plasmids and restriction endonuclease enzymes. (*a*) A "foreign" DNA molecule and a plasmid vector are selected, both carrying recognition sites that can be cleaved by the same restriction endonuclease. (*b*) Cleavage produces one or more fragments of the foreign DNA, and opens the plasmid vector. (*c*) The cuts produced by the restriction enzyme are staggered so that complementary base pairing can occur between single-strand ("sticky") ends of the foreign DNA fragments and the opened plasmid DNA. A DNA ligase enzyme then covalently bonds the two DNAs into a recombinant DNA molecule. (*d*) The plasmid carrying the foreign DNA is inserted into a host cell. (*e*) A selective process isolates cells carrying the recombinant DNA molecule. (*f*) Each isolate is further propagated, thereby generating a clone carrying the particular foreign DNA fragment that was incorporated in step (*c*). (*Adapted from Strickberger.*)

Year	Advance
1943	Evidence that DNA is the genetic material
1953	Proposal for the double helix structure of DNA
1958	Isolation of the first DNA replicating enzyme
1959	Recognition of the Down syndrome chromosome abnormality
1960	Discovery of messenger RNA
1966	Establishment of the complete genetic code used in translating codons into amino acids
1967	Development of amniocentesis and cytogenetic analysis
1970	Isolation of a restriction enzyme that cuts DNA at specific sites
1970	Discovery of the retrovirus replication cycle
1972	Production of the first recombinant DNA molecules
1973	Method for introducing recombinant plasmid vectors into bacteria for cloning
1975	Use of radioactive probes and autoradiography to identify particular DNA fragments
1975	Cloning of hemoglobin DNA copied from messenger RNA (cDNA)
1976	Elucidation of molecular diversity in immune system immunoglobulins
1976	First clinical use of recombinant DNA (prenatal diagnosis of alpha-thalassemia)
1977	Invention of laboratory techniques for sequencing DNA
1978	Synthesis of the proinsulin peptide using recombinant DNA
1978	Mapping of the sickle cell mutation using its linkage to DNA ("RFLP") markers
1981	Isolation of a cancer-causing gene ("oncogene")
1983	Direct detection of the nucleotide sequence of a specific mutant allele (sickle cell)
1983	Identification of the HIV retrovirus in patients with AIDS
1983	Use of retrovirus as a vector to replace deficient human genes in cell culture
1984	Targeting of mutations to specific genes in cell culture
1985	Invention of the polymerase chain reaction for amplifying nucleotide sequences
1986	Reverse genetics: characterization of a gene with an unknown protein product (chronic granulomatous disease) by finding its chromosomal location
1987	Production of a vaccine (anti-hepatitis B) based on recombinant DNA technology
1988	Incorporation and expression of genes inserted into mouse cells using retroviruses
1989	Clinical transfer of a marker gene to humans
1990	Replacement of defective adenine deaminase gene in somatic cells of human patients
1992	First intensive linkage map of the entire human genome using 400 gene markers
1994	Improved mapping of the human genome incorporating more than 2,000 microsatellite markers
1995	DNA cloning of a major component in human chromosome end caps (telomeres) that can be used to produce an artificial human chromosome
1996	Increased human genetic map density to more than 5,000 gene markers
1996	A physical map of more than 16,000 genes
1997	Creation of the first artificial human chromosome possessing a centromere and telomeres
1998	A physical map of more than 30,000 human genes

Source: Based on data from Caskey, Anderson, and others.

employers, insurance companies, and governmental bodies who lack socially altruistic policies might exploit and misuse genetic information to callously stigmatize individuals as defective or "inferior" because of narrow selfish motives. On the other hand, genetic information can be beneficial to parents in learning whether they have serious medical traits they can transmit to their children, and to medical scientists in learning how to treat individuals who have such traits.

BOX 25–3 SENDING IN THE CLONES

Cloning: Reproductive methods that produce offspring genetically identical to the parent.

Cloning is fairly common in nature and is used by bacteria and even many eukaryotes, especially plants (Chapter 11). Considering its prevalence, it is appropriate to ask: What may account for it?

Given the constant environmental conditions in which clones are usually found, their advantages appear relatively straightforward. Compared to sex, in which genetic variability is repeatedly produced through meiotic chromosome assortment and recombination—even when unnecessary—clones can maintain identical genotypes. Clones perform this role economically, since other than occasional mutations, their offspring are always the same. They also circumvent problems of sex that can interfere with reproductive success, such as producing males, finding mates, exposing mating pairs to a hostile environment, and other factors described in Chapter 23.

Cloning disadvantages arise because clones forego the genetic variability that sexual organisms use to accommodate many oncoming environmental changes. Also, when homozygous, clones may lack the better buffering of genetic heterozygotes ("hybrid vigor") and can accumulate sets of detrimental genes by accidental processes ("Muller's ratchet", p. 564).

In general, cloning offers a population genetic uniformity every generation, which can be advantageous when organisms face the same conditions for lengthy periods. In contrast, sex is a comparatively expensive genetic gamble that has selective advantage for organisms facing changeable and unpredictable future conditions. Also, once sex is adopted, the sexual reproductive apparatus may be too complex to dismantle, and reversion to asexual cloning may be difficult or unattainable.

EXPERIMENTS ON CLONING

Experimental cloning can be said to have begun with Spemann and others who showed that early embryonic cells of sea urchins and salamanders (at 2-, 4-, or 16-cell stages) were individually capable ("totipotent") of producing an entire embryo. However, cells taken from later embryonic stages produced abnormalities. Such developmental problems were apparently caused by inability of cells from later developmental stages to dedifferentiate (become "reprogrammed") so they could function at earlier developmental stages.

Invention of techniques in which cellular nuclei could be removed and directly transplanted into eggs began in the 1950s. These procedures had some success when embryonic donor nuclei were transferred to enucleated recipient eggs, but the transfer of post-embryonic cell nuclei showed loss of totipotency. In frogs, transplanted adult nuclei usually developed no further than the tadpole stage (Briggs and King), although by the 1960s and 1970s, techniques had advanced to show that some tadpole nuclei could produce adults (Gurdon).

These amphibian techniques were then developed and extended to mammals, so that by the 1990s, successful cloning had been shown to occur in mice, cows, rhesus monkeys, and sheep, all dependent on nuclei from very early embryos. Nuclei from later embryos—and certainly from adults—appeared ineffective, because they had presumably undergone too many changes to preserve their totipotency. (Among such changes, we do know that methylation and acetylation modify nuclear DNA and proteins as the organism matures.)

A major innovation by Wilmut and coworkers in 1997 was to "reprogram" donor nuclei taken from differentiated tissue, letting them lose their differentiated state and provide early embryonic function in the oocyte. As a result, adult nuclei could be made totipotent, and the sheep "Dolly" was then produced from a cellular nucleus taken from the udder of a 6-year-old ewe.[11]

[11] A brief summary of the transplantation procedure through which "Dolly" was born is as follows:

- Enucleated oocytes from Scottish black-faced strains were fused to nuclear donors from Welsh mountain white-faced strains, using an electric current.
- Fused "couplets" were grown for six days in ligated oviducts of temporary recipient ewes, and were then recovered and observed for embryonic development.
- Those embryos that had reached the morula/blastula stage were then transferred to the uteri of final surrogate mothers (Black-faced strain).
- Gestation for viable fetuses was about 10 days longer than the usual 150 days.

MAKING USE OF CLONING

Advantages in genetically changing organisms to produce desirable products dates back to the first human domestication of plants and animals. In the thousands of years since, we have created a world of agriculture, pharmaceuticals, and even industries such as forest products, which come from manipulating organismic genes and genomes, most commonly through selection. The prospect of using cloning now adds considerably to the range of advantages that genetic technology can offer:

1. We can mass-produce strains of agricultural plants and animals whose genomes confer parasite resistance.
2. We can mass-produce high-yield agricultural strains that maintain heterozygous hybrid vigor without the expense of producing deleterious homozygotes, and which can also be replicated without suffering hybrid sterility. Many botanists feel that cloning will lead to large increases in agricultural production, even greater than the "Green revolution" of the last few decades.
3. Animal clones can be established from donor nuclei derived from tissue-cultured cells in which particular genes have been added or deleted ("gene targeting"). Such **transgenic** animals, which now include sheep, cows, goats, and pigs, can secrete medically useful proteins in their milk, such as human clotting factors (absent in

hemophiliacs), fibrinogens (for traumatic wounds), and other special products.

4. Clones can provide animal models in the search for the cause and treatment of degenerative diseases, autoimmune diseases, cancer, and viral bacterial infectious diseases.

5. Clones can be used to produce immunologically compatible tissues or organs for transplantation to humans.

THE REACTION TO CLONING: MORALITY AND ETHICS

When successful cloning of the first adult mammal ("Dolly") became known in February 1997, it seemed to many that biotechnology was advancing at a rate that would easily lead to cloning humans. The main author of the "Dolly" paper, Ian Wilmut, told a British Parliament committee soon after, that he expected a similar technique could be used to clone humans "within two years."

The impact of such possibilities led to reactions by various governmental figures and agencies on the ethics of human cloning; practically all indicating they would ban such procedures. In the United States, for example, President Clinton immediately banned all federally funded human cloning research, also asking for a moratorium on non-federally funded research, and ordered the National Bioethics Advisory Commission to conduct immediate hearings. Since then, mice and cows have been cloned from nuclei of adult tissues, and it is likely that more species cloning is on the way, including claims by some researchers that they intend human cloning as well.

These prospects have generated considerable discussion on ethics and morality, extending to whether humans have a right to change or intervene in nature at all. Many claims have been made that certain procedures, activities, or behaviors involving departures from "natural" processes are wrong, no matter how their benefits are evaluated. Included among these are abortion, homosexuality, gene manipulation, and human cloning—all of which have been placed in the category of "unnatural," where "unnatural" is used to signify "contrary to the laws and course of nature."

In regards to gene manipulation, some opponents suggest that *any* human-directed gene change in *any* organism is "unnatural":

> Genetic engineering makes it possible to breach the genetic boundaries that normally separate the genetic material of totally unrelated species. This means that the *telos,* or inherent nature, of animals can be so drastically modified (for example by inserting elephant growth hormone genes into cattle) as to radically change the entire direction of evolution, and primarily towards human ends at that. Is that aspect of the animal's *telos* we refer to as the genome and the gene pool of each species not to be respected and not worthy of moral consideration? (Fox)

The reply from an evolutionary view is simply that the three and a half billion years of species evolution show no ethics or morality. From all we can see, evolution is strictly opportunistic, based on survival—the transmission of genes that enable reproductive success. This means that the genome of any species is neither static, nor moral, nor immoral. All genomes are the result of genetic change that replaced genomes in the past—their survival as "elephant" or "cattle" genomes depending on change with no discernable "purpose" other than survival.

Furthermore, no matter where we search, there is no evidence of a "natural" law that forbids changing organismic genotypes, or that censures the many such changes humans have made in agricultural history. No only do species themselves continuously change genes through mutation and recombination, but there is considerable evidence that species barriers have continually been crossed by viruses, transposons (mobile self-replicating DNA sequences), and plasmids carrying genes from one species to another. Mechanisms using plasmids are, in fact, a common method of gene transfer in many bacteria, and many instances are known where "resistance factors" can transfer antibiotic resistance to different bacterial strains (p. 225).

It is also clear that the physical/biological environment we consider "natural" is not constant or sacred. Organisms have regularly affected and changed the environment, and they continue to do so. Humans have long engaged in a succession of interventions into "nature" that has already changed its form, function, and substance in many localities and for many organisms. For example, stone tools, animal traps, weapons, farms, agriculture, clothing, houses, roads, ships, planes, dams, electricity, trolley cars, medicine, dentistry, engineering, computers, batteries, toothpaste, and gasoline motors, are all "unnatural" interventions by humans into a nonhuman, so-called "natural," world.

It seems a grievous error to insist that we cannot intervene into "nature" because its rules are mystical and unchangeable—that we must retain the mistaken concepts that "*what is in nature, is or should be in human culture,*" and "*what is not in nature, is immoral and unethical.*" Clearly, practices that can be deemed "unnatural" are really essential parts of the way humans live—the evolving "natural" world of humans. By this notion, gene therapy and cloning are no more "unnatural" or "unethical" than other human interventions into nature. We make our own ethical and moral rules, although these rules are neither simple nor uncomplicated.

For example, using our own cultural, ethical, and moral concepts, the arguments in the table on the following page show how difficult it can be to characterize all human cloning as immoral or unethical.

HOW CAN WE JUSTIFY INTERVENING IN NATURE?

In biotechnology, like in medicine, many people would agree that we can justify intervention by the principle

Arguments on the morality of human cloning	
Against Human Cloning	For Human Cloning
Human dignity is based on regard for one's uniqueness—an individual differs from others in ego, personality, and genetic identity. Clones lack individuality.	Human individuality and genetic uniqueness do not derive from the same source: identical twins are genetic clones, yet are considered by society and by themselves as individuals.
Cloning individuals is unethical because it is primarily narcissistic.	Cloning can be a compassionate solution for parents whose child is dying because of a fatal injury. Cloning may also be a solution to some problems of sterility.
Parents would be tempted to influence a child that is an exact genetic copy to attain parental goals rather than allow the child to attain its own goals.	Parents may impose their own goals even when the child is sexually produced, and not a genetic clone. Also, environmental influences can modify directions and goals among individuals despite their genetic identity.
Cloning can lead to militaristic and antisocial genetic choices made by political authorities.	Political decisions on matters of consequence to society, such as the use, frequency, and goals of cloning, should be made democratically. Cloning itself is neither democratic nor authoritarian. Non-clonally reproduced humans have been shown quite capable of making militaristic and antisocial choices.
Cloning techniques for humans offer no medical benefits to nonparticipants.	The technology used in cloning may enable nuclei in an adult human to be "reprogrammed" to function at an earlier embryonic stage, thus allowing malfunctioning organs, such as an aging heart or brain, to regenerate.

"Above all, do no harm"—to be beneficial or, at least, benign. Unfortunately, trade-offs are often necessary, since intervention for beneficial goals is not always benign or risk-free. For a medical example, some interventions involve surgery, radiation, catheterization, and other potentially damaging procedures. General agreement prevails that medical intervention should take into account the risk of damage and the quality of life: in other words, to evaluate the means and not just the ends.

Extending these ideas to gene manipulation and cloning, some people raise the question whether attaining improvement in human existence is worth relinquishing traditional methods of reproduction; that is, substituting "unnatural" for "natural" means. As indicated above, from all we understand, "unnatural" activities are not necessarily wrong or injurious simply because they replace earlier "natural" processes. Of more serious concern are unpredictable consequences in adopting new practices which may inadvertently lead to disasters. In the words of Rifkin, a critic of gene manipulation:

> Whenever a genetically engineered organism is released there is always a small chance that it too will run amok because, like exotic organisms, it is not a naturally occurring

conferred such status, but because we have. If we can fulfill this cultural need, it is likely that we can progress to one great prosperous, humane, and rational world. If we cannot fulfill this need, the prospect appears to be one of mutual destruction and a fall into unimaginable suffering and barbarism.

In short, because of our intellects and technologies, humans are graduating from being subjects of evolution to being coauthors of evolution. An understanding of the materials and mechanisms of change on the biological and cultural levels can provide us with the freedom to control evolution on these levels. We have in our hands our own destiny as well as that of many other creatures on earth.

SUMMARY

Unlike most other animals, humans transfer information in the form of culture (as well as hereditary information) life form. It has been artificially introduced into a complex environment that has developed a web of highly synchronized relationships over millions of years. Each new synthetic introduction is tantamount to playing ecological roulette. That is, while there is only a small chance of its triggering an environmental explosion, if it does, the consequences can be thunderous and irreversible.

Rifkin thus expresses the fear that genes introduced into genetically improved agricultural strains will escape control and cause catastrophic infections, limitless expansions, and extinctions—man-made "Andromeda strains." This fear can be extended to practically all recombinant DNA research. How can any type of gene manipulation—whether gene therapy, gene cloning, gene targeting, and so on—face up to the threat of such "runaway" disaster? How can we regulate risk and safety?

Difficult as it may be to ensure against all risk, we can evaluate potential difficulties and exercise precautions:

REGULATING RISK AND SAFETY

1. IT IS IMPORTANT TO REALIZE THAT THERE ARE RISKS IN ALMOST ANY EXPERIMENTAL PROCEDURE For example, the by-product of a drug that reduces the effect of colds or headaches or athlete's foot may be a potent carcinogen. Or an experimental strain of wheat may expand limitlessly, devastating all agricultural plants, or all trees, and so forth. At the same time, we should keep in mind that "natural" catastrophes have far exceeded those of experimental procedures: witness the Black Plague, HIV infections, Influenza epidemics, Pine wilt disease, Gypsy moth infestations, and so on.
2. THERE IS NO WAY OF PROVING THE ABSENCE OF RISK IN AN EXPERIMENTAL PROCEDURE The ultimate effect of any activity, even something as trivial as the flight of a bumblebee, is unpredictable because we cannot follow all its future interactions. Shall we then ban all beneficial technologies because there may be potential misuse or a possible unexpected deleterious effect in the future? In fact, had experiments in genetic engineering been banned, it would have excluded even our moderate success in treating cancer, AIDS, and other serious diseases, let alone the realistic possibilities we now have of curing them. Moreover, it would seem foolhardy to restrict the benefits of agricultural improvement, practically all of which now involve gene manipulation. The alternative to "risky unnatural" molecular genetics—to live in a so-called "natural" world with its own costly risks of famine, disease, and pain—seems hardly a rational choice for intelligent, ethical, and compassionate humans.
3. WHEN WE DO EVALUATE RISKS, IT SEEMS INCUMBENT THAT WE BASE THESE RISKS ON MORE IMMEDIATE AND FORESEEABLE CONSEQUENCES Again, the medical model seems appropriate: to restrict experiments to tissue cultures, animal models, clinical trials, and/or enclosed agricultural plots. Such methods allow main as well as side effects to be closely observed before using experimental treatments widely.
4. FOR THE MOST PART, SUCH RISK EVALUATIONS HAVE FUNCTIONED SUCCESSFULLY For example, most pharmaceutical drugs on the market are effective for the purposes they were designed, and their side effects can be taken into account during treatment. We should be aware that strict regulation by politically independent governmental agencies has been responsible primarily for averting pharmaceutical disasters. Since these genetic technologies introduce matters that affect everyone, decisions should be publicly based, especially since molecular biology has been a publicly supported enterprise.

* * * * *

Notwithstanding the benefits or advantages of human cloning, we should realize that human performance is a matter of both nature and nurture. For example, musical genius, like agricultural yield, not only depends on possessing a specific genotype, but on being nurtured in an appropriate environment. We can certainly benefit our species by improving conditions that allow environmentally stunted genotypes to flourish.

> To improve our species, no biologic sleight of hand is needed. Had we the moral commitment to provide every child with what we desire for our own, what a flowering of humankind there would be. (Eisenberg)

from one generation to another. The tempo of cultural evolution has been so much more rapid than human biological evolution that each person gathers new experience at a rate many times that of his or her ancestors. While cultural evolution is Lamarckian in that what is acquired can be transmitted and directed, biological evolution depends on randomly occurring genetic variation.

In the nineteenth century, proponents of Social Darwinism believed that cultural differences evolved primarily by natural selection, as embodied in the concept of "survival of the fittest." This belief justified many social inequities, and was based on the erroneous supposition that society, which often incorporates nonbiological goals and value systems, is governed by the same laws as biological evolution.

A more sophisticated approach is that of sociobiology, the advocates of which believe that there is a biological basis for much of human culture and that natural selection favors genotypes that are predisposed toward cultural development. However, although there is a biological component in many social patterns, they are often modified by their cultural context. Cultural change may occur without biological input and cannot be explained by biological laws.

Humans want to control all aspects of their evolution but are restricted by the incompatibility between cultural and biological fitness. They wish to lengthen life span, but natural selection cannot act on postreproductive individuals. They would like lower fertility, but biologically high fertility has had selective value, although socially it has led to population and ecological crises. The question remains whether or not we can direct biological evolution by cultural means and, if so, what goals would be desirable.

A large proportion of people are or will be affected by deleterious genes that persist because of forces that maintain genetic variability or because of the perpetuation by medicine of the lives of individuals suffering from genetic disorders. Eugenics represents an effort to improve the human gene pool either by reducing the frequency of such genes or by augmenting the frequency of favorable ones. Because the carriers of harmful alleles often cannot be detected, they cannot be eliminated by negative eugenics. Selecting for beneficial traits is also difficult since they are frequently multigenic, and fitness may decline in selected lines. A variety of methods, such as parthenogenesis, genetic surgery, and manipulation or artificial insemination with sperm from extraordinary individuals might be attempted in the future to improve the gene pool.

In any case, improvement of the human gene pool, intelligent direction of human evolution, and survival of the human lineage depend on eliminating anachronistic provincial, exploitative, and chauvinistic barriers and practices.

KEY TERMS

biological limitations
cloning
clutch size
culture
deleterious effects
eugenics
eutelegenesis
extragenetic system
genetic engineering
genetic system
germinal choice
hierarchy
horizontal transmission
Lamarckian mode of cultural evolution
learning
lethal equivalents
longevity
negative eugenics
parthenogenesis
population explosion
positive eugenics
provincialism
recombinant DNA
reductionism
Social Darwinism
sociobiology
transformation
transgenic
vector
vertical transmission
vitalism

DISCUSSION QUESTIONS

1. Why is biological evolution considered Darwinian, and cultural evolution considered Lamarckian?
2. What are advantages of Lamarckian evolution compared to Darwinian evolution?
3. Social Darwinism
   a. To what extent, if any, can human social structures and social changes be explained by the Darwinian laws that govern biological evolution?
   b. Why is Social Darwinism no longer an acceptable concept?
4. Sociobiology
   a. Which human social behaviors would you consider to have a biological basis? Explain.
   b. To what extent can such biologically based social behaviors be culturally modified?
   c. Would you consider culture as merely an expression of underlying biology? Explain.
5. For what human features and characteristics is there an apparent conflict between cultural demands and biological endowments?
6. What explanations have geneticists offered to account for the frequency of traits caused by deleterious genes? Can all such traits be eliminated? Should they be eliminated? Why or why not?
7. Is there evidence that human racial or social groups can be divided into "superior" and "inferior" categories by scientifically acceptable methods? If not, why then are such distinctions popular among some individuals and some societies?
8. What prospects are there for the success of negative eugenics?
9. What techniques have been proposed to implement positive eugenics, and what is the likelihood that they will be accepted and used?
10. What obstacles presently confront the possibilities for improving the human condition and future human evolution?

EVOLUTION ON THE WEB

Explore evolution on the web! Visit the accompanying web site for *Evolution,* 3/e at www.jbpub.com/evolution for web exercises and links relating to topics covered in this chapter.

REFERENCES

Alexander, R. D., 1979. *Darwinism and Human Affairs.* University of Washington Press, Seattle, WA.

———, 1987. *The Biology of Moral Systems.* Aldine de Gruyter, New York.

Anderson, W. F., 1992. Human gene therapy. *Science,* **256,** 808–813.
———, 1998. Human gene therapy. *Nature,* **392** (suppl.), 25–30.
Bajema, C. J. (ed.), 1976. *Eugenics: Then and Now.* Dowden, Hutchinson & Ross, Stroudsburg, PA.
Bannister, R. C., 1979. *Social Darwinism: Science and Myth in Anglo-American Social Thought.* Temple University Press, Philadelphia.
Barlow, G. W., and J. Silverberg (eds.), 1980. *Sociobiology: Beyond Nature/Nurture?* Westview Press, Boulder, CO.
Boyd, R., and P. J. Richerson, 1985. *Culture and the Evolutionary Process.* University of Chicago Press, Chicago.
Briggs, R., and T. J. King, 1959. Nucleocytoplasmic interactions in eggs and embryos. In *The Cell,* vol. I, J. Brachet and A. E. Mirsky (eds.). Academic Press, New York, pp. 537–617.
Buss, D. M., 1994. *The Evolution of Desire: Strategies of Human Mating.* Basic Books, New York.
Caplan, A. L. (ed.), 1978. *The Sociobiology Debate.* Harper & Row, New York.
Caskey, C. T., 1992. DNA-based medicine: Prevention and therapy. In *The Code of Codes: Scientific and Social Issues in the Human Genome Project,* P. J. Kevles and L. Hood (eds.). Harvard University Press, Cambridge, MA, pp. 112–135.
Cavalli-Sforza, L. L., P. Menozzi, and A. Piazza, 1993. Demic expansions and human evolution. *Science,* **259,** 639–646.
Cohen, J. E., 1995. *How Many People Can the Earth Support?* Norton, New York.
Comfort, A., 1963. Longevity of man and his tissues. In *Man and His Future,* G. Wolstenholme (ed.). Little Brown, Boston, pp. 217–229.
Crow, J. F., 1961. Mechanisms and trends in human evolution. *Daedalus,* **90,** 416–431.
Crystal, R. G., 1995. Transfer of genes to humans: Early lessons and obstacles to success. *Science,* **270,** 404–410.
Daily, G., et al., 1998. Food production, population growth, and the environment. *Science,* **281,** 1291–1292.
Davenport, C. B., 1911. *Heredity in Relation to Eugenics.* Holt, New York.
Devlin, B., M. Daniels, and K. Roeder, 1997. The heritability of IQ. *Nature,* **388,** 468–471.
Diamond, J., 1997. *Guns, Germs, and Steel: The Fates of Human Societies.* Norton, New York.
Dobzhansky, Th., 1962. *Mankind Evolving.* Yale University Press, New Haven, CT.
Dupré, J., 1993. *The Disorder of Things: Metaphysical Foundations of the Disunity of Science.* Harvard University Press, Cambridge, MA.
Eisenberg, L., 1999. Would cloned humans really be like sheep? *New Engl. J. Med.,* **340,** 471–475.
Epstein, C. J., and M. S. Golbus, 1977. Prenatal diagnosis of genetic diseases. *Amer. Sci.,* **65,** 703–711.
Fedoroff, N. V., and J. E. Cohen, 1999. Plants and population: Is there time? *Proc. Nat. Acad. Sci.,* **96,** 5903–5907.
Fetzer, J. H. (ed.), 1985. *Sociobiology and Epistemology.* Reidel, Dordrecht, Netherlands.
Flavell, R., 1993. Molecular genetics and new plants for agriculture. In *Genetics and Society,* B. Holland and C. Kyriacou (eds.). Addison-Wesley, Wokingham, England, pp. 87–101.
Fox, M., 1990. Transgenic animals: Ethical and animal welfare concerns. In *The Bio-Revolution: Cornucopia or Pandora's Box,* P. Wheale and R. McNally (eds.). Pluto Press, London, pp. 31–45.
Freedman, D. G., 1979. *Human Sociobiology: A Holistic Approach.* Free Press (Macmillan), New York.
Galton, F., 1869. *Hereditary Genius: An Inquiry into Its Laws and Consequences.* Macmillan, London.
Ghent, W. J., 1902. *Our Benevolent Feudalism.* Macmillan, London.
Glover, J., 1984. *What Sort of People Should There Be?* Penguin Books, Harmondsworth, Middlesex, Great Britain.
Gould, S. G., 1977. *The Mismeasure of Man.* Norton, New York.
Gowlett, J. A. J., 1993. *Ascent to Civilization: The Archaeology of Early Humans,* 2d ed. McGraw-Hill, New York.
Gregory, M. S., A. Silvers, and D. Sutch (eds.), 1978. *Sociobiology and Human Nature.* Jossey-Bass, San Francisco.
Gurdon, J. B., 1974. *The Control of Gene Expression in Animal Development.* Clarendon Press, Oxford, England.
Haller, M. H., 1963. *Eugenics: Hereditarian Attitudes in American Thought.* Rutgers University Press, New Brunswick, NJ.
Hawkins, M., 1997. *Social Darwinism in European and American Thought: 1860–1945.* Cambridge University Press, Cambridge, England.
Herrnstein, R. J., and C. Murray, 1994. *The Bell Curve: Intelligence and Class Structure in American Life.* Free Press, New York.
Hofstadter, R., 1955. *Social Darwinism in American Thought.* Beacon Press, Boston.
Holzmüller, W., 1984. *Information in Biological Systems: The Role of Macromolecules.* Cambridge University Press, Cambridge, England.
Hull, D., 1974. *Philosophy of Biological Science.* Prentice Hall, Englewood Cliffs, NJ.
Huxley, T. H., 1894. *Evolution and Ethics, and Other Essays.* Macmillan, London.
Jones, G., 1980. *Social Darwinism and English Thought.* Harvester Press, Brighton, Sussex, Great Britain.
Kaye, H. L., 1986. *The Social Meaning of Modern Biology.* Yale University Press, New Haven, CT.
Kevles, D. J., 1985. *In the Name of Eugenics: Genetics and the Uses of Human Heredity.* Knopf, New York.
Kevles, D. J., and L. Hood (eds.), 1992. *The Code of Codes: Scientific and Social Issues in the Human Genome Project.* Harvard University Press, Cambridge, MA.
Kingdon, J., 1993. *Self-Made Man and His Undoing.* Simon and Schuster, New York.
Kitcher, P., 1985. *Vaulting Ambition: Sociobiology and the Quest for Human Nature.* MIT Press, Cambridge, MA.
Lack, D., 1954. *The Natural Regulation of Animal Numbers.* Oxford University Press, Oxford, England.
Lerner, I. M., 1958. *The Genetic Basis of Selection.* Wiley, New York.
Lerner, R. M., 1992. *Final Solutions: Biology, Prejudice, and Genocide.* Pennsylvania State University Press, University Park.
Maryanski, A., and J. H. Turner, 1992. *The Social Cage: Human Nature and the Evolution of Society.* Stanford University Press, Stanford, CA.
May, R. M., 1983. Parasitic infections as regulators of animal populations. *Amer. Sci.,* **71,** 36–45.
Montagu, A. (ed.), 1980. *Sociobiology Examined.* Oxford University Press, Oxford, England.
Morton, N. E., J. F. Crow, and H. J. Muller, 1956. An estimate of the mutational damage in man from data on

consanguineous marriages. *Proc. Nat. Acad. Sci.*, **42,** 855–863.

Muller, H. J., 1963. Genetic progress by voluntarily conducted germinal choice. In *Man and His Future,* G. Wolstenholme (ed.). Little, Brown, Boston, pp. 247–262.

Müller-Hill, B., 1988. *Murderous Science: Elimination by Scientific Selection of Jews, Gypsies, and Others, Germany 1933–1945.* Oxford University Press, Oxford, England.

Mulligan, R. C., 1993. The basic science of gene therapy. *Science,* **260,** 926–932.

Murphy, T. F., and M. A. Lappé (eds.), 1994. *Justice and the Human Gene Project.* University of California Press, Berkeley.

Nabel, G. J., 1999. Development of optimized vectors for gene therapy. *Proc. Nat. Acad. Sci.,* **96,** 324–326.

Neisser, U. (ed.), 1998. *The Rising Curve: Long-Term Gains in IQ and Related Measures.* American Psychological Association, Washington, DC.

Packer, C., M. Tater, and A. Collins, 1998. Reproductive cessation in female animals. *Nature,* **392,** 807–811.

Palese, P., 1998. RNA virus vectors: Where are we and where do we need to go? *Proc. Nat. Acad. Sci.,* **95,** 12750–12752.

Peel, J. D. Y., 1971. *Herbert Spencer: The Evolution of a Sociologist.* Heinemann, London.

Pettigrew, T. F., 1971. Race, mental illness and intelligence: A social psychological view. In *The Biological and Social Meaning of Race,* R. H. Osborne (ed.). Freeman, San Francisco, pp. 87–124.

Pimentel, D., and M. Giampetro, 1994. Global population, food and the environment. *Trends in Ecol. and Evol.,* **9,** 239.

Price, D. J. da S., 1963. *Little Science, Big Science.* Columbia University Press, New York.

Proctor, R., 1988. *Racial Hygiene: Medicine Under the Nazis.* Harvard University Press, Cambridge, MA.

Rifkin, J., 1985. *The Declaration of a Heretic.* Routledge and Kegan Paul, London.

Rose, M. R., 1991. *Evolutionary Biology of Aging.* Oxford University Press, Oxford, England.

Rosenberg, A., 1994. *Instrumental Biology or the Disunity of Science.* University of Chicago Press, Chicago.

Ruse, M., 1979. *Sociobiology: Sense or Nonsense?* Reidel, Dordrecht, Netherlands.

Sahlins, M., 1977. *The Use and Abuse of Biology.* University of Michigan Press, Ann Arbor.

Shipman, P., 1994. *The Evolution of Racism: Human Differences and the Use and Abuse of Science.* Simon and Schuster, New York.

Spemann, H., 1938. *Embryonic Development and Induction.* Yale University Press, New Haven, CT.

Strickberger, M. W., 1985. *Genetics,* 3d ed. Macmillan, New York.

Sumner, W. G., 1883. *What Social Classes Owe to Each Other.* Harper & Brothers, New York.

Waddington, C. H., 1978. *The Man-Made Future.* Croom Helm, London.

Ward, L., 1893. *The Psychic Factors of Civilization.* Ginn, Boston.

Watson, J. D., M. Gilman, J. Witkowski, and M. Zoller, 1992. *Recombinant DNA,* 2d ed. Scientific American Books, New York.

Wegman, M. E., 1994. Annual summary of vital statistics—1993. *Pediatrics,* **94,** 792–803.

Weir, R. F., S. C. Lawrence, and E. Fales (eds.), 1994. *Genes and Human Self-Knowledge: Historical and Philosophical Reflections on Human Genetics.* University of Iowa Press, Iowa City.

Weiss, S., 1987. *Race Hygiene and National Efficiency: The Eugenics of William Schallmayer.* University of California Press, Berkeley.

Whiten, A., et al., 1999. Cultures in chimpanzees. *Nature,* **399,** 682–685.

Wiggam, A. E., 1923. *The New Decalogue of Science.* Bobbs-Merrill, Indianapolis, IN.

Williams, N., 1997. Biologists cut reductionist approach down to size. *Science,* **277,** 476–477.

Wilmut, I., A. E. Schnieke, J. McWhir, A. J. Kind, and K. H. Campbell, 1997. Viable offspring derived from fetal and adult mammalian cells. *Nature,* **385,** 810–813.

Wilson, E. O., 1975. *Sociobiology: The New Synthesis.* Harvard University Press, Cambridge, MA.

———, 1977. Biology and the social sciences. *Daedalus,* **106**(4), 127–140.

———, 1978. *On Human Nature.* Harvard University Press, Cambridge, MA.

———, 1984. *Biophilia.* Harvard University Press, Cambridge, MA.

Wivel, N. A., and L. Walters, 1993. Germ-line modification and disease prevention: Some medical and ethical perspectives. *Science,* **262,** 533–538.

Glossary

A

Abduction Movement of an appendage or body structure in a direction away from the midline (median sagittal) plane (for example, extending an arm laterally).

Abiotic Substances that are of nonbiological origin, or environments characterized by the absence of organisms.

Acidic A compound that produces an excess of hydrogen ions (H^+) when dissolved in water. Using quantitative hydrogen ion measurements, such solutions have a pH value less than 7.0. (*See* **pH scale**.)

Acoelomate An animal that lacks a coelom (internal body cavity).

Acritarchs Single-celled eukaryote-type microfossils of Precambrian age whose biological relationships are uncertain.

Acrocentric Chromosomes whose centromeres are near one end (between metacentric and telocentric locations).

Active sites Specific regions of an enzyme that bind substrates on which the enzyme acts.

Active transport Biochemical transport that requires the input of energy (for example, hydrolysis of ATP).

Adaptation This notion derived from the typical relationship between structure and function: that an organism's structures seem suitable ("adapted") for their tasks. Until Darwin, the cause for adaptation was commonly ascribed to intelligent (divine) guidance. Darwinians replaced this view by proposing that an adaptation is any trait that replaces other variants because of selection for greater reproductive success (*see* **Fitness**). An adaptation is a trait whose presence enhances survival or fertility. It is selection rather than intelligent design that produces and/or maintains the correlation between structure and function. The complexities of evolution, however, shroud the Darwinian concept with many qualifications. For example, should selection cease or reverse its direction, as occurs for traits that become vestigial, then the trait is no longer an adaptation, although it may have been in the past. Traits that are not maintained by selection (that is, not related to reproductive success) are generally considered "nonadaptive." Such traits may be introduced or persist in a population through mutation, random genetic drift, the accidental extinction of adaptive varieties, developmental constraints that now impede their elimination, close linkage with genes selected for other functions (*see* **Hitchhiking**), or as one of the multiple phenotypic effects of a selected gene (*see* **Pleiotropy**). Also, not all selected traits are necessarily beneficial to a population, since some may increase the reproductive success of genes or individuals but not benefit (or even decrease) population fitness (*see* **Segregation distortion, Sexual selection**). Even when selected traits are unquestionably adaptive, they often involve "trade-offs" in other traits that can lose adaptive advantages. (For example, trees that grow competitively taller put more resources into wood production than seed production.) In addition, earlier selected stages of an adaptation may have been for a function different from that of a later stage (*see* **Preadaptation**). In general, since it is quite difficult to examine historical circumstances leading to a particular trait, it can be difficult to determine how or to what extent it is an adaptation. Mostly, such determinations depend on evaluating functional utility ("optimality") for reproductive success, based on the reasonable assumption that a useful trait generally replaces or has replaced less useful variants. Unfortunately, since it can also be challenging to establish functional utility—to uncover a trait's many possible variations, and to compare their relative reproductive success—identifying adaptations can be controversial. Although selection may not be obvious, it is difficult to accept that any prominent nonadaptive trait can long persist without being affected by selection in some way and to some degree (see, for example, p. 583 footnote 1). The term is also frequently used for the process that produces adaptations (natural selection). However adaptations are defined, it is the genetic transmission of traits whose structure and function let their carriers interact successfully with the environment that drives evolution and makes biology unique and historical.

Adaptive landscape A model originally devised by Sewall Wright that describes a topography in which high fitnesses correspond to peaks and low fitnesses to valleys; each position potentially occupied by a population bearing a unique and frequent genotype.

Adaptive radiation The diversification of a single species or group of related species into new ecological or geographical zones to produce a large variety of species and groups. Such events may include the following:

1. Survivors of a catastrophe (for example, mammals) invade the adaptive zones that were abandoned by extinct species (for example, dinosaurs).
2. One or a few colonizers enter a new habitat in which competing species are absent (for example, the Hawaiian Drosophilidae).
3. One group of species (for example, pollinating insects) evolves in step with the adaptive radiation of another group (for example, angiosperms), or parasites (for example, viruses and bacteria) evolve new strains in concert with proliferation of their hosts (for example, humans).
4. A preadaptive feature (for example, the shelled reptilian amniotic egg) allows invasion into a previously inaccessible ecological zone (for example, terrestrial habitats).
5. A new morphological or physiological character (for example, pharyngeal jaw innovations in cichlid fishes) causes divergent evolution by partitioning the environment into different niches.

These various historical contingencies promote genetic diversity by letting selection—the primary biological force changing organisms through time—channel evolution in new directions.

Adaptive value The relative reproductive success (relative fitness) of an allele or genotype as compared to other alleles or genotypes. (*See also* **Fitness.**)

Adduction Movement of an appendage or body part toward the midline (median sagittal) plane, for example, bringing a laterally extended arm to the side of the body.

Adenosine triphosphate (ATP) An organic compound commonly involved in the transfer of phosphate bond energy, composed of adenosine (an adenine base + a D-ribose sugar) and three phosphate groups.

Aerobic The use of molecular oxygen for reactions that provide growth energy from the oxidative breakdown of food molecules.

Aerobic respiration An electron transport system in which oxygen serves as the terminal electron acceptor.

Algae Photosynthetic members of the eukaryotic kingdom of Protista.

Allele One of the alternative forms of a single gene (that is, a particular nucleotide sequence occurring at a given locus on a chromosome).

Allen's rule The generalization that warm-blooded animals (mammals) tend to have shorter extremities (for example, ears and tail) in colder climates than they have in warmer climates.

Allometry Differential growth rates of different body parts; during development one feature may change at a rate different from that of another feature, resulting in a change of shape. (For a change in developmental timing, *see* **Heterochrony.**)

Allopatric Species or populations whose geographical distributions do not contact each other.

Allopatric speciation Speciation between populations that are geographically separated.

Allopolyploid An organism or species that has more than two sets (2n) of chromosomes (that is, 3n, 4n, and so on) that derive from two or more different ancestral groups.

Allozyme The particular form (amino acid sequence) of an enzyme produced by a particular allele at a gene locus when there are different possible forms of the enzyme (different possible amino acid sequences), each produced by a different allele.

Alternation of generations Life cycles in which a multicellular haploid stage (1n) alternates with a multicellular diploid stage (2n).

Altruism Behavior that benefits the reproductive success of other individuals because of an actual or potential sacrifice of reproductive success by the altruist.

Amino acids Organic molecules of the general formula $R—CH(NH_2)COOH$, possessing both basic (NH_2) and acidic (COOH) groups, as well as a side group (R) specific for each type of amino acid. Normally 20 different types of amino acids are used in cellularly synthesized proteins.

Amino group An $-NH_2$ group.

Amniotic egg The type of egg produced by reptiles, birds, and mammals (Amniota), in which the embryo is enveloped in a series of membranes (amnion, allantois, chorion) that help sustain its development.

Anaerobic Growth (energy obtained from the oxidative breakdown of food molecules) in the absence of molecular oxygen.

Anaerobic respiration An electron transport system in which substances other than oxygen serve as the terminal electron acceptor (for example, sulfates, nitrates, methane).

Anagenesis The evolution of new species that takes place progressively over time within a single lineage (branch), as opposed to *cladogenesis* where a group diverges into two or more branches. (*See also* **Phyletic evolution.**)

Analogy The possession of a similar character by two or more quite different species or groups that arises from a developmental pathway unique to each group; that is, the similarity is caused by factors other than their distant common genetic ancestry. (*See also* **Convergence.**)

Aneuploidy The gain or loss of chromosomes leading to a number that is not an exact multiple of the basic haploid chromosome set (n) (for example, n + 1, 2n + 1, 2n − 1, 2n − 2, 2n + 3, and so on).

Angiosperms The flowering plants, an advanced group of vascular plants with floral reproductive structures and encapsulated seeds.

Angstrom (Å) A length one-ten billionth (10^{-10}) of a meter.

Antibody A protein produced by the immune system that binds to a substance (antigen) typically foreign to the organism.

Anticodon A sequence of three nucleotides (a triplet) on transfer RNA that is complementary to the codon on messenger RNA that specifies placement of a particular amino acid in a polypeptide during translation. (*See also* **Codon.**)

Antigen A substance, typically foreign to an organism, that initiates antibody formation and is bound by the activated antibody.

Apomixis (apomictic) Reproduction without fertilization; offspring produced from unfertilized eggs in which meiosis has been partially or completely suppressed. (*See also* **Parthenogenesis.**)

Apomorphy A character that has been derived from, yet differs from, the ancestral condition. (*See also* **Synapomorphy.**)

Aposematic Conspicuous warning coloration in potential prey species that advertises their toxicity or distastefulness to predators. Aposematic patterns usually contain bright colors or shades such as those found among wasps, monarch butterflies, coral snakes, skunks, and poisonous salamanders.

Arboreal Living predominantly in trees.

Archaebacteria Prokaryotes that, unlike eubacteria, do not incorporate muramic acid into their cell walls and possess other distinguishing characteristics. They are considered to represent one of the early cell forms.

Archetype The concept of an ideal primitive plan ("*Bauplan*") on which organisms, such as vertebrates, are presumably based. Called by Richard Owen the "primal pattern" and "divine idea."

Artificial selection Selection process in which humans are the selective agents. (*See* **Selection.**)

Asexual reproduction Offspring produced by one parent in the absence of sexual fertilization or in the absence of gamete formation.

Assortative mating Mating among individuals on the basis of their phenotypic or genotypic similarities (*positive assortative*) or differences (*negative assortative*) rather than mating among all individuals on a random basis.

Autocatalytic reaction Instances in which the agent that promotes (catalyzes) a reaction is formed as a product of the reaction.

Autopolyploid A species or organism that has more than two sets of chromosomes (polyploid) derived from one or more duplications in a single ancestral source.

Autosome A chromosome whose presence or absence is ordinarily not associated with determining the difference in sex (that is, a chromosome other than a sex chromosome).

Autotroph An organism capable of synthesizing complex organic compounds needed for growth from simple inorganic environmental substrates: *photoautotroph*, an organism that can use light as an energy source and carbon dioxide as a carbon source; *chemoautotroph* (*chemolithotroph*), an organism that obtains energy for growth by oxidizing inorganic compounds such as hydrogen sulfide.

B

Bacteriophage A virus (phage) that parasitizes bacteria.

Balanced genetic load The decrease in overall fitness of a population caused by defective genotypes (for example, homozygotes for deleterious recessives) whose alleles persist in the population because they confer selective advantages in other genotypic combinations (for example, heterozygote advantage).

Balanced polymorphism The persistence of two or more different genetic forms through selection (for example, heterozygote advantage) rather than because of mutation or other evolutionary forces.

Banded iron formation An iron-containing laminated sedimentary rock, often composed of layers of tiny quartz crystals (chert).

Basalt A fine-grained igneous rock found in oceanic crust and produced in lava flows.

Base (nucleotide) The nitrogenous component of the nucleotide unit in nucleic acids, consisting of either a purine (adenine, A, or guanine, G) or pyrimidine (thymine, T, or cytosine, C, in DNA; uracil, U, or cytosine, C, in RNA). (*See also* **Purine, Pyrimidine.**)

Base pairs *See* **Complementary base pairs.**

Basic (alkaline) A compound that produces an excess of hydroxyl (OH^-) ions when dissolved in water. Using quantitative hydrogen ion measurements, such solutions have a pH value greater than 7.0. (*See* **pH scale.**)

Batesian mimicry The similarity in appearance of a harmless species (the mimic) to a species that is harmful or distasteful to predators (the model), maintained because of selective advantage to the relatively rare mimic.

Bauplan Structural body plan that characterizes a group of organisms. (*See also* **Archetype.**)

Benthic Refers to the floor of a body of water (for example, ocean bottom, riverbed, lake bottom) and to organisms that live in it, on it, or near it.

Bergmann's rule The generalization that animals living in colder climates tend to be larger than those of the same group living in warmer climates.

Big bang theory The concept that the universe was born in a gigantic explosion about 10 to 20 billion years ago.

Bilateral symmetry Instances in which the left and right sides of a longitudinal (sagittal) plane that runs through an organism's midline are approximately mirror images of each other.

Binary fission Replication of an organism by its division into two mostly equal parts; the common form of asexual reproduction in prokaryotes and protistan eukaryotes.

Binomial expansion The binomial $(a + b)$ raised to a power n $[(a + b)^n]$ where a and b represent alternative states whose sum equals the probability of 1.

Binomial nomenclature The Linnaean principle of designating a species by two names: the name of a genus followed by the name of a species.

Biogenetic law (Haeckel) The concept that stages in the development of an individual (ontogeny) recapitulate the evolutionary history (phylogeny) of the species. (*See also* **Heterochrony**.)

Biogeography The study of the geographical distributions of organisms. A biogeographical realm is a region characterized by a distinctive biota.

Biological species concept The view that the primary criterion for separating one species from another is their reproductive isolation.

Biosphere That part of the earth containing all living organisms.

Biota All organisms, including animals (fauna) and plants (flora) of a given region or time period.

Biotic Relating to or produced by biological organisms.

Bipedal A term used mostly to describe terrestrial tetrapod locomotion that is restricted to the hind limbs when these two limbs move alternately (for example, human walking) rather than together (for example, kangaroo jumping).

Blastocoele The cavity of a blastula.

Blastopore The opening formed by the invagination of cells in the embryonic gastrula, connecting its cavity (archenteron) to the outside. In protostome phyla the blastopore is the site of the future mouth, whereas in deuterostomes the blastopore becomes the anus and the mouth is formed elsewhere.

Blastula A hollow sphere enclosed within a single layer of cells, occurring at an early stage of development in various multicellular animals.

Blending inheritance The abandoned concept that offspring inherit a dilution, or blend of parental traits, rather than the particles (genes) that determine those traits.

Bottleneck effect A form of genetic drift that occurs when a population is reduced in size (population crash) and later expands in numbers (population flush). The enlarged population that results may have gene frequencies that are distinctly different from those before the bottleneck. (*See also* **Founder effect**.)

Brachiation Apelike locomotion through trees: hanging from branches and swinging alternate arms (left, right, left,...) from branch to branch, accompanied by a rotation of the body during each swing.

Brackish Water whose salt content (salinity) is intermediate between fresh water and sea water; usually at the mouths of rivers that empty into the ocean (estuaries).

Bradytelic A relatively slow evolutionary rate.

Bryophyte Mosses and liverworts, small "primitive" land plants.

Buccal Pertaining to the inside of the mouth; side of a tooth closest to the cheek.

Burrowing animal In aquatic forms, a bottom-dweller that moves through soft benthic sediments.

C

Calorie The amount of heat necessary to raise the temperature of 1 gram of water by 1 degree centigrade at a pressure of 1 atmosphere.

Calvin cycle A cyclic series of light-independent reactions that accompany photosynthesis and that reduce carbon dioxide to carbohydrate.

Cambrian period The interval between about 545 and 505 million years before the present, marking the plentiful appearance of fossilized organisms with hardened skeletons. It is considered the beginning of the Phanerozoic time scale (eon) and is the first period in the Paleozoic era.

Carbohydrate A compound in which the hydrogen and oxygen atoms bonded to carbons are commonly in a ratio of 2/1 (for example, glucose ($C_6H_{12}O_6$), starch $(C_6H_{12}O_6)_n$, and cellulose, $(C_6H_{10}O_5)_n$).

Carbonaceous Possessing organic (carbon) compounds.

Carbonaceous chondrites Meteorites containing carbon compounds.

Carboxylic acid An organic compound that has an acidic group consisting of a carbon with a double-bond attachment to an oxygen atom and a single-bond attachment to a hydroxyl group (O=C–OH).

Carnivores Flesh eaters; organisms (almost entirely animal, rarely plant) that feed on animals.

Carrying capacity The theoretical maximum number of organisms in a population, usually designated by *K*, that can be sustained in a given environment.

Catalyst A substance that lowers the energy necessary to activate a reaction but is not itself consumed or altered in the reaction.

Catastrophism The eighteenth- and nineteenth-century concept that fossilized organisms and changes in geological strata were produced by periodic, violent, and widespread catastrophic events (presumably caused by capricious supernatural forces) rather than by naturally explainable events based on laws that act uniformly through time. (*See also* **Uniformitarianism**.)

Cell wall The rigid or semirigid extracellular envelope (outside the plasma membrane) that gives shape to plant, algal, fungal, and bacterial cells.

Cenozoic era The period from 65 million years ago to the present, marked by the absence of dinosaurs and the radiation of mammals. This is the third and most recent era of the Phanerozoic eon and is divided into two major periods, the Tertiary and Quaternary.

Centigrade scale (°C) A scale of temperature in which the melting point of ice is taken as 0° and the boiling point of water as 100°, measured at 1 atmosphere of pressure.

Centrifugal selection *See* **Disruptive selection.**

Centripetal selection *See* **Stabilizing selection.**

Centromere The chromosome region in eukaryotes to which spindle fibers attach during cell division.

Character A feature, trait, or property of an organism or population. If possible, the description of a character should include the conditions under which it is observed.

Character displacement Divergence in the appearance or measurement of a character between two species when their distributions overlap in the same geographical zone, compared to the similarity of the character in the two species when they are geographically separated. When common resources are limited, it is presumed that competition between overlapping species leads to divergent specializations and therefore to divergence in characters that were formerly similar.

Cheek teeth Mammalian premolar and molar teeth.

Chemiosmosis Linkage between a chemical process (electron transport chain) and a proton pump that causes protons (H^+) to be transferred across a membrane. This creates a proton gradient that drives a membrane-bound enzyme, ATP synthetase, to catalyze the reaction ADP + $P_i \rightarrow$ ATP.

Chemoautotroph (chemolithotroph) *See* **Autotroph.**

Chert A sedimentary rock composed largely of tiny quartz crystals (SiO_2) precipitated from aqueous solutions.

Chloroplast A chlorophyll-containing, membrane-bound organelle that is the site of photosynthesis in the cells of plants and some protistans. These organelles contain their own genetic material (circular DNA without histones) and are believed to be descendants of cyanobacteria that entered eukaryotic cells via endosymbiosis.

Chromatid One of the two sister products of a eukaryotic chromosome replication, marked by an attachment between the sister chromatids at the centromere region. When this attachment is broken during the mitotic anaphase stage, each sister chromatid becomes an independent chromosome.

Chromosome A length of nucleic acid comprising a linear sequence of genes that is unconnected to other chromosomes. In eukaryotes, histone proteins are bound to nuclear chromosomes, and this protein-nucleic acid complex can be made microscopically visible as deeply staining filaments.

Chromosome aberration A change in the gene sequence of a chromosome caused by deletion, duplication, inversion, or translocation.

Citric acid cycle *See* **Krebs cycle.**

Clade A cluster of taxa derived from a single common ancestor.

Cladistics A mode of classification based principally on grouping taxa by their shared possession of similar ("derived") characters that differ from the ancestral condition.

Cladogenesis "Branching" evolution involving the splitting and divergence of a lineage into two or more lineages.

Cladogram A tree diagram representing phylogenetic relationships among taxa.

Class A taxonomic rank that stands between phylum and order; a phylum may include one or more classes, and a class may include one or more orders.

Classification The grouping of organisms into a hierarchy of categories commonly ranging from species to genera, families, orders, classes, phyla, and kingdoms, each category reflecting one or more significant features. In practice, the decision as to the species in which to place an organism, or the genus in which to place a species, and so forth, is most often based on phenotypic similarity to other members of the group: organisms in a species are more similar to each other than they are to organisms in other species of the same genus, species in a genus are more similar to each other than they are to species in other genera of the same family, and so forth.

Cline A gradient of phenotypic or genotypic change in a population or species correlated with the direction or orientation of some environmental feature, such as a river, mountain range, north–south transect, or altitude.

Clone A group of organisms derived by asexual reproduction from a single ancestral individual.

Cloning (**gene**) Techniques for producing identical copies of a section of genetic material by inserting a DNA sequence into a cell, such as a bacterium, where it can be replicated.

Coacervate An aggregation of colloidal particles in liquid phase that persists for a period of time as suspended membranous droplets.

Coadaptation The action of selection in producing adaptive combinations of alleles at two or more different gene loci.

Coalescence A statistical term used to describe relationships among different gene sequences that are all descended from a common ancestral sequence (the "coalescent"). Instead of a genealogical tree expanding each generation from its apex toward its base, the tree is conceived as collapsing (coalescing) by proceeding in reverse, from base to apex.

Coarse-grained environment A heterogeneous environment in which individuals in a population are exposed to conditions different from other individuals.

Codominance The independent phenotypic expression of two different alleles in a heterozygote (for example, genotypes carrying both *M* and *N* alleles of the *MN* blood group show the MN blood type).

Codon The triplet of adjacent nucleotides in messenger RNA that codes for a specific amino acid carried by a specific transfer RNA or that codes for termination of translation (STOP codons). Placement of the amino acid is based on complementary pairing between the anticodon on tRNA and the codon on mRNA. (*See also* **Anticodon**.)

Coelom An internal body cavity, lined in eucoelomates (true coelomates) with mesodermal tissue that may contain organs such as testes and ovaries.

Coenzymes Nonprotein enzyme-associated organic molecules (for example, NAD, FAD, and coenzyme A), that participate in enzymatic reactions by acting as intermediate carriers of electrons, atoms, or groups of atoms.

Coevolution Evolutionary changes in one or more species in response to changes in other species in the same community.

Cofactor A small molecule, which may be organic (that is, a coenzyme) or inorganic (that is, a metal ion), required by an enzyme in order to function.

Cohort Individuals of a population that are all the same age.

Commensalism An association between organisms of different species in which one species is benefited by the relationship but the other species is not significantly affected.

Competition Relationship between organismic units (for example, individuals, groups, species) attempting to exploit a limited common resource in which each unit inhibits, to varying degrees, the survival or proliferation of another unit by means other than predation.

Complementary base pairs Nucleotides on one strand of a nucleic acid that hydrogen bond with nucleotides on another strand according to the rule that pairing between purine and pyrimidine bases is restricted to certain combinations: A pairs with T in DNA, A pairs with U in RNA, and G pairs with C in both DNA and RNA.

Complexity A state of intricate organization caused by arrangement or interaction among different component parts or processes: presumably, the greater the number of interacting parts, the greater the complexity. The term *levels of complexity* describes gradations in which complex organizations are included (nested) within others. (*See also* **Hierarchy**.) Attempts to compare the degrees of complexity among organisms have used numbers of their different kinds of structures, organs, tissues, cells, genes, and proteins. However, such numbers do not always change in a consistent fashion, and McShea points out (p. 451), "Something may be increasing [in evolution]. But is it complexity?"

Concerted evolution The process by which a series of nucleotide sequences or different members of a gene family remain similar or identical through time.

Condensation (by dehydration) The formation of a covalent bond between two molecules by removal of H_2O.

Condylarths A mammalian order that became extinct during the Miocene period of the Cenozoic era but whose first occurrences are in the late Cretaceous period of the Mesozoic. It includes a diversity of early herbivorous placental mammals and the ancestors of all later herbivores.

Constraint Constraint has been used in biology to describe factors that limit character variation or evolutionary direction. According to some authors, the term applies to traits molded primarily by physical agents and laws, such as crystallization, friction, gravity, and surface tension, but unaffected by historical contingencies, such as selection. Others argue that excluding biological factors does not help explain distinctions among phenotypic variants, nor the causes for differences among lineages. A biological concept of constraint therefore seems more useful to evolutionists—phenotypic channeling and evolutionary trends caused mostly by processes involving adaptation. That is, because adaptation depends on the availability of appropriate genes, adaptive constraints are mostly tied to organismic histories: constraints are affected by genes that evolved and were selected previously (p. 357). Among such constraining forces are directional selection (p. 542), stabilizing selection (p. 542), canalization (p. 358), and factors that may limit or direct genetic mutability (pp. 225 and 570), as well as developmental innovation (pp. 360 and 565). Some constraints may focus phenotypic trends in adaptive directions, but others (limited genetic variation and restricted developmental ability) may also limit response to new environmental challenges, leading even to extinction (pp. 358 and 452). Although constraints can provide reasonable explanations for trends and attributes, postulating which particular constraints were in force seems highly conjectural in the absence of detailed historical-phylogenetic information. For a term that can apply to so many different phenomena, *constraint's* usage in specific instances needs to be defined.

Continental drift The movement, over time, of large landmasses—tectonic plates—on the earth's surface relative to each other. (*See also* **Paleomagnetism, Sea floor spreading, Tectonic plates.**)

Continuous variation Character variations (such as height in humans) whose distribution follows a series of small nondiscrete quantitative steps from one extreme to the other. (*See also* **Quantitative character.**)

Convergence (also called *Homoplasy*) The evolution of similar characters in genetically unrelated or distantly related species, mostly because they have been subjected to similar environmental selective pressures. (*See also* **Analogy.**)

Cope's rule The generalization (not always confirmed) that body size tends to increase in an animal lineage during its evolution.

Correlation The degree to which two measured characters tend to vary in the same quantitative direction (positive correlation) or in opposite directions (negative correlation).

Cosmology Study of the structure and evolution of the universe.

Covalent bond A strong chemical bond that results from the sharing of electrons between two atoms.

Creationism The belief that each different kind of organism was individually created by one or more supernatural beings whose activities are not controlled by known physical, chemical, or biological laws.

Creodonts An extinct order of early Cenozoic placental mammals that were the dominant carnivores until replaced by the modern order Carnivora during the Oligocene period.

Crepuscular A lifestyle characterized by activity mostly during the hours around dawn and dusk.

Crossovers (chromosome) Results of a process (crossing over) in which the chromatids of two homologous chromosomes exchange genetic material. (*See also* **Recombination.**)

Cryptic A feature that is normally not visible.

Culture (social) The learned behaviors and practices common to a social group.

Cursorial Adapted for running on land.

Cusps (teeth) Elevations on the crowns of premolars and molars. The number, shapes, and positions of cusps are inherited characters that can provide useful phylogenetic information.

Cyanobacteria Photosynthetic prokaryotes possessing chlorophyll *a* but not chlorophyll *b*. Many are photosynthetic aerobes (oxygen producing) and some are anaerobes (not oxygen producing). Formerly called *blue-green algae*, their color caused by a bluish pigment masking the chlorophyll.

Cytochromes Proteins containing iron–porphyrin (heme) complexes that function as hydrogen or electron carriers in respiration and photosynthesis.

Cytology The study of cells—their structures, functions, components, and life histories.

Cytoplasm All cellular material within the plasma membrane, excluding the nucleus.

D

Darwinism The concept, proposed by Charles Darwin, that biological evolution has led to the many different highly adapted species through natural selection acting on hereditary variations in populations.

Deciduous (teeth) Teeth that are replaced during development by permanent teeth.

Deficiency *See* **Deletion.**

Degenerate (redundant) code The type of genetic code used by existing terrestrial organisms, for which there is more than one triplet codon for a particular amino acid but a specific codon cannot code for more than one amino acid. Thus, the 20 different amino acids translated in protein synthesis are coded by 61 of the 64 possible different triplet codons, some by as many as six different "synonymous" codons. (*See also* **Genetic code.**)

Dehydrogenase An enzyme that catalyzes the removal of hydrogen from a molecule (oxidation).

Deleterious allele An allele whose effect reduces the adaptive value of its carrier when present in homozygous condition (recessive allele) or in heterozygous condition (dominant or partially dominant allele).

Deletion An aberration in which a section of DNA or chromosome has been lost.

Deme A local population of a species (in sexual forms, a local interbreeding group).

Density dependent The dependence of population growth and size on factors directly related to the numbers of individuals in a particular locality (for example, competition for food, accumulation of waste products).

Density independent The dependence of population growth on factors (climatic changes, meteorite impacts, and so on) unrelated to the numbers of individuals in a particular locality.

Dentin The hard inner layer of a tooth that surrounds the tooth pulp. It is covered by even harder enamel at the crown, and by softer cement at the root.

Derived character A character whose structure or form differs (apomorphic) from that of the ancestral stock.

Deuterium An isotope of hydrogen containing one proton and one neutron, giving it twice the mass of an ordinary hydrogen atom.

Deuterostomes Coelomate phyla in which the embryonic blastopore becomes the anus.

Development The recurrent sequence of progressive changes in organisms from inception to maturity.

Dicotyledons Flowering plants (angiosperms) in which the embryo bears two seed leaves (cotyledons).

Differentiation Changes that occur in the structure and function of cells and tissues as the development of the organism proceeds. Generally, the change from an immature embryo to a more complex mature organism.

Dimorphism Presence in a population or species of two morphologically distinctive types of individuals (for example, differences between males and females, pigmented and nonpigmented forms).

Dioecious Organisms in which the male and female sex are in separate individuals.

Diploblastic An animal that produces only two major types of cell layers during development, ectoderm and endoderm (for example, Cnidaria).

Diploid An organism whose somatic cell nuclei possess two sets of chromosomes (2n), providing two different (heterozygous) or similar (homozygous) alleles for each gene.

Directional selection Selection that causes the phenotype of a character to shift toward one of its phenotypic extremes.

Discontinuous variation Character variations that are sufficiently different from each other that they fall into nonoverlapping classes.

Disruptive selection Selection that tends to favor the survival of organisms in a population that are at opposite phenotype extremes for a particular character and eliminates individuals with intermediate values (centrifugal selection).

Diurnal A lifestyle characterized by activity during the day rather than at night (nocturnal).

Divergent evolution Change leading to differences between lineages.

DNA (deoxyribonucleic acid) A nucleic acid that serves as the genetic material of all cells and many viruses; composed of nucleotides that are usually polymerized into long chains, each nucleotide characterized by the presence of a deoxyribose sugar.

DNA ligase An enzyme that joins sections of DNA together.

Domain In molecular biology, an amino acid sequence within a polypeptide chain that performs a particular subfunction in the protein (Fig. 9–14). The term has also been used in systematics to provide a tripartite division of organisms—Archaea, Bacteria, Eucarya—as a substitute for the rank of superkingdom, which commonly designates prokaryotes and eukaryotes (for example, see Fig. 11–10).

Dominance (allele) Instances in which the phenotypic effect of a particular allele (for example, *A*, the dominant) is expressed in both the heterozygote (*Aa*) and homozygote (*AA*), but the phenotypic effect of the other allele (for example, *a*, the recessive) is not expressed in heterozygotes but only in homozygotes (*aa*).

Dominance (social) Relations within a group in which one or more individuals, sustained by aggression or other behaviors, rank higher than others in controlling the conduct of group members.

Doppler effect The shift in wavelength of light or sound that is perceived as the emitting body moves toward us (shorter wavelengths, for example, blue-shifted) or away from us (longer wavelengths, for example, red-shifted).

Dorsal The back side or upper surface of an animal; opposite of ventral. (In vertebrates, the surface closest to the spinal column.)

Dosage compensation A mechanism that compensates for the difference in number of X chromosomes (or Z chromosomes) between males and females so the metabolic activities (gene expression) of their X-linked genes are equalized. Although dosage compensation is widespread among animals, the mechanism by which it is accomplished varies. In species with XY males and XX females such as *Drosophila,* male X-linked genes show increased gene expression, whereas in mammals only one X chromosome in each sex is metabolically active and any additional X chromosomes are inactivated.

Double fertilization A distinctive feature of angiosperm plants in which two nuclei from a male pollen tube fertilize the female gametophyte, one producing a diploid embryo and the other producing polyploid (usually triploid) nutritional endosperm.

Duplication Instances in which a particular section of DNA or visible chromosome segment occurs more than once.

E

Ecogeographical rules Generalizations that correlate adaptational tendencies of species with environmental factors such as climate. (*See* **Allen's rule**, **Bergmann's rule**, **Gloger's rule**.)

Ecological niche The environmental habitat of a population or species, including the resources it uses and its interactions with other organisms. Since resources and interactions are rarely constant, populations remain continually subject to selective pressures for adaptational change. A particular organism's ecological niche is commonly reflected in its adaptations when these can be specified. (*See* **Adaptation**.)

Ecology The study of the relations between organisms and their environment, in terms of their numbers, distributions, and life cycles.

Ecotype A phenotypic and genotypic variant of a species associated with a particular environmental habitat. (*See also* **Race**.)

Ectoderm The outermost layer of cells that covers the early animal embryo, from which nerve tissues and outermost epidermal tissues are derived.

Ectothermic "Cold-blooded": a body temperature primarily determined by the ambient (environmental) temperature.

Ediacaran strata Geological formations containing soft-bodied invertebrate fossils found in South Australia and other places, dating to a Precambrian period lasting about 60 or more million years.

Electron carrier In oxidation–reduction reactions, a molecule that acts alternatively as an electron donor (becomes oxidized) and as an electron acceptor (becomes reduced).

Electrophoresis A technique that separates dissolved particles subjected to an electrical field according to their mobility. Given a particular medium through which a particle moves, electrophoretic mobility depends on the size of the particle, its geometry, and electrical charge.

Endemic A species or population that is specific (indigenous) to a particular geographic region.

Endocytosis Cellular engulfment of outside material, followed by its transfer into the cellular interior encapsulated in a membrane.

Endoderm The layer of cells that lines the primitive gut (archenteron) during the early stages of development in animals, and later forms the epithelial lining of the intestinal tract and internal organs such as the liver, lung, and urinary bladder.

Endonucleases Enzymes that fragment DNA chains. (*See also* **Restriction enzymes**.)

Endosymbiosis A relationship between two different organisms in which one (the endosymbiont) lives within the tissues or cell of the other, benefiting one or both. It is now generally thought that some eukaryotic organelles, such as mitochondria and chloroplasts, had an endosymbiotic prokaryotic origin.

Endothermic "Warm-blooded": a body temperature maintained by internal physiological mechanisms at a level independent of the ambient (environmental) temperature.

Enhancer A nucleotide sequence that allows gene transcription to increase even though the gene may be quite distant. It does this by changing the configuration of the intervening nucleotide sequence, making the gene's promoter sequence more available for transcription.

Entropy The measure of disorder of a physical system. In a closed system, to which energy is not added, the second law of thermodynamics essentially states that entropy, or energy unavailable for work, will remain constant or increase but never decrease. Living systems, however, are open systems, to which energy is added from sunlight and other sources, and order can therefore arise from disorder in such systems, that is, energy available for work can increase and entropy can decrease.

Environment The complex of external conditions, abiotic and biotic, that affects organisms or populations. It provides the facilities and resources that enable hereditary data (genotypes) to produce organismic features (phenotypes).

Enzyme A protein that catalyzes chemical reactions.

Eon A major division of the geological time scale, often divided into two eons beginning from the origin of the earth 4.5 billion years ago: the Precambrian or Cryptozoic (rarity of life forms) and the Phanerozoic (abundance of life forms).

Epigamic selection Selection for mating success based on appearance or behavior during courtship.

Epigenesis The concept that tissues and organs are formed by interaction between cells and substances that appear during development, rather than being initially present in the zygote (preformed). (*See also* **Preformationism**.)

Epistasis Interactions between two or more gene loci that produce phenotypes different from those expected if each locus were considered individually. In statistical population studies that evaluate the causes for phenotypic differences, the term *epistasis* is commonly used for all phenotypic variation caused by interaction between nonallelic genes.

Epoch One of the categories into which geological time is divided; a subdivision of a geological period. For periods divided into three epochs, they are often named Early, Middle, and Late; for example, Early Cambrian, ...

Equilibrium (genetic) The persistence of the same allelic frequencies over a series of generations. Equilibria may be stable or unstable. In a stable equilibrium (for example, when the heterozygote is superior in fitness to the homozygotes), the population returns to a particular equilibrium value when the allelic frequencies have been disturbed. In an unstable equilibrium (for example, when the heterozygote is inferior in fitness to the homozygotes), such disturbances are not followed by a return to equilibrium frequencies.

Era A division of geological time that stands between the eon and the period: the Phanerozoic eon is divided into Paleozoic, Mesozoic, and Cenozoic eras; and each era is divided into two or more periods.

Estrus The interval during which female mammals exhibit maximum sexual receptivity, usually coinciding with the release of eggs from the ovary.

Eubacteria Prokaryotes, other than archaebacteria, marked by sensitivity to particular antibiotics and by the incorporation of muramic acid into their cell walls.

Euchromatin Normally staining chromosomal regions that possesses most of the active genes. (*See also* **Heterochromatin.**)

Eucoelomates *See* **Coelom.**

Eugenics The concept that humanity can be improved by altering human genotypes or their frequencies.

Eukaryotes Organisms whose cells contain nuclear membranes, mitochondrial organelles, and other characteristics that distinguish them from prokaryotes. Eukaryotes may be unicellular or multicellular and include protistans, fungi, plants, and animals.

Euploidy Variations that involve changes in the number of entire chromosome sets (n) (for example, 3n, 4n, 5n).

Eutelegenesis The use of artificial insemination to improve genetic endowment.

Eutely Constancy in the numbers of cells or nuclei from the larval stage to the adult stage.

Eutheria *See* **Placentals.**

Eutrophication The process in which an aquatic system becomes overloaded with nutrients, thereby increasing its organic productivity and causing an accumulation of debris.

Evolution Genetic changes in populations of organisms through time that lead to differences among them.

Evolutionary (molecular) clock The concept that the rate at which mutational changes accumulate is constant over time. To which genes or genomes this clock may apply, and whether it is really constant, are disputed.

Exon A nucleotide sequence in a gene that is transcribed into messenger RNA and spliced together with the transcribed sequences of other exons from the same gene. The continuous RNA molecule formed is then transferred to the ribosome and forms the template used in polypeptide synthesis. Exons ("expressed sequences") are separated from other exons in the same gene by intervening nontranslated sequences (*see* **Intron**) that are removed from the mRNA. Such intron–exon split genes are commonly found in eukaryotes but are almost entirely absent in prokaryotes.

Extant Currently in existence.

Extension Movement of an appendage so that the angle of the joint increases.

Extinction The disappearance of a species or higher taxon.

F

F (inbreeding coefficient) *See* **Inbreeding coefficient.**

Family A taxonomic category that stands between order and genus; an order may comprise a number of families, each of which contains a number of genera.

Fauna All animals of a particular region or time period.

Fecundity A measure of potential fertility, often calculated in terms of the quantity of gametes produced while sexually mature.

Feedback When the products of a process affect its own function.

Fermentation The anaerobic degradation of glucose (glycolysis) or related molecules, yielding energy and organic end products.

Fertility A trait measured by the number of viable offspring produced.

Filter feeder An animal that obtains its food by filtering suspended food particles from water.

Fine-grained environment A heterogeneous environment whose varied conditions can normally be experienced by a single individual during its lifetime.

Fitness Central to evolutionary concepts evaluating genotypes and populations, fitness has had many definitions, ranging from comparing growth rates to comparing long-term survival rates. The basic fitness concept that population geneticists commonly use is *relative reproductive success,* as governed by selection in a particular environment; that is, the ability of an organism (genotype) to transmit its genes to the next reproductively fertile generation, relative to this ability in other genotypes in the same environment ("relative fitness"). Since there are forces other than selection that influence genotype frequencies (for example, mutation, random genetic drift, migration), fitness is not the only way of characterizing short-term populational genetic changes. Nevertheless, because reproductive success, sooner or later, affects most variation, fitness and selection enter into practically all enduring organismic-environmental interactions, with adaptations their phenotypic manifestations.

Fixation Achievement of a frequency of 100 percent (monomorphism) by an allele or genotype that begins in a population at a lesser frequency (polymorphism).

Fixity of species A concept held by Linnaeus and others that members of a species could only produce progeny like themselves, and therefore each species was fixed in its particular form(s) at the time of its creation.

Flexion Movement of an appendage so that the angle of the joint decreases.

Flora All plants of a particular region or time period.

Fossils The geological remains, impressions, or traces of organisms that existed in the past.

Founder effect The effect caused by a sampling accident in which only a few "founders" derived from a large population begin a new colony. Since these founders carry only a small fraction of the parental population's genetic variability, radically different gene frequencies can become established in the new colony. (*See also* **Bottleneck effect.**)

Frequency-dependent selection Instances where the effect of selection on a phenotype or genotype depends on its frequency (for example, a genotype that is rare may have a higher adaptive value than when it is common).

Frozen accident The concept that an accidental event in the distant past was responsible for the presence of a universal feature in living organisms. Such events may include an accident in which the present genetic code was used by a group of early organisms that managed to survive some populational bottleneck, thereby conferring this particular code on later organisms.

Fundamentalism (religious) The belief that creation stories and the many events and rules given in religious documents (for example, the Judeo-Christian Bible, the Moslem Koran) are to be taken literally.

G

Galaxy A system of numerous stars such as the Milky Way (150 billion stars, 100,000 light-years across) held together by mutual gravitational effects. Galaxies, in turn, are grouped into clusters and superclusters. Our own supercluster, centered on Virgo, contains many thousands of galaxies and is more than 100 million light-years across.

Gamete A germ cell (usually haploid) that fuses with a germ cell of the opposite sex to form a zygote (usually diploid) in a process called *fertilization.*

Gametophyte The haploid gamete-producing stage of plants that have alternating generations (haploid gametophyte and diploid sporophyte). The gametophyte is produced by meiosis in the sporophyte, and its gametes are produced by mitosis.

Gamma ray A high-frequency, highly penetrating radiation emitted in nuclear reactions.

Gastrula A cuplike embryonic stage in multicellular animals that follows the blastula stage. Its hollow cavity (archenteron) is lined with endoderm and opens to the outside through a blastopore. (*See also* **Haeckel's gastrula hypothesis.**)

Gene A unit of genetic material composed of a sequence of nucleotides that provides a specific function to an organism, either by:

- Coding (via transcription into messenger RNA) for a polypeptide chain ("cistron") in a protein
- Being transcribed into a sequence of ribonucleotides used as ribosomal RNA or transfer RNA
- Possessing recognition sites for protein attachment (for example, *see* **Enhancers**, **Promoters**) that regulate processes such as replication and transcription of other genes

The position a gene occupies on a chromosome is called a *locus,* and each different nucleotide sequence of a gene is called an *allele.*

Genealogy A record of familial ties and ancestral connections among members of a group.

Gene family Two or more gene loci in an organism whose similarities in nucleotide sequences indicate they have been derived by duplication from a common ancestral gene (for example, the β-globin gene family, which includes β, γ, δ, and ε genes).

Gene flow The migration of genes into a population from other populations by interbreeding.

Gene frequency The proportion of a particular allele among all alleles at a gene locus. (Also called *allele* or *allelic frequency.*)

Gene locus The chromosomal position (nucleotide sequence) occupied by a particular gene.

Gene pair The two alleles present in a diploid organism at a specific gene locus on two homologous chromosomes.

Gene pool All the genes present in a population during a given generation or period.

Gene therapy Human-directed repair or replacement of genes that cause inherited diseases. When confined to somatic (body) cells rather than to sex cells (sperm or eggs), such gene repairs are not passed on to future generations.

Genetic code The sequences of nucleotide triplets (codons) on messenger RNA that specify each of the different kinds of amino acids positioned on polypeptides during the translation process. With few exceptions, the genetic code used by all organisms is identical: the 20 amino acids are each specified by the same codons (total: 61 codons), and the same three triplet codons are used to terminate polypeptide synthesis. (*See also* **Degenerate redundant code, Universal genetic code.**)

Genetic crossing over *See* **Crossovers (chromosomes), Recombination.**

Genetic death The inability of a genotype to reproduce itself because of selection.

Genetic distance A measure of the divergence among populations based on their differences in frequencies of given alleles.

Genetic drift *See* **Random genetic drift.**

Genetic engineering Manipulation of genetic material from different sources to produce new combinations that are then introduced into organisms in which such genetic material does not normally occur.

Genetic load The loss in average fitness of individuals in a population because the population carries deleterious alleles or genotypes. (*See also* **Balanced Genetic Load, Mutational Load.**)

Genetic polymorphism The presence of two or more alleles at a gene locus over a succession of generations. (Called *balanced polymorphism* when the persistence of the different alleles cannot be accounted for by mutation alone.)

Genome The complete genetic constitution of a cell or an individual.

Genotype The genetic constitution of cells or individuals, often referring to alleles of one or more specified genes. Provides the hereditary information necessary for phenotypic development.

Genus (plural, **genera**) A taxonomic category that stands between family and species: a family may comprise a number of genera, each of which contains a number of species that are presumably related to each other by descent from a common ancestor. In taxonomic binomial nomenclature, the genus is used as the first of two words in naming a species; for example, *Homo* (genus) *sapiens* (species).

Geographic isolation The separation between populations caused by geographic distance or geographic barriers.

Geographic speciation *See* **Allopatric speciation.**

Geological strata A series of layers of sedimentary rock.

Geological time scale The correlation between rocks (or the fossils contained in them) and time periods of the past.

Germinal choice *See* **Eutelegenesis.**

Germ plasm Cells or tissues in a multicellular organism that are exclusively devoted to transmitting hereditary information to offspring, either asexually or by means of gametes (sex cells). These "germ-line" cells are in contrast to the somatic cells that produce the nongerm-line body tissues.

Gloger's rule The generalization that warm-blooded (endothermic) animals tend to have more pigmentation in warm, humid areas than in cool, dry areas.

Glycolysis The energy-producing conversion of glucose to pyruvate under anaerobic conditions (fermentation). Subsequent steps may yield lactic acid or ethanol.

Gondwana The supercontinent in the Southern Hemisphere formed from the breakup of the larger Pangaea landmass about 180 million years ago. Gondwana was composed of what is now South America, Africa, Antarctica, Australia, and India.

Grade A level of phenotypic organization or adaptation reached by one or more species. Distantly related or unrelated species that reach the same grade are considered to have undergone parallel or convergent evolution.

Gradient Changes in the amount of a substance as it is displaced from its source.

Granite A coarse-grained igneous rock commonly intruded into continental crust.

Great Chain of Being The eighteenth-century concept that instead of a static universe, there is a continuous progression of stages leading to a superior supernatural being; the transformation of the "Ladder of Nature" into a succession of moving platforms.

Grooming Body surface cleaning by use of mouth, fingers, or claws.

Group selection Selection acting on the attributes of a group of related individuals in competition with other groups rather than only on the attributes of an individual in competition with other individuals. For example, altruism may not be beneficial to the individual altruist but can be quite beneficial to a group containing altruists. The fitness of an individual in such a group is thus, at least partially, associated with the properties of the group.

Gymnosperms A group of vascular plants with seeds unenclosed in an ovary (naked); mainly cone-bearing trees.

H

Habitat The place and conditions in which an organism normally lives. (*See also* **Environment.**)

Haeckel's gastrula hypothesis The concept that metazoans developed from swimming hollow-balled colonies of flagellated protozoans that evolved an anterior–posterior orientation in searching for food. The anterior cells, specialized for digestion, invaginated through a circular blastopore to form a digestive archenteron, and this bilayered cup, called a *gastrula* or *gastraea,* was, according to Haeckel, the progenitor of the gastrula developmental stage found in some present-day metazoans.

Half-life (radioactivity) The time required for the decay of one-half the original amount of a radioactive isotope: a period of one half-life reduces the isotope amount by one half so that a length of two half-lives leaves a remainder of one-quarter, three half-lives, a remainder of one-eighth, and so on. Each radioactive isotope has a distinctive half-life period, which remains constant over time.

Haplodiploidy A reproductive system found in some animals, such as bees and wasps, in which males develop from unfertilized eggs and are haploid, while females develop from fertilized eggs and are diploid.

Haploid Cells or organisms that have only one set (1n) of chromosomes, meaning the presence of only a single allele for each gene.

Haplotype A sequence of nucleotides, restriction sites, or marker genes inherited as a linked unit from one parent. Since more than a single genetic locus may be involved, a haplotype may be composed of a string of alleles.

Hardy–Weinberg principle The conservation of gene (allelic) and genotype frequencies in large populations under conditions of random mating and in the absence of evolutionary forces, such as selection, migration, and genetic drift, which act to change gene frequencies.

Hemizygous Genes, such as those on the X chromosome in a male mammal (hemizygote), which are unpaired in a diploid cell.

Herbivores Animals that feed mainly on plants.

Heritability In a general sense, the degree to which variations in the phenotype of a character are caused by genetic differences; traits with high heritabilities can be more easily modified by selection than traits with low heritabilities. (One measure of the heritability of a trait is the ratio of its genetic variance to its phenotypic variance—"broad sense heritability.") Obtaining a trustworthy heritability estimate demands considerable experimental control and is often valid for extremely limited conditions; that is, for specific genotypes in specific environments.

Hermaphrodite An individual possessing both male and female sexual reproductive systems. (*See also* **Monoecious.**)

Heterocercal Fish tail in which the vertebral axis is curved (usually upward).

Heterochromatin A region of the eukaryotic chromosome that stains differently from normal-staining "euchromatin" because of its tightly compacted structure. Compared to euchromatin, it is also characterized by possessing very few active genes and many more repetitive DNA sequences. It constitutes about 15 percent of the human genome and about 30 percent of the *Drosophila* genome, much of it located on either side of chromosome centromeres. In *Drosophila,* it also constitutes almost the entire Y chromosome.

Heterochrony A term Haeckel originally proposed to describe changes in timing of an organ's development during evolution. Such changes were used to explain departures from the "recapitulation" of phylogeny expected during ontogeny of descendant species. (*See* **Biogenetic law.**) Its present usage varies but still hinges on a phylogenetic change in developmental timing, whether of one organ relative to other organs, or of one organ relative to the same ancestral organ. (For organs whose growth rate changes relative to other organs, *see* **Allometry.**) Among the consequences of heterochrony are shifts in relative development of reproductive and nonreproductive tissue (*see also* **Paedomorphosis**). Such changes can cause an organism: (a) to appear more juvenile because its nonreproductive tissues develop more slowly (*neoteny*), or (b) reach sexual maturity earlier because its reproductive tissues develop more rapidly (*progenesis*). Some authors add other terms to describe degrees of developmental contraction or extension.

Heterodont An organism with structural and functional differences among its teeth.

Heterogametic The sex that produces two kinds of gametes for sex determination in offspring, one kind for males and the other for females. The heterogametic sex is the male in mammals and the female in birds. (*See also* **Sex chromosomes.**)

Heterosis (hybrid vigor) The increase in vigor and performance that can result when two different, often inbred strains are crossed. Since each inbred parental strain may be homozygous for different deleterious recessive alleles (for example, $a^1a^1 \times a^2a^2$), the cause for heterosis has been ascribed by some authors to the superiority of heterozygotes (for example, a^1a^2). (*See* **Heterozygote advantage.**)

Heterotroph An organism that cannot use inorganic materials to synthesize the organic compounds needed for growth but obtains them by feeding on other organisms or their products, such as a carnivore, herbivore, parasite, scavenger, or saprophyte.

Heterozygote A genotype or individual that possesses different alleles at a particular gene locus on homologous chromosomes (for example, *Aa* in a diploid).

Heterozygote advantage (superiority) The superior fitness of some heterozygotes (for example, A^1A^2) relative to homozygotes (for example, A^1A^1, A^2A^2). (*See also* **Heterosis, Overdominance.**)

Hierarchy A term used in some evolutionary and developmental studies to designate increasing levels of complexity or organization.

Histones A family of small acid-soluble (basic) proteins that are tightly bound to eukaryotic nuclear DNA molecules and help fold DNA into thick chromosome filaments.

Hitchhiking When a gene persists in a population, not because of selection, but because of close linkage to one or more selected genes. (*See also* **Linkage disequilibrium.**)

Homeostasis The term was classically defined by W. B. Cannon to denote the tendency of a (physiological) system to react to an external disturbance so that the system is not displaced from normal values. Probably its most common use applies to traits that measure or perform at constant values in the face of disturbing forces. An example is the persistence of a specific phenotype although confronted with genetic or environmental differences (for example, *canalization,* p. 358).

Homeotic mutations (homeoboxes and **homeodomains)** Homeosis was originally defined by William Bateson as "something [that] has been changed into the likeness of something else." In modern genetic usage, homeotic mutations cause the development of tissue in an inappropriate position; for example, the *bithorax* mutations in *Drosophila* that produce an extra set of wings (Fig. 10–28). In *Drosophila,* homeotic genes are clustered into two chromosomally separate groups, the antennapedia and bithorax complexes, in which each cluster contains several independently functioning homeotic genes. Each of these homeotic genes contains a nucleotide sequence (*homeobox*) coding for a DNA-binding polypeptide (*homeodomain*) involved in embryonic development along the animal's anterior–posterior axis. Homologous homeobox sequences are found throughout metazoan phyla, and the chromosomal organization of their homeobox-containing genes often follows the linkage order noted in *Drosophila.* The overall conservation of homeobox sequences, of the genes containing them, and of their linkage orders, indicate common developmental functions in different phyla preserved for many hundreds of millions of years, extending back to Precambrian times.

Hominid A member of the family Hominidae, which includes humans, whose earliest fossils can now be dated to about 4 million years ago (genus *Australopithecus*). Only a single hominid species (*Homo sapiens*) presently exists.

Hominoids A group (superfamily Hominoidea) that includes hominids (Hominidae), gibbons (Hylobatidae), and apes (Pongidae).

Homogametic The sex that produces only one kind of gamete for sex determination in offspring, thus causing sex differences among offspring to depend on the kind of gamete contributed by the heterogametic sex. The homogametic sex is the female in mammals and the male in birds. (*See also* **Sex chromosomes.**)

Homologous chromosomes Chromosomes that pair during meiosis, each pair usually possessing a similar sequence of genes.

Homology A common use of this term is to characterize the similarity of biological features in different species or groups *because of their descent from a common ancestor.* Homologous features may include those found in development, structure, and morphology, although similarity on the genetic level probably provides a more reliable estimate of common descent (pp. 243–244). Since such conservative features can sometimes be quantified, especially for amino acid sequences in protein or base sequences in nucleic acids, homology has also been defined as the *extent* to which two species share an ancestral character (that is, homology = *degree* of ancestral similarity), and the value obtained can then be used to help establish phylogenetic relationships among species. (*See also* **Coalescence.**) Thus, if species A and B are homologous for 70 percent of a particular ancestral protein whereas they share only 40 percent homology with species C, the assumption can be made that species A and B are more closely related to each other for this protein (have a more recent common ancestor) than to species C. We should keep in mind that homologous genes do not necessarily produce the same features, since an ancestral gene may be recruited for different functions in different lineages (p. 360). Also, as is obvious from convergence phenomena, functional, morphological, and developmental similarities can be produced by nonhomologous genetic elements of independent evolutionary origin. Emphasis on genetic homology can lead to a different, yet more realistic, phylogeny than the use of other features. Among other definitions are those that consider homology strictly qualitatively—for example, two structures in different species are or are not homologous (derived from a common ancestor)—and omit any quantitative comparative considerations as to the degree of homology.

Homoplasy Character similarity that arose independently in different groups whether through parallelism or convergence.

Homozygote A genotype or individual that possesses the same alleles at a particular gene locus on homologous chromosomes (for example, *AA* or *aa* in a diploid).

Horotelic Evolving at a comparatively average rate.

Hubble constant A ratio [(speed of galactic recession)/(distance from earth)] that indicates the rate at which the universe is expanding. Although many astronomers agree on the speed of galactic recession as determined by the red shift (*see* **Doppler effect**), they still debate galactic distance from earth, which is based on the brightnesses of celestial bodies. Estimates of the Hubble constant have ranged from about 50, signifying an age for the universe of about 15 billion years, to as much as 100, indicating a more rapid expansion, and therefore a younger age of about 7 or 8 billion years. Some newer distance measurements, using the brightnesses of supernovae, indicate a Hubble constant of 55 to 65, or a universe about 12 billion years old—a value more in accord with age estimates of the oldest stars.

Hybrid breakdown, inviability, sterility Hybrids that suffer from loss of fitness and reproductive failure.

Hybrids (hybridization) Offspring of a cross between genetically different parents or groups.

Hybrid vigor *See* **Heterosis.**

Hydrogen bond A weak, noncovalent bond between a hydrogen atom and an electronegative atom such as oxygen.

Hydrogen ion A proton (H^+) that in aqueous solution exists only in hydrated form (H_3O^+, hydronium ion).

Hydrolysis Splitting of a molecule by the addition of the three atoms from a water molecule (H_2O).

Hydrophilic A compound (for example, charged molecule) or part of a compound (for example, polar group) that has an affinity for water molecules.

Hydrophobic Compounds such as lipids that do not readily interact with water but tend to dissolve in organic solvents.

Hydrostatic pressure The pressure exerted by a liquid. When the liquid is in an elastic, muscularly controlled container (for example, the coelom of a worm), changes in shape of the container can be effected by muscularly generated hydrostatic pressure.

Hypoxia The reduction of oxygen supply to tissues.

I

Idealism The philosophy that the universe is constituted of nonmaterial ideas.

Igneous rock A rock such as basalt (fine-grained) and granite (coarse-grained), formed by the cooling of molten material from the earth's interior.

Implantation (mammals) The attachment of the embryo to the uterine wall.

Inbreeding Mating between genetically related individuals, often resulting in increased homozygosity in their offspring.

Inbreeding coefficient (F) The probability that the two alleles of a gene in a diploid organism are identical because they originated from a single allele in a common ancestor.

Inbreeding depression Decrease in the average value of a character, or in growth, vigor, fertility, and survival, as a result of inbreeding.

Inclusive fitness The fitness of an allele or genotype measured not only by its effect on an individual but also by its effect on related individuals that also possess it (kin selection).

Independent assortment A basic principle of mendelian genetics—that a gamete will contain a random assortment of alleles from different chromosomes because chromosome pairs orient randomly toward opposite poles during meiosis.

Industrial melanism The effect of soot and pollution in industrial areas in increasing the frequency of darkly pigmented (melanic) forms perhaps because of selection by predators against nonpigmented or lightly pigmented forms.

Inheritance of acquired characters The concept used by Lamarck to explain evolutionary adaptations—that phenotypic characters acquired by interaction with the environment during the lifetime of an individual are transmitted to its offspring.

Insectivore An animal that feeds primarily on insects.

Instinct An inherited (innate), relatively inflexible behavior pattern that is often activated by one or several environmental factors (releasers).

Intrinsic rate of natural increase The potential rate at which a population can increase in an environment free of limiting factors.

Introgressive hybridization The incorporation of genes from one species into the gene pool of another because some fertile hybrids are produced from crosses between the two species.

Intron A nucleotide sequence in a split gene that intervenes between two exons. The intron sequences are removed from messenger RNA, and only the exon sequences translate into polypeptides.

Intrusion Igneous rock that is inserted within or between geological strata rather than on the earth's surface.

Inversion An aberration in which a section of DNA or chromosome has been inverted 180 degrees, so that the sequence of nucleotides or genes within the inversion is now reversed with respect to its original order in the DNA or chromosome.

Ion An atom or molecule carrying a positive or negative electrostatic charge.

Isolating mechanisms Biological mechanisms that act as barriers to gene exchange between populations. These are generally divided into two groups: premating isolating mechanisms that inhibit cross-fertilization (for example, behavioral differences in courtship) and postmating isolating mechanisms that interfere with the success of the gamete or zygote even when cross-fertilization has occurred (for example, hybrid inviability or sterility).

Isomerase An enzyme that catalyzes the rearrangement of atoms within a molecule.

Isotope One of several forms of an element, with a distinctive mass based on the number of neutrons in the atomic nucleus. (The number of protons and electrons is the same in different isotopes of an element.) Radioactive isotopes decay at a rate that is constant for each isotope and release ionizing radiation as they decay. (*See* **Half-life**, **Radioactive dating**).

K

Karyotype The characteristic chromosome complement of a cell, individual, or species.

Kelvin scale (°K) A scale of temperature in which absolute zero (the point at which molecules oscillate at their lowest possible frequency, −273°C) is designated as 0°K, and the boiling point of water as 373°K.

Kilocalories (kcal) Units of 1,000 calories. (*See* **Calories**.)

Kingdom The highest inclusive category of taxonomic classification. Each kingdom includes phyla or subkingdoms. The most common presently used classification system proposes five kingdoms: Monera (prokaryotes), Protista, Fungi, Animalia, and Plantae, although some authors emphasize a tripartite division of organisms—Archaea, Bacteria, Eucarya (*see* **Domain**).

Kin selection Selection effects (for example, altruism) that influence the survival and reproductive success of genetically related individuals (kin). This contrasts with selection confined solely to an individual and its own offspring. (*See also* **Inclusive fitness**.)

Knuckle-walking Quadrupedal gait of chimpanzees and gorillas, performed by curling the fingers toward the palm of the hand and using the backs (dorsal surfaces) of the knuckles to support the weight of the front part of the body.

Krebs cycle The cyclic series of reactions in the mitochondrion in which pyruvate is degraded to carbon dioxide and hydrogen protons and electrons. The latter are then passed into the oxidative phosphorylation pathway to generate ATP.

***K*-selection** Selection based on a population being maintained at or near the limit of its carrying capacity; selection is theoretically for improved competitive ability rather than for rapid numerical increase.

***K* value (carrying capacity)** *See* **Carrying capacity**.

L

Lactation Formation and secretion of milk in maternal mammary glands for nursing offspring, a distinctive characteristic of mammals.

Ladder of Nature A concept based on Aristotle's view (the Scale of Nature) that nature can be represented as a succession of stages or ranks that leads from inanimate matter through plants, lower animals, higher animals, and finally to the level of humans. (*See also* **Great Chain of Being**.)

Lamarckian inheritance The concept that the phenotype of an organism is itself hereditary: that characters acquired or lost during the life experience of an organism, as well as characters that organisms attempt to acquire in order to meet environmental needs, can be transmitted to offspring. Lamarck proposed that it is through such means that changes in organisms (evolution) takes place. (*See also* **Inheritance of acquired characters**, **Use and disuse**.)

Language A structured system of communication among individuals using vocal, visual, or tactile signs to describe thoughts, feelings, concepts, and observations. Rather than communication, some writers emphasize the representational nature of language, defining it as a symbolic system used to store and retrieve information about experiences and concepts.

Larva A sexually immature stage in various animal groups, often with a form and diet distinct from those of the adult.

Laurasia The supercontinent in the Northern Hemisphere (comprising what is now North America, Greenland, Europe, and parts of Asia) formed from the breakup of Pangaea about 180 million years ago.

Learning Acquisition of a behavior through experience.

Lethal allele An allele whose effect prevents its carrier from reaching sexual maturity when present in homozygous condition for a recessive lethal or in either heterozygous or homozygous condition for a dominant lethal.

Life The capability of performing various organismic functions such as metabolism, growth, and reproduction of genetic material.

Life cycle The series of stages that takes place between the formation of zygotes in one generation of a species and the formation of zygotes in the next generation. (Also *life history:* the series of stages experienced by an individual of a species, from birth to death.)

Light-year The distance traveled by light, moving at 186,000 miles a second, in a solar year; approximately 6×10^{12} miles or 9.5×10^{12} kilometers.

Lineage An evolutionary sequence, arranged in linear order from an ancestral group or species to a descendant group or species (or *vice versa*).

Lingual Side of a tooth closest to the tongue.

Linkage (gene) The occurrence of two or more gene loci on the same chromosome.

Linkage disequilibrium The absence of linkage equilibrium (that is, the presence of nonrandom associations between alleles at different loci). (*See also* **Hitchhiking.**)

Linkage equilibrium The attainment of genotypic frequencies in a population that indicates that recombination between two or more gene loci has reached the point at which their alleles are now found in random genotypic combinations. For example, when an allele at one locus (for example, A_1) and an allele at another locus (for example, B_1) are found in combination at a frequency (fA_1B_1) equal to the product of their individual frequencies ($fA_1 \times fB_1$).

Linkage map The linear sequence of known genes on a chromosome obtained from recombination data.

Lipids Organic compounds such as fats, waxes, and steroids that tend to be more soluble in organic solvents of low polarity (for example, ether, chloroform) than in more polar solvents (for example, water).

Living fossil An existing species whose similarity to ancient ancestral species indicates that very few morphological changes have occurred over a long period of geological time.

Locus (plural **loci**) Strictly defined, it is the site (nucleotide sequence) on a chromosome occupied by a specific gene. Some researchers use it more broadly as a synonym of *gene*.

Logistic growth curve Population growth that follows a sigmoid (S-shaped) curve in which numbers increase slowly at first, then rapidly, and finally level off as the population reaches its maximum size or carrying capacity for a particular environment.

Longevity The average life span of individuals in a population.

M

Macroevolution Evolution of taxa higher than the species level (for example, genera, families, orders, classes), commonly entailing major morphological changes. This concept is often associated with the school of thought proposing that evolutionary events different from those responsible for changes in populations or the origin of species have caused the origin of higher taxa. (*See* **Punctuated equilibrium.**)

Macromolecules Very large polymeric molecules such as proteins, nucleic acids, and polysaccharides.

Macromutation A concept that attempts to explain the origin of a new species or an even higher taxonomic category by a *single* large mutation rather than by selection acting on many mutations. Although most geneticists agree that mutations can produce major as well as minor developmental changes, no single mutation is yet known that can cause an instantaneous speciation event, probably because such a sudden large radical change would dislocate normal genetic and developmental processes.

Malthusian parameter *See* **Intrinsic rate of natural increase.**

Mammary glands One or more pairs of ventrally placed glands used by mammalian females for nursing offspring. (*See also* **Lactation**).

Marsupials Mammals of the infraclass Metatheria possessing, among other characters, a reproductive process in which tiny live young are born, and then nursed in a female pouch (marsupium).

Meiosis The eukaryotic cell division process used in producing haploid gametes (animals) or spores (plants) from a diploid cell. Meiosis is characterized by a reduction division that ensures that each gamete or spore contains one representative of each pair of homologous chromosomes in the parental cell.

Meiotic drive *See* **Segregation distortion.**

Mendel's laws *See* **Independent assortment, Segregation.**

Mesoderm The embryonic tissue layer between ectoderm and endoderm in triploblastic animals that gives rise to muscle tissue, kidneys, blood, internal cavity linings, and so on.

Mesozoic era The middle era of the Phanerozoic eon, covering the approximately 180-million-year interval between the Paleozoic (ending about 245 million years ago) and Cenozoic (beginning about 65 million years ago). It is marked by the origin of mammals in the earliest period of the era (Triassic), the dominance of dinosaurs throughout the last two periods of the era (Jurassic and Cretaceous), and the origin of angiosperms.

Messenger RNA (mRNA) An RNA molecule produced by transcription from a DNA template, bearing a sequence of triplet codons used to specify the sequence of amino acids in a polypeptide.

Metabolic pathway A sequence of enzyme-catalyzed reactions that convert a precursor substance to one or more end products.

Metabolism A network of enzyme-catalyzed reactions used by living organisms to maintain themselves.

Metacentric A chromosome whose centromere is at or near the center.

Metamerism Division of the body, or a major portion of the body, into a series of similar segments along the anterior–posterior axis. (*See also* **Segmentation**.)

Metamorphic rock Rock that has been subjected to high but nonmelting temperatures and pressures, causing chemical and physical changes.

Metamorphosis The transition from one form into another during development (for example, a larva into a different adult form).

Metatheria *See* **Marsupials**.

Metazoa Multicellular animals.

Microevolution Evolutionary changes of the kinds usually responsible for causing differences between populations of a species (for example, gene frequency changes and chromosomal variations). Many evolutionists suggest that accumulations of such changes over time are sufficient to explain the origin of most or all taxa.

Microsatellites Tandem repeats of short di-, tri-, and tetranucleotide sequences such as cytosine–adenine–cytosine–adenine–cytosine–adenine, and so on. Such loci are abundant (humans are estimated to possess at least 35,000 loci for repeats of the C–A sequence) and mutate (change in sequence number) at a relatively high rate.

Microspheres Microscopic membrane-bound spheres formed when proteinoids are boiled in water and allowed to cool. Some cell-like properties, such as osmosis, growth in size, and selective absorption of chemicals, have been ascribed to them.

Migration The transfer of genes from one population into another by interbreeding (gene flow). (Also used to indicate movement of a population to a different geographical area or its periodic passage from one region to another.)

Mimicry Resemblance of individuals in one species (mimics) to individuals in another (models) because of selection. (*See also* **Batesian mimicry**, **Müllerian mimicry**.)

Mitochondrion An organelle in eukaryotic cells that uses an oxygen-requiring electron transport system to transfer chemical energy derived from the breakdown of food molecules to ATP. Mitochondria have their own genetic material (circular DNA without histones) and generate some mitochondrial proteins by using their own protein-synthesizing apparatus. (Most of the mitochondrial proteins are coded by nuclear DNA and produced on cytoplasmic ribosomes.)

Mitosis The mode of eukaryotic cell division that produces two daughter cells possessing the same chromosome complement as the parent cell.

Modern synthesis (evolution theory) *See* **Neo-Darwinism**.

Modifier (gene) A gene whose effect alters the phenotypic expression of one or more genes at loci other than its own.

Molecular clock *See* **Evolutionary clock**.

Monocotyledons Flowering plants (angiosperms) in which the embryo bears one seed leaf (cotyledon).

Monoecious An individual bearing both male and female organs. (*See also* **Hermaphrodite**.)

Monomers The subunits linked together to form a polymer (for example, nucleotides in nucleic acids, amino acids in proteins, sugars in polysaccharides).

Monomorphic A population or species that shows no genetic or phenotypic variation for a particular gene or character.

Monophyletic Derivation of a taxonomic group from a single ancestral lineage.

Monotremes Egg-laying mammals, presently restricted to Australasia; the platypus (*Ornithorhyncus*) and echidna spiny anteater (*Tachyglossus, Zaglossus*).

Morphology Study of the anatomical form and structure of organisms.

Müllerian mimicry Sharing of a common warning coloration or pattern among a number of species that are all dangerous or toxic to predators; resemblances maintained because of common selective advantage.

Multigene family *See* **Gene family**.

Mutation A change in the nucleotide sequence of genetic material whether by substitution, duplication, insertion, deletion, or inversion.

Mutational load That portion of the genetic load caused by production of deleterious genes through recurrent mutation.

Mutualism A relationship among different species in which the participants benefit.

N

Natural selection Differential reproduction or survival of replicating organisms caused by agencies that are not directed by humans (*see* **Artificial selection**). Since such differential selective effects are widely prevalent, and often act on hereditary (genetic) variations, natural selection is a common major cause for a change in the gene frequencies of a population that leads to a new distinctive genetic constitution (evolution). (*See also* **Adaptation, Fitness, Selection.**)

Negative eugenics Proposals to eliminate deleterious genes from the human gene pool by identifying their carriers and restraining or discouraging their reproduction.

Neo-Darwinism The theory (also called the **Modern synthesis**) that regards evolution as a change in the frequencies of genes introduced by mutation, with natural selection considered as the most important, although not the only, cause for such changes.

Neoteny The retention of juvenile morphological traits in the sexually mature adult. (*See also* **Heterochrony, Paedomorphosis.**)

Neutral mutation A mutation that does not affect the fitness of an organism in a particular environment.

Neutral theory of molecular evolution The concept that most mutations that contribute to genetic variability (genetic polymorphism on the molecular level) consist of alleles that are neutral in respect to the fitness of the organism and that their frequencies can be explained in terms of mutation rate and random genetic drift.

Niche *See* **Ecological niche.**

Nocturnal A lifestyle characterized by nighttime activity.

Non-Darwinian evolution *See* **Neutral theory of molecular evolution.**

Nondisjunction The failure of homologous chromosomes (or sister chromatids) to separate ("disjoin") from each other during one of the two meiotic anaphase stages and go to opposite poles. Because of nondisjunction, one daughter cell will receive both homologues (or sister chromatids) and the other daughter cell will receive none, leading to an increase or decrease, respectively, in chromosome number.

Nonrandom mating *See* **Assortative mating.**

Nonsense mutation A mutation that produces a codon that terminates the translation of a polypeptide prematurely. Such codons were previously called "nonsense" codons but are now generally called *stop codons* or *chain termination* codons.

Nucleic acid An organic acid polymer, such as DNA or RNA, composed of a sequence of nucleotides.

Nucleotide A molecular unit consisting of a purine or pyrimidine base, a ribose (RNA) or deoxyribose (DNA) sugar, and one or more phosphate groups.

Nucleus A membrane-enclosed eukaryotic organelle that contains all the histone-bound DNA in the cell (that is, practically all the cellular genetic material).

Numerical (phenetic) taxonomy A statistical method for classifying organisms by comparing them on the basis of measurable phenotypic characters and giving each character equal weight. The degree of overall similarity between individuals or groups is then calculated, and a decision is made as to their classification.

O

Olfactory Referring to the sense of smell.

Omnivores Animals that feed on both plants and animals.

Ontogeny The development of an individual from zygote to maturity.

Operon A cluster of coordinately regulated structural genes. In prokaryotes and in a few eukaryotes, this cluster is transcribed as a unit into a single long ("polycistronic") messenger RNA molecule which is then translated into a sequence of individual gene products that often function together.

Order A taxonomic category between class and family: a class may contain a number of orders, each of which contains a number of families.

Organelles Functional intracellular membrane-enclosed bodies, such as nuclei, mitochondria, and chloroplasts.

Organic Carbon-containing compounds. Also refers to features or products characteristic of biological organisms.

Organism A living entity. (*See* **Life.**)

Orthogenesis The concept that evolution of a group of related species proceeds in a particular direction (for example, an increase in size) because of unknown internal or vitalistic causes rather than because of nonmystical factors such as selection.

Orthologous genes Gene loci in different species that are sufficiently similar in their nucleotide sequences (or amino acid sequences of their protein products) to suggest they originated from a common ancestral gene.

Overdominance Instances when the phenotypic expression of a heterozygote (for example, A_1A_2) is more extreme than that of either homozygote (for example, A_1A_1 or A_2A_2). Overdominance has been considered a cause for hybrid vigor. (*See* **Heterosis**).

Oviparous Females that lay eggs that develop outside the body.

Oxidation–reduction Reactions in which electrons are transferred from one atom or molecule (the reducing agent that is oxidized by the loss of electrons) to another (the oxidizing agent that is reduced by the gain of electrons). For example, we can diagram an atom or molecule as A or B, each carrying a proton (p^+) and an electron (e^-), as follows:

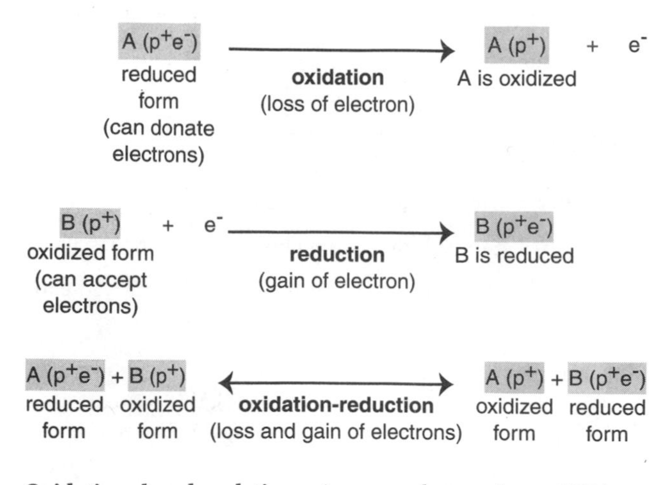

Oxidative phosphorylation A process that produces ATP by transferring electrons to oxygen.

P

Paedomorphosis The incorporation of adult sexual features into immature developmental stages. Causes for such effects are changes in developmental speed of sexual tissues relative to nonsexual tissues (*see* **Heterochrony**). In progenesis, sexual development is so rapid that the sexually mature form may remain quite small in size (p. 397 footnote 4). In *neoteny*, sexual maturation proceeds normally but somatic development slows down, producing a full-sized sexual adult with juvenile appearance. A classic neotenous form is the Mexican axolotl, an amphibian salamander that can retain its gills in tadpole form even when sexually mature.

Paleomagnetism The magnetic fields of ferrous (iron-containing) materials in ancient rocks. Among other applications, paleomagnetism provides information on the position of landmasses and continents relative to the earth's magnetic poles at the time that the rocks were formed and thus can be used to describe the historical movement of continents relative to each other (continental drift).

Paleontology The study of extinct fossil organisms.

Paleozoic The first era of the Phanerozoic eon, extending from 545 to about 250 million years ago.

Pangaea A very large supercontinent formed about 250 million years ago comprising most or all of the present continental landmasses. (*See* **Gondwana, Laurasia.**)

Pangenesis The concept of heredity, held by Darwin and others, that small, particulate "gemmules," or "pangenes," are produced by each of the various tissues of an organism and sent to the gonads where they are incorporated into gametes. The increase or decrease of specific gemmules during the use or disuse of organs was proposed in order to explain the Lamarckian concept of inheritance of acquired characters.

Panmixis (panmictic) *See* **Random mating.**

Panspermia The concept that life was introduced on earth from elsewhere in the universe.

Parallel evolution The evolution of similar characters in related lineages whose common ancestor was phenotypically different. (*See also* **Convergence.**)

Paralogous genes Two or more different gene loci in the same organism that are sufficiently similar in their nucleotide sequences (or in the amino acid sequences of their protein products) to indicate they originated from one or more duplications of a common ancestral gene. (*See also* **Gene family.**)

Parapatric Geographically adjacent species or populations whose distributions do not overlap but are in contact at one or more of their mutual boundaries.

Paraphyletic A taxonomic grouping which includes some descendants of a single common ancestor, but not all.

Parasitism An association between species in which individuals of one species (the parasite) obtain their nutrients by living on or in the tissues of another species (the host), often with harmful effects to the host.

Parental investment Parental provision of resources to offspring that increase the offspring's reproductive success at a cost of further reproductive success of the parents.

Parsimony method Choice of a phylogenetic tree that minimizes the number of evolutionary changes necessary to explain species divergence.

Parthenogenesis Development of an individual from an egg that has not been fertilized by a male gamete. Diploid eggs may arise during the meiotic process—called by some, *automictic parthenogenesis*—when the number of chromosomes in the maternal oocyte doubles, or when two of the four haploid meiotic products unite. By combining identical homologous chromosomes in a single egg nucleus, automixis can produce homozygotes for all chromosomal loci. Parthenogenesis may also occur in the absence of meiosis, when eggs are produced mitotically—called by some, *apomictic parthenogenesis.* In apomixis, a parent—whether homozygous or heterozygous—transmits its own genotype to its offspring.

Partial (incomplete) dominance Instances where two different alleles of a gene in a heterozygote (for example, A_1A_2) produce a phenotypic effect intermediate between the effects produced by the two homozygotes (for example, A_1A_1, A_2A_2).

Pelagic Refers to an entire body of water and the organisms within it, excluding the bottom (benthic) zone.

Peptide (polypeptide) An organic molecule composed of a sequence of amino acids covalently linked by peptide bonds (a bond formed between the amino group of one amino acid and the carboxyl group of another through the elimination of a water molecule).

Period (geological) A major subdivision of an era of geological time distinguished by a particular *system* of rocks and associated fossils. The Cretaceous period is named after the abundance of chalk (Latin, *creta*), and the Carboniferous period for the abundance of coal (Latin, *carbo*), in their component rock systems.

Peripatric Populations that bud off from the geographic periphery of a parental population and become genetically isolated.

Phagocytic Cellular engulfment of external material. (*See also* **Endocytosis.**)

Phanerozoic eon A major division of the geological time scale marked by the relatively abundant appearance of fossilized skeletons of multicellular organisms, dating from about 545 million years ago to the present.

Phenetic Referring to phenotypic characters that can be described or measured. Also, a system of classification that groups taxa by their degree of similarity for measured or numerically evaluated characters. (*See also* **Numerical taxonomy**).

Phenotype The characters that constitute the structural and functional properties of an organism. Phenotypic features result from interaction between the genotype, which provides developmental information, and the environment, which provides developmental facilities.

Phosphorylation The addition of one or more phosphate groups (HPO_4^-) to a compound (for example, the phosphorylation of ADP to ATP).

Photoautotroph *See* **Autotroph.**

Photosynthesis The synthesis of organic compounds from carbon dioxide and water through a process that begins with the capture of light energy by chlorophyll.

pH scale [The negative logarithm of the hydrogen ion (H^+) concentration in an aqueous solution.] A scale used for measuring acidity (pH less than 7) and alkalinity (pH greater than 7), given that pure water has a neutral pH of 7.

Phyletic evolution Evolutionary changes within a single nonbranching lineage. Although new species are produced by this lineage over time (*chronospecies*) there is no increase in the number of species existing at any one time. (*See also* **Anagenesis.**)

Phylogenetic evolution Evolutionary changes that produce two or more lineages that diverge from a single ancestral lineage. (Also called "branching evolution," *see* **Cladogenesis.**)

Phylogeny The evolutionary history of a species or group of species in terms of their derivations and connections. A phylogenetic tree is a schematic diagram designed to represent that evolution—ideally, a portrait of genetic relationships.

Phylotype (**phylotypic**) A proposed stage in embryonic development that characterizes some basic features in the body plan of a phylum (for example, see Fig. 3–10). Further evolved taxa in a lineage can then use this stage as a foundation for developing their own unique derived features.

Phylum (plural, **phyla**) The major taxonomic category below the level of kingdom, used to include classes of organisms that may be phenotypically quite different but share some general features or body plan.

Placenta A mammalian organ formed by union between the female uterine lining and embryonic membranes that provides nutrition to the embryo, allows exchange of gases, and aids elimination of embryonic waste products.

Placentals Mammals of the infraclass Eutheria, possessing, among other features, a reproductive process that uses a placenta to nourish their young until a relatively advanced stage of development compared to other mammalian groups (monotremes and marsupials).

Planula hypothesis The concept that metazoans evolved from small primitive organisms that consisted of solid balls of cells (planulae) similar to embryonic stages of sponges and Cnidaria.

Plasma membrane The boundary membrane consisting of phospholipids and proteins surrounding the cytoplasm of a cell.

Plasmid A self-replicating cellular DNA element that can exist outside the host chromosome. There are various kinds, some maintaining more than one copy per cell.

Plate tectonics The concept that the earth's crust is divided into a number of fairly rigid plates, whose movements (tectonics) relative to each other are responsible for continental drift and many crustal features. (*See also* **Tectonic plates.**)

Pleiotropy Instances when a single gene produces phenotypic effects on more than one character.

Plesiomorphy Instances when a species character is similar to that character in an ancestral species.

Polygene A gene that interacts with other polygenes to produce a quantitative phenotypic effect on a character.

Polymer A molecule composed of many repeating subunits (monomers) linked together by covalent bonds.

Polymerase An enzyme that catalyzes the synthesis of a polymer by linking together its component monomers.

Polymerase chain reaction (**PCR**) A laboratory technique that can replicate a sequence of DNA nucleotides into millions of copies in a very short time.

Polymorphism The presence of two or more genetic or phenotypic variants in a population. Usually refers to genetic variations where the frequency of the rarest type is not maintained by mutation alone. (*See also* **Balanced polymorphism.**)

Polypeptide *See* **Peptide.**

Polyphyletic The presumed derivation of a single taxonomic group from two or more different ancestral lineages through convergent or parallel evolution.

Polyploidy Variations in which the number of chromosome sets (n) is greater than the diploid number (2n). For example, triploidy (3n) and tetraploidy (4n).

Population A geographically localized group of individuals in a species that, in sexual forms, share a common gene pool. (*See also* **Deme.**)

Preadaptation A character that was adaptive under a prior set of conditions and later provides the initial stage (is "co-opted") for the evolution of a new adaptation under a different set of conditions.

Precambrian eon A major division of the geological time scale that includes all eras from the origin of the earth about 4.5 billion years ago to the beginning of the Phanerozoic eon, about 545 million years ago. The Precambrian (also known as the Cryptozoic) is marked biologically by the appearance of prokaryotes about 3.5 billion years ago and small, nonskeletonized multicellular organisms in the Ediacarian period about 50 or 60 million years before the Phanerozoic.

Predation The killing and consumption of one living organism (prey) by another (predator).

Preformationism The concept that an organism is preformed at conception in the form of a miniature adult and development consists of enlargement of the already preformed structures.

Progenote A hypothetical ancestral cellular form that gave rise to prokaryotes (archaebacteria, eubacteria) and eukaryotes.

Prokaryotes Organisms such as bacteria and cyanobacteria that lack histone-bound DNA, endoplasmic reticulum, a membrane-enclosed nucleus, and other cellular organelles found in eukaryotes.

Promoter A DNA nucleotide sequence that enables transcription (RNA synthesis) by binding the enzyme RNA polymerase. (*See also* **Enhancer.**)

Protein A macromolecule composed of one or more polypeptide chains of amino acids, coiled and folded into specific shapes based on its amino acid sequences.

Proteinoids Synthetic polymers produced by heating a mixture of amino acids. Some show proteinlike properties in respect to enzyme activity, color test reactions, hormonal activity, and so on.

Protista One of the four eukaryotic kingdoms; includes protozoa, algae, slime molds, and some other groups. (Called "Protoctista" by some authors.)

Protoplasm Cellular material within the plasma membrane.

Protostomes Coelomate phyla in which the embryonic blastopore becomes the mouth.

Prototheria *See* **Monotremes.**

Pseudocoelomates Organisms that have a coelom derived from a persistent embryonic blastocoel, largely unlined with mesodermal tissue.

Punctuated equilibrium The view that evolution of a lineage follows a pattern of long intervals in which there is relatively little change (stasis, or equilibrium), punctuated by short bursts of speciation and macroevolutionary events during which new taxa arise.

Purine A nitrogenous base composed of two joined ring structures, one five-membered and one six-membered, commonly present in nucleotides as adenine (A) or guanine (G).

Pyrimidine A nitrogenous base composed of a single six-membered ring, commonly present in nucleotides as thymine (T), cytosine (C), or uracil (U).

Q

Quadrupedal Tetrapod locomotion using all four limbs.

Quantitative character A character whose phenotype can be numerically measured or evaluated; a character displaying continuous variation.

Quantum evolution A rapid increase in the rate of evolution over a relatively short period of time.

R

Race A population or group of populations in a species that share a geographically and/or ecologically identifiable origin and have unique gene frequencies and phenotypic characters that distinguish them from other races. Because of the large amount of genetic and phenotypic variability in most species, the number of racial distinctions that can be made is often arbitrary. (*See also* **Subspecies.**)

Racemic mixture A mixture of two kinds of molecules whose structures are similar but differ in that they are mirror images of each other (one kind cannot be superimposed on the other). Each of the two molecular forms rotates the plane of polarized light in a particular direction, but the racemic mixture is optically inactive.

Radiation (phylogenetic) *See* **Adaptive radiation.**

Radioactive dating The dating of rocks by measuring the proportions of a radioactive element in an igneous intrusion and the isotopes produced by its radioactive decay. Since the rate at which a particular radioactive element decays is constant, these proportions provide an estimate of the age of the rock, which can often be confirmed by dating with other radioactive elements.

Radioactivity Emission of radiation by certain elements as their atomic nuclei undergo changes.

Random genetic drift The random change in frequency of alleles in a population. These can be caused by sampling errors that are of greater magnitude in small populations or by bottlenecks (founder effects), when population size is suddenly reduced to a few individuals. As population size increases, random drift becomes less important and selection more important in causing gene frequency changes.

Random mating Mating within a population regardless of the phenotype or genotype of the sexual partner (panmixis).

Range The geographical limits of the region habitually traversed by an individual or occupied by a population or species.

Recessive allele An allele (for example, *a*), which has no obvious phenotypic effect in a heterozygote (for example, *Aa*), producing its phenotypic effect only when homozygous (for example, *aa*).

Recessive lethal An allele whose presence in homozygous condition causes lethality. (*See also* **Lethal allele.**)

Reciprocal altruism A mutually beneficial exchange of altruistic behavioral acts between individuals. (*See also* **Altruism.**)

Recombinant DNA A DNA molecule composed of nucleotide sequences from different sources.

Recombination A chromosomal exchange process (*see* **Crossovers**) that produces offspring that have gene combinations different from those of their parents. (Also used by some authors to describe the results of independent assortment.)

Red Queen hypothesis The view that adaptive evolution in one species of a community causes a deterioration of the environment of other species. As a consequence, each species must evolve as fast as it can in order "to stay in the same place" (to survive).

Reductionism The concept that explanations for events at one level of complexity can or should be reduced to explanations at a more basic level. For example, that all biological events should be explained in the form of chemical reactions.

Regulator gene A gene that controls the rate at which other genes, adjacent or distant, will synthesize their products.

Repetitive DNA DNA nucleotide sequences that are repeated many times in the genome.

Repressor protein A regulator gene product that binds to a particular nucleotide sequence and prevents transcription.

Reproductive isolation The absence of gene exchange between populations. (*See* **Isolating mechanisms.**)

Reproductive success The proportion of reproductively fertile offspring produced by a genotype relative to other genotypes. (*See also* **Fitness.**)

Restriction enzymes Enzymes that recognize particular nucleotide sequences and cut DNA molecules at or near those sequences. (*See also* **Endonucleases.**)

Restriction fragment length polymorphisms (RFLPs) Differences between individuals in the size of DNA fragments for a particular DNA section cut by restriction enzymes. These are inherited in mendelian fashion, and furnish a basis for estimating genetic variation. They also provide linkage markers used to track mutant genes between generations.

Ribosomal RNA (rRNA) RNA sequences that are incorporated into the structure of ribosomes.

Ribosomes Intracellular particles composed of ribosomal RNA and proteins that furnish the site at which messenger RNA molecules are translated into polypeptides.

Ribozymes Sequences of RNA nucleotides that can perform catalytic roles.

RNA (ribonucleic acid) A typically single-strand nucleic acid, characterized by the presence of a ribose sugar in each nucleotide, whose sequences serve either as messenger RNA, ribosomal RNA, or transfer RNA in cells, or as genetic material in some viruses. In contrast to the base composition of DNA, RNA usually bears uracil instead of thymine.

RNA editing Information changes in RNA molecules by the addition, deletion, or transformation of ribonucleotide bases after these molecules have been transcribed from their DNA templates.

RNA splicing The joining of exons by the excision of introns.

RNA world The concept that RNA nucleotide sequences possessing catalytic and self-replicating capabilities predated catalytic protein systems in prebiological times.

***r*-selection** Selection in populations subject to rapidly changing environments with highly fluctuating food resources. Theoretically, selection in such populations emphasizes adaptations for rapid population growth rather than for the competitive ability experienced in *K*-selected populations.

S

Saltation The concept that new species or higher taxa originate abruptly because of macromutations, or because of sudden unknown causes.

Sampling error (gene frequencies) Variability in gene frequencies caused by the fact that not all samples taken from a population have exactly the same gene frequency as the population itself.

Saprophyte An organism that feeds on decomposing organic material.

Scavenger An organism that habitually feeds on animals who died naturally or accidentally or were killed by another carnivore.

Sea floor spreading Expansion of oceanic crust through the deposition of mantle material along oceanic ridges. (*See also* **Continental drift**, **Tectonic plates**.)

Sedimentary rock Rock formed by the hardening of accumulated particles (sediments) that had been transported by agents such as wind and water. The prime source of fossils. (*See also* **Geological strata**.)

Segmentation The repetition of body structures along an animal's anterior–posterior axis, as found generally in annelids, arthropods, and chordates. (*See also* **Metamerism**.)

Segregation The mendelian principle that the two different alleles of a gene pair in a heterozygote segregate from each other during meiosis to produce two kinds of gametes in equal ratios, each bearing a different allele.

Segregational genetic load *See* **Balanced genetic load**.

Segregation distortion (meiotic drive) Aberrant segregation ratios among the gametes produced by heterozygotes because of the presence of certain alleles (segregation distorters).

Selection A composite of all the forces that cause differential survival and differential reproduction among genetic variants. When the selective agencies are primarily those of human choice, the process is called *artificial selection;* when the selective agencies are not those of human choice, it is called *natural selection.* (*See also* **Adaptive value**, **Fitness**.) Although evolutionary biologists recognize other factors that contribute to genetic change, and therefore to evolution (for example, **Mutation, Random genetic drift**), selection remains the most commonly accepted cause to account for organismic adaptive features. However, selection does not have the foresight nor can development supply the means (p. 452) to enable a single population to face every eventuality. That is, although selection is a cause for evolutionary change, the amount and direction of change is limited by an organism's past history (p. 357). The regularity of extinction, embracing many lineages and practically all fossil species, indicates such limitations. That evolution proceeds continuously in the face of successive environmental contingencies is because selection is exercised in *different* populations, some of which possess adaptations that replace other populations which lack them. Among other "adaptive" hypotheses that have generally been rejected are unknown or mystical causes believed to guide evolutionary changes in non-selective yet adaptive directions, such as directed mutation, directed responses to environmental needs (Lamarckianism), orthogenesis, and saltation.

Selection coefficient (symbol s) A relative measure of the effect of selection, usually in terms of the loss of fitness endured by a genotype, given that the genotype with greatest fitness has a value of 1.

Self-assembly The spontaneous aggregation of macromolecules into biological configurations that can have functional value.

Selfish DNA The concept that the persistence of DNA sequences with no discernible cellular function (for example, various repetitive DNA sequences) arises from the likelihood that, once present in the genome, they are impossible to remove without the death of the organism—that is, they act as "selfish," or "junk" DNA, which the cell has no choice but to replicate along with functional DNA.

Semipermeable membrane A membrane that selectively permits transmission of certain molecules but not others.

Senescence The process of aging.

Septum (plural, **septa**) A dividing wall or partition between sections of an organism.

Serial ("iterative") homology Similarities between parts of the *same* organism, such as the vertebrae of a vertebrate or the different kinds of hemoglobin molecules produced by a mammal. The genetic basis for such homology can often be ascribed to gene duplications that have diverged over time but still produce somewhat similar effects. (*See* **Gene family**, **Paralogous genes**).

Sessile Attached to a substrate. An organism whose behavior is mostly nonmotile.

Sex chromosomes Chromosomes associated with determining the difference in sex. These chromosomes are alike in the homogametic sex (for example, XX) but differ in the heterogametic sex (for example, XY).

Sex linkage Genes linked on a sex chromosome. The results of such linkage may be differences between the sexes in the appearances of certain traits. For example, a recessive allele on the X chromosome is not expressed in the XX homogametic sex if that individual has a dominant allele on its second X chromosome but can be expressed in the

XY heterogametic sex, which has only one X chromosome.

Sex ratio The relative proportions of males and females in a population.

Sexual reproduction Zygotes produced by the union of genetic material from different sexes through gametic fertilization.

Sexual selection Selection that acts directly on mating success through direct competition between members of one sex for mates (intrasexual selection), or through choices made between them by the opposite sex (epigamic selection), or through a combination of both selective modes. In any of these cases, sexual selection may cause exaggerated phenotypes to appear in the sex on which it is acting (large antlers, striking colors, and so on).

Sibling species Species so similar to each other morphologically that they are difficult to distinguish but that are nevertheless reproductively isolated. (Sometimes called "cryptic species.")

Sister group A term used commonly in cladistic systematics to designate the most closely related group to a particular taxon. It derives from the concept that each significant evolutionary step marks a dichotomous split that produces two sister taxa equal to each other in rank.

Social Darwinism The concept that social and cultural differences in human societies (political, economic, military, religious, and so on) arise through processes of natural selection, similar to those that account for biological differences among populations and species.

Sociobiology The study of the biological basis of social behavior.

Somatic cells (or **tissues**) Body cells other than the germ-line tissues that produce sperm or eggs (*see* **Germ plasm**).

Speciation The splitting of one species into two or more new species (*see* **Cladogenesis, Phylogenetic evolution**) or the transformation of one species into a new species over time (*see* **Anagenesis, Phyletic evolution**).

Species A basic taxonomic category for which there are various definitions (Chapter 11). Among these are an interbreeding or potentially interbreeding group of populations reproductively isolated from other groups (the biological species concept) and a lineage evolving separately from others with its own unitary evolutionary role and tendencies (Simpson's evolutionary species concept). Employing the terms of population genetics, some definitions can be combined into the concept that a species is a population of individuals bearing distinctive genes and gene frequencies, separated from other species by biological barriers preventing gene exchange.

Split gene A gene whose nucleotide sequence is divided into exons and introns.

Spontaneous generation An early concept that complex organisms can appear spontaneously from inert materials without biological parentage.

Sporophyte The diploid spore-producing stage of plants that have alternating generations (haploid gametophyte and diploid sporophyte). The sporophyte arises from the union of gametophyte gametes and produces its haploid spores (which become gametophytes) by meiosis.

Stabilizing selection Selection that favors the survival of organisms in a population that are at an intermediate phenotypic value for a particular character and eliminates the extreme phenotypes. (Also called *centripetal,* or *normalizing, selection.*)

Stasis A period of equilibrium during which change appears to be absent, for example, in the concept of punctuated equilibrium. (*See also* **Stabilizing selection**.)

Stop codon One of the three messenger RNA codons (UAA, UAG, UGA) that terminates the translation of a polypeptide. (Also called *chain-termination codon* or *nonsense codon.*)

Strata *See* **Geological strata.**

Stromatolites Laminated rocks produced by layered accretions of benthic microorganisms (mainly filamentous cyanobacteria) that trap or precipitate sediments.

Structural gene A DNA nucleotide sequence that codes for RNA or protein. Some definitions restrict this term to a protein-coding gene.

Subspecies A taxonomic subdivision of a species often distinguished by special phenotypic characters and by its origin or localization in a given geographical region. Like other species subdivisions (*see* **Race**), a subspecies can still interbreed successfully with the remainder of the species. However, in some cases, interbreeding capabilities are unknown, and subspecies designations (for example, *Homo sapiens neanderthalensis* and *Homo sapiens sapiens*) are based entirely on phenotype.

Survivorship The proportion of individuals born at a given time (cohort) who survive to a given age.

Symbiont A participant in the interactive association (symbiosis) between two individuals or two species. This term is often restricted to mutually beneficial associations (mutualism).

Sympatric Species or populations whose geographical distributions coincide or overlap.

Sympatric speciation Speciation that occurs between populations occupying the same geographic range.

Synapomorphy The possession by two or more related lineages of the same phenotypic character derived from a different but homologous character in the ancestral lineage.

Synteny The retention of homologous genes on the same chromosome in different species, irrespective of their linkage order.

Synthetic theory of evolution *See* **Neo-Darwinism.**

Systematics Although defined by Simpson as the study of the diversity of organisms and all their comparative and evolutionary relationships, it is often used interchangeably with the terms *classification* and *taxonomy.*

T

Tachytelic A relatively rapid evolutionary rate.

Tautomer A nucleic acid base in which a hydrogen atom has moved from one position to another leading to a change in its pairing relationships.

Taxon (plural **taxa**) A named taxonomic unit consisting of a distinctive group of organisms placed in a taxonomic category, whether the unit is that of a species, genus, family, order, and so on.

Taxonomy The principles and procedures used in classifying organisms.

Tectonic plates The fairly rigid plates composing the earth's crust whose boundaries are marked by earthquake belts and volcanic chains. In oceanic regions, accretions to these plates occur at midoceanic ridges (sea floor spreading), and they are subducted under other plates at the deep oceanic trenches. Continental masses ride on some of these plates, accounting for continental drift and such processes as the mountain building that occurs when these plates collide.

Teleology The concept that natural processes such as development or evolution are guided by their final stage (*telos*) or for some particular purpose. Various philosophers separate "external" from "internal" (or "immanent") teleology. The former term is used to indicate guidance of a process toward some specified end decided by an external mystical or sagacious source; for example, "the ultimate purpose why the world was created [by God] was for the benefit of man." Thus, in contrast to science, which proposes that cause precedes effect, external teleological cause (for example, supernatural "design") arises *after* an effect's function or purpose has been foreseen or visualized. In contrast, internal teleology is generally used to indicate the end point of a process that has an understandable materialistic basis that develops from the process itself; for example, "the reason plants engage in photosynthesis and animals seek food is for survival, and the ultimate purpose of survival is for reproductive success." Among other factors, the absence of external teleology in biological evolution differentiates it from cultural evolution, into which humans have introduced external teleology by consciously designing many of its features.

Telocentric Chromosomes whose centromeres are located at one end.

Terminal electron acceptor The molecule that is the final acceptor of electrons in a metabolic pathway (for example, in aerobic respiration, oxygen is the terminal electron acceptor).

Terrestrial On the ground; also on or of the planet Earth.

Tetrapod Literal meaning, "four-footed." Commonly used to specify a member of the land-evolved vertebrate classes: amphibia, reptiles, and mammals.

Therapsids An order of synapsid mammal-like reptiles, composed mainly of fairly large herbivorous and carnivorous forms, which were dominant reptilian stocks during the Permian and Triassic periods. During the Triassic, some or many of these stocks were probably already endothermal, and one or more groups (cynodonts—"dog teeth") made the transition to smaller mammalian forms such as the morganucodontids.

Theria The viviparous mammalian subclass, consisting of marsupials and placentals.

Tissue A group of cells, all performing a similar function in a multicellular organism.

Trait *See* **Character.**

Transcription The process by which the synthesis of an RNA molecule (for example, messenger RNA) is initiated and completed on a DNA template by RNA polymerase enzyme.

Transfer RNA (tRNA) Relatively small RNA molecules (about 80 nucleotides long) that carry specific amino acids to the ribosome for polypeptide synthesis. Each kind of tRNA has a unique anticodon complementary to messenger RNA codons that specify the placement of particular amino acids in the polypeptide chain.

Transformation The process by which a cell takes up introduced DNA.

Translation The protein-synthesizing process that takes place on the ribosome, linking together a particular sequence of amino acids (polypeptide) on the basis of information received from a particular sequence of codons on messenger RNA.

Translocation An aberration in which a sequence of nucleotides is moved to a different position in the genome.

Transposons (transposable elements) Nucleotide sequences that produce enzymes to promote their own movement from one chromosomal site to another and may carry additional genes such as those for antibiotic resistance.

Triploblastic An animal that produces all three major types of cell layers during development—ectoderm, endoderm, and mesoderm.

Tritium An isotope of hydrogen that has three times the mass of an ordinary hydrogen atom.

Typology The study of organic diversity based on the principle that all members of a taxonomic group conform to a basic plan, and variation among them is of little or no significance. (*See also* **Archetype**.)

U

Ultraviolet radiation Electromagnetic radiation at wavelengths between about 4 and 400 nanometers, shorter than visible light but longer than X-rays. It is absorbed by purine and pyrimidine ring structures and is therefore quite damaging to nucleic acid genetic material.

Unequal crossing over The result of improper pairing between chromatids, causing their crossover products to differ from each other in the amounts of genetic material.

Ungulate A hoofed mammal.

Uniformitarianism A concept, popularized by Lyell in geology, that none of the forces active in past Earth history were different from those active today.

Universal genetic code The use of the same genetic code in all living organisms. (A few codons differ from the universal code in mitochondria, mycoplasmas, and some ciliated protozoa.)

Use and disuse A concept used by Lamarck to explain evolution as resulting from the transmission of characters that became enhanced or diminished because of their use or disuse, respectively, during the life experience of individuals. (*See also* **Lamarckian inheritance**.)

V

Variation A term commonly used to indicate differences in the qualitative or quantitative values of a character among individual members of a population, whether molecules, cells, or organisms.

Vascular plants Plants that have special water- and food-conducting vessels and tissues (xylem and phloem).

Vector A vehicle (for example, plasmid or virus) used to carry a genetically engineered DNA sequence into a cell.

Ventral The belly or downward surface of an animal. (In vertebrates, the surface opposite the spinal column.)

Vestigial organs Organs or structures that appear to be small and functionless but can be shown to be homologous with ancestral organs and structures that were larger and functional.

Virus A small intracellular parasite, often composed of little more than nucleic acid and a few proteins, that depends on the host cell to replicate its genetic material and to synthesize its proteins.

Vitalism The concept that the activities of living organisms cannot be explained by any underlying physical or chemical principles but arise from mystical or supernatural causes.

Viviparous Mode of reproduction in which eggs develop into live young before leaving the maternal body.

W

Wild type The most commonly observed phenotype or genotype for a particular character. Variations from wild type are considered mutants.

X

X chromosome The name given in various groups to a sex chromosome usually present twice in the homogametic sex (XX) and only once in the heterogametic sex (XY or XO).

X-linked genes Genes present on the X chromosome. (*See* **Sex linkage**.)

Y

Y chromosome A sex chromosome present only in the heterogametic sex (XY).

Z

Zootype A proposed stage in development characterized by the expression of a particular set of genes (Fig. 15–9), that governs spatial development in multicellular animals.

Zygote The cell formed by the union of male and female gametes.

Author Index

A page number in *italics* indicates pages listing complete reference citations.

C

D

E

F

G

H

I

J

K

Q

R

T

U

V

W

X

Y

Z

Subject Index

A page number in **boldface** indicates mention of the subject in a figure or figure legend.

A

B

C

D

E

F

G

H

I

J

K

M

N

O

P

Q

R

S

T

U

V

W

X

Y

Z

Part and Chapter Opener Photo Credits:

Cover: Rainbow Lorikeet: Corbis Westlight, © Australian Picture Library/Bob Walden; Dinosaur Fossil: Photo Researchers, © Francois Gohier. Part 1: shells, PhotoDisc; Part 2: stromatolites, Martin Miller/Visuals Unlimited; Part 3: ragweed pollen, David Scharf/Peter Arnold, Inc.; Part 4: peacock feather, PhotoDisc; Chapter 1: ocean waves, PhotoDisc; 2: Galapagos birds, PhotoDisc; 3: mammoth fossil, PhotoDisc; 4: watch gears, PhotoDisc; 5: crab nebula, WIYN Observatory image courtesy Eric Wilcots and Jay Gallagher/University of Wisconsin-Madison; 6: erupting volcano, PhotoDisc; 7: microspheres, Sidney Fox/Visuals Unlimited; 8: DNA model, PhotoDisc; 9: chloroplast, courtesy M. Gillott, from *Electron Microscopy,* Second Edition by John J. Bozzola and Lonnie D. Russell, reprinted with permission from Jones and Bartlett Publishers; 10: chromosomes, Biophoto Associates/Photo Researchers; 11: leaf veins, PhotoDisc (also shown on title page and preface); 12: cichlid fish, courtesy Patrick Danley/University of New Hampshire; 13: hanging Heliconia flower, PhotoDisc; 14: trilobite fossil, Alex Kerstitch/Visuals Unlimited; 15: sea urchin embryos, 2-cell stage, courtesy Ana Egaña/Tufts University; 16: sand dollars, PhotoDisc; 17: tunicates, Jane Shaw/Bruce Coleman, Inc.; 18: frog skeleton, PhotoDisc; 19: sea otter, PhotoDisc; 20: human skull, PhotoDisc; 21: flock of birds, PhotoDisc; 22: Queen butterfly, courtesy Roger Angel; 23: honey bee in flower, PhotoDisc; 24: bull elk, PhotoDisc; 25: Louvre Museum pyramid, Paris, courtesy Stephanie Torta.